RED DATA BOOK

of the Flowering Plants of Cameroon:

IUCN Global Assessments

Jean Michel Onana and Martin Cheek

Royal Botanic Gardens, Kew
IRAD-National Herbarium of Cameroon

ORCHIDACEAE

by **Benedict John Pollard**

Kew Publishing
Royal Botanic Gardens, Kew

Kew
ROYAL BOTANIC GARDENS

First published in 2011 by
Royal Botanic Gardens, Kew,
Richmond, Surrey, TW9 3AB, UK
www.kew.org

ISBN 978-1-84246-429-8

British Library Cataloguing in Publication Data
A catalogue record for this book is available from the British Library

Typesetting by Yvette Harvey
Cover Design by Christine Beard
Royal Botanic Gardens, Kew

Front cover. Main picture: Forest at Dom in the Bamenda Highlands (NW region) recently cleared for planting yams, February 2006. Photographer Yvette Harvey. From left to right: *Afrothismia winkleri* **Endangered** (Burmanniaceae), Kupe village (SW region) May 1996, *Cheek* 8354, photographer Martin Cheek; *Impatiens letouzeyi* **Endangered** (Balsaminaceae), Kodmin (SW region) Nov. 1998, photographer Benedict Pollard; *Allanblackia gabonensis* **Vulnerable** (Guttiferae), Fosimondi (SW region), Feb. 2006, photographer Barthélemy Tchiengué; *Pentas ledermannii* **Vulnerable** (Rubiaceae), Kupe summit (SW region), Oct. 1995, *Cheek* 7573, photographer Martin Cheek.

Back cover. The 815 global Red Data plant species of Cameroon occur in each of the 10 regions of Cameroon, but the macro hot-spots (>100 species per 25 km × 25 km or quarter degree square) are concentrated in SW region, shown in earlier studies to have tropical Africa's highest plant species diversity per degree square and to be home to tropical Africa's two top centres of plant diversity at Mt Cameroon, and at Mt Kupe and the Bakossi Mts. Correlation of concentrations of Red Data plant species with national parks is low since these have almost always been created for large animals and not plants. The macro hot-spot in S region is the Ngovayang Massif near Bipindi which is unprotected but which is here highly recommended as a national park; a new national park for the Yaoundé area is also advised to help protect the meso hotspots there.
S. Bachman, J.-M. Onana, C. Cole & M. Cheek.

Printed in the UK by Marstons Book Services Ltd

For information or to purchase all Kew titles please visit
www.kewbooks.com or email publishing@kew.org

Kew's mission is to inspire and deliver science-based plant conservation worldwide, enhancing the quality of life.

Kew receives half of its running costs from Government through the Department for Environment, Food and Rural Affairs (Defra). All other funding needed to support Kew's vital work comes from members, foundations, donors and commercial activities including book sales.

CONTENTS

LIST OF FIGURES

FOREWORDS

It is good to see that this book has been finished and will now be available. The support of the Darwin Initiative of the UK Government for the Cameroon Plant Red Data Book Project has now come to fruition. As the former National Focal Point contact person for the Convention on Biological Diversity (CBD) in Cameroon within the Ministry of Environment and Forests and later the Ministry of Environment and Protection of Nature, Yaoundé, I encouraged this British Initiative to support plant conservation in Cameroon, especially since it is a collaborative effort between the IRAD-National Herbarium of Cameroon and the Royal Botanic Gardens, Kew. May this book result in all of Cameroon's plant species being protected and preserved from extinction, so that in future their uses and purpose may be worked out to the benefit of people and the environment.

I equally hope that readers from all walks of life will use and share the knowledge contained in the book and its transmission will thus contribute to the sustainable management of Cameroon's rich biological resources.

Mrs. Mary Fosi Mbantenkhu
CBD focal point Cameroon (retired)

It gives me great pleasure to provide an introduction for this ground-breaking volume, the first Red Data Book for a tropical African country.

I had the privilege, with Hugh Synge and a very small team of researchers and correspondents, to publish the first of the IUCN Red Data Books for Plants in 1978, based on the preliminary work of Ronald Melville. The Director of the Royal Botanic Gardens, Kew, John Heslop Harrison in the early 70's had the foresight to see that there was a real need to match the data gathering that was well under way for animal groups, with a similar effort for plants! After a key meeting with Sir Peter Scott, I was asked to form the Threatened Plants Committee (TPC) of IUCN based at Kew but with financial support from WWF. TPC reported to the Species Survival Commission whose Chairman was Sir Peter. Under both these men's constant and very active support the Red Data Book Unit grew with plant data gathering based at Kew and a new 'Animals Unit' based in Cambridge in very large portacabins set in a field on the outskirts of the town. Here the now very fine headquarters of the World Conservation Monitoring Centre, WCMC, houses all the elements of Red Data Book research work, currently led by IUCN's Craig Hilton-Taylor, himself a botanist who was once based at the Kew Herbarium! However, Kew itself for many years housed the WCMC plant team, developing data both for the international Red Data Books but also helping to support the creation of many National Red Data Lists and Books, as well as specifically researching 'Trade threats' for CITES.

It has been a pleasure to see how many of these works now exist and how they have provided the drive to have appropriate legislation enacted to help protect the rare and endangered plants in the wild. Another side and of equal pleasure has been to see many of the young botanists I had the pleasure to appoint and work with when I was Keeper, many moons ago, have taken up the taxonomic baton but also have maintained the steady conservation drive as they have progressed in their careers. Martin Cheek is such an example, initially appointed to understudy Nigel Hepper, the head of the central and West African research activities and who had recently completed the rewriting and revision of the Flora of West Tropical Africa. Martin's enthusiasm and commitment to research into the Cameroonian flora soon became obvious, with a very large number of new species and records appearing over the years through his close and happy working relationship with all his Cameroonian colleagues. So it is no surprise to see this superb volume appear with the active support of one of the country's own leading botanists, Jean Michel Onana. They are to be congratulated for producing this vital work that will hopefully be a key to producing a robust and practical plan for the conservation of this wonderful, rich and unique tropical flora!

I started by talking of the past, but this volume allows the whole community of Cameroon to focus on the future protection of their flora, with its wide potential to enhance and maintain the resources and quality of life for everyone. As we now have the knowledge gleaned from the plants themselves, to formulate the legislation and actions necessary to ensure the maintenance and rehabilitation of this remarkable rich plant kingdom, we must ensure this fine work now stimulates those with the power to make the recommendations come to fruition. My congratulations to the team that have given us all this fine tool for the future.

Gren Lucas OBE
Keeper of the Herbarium and Library (1984–1995)
Royal Botanic Gardens, Kew

PREFACE

Not all natural areas can be protected. For a nation to progress, some development of timber production, plantation agriculture, mineral extraction, urbanisation and infrastructure, has to go forward. However, this need not result in the extinction of Cameroon's exceptionally numerous, localised and unique plant species.

In this book, 815 flowering plant species, from grasses to tall forest trees, are assessed as being globally threatened with extinction (Red Data species), most for the first time, using the IUCN 2001 standard. This is nearly a ten-fold increase on the 82 threatened flowering plant species documented for Cameroon in 1998 (Walter and Gillett, 1998). Each species in this book is described so that they can be identified, and for each, management suggestions are given so that their chances of surviving can be improved. The species are also individually and collectively mapped so that, in future, those co-ordinating development projects in Cameroon can better avoid known sites for threatened species when planning the footprint of those projects. Hotspots for threatened species in Cameroon are mapped on the rear cover.

Many plant families in Cameroon have yet to be worked out for the Flore du Cameroun. When these come to pass, or when these families are otherwise revised, additional Red Data species will emerge. Similarly, many areas of Cameroon have never been surveyed for plants, or only very incompletely surveyed, so may [illegible] contain further threatened species, including those new to science. Therefore it is important to conduct thorough, specimen-based surveys with competent botanists to check for rare plant species in any area before plans are made to clear it.

It is hoped that this book will enable those national authorities responsible for managing the environment, managers of National Parks, members of environmental NGOs, and all those interested in identifying plants, to safeguard the most threatened species of Cameroon's incredible plant diversity. It should not be forgotten that Cameroon's plant diversity, with 7850 species (Onana 2011) far exceeds those of its neighbours, such as Gabon (4710 species fide Sosef *et al.* 2006) and even exceeds that of West Africa from Senegal to Nigeria inclusive, which has rather less than the 7800 species documented for the Flora of West Tropical Africa (Keay & Hepper 1954–1972).

We plan, through this project, to make the information in this book available in condensed form to secondary school teachers, so that future generations of Cameroonians will be aware of the importance of their wild plant patrimony and be encouraged to support its protection.

Martin Cheek
Royal Botanic Gardens, Kew

Jean Michel Onana
IRAD-National Herbarium of Cameroon

ACKNOWLEDGEMENTS

Firstly we acknowledge with respect all those botanists who collected the plant specimens on which our knowledge and on which this book is based, from Kew's Gustav Mann in 1861, to the National Herbarium's Olivier Sene in 2010. Many of the earlier botanists, particularly the Germans, such as Conrau and Staudt, never returned to their home country but died young and lie buried in Cameroon.

We thank the Darwin Initiative of the UK Government, for financial support and especially Eilidh Young for constructive advice during the later stages of this project.

We thank Mary Fosi, until recently CBD Focal Point person for Cameroon, who supported this project from the start; also Paul Mzeka of ANCO, Bethan Morgan of the Ebo Forest Project, Louis Nkembi of ERUDEF, Chris Mitchell and Rachel Hogan of the former CWAF, now AAA, and Bryan Carroll of Bristol Zoo, for supporting surveys our surveys in their, previously botanically unknown, proposed protected areas. Earthwatch provided support for the support at the last of these areas, Mefou.

In London, Janis Shillito typed most of the text in this book, the species treatments, and Seren Thomas most of the introductory chapters, while the gigantic and lengthy task of databasing specimens and specimen records at Kew was done principally by Bate Oben, helped by Emma Fenton in 2006, later with Tamsyn Thring and Katie Bermingham, who also produced the draft maps for checking. George Gosline is thanked for maintaining our Cameroon database, and adding the mapping functions for this project. Final maps as seen in this book, and the automated provisional assessments from which the EOOs were taken, were produced by Steve Bachman assisted by Charlotte Cole.

Craig Hilton-Taylor and Caroline Pollock at IUCN Cambridge kindly received our species assessments in 22 batches in 2010-2011.

Tivvy Harvey is owed tremendous and unending thanks. She dedicated months to formatting, copy-editing and setting up each one of the often unevenly typed pages for the black and white version of this book, risking her sanity in efforts to keep the text standardised. In particular, she standardised Cameroon place names to:http://geonames.nga.mil/ggmagaz/ (the National Geospatial-Intelligence Agency 'GeoNames Search') and checked for correct assignment to Regions. Seren Thomas arranged the blocks of maps under Tivvy's supervision.

Brendan O'Hanrahan is thanked for supplying the species and site monitoring forms devised by Lynne Farrell who kindly agreed to their reproduction here. These are widely used in Britain and may be useful in improving Red Data plant species conservation in Cameroon.

Kew Publishing, in particular Chris Beard and Lloyd Kirton, are thanked for designing the excellent cover, while Gina Fullerlove and Lloyd Kirton devised the page template used to set up the pages for printing.

Finally, Eimear Nic Lughadha, Christine Leon and Steve Davis kindly reviewed draft versions of the manuscript and made many useful suggestions.

Martin Cheek
Royal Botanic Gardens, Kew

Jean Michel Onana
IRAD-National Herbarium of Cameroon

NEW NAMES

MARTIN CHEEK

The following are published in this volume for the first time:

Caryophyllaceae: ***Polycarpaea rheophytica*** Cheek nom. nov...p. 110

Umbelliferae: ***Lefebvrea camerunensis*** (Jacq.-Fél.) Cheek & I. Darbysh. **comb. nov.** .... p. 340

Umbelliferae: ***Lefebvrea kupense*** (I. Darbysh. & Cheek) Cheek & I. Darbysh. **comb. nov**. .... p. 341

WHAT IS A RED DATA BOOK?

MARTIN CHEEK

A Red Data book gives details, species by species, of those species in a country which are threatened with extinction, that is, Red Data Species. By drawing attention to species which are at risk from extinction, a Red Data Book provides a focus for conservation efforts. Typically each species is described and facts about its population are documented, such as where it occurs, and if known, how many individuals there are. Threats to the species are detailed and conservation actions or efforts are listed. This book follows this model and, in addition, offers management suggestions for each species.

The first Red Data Book for plants was produced at Kew in 1978, edited by Hugh Synge and Grenville Lucas, following earlier work by Ronald Melville. Its scope was the world. Although far from complete (it included only a portion of the world's threatened plant species) it succeeded in drawing attention to the fact that despite mankind's dependence on plant species, many species are either extinct or fast approaching extinction, and that action needs to be taken to avoid this. In the decades that followed, most developed countries developed their own national Red Data books for plants, and updated them every few years with new editions as additional data became available, or as conservation actions changed the threatened status of the species. Japan for example, is on its fifth edition of a Red Data book, and South Africa is on its fourth. In some cases, the quantity of data has become so voluminous that national Red Data book information is housed in databases and kept on the web, and only summaries are published as hard copy as 'Red Data Lists', with very abbreviated species treatments.

Red Data books follow the standards set under the co-ordination of IUCN, the International Union for the Conservation of Nature, a United Nations body based in Switzerland but with its Red Listing operations at Cambridge, UK. Many countries in the developing world, although vastly important for plant diversity and threatened species, lack a Red Data book. We believe this book, for Cameroon, to be the first for a Tropical African nation.

THE EVOLUTION OF THIS RED DATA BOOK

MARTIN CHEEK

The path towards this Red Data book began in 1998 with the writing of the Red Data chapter for the Conservation Checklist of the Plants of Mount Cameroon, funded by DfID (Cable & Cheek 1998: xxviii-lxxvi). Here 116 species were considered to be threatened among the 2435 taxa listed for the area in that work. Each of these 116 species was assessed according to the criteria standard at the time: IUCN (1994). For each species, following the name and the IUCN rating, notes were given under the headings 'Range:', 'Habitat:', 'Threats:', and 'Managment suggestions:'. Those assessments were reviewed and largely accepted by IUCN and appeared on the IUCN Red List website. This model proved popular and requests were made by several conservation NGOs in Cameroon to do similar books for other areas considered to be important for conservation.

Three more Conservation Checklists then followed, one in 2000 and two in 2004, all co-authored by staff of [illegible] with Jean Michel Onana of IRAD-National Herbarium of Cameroon. More details of this series are given in the next chapter. In early 2005 Cheek and Onana were considering a future research project that might best take forward the conservation actions of our institutes in Cameroon. Why not expand our geographic coverage for Red Data assessments to cover the whole of Cameroon and produce a Red Data book for Cameroon? Onana had already been compiling a manuscript of the rarest plant species of Cameroon that would be helpful to such a project (more of which below). After discussing the idea with the CBD focal point for Cameroon, Mary Fosi, and also with leaders of conservation NGOs in Cameroon, a proposal was submitted to the Darwin Initiative of the UK Government for funding in August 2005 and in May 2006 we got the news that we were successful. The initial plan had been to screen a national checklist of plant species for those with globally small ranges, at ten or less locations following the template used in our Conservation Checklists for potential protected areas. Such a national checklist had been prepared by Jean-Paul Ghogue of IRAD-National Herbarium for a research project led by Peter Linder at the Univ. Zurich and we obtained permission to use it for the Red Data project. However early in the life of the project, when reviewing that Checklist, it became clear that,

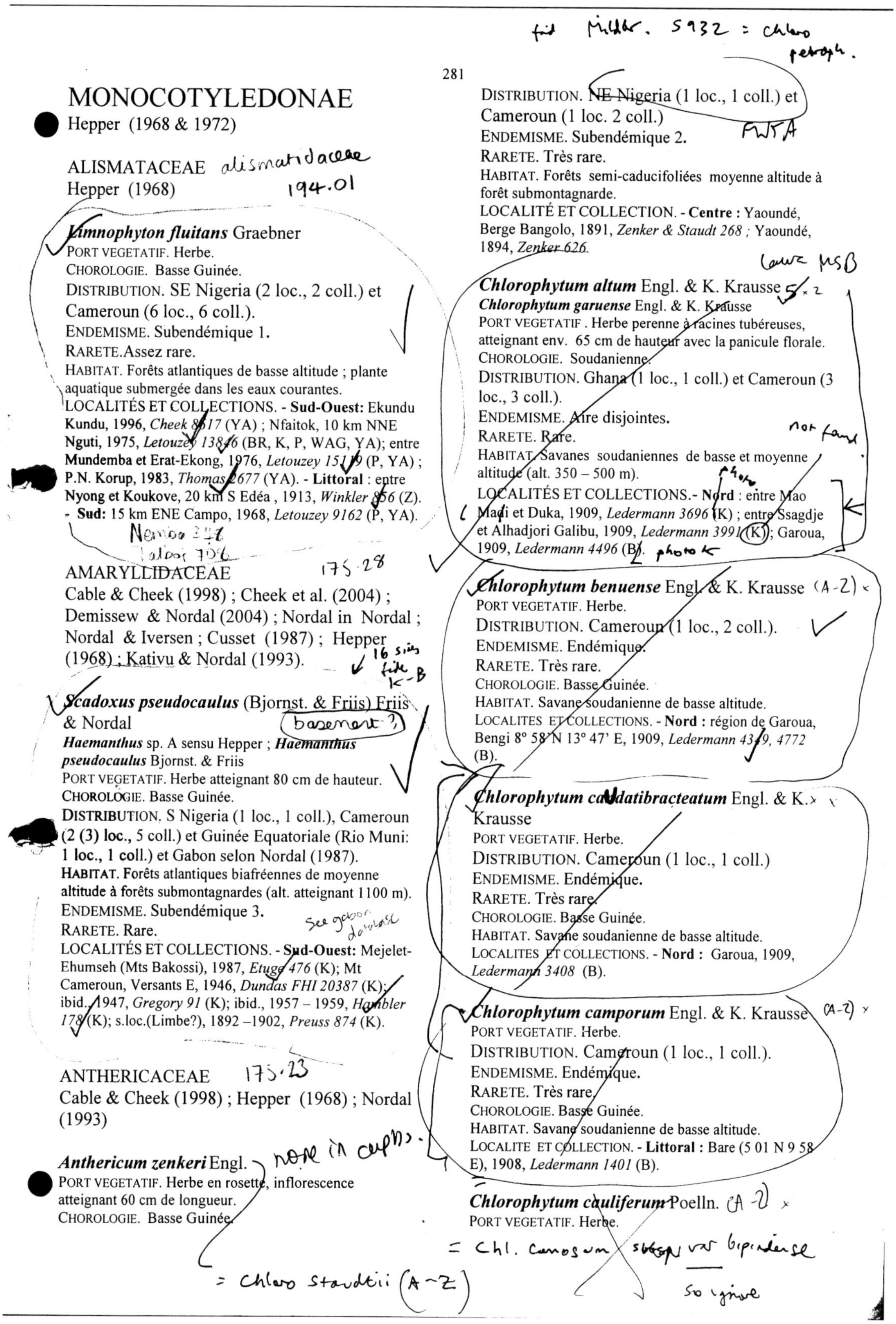

281

MONOCOTYLEDONAE

Hepper (1968 & 1972)

ALISMATACEAE
Hepper (1968)

Limnophyton fluitans Graebner
PORT VEGETATIF. Herbe.
CHOROLOGIE. Basse Guinée.
DISTRIBUTION. SE Nigeria (2 loc., 2 coll.) et Cameroun (6 loc., 6 coll.).
ENDEMISME. Subendémique 1.
RARETE. Assez rare.
HABITAT. Forêts atlantiques de basse altitude ; plante aquatique submergée dans les eaux courantes.
LOCALITÉS ET COLLECTIONS. - **Sud-Ouest:** Ekundu Kundu, 1996, *Cheek 8617* (YA) ; Nfaitok, 10 km NNE Nguti, 1975, *Letouzey 13846* (BR, K, P, WAG, YA); entre Mundemba et Erat-Ekong, 1976, *Letouzey 15119* (P, YA) ; P.N. Korup, 1983, *Thomas 2677* (YA). - **Littoral** : entre Nyong et Koukove, 20 km S Edéa , 1913, *Winkler 856* (Z). - **Sud:** 15 km ENE Campo, 1968, *Letouzey 9162* (P, YA).

AMARYLLIDACEAE
Cable & Cheek (1998) ; Cheek et al. (2004) ; Demissew & Nordal (2004) ; Nordal in Nordal ; Nordal & Iversen ; Cusset (1987) ; Hepper (1968) ; Kativu & Nordal (1993).

Scadoxus pseudocaulus (Bjornst. & Friis) Friis & Nordal
Haemanthus sp. A sensu Hepper ; ***Haemanthus pseudocaulus*** Bjornst. & Friis
PORT VEGETATIF. Herbe atteignant 80 cm de hauteur.
CHOROLOGIE. Basse Guinée.
DISTRIBUTION. S Nigeria (1 loc., 1 coll.), Cameroun (2 (3) loc., 5 coll.) et Guinée Equatoriale (Rio Muni: 1 loc., 1 coll.) et Gabon selon Nordal (1987).
HABITAT. Forêts atlantiques biafréennes de moyenne altitude à forêts submontagnardes (alt. atteignant 1100 m).
ENDEMISME. Subendémique 3.
RARETE. Rare.
LOCALITÉS ET COLLECTIONS. - **Sud-Ouest:** Mejelet-Ehumseh (Mts Bakossi), 1987, *Etuge 476* (K); Mt Cameroun, Versants E, 1946, *Dundas FHI 20387* (K); ibid., 1947, *Gregory 91* (K); ibid., 1957 – 1959, *Hambler 178* (K); s.loc.(Limbe?), 1892 –1902, *Preuss 874* (K).

ANTHERICACEAE
Cable & Cheek (1998) ; Hepper (1968) ; Nordal (1993)

Anthericum zenkeri Engl.
PORT VEGETATIF. Herbe en rosette, inflorescence atteignant 60 cm de longueur.
CHOROLOGIE. Basse Guinée.

DISTRIBUTION. NE Nigeria (1 loc., 1 coll.) et Cameroun (1 loc. 2 coll.)
ENDEMISME. Subendémique 2.
RARETE. Très rare.
HABITAT. Forêts semi-caducifoliées moyenne altitude à forêt submontagnarde.
LOCALITÉ ET COLLECTION. - **Centre :** Yaoundé, Berge Bangolo, 1891, *Zenker & Staudt 268* ; Yaoundé, 1894, *Zenker 626*.

Chlorophytum altum Engl. & K. Krausse
Chlorophytum garuense Engl. & K. Krausse
PORT VEGETATIF . Herbe perenne à racines tubéreuses, atteignant env. 65 cm de hauteur avec la panicule florale.
CHOROLOGIE. Soudanienne.
DISTRIBUTION. Ghana (1 loc., 1 coll.) et Cameroun (3 loc., 3 coll.).
ENDEMISME. Aire disjointes.
RARETE. Rare.
HABITAT. Savanes soudaniennes de basse et moyenne altitude (alt. 350 – 500 m).
LOCALITÉS ET COLLECTIONS.- **Nord** : entre Mao Madi et Duka, 1909, *Ledermann 3696* (K) ; entre Ssagdje et Alhadjori Galibu, 1909, *Ledermann 3991* (K); Garoua, 1909, *Ledermann 4496* (B).

Chlorophytum benuense Engl. & K. Krausse
PORT VEGETATIF. Herbe.
DISTRIBUTION. Cameroun (1 loc., 2 coll.).
ENDEMISME. Endémique.
RARETE. Très rare.
CHOROLOGIE. Basse Guinée.
HABITAT. Savane soudanienne de basse altitude.
LOCALITES ET COLLECTIONS. - **Nord :** région de Garoua, Bengi 8° 58' N 13° 47' E, 1909, *Ledermann 4349, 4772* (B).

Chlorophytum caudatibracteatum Engl. & K. Krausse
PORT VEGETATIF. Herbe.
DISTRIBUTION. Cameroun (1 loc., 1 coll.)
ENDEMISME. Endémique.
RARETE. Très rare.
CHOROLOGIE. Basse Guinée.
HABITAT. Savane soudanienne de basse altitude.
LOCALITES ET COLLECTIONS. - **Nord :** Garoua, 1909, *Ledermann 3408* (B).

Chlorophytum camporum Engl. & K. Krausse
PORT VEGETATIF. Herbe.
DISTRIBUTION. Cameroun (1 loc., 1 coll.).
ENDEMISME. Endémique.
RARETE. Très rare.
CHOROLOGIE. Basse Guinée.
HABITAT. Savane soudanienne de basse altitude.
LOCALITE ET COLLECTION. - **Littoral :** Bare (5 01 N 9 58 E), 1908, *Ledermann 1401* (B).

Chlorophytum cauliferum Poelln.
PORT VEGETATIF. Herbe.

Fig. 1 Sample page of the Onana manuscript (2006 version) on which this Red data book is based. The annotations derive from successive Red data project workers cross-checking records against data-sources

owing to the process of its compilation, a large number of the endemic and near endemic species of Cameroon had been omitted, reducing its value for the project since, of course, such species are most likely to be threatened. Accordingly that checklist was set aside. The project had no resources, nor time, at that point, to produce a replacement checklist. Therefore Jean Michel Onana's draft manuscript "of the Endemic, Near Endemic and Threatened Plant Species of Cameroon" (hereafter the Onana ms) became completely essential to the completion of the Red Data book. This work, compiled over years as a personal project, is a result of combining the scientific literature, ancient and modern, for references to the rare plant species of Cameroon. It is the foundation of this Red Data book.

CAMEROON – AN OVERVIEW

MARTIN CHEEK

Geography, Geomorphology, Climate and Vegetation

Cameroon, 475,442 km^2, is situated in west-central Africa between latitudes 1° and 13° N and longitudes 8° and 17° E, extending from the Atlantic coastline in the S, to Lake Chad in the extreme North at the edge of the Sahara. It is divided into 10 Regions, formerly known as Provinces.

Two major geomorphological features dominate Cameroon:

1. The S Cameroon Plateau, 500–800 m altitude, which occupies about a third of the country, namely most of E, S and Central Regions.

2. The Cameroon Fault, a failed Atlantic rift, along which mountain building has occurred in four separate epochs, runs inland from the Atlantic, nearly NE–SW, forming the Cameroon Highlands (SW, NW and W Regions) before arcing and aligning W–E, forming the Adamawa Highlands (mainly 1000-1500m alt.) in Adamawa Region, bisecting the N and S parts of Cameroon, and extending into the Central African Republic (CAR) to the E.

The Cameroon Highlands form an interrupted line, which contains Africa's highest mountains W of the Albertine Rift, namely Mt Cameroon (4095 m alt.), Bamboutos Mts (2740 m alt.), Mt Oku (3011 m alt.) and Tchabal Mbabo (2640 m alt.). The Highlands continue off-shore to the SW, in a line of islands, most notably Bioko (Equatorial Guinea), almost a twin of Mt Cameroon. The Cameroon Highlands are a mixture of isolated peaks, such as Mt Cameroon (the only active volcano in continental Africa W of the Albertine Rift) and plateau areas such as the Bamiléké Plateau and the Bamenda Highlands, as much as 75–100 km wide.

North of the Adamawa Plateau, are N and Extreme N Regions, with altitudes of only c.300 m and which are drained by the Benue, flowing W into Nigeria and the Niger River, or into Lake Chad.

The coastal plain is 75–200 km wide, falling in SW, Littoral and S Regions, and has an average elevation of 90 m.

Rainfall is generally 2–4 m per annum in the coastal plain, supporting evergreen rainforest. The Cameroon Highlands increase rainfall by forcing up clouds from the SW monsoon winds in the period April-October, the main wet season. In the drier season, November-March, the desiccating Harmattan wind from the N prevails, and visibility is reduced throughout the country due to dust particles from the Sahara. In S Region, free of the influence of the Cameroon Highlands, a short cool dry season occurs in June-July. In Central and E Regions, large areas with only 1.5 m rain per annum support semi-deciduous forest, which generally replaces evergreen forest when rainfall is below 2 m per annum. Semi-deciduous forest is generally richer in timber species and has less leached and so more fertile soils than those of evergreen forest areas. North of the Adamawa Highlands, rainfall drops to 0.6m per annum or lower and Sudanian woodland and wooded grassland dominate, changing into Sahelian grasslands in the Extreme N.

The vascular plant species diversity of Cameroon is high considering its surface area, with 7850 species, including over 100 endemic species. This is accounted for by the diversity of its geomorphology, climate and habitats (Cameroon is often termed "Africa in miniature"). Cameroon also has Tropical Africa's greatest concentration of plant diversity in terms of species per degree square (see the Hot Spots chapter). These high figures are also supported by the existence of Pleistocene Forest refugia in the S of the country (see Climate Change, below)

Human Impacts

The human population of Cameroon is c. 19.5 million. Over 200 tribes and languages occur. Officially Cameroon is bilingual in French and English. The greatest concentrations of people are in the capital city of Yaoundé (Central Region) and the port and industrial city of Douala, each with over 1 million people. Apart from these two cities, the most densely populated areas are the Bamiléké Plateau and Bamenda Highlands, with over 200 people per km^2, owing perhaps to their richer soils, cool, moist climes. Outside these areas population density is usually lower, often very much lower, with blocks in each Region of only 1–3 persons per km^2 or lower.

Food crops grown in the forest areas are macabo, yams and manioc, while maize, millet and sorghum prevail in the drier northern half of Cameroon.

Livestock production is concentrated in the grasslands of the N, zebu cattle, sheep and goats being numerous. Pastoralists from the N moved S in the 19th century to the montane grasslands of the Cameroon Highlands and the consequent impacts on those montane habitats is a cause for concern.

Major threats to natural vegetation, and to rare plant species, have been forest clearance for timber followed by agriculture in the southern half, and in the northern half, which is generally less well studied and understood botanically, overgrazing and trampling by livestock, followed by over-frequent forest fires to produce fresh grass growth for stock.

Slash and burn agriculture is widely practised and is considered a major threat to rare species of plant in forest areas. Farmers can, by cutting smaller trees in intact forest, and setting fires to the resultant debris in the dry season, clear land and cultivate crops for one or two seasons on temporarily weed-free, fertile soils, before they move on to repeat the process in another area. This process can potentially eliminate or severely deplete the population of any localised endemic species of herb, shrub, liana or small tree, that might, unknown to the farmer, be present.

Cameroon's main exports are timber and petroleum oil products. Most oil is produced off-shore, but exploration drilling is current in the coastal plain, such as around Mt Cameroon.

Much of the forest outside of protected areas has been granted as logging concessions. According to mongabay.com (accessed March 2011), total forest area is 212,450 km^2 or c. 45.6% of the land area. Deforestation rates 2000–2005 were -2,200 km^2, that is -1% over five years. Total forest loss since 1990 was reported as -13.4% (?annual?).

Wood removal in 2005 amounted to 3,211,000 m^3 for industrial roundwood and 16,561,000 m^3 for woodfuel. Logging, if properly managed, at low densities, perhaps one tree per hectare, and if arranged so that minimal damage is done to the forest, with liana cutting and best-practice extraction methods, is not necessarily inimical to conservation. We have surveyed such logged areas in Cameroon and found them still to have high conservation value. However, all too often such management does not occur and as a result, threatened species can be heavily depleted. Asian logging companies have a poor reputation in this respect. All too often, logging roads allow access to farmers practising slash and burn agriculture, attracted by easy access to new land and of getting crops to market, due to the existence of the logging roads.

Industrial crop plantations were mainly established during the German colonial rule 1884–1914. Palm oil is indigenous, but commercial strains of rubber, bananas, tea, coffee and cocoa were all introduced during the colonial epoch, often via Victoria (now Limbe) Botanic Garden. These now earn important export income. The last two species are mainly grown by smallholders. The greatest concentrations of plantings are associated with the fertile volcanic soils of parts of the Cameroon Highlands, mainly in the evergreen forest belt, especially in the neighbourhood of Mt Cameroon. If these plantations are expanded, this will have a severe impact on the many endemic and threatened species in the lowlands around the mountain. In the drier North of Cameroon, cotton is an important plantation crop. Cameroon exports market garden produce (e.g. from Mt Cameroon) to Gabon which has poorer soils, and wild-harvested forest food and other products, such as eru (*Gnetum*) and chewsticks (*Garcinia*) to Nigeria, which only has a tiny fraction of its original forest remaining.

New and Increased Threats

The threats referred to above have been in place for decades. But now, a new swathe of threats is looming, principally from the extractive sector. Mineral extraction, not previously a concern, is set for a massive expansion in Cameroon, due partly to increased global commodity prices in this sector. Mining projects for open-cast extraction of iron-ore, cobalt and copper, for example, are all being planned in the forest belt of S, E and Central Regions. All too often, these can occur in poorly known areas for plants, and impact assessments for rare plant species within project footprints are usually far from exhaustive, so there is a real probability of species becoming globally extinct. In S Region, Massif des Mammelles and Mont Elephant are both mooted for iron ore extraction, while supporting numerous threatened plant species. Several plant species are unique to Mont Elephant. The mines themselves are not the sole environmental threat since the wider footprint of such projects includes transport corridors and port facilities, the last of which may require large areas for stockpiling or grading. In Cameroon this is a concern since such projects have converged on the port of Kribi, or adjoining sites, which has some of the most diverse and threatened species rich lowland forest in Cameroon along its coast.

Hydro-electric (HE) projects are on the increase in Cameroon. While providing much-needed sustainable energy sources for communities, and often essential for extractive sector projects, they pose two classes of concern to threatened plant species.
loss of large areas of natural habitat which become submerged when the watercourse is dammed-these areas may conceivably contain localised plant species.
loss of rheophytic species. HE projects are usually placed at natural waterfalls or rapids to maximise return for investment. Rheophytic species are highly specialised, being restricted to such habitats are threatened by changes in waterflow regimes. A very high proportion of such species (see the family treatment for Podostemaceae (Dicots)) are restricted to only one or two locations and are highly threatened. Cameroon has the highest number of rheophytic plant species of all African countries.

More information on threats is given in each of the 815 species treatments that form the bulk of this book.

Climate Change

During the Pleistocene period, which concluded 10,000–12,000 years ago, the world saw massive climate change over a period of more than a million years, in which alternating warm and glacial periods in the N hemisphere corresponded with major changes in global sea-levels, temperature, and rainfall. During the glacials, Africa saw huge shrinkage in forest extent, to a few patches or "Pleistocene Refugia" which have been documented and studied by researchers over the last 20 years, most notably by Jean Maley. Cameroon and gabon have Tropical Africa's greatest concentration of postulated refugia, often associated mountains, since presumably these help create their own (orographic) rainfall, and have the possibility for plant populations, over time, to move up or down in altitude to adjust to changes in ambient temperature. Logically, these former refuges may well be important also as refuges for forest species in the future, as climate change returns to our planet. Thus it is suggested that priority be given to reviewing these refuge areas and their current protection, and enhancing their protection-levels and/or extent when needed.

Comparisons with Nigeria and Gabon

Flanked by Nigeria to the W and Gabon to the S, Cameroon is strongly differentiated from both.

Nigeria (923,786 km^2, population in 2010 c. 152.2 million people) has seen far greater forest loss than Cameroon due to conversion to timber and agriculture over many decades, and is now a net importer of timber. Currently forest area is given as 90,410 km^2, or only 9.9% of the country, but of this only 38,200 km^2 or 4.2% of the total, is rated as primary forest, surviving mainly in the SE corner. Forest losses between 1990 and 2010 were reported as 81,930 km^2, or 47.5% overall, averaging about 2.38% per annum (mongabay.com). Many threatened plant species are shared by Cameroon and Nigeria, but often there have been no records from Nigeria in more than 50 years, and former locations for threatened forest species have since been reported to have been cleared of forest, including some forest reserves, leading to concerns that they are nationally extinct. Sadly, the Forest Herbarium of Ibadan, formally the major hub for surveys throughout the country, and even into modern Cameroon, has declined as its funding has been reduced. Nevertheless, Nigeria's most species and diverse rainforest, in the SE, adjoining Cameroon, such as the Cross River National Park, is reported relatively well protected at the moment.

Gabon (267,667 km^2, population 1.2 million people) is reported as having 82% forest cover. While logging is an important source of export revenue, the range of species cut is small being mainly (76%), just one, *Aucoumea klaineana* (Burseraceae), which grows fairly rapidly and is not of conservation concern. Forest habitat loss, conversion to agriculture and slash-and-burn farming are much less of a concern than in Cameroon, due to an enviable combination of good management practice, poor soils and low population density. Gabon's network of National Park's appear better aligned to the conservation of known endemic plant hotspots than is the case in Cameroon. Gabon, like Cameroon, has numerous important Pleistocene Refuge sites, and numerous endemic forest plant species. While some threatened plant species are shared between the two countries, a natural climatic divide appears to separate them. Some of Gabon's best known species do not occur in Cameroon, or only just cross the border into S Region, for example *Aucoumea klaineana*. While possibly equally rich in lowland forest endemic species, Gabon's total number of 4710 documented plant species, Sosef *et al.* (2006) is far lower than those of Cameroon (7850 species, Onana 2011). This is probably because Gabon, compared with Cameroon, has very little ecological amplitude, with a far smaller proportion of grassland and woodland habitat overall, and of habitat types. More importantly in terms of numbers of threatened species, Gabon also has comparatively little altitudinal amplitude, its highest mountains only reaching a little over 1000 m altitude, and supporting relatively little submontane forest for this reason. In comparison, the Cameroon highlands commonly exceed 2000 m alt, the highest point reaching 4095 m, and there are large tracts of submontane forest (c. 800–2000 m alt.) which is so rich in narrowly endemic species.

CAMEROON RED DATA SPECIES IN 1998

MARTIN CHEEK

This book is not the first to cover Cameroon's Red Data plant species. Walter & Gillett (1998) documented 89 species for Cameroon in their monumental IUCN Red List of Threatened Plants. This work remains unmatched for global coverage of the world's plant species. Even though the www.redlist.org carries the latest assessments using the newer and more exacting post-1994 IUCN standards (e.g. IUCN 2001), the world's redlisters contributing to this system have still to catch up with the number of species covered in the earlier work. Below is an extract of the Cameroonian species from Walter & Gillett, with all families Dicots, Monocots and ferns, arranged alphabetically in one sequence. Although the threat nomenclature differs, and some plant names have been superseded, it is notable that most of the 82 angiosperms in table below still feature as threatened in our book. We are grateful to Harriet Gillett for providing the data which the table below summarises.

FAMILY	NAME	WORLD STATUS	
ACANTHACEAE	Isoglossa nervosa C.B. Clarke	Indeterminate	
ACANTHACEAE	Justicia preussii (Lindau) C.B. Clarke	Indeterminate	
ANNONACEAE	Uvariodendron occidentale Le Thomas	Vulnerable	Hawthorne, W. 1995. Categories of conservation priority and Ghanaian tree species. WCMC Working Document 4. 1-38 pp.
ARACEAE	Amorphophallus preussii (Engl.) N.E. Br.	Indeterminate	
ARACEAE	Amorphophallus staudtii (Engl.) N.E. Br.	Indeterminate	
ARACEAE	Culcasia panduriformis Engl. & K. Krause	Vulnerable	
ASPHODELACEAE	Kniphofia reflexum Hutch. ex Codd	Vulnerable	
ASPLENIACEAE	Asplenium adamsii Alston	Rare	
BALSAMINACEAE	Impatiens grandisepala Grey-Wilson	Indeterminate	
BALSAMINACEAE	Impatiens letouzeyi Grey-Wilson	Indeterminate	
BURMANNIACEAE	Afrothismia pachyantha Schltr.	Indeterminate	
BURMANNIACEAE	Oxygyne triandra Schltr.	Indeterminate	
CARYOPHYLLACEAE	Silene biafrae Hook.f.	Indeterminate	
CHRYSOBALANACEAE	Acioa cinerea Engl. ex De Wild.	Indeterminate	
COMBRETACEAE	Combretum batesii Exell	Indeterminate	
COMPOSITAE	Bafutia tenuicaulis C. Adams	Vulnerable	
COMPOSITAE	Helichrysum biafranum	Indeterminate	

	Hook.f.		
CYPERACEAE	Scleria sheilae Raynal	Vulnerable	
DIPSACACEAE	Dipsacus narcisseanus Lawalree	Vulnerable	
DIPSACACEAE	Succisa trichotocephala Baksay	Vulnerable	
EBENACEAE	Diospyros barteri Hiern	Vulnerable	Hawthorne, W. 1995. Categories of conservation priority and Ghanaian tree species. WCMC Working Document 4. 1-38 pp.
EUPHORBIACEAE	Amanoa strobilacea Muell.Arg.	Vulnerable	Hawthorne, W. 1995. Categories of conservation priority and Ghanaian tree species. WCMC Working Document 4. 1-38 pp.
EUPHORBIACEAE	Clutia kamerunica Pax	Vulnerable	
EUPHORBIACEAE	Crotonogyne manniana Muell. Arg.	Vulnerable	Hawthorne, W. 1995. Categories of conservation priority and Ghanaian tree species. WCMC Working Document 4. 1-38 pp.
EUPHORBIACEAE	Euphorbia letestui Raynal	Indeterminate	
EUPHORBIACEAE	Pseudagrostistachys africana (Muell. Arg.) Pax & K. Hoffm.	Vulnerable	Hawthorne, W. 1995. Categories of conservation priority and Ghanaian tree species. WCMC Working Document 4. 1-38 pp.
FLACOURTIACEAE	Caloncoba lophocarpa (Oliver) Gilg	Indeterminate	
GRAMINEAE	Eragrostis raynaliana Lebrun	Vulnerable	
GRAMINEAE	Hypseochloa cameroonensis C.E. Hubb.	Indeterminate	
GRAMMITIDACEAE	Xiphopteris villosissima (Hook.) Alston var. laticellulata Benl	Rare	
HOPLESTIGMATACEAE	Hoplestigma pierreanum Gilg	Indeterminate	
ICACINACEAE	Desmostachys vogelii Stapf	Vulnerable	Hawthorne, W. 1995. Categories of conservation priority and Ghanaian tree species. WCMC Working Document 4. 1-38 pp.
ICACINACEAE	Pyrenacantha grandifolia Engl.	Indeterminate	
ICACINACEAE	Stachyanthus cuneatus Sleumer	Indeterminate	
IRIDACEAE	Hesperantha alpina (Hook.f.) Pax ex Engl.	Indeterminate	
LABIATAE	Plectranthus dissitiflorus (Gurke) J.K. Morton	Indeterminate	
LEGUMINOSAE	Craibia atlantica Dunn	Vulnerable	Hawthorne, W. 1995. Categories of conservation priority and Ghanaian tree species. WCMC Working Document 4. 1-38 pp.
LEGUMINOSAE	Didelotia idae Oldem., De Wit. & Leon.	Vulnerable	Hawthorne, W. 1995. Categories of conservation priority and Ghanaian tree species. WCMC Working Document 4. 1-38 pp.
LEGUMINOSAE	Didelotia unifoliolata Leonard	Vulnerable	Hawthorne, W. 1995. Categories of conservation priority and Ghanaian tree species. WCMC Working Document 4. 1-38 pp.
LEGUMINOSAE	Gilbertiodendron pachyanthum (Harms) Leonard	Indeterminate	
LEGUMINOSAE	Julbernardia letouzeyi J-F.Villiers	Rare	
LEGUMINOSAE	Microberlinia bisulcata A. Chev.	Vulnerable	Songwe, C. 1990. Revised preliminary list of timbers of Cameroon with conservation categories.
LEGUMINOSAE	Monopetalanthus hedinii (A. Chev.) Pellegrin	Indeterminate	
LEGUMINOSAE	Pellegriniodendron diphyllum (Harms) J.Leonard	Vulnerable	Hawthorne, W. 1995. Categories of conservation priority and Ghanaian tree species. WCMC Working Document 4. 1-38 pp.
MEDUSANDRACEAE	Medusandra richardsiana Brenan	Indeterminate	
MELASTOMATACEAE	Amphiblemma letouzeyi Jacq.-Fel.	Rare	
MELASTOMATACEAE	Calvoa calliantha Jacq.-Fel.	Rare	

MELASTOMATACEAE	Calvoa stenophylla Jacq.-Fel.	Indeterminate	Aubreville, A. & J.-F. Leroy 1961-1995. Flore du Gabon. Published by the Gabon Government through Museum National d'Histoire Naturelle, Paris. Incomplete; 34 volumes by 1996.
MELASTOMATACEAE	Cincinnobotrys letouzeyi Jacq.-Fel.	Rare	
MELASTOMATACEAE	Warneckea memecyloides (Benth.) H.Jacques-Felix	Indeterminate	Hawthorne, W. 1995. Categories of conservation priority and Ghanaian tree species. WCMC Working Document 4. 1-38 pp.
MORACEAE	Dorstenia poinsettiifolia Engl. var. angularis Hijman & C. Berg	Indeterminate	
MORACEAE	Ficus oreslia C. Berg	Indeterminate	
ORCHIDACEAE	Bulbophyllum teretifolium Schltr.	Vulnerable	
ORCHIDACEAE	Diaphananthe bueae (Schltr.) Schltr.	Vulnerable	
ORCHIDACEAE	Disperis kamerunensis Schltr.	Vulnerable	
ORCHIDACEAE	Gastrodia africana Kraenzl.	Vulnerable	
ORCHIDACEAE	Genyorchis macrantha Summerh.	Vulnerable	
ORCHIDACEAE	Genyorchis platybulbon Schltr.	Vulnerable	
ORCHIDACEAE	Habenaria maitlandii Summerh.	Indeterminate	
ORCHIDACEAE	Polystachya albescens Ridley ssp. angustifolia (Summerh.) Summerh.	Vulnerable	
PODOSTEMACEAE	Inversodicraea musciformis G. Taylor	Indeterminate	
POLYGALACEAE	Polygala tenuicaulis Hook.f.	Indeterminate	
PTERIDACEAE	Pteris barombiensis Hieron.	Rare	
PTERIDACEAE	Pteris ekemae Benl	Rare	
RUBIACEAE	Anthospermum cameroonense Hutch. & Dalziel	Vulnerable	
RUBIACEAE	Pausinystalia brachythyrsum (K.Schum.) W.Brandt	Indeterminate	Stoffelen, P., E. Robbrecht & E. Smets 1996. A revision of _Corynanthe_ and _Pausinystalia_ (African Rubiaceae-Coptosapelteae). _Botanical Journal of the Linnean Society_ 120(4):287-325.
RUTACEAE	Balsamocitrus camerunensis Letouzey	Rare	
RUTACEAE	Oricia trifoliolata (Engl.) I. Verd.	Indeterminate	
RUTACEAE	Oriciopsis glaberrima Engl.	Rare	
RUTACEAE	Vepris heterophylla (Engl.) Letouzey	Endangered	Hawthorne, W. 1995. Categories of conservation priority and Ghanaian tree species. WCMC Working Document 4. 1-38 pp.
SAPINDACEAE	Deinbollia saligna Keay	Indeterminate	Hawthorne, W.D. 1995. Ecological profiles of Ghanaian forest trees. Oxford Forestry Institute. 345 pp. Tropical Forestry Papers 29.
SAPINDACEAE	Lecaniodiscus punctatus J.B. Hall	Endangered	Hawthorne, W. 1995. Categories of conservation priority and Ghanaian tree species. WCMC Working Document 4. 1-38 pp.
SAPOTACEAE	Gluema ivorensis Aubrev. & Pellegrin	Vulnerable	Hawthorne, W. 1995. Categories of conservation priority and Ghanaian tree species. WCMC Working Document 4. 1-38 pp.
SCROPHULARIACEAE	Aeginetia mpomii Letouzey	Rare	
SCYTOPETALACEAE	Rhaptopetalum beguei Mangenot	Vulnerable	Hawthorne, W. 1995. Categories of conservation priority and Ghanaian tree species. WCMC Working Document 4. 1-38 pp.
SELAGINELLACEAE	Selaginella raynaliana Tard.	Rare	
SIMAROUBACEAE	Nothospondias staudtii Engl.	Vulnerable	Hawthorne, W. 1995. Categories of conservation priority and Ghanaian tree species. WCMC Working Document 4. 1-38 pp.

UMBELLIFERAE	Peucedanum camerunensis Jacq.-Fel.	Indeterminate	
VITTARIACEAE	Antrophyum annetii (Jeanp.) Tard.	Rare	
VITTARIACEAE	Vittaria schaeferi Hieron.	Rare	
ZINGIBERACEAE	Aframomum longiligulatum Koechlin	Rare	
ZINGIBERACEAE	Aulotandra kamerunensis Loes.	Rare	

CAMEROON PLANT CONSERVATION CHECKLISTS, THE SERIES SO FAR

MARTIN CHEEK

The forerunner of this book is a series of so far, seven Conservation Checklists for different natural areas in Cameroon. These books have each been produced at the request of different conservation NGOs that are seeking to protect and manage the area concerned.

What is a conservation checklist?

'Normal' plant checklists seek to catalogue all the plant species present in an area, citing specimens as evidence, often with introductory information detailing the history, vegetation types etc. A conservation checklist is so named because it includes a Red Data chapter resulting from screening all the species present for their conservation status. We have used the IUCN (1994) standard in our first two checklists, and the IUCN 2001 standard for later volumes.

1. Cable, S. & Cheek, M. (eds) (1998). The Plants of Mount Cameroon: A Conservation Checklist. Royal Botanic Gardens, Kew, UK. lxxix + 198 pp.

2. Cheek, M., Onana, J.-M. & Pollard, B.J. (eds) (2000). The Plants of Mount Oku and the Ijim Ridge, Cameroon: A Conservation Checklist. Royal Botanic Gardens, Kew, UK. iv + 211 pp.

3. Harvey, Y., Pollard, B.J., Darbyshire, I., Onana, J.-M. & Cheek, M. (eds) (2004). The Plants of Bali Ngemba Forest Reserve, Cameroon: A Conservation Checklist. Royal Botanic Gardens, Kew, UK. iv + 154 pp.

4. Cheek, M., Pollard, B.J., Darbyshire, I., Onana, J.-M. & Wild, C. (eds) (2004). The Plants of Kupe, Mwanenguba and the Bakossi Mountains, Cameroon: A Conservation Checklist. Royal Botanic Gardens, Kew, UK. iv + 508 pp.

5. Cheek. M., Harvey, Y. & Onana, J.-M. (eds) (2010). The Plants of Dom, Bamenda Highlands, Cameroon: A Conservation Checklist. Royal Botanic Gardens, Kew, UK. iv + 162 pp.

6. Harvey, Y., Tchiengué, B. & Cheek, M. (eds) (2010). The Plants of Lebialem Highlands (Bechati-Fosimondi-Besali), Cameroon: A Conservation Checklist. Royal Botanic Gardens, Kew, UK. 170 pp.

7. Cheek. M., Harvey, Y. & Onana, J.-M. (2011). The Plants of Mefou proposed National Park, Yaoundé, Cameroon: A Conservation Checklist. Royal Botanic Gardens, Kew, UK. 252 pp.

Each entry for a species in our conservation checklists is designed to include information useful to those managing the survival of the species. This includes the range of the species as evidenced by specimens, general information including the assessment, often including a short description to facilitate identification. Habitat and threats are also detailed. Management suggestions are the final entry. Here we list recommended actions to facilitate management of the species for its long-term survival.

Over the years improvements has been made. In the first volume, species descriptions in the main checklist were usually reduced to the habit, or absent, as is usual in a checklist. However, beginning with the second volume, we included brief, hopefully diagnostic descriptions, of one to several lines. In the first two volumes only species assessed as threatened were distinguished. In later volumes all species [illegible] IUCN status.

METHODOLOGY

MARTIN CHEEK

As a first step in the preparation of the Red Data book, the Onana ms (see The Evolution of this Red Data Book) was screened family by family and species were ruled out which appeared to have more than 10 locations or which appeared not to have a restricted range, or in which a population reduction of more than 30% could not be demonstrated

1. Data-sources

Primary data sources
For each of the remaining species in each family, the Kew herbarium was searched for specimen records which were databased and georeferenced mainly by Bate Oben and Emma Fenton in 2006–2007. These were later supplemented by secondary data sources- literature and publicly available databases that were trawled through for additional records that could be added to the project database and also to be georeferenced (see below). Pioneers in this work were Tamsyn Thring and Aline Horwath. Meanwhile, at IRAD-National Herbarium of Cameroon, Olivier Sene was employed to do the same there for reliably identified specimens of those taxa that were not present at RBG, Kew.

An Access database was used, known at Kew as the 'Cameroon Database'. Constructed by George Gosline to replace a BRAHMS system used previously, this combined specimen and species database had been used to generate several of the Conservation Checklists (see preceding chapter) that had been published in recent years and so had proven robust, user-friendly and reliable. In 2008, George installed into his database-system Diva software that allowed easy dot-mapping of any chosen taxon, and in 2010 a link to Google Earth allowing the same with satellite imagery.

Family by family, the species treatments that form the bulk of this book were written. The specimen records compiled on the database by Bate and Emma (above) were used to output maps (initially by Bate Oben, later by Katie Bermingham and Tamsyn Thring). These were checked for erroneous placement of location points.

Secondary data-sources.
Genus by genus, literature and specimens were searched for additional information on species that might not have been available to Onana when writing his ms. Sometimes this led to the discovery of enough additional locations for a species to show that it could no longer be considered threatened, or that the range of the species was much wider than had previously been believed, with the same result.

Literature:

Flore du Cameroun (1963-present) compiled principally in Paris was the most useful single literature source for this Red Data Book project. Its 37 volumes are the most authoritative treatments for many important plant families in Cameroon. Crucially, they cite lengthy and comprehensive specimen details for all species included, allowing databasing and georeferencing even where the actual specimens are located in Paris and have not been easily accessible due to building work recently. Sadly many important families in Cameroon still lack Flore du Cameroun treatments, principally Rubiaceae, Euphorbiaceae and Compositae, and some of the earlier treatments are now a little outdated, such as those for Sapotaceae and Leguminosae-Caesalpinoideae.

Flore du Gabon (Fl. Gabon, 1961- present) was compiled in parallel to the Flore du Cameroon at Paris until the early 21st Century when these two projects diverged in management and format. Sadly, specimen data is not included in the new format, reducing its value hugely for the current project. However family treatments to about 2005 keep to the Flore du Cameroun format (see above). This series was therefore extremely useful in providing specimen data on species threatened in Cameroon which extend into Gabon, as is often the case for rare species of S.Region. To a large extent the utility of the Fl. Gabon series for our project was superseded by the Gabon database (see below) when that became available to us in 2009.

Flora of West Tropical Africa (FWTA), the second edition (1954-1972) is still the world's only completed Tropical Regional Flora although now a little outdated. Compiled principally at Kew, it is useful for covering the SW and NW Regions of Cameroon (formerly Southern British Cameroon) and also for providing data on locations in Nigeria, and further West, for the many threatened species that occur not just in western Cameroon but in those countries also. Being of concise format, no more than three specimen citations are given per country for each taxon, but for rare species this is not usually a difficulty.

Thomas *et al.*, 2003 (Tree species of Southwestern Cameroon; Center for Tropical Forest Science, Smithsonian Tropical Research Institute, Washington) documents the c. 492 tree species that occur in the Smithsonian 50 ha plot in South–Central Korup National Park in SW Region. Many of these are threatened. This book contains authoritative taxon records but sadly lacks specimen data. Importantly, it provides the best population data in Cameroon for each included species. For those threatened species of tree occurring within the plot, the Smithsonian project provides the reassurance that they will probably be the best protected and best monitored of all those in Cameroon.

Tchouto (2004) Plant Diversity in a Central African Rain Forest. Ph.D. thesis, Dept. Plant Sciences, Uni. Wageningen. This volume is focussed on the plant species of the Campo-Ma'an National Park in S Region, especially those which are threatened,. Among the many sensational achievements by Tchoutou in the course of this work was the rediscovery of the long-lost enigmatic tree genus of the Euphorbiaceae, *Afrotrewia*. Several new, narrowly endemic and Threatened species came to light from Tchouto's surveys, such as the bizarre *Cassipourea alternifolia* Breteler and the notable *Tapura tchoutoi* Breteler. Regrettably, full specimen data are omitted from this work, and tracking down the specimens which underpin the results presented is difficult. Fortunately most species are given locations allowing databasing and georeferencing. While some specimens are at WAG, the majority are thought to be at the old TROPENBOS herbarium at Kribi. Therefore it has not been possible to verify these records, although most must be beyond doubt, having been identified at WAG. Tchouto's book is critical in providing data on rare plants in one of Cameroon's most important locations for conservation of plant diversity. He gives details of 141 species which are either endemic or near endemic to the broader Campo-Ma'an area.

Flore D'Afrique Centrale, formerly the Flore du Congo Belge et du Ruanda-Urundi (1948- present) is the highly authoritative series of volumes covering most of the vascular plant families of the immense Congo (Kinshasa) currently known as Democratic Republic of Congo or DRC, and formerly Zaire. Rubiaceae is the single most important family that is still lacking. This excellent work cites detailed specimen data for each species, allowing databasing and georeferencing even although the specimens concerned are often located only in the Meise Herbarium (BR) near Brussels, or at Yangambi. This work has been the major source of Congo [illegible] Cameroonian species that extend [illegible]

Flora of Tropical East Africa (FTEA, 1949-present) has been useful for the small percentage of threatened or potentially threatened species of Cameroon that are widespread in Tropical Africa, such as Umbelliferae and Apocynaceae-Asclepiadaceae. Compiled at RBG, Kew, it includes a list of all countries in Tropical Africa in which each species occurs, in addition to detailed specimen reference data for up to four records for each country within its coverage (Uganda, Kenya and Tanzania).

Exell (1973) Angiosperms of the islands of the Gulf of Guinea. Bull. Mus. (Nat. Hist.) Botany 4(8): 325–411. This checklist covers all the Gulf of Guinea islands and cites basic specimen records allowing mapping of those rare Cameroonian species which extend into the Gulf of Guinea. Equatorial Guinea (Bioko (Fernando Po)) in particular, shares a large number of rare species with SW Region Cameroon.

Electronic data sources

Late in 2009, Marc Sosef, Head of Herbarium Vadense (WAG) at Wageningen University, Netherlands, provided a link (http://dps.plants.ox.ac.uk/bol/Gabon/Search/Fullsearch?searchType=Collection) to the Gabon database. Wageningen, long-partnered with the National Herbarium of Gabon (LBV) have brought about a spectacular and very useful achievement in using Denis Filer's BRAHMS in databasing and georeferencing 90% of specimens ever collected in Gabon. This facility is regrettably not available for any other country in Tropical Africa. Using this link it was possible to see if additional records in Gabon were available for any of our potentially threatened species. Gabon shares a border with Cameroon and is c. 90% covered with evergreen forest – the vegetation type in Cameroon which is most rich in threatened species. Despite their immediate proximity, there are great differences between the species of Gabonese and Cameroonian forest, particularly as far as Red Data species are concerned. However, there are also many cases of rare species being common to both countries, so the availability of the Gabon database has been very welcome. Records from the Gabon database were not added to our baseline data for the Cameroon Red Data book uncritically. Records for families such as Dichapetalaceae, Connaraceae, Apocynaceae, Begoniaceae and Loganiaceae were added without question since researchers at WAG have led the world in monographing these. Regarding other families, these were added on a case-by-case basis, and, when records were considered to be questionable, this has been raised in the appropriate species treatment, whether the record was mapped or not. This procedure was also used for records from TROPICOS, the publicly available database the Missouri Botanical Garden (MO), USA, which hosts the records of Duncan Thomas who was employed by MO up to about 1987 and who made many important contributions to Cameroonian plant knowledge. An example of such a questionable record might be one identified by a non-specialist from lowland Gabon or Cameroon, geographically distinct from a species otherwise resticted to the high altitudes in the Cameroon Highlands. Another example might be a specimen of a taxon viewed and found to be morphologically discordant with authenticated specimens.

2. Global Conservation Assessments Of Species Data

A concise summary of the IUCN 2001 Criteria for assessing threatened species is given in table form at the end of this chapter, reproduced from that devised by the IUCN Cambridge office, with permission.

Criterion B

Because of the nature of the data available, most conservation assessments of plant species have employed Criterion B, which depends on geographic range, principally the number of locations for a species, supported with data on either or both of AOO (area of occupancy) and EOO (extent of occurrence) together with information on direct threats to the species.

Location records are based on verifiable specimen records since observation-based (sight) records are unreliable in a country where there is no comprehensive illustrated species identification reference work (in contrast with the situation for bird species) and where the total number of species is so large, approaching 10,000 species. This is the position in most tropical countries. Over time the definition of a 'location' for IUCN assessment purposes has both changed and solidified. This has been reflected in the assessments in our conservation checklists 1998–2011. Initially we counted as separate locations any specimen records separated by more than one or two kilometres. However, in our more recent conservation checklists (Cheek *et al.* 2010, Harvey *et al.* 2010) we have followed the more detailed and precise IUCN (2003) guidelines and similar recent IUCN guidance in a broader definition of location which is defined not as a point, but as an area where a single common threat might destroy all individuals present. For example in the past, three widely geographically separated specimens in a forest reserve might have been treated as three separate locations, but now they would be treated as three sites at a single location since, should the reserve be granted as a logging concession, all three are equally at risk of being eliminated. It must be noted that not all logging activity necessarily poses a high level of threat to all rare plant species. Nonetheless, this broader definition of location has resulted in several species that we had assessed in our earlier checklists as Vulnerable, now being assessed as Endangered.

Criterion B2: AOO (area of occupancy) implies the actual area occupied on the ground by a species, calculated by combining the areas occupied at all known locations. In many cases this can be difficult to record unless a labour-intensive survey is executed just for this purpose for a particular species. In other cases, such as *Eriocaulon parvulum*, where all locations have been recorded by us, and where the species is restricted to a particular microhabitat, easily located and measured on the ground (seasonally wet, seeping basalt slopes), we can calculate this actual AOO quite accurately. In many cases in our earlier conservation checklists, we judged that using an arbitrary 1 km^2 cell per location was a reasonable upper limit for the actual area of occupancy. A third approach involves calculating the AOO of a species from the GIS- logarithmic approach of Willis *et al.* (2003) (see below). Instead we chose to adopt a 4 km^2 cell size simply because it is currently firmly favoured by IUCN (IUCN 2003). In practice, in many cases, this allows AOO to be calculated by multiplying the number of locations by this cell size. Exceptions occur when, in one broad location, numerous sites are recorded which if plotted, would occupy more than a single 4 km^2 grid cell. Further, aquatic species, recognised by IUCN as a special case, have been treated with a 1 km^2 cell as recommended by IUCN.

Criterion B1: EOO was extracted (where available) from the automated assessments generated using Willis *et al.* 1997 methodology by Steve Bachman (see below) and then inserted in our species treatments, often allowing species previously assessed using B2 to be attributed B1 status also

Threats under Criterion B. To qualify for Criterion B, in addition to the geographic factors of location and/or AOO/EOO being met, direct threat must also be shown to the species being assessed. In assessing and ascribing these threats, we, the authors, depended mainly on our personal knowledge of factors on the ground during years of fieldwork in different regions of the country, and from information from contacts established during that time, and also from data presented at the IUCN workshop in Yaoundé in July 2004 to evaluate conservation assessments of aquatic organisms, which we both attended. At that workshop data on threats was compiled and presented by co-ordinators from IUCN and WWF. It was intended that this might be published as a 'Threat Map' and be included in this volume. Sadly, this has not yet come to pass but hopefully it will be possible in future editions.

Species with four locations and/or low AOO/EOO yet without evidence of direct threats do not satisfy Criterion B and must be assessed instead under Criterion D (see below).

GIS-based automated conservation assessment. Willis *et al.* (2003) http://www.springerlink.com/content/m7588r021

2585577/ sets out a GIS-based system that allows automated production of provisional conservation assessments by plotting site data, from which the EOO is calculated, and, using an algorithm, the AOO, thence the provisional conservation assessment using Criterion B. More information about this system is available at Moat, J. (2007). Conservation assessment tools extension for ArcView 3.x, version 1.2. GIS Unit, Royal Botanic Gardens, Kew. Available at: http://www.rbgkew.org.uk/gis/cats. The assessments are provisional since they do not take into account threats. The virtue of this system is that once all specimens of a species have been databased and georeferenced, many hundreds of fully objective provisional assessments can calculated rapidly, and far less laboriously than by any other for any method. The system does have limitations. It is restricted to Criterion B. Taxa with just one or two locations cannot be dealt with since a polygon cannot be constructed. Local knowledge of threats on the ground is needed before the assessment can be converted from provisional to final. All taxa in this Red Data book were processed through this GIS system by Steve Bachman and his assistant Charlotte Cole. All EOOs included in our species treatments derive from their work.

Criterion A . Population reduction. This criterion is used where evidence of population loss is available, whether in the past, present, future or a combination of these (subcriteria A1-A4). Most commonly for Cameroonian species criterion A2c is used at the Vulnerable level, since reduction of more than 30% within three generations (up to 100 years) can reasonably be argued in two cases:
Cases where more than 30% of the locations of a specimen have been destroyed. For example a tree species known only from six historic locations, Libreville, Kumba and four locations in S Region, can be inferred to have lost one-third (33.3%) of it population due to vast urbanisation in the last century at Libreville and Kumba, especially if there are no recent records from those locations despite their being important centres for botanical surveys and research in recent decades.

Cases where a species is restricted to a particular habitat which has seen reduction, where this reduction has been quantified. For example there are two GIS studies using satellite imagery in the Bamenda Highlands section of the Cameroon Highlands, which had already seen an estimated 96% loss of its original submontane and montane forest (Cheek *et al.* 2000: 49–50). The first study (Moat in Cheek *et a.l* 2000: back cover) [illegible] surviving submontane/montane forest [illegible] Ijim Ridge part of the Bamenda Highlands between January 1987 and January 1995 (eight years). The second study shows a c. 50% reduction in forest habitat further N in the Bamenda Highlands, during the 15 years between Feb. 1988 and Jan. 2003 (Baena in Cheek *et al.* 2010: back cover). Thus a submontane forest tree species which is restricted to a section of the Cameroon Highlands which includes the Bamenda Highlands, and which can be shown to be more or less equally frequent throughout its range, can be logically argued, by averaging, to have lost more than 30% of its population in the last 100 years. This is because losses in the Bamenda Highlands are well documented due to the studies referred to even if populations of the species in other parts of the Cameroon Highlands, such as Mt Kupe, are relatively secure. *Allanblackia gabonensis* is an example of such a submontane tree species.

It is to be hoped that more studies quantifying forest habitat loss over time will be carried out in other parts of Cameroon in future, so that Criterion A can be used more widely to assess species.

Criterion C Low population size with threats. This criterion has been used for only a small proportion of the species in this Red Data Book because it requires firstly, detailed knowledge of the numbers of mature individuals of a species combined with precise information on losses over time, or the breakdown between subpopulations of the numbers of individuals. In future, especially if basic population surveys are executed as has been recommended for most species in this book, and if monitoring takes place at yearly or 5-yearly intervals, as also advocated, then data will become available enabling increased use of this criterion.

Criterion D Very low population size and (VU D2) low AOO or numbers of locations, in the absence of direct threats.
This Criterion is the only one of those considered so far which does not need demonstrated a direct threat or reduction of the population of a species in order to qualify for threatened status. Logically, it requires much lower thresholds for numbers of mature individuals for a species to qualify (e.g. less than 50 individuals for CR status).

In this book CR D and EN D status has been invoked in few cases, especially in Leguminosae-Caesalpinioideae, where there is data on numbers of individuals of a species. Often this pertains to recently discovered species where numbers of individuals seen at each known location have been recorded or have been obtained by questioning the botanist-collector. As with criterion C, it is likely in future that this Criterion will be used more frequently if surveys of basic population data are executed as has been extensively advised throughout this book.

VU D2 has been used throughout this book for species that appeared to qualify under Criterion B (see above) on the basis of low AOO and numbers of locations but for which no direct threat is so far known. Examples are species restricted to the summit areas of Mt Cameroon or Mt Kupe which may be very rare but, for the present, face no threat from man.

From time-to-time rare species were uncovered that did not appear in the Onana 2006 ms. These were added to the database for assessment.

A fair proportion of the species treated below, about one third, had already been assessed for their IUCN status in our 'Conservation Checklist' series of books for different parts of Cameroon (1998-present), see preceding chapter. These treatments were reviewed. In many cases they needed revision before they could be included here. In some instances sufficient additional locations had since come to light that the species no longer qualified as threatened but might instead be rated as NT (Near Threatened). Such cases are noted in the appropriate family introduction. In other cases new data, or a re-assessment of re-existing data, necessitated an increase in the threat assessment. Thus a species previously assessed as VU (Vulnerable) might become EN (Endangered). In all cases (unless overlooked in error), earlier assessments of the taxon are referred to in the species treatment concerned, and the reasons for the change (if any) noted.

THE 2001 IUCN RED LIST CRITERIA (VER. 3.1) FOR CRITICALLY ENDANGERED, ENDANGERED AND VULNERABLE

A. Reduction in population size

A. Reduction in population size	Critically Endangered (CR)	Endangered (EN)	Vulnerable (VU)
A1. An observed, estimated, inferred or suspected population size reduction of ...			
	≥90%	≥70%	≥50%
over the last 10 years or 3 generations whichever is the longer, where the causes of the reduction are: clearly reversible AND understood AND ceased, based on (and specifying) any of the following: (a) direct observation (b) an index of abundance appropriate for the taxon (c) a decline in area of occupancy, extent of occurrence and/or quality of habitat (d) actual or potential levels of exploitation (e) the effects of introduced taxa, hybridisation, pathogens, pollutants, competitors or parasites.			

A. Reduction in population size	Critically Endangered (CR)	Endangered (EN)	Vulnerable (VU)
A2: An observed, estimated, inferred or suspected population size reduction of ...			
	≥80%	≥50%	≥30%
over the last 10 years or three generations, whichever is the longer, where the reduction or its causes may not have ceased OR may not be understood OR may not be reversible, based on (and specifying) any of the following: (a) direct observation (b) an index of abundance appropriate for the taxon			

(c) a decline in area of occupancy, extent of occurrence and/or quality of habitat (d) actual or potential levels of exploitation (e) the effects of introduced taxa, hybridisation, pathogens, pollutants, competitors or parasites.

A. Reduction in population size	Critically Endangered (CR)	Endangered (EN)	Vulnerable (VU)
A3: A population size reduction of ….			
	≥80%	≥50%	≥30%
projected or suspected to be met within the next 10 years or 3 generations whichever is the longer (up to a maximum of 100 years), based on (and specifying) any of the following: (b) an index of abundance appropriate for the taxon (c) a decline in area of occupancy, extent of occurrence and/or quality of habitat (d) actual or potential levels of exploitation (e) the effects of introduced taxa, hybridisation, pathogens, pollutants, competitors or parasites.			

A. Reduction in population size	Critically Endangered (CR)	Endangered (EN)	Vulnerable (VU)
A 4. An observed, estimated, inferred, projected or suspected population size reduction of ..			
	≥80%	≥ 50%	≥30%
over any period of.10 years or 3 generations whichever is longer (up to a maximum of 100 years), where the time period includes both the past and the future, and where the decline or its causes may not have ceased OR may not be understood OR may not be reversible, based on (and specifying) any of the following: (a) direct observation (b) an index of abundance appropriate for the taxon (c) a decline in area of occupancy, extent of occurrence and/or quality of habitat (d) actual or potential levels of exploitation (e) the effects of introduced taxa, hybridisation, pathogens, pollutants, competitors or parasites.			

B: Geographic range

B: Geographic range ...	Critically Endangered (CR)	Endangered (EN)	Vulnerable (VU)
... in the form of either B1 (extent of occurrence) OR B2 (area of occupancy) OR both:			
B1: Extent of occurrence estimated to be (km^2), and estimates indicating any two of a-c:	<100	<5,000	<20,000
B2: Area of occupancy estimated to be (km^2), and estimates indicating any two of a-c:	<10	<500	<2,000
a: Severely fragmented or known to exist at	only 1 location	≤ 5 locations	≤ 10 locations
b. Continuing decline, observed, inferred or projected, in any of the following: (i) extent of occurrence (ii) area of occupancy (iii) area, extent and/or quality of habitat (iv) number of locations or subpopulations (v) number of mature individuals. c. Extreme fluctuations in any of the following: (i) extent of occurrence (ii) area of occupancy (iii) number of locations or subpopulations (iv) number of mature individuals.			

C: Population size

C: Population size ...	Critically Endangered (CR)	Endangered (EN)	Vulnerable (VU)
... estimated to number fewer than (mature individuals) and either	<250	<2,500	<10,000
C1. An estimated continuing decline of at least	25%	20%	10%
in (years)	3	5	10
or (generations)	1	2	3
whichever is longer (up to a maximum of 100 years in the future) OR			
C2: A continuing decline, observed, projected, or inferred, in numbers of mature individuals AND at least one of the following: (a-b)			
a) Population structure in the form of one of :			
(i) no subpopulation estimated to contain more than (mature individuals), OR	50	250	1,000
(ii) at least (%) of mature individuals are in one subpopulation	90%	95%	All (100%)
(b) Extreme fluctuations in number of mature individuals.			

D: Population size

D1: Population size	Critically Endangered (CR)	Endangered (EN)	Vulnerable (VU)
estimated to number fewer than (mature individuals)	<50	<250	<1,000

VU D2. Population with a very restricted area of occupancy (typically less than 20 km^2) or number of locations (typically 5 or less) such that it is prone to the effects of human activities or stochastic events within a very short time period in an uncertain future, and is thus capable of becoming Critically Endangered or even Extinct in a very short time period

E: Quantitative analysis

E. Quantitative analysis	Critically Endangered (CR)	Endangered (EN)	Vulnerable (VU)
showing the probability of extinction in the wild is at least	50%	20%	10%
within (years)	10	20	100
or (generations)	3	5	-
whichever is the longer (up to a maximum of 100 years).			

LIMITATIONS: FUTURE WORK NEEDED

MARTIN CHEEK

This Red Data book is very far from perfect, we realise. Here we give details of some important shortcomings that we hope will be addressed in future editions.

1. Limitations of descriptions and illustrations. Ideally, each species treatment would have an illustration, description, and supporting notes detailed enough to enable unambiguous identifications of the species in the field. Unfortunately, due to time constraints, some species descriptions are overly long and technical (usually taken directly from their protologues) while others are overly short (often taken from a conservation checklist). In some cases, such as Orchidaceae, the author, Benedict Pollard, ran short of time and there are no descriptions. Instead the reader is referred to the Flore du Camer[illegible] family. 'A picture speaks a thousand [illegible] threatened species have never [illegible] few photographs exist of them. To commission illustrations for each threatened species was beyond the scope of this project and would have cost more than a million euro. Similarly, the re-use cost of including existing illustrations within copyright has restricted the degree to which such figures could be included. We hope that future editions of this book will be better illustrated.

2. Limitations of mapping accuracy. Georeferencing of historic specimens included in this work is often accurate only to within a ± 5–10 km radius. In order to complete georeferencing of specimens expeditiously, historic specimens, for example those from the German period (c 1890–1914), which often only give the name of the nearest town, (e.g. Zenker: Bipindi) were referenced to the town centre. Vastly more time would have been needed to research precisely exactly where the collector might have obtained the specimen. In future this should be addressed. One approach might be to reference such specimens to areas with natural habitat nearest to such towns, as observed with aerial/satellite imagery. However this would involve making some assumptions as to where the collections were actually made.

3. Limitations of taxonomic impediments. Many plant groups were not assessed, or only partly assessed, because there is no means to assess which of their species are 'real' and which are not, or no means to reliably identify plants as belonging to a particular species. This taxonomic impediment has usually been caused by the lack of a Flore du Cameroun account, and/or, the absence of any revision of the group for Tropical Africa. It is hoped that this impediment will be reduced by the publication of more such works in future.

4. Restrictions of groups included. This Red Data book does not cover all plant groups. Ferns, bryophytes, lichens and algae, have not been included. Currently neither Kew nor the National Herbarium of Cameroon has specialists to provide expert identifications in these areas. It is hoped that other researchers at other institutes will be able to fill this gap.

5. Infra-specific taxa have been omitted from this book in order to reduce costs. The exceptions are such taxa that had already been assessed as threatened in our conservation checklists (1998–2011). IUCN (2001, 2003) rules are that infra-specific taxon assessments are only accepted if the species itself has also been assessed. In future editions it is hoped to address this.

6. Limitations due to lack of surveys. Undoubtedly many additional threatened species to those in this book, both new to science, and new to Cameroon, will come to light when unsurveyed or under-surveyed areas of Cameroon have been visited by botanists. These areas can be identified from studying the maps in Onana 2011 (in press) which map the location of all collections at the National Herbarium of Cameroon.

7. GIS habitat-loss studies. More assessments could be made under Criterion A if the shortage of GIS-based vegetation change studies could be addressed. Further studies of habitat loss over time, such as those by Moat in Cheek *et al.* 2010, and Baena in Cheek *et al* 2010, are needed, especially in key areas in the forest belt of Cameroon.

8.Scarcity of population data. For many threatened species we do not know how many individuals are present at a location since it is not recorded on the specimen label. Ground surveys to collect basic population data have been recommended in most of the species treatments in this book, in part so that in future more assessments will be possible under criteria C and D. Further benefits of collecting such data are that they will form a baseline for future monitoring, and better management planning for the survival of the species in the wild.

ANALYSIS OF RESULTS

MARTIN CHEEK

1. Statistical summary

815 Cameroonian vascular plant taxa have been treated in this book as being globally threatened according to IUCN (2001). This amounts to 10.38% of the 7850 indigenous and naturalised taxa recognised for Cameroon by Onana (2011). This percentage could be higher but for the following factors:

A. The assessments in this book are done on a global basis and not on a purely national or regional basis as is often the case in Red Data books.
B. Some plant groups included in Onana (2011), such as Pteridophytes, were not assessed in this volume.
C. Due to the taxonomic impediment caused by a lack of revisions and/or Flora accounts for identifying several major groups such as Compositae, many taxa could not be assessed reliably and so were not treated as threatened.
D. Due to lack of data on habitat loss in many areas of Cameroon, the opportunity to assess species using Criterion A was diminished, so that some species that might qualify as threatened were not so treated.

It is no surprise that in the main, the families with the greatest number of species in total also have the highest number of threatened species, at least in the top five cases of families with the greatest number of species (compare Table 1 with Table 2.). However some smaller families with exceptionally high proportions of threatened species such as Podostemaceae also figure in the top 10.

Table 1. The top 10 families, having the greatest number of threatened species in Cameroon. Total family species numbers derive from Onana 2011.

	Family	**Number of Threatened Species/Total No. Species**
1	Rubiaceae	136/789
2	Orchidaceae	112/436
3	Leguminosae	71/791
4	Acanthaceae	34/184
5	Euphorbiaceae	30/171
6	Podostemaceae	29/39
7	Melastomataceae	29/116
8	Sapindaceae	24/85
9	Celastraceae	21/112
10	Begoniaceae	20/50

Table 2. The top 10 families, having the greatest number of species in Cameroon. Numbers derive from Onana 2011.

	Family	**Total number of species**
1	Leguminosae	791
2	Rubiaceae	789
3	Gramineae	479
4	Orchidaceae	436
5	Compositae	257
6	Acanthaceae	184
7	Lamiaceae	174
8	Euphorbiaceae	171
9	Apocynaceae	143
10	Sterculiaceae	142

The proportion of threatened species within a family varies widely from family to family. In some families it is far lower than the average of 10.38%. For example , in Combretacae, in which most species occur in drier, semi-deciduous forest, only two of the 62 species in Cameroon are treated as threatened (3.2%). Some well-known families, such as Irvingiaceae, appear to have no threatened species at all. Conversely, other families have a high proportion of threatened species. These are tabulated below (Table 3) excluding families with less than ten species in Cameroon. Top of the list is Podostemaceae, restricted to rapids and waterfalls, with many species restricted to less than five locations, often on a single river, and often threatened by hydro-electric projects. The families listed, being rich in threatened species, should be prioritised and targeted in surveys of unknown areas, if species of conservation priority and indicators of high conservation value, are being sought.

Table 3. Top 10 families having the highest proportion of threatened species excluding families with <10 species in Cameroon. Total number species figures derive from Onana (2011)

	Number threatened species	Total number species	% of threatened species
Podostemaceae	29	36	80.5
Burmanniaceae	9	16	56
Myrsinaceae	16	33	48
Begoniaceae	20	50	40
Balsaminaceae	8	25	32
Scytopetalaceae	5	26	31
Sapindaceae	24	85	28
Menispermaceae	9	34	26
Apocynaceae-Asclepiadaceae	15	65	23
Myrtaceae	6	26	23

The taxonomic impediment is still high for Cameroon, and has undoubtedly skewed our results. Family rankings for numbers and proportions of threatened species will undoubtedly change once Flore du Cameroun accounts or Tropical African revisions are available for those many families that currently lack them. Those families most in need of taxonomic revision for Cameroon, so that their species can be properly assessed for conservation status are listed in Table 4. Once these impediments are removed, additional threatened species will undoubtedly be revealed for Cameroon.

Table 4. Families in Cameroon most in need of taxonomic research so that their species can be properly assessed for conservation status. Species numbers are from Onana 2011.

Euphorbiaceae 171	Violaceae 58
Compositae 257	Lauraceae 47
Leguminosae-Papilionoideae 472	Myrsinaceae 33
Cyperaceae 217	Sterculiaceae 142
Labiatae 174	Zingiberaceae 50

HOTSPOTS OF THREATENED SPECIES IN CAMEROON

MARTIN CHEEK, JEAN MICHEL ONANA & STEVE BACHMAN.

Locations within Cameroon for all 815 species treated as threatened in this book were plotted on a map of Cameroon and are displayed on a quarter degree square grid (see map on rear cover and caption on history page). The sum of threatened species for each quarter degree grid cell (approx. 28 km × 28 km) was calculated and is displayed as a red dot at the centre of each cell using a proportional scale. Owing to the scale of the grid, allowance has to be made for the slight mismatch between the actual location of threatened species, and the red spot that represents them. In an extreme case, such as Mone FR in SW Region, four cells appear to meet in the centre of the reserve, which is approximately 25 km × 25 km in size, so that the threatened species of the reserve are all mapped as spots just outside the four corners of the Reserve and none are mapped in the reserve itself!

Spot size correlates directly to the number of threatened species per cell, whether 1, 10–49, 50–99 or 100+, using a continuous scale. Threatened species occur in every Region of Cameroon but the highest concentrations, represented by the larger spots, are not evenly distributed (see rear cover). Protected areas have been superimposed on the map so that density of threatened species can be visualised along with the distribution of protected areas.

Macro-Hotspots of >100 Threatened species

That South West Region has the greatest concentrations of threatened species in Cameroon is to be expected since it also has the highest diversity of plant species per degree square in Tropical Africa (Barthlott, W., Lauer, W., & Placke, A. (1996). Global distribution of species diversity in vascular plants: towards a world map of phytodiversity. Erkunde Band 50: 317–328). The South West Region also has Tropical Africa's most species-diverse centres of plant diversity, both with more than 2300 species (Mt Cameroon, Cable & Cheek 1998, and Kupe-Bakossi, Cheek *et al.* 2004). Therefore it is no surprise that five of the six Cameroonian cells with more than 100 threatened species are to be found in SW. Two of the cells correspond to the lower slopes of Mt Cameroon and two to the Kupe-Bakossi area. The fifth corresponds to the southern part of Korup National Park. All three of these locations have one or more protected areas, but the correlation between the location of the threatened species and the location and status of the associated protected area is variable;

Korup National Park. In this case there is an excellent match. Most of the threatened species occur inside the secure boundaries of this well-managed and world-renowned National Park

Mt Cameroon. Although a National Park has been newly created (2010), it mainly occupies the high altitude areas and does not cover any of the areas immediately around the massif, which host most of the threatened species of the mountain. These areas are incompletely covered by a network of Forest Reserves set up for timber production, but some have been partly converted to agriculture, a particular threat since due to volcanic activity on Mt Cameroon, soils are often fertile. [illegible] have already been [illegible] rubber and oil palm (see Cable & Cheek 1998).

Mt Kupe and the Bakossi Mts; the Bakossi National Park includes only a fraction of the Bakossi Mts. Mt Kupe, with its many unique species, has no formal protection but depends on the protection of the communities that surround it. The Mt Kupe cell has the greatest concentration of threatened species in Cameroon.

South Region, Cameroon; Bipindi and the Ngovayang Massif. This is the only location outside SW with a cell with > 100 threatened species. There is no protected area in the vicinity.

Meso Hotspots of 50–100 threatened species

These hotspots encompassing 50–100 threatened species are also concentrated (6 spots) in SW Region, adjoining the macro hotspots (above), but also occur in S Region (two spots), C Region (1 spot), W Region (1 spot) and NW Region (3 spots).

South West Region:

1. Bimbia-Bonadikombo (formerly Mabeta-Moliwe)-eastern foothills of Mt Cameroon
2. Lake Barombi Mbo (Kumba) and S Bakunda FR-northeastern outliers of Mt Cameroon
3. Mokoko FR northwestern outliers of Mt Cameroon
4. & 5 W Bakossi (N & S)
5. Bakaka FR

South Region:

1. Campo Ma'an National Park
2. Lolodorf-Ngovayang area

Central Region:

1. Yaoundé area; inselbergs with submontane forest

West Region:

1. Bamboutos Mts

North West Region:

1. Bali Ngemba Forest Reserve
2. Ijim Ridge
3. Mt Oku

Micro hotspots (<50 threatened species)

These are scattered very thinly through most regions except Extreme North but are moderately dense in SW, NW, S, C and Littoral Regions. They are not detailed individually here, being so numerous.

No correlation: Red Data plants and animals?

Most National Parks were set up for big animals, not for the conservation of threatened plant species, so it is not surprising that there is almost no correlation between areas of documented importance for the conservation of Red Data plant species and National Parks in most of Cameroon (see rear cover of this book where hotspots and protected areas are mapped). The exceptions are those protected areas in the wet evergreen, mainly lowland forest areas towards the coast. Since this habitat is so rich in narrowly distributed plant species, it is to be expected that National Parks in these areas, while originally set up for animals, are also important for plant conservation: Korup National Park, Campo-Ma'an National Park, and, the exception in being set up mainly on the basis of plants, the Bakossi National Park.

THE WAY FORWARD: PLANT SPECIES CONSERVATION IN CAMEROON

MARTIN CHEEK

For the first time, we have identified, described and mapped most of Cameroon's threatened plant species. We now need to work out how to protect and manage them so that they survive for future generations. If this is not done, they risk being accidentally destroyed: many are known from just a single location. Over time, new uses are being found for plant species previously unknown to have any use whatsoever. If species become extinct, such uses cannot be discovered or exploited for the benefit of Cameroonians or for the world.

It is to be hoped that despite its limitations, this book will form the basis, or a key reference for, a future Cameroon National Plant Conservation Strategy, and that it will promote a review of the existing protected area network, so that it better protects globally threatened plant species.

Protected areas: creating new and uprating old.
Protection of wild plant species is carried out around the world by creating protected areas such as National Parks. There is no doubt that creating more such parks, and upgrading some production forest reserves to National Park status, would be a positive move to enhance the survival of threatened species. It is recommended that all six of the super hotspots (>100 threatened species per 25 km × 25 km) and 13 hotspots (50–99 threatened species) identified in the previous chapter are adequately protected in this manner, where they are not already.

The single largest gap in protecting threatened plant species, as revealed in this project, is the lack of a protected area at the super hotspot that is the Ngovayang Massif between Bipindi and Lolodorf in S Region. We strongly recommend that this is addressed by the creation of a new National Park at this location.

We also recommend that the newly created Bakossi and Mt Cameroon National Parks be extended to include additional sites for threatened plant species, and that several Forest Reserves and community forests which are hotspots for threatened species are considered for upgrading to more fully protected areas such as National Parks. These are: Bimbia-Bondikombo; Lake Barombi Mbo and S Bakundu FR; Mokoko FR; Bakaka FR; inselbergs in the Yaounde area such as at Mt Eloumden or N'kolbison and Ototomo Reservoir; Bamboutos Mts; Bali Ngemba FR; Mt Oku and the Ijim Ridge.

Micro-protected areas for plants, and community management.
There are fundamental differences between the needs for protected areas for plant versus large animals. Protected areas for individual plant species do not need to be as vast as those that have been created for individual large animal species which often require very large ranges for groups of families. In contrast, individual plants do not move (of course) and rare species often have quite small populations. Extreme examples are to be found in Cameroon with the family Burmanniaceae/Thismiaceae in which each of the threatened species occupies no more than a few metres in total. So long as these few m^2 are protected, together with enough of the surrounding forest to maintain the microclimate, those species can be considered reasonably secure. However, if these few m^2 of forest are cut, burned, and cultivated, as can happen all to easily with slash and burn and other forms of agriculture, those species can easily become globally extinct, as has almost certainly happened already with *Oxygyne triandra* (Burmanniaceae).

Even gregarious tree species which occupy much larger areas than do the Burmanniaceae mentioned, need much smaller protected areas than do large animals. In Malaysia, remnant populations of rare dipterocarp timber species have recently been formally protected in Parks of only 3 km^2 in extent, apparently adequate to protect the continued genetic viability of the species (P. Ashton pers. comm. 2011). Therefore to protect Cameroon's large numbers of threatened plant species, large numbers of micro-national parks, or plant sanctuaries, equivalent to Sites of Special Scientific Interest (SSSIs) in the UK, might be appropriate. These might correspond with the any currently unprotected minor and meso hotspots (see preceeding hot spot chapter) as mapped on the rear cover of this book.

There are alternatives to formal, government-managed protected areas. Cameroon already has pioneered Community Forests, officially recognised protected areas proposed and managed by local communities. Perhaps the best model for protecting sites with small areas with threatened plant species or even individual threatened trees, outside national parks, might again be direct local management, either through community elders or through community-recognised individual land-owners. In some cases incentives might be appropriate to provide motivation for continued protection. In the Bakossi area we have pioneered such a scheme over the last ten years or more by renting small areas of forest from local land managers, with globally threatened species that would otherwise be cleared to cultivate cocoa, coffee or pineapples.

Monitoring Sites and Populations of Threatened Plant Species in Cameroon

If Cameroon's 815 threatened (Red Data) plant species are to survive long-term, it is important that both their populations and the sites that contain them are monitored on a regular basis. This practice is observed in many countries outside of Africa. By this means, authorities in Cameroon, and thence the World, can be assured of the survival of such species, and if trends are negative, consider taking action to avoid extinction of the species and to assist it to proliferate again, by appropriate management.

The following pages contain site and population monitoring advice (Annex A) and monitoring forms and supporting notes (Annex D) together with a population location form (Annex C). These forms are intended to offer assistance to those in cameroon who are responsible for or interested in monitoring sites and populations of globally threatened plant species; whether Ministry of the Environment staff, NGO staff and volunteers, Community Forest Management committees or National Herbarium staff.

The basic population data that would be gathered when using this system in the first instance would form the baseline for all future monitoring of the populations of threatened species at the sites surveyed.

If any item on the forms proves difficult to record or grasp, the recorder should move to the next item and gather what data is available at the time.

The forms, advice and notes given below have been modified to suit Tropical Africa, but are still heavily based on the Rare Plant Monitoring forms and Site Condition Monitoring forms designed and developed by official government conservation agencies in Britain (England, Wales, Scotland and N Ireland) and by the Botanical Society of the British Isles (BSBI), principally by Lynne Farrell (now retired) but input also from Robin Payne at Scottish natural heritage and from Jim McIntosh, BSBI Scottish officer. The authors of this book are grateful for their permission in allowing these forms to be used in modified form here.

ANNEX A

SITE CONDITION MONITORING OF THREATENED VASCULAR PLANTS

Objective
Locate populations of specified target species at selected sites and report on location, population size, regeneration and environmental parameters, especially threats.

Methods

1. Visit site and search appropriate habitat for target species. On large sites examine only the suitable habitat in the target 1-km. sq. centred on past population records. It is important to search for new locations of the target species in the square as well as current records but since species tend to cluster in favoured habitats it is usually possible for a single habitat area to be searched for several species at the same time.
2. Record the location of those populations in detail on the standard recording form adding map and annotated photograph(s) to identify the location precisely. Use a Geographical Positioning System accurate to at least 10m to obtain the grid reference. Hand held GPS units such as the Garmin Etrex can create waypoints that can be downloaded to a GIS on a PC. Each waypoint is given a Unique Identifier UID (a number) and the grid reference of the site. Field notes or descriptions can be recorded separately and linked to the waypoints using the UID. If a recording form already exists it is only necessary to check the information and correct if necessary.
3. Establish the extent of each population located and count number of plants in small populations up to about 100 plants. For larger populations obtain an estimate by defining extent, counting samples and then extrapolating from these.
4. Quantify evidence of successful regeneration e.g. seed production, evidence of a range of age structure (especially seedlings and young plants) and evidence of vegetative expansion.
5. Briefly record the environmental parameters that have been identified for indirect monitoring on habitat occupied by the species (or which should be occupied by the species where it has already been eliminated). These environmental parameters are identified in tables which will be provided.
6. Note any damaging activities or factors of conservation concern such as overgrazing, poaching, signs of eutrophication, etc.

Outputs

For each population:
A location recording form (Annex C) to which should be added supplementary maps and annotated photographs to provide more detail than the grid reference for relocation of the record. Extracts of printed maps are preferred to sketch maps. Ideally photographs should show the location from the middle distance so as to include landscape features that help relocation. Photographs are most useful in the uplands where maps often lack accurate detail but it is accepted that there are locations (and weather conditions) where photography will not be possible.

For each site:
i) a site monitoring form (Annex D) for each species to provide standard assessment with information on population size (including number estimates), regeneration stages, environmental records (including management) and remedial management required (if any);

For the survey as a whole:
A brief final report. This should headline key findings in site order (e.g. new populations of interesting species, declines or increases in estimations for particular populations, problems with site management, etc.). It should also discuss any unresolved problems with the methodology (preferably with proposals for practical solutions). It would be particularly useful to have comments on the environmental parameters proposed for indirect monitoring.

Dissemination
Site reports will be copied to local authorities; the National Herbarium of Cameroon; CBD focal point person of the Ministry of the Environment. Population location data will be collated as part of the threatened plants in Cameroon dataset by the National Herbarium of Cameroon and used to prioritise conservation efforts and to refine future Cameroon Plant Red Data books.

Timescale
In general the best time for monitoring is during the flowering period to permit confident identification of species and the collection of voucher herbarium specimens 9if this doers not further threaten the plant species concerned) so as to allow the identification to be checked by the national Herbarium, and if required, International Specialists for the group in which the species falls.

Access permission for survey

Contractors should not enter land without prior consent to do so from owners and occupiers.

Contractors are required to liaise with owners, land managers, farmers, as appropriate, in order to avoid causing disturbance to land management or susceptible conservation interests. Contractors must adhere to all reasonable measures requested by owners and occupiers.

VASCULAR PLANT - POPULATION LOCATION FORM

Species:	Date:
Recorder(s):	GridRef: GPS used?
Location name:	Region.

Habitat type:

ASSOCIATED SPECIES

POPULATION SIZE (TICK BOX OR ENTER COUNT)

	No.	<10	10 - 100	101-300	301-1000	1001-3000	30001-10000
mature plants							
flowering or seeding plants							
juvenile plants							
Seedlings							
TOTAL - note unit counted (e.g. rosette, vegetative shoot, clump, flowering shoot, etc)							
Area occupied:							

REGENERATION OBSERVED (tick box)

flowers	seedheads	seedlings	juveniles	vegetative expansion

DAMAGED PLANTS (note nature of damage, number of plants affected and assumed cause)

HABITAT CONDITION COMMENTS e.g. sward height, grazing of target or neighbouring plants, physical disturbance, presence of other species indicating longer-term disturbance, proportion of open ground, shade, canopy structure, visible pollution, water clarity, water depth

OTHER PERCEIVED THREATS (note any evidence)

SCM VASCULAR PLANTS - SITE MONITORING FORM

SITE NAME	SPECIES
SURVEYOR(S)	DATE(S)

1. LOCATIONS

Re-locate populations of target species on site using past recording forms where available.

LOCATION SUMMARY ✓ TABLE

NO OF LOCATIONS		
1		UNFAVOURABLE
2 or more		FAVOURABLE

2. POPULATION SIZE

Instructions. Count no. of plants in populations of target species. Record counting unit.

POPULATION SIZE WORKING ✓TABLE

Location ↓	<10	11-25	26-100	101-300	301-1,000	1001-3000	3001-10,000	>10,000

POPULATION SIZE SUMMARY ✓ TABLE

PERENNIAL		ANNUAL/BIENNIAL	
<25	>25	<100	>100
UNFAVOURABLE	FAVOURABLE	UNFAVOURABLE	FAVOURABLE

3.-DECLINE IN POPULATION SIZE

Instructions. Ensure that the method used to count the plants on current visit is similar to that used on previous occasions. Compare the numbers obtained during the current visit with those from the past two visits.

DECLINE WORKING ✓TABLE

POPULATION SIZE TODAY	POPULATION SIZE ON PREVIOUS OBSERVATION	POPULATION SIZE ON PREVIOUS OBSERVATION	OVERALL CHANGE (ORDER OF MAGNITUDE)

DECLINE SUMMARY ✓ TABLE

all spp.	NO DECLINE	FAVOURABLE
Annual spp.	DECLINE 1-2 ORDER OF MAGNITUDE	FAVOURABLE
Perennial spp.	DECLINE 1 ORDER OF MAGNITUDE	WARNING
Perennial spp.	DECLINE 2 OR 2+ ORDERS OF MAGNITUDE	UNFAVOURABLE
Annual spp.	DECLINE 3 OR 3+ ORDERS OF MAGNITUDE	UNFAVOURABLE
all spp.	STEADY DECLINE OF 1 ORDER PER VISIT FOR 2 VISITS	UNFAVOURABLE

4. REGENERATION

Instructions. Examine target populations closely and record evidence of regeneration.

REGENERATION WORKING ✓ TABLE

	LOCATION:						
DYNAMIC POPULATION STRUCTURE	(i.e. CLEAR RANGE OF PLANT SIZE)						
EVIDENCE OF VEGETATIVE REGENERATION	(i.e. PLANTS CONNECTED BY RUNNERS ETC.)						
EVIDENCE OF SEXUAL REGENERATION	SEEDLINGS						
	SEED HEADS						
	FLOWERS						

REGENERATION SUMMARY ✓ TABLE

	ABSENT	PRESENT
Age structure		
Sexual		
Vegetative		
	UNFAVOURABLE	FAVOURABLE

Presence of any category permits favourable status although absence of any evidence of a range of age structure and seedling stage of sexual regeneration should be taken as a WARNING of potential unfavourable status.

5. HABITAT CONDITION

Instructions. Use the parameters identified in the relevant table for the species.

ENVIRONMENTAL PARAMETERS WORKING TABLE

Parameter		Notes
e.g. Niche availability (is there suitable but unoccupied habitat present for the target species, if so how much?)		
e.g. Physical integrity (is the habitat undisturbed or has it been damaged to some extent and if so, by what and when?		
e.g. Competition from Invasive species? (What other species if any appear to be threatening the habitat in which the target species sits?)		

OVERALL SUMMARY TABLE

	UNFAVOURABLE	FAVOURABLE
POPULATION NUMBER	1	2+
POPULATION SIZE	1-25 plants in each population	>25 plants in each population
POPULATION DECLINE	-2 orders of magnitude	0-1 orders
EVIDENCE OF REGENERATION	absent	present

6. ACTIVITIES OR EVENTS AFFECTING THE SPECIES

Activity Assessment

For all features, which of the following types of activity or event are having a positive or negative effect on the condition of the feature? Identify **no more than three positive (+) and three negative (-)** activities (on or off the site) affecting each feature, by putting a +/- in the box.

	Identify species if specific							
1. Agricultural operations (e.g. cutting trees and shrubs, burning, tilling)								
2. Over-grazing (including wild animal browsing)								
3. Under-grazing								
4. Burning								
5. Game or fisheries management (e.g. introduction of stock, cutting of river banks)								
6. Water management (including dams, drainage, dredging or alterations to the water table).								
7. Water quality (including silt, water pollution (direct or diffuse), run-off, nutrient enrichment, eutrophication etc)								
8. Forestry operations (including logging)								
9. Recreation / disturbance (including clearing of habitat; trampling)								
10. Development carried out at local level (including construction of new houses)								
11. Development carried out at National or Regional level (e.g. major roads; bridge building; pipelines)								
12. Invasive species (including bracken or scrub or grassland)								
13. Earth Science feature obscured / eroded (e.g. coastal erosion; cliff falls; overshading of rockfaces) / modified (e.g. cave entrances)								
14. Dumping / spreading / storage of materials								
15. Other (specify)								

6. SITE MANAGEMENT

		Code	**Comments**
Are existing management measures if present likely to safeguard the species?	[1] yes [2] yes, but could be improved [3] uncertain [4] no		
Type(s) of management activities if present	[1] all are appropriate [2] 1 or more uncertain [3] 1 or more inappropriate		
Intensity & timing of management activities if present	[1] all are appropriate [2] 1 or more uncertain [3] 1 or more inappropriate		

SITE CONDITION MONITORING - VASCULAR PLANTS
ADDITIONAL NOTES TO BE READ IN ASSOCIATION WITH RECORDING FORM

OVERALL CONSIDERATIONS

Before visiting the site you should carry out the following.

1. Obtain past recording forms if available to speed your relocation of target plant populations and make sure that you re-locate the same population that was recorded previously. (If there is not an existing population form, create one by completing form in Annex C and photographing site during your monitoring visit.)

2. Identify past methodology used to count population. If this is the first monitoring visit it is particularly important to describe for future workers the unit of vegetation taken to represent a single plant for the purposes of the count (see 'population size' below).

3. Most rare plants are easy to identify once you have seen them. If you do not know the species, look at pictures and descriptions before visiting the site. Then visit a known population with a clearly recorded location to see the species for the first time. Spend a little time making sure that you have seen its full range of appearance there.

Timing of visits. It is best to monitor most plants when in flower or fruit (see Annex A). Approximate times for flowering are provided in floras e.g. FWTA and Flore du Cameroun and on specimens housed in the National Herbarium Yaoundé or the herbarium at the Limbe Botanic Garden, but precise information for the site may be best obtained from past site notes although flowering times may vary by several weeks between years.

1. NOTES ON COUNTING LOCATIONS

Explanation of target. A single isolated plant population is vulnerable to chance losses and genetic change. .

Instructions. Use past species recording forms if they exist, to relocate known locations. Count separate micro-sites as any that are more than 40m apart. If a population appears to be extinct it is very important to search adjacent suitable habitat for the presence of other populations as a replacement population may have taken its place.

Exceptions. A few species occur in isolated, very large populations i.e. more than 10,000 plants. These should not be recorded as unfavourable indicators of site condition as they probably act in the same way as a cluster of smaller populations.

2. RECORDING POPULATION SIZE

Explanation of target. Small plant populations are more vulnerable to chance losses as a result of natural (or unnatural population fluctuations) and genetic change.

Instructions. First spend some time establishing the true limit of the population. It is very useful to mark this on the ground. Count plants in small populations <100 plants. It can be very helpful to mark plants as they are recorded with pins, beads, canes or (least damaging) on a photograph. A Polaroid camera is useful to provide an instant print. Estimate the number of plants in populations >100 plants on the log scale. For these populations it is important to make some careful sample counts and multiply by the area of the population bearing in mind that you are only seeking a log scale estimate. Make systematic decisions as you estimate e.g. are there more or less than 25, than 100, than 1000 or than 10,000.

It is difficult to count many plants because of vegetative regeneration and clumping, but rarely impossible. Look at clumps or dense stands and decide on a functional or even arbitrary unit. It is important to record the chosen unit for posterity. Extrapolate from them to obtain an estimate where cover otherwise appears to be continuous. Flowers or flowering shoots are a good measure of population size but be aware that unrecorded vegetative plants are likely to be present and should be located and included in the count

3.-RECORDING DECLINE IN POPULATION SIZE

Explanation of target. Decline may be an important indicator of unfavourable management conditions but natural fluctuations also occur. Conclusions are vulnerable to any change in methodology.

Instructions. Compare the results from the current visit with those from the past two visits if data exists. Ensure that the method used to count the plants on current visit was similar to that used on previous occasions. Annual plants naturally fluctuate significantly between years hence higher limit. For perennial plants a change of 2 orders of magnitude should be considered unfavourable. A change of 1 order of magnitude is significant where this has occurred in the two periods between 3 successive visits (i.e. a decline of 2 orders in total).

Exceptions.

Perennial plants would not normally be expected to vary by an order of magnitude over a single 6 year period and so this should be considered to be a WARNING of unfavourable conditions. A decline of this magnitude should not be recorded as unfavourable but detailed annual monitoring should be introduced to establish the true trend.

4. RECORDING REGENERATION

Explanation of target. Successful regeneration is crucial to the survival of populations and is a valuable indicator of future trends. Evidence of a range of young and old plants is clearly favourable but is not obvious in all species, requiring you to fall back on indications of the success of stages of regeneration.

Instructions. Examine small areas of open ground closely to detect very young plants. Note that the stages of sexual regeneration identified in the form are in order of significance i.e. seedlings are much more valuable than flowers.

A dynamic population structure is one that shows a range of plants that by size or appearance indicate a spread of different age classes. Vegetative regeneration can be detected by runners or clumps with older plants at the centre and younger plants at the outside. Flowers are the weakest evidence of successful sexual regeneration since many flowers do not produce seed. Full seed heads are potentially more significant than flowers, or empty seed heads. Seedlings are often very difficult to locate and identify as young seedlings are often very unlike the parent plant. Ideally seedlings should show cotyledons but small plants that are well isolated from mature plants can be accepted.

5.-RECORDING ENVIRONMENTAL PARAMETERS

Use the parameters identified in the relevant table for the species. Establish which table to use by referring to table of spp/habitat affinities. The parameters for Suite 13 (**HIGH-ALTITUDE SCREES AND CLIFFS) are given below.**

Niche availability	**Map habitat area that should be occupied by species (either area or length)**	Favourable condition should be that all suitable habitat should be occupied by species.
Physical integrity (physical damage to the vegetation)	**Visual assessment (signs of grazing by mammals or of cutting or clearing of forest or grassland)**	Favourable condition requires that signs of forest clearance, or cutting; or stock grazing are absent and that signs of grazing or browsing by wild mammals no more than rare.
Competition from alien flora	**Visual assessment (vegetation composition)**	Associated vegetation should not include competitive alien species (e.g. *Chromolaena odorata*).

6.-RECORDING ACTIVITIES AND EVENTS

Identify only the three main 'activities' directly affecting the species and indicate whether the effect is positive ('+'), negative ('-') or neutral ('0') i.e. do not only record activities that are actually causing harm. Under comments identify the precise activity (e.g. grazing by elephants, road building, etc.) and whether the activity was one-off or is continuing. Species can be grouped together if appropriate so that in practice a single form may suffice for all the species on a site.

7.-RECORDING MANAGEMENT

This item is the most judgmental and the reason for an unfavourable rating here should be clearly visible in the earlier sections. Identify the unfavourable factors under comments and the remedial management that in your view is required. Again species can be grouped.

FAMILY ENTRIES EXPLAINED

MARTIN CHEEK

The species treatments in this book are grouped alphabetically within genus, and the genera, then arranged alphabetically, within families. Plant families each have a set of characteristics shared by all or most species and genera that they contain. These are summarised in each family treatment so as to assist identification or confirmation of identification. References are given, such as Flore du Cameroun family treatments, where these exist, or recent revisions or other scientific references. Such references should be consulted for more extensive descriptions and the most complete botanical data on each species. In many families or genera such references are lacking, making it very difficult to assess the conservation status of the species in that group since unless a species can be correctly identified, it cannot be assessed. Such cases are highlighted in the family treatments in the hope that this will encourage others to fill such gaps in our knowledge.

Family names are in a state of flux at the time of writing. Recent molecular-cladistic methods have shown that some families and genera traditionally recognised can no longer be maintained as they were formerly. In other cases the evidence for change is less conclusive and although changes in classification have been proposed in one year, subsequently they have been retracted. For example *Harungana* was united with *Psorospermum* (Guttiferae/Hypericaceae) recently, but a later reanalysis, with additional data, showed this to have been an error, and the original proponent of uniting the genera now maintains them as separate entities. Therefore, some caution is needed before accepting all proposed changes outright and immediately. In the main, in this book, 'traditional' families are maintained in order to be consistent with the treatments in our seven Conservation Checklists for Cameroon published 1998–2011. However in cases where family limits are being altered, reference is made to this, for example the family heading Guttiferae is followed by Clusiaceae/Hypericaceae in brackets, signalling that some authors now accept the latter two families in place of the first and that in subsequent editions of this book it is likely that the second alternative will be taken. Heywood, V.H., Brummit, R.K., Culham, A., & Seberg's Flowering Plant Families of the World, RBG, Kew, 2007, should be consulted for more information on plant family delimitation. This excellent work has been much used in the writing of the family treatments in this book.

In some large families with many genera, such as Rubiaceae, short descriptions of the genera and published revisions or other references on the taxonomy of the genus, are given separately, before each genus, rather than in the family treatment, so that all the information on each genus can be found on one page.

General material relating to the global range of the family, and often the number of genera and species, together with notes on any species which might be conspicuously useful to man, are given, as are notes on the habitats in which the species of the family are usually found in Cameroon.

The family accounts usually conclude with notes on species that are not treated as definitively threatened but which either were formerly considered so, or which might be considered so in future. In the first case are species which were assessed as threatened in our seven Cameroon Conservation Checklists 1998–2011, but which new data have shown for example to have additional locations or larger ranges, so taking these beyond the threshold for Vulnerable status into Near Threatened status.

In the second case are listed species which i) came to light only recently in the compilation of this book, where there was insufficient time to research their global range and locations but which look to be potentially threatened; ii) are poorly known species where delimitation or identification is unclear because there is no Flore du Cameroun account or modern revision, for example, some of the many German names of species published about a century ago based on specimens now most in Berlin. Should a revision or Flore du Cameroun account become available for such families in future, many of these names might be resolved and some might be shown to be additional threatened species.

SPECIES ENTRIES EXPLAINED

MARTIN CHEEK

This book is intended as one of reference and is not meant to be read from cover to cover. The worksheet approach of the original Plant Red Data Book project of Ronald Melville has been maintained. Thus every species treatment is intended to stand by itself (although some useful data is contained in the preceeding family treatment). As a result there is a certain amount of repetition between species treatments. Similar species, with similar conservation status and geography often have similar suggestions for management.

The entry for each species follows the template established in the Cameroon Conservation Checklist series (Cable & Cheek *et al.* 1998, Cheek *et al.* 2000, Cheek *et al.* 2004, Harvey *et al.* 2004, Cheek *et al.* 2010, Harvey *et al.* 2010, and Cheek *et al.* 2011).

The accepted scientific species name is given followed by the standard author abbreviation. Synonyms in recent use, if any are then given in brackets. Local names as used in Cameroon are then cited. However, in practice, most rare species are unknown to people in the surrounding area and so have no such names. The level of threat is then given, following the IUCN (2001) system: VU=Vulnerable, EN=Endangered and CR= Critically Endangered, followed by the category code which summarises the basis for the assessment e.g. VU B2a,b(iii).

Range: countries in which the species occurs are listed from W to E and from N to S. In each country, locations are given, usually with numbers of collections where there are several from one location, to give an indication of the frequency of the species. Locations for Cameroon are cited more rigorously than those for other countries, and are cited under Region, usually abbreviated to a single letter, e.g. SW for South West Region.

A paragraph then follows, justifying the threat assessment in full. Usually this is based on the low number of locations, AOO (area of occupancy) and presence of one or several threats to the species (Criterion B). More extensive information on methods used in generating the assessments is given in the Methodology chapter. This should be read by these unfamiliar with the IUCN (2001) system which is followed throughout this book. A short description of the species has been included to assist identification in the field. Should more detailed descriptions be needed, references are given in either the species or family entry to where these are to be found, if they exist. Descriptions follow the standard pattern and are ordered by habit, stem, leaf, inflorescence, flower, fruit and seed, where these are known. Efforts have been made to reduce the technical content of the descriptions to make them more user-friendly, but non-botanists may still need to use a Glossary to help comprehension (The Plant Glossary, H. Beentje, RBG, Kew, 2010 is recommended). Where possible, reference is made to features that enable the species in question to be distinguished from similar species with which it might be confused.

General material relating to the discovery or botanical or cultural interest of the species is included where possible.

Threats: are described where known, usually by reference to the clearance or decline of the habitat in which the species occurs, since very few species are themselves directly targeted by man, the exception being some timber species. Competitive invasive plant species, and introduced fungal diseases, which are major threats to rare species in other parts of the world, are not known to be threats in Cameroon. Threats, or population reduction due to threats are an integral part of most IUCN 2001 categories, Criterion D (low numbers of individuals) being the exception.

Management Suggestions: these notes are not part of a standard IUCN 2001 assessment but are intended to offer ideas to those who are responsible, or who might take on responsibility, for the survival of the species here assessed as threatened. In the case of species known from several (6–10) recent records, the advice is primarily to gather basic population data which can form a baseline for future monitoring, so that such managers might be Ministry of Environment staff responsible for individual protected areas such as National Parks, NGO staff involved with protected areas or proposed protected areas, elders or council members of communities managing community forests or individuals interested in helping to protect the national patrimony of Cameroon.

Maps explained

The maps of globally threatened species are printed in a block at the end of this book, six per page, ordered in the same sequence as the species treatments, that is, alphabetically by family, then by genus, then by species, with Dicotyledon families treated first, followed by Monocotyledon families. Each species is captioned with its scientific name, abbreviated IUCN threat code (VU, EN or CR) and the family in which it is placed in this book.

Map width is standard throughout, showing Ghana in the W and in the E, the western parts of Central African Republic (CAR) and Congo (Kinshasa). While this window shows the global distributions of c. 95% of the species treated, the remainder, with wider distributions, do not have their global distributions shown if they occur outside this window. We apologise for this limitation, brought about by the need to economise with time. To produce, size and fit 815 maps one-by-one, and not in an automated system as used here, would have further delayed production of this book. If there is any doubt as to whether you are viewing a complete global distribution or not, please refer to the species treatment, which will clarify the matter.

BIBLIOGRAPHY FOR INTRODUCTORY CHAPTERS

MARTIN CHEEK

Beentje, H. (2010). The Plant Glossary, RBG, Kew.

Cable, S. & Cheek, M. (eds) (1998). The Plants of Mount Cameroon: A Conservation Checklist. Royal Botanic Gardens, Kew, UK. lxxix + 198 pp.

Cheek, M., Onana, J.-M. & Pollard, B.J. (eds) (2000). The Plants of Mount Oku and the Ijim Ridge, Cameroon: A Conservation Checklist. Royal Botanic Gardens, Kew, UK. iv + 211 pp.

Cheek, M., Pollard, B.J., Darbyshire, I., Onana, J.-M. & Wild, C. (eds) (2004). The Plants of Kupe, Mwanenguba and the Bakossi Mountains, Cameroon: A Conservation Checklist. Royal Botanic Gardens, Kew, UK. iv + 508 pp.

Cheek. M., Harvey, Y. & Onana, J.-M. (eds) (2010). The Plants of Dom, Bamenda Highlands, Cameroon: A Conservation Checklist. Royal Botanic Gardens, Kew, UK. iv + 162 pp.

Cheek. M., Harvey, Y. & Onana, J.-M. (2011). The Plants of Mefou proposed National Park, Yaoundé, Cameroon: A Conservation Checklist. Royal Botanic Gardens, Kew, UK. 252 pp.

Harvey, Y., Pollard, B.J., Darbyshire, I., Onana, J.-M. & Cheek, M. (eds) (2004). The Plants of Bali Ngemba Forest Reserve, Cameroon: A Conservation Checklist. Royal Botanic Gardens, Kew, UK. iv + 154 pp.

Harvey, Y., Tchiengué, B. & Cheek, M. (eds) (2010). The Plants of Lebialem Highlands (Bechati-Fosimondi-Besali), Cameroon: A Conservation Checklist. Royal Botanic Gardens, Kew, UK. 170 pp.

Heywood, V.H., Brummit, R.K., Culham, A., & Seberg, O.(2007). Flowering Plant Families of the World, RBG, Kew.

Gabondatabase online. (http://dps.plants.ox.ac.uk/bol/Gabon/Search/Fullsearch?searchType=Collection)

IUCN. 1994. *IUCN Red List Categories*. Prepared by the IUCN Species Survival Commission. IUCN, Gland, Switzerland.

IUCN. 2001. *IUCN Red List Categories and Criteria : Version 3.1*. IUCN Species Survival Commission. IUCN, Gland, Switzerland and Cambridge, UK.

IUCN. 2003. *Guidelines for Application of IUCN Red List Criteria at Regional Levels: Version 3.0*. IUCN Species Survival Commission. IUCN, Gland, Switzerland and Cambridge, UK.

Keay, R. & Hepper, F.N., eds. (1954-1972). The Flora of West Tropical Africa. 3 Vol., Crown Agents, London.

Lucas, G. & Synge, H. (1978). The IUCN Plant Red Data Book, IUCN, Gland.

Moat, J. 2007. Conservation assessment tools extension for ArcView 3.x, version 1.2. GIS Unit, Royal Botanic Gardens, Kew. Available at: http://www.rbgkew.org.uk/gis/cats.

Onana, J.-M. (2011 in press). The Vascular Plants of Cameroon, a Taxonomic Checklist, IRAD-National Herbarium of Cameroon, Yaoundé

Sosef, M.S.M. *et al.* (2006). Checklist of Gabonese Vascular Plants. Scrpta Botanica Belgica 35, National Botanic Garden of Belgium.

Walter, K.S. and Gillett, H.J. [eds] (1998). *IUCN Red List of Threatened Plants*. Compiled by the World Conservation Monitoring Centre. IUCN – The World Conservation Union, Gland, Switzerland and Cambridge, UK. lxiv + 862pp.

Willis, F., Moat, J. & Paton, A. (2003). Defining a Role for herbarium data in Red List assessments: a case study of *Plectranthus* from eastern and southern tropical Africa. Biodiversity and Conservation 12(7): 1537–1552.

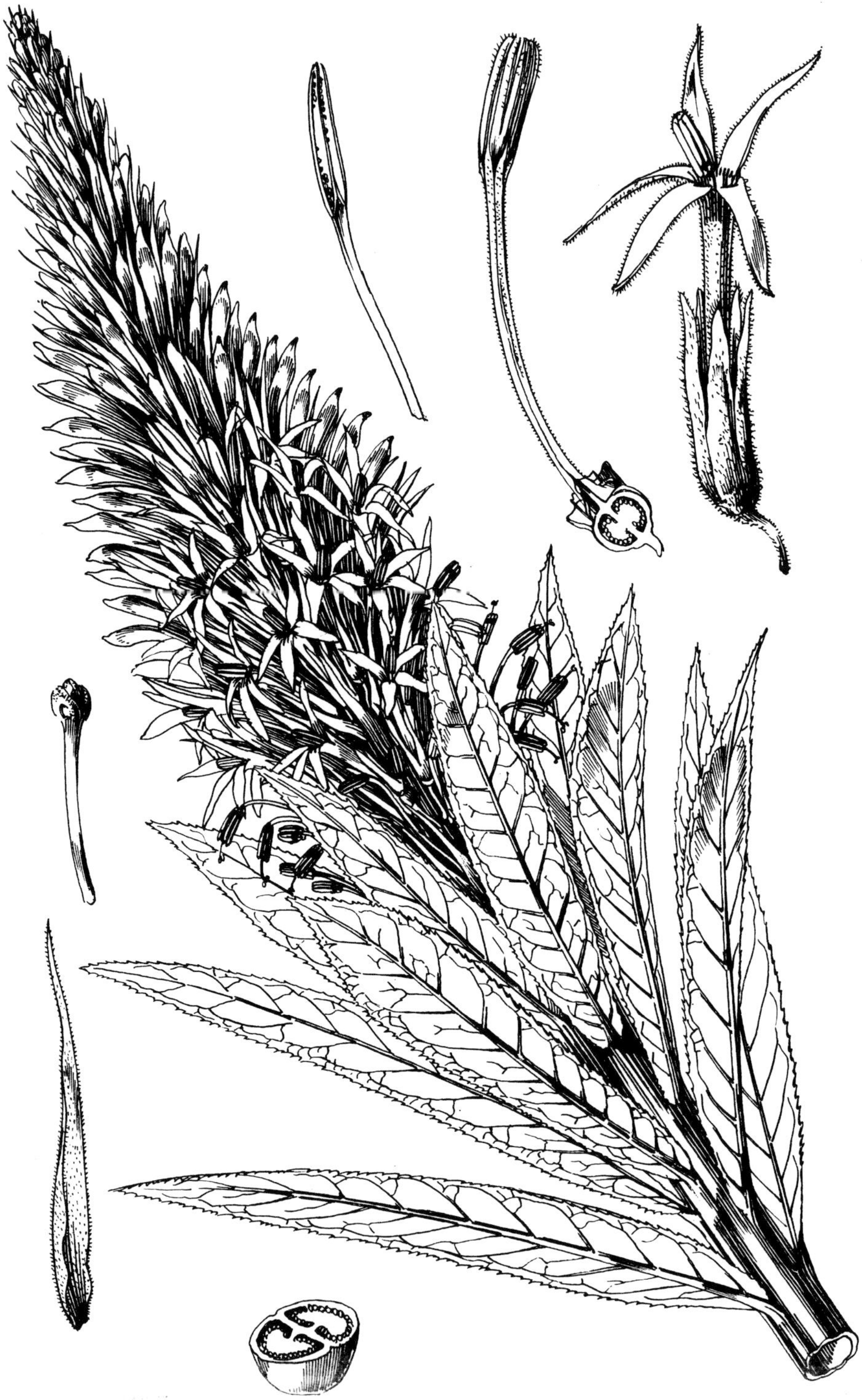

Fig. 2 *Lobelia columnaris* (Campanulaceae) by W.E. Trevithick

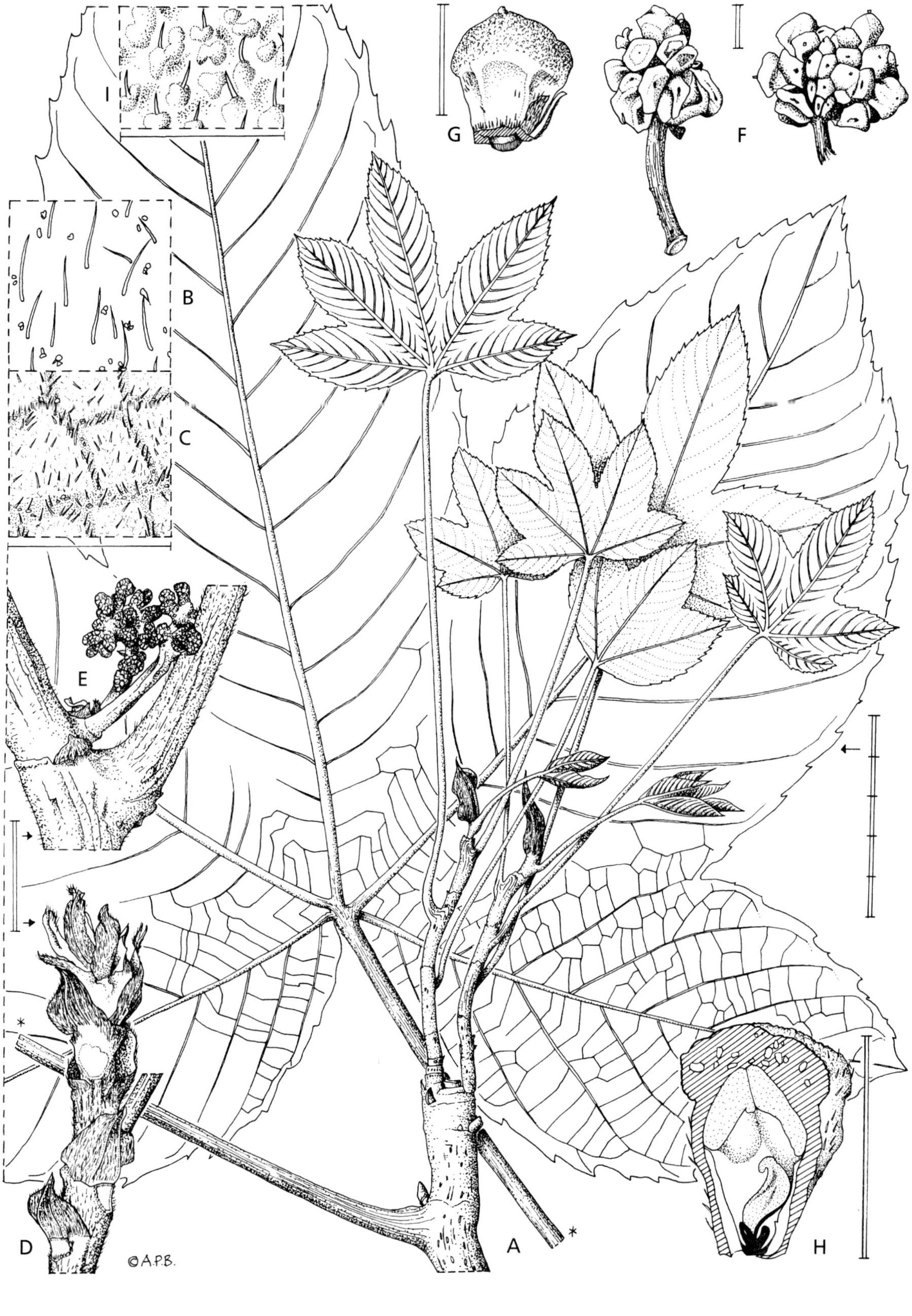

Fig. 3 *Myrianthus fosi* (Cecropiaceae) by A. Brown

Fig. 4 *Helichrysum cameroonense* (Compositae) by W. Fitch

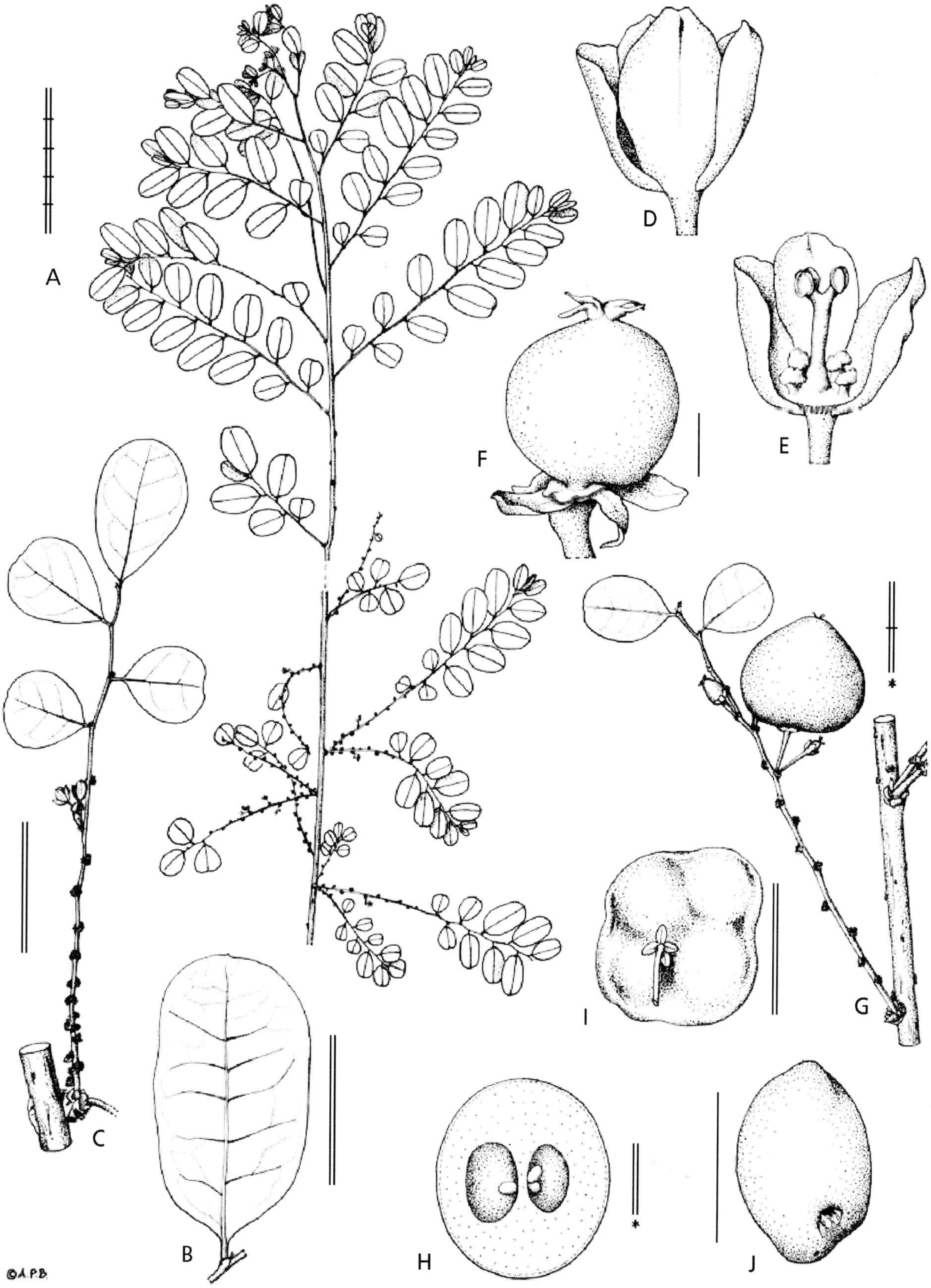

Fig. 5 *Phyllanthus kidna* (Euphorbiaceae) by A. Brown

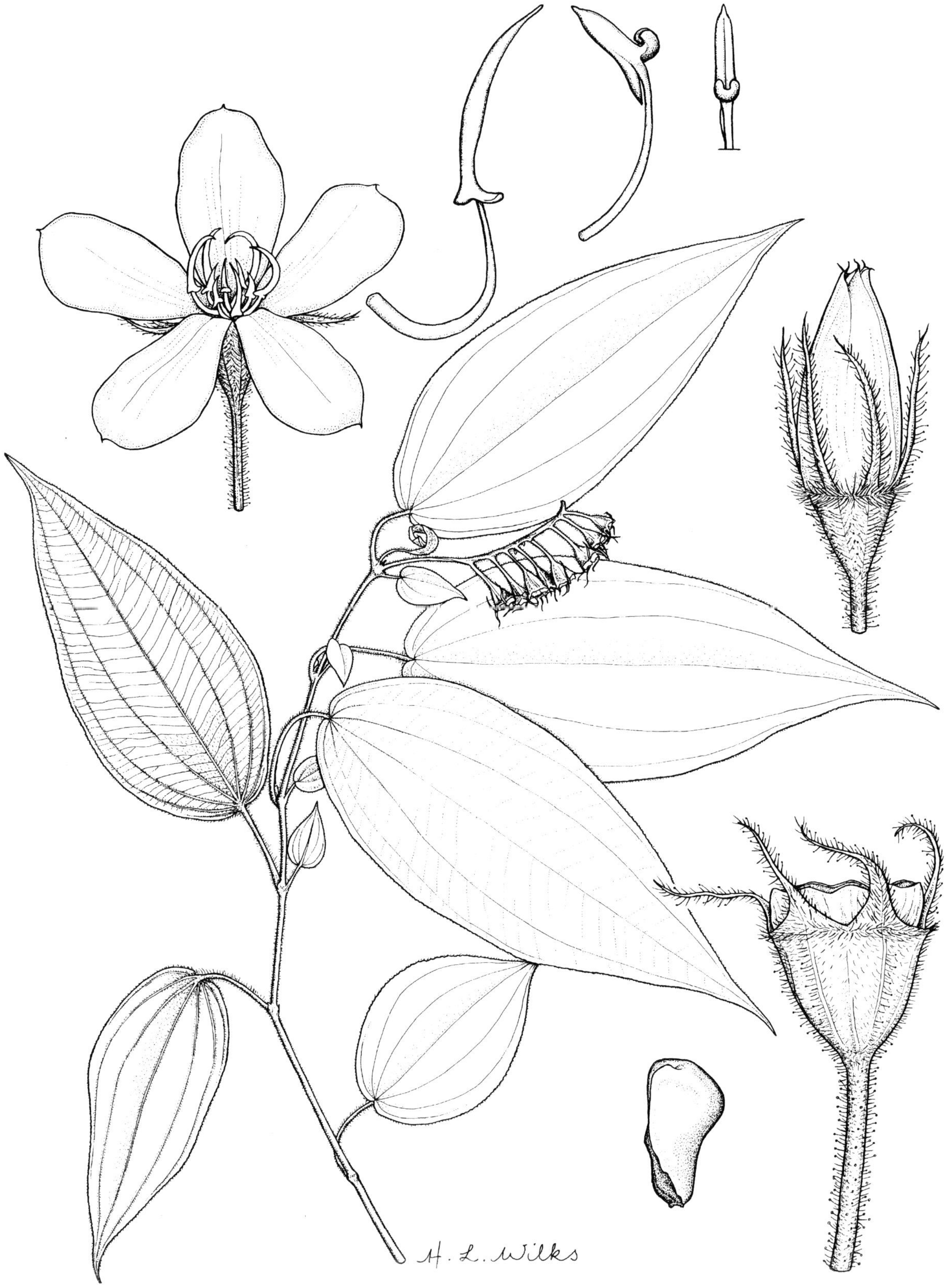

Fig. 6 *Amphiblemma monticola* (Melastomataceae) by H.L. Wilks

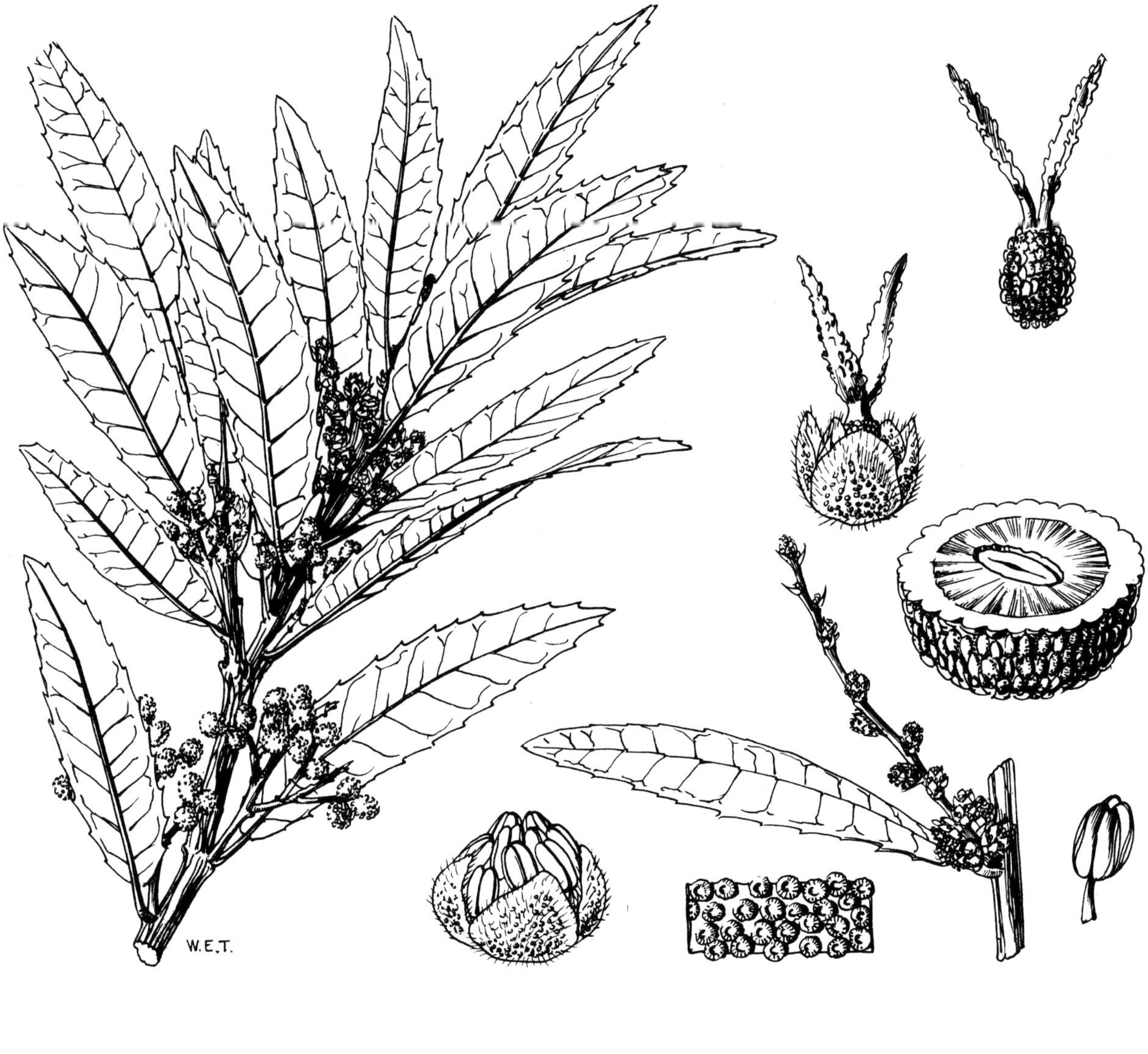

Fig. 7 *Morella arborea* (Myricaeae) by W.E. Trevithick

Fig. 8 *Ardisia dom* (Myrsinaceae) by A. Brown

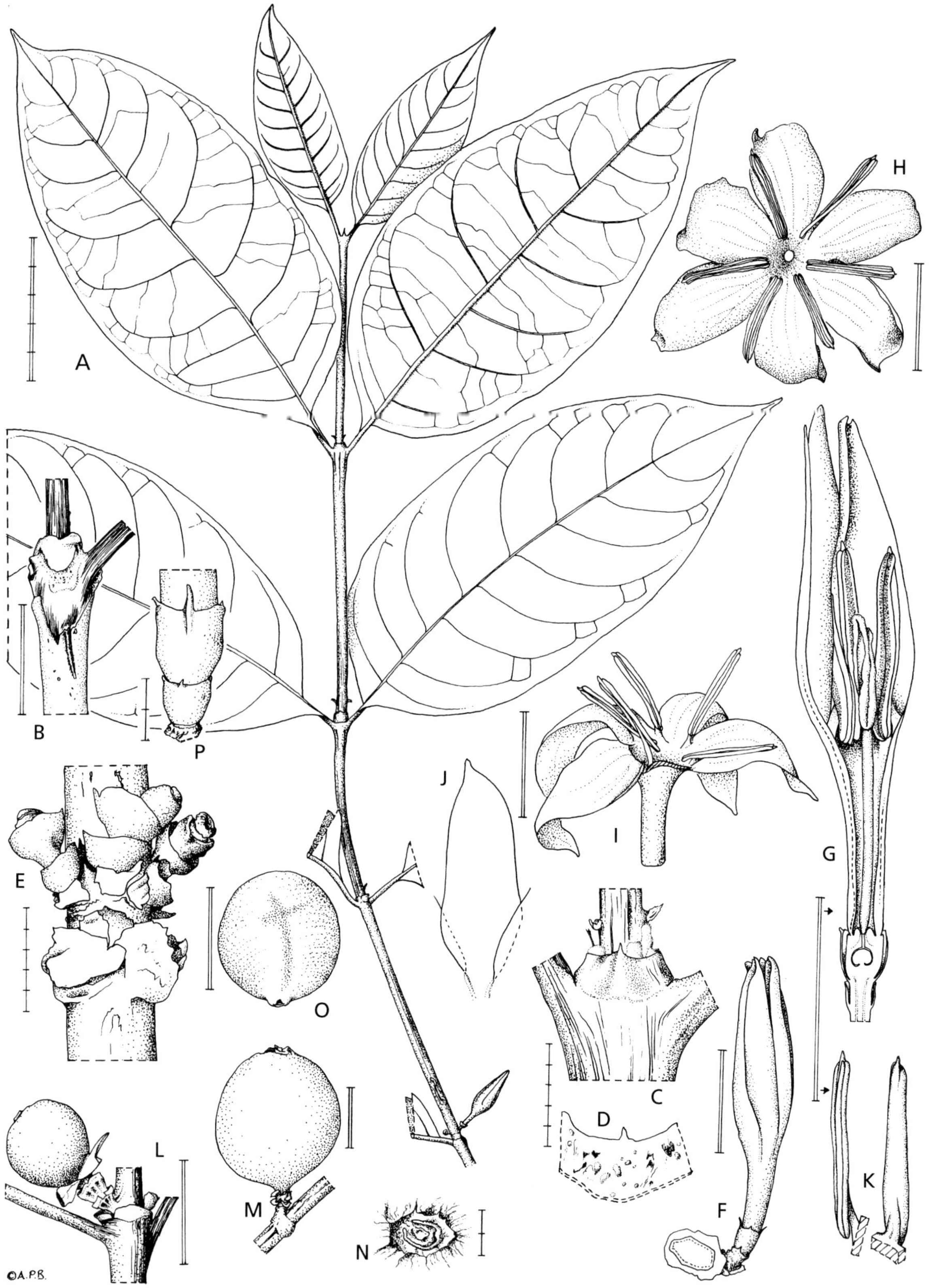

Fig. 9 *Argocoffeopsis fosimondi* (Rubiaceae) by A. Brown

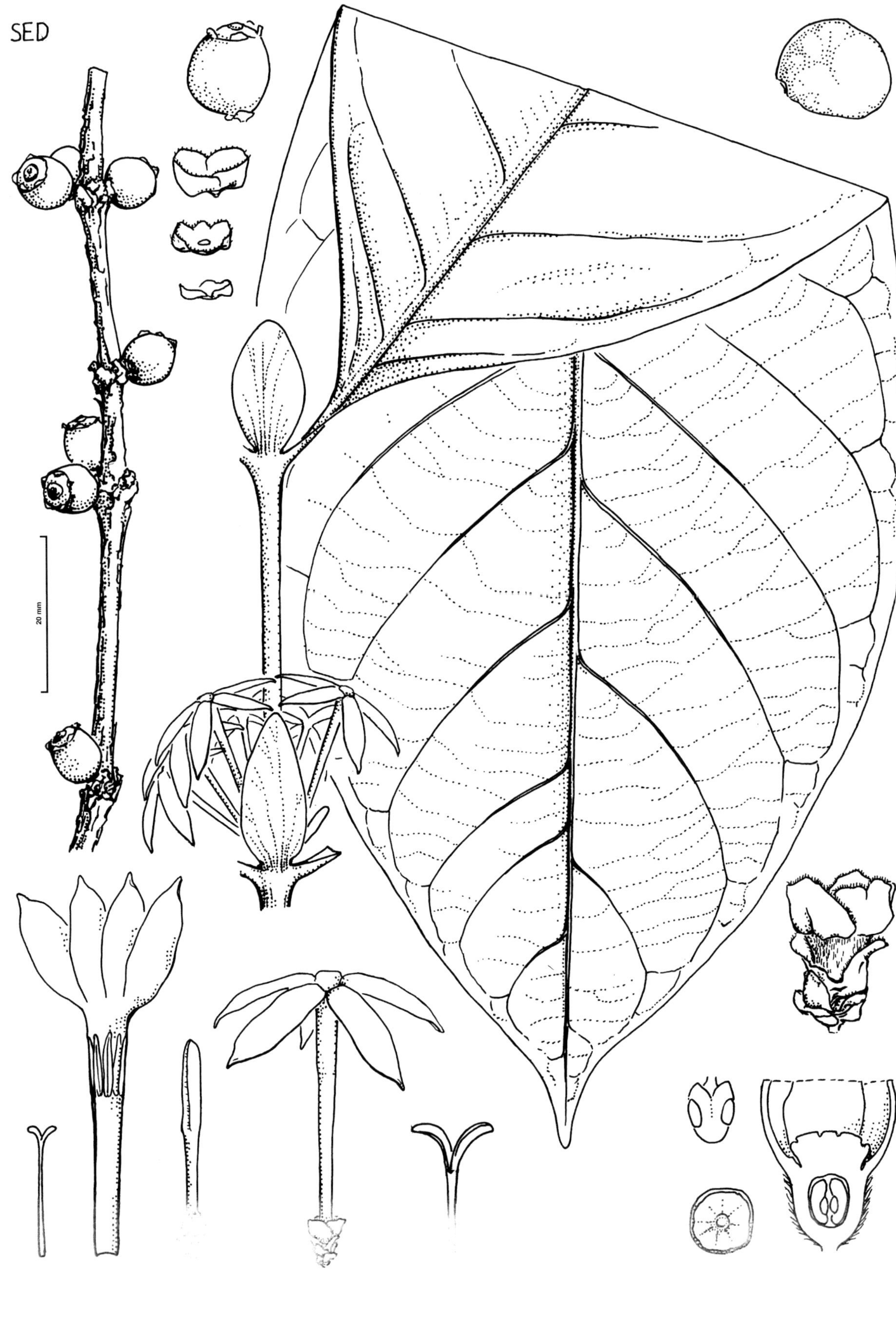

Fig. 10 *Belonophora ongensis* (Rubiaceae) by Sally Dawson

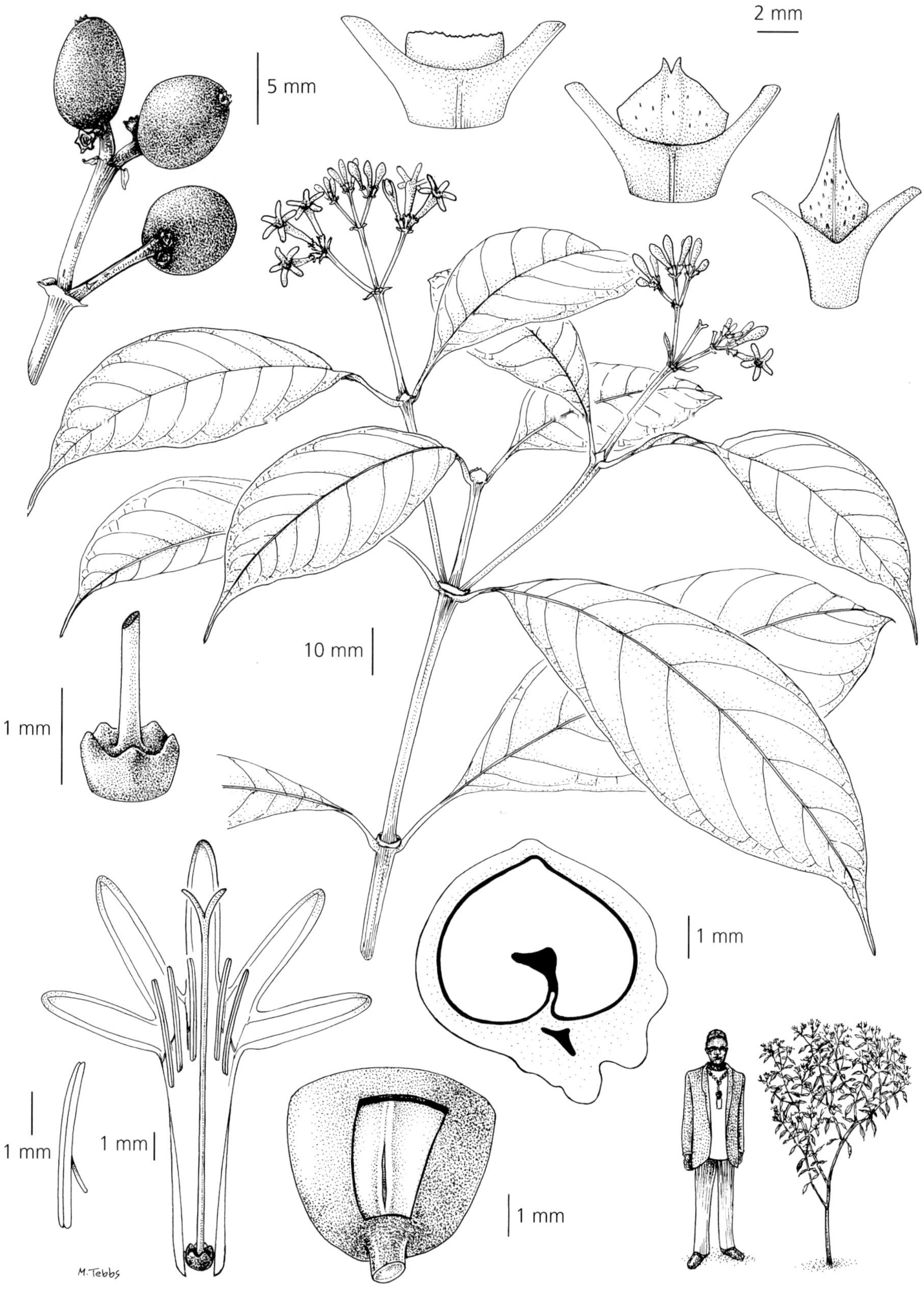

Fig. 11 *Chassalia laikomensis* (Rubiaceae) by M. Tebbs

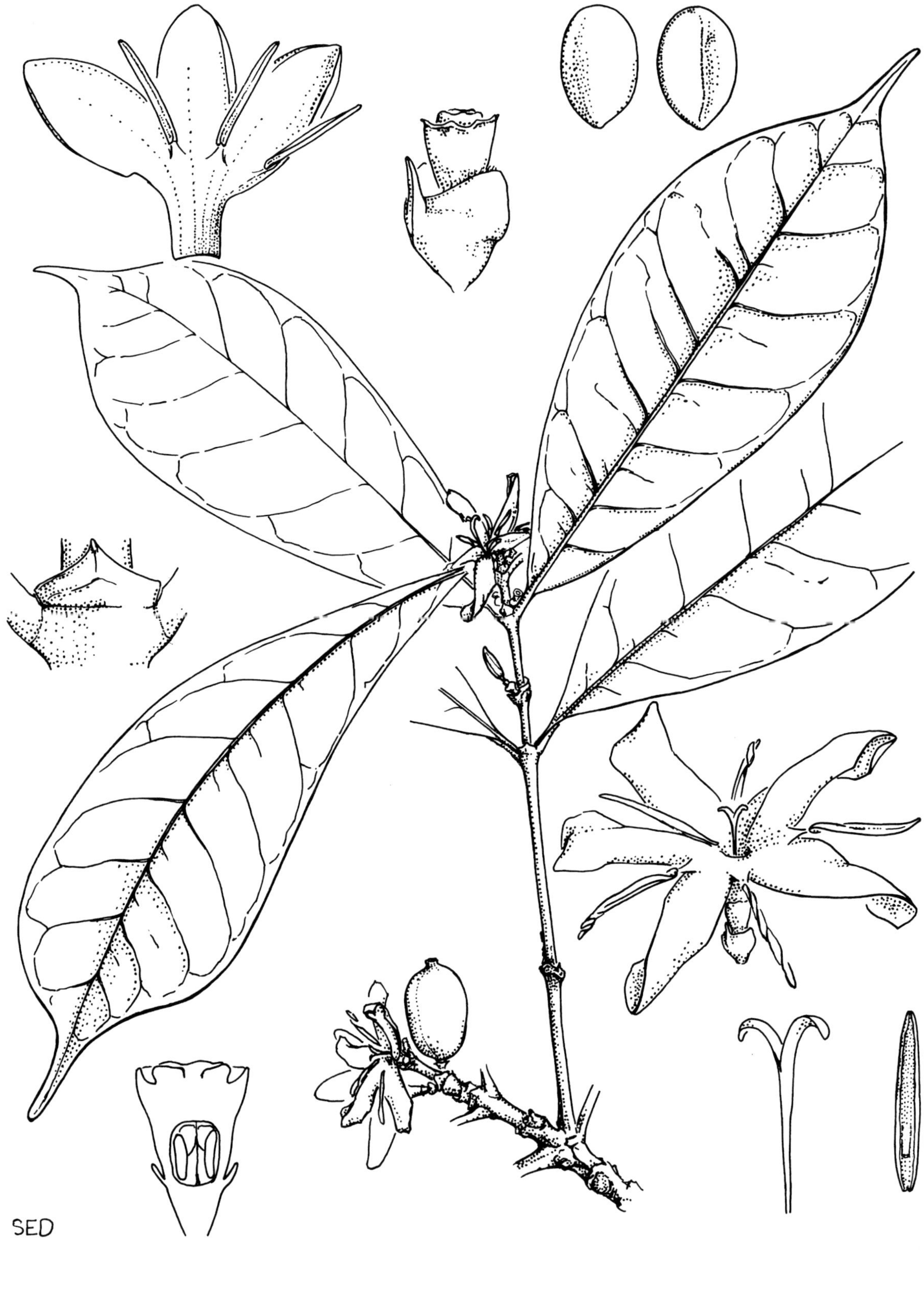

Fig. 12 *Coffea montekupensis* (Rubiaceae) by S. Dawson

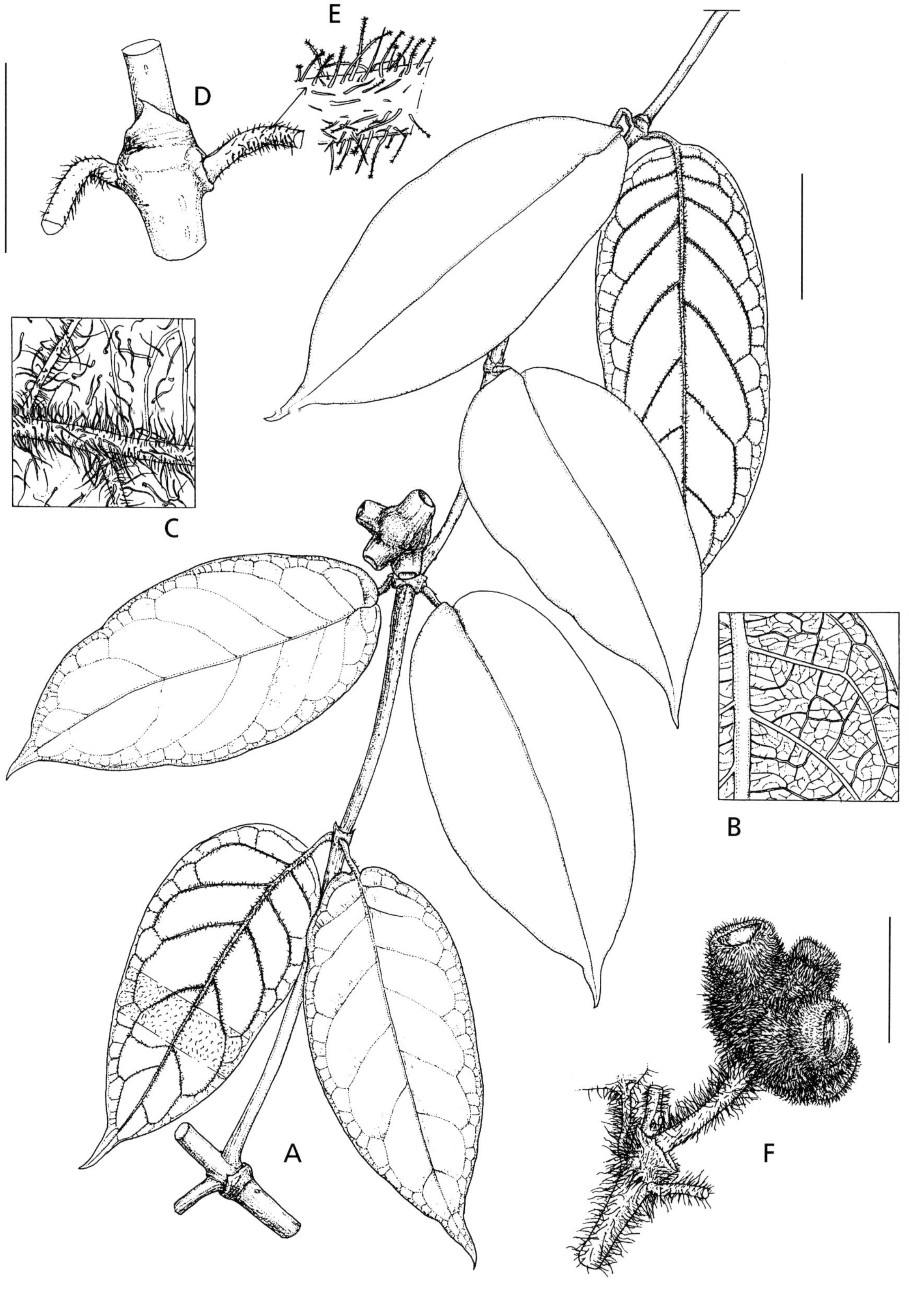

Fig. 13 *Morinda mefou* (Rubiaceae) by Juliet Williamson

Fig. 14 *Oxyanthus okuensis* (Rubiaceae) by Margaret Tebbs

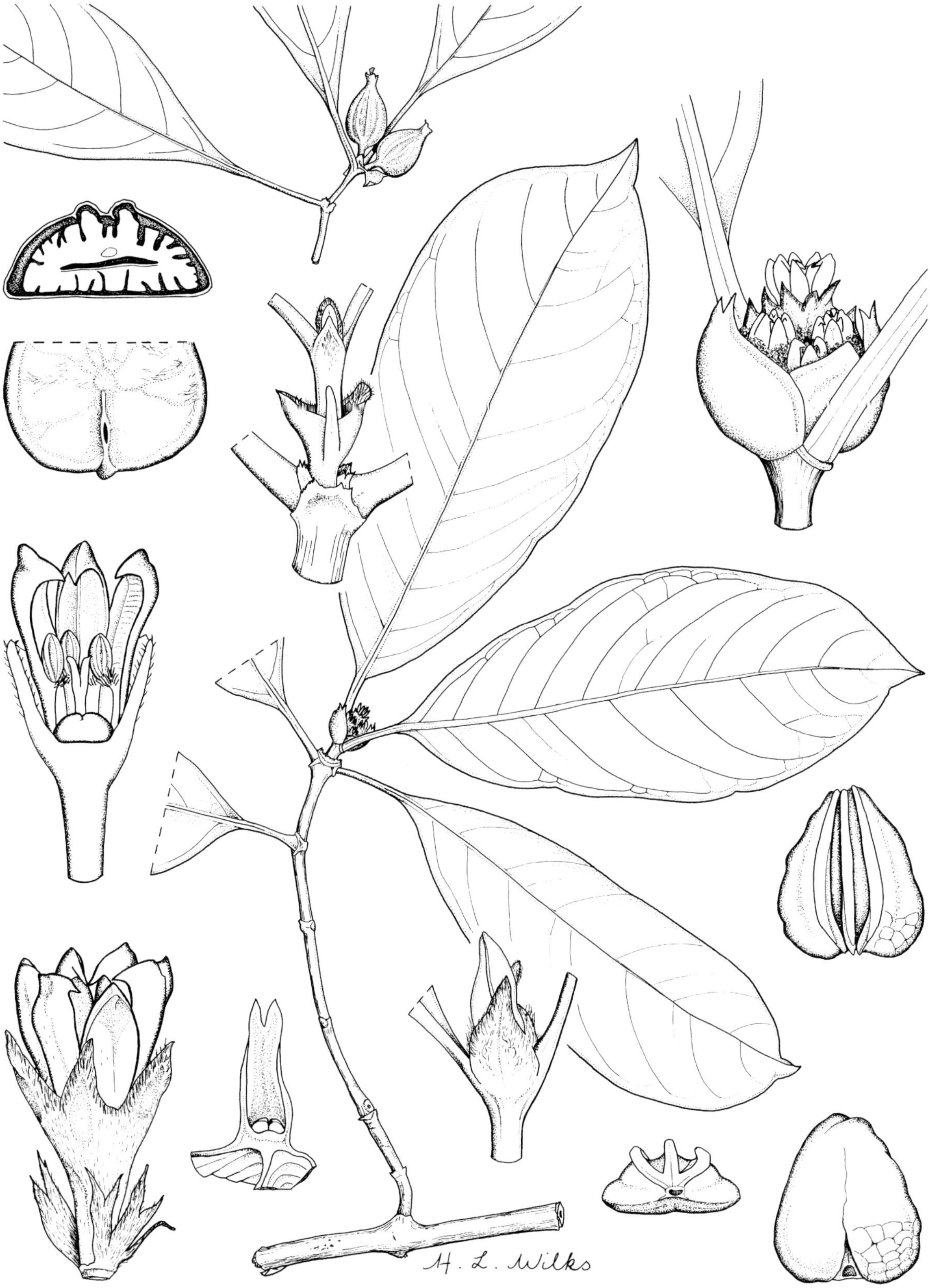

Fig. 15 *Psychotria bakossiensis* (Rubiaceae) by H.L. Wilks

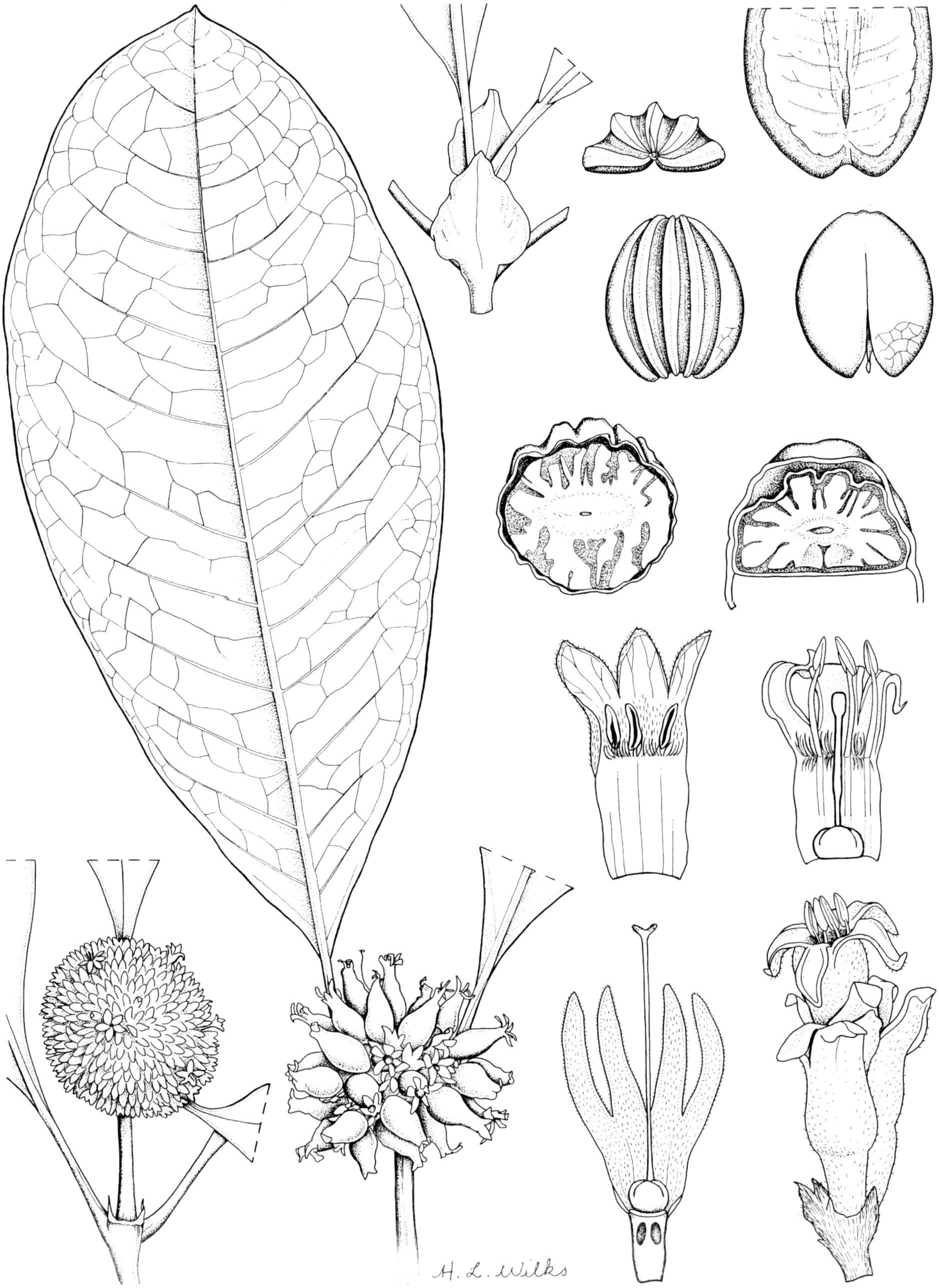

Fig. 16 *Psychotria geophylax* (Rubiaceae) by H.L. Wilks

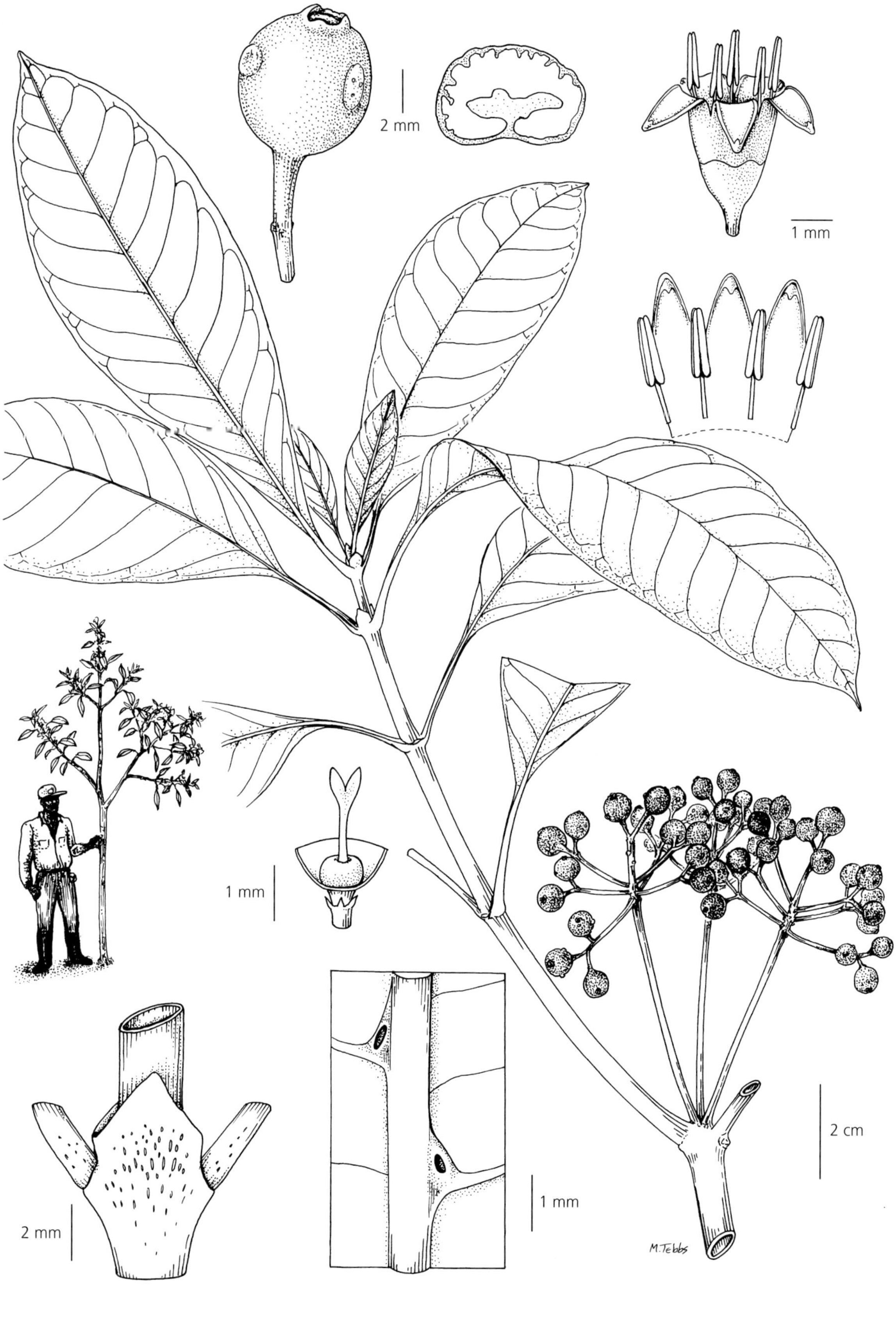

Fig. 17 *Psychotria moseskemei* (Rubiaceae) by M. Tebbs

Fig. 18 *Stelechantha arcuata* (Rubiaceae) by S. Dawson

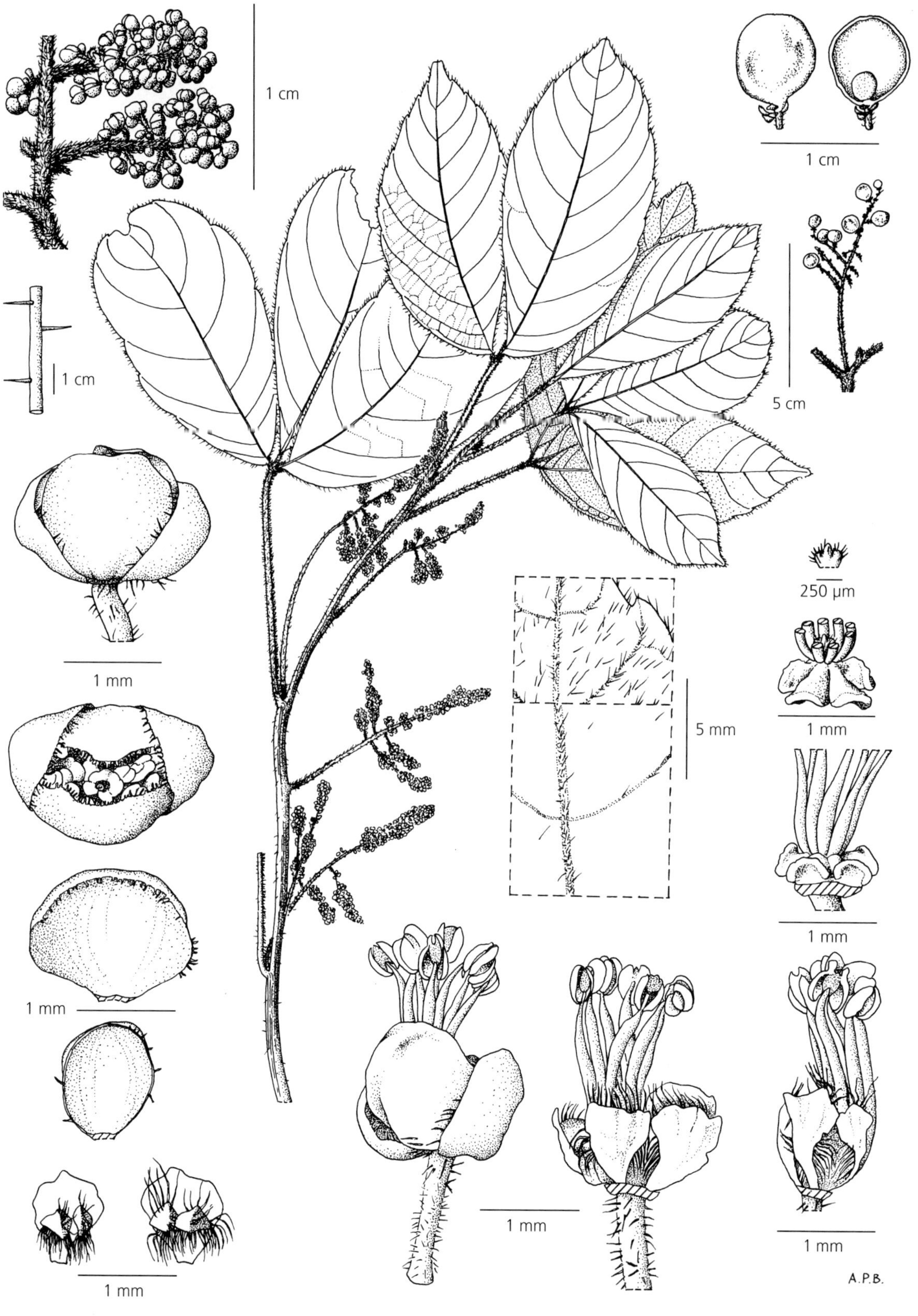

Fig. 19 *Allophylus ujori* (Sapindaceae) by A. Brown

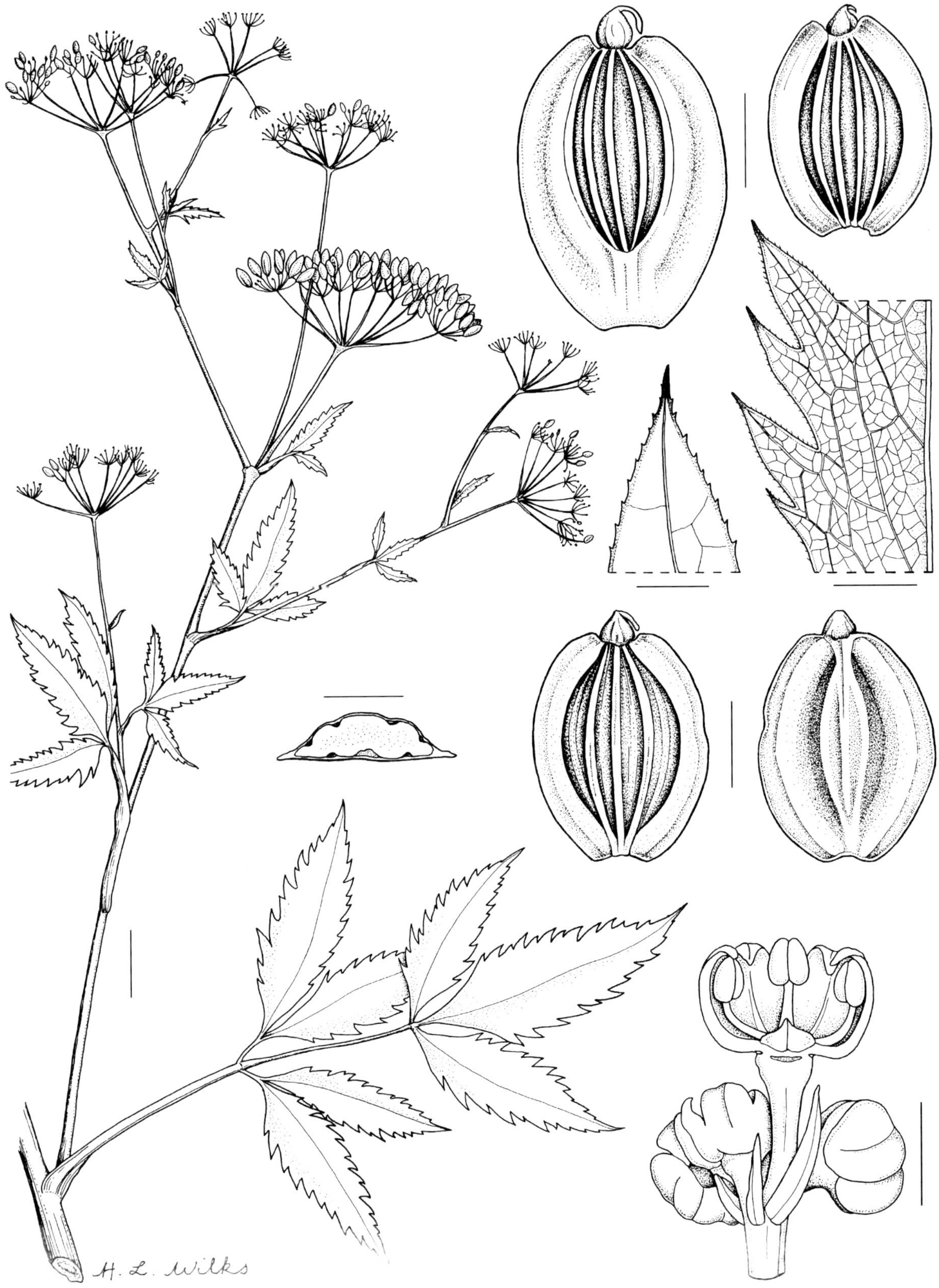

Fig. 20 *Lefebvrea kupense* (Umbelliferae) by H.L. Wilks

Fig. 21 *Coleochloa domensis* (Cyperaceae) by M. Tebbs

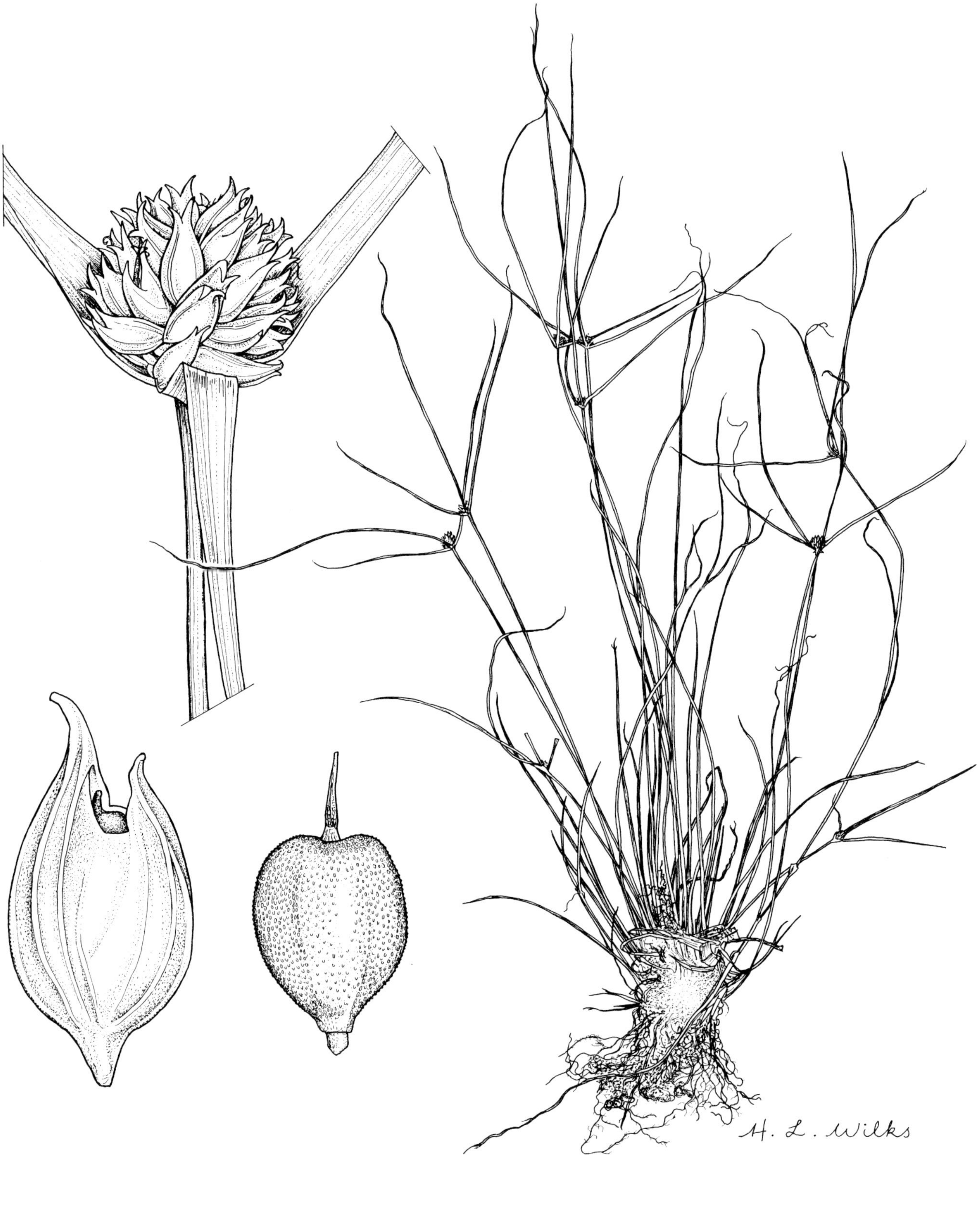

Fig. 22 *Cyperus rheophytorum* (Cyperaceae) by H.L. Wilks

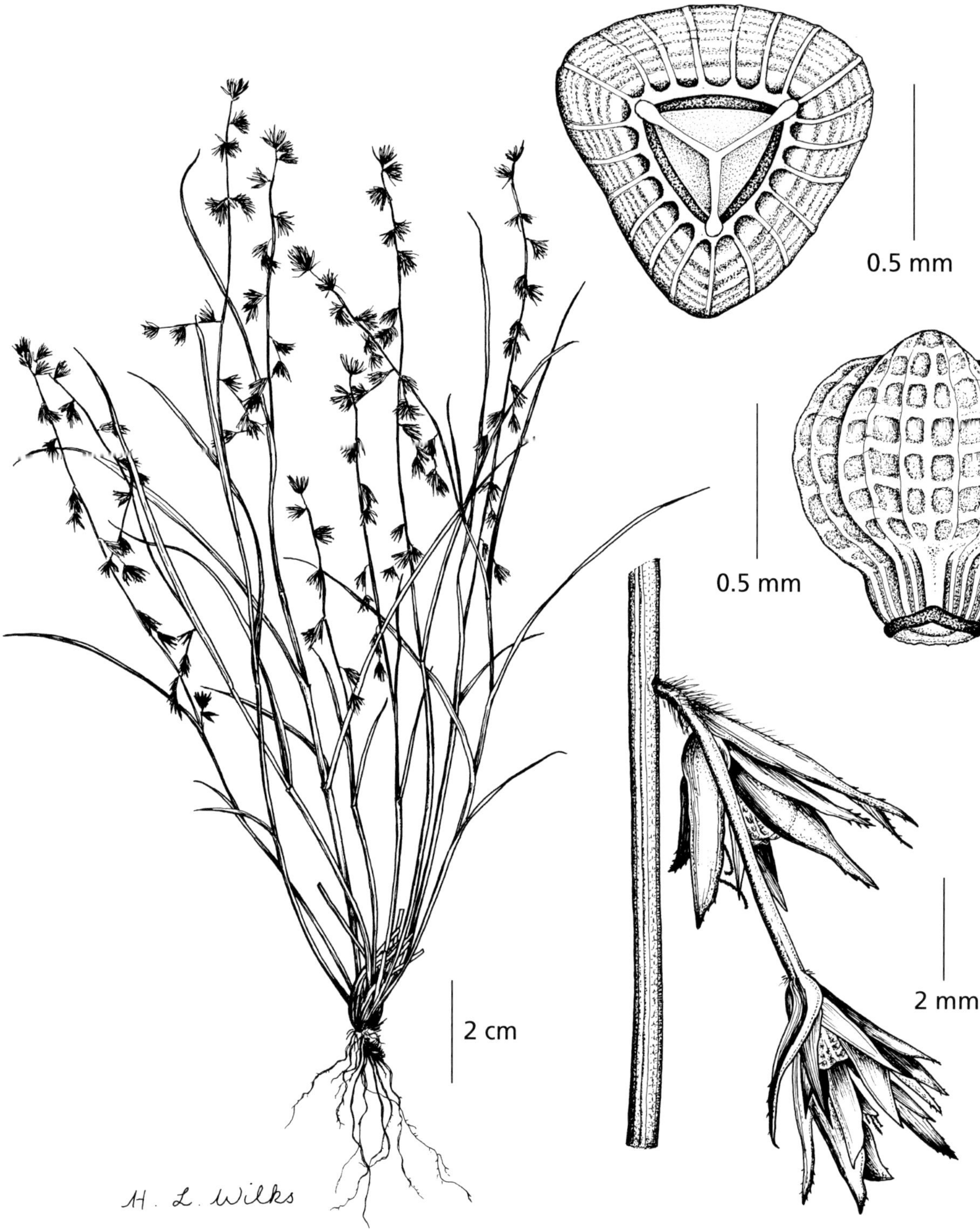

Fig. 23 *Scleria afroreflexa* (Cyperaceae) by H.L. Wilks

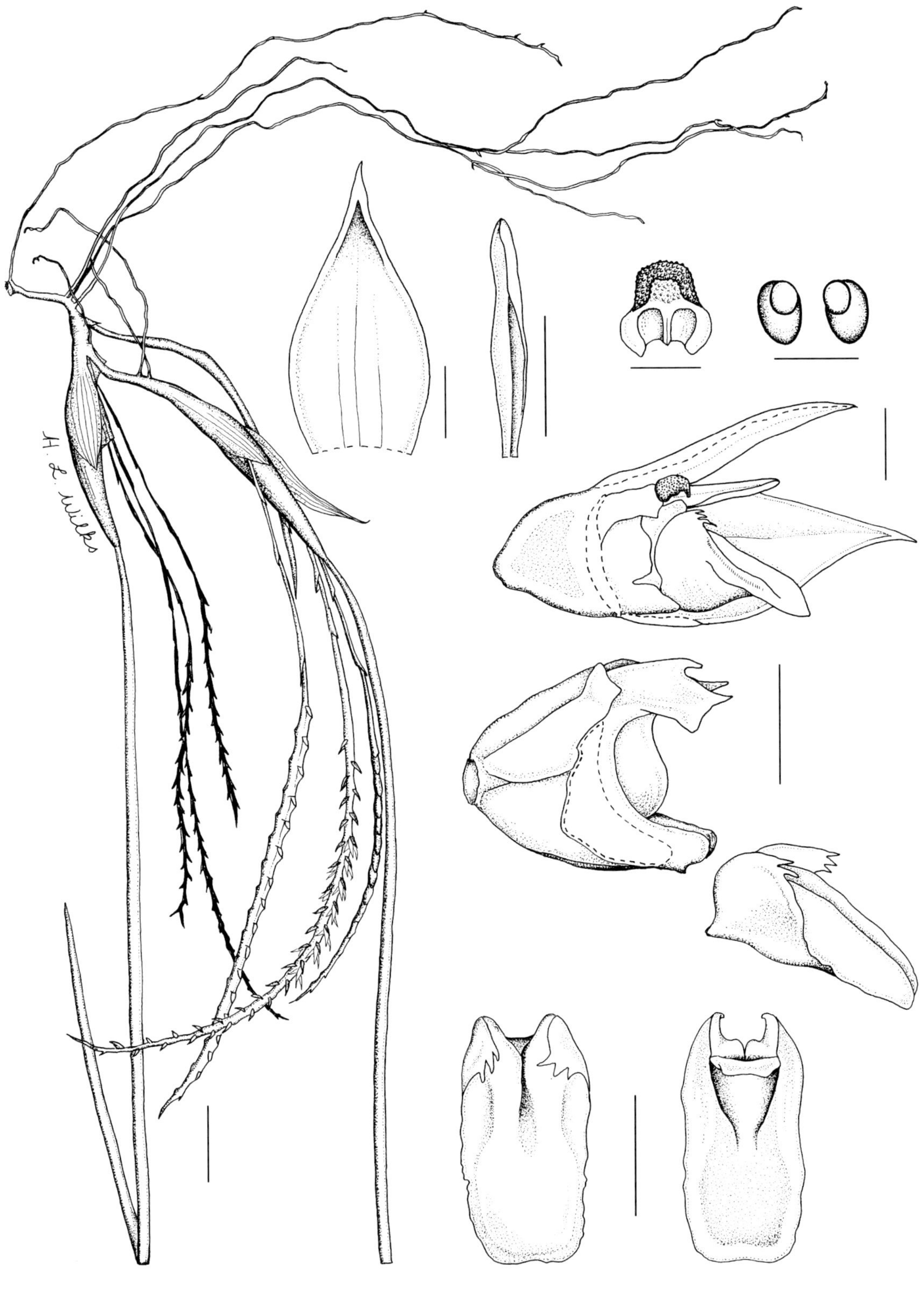

Fig. 24 *Bulbophyllum teretifolium* (Orchidaceae) by H.L. Wilks

DICOTYLEDONAE

ACANTHACEAE

assessed by Iain Darbyshire (ID) and Martin Cheek (MC)

Although no Flore du Cameroun exists for this family, much research work on delimiting taxa by Darbyshire (various genera), Vollesen (various genera), Champluvier (*Brachystephanus* with ID), Pearce (*Isoglossa* with ID) and Sidwell (*Brillantaisia*) means that species in many genera are generally well delimited, see e.g. Champluvier and Darbyshire (2009) A revision of the genera *Brachystephanus* and *Oreacanthus* (Acanthaceae) in tropical Africa. Syst. & Geogr. Pl., 79: 115-192.

A peculiarity of several forest Acanthaceae species is that they flower en masse only at intervals of 7, 9 or more years when they can be very conspicuous for several weeks. In the intervening years they flower not at all, or only very sparsely, and so cannot be reliably identified. After flowering, the whole population dies, regenerating from seed, therefore qualifying as showing massive fluctuations in population under Criterion Bc, otherwise rarely applicable to forest plants. While mass-flowering of some species is well known (e.g. in the genera *Acanthopale*, and *Brachystephanus*), due to the scarcity of long term field studies, the status of some species as mass-flowering can only be guessed at.

Dicliptera silvestris Lindau previously assessed as VU A2c by ID in Cheek *et al.* (2004: 140) is not maintained here since the species is now known to be a synonym of the more widespread and common (LC) species *D. maculata* Nees subsp. *maculata* (ID pers. comm. 2008).

Brachystephanus nimbae Heine previously assessed as VU B2ab(iii) by ID & MC was formerly recorded from Mt Kupe (Cheek *et al.* (2004: 139)) but this has now been shown to be due to a misidentification: "I am currently completing the revision of *Brachystephanus* written by Dominique Champluvier. Having looked again at *B. nimbae*, I conclude that this species does not occur in Cameroon, the specimens from Mt Kupe being a slender form of *B. jaundensis*. Unfortunately, this was a Red Data species for Cameroon whilst *B. jaundensis* is of Least Concern. Following this re-delimitation, *B. nimbae* is now restricted to Upper Guinea, B. jaundensis to Lower Guinea and the closely related *B. nemoralis* to Congolian forest." (I. Darbyshire pers. comm. to Onana, Sept. 2008)

Acanthopale decempedalis C.B.Clarke assessed by ID
VU B2b(iii)c(iv)

Range: Nigeria (Cross River State (2 coll.)), Equatorial Guinea (Bioko (5 coll.)) and Cameroon (SW: Kupe-Bakossi (9 coll.), Mt Cameroon (9 coll.). NW: Bamenda Highlands (Mt Oku (1 coll.); elsewhere (3 coll.))).

This taxon forms gregarious communities together with *Mimulopsis solmsii*, *Brachystephanus giganteus* (both Acanthaceae) and *Plectranthus insignis* (Labiatae), flowering together en masse on 7-9 year cycles followed by die-off and subsequent regeneration from seed (Cheek *et. al.* 2000: 21, Cheek *et al.* 2004: 60–61), so showing extreme fluctuations in the number of mature individuals. *M. solmsii* is a widespread Afromontane taxon but the remainder are restricted to the west Cameroon uplands phytochorion. The dead woody stems of these taxa following mass-flowering years, provide a source of firewood for local populations in the Bamenda Highlands. Kupe-Bakossi appears the most important area for *A. decempedalis*, being recorded at 6 places there. The assessment of this taxon made in Cheek et al. (2004: 138) is maintained here since no new data are available. This taxon would probably also rate as VU under criterion A3c. EOO is 39473 km^2.

Habitat: Cameroon Highlands, closed canopy submontane and montane forest understorey, occasionally in disturbed forest; (700–)1000–2300 m alt.

Threats: much of the submontane and montane forest in SW remains relatively undisturbed, but clearance for agriculture has been widespread in the Bamiléké Plateau/Bamenda Highlands, threatening these populations where they have not already been wiped out. Furthermore, the cyclical mass-flowering habit of this species results in large fluctuations in mature populations, making it susceptible to short-term stochastic change, for example local fire events or landslides which could decimate seedling populations.

Management suggestions: current population sizes in terms of both number of individuals and area of occupancy should be assessed as this is not clear from the specimen data. A better understanding of the flowering cycle of this community should be gained, including population fluctuations between cycles. Bamenda Highlands populations outside protected areas may provide valuable information on how immature populations respond to increased anthropogenic pressures.

Afrofittonia silvestris Lindau assessed by ID
VU A2c

Range: Nigeria (Cross River State (5 coll.)), Equatorial Guinea (Bioko (1 coll.)) and Cameroon (SW: Bakossi (2 coll.); Korup (7 coll.); Mt Cameroon (14 coll.); Banyang-Mbo (1 coll.)).

Predominantly a lowland forest taxon, this prostrate perennial forest floor herb is recorded as "locally abundant" in Korup and widespread on Mt Cameroon (Cable & Cheek 1998: xxxviii, where it was first assessed as VU), but appears rare in Bakossi and absent from Mt Kupe. The assessment of this taxon made by ID in Cheek *et al.* (2004: 138) is maintained here. Since that date forest destruction near Mt Cameroon at Bimbia-Bonadikombe (Mabeta-Moliwe) has been reported (Alicha pers. comm. 2008) likely rendering the species extinct there. EOO is calculated as 36459 km^2.

Habitat: lowland closed-canopy forest, often growing in deep shade on forest floor; 0–500 m alt. Rarely in mid-elevation forest; 500–1400 m alt. (Bakossi).

Threats: continued forest clearance of low altititude evergreen rainforest habitat within its range, particularly on the lower slopes and foothills of Mt Cameroon, which is considered to have exceeded 30% in the last 100 years. However, higher altitude sites discovered in Bakossi are less threatened.

Management suggestions: since the foothills of Mt Cameroon are the main stronghold for this species, these should be monitored to ensure that the species survives and that protected areas are being respected. There are concerns that habitat loss here in the last few years, since the closing of the Mt Cameroon Project at Limbe, is set to continue. The high altitude sites in Bakossi may prove important for the conservation of this taxon and so should be assessed more fully; atypically for *Acanthaceae* taxa, this species is readily identifiable when sterile, aiding population assessments.

Anisotes zenkeri (Lindau) C.B.Clarke assessed by MC
EN B1ab(iii)+B2ab(iii)

Range: Cameroon (C: N'kolbison (1 coll.); Eloumden (1 coll.); "Yaunde-Station" (2 coll.). E: Nanga Eboka (1 coll.)).

This is a spectacular 2–3 m tall shrub with dense clusters of pink or red flowers each 5 cm long, probably bird-pollinated. The leaf blades reach 30 × 13 cm. It was first collected in 1890-94 by the great German collector Zenker from "Yaunde-Station, 800 m". In Feb. 1956 it was recollected by Benoit Mpom (2280SRFK) at 1500 m on Eloumden, one of the seven hills of Yaoundé. The Dutch botanists Birgitte and Willi de Wilde refound it at N'kolbison, 600 m alt. (1644, 31 Dec. 1963) and it was last recorded 35 km from Nanga Eboka to Bertoua in March 1969 by the American Bill Sanford (6130). Here it occupied a forest remnant by a stream at 600 m alt. Assessed here as Endangered since threatened (see below) at all three sites. AOO is estimated as 12 km^2 using 4 km^2 cells while EOO is calculated as 845 km^2.

Habitat: submontane evergreen forest understorey; 600–800(–1500) m alt.

Threats: forest clearance for small-scale farming at N'kolbison and Eloumden (pers. [illegible]–2007) and at Nanga Eboka (specimen data above).

Management suggestions: since so many species are found only on the Yaoundé hills, it is imperative that one or several be protected so that their natural vegetation and species survive the expansion of the city. *Anisotes zenkeri* has great potential as an ornamental plant for gardens. Seed should be collected from the wild to enable its introduction to horticulture as a safety net should it become extinct in the wild.

Asystasia glandulifera Lindau assessed by ID
VU B2ab(iii)

Range: Nigeria (Mambilla, Adamawa, Mayo Ini (1 coll.)) and Cameroon (SW: Kupe-Bakossi (6 coll. at 3 sites). E: Bertoua (1 coll.); Betare Oya (1 coll.). W: Bafoussam-Foumbot (1 coll.). NW: Bamenda (1 coll.)).

Previously known from 5 scattered locations, inventory work in Kupe-Bakossi considerably increased the known range, with 3 additional sites. The assessment of this taxon made in Cheek et al. (2004: 138) is maintained here since no new data is available. EOO is calculated as 118295 km^2.

Habitat: mid-elevation to submontane forest, preferring forest margins and clearances and lingering in farmbush; 800–1550 m alt.

Threats: intensification of land use at lower elevations, particularly the conversion of lower forest margins to arable agriculture; this is a significant threat around larger habitations such as Nyasoso in Bakossi, where this taxon has been collected 4 times, and around Bafoussam, a very large town in W Region around which almost all natural habitat has long been lost.

Management suggestions: monitoring of the populations above Nyasoso should provide further information on the habitat requirements of this species, particularly the extent to which it tolerates anthropogenic disturbance. Active protection of threatened sites may be required.

Barleria bornuensis S.Moore assessed by MC
VU B2ab(iii)

Range: N Nigeria (Sokoto Prov., Keana (1 coll.); Katsina (Bakori, Dutsin Makurdi) (1 coll.); Lake Chad, Bornu (1 coll.); near Zuru (1 coll.)) and Cameroon (N: from 100 km S Maroua (1 coll.); 4 km N Baikoua and Bidzar (Garoua-Maroua) (1 coll.)).

This spiny 15–45 cm tall shrub has sky-blue flowers 3–4 cm long. It is reported to be very local. Since it is only known from 8 sites with the threats described below (AOO 32 km^2 with 4 km^2 cells) it is assessed here as Vulnerable. EOO is calculated as 239007 km^2.

Habitat: savanna, 300 m alt.and above.

Threats: over-grazing by stock.

Management suggestions: if further survey work does not show this species to have a greater range or number of sites than warrants threatened status, then measures should be taken to include this plant in a protected area and a management protocol researched and applied.

Brachystephanus giganteus Champl. assessed by ID (syn. *Oreacanthus mannii* Benth. FWTA 2: 432 (1963). VU A3c+ B1ab(iii)+B2b(iii)c(iv)

Range: Equatorial Guinea (Bioko (2 coll.)), Cameroon (SW: Kupe-Bakossi (2 coll.); Rumpi Hills (1 coll.); Mt Cameroon (13 coll.); Lebialem-Bamboutos (1 coll.). NW: Bamenda Highlands (Mt Oku (3 coll.), Bafut Ngemba (1 coll.))).

A species of the gregarious, mass-flowering herbaceous community of upland forest, it is more restricted in altitudinal range than *Acanthopale decempedalis*. It appears locally common on Mt Cameroon but is scarce elsewhere, being restricted to Mwanenguba and on Mt Kupe. Both are pre-1980 collections, and this species was not recorded during extensive inventory work at these sites in the 1990s, perhaps because these surveys did not coincide with mass-flowering years. The assessment of this taxon made in Cheek *et al.* (2004: 139), based on its habitat loss exceeding 30% in c. 30 years, is maintained here since no new data are available apart from the Lebialem record that does not change the assessment. Seven sites and an estimated AOO of 56 km² using 4 km² cells (but allowing for larger AOO at Mt Cameroon and Mt Oku where there is a more extensive area), with threats allows the assessment under B2. EOO is calculated as 11930 km².

Previously named *Oreacanthus mannii*, this taxon has been reassigned to *Brachystephanus* by D. Champluvier (Champluvier & Darbyshire op. cit.).

Habitat: closed canopy montane forest; (1300–)2000–2600 m alt.

Threats: as for *Acanthopale decempedalis*.

Management suggestions: as for *Acanthopale decempedalis*. In addition, a survey of this taxon should be carried out during the next mass-flowering year on Mt Kupe and Mwanenguba.

Brachystephanus kupeensis Champl. assessed by ID CR B1ab(iii)

Range: Cameroon (SW: Mt Kupe, above Nyasoso (2 coll.)).

Unknown until the inventory work of the 1990s, this taxon has been collected at two sites within 1 km of each other from the lower slopes of Mt Kupe near Nyasoso; no record of the abundance of this taxon at these sites has been made. The assessment of this taxon made in Cheek *et al.* (2004: 139) is maintained here since no new data are available.

Habitat: submontane forest and secondary regrowth forest; 900–1100 m alt.

Threats: the continued expansion of Nyasoso town has resulted in increased agricultural activity on Mt Kupe up to approx. 1000 m alt., with further encroachment inevitable in the near future. The locations for this species are within 100-200 m of intensive farming activity.

Management suggestions: it is imperative that the collection sites are revisited to assess current population levels; care should be taken to separate this taxon from similar *Brachystephanus* spp. using the key in Champluvier & Darbyshire (op.cit.). A better understanding of its habitat requirements should be gained, in particular the extent to which it tolerates disturbance. Notifying the Nyasoso population of its existence and rarity, perhaps through issuing a conservation poster on this species to local schools and community halls, may help in the protection of these sites. Other potential sites on the mountain should be surveyed for new populations.

Brachystephanus longiflorus Lindau assessed by MC VU B2ab(iii)

Range: Nigeria (Ogoja (1 coll.)), Equatorial Guinea (Bioko (3 coll., 1 site)) and Cameroon (SW: Mt Kupe (2 coll.); Mt Cameroon (2 coll., 1 site); Mbango, Ndikinimeki, Eséka, Ngambe (1 coll./site each)).

Despite the addition of four new sites and an expanded range from the Champluvier & Darbyshire revision (op. cit.), the assessment of this taxon made in Cheek *et al.* (2004: 139) is maintained here since only 10 sites are known (AOO 40 km² with 4 km² cells). Addition of further sites will disqualify the taxon from criterion B, but it may then yet qualify under criterion A. EOO is calculated as 30525 km².

The paucity of collections from Cameroon despite extensive survey work (e.g. not refound at Mt Cameroon despite intensive surveys in the 1990s), and its absence from such areas as the Bakossi Mountains and the Rumpi Hills, suggest that this species is restricted in range and genuinely rare.

Habitat: submontane and montane forest undergrowth; (1000–)1300–1550 m alt.

Threats: forest clearance for wood and agriculture threatens the Nigerian population and also that at Mt Cameroon (Buea).

Management suggestions: so long as forest remains largely undisturbed above 1000 m, no intervention is required in Bioko or Cameroon; confirmation of its continued existence in Nigeria is required.

Brachystephanus oreacanthus Champl. assessed by MC

EN B2ab(iii)

Range: Guinea (Conakry) (Simandou (2 coll.)) and Cameroon (NW: Metchum R (1 coll.)).

This 1 m forest herb is interesting as it is intermediate in several characters between *Brachystephanus* and *Oreacanthus*, which is why Champluvier sunk the latter into the former (Champluvier & Darbyshire op.cit.). The purple flowers each 1.5 cm long are borne in loose terminal spikes c. 30 cm long. It is known in Cameroon only from *Letouzey* 13412 (fl. 3 Dec.1974), from Metchum R near Obang, 15 km S of Wum, along a logging road. Here *Brachystephanus oreacanthus* is assessed as endangered since only two sites

are known (AOO 8 km² with 4 km² cells) with threats as detailed below. The actual AOO is far smaller and has been measured at the Guinea site, where a detailed survey specifically to study this species occurred in 2010, to be only a few m².
See the revision of *Brachystephanus* in Syst. & Geogr. Pl. 2009: 148 (Champluvier & Darbyshire).
Habitat: submontane forest; c. 800 m alt..
Threats: in Cameroon, forest clearance for agriculture (pers. obs.); in Guinea fires set in lowland grasslands have begun to enter submontane forest near this species (Haba and Haba pers. comm. 2009). The Guinea site is near a proposed iron mine but outside the proposed footprint and efforts are being made to protect the location.
Management suggestions: the forest in Cameroon in which this species survives should be rediscovered, made known to the local population and officials, and protected so that the species might survive. The usual basic populational data (density, range , regeneration levels, threats) should be gathered as a baseline for future monitoring and to gauge needs for future management intervention. It is known that some species of *Brachystephanus* (e.g. Cheek *et. al.* 2000: 21, Cheek *et al.* 2004: 60–61) flower en masse but only at intervals of many years, so this species may have been passed by when sterile by other botanists if it behaves in this way. However, at the one site still known in Guinea, at Simandou, the species has flowered at the end of both 2008 and 2009. Seed was collected in early 2010 with a view to safeguarding and multiplying the species for possible future wild plantings.

Brillantaisia lancifolia Lindau assessed by ID
VU B2ab(iii)
Range: Nigeria (Cross River State (2 coll.)), Cameroon (SW: Bakossi (3 coll.); Mt Cameroon (2 coll.)) and Gabon (2 coll.).
This taxon is apparently obligately rheophytic. It is easily recognised from other species in the genus by its narrowly elliptic leaves. Although having a large extent of occurrence, this species is rare across its range, being known from only 10 locations, though recorded as "locally abundant" at both Ndile waterfall, Ngomboaku (*Mackinder* 335) and on the Onge River (*Watts* 930). As an aquatic species, IUCN currently accept a grid cell of 1 km² per site, giving an AOO of 10 km².). EOO is calculated as 38824 km². The assessment of this taxon made in Cheek *et al.* (2004: 140) is maintained here. The reference to Rio Muni has been removed since unsubstantiated.
Habitat: rocky river margi[illegible] to submontane forest; 0–1350 m alt.
Threats: forest clearance in upstream sites may result in loss of populations through increased run-off causing flooding or silting of its favoured habitat. This is a potential threat at, for example, the Mungo R in Bakossi, as upriver activity is high. Increased use of rivers by humans, for example providing water for livestock, may cause excessive disturbance through trampling at the water's edge.
Management suggestions: monitoring of populations where upstream disturbance is occurring to assess the sensitivity of this species, for example along the Mungo R. Prevention of forest clearance in catchments where this species is recorded.

Chlamydocardia subrhomboidea Lindau assessed by MC
EN B2ab(iii)
Range: Cameroon (SW: Loum FR (2 coll.); Mungo R FR (1 coll.). C: Nguila-Ntui (1 coll.). Moliki (1 coll.) and Congo (Kinshasa) (Equateur Prov. at Banzyville, Businga (1 coll.)).
Erect, sparsely branched, glabrous forest herb 30–60 cm tall, stems terete with 2 lines of hairs. The terminal spikes are 3–10 cm long clothed in green elliptic bracts 1–1.5 cm long. The pale yellow green, 4-lobed corollas are inconspicuous and produced in Nov. & Dec. Several specimens formerly identified as *C. subrhomboidea* in Cameroon have been renamed as *C. buettneri* but "tending towards *C. subrhomboidea*" by Darbyshire at K. The two taxa were united in FWTA but separated in Fl. Gabon. *C. buettneri* has heart-shaped bracts.
Here assessed as Endangered since five sites are known (AOO 20 km² with 4 km² cells) and since threats are as below. EOO is calculated as 17889 km².
Habitat: Lowland semi-deciduous forest; 150-600m alt.
Threats: the sites at Loum and Mungo R are threatened by clearance of the forest understorey for farming of coco-yams and cocoa (pers. obs. MC). Threats at the other sites are unknown.
Management suggestions: examination of specimens at BR may well reveal further sites in Congo (Kinshasa), reducing the threat assessment level of this species. Further surveys might also yield new sites if this inconspicuous species is targeted. Areas of the Loum and the Mungo R FR which have *C. subrhomboidea* could be protected as a nature reserve and the understorey maintained intact, to secure the future of this plant. Monitoring would then be advisable.

Crossandra obanensis Heine assessed by MC
EN B1ab(iii)+B2ab(iii)
Range: Nigeria (Oban Rock at Oban (2 coll.) and Cameroon (C: N'kolbison (1 coll.))..
This single-stemmed shrublet reaches 50 cm tall, the 10 cm terminal spike bearing 2 cm yellow flowers among a rosette of leaves. One of the Nigerian collectors reports seeing but a single plant, so it appears scarce even at the two known sites. Since it is threatened (see below) and since only two sites are known (AOO 8 km² with 4 km² cells), it is here assessed as Endangered. EOO is calculated as 194 km².

Habitat: wet areas, on rock or in swamps, in darkest parts of evergreen forest; 300–600 m alt.
Threats: at N'kolbisson forest continues to be cleared for agriculture (pers. obs.)
Management suggestions: establishment of a protected area at N'kolbisson would maintain the forest habitat of this rare taxon. Further surveys might discover new sites for this plant.

Dicliptera alternans Lindau assessed by MC
VU B2ab(iii)
Range: Cameroon (SW: Buea, Mt Cameroon (1 coll.); Mundame (1 coll.); N'kolbison (1 coll.)), Congo (Brazzaville) (Boudika (1 coll.)), Congo (Kinshasa) (Lodja-Sankura (1 coll.)) and W Uganda (Kibale NP (1 coll.)).
This 0.15–1.5 m weak, probably short-lived perennial herb of forest understorey has ovate to elliptic leaves 2.5–10 cm long. The loose terminal spikes are 3–8 cm long, conspicuous for the paired ovate 7 – 8 mm long sepals which are white with a green edge and midrib. The corollas are pale white, fall early and are seldom seen.
While widespread (EOO is calculated as 1399400 km²), *D. alternans* is very rare, with only six sites globally (AOO 36 km² with 4 km² cells). Even in intensively botanised sites such as Mt Kupe, it has not been picked up, and at Mt Cameroon, not since 21 Jan. 1891 (*Preuss* 604, type), despite intensive botanical surveys there in the 1990s. Given the threats (below) it is here assessed as Vulnerable.
Habitat: lowland and submontane forest; 50 – 1500 m alt.
Threats: while the Ugandan site is in a National Park, most other sites are insecure. At both Buea and N'kolbisson, the forest habitat has suffered and will continue to be lost to housing and agriculture as these large towns continue to expand.
Management suggestions: fortunately the Ugandan site is protected, but it would be good to obtain protection of at least one other site, e.g. at Mundame, Buea or N'kolbisson. Before this can be done, the plants need to be refound in the wild, supposing they survive at these sites. Public education of the importance of conserving these plants at the latter two sites would be valuable.

Dischistocalyx champluverianus Lejoly & Lisowski assessed by MC
EN B1ab(iii)+B2ab(iii)
Range: Equatorial Guinea (Rio Muni (two sites)) and Cameroon (S: Zingui (two sites)).
Known as Lékwajo in the language bagielli (*Letouzey* 9103).
This epiphyte grows on bushes and is 0.5–1 m tall with unequal leaves, obovate bracts and white or very pale violent flowers 2–3 cm long. In Cameroon known from two sites: 10 km SSE of Zingui (*Letouzey* 9006, 13 March 1968) and 15–25 km SW of Zingui [45 km SSE of Kribi, near Campo] (*Letouzey* 9103, 22 March 1968). Biafran forest with *Sacoglottis, Lophira & Caesalpinoids* on sand. Two further sites are known in Rio Muni; estuaire du Rio Muni, village Mayang, au bord la Mitong R (*Lisowski* 1242) and route de Nco vers Misoborg, au grande cascade (*Nguema & Lejoly* 190) fide Bull. Jard. Bot. Natl. Belg. 67:107 (1999).
Here the species is assessed as Endangered since only four sites are known (AOO 16 km² with 4 km² cells) with threats as below. EOO is calculated as 1667 km².
Habitat: understorey of wet evergreen forest, altitudinal range unknown.
Threats: *Lisowski* 1242 was recorded near Mayang village making it vulnerable to forest clearance from slash and burn agriculture.
Management suggestions: further exploration may provide further sites for this taxon. If not included in a protected area, this should be rectified.

Dischistocalyx rivularis Bremek. assessed by MC
CR B2ab(iii)
Range: Cameroon (S: Lolodorf (1 coll.)).
Known only from the type (Lolodorf, *Zenker* 4056 coll. 1911) according to Bremenkamp (Bot. Jahrb. 73:140, 1943). The isotype at K has Bipindi printed on the label.
This *Dischistocalyx* appears unremarkable within the genus. Here the species is assessed as Critically Endangered since only one site is known (AOO 4 km² with 4 km² cells) with threats as below.
Habitat: probably lowland or submontane evergreen forest (not recorded).
Threats: Lolodorf has expanded since 1911, possibly threatening the site of this species due to urbanisation and slash and burn agriculture. The exact site of the location remains unknown.
Management suggestions: a badly needed revision of this genus might alter species delimitations, hence assessments. Efforts should be made to rediscover this poorly known species at Lolodorf and secure its protection.

Hygrophila mediatrix Heine assessed by MC
EN B2ab(iii)
Range: Cameroon (Adamawa: Tourningal (1 coll.)), CAR (Bamingui-Bangoran (1 coll.) and Chad (Marou and Aloua (2 coll.)).
Known in Cameroon only from the type specimens: km 45 route de Ngaoundéré à Belel: mare de Tourningal (*Jacques-Felix* 8821, 24 Oct. 1967); and in Chad: Marou, bord d'une petite digue en zone inondée (*Fotius* 1255, 4 Dec. 1968) and from Aloua, large mare temporaire peu profonde (*Fotius* 1834, 9 Nov. 1969). The CAR reference derives from Missouri data and has not been confirmed (*Fay* 15677).
Aquatic herb with long, leafy, tubular, ?rooting underwater stem, the apical 10 cm emerging above the surface, aerial

leaves 2 cm, feathery-pinnate, flowers 2 cm, white, falling in the course of the afternoon.
Heine (Adansonia 11: 653 (1971). Here the species is assessed as Endangered since only three sites are known (AOO 12 km² with 4 km² cells) with threats as below. EOO is calculated as 196782 km².
Habitat: seasonal lakes in savanna habitats.
Threats: trampling of the wetland habitat for watering stock of planting of short cycle crops to benefit from the humid soils are likely to threaten seed production of this probable annual. Extreme fluctuations in population are likely due to seasonal variation in rains.
Management suggestions: if not already secure in a protected area, this should be rectified. Further survey work might produce more sites for *Hygrophila mediatrix*.

Isoglossa dispersa I.Darbysh. & L.J.Pearce (syn. *Isoglossa* sp. A of FWTA) assessed by MC
VU B2ab(iii)
Range: Cameroon (NW: Bamenda Highlands at Nkambe (1 coll.); Metchum (1 coll.); Nkom-wum (1 coll.). W: Mbo Plain (1 coll.), Sierra Leone (Loma Mts (2 coll.)) and Guinea (Conakry) (to be confirmed).
This taxon is a herb 0.4–2.5 m tall with a 30 cm panicle of 1 cm long pink-white flowers. Montane Cameroon *Isoglossa* are being revised by Pearce and Darbyshire. This taxon is known from 6 locations, with an AOO of 24 km² using 4 km² cells. EOO is calculated as 240902 km².
Habitat: lowland and submontane forest, often in damp areas; 300 – 1000 m alt.
Threats: apart from one record in Sierra Leone and one likely one in the Guinea Highlands, this species is restricted to the Cameroon Highlands where it is vulnerable to forest clearance for agriculture. None of the sites are known to be protected in national parks or included in areas managed for nature conservation.
Management suggestions: the species should be refound in the wild, and one of its forest sites protected. The site at Metchum is a strong contender, if it survives there, since it is also the only known site for *Brachystephanus oreacanthus*.

Isoglossa nervosa C.B.Clarke assessed by MC
VU B2ab(iii)+c(iv)
Range: Equatorial Guinea (Bioko: Pico Basile), Cameroon (SW: Mt Cameroon (Mann's Spring and Bokwango); Bamboutos. NW: Mt Oku. Adamawa: Mt Daland) and Nigeria (Mabile-Ngelyaki).
Previously assessed as VU D2 (Cable & Cheek 1998: xxxviii) when known only from Mann's Spring. The re-assessment of Cameroon montane *Isoglossa* by Pearce and Darbyshire (in prep.) has extended the range of this taxon – it is now known from 7 sites along the Cameroon Highlands, with an AOO of 28 km² using 4 km² cells. EOO is calculated as 21960 km². Mt Cameroon remains the focal area for this taxon, with 7 of the 14 known collections. Believed to be mass-flowering.
Habitat: montane forest; 2000 – 2950 m alt.
Threats: while the Mann's Spring and Mt Oku sites appear secure from disturbance, those at Bokwango, Banboutos and Mabila are under great threat of clearance for agriculture, if not already lost. The status of the Basile and Adamawa sites is not known.
Management suggestions: monitoring at the secure sites and a recensus at the remaining locations is advisable.

Justicia camerunensis (Heine) I.Darbysh. assessed by ID
VU A2c
Range: Nigeria (Gembu Dist. (1 coll.)) and Cameroon (SW: Mt Kupe, Nyassoso (8 coll.); Mt Cameroon, Bova (1 coll.); Kumba (1 coll.). NW: Bafut-Wum (1 coll.). C: nr. Yaoundé (2 coll.); Bertoua (2 coll.)).
In Cameroon this taxon has a patchy distribution, being absent from several seemingly suitable areas including the Bakossi Mountains and much of Mt Cameroon. The assessment of this taxon made in Cheek *et al.* (2004: 140) is maintained here since no new data are available. EOO is calculated as 108578 km².
Habitat: understorey of closed canopy mid-elevation forest; (600–)900–1400 m alt.
Threats: throughout its range, it is threatened by extensive forest clearance. At more than 30% of the known sites, its habitat either has or is being lost. The Gembu and NW population is highly threatened as closed-canopy forest is scarce here today and still being lost; it was noted in the collection of 1975 that the forest patch in which it was found was "now in exploitation by local people with handsaws" (*Leeuwenberg* 8648); this site is therefore likely to be lost. At Mt Kupe, it was recorded most often above Nyasoso, at around 1000m alt., where agricultural encroachment is causing significant losses of forest. Sites around Yaoundé and at Bova are also likely to have been lost. The status at Bertoua and Obang is unknown.
Management suggestions: protection of existing mid-elevation forest where possible.

Justicia leucoxiphos Vollesen, Cheek & Ghogue assessed by ID
EN B1ab(iii)
Range: Cameroon (SW: Bakossi Mountains and Rumpi Hills (10 coll.)).
First collected in the 1980s. this species is now well known from the Bakossi Mountains following extensive survey work in the 1990s. It is recorded from 5 separate locations; Kodmin, Mejelet, Mwambong-Jide, Ngomboaku and Mbu Bolomi (Rumpi Hills), having an area of occupancy of less

than 5000 km^2. EOO is calculated as 450 km^2. It is locally not uncommon, though its range does not extend to Mt Kupe or Mwanenguba in the east. The assessment above is taken from that by Cheek in the protologue (Vollesen *et al.*, Kew Bull. 59, 2004). Part of the Bakossi Mts was gazetted as a National Park in 2007. The assessment of this taxon made in Cheek *et al.* (2004: 140) is maintained here.
Habitat: closed canopy forest understorey, rarely secondary forest; 900–1500 m alt.
Threats: an inferred decline in habitat quality due to illegal logging activity which begun within its area of distribution in 2002, but which has recently reduced following efforts from local conservation groups and communities.
Management suggestions: as a striking and easily recognisable species, *J. leucoxiphos* could contribute to the promotion of community-based forest protection. Conservation posters on the rarity and uniqueness of this species could be distributed to schools and community centres, encouraging local communities to protect areas where it is found. Further inventory work in the Bakossi Mountains and Rumpi Hills is likely to uncover new locations.

Justicia orbicularis (Lindau) V.A.W.Graham assessed by ID
VU A2c
Range: Nigeria (Cross River State (3 coll.)), Equatorial Guinea (Bioko (1 coll.)) and Cameroon (SW: Bakossi (3 coll.); Kumba (1 coll.). C: nr. Yaoundé (2 coll.). S: Ébolowa-Amban (1 coll.). Littoral: Bafang-Yabassi (1 coll.)).
A highly distinctive species known from just 12 locations. Populations appear disjunct, being absent in several seemingly suitable areas. In Bakossi, it is only known in the west, though further exploration of the lowlands of southern Bakossi may reveal further populations. It is also absent from the lowlands around Mt Cameroon, and has not been recorded in extensively collected forest areas around Bipinde and Kribi in S Cameroon. The assessment of this taxon made by ID in Cheek *et al.* (2004: 141) is maintained here, it being estimated that more than 30% of the population has been lost in the last three generations, here estimated as amounting to 30 years. EOO is calculated as 73107 km^2.
Habitat: Lowland forest, often along streams; 0–700 m alt.
Threats: throughout its range, deforestation has been extensive in the lowlands, and this continues outside protected areas.
Management suggestions: as the species is readily identifiable, population census data should be possible in protected areas, and an assessment of its habitat requirements made. Further inventory work in suitable lowland sites, such as the Loum FR, may provide additional sites.

Justicia telloensis Hedrén assessed by MC
EN B2ab(iii)
Range: Cameroon (Adamawa: Adamawa Highlands (2 sites)).
This taxon is a straggling herb to 1 m tall. The leaves are narrowly ovate, coriaceous, c. 2.5 × 1 cm, apex and base rounded. The flowers are purple, c. 0.5 cm long. It was identified as a possible new species in Cheek *et al.* (2000: 55, 112) where it was listed in the red data account as *Justicia* sect. *Harnieria* sp. nr. *J. heterocarpa* subsp. *praetermissa*. It was only later that Kaj Vollesen brought to light that the species had already been published by Hedren (Nordic. J. Bot. 10: 269 (1990).
It is thus far known only from from two sites: 1.Mbi Crater (*Etuge* 4580, 9 Dec. 1998). 2. Near falls in Tello R, about 47 km E of Ngaoundéré (*W.J.J.O. de Wilde* 4289, 27 Nov. 1964 and *Jacques-Felix* 8831, 24 Oct. 1967). In view of the threat cited below, *Justicia telloensis* is here assessed as Endangered.
Habitat: swampy grassland near rivers or in crater lake; 1200-1750 m alt. The vegetation of the Mbi Crater swamp is documented in Cheek *et al.* (2000: 18).
Threats: the status of the site at Tello is unknown. While the Mbi Crater is an official wildlife sanctuary, it is surrounded by an intensive agricultural estate (wheat and stock raising) and so not easily accessed by the local government wildlife officer based at Belo. At Mbi the plant is not very common and can be considered under threat from the possibility of trampling by farm stock.
Management suggestions: the Tello site should be revisited and monitored. The wildlife officer responsible for Mbi should be provided with a motorbike and fuel so as to access and monitor this important site for plant conservation.

Justicia tenuipes S.Moore assessed by MC
EN B2ab(iii)
Range: Nigeria (Oban) and Cameroon (SW: Korup).
This creeping perennial herb of evergreen rainforest on sand is distinctive. The elliptic-oblong leaves, c. 8–10 × 3–4 cm, gather soil on their lower surface and are held flat to the ground, forming a loose rosette from which 1–3 scapes, each 10–15 cm, arise in Dec.–Jan. Branched 2– 3 times, the loose panicle bears c. 10 white and pink slender corollas 1.5 cm long, surrounded by filamentous sepals. At Korup this species is locally common in the S part, where 4 specimens are known.
Since only two sites are known, one threatened (see below), *J. tenuipes* is here assessed as Endangered since only two sites are known (AOO 8 km^2 with 4 km^2 cells) with threats as below. EOO is calculated as 86 km^2.
Habitat: shaded forest floor on sand, 3–4 m rainfall p.a.; alt. 50–400 m.

Threats: the type locality at Oban is possibly outside the Cross River NP and may already have been destroyed by town expansion and forest clearance. The Korup site is securely protected.
Management suggestions: the sites for this species at Korup should be monitored to check that the population remains stable.

Physacanthus talbotii S.Moore assessed by MC
EN B1ab(iii)+B2ab(iii)
Range: Nigeria (Stubbs Creek, Eket; Oban) and Cameroon (SW: Korup NP).
This perennial, mat-forming, creeping herb has dark green ovate, round-based leaves, c. 5 cm long. The flower arises vertically 6 cm tall, from the stem apex. The slender corolla tube is sheathed at the base in a green, stout, corrugated, cylindric calyx 2.5 cm long. The five equal blue petals are held flat, facing the sky, each blue-mauve with a purple centre. This species appears infrequent even at its three known sites. In view of the threats below it is here assessed as Endangered since only three sites are known (AOO 12 km² with 4 km² cells) with threats as below. EOO is calculated as 2750 km².
Habitat: lowland evergreen forest on sandy soils in high rainfall areas.
Threats: while the site in Cameroon is well protected, those in Nigeria are threatened. Here forest is under great pressure, and Stubbs Creek, Eket, approaching the oil rich Niger Delta, is thought to have an Exxon installation.
Management suggestions: the site(s) at Korup should be monitored to ensure that populations are stable. At Oban, the type locality (*Talbot* 972, 1911) an effort should be made to refind this plant and to determine whether it occurs in the Cross River NP.

Pseuderanthemum dispersum Milne-Redh. assessed by ID
VU B2ab(iii)
Range: Nigeria (Cross River State (3 coll.)) and Cameroon (SW: Bakossi (1 coll.); Mt Cameroon (1 coll.); Babenga nr. Limbe (1 coll.). S: Bipindi (1 coll.)).
This robust herb or shrub appears rare throughout its range, and was recorded only once during the extensive inventory work in SW Cameroon, during the 1980s and 1990s, a collection by D.W.Thomas at Bangem in north Bakossi. It has never been recorded in central or southern Bakossi or Mt Kupe and has not recently been found on Mt Cameroon. The assessment of this taxon made by ID in Cheek *et al.* (2004: 141) is maintained here. EOO is calculated as [illegible] km².
Habitat: undergrowth of mid- to low-elevation forest; 0–1000(-1400) m alt.
Threats: a continued decline in habitat is inferred by clearance of extensive areas of lowland and mid-elevation forest throughout its range for plantation agriculture and timber.
Management suggestions: attempts to rediscover this species on Mt Cameroon should be made. The collection from Bangem should be verified as the specimen was not available at Kew.

Sclerochiton preussii (Lindau) C.B.Clarke assessed by ID
EN B1ab(v)
Range: Nigeria (Obudu Plateau (1 coll.) and Cameroon (SW: Mt Cameroon (2 coll.); Mt Kupe (3 coll.)).
First collected "near Buea" in Nov. 1891, it was rediscovered on Mt Cameroon 101 years later in the saddle of Mt Etinde–Mt Cameroon. It was subsequently found in the 1960s at Obudu, Nigeria. The Mt Kupe collections from the 1990s, from above Kupe Village and Nyasoso, represent a significant expansion in the area of occupancy, but it has not been recorded from elsewhere in Bakossi. The assessment of EN (since only three sites are known (AOO 12 km² with 4 km² cells) with threats as below) which was first made in Cable & Cheek (1998: xxxviii) and revalidated by ID in Cheek *et al.* (2004: 141), is maintained here since no new data are available. EOO is calculated as 7327 km².
Habitat: an understorey shrub of closed canopy submontane and montane forest; 1100–1400 m alt.
Threats: stems cut for trap "springing sticks" on Mt Kupe (Cheek, *pers. obs.*); this is likely to have an impact upon the population of mature individuals at the two Kupe sites.
Management suggestions: as stated in Cable & Cheek (1998: xxxviii), an evaluation of the threat posed by exploitation for trapping is advised, and local trappers should be encouraged to use alternative, more common taxa.

Staurogyne bicolor (Mildbr.) Champl. assessed by ID
VU B1ab(iii)+B2ab(iii)
Range: Cameroon (SW: Kupe-Bakossi (11 coll.); Banyang Mbo-Ntale, (1 coll.); Mbu-Bakundu (1 coll.). S: 15km north of Kribi (1 coll.); Bipindi-Ébolowa (1 coll.)).
Previously known from only 3 disjunct locations, the botanical inventory work of the 1980s–90s identified the Kupe-Bakossi as highly important for this species, with 6 new locations from Bangem in the north to Nyale in the west and Mt Kupe in the east. It has not so far been collected on the slopes of Mt Cameroon or in neighbouring Nigeria and Bioko. The assessment of this taxon made by ID in Cheek *et al.* (2004: 141) is maintained here since only six sites are known (AOO 24 km² with 4 km² cells) with threats as below. EOO is calculated as 18004 km².
Habitat: an understorey herb of mid-elevation to submontane forest, often near water; 50–1500 m alt.
Threats: as this species extends to forest below 1000 m, it is threatened by widespread deforestation.

Management suggestions: protection of the remnant mid-elevation forest in Bakossi appears imperative for the survival of this species. Attempts should also be made to rediscover the disjunct populations in S.

Staurogyne kamerunensis **(Engl.)** Benoist subsp. ***calabarensis*** Champl. assessed by MC
VU B2ab(iii)

Range: Nigeria (Oban; Orem; and Aboabam at Ikom Distr.) and Cameroon (SW: Korup (Mundemba; Ikenge village; and S of Mt Juahan)).

A stoloniferous, densely puberulent forest floor perennial herb with ascending flowering stems 15 – 20 cm tall. The terminal, broad flowering spikes reach 10 cm long and are densely covered in elliptic green bracts c. 2 cm long among which the white flowers of about the same size emerge. The EOO of this taxon is highly restricted but within some parts of its range, at least it is locally common. Over four weeks in April 1996, c. 6 specimens were made in a survey area several km^2 S of Mt Juahan in the centre of Korup NP, suggesting that a grid of more than 2 km × 2 km may be appropriate for this taxon. This subspecies, being known from only 6 sites (AOO 24 km^2 with 4 km^2 cells) within a small EOO (calculated as 4379 km^2) is assessed here as Vulnerable due to the threats stated below.

Habitat: lowland evergreen forest in 3–4 m rainfall area, mostly on poor rocky soils, c. 300 m alt.

Threats: while the sites in Korup are well protected, several of those in Nigeria (Oban, Orem) lie along a road and their forest habitat is vulnerable to forest clearance for timber and agriculture.

Management suggestions: monitoring of the sites at Korup is advisable to ensure the future well-being of this species.

Staurogyne pseudocapitata Champl. assessed by MC
EN B2ab(iii)

Range: Cameroon (SW: 20 km W Mamfe. S: 35 km S Bengbis; Asseng, Nyabéssan. E: près Lomie) and Gabon (Cristal Mts).

This small creeping herb of swampy forested areas was published by Champluvier in 1991 (Bull. Jard. Bot. Nat. Belg. 61:123). The ovate leaves are 1.5–4.5 × 1–3.2 cm, the terminal flower heads 1.5 cm diam. with inconspicuous tubular white flowers 6.5–8 mm long appearing among bracts of the same length.

Staurogyne pseudocapitata is here assessed as Endangered given that it is only known from five sites (AOO 12 km^2 with 4 km^2 cells) and is threatened (see below). EOO is calculated as 121250 km^2.

Habitat: *Raphia* swamps and periodically inundated areas in shade of low altitude forest; forming carpets.

Threats: logging operations in forest may threaten the species by removing shade, churning up muddy areas and altering drainage patterns.

Management suggestions: further survey work is likely to produce more sites for this inconspicuous species, possibly revealing that its threat status should be downgraded. This aside, at least one site should be included in a protected area, monitored and managed to ensure the survival of this species.

Stenandrium thomense (Milne-Redh.) Vollesen assessed by MC
EN B2ab(iii)

Range: São Tomé & Príncipe (São Tomé (1 site)) and Cameroon (NW: Metchum R in Wum Distr. SW; Etinde. S: Akom II; Dipikar Is.)

First discovered on São Tomé and described as *Crossandra thomensis*, it was found in Cameroon at Metchum in June 1951 (*Ujor* in FHI 29253), then again in 1993 at Mt Etinde (*Williams* 74). Tchouto (2004) has reported it from two sites in Campo-Ma'an but we have not seen the specimens. EOO is calculated as 92404 km^2.

A prostrate brown-hairy forest floor herb with obovate leaves to c. 10 × 5 cm. The flower spike is 2–10 cm long with acute bracts 6 mm long and calyx segments 12 mm. The flowers are purple, 3 cm diam. It is here assessed as Endangered since only five sites are known (two of which need verification; AOO 20 km^2 with 4 km^2 cells) and due to its threats (below).

Habitat: lowland to submontane forest in shade; 400–600(–1100) m alt.

Threats: forest clearance for agriculture, e.g. at the Metchum site (see comments under *Brachystephanus oreacanthus*) and especially at Etinde where plans are advanced to clear the forest for plantations for oil-palm and other crops. Since 1993 much clearance has already occurred, possibly extirpating the only known site for this taxon in the Mt Cameroon area.

Management suggestions: the Tchouto voucher specimens, presumably at WAG, should be verified since they were probably made without reference to authentic specimens. Confusion with *S. guineensis* is possible. Measures should be taken to refind the plants at Etinde and at Metchum and to secure their long-term protection. A programme to multiply plants at Limbe Botanic Garden for reintroduction to safe sites would be appropriate.

Thunbergia rufescens Lindau assessed by MC
EN B2ab(iii)

Range: Nigeria (Ondo Prov.: Akure FR and Aiyetoro Owenna), Cameroon(SW: Barombi-Mbo to Kake at Kumba)..

A red stellate-hairy climbing shrub to 6 m in forest, with opposite elliptic leaves c. 12 × 7 cm. The white flowers are 5–7 cm long. First discovered in the 1890's by Preuss near lake Barombi Mbo (Kumba), it has not been refound in Cameroon despite much botanising at this site by Duncan Thomas and associates in the 1980's when he lived at Kumba.

Thunbergia rufescens is anomalous, having several characters of the genus *Phellocalyx*: stellate hairs on bracteoles and corolla tube base, entire leaves (Heine in FWTA 2: 402 (1963)). In view of the severe threats to this species (below) and being known from three (AOO 12 km² with 4 km² cells) with threats as below, it is here assessed as Endangered.

Habitat: lowland evergreen forest, flowering in Sept.

Threats: Nigeria's forests, including forest reserves, have seen widespread loss of canopy over recent decades. Forest inside the crater of Barombi has been at least partly cleared, and some crops planted (pers. obs.)

Management suggestions: it is a matter of urgency to rediscover *Thunbergia rufescens*, if it is not already extinct. Forest clearance around Barombi Mbo should cease and the remaining forest be protected and restored to its former extent, in view of the importance of this site for plant conservation, and to protect the water quality of the lake, which supplies drinking water to Kumba. Should this *Thunbergia* be refound, it could be propagated and incorporated in the forest restoration around the lake. Cultivation of the species at Limbe Botanic Garden might also safeguard it.

Whitfieldia preussii (Lindau) C.B.Clarke assessed by ID

VU B1ab(iii)+B2ab(iii)

Range: Equatorial Guinea (Bioko (1 coll.)) and Cameroon (SW: Bakossi (1 coll.); Matoh (1 coll.); Kumba (1 coll.); Barombi (1 coll.). S: Bipindi (3 coll.)).

This taxon appears rare throughout its range, although at Malabo-Luba, Bioko in 1989 it was recorded as "frecuente en colonias reducidas". In Bakossi, it is known only from the Mungo R FR. The Bipinde collections are from the early 1900s; no subsequent records are known from S. The assessment of this taxon made by ID in Cheek *et al.* (2004: 142) is maintained here since only six sites are known (AOO 36 km² with 4 km² cells) with threats as below. EOO is calculated as 17967 km².

Habitat: an understorey shrub of lowland forest, rarely secondary scrub; 0–1350m alt.

Threats: forest clearance throughout its range. However, at Kumba this species was collected from a road verge, suggesting that it can tolerate some disturbance.

Management suggestions: further data is needed on the range and habitat requirements of this species, particularly the extent to which it can tolerate disturbance. The continued existence at Bipindi should be confirmed during future botanical inventory work at this site.

AMARANTHACEAE

assessed by Iain Darbyshire

Achyranthes talbotii Hutch. & Dalziel

VU B2ab(iii)

Range: Nigeria (Cross River State (2 coll.)) and Cameroon (SW: Mt Cameroon (8 coll.); S Bakundu FR (1 coll.); Bakossi (2 coll.); Banyang-Mbo (1 coll.). NW: Bu (1 coll.). Littoral: nr. Nkongjok (1 coll.); nr. Yingui (1 coll.)).

This taxon was treated in Cable & Cheek (1998: xxxix) as Endangered, being then recorded from only 3 sites, the type collection from Oban District, Nigeria and the Mt Cameroon collections on the Onge and Joke rivers. However, with the additional collections from Baseng in the Jide Valley, Nyandong in western Bakossi, the Banyang-Mbo Wildlife Sanctuary and from Bu, NW, together with the records of this taxon from two rivers in Littoral previously omitted, it is was downgraded to Vulnerable by Darbyshire in Cheek *et al.* (2004: 142) since known from 10 locations (equating to an AOO of 40 km² using 4 km² cells). This assessment is maintained here with the note that, with further survey work more sites are likely to be found, when the species will need to be downgraded further, to NT unless it can be assessed under another criterion. EOO is calculated as 32555 km².

Habitat: fast-flowing rivers, growing half submerged, rooting among boulders; often around waterfalls; 0–750 m alt.

Threats: the proposed conversion of lowland forest around Mt Cameroon to plantation is likely to threaten the plant communities of the rivers that drain the area, such as the Onge, an important site for this taxon, through flooding and excessive silting of their habitat. Illegal logging for timber in lowland Bakossi is likely to have a similar impact. Potentially, hydro-electric projects might threaten this species by reducing water course flow rates.

Management suggestions: botanical surveys of other suitable sites in SW and Regions should reveal further populations of this taxon; monitoring of the riparian plant communities in response to increased upstream run-off should be carried out.

Cyathula fernando-poensis Suess. & Friedrich

VU D2

Range: Equatorial Guinea (Bioko (3 coll.)) and Cameroon (SW: Mt Kupe (1 coll.); Bakossi Mts (1 coll.)).

This taxon was previously known from only 3 collections made in 1951 at El Pico, Bioko by A.S. Boughey, with no subsequent collections from the island. The collection on Mt

Kupe above Nyasoso (*Lane* 242) represented the first record for continental Africa; this was followed by discovery of a second site on the trail from Kodmin to Mwanzum (Muahunzum) in north Bakossi. There are currently no records of abundance of this species at its 3 known localities. The paucity of collections of this taxon is perhaps due in part to the inconspicuous nature of this species, although it is clearly rare throughout its range. This species was assessed as Vulnerable by Darbyshire in Cheek *et al.* (2004: 142) since so few sites are known, and that assessment is maintained here. No new data are available. No direct or immediate threats are known at any of the sites, but were these to be discovered the species should be reassessed as Endangered (EN B2ab(iii)).
EOO is calculated as 1198 km^2.
Habitat: understorey of closed canopy submontane and montane forest; 1330–2100 m alt.
Threats: the forests of Bioko and west Cameroon over 1000 m alt. are currently under only minimal anthropogenic pressure. However, the low number of locations, and their apparent isolation, render this species vulnerable to stochastic events such as localised landslide or overgrazing by native mammals. It is possible that following the exhaustion of lowland forest resources, there will be increased human encroachment into the montane forests in future unless protected by law.
Management suggestions: a survey of the current known populations should be undertaken, including its rediscovery on Bioko. This species should be sought during any future botanical inventory work in additional montane sites such as the highest parts of the Rumpi Hills.

ANACARDIACEAE
assessed by Martin Cheek

Among the trees and shrubs of Cameroon with alternate pinnate leaves, Anacardiaceae can be identified by strong smell of mango when the stem is cut, and by a clear or watery white exudate from the wound which oxides and dries black. This exudate causes skin allergies in some people. The best known species, the mango, *Mangifera indica*, is native of India, but many Cameroonian species also have edible fruits, such as the bush bon-bon, *Trichoscypha acuminata*. As in the mango, many species of the family have the style arising asymmetrically on the ovary, or if more than one style per fruit (*Trichoscypha*) separately from each other.

There is no Flore du Cameroun account, but Breteler has revised the two main genera for W-C Africa, only one of which contains threatened taxa (see below).

In previous Red Data publications in Cameroon, two species of *Trichoscypha* were assessed by us as threatened (*T. bijuga* Engl. and *T. camerunensis* Engl., in Cable & Cheek 1998). This genus has since been revised by Breteler (Adansonia 26: 97–127 (2004)). Species delimitation has changed greatly, with many formerly recognised being sunk into larger more widespread species that in most cases merit 'Least Concern' status. The same applies to the genus *Sorindeia. S. mildbraedii* was listed as VU by WCMC [World Conservation Monitoring Centre] (1997), but is now regarded as a synonym of *S. winkleri*. This is explained in detail in Cheek *et al.* (2004: 143). Only two rare/restricted range species of the family are now known to occur in Cameroon:

Trichoscypha hallei Breteler
EN B2ab(iii)

Range: Cameroon (Akom II) and Gabon (Bélinga).
This rain forest understorey shrub has 15–17 leaflets, velvety fruits and 4-petalled flowers with stalk less than 1mm long. According to Breteler (loc.cit. 2004: 111) it is known from two sites. One in Gabon at Bélinga (*Cours* 7011 and *Hallé* 2822), the other in Cameroon at Akom II, Nkol Dangueng (*Van Andel et al.* 3717). With an AOO of 8 km^2 using 4 km^2 cells and threats as below, it is here assessed as endangered.
Habitat: rainforest; 500–900 m alt.
Threats: at Bélinga iron ore extraction is now being resumed by Chinese interests (Breteler pers. comm. 2008). Establishment of open cast mining and its infrastructure result in habitat clearance. No threats are currently known to the Akom II site.
Management suggestions: the forest at Bélinga has several rare and unique species and a representative portion should be protected. A survey to rediscover this species in the wild, followed by public education, and discussions with forest managers would facilitate the survival of this rare species.

Trichoscypha engong Engl. & Brehmer
VU B2ab(iii)

Range: Cameroon (E: Lomié (1 site)), Equatorial Guinea (Rio Muni: Mt. Alen/Moka and Nkolentanga (2 sites)) and Gabon (Assok, Oveng, Makokou and Kougouleu (4 sites)).
This evergreen forest tree reaches 25 m tall, with a diameter at breast height of 1 m. The leaves can have up to 25, pinnately-arranged oblong leathery leaflets each 16-20 × 6-9 cm, with 13-20 pairs of lateral nerves. The inflorescences are borne on the thicker branches, bearing edible glabrous fruits 4 cm long (Breteler loc. cit. 2004: 109-110). *Trichoscypha engong* here is assumed to be vulnerable since known from 6 sites (AOO 24 km^2) with habitat threats as described below. EOO is 78595 km^2.
Habitat: evergreen forest up to c. 800 m alt.

Threats: at least one site, that near Assok, is on a road, near a village and so vulnerable to forest clearance for agriculture.
Management suggestions: since this species has edible fruits (known as Engong in Gabon), it could be known to local communities and possibly even planted in a small-scale. This could be encouraged. Further surveys may well show this species to have more sites and deserve removal from the threatened list. Until this is proven, sites for *T. engong* should be protected and monitored.

ANCISTROCLADACEAE

assessed by Martin Cheek

The single genus of this family *Ancistrocladus,* are all woody forest climbers, distinctive in their short side branches (spur shoots) which each bear 1 or 2 coiled woody hooks and end in funnels of leathery simple leaves in which are borne the panicles of pale flowers with petals only 2–6 mm long. These are succeeded by 1cm leathery-dry wind-dispersed fruits, the sepals developed into wings. Thirteen species were recognised for Tropical Africa by Cheek (Kew Bull. 55: 871–882 (2000)). The genus was revised by Taylor *et al.* in 2005 (Ann. Missouri Bot. Gard. 92: 360 399). With the benefit of a larger number of specimens being available for their study, Taylor *et al.* 2005 (op.cit.) modified many of the assessments made in Cheek (2000 op.cit.). Those relating to Cameroon are as follows:

Ancistrocladus guineensis Oliv. now known from Nigeria to Gabon, having been assessed as Endangered A2c (Cheek 2000 op.cit.) is assessed as LC by Taylor *et al.* (2005 op.cit.); *A. letestui* had been assessed by Cheek in Cable & Cheek (1998: xl) as VU B1+2a and in Cheek *et al.* 2000 op. cit. as DD, Taylor *et al.* 2005 treat it as LC. Two species are here accepted as threatened in Cameroon.

Ancistrocladus grandiflorus Cheek
VU B1ac + B2ab(iii)

Range: endemic to SW Region Cameroon (SW: Bimbia-Bonadikombe Forest near Limbe; Mundemba; Mbu Bakundu; Dikome Balue; Banyang Mbo; Masaka Batanga).
This species has larger flowers than any other in the genus (petals 9–12 mm long) and also bears very large, long leaves 16–35 cm long on spur branches, 49–56 cm on the main stem.
First assessed as CR A1c+2c, C1+2a (Cheek in Cable & Cheek 1998: xl as *Ancistrocladus* sp. nov.) when known only from Mt Cameroon, but reassessed by Taylor *et al.* 2005 as VU B1ac on the evidence of many additional collections between Mundemba and Mt Cameroon. Six sites can now be accepted (AOO 24 km² with 4 km² cells) and Taylor *et al.* calculated an EOO of >5000 km² (we calculate EOO 5405 km²) which with the threats indicated below allow the assessments indicated above.
Habitat: lowland evergreen forest; 60–940 m alt.
Threats: clearance of forest for agriculture and urban expansion. The site at Bimbia has almost certainly been lost since reserves for the species there in 2009 (Marcella Corcoran) and 2008 (David Alicha) with Limbe BG staff including Paul Ndumbe who had helped collect the type material, failed to find it. The site at Mundemba is unprotected and at risk from urban expansion.
Management suggestions: the Rumpi Hills offer the best hope for protecting this species since as a forest reserve, there is some protection and forest in the area is at low risk from threats, while most records of the species derive from that site. Surveys to rediscover and put in place plans with local communities to protect this species are advisable. Data on growth rates, population density and regeneration levels should be gathered. Cultivation at Limbe BG would allow the possibility of multiplication for reintroduction.

Ancistrocladus korupensis D.W.Thomas & Gereau
EN B2ab(iii)

Range: Nigeria (Oban Hills, put possibly extinct there) and Cameroon (SW: Korup NP).
This species excited much interest when discovered to have anti-AIDS potential, much reported in New Scientist and other newspapers. However the active ingredient Michellamine B was later found to be highly toxic to human cells.
Liana with spur-shoot leaves oblanceolate, (8–)14.5-34.2 × (2.9–)4.6–11 cm, flowers with petals 2–3 mm long, exceeded by sepals, sepals highly unequal in flower and fruit, in fruit sepal wings are spathulate; largest 34–48 mm long; styles 2 or 3; peduncle 4–13 cm long.
First assessed LR/cd i.e. Lower Risk, but conservation dependent (Cheek 2000 op.cit.) this taxon was reassessed by Taylor *et al.* (2005 op.cit.) as Vulnerable (VU B1ac) citing an EOO of less than 5000 km² and being potentially threatened outside of the two protected areas cited above. Here the taxon is reassessed as Endangered, since it is only known at two locations, at one of which it is threatened if it still survives (Oban, see below). At the other it occurs over a large but, probably continuous area with an AOO estimated here of 8 km², giving a total of 12 km² for AOO if the Oban site is given a cell size of 4 km². Several large parts of Korup have been intensively studied yet the species is unknown from there, hence its highly localised distribution is not an artefact.
The map accompanying this text shows a third locality to the NE of Korup. This is based on *Thomas* 9999, on the Babong–Baro Rd from Nguti (Missouri data). However, this is stated to be a sapling 1–2 m tall and thus must be regarded as only a tentative identification since species characteristics

are not known at this stage. Therefore this is not accepted as a location in our assessment as a record of the species, although it has been included and so has distorted the automated EOO calculation (2560 km²).

Habitat: lowland evergreen forest c. 200 m alt.

Threats: inside the Korup NP this taxon is quite secure being locally common and well-protected. However, if the record from Oban is correct (Duncan Thomas pers.comm. reports that only a sterile specimen exists), then the species is threatened there since although a National Park is gazetted, there are difficulties with agricultural incursions (Cheek pers.obs. 1993) and it is not certain that the record derives from the NP. It is perfectly possible that the taxon has been lost at the Oban locality, and is possibly lost to Nigeria also.

Management suggestions: the Korup NP is the best hope for the survival of this species and may, in fact, be the only site where it now survives. Given the well-known local abundance of the taxon no immediate management appears to be needed. However, a census of the population size and range, and regeneration levels would be useful for understanding this species better and dealing with any future stochastic events, such as cataclysmic storm-driven tree-falls. This species is already cultivated at Limbe BG where it has thrived for many years.

ANNONACEAE

assessed by Martin Cheek, advised by George Gosline

This is a large pantropical woody family, most species being small or medium-sized trees of lowland evergreen forest. A few genera (e.g. *Monanthotaxis*) are lianas. Generally, Annonaceae are indicators of good quality forest – few are pioneers. They are easily recognised by having mainly distichous, alternate, simple, entire leaves folded along the midrib when in bud and lacking stipules. The apical buds often have appressed coppery hairs, and most distinctively of all, when stems are broken, they issue a strong spicy smell otherwise rarely encountered in woody African forest groups (the others being Lauraceae, which dry brown; Myristicaeae, which have whorled branching; and Monimiaceae/Siparunaceae, which are very few in number). In flower Annonaceae have sepals and petals in whorls of three, usually free; the anthers are numerous, spirally arranged in a compact convex head, the ovaries are free (except *Isolona* and *Monodora*) also spirally arranged, and held at the apex of the staminal head. Fruits are apocarpous (except *Monodora* and *Isolona*).

Annonaceae are not among the major timber species, but they are important sources of spices in Cameroon. No market in the forest zone lacks dried fruits of *Xylopia aethiopica* (peppery spice) or seeds of *Monodora myristica* (spices). Bark of *Annickia (Enantia) chlorantha* is highly regarded in the belief that it treats malaria, yellow fever and viruses.

There is no Flore du Cameroun account for Annonaceae, and most African genera have not been revised in 100 years or more. Annick Le Thomas, who wrote the excellent Fl. Gabon account (1969) drafted a volume for Cameroon (not seen by us) which has been in the process of being completed by Richardson and Couvreur these last several years. When this becomes available species ranges and delimitation (and generic attribution) are likely to be radically altered from that available to us now rendering any existing Red Data assessments immediately out-of-date. Until that day, this family in Cameroon must be regarded as suffering from taxonomic impediment, excepting the genera *Isolona* and *Monodora* recently monographed by Couvreur (Systematic Botany Monographs 87: 1-50 (2009)). Couvreur recognised 20 and 14 species in these genera respectively. He also assessed the conservation status of each of these taxa using IUCN 2004 guidelines and calculating AOO with a cell size of 9.98 km² (Couvreur loc. cit. 4). However, he was not explicit in stating the parameters of the assessment for most of the species he assessed, therefore this is addressed below, as far as possible. Nevertheless Couvreur's work is extremely helpful in reaching conservation assessments. We have not repeated Couvreur's excellent species distribution maps here.

Couvreur has sunk the Cameroon Highland endemic *Isolona maitlandii* under the more widespread Congolian *I. congolana*, which we must accept.

We agree with most of Couvreur's IUCN assessment ranks for Cameroon species except that *Isolona dewevrei* (De Wild. &T.Durand.) Engl. & Diels assessed as LC has only nine spot/clusters hence is probably better placed as NT or VU B2ab(iii). Meanwhile, *I. thonneri* (De Wild. & T.Durand) Engl. & Diels, with 13 sites is assessed as VU B2ab(iv) when NT seems more appropriate.

The account below, apart from covering the genera revised by Couvreur, also includes some species of *Uvariopsis* since this genus has been the subject of attention from workers such as Kenfack, Gereau, Schatz and Gosline (papers cited below under those species). Apart from these we include a few other species which have been studied by us in connection with our series of conservation checklists (Cable & Cheek 1998, Cheek *et al.* 2000, Harvey *et al.* 2004, Cheek *et al.* 2004, Harvey *et al.* 2010, Cheek *et al.* 2010, Cheek *et al.* 2011). Here their treatments are revised and updated.

ANNONACEAE

Several species referred to in our Conservation Checklist for Mt Cameroon (Cable & Cheek 1998) as being possibly to be assessed in future as threatened (e.g. *Artabotrys sp. aff. rhopalocarpus, Cheek, Elad and Ndumbe* 3470) have not yet been taken forward. Several other species that were assessed as threatened in that volume but had not been formally published remain unpublished and therefore cannot be accepted by IUCN so are not included here. Hopefully they will be resolved in future and appear in a future addition:
Isolona sp. nov. (CR A1c+2c), based on *Watts* 687 (Bambuko FR) and *Wheatley* 501 (Etinde).
Piptostigma sp. nov. aff. glabrescens Oliv. (CR A1c+2c) known from three collections in Onge and one in Mokoko.
Polyceratocarpus sp. nov.(CR A1c+2c) known only from the Mokoko/Bambuko boundary (*Watts* 627, March 1993, Mundongo).

In our book on the Plants of Kupe, Mwanenguba and Bakossi Mts (Cheek *et al.* 2004), *Isolona zenkeri* Engl. was assessed as VU B2ab(iii). However, the revision by Couvreur (2009: 87) has shown that it has so many sites in Gabon, extending also to Congo (Brazzaville), in total 10-15 localities, that he lists it as LC. This is accepted here although NT is probably more appropriate. Similarly, *Uvariodendron giganteum* (Engl.) R.E.Fr. was assessed as VU B2ab(iii). We now know that there are 11 sites for the species in Cameroon and Gabon, some with several collections, so it is here downrated to NT. However, with better habitat loss data, in future it might be reassessed as threatened again under criterion A. This also applies to *Cleistopholis staudtii* Engl. & Diels, assessed as VU B2ab(iii) but now known from more than 10 sites.

49 taxa are listed by Onana (ms) as being endemic, near endemic or of conservation concern. Those not accounted for here, with their provisional assessments by Onana (ms) are:

Anonidium friesianum Exell VU B2ab(iii)
Artabotrys jollyanus Pierre ex Engl. & Diels CR B2ab(iii)
Artabotrys congolensis De Wild. & T.Durand EN B2ab(iii)
Balonga buchholzii (Engl. & Diels) Le Thomas EN B2ab(iii)
Duguetia dilabens Chatrou & Repetur EN B2ab(iii)
Friesodielsia discostigma (Diels) Steenis CR B2ab(iii)
Letestudoxa bella Pellegr. NT
Meiocarpidium lepidotum (Oliv.) Engl. & Diels NT
Monanthotaxis angustifolius (Exell) Verdc. (syn. *Enneastemon*) CR B2ab(iii)
Monanthotaxis congoensis Baill. VU D2
Monanthotaxis elegans (Engl. & Diels) Verdc. VU B2ab(iii)
Monanthotaxis oligandra Exell EN B2ab(iii)
Piptostigma glabrescens Oliv. var *glabrescens* VU Bab(ii,iii)
Piptostigma multinervium Engl. & Diels VU B2ab(iii), D2
Polyceratocarpus microtrichus (Engl. & Diels) Ghesq. ex Pellegr. EN B2ab(ii,iii,iv)
Uvaria bipindensis Engl. ex Engl. & Diels EN B2ab(iii)
Uvaria heterotricha Pellegr. NT
Uvaria obanensis Baker f. VU A2c; B2ab(iii)
Uvaria comperei Le Thomas VU B2ab(iii)
Uvariastrum zenkeri Engl. & Diels VU A1c, B1+2c (IUCN 2002/WCMC 1998) = VU A2c; B2ab (iii)
Uvariodendron connivens (Benth.) R.E.Fr. NT
Uvariopsis bakeriana (Hutch. & Dalziel) Robyns & Ghesq. VU B2ab(ii,iii)
Uvariopsis dioica (Diels) Robyns & Ghesq. VU D2
Uvariopsis sessiliflora (Mildbr. & Diels) Robyns & Ghesq. CR B2ab(ii,iii)
*Uvariopsis solheidii (*De Wild.) Robyns & Ghesq. EN B2ab(iii)
Uvariopsis zenkeri Engl. CR A2c+3c
Xylopia aurantiiodora De Wild. & T.Durand VU B2ab(iii)
Xylopia flamignii Boutique VU D2
Xylopia gilbertii Boutique VU B2ab(iii)
Xylopia letestui Pellegr. EN B2ab(iii)
Xylopia mildbraedii Diels EN B2ab(iii)
Xylopia phloiodora Mildbr. NT

Boutiquea platypetala (Engl. & Diels) Le Thomas EN B2ab(iii)

Range: Cameroon ((coastal) SW: Idenao. S: Bipindi; Kribi) and Equatorial Guinea (Bioko).
This monotypic genus is close to, and may yet be united with *Neostenanthera* Exell. It was erected by Le Thomas (Adansonia sér. 2, 5: 531 (1966)) and is known from eight specimens, but only five localities. The type is from Bipindi, and the species is also known from Kribi besides Bioko. On Mt Cameroon, it is known only from *Nkeng* 69 (June 1992, Idenao). A shrub or tree 3–5(–10) m tall, this species has conspicuous flowers and fruits, and is not easily overlooked when fertile. Consequently, the sparse and scattered collections probably indicate rarity, rather than undercollection. The text here is taken from the treatment by Cheek in Cable & Cheek (1998: xl) where it was assessed as EN A1c+2c under the 1994 criteria. Here EN status is maintained, but it is more appropriate to assess it now under criterion B, since 5 sites are known (AOO 20 km^2 using 4 km^2 cells) and threats are as below. EOO is calculated as 15109 km^2. It is notable that in more than ten years of active surveys in Cameroon since the original assessment was made, no further records have come to our attention.
Habitat: lowland evergreen forest.
Threats: clearance of forest for timber extraction, followed by agriculture and particularly plantations. Plantations in the Idenao area are believed to be scheduled for expansion soon. Lowland forest is probably the most heavily threatened vegetation type on Mt Cameroon. At Kribi there has been

clearance of coastal forest for holiday homes and new development projects (Cheek pers. obs.).

Management suggestions: an attempt to be made to rediscover this tree at Idenao, where it was most recently recorded (see above). Seed should be collected for propagation at Limbe BG and subsequent replanting in secure habitats in the wild. Data on growth rates, regeneration levels in the wild, and population data would be valuable in managing the survival of this species.

Isolona pilosa **Diels** (mapped in Couvreur 2009: 79, fig. 44)

VU B2ab(iii,iv)

Range: Cameroon (E: 21 km N Moloundu), Gabon (Lastoursville (2 sites)), Congo (Brazzaville) (Odzala NP; Niari; Mayombe), Congo (Kinshasa) (Luki; Kisangani; Kasai Oriental) and Angola (Cabinda).

Tree to 13 m tall, distinct from other *Isolona* in the densely hairy adaxial midrib, pedicels and corolla lobes.

Couvreur's assessment can be justified since ten sites are known (see above and Couvreur loc.cit. fig. 44, p.79) and since threats are as below. If 4 km² cells are used an AOO of 40 km² can be accepted using the conventions of this volume.

Habitat: lowland evergreen forest.

Threats: slash and burn agriculture for *Manihot* cultivation in Mayombe and Niari sites (Moutsambote pers.comm. 2009).

Management suggestions: since the only collection of this species in the last 30 years (made in 1998) is from the Odazala NP, this may be the best place to focus conservation activity. Efforts should be made to rediscover the species there and to gather the usual data needed to inform management decisions.

Isolona pleurocarpa Diels (mapped in Couvreur 2009: 81, fig. 45)

VU B2ab(iii)

Range: SE Nigeria (Akamkpa Rubber Estate, Calabar) and Cameroon (SW: Korup NP; S Bakundu FR. S: Bipindi-Mimfia; 20 km NE of Kribi at Mt Elephant; 20 km SE Kribi at Colline Nkolo Manga)).

Couvreur (loc.cit.) rated this species as EN B2ab(iii) but since six sites are known (AOO 24 km² with 4 km² cells) with threats as below, *Isolona pleurocarpa* is here reassessed as VU B2ab(iii). It can be distinguished from other species by the rounded and ribbed fruit, the glabrous young leaves, lamina inserted laterally on the petiole with a decurrent base. The corolla lobes are ovate to narrowly ovate with a narrowed base and an acute apex.

Habitat:lowland evergreen coastal forest.

Threats: forest in Nigeria has seen dramatic losses in recent decades and this species may now even be endemic to Cameroon. S Bakundu FR has seen clearance by Nigerians for yam farming, while Mt Elephant is threatened by mineral extraction.

Management suggestions: the most securely protected site for this species is probably the Korup NP which may be the best focus for conservation efforts. It is possible that the two large plot projects there may have data on population data and regeneration levels for the species to assess needs for management intervention.

Monodora zenkeri **Engl.** (mapped in Couvreur 2009: 132, fig 71)

VU B2ab(iii)

Range: Cameroon endemic (C: Yaoundé; 38 km N Bafia; 4 km NE Otélé. E: Bertoua-Batouri 5 km; 5 km S Nguélémendouka; Deng Deng. S: Bibye; Oveng)

While Couvreur rated *M. zenkeri* as EN B2ab(iv,iii), we have reassessed it here as VU since eight sites (above, AOO 32 km² with 4 km² cells) are known, with threats as below. EOO is calculated as 31650 km².

Couvreur states (2009: loc. cit. 132) that it differs from other W-C African species in the straight margins to the petals and in the inner petals connivent at the centre, not at the margins.

Habitat: disturbed lowland semi deciduous to evergreen forest; 600-700 m alt.

Threats: *Monodora zenkeri* grows in an area rich in semi-deciduous forest favoured by commercial timber species and so heavily affected by logging, often followed by cultivation for *Theobroma cacao* and/or *Manihot* (pers. obs.).

Management suggestions: every effort should be made to determine if *M. zenkeri* occurs in the Dja NP which is within its natural range and would offer possibly the best hope for conservation efforts. Data on population density and regeneration levels are needed to assess needs for management intervention.

Piptostigma calophyllum Mildbr. & Diels

VU B2ab(iii)

Range: Cameroon (SW: Bakossi (3 coll.), Rumpi Hills (1 coll.). Littoral: Edéa (1 coll.). S: Ébolowa (1 coll.); Sangmélima (1 coll.)) and Gabon (2 coll.).

Spectacular for its large and beautiful leaves, this small cauliflorous tree is not easily overlooked. It was first published in 1915, based on *Mildbraed* 5791 from Ebolowa. This species is only known from seven sites (taking the three Bakossi records as one site), equating to an AOO of 28 km² with 4 km² cells, and with threats as below. Therefore the assessment of this species in Cheek *et al.* 2004 is maintained here. No new data has come to light since that publication excepting the EOO, calculated here as 113804 km².

Habitat: pristine submontane evergreen forest; 900–1320 m alt.

Threats: forest clearance for logging and agriculture (pers. obs. Cheek).

Management suggestions: the three known trees in Bakossi constitute over a third of the global total suggesting that this locality might be the best focus of conservation efforts. None of these three seem under immediate threat, although this cannot be said of those elsewhere. Education, perhaps through a conservation poster, might facilitate protection of this charming tree. So far as is known, no individuals are protected in a National Park.

Uvariopsis korupensis Gereau & Kenfack

EN B2ab(iii)

Range: Cameroon (SW: Mt Cameroon; Korup; Bakossi; Takamanda).

This small tree bears its flowers from a woody boss that circles the trunk at ground level. The orange and cream, fleshy flowers form a continuous cushion around the trunk. Kenfack *et al.* distinguish the species in having a leaf blade 30–52 × 9–14 cm, sepals 2-5 mm long; petals 10–35 × 5–12 mm (Adansonia sér. 3, 22(1): 41 (2000)).

Described in 2000, the Takamanda record is newly added here. Two further new records derive from Missouri data so have not been seen by us but since collected by Kenfack and his assistant must be accepted: *Kenfack* 1026 and *Sainge* 326. Although we did not have access to their geographic data, it is most likely that they also derive from Korup where these botanists have been based.

Uvariopsis korupensis is here assessed as Endangered since four sites are known (AOO 16 km² with 4 km² cells and threats as below). EOO is calculated as 9016 km². When in flower in January it is extremely conspicuous yet even in Korup, where most collections derive, it is rarely encountered. Despite spending several weeks in Korup, Cheek only saw a single individual of this species. At all other sites only a single individual is known, it is believed

Habitat: lowland evergreen forest in hilly terrain.

Threats: secure at all sites apart from stochastic events, apart from at Mt Cameroon where threatened by forest clearance for timber and agriculture (Cheek pers.obs.)

Management suggestions: this tree is especially secure at Korup. Population and regeneration data may be available from the two large plots there, otherwise should be gathered to inform management decisions.

Uvariopsis submontana Kenfack, Gosline & Gereau

EN B2ab(iii)

Range: Cameroon (SW: Mt Kupe (12 coll.). Littoral: Yabassi-Yingui (1 coll. *Kenfack* 1602)).

In January, the gnarled trunks of this medium-sized canopy tree are largely covered in big pink, fleshy, velvety-hairy flowers in the males. The 12 collections on Mt Kupe were made at three sites although they could be taken as one since proximity is close. Taken as a global total of four sites, AOO is 16 km² with 4 km² cells. EOO is calculated as 393 km². The site near Kupe village occurs very close to the upper level of cultivation and may be decimated if this level is elevated further. The absence of this species from the Bakossi Mts is notable. The original assessment from Cheek *et al.* (2004: 144) is maintained here.

This species was published in Novon 13(4): 444 (2003).

Habitat: submontane evergreen forest, 840–1200 m alt.

Threats: clearance of forest for agriculture, especially at the lower parts (below 1000m) of its altitudinal range.

Management suggestions: public education of the importance of this tree is advisable to assure its continued existence at its stronghold on Mt Kupe; a poster explaining this should be produced for local use.

Uvariopsis vanderystii Robyns & Ghesq.

EN B2ab(iii)

Range: Cameroon (SW: Takamanda (1 coll.); Mt Kupe (2 coll.)), Gabon (1 coll.) and Congo (Kinshasa) (1 coll.).

Flowers in this small tree are only produced at ground level and radiate from the trunk on pedicels 4–5 cm long. Although moderately widespread, it is both patchy and rare: despite intensive inventories at Korup and Mt Cameroon it has not been recorded there. We have no data on threats (or its continued existence) at sites in Gabon and Congo (Kinshasa), but the prospects of continued forest quality at Takamanda have not been enhanced by construction of a new road there; the site at Mt Kupe, above Kupe village, is within 100 m of cultivated land and has been earmarked for expansion. However, the small plot containing this *Uvariopsis* has been rented for conservation purposes by us from the farmer and monitored since 1995 and has seen no deterioration in that time. With four sites, AOO is 16 km² with 4 km² cells. EOO is calculated as 15109 km². The original assessment from Cheek *et al.* (2004: 144) is maintained here.

Habitat: lowland and submontane evergreen forest; 780–1000 m alt.

Threats: forest clearance for agriculture.

Management suggestions: at Mt Kupe continued monitoring and rental payments are needed to ensure the continued survival of this species. The Takamanda site is partly protected as a wildlife reserve but the tree(s) of *U. vanderystii* should be refound and monitored. Consideration should be given to educating the inhabitants of the area as to the importance of this plant. At the other locations, trees should be refound (if still existing), censused and considered for protection by the appropriate authorities.

Xylopia africana (Benth.) Oliv.

VU A2c

Range: São Tomé & Príncipe (São Tomé (1 coll.)), SE Nigeria (Obudu (1 coll.)) and Cameroon (SW: Mt Cameroon

(4 coll.); Bakossi (10 coll.); Rumpi Hills (1 coll.); Fosimondi (1 coll). NW: Bali Ngemba (numerous coll.)).
A canopy tree restricted to the submontane forests of the Cameroon Highlands, it can be distinguished from all other *Xylopia* in the area (apart from the lowland *X. staudtii*) by the globose, not long-beaked, flower buds. Of all the forests that we have surveyed, *Xylopia africana* has been found in greatest density at the Bali Ngemba FR in NW Region. This, now the largest remnant of the forest that cloaked the Bamenda Highlands at the 1300–1900 m range, is only 100 ha in extent and shrinking fast due to illegal clearance for farming.
Presumably this species was once common throughout the Bamiléké Plateau and Bamenda Highlands where it is now all but extinct. While there are no figures for rates of forest loss in the Bamenda Highlands as a whole, in one area which has been studied, the Kilum-Ijim area, forest loss of 25% over 8 years in the 1980s–1990s has been recorded (Moat in Cheek *et al.* 2000: back cover) and in another 50% in 15 years (Baena in Cheek *et al.* 2010: back cover). Past and ongoing forest loss in the Bamenda Highlands is therefore the main basis for the threat to *Xylopia africana*. On Mt Cameroon it appears rare, being found only twice in the surveys of 1992–1994. Elsewhere in the mountains of the Cameroon line it is also known from the extension into Nigeria: the Obudu Plateau where it is also threatened due to forest clearance, if indeed, it is still extant there. It is also known from São Tomé in the Gulf of Guinea. Strangely, it is not known from Bioko. Bali Ngemba, Mt Kupe and the Bakossi Mts probably now support the largest single subpopulations of *Xylopia africana*. The original assessment from Cheek *et al.* (2004: 144) is maintained here.
Habitat: submontane and lower montane forest; 800–2000 m alt.
Threats: clearance of forest for timber and agricultural land.
Management suggestions: if proposals to protect forest above 1000 m alt. in much of Bakossi are enacted and respected, this subpopulation seems secure.

APOCYNACEAE

assessed by Martin Cheek advised by David Goyder

This family have been recently enlarged to include the Asclepiadaceae since the two are now known to be the halves of one spectrum. The taxa listed below were those traditionally placed in Apocynaceae in the narrow sense – maintained here purely for convenience and for continuity with earlier taxonomic literature such as FWTA. As with Asclepiadaceae in the traditional sense, traditional Apocynaceae in Tropical Africa have opposite, simple leaves, actinomorphic flowers with tubular corollas didymous stamens and superior bilocular ovary-fruits (except *Pleiocarpa*). Apocynaceae however, generally have white (except e.g. *Strophanthus*), not translucent exudate, and more simple flowers, with, for example, free stamens, and a simpler style.

Apocynaceae in this sense are mainly pantropical. In tropical Africa, certainly in Cameroon, ecologically they are almost entirely restricted to forest habitats. Several genera are trees, e.g. *Alstonia, Picralima, Tabernaemontana, Rauvolfia* or shrubs, but most species diversity occurs in liana genera, above all *Landolphia* which as c. 50 species in Tropical Africa.

Apocynaceae species often contain cardiac glycosides and so are often both toxic and medicinally important. Their white exudate contains rubber polymers and *Funtumia elastica,* a wild Cameroonian forest species was commercially extremely important in providing commercial rubber for the Western market in the late nineteenth and early twentieth centuries. Later it was replaced by plantation rubber from *Hevea brasiliensis* introduced from Brazil via Britain. Ecologicically Apocynaceae are important in providing large quantities of nutritious fruit to primates and birds.

While there is no Flore du Cameroun account for Apocynaceae, there is little difficulty delimiting species due to the excellent work of Leeuwenberg and his students at Univ. Wageningen in the Netherlands. They have published revisions of each and every African genus of traditional Apocynaceae. However, there is no modern key to the Tropical African genera. This treatment depends heavily on the revisions which are cited under the relevant species acounts.

Apart from the species assessed below as threatened , several others are known from only slightly more than ten locations (NT) and with better data on population reduction or number of individuals, might well in future be assessed as threatened, these being: *Ancylobothrys pyriformis, Callichilia basileis, Landolphia glandulosa, L. jumellei, L. leptantha, L. letestui, L. stenogyna* and *Tabernaemonanta contorta*. When the Cameroon submontane endemic *Tabernaemontana* is finally delimited, it is likely to be assessed as VU.

Alafia whytei Stapf

VU B2ab(iii)

Range: Liberia (Sinoe R basin; Jaurazon), Ivory Coast (Grabo-Fété Rd 2 km; Téké Forest; Sanvi), Ghana (Subri R FR) and Cameroon (SW: Limbe).
Here *Alafia whytei* is assessed as Vulnerable since nine locations are known (AOO 36 km^2 with 4 km^2 cells) with threats as below. EOO is calculated as 141317 km^2. Specimen and descriptive data are taken from the revision of *Alafia* by Leeuwenberg (Kew Bull. 52(4): 769-839 (1997).

Liana to 5 m or more, exudate white, stems dark brown with paler lenticels; leaves papery, obovate or elliptic 6–18.5 × 2.5–8.2 cm, acuminate, sparsely hairy below, lateral nerves 4-7 pairs; tertiary nerves conspicuous; inflorescence dense, subglobose, 2–5 cm diam., subsessile, puberulous; pedicels 1-3mm; sepals ovate 1–2 mm; corolla dark red, puberulent, tube 11–13 × 1.5–5 mm, lobes suborbicular, 4–7 mm, spreading; fruit follicles two, 15–50 × 1 cm, puberulous.
Habitat: lowland evergreen forest.
Threats: clearance of habitat for timber extraction followed by agriculture for e.g. Cocoa cropping (Ivory Coast, Ghana); urban and extractive industrial expansion (Limbe).
Management suggestions: more exact georeferencing of known sites and mapping against surviving habitat is likely to show that this species has already been lost at several historic sites (see threats). Selection of which known sites are secure is needed before surveys to gather basic populational data are mounted and local community leaders are advised and supported in protecting *Alafia whytei*.

Baissea ochranta K.Schum. ex Stapf
EN B2ab(iii)
Range: Cameroon (S: Bipindi; Kribi-Edéa 40 km) and Gabon (SE Sindara).
Here *Baissea ochranta* is assessed as Endangered since effectively there are only two known locations (AOO 8 km² with 4 km² cells) and threats as below. Location Data are taken from the revision of *Baissea* by van Dilst (Bull. Jard. Bot. Nat. Belg. 64: 89-178 (1995)). Liana to 5 m, glabrous, stems grey-green, smooth; leaves ovate or oblong 3.4–12.3 × 1.6–5.7 cm, apiculate or acumen 5–16 mm, secondary nerves 3–5 pairs, almost forming a submarginal nerve, tertiary nerves scalariform, domatia present. Inflorescence of axillary cymes 1–10-flowered, peduncles 11–46 mm, pedicels 6–14 mm; flowers with sepals ovate 1–1.5 mm; corolla tube white, yellow or green, cylindrical 5.2–6.8 mm, lobes ovate, 6–9 × 2–2.5 mm; fruit follicles each 50 cm, pustulate, glossy.
Habitat: lowland evergreen forest; 30–40 m
Threats: slash and burn agriculture in Gabon and Bipindi areas, but development and extractive infrastructure are a risk at the Kribi-Edéa site.
Management suggestions: rediscovery of this species at one or more of its known sites; surveys to provide basic populational data for future monitoring and intervention assessments; advising leaders of local communities nearby, to seek their support in protecting this species. Care should be taken not to confuse this species with others of the genus in Cameroon, for which van Dilst 1995 should be referred to.

Callichilia monopodialis (K.Schum.) Stapf
EN B2ab(iii)
Range: Cameroon endemic (C: Mt Kala; Ayos; N'kolbison; Bot. Makak; Yaoundé; Eloumden; Ototomo FR. S: Bipindi; 25 km NW Eséka; Campo-Ma'an (Tchoutou 2004)).
Shrub 0.5–2 m, pale brown; leaf-blades elliptic to obovate 5–27 × 3–12cm, acumen to 1.8 cm, cuneate, lateral nerves 5–16 pairs; inflorescence peduncle 5–50 mm, bracts dense, each 2–5 mm; pedicel 3–10 mm; sepals 5–20 × 1.5–3 mm; corolla curved 42–63 mm, lobes overlapping to left, 17–30 × 20 mm; fruit carpels narrowly ellipsoid, c. 35 × 7 mm, orange.
Here *Callichilia monopodialis* is assessed as EN since the sites equate to five locations, taking the numerous collections in and around Yaoundé as one location. AOO 20 km² with 4 km² cells, threats as below. EOO is calculated as 11185 km². Data are taken from the revision of *Callichilia* by Beentje (Meded. Landbou. Wag. 78-7: 1-32 (1978)).
Habitat: lowland to submontane forest; to 950 m alt.
Threats: slash and burn agriulture; additionally, in the Yaoundé area, urban expansion is a major threat.
Management suggestions: rediscovery of this species at one or more of its known sites; surveys to provide basic populational data for future monitoring and intervention assessments; advising leaders of local communities nearby, to seek their support in protecting this species. Since most records occur in and around Yaoundé it is logical to consider setting up a protected area here (many other threatened species also occur in and around Yaoundé).

Dictyophleba setosa B.de Hoogh
VU B2ab(iii)
Range: Cameroon (S: Kribi-Ébolowa, km 16; Kribi-Mpolongwe, km 15: Kribi-Lolodorf, km 19; Campo, 35 km E, Mvini; Kribi 40 km S, Ebodié; Campo-Kribi, 6 km. C: Mefou proposed NP) and Gabon (Ngounié at Fougamou).
Here *Dichtyophleba setosa* is assessed as Vulnerable since eight locations are known (see range, above, AOO 32 km² with 4 km² cells, threats as below). EOO is calculated as 48839 km². Data are taken from de Hoogh's revision of *Dictyophleba* (Bull. Jard. Bot. Natl. Belg. 59: 207–226 (1989)).
Liana to 15 m, trunk rough; stems hirsute, hairs red-brown; leaves narrowly obovate 8–21 × 3.5–105 cm, acuminate-cuspidate, base cordate, lateral nerves 9–13 pairs, scabrid-hirsute below; petiole 0.5–2 cm; stipule interpetiolar, laciniate, black, 5–20 mm; inflorescence 10–32 x 1–4 cm, cymes alternate, scabrid-hirsute, sepals narrow oblong 5–6 × 1 mm, bristles; corolla white, tube 4.1–6 mm; lobes oblong 5–11 × 1–2 mm, acute; fruit 1.5–5 cm.
Habitat: coastal and riverine evergreen lowland forest.
Threats: apart from slash and burn agriculture, known threats are few at all sites apart from the Kribi area where extractive industry infrastructure represent a clear threat.

Management suggestions: rediscovery of this species at one or more of its known sites; surveys to provide basic populational data for future monitoring and intervention assessments; advising leaders of local communities nearby, to seek their support in protecting this species. Care should be taken not to confuse this species with others of the genus in Cameroon, for which de Hoogh (1989) should be referred to. Gazettement of the proposed Mefou NP would help secure the protection of this species.

Isonema buchholzii Engl.

VU B1+2ab(iii)

Range: Nigeria (Cross River state, Imo R; Ikot; Oron-Eket, km 50; Calabar; Ikom) and Cameroon (SW: Kumba. Littoral: Abo R; 5 km NE of Douala).

Here *Isonema buchholzii* is assessed as Vulnerable since eight locations (above, AOO 32 km² with 4 km² cells, threats as below). EOO is calculated as 19849 km². Data are taken from the revision of *Isonema* and *Pycnobotrys* by van der Ploeg (Meded. Landbou. Wag. 83-4: 1–20 (1983)).

Shrub or climber 3–6 m, stems dark purple, puberulous when young; leaves elliptic-obovate 4.8–10 × 3.2–5.3 cm, acuminate, glabrous apart from domatia, secondary nerves 4–5 pairs, margin revolute; inflorescence lax, thryrsoid, 11.5–19 × 9–16 cm, pubescent; sepals ovate c. 2 × 1 cm puberulent; corolla pink, white, or yellow; tube 8.5–10 mm, lobes 4–7.5 × 1.5–3 mm, minutely pubescent.

Habitat: riverside and swamp, lowland evergreen forest.

Threats: nearly 90% of Nigeria's original forest has been lost and losses are ongoing. Since 5 of the 8 locations are in Nigerian forest, the main threats are clear: forest clearance for timber and agriculture.

Management suggestions: rediscovery of this species at one or more of its known sites; surveys to provide basic populational data for future monitoring and intervention assessments; advising leaders of local communities nearby, to seek their support in protecting this species. Care should be taken not to confuse this species with others of the genus in Cameroon, for which van der Ploeg (1983) should be referred to.

Landolphia flavidiflora (K.Schum.) J.G.M.Pers.

VU B2ab(iii)

Range: Cameroon endemic (SW: Kumba; Limbe. S: Kribi-Lolodorf, 20 km; Bipindi. C: Yaoundé; Makak; 50 km S Yaoundé).

Here *Landolphia flavidiflora* is assessed as Vulnerable since seven locations are known (AOO 28 km² with 4 km² cells, threats as below). EOO is calculated as 20929 km². Data are taken from the revision of the African species of *Landolphia* by Persoon (Wag. Agric. Univ. Papers 92-2: 1–232 (1992)).

Liana to 35 m, tendrils to 20 cm; stems lenticellate, glabrous; leaf-blade ovate to obovate, 6.5–14.5 × 3–6.5 cm, glossy, leathery, caudate, cuneate, glabrous, 6–10 pairs lateral nerves; reticulate; inflorescences 1–7 cm, 1–8-flowered; sepals triangular-ovate, 1.4–2 mm, acute; corolla cream, tube 1.4–2.6 mm, lobes ovate-oblong, 10–13 × 2.5–4.5 mm, spreading.

Habitat: lowland evergreen forest; 0–1000 m alt.

Threats: urban expansion (Limbe, Yaoundé, Kumba), slash and burn agriculture; conversion of forest to timber and agriculture.

Management suggestions: rediscovery of this species at one or more of its known sites; surveys to provide basic populational data for future monitoring and intervention assessments; advising leaders of local communities nearby, to seek their support in protecting this species. Care should be taken not to confuse this species with the numerous other species of the genus in Cameroon, for which Persoon (1992) should be referred to.

Landolphia maxima (K.Schum. ex Hallier f.) Pichon

VU B2ab(iii)

Range: Nigeria (Oban), Cameroon (SW: Kumba; Korup. Littoral: Eboné-Yabassi km 10. S: Lobé R–Batanga; Bipindi) and Gabon (Mondah FR at Cap Esterias; Makokou).

Liana 15 m or more, trunk with longitudinal crests; leaf-blade ovate to obovate, 21.5–40 × 10–22 cm, acumen to 2 mm, lateral nerves 7-13 pairs, scalariform, lower surface glabrous often black dotted beneath; petiole 9–42 mm; inflorescence 3–7 × 4–10 cm, 1–20-flowered, subfasciculate, sepals 1.7–3.4 mm hairy; corolla white, tube 11–14 mm, lobes lanceolate, 14–25 mm.

Here *Landolphia maxima* is assessed as Vulnerable since eight locations are known (see range, above; AOO 32 km² with 4 km² cells) with threats as below. EOO is calculated as 115025 km². Data are mostly derived from Persoon (1992 loc.cit.)

Habitat: lowland evergreen forest.

Threats: forest clearance for timber and agricultural land magnified by proximity to large population centres (Oban, Kumba, Mondah, Kribi locations). Slash and burn agriculture.

Management suggestions: this species is protected and secure at the Korup so this might be the focus of conservation activities for the species and here surveys to provide basic populational data might be conducted, while providing the means to identify the species to the Park Managers. Care should be taken not to confuse this species with the numerous other species of the genus in Cameroon, for which Persoon (1992) should be referred to.

Landolphia uniflora (Stapf) Pichon

EN B2ab(iii)

Range: Nigeria (Oban), Cameroon (S: Lokoundjé falls at Ebéa) and Gabon (Libreville – Sibang).

Here *Landolphia uniflora* is assessed as Endangered since three locations are known (listed above, AOO 12 km² with 4 km² cells, threats as below). EOO is calculated as 33159 km².
Liana, at least 5 m long, stems to 3 cm diam., yellow-brown, fissured, lenticels 2 mm diam.; leaf-blades narrowly elliptic or obovate, 9–26 × 2.5–9.5cm, acumen 0.3–2.8 mm, secondary nerves 4–7 pairs, forming a strong looped marginal nerve, pilose on midrib and nerves below; petiole 4–18 mm; inflorescence 1–2-flowered, pedicel 1.5–5 mm, pilose, bracts 4–8, sepal-like; sepals 2.6–5 mm; corolla white or purple, tube 13–18 mm, lobes slender, 13.5–23 × 4–5.5 mm, acute; fruit yellow, ellipsoid or ovoid, 4–8 × 2–4 cm. The data presented here is mostly derived from Persoon 1992 loc.cit.).
Habitat: lowland evergreen forest; 0–100 m.
Threats: urbanisation has probably extinguished this species at Sibang while forest clearance for timber and agricultural land is a major threat in Nigeria.
Management suggestions: rediscovery of this species at Lokoundjé falls at Ebéa in S Region; surveys to provide basic populational data for future monitoring and intervention assessments; advising leaders of local communities nearby, to seek their support in protecting this species. Care should be taken not to confuse this species with others of the genus in Cameroon, for which Persoon 1992 and the description above should be referred to.

Malouetia barbata J.Ploeg
EN B1+2ab(iii)
Range: Cameroon endemic (S: Kribi, S bank Kienké R; Campo-Ma'an: Tchoutou 2004).
Shrub to 3 m with white exudate, glabrous, branches dark brown; leaf-blade elliptic 6.7–18 × 2.2–6.7 cm, acumen slender, base obtuse, lateral nerves c. 7–9 pairs, domatia pits glabrous; petiole 3–6 mm; inflorescences 5–8 cm, 1–4-flowered in Jan. & Feb., peduncle 2–3 mm, pedicels c. 3 cm; sepals 3 mm; corolla tube greenish white, 3–3.8 cm, lobes ovate-acuminate, spreading, 12–17 × 6–8 mm, upper surface hairy; fruit follicles 12 × 0.5 cm; seeds 1.5 cm, pilose at base and apex.
Here *Malouetia barbata* is assessed as Endangered since two locations are known (listed above; AOO 8 km² with 4 km² cells, threats as below). The data here is mainly taken from van der Ploeg's Revision of the Genera *Cyclobotrya* Stapf, *Dewevrella* De Wild, and of the African species of the genus *Malouetia* A. DC (Apocynaceae), Agricultural University Wageningen Papers 85-2 (1985).
Habitat: periodically [illegible] forest near sea-level.
Threats: at Kribi urbanisation and infrastructure for extractive industries are threats to habitat as is development for tourism and retirement accommodation.

Management suggestions: Peguy Tchoutou should be engaged to rediscover his Campo-Ma'an NP location so that surveys to collect basic populational data for future monitoring and management can be put in place. Park Managers should be helped to identify and manage the species.

Petchia africana Leeuwenb.
EN B2ab(iii)
Range: Cameroon endemic (S: 8 km N of Kribi; Lolodorf; Bipindi; Campo-Ma'an – las + three fide Tchoutou 2004).
The genus *Petchia* has eight species all found in Madagascar – Comores except one species endemic to Sri Lanka and another in S Region of Cameroon (Leeuwenberg in *Craspidospermum* Boj. ex A. DC., *Gonioma* E. Mey., *Mascarenhasia* A.DC., *Petchia* Livera, *Plectaneia* Thou. and *Stephanostegia* Baill, Wageningen Agricultural University Paper 97-2, 1997).
Shrub 1.5–2.5 m, glabrous, white exudate, branches green; leaf-blades narrowly elliptic, 14–21 × 4.5–7.5 cm, apex acuminate-caudate, 1–2 cm, base cuneate, secondary nerves straight, 12–20 pairs, petiole 6–10 mm long; inflorescence terminal, 2 × 2 cm, peduncle 0.5–1 mm; pedicels 4 mm; sepals slender, acuminate 2–3.5 × 0.5 mm; corolla pale yellow, tube 13–15 × 2mm, lobes spreading, elliptic, 6 × 3 mm apex rounded; fruit follicles orange, presumed fleshy, moniliform, 27 × 2 cm, of 9–11 units.
Here *Petchia africana* is assessed as Endangered since four locations are known (see range above; AOO 16 km² with 4 km² cells, threats as below). EOO is calculated as 3094 km². Only known from the type location near Kribi until Tchoutou (2004 loc.cit.) reported three more locations.
Habitat: lowland evergreen forest.
Threats: slash and burn agriculture. At Kribi, urbanisation and infrastructure from extractive industries are major threats.
Management suggestions: Peguy Tchoutou as the most knowledgeable person in recognising this taxon should be involved in surveys to rediscover and present it with the support of local officials and traditional leaders. Basic populational Data are needed for future monitoring and management planning.

Pleioceras zenkeri Stapf
VU B2 ab(iii)
Range: Nigeria (Oron-Eket Rd), Cameroon (SW: S. Bakundu FR. S: Bipindi; 58 km E Kribi at Fenda), Gabon (Wolou-Ntem, Oyem; Abang-Estuaire) and Congo (Brazzaville) (Placongo, tentative identification).
Shrub or liana 3–4 m, to 9 cm; leaf-blade elliptic or obovate, 4.5–13 × 1.2–7 cm, acuminate, secondary nerves 6–8 pairs, nerves pubescent below; petiole 3–7 mm; inflorescence 4.5–9 × 6.5–11.5 cm, many-flowered, peduncle 2.5–4.5 cm;

pedicels 3–7 mm; sepals ovate 1.6–2.2 mm; corolla yellow, apex red or purple, 6–10 mm, lobes 3–5.5 mm minutely pubescent; corona-like appendages from throat, filiform, broom-like, 3–5.8 mm long; fruit follicles dark yellow to brown, 50–75 × 0.5–1cm, seed body 2 cm, hair tuft 8 cm.
Here *Pleioceras zenkeri* is assessed as Vulnerable since seven locations (above; AOO 28 km² with 4 km² cells, threats as below). EOO is calculated as 123308 km². The data here mostly derives from Barink's Revision of *Pleioceras* Baill., *Stephanostema* K. Schum. And *Schizozygia* Baill. (Apocynaceae), Meded. Landbouwhogeschool Wageningen 87-3 (1983).
Habitat: forest; 90-200m.
Threats: destructive logging (Placongo), logging followed by yam cultivation (S. Bakundu and Nigeria); slash and burn agriculture (elsewhere).
Management suggestions: rediscovery of this species at one or more of its known sites; surveys to provide basic populational data for future monitoring and intervention assessments; advising leaders of local communities nearby, to seek their support in protecting this species.

Tabernaemontana hallei (Boiteau) Leeuwenberg
VU B2ab(iii)
Range: Cameroon (S: Nyabéssan; 55 km W Ébolowa; Nkoemvom) and Gabon (25 km NE Asok; 29 km SW Makokou; Bissok; Bitam; Adzabilone).
Here *Tabernaemontana hallei* is assessed as Vulnerable since eight locations (see above; AOO 32 km² with 4 km² cells, threats as below) are known. EOO is calculated as 56945 km². Data are derived from Leeuwenberg's Revision of Tabernaemontana – The Old World Species, R.B.G. Kew, 1991. Here he delimits 55 species most of which occur in Africa.
Shrub 0.6–2.5 m, glabrous; branches with large lenticels; white exudate; leaves in a pair equal or one up to 3 × larger and narrower; elliptic or obovate, 7–26 × 1.5–9.5 cm, apex rounded then abruptly acuminate, cuneate, lateral nerves 7–12 pairs, black dotted below; inflorescence peduncle 1-2 mm; flowers 4, congested, pedicels 3–7 mm; bracts caducous; sepals ovate, 5–6 mm; corolla white, 7–12 cm; lobes elliptic, 3–4.5 × 1–2 cm; fruit of two separate mericarps, each normally oblong, 5–7.5 × 1.5 × 1.5 cm.
Habitat: swampy river banks in evergreen forest; 0–700 m alt.
Threats: slash and burn agriculture.
Management suggestions: rediscovery of this species at one or more of its known sites in S Region; surveys to provide basic populational data for future monitoring and intervention assessments; advising leaders of local communities nearby, to seek their support in protecting this species. Care should be taken not to confuse this species with the numerous other species of the genus in Cameroon, for which Leeuwenberg 1991 should be referred to.

ARALIACEAE

assessed by Martin Cheek

A family of trees, shrubs, epiphytes and lianas, mainly of evergreen forest but with a species of *Cussonia* in the subsahelian wooded grassland belt. Essentially the woody Umbelliferae, they share umbellate inflorescences, 2-styled ovaries in inconspicuous flowers, similar aromatic compounds, mainly palmately compound alternate leaves.

No Flore du Cameroun treatment exists, but the few species in W Africa have been well delimited by David Frodin (unpublished) who has spent decades working on the taxonomy of this family.

Polyscias fulva, a widespread pioneer submontane tree is traditionally important for making drums and carvings in the Cameroon Highlands, but is not threatened. *Cussonia bancoensis* is a very rare but widespread (Ivory Coast to Cameroon) tree that could be rated as threatened in future but for the present must be treated as NT.

Schefflera hierniana Harms
VU B2ab(iii)
Range: Equatorial Guinea (Bioko) and Cameroon (NW: Belo to Lake Oku (1 coll.). SW: Rumpi Hills (1 coll.); Mt Cameroon (2 coll.); Bakossi Mts (2 coll.); Lebialem Highlands at Fosimondi (2 coll.)).
A strangling epiphytic shrub of cloud forest resurrected by David Frodin from synonymy with the much commoner *S. barteri*, this rare species is known from only six locations (AOO 24 km² with 4 km² cells, threats below) along the wetter, submontane forested parts of the Cameroon Highland chain. It was assessed as Vulnerable in Cheek *et al.* (2004: 145) and that assessment is maintained here although Tchiengué has since found it in the Fosimondi area, where forest is extremely threatened. EOO is calculated as 8611 km². In future it is likely that this species will be rated as Endangered since it seems to be genuinely rare within its range. It is separated from *S. barteri* by having a slender 1–2 cm acumen, persistent fleshy-leathery bracts at the base of the inflorescence which are most easily seen in immature inflorescences, and congested densely brown-scurfy pubescent inflorescence axes, partial-peduncles 2–3 mm long, pedicels 2 mm (Frodin & Cheek in Cheek *et al.* 2004: 246)
Habitat: evergreen forest; 900–1400(–?2100 m) alt.
Threats: forest clearance for agriculture and wood. It has not recently been seen at the Belo to Lake Oku site despite intensive surveys in the late 1990s (Cheek *et al.* 2000) and

since forest has almost disappeared between these two locations, it is probably extinct there. Habitat degradation has been steady on some parts of Mt Cameroon at these altitudes and the species was not refound there in intensive surveys of the early 1990s. At Fosimondi forest is disappearing very rapidly to create farms (Cheek in Harvey *et al.* 2010).

Management suggestions: the best hope for the survival of this species may be the Bakossi Mts at the Kodmin and Nzee Mbeng sites since pressure here is relatively low and since the species was seen here recently (1998). Continued surveys of the unexplored parts of the Bakossi Mts might well reveal new sites. Further surveys at its other sites, if focussed on this species, might yet reveal that it survives so long as significant areas of forest exist.

Schefflera mannii (Hook.f.) Harms
VU A3c

Range: Equatorial Guinea (Annobón (1 coll.). Bioko (10 coll. at 4 sites)), São Tomé & Príncipe (São Tomé (1 coll.)). SE Nigeria (Obudu (1 coll.); Mambilla (3 coll.)), Cameroon (SW: Mt Cameroon (7 coll.); Kupe and Mwanenguba Mts (6 coll. at 4 sites). NW: Mt Oku (7 coll.); Bamboutos and Bafut-Ngemba (each 1 coll.)).

One of the very few montane (above 2000 m alt.) trees that are endemic to the Cameroon uplands (another is *Morella arborea*), this evergreen canopy tree begins life as an epiphytic shrub. Only one other montane *Schefflera* occurs in montane (above 2000 m) Cameroon Highlands, this being *S. abyssinica*, also a tree, it is easily distinguished by having deciduous, papery, long-acuminate leaves (not evergreen, thickly leathery, non-acuminate leaves).

Forest clearance for agriculture and wood has reduced the habitat of this species by an estimated 30% or more over its whole range due principally to loss in the Bamiléké Plateau-Bamenda Highlands, which, having the largest area above 2000 m in the Cameroon uplands, was probably once the stronghold for this species. EOO is calculated as 98304 km^2

Habitat: montane forest; (1400–)2000–2400 m alt.

Threats: Between 1987 and 1995, 25% of forest was lost in one area of the Bamenda Highlands (Moat in Cheek *et al.* 2000: back cover). This tree was first assessed as vulnerable in Cheek *et al.* (2004: 145) on the basis of the data above and this assessment is maintained here. No new data have been acquired except that a further study of forest loss in the Dom area of the Bamenda Highlands shows that 50% was lost in 15 years recently (Baena in Cheek *et al.* 2010: back cover). Extensive losses of habitat have also occurred at Mwanenguba, Obudu, Bamboutos and Bafut-Ngemba. Tree clearance for timber, firewood and agricultural land are the main threats.

Management suggestions: the status of *S. mannii* trees on the Gulf of Guinea islands such as Annobón requires elucidation, perhaps by survey. Are they truly conspecific given the morphological differences? The record from Príncipe derives from a note in FWTA and no specimen has been seen. At Mts Kupe, Cameroon and Oku (Kilum-Ijim) the species seems currently secure.

ARISTOLOCHIACEAE

assessed by Martin Cheek

These are mainly stem-twining lianas, creepers or subshrubs with alternate, simple, often cordate leaves, lacking stipules, the stems lacking coloured exudates or scent. The flowers are often large and conspicuous, and scented of dead animals to attract fly pollinators. The calyx is three-lobed and is the showy part since the petals are minute or absent. The stamens and style are held in a column above the inferior ovary. Without major use to man apart from various traditional medicinal uses and some species cultivated for their ornamental flowers.

There is no Flore du Cameroun account for the family. This pantropical family of 7–12 genera and c. 600 species is represented in Cameroon by two genera. *Aristolochia* occurs in semi-deciduous forests and woodland and has dry fruits which open like an upside-down parachute, releasing winged seeds. With nearly 500 species it is the largest genus in the family and none of the species in Cameroon are known to be threatened. *Pararistolochia* species are restricted to the forest of W-C Africa (revised by Odile Poncy, Le genre *Pararistolochia*, Aristolochiaceae D'Afrique Tropicale, Adansonia Sér. 2, 17(4) 465-494 (1978)) and New Guinea. They have similar flowers to the preceeding but fruits which are fleshy, resembling cucumbers. They can be identified to genus in forest by their stems which are figure-of -8-shaped in section. Several species in Cameroon are range-restricted and/or rare. Favio Gonzalez of New York is actively researching this genus.

Pararistolochia goldieana (Hook.f.) Hutch. & Dalziel is a climber which almost certainly has the largest flowers (c. 30 × 30 × 30 cm) of any African species of flowering plant. It is the African equivalent of the SE Asian *Rafflesia*, matching it in habitat, colour, size and scent of the flowers. It was previously assessed in Cheek *et al.* (2004: 146) as VU A2c. Although widespread it is relatively rare: Sierra Leone (2 coll.); Nigeria (6 coll.); Equatorial Guinea (Bioko (1 coll.)); Cameroon (several. coll). It is here reassessed as NT since it appears to survive happily in disturbed or regenerating forest and has more than ten locations now known.

Pararistolochia ceropegioides (S.Moore) Hutch. & Dalziel

VU A2c+3c

Range: Cameroon (SW: Mt Kupe (19 coll.). NW: Dom (1 coll.); Littoral: Mt Nlonako. C: Metet near Mefou proposed NP; Mékoassi. S: 70 km ESE Ébolowa; Bitye. E: 25 km NE Moloundou) and Gabon (Belinga; Massif du Chaillu; Mts Doudou; Rabi E).

This canopy climber is easily spotted when fertile since the orange, bijoux flowers are produced in clusters at eye-level from the figure-of-eight shaped stems. It is widespread but fairly rare in semi-deciduous and submontane forest in the forest belt of Cameroon and Gabon. In Bakossi, although common (19 coll.), that part of its population occurring below 1000 m alt. is under threat of forest clearance, while there has been extensive forest loss at some of its known locations elsewhere, e.g. at Nlonako and Yaoundé.

Its presence at Dom is a range extension, so collection of flowering material is needed to confirm the identification.

Overall, 30% habitat loss for this species over the past three generations seems conservative given that it is apparently a long-lived perennial with generation times possibly exceeding 10–20 years. Forest in the Dom area declined by c. 50% in 15 years recently and is ongoing. However, habitat in other parts of the range, such as Gabon, tends to be much less threatened. Data on collections cited under range for this species is taken from Poncy (1978) Adansonia 17 (4): 465–494.

The assessment above is updated and modified from those in Cheek *et al.* (2004: 145–146 and 2010: 94–95).

Climber 5–15 m; stem 8-shaped in section, 1–2 cm diam., flexible; leaves leathery, ovate, c. 13 × 8 cm, 3-nerved, acuminate, rounded-truncate; inflorescence cauliflorous, 1–2 m from ground, 3–8-flowered; flowers 5–7 cm; perianth tube pouched at base, pale-brown outside; tepals 3, equal, 1–2 cm, slightly splayed, throat orange; fruit fleshy, ellipsoid, c. 5–6 × 3–4 cm, 5-ridged, base and apex truncate; pedicel 3–4 cm.

Threats: forest clearance for timber and agriculture.

Habitat: semi-deciduous and submontane forest; 600–1600 m alt.

Management suggestions: research is needed to decide whether as currently delimited, more than one taxon is included, since this appears a very broadly defined species at present.

Pararistolochia preussii (Engl.) Hutch. & Dalziel
EN B2ab(iii)

Range: Cameroon endemic: (SW: Mt Cameroon area, 4 coll. at S Bakundu and Mabeta-Moliwe. S: Ambam-Ébolowa; Kribi-Campo; Ile de Dipikar (Tchouto 2004)).

This species was only known from a leafless specimen, *Preuss* 108 (type, Barombi Mbo, Kumba, probably extinct at this location, specimen assumed destroyed at Berlin), until a collection attributed to this species by Keay was found in the Mt Cameroon–S Bakundu area at Banga (*Brenan* 9484, collected in March). This species was recollected in the foothills of Mt Cameroon in Mabeta-Moliwe (now Bimbia-Bonadikombo), on the southern base line, at 400 m from the origin (*Sunderland* 1197, fl. 6 April 1992 and *Wheatley* 137, fl. 9 April 1992).

According to Gonzalez (pers. comm.) a further specimen of this species is known from the Kribi-Campo area (*de Wilde* 8011, WAG) and Poncy in Adansonia 17: 484, 1978 lists in addition a collection from Ébolowa: *Letouzey* 10053 (P). This species is undoubtedly very rare.

Assessed by me as CR A1c+2c C2a in Cable & Cheek (1998: xlii), it is here reassessed as Endangered since there are probably five surviving locations for the species (see above; AOO of 20 km² with 4 km² cells) with threats as below.

Liana to 15 m high; leaves papery, dark green, glossy above, ovate-oblong, 13–16 × 7–9 cm, acuminate, entire; perianth pale dull brown outside with dull purple markings, ovary 2 cm, perianth 7–10 cm, inflated part 2–3 cm, tube 2.5–3.5 cm, throat 1–1.5 cm wide, lobes equal, triangular, 1–1.5 cm wide at base, narrowing abruptly, with filamentous apices 3–4 cm long, concolorous within, throat deep purple; stamens 8. Inflorescence borne on old wood

Habitat: disturbed lowland forest.

Threats: forest clearance. All known sites of this species in the Kumba-Mt Cameroon locations are under pressure or have been lost to the species due to timber extraction, or forest clearance for agriculture and plantations. Slash and burn agriculture is proably the main threat in the S Region locations.

Management suggestions: this ornamental species is a prime candidate for cultivation and multiplication at Limbe BG, with a view to reintroduction to the wild at secure sites.

Several collections from submontane forest of what appeared to be a new orange-flowered species of *Pararistolochia* from Mt Cameroon and Mt Kupe (*P. sp. aff. ceropegioides* (S.Moore) Hutch. & Dalziel), were first reported in Cable & Cheek (1998: xlii) but have since been confirmed as falling within the range of variation of *P. ceropegioides*.

ASCLEPIADACEAE (APOCYNACEAE)

assessed by Martin Cheek and advised by David Goyder

Asclepiadaceae are now generally included under Apocynaceae with which they form a continuum. They have simple, opposite leaves which lack stipules but may have an interpetiolar ridge, sometimes with appendages. The flowers are regular, actinomorphic, the petals united at the base, the stamens united in a tube, often with specialised appendages, around the style-stigma. All species have white exudates when wounded except *Brachystelma*, *Ceropegia*, *Neoschumannia*, *Tylophora* and some species of *Marsdenia* which all have clear exudates. Most species have plumed seeds capable of long-distance dispersal. In some species the plumes are valued for making luxury papers. Many species have medicinal properties. Several genera are cultivated by hobbyists for their ornamental, often unusual, flowers.

The Red Data species of this group fall in two ecological parts:

1. Climbers of evergreen forest in the south of Cameroon. Such Asclepiad species are often widespread in tropical Africa from W to E, but very rare, often with only a single record per country. These are vulnerable to forest clearance.
2. Geophytes of grasslands and savanna in the north of Cameroon. These have underground rootstocks, producing annual short stems above ground before they flower in April–June, sometimes after fires.

There is no modern taxonomic account of the Asclepiads of Cameroon. The species recognised here follow the guidance of Dr David Goyder at RBG, Kew, who is working towards completing a Flore du Cameroun account.

Batesanthus purpureus N.E.Br. has previously been listed as VU (Cheek *et al.* 2004: 247). At that time this species was considered by H.J.T. Venter (BLFU) to be distinct from *B. intrusus* S.Moore and *B. parviflorus* C.Norman. Later he revised his species concept and now unites all three taxa under the first, which he treats as highly polymorphic (H.J.T. Venter & R.L.Verhhoeven, 2009, Morphology and taxonomy of *Baseonema* and *Batesanthus* (Apocynaceae: Periplocoideae, S. African J. Bot. 75: 445-455.). As a result, the number of sites for *B. purpureus* has risen to exceed the threshold of 10 used in Criterion B of IUCN (2001). For example, Sierra Leone holds one site, Nigeria two and Cameroon nine. The species also occurs in CAR, Congo (Kinshasa), Gabon, Angola and Liberia. Therefore it is here reassessed as Near Threatened.

Secamone racemosum (Benth.) Klack., formerly *Rhynchostigma racemosum* Benth., was previously assessed as VU (Cheek *et al.* 2004: 146, 249). We here reassess this species as Near Threatened since more than 10 sites are known (Criterion B) and data to assess the taxon under Criterion A is considered inadequate.

Tylophora urceolata Meve was formerly assessed as VU (Cable & Cheek 1998: xliv) but is now united with *T. anomala* N.E.Br. by Goyder (pers. comm. 2008) under this last name. Consequently the range and number of sites is so expanded that it is here assessed as Least Concern.

Asclepias kamerunensis Schltr. (syn. *Gomphocarpus kamerunensis* (Schltr.) Bullock)

EN C2a(i)+D

Range: Ghana (Shiare, Bueni-Krachi Distr.), Nigeria (Onitscha Prov., Nsuka; Zaria Prov., 3 m E Mando) and Cameroon (NW: Bamenda-Ngong; Mbaw Plain. E: Layé, 30 km W Batouri. Adamawa: Ngom. Extreme N: Kongola).

A geophyte 60–90 cm tall from a knobbly tuber c. 3 cm diam. The cylindrical glabrous stems bear linear-oblong leaves c. 10 × 0.5–1 cm, in pairs. The flowers are dull green and yellow, each 1 cm across in umbels of 4–8, on 2 cm pedicels. Known from eight sites (AOO 32 km² with 4 km² cells, with threats as below and lack of any recent collections allowing a continuing decline to be inferred). Only 2–3 plants are generally noted at each site, so no more than 500 individuals, and as little as 24, can be estimated from known observations. Accordingly, *Asclepias kamerunensis* is here assessed as Endangered. EOO is calculated as 632485 km².

Habitat: among rocks on steep slopes; 800–1000 m alt.; fl. April–June.

Threats: as for *Brachystelma exile*.

Management suggestions: as for *Brachystelma exile*.

Brachystelma exile Bullock

EN B2ab(iii)+D

Range: Nigeria (Ancho; Zaria, Birnin Gwari at Mando FR), Cameroon (Adamawa: Mbousa) and CAR (Bouar; Parc Manovo Gounda, St. Floris)

"A geophyte with a bun-like tuber up to 5 cm diam. and 2 cm thick. Stem erect, to 40 cm high. Corolla lobes purple within, greenish without. Corona greenish-yellow." (*Keay* FHI 25848, holotype). The stems bear 2–4 slender flowers, each c. 5 cm long and divided to the base into 5 linear, long-hairy lobes. The leaves at anthesis are sparse, linear, c. 10 × 0.5 cm.

Here assessed as Endangered, being known from five sites (AOO 20 km² with 4 km² cells), with threats as below. *Fay* (7127) records it as being uncommon. Less than 20 individuals are known from specimens, so we estimate that less than 250 individuals are known in the wild. EOO is calculated as 306190 km².

Habitat: *Hymenocardia–Crossopteryx* savanna, in sparse grassland over thin soils; 600–1000 m alt.; fl. May & June, possibly after fires.

Threats: agriculture. The type specimen, from Zaria Province, was in the neighbourhood of farm fallow.

Management suggestions: an effort should be made to refind this species in the wild, e.g. at Mbousa, and to incorporate it within a protected area.

Brachystelma omissum Bullock

VU B2ab(iii)
Range: Nigeria (Sardauna Prov., Yelwa Dist.; Mambilla Plateau) and Cameroon (NW: near Mankon; Jua. W: Ndop Plain near Bamessi. Adamawa: Kunre; Mbousa. C: Mbala–Yaoundé).
A geophyte with tuber c. 8 cm diam. Producing a robust, erect, glabrous stem c. 30 cm tall, and pairs of oblanceolate leaves c. 10 × 2 cm. The purple-grey-green slightly lobed flowers are 2–3 cm diam., held in large terminal umbels. Seemingly restricted to drier parts of the Cameroon-Adamawa Highlands, this plant appears not to flower after burning but to appear with the early rains. The geographically anomalous specimen from Central Region has been verified by Goyder. Here assessed as Vulnerable given the threats below and since only seven sites are known (AOO 28 km² with 4 km² cells). It is never described as common. EOO is calculated as 70853 km².
Habitat: montane grassland and savanna, withstanding heavy cattle-grazing; 900–1300 m alt; fl. April & May.
Threats: tilling of savanna and grassland for crop-production e.g. at Mbala.
Management suggestions: as for *Brachystelma exile*.
Note: this species is closely similar to *B. togoense* (Goyder pers. comm. 2008).

Ceropegia ledermannii Schltr.
EN B2ab(iii)+D
Range:Benin (Borgou-Papané), Nigeria (near Oyo; Vom-Bauchi Plateau; Sardauna Prov., Mayo Selbe; Takwara [not located]); Cameroon (N: Rey Bouba).
First discovered at Rey Bouba in Cameroon in 1908/09 by Ledermann (Bot. Jahrb. Syst. 51:154 (1913)), it has not been seen there since, and his specimen was destroyed at Berlin.
Growing from a tuber c. 15 × 15 × 5 cm, the erect stem reaches 15–45 cm. The leaves are paired, patent, linear, c. 10 × 1 cm. The erect, purple spotted white flowers are 10–15 cm high with the corolla tube dilated at the base, then constricted above, before widening gradually to the 1.5 cm wide mouth, over which the 5–8 cm long erect filamentous lobes stand guard. Known from only five sites (AOO 20 km² with 4 km² cells) with the threats below, usually with a single plant at each site, therefore a total of less than 250 plants are known in the wild. EOO is calculated as 208804 km².
Habitat: savanna; 300–1000 m alt.; fl. April–June.
Threats: tilling of savanna for agriculture.
Management suggestions: as for *Brachystelma exile*.

Ceropegia rhynchantha Schltr.
VU D
Range: Senegal (1 coll.), Guinea (Conakry) (1 coll.), Mali (1 coll.), Ghana (Sobaare Hill, 1 m S of Lawra), Nigeria (Ilorin-Jebba, 8 m; Naraguta; Panshanu Pass; Zaria, Igabi Distr., Anara FR) and Cameroon (N: Kokoumi on Bénoué R. Extreme N: 10 km E Mokolo at Souledé; 10 km SE Mokolo). First collected in 1909 at Kokoumi on the Bénoué R by the German *Ledermann* (4732), this twining herb is restricted to rocky grassland. The corolla tube is purple at base, then with a white belt, then purplish again and finally yellow, on top. It is globose, 1 cm diam. at the base, then constricted tightly before dilating at the mouth. The corolla lobes are united in the upper part into an erect rod with capitate apex. Here assessed as Vulnerable: although 11 locations are known with threats as below, less than 500 individuals are estimated since only one individual appears to be present at one site. EOO is calculated as 981600 km².
Habitat: rock outcrops in grassland and savanna, rarely in dry forest; 300–1000 m; fl. July–Sept.
Threats: as for *Ceropegia ledermannii*.
Management suggestions: as for *Brachystelma exile*.

Epistemma decurrens H.Huber
EN B1+B2ab(iii)
Range: Cameroon (Adamawa: Gotel Mts, slopes of Tchabel Ouade. NW (Bamenda Highlands): Dom; Mt Oku; Bali Ngemba FR).
An epiphyte, known with certainty from only two flowering gatherings (*Letouzey* 8606 and *Darbyshire* 378), but with two others in fruit almost certainly conspecific. The leaves resemble those of *E. rupestre*, while its flowers are 1 cm wide, yellow. Here it is assessed as Endangered since only four locations are known (AOO 16 km² with 4 km² cells) and at two of which it is acutely threatened. EOO is calculated as 1589 km².
Habitat: montane forest 2000 m alt., flowering in June.
Threats: its forest habitat (e.g. at Bali Ngemba pers. obs. Cheek (Harvey *et al.* 2004: 28)) is under threat due to felling for timber, agricultural land and fires set by pastoralists. Most montane forest in the Cameroon Highlands has been lost and that which remains is disappearing fast.
Management suggestions: an attempt should be made to refind this species at its known localities and to protect it, if it survives, working with local communities. Further survey work in the Bamenda Highlands may reveal more sites. Several fruiting collections (made in Nov.) of *Epistemma* from near Lake Oku are known (*Pollard* 41, *Cheek* 8767) and from Dom (*Cheek* 13648) which either represent this species (very likely) or another, possibly undescribed species of *Epistemma*. These needed to be re-collected when in flower (probably June) to secure a specific identification.

Epistemma rupestre H.Huber
EN B1+B2ab(iii)
Range: Cameroon (C: N'kolbison; Mt Yangba, 42 km NNE Bafia. S: Akoakas Rock, 25 km SE Ébolowa).

This shrub of granite rock outcrops is unique in the genus in being epilithic, not epiphytic. Reaching 1 m tall, the purple stems with white exudate are puberulent at first, bearing elliptic leaves 12 × 7 cm. The terminal flowers are purple, 1–1.2 cm wide. The 2-fruiting carpels are linear, 12 × 1.5 cm with plumed seeds. Known from only three sites (AOO 12 km² with 4 km² cells), with the threats below, it is here assessed as Endangered. EOO is calculated as 500 km².

Habitat: scrub on granite, with *Nephrolepsis biserrata*, *Clappertonia polyandra* and *Euphorbia kamerunica*; 750–1200 m alt., flowering in June.

Threats: rock outcrops in the N'kolbison area are threatened with granite extraction for building; soil pockets are cultivated for *Manihot* production (pers. obs. Cheek 2006).

Management suggestions: further surveys of suitable habitat may find this species to be more widespread than it appears. One or more of its sites should be protected, especially since its habitat supports other rare species such as *Tricarpelema africana*.

Maclaudia felixii Venter & R.L.Verh.
VU B2ab(iii)

Range: Guinea (Timbo; Mamou; Kindia-Benna; Gangan), Sierra Leone (Musaia; Njala; Sandra Tenraran), Ivory Coast (Talopa; Bouaké); Cameroon (NW: Njinikom).

Venter & Verhoven described this monotypic genus, a purple-flowered woody climber, in Bot. J. Linn. Soc. 115:57–63 (1994). The elliptic leaves to 10 × 4 cm are cuspidate, cordate and purple veined, the flowers 5 mm wide in axillary panicles. Although widespread, it appears rare. In Cameroon, despite tens of thousands of survey specimens collected in recent decades, it has not been refound. Globally, it has not been seen since 1963. Here it is assessed as Vulnerable since known from only 10 sites and due to the threats reported below. It has never been reported as common. Four local names are reported from Sierra Leone (in litt. Herb. Kew) so it may have local uses. The Talopa locality name has not been traced so is not mapped.

Habitat: orchard bush savanna along streams and forest edge; 400–1500 m alt., fl. April–Dec.

Threats: clearance of natural vegetation for agriculture (e.g. Njinikom) and mining infrastructure (Kindia-Benna, Mamou).

Management suggestions: a targeted search for this species at Njinikom might rediscover this plant, last seen there in 1931. Further survey work in Sierra Leone and Guinea (where it appears most common) may produce new sites for this taxon and allow it to be downrated from VU status, however it should first be verified that it survives at its other sites. It is recommended that sites are given protection.

Marsdenia magniflora P.T.Li (syn. *M. grandiflora* C.Norman)
VU B2ab(iii)

Range: Sierra Leone (Njala & Rowala), Liberia (Nimba–Mt Detton), Ivory Coast (Banco FR), Nigeria (Ife-Ife), Cameroon (SW: Bakossi), Congo (Kinshasa) (Yangambi) and Tanzania (Kagesa).

This brown hairy forest climber is sometimes known as the African *Hoya* due to its spectacular large purple-blue flowers each 1.5–2 cm across (Goyder pers. comm. 2008). The oblong-elliptic, acuminate-cordate leaves reach 10 × 5 cm. The plant has a wide but very sparse distribution, being extremely rare across its range, despite being so conspicuous. Here assessed as Vulnerable since known from 8 sites (AOO 32 km² with 4 km² cells), with the threats below. EOO is calculated as 1101230km².

Habitat: mainly submontane evergreen forest; c. 500 – 1500 m; fl. June (Cameroon).

Threats: forest clearance for agriculture e.g. in Bakossi (pers. obs. Cheek 2003).

Management suggestions: this plant should be introduced to cultivation so as to allow its inclusion in revegetation projects. Inclusion in protected areas is advisable. So far it is included only in one forest reserve.

Neoschumannia kamerunensis Schltr.
CR D

Range: Cameroon (SW: Mt Cameroon (2 coll.); Mt Kupe), Ivory Coast (1 coll.) and CAR (1 coll.).

The following text is modified from the assessment in Cheek *et al.* (2004: 146). Discovered at Man O War bay near Mabeta-Moliwe in 1899 by Schlechter, it has not yet been rediscovered at that site despite repeated searches in 1992, 1993 and 1995. One specimen was located in the 1960s at Mt Tonkui in Ivory Coast (probably exterminated there according to data from Gautier), it was not until March 1995 that an Earthwatch sponsored expedition at Mt Cameroon rediscovered this extremely rare vine above Likombe, in farmbush. Subsequently, one of the original discoverers, Etuge (the other being Meve, an Asclepiad specialist) found a population of seven plants at Mt Kupe in forest at the edge of a farm. Since Oct. 1995 the area at Mt Kupe has been rented from the local farmer in order to protect the plant. Another possible site was later found by Etuge in nearby northern Bakossi but the specimen has still to be verified. Finally, David Harris discovered this species in the Dzanga-Sangha reserve in southern CAR in 1996. The assessment of Critically Endangered is maintained for this taxon, on the basis of less than 50 individuals being known in the wild. In addition it is noteworthy that its geographic range is severely fragmented, its area of occupancy being less than 12 km² (given three probable extant sites, an area of occupancy of 12 km² is calculated using 4 km² cells), and continued decline of habitat quality being projected. EOO is calculated as 178009 km².

Habitat: lowland to submontane forest, withstanding and perhaps benefiting from some disturbance; 0–1000m alt.
Threats: although known to withstand some disturbance, presumably being able to regenerate after cutting, there is no doubt that intensive agriculture and tree clearance would destroy this rare species. The population in Bakossi is at the edge of a coffee farm which if extended would destroy the *Neoschumannia* plants. However a rent is paid so long as the plants remain undisturbed. The Likombe plant in farm bush depends on the goodwill of a local farmer for survival and has not been verified for several years.
Management suggestions: the northern Bakossi site requires verification. The now well-known plant above Likombe should be protected by formal agreement with the local landowner. A search for more individuals in the same area was carried out without success in 1995. This species may be a candidate for propagation at Limbe Botanic Garden for reintroduction to a protected, managed area or areas. The site at Mt Kupe remains the primary hope for the survival of this species in the wild. Regular monitoring by Martin Etuge and the author over the last 12 years has shown the population at this site to be stable.

Pachycarpus medusonema Bullock
EN B1+B2ab(iii)
Range: Cameroon (Adamawa: Ngaoundéré *Mildbraed* 10127; Meïganga, *Surville* 302; 80 km E of Ngaoundéré *Piot* 404; Wakwa *Piot* 94).
A brown felty erect herb 30–60 cm tall from an unknown rootstock. The paired 10 × 2.5 cm oblong leaves bear more than 20 pairs of lateral nerves and are spaced at 10 cm intervals. The terminal umbel holds 6–8 white flowers, each 2 cm across and on 2 cm pedicels. The 5 petals are divided to the base and held partly splayed, with the thread-like corona lobes slightly protruding. Endemic to Cameroon and known from four collections at effectively three sites (AOO 12 km^2 with 4 km^2 cells), it is here assessed as Endangered due to the threats listed below. EOO is calculated as 3170 km^2.
Habitat: grassland on basalt or granite; c. 1200 m alt.; fl. Jan., March or August.
Threats: town expansion at Ngaoundéré and Wakwa.
Management suggestions: efforts should be made to refind this plant around Ngaoundéré and to include it in a protected site.

Pentarrhinum ledermannii (Schltr.) Goyder & Liede (*syn. P. abyssinicum* Decne subsp. *ijimense* Goyder)
VU B2ab(iii)
Range: Cameroon (N: Kokoumi on Bénoué R (1 coll.). NW: Mt Oku and the Ijim Ridge (5 coll.); Bali Ngemba FR (1 coll.)), Burundi (Bubanza, Mugomero, 1 coll.) and Tanzania (Kilimanjaro; Uluguru FR).

Herbaceous climber to c. 6m tall, exudate clear. Stems c. 2mm diam., glabrous, internodes on flowering stems c. 15–20cm long. Leaves opposite, membranous, glabrous; blades ovate, c. 9 × 6cm, apex acuminate, base broadly cordate; petiole 4–5cm long; stipule-like subsidiary leaves ovate, c. 1 × 0.6cm. Inflorescence axillary, peduncle as long or longer than the subtending petioles, bearing a cluster of 8–15 flowers; pedicels c. 5mm long. Flowers 5-merous, c. 5 mm diam., white flecked with pink. Fruits with two follicles arranged in a line, follicles cylindrical, smooth, c. 4 × 0.6 cm, apex acute.
This subspecies can be distinguished from *P. abyssinicum* subsp. *angolense*, which also occurs in our area (e.g. *Thomas & McLeod* 5984 from forest above Oku village) by its smaller flowers which have a shorter corona bearing shorter teeth and have a more clearly stipitate gynostegium (Goyder *pers. comm.*).
This forest climber was discovered to be a potential novelty by David Goyder in 1998 while identifying specimens of Asclepiadaceae from the 1996 expedition to the Ijim Ridge. At that time only a single specimen was known. *Etuge* 3565 was collected near Tum at Ijim on 21 November 1996. A second specimen from Kilum, collected slightly earlier (*Buzgo* 798) lacks collecting data and so had been passed over. A new site was discovered for this plant by Pollard at the medicinal forest at Laikom in December 1998 (*Pollard* 368). In November 1999 Etuge led a large party of botanists back to the Tum forest patch which had yielded his specimen in 1996. Here about a dozen plants were counted flowering and fruiting profusely on both sides of the stream that bisects the forest. On the west side of that stream this climber is locally abundant over several square metres of *Solanecio mannii*, to the extent that it is difficult to distinguish one individual from another. On the east side of the stream, three plants were seen, each separated by 10–15 m. New material, including flowers in spirit, was collected (*Cheek* 9943) so that the taxon could be formally described in the Plants of Mt Oku and the Ijim Ridge (2000). This taxon appeared on the front cover of that book. Sadly Etuge (*pers. comm.*) reported that the Tum forest patch had been reduced by about two–thirds in the three years since he was there last. A third site for this rare plant was discovered by the Ghanaian botanist Amponsah at "Back valley", Mbingo on 10 November 1999 (*Cheek* 10063). In October 2001, Etuge (*Etuge* 4267) found this taxon at a fourth site, Bali Ngemba FR.
Liede and Goyder discovered this taxon from E African specimens in recent years, and found that it had first been published in Bot. Jahrb. 51: 155 (1913) by Schlechter, as a *Cynanchum* based on a Ledermann specimen from Kokumi in the Bamenda Highlands. The name transfer was formally made by Goyder and Liede-Schumann in 2008/2009 (Notes on *Cynanchum* and *Pentarrhinum* (Apocynaceae: Asclepioideae) in Tropical Africa, Kew Bull. 63: 463-466). That publication also adds new records from Tanzania.

The CR A1c assessment made in Cheek *et al.* (2000: 55) and maintained in Harvey *et al.* (2004: 60), is no longer appropriate. In the first place the causes of the population reduction have not ceased as is required for A1 qualification. In the second we can no longer say that more than 90% of the population has been lost in the last 3 generations/10 years, given the three newly discovered locations E of the Congo basin about which we know very little at present regarding the levels of threats or protection, or of population reduction.

Since six locations are now known for the species (see range above, AOO 24 km² with 4 km² cells, threats as below), under criterion B we can now assess *P. ledermannii* as VU B2ab(iii). EOO is calculated as 934729 km². However, when better data on the locations and subpopulations in E Africa is known, it is likely that we will be able to re-assess this taxon as EN.

This taxon is completely unprotected in Cameroon except at Bali Ngemba where it appears rare and protection levels are low. In Tanzania it occurs in the Uluguru South catchment Forest reserve, but it is uncertain whether the species is protected there or not.

Habitat (in Cameroon): submontane forest with *Garcinia smeathmannii, Solanecio mannii, Cuviera longiflora,* and *Pouteria altissima*; 1750–1900 m alt.

Threats: clearance of forest for cultivation of crops. It is considered that over 50% of the forest habitat of this species has been lost in the last 100 years in Cameroon and that this loss is ongoing.

Management suggestions: effort should be made to protect the surviving forest patch at Tum (site of the largest known sub-population of this taxon but outside the present protected area boundary) if it survives there, and to establish whether other sites for this variety exist apart from the five listed. It should also be established whether or not the species survives at Kokumi and at the E African sites.

Raphionacme keayii Bullock

EN D

Range: Nigeria (Mando FR, E of Kargi in Zaria Prov.) and CAR (Beretum). Not known but likely in Cameroon.

A brown hairy erect herb 0.3–1 m tall from an underground tuber c. 3 × 7 cm (dried). It produces white exudate. The leaves are c. 10 × 2 cm and the spike-like inflorescence, c. 30 cm long, bears green and red flowers c. 1 cm wide. First collected in 1914 by Mildbraed, it was refound in Nigeria by Keay in 1950 and has not been seen since. Here it is assessed as Endangered since known from only two, widely separated sites despite being a conspicuous, easily recognised plant, with less than 250 individuals being estimated.

Habitat: *Isoberlinia–Uapaca* open woodland; c. 800 m alt.; fl. June–July.

Threats: natural habitat in Nigeria has seen great depredations in recent decades with incursions of agriculture into forest reserves.

Management suggestions: every effort should be made to refind this species and protect it before it becomes extinct. Beretum in CAR offers the best prospect since CAR has seen less habitat destruction than Nigeria. However a watch should be kept for *R. keayi* in intervening Cameroon. It may well be found there.

Secamone letouzeana (H.Huber) Klack.

VU B2ab(iii)

Range: Cameroon (SW: Bakossi Mts. W: Bangangté. E: 20 km W Deng Deng. C: N'kolbison), CAR (50 km E Bambari), Congo (Kinshasa) (Nsélé-Maluku, 4 coll.), Congo (Brazzaville) (Lékoumou: near Simonbondo, Ogooué-Lékéti proposed National Park; Doumai) and Gabon (Haut Ogooué; 15 km W Lope; Mont du Casque).

Shrubby climber, stems with retrorse red hairs; leaf-blades 7–9 × 2–3 cm, acuminate, base cuneate, cymes extra-axillary, shorter than leaves; corolla greenish white, sweet scented, lobes 3–5 × 0.5–0.9 mm.

Known from 10 sites (AOO 36 km² with 4 km² cells), with the threats detailed below, *S. letouzeyana* is here assessed as Vulnerable. EOO is calculated as 839262 km².

Habitat: evergreen forest; 600–1300 m alt.; fl. Jan., March and July.

Threats: much habitat has been cleared at N'kolbison for city expansion since the species was collected there (de Wilde pers. comm. 2003). Almost all natural forest habitat in Bangangté has been cleared for agriculture.

Management suggestions: surveys are needed to rediscover and protect this species.

BALANOPHORACEAE

assessed by Martin Cheek

A strange family of root parasites which spend most of their existence out of sight below ground, appearing above ground only to flower and fruit, when such is their strange appearance, that they are often taken for fungi. *Thonningia sanguinea* is widespread in tropical Africa, and fairly common in Cameroon, its 2-3cm wide inflorescences being found in forest floor litter. These resemble a *Protea* flower, being surrounded by petal-like red-pink bracts, the central mass comprising hundreds of reduced flowers. The family has no commercial value. The second species of the family in Cameroon, treated below, is very rare.

Chlamydophytum aphyllum Mildbr.

EN B2ab(iii)

Range: Cameroon (E: Deng Deng), Congo (Brazzaville) (Madingo-Kayes) and Congo (Kinshasa) (Taketa, terr. Oshwe, R Lukenie; Kokolopor).

Most of the information in this note is taken from the superb paper by N. Hallé in Adansonia 17 (3): 249-261 (1978), documenting three locations. A fourth location was recently found by Georgieu *et al*: African Journal Ecology 48 (3): 849-852 (2010).

Here *Chlamydophytum aphyllum* is assessed as Endangered since only four locations are known (above; AOO 16 km² with 4 km² cells) and since threats are as below.

Leafless herb emerging above ground only to flower; parasitising the roots of *Tessmannia*; inflorescence when juvenile emerging in an obovoid 'volva' to 18–45 × 8–12.5 cm, the surface grey-brown, leathery; at length rupturing to reveal the many-flowered inflorescence which resembles a dirty crimson cauliflower 10–21 cm diam, the fleshy central axis 4–5 cm diam. at the base; the individual flowers fleshy, 5–15 mm diam. Flowering in Feb. and Dec. (Congo (Brazzaville)).

Habitat: evergreen forest on white sand (Congo (Brazzaville)).

Threats: habitat destruction for tourism and development (Congo (Brazzaville)). Trees of *Tessmannia* (on which *Chlamydophytum* depends) are logged for timber.

Management suggestions: since this plant is so rare, and mostly inconspicuous, it might easily be destroyed at its few locations by people understandably ignorant of its existence and importance. Therefore efforts should be made to pinpoint its locations (surveying in the flowering season) and to engage leaders of local communities in its protection and monitoring. The host trees should also be protected in order to continue sustaining their parasites.

BALSAMINACEAE

assessed by Martin Cheek

Impatiens etindensis Cheek & Eb. Fisch (syn. *Impatiens* sp. nov. 2 aff. *grandisepala* in Cable & Cheek 1998)

EN B2ab(iii)

Range: Cameroon endemic (SW: Mt. Cameroon at Mt. Etinde & Njonji (2 sites, 5 coll.)).

A tuberous canopy epiphyte with dramatic red and yellow flowers, this species is conspicuous near the summit of Mt. Etinde in the wet season, when in flower. No other Cameroonian species is known to have a tuber. Published in 1999 (Cheek & Fischer in Kew Bull. 54: 471-475), *I. etindensis* was previously named as *I.* sp. nov. 2 aff. *grandisepala* and assessed as VU D2 in Cheek & Cable (1998: xlv, 21). The assessment here of *I. etindensis* as endangered is taken from Cheek & Fischer loc. cit. and is maintained here since no new data are available. AOO is 8 km² using 4 km² cells, threats as below. Further work might prove the two sites contiguous, in which case CR would be appropriate. EOO is 48 km².

Habitat: submontane evergreen forest; (450-)1600-1720 m alt.

Threats: forest above Njonji is threatened by expansion of oil palm plantations.

Management suggestions: monitoring of the plants in the summit area of Mt. Etinde is advisable. Protection of forest above Njonji should be considered if the record site for this species is to survive. Cultivation at Limbe BG is likely to prove difficult due to the altitudinal disparity.

Impatiens frithii Cheek (*Impatiens* sp. nov. 1 of Cable & Cheek 1998)

EN B2ab(iii)

Range: Cameroon endemic (SW: Bakossi Mts, Kodmin area (4 coll.) and Mt Etinde (4 coll.). Littoral: Ebo (1 coll.)).

First referred to as *Impatiens* sp. nov. 1 and assessed as VU D2 (Cable & Cheek 1998: xliv), the species was formally named in 2002. The slender, hairpin-shaped flowers of this epiphytic species are flame-red. They are produced in the wet season from stems often only 15cm long which scramble along mossy branches of low trees and shrubs. Assessed as vulnerable (Cheek & Csiba 2002, see reference below), *Impatiens frithii* is here reassessed as endangered given an estimated area of occupancy of 15 km^2, being known from only two broad but very fragmented (100 km interval) sites and the threats mentioned below, referred to in detail in Cheek & Csiba (A new epiphytic species of *Impatiens* (*Balsaminaceae*) from western Cameroon. Kew Bull. 57(3): 669-674 (2002)). This assessment is updated from that made in Cheek *et al.* (2004: 147), since a third location has been added, at Ebo. EOO is calculated as 6840 km².

Habitat: moss forest; (800–)1100–1700m alt.

Threats: planned plantation expansion (Mt Etinde) and reservoir scheme (Bakossi Mts).

Management suggestions: see *Impatiens letouzey and I. etindensis*.

Impatiens gongolana N.Hallé

EN B2ab(iii)

Range: Cameroon (S: Campo Ma'an (1 coll.)) and Gabon (Cristal Mts., 3 coll.).

This relative of *I. filicornu* is distinguished by slender leaves that are hairy above. It is a terrestrial herb that was believed endemic to the Cristal Mts. until Tchouto (2004) identified it in Campo Ma'an. We rely on the identification of the voucher at WAG being reliable. We here assess *I. gongolana* as endangered in view of the threat to its habitat and the four sites as detailed above. AOO is 8 km² using 4 km² cells, threats as below.

Habitat: evergreen lowland and lower submontane forest; alt. unknown.
Threats: logging followed by agriculture. Campo Ma'an NP has seen incursions by rubber plantations in recent years (UNEP 2008).
Management suggestions: sites for *I. gongolana* should be pinpointed and at least one should be protected by inclusion in a protected area or by other means.

Impatiens grandisepala Grey-Wilson
CR B2ab(iii) + D
Range: Cameroon endemic (SW: Mt Cameroon at Efolofo (1 pre-1988 collection)).
Known only from a single collection (*Satabié* 303, YA, P) made in June 1979 from Efolofo, 30 km NW of Muyuka on the N. side of the mountain. "*Impatiens grandisepala* is an extraordinary species known only from a single gathering in the Cameroun. The most obvious features of this plant are the very large and conspicuous lateral sepals which droop downwards, partially or wholly concealing the lower sepal and spur, a characteristic quite unlike any other African species of *Impatiens*." (Grey-Wilson in Impatiens of Africa (1980: 216)). Unsatisfactorily assessed as DD by Cheek in Cable & Cheek (1998: 20). AOO is 4 km^2 using 4 km^2 cells, threats as below. Less than 50 individuals were recorded. Here assessed as CR.
Habitat: epiphytic in deeply shaded montane forest; 1150 m alt.
Threats: forest clearance for agriculture.
Management suggestions: this species should be looked for in the wet season, ideally with the help of Dr Satabié (retired head of the National Herbarium) and the state and extent of the population assessed.

Impatiens hians Hook, F. *var.* ***bipindensis*** (Gilg.) Grey-Wilson
VU B2ab(iii)
Range: Cameroon (SW: Korup-Mundemba; Mt. Cameroon at Onge FR, Idenau and Etinde. S: Kribi–Zingui; Bipindi–Lokundjé) and Gabon (3 sites in Cristal Mts).
This taxon probably deserves specific status. It is an epiphyte with straggling stems, resembling *I. frithii* but with much larger flowers. Other differential characters are listed in Cheek & Csiba 2002 (Kew Bull. 57: 669-674). It is here assessed as vulnerable: AOO is 32 km^2 using 4 km^2 cells (8 locations), threats as below. EOO is calculated as 44177 km^2.
Habitat: evergreen lowland and lower submontane forest; c.200–1000m. alt.
Threats: habitat clearance on the lower slopes of Mt. Etinde/Mt. Cameroon for plantation expansion. In the Cristal Mts. plants are in the path of mineral prospection operations (Cheek pers. obs. 2007).

Management suggestions: further survey work might well locate extra sites for this taxon, and enable de-listing from the threatened list. Unless and until this occurs this taxon should be protected. Subpopulations in Gabon are believed to be within the Cristal Mts NP of that name.

Impatiens kamerunensis Warburg subsp. ***obanensis*** (Keay) Grey-Wilson var. ***parviflora*** Grey-Wilson
CR B2ab(iii); D
Range: Cameroon (SW: Mamfe (1 site)).
A pink-flowered 20cm tall terrestrial herb known from a single record made in 1962 and not seen since. Here assessed as critically endangered. This probable annual is here assessed as critically endangered since it is known from a single site near a major town so habitat destruction is a threat. By extrapolation from other related species, less than 500 individuals are likely to be present at the site, if it survives.
Habitat: on rock in evergreen forest belt; c.400m alt.
Threats: see above.
Management suggestions: the type locality should be surveyed to discover if the taxon survives and if so, the threats it faces. The taxonomic status of the plant should be reviewed in the light of additional material. If the plant is re-found a management plan should be executed to protect the species.

Impatiens letouzeyi Grey-Wilson
EN B2ab(iii)
Range: Cameroon endemic (SW: Bakossi Mts (5 coll.) and Lebialem Highlands-Fosimondi (1 site)).
This robust, epiphytic herb has the largest flowers of all Cameroonian, and perhaps African, *Impatiens*. It is easily recognisable due to the tightly inwardly coiled spur, and toothed lateral sepals. Described in 1981 from a single specimen in western Bakossi collected in the 1970s (*Letouzey* 15353), it was rediscovered by an Earthwatch team led by George Gosline in 1998 and subsequently found to be fairly common in the Kodmin-Edib area. At the end of the wet season, plants are readily spotted by their fallen flowers under the trees on which they grow. This plant has not been found on Mt Kupe despite searches in all months over several years. It is arguable as to whether the Bakossi records should be taken as two or four sites. Opting for the second gives a total number of three sites (AOO 12 km^2 using 4 km^2 cells, threats as below). The assessment above is maintained from that given in Cheek *et al.* (2004: 147). The additional record from Lebialem both extends the range and increases the threat level to the species but does not take the species out of the EN status. EOO is calculated as 1566 km^2.
Habitat: epiphytic in crown of trees 4–6 cm from ground, in shrubs over streams or very rarely, terrestrial, on lakeside *Sphagnum* blanket; 1200–1350m alt.

Threats: the planned reservoir scheme near Kodmin may threaten part of the population of this species. At Fosimondi the trees on which this species is growing are being cut down to plant crops.
Management suggestions: a detailed study of this species at Kodmin-Edib where the species is most secure and where conservation efforts might be focussed, would provide more precise data on population density and demography, as well as assisting in the placement of future development schemes. Botanical surveys in other parts of the Bakossi Mts, or Rumpi Hills, might discover more sites which would reduce the threat assessment of this taxon.

Impatiens sakeriana Hook.f.
VU A3c
Range: Equatorial Guinea (Bioko) and Cameroon (SW: Mt Cameroon (6 coll.); Mwanenguba (2 coll.); Fosimondi; Bamboutos Mts (1 coll.). NW: Bamenda Highlands (1 coll.); Bali Ngemba FR (3 coll.); Mt Oku (12 coll.); Dom (1 coll.)). This often robust, presumed perennial, locally common terrestrial herb has the highest altitudinal range of all Cameroonian *Impatiens*. Secure on Mt Cameroon, and probably Bioko, forest at Mwanenguba, the Bamenda Highlands and Bamboutos Mts has been under pressure from grassland fires set by graziers and for clearance for agriculture; forest in the Kilum-Ijim area outside the protected area having seen a reduction in cover of c. 25% in eight years of the 1980s and 1990s (Moat in Cheek *et al.* 2000: back cover). In the Dom area 50% forest loss has occurred 1988–2003 (Baena, in Cheek *et al.* 2010: back cover). Assuming a generation time of ten years, it is estimated that about 30% habitat loss may have occurred over its whole range in the last 30 years. EOO is calculated as 27837 km².
The assessment above is updated from that in Cheek *et al.* (2004: 147) and Cheek in Harvey *et al.* (2004: 61).
Habitat: understorey of montane forest; 2000-3000m alt.
Threats: see above.
Management suggestions: enforcement of existing protected area boundaries. Demographic studies are needed to elucidate generation time and ecological requirements of this taxon.

REFERENCES

Cheek & Csiba (2002 A new epiphytic species of *Impatiens* (*Balsaminaceae*) from western Cameroon. Kew Bull. 57(3): 669-674.).
UNEP 2008

BEGONIACEAE

assessed by Martin Cheek

Despite the absence of a Flore du Cameroun account, most of the Cameroonian species of *Begoniaceae* can be identified using the research publications of Wageningen Univ., Netherlands. For decades this institute has specialised in African *Begonia* taxonomy. The species of sections *Squamibegonia* (de Wilde, J.J.F.E. & J.C. Arends (1980). *Begonia* section *Squamibegonia* Warb. a taxonomic revision. Miscellaneous Papers Landbouwhogeschool Wageningen 19: 377–421), *Tetraphila* (de Wilde, J.J.F.E. (2002). Studies in Begoniaceae VII. Wageningen University Papers. Backhuys, Leiden), *Loasibegonia* and *Scutobegonia* (Sosef, M. (1994) Studies in Begoniaceae V. Wageningen University Papers. Backhuys, Leiden.) have all been treated. Readers are referred to these works for illustrations and diagnostic characters. Since these revisions, the ranges of some species and numbers of sites has increased due to recent surveys, such as those of Tchoutou (2004) in Kribi-Campo and our own in western Cameroon. The last two sections mentioned comprise the 'refuge begonias'. These are usually terrestrial forest-floor herbs, yellow flowered, that are of great interest for conservation since several lines of evidence suggest that they are indicators of ancient forest refuge areas (Sosef 1994 loc. cit.). Many of these species are highly restricted in range and so are assessed here as Red Data species.

All the species of *Begonia* treated here are dependent on shaded habitats, generally of forest. They sometimes have very specific microhabitats within forest, such as rock faces with proximity to waterfalls. In general they are fairly conspicuous and not easily overlooked in surveys since they occur near the forest floor and have brightly coloured flowers borne over a long flowering season.

Begonia adpressa Sosef (Sect. *Loasibegonia*)
EN B2ab(iii)
Range: Cameroon (SW: Kupe-Bakossi; Rumpi Hills; S Bamboutos; Lebialem Highlands. Littoral: Mt Nlonako. (8 pre-1995 coll.)).
Described by Sosef in 1992, when only eight collections were known (Sosef 1994: 150), all of which were from W of Bangem in in the northern Bakossi Mts apart from three collections immediately to the north in the Bamiléké area, one to the west, in the Rumpi Hills, and one immediately to the east, at Mt Nlonako. To these we added six further collections made since 1995, from three sites in the Bakossi Mts one from Kupe village, one from Nyasoso, the others all from Kodmin (Cheek *et al.* 2004). Previously (in Cheek *et al.* 2004: 147, Harvey *et al.* 2010: 70) we held that these constituted nine locations and an area of occupancy of 14 km² calculated at 1 km² per collection, but here a wider concept of "location" is used and five are accepted (see under "range" above). Therefore *B. adpressa* is here reassessed as Endangered (five locations, AOO 20 km² with 4 km² cells

and threats below). The habitat quality at the Bakossi sites is projected to remain stable if proposed conservation outcomes come to pass, but all other sites except the Rumpi Hills, natural forest at this taxon's altitude range is under severe pressure, or have been eliminated.

Habitat: terrestrial and on rocks in forest; 1000–1750 m alt.

Threats: forest clearance for timber, followed by agriculture, especially at Nlonako, Bamboutos and Lebialem sites where active and ongoing.

Management suggestions: the future for *Begonia adpressa* seems bleak outside of the Rumpi Hills and Bakossi Mts. However, at these last sites it seems secure if proposed new protected areas are implemented and respected.

Begonia bonus-henricus J.J.de Wilde (Sect. *Squamibegonia*)

EN B2ab(iii)

Range: Cameroon (SW: Rumpi Hills, (1 coll.); Mt Kupe and Bakossi Mts (17 coll.). S: Ébolowa (1 coll.))

First described in 1980, this species is unique in sect. *Squamibegonia* Warb. (characterised by peltate scales, indehiscent globose berries, inflorescences enclosed in sheathing bracts) in having long (c. 1 m) slender (1.5 mm diam.) pendulous or rooting stems. It was then known from just five collections at three sites: c 15 km SE of É bolowa, Rumpi Hills and, in Bakossi, Ngomboaku and Ngussi (coll. from last two not seen). This geographical range is very similar to that of *Newtonia duncan-thomasii*, although the *Begonia* also occurs on Mt Kupe and has a slightly lower altitudinal range. We have added fifteen more collections in Bakossi to those cited by J.J. De Wilde & Arends (1980 loc. cit.). Excellent illustrations, notes and a description are given by those authors. The conservation assessment of this species in Cheek *et al.* (2004: 148) of VU A3c is raised here to EN since three locations (see under range above) are known (AOO is estimated as 24 km^2 with 4 km^2 cells, allowing 12 km^2 for the Bakossi area which has numerous collections from adjoining sites

Habitat: hemiepiphytes in forest; 660–1300 m alt.

Threats: clearance of forest for timber and agriculture at lower altitudes (below 1000 m alt.) seems unfortunately inevitable in the Bakossi area.

Management suggestions: assuming that proposed conservation plans in the Bakossi area are enacted, this locally common species seems fairly secure there above the 1000 m contour and no other conservation actions seems necessary.

Begonia duncan-thomasii Sosef (Sect. *Loasibegonia*)

EN B2ab(iii)

Range: Cameroon (SW: Kupe-Bakossi and Rumpi Hills).

Described by Sosef in 1992 and known only from four pre-1995 collections (Sosef 1994: 156), all but one of which are in Kupe-Bakossi, the exception being *Satabié* 259 from the Rumpi Hills; the remainder were collected by Duncan Thomas. These three sites are at the summit and on the west side of Mt Kupe, and west of Bangem. To these in 1998, we have added only two more collections from a fifth site, near Kodmin at Abo'h (*Biye* 31 and *Etuge* 4025). The conservation assessment of this species in Cheek *et al.* (2004: 148) was VU B2ab(iii), but here we adopt the current IUCN concept of location and (see range) accept two, for an assessment of EN (AOO 8 km^2 with 4 km^2 cells, threats below).

Habitat: *B. duncan-thomasii* appears to be restricted to shady, damp, vertical rock-faces in forest, sometimes with the rare *Nodonema lineatum*; 1550–2000 m alt.

Threats: see *Begonia schaeferi.*

Management suggestions: the sites of this species merit revisiting and regular monitoring for numbers of individuals, regeneration levels and threats.

Begonia furfuracea Hook.f. (Sect. *Tetraphila*)

EN B2ab(iii)

Range: Equatorial Guinea (Rio Muni (1 coll.); Bioko (4 coll., 2 sites)) and Cameroon (SW: Mt Kupe and the Bakossi Mts (8 coll.). NW: Bamenda, Kondo (1 coll.)).

This epiphytic species of *Begonia* sect. *Tetraphila* was recently revised by J.J.F.E. de Wilde (2002). He cites evidence that the subpopulation on Bioko may have been effected by cultivation of cocoa, coffee and bananas, for which most of the original vegetation below 700 m was cleared. *Begonia furfuracea* was first collected there by Mann in 1860, recollected many decades later by Mildbraed and then relatively recently (c. 1960s) by Sanford and Carvalho.

The Cameroonian subpopulation first came to light in the 1970s at Kondo, 40 km NW of Bamenda (*Letouzey* 14186) an area where most forest has now been extirpated. The species probably now has its stronghold at Kupe-Bakossi. De Wilde cites three collections from the area (de Wilde 2002: 89). To these can be added a further five (see Cheek *et al.* (2004: 252)). The conservation assessment of this species in Cheek *et al.* 2004: 148 was VU B2ab(iii), but here we adopt the current IUCN concept of location and (see range) accept five, for an assessment of EN (AOO 20 km^2 with 4 km^2 cells, threats below). The Bata [Rio Muni] site is newly added here, based on a determination by Plana.

Habitat: evergreen forest, 1000–1300 m alt.

Threats: assuming that proposed conservation plans in Bakossi are enacted this species seems secure there, its stronghold. Some habitat destruction occurs at its known sites, but levels are low. In the Bamenda Highlands 96.5% of forest was estimated as lost (Cheek *et al.* (2000: 6)) and that which survives is disappearing fast (50% lost 1988-2003 in the Dom area fide Baena in Cheek *et al.* 2010: back cover).

Management suggestions: yearly monitoring of plants to assess population trends seems advisable. An attempt should be made to rediscover and assess the status of the Bioko subpopulation.

Begonia heterochroma Sosef (Sect. *Loasibegonia*)
EN B2ab(iii)

Range: Cameroon (S: 13-15 km N Kribi and Campo-Ma'an) and Gabon (Cristal Mts (4 coll.)).
This is a terrestrial yellow-flowered refuge *Begonia* (Sosef loc. cit. 1994: 160-162). The peltate leaves are often kidney-shaped, the apex rounded, the upper surface hirsute, with erect red hairs. *Begonia heterochroma* is here assessed as EN (three locations, six site-records for an AOO of 24 km^2 with 4 km^2 cells and in view of the threats detailed).
Habitat: evergreen forest, 80–360 m alt.
Threats: currently the Cristal Mts and Campo-Ma'an habitat seem secure, but coastal forest near Kribi has seen clearance for tourist activities largely and extensive expected gas and oil related installations will destroy more habitat.
Management suggestions: the forest site N of Kribi should be protected from clearance, otherwise the species will be lost both from its type locality in Cameroon. In Gabon the correspondence between known sites and boundaries of the Cristal Mts National Park should be examined. Sites for the species outside the boundary should be considered for protection.

Begonia mbangaensis Sosef (Sect. *Scutobegonia*)
EN B2ab(iii)

Range: Cameroon endemic (S: Kribi–Ébolowa km 81 near Mbanga; Nkolesesan Hill; near Lolodorf and Bipindi (Tchoutou (2004))).
This terrestrial refuge species has elliptic, peltate leaves (4.8 -10.5 x 3.5 – 8.8 cm) with a rounded apex. Originally known from three collections, all possibly from the same site at "Mt Nkolesesan" (data Sosef loc. cit. 1994: 259). Tchoutou (2004) cited two new sites for the species. *Begonia mbangaensis* is here assessed as Endangered: with three localities (AOO 12 km^2 with 4 km^2 cells) and proximity to a village posing a threat to its habitat.
Habitat: evergreen forest; alt. unknown.
Threats: see above
Management suggestions: the authorities of Mbanga should be apprised of the existence of this rare plant and efforts made to rediscover, monitor and protect the site, also those discovered by Tchouto. *Begonia mbangaensis* might be found at additional sites if further surveys are mounted.

Begonia microsperma Warb. (Sect. *Loasibegonia*)
EN B2ab(iii)

Range: Cameroon (SW: Bai fall on Meme R SE of Mbongé. S: Mt Nkolumbembe, Ébolowa–Campo–Kribi, Ebianemeyong & Ma'an in Campo-Ma'an NP).
An epilithic or epiphytic yellow-flowered refuge *Begonia*. The peltate leaves are ovate, coarsely bullate, with bristle-like hairs above. Unusually the stipules lack fimbriae. Since only three sites are known (AOO 12 km^2 with 4 km^2 cells) and due to the threats noted, *B. microsperma* is here assessed as Endangered. A Thomas collection from Takamanda appears as this species on the TROPICOS database but has been redetermined by Sosef (1994 loc. cit.) as *B. staudtii*.
Habitat: on spray-soaked surfaces in evergreen forest habitats at low altitude.
Threats: coastal Mts in the Kribi–Campo area are threatened by iron mining.
Management suggestions: known sites should be revisited and surveyed for the size of the population, and regeneration levels recorded. The possibility of protecting both sites outside the Campo–Ma'an NP should be considered and public education used to raise awareness of the importance of the plant. The Bai waterfall may have potential as a tourist location which might provide a basis for protecting this site.

Begonia minuta Sosef (Sect. *Loasibegonia*)
CR D

Range: Cameroon endemic (S: c. 20 km NW of Bipindi at Nkol Tsia ridge (1 site/location)).
Named for its minute (1.2–3.9 × 0.5–1.5 cm) lanceolate-peltate leaves. This terrestrial refuge *Begonia* is known from a single site. The plants grow on a flat, wet, mossy rock face in an evergreen forest area. Refuge begonias were revised by Sosef (loc.cit 1994) from which source data here cited are derived. The species is here assessed as Critically Endangered since no more that 500 individuals are known to exist at the only known site.
Habitat: see above.
Threats: exposed rock outcrops with road access are vulnerable to exploitation for aggregate for civil engineering projects, such as those associated with the extractive industries expected in the Kribi area. In addition, a single rock fall or quarrying event could wipe out the species.
Management suggestions: the importance of this site should be drawn to the attention of the traditional and governmental local authorities so as to protect the plants. It is to be hoped that new sites might be found for this strange *Begonia*. Further survey work might help with this.

Begonia montis-elephantis J.J. de Wilde (Sect. *Scutibegonia*)
CR B2ab(iii); D

Range: Cameroon endemic (S: 20 km SE Kribi at Mt Elephant (1 site/location)).

A creeping herb of wet rock-faces, this is a recently discovered yellow-flowered refuge *Begonia* (De Wilde 2002, Wageningen Univ. Papers 01.2: 259-266). *Begonia montis-elephantis* is distinct for its very narrowly elliptic pendant leaves, held at 90° from the petiole (7–17.5 × 1–2.5 cm), laciniate stipules and c-shaped stigma lobes. It is here assessed as Critically Endangered since known from a single site (AOO 4 km^2) with threats as below and, where less than 500 individuals can be seen (photos in De Wilde 2002 loc. cit.)

Habitat: near vertical, wet rock faces in evergreen forest belt; 270 m alt.

Threats: Mt Elephant is reputed to be threatened by iron ore mining (pers. comm. At IUCN Workshop May 2008, Yaoundé).

Management suggestions: this species is known from a small site. If iron mining goes ahead at Mt Elephant, all efforts should be made to protect the rock-face on which it grows. It is to be hoped that survey work might discover new sites for *B. montis-elephantis*.

Begonia oxyanthera Warb. (Sect. *Tetraphila*)

VU A3c

Range: Nigeria (Obudu; Mambilla), Equatorial Guinea (Bioko (6 coll. from 2 sites)) and Cameroon (SW: Mt Cameroon (4 coll.); Mt Kupe and the Bakossi Mts (5 coll.); Rumpi Hills (1 coll.); Mwanenguba (1 coll.); Lebialem Highlands/Bamboutos Mts (2 coll.). NW: Bamenda Highlands (17 coll.)).

The range data above are taken from the account of the species in the recently published revision of *Begonia* sect. *Tetraphila* (de Wilde 2002). In previous Red Data assessments (Cable & Cheek 1998: xlv and Cheek *et al.* 2000: 56) this taxon was listed as LR nt. Here it is reassessed as Vulnerable because of habitat destruction in what appears to be its main subpopulation in the Bamenda Highlands from which most of the c. 30 specimens listed by de Wilde derive. Moat (in Cheek *et al.* 2000: back cover) records forest loss in the Kilum-Ijim protected area itself. Since the 'generation' duration of this taxon might easily be five years, it is estimated that habitat loss for the species over its area of occupancy as a whole is likely to have been over 30% in the last 15 years. The conservation assessment of this species in Cheek *et al.* 2004: 148–149 is maintained here since no additional data are available apart from its discovery at Lebialem, where its habitat is very severely threatened.

The increasing density of this species, moving northwards (Mt Cameroon and Bioko only c. 2 sites each; Bamenda Highlands numerous sites and [illegible] reflection of a longer dry season [illegible]

Habitat: submontane and montane forest, 1200–2200(–2400) m alt.

Threats: forest clearance for wood and agriculture (mainly in the Lebialem/Bamboutos/Bamenda Highlands).

Management suggestions: subpopulations in SW Province, and probably also in Bioko, are fairly secure, lacking threats. In NW Province the substantial subpopulation in the well protected Kilum-Ijim site (Mt Oku and the Ijim Ridge) may be the only locality where the taxon will survive, unfortunately, thus protection here is important.

Begonia pelargoniiflora J.J.de Wilde & J.C.Arends (Sect. *Tetraphila*)

EN B2ab(iii)

Range: Equatorial Guinea (Bioko (2 coll.)) and Cameroon (SW: Bakossi Mts (1 coll.). Littoral: Mt Nlonako (1 coll.)).

First published in 1992, an updated account was included in de Wilde's revision of *Begonia* sect. *Tetraphila* (2002: 183). *B. pelargoniiflora* shows a similar disjunct distribution to that of *B. furfuracea*, that is, Bioko, northwards to the Bakossi Mts but omitting the very well collected Mt Cameroon. However, whereas de Wilde records eight collections for *B. furfuracea*, there are only four known for *B. pelargoniiflora*, two from Bioko and one each from the Bakossi Mts and the adjoining Mt Nlonako. The fact that it has not been recollected in the last 10 years despite intensive botanical inventory work e.g. in Bakossi, illustrates that this species really is very rare. By comparison, 30 specimens of the related and similar *B. longipetiolata* were made between 1995–2003 in Kupe-Bakossi. On the basis of an area of occupancy of 4 km^2 per site, the total area of occupancy is calculated as just 16 km^2 for this taxon. Its range was considered severely fragmented on the basis of its absence from Mt Cameroon and so it was assessed as CR in Cheek *et al.* 2004: 149. However since current IUCN thinking discourages this use in such cases its rating is here lowered to Endangered. No additional data are available since the 2004 assessment.

Habitat: forest; 1000–1500 m alt.

Threats: forest clearance for wood and agriculture, e.g. ongoing at Mt Nlonako, the type locality.

Management suggestions: an effort should be made to rediscover this taxon and to assess the size of the subpopulation and its demography in detail. In view of its rarity this species might be suitable for propagation and reintroduction in suitable localities.

Begonia preussii Warb. (Sect. *Tetraphila*)

VU A3c

Range: Nigeria (3 coll.), Equatorial Guinea (Bioko (1 coll.)) and Cameroon (SW, S & Littoral: Kribi-Edea-Korup-Lebialem ([illegible])).

This epiphytic lowland *Begonia* was previously assessed as Endangered (Cable & Cheek 1998: xlv-xlvi) but is here downrated to Vulnerable in the light of the taxonomic revision by de Wilde (2002, source of the range data above) showing it to be more common and widespread in Cameroon than was previously thought. In addition, the discovery of eight further specimens at five sites in Mt Kupe-Bakossi Mts confirms the foregoing. Nonetheless, due to the altitudinal range of this taxon, ongoing habitat loss of about 30% over the past 15 years (estimated as equating to three generations) can be postulated. The conservation assessment of this species in Cheek *et al.* 2004: 149 is maintained here since no additional data are available apart from a new record at Lebialem which needs further confirmation since it extends the range of the species N considerably, and has a much higher alt. (1486 m) than all others. Its forest habitat at this site is extremely threatened.

Habitat: evergreen forest; sea-level–1000(–1486m?) m alt.

Threats: clearance of forest for wood and agriculture, both small-holder and plantations, is a major threat throughout its range and probably accounts for the lack of collections from Bioko in the last century (where forest was largely cleared below 1000 m alt.). More than 30% of its historical sites have seen severe habitat degradation or total loss due to plantations of rubber, bananas, oil palm, tea and coffee, and those near towns, urbanisation.

Management suggestions: this taxon is most likely to survive in lowland forest reserves such as the Bakossi FR and Mokoko FR, but only if their boundaries are respected and they are not scheduled for logging as is feared. Resources need to be found to upgrade such sites to National Parks.

Begonia prismatocarpa Hook. subsp. ***delobata*** Sosef (Sect. *Loasibegonia*)
EN B2ab(iii)

Range: Cameroon endemic (SW: Korup NP(1 coll.); Nta-ali (1 coll.); Rumpi Hills (2 coll.); Kupe-Bakossi (2 pre-1995 coll.)).

First recognized in 1994 (Sosef 1994: 179), this subspecies was first collected in October 1946 in the northwest part of Kupe-Bakossi (*Dundas* in FHI 15329). It seems restricted to lowland and submontane sites bordering the upper reaches of the Mungo valley, extending eastwards down the Mbu valley (*Thomas* 5351). The fact that in our inventory of Bakossi plants in the late 1990s only 1 in c. 9,000 specimens (*Cheek* 9645) belong to this taxon indicates its rarity, and also its restriction to western Bakossi, which we have sampled sparingly. The conservation assessment of this species in Cheek *et al.* 2004: 149–150) was VU B2ab(iii), but here we adopt the current IUCN concept of location and (see range) accept four, for an assessment of EN (AOO 16 km^2 with 4 km^2 cells, threats below).

Habitat: terrestrial on soil, or on rocks, or epiphytic on the base of tree trunks, in forest; 400–1100 m alt.

Threats: forest clearance for timber, followed by agriculture. Being restricted to forest at lower altitudes, this rare plant is especially vulnerable to illegal logging, which is currently so prevalent throughout its range (*pers. obs.*).

Management suggestions: the low altitudinal range of this taxon leaves it outside of protected areas within its range apart from the Korup NP. Therefore public education as to the advisability of conserving this taxon is recommended. The site at Nyale Rock is remarkable for the lack of pressure and the possibilities of long term conservation.

Begonia pseudoviola Gilg (Sect. *Loasibegonia*)
EN B2ab(iii)

Range: Cameroon endemic (SW: Mone FR; Lebialem/S Bamboutos Mts (6 coll.); Kupe-Bakossi (3 coll.). Littoral: Mt Nlonako (3 coll.)).

Described by Gilg in 1904 from *Conrau* 10 collected in Nov. 1898 between Banti and Babesong at 600–700 m alt., this beautiful species is easily recognised by its diminutive size and its often purple-black leaf-blades which bear long, white, patent-curved hairs on their upper surfaces.

It appears restricted to submontane forest areas along a short section of the Cameroon Highlands from northern Bakossi to southern Bamboutos Mts/Bamléké plateau. The nine collections recorded by Sosef (1994: 182) are immediately to the NE (Nkongsamba–Mt Nlonako) or N (S and W Bamboutos Mts) of Kupe-Bakossi. Until we worked at Ngomboaku in 1999, recording three sites for the species in a fairly small area east of the village, *B. pseudoviola* was entirely unknown from Kupe-Bakossi. Owing to its relatively low altitudinal range, *B. pseudoviola* is extremely vulnerable to forest clearance (see discussion at beginning of this chapter). For the purposes of this assessment we rate this species as EN under criterion B, being known from four broad locations and having an area of occupancy of 48 km^2 using 4 km^2 cells for each collection site. The prognosis for habitat destruction for this taxon is high. This taxon might be better assessed under criterion A, but lack of data on the state of sites west of the Bamboutos Mts made this difficult to apply until recently. In 2009 observations at Lebialem proved the species to be extremely threatened west of the Bamboutos. However specimens from that location are slightly anomalous, having green not black leaves and having a lower range (250–400 m) so need further scrutiny. The conservation assessment of this species in Cheek *et al.* 2004: 150 of VU is therefore raised here, using the current IUCN concept of location.

Habitat: wet mid-elevation forest; (250–)600–850 m alt.

Threats: see above.

Management suggestions: revisting the surviving sites to assess the population size, regeneration and threats should be done to obtain baseline data for future monitoring. None of the known sites has formal protected status as far as is known, this should be addressed if possible.

Begonia quadrialata subsp. **_dusenii_** (Warb.) Sosef (Sect. *Loasibegonia*)
CR A3c

Range: Cameroon (SW: strictly endemic to the lower slopes of Mt Cameroon between Bakingili and Njonji).

Sosef (1994: 194), in his monumental revision of the 73 yellow-flowered African "refuge begonias", lists four collections of this taxon, including the type (*Dusen* 90, probably collected in 1891/2). *Watts* 501 is the only collection of this species made in the 1990's inventories. He collected it in Oct. 1992, at Njonji Lake, alt. 500 m, on a small tree, at about 2 m from the ground. Only two of the earlier collections have detailed locality data, and both are from Bakingili, so it appears that species may be restricted to a belt between 500-800 m alt. between Bakingili and Njonji. After several failed efforts to refind this plant, staff at Limbe Botanic Garden (LBG) (Paul Ndumbe and Yves Nathan) and RBG, Kew (Marcella Corcoran) finally succeeded in Dec. 2009, the identification being confirmed by Marc Sosef. Live material was transferred to LBG for cultivation, and hopefully multiplication and reintroduction to the wild.

This taxon is one of seven yellow-flowered begonias on the mountain, one of largest conglomerations in Africa and perhaps indicative (see Sosef 1994) of its importance as a refuge area. The original assessment of this species made in Cable & Cheek 1998: xlvi is maintained here since no new data are available.

Habitat: on tree trunks or wet rocks in deeply shaded undergrowth of forest; 500–800 m alt.

Threats: forest disturbance for timber extraction and clearance for agriculture, particularly plantations.

Management suggestions: now that this subspecies has been rediscovered in the wild, its range and population density, regeneration levels and threats should be assessed. Care should be taken not to confuse it with the six other yellow-flowered *Begonia* taxa on the mountain!

Begonia rubromarginata Gilg (Sect. *Tetraphila*)
EN B2ab(iii)

Range: Nigeria (Erin-odo-oke waterfalls in Ilesha Distr.) and Cameroon (W: Bangwa. SW 35 km W of Nguti).

A terrestrial or epiphytic scrambling or scandent herb, stems to 1m long initially scaly-scurfy, otherwise glabrous. The stipules are leafy and persistent, 12-18x4-7 mm and the cylindrical ovary densely hairy. *Begonia rubromaginata* is here assessed as Endangered since known from 3 sites (AOO 12 km^2 with 4 km^2 cells). Less than 500 individuals are known. Data here is taken from De Wilde (2002: 202-207, loc. cit.)

Habitat: Evergreen forest; 400-1000m alt.

Threats: Bangwa (a Conrau site) from c. 1900 is believed to be W of Bangangté and E of Fontem in the Bamiléké Highlands which are now almost completely deforested due to agricultural operations. This site is probably lost now from that original site.

Management suggestions: the continued existence of this species in Nigeria needs verification, assessment and monitoring. The site W of Nguti may fall in northern Korup NP in which case the species should be incorporated into the management plan. Otherwise communities near to the sites should be consulted and advised on the importance of this plant. Ex situ cultivation at Limbe Botanic Garden might allows this species to be used in forest restoration projects within its range. Further surveys might locate new sites and decrease the threat status of *B. rubromaginata*.

Begonia scapigera Hook.f. subsp. **_scapigera_** (Sect. *Loasibegonia*)
EN B2ab(iii)

Range: Nigeria (Erin-Odo and –Ilesha waterfalls, Western State: 1 site), Cameroon (SW: Buea, Etinde, and Neu-Tegel at Mt Cameroon. C: Ndikiniméki (4 sites)) and Gabon (Oveng near Mitzic (1 site)).

This is a robust yellow-flowered terrestrial refuge *Begonia* with petioles 6–30 cm long and black rimmed glabrous peltate leaf-blades 6–20 × 4–14 cm. Assessment of this taxon was overlooked in the Plants of Mount Cameroon (Cable & Cheek 1998) but with only four locations as listed under range above (AOO 28 km^2 with 4 km^2 cells per site-record) and the threats below. *B. scapigera* subsp. *scapigera* is here assessed as Endangered. Reference to this taxon in FWTA is based on *Johnston* s.n. from Rio del Rey subsequently identified by Sosef (1994) with doubt as *B. hirsutula* .

Habitat: mainly submontane evergreen forest near stream, 560–1250 m alt.

Threats: while Mt Cameroon is the stronghold for this species, it is also here that most threats occur. Franke *et al.* (see Burmanniaceae) and Google Earth suggest that habitat at Neu-Tegel has been lost to Oil Palm plantation. Forest along Buea has been lost as the town expands. The oil palm plantations that flank Etinde are being pushed upslope, destroying more of the habitat of this and other rare plants.

Management suggestions: submontane forest at Mt Cameroon at Etinde–Buea should be protected to protect this species which there is comparatively common locally, with 8 records from the Etinde area alone (Cable & Cheek 1998: 23). Further survey work may extend the range of the taxon but it is apparently absent from some areas where it might be expected, such as Kupe-Bakossi: (Cheek *et al.* 2004). It is possible that plants can be verified as locally extinct at Neu-Tegel and Buea.

Begonia schaeferi Engl. (Sect. *Loasibegonia*)
VU D2

Range: Nigeria (Obudu Plateau (1 coll.)) and Cameroon (SW: Mwanenguba (1 coll.), Lebialem Highlands/Bamboutos Mts (3 coll.). Littoral: Mt Nlonako (1 coll.); Kongoa Mts (1 coll.). NW: Bamenda Highlands (3 coll.)).

First collected in Nov. 1900 at Bare, Mwanenguba, by Schaefer and described in 1921, the species has not since been seen in the Kupe-Bakossi area and may be extinct there. Personal observations of *B. schaeferi* in the Bamenda Highlands shows that this species is demanding in its habitat requirements and likely to occur at extremely few sites within a given area (Cheek *et al.* 2000: 56–57, 117).

The assessments of this species as VU D2 in Cheek *et al.* (2000: 56–57; 2004: 150) are maintained here: no new data are available on the taxon apart from a recent record from Lebialem/Bamboutos where its habit is extremely threatened (Harvey *et al.* 2010: 72–73). The information presented below is mainly taken from those works. This taxon would also qualify as VU under criteria A3c and B2ab(iii).

Habitat: on rocks and vertical rock faces in moist to comparatively dry places in primary submontane to montane forest, the latter sometimes with trees not taller than 6–12 m; c. 1500–2300 m alt. (Sosef 1994).

Threats: while cliff faces generally are unlikely to be disturbed except for extraction of rock aggregate for civil engineering, if easily accessible by road, clearance of adjoining forest for fuel and agriculture could endanger this species by removing the shade necessary for its survival.

Management suggestions: during surveys of cliff spaces (in forest) this species should be looked for and, if located, the number of plants and locality recorded as a baseline for future monitoring.

Begonia stellata Sosef (Sect. *Loasibegonia*)

CR B2ab(iii); D

Range: Cameroon (SW: Takamanda (1 site)).

This extremely rare and poorly known terrestrial refuge *Begonia* is unusual in having pink, not yellow flowers and in its stellate-scale indumentum, oblong, subcordate, non-peltate, bidentate leaves. It is only known from the type collection "forested ridges above Mbilishe village, canopy low (5–10m), some trees to 30m, ... lots of epiphytes 6° 13'N, 9° 28'E, 1000-1200 m" (7 May 1987–*Thomas, Fay & Doumenge* 7437). This data is taken from Sosef (1994: 214-215, loc. cit.). *Begonia stellata* is here assessed as Critically Endangered since known from a single site (AOO 4 km^2) with less than 500 individuals presumed and being threatened by proximity to a village and the usual threat from agriculture that is entailed.

Habitats: evergreen forest; 1000–1200 m alt.

Threats: see above

Management suggestions: *Begonia stellata* should be incorporated with the management plan of Takamanda NP and the elders and people of Mbilishe village informed of the importance this very rare plant. Further surveys might expand the range of *B. stellata* and provide data to further survival of the species.

Begonia zenkeriana L.B.Smith & Wasshausen (Sect. *Scutobegonia*)

VU B2ab(iii)

Range: Cameroon (S: Kribi–Campo–Ébolowa–Lolodorf area (10 sites)).

This robust refuge Begonia grows 0.3 m tall, with erect, very narrowly elliptic, ± glabrous, non-peltate leaves 11.5–19 × 4.4–8.3 cm. The leaf apex is acute, the base obtuse, slightly asymmetrical, the margin wavy. The petals are white, marked with red. *Begonia zenkeriana* is assessed as Vulnerable since known from ten locations (AOO 40 km^2) in an area where it is threatened (see below). To the seven sites listed by Sosef (1994), three more were added by Tchoutou (2004), namely Mammelles, Bipindi and Dipikar island.

The site-records are in such close proximity to each other that it could be contended that less than ten locations exist.

Habitat: lowland evergreen forest.

Threats: forest habitat within the range of this species is threatened by logging, future oil and gas installations, iron-ore mining and touristic developments.

Management suggestions: a stronghold for *B. zenkeriana* is the Kienke Forest and Kienke R. Considerations should be given to upgrading the protected status of this FR or of protecting another area where this species survives (e.g. Campo-Ma'an NP, if confirmed there), and *B. zenkeriana* should be incorporated into the management plan.

Franke *et al.* (see Burmanniaceae)

BIGNONIACEAE

assessed by Martin Cheek

The Cameroonian species are well worked out thanks to Flore du Cameroun 27 by Gentry (1994). *Fernandoa* was revised by Bidgood in Kew Bull. 49: 381 (1994).

The Bignoniaceae are pantropical, mainly trees, and especially in the neotropics, also climbers. Amongst the trees with pinnately compound leaves they are easily identified, even when sterile, by their opposite leaves.

Their large showy flowers have tubular corollas coloured bright red (e.g. *Spathodea campanulata*), yellow (*Markhamia lutea*) or purple (several species of *Stereospermum*). Generally they are fast growing with soft wood of little value, but they are often planted as ornamental trees.

Fernandoa ferdinandi (Welw.) K. Schum
VU B2ab(iii)

Range: Cameroon (S: 25 km S Djoum. E: 36 km NE Moloundou. C: Mefou proposed NP), Gabon (Minkebe; and Lastoursville-Poungou; ibid-Bounzoco; Moila-Mounabounou) and Angola (Golungo Alto, type).

Tree to 30 m, puberulent; leaves deciduous, opposite pinnate, with 5–11 leaflets, , each 2–7 × 1–3.5 cm, entire, puberulent bronze below. Inflorescence reduced, ramiflorous, 1–3-flowered; corolla orange, 6–9 × 3–4 cm; anther anthecae 3–4 mm.

This tree, known from only the eight sites above (AOO 32 km² with 4 km² cells) and with the threats below is here assessed as vulnerable. EOO is calculated as 372113 km².

Habitat: semi-deciduous and riverine-evergreen forest mosaic; c. 700 m; fl. April and Nov.

Threats: the Djoum and Moloundou sites are believed to be in logging areas and at the Mefou site tree clearance for agriculture continues (pers.obs. 2006).

Management suggestions: upgrading the Mefou proposed NP to official status, and cessation of clearance for agriculture within its boundaries would provide a secure site for this rare tree. Clarification on the status of the E and S Region sites by field survey is advisable. Information on the threat status of *F. fernandi* in Gabon is needed.

Stereospermum zenkeri K.Schum. ex De Wildeman
Edzodzongui, Evoula.
CR B1+B2ab(iii)

Range: Cameroon (C: Yaoundé).

Tree, puberulent. Stems 4-angled to cylindric, lenticels absent. Leaves opposite, pinnately 9–13-foliolate; leaflets oblong-ovate 5–13 × 2–5cm, acuminate, entire, lower surface with punctae. Inflorescence terminal, paniculate; calyx 10–15 × 7 mm; corolla tubular, 3.5–5 × 1 cm.

This tree is known from four collections, all from Yaoundé, but without further localisation apart from *Service Forestier* 97, given as Ototomo. Here *Stereospermum zenkeri* is assessed as Critically Endangered given a single site (AOO 4 km² with 4 km² cells) and the threats below. EOO is calculated as <100 km².

Habitat: not known, possibly submontane evergreen forest on hills.

Threats: urbanisation. Yaoundé, capital of Cameroon, has expanded steadily, and inevitably, natural habitat has been lost, threatening the species.

Management suggestions: plant surveys in and around Yaoundé should [illegible] encountered in case they should encounter this plant. If found, basic population data should be gathered, and plants marked so that seed can be collected for propagation.

BORAGINACEAE

assessed by Martin Cheek

This is such an extremely varied worldwide family that it is difficult to characterise. It is hard to believe that genera so dissimilar as the Cameroonian lowland trees *Cordia* and *Ehretia* (with juicy ovoid fruits in a cupular calyx arising from flowers with an open corolla) are in the same family as *Cynoglossum* and *Myosotis* which are montane grassland herbs having four dry fruitlets or nutlets and a gynobasic style following flowers in which the corolla tube is closed by a corona. Only a single species in Cameroon is treated as threatened here, *Myosotis*, a genus widespread in temperate areas from Europe to New Zealand and in Tropical Africa only on high mountains. With its scorpoid cymes, alternate simple leaves, its rough indumentum (due to stout simple hairs arising from a raised multicellular base), tubular, radially symmetric corollas, *Myosotis* are only likely to be confused with the genus *Cynoglossum* which occur in the same habitat and are much more common in Cameroon, with numerous varieties (some quite rare) in several species. However *Cynoglossum* have four large flattened fruitlets that exceed the calyx and are hooked projections, while *Myosotis* has four small globose, smooth nutlets concealed inside the calyx.

Regrettably, there is no Flore du Cameroun account available for the family at the moment.

Myosotis cameroonensis Cheek & R.Becker (syn. *Myosotis* sp. nr. *vestergrenii* Heine, FWTA 2 (1963: 325); Cable & Cheek, (1998: xlvi, 25).
VU D2

Range: Cameroon (SW: strictly endemic to Mann's Spring at Mt Cameroon).

Presumed perennial, herb 20–30 cm tall, stem decumbent, up to 1.05 m long, basal part of stem trailing along the ground, producing numerous side shoots, apical part ascending, sometimes dull purplish on exposed side, sparsely pubescent with stout white, patent hairs 0.4–0.6 mm long with swollen bases. Leaves sessile, heteromorphic, oblanceolate-spatulate 2.5–5.5 × (0.7–)0.8–1.0 cm, the narrowed, subpetiolar part (1–)3–4 mm wide, apex acute, apiculum 0.3 × 0.3 mm, base cuneate, fairly densely patent-hairy, the base of each hair surrounded by a ring of 10–15 cystoliths, these most conspicuous towards the leaf apex on the lower surface; distal 4–6 leaves sessile, narrowly ovate or elliptic, 1.5–3 × 0.5–1 cm, apex acute, base clasping the stem for about half its circumference. Inflorescences 1–3, at stem apex, the main inflorescence terminal, unbranched, 8.5–16 cm long, 6–17-flowered, with 1-2 slightly shorter inflorescences arising from the adjoining distal-most axils, indumentum as the stem; peduncle 4–10 cm long, basal internodes 0.6–2.5(–3) cm long, bracts absent, pedicels 4–6 mm long in fruit. Flowers 3–5 mm across, calyx 2–3 mm long, basal part campanulate, lobes oblong-triangular, 0.87–1.5 mm long. Corolla bright sky blue with yellowish white eye; tube 1.25–1.6 mm long; faucal appendages transversely semi-circular, papillate, c. 0.2 × 0.4 mm; corolla lobes patent, orbicular or transversely elliptic, 1.2–1.45 × 1.25–1.6 mm, apex rounded or irregularly notched.

This species was described by Cheek and Becker in "A New Species of Myosotis L. (Boraginaceae) from Cameroon, with a Key to the Tropical African Species of the Genus" Kew Bull. 59: 227–231(2004). The conservation assessment was maintained from that used (under the synonym) by Cheek in Cable & Cheek (1998: xlvi) which is also maintained here, no new data being available. *Myosotis abyssinica* also occurs on Mt Cameroon but is a small erect annual with calyx longer than the fruiting pedicel and not vice versa. These are the only two species of the genus in Africa W of the Albertine Rift.

The range of this species appears to be very small. Not only is *Myosotis cameroonensis* restricted to Mt Cameroon, it also appears to be restricted to one small part, W of Mann's Spring, the only perennial water source on the mountain above c. 800 m altitude. It may be that, like many other species of the genus, this species prefers habitats with a constantly high water table, and that this is only to be found on the mountain, within the required altitudinal range, at the Mann's Spring location. The lack of permanent surface water on Mt Cameroon results from it being a mass of lava flows of different ages; the cinder-like soils are so free-draining that rainfall is quickly absorbed. Apart from Mann's Spring, the only other location for the species on the mountain is "several miles west of Mann's Spring" (*Morton* 875), so that the species has a linear range of at least 5 km. If the precise itinerary of Johnston on the mountain was known, it might be possible to identify the locality of his specimen. The location of "Ukele Camp" (*Maitland* 1342) is unsure, but this is probably an orthographic variant for Wikile, a permanent hunters camp a few km west of Mann's Spring which is also supplied by springs, although none are so reliable as Mann's Spring (Cheek *pers. obs.*). It might be thought the restriction of *Myosotis cameroonensis* to the Mann's Spring area is an artefact caused by a concentration of collections at this site due to its convenience for camping. In fact, two other parts of the mountain where the forest:grassland interface occurs at this altitude are also well collected; these are the route to the summit from Buea and the route from Bonakanda to the antennae at the northern tip of the massif. At neither of these locations has *Myosotis* been collected (Cheek pers. obs.). In the Red Data assessment for this taxon in Cable & Cheek (1998: xlvi), it was incorrectly stated that it had not been collected recently, and elsewhere (op.cit. xxxiii) that it had not been collected since 1952 and so was a cause for concern. Cable & Cheek (1998: 25) also list *Wheatley* 334, but this seems to be an error for *Mbatcho* 334 (*Wheatley* 334 is actually *Sabicea calycina, Rubiaceae*), also from west of Mann's Spring. The most recent collection of the taxon, *Cheek* 5341, was unaccountably passed over in Cable & Cheek (1998).

Habitat: at boundary of montane grassland with forest; alt. 2100–2700 m.

Threats: the location for this species is under no threat from human activities (Cheek *pers. obs.*), but is vulnerable to stochastic events, such as lava flows from volcanic eruptions. Another taxon is also, so far as is known, restricted to the Mann's Spring area. This is *Angraecopsis cryptantha* P.J.Cribb (*Orchidaceae*) (Cable & Cheek 1998: 169). *Isoglossa nervosa* C.B. Clarke (*Acanthaceae*) formerly thought to be similarly restricted is now known to be more widespread.

Management suggestions: basic population data on this species, such as its exact range, frequency and regeneration levels, and on its life-cycle, (e.g. whether it is indeed a perennial or not) are needed.

BURSERACEAE

assessed by Jean Michel Onana

Burseraceae are a pantropical family consisting of shrubs and trees known for their frankincense and myrrh. Individuals are identified in the field by the resin that flows from the bark when making a cut, and especially the smell of turpentine, due to the presence of amyrin, a substance that allows them to be separated from other families that also secrete resins. The morphological characters are varied. Nevertheless the Burseraceae are recognised by their alternate, compound imparinnate leaves (rarely paripinnate), often grouped toward the apex of the stems, with from opposite or alternate (1–) 2 to 12 leaflets. Inflorescences are panicles of biparous cymes, flowers pedicellate or not, with ovary bilocular, the 2-locules biovulate, or trimerous. Fruits are either fleshy, and edible by humans and/or animals, or winged and dispersed by wind. Some are timber species such as Okoumé (*Aucoumea klaineana* Pierre) Aiele (*Canarium schweinfurthii* Engl.) or Ozigo (*Dacryodes buettneri* (Engl.) H.J.Lam) and are subject to commercial exploitation; others are sources of vegetable fat, such as the common African plum (*Dacryodes edulis*), or the "black fruit" (*Canarium schweinfurthii*), or sources of sugar, such as Atom (*Dacryodes macrophylla* (Engl.) H.J.Lam). The volume of the "Flore du Cameroun" for Burseraceae is not yet published, but studies have been conducted to describe the species in Cameroon (Onana, 1998, Onana, 2003) and Gabon (Aubreville, 1963), a large part of the species of the last being common to both countries. Moreover the genera *Dacryodes* and *Santiria*, which are the most common in moist forests and which have long posed problems of identification, have recently been revised for Africa (Onana, 2008, 2009).

Assessments for some species were already accepted by IUCN (2004) (*Aucoumea klaineana*: VU A1cd; *Dacryodes igaganga*: VU A1cd + 2c) or have had their conservation status proposed in previous works (Onana, 2003, 2008, 2009). Further species of *Dacryodes*, due to their restricted range, small number of collections and imperfectly known, taxonomic status, are additional candidates for threatened status. The clearance of forest habitat for agriculture is the main threat because the subpopulations of such species are becoming increasingly fragmented and their areas of occupancy becoming increasingly reduced.

Dacryodes camerunensis Onana

VU B2ab(iii)

Range: Cameroon (S: Ébolowa (1 coll.); Nkoemvone (1 coll.)), Gabon (Lastoursville (3 coll.); Mbolzore (1 coll.); Oyem (1 coll.); Oveng (1 coll.)) and Congo (Brazzaville) (Brazzaville (1 coll.)).

For a long time this species was confused with *Dacryodes klaineana* due to the similarity of their inflorescences (e.g. Aubreville, 1963). This species of equatorial rainforest has 3-7 leaflets, flowers reddish-brown tomentose, disc and ovary tomentose, and globose fruit covered with stellate reddish-brown hair. According to Onana (2006 (published in 2007), 2008), the species is distributed over seven sub-populations distributed in three countries. With a total of 28 km² AOO on the basis of 4 km² per site, and threats as indicated below, it is confirmed as Vulnerable as previously evaluated (Onana, 2003). Population fragmentation is a concern.

Habitat: lowland mixed evergreen and moist semi-deciduous forest; 400–550 m alt.

Threats: deforestation due to agriculture is the main threat to this species. It appears to be unknown to local people (no local names are recorded). The species is generally distributed in areas that are not under protection, and this exposes trees to be felled.

Management suggestions: there is a need for research into the known locations for this species to ensure the continued existence of individual trees, given that all the individuals we know to date are in populated and not in protected areas. Surveys in the surrounding forests in neighbouring protected areas must be carried to find other individuals, since this will help to protect this species. An introduction into the neighbouring NP of Campo-Ma'an in Cameroon will ensure the survival of this species. Management and traditional authorities should be provided with the information needed to identify and protect the species.

Dacryodes igaganga Aubrév. & Pellegr.

VU A1cd + 2c + B2(ii, iii, iv)

Range: Angola (Cabinda (2 coll.)), Cameroon (S: Ebom (2 coll.)) and Gabon (Haute Ngounyé (1 coll.); Mbelale (1 coll.); Lope (1 coll.); Ikoy station (1 coll.); Divangui (1 coll.); Rabi-Kounga (2 coll.)).

This species was for a long time confused with the well-known and more widespread *Dacryodes edulis*, but it can be distinguished by its leaves with petioles, winged (not winged in *D. edulis*), 3–5 pairs of leaflets, lacking pseudostipules (versus present in *D. edulis*), flowers with glandular hairs inside the petals, the disc glabrous, ovary tomentose-stellate, and in the globular fruit with a hard endocarp bearing a conspicuous scutellum According to Aubréville (1962) and Onana (2008), the species is known from eight locations. The AOO is thus 32 km² on the basis of 4 km² cells. However, the EOO is more than 20000 km^2, being distributed from S Cameroon to Cabinda, this let Onana (2003) to propose that the species is not threatened (NT). But the species was assessed VU A1cd + 2c (IUCN (2004) citing an estimation of the reduction of the population of >30%. Here, that assessment of IUCN is confirmed because the population is indeed being reduced due to the degradation of its forest habitat. In addition, criterion B is also triggered considering the number of locations and AOO, and so the threatened status is revised as above.

Habitat: lowland evergreen forest; 500 m alt.

Threats: this species is distributed in forests subjected to farming. Deforestation is active, but protective measures have been undertaken by the establishment of protected areas in Gabon, the main country, where the species is widespread. In Cameroon, the species is at its northern limit, and it is known from a single locality in an area that is not protected and so is isolated from the subpopulations in Gabon.

Management suggestions:in Gabon this species is distributed partly within protected areas (e.g. Lope National Park) so will potentially benefit from management plans of these areas for its survival. But in Cameroon, where the species is rarer and not located in protected areas, research to find this species must be undertaken near the known locality of Ebom in protected areas such as the Campo-Ma'an NP. An introduction using the seed collected from Ebom into the national park is an appropriate measure to maintain the genetic heritage of this subpopulation, although it is not normally allowed to introduce a species into a national park.

REFERENCES

Aubréville, A. (1962). Burséracées. In: Aubréville, A. (ed.) Flore du Gabon 3: 53–97. Muséum National d'Histoire Naturelle. Paris.

Onana (1998). Les Burseracées du Cameroun: Taxonomie, biométrie foliaire et biologie. Thèse de 3e cycle. Université de Yaoundé I.

Onana, J.-M. (2003). Inventaire synoptique et évaluation du statut de conservation des Burseraceae camerounaises. Etat des connaissances. Ann. Fac. Sci. Yaoundé I, sér. Sci. Nat. Vie 32(2): 44–56.

Onana, J.-M. (2006, published 2007). *Dacryodes camerunensis* (Burseraceae) a new species from central Africa. Kew Bull. 61(4): 579–584.

Onana, J.-M. (2008a). A synoptic revision of *Dacryodes* (Burseraceae) in Africa, with a new species from central Africa. Kew Bull. 63(3): 385–400.

Tchouto, M.G.P. (2004). Plant diversity in a central African rain forest: implications for biodiversity conservation in Cameroon. Ph D. thesis. Department of Plant Sciences, Biosystematics Group, Wageningen University, the Netherlands.

IUCN (2004). 2004 IUCN Red List of Threatened Species.www.redlist.org. Downloaded on 09/08/2005.

Wilks, C.M. & Issembé, Y.A. (2000). Guide pratique d'identification: les arbres de la Guinée Equatoriale. Région continentale. Projet Curef, Bata, Guinée Equatoriale. 546 p.

CAMPANULACEAE INCLUDING LOBELIACEAE

assessed by Martin Cheek, advised by Mats Thulin

A cosmopolitan family of about 2000 species, mainly of herbs, most diverse in temperate climes and with few species in Cameroon, none of which are endemic. The leaves are simple, alternate, lacking stipules, the flowers with an inferior ovary and five petals united in a tube, and stamens with anthers all united or held together. There is no Flore du Cameroun account, but Fl. Gabon 40 (2010) includes treatments by Thulin of two of the species below. Either one family with two main subfamilies can be recognised, or as in Fl. Gabon, two separate families, separated as follows:

Campanuloideae: flowers radially symmetrical, anthers held together (e.g. *Wahlenbergia*)
Lobelioideae: flowers zygomorphic, anthers connate (e.g. *Dielsantha, Lobelia*)

Lobelia is the largest genus of the family in Cameroon, and often confused with *Lindernia* (Scrophulariaceae) which has similar shaped and sized flowers, but differs in having superior ovaries and anthers free from each other and included in the corolla tube.

While the genus *Wahlenbergia* was revised by Thulin (Symbol. Bot. Upsal. 21: 1 (1975), there is no modern revision of *Lobelia*. Thulin has written Flora accounts, and published new species in that connection, for Flora of Tropical East Africa, Flore D'Afrique Centrale and Flora Zambesiaca, but Cameroon remains untreated as a whole since the revisions and updates of Wimmer in 1953 which are listed by Thulin in Nordic J. Bot. 3 (1983). These we have not been able to screen for potentially threatened Cameroonian taxa.

Dielsantha galeopsoides (Engl. & Diels) E.Wimm. (syn. *L.sylvicola* Lejoly & Lisowski)
VU A2c

Range: SE Nigeria, Equatorial Guinea (Bioko and Rio Muni (1 coll. each)), Cameroon (SW: Korup; Barombi Mbo (1 coll. each); Kupe-Bakossi (3 coll. in 1990s). Littoral: Ebo. S: Bipindi; N'kolandom), Gabon (Cristal Mts; Moumba) and Congo (Brazzaville) (Odzala).

This sprawling perennial herb resembles a blue-flowered *Lobelia*, to which it is closely related. It is placed in its own genus, differing from *Lobelia* in the peculiar, longitudinally-dehiscent, papery fruit with brown splayed valves and in its shade-demanding ecology, being restricted to forest while *Lobelia* need open areas. Discovered in the late 19th century at Barombi near Kumba, it appears not to have been collected at any of its three previously known locations for over sixty years. Despite intensive surveys in the 1990s this taxon remains unknown in the Mt Cameroon area. Even in areas where it occurs, it is never common, often only one plant being at a location (Cheek pers. obs.)

Previously assessed as EN A3c in Cheek *et al.* (2004: 150–151) on the basis that a population reduction of two-thirds over the next ten years was projected. New data has extended the number of locations and the range to Littoral and S Regions of Cameroon and to Gabon, Rio Muni and Congo (Brazzaville), where forest loss is low, so that population reduction is considered much lower. However, of the 12 locations now known (see range, above; AOO 48 km^2 with 4 km^2 cells; EOO is calculated as 114853 km^2), it is considered likely to be extinct or to become extinct shortly, at four of these, due to clearance of its forest habitat (see details below), therefore a population reduction exceeding 30% in the last c. 10 years can be estimated.

Habitat: lowland and lower submontane evergreen forest; 650–1000 m alt.

Threats: forest clearance for agriculture: suitable habitat at Bioko was destroyed for cocoa plantations, habitat at Lake Barombi Mbo was reported destroyed in 2002 (A. Davies. *pers. comm.*) and Nigeria has seen c. 90% loss of original forest. The species has been seen at neither of these localities, nor at Bipindi, in the last 100 years despite recent botanical surveys at all these locations. In Bakossi *Dielsantha galeopsoides* is widespread but rare. Known from only three sites, it is endangered by farm expansion in two of these: Nyasoso Nature Trail and Manehas FR. At Nyale it is probably fairly secure owing to its remoteness and low population.

Management suggestions: Bakossi, Ebo and the Cristal Mts are the most important locations for this genus according to available data, having the greatest concentration of recent records. Research in population density, demography and recruitment would best be conducted at Nyasoso since the plant exists on the very edge of the town in the 'Nature Walk' above the school. This might also be the best location to propagate the species for reintroduction to safe sites. Verification of the losses of this species at the four locations mentioned above is needed and if confirmed, in future this species should be reassessed under criterion B2.

Lobelia columnaris Hook. f.

VU A3c+B1ab(iii)

Range: Equatorial Guinea (Bioko) and Cameroon (SW: Mt. Cameroon; Lebialem Highlands. W: Bamboutos. NW (Bamenda Highlands): Bali Ngemba; Mt Oku; Ndu; Dom).

Herb 2 m tall, unbranched. Stem 2 cm diam. at base, glabrous. Leaves oblanceolate-oblong, c. 30 × 8 cm at base of stem, gradually diminishing in size towards the apical inflorescence, apex acute, margin inconspicuously serrate, softly hairy. Inflorescence occupying the apical half of the plant, densely flowered, unbranched spike. Flowers with corolla pale blue, 2–3 cm long. It is difficult to confuse this species with any other in the Cameroon Highlands.

This Giant Lobelia, with 2 m spikes of blue flowers, is the only representative in continental W Africa of a group for which E African mountains are famous. It thrives at the montane forest: grassland ecotone. Previously assessed as LR/nt in Cheek *et al.* (2000: 57), its threat rating is here revised upwards for the reasons indicated below.

Although there are numerous collections from Mount Cameroon, it proved extremely rare during the 1992 survey there, when only one or two specimens were seen (Cable & Cheek (1998: 25). In contrast, it is abundant in many localities at Mt Oku (pers. obs. 1996, 1998, 1999, where it is gregarious, 10–50 plants being commonly found at a site). It also appears commonly recorded elsewhere in the Bamenda Highlands and at Bioko. However it is here estimated that more than 30% of its habitat has been lost due to forest reduction over the last 15 years (estimated duration of three generations) over its range: see threats below. EOO is calculated as 18989 km².

Habitat: boundary between montane forest & grassland; 2000–2200 m alt.

Threats: fires from grassland moving into forest edges; forest loss due to timber extraction and clearance for agricultural land. 96% of original forest is calculated as being lost in the Bamenda Highlands (Cheek *et al.* 2000: 6) and in the Dom area, c.50% of what survived was lost in the 15 years between 1998-2003 (Baena in Cheek *et al.* 2010: back cover).

Management suggestions: better protection of montane forest, especially in the Bamenda Highlands which appears to be the critical location for the species. Informing managers of protected areas, including elders managing community forests, of the threatened status of this species is advisable. Evaluation of the threats in Bioko to this species would also be useful. A study of the life-cycle of this species would inform future monitoring, management interventions, and threat assessments.

Lobelia gilletii De Wild.

EN B2ab(iii)

Range: Cameroon (C: Mefou proposed NP), Gabon (Billagore [Mbilagone]; Doudou Mts), Congo (Brazzaville) and Congo (Kinshasa) (Kasai, vallée de la Djuma).

Prostrate, mat-forming herb; stems glabrous, rooting; leaf-blade ovate, 5–10 × 5–10 mm, apex rounded, base cordate, margin dentate or crenate, sparsely pubescent on the upper face; petiole 2–7 mm, glabrous; flowers simple, axillary; pedicel, 5–12 mm; calyx 1–2 mm; corolla white, 7–10 mm; fruit obovoid, 3–5 mm, indehiscent.

Here *Lobelia gilletii* is assessed as Endangered since five locations (see range, above; AOO 20 km^2 with 4 km^2 cells) are known, with threats below. EOO is calculated as 588081 km². Three specimens are known from Congo (Kinshasa) but only one location. All material known has been identified by Mats Thulin (UPS) author of Flore D'Afrique Centrale Lobeliaceae (1985) from which the data cited is taken – apart from the Doudou Mts and Congo (Brazzaville) locations which derive from Olivier Lachenaud (G. Walters pers. comm to Cheek, May 2010).

Habitat: lowland swamps in forest areas; c. 700 m alt.

Threats: agriculture at Mefou – a small local community, only 100 m distant from the swamp location, use the habitat for *Raphia* cultivation and extraction.

Management suggestions: the ecological requirements of this species should be researched to ensure it can be managed for survival. Some disturbance may be beneficial for the species. Basic population data should be gathered to form a baseline for future monitoring.

Wahlenbergia ramosissima (Hemsl.) Thulin subsp. ***ramosissima***

VU B2ab(iii)

Range: Nigeria (Chappal Waddi, Mambilla Plateau (2 coll.)) and Cameroon (SW: Mt Cameroon (5 coll.); Mwanenguba (2 coll.). NW (Bamenda Highlands) Bali Ngemba; Sabga; Ijim Ridge (5 coll.). Adamawaa: Mts Poli, Tchabal Mbabo; Nganha (1 coll each)).

Annual herb 4–15 cm tall, erect, unbranched in the lower half. Stem filiform, bearing evenly sized and shaped leaves at internodes of 5–10 mm. Leaves elliptic, c. 5 × 3 mm, apex rounded to acute, each side bearing 3–5 crenate teeth; petiole c. 1 mm long. Inflorescence 1–10-flowered, diffuse; bracts linear, 3 mm long; pedicels 7–30 mm long. Flowers c. 4 mm diam., petals blue, marked white at base.
This herb probably germinates and grows through the wet season, flowers and fruits at the beginning of the dry season, perennating as seed in the soil.
Previously assessed (Cable & Cheek 1998: xlvii; Cheek *et al.* 2000: 57–58; Cheek *et al.* 2004: 151) as Vulnerable. This rating is maintained here since although the number of known locations has increased, it is still known from only 17 specimens at nine mountain locations (see range, above; AOO 36 km^2 with 4 km^2 cells, threats as below) along the Cameroon Highland line. EOO is calculated as 71389 km^2.
Habitat: montane grassland, forest ecotone, on bare ground between *Sporobolus* grass tussocks and on shady banks with *Radiola*; 1500–2600 m alt.
Threats: trampling by cattle during the wet (growing) season is a threat at all sites excepting Mt Cameroon where no livestock are grazed in montane grassland.
Management suggestions: assessing the size of the subpopulations, ideally when the plants are in flower and most conspicuous, is advisable. Monitoring variation in numbers of individuals present from year to year would also help better assess the threat to this taxon.

CAPPARACEAE

assessed by Martin Cheek

Thirty-eight species in nine genera are recognised for Cameroon by Kers (Flore du Cameroun 29 (1986)). Related to the Cruciferae, they similarly produce piquant oils but are woody shrubs, often with 3-foliolate leaves. They are most easily recognised in flower or fruit, since the ovary, and later the fruit, is produced on a long stalk, the gynophore, among the usually numerous free petals and stamens.

Species of this family are most easily encountered in the dry wooded grasslands of the north of Cameroon, but those species are widespread, whereas in forest several of the species are quite rare, in fact one can spend several days surveying evergreen forest before encountering a Capparaceae.

Ritchiea macrantha Pax & Gilg
VU B2ab(iii)
Range: Nigeria (Cross River), Cameroon (SW: Kumba to Ikiliwindi. Littoral: 30km N Eséka near Boga, Mambe Forest) and Gabon (Kessipoucou; Ibanga; Monts de Cristal).
Shrub or small tree to 5 m, glabrous; highly lenticillate; leaves 3-foliolate, long-petiolate, not clustered at stem apex, central leaflet broadly lanceolate to obovate, 12.5–20 × 4-7.5cm, long acuminate, base cuneate, lateral nerves 3–5 on each side of the midrib, lateral leaflets asymmetric, ovate to lanceolate, slightly smaller; inflorescence terminal on leafy branches; petals numerous, 9–12 cm, stamens numerous, 4.5 cm.
Here *Ritchiea macrantha* is assessed as Vulnerable since six sites (above; AOO 24 km^2 with 4 km^2 cells) are known with threats as below.
Habitat: lowland evergreen forest.
Threats: forest clearance for housing and agriculture, especially in the Kumba area. Threats at the Gabon sites are not known.
Management suggestions: searching at the most recently discovered Cameroon sites, near Eséka is advised as being most likely to be successful. Basic populational data should be collected and local leaders advised if the plant is found so that future monitoring and protection might be possible.

CARICACEAE

assessed by Martin Cheek

Cylicomorpha solmsii (Urb.) Urb.
VU B2ab(iii)
Range: Cameroon endemic (SW: Mt Cameroon-Buea-Bulifambo (2 coll.); Etome; Mbonge-Bakosso; Korup; Kumba (Barombi, 3 coll.); Kupe-Bakossi (c. 15 coll.). NW: Wum, Bu (3 coll.); Mujung (1 coll.). E: 80 km SSW Yokodouma (1 coll.)).
Assessed as Vulnerable in Cheek *et al.* (2004: 151). That assessment is maintained here despite the addition of some new records. Known from seven broad, mainly submontane forest locations (above; AOO 28 km^2 with 4 km^2 cells, EOO 93395 km^2, threats as below), this dramatic dioecious tree, one of only two *Caricaceae* on the African continent (the second species occurs in E Africa), is unmistakable even when sterile due to the similarity with *Carica papaya,* yet being a forest tree with broad (to 1 m or more diam.), soft, spiny trunks which yield white exudate, and having papaya-like digitately lobed leaves. It appears to be a pioneer, but is always very local, and usually gregarious. The hollow trunks were formerly used for imprisoning miscreants in Bakossi.
This species is mainly restricted to the lower slopes of the Cameroon Highlands. The single record from E Region from semi-deciduous forest with Ulmaceae-Sterculiaceae (*Letouzey* 12237) seems unlikely but the specimen has been verified as *Cylicomorpha* by MC at K. *Buckner* 381 (Missouri data) from savanna in CAR requires verification before it can be accepted. In the absence of a specimen I have not included my own sight record from Yaoundé mentioned in Cheek *et al.* 2004: 151.

Habitat: lower submontane forest, usually; (400–)1000–1400 m.
Threats: forest clearance for agriculture and wood especially in the Bamenda Highlands area. It may now be extinct at Barombi, Kumba since no records have been found in nearly 100 years.
Management suggestions: the demography of this taxon is not documented and requires study. Nyasoso at Mt Kupe would be a good base for such research, and as a centre for conservation of the species since a sizeable population exists on the main trail to the summit at 1100 m alt. and other subpopulations occur elsewhere in Bakossi e.g. near Kodmin. A good population also exists at Nkom-Wum. In the Bakossi Mts, traditions attached to these trees are likely to help protect them. Verification of the *Buckner* specimen is required, also of occurrence in Yaoundé.

CARYOPHYLLACEAE

assessed by Martin Cheek

This family of herbs, less usually subshrubs, are characterised by simple, opposite leaves which lack stipules (except *Drymaria* and *Polycarpaea*). Eight genera occur in Cameroon. The actinomorphic flowers are held in terminal cymes with five free sepals (united in a tube in *Silene*) and petals, stamens ± free, 5 or 10, ovary superior with 3 to 5 styles.

Caryophyllaceae usually demand high light and so are absent or rare in forest. Most Cameroonian genera occur in grassland or grassland: forest edges on high mountains (*Cerastium, Stellaria, Sagine* and *Silene*). These genera are wild also in temperate areas such as Europe. However, *Polycarpaea* occur at low altitudes, for example in wooded grassland of C and Adamawa regions of Cameroon.

Polycarpaea garuensis J.P.Lebrun
EN B2ab(iii)
Range: Nigeria (40 miles SE Gambe) and Cameroon (N: Garoua; Mt Tinguelin; Lamoudan, 30km S Garoua; Sanguéré, 14km SE Garoua).
We here assess *Polycarpaea garuensis* as endangered since it is known from only four sites (above, AOO 16 km² with 4 km² cells, EOO 1493 km².) and threats as below. Lebrun states that it is known only from a diameter of 40 km² around Garoua, citing it as a remarkable case of micro-endemism. The description below is taken from the protologue of Lebrun (Adansoni[illegible] (1977)) since we have not had acc[illegible] from the Nigerian record identified by Lebrun in [illegible] and [illegible]pearing to us as possibly anomalous.
Tufted, silvery hairy subshrub, stems to 17 cm tall, clothed in linear leaves, 8–10 × 0.6 mm these shorter and much denser at base of stem, towards the apex spaced at up to 2.5 cm, opposite, covered in long white hairs, mucronate; stipules lanceolate, 4 mm long, scale-like, margins hairy; inflorescence capitate, terminal, 8–10 mm diam.; sepals 5, bright white, 2 × 0.6 mm, petals 5, red, half as long as sepals. Capsules 1.5 mm, seeds 5.
Habitat: bare, sandstone rock (in hills around Garoua) c. 400 m alt.
Threats: grazing and trampling by livestock.
Management suggestions: further studies are needed at Garoua to determine whether this is annual or perennial, the level of regeneration, detailed threats, the possibility of protecting one of its sites. Although Lebrun (op.cit.) gives some data on population density, this needs to be supplemented. Flowering occurs April–June when surveys should take place. It is possible that further survey work may show that this species is more common and widespread than currently understood, enabling it to be downrated to VU.

Polycarpaea rheophytica Cheek nom. nov. replacing *Polycarpaea rupicola* Lebrun & Stork nom. illeg. in Adansonia sér. 4, 9(1):39-42 (1987).
Polycarpaea rupicola Lebrun & Stork is a later homonym, (and therefore illegitimate according to the code) for *P. rupicola* Pomel in Nouveaux Matérieux pour la Flore Atlantique, Paris. Vol 1: 202 (1874). Described from Algeria. Type: Cameroon, Sanaga R, 70km SSW Bafia on route Douala-Bafia, fl. fr. 3 Jan. 1971, *Letouzey* 9792 (holotypus P; isotypus. A.L.F.)
EN B2ab(iii)
Range: Cameroon endemic (Sanaga R: Douala-Bafia, 70km SSW Bafia at port Kikot; Nkongnok. Littoral: Nkam R near Sahé, Bafang-Yabassi. NW: River Katsina at Kimbi Bridge)
Polycarpaea rheophytica is here assessed as endangered since it is known from four sites (above, AOO 16 km² with 4 km² cells) and threats as below.
Rheophytic subshrub with several slender, hairless, probably annual stems 20 cm tall from a stout (1–2 cm diam.) vertical rootstock. Leaves in whorls of 2–6, oblanceolate, 10–15 × 1–2.8 mm, stipules ovate 1–2 mm long, scale-like. Inflorescence a terminal cyme, 9–12-flowered, 4 × 4 cm; sepals 5, lanceolate 2.3 × 0.87 mm, silver-white; petals 5, 1 mm, spotted red; capsule ovoid 1.5 mm; seeds reniform, 0.8 mm.
Habitat: between boulders at edge of rivers, almost certainly submerged in the wet season, flowering and fruiting in February.
Threats: hydroelectric dams have been proposed for several large rivers in Cameroon including the Sanaga. By permanently flooding some areas of river and changing the

flow regime in others, they may have a deleterious effect on rheophytic species such as this.
Management suggestions: further survey work along the river habitat of this species might yield more sites and provide data on density and regneration levels. February would be the best month for such work, when plants are in flower.

Silene biafrae Hook.f.
VU D2
Range: Cameroon endemic (SW: Mt Cameroon).
This erect, probably annual herb, 20–30(–50) cm high, with 1–7 brownish-pink flowers in October and November, is common at about 3000 m alt., scattered thinly, between Mann's Spring and Bokwango Hut 3 (pers. obs, *Cheek* 3661 and *Thomas* 9326). It favours sparse grassy areas, often on cinder slopes. Its closest relative may be *S. burchellii* Otth, a species of montane eastern Africa including E Congo (Kinshasa), Rwanda and Burundi. However *S. lynesii* Norman is less geographically distant, occurring in the Tibesti as well as in S Sudan and is also noted to occur on volcanic soils. No other member of the genus occurs in Cameroon, nor in W Africa. 10 historic specimens are known, the earliest being the type, *Mann* 2034, collected at Mann's Spring, 8–10,000 ft in November 1862. It has also been collected above Buea at Huts 2 and 3, and so appears to be distributed around a good swathe of the mountain, perhaps around its whole circumference. The assessment here, by Cheek in Cable & Cheek (1998: xlvii–xlviii), is maintained since no new data are available. EOO is calculated as 41 km², but accuracy needs reviewing.
Habitat: lava and cinder slopes, grassland; 2700–3600 m alt.
Threats: none known.
Management suggestions: none. This species appears in no immediate danger of extinction, but a survey to record basic populational data would be useful as a baseline for future monitoring.

CECROPIACEAE

assessed by Martin Cheek

The c. 200 species and five genera of this family were formerly included in either Urticaceae, or, most recently, Moraceae, until Berg segregated them as a separate family in 1978. They occur in the tropics of both America and Africa and are mainly pioneer trees or soft stemmed treelets with numerous minute flowers, as occur in both families mentioned. However generally they have digitately compound or lobed leaves, and individuals are either male or female, with radically differing inflorescences.

The Flore du Cameroun account for Moraceae by Berg *et al.* (1985: 270), includes Cecropiaceae. De Ruiter (1976) revised *Myrianthus* P. de Beauvois and *Musanga* R.Br., the two African genera of the Cecropiaceae, both endemic. De Ruiter recognised seven species of *Myrianthus* which are all trees or shrubs (one is a liana) of evergreen forest.

Musanga cecropioides is known to everyone in the forest belt of Cameroon since it is a the commonest fast-growing tree which colonises disturbed areas in lowland evergreen and semi-deciduous forest, including farm-fallow. Its seeds only germinate when soil is exposed to direct sunlight so it does not regenerate inside the forest canopy. Trees grow very fast but only live for 20 or so years before they die.

Myrianthus arboreus, the bush pineapple, is also a pioneer. It has edible orange fruit about the size of a pineapple and with similar taste.

Apart from the species treated below, *Myrianthus preussii* is also characteristic of submontane forest, occurring in Nigeria, Cameroon and Gabon and is here assessed as Near Threatened. In future it might be rated as Vulnerable due to population reduction in the Cameroon Highlands.

Myrianthus fosi Cheek (syn. *Myrianthus* sp. 1, Cheek & Darbyshire p. 260 in Cheek *et al.,* (2004)). [Boussombossomb (Banen)]
VU B2ab(iii)
Range: Cameroon endemic (SW: Bakossi Mts; Rumpi Hills; Mamfe area; Lebialem Highlands. W: 12 km N Bafang. Littoral: Ebo Forest. C: Ndikiniméki, 40 km S; ibid 40 km NW).
Myrianthus fosi was assessed as Vulnerable by Cheek and Osborne in Harvey *et al.* (2010: 62, 73) and this is maintained here. Only eight locations are known equating to an AOO of 32 km^2 if 4 km^2 cells are used. Threats are given below.
Named (noun in apposition) in honour of Mrs Mary Fosi, former National Focal Point person for the Convention on Biological Diversity with the Ministry of the Environment and Forests of Cameroon who has long supported the project for a Red Data book for the Plants of Cameroon and who is herself, like this species, derives from the Cameroon Highlands.
This appears to be a very sparsely distributed species even within its submontane forest habitat in the Cameroon Highlands. For example, although c. 9000 specimen numbers were made at Kupe-Bakossi, only the single specimen cited above was made of this species. However at a very local level it can be common since *Letouzey* 11208 remarked 'abundant' and *Letouzey* 14532 'frequent'. It may qualify as a rare pioneer since some collections derive from habitat which is disturbed, e.g. *Satabié* 105 from a young coffee plantation. Only at the Ebo site is more than one

specimen known which may suggest that it occurs most densely here.

Tree 5–15 m tall, probably evergreen; bark smooth, trunk c.15 cm dbh; stems of leafy shoots scented of antiseptic or 'deep heat' when cut, terete, fistular, 8–9(–12) mm diam., internodes 7–20 mm long when mature, outer surface matt red brown, with scattered raised concolorous elliptic lenticels 1.5–2 mm long, indumentum dense of appressed simple, soft, straight golden-brown hairs 0.1–0.5 mm long, soon glabrescent; leaves alternate, simple, chartaceous to coriaceous, ± orbicular in outline, 15–30 cm long, 11–36 cm broad, base truncate to cordate, 3–5(–7)-lobed, lobing by 1/3–1/2(–2/3) of the radius of the leaf as measured between the outermost and its neighbouring lobe, lobes oblong-elliptic, apex broadly acute to rarely long acuminate, central lobe with 6–14 secondary nerves on each side of the midrib, nerves arching near margin towards leaflet apex, not terminating in a marginal tooth, lower surface of the blade brownish white, tertiary nerves scalariform, nerves brown, interstitial areas brownish white, matt, with arachnoid hairs mixed with straight, transparent, patent 0.05–0.2 mm long hairs, upper surface subglossy, dark green, hairs sparse, straight, translucent (0.2–)0.3–0.4 mm long, margin serrate-dentate, teeth irregular, slightly hooked, apex mucronate, c. 1.5 teeth per cm, each 3–11 mm long; petiole 6–28 cm long, both long and short on the same stem at adjacent nodes, longitudinally ribbed, matt dark brown, indumentum as stem; stipules intrapetiolar, at length deciduous, subpersistent, clasping the stem by ¾ to 9/10 of the circumference, ovate, 0.7–0.8 × 0.9–1.2 cm, midrib prominent, hairs appressed, moderately dense. Staminate inflorescences immature (including the peduncle) 1.1.5–1.5 × 0.4–0.8 cm with 2–4 primary branches, each one itself branching dichotomously; ultimate flowering units bearing c. 20 flowers (in bud only), ellipsoid, c. 2.5 × 2 mm; peduncle c. 1.3 cm long, 1–2 mm thick, indumentum of translucent patent simple hairs 0.1–0.3 mm long, the surface moderately densely covered with brown ellipsoid structures 0.02–0.03 mm long; female inflorescence unknown; infructescence with peduncle 30 × 5 mm, fruiting head 3–3.5 cm diam., with 13–22(–30) fruits; bracts oblong, c. 4 × 2.5 mm; fruits edible, ripening from green to yellow (*Osborne* 46), obovoid, 12 × 12 mm. Local name: Boussombossomb (Banen *fide* L. Beheng), fruit edible, also eaten by monkeys, chimps and gorillas (*Osborne* 46).

Habitat: submontane forest with *Santiria trimera* (Burseraceae), *Allanblackia gabonensis* (Guttiferae), *Syzygium staudtii* (Myrtaceae): 832–1400 m alt.

Threats: the species is threatened at Lebialem Highlands since its forest habitat has been observed to have been cut down for farms over several years of study. While the sites in Rumpi Hills and Bakossi Mts are likely to be unthreatened for the moment due to low population pressure and remoteness from infrastructure, the status of those near Bafang, Mamfe and Ndikiniméki is unknown and requires investigation.

Management suggestions: since there are prospects of formal protected status at Ebo, and a long term conservation project there it seems logical that further research and conservation efforts should be focussed here. However efforts should be made to educate conservation managers and local communities at other sites to recognise and protect *Myrianthus fosi*. Basic populational data should be recorded as a baseline for monitoring. Consideration should be given to seed collection both as a safeguard against extinction in the wild and to multiply young plants for augmenting the wild population in safe sites within the range of the species.

REFERENCES

Berg, C. C., Hijman, M.E.E. & Weerdenburg, J.C.A.(1985). Moracées. Flore du Cameroun 28. MESRES, Yaoundé.

Cheek, M., Pollard, B.J, Darbyshire, I., Onana, J.-M. & Wild, C. (eds) (2004). The Plants of Kupe, Mwanenguba & the Bakossi Mts, Cameroon: A Conservation Checklist. Royal Botanic Gardens, Kew, UK. iv + 508 pp.

de Ruiter, G. (1976). Revision for the genera *Myrianthus* and *Musanga* (Moraceae). Bull. Jard. Bot. Nat. Belg. 46: 471–510.

CELASTRACEAE

assessed by Martin Cheek

A cosmopolitan family represented in Cameroon by over 100 species in three groups.

Mainly lianas which climb mainly by the twining lateral stems. The leaves are simple and opposite and on the same stem, distinctively, subopposite (nearly but not quite perfectly opposite). Stipules are usually absent except in some Hippocrateoids. The flowers are radially symmetrical and flat, often yellow or orange, usually with a ring – or cylindrical nectar disc next to the whorl of (2–)3–5 curved stamens and the commonly 3-loculed pistil. Several species have a rubbery resin but this is often difficult to view. Fortunately the family are treated by Flore du Cameroun from whence the descriptions below are mainly derived. No new taxa are known to have been published in Cameroon since those publications.

Potentially possible to confuse with Guttiferae or Apocynaceae (but lacking coloured exudate) or Rubiaceae (but lacking a proper, **entire** interpetiolar stipule, although an interpetiolar ridge, sometimes with a fringe of colleter hairs, is present in many Hippocratoids).

1. *Celastroids.* Mainly shrubs, with the disc placed between stamens and pistil, fruits fleshy, 1-seeded except the largest genus *Maytenus,* unusual in dehiscent capsules with arillate seeds, also unusual in that leaves are often alternate, stems often spiny. Treated in Flore du Cameroun 19: 3-32 by Villiers (1975).

2. *Hippocratoids.* Forest canopy flowering lianas. Formerly treated as the genus *Hippocratea* but now, following Loesener's classification, regarded as 14 genera in Flore du Cameroun 32 (Hallé 1990). All species share the same distinctive fruits resembling a 3-bladed propeller, the flattened blades being oblong mericarps with woody-leathery walls splitting down the midline into two parts and releasing numerous winged seeds. Three tribes have been recognised:
Hippocrataeae. Extra-staminal disc, seeds glabrous, indumentum simple. (*Hippocratea* and nine satellite genera).
Campylostemoneae. Disc absent, otherwise as above (*Campylostemon, Bequaertia, Tristemananthus*).
Helictonemeae. As first tribe, but seeds hairy; marginal raphe; hairs stellate (*Helictonema*).

3. *Salacioids.* Understorey forest lianas, rarely small trees or shrubs. Fruits orange or red, leathery, indehiscent, containing several seeds in translucent, sweet pulp. Also treated by Hallé (1990, loc.cit.) (*Thyrosalacia*, terminal, paniculate, inflorescence, 3 species; *Salacighia* flowering when leafless, 1 species; *Salacia* axillary, contracted inflorescences when stems are leafy, c. 50 species in Cameroon).

Two new species awaiting description were assessed as threatened in Cable & Cheek 1998, both of which have not yet been formally published so cannot be maintained as threatened until this is rectified, following IUCN guidance:

Salacia sp. nov. 1 (sp. A of FWTA 1: 632) CR A1c+2c. Range: Mt Cameroon

Salacia sp. nov. 2 CR A1c+2c. Range: Mt Cameroon.

Campylostemon mitophorum Loesener
VU B2ab(iii)

Range: Cameroon (SW: Mt Cameroon, Bimbia. Littoral: Ndokmen II, Yingui. S: Bipindi; Ébolowa-Minkok), Equatorial Guinea (Bioko (fide Flore du Cameroun)) and Gabon (Mebomen).
Here *Campylostemon mitophorum* is assessed as Vulnerable since six sites are known (AOO 24 km^2 with 4 km^2 cells) and threats are as below. EOO is calculated as 32483 km^2.
Hippocratoid liane, 20 m, resin, glabrous; branches red regulose; blade drying brown, elliptic-oblong, 8–16 × 4–8 cm, acumen <1 cm, base acute to rounded, margin entire to slighly irregular; midrib not raised above; lateral nerves 6-8 pairs, intersecondary conspicuous, perpendicular; petiole 9–14 mm; cymes in axillary pairs, 3.5–9 cm, drying red brown, peduncle 1.5–4.2 cm; flower buds globose, 2–3 mm; open flowers 6–7 mm wide, perfumed, white turning yellow, stamens 5, introrse, disc absent, stigma trilobed, subsessile (N. Hallé Flore du Cameroun 32: 225 (1990)).
Habitat: lowland evergreen forest.
Threats: unknown for most sites, but both Bimbia and lower altitude Bioko have seen heavy losses of the habitat of this species in recent years, which losses are ongoing.
Management suggestions: the community forest at Bimbia should be surveyed in case the species survives there. If refound, basic populational data should be gathered as a baseline for future monitoring, the forest management council consulted on the story and identification of the plant. Propagation at Limbe Botanic Garden is advisable.

Elaeodendron kamerunense (Loesener) Villiers (syn. *Cassine glauca* var. *kamerunense*)
EN B2ab(iii)

Range: Cameroon (NW: Bamenda Highlands) and Congo (Kinshasa) (Lake Albert, Nioka).
Here *Elaeodendron kamerunense* is assessed as Endangered since only two sites (AOO 8 km^2 with 4 km^2 cells) and with threats as below.
A celastroid glabrous shrub with quadrangular branches; axillary bud minute; leaves opposite; stipule triangular; leaf-blade elliptic 5–9 × 1.9–4.5 cm, subacuminate, base cuneate-acute, margin strongly dentate, lateral nerves 8–11 pairs; petiole 1–1.4 cm; inflorescence of repeatedly dichotomous cymes c. 24-flowered; peduncle 0.6–2.5 cm; flowers with petals oblong-elliptic, 1.75 mm; stamens 5, 1.25 mm long; intrastaminal disc 5-lobed; ovary superior; drupe assumed symmetric.
Described in 1932 as *Elaeodendron glaucum* pers. var. *kamerunense* Loes. based on *Ledermann* 5796 and 5958 from the Bamenda Highlands. Subsequently a further specimen from eastern Congo (Kinshasa) was identified by Wilczek in Flore du Congo Belge 9: 131 (1960), but transferred to *Cassine.* Villiers raised the variety to species rank in *Elaeodendron*, neotypifying on the Congo (Kinshasa) material since the Cameroon specimens were destroyed. (Flore du Cameroun 19: 10, 1975).
Habitat: montane forest or scrub (deduced).
Threats: not seen in Cameroon since the type collections were made in 1907–1908 in the Bamenda Highlands where massive forest losses have occurred of that forest which survives – in one area. 50% loss has occurred in 15 years recently (Baena in Cheek *et al* 2010: back cover).

Management suggestions: Ledermann's itinerary should be studied to pin down where in the Bamenda Highlands he may have collected it. Efforts should be made to refind it and if successful, basic populational data should be collected as a baseline for future monitoring. Material should be collected for propagation so that it can be introduced to reforestation projects. Leaders of settlements adjacent should be informed and assisted in identifying this species.

Loeseneriella camerunica (Loesener) Hallé
EN B2ab(iii)
Range: Cameroon (S: Bipindi) and Gabon (Tsengué; Lastoursville; Lagune de Mayomba).
Here *Loeseneriella camerunica* is assessed as Endangered (AOO 16 km² with 4 km² cells; sites four listed above; threats below). EOO is calculated as 55791 km².
Hippocratoid woody liana, resin absent; stems and leaves glabrous; stems finely rugulose and lenticellate; leaf-blade leathery, drying red brown, elliptic-oblong, 5–20 × 3–10cm, acumen 8–13 mm, base acute to rounded, margin indistinctly denticulate, secondary nerves 6–8 pairs, nervelets indistinct; petiole 5–13 mm; stipules a thin transverse fringe; cymes 4–8 cm, axes finely red puberulent; buds ovate, conic, 5–6 mm, densely puberulent; flower orange-yellow 10–12 mm wide, petals long-triangular 4–6 × 2 mm minutely puberulent; disc 2.5–3 mm diam. cup-shaped, inner surface hairy, difficult to view; stamens 3 reflexed, curved, on androphae; fruit valve 89 × 15 mm.
Described from Zenker's specimens at Bipindi, all subsequent sites discovered have been in Gabon.
Habitat: lowland evergreen forest.
Threats: slash and burn agriculture. Further threats are unknown.
Management suggestions: Bipindi should be the focus of surveys to refind this species in Cameroon. Basic populational data should be collected as a baseline for future monitoring if the search succeeds, and local leaders informed of the location and identity of the plants. Currently no protected areas are known in the Bipindi area despite its importance for plant conservation.

Pleurostylia serrulata Loesnener
EN B2ab(iii)
Range: Cameroon endemic (Littoral: Douala-Edea. C: Yaoundé-Bafia 110km. E: 35km NNE Moloundou. S: Massif de Nyolé; NW Banda).
Here *Pleurostylia serrulata* is assessed as Endangered (four locations, AOO 16 km² with 4 km² cells, threats as below). EOO is calculated as 157[illegible] km².
Celastroid shrub 2 m. [illegible] with stellate and simple, white hairs; lea[illegible] × 1.6–3 cm, acuminate, base rounded-obtuse, margin toothed, midrib pubescent above and below; secondary nerves 6–7 pairs, uniting in a marginal nerve, yellow; inflorescences axillary, peduncle 1.3–2.5 cm; flowers orange, 4-merous, pedicel glabrous; sepals orbicular 1 mm; petals free 1 mm; drupe elliptic, asymmetric, 1 × 0.6 cm, orange-red, style excentric.
Habitat: lowland evergreen forest.
Threats: slash and burn agriculture.
Management suggestions: surveys to refind this species are needed in Cameroon. Basic populational data should be collected as a baseline for future monitoring if the search succeeds, and local leaders informed of the location and identity of the plants. Currently no protected areas are known for any of the locations.

Pristimera biholongii Hallé
CR B2ab(iii)
Range: Cameroon endemic (C: N'kolbison).
Named for the Cameroonian specimen collector Biholong. Here *Pristimera biholongii* is assessed as Critically Endangered (a single location, cited above, AOO 4 km² with 4 km² cells, threats as below).
Hippocratoid liana >5 m, stems cylindric, red-brown, lentricels white, leaves 5–9 × 2.2–4.5 cm, acumen 4–10 mm, base cuneate to rounded, margin toothed, secondary nerves 5–7 pairs, finely reticulate; petiole 3–6 mm; cymes axillary on leafy stems, 1.5–2 cm, puberulent; flower buds ovate 1.5–2 mm; flowers pale green, 1.5–2 mm diam., cup-shaped; petals triangular 1.2–1.6 mm, glabrous; disc cupular 0.3–1 mm anthers 3, subsessile, style-stigma entire. Mericarps 5–6.5 × 2 cm.
Known only from two collections, made in April 1962 (fl.) and Dec. 1963 (fr.).
Habitat: submontane forest and scrub on inselbergs; 700 m.
Threats: N'kolbison, close to Yaoundé, has seen clearance of natural habitat due to urbanisation and small-holder agriculture in recent years (pers. obs.).
Management suggestions: N'kolbison and similar inselberg areas around Yaoundé should be targetted for surveys to refind this species. Since *Talbotiella breteleri (*Leg.-Caesalp.) has been refound in the Yaoundé area recently, we must be hopeful that *Pristimera biholongii* will also be rediscovered. Seeds should be banked and material taken for propagation and multiplication if this does not compromise the survival of the plants in the wild.

Pristimera breteleri N.Hallé
CR B2ab(iii)
Range: Cameroon endemic (C: N'kolbison)
Here *Pristimera breteleri* is assessed as Critically Endangered (a single location, cited above, AOO 4 km² with 4 km² cells, threats as below).
Hippocratoid liana >5 m, similar vegetatively to *P. biholongii* and collected in flower by the same collector (Breteler), on the same day, in the same locality. Differing in the 2–3.5 cm, glabrous cymes; disc annular 0.01–0.9 mm, stigma trilobed.

Known only from the type collection.
Habitat: submontane forest and scrub on inselbergs; 700 m.
Threats: N'kolbison, close to Yaoundé, has seen clearance of natural habitat due to urbanisation and small-holder agriculture in recent years (pers. obs.).
Management suggestions: N'kolbison and similar inselberg areas around Yaoundé should be targetted for surveys to refind this species. Since *Talbotiella breteleri (*Leg.-Caesalp.) has been refound in the Yaoundé area recently, we must be hopeful that *Pristimera breteleri* will also be rediscovered. Seeds should be banked and material taken for propagation and multiplication if this does not compromise the survival of the plants in the wild.

Salacia conraui Loesener
CR B2ab(iii)
Range: Cameroon endemic (SW: 40 km NW Nguti at Tali).
Known only from the original specimen, *Conrau* 140, location above, collected 1898–1899, *Salacia conraui* is here assessed as Critically Endangered (AOO 4km² with 4km² cells, threats below).
Shrub(?), glabrous; young branches densely minutely lenticellate; leaves grey-brown, glossy above; weakly discoloured opposite, elliptic-ovate or oblong, 10–14 × 3.5–6 cm, shortly acuminate, base acute-obtuse, entire or subdentate, secondary nerves 5–7 pairs; petiole 3–5 mm; inflorescence axillary, sessile, 6-flowered; pedicels 10–15 mm; flowers 18 mm diam., petals clawed 6–8 × 8–10 mm; disc 4–5 mm diam., stamens 2 mm; pistil 3-angled, stigma unlobed.
Known only from the type collection, *Conrau* 104, now destroyed at Berlin.
Habitat: submontane forest (deduced).
Threats: slash and burn agriculture is prevalent in the region.
Management suggestions: research on Conrau's itinerary in Cameroon would aid pinpointing the original locality of this plant and assist planning for where to redicover it. If this hope is realised, basic population data and propagation material for multiplication should be gathered. Elders of communities should be apprised of the existence and need to protect this liana.

Salacia dimidia Hallé
VU B2ab(iii)
Range: Cameroon (SW: Mt Cameroon foothills: Onge-Idenau, several coll; Mamfe-Kendem. Littoral: Edéa; Kompina. C: Ndoknabao; N'kolbison. S: Nloubessa Boulou) and Gabon (Ngoiné; Monkalaba Reserve).
Here *Salacia dimidia* is assessed as Vulnerable (ten locations listed above; AOO 40 km² with 4 km² cells, threats below). EOO is calculated as 139474 km².
Shrub 0.3–1 m, glabrous, stem 4-lined; leaves elliptic or oblong, 9–21 × 4–9 cm, acuminate, base rounded-cordate or obtuse, weakly toothed, secondary nerves 8–16 pairs, finer nerves indistinct; inflorescences axillary, 2–6-flowered, fasciculate; bud globose, 3–4 mm; pedicel 4–7 mm; flower 9–15 mm diam., petals orange-pink orbicular, 4–5 mm; disc cupular, rim sinuate, 2.8–4 mm diam., stamens 3, 3 mm; pistil unlobed; fruit submoniliform, beaked, 6–8 × 2.5 cm, orange-red, seeds several, in fleshy pulp.
Habitat: lowland evergreen forest.
Threats: forest clearance for logging followed by agriculture, especially in the coastal foothills of Mt Cameroon at Idenau and Onge. For threats at N'kolbison see *Pristimera breteleri.*
Management suggestions: *Salacia dimidia* occurs in protected areas in Gabon, but not in Cameroon, excepting forest reserves which are likely to be commercially logged soon. Thought should be given to creating a new National Park at one of its Cameroon locations, ideally one where many other threatened species occur, such as at Onge or N'kolbison. Basic populational data should be gathered and local communities educated to identify and protect this species formally unprotected sites.

Salacia fimbrisepala Loesener
EN B2ab(iii)
Range: Ghana (Kwahu) and Cameroon (SW: Mt Cameroon, near Buea, along river).
Not seen in Cameroon since the type and only gathering a century ago at the single known location (above). Previously assessed as CR A1c+2c in Cable & Cheek (1998: xlviii). Here assessed as Endangered (AOO 8 km², 4 km² cells, threats as below).
Liana, glabrous; leaves opposite; elliptic, 12–18 × 5–7.3 cm, obtusely acuminate; cuneate, dentate, nerves finely reticulate, petiole 8–13 mm; fascicles few-flowered; flowers large, brown-yellow, drying black, pedicel 5–6 mm, sepals deltoid, margin lacerate-fimbriate, inner 3 larger; petals 5–6 mm; disc 4 mm, pentagonal.
Only known from the type collection *Lehmbach* 228 (fl. April) destroyed at Berlin. A second collection, from Ghana, not seen at K, is cited in FWTA 1: 632 (1958) but is not referred to by Hallé (1990 loc.cit.).
Habitat: submontane forest, along river; 1000 m alt.
Threats: Buea, now a large University town and regional centre, has expanded greatly in the last 100 years and will continue to do so. The habitat of this species, if not already lost, is threatened from urbanisation and associated agriculture.
Management suggestions: since this species has not been recollected recently, despite several major surveys on Mt Cameroon in the 1990s, there is cause for concern regarding its survival, especially since Buea, which is given as the

locality, has lost so much of its natural forest vegetation. Surveys are needed to rediscover this species. If successful, basic populational data should be collected and the species propagated at Limbe Botanic Garden for reintroduction.

Salacia lebrunii Wilczek

VU B2ab(iii)

Range: Cameroon (SW: Bechati-Lebialem), Gabon (33 km E of Lastourville; Belinga iron mine), Congo (Brazzaville) (Chaillu) and Congo (Kinshasa) (Kinshasa, Lac Leopold II; Lodja-Kole; Yangambi).

Known from six localities (AOO 24 km^2 with 4 km^2 cells) and with threats as below, *Salacia lebrunii* is here assessed as Vulnerable, following the assessment in Harvey *et al.* (2010: 73–74). EOO is calculated as 671124 km^2.

This liana has 4-ridged stems and elliptic leaves c. 7–9 × 3.2–5 cm, including a 0.5 cm acumen, lateral nerves are 3–6 pairs. The flowers are single, petals light orange, 2 × 1 mm, stamens only 2 (usually 3 in *Salacia*); fruit subglobose, hard, orange, 2–3 cm diam.

Habitat: lowland forest; 100–400 m alt.

Threats: Belinga is set to become a Chinese owned iron mine. At Lebialem–Bechati, forest clearance for agriculture in ongoing.

Management suggestions: in Chaillu threats are few and Ogooué-Leketi proposed NP (the site of this species) may be the best focal point for conserving the species. The status at the other sites is unknown. The elders at Bechati should be informed of the presence and conservation importance of this species.

Salacia lehmbachii Loes. var. ***pes-ranulae*** N.Hallé (assessed by Iain Darbyshire & Benedict Pollard modified by Martin Cheek)

VU B2ab(iii)

Range: Nigeria (Cross River State: Oban (1 coll.)) and Cameroon (SW: Mt Cameroon at Bimbia, 2 coll.); Bakossi FR (1 coll.), Lebialem (1 coll.), 15 km S of Akwaya (1 coll.). S: S of Ébolowa (3 coll.)).

Salacia lehmbachii sensu lato is distributed from Sierra Leone to Tanzania; the complex comprises seven varieties of which var. *pes-ranulae* is both the most localised in distribution and the most distinct morphologically. It probably merits recognition at species rank (Gosline pers. comm.)

Salacia lehmbachii in all its several varieties is distinctive in being a shrub or small tree, glabrous, with orange-red flowers 6–8.5 m[illegible]-peduncled axillary inflorescences. T[illegible]og foot) is named for the alternate [illegible] of the peduncle believed to resemble a frog's foot. In addition the peduncle often lies in the groove of the leaf-blade's midrib so that the flowers appear to emerge directly from the middle of the blade.

First described in 1986, all previous collections of this taxon were made prior to 1980, when lowland forest south of Ébolowa appeared to be the centre of its distribution. These forests have experienced significant reductions following expansion of lowland plantation agriculture. The discovery of this taxon in the protected Bakossi FR is of significance to its future conservation. This assessment was originally made in Cheek *et al.* (2004: 151) and was maintained in Harvey *et al.* (2010: 74), followed here, since only one extra record (Lebialem) has been added since. Six sites are known (AOO 24 km^2 with 4 km^2 cells). EOO is calculated as 45656 km^2.

Habitat: dense lowland forest understorey; 450–700 m alt.

Threats: continued loss of lowland forest throughout the species range, particularly in Cross River State, Nigeria and the forests of S and SW Regions, Cameroon. Illegal encroachment of agriculture into the Bakossi FR, facilitated by European Community road building, threatens this population.

Management suggestions: heightened protection of remaining forest at the Bakossi FR may help to preserve this taxon at Bakossi. Further botanical inventory work in lowland sites, such as the neighbouring Loum FR, may reveal further populations; care should be taken to accurately identify all future collections of *S. lehmbachii* to variety level.

Salacia lenticellosa Harms.

VU B2ab(iii)

Range: Nigeria (Oban); Cameroon (SW: Mt Cameroon-Mokoko; Nta Ali. E: Lomie; Yaoundé-Deng-Deng 90km) and Gabon (Medouneu-Simanguen).

Here *Salacia lenticellosa* is assessed as Vulnerable since six locations (see above; AOO 24 km^2 with 4 km^2 cells) are known, with threats as below. EOO is calculated as 135471 km^2.

Shrub 2 m or liana 25 m, glabrous; stems brown and lenticellate; leaves opposite, blade oblong or elliptic 10–25 × 5–12 cm, acutely acuminate, base rounded to acute, margin entire or weakly toothed, lateral nerves 7–11 pairs; petiole 7–13 mm; fascicles axillary, peduncle 0–4 mm; pedicels 5–10 mm, flowers >8 mm wide, disc cupulate, 5-lobed, rim thickened, stamens 3, pistil entire; fruit subglobose 3.5–4 cm.

Habitat: lowland evergreen forest.

Threats: Nigeria has lost nearly 90% of its original forest (mongabay.com) and surviving forest, largely in the SE near Oban is under great pressure for timber extraction and agricultural land. Mokoko FR is likely to be logged in the near future.

Management suggestions: within Cameroon the most recent observations of this species have been at Mt Cameroon-

Mokoko, therefore it makes sense to set aside part of this forest reserve containing this and other threatened species, for protection as a nature reserve. Whichever location is adapted to protect this species should be surveyed for basic populational data as a baseline for future monitoring. If population size is low, propagation for multiplication and reintroduction could be arranged at Limbe Botanic Garden potentially.

Salacia letouzeyana Hallé
VU B2ab(iii)

Range: Nigeria (Oban) and Cameroon (SW: Ekondo Titi/Masore; Abakpa, SW Mamfe. S: Kribi; Ébolowa. C: 50 km S Ndikiniméki at Kikot bridge).

Here *Salacia letouzeyana* is assessed as Vulnerable since there are seven sites (AOO 28 km² with 4 km² cells) and threats as below. EOO is calculated as 48753 km².

Liana 2.5–10 m, glabrous; stems pale orange brown, glabrous; leaf-blade 6–14 × 2–6 cm, acuminate, base acute, subentire to denticulate, secondary nerves 7–10 pairs, lower surface yellow-green; petiole 5–15 mm; glomerules axillary, 1–few flowers; flowers orange-pink, 3.5–4.8 mm diam; disc circular 2 mm, stamens 3, pistil entire, fruit orange, subglobose 4–6 × 3–4 cm, smooth or warty.

Habitat: lowland evergreen forest.

Threats: unknown at most sites but forest habitat has been lost at Oban and in Kribi, the latter due to increased urbanisation and with infrastructure from extractive industries expected to destroy habitat in future. Slash and burn agriculture is widespread in the forest belt.

Management suggestions: rediscovery of this species should be attempted and efforts made to include it in a protected area; basic populational data should be gathered as a baseline for future monitoring. Local communities should be assisted to identify the plant so as to protect it.

Salacia lucida Oliv.
VU B1+B2ab(iii)

Range: Nigeria (Calabar) and Cameroon (SW: Mundemba. S: Kribi; 25 km NNE Kribi; Bipindi; 60 km SW Eseka).

Here *Salacia lucida* is assessed as Vulnerable since six sites (above) are known (AOO 24 km² with 4 km² cells, threats below). EOO is calculated as 14481 km².

Liana 5–20 m, glabrous, leaf-blade elliptic 4.5–13.5 cm × 2–7.5 cm, acumen 5–11 mm; base acute; denticulate to dentate; secondary nerves 7–9 pairs; petiole 7–8 mm; inflorescence dichasial, axillary, pedunculate, pedicels 4–5 mm, flower 6 mm diam. pale green to orange; sepals slender, hooded at apex; disc cylindrical, papillose, enclosing pistil base and giving rise to 3 sagittate stamens; pistil entire.

Habitat: littoral forest and rear of mangrove.

Threats: Hallé (loc.cit.65) reports that this species can survive in farm bush (broussoilles postculturales).

Management suggestions: as for *S. letouzeyana.*

Salacia mamba N.Hallé (assessed by IainDarbyshire and Benedict Pollard, modified by Martin Cheek)
VU B2ab(iii)

Range: Cameroon (SW: Mt Kupe (3 coll.). Littoral: Bakaka/Nkongsamba-Loum (3 coll.). C: Otélé, nr. Yaoundé (1 coll.). S: 10km SW of Amban (1 coll.)), Gabon (nr. Makokou (3 coll.)) and Congo (Brazzaville) (Mayombe (1 coll.); Komono to Mbila (1 coll.)).

Salacia mamba is a 10–20 m glabrous liana, the stems are red-brown, lenticellate with leaves glossy, as though varnished above, elliptic, 8–14 × 3.5–6 cm, acumen short, base acute-obtuse, margin entire, secondary nerves 5–7 pairs; petiole 6–9 mm; glomerules axillary, pedicels 3–4 mm; flower orange, 4 mm diam., stamens 3, erect, concealing the 3 lobed pistil; fruit globose, 2 cm, orange.

The assessment made in Cheek *et al.* 2004:152 is maintained here, since there are seven (above) localities (AOO 28 km² with 4 km² cells). EOO is calculated as 147586 km². Although relatively widespread, it is clearly rare throughout its range and absent from many seemingly suitable sites. The discovery of this species on the western slopes of Mt Kupe extended its western range limit; it is also known from the lowlands adjacent to this mountain to the east, although the expansion of settlements along the Loum-Nkongsamba Rd may have caused local population losses since its collection there in the 1970s.

Habitat: dense riverine, lowland and mid-elevation forest; 500–1000 m alt.

Threats: continued loss of lowland forest for timber and agriculture in Cameroon; the Gabonese and Congo (Brazzaville) sites are likely to have remained relatively undisturbed.

Management suggestions: protection of existing lowland forest sites invoking the interest and support of local communities where possible. Care should be taken to separate this taxon from other lowland *Salacia* species in future botanical inventory work in Lower Guinea.

Salacia talbotii E.G. Baker
EN B2ab(iii)

Range: Nigeria (Calabar-Mamfe Rd at Oban and Osomba; Ikom 15 km SE) and Cameroon (SW: Lake Eghaham; Mt Cameroon foothills at Onge and Mokoko) possibly in Gabon. Distinctive in the spiralled sepals and petals merging in a single series of 15 units, the margins thin, pale brown, splitting.

Liane >5 m, glabrous, stem 4-angled, grey-green, pale, smooth, leaf-blade elliptic, 8–22 × 4–8.5 cm, acumen 4–9 mm, base acute, secondary nerves 6–8 pairs, a fine venation visible on both sides; petiole 7–14 mm; glomerules axillary to 10-flowered, pedicel 4–7 mm; flower 12–16 mm diam.

greenish white to orange, petals and sepals with brown, thin, splitting margin; disc shortly cylindric, stamens 3, pistil entire; fruit orange, elliptic, 3-ribbed, ribs warty, rostrate 7–10 × 5 cm.

Here *S. talbotii* is assessed as Endangered given the four broad locations listed above (AOO 16 km² with 4 km² cells) and threats below. EOO is calculated as 6289 km². Although specimens are cited on the Gabon database as this species, they are here excluded since they have dark brown pustulate (not pale grey smooth) stems and lack a fruit rostrum.

Habitat: along rivers in lowland evergreen forest (*De Wilde* 9121 at K).

Threats: Nigeria has lost nearly 90% of its original forest cover and that which remains is under great pressure for timber extraction and agricultural conversion, including that in SE Nigeria.

Management suggestions: the densest records of this species occur in the foothills of Mt Cameroon in Onge and Mokoko. This would be a logical zone in which to place a National Park or other permanent protected area to ensure the survival in the wild of this plant. Whichever location is adapted to protect this species should be surveyed for basic populational data as a baseline for future monitoring. If population size is low, propagation for multiplication and reintroduction could be arranged at Limbe Botanic Garden potentially.

Salacia volubilis Loesener & Winkler

EN B2ab(iii)

Range: Cameroon (SW: Korup; Mt Cameroon at Onge and Buea; Limbe; Okola-Mua-Evodula; N'kolandom 10 km S Ébolowa) and Gabon (Doudou Mts).

Here *Salacia volubilis* is assessed as Endangered since five broad locations (above) are known. At one of these, Mt Cameroon, it occurs at three sites, so an AOO of 28 km² is calculated using 4 km² cells. Threats are as below. EOO is calculated as 106522 km².

Liana 1–15 m, glabrous; young stems 4-angled; leaf-blades ovate-elliptic 6–15 × 3–7 cm, acuminate, base rounded to subcordate, dentate, lateral nerves 8–11 pairs; petiole 2–5 mm; glomerules axillary, 1–5-flowered; pedicel 8–20 mm; flower orange 5–8 mm diam; petals oblong with 2–3 concavities at inner apex; disc inconspicuous, stamens 3; fruit globose, 2–3 cm, smooth.

Habitat: lowland evergreen forest.

Threats: while the lower slopes of Mt Cameroon have the greatest density of this species, it is here that the greatest known threats occur. The type [illegible] already be lost to this species due to urbanisation [illegible] infrastructure for extractive industries (oil, timber) and plantation agriculture (oil palm, rubber, bananas). Buea has seen urbanisation, while the Onge forest location has been targetted for logging.

Management suggestions: surveys to rediscover this species should be mounted initially in the Mt Cameroon area (greatest density). Basic populational data should be collected, and, if plants are in highly threatened sites, consideration should be given to their cultivation at Limbe BG for multiplication and reintroduction at safe sites.

Simirestis staudtii (Loesener) Hallé (*Hippocratea staudtii* Loesener)

EN B2ab(iii)

Range: Sierra Leone (Kenema-Sedafu) and Cameroon (C: Yaoundé).

Here *Simirestis staudtii* is assessed as EN since there are two locations (AOO 8 km² with 4 km² cells, threats as below).

Hippocratoid liana, 10 m, stems glabrous, finely ridged; blade ovate-oblong 6–17 × 3–8 cm, feebly acuminate, base rounded to subacute, entire, 6–7 pairs lateral nerves; petiole 8–14 mm; cymes axillary, slender, forking up to 6 times, peduncle 2–4 cm; flower green, 5–7.5 mm diam., sepal margin ciliate, disc-androphote cylindric, stamens 3, pistil unbranched.

Habitat: lowland to submontane evergreen forest (deduced); 800 m alt.

Threats: only collected once in Cameroon at Yaoundé in 1890–1894, fl. May – *Zenker & Staudt* 325. In the last 100 years this capital city has expanded greatly at the expense of natural habitat. This urbanisation continues.

Management suggestions: searching surviving forest and scrub on the hills around Yaoundé in the flowering season of May may yet rediscover this lost species. If refound, basic populational data should be gathered to aid future monitoring and assessment of intervention needs; settlements near plants should be informed. Seed should be obtained for banking since likely to be orthodox, and seedlings raised for introduction in safe sites.

Thyrosalacia pararacemosa Hallé

EN B2ab(iii)

Range: Cameroon (Littoral: Yabassi-Yingui; Ndikiniméki. S: Eséka 30 km N; 30 km WNW Eséka) and Gabon (Makoukou, 5-10 km S).

Salacioid liana to 6 m or shrub c. 1 m, bark folded and cracked; blade elliptic-obovate, 8–23 × 3–9.5 cm, acuminate, entire, secondary nerves 7–12 pairs; petiole 5–12 mm; raceme c. 6 cm, 'pedicels' 1.2–2.5 cm, flowers in Dec./Jan., pale green or yellow, 15–22 mm diam; disc double; stamens 3; pistil unbranched; fruit ellipsoid, 3-ridged at base, warty 30 × 22 mm, 1-seeded.

Here *Thyrosalacia pararacemosa* is assessed as Endangered (five locations, AOO 20 km² with 4 km² cells, threats as below). EOO is calculated as 22114 km².

Habitat: lowland, evergreen forest.

Threats: slash and burn agriculture at most forest locations; logging followed by agriculture in Yabassi-Yingui area (pers. obs.).
Management suggestions: while none of the sites is known to be within a National Park, the Yabassi-Yingui site is near the proposed Ebo NP and so there is a possibility that the species may occur there. If rediscovered, basic populational data should be gathered as a baseline for future monitoring.

Thyrosalacia racemosa (Loes. ex Harms) N.Hallé (assessed by Iain Darbyshire and Benedict Pollard modified by Cheek)
CR A2c
Range: Cameroon (Littoral Province: Yabassi (1 coll.); Babong (1 coll.); Lom (= Loum) (2 coll.)).
Resembling *T. pararacemosa* in all respects except the disc, which is simple, not double. Hallé awaited the opportunity to dissect the flowers of this species before concluding whether the two taxa should be united or not.
This species is only known by 4 Ledermann collections from 1908 and 1909 in the lowland forests around the Yabassi-Loum area. It was assessed in Cheek *et al.* (2004: 152) as Critically Endangered on the basis that 80% of the population has been lost in the last 100 years using habitat destruction as a proxy (see notes below). Since there are three locations (above) AOO is calculated as 12 km^2 with 4 km^2 cells. EOO is calculated as 527 km^2.
Habitat: lowland forest, c. 200 m.
Threats: as a climber, this species is particularly sensitive to understorey clearance, and from personal observation at the Loum FR, an estimate of 80% reduction of population size over three generations is made here, based on the continuing development of extensive cocoa and banana smallholdings within the reserve. The Babong subpopulation is likely to have disappeared due to urban development, and probably only the Yabassi sub-population has survived the last 100 years.
Management suggestions: protection of existing lowland forest sites. Surveys in the Yabassi area would be most welcome, and this species made one of the targets for re-collection and population size monitoring. It is to be hoped that future survey work will reveal this species to occur in a protected area such as the Ebo proposed National park, which is E of Yabassi.

CHRYSOBALANACEAE

assessed by Martin Cheek, advised by Ghillean Prance).

A pantropical family of trees, or less usually shrubs, mainly in evergreen lowland forest but with some species in drier savanna woodland. Formerly included with Rosaceae due to their floral structure, recent evidence has shown that their closest relatives are Dichapetalaceae. Young, 1–2 m tall individuals of Chrysobalanaceae are difficult to separate from Dichapetalaceae in the forests of Cameroon, both having alternate, stipulate, simple leaves with similar dark gland spots. Most species of the last family are lianas when adult, and have glands dispersed in patterns throughout the leaf, visible in transmitted light while in Chrysobalanaceae the glands occur more usually on the surface, near the petiole and the adults are small to medium sized trees.

The Flore du Cameroun account by Letouzey and F. White (Vol. 20 (1978)) is excellent for species identification and delimitation (35 species in six genera). Since that time, Ghillean Prance has published numerous revisions, in the course of which all African *Acioa* species have been transferred to *Dactyladenia*. The definitive global account for the family is volume 9 (parts 1 & 2) of Flora of the World (Prance & Sothers 2004, Species Plantarum, IOPI, Canberra).

Recently three new *Magnistipula* taxa endemic to Cameroon have been discovered and published (Pollard *et al.* in Harvey *et al.* 2004: 51; van der Burgt, in press). In addition to the taxa assessed as threatened below, several other Cameroonian species: *Dactyladenia lehmbachii, D. letestui* and *Magnistipula glaberrima* only just exceed 10 sites so are here assessed as NT. However, if better data on habitat loss were available it is likely they would be assessed as threatened under Criterion A. If populational studies confirmed that these species occur at extremely low densities (this might be done by extrapolation from van der Burgt's work in Korup perhaps) then they might rate as highly threatened under Criteria C and/or D.

Almost all of the specimens cited have been identified or confirmed in identification by Ghillean Prance.

Dactyladenia are recognised by the stalked glands on the edge of stipules and of floral bracts and long exserted stamens united most of their length. In *Maranthes* and *Magnistipula* these features are lacking, the stamens being free almost to the base, and little exserted. In *Magnistipula* the flower is strongly zygomorphic, the stamens c. 7, included, the fruits indehiscent. In *Maranthes* the zygomorphy is weak, stamens 20–60, fruits with lateral valves in the endocarps. The descriptions below are either taken from our checklists (Harvey *et al.* 2004; Cheek *et al.* 2004) or from Letouzey and White (1978).

Dactyladenia cinerea (Engl. ex De Wild.) Prance & F. White
EN B2ab(iii)

Range: Cameroon endemic (SW: Mokoko at Mt Cameroon. S: Bipindi-type specimen; Grand Batanga (Tchoutou 2004)). Here *Dactyladenia cinerea* is assessed as Endangered since only three sites are known (see above, AOO 12 km^2 with 4 km^2 cells) with threats as below. At each site apart from Bipindi only a single record exists suggesting that even locally, it is rare. The Grand Batanga record has not been verified by us. EOO is calculated as 7719 km^2.

Shrub or tree to 15 m, young stems and inflorescences densely grey appressed hairy; leaf-blade elliptic, 12–14 × 6–7 cm, acumen 1 cm, cuneate, lateral nerves 6(–9) pairs rising at 45°, curved upwards, glabrous; petiole 6 mm, stipules triangular 6–8 mm, base clawed, subcordate, palmate-nerved; inflorescences subracemose 8–10 cm, in distal axils; bracts as stipules but 4–5 mm, margin glandular; receptacle 10–15 mm, sepals 5 mm, both laxly appressed white hairy, sepals 5 mm.

Habitat: lowland evergreen forest.

Threats: commercial logging is going ahead at Mokoko. There is a high risk of subsequent forest degradation. Grand Batanga, close to Kribi, risks coastal development for tourism and leisure homes.

Management suggestions: rediscovery of this species should be attempted, and basic populational data recorded to inform future monitoring and management interventions. Local communities and authorities should be informed of the existence of this tree and assisted to identify it since they will be responsible for protecting it. Seed should be collected for raising in Limbe Botanic Garden and subsequent introduction to the wild at safe sites, possibly as part of forest restoration projects.

Dactyladenia johnstonei (Hoyle) Prance & F.White

[Tileelee (Bamenda; *Johnstone* 74/31)].

CR A3c

Range: Nigeria (Obudu (1 coll.)) and Cameroon (NW: Bamenda Highlands (5 coll.). SW: Mt Kupe (1 coll.)).

Shrub 3 m to tree 10 m or more, young stems hispid, hairs dispersed, 2–3 mm; leaves ovate-elliptic, 6–12 × 2–5 cm subacuminate, base rounded to obtuse, lateral nerves 6 pairs; scabrid above, caducous long hairy below; petiole 5–6 mm, rugulose; stipules long triangular, 4 mm, margin stalked glandular; inflorescence pseudo-racemose, 10 cm, 1–2-branched; hispid-pilose; receptacle 10 mm; sepals and petals 5 mm.

This rare, hispid tree was first collected in the Bamenda Highlands in 1931 (*Johnstone* 74/31). Since that time four other collections from [illegible] in the Bamenda Highlands have been [illegible] Nkambe and Fonfuka, as well as [illegible]nt Nigeria. EOO is calculated as 14747 km^2. Our re[illegible]n Mt Kupe (*Cheek* 10158) is the first from SW Region. Given the threats below and an estimated generation time of 30 years, with past and projected population reduction due to forest loss at these sites of at least 80% over 90 years (see below), this taxon is here assessed as Critically Endangered under criterion A3. This assessment is maintained and modified from Cheek *et al.* (2004: 152). The absence of the species from areas of apparently suitable habitat within its range that have recently been surveyed (see below) points out how rare it is.

Habitat: forested valley slopes, understorey, or forest edges; 950–1600 m alt.

Threats: felling for timber and clearance for agricultural land: these problems are exacerbated in NW Region by a shortage of both commodities for a dense human population. Forest at this altitude is particularly under threat and may well no longer survive at the localities where this taxon was previously collected. A 25% reduction in forest area over eight years was recorded by Moat (in Cheek *et al.* 2000: back cover) near Mt Oku using GIS techniques and a 50% loss in 15 recent years in the Dom area (Baena in Cheek et al. 2010: back cover). A similar situation is recorded in Cross River State, Nigeria. Forest at Mt Kupe is less threatened, at least above 1000 m alt. However the Kupe material is recorded at 950 m and is thus vulnerable to agricultural pressure from farmers in the valley bottom, particularly for expansion of small-holder *Coffea canephora* (robusta coffee).

Management suggestions: efforts should be made to rediscover this taxon in NW Region and to protect it. Individuals may yet survive in royal, sacred or village forests and might be protected there. It appears to be unknown at Bali Ngemba FR the only extant formally protected area covering the altitudinal range of this species in the Bamenda Highlands (Harvey *et al.* 2004). Notably it is also absent from two other fairly well surveyed areas within its range: at Dom and at Lebialem Highlands (Cheek *et al.* 2010, Harvey *et al.* 2010).

At Mt Kupe the one individual known should be rediscovered and an effort made to find others. Education of the local population is suggested regarding the importance of this species for conservation. Basic populational data should be gathered as a baseline for future monitoring and management planning.

Dactyladenia mannii (Oliv.) Prance & F.White

CR A3c

Range: Equatorial Guinea (Bioko (possibly extinct)) and Cameroon (SW: the foothills of Mt Cameroon (four pre-1960 records, one modern)).

The type collection, and the only known record from Bioko (*Mann* 1427) refers to a "climbing shrub 15–20'". However, Maitland records it (*Maitland* 467, March 1929, nr Buea 2800') as "a small tree of 30 ft on edge of forest").

Scandent shrub 4–6 m young stems white appressed hairy but rapidly glabrous and sparsely lenticellate; leaf-blade elliptic, 12–18 × 5–9 cm, acumen 1 cm, base acute-obtuse,

lateral nerves 6–7 pairs ascending, glabrous, petiole 8 mm; stipules asymetric, ovate, 6 × 3–4 mm, strongly nerved; inflorescence pseudoracemose, 8–12 cm, in distal axils, bracts ovate, 2.3 × 1 mm, outer surface long hairy, bracteoles smaller inserted halfway up the pedicel; receptacle 14–20 mm, sepals 6–7 mm, petals obovate 4 mm.
Two other historic collections are also from near Buea (*Maitland* s.n.) and from Kumba, (*Staudt* 513) or 15 km SW of Kumba (*Lobe Babute* 9), but the only modern collection (*Wheatley* 790) is from the Onge forest. Known from a total of only 7 collections. EOO is calculated as 2316 km^2.
This species was assessed as CR A1c+2c by Cheek in Cable & Cheek (1998: xlix). This is maintained here but modified according to the current criteria of IUCN (2001), on the basis that it is estimated that more than 80% of the lowland forest habitat of this species has been lost in the last 100 years, mainly converted into cocoa (Bioko) banana, oil palm, tea, or rubber plantations, or small-holder market-gardening, particularly at the very spots where it was previously recorded and from whence it has not been refound despite recent surveys at those sites, also that these losses are ongoing, for example the Onge forest, source of the only collection known in the last 80 years and still intact when surveyed in 1993, is likely to be logged soon.
Habitat: lowland and submontane forest, sometimes along rivers, flowering in March; up to 800 m alt.
Threats: forest clearance for housing, timber and agriculture: see above.
Management suggestions: since all of the sites from which this species was known are under heavy threat or have already been lost to forest, it is advisable to assess the population of this species most likely to survive, that is, at Onge and develop a means of protecting it with forest managers, if not already too late. Seed should be obtained for introduction into Limbe BG, so that plantlets can be raised for introduction to safe sites.

Magnistipula butayei De Wild. subsp. ***balingembaensis*** Sothers, Prance & B.J.Pollard
CR A3c; B1ab(iii)+2ab(iii).

Range: Cameroon endemic (NW: Bali Ngemba FR)
Tree to 15 m tall. Branches glabrous, lenticellate. Leaves narrowly-ovate, elliptic or elliptic-oblong, 9.9–13.4(–18.5) × 3.4–5.4(–6.8) cm; lamina glabrous throughout, drying pinkish above and on the veins beneath; apex acute to (long-) acuminate; base rounded to cuneate (rarely subcordate), occasionally almost shortly-attenuate; glands on lower leaf surface usually in 1–4(–several) basal pairs either side of midrib, rarely with a basal pair on leaf-margin, with some also scattered randomly in apical half of blade; secondary veins 7–10(–16) pairs, glabrous; petiole glabrous, woody, articulate, 5–6(–10) mm long, pulvinate. Stipules caducous (not seen). Inflorescences terminal or subterminal panicles; branches horizontal or patent, to 9 cm. Flowers 7–8 mm long. Receptacle tube 3.5–4.0 mm long, ventrally slightly gibbous, puberulous outside, occasional glands seen on exterior, usually nearer the base. Sepals 3–4 mm long. Petals white, ciliolate at apex. Stamens 7, on one side, connate for a third of their length.
"*Magnistipula butayei* De Wild. subsp. *balingembaensis* Sothers, Prance & B.J.Pollard is only known from four collections inside the reserve (AOO, and EOO, c. 8 km^2). Our estimate of generation time for this subspecies is c. 20 years, so that within three generations (60 years) from now, we suspect c. 80% loss of forest cover, based on observations of current habitat loss (Cheek, Etuge, Onana, Pollard & Zapfack pers. obs. 2000–2002). It is therefore here assessed as Critically Endangered, following the guidelines of IUCN (2001)." (Pollard *et al.* in Harvey *et al.* 2004: 51–52 , where the tree was first published). This assessment is maintained here since no new data are available except that rates of loss of surviving forest are higher than previously thought (e.g. 50% in 15 years in the Dom area of the Bamenda Highlands 1988–2003, Baena in Cheek *et al.* 2010: back cover).
Habitat: submontane forest understorey and riverine forest; (?1310–)1700–1950 m alt.
Threats: within the boundaries of the reserve, there is widespread felling of canopy trees (e.g. *Synsepalum msolo* (Engl.) T.D.Penn. (Sapotaceae)), large emergent species (e.g. *Pterygota mildbraedii* Engl. (Sterculiaceae)), and many smaller tree species, as well as extensive clearance of shrubby and herbaceous undergrowth vegetation (Pollard pers. obs. 2000–2002). This is the work of local farmers, particularly those from the far side of the higher altitude grassland ridge towards Pinyin (c. 2000 m), who are increasing the number and scale of their agricultural small-holdings to increase productivity of various crops, notably coco-yams and potatoes. *Ghogue* 1081 was collected from 'disturbed plantation area in grassland', which alludes to the fact that the natural habitat of this taxon is being altered by farming practices at the higher elevations of the reserve, towards the grassland ridge. All of the reserve's native vegetation is thus under threat, and there are currently no efficient protective measures in place to monitor or prevent such activities. Without conservation-directed intervention, it is likely that this decline in habitat quality will continue unchecked and could contribute towards the extinction of several threatened plant taxa known to occur within the reserve (Pollard *et al.* in Harvey et al. 2004: 51–52)
Management suggestions: this tree should become a priority target for seed-collection and cultivation for introduction at other suitable sites nearby, especially in reforestation programmes, and could be used to highlight the urgent need for *in-situ* conservation measures. It has been suggested as a 'flagship' taxon for the conservation of the plants of the Bali Ngemba FR, which may be the last remaining patch of submontane forest in the whole of NW Region, 96.5% of the forest having been lost in the Bamenda

Highlands region (Cheek *et al.* 2000: 6). The forgoing data is modified slightly from Pollard *et al.* in Harvey *et al.* 2004: 51–52.

Magnistipula butayei De Wild. subsp. ***korupensis*** Burgt

CR D

Range: Cameroon endemic (SW: Korup NP).

The data in this species account is entirely taken from van der Burgt (2010 in press, Two new taxa in *Magnistipula* (Chrysobalanaceae) from Korup National Park, Cameroon, in Plant Ecology and Evolution).

Tree, to c. 30 m high, dbh to 74 cm; bole cylindrical or slightly fluted, to c. 15 m long; buttresses to 3 m high and extending to 0.5 m from the stem; bark brown, with rough, brittle flakes; stems hairy when young, hairs to 1.5 mm long; older stems glabrescent, with light grey lenticels; stipules in pairs, free, persistent; lanceolate, to 28 × 6 mm, base shortly petiolate, apex acute, margin entire but sometimes with a single lobe to 3 mm long, placed on the proximal side at the base; venation leaf-like, a single primary vein running nearer to the distal side, up to 5 glands evenly distributed across the stipule blade, sessile, 0.15 mm diam.; outer surface glabrescent, margins and veins of inner surface hairy with hairs to 1 mm long; leaves alternate, elliptic to slightly obovate, base (sub-)cordate, margin entire, apex acuminate; 8–15 × 3–7 cm, juvenile leaves to 25 × 9 cm with base auriculate; petiole 2–3 mm, hairy; both sides medium glossy when dried; 8–11 pairs of secondary veins, usually 10–15 mm apart in the central part of the leaf; hairs simple, to 1.5 mm long, upper surface of young leaves with a few erect hairs, becoming glabrous, the hairs leaving behind c. 3–6 small gland-like scars per mm^2; lower surface with hairs mostly on the veins, persistent; midvein impressed and hairy above, prominent and hairy below; glands 23–26 per leaf, distributed across the leaf blade but more common near the midrib, sessile; inflorescence a panicle, terminal or axillary, 4–13 cm long, axis with dense ferruginous hairs c. 0.2 mm long and sparse hairs to 0.8 mm long; pedicel 1–2(–4) mm long, receptacle 4 mm long, densely hairy, hairs to 0.2 mm long; sepals 5, triangular, hairy outside and inside, c. 2.5 × 1.5 mm, slightly unequal in size. Flowering in March.

This species was discovered in the "P transect plots" of the southern Korup NP during several years work there recently by van der Burgt. The total size of the plots is 155.75 ha. All trees over 50 cm dbh have been registered and identified. Of the total 3181 registered trees over 50 cm dbh, nine trees to *M. butayei* subsp. [illegible] dbh of between 10 and 50 [illegible] ha each (total [illegible] randomly [illegible] the P transect plots. Of the [illegible] registered tr[illegible] 10 and 50 cm dbh, five trees were identified as *M. butayei* subsp. *korupensis*. One of these 14 died, but two more trees were found just outside the plots. Globally therefore, only 15 individuals are known of this species, enabling the CR D assessment.

Habitat: rainforest rich in tree species from the Legume subfamily Caesalpinioideae, on well-drained sandy soil; 100 m asl.

Threats: none are known.

Management suggestions: the Korup Park Managers should be informed of the rarity of this tree and assisted in identifying it since they are responsible for its long-term protection.

Magnistipula conrauana Engl.

EN A3c

Range: Cameroon (SW: Lebialem-Bamboutos Mts (5 coll.); Mwanenguba (1 coll.); Kupe-Bakossi (4 coll. at 4 sites)).

The type species of the genus, presumable whence the generic name *Magnistipula* (big stipules) arises. Discovered near Fontem at Bangwa by Conrau in the late 19th century.

Shrub or tree to 12 m, branches glabrous, lenticellate; stipules foliaceous, ovate, base oblique, 3–4 × 2–3 cm; leaves ovate to oblong-elliptic, 22–25 × 10–12 cm, base acute to rounded, decurrent, glabrous, numerous glands towards base and under acumen, lateral veins 6–8 pairs; panicles spreading, terminal and axillary, to 30 cm, often subtended by foliaceous bracts, rachis glabrous; normal bracts 3 × 2.5 mm, triangular, margins glandular; bracteoles 1 mm, eglandular; pedicels 4–5 mm, receptacle obliquely campanulate, curved, 6–7 mm, glabrous outside, deflexed-villous within, sepals 4 × 2.5 mm, ciliate, tomentellous within, petals white to pale violet, 2–4 mm, stamens 7, 4–6 mm long, staminodes c. 7, tooth-like; ovary glabrous, style curved, 3–5 mm, fruit ovoid, 20 × 35 × 55 mm.

Letouzey and White (Fl. Cameroun 20 (1978)) record the collections for this taxon as restricted to the Bamboutos-Mwanenguba area. During the late 1990s we found the taxon also to occur, rarely although fairly widespread, at Mt Kupe and the Bakossi Mts. It is estimated here that this canopy tree will become extinct in the wild at Bamboutos and Mwanenguba (given the threats indicated below) in the next 100 years but has a good possibility of it surviving in the newly discovered southern part of its range at Kupe-Bakossi. EOO is calculated as 6227 km^2.

The assessment here, made in Cheek *et al.* (2004: 152–153), based on an estimated >50% population ongoing loss due to forest habitat loss in the last 90 years, is maintained since no new data are available apart from the record at Lebialem. A Missouri record from Korup (*Gentry* 52789) is regarded as very unlikely, being far outside the altitudinal range of the species. Two records from the Rumpi Hills by Thomas, are more likely, but these are based on field determinations only and need verification before they can be accepted.

Habitat: submontane forest, 1000–1500 m alt.

Threats: the Bamboutos Mts are densely populated and intensively cultivated, for such crops as *Coffea arabica*. What few fragments of forest remain are under great pressure to supply wood and for further agricultural land. The rate of forest area loss of 25% in eight years for the Oku area cited elsewhere is probably exceeded in the Bamboutos area, and this rate of loss appears to be continuing unabated.
Management suggestions: Letouzey & White (1978 *loc. cit.*) mention this plant occurring in hedges in farmland in Bamboutos. Encouragement of this practice, and also evaluating sacred forests there for the existence of this species, might be the means to survival in the northern part of the range. In the southern part of its range *M. conrauana* seems likely to survive if existing conservation plans go ahead for the Bakossi NP.

Magnistipula cuneatifolia Haum.

EN B2ab(iii)

Range: Cameroon (SW: Bimbia-Bonadikombo at Mt Cameroon; S: Abong Mbang), and N Gabon.
Large tree to 80cm diam. at base, leafy stems glabrous, slightly angular; leaf-blade oblanceolate, to 20 × 8 cm, acumen to 1.2 cm, base cuneate, lateral nerves 5–6 pairs ascending, glabrous; petiole 5–12 mm, transversely ridged; stipules falcate, 6–8 × 1 mm appressed hairy, then glabrous; panicles terminal, 20–25 cm, golden brown tomentellous, receptacle 6mm; sepals and petals 4–6 mm.
This tree (to 80 cm diam.) is known from only three sites, N Gabon (*Le Testu* 9376, type), Abong Mbang (*Letouzey* 3961) and a plot voucher specimen from the Mt Cameroon foothills at Mabeta-Moliwe, now known as Bimbia-Bonadikombo (*Cheek* in W 831, transect B, plot -1, June 1992, identified by Ghillean Prance). This tree seems extremely rare though fairly widespread.
A specimen record (*Ewango*) from Orientale Prov., Congo (Kinshasa) identified by the collector as this species is to be treated with caution since it differs in several features from other material of the species.
This species was assessed as CR A1c+2c by Cheek in Cable & Cheek (1998: xlix). This is modified here since although the site at Mt Cameroon has probably been lost due to smallholder market-gardening according to reports of Corcoran and Alicha in recent years, we do not have detailed information on threats at the other sites. Therefore the original assessment seems inappropriate and the species is now assessed as Endangered under criterion B since threats and three sites (AOO 12 km^2 with 4 km^2 cells) are known. EOO is calculated as 88976 km^2.
Habitat: lowland evergreen forest.
Threats: forest clearance for timber extraction and agriculture.
Management suggestions: although habitat loss in recent years in the Bimbia area is not encouraging, a further effort should be made to retrace this plant at the plot in which it was found in Mabeta-Moliwe (TB-1) in 1992, and thence to acquire the information necessary to protect this species at that site and to introduce it to Limbe Botanic Garden for multiplication and reintroduction at secure sites.

Magnistipula multinervia Burgt ined.

CR D

Range: Cameroon endemic (SW: Korup NP).
The data in this species account is entirely taken from van der Burgt (2010, in press Two new taxa in *Magnistipula* (Chrysobalanaceae) from Korup National Park, Cameroon, in Plant ecology and Evolution).
Tree, to 41 m high, dbh to 57 cm; stems hairy when young, hairs to 1 mm long; older stems glabrescent, with light grey lenticels; bud scales brown, up to c. 6, caducous; stipules in pairs, free, caducous, light green; oblong to ovate, to 12 × 4 mm, base 1.5–2 mm wide, apex acute, margin dentate, 10–20 parallel, longitudinal veins, the outer veins ending at the teeth on the margin; 0–2 sessile glands of 0.25 mm diam. at the edge near the base; outer surface, margins and base of inner surface densely hairy with hairs to 1 mm long, inner surface with hairs to 0.1 mm long. Stipules on lower part of the shoot c. rectangular; stipules on upper part of the shoot c. triangular. Leaves alternate, elliptic, base cuneate to obtuse, margin entire, apex acuminate; 7–10 × 2.5–4 cm; petiole 4–9 mm, hairy; both sides medium glossy when fresh, dull when dry; 8–13 pairs of secondary veins, usually 5–7 mm apart in the central part of the leaf; hairs simple, to 1.5 mm long, upper surface of young leaves with many erect hairs, becoming glabrous. Inflorescence a panicle, terminal or axillary, 8–12 cm long, 2–6 cm wide; densely hairy with ferrugineous hairs to c. 1 mm long. Flower pedicel free part 2–5 mm long, part fused with receptacle 4 mm long, densely hairy, flowers opening late Feb.
This species was discovered in the "P transect plots" of the southern Korup NP during several years work there recently by van der Burgt. The total size of the plots is 155.75 ha. All trees over 50 cm dbh have been registered and identified. Of the total 3181 registered trees over 50 cm dbh, 3 trees were identified to *M. multinervia*. Trees with a dbh of between 10 and 50 cm were registered in 56 subplots of 0.25 ha each (total size 14 ha), randomly located within the P transect plots. Of the 5755 registered trees between 10 and 50 cm dbh, 1 tree was identified to *M. multinervia*. Globally therefore, only four individuals are known of this species, enabling the CR D assessment.
Habitat: rainforest rich in tree species from the Legume subfamily Caesalpinioideae, on well-drained sandy soil; 100 m asl.
Threats: none are known.
Management suggestions: the Korup Park Managers should be informed of the rarity of this tree and assisted in

identifying it since they are responsible for its long-term protection.

Maranthes sanagensis F. White.

EN B1+B2ab (iii)

Range: Cameroon endemic (C: banks of the Sanaga at Nachtigal, 60 km SE Bafia; also 7 km N Olembé at 60 km N Yaoundé; Kikot bridge; banks of the river Ndjim 25 km ENE Bafia; confluence of Asamba and Sanaga)).

Small riverine tree 6–8 m, the trunk often twisted and prostrate; leaf-blade ovate to elliptic 8.5–15 × 4.5–9.5 cm apex acute-obtuse, abruptly cuspidate, base subcordate, lateral nerve 6–10 pairs, fine nerves reticulate, prominent below glabrous; petiole 5–8 mm; stipules not recorded; inflorescence c. 20-flowered, of simple or branched racemes with cymes 1–3-flowered, receptable obconic 8–14 × 4–5 mm; sepals suborbicular, 6–10 mm; petals 18 × 13 mm; stamens 50–60, ± free.

Here *Maranthes sanagensis* is assessed as Endangered. Although eight specimens are known from six sites all but one of these are along the Sanaga R so that a single event affecting a section of the middle course of that river could threaten all but one of the sites. Therefore two locations can be admitted, taking the Sanaga sites as one. AOO is 6 km² with 1 km² cells as recommended currently by IUCN for aquatic species. EOO is calculated as 4704 km².

Habitat: evergreen riverine forest; 400–500 m alt.

Threats: the species is not immediately threatened with extinction, but some habitat loss has occurred and is ongoing due to sand extraction from the Sanaga for construction in Yaoundé (Cheek pers. obs. 2006).

Management suggestions: basic populational data should be gathered as a baseline for future monitoring and local communities such as fishermen and sand extractors, educated as to the existence and rarity of this tree.

COMBRETACEAE

assessed by Martin Cheek, advised by Carel Jongkind

The Combretaceae are characterised by their hypanthoid flowers, the receptacle forming a cup or tube above the inferior ovary, to the lip of which the free petals and calyx lobes are attached. In the absence of flowers they can sometimes be recognised by their winged fruits or their opposite and subopposite leaves which often have minute glands on the lower surface. However there are taxa with alternate leaves, and those lacking glands.

The Flore du Cameroun account (vol. 25 (1983), Liben) recognises nine genera: *Strephonema*, *Guiera*, *Pteleopsis*, *Combretum*, *Quisqualis*, *Terminalia*, *Laguncularia*, *Conocarpus* and *Anogeissus*. The last three are salt-tolerant and mainly associated wth mangrove, while the first is usual in freshwater swamp and riversides. The remainder grow mainly in dry habitats such as woodland or semi-deciduous forest. They are less usual in evergreen forest. *Terminalia ivorensis* and *T. superba* are timber trees of forest but otherwise there are few uses for the family in Cameroon. With 35 species, *Combretum* is the largest genus in Cameroon and is the only one with threatened species. Most *Combretum* are lianas, usually hairy, often with scales or glands on the lower surface of the subopposite leaves which often articulate above the petiole base which, when the blade falls, often grows into a spine. The two species here do not appear in Flore du Cameroun, being described later, in the 1990's, by Carel Jongkind of Wageningen, who has a long-term project to monograph the African species.

Combretum exellii Jongkind

EN B2ab(iii)

Range: Cameroon (SW: Mt Cameroon area, Barombi Kang; 5 km S Ediki; S, Bakundu FR) and Gabon (Cristal Mts; Bellvue; NNW N'djolé).

The six sites equate to three locations (AOO 24 km² with 4 km² cells for each site) and since threats are known, *Combretum exellii* is here assessed as Endangered.

Large liana, leaves oppposite, with scales, blade to 16 × 8 cm, acuminate, leathery, above matt dark green, below covered with a continuous yellow indumentum, lateral nerves 7–9 pairs, tertiary nerves scalariform highly prominent; panicles axillary, 8 cm, peduncles 0.5–1.5 cm, bracts inconspicuous; flowers sessile, 4-merous, outer surface with scales and hairs, ovary-receptacle 1.5–3 mm; hypanthium saucer-shaped; petals obcordate, 1.5 mm, glabrous, yellow; fruits 4-winged.

The indumentum on the lower leaf-blade is distinctive.

Habitat: lowland evergreen to semi-deciduous forest.

Threats: forest clearance in the lowlands between Kumba and Limbe has been extensive to provide timber and agricultural land for small-holders, cocoa, yams, bananas, rubber and oil palm.

Management suggestions: the prospects of rediscovering this species in Cameroon do not look promising since all known sites are around Kumba which has seen great losses and degradation of forest (above) and no recent records have been seen. However it is possible that it survives in some remaining forest and so searches should be made. Basic populational data should be collected to aid future monitoring. Propagation from seed for introduction to safe sites is advised since *Combretum* species commonly produce large numbers of fruit.

Combretum echirense Jongkind

EN B2ab(iii)

Range: Cameroon (S: Bipindi) and Gabon (Rabi-koung. Asanta (3 coll.); 10-20 km N N'djolé).

Here *Combretum echirense* is assessed as Endangered since effectively, three locations are known (AOO 12 km² with 4 km² cells) and with threats as below.
Large liana with scales, stems grey, with sparse short hairs; leaves opposite, elliptic, to 13 × 6 cm, papery to leathery, mat, glabrous, lateral nerves 3-6 pairs, often with pocket domatia, acuminate, base attenuate; flower-spikes to 8 cm axillary or terminal, bracts inconspicuous; flowers 4-merous, receptacle 4 mm, shortly hairy; upper receptacle bell-shaped, 1.5 mm wide.
Closely similar to *C. fuscum*, differing in the free edge to the floral disc.
Habitat: lowland evergreen forest.
Threats: slash and burn agriculture; infrastructure for petroleum extraction (Gabon).
Management suggestions: Bipindi forest should be searched for this taxon, taking care not to confuse it with the apparently similar *C. fuscum*. Although not seen at Bipindi in 100 years it may well survive there yet since good forest still exists around the town and little general collecting has occurred there in intervening years. Basic populational data should be collected and local leaders advised if the species is rediscovered, so that future monitoring and protection can be facilitated.

COMPOSITAE (ASTERACEAE)

assessed by Martin Cheek

One of the world's most important plant families in terms of numbers of species, the Compositae rival legumes, orchids, grasses and euphorbs in this respect. Composites are easily recognised by their capitula – heads of flowers that at first resemble a single flower but are made up of tens or hundreds of reduced flowers known as florets. Florets at the outer edge of the capitulum may have their corollas vastly expanded as a rectangular or strap-like, brightly coloured structure (a `ligule') which gives the capitulum the appearance of having free petals. These are known as 'ray-florets' as opposed to the regular, more centrally placed 'disc-florets'. The inferior ovary gives rise to a single-seeded nutlet or achene which is often distributed by wind due to the whiskery calyx which extends as the fruit develops into a feathery tuft in many species; in others the calyx is barbed and the fruits are dispersed by latching onto passing animals, e.g. in *Bidens*. The leaves of composites are generally simple, alternate, lacking stipules, and often scented of spice or resin, and are densely hairy. Some species produce a white exudate when wounded, e.g. *Lactuca.*

Ndolé, an important Cameroonian food, is made from the pounded leaves of a native, but usually cultivated species, *Vernonia amygdalina*, or "bitter leaf". Lettuce, *Lactuca sativa*, a European import, is also cultivatedas a salad food, as are many ornamental species, commonly known as marguerites or daisies. *Pyrethrum* is the source of important insecticides of that name, while many other genera provide locally important medicines or food flavourings.

Cameroon and West Africa's most noxious weed of cultivation, *Chromolaena odorata*, a native of S America, is a composite. Several other important weeds of farmland likewise belong to this family.

Composites are entirely absent from intact evergreen lowland forest appearing only in disturbed or edge areas of this habitat in Cameroon. They are at their most diverse in grasslands, such as the montane grassland of the Cameroon highlands. Composites are also numerous in the subsahelian grasslands of the N of Cameroon.

Fortunately many of the rare Cameroon Highland species of the family were worked out by Charles Dennis Adams in the 1960s, and his FWTA account in volume two of that work, in 1963, is very useful. Regrettably there is no Flore du Cameroun account for this important family, so areas of Cameroon outside SW and NW regions have no guide to identifying species, nor are there recent or readily available revisions for most of the genera in Cameroon, and so identifications of specimens of all but the commonest species, is difficult without the help of composite specialists and reference material available in Europe.

Aedesia engleriana Mattf. is believed to be described from Cameroon but is not included in FWTA, nor have we seen any material or data relating to it, here assessed as DD.

The following taxa appear restricted to the Adamawa/Cameroon Highland areas in Nigeria and Cameroon and are here assessed as NT since slightly more than 10 locations are known, with threats: *Echinops lanceolatus* Mattf., *Echinops mildbraedii* Mattf., *Echinops giganteus* A.Rich. var. *lelyi* (C.D.Adams) C.D.Adams and *Vernonia iturensis* Muschl. var. *occidentalis* (C.D.Adams) C.Jeffrey

Crassocephalum bauchiense (Hutch.) Milne-Redh. previously assessed as VU B2ab(iii) in Cheek *et al. 2004*, Harvey *et al.* (2004: 62), Cheek *et al.* (2010: 96) is here reassessed as NT since its range is now believed to extend to Kivu and Uganda.

Vernonia acrocephala Klatt is a submontane pyrophyte of Guinea (Conakry), Sierra Leone, the Cameroon Highlands (including Nigeria), Angola, Congo (Kinshasa), Tanzania, Zambia and Malawi. Despite the wide range it is rare. Since more than 10 locations are known, for the moment it is here assessed as being NT.

Vernonia calvoana (Hook.f.) Hook.f. subsp. *calvoana* var. *calvoana* was assessed as VU D2 in Cheek *et al.* (2004: 154) being known from only three locations in the Cameroon Highlands and there is not new data available to change this rating. However, varieties of subspecies are not being considered in this treatment.

Vernonia guineensis Benth. var. *cameroonica* C.D.Adams was assessed in Harvey *et al.* (2004: 62) as VU B2a,b(iii) on the basis of its cited range being Nigeria (Gashaka, 1 coll.) and Cameroon (NW: Ndop; Bamenda-Banso; Bamenda; Bambuluwe; Nchan; Bali Ngemba; Bambili (1 coll. each). C: Ndikiniméki (1 coll.). W: Bangwa (1 coll.). This is a pyrophyte is restricted to the Bamenda Highlands of NW Province, Cameroon, apart from a few outlying areas. However difficulty was experienced identifying to variety material of this species in the course of this project and so until this is resolved, the taxon is not included in the treatments below.

Aedesia spectabilis Mattf.
VUB2ab(iii)

Range: Nigeria (Mambilla Plateau), Cameroon (NW: Ndop; Mbaw; Mbaw-Mbiami. W: Lacs Paponoun. Adamawa: Meiganga-Betaré) and CAR (Bouar; Boheng; Koumbala):
A 60 cm tall erect, few stemmed subshrub from a cluster of carrot-like root tubers. The terminal capitulum is 5-7 cm diam. with white-felted bracts. The numerous flaccid hairy, cauline oblong-oblanceolate leaves, c. 30 × 3 cm, have 3 mm marginal teeth. Discovered by Mildbraed in 1914 at Bouar, this spectacular, bromeliad-like plant is here assessed as Vulnerable since nine locations (AOO 36 km^2 with 4 km^2 cells) are known, with threats as below. *Fay* 4864 records the plant as used in CAR for children's constipation (ngendekoro in Suma).
Habitat: grassland and savanna; 600–1200 m; fl.fr. May–July(–Nov)
Threats: cattle trampling, grazing, and pasture management, e.g. frequent burning, are threats to this species, it is believed.
Management suggestions: none of the known locations appears to be in a protected area so, near locations for the species it is important to help community elders to understand the rarity and need to protect this plants, and to help them identify it. Data on local range, density, regeneration levels and life-cycle should be acquired to form a baseline for monitoring. Cattle exclusion experiments would help better gauge the threat from this source.

Bafutia tenuicaulis C.D.Adams subsp. ***tenuicaulis***
VU B2ab(iiii)

Range: Nigeria (Mambilla Plateau (3 coll.)) and Cameroon (W: Bamboutos Mts (2 coll.). NW: Bamenda Highlands (several locations, 5 coll. pre-1998); Mbam (1 coll.)).
This annual herb is the only species of a genus (named for the Bafut-Ngemba Reserve near Bamenda where it was first discovered) restricted to the Bamenda Highlands and their outliers. It seems fairly specific to rock outcrops. Within Mt Oku and the Ijim Ridge (Cheek *et al.* 2000), most of our specimens were collected at Oku-Elak during November of 1996: *Munyenyembe* 825, 878, *Zapfack* 1199, near the summit (*Zapfack* 1130), at Ewook Etele Mbae (Oku-Kumbo 10 km *Zapfack* 1220, 1224), or at Shambai near Oku Elak (*Munyenyembe* 878). One of these collections was from farm fallow, showing that the species is capable of adapting to disturbance. At Ijim, *Bafutia* was recorded from an outcrop near the "Mbesa Swamp" in December 1998. According to Pollard (*Pollard* 358) 20 plants occurred at this site where it is threatened by cattle trampling and the taxon is otherwise extremely rare there. A large population of c. 60 plants was discovered on an inselberg at Mbingo in November 1999 (*Cheek* 10060).
This taxon was assessed as LR/nt in Cheek *et al.* (2000: 58) but it is here reassessed as Vulnerable since less than ten locations are known (above; AOO 36 km^2 with 4 km^2 cells), with threats as below.
Annual herb 3–15(–30) cm tall, glabrous. Stems c. 1 mm diam., terete. Leaves sessile, mostly basal, linear-oblong, up to 4 × 0.4 cm, apex rounded, margin subentire. Inflorescence a diffuse corymb forming the apical half of the plant, capitula 3–15; peduncles filiform, 2–3 cm long, lacking bracts. Capitula 3–5 mm wide; involucre with 5–8 bracts united in a cup with 5–8 acute, erect lobes, c. 1 × 1 mm. Ray florets absent, disc florets 12–15, purple.
Habitat: rock outcrops; 1800–2800 m alt.
Threats: trampling and grazing by cattle (see above)
Management suggestions: a survey is needed to find all the sites for this taxon at Kilum-Ijim, and to confirm that it is indeed specific to rock outcrops. Information on its palatability to livestock would be useful. A survey of basic populational data would serve as a baseline for future monitoring.

Bafutia tenuicaulis C.D.Adams subsp. ***zapfackiana*** Beentje & B.J.Pollard
EN B2ab(iii)

Range: Cameroon endemic: (NW: Mt Oku, summit and lower slopes above Elak).
Previously assessed as VU D2 (Cheek *et al.* 2000: 58), it is reassessed here as Endangered since only two closely proximal locations are known at each of which only a single

collection is known (above; AOO 8 km^2 with 4km^2 cells), with threats as below.

Differing from they type subspecies (see above) in that the stem, leaves and especially pedicel and involucre are pilose to pubescent, noty glabrous; the involucre base is much larger (2.5-6.5 mm), there are 8–14 involucral lobes (as opposed to 8), with paler zones of tearing below sinus (no zones) and the pappus setae are 0.5-0.8 mm long as opposed to 0.1-0.5 mm. (Beentje *et al.* in Cheek *et al.* 2000: 121).

The taxon seems so distinctive that it might be raised to species level.

Habitat: rock outcrops, fallow; 2000–2800 m alt.

Threats: trampling and grazing by cattle, goats and sheep in the summit area of Mt Oku, clearance for agriculture at lower altitudes.

Management suggestions: a survey is needed to determine is there are other sites for this taxon at Kilum-Ijim, as is likely, and to confirm its ecological requirements. A survey of basic populational data would serve as a baseline for future monitoring. In view of its apparent rarity, seedbanking, and public education to explain the need for the protection of this plant is advised.

Bidens mannii T.G.J.Rayner (syn. *Coreopsis monticola* (Hook.f.) Oliv. & Hiern)
VU B2ab(iii)

Range: Cameroon endemic (SW: Mt Cameroon (10 coll.); Mwanenguba; Kupe Rock. W: Bamboutos (1 coll.); Lake Aweng (1 coll). NW: Santa Mt (1 coll.); Mt Oku (2 coll.)).

The assessment of this species in Cheek *et al.* (2004: 153) is maintained here since only seven locations are known (above; AOO 28 km^2 with 4 km^2 cells), with threats as below.

This floriferous, robust herb looks like a garden escape, and is certainly worthy of horticultural attention. It is abundant, common and secure on old lava flows on Mt Cameroon, but elsewhere it appears rare and precarious (see threats below). In several fairly intensively investigated areas of the Cameroon Highlands this species has not been recorded and is apparently absent (Harvey *et al.* 2004, Harvey *et al.* 2010, Cheek *et al.* 2010).

Herb to 2 m tall; young stems and leaves reddish; cauline leaves deeply 3(–5)-lobed, to 6 × 4 cm, the lobes deeply toothed to lobed; capitula pale-yellow, c. 4.5 cm diam., rayed, rays elliptic, 2 × 1 cm; involucre of 5 inner broadly triangular hairy bracts, c. 1 cm long, and 5 outer narrowly oblong glabrous bracts, c. 2 cm long.

Habitat: grassland, near forest, often at well-drained sites; (1000–)2000m alt.

Threats: cultivation of habitat for crops e.g. at Lake Aweng, Santa and Bamboutos, is a threat.

Management suggestions: this species is probably secure at Mt Cameroon where human pressures in montane grassland are currently low and where these species appears most frequent. Conservation efforts might best be focussed here. A survey to quantify population structure here is advised to form a baseline for future monitoring.

Crassocephalum bougheyanum C.D.Adams
VU B2ab(iii)

Range: Equatorial Guinea (Bioko (3 coll.)) and Cameroon (SW: Mt Cameroon (14 coll.). SW/W: Lebialem/Bamboutos Mts (1 coll.). NW: Mt Oku and Ijim Ridge (7 coll.); ?Dom (photo. record only)).

Crassocephalum species in the Cameroon Highlands appear to be annuals which grow through the wet season and flower and fruit at the beginning of the dry season. Intermittent forest disturbance favours these species by creating "light gaps" in which they flourish. However, they do not survive in open grassland or scrub, and so are forest-dependent species. *Crassocephalum bougheyanum* is distinguished from its congeners by the stems bearing usually just a single, very large capitulum. It was named by C.D.Adams in 1957 in honour of Prof. Boughey of Ghana, a pioneer of the study of the flora of upland Cameroon. Segregated from *C. vitellinum* (Benth.) S.Moore by Adams, *C. bougheyanum* is a larger plant, 90–180 cm tall, with a more restricted distribution. *C. montuosum* is another closely related species and care should be taken to avoid confusion in identification. Locally it appears a relatively common species in the few locations in which it occurs, but it appears absent from some parts of the Cameroon Highlands which have been fairly intensively investigated, not being recorded from Bali Ngemba for example (Harvey *et al.* 2004).

Previously assessed as LR/nt (Cheek in Cable & Cheek 1998: xlix, Cheek et al. 2000: 58–59) it is here reassessed as Vulnerable since only six locations (above; AOO 24 km^2 with 4 km^2 cells) are known, with threats as below.

Annual herb 90–180(-250) cm tall, glabrous, lacking exudate. Stems c. 8 mm wide at midheight, internodes 5–15 cm, unbranched or very sparsely branched. Leaves membranous, those of the mid–stem ovate-trullate, 5–9 × 2–5.5 cm, apex acute, base obtuse-decurrent, unequal, margin bidentate; petiole 2–3 cm. Inflorescence of a single (rarely 2–3), terminal capitulum c. 2.8 cm wide, orange-yellow.

Habitat: clearings and at the edge of montane forest; 1500–2500 m alt.

Threats: forest clearance for agriculture, c. 50% of surviving forest was lost in 15 years recently (1998-2003) in the Dom area of the Bamenda Highlands (Baena in Cheek et al. 2010: back cover).

Management suggestions: this species is dependent on forest protection, which seems best assured at Mt Cameroon, Mt Oku and Ijim, and if confirmed as being present, at Dom. Some disturbance is beneficial to the species. So long as these conditions are maintained the species is likely to persist, however baseline population studies against which future monitoring can be done are suggested.

Helichrysum biafranum Hook.f.

CR B2ab(iii)

Range: Cameroon endemic (SW: Mt Cameroon, 2 coll.).

Known from only two collections. The type, *Mann* 1934 *pro parte* (a mixed collection) is recorded as "10' tall, 7,000', Dec. 1862" and is probably somewhere in the vicinity of Mann's Spring, since his letters to Hooker in the Kew archives show that he was camped there then. The second collection, *Boughey* in GC 12656), was made on 11 Dec. 1952: "Woody herb in second tongue of forest from Mann's Spring". Recent inventories have not found this plant, but since they have not been conducted in this habitat in December (when *H. biafranum* seems to flower) insofar as is known, the species may have been overlooked.

This species was assessed as VU D2 by Cheek in Cable & Cheek (1998: xlix–l) but is here reassessed as Critically Endangered since known from a single location (above; AOO 4 km² with 4 km² cells) are known, with threats as below.

An erect, slender, branched herb to 3 m high with narrowly winged aerial stems bearing linear leaves, less than 2 cm wide, subscabrid below, inflorescence branches bearing few or no leaves and numerous bright yellow flower heads (including bracts) 6-7 mm diam.

Habitat: edge of montane forest; 2100–2200 m alt.

Threats: a lava flow over Mann's Spring could destroy this species, if it is as localised as it seems.

Management suggestions: high priority should be given to finding this species, preferably searching in December, in the forest:grassland ecozone. An attempt to investigate the range and numbers of individuals of this species (is it really restricted to only Mann's Spring?) is important for conservation planning. If the species does occur at this ecozone, it is to be hoped that its range extends to cover a larger part of the grassland-forest perimeter than just that near Mann's Spring.

Helichrysum cameroonense Hutch. & Dalz.

EN B2ab(iii)

Range: Nigeria (Chappal Waddi (3 coll.)) and Cameroon (SW: Mt Cameroon (10 pre-1988 coll.). SW/W: Lebialem/Bamboutos Mts (1 coll.). NW: Mt Oku (1 coll.); Bafut-Ngemba (1 coll.), several other locations in Bamenda Highlands. Apparently absent from Bioko (Equatorial Guinea).

Assessed as LR/nt by Cheek in Cable & Cheek (1998: l) and Cheek *et al.* (2000: 59), it is here reassessed as Endangered since known from a six locations (above; AOO [illegible] with 4 km² cells) are known, with threats as below.

H. cameroonense appears to be a biennial. In the first year it produces a rosette resembling a small *Dendrosenecio*. In the second year, after vertical growth, the stem produces in a dense mass up to 50 capitula, each about 2.5 cm diam. These large plants are highly conspicuous and in local areas relatively common scattered in sparse grassland at about 2000 m alt., e.g. on Ijim Ridge between Anajua and Jikijem. This species appears to recolonise bare, rocky ground (see e.g. account in Cable & Cheek 1998: l). The fact that only two specimens of this species were gathered in the 1996 and 1998 surveys of Mt Oku may reflect the fact that this plant is time-consuming and awkward to collect, being so bulky, rather than rarity.

Biennial (?) herb 1–2 m tall. Stem woody at base, 2.5 cm diam., unbranched, densely clothed by leaves. Leaves at mid-stem level sessile, ovate-lanceolate, c. 14 × 3 cm, apex acute, margin finely serrate, aromatic, viscid white-woolly below. Inflorescence a dense, terminal, fairly flat assemblage of 20–50 capitula forming an aggregation 10–20 cm diam. Capitula 2–3 cm diam., disc dull orange, ray involucral bracts glossy pale yellow.

Distinguished by being larger in most parts from the similar *H. foetidum* which has stem diam. c. 5 mm, leaves c. 1.8 cm wide, capitula c. 1.5 cm diam.

Habitat: montane grassland, particularly on broken rocky ground; 1900–3000 m alt.

Threats: lava flows on Mt Cameroon; cattle grazing and trampling, and burning associated with cattle management, at all locations other than Mt Cameroon where grazing by domestic stock is absent.

Management suggestions: gathering of basic populational data at the known sites would facilitate future monitoring, acting as a baseline.

Helichrysum mannii Hook.f.

EN B2ab(iii)

Range: Equatorial Guinea (Bioko (2 coll.)) and Cameroon (SW: Mt Cameroon (11 pre-1988 coll.)).

Previously assessed as LR nt by Cheek in Cable & Cheek (1998: l), reassessed here as Endangered since only two locations (above; AOO 8 km² with 4 km² cells) are known, with threats as below.

The characters distinguishing this species from the similar *H. cameroonense* are listed there. This pair may be altitudinal vicariants. Certainly *H. mannii* occurs at much higher altitudes than *H. cameroonense*, e.g. *Dundas* in FHI 20362: "within 50 feet of the mountain top". The lower altitudinal limit of *H. mannii* is not completely clear from the data available. Mann certainly confused the two species. Vegetatively they may be almost indistinguishable although the differences in indumentum should help discriminate between them. The absence of specimens of *H. mannii* in the 1992 inventory of Mt Cameroon probably indicates the low coverage of altitudes above 2400 m. Hannah Banks (Lava-flow expedition 1995) collected this species above Buea "three-quarters of the way to Hut 3, in rocky places".

A stout herb 30–60 cm high with a pithy stem and crowded sessile, amplexicaul, ovate-lanceolate 12 × 3 cm, leaves, scabrid above, woolly-tomentose below, flower heads clustered, involucral bracts broadly ovate, subacute, silvery-white, heads 1.5–2 cm diam., florets cream to yellow.

Habitat: montane grassland, particularly on cinders at high altitude; 2400–4000 m alt.

Threats: lava flows on Mt Cameroon: eruptions occur every 20 years, roughly, and are particularly prevalent in the area occupied by this species.

Management suggestions: a survey to record basic populational data would form a baseline for future monitoring.

Mikaniopsis maitlandii C.D.Adams

EN B2ab(iii)

Range: Nigeria (Chappal Waddi (2 coll.)), Equatorial Guinea (Bioko (1 coll.)) and Cameroon (SW: Mt Cameroon (5 coll.); Bakossi-Kupe (2 coll. at 2 sites)).

This climber of gaps in cloud forest was previously assessed (Cable & Cheek 1998: li) as LR nt and in Cheek *et al.* (2004: 153–154) as VU B2ab(iii). Here it is reassessed as Endangered since only five locations (above; AOO 20 km² with 4 km² cells) are known, with threats as below. On Mt Cameroon it was not detected during the intensive surveys of the early 1990s, so is at least very rare there.

Shrubby climber with purplish stems; leaves fleshy, ovate and cordate, to 8 cm long; 5–7-veined from base; capitula in dense clusters, without rays; florets yellow.

Habitat: montane forest gaps, sometimes with *Mimulopsis solmsii*; 1000–2200 m alt.

Threats: forest clearance for agriculture and wood, particularly likely at the lower part of its altitudinal range at sites such as Chappal Waddi and Mt Cameroon (plantation expansion to the 1000 m contour is a major threat).

Management suggestions: this rarely-seen species is perhaps most easily refound and studied at the sites where it has most recently been recorded, both in Bakossi. Data on demography and population size would better inform management decisions. In the meantime, education of the public by means of e.g. conservation posters might help reduce pressure on subpopulations that survive.

Mikaniopsis tedliei (Oliv. & Hiern) C.D.Adams

VUB2ab (iii)

Range: Guinea (Conakry) (Simandou), Sierra Leone (Loma; Tingi Hills), Ghana (Hemang, Pra R (numerous coll)), Cameroon (SW; Mt Cameroon (numerous coll.) at Mann's Spring and 'Litoka'), Congo (Kinshasa) (Kimoha; Bubulu; Libulu) and Uganda (Bugoma, Masaka Distr.). A false record for Togo appears on our map.

Assessed in FTEA Compositae part 3 as LC ("Widespread, least concern") it is reassessed here since although widespread, only nine locations are known (above, AOO 36 km² with 4 km² cells) with threats as below.

Woody climber, branches striate, puberulous and glandular when young, then glabresent; leaves fleshy, alternate, c. 6 × 5 cm, acuminate, margin repand-dentate, petiole 3–6 cm; capitula numerous in 10–15 cm, slender, axillary and terminal corymbs; capitula scattered, white; involucral bracts of the inner series c 12, 1 cm long, long-acute, slightly exceeded by the florets.

Habitat: submontane or montane forest: grassland transition; 900–2000 m alt. (Cameroon)

Threats: unknown in Cameroon, but an open cast iron ore mine threatens the only known location in Guinea (Conakry).

Management Suggestions: further investigation on the delimitation of this species is needed since the Ugandan material is rather different to that of W Africa in several characters (leaves cordate at base, larger, tormentose below). Population data should be recorded to enable future monitoring at Mann's spring. Plants are most frequent at Pra R in Ghana and 200 m on Mt Cameroon, using specimen records as an indicator. Conservation efforts therefore might best be focussed in these areas.

Mikaniopsis vitalba (S.Moore) Milne-Redh.

VU B2ab(iii)

Range: Cameroon (SW: Nyasoso, Mt Kupe (1 coll.). E: Diang, 40 km W of Bertoua (1 coll.). C: Yaoundé (2 sites), Uganda (2 sites), Gabon, Angola and Congo (Kinshasa) (1 coll. each).

The assessment in Cheek *et al.* (2004: 154) is maintained here since no new data are available.

This rare climber, although very widespread, is known only from nine locations scattered over five countries. More work is needed, ideally a revision of the genus, to confirm that the specimens concerned all do belong to the same taxon.

Climber to 8 m, glabrous; stems terete, ridged; leaves ovate, acumen 0.5 cm, base oblique, entire, 3-nerved from base; petioles 10 cm; racemes 15 cm; pedicels 1.5 cm; capitula without rays; florets white, 7 mm.

Habitat: lowland gallery, swamp, submontane forest or savanna; 700–1600 m alt.

Threats: forest clearance for agriculture and wood, and (e.g. Yaoundé area) urban expansion.

Management suggestions: more material is needed of this taxon from Nyasoso (*Etuge* 2177) to confirm that it is indeed *M. vitalba*. Although in *Mikaniopsis* it is best placed here, it differs from the specimens at other sites in having leaves rounded (not cordate) at the base and more than twice as long as broad, and elongated, not congested, racemes, and in growing at 1600 m, not 700–750 m alt. Site surveys are required to investigate subpopulations and to gather data on demography and recruitment. This taxon should be looked for in ongoing surveys at the proposed National Park at

Mefou, which might form a secure base for the taxon. The Nyasoso site is probably secure.

Senecio burtonii Hook. f.

VUB2ab (iii)

Range: Cameroon endemic (SW: Mt Cameroon, 29 coll. W: Bamboutos Mts (1 coll.). NW: Bamenda Highlands at Mba Kokeka (1 coll.); Mt Oku (4 coll.); Ijim (2 coll.); Santa (1 coll.)).

Named for the adventurer Richard Burton by Joseph Hooker, this species is here assessed as Vulnerable since there are six locations (above; AOO 24 km^2 cells) and threats as below.

Herb to 1.5 m, stem hard with longitudinal ridges, silvery green and dark red, leaves alternate, simple, clasping stem, blade white below, margin revolute; capitula terminal and axillary, yellow; ligules c 13; bracts green; flowering Oct.–Jan. and April.

Habitat: montane forest-grassland transition; 2100–3000 m alt.

Threats: clearance of montane forest, especially in the Bamenda Highlands where 96.5% is estimated as lost already (Cheek *et al.* 2000: 6). In addition cattle trampling and grazing, and fires set by pastoralists may explain why the species is so rare at locations where such grazing occurs, while so common where it does not (this being Mt Cameroon with 29 of the 42 records of the species).

Management suggestions: montane forest protection in W and NW Regions would help protect this species. A cattle exclusion study might resolve the impacts of this regime on the species. Data on life-cycle duration, regeneration, frequency and local range would help form a baseline for future monitoring. Since Mt Cameroon has the densest population (mainly at Hut 2 Buea), preservation efforts might best be followed there.

Vernonia bamendae C.D. Adams

VU B2ab(iii)

Range: Nigeria (Mambilla Plateau, Chappal Waddi (1 coll.)) and Cameroon (NW: Bamenda Highlands (6 coll.). W: Foumban).

Previously assessed as VU D2 (Cheek *et al.* 2000: 59–60) here it is reassessed as VU B2ab(iii) since six locations are known (above; AOO 24 km^2 with 4 km^2 cells) are known, with threats as below.

Dennis Adams described *Vernonia bamendae* citing *Maitland* 1514 (Laikom, on grassy hill slopes, 6000', May 1931) as type specimen (Adams 1957). This remains the only specimen known from the Kilum-Ijim area. Earlier in his number series, Maitland had collected two other specimens of this species from other parts of the Bamenda Highlands. *Maitland* 1455 is labelled "Bamenda, Basenako, 5,000', on stony grassy slopes". *Maitland* 1457 is labelled "Nchan, June 1931". *Keay* in FHI 28410 collected the fourth specimen from 7,400' on the NW slopes of Mba Kokeka near Bamenda. *Boughey* in GC 11013, from Santa Mt. at 2500 m followed on 25 Dec. 1952. *Daramola* in FHI 46649 was the last coll. from the Bamenda area (Bamenda, 27 Jan. 1957). The first and only collection from Nigeria was made from Chappal Wadi on the border with Cameroon (*Tuley* 2041, 7,500', 19 November 1969). The most recent collection of *Vernonia bamendae* was on 1 November 1974 from Mbam, 2335 m alt., 35 km NW of Foumban in W Region (*Letouzey* 13100).

It is curious that despite collecting expeditions in 1996, 1998 and 1999, we have not rediscovered this species from its type locality in the Ijim area; nor has it been discovered at any of our surveys in the Bamenda Highlands generally (1996–2006). This species seems to flower throughout the dry season, from early November until June, during part of which (Nov. and Dec.) we have conducted fieldwork in the area. The explanation for our not rediscovering the species may be that it is not only narrowly endemic, but, within its small range, extremely rare and possibly declining.

Stout herb (1–)1.5(–2) m tall, probably perennial. Stems c. 1 cm diam., densely brownish tomentose, finely ribbed. Cauline leaves triangular, c. 12 × 5 cm, apex rounded, base cordate or cordate-auriculate, lower surface with raised reticulate veins, pale grey-brown pubescent. Inflorescence spreading. Capitula c. 1.5 cm wide, florets mauve or purple.

Habitat: grassy slopes, sometimes near forest boundary or in wet hollows; 1500–2500 m alt.

Threats: grazing, trampling by cattle and fire set by graziers are likely to be important factors in threatening this species.

Management suggestions: this species still awaits rediscovery despite numerous surveys in the last 10–15 years within its range. It may be extinct in much of its range, in which case consideration might be given to its reintroduction, providing that a source of seed-bearing plants can be found.

Vernonia chapmanii C.D.Adams

VU B2ab(iii)

Range: Nigeria (Mambilla Mts) and Cameroon (Adamawa: 32 km SSE Kontcha; 2km W Banyo; 50km SW Banyo; 30km NNE Meïganga; Niadaba, Yoko; Djouroum 13km S, Ngaoundéré).

Here *Vernonia chapmanii* is assessed here as Vulnerable since only seven locations (above; AOO 28 km^2 with 4 km^2 cells) are known, with threats as below.

A perennial herb from a woody rootstock, stems leafy, 30–50 cm tall, leaves leathery, suberect, reticulate; capitula subsessile, solitary, partly hidden in uppermost 5–6 leaves, involucral bracts containing purple florets, achenes pubescent.

Habitat: grassland or wooded grassland (savanna) with *Hyparrhenia* and *Terminalia glaucescens*; 100–1600 m alt.

Threats: grazing, trampling and frequent fires resulting from graziers and their animals; cultivation of habitat.
Management suggestions: inclusion of this species in a National Park is desirable, but if this is not possible, elders in communities near locations for the species should be informed of the rarity of the species and advised to protect it. A survey of basic populational data would serve as a baseline for future monitoring. In view of its rarity, seedbanking, and public education to identify the species and explain the need for the protection of this plant is advised.

CONNARACEAE

assessed by Martin Cheek

'The Connaraceae, a Taxonomic Study with emphasis on Africa' by Jongkind and Lemmens (Agric. Univ. Wageningen Papers 89-6, 1989) is the basis for the treatment below. Jongkind and Lemmens reduced the number of species recognised in Africa from the 154 in 16 genera previously accepted, to 49 species in 10 genera.

In Cameroon most species are woody lianas of forest although *Jollydora* is a small tree), with pinnately compound leaves (except *Agelaea* which is trifoliolate). They resemble Leguminosae-Papilionoideae when flowers or fruits are absent, but lack the stipules of that last group, so with careful examination, can be distinguished. Unlike papilionoid legumes they have several, not one, carpels, and actinomorphic, not zygomorphic flowers.

The absence of a Flore du Cameroun account is mitigated by the work above.

Cnestis macrantha Baill.

EN B2ab(iii)

Range: Nigeria (Oban FR 9 coll.; Old Calabar R), Cameroon (SW: Mamfe; Korup NP) and Gabon (Mondah Forest near Libreville).
Liana, stems densely yellow-brown pilose; leaves 10–17-jugate, petiole 0.5–4.5 cm, rachis 11–22; leaflets papery, opposite, elliptic or oblong, 1–5 × 1–1.5 cm, rounded, truncate (-subcordate) at base; petiolules to 1 mm; racemes or panicles single in young leaf axils near branch apex, 5–24 cm long, to 40-flowered, densely yellowish pilose; sepals 5–6.5 mm; petals obovate or obcordate, 2–3 × 1.3–2 mm; fruit follicles 1–4, shortly hairy, narrowly ovoid, 4–6 × 1–1.5 cm.
Here *Cnestis macrantha* is assessed as Endangered since only five locations are known (above; AOO 20 km^2 with 4 km^2 cells) and threats as below.
Several other, more common *Cnestis* species are distinguished by differences in inflorescence position, leaflet number, and petal shape.
Habitat: secondary forest.
Threats: forest clearance for fuel-wood, timber and agricultural land is a threat to this species at all locations except Korup NP which is secure.
Management suggestions: rediscovery of this species, ideally at Korup, where secure, is advisable. Basic populational data should be collected to form a baseline for monitoring. Characterisation of the stem would aid survey work since leaves and flowers/fruit are likely to be mainly inaccessible in the forest canopy.

Cnestis macrophylla Gilg ex Schellenb.

EN B2ab(iii)

Range: Nigeria (Ogun State, Omo FR near Oloji), Cameroon (S: Bipindi) and Gabon (Ogooué).
Here *Cnestis macrophylla* is assessed as Endangered since only three locations are known (above; AOO 12 km^2 with 4 km^2 cells), with threats as below.
Liana, stems brown-pubescent, soon glabrescent; leaves 2–4-jugate, petiole with black articulation at base, 8–13.5 cm, rachis 7.5–19 cm, glabrous; leaflets papery, lateral ones opposite, elliptic or obovate, 7–19 × 4–8 cm acuminate, cuneate or rounded at base, lateral nerves 6–8 on each side, tertiary reticulate, both surfaces glabrous; petiolules 8–11 mm; racemes 3–5 together on stem below the leaves, each up to 16.5 cm, to 30-flowered, brown pubescent; sepals 5.5–7.5 mm; petals narrowly oblong to linear 8 × 0.6–1 mm; follicles 1–3 infruit, ellipsoid, oblique, 3.5–4.4 cm with stinging red hairs outside.
Separated from the very common *C. corniculata* only by the black, very long (8–11 mm) petiolules.
Habitat: lowland evergreen forest.
Threats: forest clearance for timber and agricultural land in Nigeria.
Management suggestions: as for *C. macrantha* above, except that efforts should be focussed at the only known Cameroon site at Bipindi and since there is no National Park there, local communitities should be involved in protection if the species can be refound.

Jollydora glandulosa Schellenb.

EN B2ab(iii)

Range: Nigeria (Obudu, Boshi-Okwangwo FR) and Cameroon (SW: Likomba, 15–35 km NE of Limbe; Kumba to Ediki. NW: 18 km S of Wum, near Obang).
Treelet to 5 m, usually unbranched; leaves alternate, pinnately compound, petiole 15–18 cm; rachis 10–45 cm; leaflets to 13, papery, oblong to obovate-elliptic, 8–43 × 2.5–10.5 cm, acumen to 1.5 cm, base cuneate, petiolule 4–10 mm, midrib with glandular hairs on both sides, lateral nerves 6–13 on each side; racemes to 9 mm, 1–3-flowered, axillary; sepals oblong, 2–6 mm, with both glandular (mainly) and non-glandular hairs; petals oblong, 1 cm; fruit ovoid-ellipsoid, to 4 × 2.5 cm, yellow, not stipitate, 1-seeded.

Only three species exist in the genus, only one of which is fairly common, *J. duparquetiana* (Cross to Congo Rivers). This differs from *J. glandulosa* in having a short stipe to the fruit and in lacking glandular hairs.
Here *Jollydora glandulosa* is assessed as Endangered since effectively only three locations (above) are known and there are severe threats (below). AOO is 12 km² with 4 km² cells.
Habitat: lowland evergreen forest.
Threats: all four sites are in areas that have been affected by forest clearance for timber and agriculture, and it may have been lost in several of these.
Management suggestions: attempts should be made to refind and protect this species in Cameroon if it still exists. Recent collections of *Jollydora* from SW and NW regions should be re-examined for glandular hairy leaflet midribs and absence of fruit stipes in case they have ben misidentified as the more common species. In this way more recent collections might be found, from sites where the species might be more likely to survive. Basic populational data should be gathered for future monitoring and seed collected for multiplication and introduction to safe sites.

CONVOLVULACEAE

assessed by Martin Cheek

A globally widespread family of stem-twining climbers, the species are characterised by their alternate, usually simple, estipulate leaves, often producing white exudate when wounded, and flowers with funnel-like corollas, producing capsules with several seeds. In Cameroon there are 16 genera, with *Ipomoea* the largest. The best known species is the potato or sweet potato, *Ipomoea batatas,* originating in the New World, cultivated for its edible tubers, white 'morning glory' (*Ipomoea purpurea*) is cultivated for its beautiful blue flowers. A third species, *Ipomoea involucrata* with beautiful pink flowers, is a weed of cultivation and fallow land. Most Convolvulaceae in Cameroon are species of disturbed areas or forest edge. The lack of a Flore du Cameroun account is an impediment to identification of species and some threatened species may have been overlooked as a result.

The threatened species thus far identified belong to two genera which are atypical of the family, being restricted to tropical, usually evergreen, forest. These are woody lianas, flowering and fruiting in the canopy. Usually only the distinctive fallen fruits are seen on the forest floor. These have a heart-shaped papery, dry, leaf-like structure, to which a small egg-shaped capsule is [illegible] side. In *Calycobolus* the leaf-like structure is a sepal and the capsule is concealed by another although smaller, expanded sepal. In *Neuropeltis* the sepals are not enlarged in fruit, and it is a bract that forms the leaf-like structure.

Calycobolus was revised by Lejoly and Lisowski (Bull. Jard. Bot. Nat. Belg. 55: 27–60 (1985)) who recognised 30 species, of which 25 occur in continental Africa. The species are mainly woody forest lianas, flowering and fruiting, like *Neuropeltis*, in the forest canopy and difficult to observe unless there are fallen trees, or if forest edges available. Many of the species were first described as *Prevostea*. Breteler is currently revising species limits in this genus.

Calycobolus gilgianus (Pilger) Heine
CR B2ab(iii)

Range: Cameroon endemic (S: Bipindi).
Woody liana, branches blackish, longitudinally striate, glabrous; petiole 10–15 mm, finely pubescent; leaf-blade elliptic-obovate, apex rounded, shortly acuminate, 11–16 × 6–9 cm, margin revolute, papery, glabrous, mat and olive-green on both sides, margin revolute; lateral nerves 7–8 pairs, reticulation prominent on both surfaces; cymes axillary, 3–8-flowered, rachis 10–15 mm, white pubescent; flowers with corolla urceolate, 16 mm long, sparsely pubescent.
Recognisable by the axillary inflorescence, urceolate corolla more than 10 mm long, and elliptic, shortly acuminate, leaf.
Here *Calycobolus gilgianus* is assessed as Critically Endangered since only a single location is known (AOO 4 km² with 4 km² cells) and since threats are as below.
Habitat: lowland evergreen forest (assumed)
Threats: slash and burn agriculture.
Management suggestions: rediscovery of *Calycobolus gilgianus* should be attempted in the Bipindi area, where it was last seen 100 years ago. Forest edge and tree-fall areas should be targeted in the flowering season and if successful the site should be demarcated and protected with involvement of local communities. Efforts should be made to characterise the stem so as to be able to conduct a census of the population as a baseline for future monitoring.

Calycobolus micranthus (U.Dammer) Heine
CR B2ab(iii)

Range: Cameroon endemic (SW: Kumba).
Liana-like shrub 2–2.5 m, young branches pubescent; petioles 5(–10) mm, blade obovate, 12 × 6 cm, long acuminate, cuneate, glabrescent except the pubescent midrib, lateral nerve c. 8 pairs; inflorescences few-flowered axillary panicles, bracts minute; corolla white, strongly urceolate, 7–8 mm long.
Distinct in the axillary inflorescences with small, urceolate corollas.
Here *Calycobolus micranthus* is assessed as Critically Endangered since only a single location is known (above; AOO 4 km² with 4 km² cells) and threats as below.
Habitat: lowland evergreen forest.

Threats: slash and burn agriculture; plantation agriculture (rubber; oil palm); urbanisation.
Management suggestions: rediscovery of *Calycobolus micranthus* should be attempted in the Kumba area, where it was last seen 100 years ago. Forest edge and tree-fall areas should be targeted in the flowering season and if successful the site should be demarcated and protected with involvement of local communities. Efforts should be made to characterise the stem so as to be able to conduct a census of the population as a baseline for future monitoring.

Neuropeltis, revised by Lejoly & Lisowski (Bull. Jard. Bot. Belg. 54: 471–482 (1984)) was recently reviewed for Africa by Breteler having described a new species and recognising in all, nine for the continent (Plant Ecology and Evolution 143 (2): 176–180 (2010)). Apart from those proposed as threatened below, *N. alnifolia* Lejoly & Lisowski, from Nigeria, Cameroon, CAR, Equatorial Guinea and Gabon is rare and merits an NT assessment. Further work might show it to be threatened.

Neuropeltis laxiflora Lejoly & Lisowski
EN B2ab(iii)

Range: Cameroon (S: 40 km ESE Kribi, S Zingui) and Gabon (Mbigou-Malinga; Libreville; Lastoursville (5coll.)).
Here *Neuropeltis laxiflora* is assessed as Endangered since only four locations are known, one of which has five sites, hence AOO is calculated as 32 km², using 4 km² cells, threats as below.
Liana with leaves glabrous; inflorescence large, to 20 cm, sparsely flowered, corolla <5 mm long, stamens included, ovary glabrous; fruit glabrous, with calyx less than half as long as fruit.
The description is extracted from Breteler's key since the revision by Lejoly and Lisowski is absent from Kew library and no specimens are present at Kew currently.
Habitat: evergreen lowland forest.
Threats: urbanisation (the location at Libreville has probably been lost) logging followed by agriculture; slash and burn agriculture.
Management suggestions: in Cameroon, rediscovery and protection at the Zingui site is suggested. Globally speaking conservation efforts would be best focussed around Lastoursville where it appears to be most common (see above). Efforts should be made to protect it with support from local communities.

Neuropeltis pseudovelutina Lejoly & Lisowski.
EN B2ab(iii)

Range: Cameroon endemic (E: 8 km NE Doumé on road to Bertoua. S: Ébolowa-Minkok; Bipindi; Campo-Ma'an (Tchoutou 2004)).
Liana brown appressed-hairy, leaves elliptic, densely tomentose, dark brown below, 7–13 × 3.5–7 cm, mucro/acumen <5 mm base obtuse or rounded, lateral nerves 8–9 pairs, nerves plane or impressed above; inflorescence of axillary spikes 6–8 cm and a terminal panicle of spikes to 18 × 6 cm, corolla 5–6 mm long; bracts 3–4.5 cm diam. in fruit.
Here *Neuropeltis pseudovelutina* is assessed as Endangered since only three locations are known (above; AOO 12 km² with 4 km² cells) and because threats are as below.
Habitat: lowland evergreen forest.
Threats: slash and burn agriculture; logging followed by agriculture.
Management suggestions: a record of this species from Gabon (*Thomas* 6557, TROPICOS) is suspicious since it does not appear on the Gabon database, nor is the species recorded in that country according to the Gabon checklist despite the work at WAG and in Gabon by Breteler, an authority on the genus. The Campo-Ma'an record has not been seen, but if identification is confirmed, this location would seem to be ideal for the conservation of the species since it is protected as a National Park. Ideally basic populational data should be collected to aid future monitoring and management decisions.

CTENOLOPHONACEAE
assessed by Martin Cheek

Apart from the single African species below, this family is restricted to SE Asia. Easily recognised from the leafy stems. No other tree family in Cameroon has opposite simple leaves with an interpetiolar stipule and lepidote scales.

Ctenolophon englerianus Mildbr.
VU B2ab(iii)

Range: SE Nigeria (Stubbs Creek; Benin-Sapoba; Degema), Cameroon (Littoral: Douala-Edéa), Gabon (four locations), Congo (Brazzaville) (Kouilou-Kayes) and Angola (Cabinda).
First collected in May 1923 in Cabinda by Gossweiler, this large tree is gregarious but restricted to freshwater swamp forest near the coast. The first records in Cameroon date from 1976 (*Waterman & McKey* 828).
Large (30–40 m) emergent tree with fluted bole, slash thick and red, masses of small red aerial roots; indumentum of stems and leaves of minute lepidote scales; leaves opposite, blade elliptic-ovate, 4.5–7.5 × 2.5–4.5 cm, leathery, glossy above, drying brown below; petiole 0.4–1 cm, articulated at base; interpetiolar stipule oblong-triangular 3 mm; inflorescence 2–3 cm, terminal, flowers white, 7–8 mm; capsule ellipsoid, 1-valved, 1.5–2 cm, seed single, aril fimbriate.
Here *Ctenolophon englerianus* is assessed as Vulnerable since it is known from ten locations (AOO 40 km² with 4

km² cells) three of which, in Nigeria, are believed to have suffered forest habitat losses in recent decades. Nigeria has lost nearly 90% of its original forest cover. If extinction at these locations can be verified, the species would qualify as VUA2c. EOO is calculated as 284508 km².
Habitat: lowland freshwater swamp forest in the coastal evergreen forest belt; sea-level to 100 m alt.
Threats: none are known apart from those in Nigeria (above, clearance for timber and, at Stubbs Creek, for oil-industry related infrastructure). Timber used for house and canoe building in Nigeria.
Management suggestions: the sole Cameroon location known so far is in the Douala-Edéa forest reserve. Consideration should be given to protecting at least part of this as a National Park since it is rich in plant diversity and contains several other threatened species.

CUCURBITACEAE

assessed by Martin Cheek

One of three fairly common Cameroonian families of plants which are herbaceous climbers with alternate simple leaves and tendrils, the other two being Passifloraceae (passion flowers) and Vitaceae (grapes). While Vitaceae have leaf-opposed tendrils, and Passifloraceae are glabrous with a pair of petiole glands, Cucurbitaceae generally have subaxillary tendrils, an absence of petiole glands (except *Lagenaria* which have such glands), and, often, a bristly, white indumentum.

The genera and species of Cucurbitaceae in Cameroon are very well resolved thanks to the work of Kerauden in Flore du Cameroun 6 (1967). She recognised 58 species in 28 genera. No further species are known to have been described since that date for Cameroon.

Egusi, an important food in Cameroon, is prepared by steaming in Marantaceae leaves from the pounded seed of *Cucurmeropsis mannii* Naudin, which is cultivated for that purpose in Tropical Africa as well as being native. *Lagenaria siceraria* is cultivated for its dried fruits used as traditional drinking vessels in Bakossi with which no man is without. *Telfairia occidentalis* is often cultivated, not for its massive, ribbed fruits, but for its edible leaves. It is also indigenous. *Cucumis melo* (melon), *Luffa cylindrica* (luffa), *Citrullus lanatus* (melon or water melon) of the New World, *Cucumis sativus,* the cucumber and *Momordica charantia,* all possibly or certainly introduced, are cultivated on a small-scale in Cameroon.

Within Cameroon the rarest Cucurbitaceae are concentrated not in the coastal evergreen forests but in the semi-deciduous-evergreen forest mosaic of the S Cameroon Plateau, especially in the area NE of Yaoundé. These are not usually a family of completely undisturbed habitats but nonetheless clearing of forest is a major threat to their survival.

Cyclantheropsis occidentalis Gilg. & Mildbr.

EN B2ab(iii)

Range: Cameroon (E: Yaoundé–Deng Deng, 90 km; Lomié. S: Djoum area, Mintom-Alati 65 km).
Here assessed as Endangered since three locations (above; AOO 12 km² with 4 km² cells) are known with threats as below.
Slender climber, stems glabrous; leaves alternate, tendrils near axillary, unbranched or bifid, blade near succulent when live, glabrous, ovate-oblong, 7–13 × 4–8 cm, subacuminate, base cordate then abruptly decurrent 3–5-nerved at base, veinlets conspicuous; panicles many-flowered, 20 cm; pedicels 1 mm; flowers minute 1.3 mm.
This is the only member west of the Congo basin of a genus remarkable for having 1-seeded winged fruit, arising from a unilocular ovary with one ovule, and anthers united in a ring.
Habitat: probably forest: grassland transition.
Threats: conversion of forest to agriculture followed by logging.
Management suggestions: rediscovery of this species, not seen since Jan. 1973, is advisable, followed by recording of basic population data to aid future monitoring and conservation management.

Momordica camerounensis Kerauden

EN B2ab(iii)

Range: Cameroon endemic (C: N'kolbison; Mefou proposed NP. S: Nkane, 27 km WNW Ambam)).
Here *M. camerounensis* is assessed as Endangered since only three locations are known (above AOO 12 km² with 4 km² cells) and threats are as below. This assessment is repeated in Cheek *et al.* 2011.
Climber; tendrils simple, unbranched; leaves leathery, ovate, 7–14 × 5–8 cm, acumen short, base cordate, entire, glabrous, tertiary nerves raised reticulate; petiole 2.5–4 cm, glabrous; racemes axillary, 8–9 cm, 3–10-flowered; bracts foliaceous, ovate, 1 mm, pubescent; pedicels 2 mm; sepals triangular, 3 mm; petals yellow, ovate, 5 × 4 mm.
Habitat: riverine forest edge; 600–700 m alt.
Threats: forest clearance for timber followed by agriculture.
Management suggestions: formal recognition of the proposed National Park would provide a protected site for this species. Comparison with the type material at Paris is needed to confirm identification of the Mefou material.

Momordica enneaphylla Cogn. (assessed by Iain Darbyshire, updated by Martin Cheek)

VU B2ab(iii)

Range: Cameroon (SW: Kupe-Bakossi (3 coll., 2 sites). C: Mefou proposed NP; 20 km NW of Eséka (1 coll.). S: Bipinde (1 coll.)), Gabon (La Mvoum, Fougamou), Congo (Kinshasa) (Forestier Central: Yangambi (3 coll.); Watsi, Boende (1 coll.); Yabwesa, Isangi (1 coll.)) and Congo (Brazzaville) (Massif du Chaillu).

First assessed as VU B2ab(iii) by Darbyshire in Cheek *et al.* (2004: 154), the assessment is maintained here and in Cheek *et al.* 2011 since it is still the case that no more than 10 locations are known (AOO 40 km^2 with 4 km^2 cells) and threats are as below.

This little known liana was first collected in Gabon, being described in 1888. Collections from Congo (Kinshasa) are all from the central Congo Basin. In Cameroon, the discovery in 2004 of 2 sites in the Kupe-Bakossi area, on Mt Kupe above Kupe Village and at Ngomboku in the Bakossi Mts, extended this species' range northwestwards.

This is a highly distinctive member of the Cucurbitaceae family, being the only bipinnate species of W Africa, with 3 ternate leaflets. *Sidwell* 443 appears to be the first collection of the currently undescribed female flowers of this species; this specimen demonstrates how this taxon could be easily overlooked, as it is flowering prior to the development of leaves. A collection of a single *Momordica* flower on a trailing leafless stem at Nyandong, western Bakossi (c. 500 m alt.) by M. Etuge in March 2003 may prove to be this species on further investigation.

Habitat: primary forest, including riverine and swamp forest; occasionally in secondary forest growth; 450–1070 m alt.

Threats: on Mt Kupe, this species is recorded in forest down to 900 m, which is below the lower limit of effective protection on this mountain, thus this population is threatened by forest clearance. The population at Mefou proposed NP is not yet secure.

Management suggestions: formal protection of Mefou as a National Park would give this species a protected area in Cameroon. If the low levels of forest clearance continue in Gabon and Congo (Kinshasa), this taxon should remain extant there, though its rediscovery in these two countries is advisable in light of the age of the known collections.

Momordica gilgiana Cogn.

VU B2ab(iii)

Range: Cameroon (C: 30 km SE Yaoundé. E: Bertoua-Deng Deng 15 km; Lomié; Deng Deng; River Dja, Ntouo-Meu rivers; Woutchaba-Samat. SW: Buea; Nyandong); Equatorial Guinea (Bioco: Pico Basile) and Gabon (Belinga and Franceville).

Here *Momordica gilgiana* is assessed as Vulnerable given nine locations (above; AOO 36 km^2 with 4 km^2 cells) and with threats as below.

Herbaceous, glabrescent climber, stems finely ridged; leaves alternate, trifolioliate, petiole 2–3 cm, with a large green gland at its insertion at the node; petiolules 6–9 mm, leaflets papery, ovate-elliptic, central arc, 8–10 × 4–6 cm, laterals slightly smaller and asymmetric, margin with a few, widely-spaced, small teeth, upper surface punctate; male inflorescence umbellate, bract absent, peduncle 6–13 cm with 2–3 large glands at apex, flowers 15–30, pedicels 1.2–2.5 cm, floral cup 3 × 6 mm, petals obovate, 2 × 1 cm; fruit obovoid, 10 × 7 cm rostrate, covered in conical tubercles of two sizes, glabrous.

Habitat: semi-deciduous forest.

Threats: logging followed by agriculture.

Management suggestions: discovery of further locations may remove this species from the threatened list. Mapping of locations against existing protected areas might produce a site for focussing conservation efforts. Failing this existing areas such as Dja NP, might be found to yield this species if surveyed. Local communities should be involved in protecting the species if possible. Basic populational data would as a baseline help future monitoring. Discovery of new locations might take the species beyond the threatened threshold.

Momordica jeffreyana Kerauden

VU B2ab(iii)

Range: Cameroon (S: Bitye near Dja; 35 km S Bengbis), Gabon (Makakou; Tchimbélé; Belinga), Congo (Kinshasa) (Tschelai, Shimbanza; Ikela: - Yalikungu; Boketa; Ile Tukutu; Yangambi) and Congo (Brazzaville (Massif du Chaillu)). The Gabon locations are not shown on our map.

Here *M. jeffreyana* is assessed as Vulnerable since nine locations (above; AOO 40 km^2 with 4 km^2 cells) are known with threats as below.

Climber, stems reaching 3–4 m and up to 1 cm diam., glabrous; leaves alternate, simple, petiole 1.5–2 cm, blade obovate-lanceolate 10–14 × 5–6 cm, subacuminate, shallowly cordate to truncate; glabrous, entire; tendrils simple; male flowers axillary, fasciculate, 3–10, enclosed in a rhombic bract 2 × 1 cm; pedicel 1–2 cm, not winged; flower bud 5 × 8 mm; petals obovate, 8 × 6 mm; fruit ovoid, 4–13 cm, 5–7 ridged, ridges spiny.

Habitat: evergreen forest edges, often near rivers where disturbed.

Threats: logging of forest followed by cultivation.

Management suggestions: mapping of locations against existing protected areas might produce a site for focussing conservation efforts. Failing this existing areas such as Dja NP, might be found to yield this species if surveyed. Local communities should be involved in protecting the species if possible. Basic populational data would as a baseline help future monitoring. Discovery of new locations might take the species beyond the threatened threshold.

Momordica obtusisepala Kerauden

CR B2ab(iii)

Range: Cameroon (E: Bertoua–Deng Deng, 16 km).

Only known from *Breteler* 1838 collected in 1961.

Here assessed as Critically Endangered since only a single location (above; AOO 4 km² with 4 km² cells) is known with threats as below.

Climber, stems to 5 mm diam., deeply ridged, glabrous; leaves alternate, petiole 4–14 cm, finely pubescent; blade papery-membranous, broadly ovate 15–16 × 15–16 cm, acute, base cordate, margin subpentagonal, minutely dentate; obscurely puberulent on both surface; 3–5-nerved at base; tendrils bifid; racemes on stems below leaves, 9–10 cm, bracts minute, hooded; pedicels 5–8 mm; floral cup 6–7 × 8–9mm; sepals ovate 5.5 × 6.5 mm, obtuse, glabrescent; petals white, obovate, 15 × 8 mm.

Habitat: semi-deciduous and evergreen forest.

Threats: logging of forest followed by agriculture.

Management suggestions: the only known location should be revisited to attempt to rediscover this species. Breteler, the original collector should be consulted in case he can give more detailed pointers on finding the site. Since formal protection of the site is unlikely even if the plant is refound, the local community responsible should be approached to protect the species. Seed banking is recommended as an insurance – and propagation to strengthen the wild population. The Yaoundé herbarium should be checked to discover if more specimen locations do not exist.

Telfairia batesii Kerauden

CR B2ab(iii)

Range: Cameroon (S: Bitye, on Dja R).

Here *T. batesii* is assessed as Critically Endangered being known from a single location (above; AOO 4 km² with 4 km² cells) with threats as below.

Climber, stem glabrous apart from the hairy nodes; leaves alternate, trifoliolate, petiole 2–4 cm, finely tomentose; leaflets obovate-elliptic, 4–7 × 2.5–3 cm, long-acuminate, base attenuate; laterals smaller, with base asymmetric, margins entire or with a few, sparse, small teeth; upper surface sparsely punctuate, lower with hooked hairs; lateral nerve 4–5 pairs; tendrils forked; racemes slender, 20 cm, pedicels 1.2–1.5 cm, each with a bract at midpoint; bract elliptic; floral cup 3 × 5 mm; sepals subulate, 4 × 1 mm, glabrous; petals ovate 10 × 5 mm; stamens 3 on filaments with a basal hair-tuft.

Differing from *T. occidentalis* in the subulate sepals which lack teeth and the secondary nerves strongly ascending.

Habitat: probably riverine forest.

Threats: logging of forest followed by agriculture.

Management suggestions: rediscovery of this species, not seen since it was discovered 50 years ago, should be attempted. If this is successful, basic populational data should be collected as a baseline for future monitoring. The local community should be informed and asked to assist in the protection of this species.

ALUKA features an image of *Dang D.* 500 (Douala-Nkongsamba, or Njombé) of what may be a second specimen-site of this species. If this can be confirmed the rating needs to be reassessed as EN. Since this is a relative of the cultivated *T. occidentalis*, it may have crop-breeding value for that species.

DICHAPETALACEAE

assessed by Martin Cheek assisted by Laura Pearce

This is mainly a pantropical family of evergreen forest, with two genera present in Cameroon: *Tapura* with three species, one of which is threatened, are all trees, while *Dichapetalum* with 42 or more species in Cameroon, are almost all stem twining lianas. Among the families with alternate, simple leaves, they are recognised when sterile by a characteristic appressed grey indumentum in many species, combined with stipules, and dark 'gland' spots within the leaf-blades, most easily visible when held up to the light. These spots are distributed irregularly, mainly on each side of the midrib and along the ends of the secondary nerves. In some species they appear on the lower surface near the petiole, as in Chrysobalanaceae, to which they are related and sometimes confused when young and sterile (see diagnostic notes under that family). The flowers, only 3–4 mm long, are usually held in tight axillary clusters, rarely diffuse; sometimes the infloresceneces arise directly from the petiole. The five petals are all equal in size in *Dichapetalum*, but unequal in *Tapura*. The family takes its name from the dichotomously bilobed petals. The fruits are fleshy, 2–5 cm, irregular, often lobed suborbicular with 1–3 seeds, the epiderm is often covered in irritant hairs.

The family contains an unusual highly toxic compound, fluoro-acetic acid, which disrupts the tricarboxylic acid respiration system of vertebrates. Locally *Dichapetalum* speces are used for poisoning a wide variety of animals.

Breteler wrote the Flore du Cameroun 37 (2001) account, based on his research on the family across Africa over many years. His work is vital for the correct identification of Dichapetalaceae species, and spans 40 years. His revision of the genus has been broken into six parts, the species ordered alphabetically, as follows: Breteler (1973), species a-b Meded. Landbou. Wag. 73-13; (1978), species c-f, Meded. Landbou. Wag. 78-10; (1979), species g-l, Meded. Landbou 79-16; (1981), species m-q, Meded. Landbou. Wag. 81-10; (1982), species r-z, Meded. Landbou. Wag. 82-8. In addition, new species found subsequent to this revision have been published separately, in various scientific journals, such as is the case for *D. korupinum*, below. It is to be hoped that all these works will eventually be united in a single publication so as to facilitate ease of use. All of the descriptions and identifications in the species treatments below have been abstracted from Breteler's works. Apart from the species assessed as threatened below, the following taxa occurring in Cameroon, while known from more than 10 locations globally, can be considered Near Threatened since they are known from either less than 20 locations and/or are have a fairly small range. Reassessment of these in future with better data (e.g. on population/habitat reuction) might elevate them to threatened status: *Dichapetalum altescandens, D. barbatum, D. crassifolium* var. *integrum, D. dewevrei, D. librevillensis, D. pulchrum* and *D. tetrastachyum.*

Dichapetalum sp. nov.? (*von Rege* 62a. SW. Transect B, Mabeta-Moliwe, Mt Cameroon, August 1993). Good flowering material is required of this plant, described as a tree 1 m tall, and having pilose stems (both unusual in this genus) before this can be confirmed as a new species by Dr Breteler. If this is done, it may rate inclusion in a future addition of this Red Data list. This note first appeared in the Red Data chapter of Cable & Cheek (1998: li). To date no further material has come to light.

Dichapetalum korupinum Breteler

CR D

Range: Cameroon (SW: Korup NP, 6 km W of Ikenge (1 coll.)).

A liana, leaf stems densely tomentose, blade elliptic, 2.5–3 times as long as wide, (8–)9–11 × 3–4 cm, acumen gradual, 1.5–2(–2.5) cm, base rounded, lateral nerves 4–5(–6) on each side of the midrib, sparsely subappressed short -hairy on the midrib and main laterals on both sides, glabrescent above; glands few, small, dispersed, often more numerous above, stipules early caducous, narrowly triangular, 2–4 × 0.5 mm, densely tomentose. Inflorescences sessile, scorpioid, simple, rarely with two branches, to 6 mm long, few-flowered. Sepals sub-erect, ± 1.5 mm; petals suberect, united with staminal filaments in a 1 mm long tube, lobes obovate, 2 mm long, split to 0.5 mm; flowering in April (Breteler in Kew Bull. 51(1): 201 (1996)).

Here *Dichapetalum korupinum* is assessed as Critically Endangered under Criterion D, since less than 50 individuals are known. The southern Korup NP has become one of the most intensively sampled forests in Cameroon, with many thousands of specimens collected since the early 1980's principally by the team of Duncan Thomas. Despite this, just a single individual of this species has been recorded (*Thomas* 7620) suggesting that it is genuinely rare.

Habitat: lowland evergreen forest.

Threats: none are known – the site is secure from man-made threats.

Management suggestions: the site should be revisited to acquire basic populational data (density and range of individuals, regeneration levels, indication of any threats although these are not suspected) for use in future monitoring and planning of management needs. Individuals of forest lianas can spread clonally over very large areas so care is needed to distinguish likely clones from genetically different individuals.

Dichapetalum letouzeyi Breteler

CR D

Range: Cameroon endemic (SW: Korup NP).

A large liana, stems with short, dense, brown pubescence extending to the stipules and petioles; leaves elliptic to obovate, 11–14 × 6–8 cm, acuminate, base rounded, lateral nerves 7–8 pairs, midrib pubescent above and below, glands dense and large; petiole 6–12(–18) mm; stipules triangular-ovate, 2–7 × 1–1.5 mm. Inflorescence diffuse, 4–5 times dichotomously branched, c. 100 flowers; peduncle 6–17 mm united with petiole; sepals reflexed, ovate 2.5–3 mm long; petals narrowly obovate, 3.5 mm, split by 1.5 mm, flowering in April (Breteler in Kew Bull. 51(1): 201 (1996)).

Here *Dichapetalum letouzeyi* is assessed as Critically Endangered under Criterion D, since less than 50 individuals are known. The southern Korup NP has become one of the most intensively sampled forests in Cameroon, with many thousands of specimens collected since the early 1980's principally by the team of Duncan Thomas. Despite this, just a single individual of this species has been recorded (*Thomas* 4701) suggesting that it is genuinely rare.

Habitat: lowland evergreen forest.

Threats: none are known – the site is secure from man-made threats.

Management suggestions: the site should be revisited to acquire basic populational data (density and range of individuals, regeneration levels, indication of any threats although these are not suspected) for use in future monitoring and planning of management needs. Individuals of forest lianas can spread clonally over very large areas so care is needed to distinguish likely clones from genetically different individuals.

Dichapetalum oliganthum Breteler

VU B1+B2ab(iii)

Range: Cameroon endemic (S: Kribi to Campo area, c. 6 locations, numerous coll.).

Liana or shrubby liana: stems densely pubescent; leaves elliptic, 7–12 × 2–4 cm, acumen 0.5–1.5 cm, base rounded; lateral nerves 5–7 pairs, glabrescent on both faces, glands small, dispersed, lower surface only; petiole 1–3 mm; stipules long persistent, narrowly triangular, often curved, 2–

5 mm long; inflorescence 3-flowered, silver pubescent; peduncle <1 mm; pedicel 1.5 mm, sepals reflexed, ovate to obovate, 1.5–2.5 mm; petals free, obovate, 1.5–2 mm, cleft 0.5–1 mm. Unusual in the few-flowered cymes.

Here *D. oliganthum* is assessed as Vulnerable since six locations (AOO 24 km² with 4 km² cells) are known, with threats as below. EOO is calculated as 2045 km².

Habitat: lowland evergreen forest near the coast.

Threats: the Kribi area faces numerous threats to its natural habitat. New infrastructure connected with extractive industries is expected to require clearance of habitat, and urban expansion, and holiday cottages along the scenic coastline are also causing habitat loss.

Management suggestions: Tchoutou should be contacted to establish the number of individuals observed by him in the field. If less than 50, this species should be re-assessed as CR D. It is advisable that efforts are made to return to the original sites and discover basic population data (density, range, regeneration levels, threats) for this species to inform monitoring and gauge needs for management intervention. Campo-Ma'an managers and community leaders near known sites should be assisted to identify and so protect this rare liana.

Dichapetalum potamophilum Breteler

EN D

Range: Cameroon (Littoral: 70 km SSW Bafia, bank of Sanaga R) and Gabon (NE quadrant, 3 loc.)

Liana, woody stems deeply lobed, lenticels arranged in five ranks; leafy stems shortly brown pubescent; leaves ovate-elliptic 4–11 × 3-5cm, obtuse-mucronate, base rounded and shortly cordate, lateral nerves, prominent above and below, 5–9 pairs, glabrescent, glands on both surfaces, dispersed, fairly distinct. Stipules caducous, 1–3 mm; petioles 3–5 mm. Inflorescences subsessile, 3–5-branched, to 3 cm; pedicels 3 mm, sepals ovate 1.5–2.5 mm, petals obovate 2–3 mm, cleft 0.5–1.5 mm (Breteler, 2001).

Here *D. potamophilum* is assessed as Endangered under Criterion D since we estimate that less than 250 individuals have been recorded. Four sites (AOO 16 km² with 4 km² cells) are known, but no direct threats are known by us at present. EOO is calculated as 51563 km².

Habitat: along rivers in lowland evergreen forest.

Threats: none are known.

Management suggestions: it is advisable that efforts are made to return to the original sites and discover basic population data (density, range, regeneration levels, threats) for this species to inform monitoring and gauge needs for management intervention. Community leaders near known sites should be assisted to identify and so protect this rare liana. None of the sites are in protected areas it appears.

Dichapetalum reticulatum Engler

VU B2ab(iii)

Range: Nigeria (Ibadan S FR; Akure; Udo, Ubiaja) and Cameroon (C: Yaoundé. S: Bipindi; Campo-Ma'an (Tchoutou)).

Liana, stems densely villous; leaves obovate-elliptic, 11–16 × 7–8 cm, acumen 0.5–1 cm, rounded to obtuse (-cordate) at base; lateral nerves 8–10 pairs, upper surface appressed hairy, persisting on the midrib; lower surface softly hairy; glands few, inconspicuous; inflorescence subumbellate, <25 flowers; peduncle 8–12 mm, pedicel 6 mm; calyx lobes 3–4 mm; petal lobes 3–3.5 mm, cleft 1 mm.

Here *Dichapetalum reticulatum* is assessed as Vulnerable since six locations (see above; 24 km² AOO with 4 km² cells) are known with threats as below. EOO is calculated as 118442 km².

Habitat: lowland evergreen to semi-deciduous forest.

Threats: all of the Nigerian sites must be considered under immediate and severe threat given the great losses that have occurred in recent decades, assuming that they survive. In Cameroon forest habitat at Yaoundé has been lost as urbanisation proceeds. Threats at Bipindi and Campo are unknown.

Management suggestions: it is advisable that efforts are made to return to the original sites and discover basic population data (density, range, regeneration levels, threats) for this species to inform monitoring and gauge needs for management intervention. Community leaders near known sites should be assisted to identify and so protect this rare liana. Only the Campo-Ma'an location is likely to be in a protected area.

Dichapetalum witianum Breteler

EN B2ab(iii)

Range: Cameroon (S: Ébolowa–Kribi, 21 km) and Gabon (four locations).

Liana, main axis 5-lobed; stems sparsely pubescent; leaf-blades obovate-elliptic, 9–15 × 4–7 cm, acumen 0.5–1.5 cm, base cuneate, lateral nerves 5–6 pairs, sparsely puberulent, glabrescent; glands inconspicuous on lower surface, along the midrib and near the petiole; stipules early caducous, 2–5 mm; petioles 4–6 mm; inflorescence 3–5-branched, 5 flowers, peduncle 5–15 mm; pedicel 2 mm, sepals oblong, 5–6.5 mm, petals oblong, 5–6 mm, cleft <1 mm.

Here *D. witianum* is assessed as Vulnerable since five sites (AOO 20 km² with 4 km² cells) are known, with threats as below. EOO is calculated as 75795 km².

Threats: slash and burn agriculture.

Management suggestions: it is suggested that efforts are made to return to the original sites and discover basic population data (density, range, regeneration levels, threats) for this species to inform monitoring and gauge needs for management intervention. Community leaders near known sites should be assisted to identify and so protect this rare liana. None of the Cameroonian sites are in protected areas it appears.

Tapura tchoutoi Breteler

EN B2ab(iii)

Range: Cameroon endemic (S: Campo-Ma'an area–Bifa (Bipindi–Mimfia; Bibabimwoto; Dipikar Island)

Published in 2002 (Breteler, Adansonia 24 (2): 267), this species is closely related to *T. neglecta* Hallé & Heine from Gabon which shares the compact sessile inflorescence which is free from the petiole of the subtending leaf (Breteler 2002 op. cit.).

Small tree, stems appressed pubescent, glabrescent; leaf-blade obovate-oblanceolate 15–24 × 5–6 cm, acumen abrupt, slender, 1.5–3 cm, base cuneate to rounded, slightly asymmetric, lateral nerves 9–11 pairs, lower surface nerves appressed pubescent, glands small, on lower surface, along midrib and margins; inflorescence sessile, to 25-flowered, pedicel articulate, to 1.5 mm; flowers 6 mm, sepals 5, unequal, oblong-elliptic, outer two 3 mm, inner three 4–5 mm; petals 5 mm, fused with stamens basally forming a 3 mm tube, large petal lobes 2, 2.5 mm, cleft 1 mm; small petal lobes 1 mm, entire.

Known only from two collections by Tchoutou at two sites in the Campo-Ma'an area collected in 2000 and 2001 (Breteler, Adansonia sér.3, 24 (2): 267-269 (2002)), *Tapura tchoutoi* is here assessed as Endangered since two sites are known (AOO 8 km² with 4 km² cells) and threats are as given below.

Habitat: lowland evergreen forest rich in Caesalpinoid legumes; 0-100 m alt.

Threats: slash and burn agriculture.

Management suggestions: Tchoutou should be contacted to establish the number of individuals observed by him in the field. If less than 50, this species should be re-assessed as CR D. It is advisable that efforts are made to return to the original sites and discover basic population data (density, range, regeneration levels, threats) for this species to inform monitoring and gauge needs for management intervention. Campo-Ma'an managers and community leaders near known sites should be assisted to identify and so protect this rare tree.

DIPSACACEAE

assessed by Martin Cheek

A small family of herbs, mainly in the N Temperate zone but with a few species extending into tropical African mountainous areas.

Closely similar to Compositae in having reduced flowers clustered in heads subtended by bracts, Dipsacaceae differ in the stamens being free, often exserted from the corolla, and not united in a cylinder around the style-stigma. Only two species occur in W and C Africa west of the Albertine Rift. Both species are of otherwise European genera and are endemic to the Cameroon Highlands, being found at the boundary between montane forest and grassland. Flore du Cameroun 21 (Lawalreé 1980).

Dipsacus narcisseanum Lawalrée (syn. *Dipsacus pinnatifidus* sensu FWTA)

EN B1+B2ab(iii)

Range: Cameroon endemic (W: Bamboutos Mts. NW: Bamenda Highlands at Kilum-Ijim)

Perennial herb with one to several apparently biennial stems from a several-crowned underground rootstock. Vegetative rosettes with numerous lax leaves c. 45 cm long, densely felty. Inflorescences presumably produced in second year. 1–4 m tall, peduncles 1 cm diam., stout, ridged, bearing 1–9 short (c. 10 cm) stalks in upper part, each bearing rigid spherical capitula c. 6 cm diam. Flowers whitish, c. 1 cm long, not very conspicuous, in spiral rows around capitula, only one row developed at one time. Fruits not seen. Apparently flowering throughout November.

Rare on Kilum-Ijim. Only 3 sites and c. 30 plants seen in 1996–1999. At Kilum, seen at two sites in 1996 (pers. obs. Cheek).

1. KA transect, in forest belt in upper sector towards grassland, on steep, thinly clad walls of ravine. Two to three plants seen.
2. KA transect, 200 m W of grassland: forest boundary, at top of cliff edge below grassland facing N. About 8 plants seen.
3. At Ijim, seen at one extended site on roadside bank, grassland-woodland boundary amongst *Ixora* bushes c. 2 m high north of Gikwang Rd, between Lake and Jikijem junction, c. 2500 m alt. About 15 inflorescences and more vegetative rosettes seen at this grid reference in November 1996. Monitoring in November 1999 showed this population to be virtually unchanged. This is the largest population known. Along the road from this population, towards Aboh, 3–4 other groups of 1–5 inflorescences were seen.

This species was previously assessed as VU D2 in Cheek *et al.* (2000: 60) but is here reassessed since it is now considered that direct threats do exist (see below). Two broad locations (above) are known, but AOO is estimated as 12 km² using 4 km² cells for each of the adjoining Kilum, Ijim and also the more distant Bamboutos subpopulations. With better census data at Kilum-Ijim and Bamboutos, it would be possible to reassess this species under criterion C or D. EOO is calculated as 607 km².

Dipsacus is genus of about 15 species restricted to Eurasia (Europe 8) and Tropical African mountains. This is the only species known in W–C Africa.

Habitat: on banks and cliffs or cliff edges, in scrub or grassland near forest; c. 2000 m alt.

Threats: previously it was stated that "there are no obvious threats to this species" (Cheek *et al.* 2000: 60). However, it is now considered that cattle trampling and goat-grazing probably are the main threat to this species together with loss of forest habitat on which the species depends. Studies by Baena in Cheek *et al.* (2010: back cover) have show c. 50% forest loss in the Dom area of the Bamenda Highlands 1988–2003. In addition the species appears to be so rare that unintentional clearance (e.g. for a cattle compound, or road diversion) at any of its sites might endanger it.

Management suggestions: the reason for the rarity of this species is unclear and needs more study, perhaps using experimental plots. The explanation may be that establishment of new colonies demands conditions rarely encountered. Seed gathering for banking and for multiplication and reintroduction should be considered. See comments below on the ecologically similar and related Succisa. Surveys are needed in the Bamboutos Mts to determine if the species still survives there despite forest loss. If refound, basic populational data of the sort recorded above for the Kilum-Ijim location should be collected as a baseline for future monitoring. The Kilum-Ijm sites should be recensused to assess trends in the health of the species there.

Succisa trichotocephala Baksay

EN B2ab(iii)

Range: Nigeria (Chappal Wadi, 1 coll.) and Cameroon (SW: Mt Cameroon (numerous coll.). W: Bamboutos Mts (two coll.). NW: Kilum-Ijim Mt Oku (several coll.); Bambili Lakes (1 coll.)).

This is the only species of the genus Succisa in Tropical Africa. The remaining three species are Eurasian. This species appears to be a perennial, with a subterranean perennating rootstock that withstands fire. It produces biennial rosettes that flower in their second year, though more observations are needed to confirm this.

Perennial herb, with one to several apparently biennial stems from a several-crowned underground rootstock. Vegetative rosettes with numerous thinly hairy, lax leaves c. 30 cm long. Inflorescence c. 0.5 m tall, lax, branching from near the base, basal branches over 30 cm long. Capitula hemispherical, c. 2.5 cm diam.. Flowers c. 1 cm long, white or mauve. Flowering abundantly in October and November. Young fruits were seen at the end of November.

Observations above [illegible], in October 1996, showed that this species survived, though not abundantly, and was flowering in previously fired grassland that was being cattle grazed. Cattle seem to avoid eating *S. trichotocephala.*

Kilum: transect KA, just above forest, about 4 plants seen thinly scattered in closely cropped Sporobolus grassland: apparently not grazed. Also reported common from near main peak (fide Munyenyembe).

Ijim: Gikwang rd: almost immediately on entering the gate into the protected area from Aboh, *Succisa* occurs intermittently as scattered individuals and sometimes in dense communities of 10–50 plants in glades in *Gnidia glauca* woodland, with light shading. This is true of a 2–3 km stretch of the road at this point. At the Kilum-Ijim Forest Project resthouse further along that same road, an apparently natural grassy area occurs in which the dominant plant during our visit in November 1996 was *Succisa.* An estimated 400–500 flowering plants in about a hectare. Mt Cameroon: during Oct 1992 and 1993, MC was able to study *Succisa*, then in flower, over several days in the area south of Mann's Spring. Here it occurs only as scattered individuals in relatively ungrazed (no domestic animals) grassland, never more than about 200 m from the forest. Plants were also recorded at Laikom Ridge in December 1998 (Pollard pers. comm.).

Hooker treated *Mann* 1309 (Jan. 1862, 10,500 ft.), the first specimen known of this species, as identical to the European 'Devil's Bit Scabious', *Succisa pratensis. Succisa trichotocephala* was only described as a separate species in 1952, before which the African plants were only varietally distinguished (in the 1930 s) as *Succisa pratensis* var. *kamerunensis* B.L.Burtt.

Previously this species was assessed in Cable & Cheek (1998: li–lii) and in Cheek *et al.* (2000: 60–61) as LR/nt since it was considered that there were "no apparent threats to this species". However, it is now considered that its forest: grassland habitat is as equally threatened as forest in its area, since, without forest, it will cease to exist. Moreover, although flowering plants have been observed in grassland that had been fired, and also that which was being grazed, this does not necessarily mean that these conditions are not deleterious in the long-term to the species.

Here *Succisa trichotocephala* is reassessed as Endangered since there are five locations (above, AOO is estimated as 20 km^2 using 4 km^2 cells) and threats are as above. Under citerion A, it would rate as vulnerable since more than 30% of the forest dependent habitat of this species is estimated as having been lost within three generations. Studies by Baena in Cheek *et al.* (2010: back cover) have shown c. 50% forest loss in the Dom area of the Bamenda Highlands 1988–2003. However at Mt Cameroon forest loss at higher altitudes is though to be low or nil. EOO is calculated as 13622 km^2.

Habitat: grassland, near boundary with forest; open glades in *Gnidia glauca* woodland; 2000–2800 m alt.

Threats: there are within the Kilum-Ijim area, beyond clearance of forest for agriculture. Grazing by horses and cattle was prevalent in one area where the species was observed to be fairly common in November 1996 (Gikwang Rd, Ijim) but appeared to cause no adverse effects to the *Succisa.*

Management suggestions: insufficient data available to suggest a management regime, but fire is possibly important at intervals in preventing shading-out by shrubs and trees, while too frequent fires, and too dense stocking rates, might threaten the species. The underground rootstock is likely to provide protection in the event of dry season fire. The large, lax leaves help minimise competition from adjoining herbs by smothering them. This species appears to be in no immediate danger of extinction, being relatively common locally for the present, at some sites. Nonetheless, if surveys at the montane forest: grassland ecotone are being conducted, the opportunity should be taken to collect data on the population size and range of this species.

DROSERACEAE

assessed by Martin Cheek

A worldwide family of carnivorous (animal-eating) herbs, the Droseraceae consists of four genera, all with a single species except *Drosera* itself, with c. 150 species, the only genus recorded in Cameroon, although the elusive free-floating aquatic, *Aldrovanda* is also likely to be found. *Drosera* can be recognised by their gland-covered leaves to which small insects are attracted, trapped, and at length, digested and absorbed. The glandular hairs each have a red head, which secretes a translucent, dew-drop-like globe of mucilage which sparkles in the sun and glues to the leaf any fly that might touch it. Tropical African *Drosera* all occur in seasonally wet, nutrient poor, open habitats from sea-level to high altitude. The commonest species in Cameroon, *Drosera madagascariensis*, is used to treat certain medical conditions. No Flore du Cameroun account is available.

Drosera pilosa Exell & Laundon

VU D2

Range: Cameroon (NW: Bamenda Highlands (2 locs)), Congo (Kinshasa) (Msipashi, 24 km SSW), Burundi (Prov. Bururi (1 coll.)), Kenya (Mau Forest; Cherangani Hill (1 coll. each)) and Tanzania (Mt Rungwe (3 coll.)).

Distinct from the more widespread *D. burkeana* in the pilose, non-glandular peduncle, this sundew is known from one or two records each in Congo (Kinshasa), Tanzania, Kenya and Burundi. In Cameroon it was only known from two collections, both in the Bamenda Highlands (Nchan and Bum) but appears not to have been recollected since 1931 until Nov. 2000, at a new site, Bali Ngemba (*Pollard* 531). *Pollard* 1044 from the same locality, but sterile, probably represents the same entity. The area of occupancy of *D. pilosa* is estimated at 8 km^2, 1 km^2 per site. The assessment of VU D2 here is maintained from that in Harvey *et al.* (2004: 62), signifying that although very few plants and few locations are recorded, threats are currently unknown. Pollard (pers comm.) notes that this species is rare, just one or two individuals being found after extensive searching. By contrast other species of the genus are often gregarious being found in dense colonies. Other apparently similar areas of habitat searched yielded no individuals of this plant.

Carnivorous acaulous (?annual) herb; leaf rosette 3 cm diam., leaves to 20, appressed, spathulate, blade subcircular 5 × 5 mm, rounded-truncate, base cuneate-decurrent; petiole 1 cm; scape 8 cm, pubescent with angled hairs, not glandular; flowers 5 mm.

Habitat: seasonally wet, open, peaty soils between grass tussocks on steep slopes; 1450 m.

Threats: extremely rare, otherwise unknown.

Management suggestions: a survey of the site at Bali Ngemba FR should be made to assess the size and demography of the subpopulation. This species may be an annual or short-lived perennial, flowering at the end of the wet season. Regular burning of grassland (reported by *Pollard* 531) may favour this plant by removing the shade inimicable to most *Drosera* (pers. obs.).

EBENACEAE

assessed by Martin Cheek, advised by George Gosline

Revised for the Flore du Cameroun 11 by Letouzey and White (1970: 36 species), several new, rare species have since been described: *Diospyros kupensis* (Gosline & Cheek, Kew Bull. 53: 461–465 (1998)) and, *D. korupensis* and *D. onanae* (Gosline, Nordic J. Bot. 27(5): 353–358 (2009)). Jongkind (Kew Bull. 62(4): 637 (2007)) has shown that what was treated as *D. barteri* in Cameroon is *D. rubicunda* Gürke.

Often known as the 'ebonies', the genus *Diospyros* are famed above all for the hard black wood produced by a minority of the species. The main C African ebony-producing species in *D. crassiflora* Hiern which occurs from Nigeria to Congo (Kinshasa) and is relatively common in the forests of Cameroon, although trees large enough to provide the commercially useful heartwood are rarely seen. Although listed by IUCN www.redlist.org (viewed Oct. 2010) as EN we disagree for the reasons stated and would list it as Near Threatened. Several other species have edible spherical juicy fruits, and a few produce dye from their bark. Douglas Gibbs of BGCI is planning to co-ordinate an assessment of the IUCN status of *Diospyros* species worldwide.

Most species occur in wet lowland evergreen forest, but a few occur in drier habitats.

Diospyros are easily recognised among other woody forest shrubs and trees with alternate simple leaves by their tendency to have verticillate branching: lateral branches arise in whorls of three or more at one point on the main stem, spreading horizontally. The stem then continues naked until another whorl is produced. As with the Annonaceae, the leaves lack stipules but confusion is avoided by smelling stem scrapings. Those of Annonaceae are spicy, while those of *Diospyros* are scentless. *Diospyros* trees are reputed to have a thin, charcoal-like underbark but this is often hard to observe. The flowers have parts in threes or fours (less usually fives), often with a persistent leafy calyx and tubular corolla.

Diospyros alboflavescens (Gürke) F.White
EN B1ab(iii) B2ab(iii)

Range: Cameroon endemic (S: Kribi–Bipindi).
Records of this species from Nigeria (e.g. *Daramola* 160, *Smith* 3, *Gentry* 32720, 32718, and from CAR (e.g. *Fay & Harris* 8593) have been redetermined as *D. iturensis* (Gürke) Letouzey & F.White by Gosline. Apart from the BipindI type specimens of Zenker from a century ago, only one modern collection is confirmed: *Bos* 4995 from 10 km NE Kribi dating from the 1970's. No specimen has been seen for a record from Zingui attributed to Tchoutou so this must be treated with caution although included on our map.
Diospyros alboflavescens is a tree to 20 m with glabrous stems and leaves, the leaves are ovate-oblong, 10–20 × 6–10 cm with a triangular, obtuse apex and rounded base. The six pairs of lateral nerves have crater-like glands in their axils and the veinlets dry red-brown. The male flowers are in dense cymes of c. 20 each with a 3 mm cup-like truncate, glabrous calyx and a 12 mm 3-lobed tubular yellow corolla.
Here *D. alboflavescens* is assessed as Endangered since only three sites are known (above, AOO 12 km² with 4 km² cells) and threats are as below. EOO is calculated as 993 km²
Habitat: lowland evergreen forest.
Threats: slash and burn agriculture is practiced throughout this area. Natural habitat in the Kribi area is threatened by development infrastructure such as transport, storage and ports associated with extractive industries.
Management suggestions: given the exceptionally high number of plant species unique to Kribi–Bipindi a National Park should be created there to protect them. Failing this trees of this species should be rediscovered and brought to the attention of traditional and governmental authorities so as to protect them for the future. Data on density, range and regeneration of this species are needed to gauge needs for intervention.

Diospyros korupensis Gosline
VU B2ab(iv)

Range: Cameroon endemic (SW: S Korup NP; Mt Cameroon at Bomana Bakweri, Bakingili, Njonji).
A tree to 25 m unusual in its large oblong leaves 25–45 × 11–15 cm with whitish cast below. The 5-merous male flowers have a 2–3 mm cupular calyx and a cylindrical white, glabrous corolla c. 15 × 4 mm with lobes more than 10 times as long as the tube.
Known only from the two areas indicated. In Southern Korup over 1000 specimens of greater than 1 cm diam. at 1.5 m from the ground were recorded in the Smithsonian 50 Ha research plot, while in the Newbery plot, 11 trees over 10 cm diam. are known in a 71 ha area. Yet 15–20 km to the N in the Ekundu-Kundu area no individuals were found in a series of plots made there (Gosline loc. cit.). At Mt Cameroon it appears to be confined to the arc of forest on the S side of the main massif being unrecorded from Onge, Mokoko, S Bakundu and Bimbia-Bonadikombo. Area of occupancy was estimated as 600 km² (loc. cit.) and given two effective locations and the threats below, his Endangered assessment (Gosline 2009, loc. cit.) seems justified and is maintained here. It seems likely that the species will be found in the intermediate area. EOO is calculated as 939 km²
Habitat: lowland evergreen forest with rainfall greater than 3 m p.a.
Threats: being inside the National Park, the Korup trees are secure, but those at Mt Cameroon are threatened by forest clearance of surviving forest to expand existing plantations of oil palm and other tropical crops.
Management suggestions: excellent data (cited above) are already available for monitoring the subpopulation at Korup. Management intervention seems unnecessary at this stage since numbers of individuals and regeneration levels are so high there. At Mt Cameroon local communities and authorities should be provided with the means to recognise and protect this tree and seedlings could be included in the Limbe BG for the purpose of public education and further studies.

Diospyros kupensis Gosline
VU B2ab(iii)

Range: Cameroon (SW: Mt Kupe (8 coll.); Nyale, central Bakossi (2 coll.); Nyandong, W Bakossi (1 coll.); Enyandong, N Bakossi (1 coll.)).
The assessment here, by Cheek & Darbyshire in Cheek *et al.* (2004: 154–155) is maintained here since no new data are available. However, if, as is credible, the several sites in Bakossi are taken as a single location, the species will need to be reassessed as CR. EOO is calculated as 308 km² which is within the EN threshold.
First collected in 1995, this species appears restricted to Mt Kupe and the Bakossi Mountains. It remains fairly numerous on the slopes of Mt Kupe above Kupe Village; a colony of 30–40 individuals was recorded in one location here in the mid 1990s. Elsewhere it is rare, though is easily overlooked when sterile. In 2003 it was recorded for the first

time in W Bakossi, referring to a single sterile specimen adjacent to the track from Nyandong to Messaka, extending the taxon's altitudinal range down to 460 m. It was also recorded for a third time at the base of Nyale Rock (Etuge *pers. obs.*).

This species appears most closely related to *D. conocarpa* Gürke & K.Schum, but it lacks the modifications for ant association of this taxon. *D. kupensis* displays an interesting form of vegetative reproduction; in many specimens the main trunk of the monopodial growth is angled at approximately 45°, with side shoots growing vertically and producing adventitious roots, which eventually produces several vertical stems from originally monopodial growth (Gosline & Cheek (1998). Kew Bull. 53: 463).

Habitat: understorey of undisturbed mid-elevation to submontane forest on well-drained slopes; 460–1300 m alt.

Threats: several populations have been recorded on Mt Kupe below 1000 m alt., which is below the limit of effective protection through restricted land use at this site; these populations are therefore threatened by agricultural encroachment. The Nyandong site is threatened by disturbance along the well-used track to Messaka.

Management suggestions: further botanical inventory work will likely reveal further locations for this taxon in the Bakossi Mountains. Active conservation of the larger populations on Mt Kupe may be required; informing the community at Kupe Village and neighbouring settlements of the rarity and uniqueness of this species has begun with a conservation poster specific to *D. kupensis*, with the intention that this will promote community-led protection.

Diospyros longiflora Letouzey & F.White
VU B2ab(iii)

Range: Cameroon (Littoral: Douala-Edéa FR. C: Eséka area. S: Kribi–Campo area) and Gabon (Cristal Mts at Mala).

A shrub reaching 7 m tall with lanceolate leaves 12–20(–38) cm apex acute, lateral nerves 5 pairs. Similar to *D. korupensis* in the white cast of the underblade, the cupular calyx and deeply 5-lobed corolla, but differing in habitat preference, and in having only 5–10 (not 30–50) flowers per fascicle, pointed, not rounded petal tips and smaller leaves.

Here *D. longiflora* is assessed as Vulnerable since seven sites are known (see accompanying map, AOO estimated as 28 km² using 4 km² cells) with threats as below. The AOO in this case is likely to be realistic since the species is reported locally abundant in its range, (Gosline 2009 loc. cit.) and therefore each specimen record may well approach an area occupied by the species of as much as 4 km². EOO is 21698 km.

Habitat: lowland riverine and swamp forest.

Threats: swamp forest is generally amongst the least threatened habitats being unsuitable for the usual forms of agriculture practiced in Cameroon. However one site, at Kribi, is vulnerable to plans to increase infrastructure for transport, storage, and export of products from extractive industries.

Management suggestions: confirmation by survey is needed that two of the sites do in fact fall within the Douala-Edéa and Campo-Ma'an protected areas as is suspected, and that the trees still exist here. Management and traditional authorities at those places should be provided with the information needed to identify and protect the species. Data on regeneration levels and population density could be gathered to aid an understanding of future intervention needs.

Diospyros onanae Gosline
EN B2ab(iii)

Range: Cameroon endemic (SW: 10 km W Banga Bekele. S: Ebolowa-Minkok 30 km)

A 4–20 m tall tree with densely brown pubescent stems and petioles. The ovate to lanceolate leaves are 4–9 × 2–5 cm, the lower surface with a whitish cast glabrous except for the white hairy nerves. The petioles are 5 mm long. The 4-merous flowers have triangular, 1 mm long calyx lobes and a tubular, 10 × 3 mm corolla with 1 mm lobes. Gosline (2009 loc. cit.) in describing the species assessed it as EN B1ab(iii) presumably an error for that above since only the AOO, and not the EOO was cited. That assessment is maintained here, being based on two sites (above) and an AOO of 10 km² with threats as below.

Habitat: lowland evergreen forest, 2–2.5 m rainfall p.a.

Threats: Gosline (loc.cit.2009) states that both known sites were being cleared of forest for timber and/or farming when the specimens were made.

Management suggestions: the first priority is to revisit the two known sites in order to determine if any trees survive and, if they do, to secure their protection by providing representatives of local communities with the knowledge to identify and protect them. Data on density, range and regeneration levels at the two sites would aid future management. Propagation material should be raised at the Limbe BG as a back up and as a means of multiplying the species for reintroduction into the wild as is likely to be needed given the available data.

Diospyros platanoides Letouzey & F.White
EN B2ab(iii)

Range: SE Nigeria (Calabar to Afi area, 3 sites), Cameroon (SW: Lake Ejagham) and Gabon (40 km SSW Lastoursville).

Distinctive and named for its trunk, the bark of which falls in plaques up to 12 cm in diam., leaving a pattern as seen in a *Platanus* or plane tree. The oblong-elliptic/lanceolate leaves, 7.5–11.5 × 3–4.5 cm, have 2 glands at the base. The 3–4-merous flowers have a 2 mm glabrous calyx with rounded

lobes and a barrel-like corolla 6 × 3.5 mm lubed by ¼ its length.
The type and only known Cameroonian specimen is from Lake Ejagham. *Diospyros platanoides* is here assessed as EN since five sites are known (above, AOO 25 km^2 with 4 km^2 cells) with threats as below. EOO is 37417 km^2.
Habitat: lowland evergreen forest.
Threats: Nigeria has seen heavy forest losses in recent decades and it is thought that only the Afi site has some protection, the other two Nigerian sites being unprotected and vulnerable to logging and agriculture, if trees survive there at all. Threats at the Cameroonian and Gabon sites are unknown.
Management suggestions: the priority is to rediscover this species at one of its sites. Ideally one which already receives some protection, and then to gather the usual populational data to assist future monitoring and management interventions. Leaders of local communities near any surviving trees should be provided with the means to identify, and so protect them.

Diospyros soyauxii Gürke & K.Schum
EN B1ab(iii) B2ab(iii)

Range: Cameroon (S: 50 km SE Kribi; Campo-Ma'an NP?) and Gabon (Libreville area: Sibange; Mondiah First; Libreville–Kango 15–17 km; Ntoum–Cocobeach 35 km).
A 2–3 m shrub, stems with appressed hairs. The leaves are elliptic or lanceolate-elliptic, up to 21 × 10 cm with a short broad acumen, often with a mucron, and a rounded to truncate base, lateral nerves are 5–8 pairs, upper fruit drying red-brown, glabrous; lower surface drying grey, with nerves hairy, and 3–5 large glands in the axils of the nerves. Petioles are 1.5 × 0.2 cm; female flowers 4-merous, 1–9 in each axil, pedicels 3–13 mm; calyx 3 mm, lobed by 2/3, tomentose; corolla narrowly conical in bud, 8 × 3 mm, lobed by ¼ tomentose, lacking staminodes.
For many years this species was only known from Libreville where it is probably extinct within the city limits due to urbansiation. However, Breteler (Gabon database) has identified three of his own collections immediately to the E and N of Libreville as this species. The only confirmed record in Cameroon is that of Letouzey from 50 km SE of Kribi. Althought the records of Tchoutou from Campo-Ma'an are mapped here, their identity requires checking, as does that of *Le Testu* 2289 from Sindara. Here *D. soyauxii* is assessed as EN since five sites are known (AOO 20 km^2 with 4 km^2 cells) and threats are as below. EOO is 2016 km^2.
Habitat: littoral low-altitude forest predominently.
Threats: probably already lost from the type locality at Libreville. Continued city expansion and needs (timber for construction etc) are likely to threaten two of the three new Breteler sites – those closest to Libreville. Threats to the Cameroonian location(s) is unknown, apart from slash and burn agriculture.
Management suggestions: verification of the Tchoutou records from Campo-Ma'an is a priority. If confirmed this National Park should be the logical focus of conservation efforts. Basic populational data should be gathered to assist future monitoring and managment interventions, if needed, and the means to identify this species passed to the park and traditional authorities. It seems very likely that further work will yield additonal locations for this species, crossing the threshold to VU status under criterion B, although with better habitat-loss imagery, this species might yet be retained as EN.

EUPHORBIACEAE SENS. LAT.
assessed by Martin Cheek

Recent molecular studies have supported the break up of Euphorbiaceae into smaller families based on sometimes rather difficult-to-observe morphological features. Phyllanthaceae and Putranjivaceae (*Drypetes* in Cameroon) have both been included in the greater Euphorbiaceae in recent decades in Flora accounts, and so are treated in this way here. However, they are separated in modern works. In future this division may well continue and still further splits could be raised to family level such as Uapacaceae and Oldfieldiaceae.

Euphorbiaceae in the wide, traditional sense usually have alternate, simple leaves with stipules, often unisexual flowers with the ovary superior and three flat, often branched stigmas leading to a distinctive 3-valved fruit in which the central column persist after the valves have fallen. Less than half the tropical African forest species have the white exudate that characterise the species that predominate in Europe and N America.

Euphorbiaceae in the wide sense occupy almost every habitat and every niche imaginable in Cameroon. However, few if any are canopy emergents of intact evergreen forest and none are epiphytes. Otherwise they range from minute herbs, to shrubs, to climbers and large trees. In the far N of Cameroon, cactus-like *Euphorbia* can be found. However, most species in Cameroon are shrubs and trees of forest often recognisable by their catkin-like inflorescences (tail-like, with many, highly reduced, flowers) and the presence (except in Phyllanthaceae) of two globe-like glands at the top of the petiole.

In Cameroon this family is of immense use to man, mainly due to species introduced from S America, such as *Manihot esculenta* (manioc, cassava, miondo – a very important food crop) and *Hevea brasiliensis* (rubber – an important plantation crop). Indigenous species are also important. *Ricinodendron heudelotii* provides seeds sold for soup mixes in every village (Njansang) and *Tetracarpidium* provides the widely sold 'noyaux'. Several species of *Euphorbia* are pantropical weeds of cultivation.

Identifying Cameroonian species of Euphorbiaceae to species is often not easy since there is no Flore du Cameroun account. Although the Flora of West Tropical Africa (FWTA) account is helpful, especially for SW and NW Regions, many species and even genera in other Regions of Cameroon are not included. In addition so many new specimens from previously unexplored areas are now available that the FWTA account is now dated. That giant of African botany Léonard recently revised many Congolian genera of Euphorbiaceae and completed parts for the family for Flore d'Afrique Centrale, but sadly did not extend his studies usually to include taxa of areas W of Congo (Kinshasa). Therefore the most recent treatments for the Euphorbiaceae of most of Cameroon are those based on the collections of German explorers a century ago, as analysed by Pax and K. Hoffman.

Apart from the species assessed as threatened below, the following can be rated as Near Threatened, being known from 11–20 locations. With better data they may in future rate threatened, e.g. under Criterion A. These are: *Bridelia speciosa* Muell. Arg., *B. ripicola* J.Léonard, *Cleistanthus bipindensis* Pax, *Cleistanthus mildbraedii* Jabl., *Crotonogyne preussii* Pax, *C. strigosa* Prain, (formerly VU in Cheek *et al.* (2004: 156)), *Drypetes molunduana* (formerly VU in Cheek *et al.* (2004: 156)), *Grossera macrantha* Pax, *G. multinervis* J.Léonard, *Macaranga occidentalis* (Muell.Arg.) Muell.Arg., *M. saccifera* Pax, *Maesobotrya bipindensis* (Pax) Hutch., *Mareya brevipes* Pax, *Oldfieldia africana* Benth. & Hook, *Pentabrachion reticulatum* Muell.Arg., *Phyllanthus diandrus* Pax, *P. mannianus* Muell.Arg., *Pycnocoma chevalieri* Beille, *Thecacoris trichogyne* Muell.Arg. and *T. viridis* (Muell.Arg.) G.L.Webster.

Several genera are in such serious need of revision that their species in Cameroon have not been included except in the case of a few well-delimited species: *Drypetes* (Putranjivaceae), *Erythrococca, Maesobotrya,* parts of *Phyllanthus,* and *Uapaca.* Species in these genera which, after revision, may be shown to be threatened are: *Drypetes armoracia* Pax & K.Hoffm., *D. calvescens* Pax & K.Hoffm., *D. celastrinea* Pax & K.Hoffm., *D. cinnabarina* Pax & K.Hoffm., *D. dinklagei* (Pax) Hutch., *D. diopa* (Hiern) Brenan, *D. fallax* Pax & K.Hoffm., *D. gracilis* Pax & K.Hoffm., *D. iturensis* Pax & K.Hoffm., *D. peltophora* S.Moore, *D. polyantha* Pax & K.Hoffm., *D. rubriflora* Pax & K.Hoffm., *D. spinosodentata* (Pax) Hutch., *D. tessmanniana* (Pax) Pax & K.Hoffm., *Erythrococca hispida* (Pax) Prain, *E. membranacea* (Muell.Arg.) Prain, *E. rivularis* (Muell.Arg.) Prain, *Maesobotrya dusenii* (Pax) Pax, *M. bipindensis* (Pax) Hutch., *M. scariosa* Pax & K.Hoffm., *Phyllanthus polyanthus* Pax, *Uapaca nymphaeantha* Pax & K.Hoffm. and *Uapaca stipularis* Pax & K.Hoffm.

In addition, for these and other genera, there is the question again of old German names based on unicates destroyed at Berlin, such as *Macaranga ebolowana* Pax & K.Hoffm., or *Euphorbia juvoklanti* Pax based on *Winkler* 490 collected in 1904 from Victoria (Limbe). Are some or all of these 'real' but 'missing' species as with the case of *Afrotrewia kamerunica* which was recently rediscovered? We may never know.

Afrotrewia kamerunica Pax & K.Hoffm.

EN B2ab(iii)

Range: Cameroon (S: Kribi–Ébolowa, 27 km; Massif des Mamelles; Boussebeliga bridge) and Gabon (Moyen-Ogooué, Missanga).

Afrotrewia, a genus with a single species, known only from Cameroon, had been lost for many decades until it was rediscovered from specimens collected by Peguy Tchouto recently (Kulju *et al.*(2008). Taxon 57(1) 137–143).

Here *Afrotrewia kamerunica* is assessed as Endangered since only four locations (above; AOO 16 km^2 with 4 km^2 cells) are known, with threats as below.

Shrubs or small trees to 10 m, puberulent, hairs pale red, stellate to simple, 0.2 mm; stipules triangular, 1–3 mm; acute; leaves alternate, simple, petiole 4–9 mm terete, apex pulvinate; blade ovate-elliptic to obovate 10–20 × 2.7–5.8 cm, acuminate, base obtuse to subcordate, marginal teeth 9–18 per side, glands scattered over surface on both sides, lateral nerves 9–13 per side, uniting in a looping marginal nerve; inflorescences terminal, paniculate, 6 cm; bracts 1–2 mm; sparsely hairy, flowers 1–3 per bract; pedicels 6–10 mm; sepals 3, hooded, 1 mm; petals absent; disc as long as sepals; male flowers with anthers 30–40; female flowers with sepals 5–6; anthers absent; ovary 3-locular, stigmas 3, each twice bifid; fruit 1 cm, woody, 3-valved; seeds 6–7 mm.

Habitat: lowland evergreen forest.

Threats: future open cast iron-ore (Massif des Mamelles); slash and burn agriculture; infrastructure for extractive industries and urbanisation (Kribi area).

Management suggestions: Peguy Tchouto should be engaged to refind this species. Data on density, range, threats, regeneration should be collected as aid future monitoring and to guide management plans. Logically, conservation efforts should be focussed in Campo Ma'an NP since half the locations known are there.

Antidesma pachybotryum Pax & K.Hoffm.

VU B2a,b(iii)

Range: CAR (Bouar (1 coll.)) and Cameroon (W: Dschang (2 coll.); Mbouda (1 coll.). NW: Bamenda; Dom. Adamawa: Tibati & Tibati-Mbanti (1 coll. each)).

Known historically from 7 locations (AOO 28 km^2 using 4 km^2 cells) in an arc extending from S to E along the Bamenda and Adamawa Highlands. The assessment by Cheek in Cheek *et al.* (2010: 96–97) is maintained here since no new data are available. It was assessed as vulnerable due to threats (see below), low number of both sites and low A00: the species appears not to be gregarious – no more than a single individual being noted at a site. It has not been recorded from any of these sites for more than 40 years, apart from at Dom (see table below). Given the pressures highland habitat has faced in Cameroon (see below) it may no longer occur at the former sites. It is not ubiquitous within its range. Despite intensive inventory work, it was not found at Bali Ngemba (Harvey *et al.* 2004), nor in the Lebialem Highlands (Harvey *et al.* 2010).

Unusual in the genus for its high alt. habitat, large leaves, and long brown tomentum.

Habitat: submontane forest, near water?; c. 1000–1630 m alt.

Threats: no data is available for the sites near Tibati or Bouar, but at Bamenda, Mbouda and Dom, all within the Bamenda Highlands, there has been extensive loss of forest, especially submontane forest, with 93–96.5% having been lost in total from the original state. Between 1987–1995 alone, 25% of forest cover was lost in the Mt Oku & Ijim Ridge areas of the Bamenda Highlands, based on satellite imagery studies by Moat (Cheek *et al.* 2000: back cover). (Cheek *et al.* 2000: 49–50). At Dschang in the Bamenda Highlands, forest loss has been more extreme.

Management suggestions: efforts should be made to rediscover the species in surviving submontane forest patches within its range and then to explore protection of these sites. The highest priority is continued protection of the community forest at Dom, the only known site for the survival of this species where a search should be made for the original plant and where efforts should be made to find more plants.

This species could be propagated and included in forest replanting schemes in the Bamenda Highlands by NGOs such as ANCO, but the preferred proximity of the species for wet ground, if confirmed, should be respected.

Note: since the original three Ledermann specimens made in 1908–09 at Dschang, Tibati, Mbanti, were destroyed at B, and no illustration is available, our identification of this taxon (by Tchiengué, [illegible] Cheek) has depended on the Mildbraed specimen [illegible] 1914, determined at B, now held at K.

Table: collections of *Antidesma pachybotryum*

Specimen	**Site**	**Date**
Ledermann 1569	Dschang	1908–1909
Ledermann 2405	Tibati	1908–1909
Ledermann 2311	Tibati-Mbanti	1908–1909
Mildbraed 9358	Buar = Bouar, c. 1000 m	Fl. May 1914
Meurillon in CNAD 325	Dschang, 1300 m	Fr. 14 June 1966
Meurillon in CNAD 940	Mbouda, Plaine de Bagam 1350 m	Fl. 5 Sept. 1967
Daramola in FHI 40496	Bamenda town, General Hospital	Fl. 20 Feb. 1959
Cheek 13469	Dom Forest, Bamenda Highlands, 1630 m	Fr. 25 Sept. 2006

Cleistanthus camerunensis J.Léonard.

CR B2ab(iii)

Range: Cameroon endemic (Littoral: Douala).

Here *Cleistanthus camerunensis* is assessed as Critically Endangered since only a single location (above; AOO 4 km^2 with 4 km^2 cells) is known, with threats as below. Several specimens from other locations have been found which may prove to be this species but require confirmation, e.g. *Louis et al.* 349 and *Letouzey* 4862.

Tree or shrub, leaves 7–13.5 × 2–3.5 cm, acumen 2 cm, base acute-obtuse; petiole 3–5 mm; male flowers with pedicels and disc white, glabrous; female flowers with disc pubescent inside; capsule 1.2 × 1.5–1.7 cm, pedicel and exocarp glabrous (from the protologue, Letouzey in Bull. J.Bot. Brux. 30: 430 (1960).

Habitat: lowland evergreen forest.

Threats: urbanisation and slash and burn agriculture.

Management suggestions: rediscovery of this taxon should be attempted in remaining forest in the Douala area, perhaps the Douala-Edéa reserve. Research on the collector's itinerary might help narrow the target area. Data on density, range, threats and regeneration levels would inform future monitoring and management planning.

Cleistanthus letouzeyi J.Léonard.

VU B2ab(iii)

Range: Cameroon (SW: Korup; Mt Cameroon–Onge. C: 40 km E Mbalmayo. E: Abong Mbang to Lomié; Bertoua 53 km) and Gabon (Camp Waka; Massif du Chaillu; Cristal Mts at Kinguele; Ngoko on Ogooué R.; Rabi NE).

Here *Cleistanthus letouzeyi* is assessed as Vulnerable since ten locations are known (above; AOO 40 km^2 with 4 km^2 cells) and since threats are as below.

Small tree, trunk 20 cm diam., adult branches glabrous, brown; leaves alternate, stipules caducous, petiole 8–13 mm, transversely rugose; blade elliptic-obovate 13–20 × 4–7.5 cm, leathery, glabrous, secondary nerves 7–8, tertiary nerves scalariform; inflorescences axillary, solitary 1–3.5 cm, red tomentose, internodes 3–13 mm; pedicels 6–8 mm; flowers

5-merous, sepals valvate, triangulate-ovate, 7–8 × 2–2.5 mm, petals obovate 1.5 × 1 mm; stamens 6.
Léonard (Adansonia ser. 2, 3: 65 (1965) from whence this description is taken) states this species to be close to *C. bipindensis* Pax but that it is separated by the longer petiole, the glabrous leaves, the 6 stamens and etc.
Habitat: lowland evergreen or mixed semi-deciduous-evergreen forest.
Threats: logging, followed by plantation or slash and burn agriculture (e.g. Mt Cameroon–Onge).
Management suggestions: fortunately one site is already in a securely protected National Park: Korup. It seems logical to focus conservation efforts there. Data on population density, range, regeneration levels and threats should be collected to form a baseline for future monitoring and for a needs assessment.

Clutia kamerunica Pax
EN A2c
Range: Nigeria (Chappal Waddi (1 coll.)) and Cameroon (NW: Bamenda to Bafut Ngemba (7 coll.); Bali Ngemba FR and Baba 2 (3 coll.). W: Bamboutos–Dschang (1 coll.); Market Singwa (1 coll.)).
First collected in the Bamenda Highlands in the early 20th Century (*Ledermann* 1658, Market Singwa) this white, downy shrub is notably absent from the Kilum Ijim area (Mt Oku and the Ijim Ridge) that has been surveyed in some detail recently (Cheek *et al.* 2000), and also appears from absent from Dom (Cheek *et al.* 2010 and Lebialem Highlands (Harvey *et al.* 2010) which are also locations where it might reasonably be expected to occur and which have also been reasonably intensively studied. Over half the known collections were made on various sites near Bamenda in or near the Bafut Ngemba FR (Mba Kokeka, Bambuluwe etc.). Once this was the largest area of natural forest in the Bamenda Highlands but has since been decimated.
This assessment is maintained from Cheek in Harvey *et al.* (2004: 63) since no new data are available.
Habitat: submontane and montane forest, forest edge or riverine forest where, usually, single isolated plants are found; 1300–2300 m.
Threats: forest clearance for wood and agriculture. Bafut Ngemba FR, by 2000, had been 99% converted to *Eucalyptus* or agriculture, forest near Dschang is under enormous pressure from agriculture. More than 50% of *Clutia kamerunica* appears to have been lost over recent decades, largely through the demise of Bafut Ngemba as a natural forest.
Management suggestions: a survey for *C. kamerunica* should be executed at Bafut Ngemba FR and the adjoining area to establish whether the species is indeed extinct there, or not. Surveys of this sort at other known locations for the species would also be valuable. Conservation efforts to protect the species are probably best focussed on the sites at Bali Ngemba FR, and at the village forest of Baba 2, since plants have been seen here in the last few years. Education of the public likely to affect the survival of the plants could be carried out by a poster campaign, ideally by such organizations as the Bamenda Highlands Forest Project who are already carrying out conservation work in the area.

Conceveiba leptostachys Breteler (syn. *Aubletiana leptostachys* (Breteler) J.Murillo
VU B2ab(iii)
Range: Cameroon (SW: Korup. S: Kribi 18 km N; ibid 26 kms; ibid. Lobé R) and Gabon (three loc.).
Here *Conceveiba leptostachys* is assessed as Vulnerable since seven locations are known (above; AOO 28 km^2 with 4 km^2 cells) and since threats are as below.
Shrub-treelet to 5 m, bark dirty white; leafy stem and petioles glabrescent, leaves alternate, stipules narrowly triangular 1–4 mm; leaves often crowded at stem apex, both long and short petioles on one stem, petioles 0–7 cm, blade elliptic to narrowly obovate (ovate in sessile leaves) 10–17 × 3–7 cm, acumen 1–2.5 cm, base obtuse to cordate (sessile leaves) or cuneate (long-petioled leaves), margin undulate-dentate in upper half, base with gland pair on upper surface, a few flat glands on both surfaces; male inflorescences erect, unbranched, axillary, interrupted spikes to 15 cm, glabrous or puberulous; flowers yellow, up to 4 together, pedicels to 0.5 mm; calyx cylindric, lobes 4, triangular, 2 mm, stamens 8–12; female inflorescence 1-flowered, peduncle 2 cm, bracts and perianth inconspicuous, ovary 2–3 mm long, 3-lobed style 5 mm, at apex 6-branched; fruit 1–2 cm, 3 lobed.
Conceveiba was only known from Asia until 1990. Now two species are known in Africa, both in Cameroon (Breteler Bull. Jard. Bot. Nat. Belg. 63: 209–217 (1994)). Recently this species was transferred to *Aubletiana* but this is not followed here.
Habitat: lowland evergreen forests.
Threats: infrastructure for extractive industries, urbanisation (Kribi).
Management suggestions: fortunately one site is already in a securely protected National Park: Korup. It seems logical to focus conservation efforts there. Data on population density, range, regeneration levels and threats should be collected to form a baseline for future monitoring and for a needs assessment.

Croton aubrevillei J.Léonard
VU A2c, B2ab(iii)
Range: Ivory Coast (Mt Nimba; Amitioro; Anoumaba; Agboville); Ghana (Kade) and Cameroon (SW: Nyasoso).
This rare tree was assessed by Hawthorne in 1997 as VU A1c, B1+2c, but listed only from Ivory Coast and Ghana (www.redlist.org (IUCN 2003)). The range was extended to

Cameroon in Cheek *et al.* (2004: 155), and the assessment was maintained there as vulnerable, but adjusted to accommodate the post-1997 modification in criterion B. The extent of occurrence of this taxon now exceeds 20,000 km^2. It is only known from six locations (above; AOO 24 km^2 with 4 km^2 cells) with threats as below.

Given continued forest loss in Ivory Coast since 1997 and the precariousness of the taxon in Cameroon, future re-evaluation is likely to rate this taxon as endangered.

Tree 15 m, stem and lower leaf surface densely carved in white, fimbriate, peltate scales; leaves drying dark brown above with minute stellate hairs, ovate, c. 8.5 × 5 cm, acuminate, base rounded, basal glands prominent, stalked, petiole 4–8 cm.

Habitat: lowland evergreen forest; to 850 m alt.

Threats: clearance of forest for agriculture.

Management suggestions: the location for the only known Cameroonian specimen, Nyasoso, 850 m alt., indicates that this species might already be lost to forest clearance. A search should be made to rediscover the tree, and a survey conducted to discover if any other trees of the species occur in the area, and whether regeneration is proceeding or not. Artificial propagation and public education might be suitable options to ensure the survival of this taxon. A record of this species from the Cristal Mts (Gabon database: *Breteler* 12930), if verified, would extend the range of the species and possibly provide a secure site for protection.

Crotonogyne impedita Prain

VU B2ab(iii)

Range: Cameroon endemic (SW: Kumba; Nta Ali FR; Konye. Littoral: Loum. C: Mt Kala; 13 km SW Eseka. S: Ngovayang; Campo-Ma'an NP (Tchouto 2004)).

A poorly known rainforest shrub, *Crotonogyne impedita* appeared to be known from only four collections at as many sites and so was assessed as CR A2c in Cheek *et al.* (2004: 155–156). Three of those sites are on the western and southern boundary of the Bakossi checklist area in SW Province. Two of these provided the original material (*Büsgen* 163 from Kumba and *Ledermann* 6397 from Loum) at the beginning of the 20th century, but there are no modern collections from either site. In the 1980s, D.W. Thomas, specialist in Cameroonian Euphorbiaceae, made the other two collections known, at Konye in western Bakossi (*Thomas* 5167) and Mt Kala (*Thomas* 3473). Recently we have discovered several other records from around Cameroon so here reassess the species as Vulnerable since there are now from nine locations known (above; AOO 36 km^2 with 4 km^2 cells) with threats as below.

Shrub 2 m, highly branched, leaves arranged in loose funnels, indumentum stellate and lepidote; leaves oblanceolate or elliptic, to 20 × 6 cm, base cuneate, drying brown, glabrous below, petiole 1–3 cm; stipules glabrous; male inflorescence with 1–3 branches; fruit with pedicel 3 cm.

Habitat: understorey of semi-deciduous and evergreen forest; to 400 m alt.

Threats: most of the understorey of Loum FR has already been lost to agriculture; much of the forest that survived at Kumba itself, e.g. Barombi Mbo, has been lost in recent decades. Recent reports are that the forest at Mt Kala has been extensively cut-over. Konye lies on the Kumba–Mamfe Rd that is in the process of being upgraded so increased deforestation along this route is predictable. Mt Kala forest is reported to be at risk from slash and burn agriculture.

Management suggestions: rediscovering this rare species, is desirable. It is suggested that the starting point be the sites of the D.W. Thomas collections. *In situ* protection at e.g Campo-Ma'an NP would be ideal, but if unrealistic, cultivation in a botanic garden, such as Limbe, may be warranted, with a view to reintroductions to secure sites in future.

Crotonogyne zenkeri Pax

VU A2c, B2ab(iii)

Range: coastal Cameroon (SW: W Bakossi. C: Nyong at Makak. S: 5 km S Kribi; Batanga; Bipindi) and Gabon (Libreville; Malibé; Rabi).

This forest shrub appears restricted to wet forest close to the Atlantic Coast or along rivers. Despite intensive recent inventories it is unknown from the Mt Cameroon area (Cable & Cheek 1998). The assessment made in Cheek *et al.* (2004: 156) is maintained here even though more locations are known since the figures are still within the VU threshold. 13 collections from seven locations are known. At least four of these sites are threatened.

Vulnerable status is merited since it is only known from seven locations (above; AOO 32 km^2 with 4 km^2 cells) with threats as below.

Shrub 2–8 m, resembling *C. impedita,* but stem and inflorescence indumentum brown stellate-pubescent, leaves to 35 × 10 cm, male inflorescence unbranched.

Habitat: wet lowland evergreen forest; to 200 m alt.

Threats: forest clearance followed by agriculture in the Kumba-Mamfe area (see notes under *Crotonogyne strigosa*), forest clearance for touristic development (Kribi-Batanga area, *pers. obs.* 2000) and for urban expansion in the Libreville area, Gabon (*pers. obs.* 2002).

Management suggestions: the western Bakossi location for this taxon should be resurveyed in order to rediscover this plant and evaluate the size and demography of its subpopulation, if extant. Surveys of the same nature are advised at other sites. This data could form the basis of a plan to protect the species.

Drypetes magnistipula (Pax) Hutch.

EN B2ab(iii)

Range: Cameroon (SW: Bakossi Mts (2 sites). S: Bipindi (2 coll.) and Gabon (Cristal Mts, Tchimbele dam (3 coll.).

This 4 m treelet is named for its large persistent ovate stipules, probably the largest of any African *Drypetes*, measuring 3.5 × 1.5 cm.

The assessment made in Cheek *et al.* (2004: 156) is maintained here since no new data are available. Three locations, comprising four sites are known (above; AOO 16 km² with 4 km² cells) with threats as below.

Treelet 4 m; stems 4-ridged, glabrescent; leaves oblong (-obovate), 30 × 11 cm, acuminate, unequally rounded-truncate, margin subentire, nerves 8 pairs, looped, weakly linked near margin, petiole 1.5 cm; stipules foliaceous, persistent, ovate-triangular, 3.5 × 1.5 cm, acute, base cordate; inflorescence few-flowered, cauliflorous; flowers not seen.

Habitat: lowland evergreen forest; 800–1000 m alt.

Threats: clearance of forest for timber and agriculture is ongoing in both host countries but, so far as is known, is not intense at any of the four sites known for the species.

Management suggestions: although distinctive, *D. magnistipula* is a poorly known species. Since neither the Flore du Cameroun Euphorbiaceae account, nor that for Fl. Gabon, are available, no recent literature is available on the taxon. It is entirely possible that more collections, representing further sites, exist in herbaria other than that of Kew, such as BR, MO, P and WAG. These should be checked. Demographic and other populational data desirable in assessing the conservation requirements of this species could be obtained if the sites of *Thomas* 5313 and *Doumenge* 519 are found intact.

Drypetes moliwensis Cheek & Radcl.-Sm. (syn. *Drypetes sp. nov. aff. similis* Hutch. in Cheek (1992). A Botanical Inventory of the Mabeta-Moliwe Forest: 32 & map 6; *Sibangea sp. aff. similis* (Hutch.) Radcl.-Sm. in Cable & Cheek 1998: lii, 54).

CR A2c+B2a,b(iii)

Range: Cameroon: (SW: Mt Cameroon at Bimbia–Bonadikombo (3 sites)).

Drypetes moliwensis had been allocated the IUCN rating CR A1+2c (Cable & Cheek 1998: lii, as *Sibangea sp. aff. similis*) on the basis of the threat to the Bonadikombo-Bimbia (Mabeta-Moliwe) forest to which, on the basis of available evidence, it appeared restricted. Almost all of this forest was stated as standing on land reserved for plantation expansion, although a more immediate threat was thought to be forest clearance from small-holder agriculturalists from the nearby town of Limbe (Cable & Cheek (1998: xxxvi)). This assessment is updated here to reflect that most of this forest had indeed been cleared in the previous ten years (site visits by Kew with Limbe BG staff in 2007–2009). Furthermore it can also be assessed as Critically Endangered since only a single location, comprising four sites is or rather was known (above; AOO 12 km² with 4 km² cells) with threats continuing as stated.

Tree 4–8 m tall; bark light grey to pale brown, smooth; stems buff to pale grey, with raised, brown, pustular, circular lenticels up to 1 mm diam.; youngest growth fulvous-pubescent, glabrescent; leaves usually markedly unequal from the base, oblong-elliptic to slightly oblanceolate, 15.4–29.6 × 7.0–10.1 cm, acumen (0.4–)0.7–1.3 cm, midrib drying black, lamina with 7–13 secondary veins on each side of the midrib, prominent above; tertiary venation reticulate, prominent on underside when mature; margin remotely and shallowly serrate to subentire, slightly revolute, coriaceous; petiole 1.4–2.0 cm long drying almost black; stipules minute, <1 mm across, soon falling; inflorescence rami-florous, flowers in fascicles of up to 8, bracts and bracteoles caducous, bracts transversely elliptic, concave, c. 2.5 mm wide, 1 mm long, subglabrous; bracteoles laterally compressed, triangular, c. 1 × 1 mm, densely pubescent; pedicels of both male and female flowers 2–4 mm long, shortly pubescent; male flowers with a deeply (4–)5-lobed, imbricate calyx, sepals ovate to oblong, 3.5–4.5 × 1–3 mm, yellowish green, inner surface densely hairy with long, appressed hairs, outer surface thinly pubescent; anthers (9–)10–15, 2–4.0 mm long, inserted in the obscure notches at the margin of a highly rugose, convex, slightly lobed disc c. 2.5 mm diam; female flowers as the male but calyx lobes ovate to more or less orbicular, 5–7 × 5–7 mm, densely pubescent on inner surface; disc convex, fleshy, undulating at the margin, light yellow, glabrous, 3–5 mm diam. having 30–40 very faint radial ridges; ovary subspherical, 2.5–3 × 1.5–2 mm, ovary wall 0.5–0.75 mm thick, unilocular, ovules two, collateral, placentation common, parietal; style stout, ±1 mm long, 1.5 mm wide, stigma dark orange brown, shallowly 3-lobed, convex, glabrous, 2.5–3.5 mm in diam.

Drypetes moliwensis differs from *D. similis* Hutch. (= *Sibangea similis* (Hutch.) Radcl.-Sm.) in having smaller leaves, but with the same number of lateral nerves and also in having pubescent sepals. From *D. aylmeri* Hutch. & Dalz. it differs in the slightly larger leaves with far fewer lateral nerves and in having non-orbicular sepals. The male flowers of *D. moliwensis* are remarkable in West Africa in combining (9-)10–15 stamens and a highly rugose disc.

D. moliwensis is named for Moliwe, the nearest village to the site of the only collections known. Another collection, *D. Thomas* 6823 (YA) from the Korup forest to the W of Mt Cameroon bears superficial similarities to *D. moliwensis* and may prove conspecific when opportunity allows a detailed examination.

Habitat: lowland evergreen forest; 80–100 m alt.

Threats: clearance of forest for agriculture.

Management suggestions: attempts should be made to refind this species at Bimbia-Bonadikombo since there is a chance that despite the loss of forest habitat at the location, it may survive in one patch or another. Staff from Limbe BG should be involved in the search since they have local knowledge of this forest. Exact locality data is available in the protologue and in the references cited under the synonym. If the species is refound, trees should be marked and revisited when fruit are available so that seedlings can be raised at e.g. Limbe BG for public education purposes and for reintroduction into the Bimbia Community Forest or other protected areas in the area, if they can be found.

Drypetes preussii (Pax) Hutch

VU B2ab(iii)

Range: SE Nigeria (Oban, 1 coll.), Cameroon (SW: Mt Cameroon at Onge and Mokoko (6 coll.); Kumba; Kumba–Mamfe (1 coll. each); Korup (5 coll,). S: Bipindi (6 coll.), Campo- Ma'an NP (Tchouto 2004)), Equatorial Guinea (Rio Muni: 21 km N Mitong) and Gabon (5 km SE of Koumémayong; 5 km from Ekorado).

Assessed by WCMC in 1997 as VU B1 +2c (www.redlist.org (IUCN 2003)), Drypetes preussii was there treated as being restricted to Cross River and adjoining forests in Cameroon. Inspection at the Kew Herbarium shows that its range extends along the coast to Gabon. Accordingly its extent of occurrence now exceeds 20,000 km^2, so was re-evaluated in Cheek et al (2004: 157) according to the new model for Criterion B (IUCN 2001). The assessment of Vulnerable is maintained here although additional locations are now known. Nine locations are known (above; AOO 36 km^2 with 4 km^2 cells) with threats as below.

Tree 10 m; stems densely puberulent, terete or striate; leaves elliptic, to 17 × 6 cm, acuminate, slightly unequally obtuse-rounded, serrate, nerves 7 pairs, strongly looped 1 cm from margin; petiole 3 mm; stipules caducous; inflorescence cauliflorous 7 mm, white, pedicels 7 mm; fruit ellipsoid, 3-lobed, 2 cm, smooth, stigmas minute.

Habitat: lowland evergreen forest; 200 m alt.

Threats: threats to lowland forest in the Mt Cameroon area are documented in Cable & Cheek (1998: xxxiv–xxxviii) and in western Bakossi in Cheek et al. (2004). Mokoko and Onge are under imminent threat of logging (2010).

Management suggestions: former locations for this taxon should be resurveyed in order to rediscover this plant and evaluate the size and demography of its subpopulation, if extant. This data could form the basis of a plan to protect the species.

Drypetes staudtii (Pax) Hutch.

VU B2ab(iii)

Range: Nigeria (Omo FR; Ikom, Calabar–Oban) and Cameroon (SW: Mt Cameroon (Mokoko FR; Bambuko; S Bakundu; Etinde); Korup (several coll.); Bakossi (2 sites). NW: Wum (1 coll.). S: Lolo[illegible]

This canopy tree has very large, glossy, leathery, finely-veined oblong leaves resembling those of *D. magnistipula* but lacking persistent stipules. Its geographic distribution is patchy. Although known from only nine locations there are indications from the number of collections at some of these that it is locally fairly common, occupying a relatively large area. At Omo there are four collections, and at the Mokoko R FR (Mt Cameroon) there are eight, with many also at Korup. Meanwhile, adjacent forest reserves such as Onge have no records of the taxon at all, so it appears patchy. In Cheek *et al.* (2004: 157), this species was assessed as VU B2ab(iii). The assessment of Vulnerable is maintained here although additional locations are now known. Nine locations are known some occupying several cells (above; AOO 80 km^2 with 4 km^2 cells) with threats as below.

Were better data on local threats available throughout the range of this taxon, it would be better assessed under criterion A and then would be likely to rate EN or CR. Extensive losses of forest areas have occurred in Nigeria and are ongoing at Wum (pers. obs.). The forest at Mokoko and S Bakundu has also been under great pressures for clearance.

Tree or shrub 15 cm dbh; stems strongly 4-angled, glossy purple-brown; leaves glossy, oblong, 24 × 9 cm, obtuse, obliquely rounded, margin finely serrate-toothed, midrib red-brown, nerves c. 12, quaternary nerves prominent; petiole 5 mm; stipules caducous; inflorescence cauliflorous, fasciculate; male flowers 15 mm diam., yellow, stamens c. 30, pedicels 15 mm.

Habitat: wet lowland evergreen forest, tending to sand substrates; 400 m alt.

Threats: clearance of forest for timber and expansion of agriculture, both large scale commercial and small-holder.

Management suggestions: investigation of unpublished forest plot data from e.g. Mokoko might give data useful for developing a plan for the management of this species. In the absence of this the Korup or Nyandong site in Bakossi could be surveyed for this data. These latter look to be secure sites for the taxon.

Drypetes tessmanniana (Pax) Pax & K.Hoffm.

EN B2 ab(iii)

Range: Equatorial Guinea (Rio Muni) and Cameroon (SW: Mt Cameroon (possibly extinct). E: Abong Mbang (to be confirmed). S: Campo Ma'an (Tchouto 2004)).

The type specimen of this species, *Tessmann* 996, was collected in Rio Muni in 1908/09 and the species had not been recollected until a specimen believed to be conspecific (*Mildbraed* 10600) was collected at Likomba in the eastern foothills of Mt Cameroon in Nov. 1928. This species appears to be a shrub, 2 m tall, so should not be easily overlooked in flower or fruit and the fact that there no further collections have been made nearby (e.g. in the Mabeta-Moliwe inventory of 1992) suggests that *D. tessmanniana* is genuinely rare. The species was assessed as CR A1c+2c by Cheek in Cable & Cheek *et al.* (1998: lii), but since this time the Abong Mbang specimen has come to light, also the reference to the species in Tchouto 2004 at Campo-Ma'an, so it is here reassessed as Endangered given that there are now four possible locations (above; AOO 16 km^2 using 4 km^2 cells) with threats as below.

Habitat: lowland evergreen rainforest.

Threats: forest clearance. In 1998 I stated that "it may be that the species is extinct in Cameroon, in view of forest destruction in the Likomba area over the last 60 years. Heavy logging of lowland forest in Rio Muni is reported at the moment." While the two possible new locations give hope that this was too pessimistic, it is still the case forest clearance threatens the existence of this species.

Management suggestions: the best hope for the survival of this species is its occurrence in Campo-Ma'an NP. This identification should be confirmed and the species rediscovered there, and information and advice given to local managers and leaders so that the species can be protected and monitored. Peguy Tchouto should be engaged to assist with this task.

Euphorbia letestui J.Raynal.

EN B2ab(iii)

Range: Cameroon (S: Akoakas, 27 km SE Ébolowa) and Gabon (Mt Courn, 45 km N of Oyem; Rocher du Chula, 15 km NE Oyem; Mt Koum, 32 km Oyem; Mt Mengong, Mitzic-Sam 45 km; Rocher du Ntan, Sam; 25 km ESE Medoneu).

This cactus-like plant is restricted to large rocks which emerge out of the forest (inselbergs). Here *Euphorbia letestui* is assessed as Endangered since the seven specimen-sites equate to four locations (above; AOO 16 km^2 with 4 km^2 cells) and threats as below.

Habitat: inselbergs in forest.

Threats: rock extraction for building aggregate for infrastructure projects is a threat for inselberg-specific species such as this (Cheek, pers. obs.).

Management suggestions: consultation with the inselberg team that are active in western equatorial Africa (Ludovic Banuk, Ingrid Parmentier, Lejoly *et al.*) is advised to determine if, as is likely, populational data on this species is available, and the extent and prevalence of threats. This data would be a baseline for monitoring and assessing management needs: if not available, it should be collected.

Grossera major Pax.

EN B2ab(iii)

Range: Cameroon (SW: Korup. S: Bipindi) and Gabon (Tchibanga).

Here *Grossera major* is assessed as Endangered since there are three locations (above; AOO 12 km^2 with 4 km^2 cells) and threats are as below. The Tchibanga location is unmapped.

Govaerts *et al.* (2000: World Checklist and Bibliography of Euphorbiaceae 3: 976) remark that this species is close to and possibly conspecific with the more common *G. paniculata* Pax.

Tree 9 m tall; 6cm diam, leaves attenuate, internodes 5 cm, glabrous; petioles 6–7.5 cm, swollen at apex; blade oblanceolate, 25–31 × 8–10.5 cm, acumen 2 cm, base cuneate, lacking gland pair; marginal teeth glandular, 1.5 cm apart; lateral nerves c. 12 pairs broadly scalariform; penicle terminal, 35 × 40 cm, branches few, lowest 24 cm long; flowers sparse, 1 cm wide, petals white, 5; stamens white, united.

Habitat: lowland evergreen forest.

Threats: slash and burn agriculture (BipindI area).

Management suggestions: verification of the validity of this species is advisable. Korup NP is secure and so the logical site in which to focus conservation efforts for this species. Once refound, data on range, density and regeneration of the population should be gathered as a baseline for future monitoring and conservation planning.

Hamilcoa zenkeri (Pax) Prain

VU B2ab(iii)

Range: Cameroon (SW: Mt Cameroon at Bomana-Bakweri–Idenao (1 coll.); S of Kumba at Barombi Kang in S Bakundu FR (8 coll.). Littoral: Loum FR (2 coll.). S: Bipindi (10 coll.); Longji (2 coll.); 30km WNW Eséka; Campo-Ma'an (Tchoutou 2004)).

An understorey shrub, vegetatively resembling a *Cola* but with slightly serrate leaves, this monotypic genus is only known from seven locations (above; AOO 28 km^2 with 4 km^2 cells) with threats as below leading to an assessment of Vulnerable, maintaining that made in Cheek *et al.* (2004: 157). At several of its locations it is known from many collections and so may well be gregarious, even locally fairly common. Its absence from all Mt Cameroon forests except for the drier S Bakundu FR (except 1 record at Idenao), and its existence in Bakossi only at the drier, semi-deciduous forest of Loum FR, both suggest that the taxon might avoid the wetter evergreen forests.

Tree 15 m or more; stems grey, glabrous, trunk cylindric, bark and slash whitish brown, slow white exsudate; leaves drying yellowish green below, papery, oblanceolate-oblong, 18 × 6 cm, acumen 1 cm, acute, secondary nerves acute, c. 6 pairs, raised, margin sparsely and acutely serrate, basal gland pair adaxial, inconspicuous, petiole variable, 1–4 cm, swollen at base and apex; stipules caducous; inflorescence axillary racemes, 4 cm; flowers 6–8, each c. 4 mm.

Habitat: evergreen and semi-deciduous forest; to 400 m alt.

Threats: much of the understorey of the Loum FR has been replaced by small-holder plots of e.g. coco-yam (*Colocasia*) and cocoa (*Theobroma*). A large part of S Bakundu FR is converted to yam cultivation and the specimens from Longji and Barombi Kang both record its presence in logged or degraded forest suggesting that it can tolerate some disturbance, including logging. No data is available on the other sites. In future this taxon might be better assessed under Criterion A if data are available.

Management suggestions: reducing incursions into government forest reserves would protect this species in SW Region. This might be assisted by a public education programme. The sites in Littoral and S Region should be assessed for the existence of *Hamilcoa*, and of threats, in particular conversion of forest to other uses.

Leeuwenbergia africana Letouzey & N.Hallé.
EN B2ab(iii).

Range: Cameroon (SW: Mt Cameroon lower slopes and foothills, Onge, S Bakundu; Rumpi Hills; Nta Ali FR), Gabon (Belinga) and Congo (Brazzaville) (Mayombe).

Named for Toon Leeuwenberg who collected the type material at Mt Cameroon in the 1960's, the genus has a single species, here assessed as Endangered, since only four locations are known (above; AOO 32 km² using 4 km² cells for each site, of which there are four at the Mt Cameroon location) with threats as below.

Tree 15 m tall, 40 cm diam., slash fibrous with red exudate; young stems, petioles and pedicels with patent 1–2 mm hairs; leaves alternate, stipules leafy 1 × 0.3 cm; petiole 18–25 cm, leaflets 5–7, the outermost about half as long as the median; median with petiolule 0.5–4.5 cm, blade elliptic or oblanceolate, 20–28 × 7.5–10.5 cm, acutely 0.2-1.5 cm, base cuneate to decurrent; entire; lateral nerves c. 12 pairs; inflorescence 1–3-flowered, 12 cm; pedicel 2 cm, stout; flower bud 1.5 cm; petals elliptic 1.5 cm.

Leeuwenbergia, holding this single species, was erected in 1974 (Adansonia ser. 2, 14 (3): 386). *Oldfieldia* of W Africa is similar in its digitately compound leaves but these are opposite.

Habitat: lowland evergreen forest.

Threats: logging followed by agriculture, both plantation (oil palm, bananas, rubber), cocoa and market gardening in the Mt Cameroon area, and iron-ore extraction (Belinga) and slash and burn agriculture (Mayombe).

Management suggestions: the forests around Mt Cameroon have the greatest density of *Leeuwenbergia* documented, suggesting that conservation efforts are best focussed here despite the fact that threats are also most severe here. The Limbe BG offers the possibility of seed being collected for propagation and introduction to safe sites in the wild.

Macaranga paxii Prain
VU B2ab(iii)

Range: Nigeria (Oban; Oban-Mamfe), Cameroon (Littoral: Bakaka FR. W: 35 km ESE Foumbam. C: Yaoundé. E: Mintomiaka 45 km S Batouri; Boumba R. 40 km SSW Yokadouma) and Congo (Brazzaville) (NW, 1 loc.).

Here *Macaranga paxii* is assessed as Vulnerable since eight locations are known (above; AOO 32 km² with 4 km² cells) and since threats are as below.

This species was accepted in Whitmore (The Genus *Macaranga*, A Prodromus, RBG Kew, 2008: 203). The type location is Yaoundé but the species has not been collected there in the last 120 years (*Zenker & Staudt* 106).

Species of *Macaranga* are recognisable by the presence of numerous bun-like glands on the lower leaf-surface.

Shrub or small tree to 10 m with spreading 8 mm spines; twigs strongly ridged, sparsely pubescent or glabrous; leaves alternate, stipules caducous, ovate, 2.5 cm, acute, keeled, papery; blades ovate-elliptic, to 23 × 11 cm, acuminate, base rounded, margin coarsely dentate towards apex, teeth gland-tipped, sparsely white hairy on midrib below; male panicles 7–10 × 3 cm, repeatedly branched, puberulous, peduncle 1–2 cm, bracteoles suborbicular, 3–4 mm crenate, hooded, larger than the flower cluster.

Habitat: semi-deciduous or drier evergreen forest; c. 600 m alt.

Threats: urbanisation (Yaoundé), logging followed by agriculture (other locations).

Management suggestions: none of the known locations appears to be protected, however *Macaranga* species generally are pioneers of forest and so some forest disturbance is needed for their survival.

Efforts should be made to refind this species at one of its known locations and then to establish its precise ecological niche. The known locations vary from semi-deciduous forest (Batouri-Yokadouma) to evergreen forest (Oban). Data on population density, range, regeneration and threats should also be collected to aid future monitoring and management planning.

Necepsia afzelii Prain subsp. ***zenkeri*** Bouchat & J.Léonard
EN B2ab(iii)

Range: Cameroon (SW: 6 km N Ekondo Titi, at Masore. S: Kribi–Ébolowa, 12 km at Kienké FR; Bipindi) and Gabon (Lastoursville).

According to Harris (1999 loc. cit.) this species also occurs in E Region at Lobéké, but this requires verification since the location is at odds with the rest of the range.

The genus *Necepsia* is easily recognised in having large glands scattered on the blade on both surfaces (also known in *Afrotrewia*), otherwise the leaves resemble those of *Alchornea hirtella*. Revised by Bouchat & Léonard, Bull. Jard. Bot. Nat. Belg 56 (1986).

Shrub or small tree 1–6 m, stems dark brown puberulent, leaves alternate, stipules triangular, 5 × 2 mm; petiole 5 mm; blade oblanceolate, 15–28 × 4.5–10 cm, acumen slender, 1 cm; base acute; serrate-glandular, lateral nerves 8 pairs, scalariform; spikes axillary 6–11 cm, interrupted, flowers several per node, buds hairy.

Here this taxon is assessed as Endangered since four locations (above; AOO 16 km² with 4 km² cells) and threats as below are known.

Habitat: lowland evergreen forest.

Threats: forest clearance for urbanisation, extractive industry infrastructure; slash and burn agriculture (Kribi).

Management suggestions: consideration should be given to upgrading the Kienke FR to National Park status given that this species not only hosts this taxon, but several other very rare plant species. Effort is needed to refind this plant in the wild, ideally in a protected area, and to record basic populational data so as to give a baseline for future monitoring. If not rediscovered at Kienke, then Bipindi and Masore locations should be attempted with support and engagement of the local communities that might be involved in protecting plants if successful.

Phyllanthus is a pantropical genus of 200+ species. It is the type and largest genus of the Phyllanthaceae, a segregrate of Euphorbiaceae. It and related genera lack white exudate and leaf-base glands. *Phyllanthus* species have distinctive 'plagiotropic' branches which closely resemble the pinnately compound leaves of papilionoid legumes, and are usually only differentiated by the presence of minute flowers, and later fruits, on the lower surface of this apparent compound leaf, hence the origin of the generic name ('flowering leaf').

The genus is commonly encountered as a weed of habitation and some of these species are medicinal; others poisonous. *Phyllanthus muellerianus* is a spiny forest climber, the bark of which is used to flavour beer widely in Cameroon. Less well-known are the rarer subshrubs of forest, some of which are very rare (*P. nyale, P. caesiifolius*). Brunel & Roux e.g. Willdenowia 14: 379–391 (1984) delimited many new and rare taxa in the 1980's, but more remain to be described. One new species on the brink of publication is *P. kidna* ined. (Challen *et al.* Syst. Bot. in press), a beautiful tree with fleshy fruits, so far known only from the Mefou proposed NP. This and other species as yet undescribed are likely to appear in future editions of this Red Data book.

Phyllanthus aspersus Brunel & Roux.
CR B2ab(iii).

Range: Cameroon endemic (S: 20 km NNE of Bipindi at Nsola).

Here *Phyllanthus asperses* is assessed as Critically Endangered since known from a single locations (above; AOO 4 km² with 4 km² cells), with threats as below.

Herb or subshrub 30–50 cm tall, base woody; branches winged, glabrous; plagiotropic branches 20–30 cm, leaves oblong to elliptic, 15–25 mm, secondary nerve pairs 10, obtuse, base rounded, petiole 1.2–1.3 mm; stipules 1.8 mm; male flowers in small groups in the middle part of the branchlet (more than 20 leaves from the base); pedicel c. 1 mm; tepals 6, oblong, 1–1.3 × 0.5–0.7 mm.

Habitat: lowland evergreen forest at the base of a waterfall among damp rocks.

Threats: slash and burn agriculture.

Management suggestions: rediscovery of this species, collected once previously, in 1976 (*Letouzey* 12681) is needed if basic populational data is to be collected as a baseline for future monitoring, and to assess needs for management planning.

Phyllanthus caesiifolius Petra Hoffm. & Cheek
CR C2a(i), D

Range: Cameroon endemic (SW: Bakossi Mts (2 loc.)).

A monopodial, stoloniferous forest shrublet, *Phyllanthus caesiifolius* is immediately recognised even when not fertile by its bicoloured leaves; the upper surface is dark green in the upper half and whitish blue in the bottom half.

The paper in which the species was published (Hoffmann, P. & Cheek, M. (2003). Two new species of *Phyllanthus* (*Euphorbiaceae*) from southwest Cameroon. Kew Bull. 58(2): 437–446) carries the conservation assessment reproduced in Cheek *et al.* (2004: 157–158) and maintained here in the absence of new data. Neither of the two known subpopulations (c. 15 km apart) contain more than 50 individuals. It appears to be vegetatively apomictic.

Monopodial shrublet 0.2–0.4 m, glabrous; lateral branches horizontal, 8–16 cm; leaves oblong to ellipsoid 1–2.5 × 0.6–1 cm, obtuse to rounded apex and base, upper surface with basal half bluish white, upper half dark green; flowers pale green, c. 1.5 × 1.5 mm, pedicels 4–5 mm.

Habitat: undisturbed submontane forest with *Gymnosiphon longistylus*, *Xylopia africana*, *Oncoba ovalis*, *Dielsantha galeopsoides*, *Pentaloncha* sp. nov.; 1000–1275 m alt.

Threats: forest clearance for agriculture. The eastern location is on a route along which forest is being cleared for farming from the E.

Management suggestions: continued surveys are advised in the altitudinal range indicated in the hope of discovering further sites for the taxon. Production of a conservation poster for use in Bakossi might help locate more sites and educate people as to the rarity of this species.

Phyllanthus caligatus Brunel & Roux
EN B2ab(iii)

Range: Cameroon endemic (W: Bamboutos Mts. Adamawa: Tello, E of Ngaoundéré).

Here *Phyllanthus caligatus* is assessed as Endangered since only two locations (above; AOO 8 km² with 4 km² cells) are known and since threats are as below.

Herb or subshrub 0.3–1.5 m, branches rooting adventitiously, plagiotropic shoots glabrous, smooth, blade oblong, 15–23 × 9–13 mm, rounded, base cuneate, lateral nerves 5–7 pairs, marginal nerve smooth, petiole 0.8–1.6 mm, stipules 1.5–3.3

mm, male flowers produced in clusters on the proximal axils of the branches, female on the distal; male pedicel 3.8–4.4 mm, tepals 6, equal, oblong, 1.3–1.7 × 0.8–1.3 mm, median nerve broad and distinct.

Habitat: not known precisely, but probably submontane forest:grassland interface or scrub

Threats: not known at Tello, but original habitat at Bamboutos Mts is under great threat from clearance for agricultural land and degradation due to grazing practices.

Management suggestions: rediscovery of this species is needed if basic populational data is to be collected as a baseline for future monitoring and to develop a management plan. Tello should be prioritised since two independent records are known at that location including the most recent for the species (*W. de Wilde* 4324 in 1964).

Several other Endangered species also occur at Tello, such as *Justicia telloensis*, so consideration should be given to protecting a location here, perhaps as a National Park.

Phyllanthus dusenii Hutch.

EN B2ab(iii)

Range: Nigeria (Oban) & Cameroon (SW: Korup, Mana R; Mamfe. S: Kribi–Ébolowa; Bipindi-Lolodorf).

Here *Phyllanthus dusenii* is assessed as Endangered since only five locations (above; AOO 20 km² with 4 km² cells) are known, and threats are as below.

The two locations in S Region derive from TROPICOS and are based on specimens determined by Grady Webster which have not been seen by the present writer, nor were they seen or cited by Brunel & Roux (op. cit.) from whom the description below is based.

Shrub to 1.5 m, perennial, orthotropic and plagiotropic stems angled and with the leaf petioles and midribs, densely puberulent; blade elliptic 8–20 × 2.5–6 mm, apex obtuse, base rounded, petiole 1mm, stipules needle-like, 2.2–3.2 mm; male flowers in clusters of proximal axils, pedicel 1 mm, tepals 6, subequal, oblong 1.3–2.5 × 0.5–0.8 mm.

Habitat: fast-flowing water on rock in the lowland evergreen forest belt.

Threats: forest habitats in Nigeria have seen massive destruction and loss in the 100 years since this species was recorded in Oban. Logging, followed by plantation and small-holder agriculture have been drivers.

Managements suggestions: while Mamfe may represent a secure location for the species, the Mana at Korup is the logical focus for conservation efforts since it has the largest number of records, is at least part protected, and has an extensive community of other rheophyte species. Basic populational data should be collected as a baseline for future monitoring and conservation planning. The S Region specimens should be re-examined to confirm their identity.

Phyllanthus kidna Challen & Petra Hoffm. sp. nov. ined.

CR B2ab(iii) + D

Range: Cameroon endemic (C: Mefou proposed NP).

Since it has been estimated that, at the moment, only five individuals of *Phyllanthus kidna* are known from the single location so far discovered (AOO 4 km^2 using 4 km^2 cells) where it is threatened by agriculture, this species was assessed as Critically Endangered by Cheek in Challen *et al.* (Syst. Bot. in press) and in Cheek *et al.* 2011 (in press).

Tree, 10–15 m; monoecious or dioecious, most likely deciduous; trunk cylindrical, dbh 15–20 cm diam., dilated at ground level; orthotropic branchlets (resembling pinnate leaves) terete, simple, with up to 29 leaves per shoot; leaves distichous, glabrous, papery, green tinged red when young; leaf-blades ovate to oblong (-orbicular), 0.9–2.3 × 0.5–1.3 cm, entire, basally rounded to obtuse, apically rounded to obtuse (mucronate), drying discolorous, dull olive adaxially, pale glaucous green abaxially, veins (5)7–9 pair; petioles 1.0–1.3 × 0.3 mm; stipules triangular, 0.5 mm; flowers solitary or fasciculate, the fascicles in the axils of lower leaves of brachyblasts, consisting of 1 pistillate and many staminate flowers, also some solitary pistillate flowers in the axils of leafy plagiotropic shoots; staminate flower pedicels 1.3 mm long, capillary; sepals 4, 0.7–1.3 × 0.5–0.8 mm, equal, suborbicular, erect, apically acute, glabrous, white, occasionally tinged red, margins hyaline; petals 0; mature fruit 2 locular, with 1–2 ovules per locule, indehiscent, 1.3 × 1.1–1.3 cm, subglobose with a fleshy pericarp, drying mid-brown; pedicels 4.5 mm; seeds irregularly ovoid, 0.6 × 0.8 cm.

Habitat: cut-over semi-deciduous forest planted with Cocoa (*Theobroma*). Other surviving indigenous tree species include *Triplochiton scleroxylon* K.Schum., *Celtis* L. spp., and *Baillonella toxisperma* Pierre.

Threats: one of the two individuals (*Cheek* 13098) seen by the author in 2006 was regrowing from a stump of a tree that had been cut in the previous 1–2 years, possibly to facilitate cocoa cultivation or to furnish firewood or poles. Much of the area of the proposed National Park has previously been logged for timber for export and subsequently been cultivated for small-holder cocoa or manioc which was ongoing during fieldwork there in 2002–2006.

Management suggestions: it is to be hoped that the species will be found elsewhere in the Yaoundé area and that in future its threatened status will be reduced. Surveys to address this are suggested by asking in communities for this tree using the Ewondo name "kidna". The surviving trees in the Park should be flagged, signposted, and made the subject of an education campaign in the area to avoid their being cut down by mistake. Seed should be collected and both seed-banked and raised and replanted in the Park so that the wild population in the proposed National Park can be strengthened.

Phyllanthus nyale Petra Hoffm. & Cheek
CR C2a(i), D

Range: Cameroon (SW: Bakossi Mts (2 sites)).

This monopodial forest shrublet is distinguished from the other six species of *Phyllanthus* in Bakossi by its winged, not terete, branchlets. It is named for the type locality, Nyale Rock, a dramatic inselberg located east of Nyandong on the footpath to Kodmin. The paper in which the species was published (Hoffmann, P. & Cheek, M. (2003). Two new species of *Phyllanthus* (*Euphorbiaceae*) from southwest Cameroon. Kew Bull. 58(2): 437–446) carries the conservation assessment reproduced in Cheek *et al.* (2004: 158) and maintained here in the absence of new data. Neither of the two known subpopulations contain more than 50 individuals. *Phyllanthus nyale* was published with *P. caesiifolius* in a paper that included the conservation assessment above (Hoffmann & Cheek, Kew Bull. 58: 437–446 (2003)).

Monopodial shrublet resembling *P. caesiifolius* but leaves not bicoloured above, lateral stems strongly winged, the wings as wide as the main stem.

Habitat: submontane forest; 1000 m alt.

Threats: forest clearance for agriculture.

Management suggestions: as for *Phyllanthus caesiifolius*.

Phyllanthus raynalii Brunel & Roux
EN B2ab(iii)

Range: Cameroon (S: Ébolowa) and Gabon (Belinga).

Here *Phyllanthus raynalii* is assessed as Endangered since only two locations are known (above; AOO 8 km² with 4 km² cells) and since threats are as below.

Shrub to 3 m, branched, glabrous, plagiotropic branches 8–10 cm, laterally ridged the ridges denticulate, leaves 16–22; blade lanceolate-ovate 9.5–27 × 4.5–11 mm, obtuse, base cordate, lateral nerves 7–10 pairs, petiole 0.7–1.2 mm, stipules 1.5-2 mm; male flowers in clusters in axils of proximal leaves, pedicel c. 1 mm; tepals 6 subequal, oblong, 1.3–1.6 × 0.5–0.7 mm, midrib 0.2–0.3 mm wide; distal axils with single female flowers.

Habitat: evergreen forest (inferred).

Threats: Belinga is a proposed iron-ore open-cast mine.

Management suggestions: rediscovery of the plants at the original site at Ébolowa so as to secure their protection with the support of the local community since there is no National Park present.

Pseudagrostistachys africana (Müll.Arg.) Pax & K.Hoffm. subsp. ***africana***
VU A2c

Range: Ghana, São Tomé & Príncipe (São Tomé), Equatorial Guinea (Bioko), SE Nigeria (Obudu Plateau (1 site)) and Cameroon (SW: Mt Etinde; Mt Kupe and Bakossi Mts; Lebialem. NW: Bali Ngemba FR; Mbaw Plain. C: Mefou proposed NP. S: Kribi).

Listed as VU A1c, B1 + 2c by Hawthorne in 1997 (www.redlist.org (IUCN 2003)), this monotypic genus, a tree restricted to submontane forest (apart from at one lowland site in Ghana and one S of Yaoundé at Mefou proposed NP: it appears also to be local in swamp forest), probably now has its largest subpopulation in Bakossi, where it is fairly secure. It was reassessed as VU A2c, B2ab(iii) in Cheek *et al.* (2004: 158) on the basis of more extensive disturbance data in Cameroon and according to the modified IUCN criteria of 2001. That assessment is maintained in Harvey *et al.* (2004: 63), Harvey *et al.* (2010: 75) and Cheek *et al.* (2011: in press), and is maintained here as far as criterion A is concerned, but the B assessment is dropped since more than 10 sites are now known, now that it has been discovered at Mefou and Lebialem. It is estimated that more than 30% of the population has been lost in the last 100 years due to submontane or swamp forest habitat destruction in major parts of its range, principally Ghana, Nigeria and, in Cameroon, the Bamboutos and Bamenda Highlands. This destruction is ongoing. This is one of several species showing habitat disjunction between lowland swamp and submontane forest. In the field it is readily recognisable by the large Irvingiaceae-like sheathing apical stipule and the long-scalariform tertiary leaf venation.

Tree 5–20 m tall; bark whitish with green spots, slash brown to red; leaves leathery, elliptic, 30 × 11–18 cm, subacuminate, base rounded then slightly decurrent with a pair of flat glands, ± serrate, nerves 21–24 pairs, venation scalariform, petiole 2.5–5 cm; stipule single, 3 cm, caducous, scar completely encircling stem.

Habitat: submontane, or rarely lowland swamp forest; 500–1500 m alt.

Threats: forest clearance for wood and agriculture (Obudu Plateau, Bamboutos/Bamenda Highlands: Lebialem and Bali Ngemba FR).

Management suggestions: the status of this taxon in São Tomé and Bioko needs more investigation. Bali Ngemba FR represents the most easily accessible and dense population of the taxon, followed by Mt Kupe and Mt Etinde; these are the more promising sites for demographic studies of the taxon that would inform management planning. It is possible that surveys in lowland swamp forest will find more sites for this species.

Thecacoris annobonae Pax & K.Hoffm.
EN A2c

Range: Equatorial Guinea (Annobón) and Cameroon, (SW: Korup (1 coll.); Mt Cameroon at Mungo R FR (1 coll.); Kumba–S Bakundu, Banga (9 coll.); Kumba–Mamfe (6 coll.); Mt Kupe at Ngussi (1 coll.).

Judging by the number of collections (nine), this species was most abundant in the Kumba–S Bakundu area. It has not

been seen in S Bakundu in decades, although a single collection was made nearby in the Mokoko FR (Cable & Cheek 1998) in the 1990s. From Kumba its range extends or extended, northwest along the valley towards Mamfe (6 collections), eastwards towards Tombel, then up the Jide trough to Ngussi (1 coll. each).
This assessment was originally made in Cheek *et al.* (2004: 158) and is maintained here since a decline of more than 50% in the population over the last 100 years can be estimated
Tree 3–10 m, no exudate; stems glabrescent, grey-wrinkled; leaves drying dark green-brown, oblong to obovate, c. 20 × 9 cm, acuminate, obtuse, entire, nerves c. 6 parts; petiole 0.5 cm; stipule caducous; male inflorescence 1 per leaf axil, interrupted spikes c. 20 cm; flowers yellow, 2 mm;
Habitat: lowland evergreen forest; to 400 m alt.
Threats: clearance of forest for wood, agriculture and urban expansion; large parts of S Bakundu have been cleared for yam plantations (Cable & Cheek 1998: xxxvi) and the Kumba–Mamfe Rd is being massively upgraded with concomitant expansion of agriculture likely. 50% loss of the population of *T. annobonae* is estimated to have occurred in recent decades and this loss is continuing.
Management suggestions: the type, and modern material of *T. annobonae* from Annobón, need to be examined in more detail to establish whether it is the same entity as the Cameroonian material (see FWTA 1: 372 (1963)). A survey in Annobón is needed to confirm survival of the species there. Although it has not been found in surveys at Bakossi in the late 1900s, lowland forest was not comprehensively investigated and it may well survive there. Once rediscovered, a plan for the management of *T. annobonae* can be made.

FLACOURTIACEAE (ACHARIACEAE/SALICACEAE)

assessed by Martin Cheek

The family traditionally known as Flacourtiaceae has been divided into two on molecular phylogenetic grounds by workers in that area. However this division has been rejected (see Brummitt in Heywood *et al.* (2007). Flowering Plant Families of the World, RBG, Kew), since there are no means of reliably separating the two groups on physical evidence, although there is a tendency to different flower sizes. The two groups are: those with a tendency to larger flowers (redefined and expanded Achariaceae) and those with a tendency to smaller flowers (redefined and expanded Salicaceae).

Flacourtiaceae have alternate, simple often pulvinate leaves, with or without evident stipules. When wounded they generally lack a distinctive scent or coloured exudate, and the leaves lack any type of gland. In the absence of flowers, they are most likely to be confused with Euphorbiaceae (usually, but not always, distinguished by the presence of leaf glands, and, sometimes, white exudate or stellate hairs) and/or Tiliaceae/Sparrmanniaceae (distinguished by a tendency to having basally 3 or 5-nerved leaves with stellate hairs). In flower or fruit, Cameroonian Flacourtiaceae are easily separable from Euphorbiaceae/Tiliaceae (traditional sense) in having three parietal placentas. The ovary is superior with a single style-stigma surrounded by numerous free stamens (but four separate style-stigmas and basally united filaments in *Trichostephanus*) and petals in a flower with radial symmetry.

No Flore du Cameroun account exists for this family, although there is a recent Fl. Gabon account (Breteler & Hul, Fl. Gabon 34 (1995)) in which 13 genera and 34 species are recognised. Consequently it is likely that one or more additional threatened species of this family have been passed over inadvertently.

One of the threatened species below was first described in the genus *Oncoba*. Subsequently segregate genera *Lindackeria*, *Caloncoba* and *Camptostylus* were erected. Recently Breteler and Hul (1995 loc. cit.) have convincingly argued for sinking these back into *Oncoba*. Interestingly, while most *Oncoba sens. lat.* fall into newly defined Achariaceae, at least one falls in newly defined Salicaceae, on molecular evidence. In addition to the species treated as threatened below, several others are likely to appear in a second edition of this book:

Casearia prismatocarpa Mast. (syn. *Casearia calodendron* Gilg) from Guinea (Conakry), Sierra Leone, Ivory Coast, Ghana, Togo, Nigeria, Cameroon, Congo (Kinshasa) appears to be borderline VU/NT with only 10 or 11 locations. Further analysis for a future edition of this book might well show that it be treated as threatened. For the time being it is treated as Near Threatened here.

Dovyalis sp. nov. of submontane forest in Kupe-Bakossi (Cheek *et al.* 2004: 295) is likely to be assessed as threatened once it is formally described.

Oncoba ovalis Oliv. (syn. *Camptostylus ovalis* (Oliv.) Chipp), a submontane tree of the Cameroon Highlands extending from Mt Cameroon, through Bakossi, before jumping to the Lebialem Highlands and possibly Nigeria (Ikom, *Keay* in FHI 2848 "*Camptostylis* sp. nov. aff. *ovalis*) was listed as LR nt in Cable & Cheek 1998: lii–liii). It may well also merit VU A3c status due to population reduction due to habitat destruction. For the time being it is treated as Near Threatened here.

Oncoba sp. nov. [Feewu a wu (Kom: Clement Toh)] was listed in Cheek *et al.* (2000: 62) as Extinct in the Wild being considered formerly endemic to Mt Oku and the Ijim Ridge, but then known only in cultivation. Subsequently a single plant was found in the wild at Bali Ngemba (Cheek in Harvey *et al.* 2004: 99) but since it had not been formally described and named, it was not possible to give it an IUCN threat status. This is still the case. Once addressed, this spiny (sub)montane forest tree is likely to be rated as CR.

Dovyalis cameroonensis Cheek & Ngolan
CR D

Range: Nigeria (Obudu (1 coll.)) and Cameroon (NW: Bamenda Highlands at Kilum-Ijim (several sites); Baba II (1 coll.)).

This taxon is probably closely related to *D. spinosissima* Gilg from similar habitat (montane lake margins) in Kivu, Congo (Kinshasa) to Malawi. It differs by the glabrous fruits (not hairy), flowers with 14–21 styles (not 8–12) and much larger flowers. This species is only known with certainty from the forest in the crater of Lake Oku (16 trees) and from a site with eight trees between Oku-Elak and Nsoh in forest above Simonkoh (Kemei and Innocent Wultoff pers. comm. November 1999) though was probably formerly more widespread. Trees sometimes grow at the very edge of the lake but also occur c. 50 m up near the rim of the crater on the S side. It was reported in forest above Oku-Elak, but becoming rare there fide Peter Wambeng (November 1996) and this has not been confirmed since. An unlocated ('Bamenda district') collection in male flower (*Daramola* in FHI 41575 at Kew and Yaoundé herbaria) made on 5 July 1959, almost certainly belongs to this species, but we have not yet compared male flowers. Oku was formerly part of Bamenda District. The foregoing notes are taken from Cheek *et al.* (2000: 61–62) were, as *Dovyalis* sp. nov., it was assessed as CR D. Subsequently the species was formally described by Cheek and Ngolan (Kew Bull. 61(4): 597–600; fig. 1, map (2007)) by which time additional data had come to light. One of the Simonkoh trees was felled, possibly for firewood or tool handles (R. Ngolan pers. obs. March 2003). Both this site and that at Lake Oku are within the boundary of the BirdLife supported Kilum-Ijim Forest. The second site has been proposed as a Plant Life Sanctuary and the former has been included in the management plans of the Mbai Community Forest wherein the *Dovyalis* trees are specifically mentioned as being important for conservation (I. Wultoff pers. comm.). A third locality was discovered in the community forest of Baba II village near Bali Ngemba FR in October 2001, where a single fruiting tree was discovered by Martin Etuge in the course of a day's inventory by 20 people (M. Cheek & R. Ngolan pers. obs.). A fourth location was discovered in a forest patch in the area of Santa Mbei. The exact number of trees is unknown, but thought to be less than five. A fifth location was discovered when examining specimens at the Kew Herbarium: *J.B. Hall* 2912 from the Obudu Cattle Ranch in SE Nigeria. However, it is not certain that this site survives since destruction of natural habitat in this area has proceeded apace since the specimen was collected in 1973 (E. Obot pers. comm. 2002). Recently (Sept. 2003) we have learnt from Wultoff (Tah in litt.) that additional localities have been discovered for the species. We are treating these with caution until we have verified either specimens or live trees in person due to the possibility of confusion with other spiny trees with similar leaves, such as *Oncoba* sp. nov. (Flacourtiaceae) and *Maytenus* spp. (Celastraceae). The potential additional localities are all in the area protected by the Kilum-Ijim project, as follows. Kilum site: Nsoh, Kevu, and the summit. Ijim site: Afua, Mbesa, Ijim plateau, and Ngemsiba.

Since a total of only 36 trees have been recorded, the CR D assessment was maintained in the protologue, as it is here, no new data being available apart from the EOO, calculated here as 3272 km^2.

Dioecious, ?deciduous tree 4 to 8 m tall, bearing spines to 4 cm long on branches near the trunk. Leaves papery, elliptic, c. 10 × 5 cm, apex acute, margin serrate. Flowers 3–4 cm diam., green and white, felty. Fruits c. 6–7 cm diam., glossy, ripening yellow, soft. Seeds numerous, elliptic, c. 8 mm long, flattened. In flower and young fruit in early November. Mature fruits seen in late November.

This species has been confused with the even rarer *Oncoba* sp. nov. with a similar distribution. Both are small, spiny trees of the Flacourtiaceae. However our species has soft (not hard) fruits and greenish flowers, not so showy as the white petalled, yellow stamened *Oncoba.*

Habitat: evergreen montane forest; c. 1900–2200 m alt.

Threats: timber extraction (heavy poles c. 15 cm diam. for building construction) was noted in the area where the first trees were seen in the Oku crater-lake forest and threatens the existence of this species, as does usage as a fuel wood and clearance for agriculture.

Management suggestions: felling of trees of this and other species in the lake Oku crater should be prevented. An attempt should be made to mark, record and monitor all trees of this species. A check should be made to study seedling regeneration. Is recruitment observable? A planting programme is advisable using seedlings raised from harvested fruit, continuing that already established with ANCO [the Apiculture and Nature Conservation Organisation] which has incorporated seedlings of this species in its reforestation plantings with the local community at the forest of Dom.

Homalium hypolasium Mildbr.

VU B2ab(iii)

Range: Equatorial Guinea (Rio Muni: Nono) and Cameroon (SW: Mt Kupe. C: Mbam at Yoko. E: Yaoundé-Deng Deng; Yokadouma; Lomié (1 coll. each)).

Described in the early 20th century from two collections (*Tessmann* 485 collected in 1908 from Rio Muni and *Mildbraed* 8508 collected in 1914 NE of Yaoundé), there appeared to have been no subsequent collection of *H. hypolasium* until 1985 (*Thomas* 5100 at Mt Kupe), consequently the species was assessed in Cheek *et al.* (2004: 159) as EN (three locations, with threats). However, three additional locations have been revealed by research at YA by Onana (see above), requiring a reassessment here to Vulnerable since there are now six locations (above, AOO 24 km^2 with 4 km^2 cells) and threats as below. EOO is calculated as 178322 km^2.

Tree 4.5–8 m; leaves papery, drying grey-brown above, green below, leaves elliptic, c. 13 × 8 cm, apex emarginate, base rounded, margin finely serrulate; lateral nerves 12 pairs, lower surface densely velvety-yellowish brown tomentulose; stipules caducous; inflorescence terminal, c. 20 × 15 cm, with 5–15 branches; petals in fruit c. 2 mm.

Habitat: lowland semi-deciduous/evergreen forest and submontane evergreen forest, to 1100 m alt.

Threats: forest clearance for agriculture and timber at Mt Kupe and probably (site visits needed to confirm) at the other two sites.

Management suggestions: Mt Kupe offers the most likely opportunity of refinding this species. It is advised that an effort be made to do this, and if successful, that further individuals be sought and censored in the area, together with levels of regeneration. If refound, conservation education, perhaps in the form of a poster, is suggested to benefit the local population. Surveys at the other sites should be conducted if possible. Further work is needed to confirm identifications of the known specimens since it seems unlikely that such a rare species should occur in such a wide range of vegetation types.

Oncoba lophocarpa Oliv. (syn. *Caloncoba lophocarpa* (Oliv) Gilg)

VU A3c+B1ab(iii)

Range: Cameroon (SW: Mt Cameroon (c. 6 coll.), Kupe-Bakossi (13 coll. at 6 sites). NW (Bamenda Highlands): Bali Ngemba (2 coll.)).

This canopy tree of submontane forest is notable for the rope-like inflorescences that can bear fried-egg-like flowers several metres from the trunk, where they emerge from the leaf litter like parasites. Previously this species had been listed as LR nt (Cable & Cheek, 1998: lii–liii) on the basis that it was restricted to Mt Cameroon where its habitat was considered unthreatened. Independently it was assessed as VU D2 under the name *Caloncoba lophocarpa* on the basis of its occurring at Banguem, Mamfe, Bakossi Mts and Mt Cameroon (WCMC 1997, based on forms by Peguy, T: www.redlist.org (IUCN 2003)). We then discovered that its range and number of sites are larger than previously thought (see above) and that it no longer qualifies as VU D2 but assessed it as VU A2c (in Cheek *et al.* 2004: 159; Harvey *et al.* 2004: 63–64) since forest losses in the Bamenda Highlands part of this extended range equate to over 30% habitat loss over the last 100 years over the species range as a whole. This last assessment is modified here to A3c since population reduction due to habitat losses are ongoing at locations such as Bali Ngemba, and B1 status is added since the EOO is calculated here as 6254 km^2.

Tree (6–)12–25 m; leaves elliptic or elliptic-ovate, c. 14 × 7 cm, long acuminate; inflorescences cauliflorous, c. 2 cm long, or on long, whip-like, mostly leafless branches radiating 3–5 m from the trunk on the ground; flowers 5–7 cm diam.; sepals 3, 1.5 cm; petals white, 8, c. 3 cm; fruit ovoid, c. 5 × 4 cm, strongly 8–12 winged.

Habitat: submontane evergreen forest; (400–)800–1950 m alt.

Threats: forest clearance for wood and agriculture. This is most significant at the lower parts of the altitudinal range (below 1000 m) at all three areas. In the Bamenda Highlands the tree is probably now restricted to the Bali Ngemba FR. These extensive highlands were probably once home to the main subpopulation of the species given the density of the species at the Bali Ngemba remnant. Elsewhere in the Bamenda Highlands, 25% loss of forest in the eight years between 1987–1995 is recorded by Moat in Cheek *et al.* (2000: back cover). With 13 collections at six sites, Kupe-Bakossi appears to be the stronghold for this species, where it is relatively secure and unthreatened.

Management suggestions: a population census, focussing upon the known sites at Kupe-Bakossi and Bali Ngemba, should be carried out and specific threats recorded. A poster campaign focussing on this striking species may serve to promote forest conservation to the local communities.

Trichostephanus acuminatus Gilg

VU B2ab(iii)

Range: Cameroon (S: Ébolowa, 15 km S, at Nkoemvom; Bipindi at Amugu) and Gabon (Lastoursville,10 km S at Micouma; 30 km SW at Roungassa; 70 km W at Lissacho; 20 km SW at Bangassou).

Here *Trichostephanus acuminatus* is assessed as Vulnerable, since there are six locations (above; AOO 24 km^2 with 4 km^2 cells) with threats as below. EOO is calculated as 62299 km^2.

Trichostephanus is a genus of two species of small trees or shrubs restricted to Gabon and Cameroon. First described a century ago, the second species was published by Breteler in 1990 (J. Bot. Belg. 60 (1-2): 143–146). This treatment is based mainly on the excellent paper by N. Hallé & J.F.F.E de Wilde (Adansonia 18(2): 167–18 (1972)).

Shrub or small tree 2–10 m, stems zig-zag; internodes pubescent when young, leaves simple, alternate, distichous, blade obovate, 9–20 × 3–9 cm, acumen 10–27 mm, base cuneate, margin entire or with 1–4 teeth to 2 mm on each side, secondary nerves 5–6 pairs, finely reticulate, petiole 2–4 mm, grooved above; stipules linear 3–20 mm; inflorescence a contracted cyme, 1–6 flowered, enveloped in spathe-like bracts, tepals 4; stamens in a single whorl, numerous, united at base in a cylinder, inner surface with a corona of staminodes; male flowers with pistil minute; female superior ovary cylindrical, styles 4, separate; fruit unknown.

Habitat: lowland evergreen forest, sometimes near *Raphia* swamps.

Threats: slash and burn agriculture; logging followed by agriculture.

Management suggestions: N. Hallé (loc. cit.) describes the population biology of this species. Repeated observations at the main Ébolowa site suggested that fruit are not found although the species dominates the understory in an area of 1000 m^2, with dozens of stems as a range of heights. He deduced that reproduction was solely vegetative and that all individuals are but a single clone, which seems credible, and relevant to planning for conservation management. A priority for action should be to protect the most recently discovered and best identified Cameroonian site- on the right bank of the river Seng (tributary of the Mila R) at the edge of the station de Recherche du Cacao de N'Koemvone [Nkoemvom] (2°49'30''N, 11°8'E).

GESNERIACEAE

assessed by Martin Cheek

A pantropical family of c. 1400 species, sharing similar flower structure to the closely allied and larger and more diverse family *Scrophulariaceae*. African Gesneriaceae are herbs which have brittle, succulent, pubescent leaves and stems, and more or less zygomorpic, (bilaterally symmetrical) tubular corollas. The two most species-diverse genera in Africa are *Streptocarpus* and *Saintpaulia* both of which are well-known in European households as pot-plants. *Saintpaulia* are E African, while *Streptocarpus* are most diverse in S and S-C Africa, with a few species extending to W Africa including two, or perhaps three, in Cameroon. Cameroon is comparatively poor in species diversity in this family, but rich in monotypic genera - *Schizoboea*, *Epithema*, *Acanthonema*, besides *Streptocarpus*, and (below) *Nodonema*, occur. *Trachystigma* is endemic to the nearby Cristal Mts in Gabon. Ecologically, this family in Cameroon prefer damp rocks or banks in the shade of submontane forest.

Flore du Cameroun 27 by B.L. Burtt in 1984 treats Gesneriaceae. *Streptocarpus muscicola* Engl. was only known from the syntypes, since destroyed at Berlin, from the Mt Oku and Banyo areas (See Flore du Cameroun 27 loc. cit. for discussion). While it may prove an endemic, threatened species, I suspect it from its range to be the widespread *S. elongatus*. Collections from the type localities should be obtained and studied to resolve this question.

Schizoboea kamerunensis (Engl.) Burtt is regarded as monotypic but occurs in three discrete areas of Africa: Mt. Cameroon and Bioko; Rwanda and Kivu; S Tanzania. Burtt (loc. cit.) states that differences between the populations probably amount to ecological forms. However it would be worth reinvestigating the subject with the aid of additional specimens and also, using molecular methods. Should the material be taxonomically separable, the entity at the first area would might well rate as threatened since only two records are known from Mt Cameroon.

Nodonema lineatum B.L.Burtt

VU B2ab(iii)

Range: Nigeria (Ogoja, Obudu Cattle Range and Boshi Extension FR (5 coll.)) and Cameroon (SW: Nta Ali (1 coll., 1 obs.); Kupe-Bakossi Mts (4 locs.)).

First collected by Jane Medler in 1973 in Nigeria, and described in 1981, this monotypic genus resembles an African Violet (*Saintpaulia*). Its habitat requirements are highly specific, but occur at several of the 'rocks' (inselbergs) in Kupe-Bakossi, notably Bime Rock, Kupe Rock and two smaller rocks above Nyasoso.

The assessment in Cheek *et al.* (2004: 159) is maintained here since no new data are available: Vulnerable since it is known from only six locations (above, AOO 24 km^2 with 4 km^2 cells) and threats as below.

Rhizomatous, rosulate herb; petioles long; leaves cordate, 3–10 × 2.5–6.5 cm, upper surface softly-pubescent, lower surface subglabrous; corolla tubular, lobed, white with purple guide stripes in tube; stamens short; fruit subglobose.

Habitat: perennially wet, mossy, vertical granite in forest shade; 800–1800 m alt.

Threats: removal of shade due to forest clearance is the most likely threat. The rock face habitat itself is not likely to be mined in the near future at the locations known in Cameroon due to current lack of road access, but plants are vulnerable to rock falls. Much forest clearance in Ogoja has occurred in recent decades and this has possibly threatened sites for the species at that location.

Management suggestions: data on the population size and demography of this species would inform better conservation management of this taxon. Such studies would best be carried out from Kupe village or Nyasoso both of which have subpopulations nearby.

GUTTIFERAE (CLUSIACEAE/HYPERICACEAE)

assessed by Martin Cheek

Variously defined as either two, or as a single family of shrubs and trees. All species share opposite, simple, exstipulate leaves and produce yellow (rarely white or orange) exudate from minute canals in the tissue, or instead, having secretory cavities separated (and not continuous as canals), appearing as block dots or dashes on the surface of the leaf or within, when the leaf is held up to the light.

The **Clusiaceae** (excluding the Hypericaceae) in Cameroon are mainly trees of evergreen forest, all producing yellow exudate when wounded. Their regular flowers often have five, fleshy, free, petals, the ovary with five short, stout, often flattened, often partly united style-stigmas. The best known species in Cameroon is probably *Garcinia kola*, which produces 'Bitter Kola' widely grown and sold for its bitter tasting, medicinally stimulating seeds. Its fruit flesh is also edible. Perhaps surprisingly it has been assessed as threatened by other workers. Several species of *Garcinia*, principally *G. mannii* are valued as sources of 'chewsticks' for cleaning teeth, and many wild trees are felled to meet this demand both inside Cameroon and for export to Nigeria. The bark and seeds of species such as *G. lucida* are used medicinally and in flavouring alcohol.

Mammea, *Pentadesma*, *Allanblackia* and *Symphonia* are the other main genera of this group in Cameroon. While present in lowland forest, they are perhaps most prevalent in submontane forest. Generally Clusiaceae are very good indicators of intact, good quality, diverse evergreen forest. There is no Flore du Cameroun or Fl. Gabon account. Revisions by Bamps for several genera in Congo are useful for Cameroon e.g. *Allanblackia* in Bull. Jard. Bot. Nat. Belg. 39: 345–372 (1969), but an African-wide revision is badly needed for *Garcinia*.

The **Hypericaceae** are typically shrubs or subshrubs of montane areas, especially *Hypericum* itself. *Psorospermum febrifugum* is a characteristic shrub of the very widespread Guinean savanna belt at a wide range of altitudes, but other species of the genus are found at the forest; grassland boundary or transition zone, while *Harungana madagascariensis* is probably the best known species of the group in Cameroon. It is a forest pioneer that rapidly colonises farm fallow and also has medicinal uses. It is the Hypericaceae that lack free-flowing yellow exudate, having instead black or translucent cavities in or on petals and leaves. The flowers have membranous petals and an ovary with five filamentous styles each with a swollen head. Again there is no Flore du Cameroun account, which can make species identification difficult for *Psorospermum*, especially.

Apart from the species treated below, several others appear Near Threatened, but when re-examined with better data may well be evaluated in future as Threatened:
Garcinia conrauana Engl. (Cameroon and Gabon)
Garcinia letestui Pellegr. (Cameroon and Gabon)
Garcinia preussii Engl. (Cameroon, Rio Muni, Gabon)
Garcinia staudtii Engl. (Nigeria and Cameroon)
Psorospermum densipunctatum Engl. (montane Sierra Leone, Nigeria and Cameroon)
Psorospermum membranaceum Wright (Cameroon, Rio Muni, Gabon)
Psorospermum staudtii Engl.(Cameroon, Congo(Kinshasa), Gabon)

Allanblackia gabonensis (Pellegr.) Bamps
VU A2c

Range: Equatorial Guinea (Bioko), Cameroon (SW: Kupe-Bakossi (11 coll.); Lebialem-Bamboutos (Fosimondi). NW: Bali Ngemba, Bamenda Highlands (several coll.). W: Mt Bana, Batcham. C: Yaoundé (1 coll.). S: Ébolowa, Sangmélima (1 coll. each)) and Gabon (Chaillu Mts: Moubighou, Moucongo and Tcyengue (1 coll. each)).
This tree is conspicuous for carpeting the submontane forest floor with its pale lemon-coloured fallen flowers (usually red in the Lebialem and Bali Ngemba subpopulations), each about 6 cm across. The largest part of its domain was probably once the Bamenda Highlands, where submontane forest is now confined to a few small parcels, the largest of which is at Bali Ngemba. Forest in these highlands is still being lost: 25% of surviving forest in one area disappeared in an eight year period up to 1995 (Moat in Cheek *et al.* 2000: back cover). In another area 50% was lost in 15 years 1988–2003 (Baena in Cheek *et al.* 2010: back cover). Overall, it is estimated that more than 30% population reduction of this species over the last 100 years has occurred due to habitat loss. For this reason the species was assessed as Vulnerable in Cheek *et al.* 2004: 160, Harvey *et al.* 2004: 64 and ibid 2010: 75–76, and these assessments are maintained here. Mt Kupe and the Bakossi Mts are now probably the stronghold for *A. gabonensis*. Elsewhere the species occurs on several of the small hills dotted through the forest belt in S and C Region, finally extending into the Chaillu Mts where it was first discovered and named (in earlier references I erroneously stated that it occurred in the Cristal Mts). Bamps (loc. cit. 1969) was first to apply the name to material from Cameroon, which had in FWTA 1 (1958) been referred to with uncertainty as to its identity. Bamps was less certain

about the specimen from Bioko being conspecific. It had been referred to as *A. monticola* Mildbr. in FWTA. The absence of this species from the Cristal Mts seems strange. EOO is calculated as 144789 km^2

Tree 10–30(–45) m; leaves obovate, c. 12 × 5 cm, acuminate, obtuse to rounded, lower surface matt; lateral nerves c. 15 pairs, resin canals inconspicuous, midrib pinkish red; petiole 1.5 cm, with axillary cup; male inflorescence terminal, 3–15-flowered; flowers pale yellow or pink, 4.5 cm diam.; petal apex rounded; staminal phalanges 5, with anthers on both upper and lower surface; central disc 5-lobed, slightly undulate; fruit ovoid c. 15 cm.

Habitat: submontane forest; 700–1500 m alt.

Threats: continued forest clearance for agriculture and wood particularly in the Lebialem/Bamboutos/Bamenda Highlands and Yaoundé areas.

Management suggestions: this species is fairly secure in the upper part of its altitudinal range in Kupe-Bakossi which should be the logical focus of conservation efforts, but enforcement of the forest reserve boundary is needed if it is to survive at Bali Ngemba. Surveys should be conducted on the forested hills from which it has been collected elsewhere in Cameroon and also in the Chaillu Mts, to determine whether it survives there and whether it can be protected at any of these sites.

Allanblackia stanerana Exell & Mendonca
EN B2ab(iii)

Range: Cameroon (S: Kribi, 50 km SE), Congo (Kinshasa) (Mayumbe, Kisafu, Maduda) and Angola (Cabinda: Mayumbe–R Lufo and Belize).

Here *Allanblackia stanerana* is assessed as Endangered since three locations (above; AOO 12 km^2 with 4km^2 cells) are known with threats as above. EOO is calculated as 12542 km^2

Based on material from the Mayumbe Mts of Cabinda (Exell and Mendonca p. 20 in J.Bot. 74 (suppl.) 1936), this species was maintained with extended range by Bamps (op.cit. 1969: 354) but remains very rare. No new records have been published since 1969.

Tree 25-30 m, trunk to 1 m diam., leaf-blades elliptic to oblanceolate, 11–18 x 3.5–6 cm, acuminate; petioles 8–10 mm; peduncles to 10 cm, sepals 6–10 × 7–8 mm (outer); petals white, 22 × 14 mm; stigmatic disc lobes 5 × 5 mm obcordate, more or less smooth; stamina phalanges 12–15 × 1.5–3 mm, both upper and lower surfaces, and apex, with anthers; fruit 7 cm long, obpyriform, 5-ridged.

Habitat: submontane evergreen forest (deduced)

Threats: slash and burn agriculture (Cameroon)

Management suggestions: rediscovery of this species at its only known location in Cameroon is advised. When refound, local community leaders should be advised to protect the tree and seed collection organised so that new plants can be propagated for introduction to safe sites within its range. Basic populational data should be collected as a baseline for monitoring.

Garcinia brevipedicellata (Bak.f.) Hutch. & Dalz.
VU B1+B2ab(iii)

Range: Nigeria (Oban numerous coll; Acharan FR at Kabba) and Cameroon (SW: Kumba area including S Bakundu; Kumba–Mamfe Rd at Konye; ibid, Weme; Manje, Ndian; Rumpi Hills nr Madie R; Masaka-Batanga). A single record from Gabon (*McPherson* 17045, Gabon database) is probably an error since it has white petals and a green stigma.

Here *G. brevipedicellata* is assessed as Vulnerable since there are effectively eight locations (above; AOO 32 km^2 with 4 km^2 cells) and threats as below. EOO is calculated as 13272 km^2

Tree 5–15 m, slash white; glabrous; stems finely ridged when dry; leaves obovate to elliptic, 9–14 × 3–6.5 cm, acumen abrupt, 1cm, base cuneate, papery, lateral nerves 6–10 pairs, arising at 60° from midrib, uniting near margin, resin canals conspicuous on lower surface, black, sinuous, running longitudinally; petals 4, yellow; staminal bundles free, much longer than ovary, stigma red-purple; fruit subglobose, 2–3 cm, 2-seeded.

Distinctive in flower due to the yellow petals and red or purple flat stigma.

Habitat: lowland evergreen and semi-deciduous forest

Threats: used as chewing sticks; its lowland evergreen forest habitat is being steadily cleared for agriculture throughout its range.

Management suggestions: surprisingly this species seems to be absent from Korup so focussing on conservation here is not an option. Rather, the Rumpi Hills reserve may be the best choice although its protected status is not certain. Basic populational data should be collected as a baseline for monitoring.

Garcinia densivenia Engl.
EN B2ab (iii)

Range: Cameroon endemic (Littoral: Douala–Edéa reserve at Lake Tissongo. S: Bipindi, possibly also Campo-Ma'an. SW: possibly Mt Cameroon at Bimbia).

Described from Zenker specimens from Bipindi the only other reliable material I have seen is that from Lake Tissongo in Douala-Edéa determined by Bamps in May 1978 (*Waterman & McKey* 869). The specimens cited from Mt Cameroon at Mabeta-Moliwe were sterile and it can be easy to misidentify sterile specimens of this genus. *Talbot* 229 from Oban was misidentified as this species in FWTA ed. 1. *Chevalier* 33148 from Douala was also misnamed as this taxon. Tchouto (2004) cites three locations for this tree in Campo-Ma'an but in view of the observations above, they should be verified before being finally accepted. Here *G. densivenia* is assessed as Endangered since there are effectively three locations (above; AOO 12 km^2 with 4 km^2 cells) and threats as below. EOO is calculated as 10463 km^2

Understorey tree, trunk to 15 cm diam. with scant, cream latex; stems 2 winged below leaf insertion, 2-ridged on remaining face, glabrous; leaves elliptic 14–20 × 5–9 cm, acumen 1 cm base obtuse-acute, lateral nerves 6–8 pairs on each side, impressed above, prominent below, resin canals not conspicuous, tertiary and quaternary nerves finely reticulate, and shining above; flowers 1–3(–10) subfasciculate, axillary; pedicels 6 m; petals 8 × 4 mm; fruit subglobose, smooth, 2.5 cm diam., stipe 0.5 cm, stigma 5-lobed.

Habitat: lowland evergreen forest.

Thrats: slash and burn agriculture (Bipindi area).

Management suggestions: verification of the evidential basis of this species occurring at Bimbia and Campo-Ma'an is needed. If these cannot be verified, the logical focus for verification efforts might be the Douala-Edéa reserve since it is the most recent record known and has some hopes as a protected area. Basic populational data should be recorded to gauge management needs and to form a baseline for monitoring. Consideration should be given to involving local leaders in protecting the species and to collecting seed for propagating new plants to augment wild populations.

Garcinia kola Heckel

VU A2cd

Range: Sierra Leone, Liberia, Ivory Coast, Ghana, Benin, Cameroon, Gabon, Congo (Kinshasa).

Tree 5–12 m; bole cylindrical; leaves elliptic, 13 × 6 cm, acuminate, obtuse, resin canals black, parallel to nerves; petiole 1.5 cm; inflorescence subumbellate, subsessile on short shoots, 10–15-flowered; flowers c. 1 cm diam.; petals concave; staminal bundles 4; fruit orange, fleshy, 10 cm diam., containing brazil-nut like shaped and sized seeds, the "bitter cola" of commerce.

The assessment above is given by Hawthorne (cited as 1997 in www.redlist.org (IUCN 2003)), on the basis that it "is probably the most important source of chewsticks. Over-exploitation has caused population declines. Seedlings are uncommon and slow-growing". That assessment is apparently based only on its occurrence in Ghana and Congo (Kinshasa) since these are the only two countries cited under the distribution. The range data cited above is taken from FWTA. In SW Region Cameroon, the main use for the species is not as a chew-stick (for which the most important species is *Garcinia mannii*), but for the comestible-medicinal seeds ('Bitter cola'), probably harvested sustainably from the fallen fruits. The seeds are marketed extensively by vendors all over at least the southern part of Cameroon. The species is at least occasionally cultivated (*pers. obs.* Kupe village) and perhaps commonly so in Cameroon. On the basis of these observations in Cameroon, and given the large range of the species, *G. kola* would not otherwise be assessed as threatened. However, it is perfectly possible that general habitat loss and felling for dental hygiene in the western part of its range is sufficient to justify Hawthorne's rating, so this is maintained here with some hesitation. The foregoing assessment is taken from Cheek *et al.* 2004: 160 and is maintained here in the absence of any new data. The species is not mapped here since it is rather common in the forest zone of much of its range, especially in Cameroon. EOO is calculated as 1067820 km^2 based on an incomplete dataset.

Psorospermum aurantiacum Engl.

VU B2ab(iii)

Range: Nigeria (Obudu Plateau (5 coll.) and Mambilla) and Cameroon (SW: Lebialem. NW: Bambui, Bamenda, Kumbo–Oku, Bafut Ngemba, Bali Ngemba (12 collections). W: Kounden (1 coll.)).

This tree or shrub is distinctive for the dense orange-brown hairs on the lower surface of the leaf, the upper surface a contrasting glossy black when dried. It appears confined to the Bamenda Highlands, with outliers in the adjoining Obudu Plateau and Lebialem/Bamboutos Mts (Kounden). Six locations can be accepted if the Bamboutos/Bamenda Highlands are taken as one large subpopulation. Clearly a 4 km^2 cell size is inadequate there. Overall an AOO of 80 km^2 is estimated. Here *G. brevipedicellata* is assessed as Vulnerable since there are effectively eight locations (above; AOO 32 km^2 with 4 km^2 cells) and threats as below. EOO is calculated as 26138 km^2

In future this species could be better assessed under criterion A2c since its distribution and habitat corresponds with that which has seen most loss in Cameroon. This assessment is maintained from that in Harvey *et al.* 2004: 64, Harvey *et al.* 2010: 76 and Cheek *et al.* 2010: 97–98.

Habitat: edge of gallery forest with grassland; 1500–1800 m alt.

Threats: dry season grassland fires, usually set by man, burn into the montane and submontane forest in the Cameroon Highlands, reducing its area. It is possible, even likely, that *P. aurantiacum* by the nature of its habitat, has some resistance to fire and may even benefit from occasional fires. However, the current frequent and intense fires may affect individuals adversely. Conversion of forest to farmland, by contrast, is an undoubted threat. Over 25% of forest in one sample area of the Bamenda Highlands was lost in the 1980s–1990s (Moat in Cheek *et al.* 2000: back cover) and 50% over 15 years 1988-2003 in another (Baena in Cheek *et al.* 2010: back cover).

Management suggestions: research to explore the effect of different fire regimes on this species is advised. In the short term, the highest priority is to re-find individuals at the known sites and to seek means to protect these. While almost all natural forest at Bafut Ngemba FR has disappeared already, Bali Ngemba FR still remains fairly intact and may be the best prospect for the conservation of *P. aurantiacum*.

HOPLESTIGMATACEAE

assessed by Martin Cheek

An obscure endemic Tropical African family of trees resembling and possibly related to *Cordia* and *Ehretia* of Boraginaceae. Only two species are known, both rare, one widespread (*H. klaineanum,* Gabon to Ivory Coast, probably Near Threatened or Vulnerable), the other apparently restricted to Cameroon, which is treated below.

The trunks of *Hoplestigma* have no distinctive scent or exudate, but when cut, rapidly oxidise from pale yellow to black. Young leaves and stems have brown simple hairs. The alternate, simple, entire leaves have pinnate nerves and lack stipules. The flowers are showy, white, 1-2 cm wide, in axillary cymes, resembling those of *Anthocleista* (Loganiaceae/Gentranaceae), having a tubular corolla with numerous rounded, overlapping lobes and exerted stamens attached to the top. The superior ovary bears two long, basally united styles and is succeeded by a fleshy fruit of bizarre internal structure that includes two bone-like endocarps. There is no Flore du Cameroun account. No uses are known. The two species are separated as follows by Hepper in FWTA 2: 16 (1963):

Leaves densely pubescent beneath at least on midrib and nerves with rather long, crisped hairs; lamina broadly obovate-elliptic, rounded and slightly apiculate, narrowed to base, 15–30 cm long, 5–15 cm broad, venulose-reticulate above; lateral nerves about 10 on each side of midrib; petiole about 2 cm long; cymes terminal, softly tomentose all over; calyx about 1 cm long; corolla 1.5 cm long; fruits broadly ellipsoid, 3 cm long, 2.6–2.9 cm broad, glabrous .... 1. **pierreanum**

Leaves beneath with small, stiff and straight appressed hairs, sometimes sparse, lateral nerves about 8 on each side of midrib; petioles 1-1.5 cm long; cymes covered with dense stiff hairs; fruits 2 cm long, 2.2-2.4 cm broad .... 2. **klaineanum**

Hoplestigma pierreanum Gilg

CR D

Range: Cameroon endemic (SW: Korup; Mt Cameroon at Mokoko and Likomba. S: Bipindi).

This tree was only known from the type collections of Zenker (2575, 3650, 2632 & 3383) at Bipindi in 1903–08 (no recent collections from this area have been seen) and from the foothills of Mt Cameroon until the 1990s. The first collection from the mountain's foothills was that from Likomba (*Mildbraed* 10774, Dec. 1928; no natural forest now survives there, it is believed), just east of Mabeta-Moliwe, where it should be looked for although forest there has also been much reduced in area. The second collection (rather poor, only a few fallen leaves) is from the western foothills, in the Mokoko forest (*Thomas* 10022) now mooted for logging and oil exploration, and I also made a collection of fallen leaves and endocarps in the Korup (*Cheek* 8800, in 1998).

I assessed *Hoplestigma pierreanum* in Cable & Cheek (1998: liii) as CR A1c+2c (Critically Endangered), before categories were updated in IUCN (2001). Here I update the assessment to CR D on the basis that only a single tree is known (pers. obs.), or probably exists, at each site so that no more than four individuals are known to exist or to have existed, with certainty. EOO is calculated as 3838 km^2.

Habitat: lowland evergreen coastal forest at or near sea-level.

Threats: forest clearance for timber and agriculture, especially the sites around Mt Cameroon which have been replaced or are at risk from small-holder agriculture and plantations of cocoa, bananas, rubber and oil palm.

Management suggestions: in view of the extreme rarity of this species and the evolutionary isolation of the genus, it is worth considerable effort to rediscover it. The best starting point would be Mokoko and Korup, where the most recent collections have been made, although since they were incomplete, their identification needs confirmation once they have been recollected. If the Korup record is confirmed, this would be the logical focus of conservation efforts since it is well-protected. It is also recommended that the site where the original collections in Bipindi were made be pinpointed and revisited to rediscover the precise type locality. An isolated tree of this genus discovered by me in a cocoa small-holding in the early 2000s in or near the Mokoko R FR in Bakossi is thought to be *H. klaineanum.* An agreement with the farmer concerned to pay a rent so long as the tree is allowed to survive has so far encouraged its survival, while fruits collected from it have produced abundant seedlings at a nursery nearby for planting in school compounds for conservation- educational purposes, and for planting in forest to augment the wild population. This approach might be taken with any trees of *H. pierreanum* once refound.

HUACEAE

assessed by Martin Cheek

This family of onion-scented shrubs and trees of evergreen forest are restricted to Guineo-Congolian Africa. All four species can be found in Cameroon and Gabon, but only one appears threatened. Known as 'bush onion', the fruits and/or bark of several species are harvested from the wild and sold in local markets as a food-flavouring.

The family are easily recognised by the unusual scent of their stems, otherwise known in Cameroon only in *Scorodophleus zenkeri* (Leguminosae subfamily Caesalpinioideae.) and *Olax subscorpoidea* (Olacaceae), among forest species. The leaves are alternate, simple, entire, with pinnate nerves and conspicuous stipules. The indumentum is stellate or lepidote. No Flore du Cameroun account exists but the Fl. Gabon treatment by Yves Issembe in 2009 covers the same species found in Cameroon.

Afrostyrax macranthus Mildbr.

VUB2ab (iii)

Range: Cameroon (S: Kribi-Campo, 44 km; Kribi-Edéa, 15 km; Kribi-Lolodorf, 2 km; Mvini, 35 km E Campo; Mt Elephant; Campo-Ma'an) and Gabon (SW Lambarene; Nkoulounga).

Shrub 3–6 m, bark white-grey; leafy stems red-brown hispid; stipules persistent, 3–5 × 1 mm; leaf-blade elliptic-oblong, 7–14 × 3–6 cm, acumen 1–2 cm, abrupt, bark rounded, margin revolute, lateral nerves 6–8 pairs, glabrous, or the nerves glabrescent; flowers single to three in axillary clusters, petals free, pale yellow, 5 mm; fruit ovoid 1–2.5 cm. Distinguished from the similar *A. kamerunensis* by having 6–8 pairs of leaf nerves and a revolute margin (not 4–5 pairs with margin flat).

Here assessed as Vulnerable since eight locations are known (above; AOO 32 km^2 with 4 km^2 cells) with threats as below.

Habitat: lowland evergreen forest.

Threats: slash and burn agriculture in forest; oil drilling (Gabon); iron ore extraction (Gabon); urbanisation and infrastructure development for extensive industry (Kribi).

Management suggestions: Campo-Ma'an NP is the logical focus for conservation of this species since it is a protected area. Basic populational data should be collected there to support future monitoring.

ICACINACEAE (INCLUDING STEMONURACEAE & LEPTAULACEAE)

assessed by Martin Cheek

This family has recently been divided up so that some Cameroonian genera once included have now been placed in other families although maintained here for the present. *Lasianthera* is now included in Stemonuraceae and *Leptaulus* in Leptaulaceae (Heywood *et al.* (2007). Flowering Plant Families of the World: op. cit.). As newly defined, this remains a pantropical family, with about 35 genera and 300 species. Stem-twisting lianas, less usually trees, they lack exudates or distinctive scent, have alternate leaves (opposite with tendrils in *Iodes*), lacking stipules, having usually pinnate nerves and an entire margin, less usually palmately lobed. A short, appressed grey indumentum is often present on the stems, sometimes becoming rough to the touch, or even irritating. The inflorescence is racemose, often slender and pendulous, with small flowers, glabrous stamens 1-locular ovary, followed by fleshy drupes. *Iodes, Chlamydocarya*, *Icacina*, *Alsodeiopsis Rhaphiostylis*, *Desmostachys*, *Lavigeria*, *Stachyanthus*, *Polycephalium* and *Pyrenacantha* are other genera in Cameroon. Flore du Cameroun 15 produced by J.-F. Villiers in 1973 treats this family and allows identification of species. Most of the following treatments depend on this work. Previous revisions for Africa were those of Engler in the 1920s, and Sleumer in the 1940s.

Apart from the species treated as threatened below, the following, for the present treated as Near Threatened, may also prove Threatened when examined in more detail:

Alsodeiopsis weissenborniana J.Braun & K.Schum. (Cameroon, Gabon & Angola)

Desmostachys brevipes (Engl.) Sleumer (Cameroon and Gabon)

Desmostachys vogelii (Myers) Stapf (Sierra Leone, Liberia, Ivory Coast and Cameroon)

Iodes kamerunensis Engl. (Nigeria, Cameroon & Gabon)

Alsodeiopsis rubra Engler

VU B2ab (iii)

Range: Cameroon (SW: Korup; Masore. C: Eséka; L: Lake Tissongo. S: Bipindi; Campo-Ma'an (Tchouto 2004)) and Gabon (Cristal Mts; Oveng; M'passa).

Here *Alsodeiopsis rubra* is assessed as Vulnerable since nine locations are known (above; AOO 36 km^2 with 4 km^2 cells) and since threats are as below. EOO is calculated as 84602 km^2.

Shrub 1–2.5 m, stem hirsute, golden; leaves alternate, petioles 1.5 mm; blade lanceolate or oblanceolate, 8.5–18 × 4.5–6 cm; apex long acute acuminate, mucronate, base rounded, asymmetric, lateral nerves 8–18 pairs, uniting near the margin and pubescent on both surfaces, tertiary nerves conspicuous, scalariform; panicles axillary, peduncle, 1–2.5 cm, yellow-hirsute; flowers c. 5 mm, yellow-orange hirsute; fruits ellipsoid, 2.8 × 1 cm, brilliant orange, hairy.

Habitat: lowland evergreen forest.

Threats: slash and burn agriculture around Bipindi, Eséka and Masore.

Management suggestions: secure sites for the species are at Korup NP, and Campo-Ma'an where conservation efforts might best be focused. Basic populational data should be collected as a baseline for monitoring, and park managers assisted in identifying the species so as to better protect it. Flore du Cameroun 15 mentions this species as occurring in Equatorial Guinea, which requires verification.

Alsodeiopsis zenkeri Engler [Lelok Onone (bakoko)]
VU B2ab(iii)

Range: Cameroon (C: Eséka; Obala, chutes Sanaga. S: Bipindi; Déhané on Nyong; chutes de Ntem near Nyabéssan; Campo Ma'an (Tchoutou 2004)) and Gabon (Lastoursville to Koulamoutou).

Here *Alsodeiopsis zenkeri* is assessed as Vulnerable since there are seven locations (above; AOO 28 km^2 with 4 km^2 cells) and threats as below. The isolated Gabonese record was identified by Villiers so must be accepted. EOO is calculated as 72045 km^2

Shrub, stems appressed yellow hairy, leaves alternate, petiole 2–3.5 mm; blade elliptic, 4–13 × 1.3–3.5 cm, acumen obtuse, mucronate base attenuate, lower surface with hairs as stem; secondary nerves 4–7 pairs, raised, pubescent, uniting near margin, domatia and nervelets conspicuous; panicle flat-topped, peduncle 0.8–1.5 cm, pubescent; flowers 5 mm, brown pubescent.

Habitat: lowland evergreen forest.

Threats: slash and burn agriculture; urban expansion (Bipindi).

Management suggestions: this species is probably secure in the Campo-Ma'an NP, whence three records derive (Tchouto 2004) although identifications should be confirmed. Basis populational data should be collected as a baseline for monitoring and the Park Managers assisted to identify the plant.

Leptaulus grandifolius Engl.
VU B2ab (iii)

Range: Cameroon (SW: S Bakunda FR (2 coll.). Littoral: Masok. S: Bipindi (3 coll.); Kribi area (6 Coll.); 45 km S Kribi (1 coll.)) and Gabon (Cristal Mts (2 coll.).

Here *Leptaulus grandifolius* is assessed as Vulnerable since six locations are known (above; AOO 24 km^2 with 4 km^2 cells) and threats are as below. EOO is calculated as 25018 km^2.

Shrub 3–4 m, leafy stems strongly horizontal, glabrous; stems longitudinally ridged; leaves alternate, petioles 1.5–2 cm; blade elliptic, 15–32 × 6.5–11.5 cm, glossy above, strongly obtuse-acuminate, base cuneate; secondary nerves 8–12 pairs, uniting near the margin; fascicles axillary, flowers subsessile, 5-merous, yellow, sepals lanceolate, 2.5 mm; corolla white, united in a tube, 1.2 cm, stamens epipetalous; drupe oblong.

Habitat: lowland to submontane evergreen forest.

Threats: forest clearance for yam farming (S Bakunda); urbanisation, infrastructure for extractive industry (Kribi area); slash and burn agriculture (Masok and Bipindi).

Management suggestions: none of the Cameroon locations appears to be within a protected area so the assistance of local community leaders should be invoked to protect plants near locations where it occurs. The first step is to rediscover the species in the wild. The Kribi area offers the best prospect, having the most numerous and recent records. Basic populational data should be recorded as a baseline for future monitoring.

Pyrenacantha cordicula Villiers (assessed by Iain Darbyshire)
EN B1+2ab(iii)

Range: Ivory Coast (60 km N of Sassandra, Davo R Gorge (1 coll.)), Ghana (Draw R FR (1 coll.)), Cameroon (SW: Ngombombeng near Nyasoso (1 coll.) and Equatorial Guinea (location unknown).

Known from the four locations above, each from a single collection, *Pyrenacantha cordicula* is here maintained as Endangered (AOO 16 km^2 with 4 km^2 cells and threats below) following the assessment by Darbyshire in Cheek *et al.* (2004: 160–161). EOO is calculated as 174 km^2.

Despite the wide range of this taxon, it appears extremely scarce throughout. It was first described by Engler under the invalid name *Chlamydocarya tessmannii* Engl., and was recorded as being found in Cameroon though no specimens were cited. J.F.Villiers formally published the taxon, originally under the illegitimate name *Pyrenacantha cordata* Villiers, following the collection of a specimen with male flowers in the Ivory Coast in 1959 (*Leeuwenberg* 2084), from where it has not subsequently been recollected. Villiers also recorded the species as occurring in Cameroon and Equatorial Guinea, but cited no specimens. The Ghana collection, made in 1974, contains male flowers. Female flowers and fruits were unknown until the collection by M. Etuge and D.W. Thomas (*Etuge* 28) at Ngombombeng, Cameroon in 1987. Data is here derived from Flore du Cameroun. Equatorial Guinea was maintained for this species (as *Chlamydocarya tessmanniii* Engl.) in an aside by Engler, but no further specimen location data were given (see Flore du Cameroun for reference).

Liana, felty-pubescent throughout; leaves elliptic, c. 11.5 × 5 cm, apex acuminate, base rounded; lateral nerves 5 pairs, nerves yellow below; petiole to 1.5 cm; female flowers in dense sessile axillary glomerules, c. 0.7 cm; male flowers in axillary fascicles of racemes 4–14 cm; ovary truncate at apex; fruit ellipsoid, 1.7 × 1.2 cm, green, felty, subrostrate.

Habitat: an understorey climber in dense humid forest or secondary forest; c. 750 m alt.

Threats: widespread and continued loss of lowland and mid-elevation forest in the Ivory Coast is likely to threaten any

extant populations of this species here. The site at Ngombombeng in Cameroon lies along the route from Nyasoso to Ngomboku, a somewhat populous area with resultant widespread loss of forest below 1000 m alt., the limit of effective forest protection on the adjacent Mt Kupe.
Management suggestions: more data are required on the distribution of this species in Cameroon and Equatorial Guinea, including previous collecting locations. The forest around Ngombombeng and adjacent forest areas should be surveyed to try to rediscover this taxon here. Formal description of the fruits of this taxon, from the *Etuge* 28 collection, should be made to aid field botanists in identification of this taxon in future.

Pyrenacantha grandifolia Engler
CR B2ab(iii)

Range: Cameroon endemic (S: Bipindi)
Here *Pyrenacantha grandifolia* is assessed as Critically Endangered since only a single location is known (above; AOO 4 km^2 with 4 km^2 cells) and with threats as below. Not seen since the type collections in 1907, now destroyed.
Liana, stems yellow, appressed red hairy; leaves alternate, petiole 1–1.5 cm, blade oblong, 20–23 × 9–10 cm, acumen 1 cm, base weakly auriculate, lateral nerves 8 pairs, uniting near the margin, nervelets densely reticulate; spike with rachis to 10 cm in fruit, petals discoid, drupes oblong, 2-ridged, 4 × 2 × 1.5 cm, glabrescent.
Habitat: lowland evergreen forest.
Threats: slash and burn agriculture
Management suggestions: rediscovery of this lost species is a priority. Its validity seems assured since all previous workers on Icacinaceae have accepted it. Surveys should be focused around Bipindi where it was last seen. If refound, basic populational data should be collected so that there is a baseline for monitoring. Support from local people should be sought to protect the species since there is no protected area known in the Bipindi area.

Pyrenacantha longirostrata Villiers
EN B2ab(iii)

Range: Cameroon (SW: Mt Cameroon-Mokoko; Kumba-Limbe; Lebialem-Bechati; Akwaya-Mamfe) and Gabon (Monts de Cristal).
Known from the five locations above, each from a single collection, *Pyrenacantha longirostrata* is here maintained as Endangered (AOO 20km^2 with 4km^2 cells and threats below) following the assessment by Cheek in Harvey *et al.* (2010: 77). EOO is calculated as 36637 km^2.
Climber, 5 m; stem twining, minutely sparse puberulent to glabrous; leaves alternate, elliptic, 10–13 × 4.5–8 cm, acumen 1 cm, base rounded to obtuse, lateral nerves 3–6 pairs, quaternary nerves raised, reticulate, with papillae in areolae; petiole 1–1.5 cm, angled near base; inflorescence 6–9 cm, spike-like, lateral; apex with 6–8 flowers; flowers 3 mm long, 4-merous; infructescence 7 cm, 0.5–1 m from ground; fruits 2, bright orange, heart-shaped, 2.5 × 1.5 cm with a beak 1–2 cm long; seeds dark brown, 15 × 14 × 8 mm
Habitat: lowland evergreen forest c. 200–600 m alt.
Threats: highly threatened at most known lowland sites in Cameroon by logging followed by agriculture, especially at Kumba-Limbe where it may already have been lost by forest clearance, and at Mokoko FR (scheduled for logging in 2010) and Bechati (pers. obs.). The threat status of the species at Monts de Cristal and Akwaya-Mamfe is unknown.
Management suggestions: there is no obvious site that might be the focus for conservation of *P. longirostrata*. One option is to seek to refind plants at the most recently discovered site, at Bechati (Sept. 2006). Here the elders of Bechati should be consulted to secure the protection of the species, and data gathered on regeneration levels, density and range of the species.

Rhaphiostylis elegans Engler
CR B2ab(iii)

Range: Cameroon endemic (S: Bipindi)
Here *Rhaphiostylis elegans* is assessed as Critically Endangered since a single location is known (above; AOO 4 km^2 with 4 km^2 cells) with threats as below.
Tree, leaves alternate, petiole 1–2 mm, leaves oblong-ovate, 5–8 × 3–5 cm, glossy on both sides, acumen 1–2 × 0.2 cm, secondary nerves 4–5 pairs, uniting at the margin; drupe carried on a slender pedicel, reniform 1.5 × 0.8 cm, apiculate. Known only from the type, *Zenker* 3615 collected in Jan 1908 and destroyed in 1943 in Berlin. Treated in Flore du Cameroun as an imperfectly known species.
Habitat: lowland evergreen forest
Threats: slash and burn agriculture
Management suggestions: rediscovery of this lost species is advised. Its validity seems likely since all previous workers on Icacinaceae have maintained it. Surveys should be focused around Bipindi where it was last seen. If refound, basic populational data should be collected so that there is a baseline for monitoring. Support from local people should be sought to protect the species since there is no protected area known in the Bipindi area.

Rhaphiostylis ovatifolia Sleumer
EN B1+2ab(iii)

Range: Cameroon endemic (S: Elabi–40 km S of Kribi; Grand Batanga; Lolabé).
Treated as an imperfectly known species in Flore du Cameroun when only known from the type, *Ledermann* 648, Elabi, fl. Sept 1908, destroyed at Berlin in 1943. Apparently rediscovered by Tchouto (2004) who cites two other locations, although specimens have not been seen.

Here *Rhaphiostylis ovatifolia* is assessed as Endangered given three locations (above A00 12 km^2 with 4 km^2 cells) and with threats as below. EOO is calculated as 3.86 km².
Liana, glabrous; leaves alternate, petiole 2–3 mm, blade ovate-elliptic 7–11 × 3.5–5.5 cm, acumen 1.5 mm, obtuse; base rounded, secondary nerves 5–6 pairs, uniting near the margin, veinlets feeble; fascicles axillary, 5–7–flowered; pedicels 1.5–2.5 cm, sepals oblong, 0.5 mm; petals white, oblong, 5 mm.
Habitat: lowland evergreen forest.
Threats: clearance of habitat along the coast for urbanisation, touristic development infrastructure for extractive industries.
Management suggestions: Peguy Tchouto's expertise should be sought to rediscover this species and to protect it (see above). If plants occur in Campo-Ma'an NP, then this is the logical focus for conservation efforts and baseline populational data should be recorded here for monitoring. Park managers should be provided with the means to identify and so better protect the species.

Rhaphiostylis poggei Engler
CR B2a(iii)
Range: Cameroon endemic (SW: Kumba at Lake Barombi Mbo).
Here *Rhaphiostylis poggei* is assessed as Critically Endangered since only a single location is known (above; AOO 4 km^2 with 4 km^2 cells). Threats are as below. An accepted species in Flore du Cameroun 15: 36. Known only from the type *Preuss* 561, Barombi, collected in the 1890s but lost in Berlin 1943. Flore du Cameroun remarks that Engler stated that it can be found in "Zaire".
Liana-shrub, glabrous, stems zig-zag, leaves alternate, petiole 3-4 mm; blade oblong, 8 × 3–4 cm, subacuminate, base obtuse, secondary nerves 5-6 pairs, uniting near margin, finer nerves forming a reticulum; fascicle axillary; pedicel 1 cm; flower buds yellow oblong, 6 mm.
Habitat: lowland evergreen forest.
Threats: clearance for agriculture (pers. obs.)
Management suggestions: since Barombi Mbo has been collected several times by botanists over the last century, the absence of further collections since 1890 suggests that this apparently very localised species might now be extinct. Surveys in Barombi Mbo should be recorded as a baseline for monitoring and land managers involved in its protection. Consideration should be given.to propagating the species from seed at Limbe BG for augmenting the wild population.

Rhaphiostylis subsessifolia Engler
EN B1+2ab (iii)
Range: Cameroon endemic (S: Grand Batanga; Mt Elephant; Ebianemeyong).
Here *R. subsessifolia* is assessed as Endangered since three locations (above; AOO 12 km ^{2} with 4 km^2 cells) are believed to exist, with risks as below. EOO is calculated as 306 km². Treated as a poorly known species in Flore du Cameroun, being only known from the type collection at [Grand] Batanga (*Dinklage* 1028, fr. Dec. 1890) destroyed at Berlin in 1943. Apparently it was refound by Tchoutou 2004 at the second and third locations cited above, although specimens have not been seen.
Shrub, stems zig-zag. Internodes 3-4 cm. Leaves alternate, petiole 2 mm, blade ovate 6–9 × 4–6 cm, acumen obtuse 1–2 cm, secondary nerves 5–6 pairs, uniting near margin, nervelets in open reticulum; drupes reniform, apiculate, sides flattened and ridged, pedicel excentric.
Habitat: lowland evergreen forest.
Threats: extractive industry (Mt Elephant); touristic development and infrastructure (Grand Batanga).
Management suggestions: Peguy Tchouto should be consulted to refind this species again and to help collect baseline populational data for monitoring, and to work with local people to help protect the species since none of its sites appears to be protected. This species may soon become Critically Endangered if mining at Mt Elephant goes ahead. It is suspected that the plants at Batanga have been lost.

Stachyanthus cuneatus Sleumer
CR B2ab (iii)
Cameroon endemic (C: Lom).
Here *Stachyanthus cuneatus* is assessed as Critically Endangered since only a single location is known, with threats as below (AOO 4 km^2 with 4 km^2 cells). An accepted species in Flore du Cameroun even though known only from *Ledermann* 6437 from Lom, collected in flower Dec. 1909, destroyed at Berlin in 1943.
Liana, stems appressed pubescent but soon glabrescent, leaves alternate, petioles 1.5–2 cm, blade oblong 10–14 cm, flowers sessile, calyx campanulate, glabrous, shortly 5-toothed, petals oblong, 5, glabrous; stamens 5.
Habitat: lowland evergreen forest with semi-deciduous elements.
Threats: conversion of forest to agriculture (pers.obs.)
Management suggestions: research is needed on Ledermann's itinerary to narrow down the location of the type record. It may be at Loum in Littoral rather than modern Lom, as mapped. A survey to refind the species is needed, if it still survives, so that basic populational data can be collected as a baseline for monitoring, and so that local managers can be helped to protect the plant. Loum FR has suffered the loss of its understorey to agriculture in many areas.

IXONANTHACEAE

assessed by Martin Cheek

A small pantropical family of trees with two genera: *Ixonanthes* in SE Asia, and *Ochthocosmus* in Tropical Africa and America, with most species in Africa. Flore du Cameroun 14 by F. Badré in 1972 treats this family, recognising three species. The present treatment derives largely from this work. Among tree families with alternate, simple leaves, Ixonanthaceae are most likely to be mistaken due to the serrate margin, for *Drypetes* (Euphorbiaceae-Putranjivaceae) but that genus has asymmetric leaf-bases, Flacourtiaceae (genera such as *Scottellia*) and Ochnaceae. Here flowers and fruits are needed to separate them reliably. The five free sepals and petals, and 5-merous 1 or 2 whorled, free stamens around a superior 1-styled pistil which produces small capsular, 1-seeded fruit are typical of the family as seen in Cameroon.

No important uses are recorded yet for the Cameroonian species.

Ochthocosmus calothyrsus (Mildbr.) Hutch & Dalz. VUB2ab (iii)

Range: Cameroon (S: 15–25 km SW Zingui; Kribi-Bipindi; Campo, 10 km ESE) and Gabon (Lopé NP; Rabi South West; Agouma; 40 km NW Doussala).

Here *Ochthocosmus calothyrsus* is assessed as Vulnerable since six locations are known (above; AOO 24 km^2 with 4km^2 cells) with threats as below. EOO is calculated as 58003 km^2.

Tree to 15 m, glabrous; leaves alternate, stipulate, petiole rugose, 2–7 mm; blade elliptic-obovate, 8–27 × 3–10 cm, base attenuate, margin acutely serrate, secondary nerves c. 6 pairs, not prominent, finer nerves indistinct; panicles in terminal clusters, 10–15 cm, pedicels 4–10 mm, sepals ovate, 1.5 mm; petals oblong, 3–4 mm; stamens 10, in two ranks of differing lengths, free.

Distinguished from congeners by the 10 stamens, pedicellate flowers, and indistinct finer nervation.

Habitat: lowland evergreen forest

Threats: clearance of forest habitat for urbanisation and infrastructure for industries (Kribi) for plantations, and for slash and burn agriculture.

Management suggestions: the best hope for protecting this species in Cameroon may be the Campo- Ma'an NP. If presence of the species can be confirmed here, basic populational data should be collected as a baseline for monitoring and Park Managers provided with the means to identify the species. Little is known of the precise ecological requirements of this species - this should be addressed if the opportunity allows.

LABIATAE (LAMIACEAE)

assessed by Martin Cheek and Benedict Pollard

A world-wide family of herbs and climbers (*Clerodendrum*), less usually trees (*Vitex*), characterised by opposite, simple leaves which, when crushed, release pungent scent due to their essential oils. Interpetiolar stipules and coloured exudates are absent in this family. The flowers are often held in terminal, spike-like inflorescences in which several or many flowers occur in contracted cymes at each node. Individual flowers have a cupular 3–5- toothed calyx, a tubular corolla with upper and lower lips well differentiated (bilateral symmetry), often with stamens exserted and superior ovaries which produce a fruit with 4 nutlets and a gynobasic style. Labiates are most likely to be confused with Acanthaceae in Cameroon, but the last lack essentials oils and so are more or less scentless. They also lack the 4 nutlets, having instead a two-valved capsule.

Until the early 1990s, woody genera, such as *Clerodendrum* and *Vitex* (and the non-native timber trees *Gmelina* and *Tectona*) were included in Verbenaceae, but the work of Cantino and colleagues realigned these two families when published in 1992, and has been widely followed. Verbenaceae are now slimmed down to far fewer genera than before, none of which are threatened in Cameroon, but including well-known species such as *Lantana camara,* an invasive weed in many parts of the world, and *Stachytarpheta,* mainly S American and also weedy in Africa. Verbenaceae are differentiated by having stamens included, the corolla less markedly differentiated into an upper and lower lip (the five lobes being more or less equal) and with the ovary not divided into four nutlets, but entire, with an apical style (see Brummitt in Heywood *et al.* (2007), Flowering Plant Families of the World, RBG, Kew).

Labiates mostly require full sunlight so are mainly found in grassland or woodland habitats, especially montane grassland areas, although several species of *Plectranthus* can grow in light shade at the edge of forest, and a few species, such as the mass-flowering *Plectranthus insignis* only occur in montane forest. *Achyrospermum* is the only genus in Cameroon which is restricted to forest shade.

The Labiates most useful to man are those of Europe or the Mediterranean in the main. Rosemary, mint, sage, thyme, basil, marjoram are all culinary herbs in the Labiate family which derive from this area. In Cameroon, a species of basil, *Ocimum*, is very commonly cultivated for this purpose. Some species of *Plectranthus* have edible tubers, and many have potential as garden plants.

There are no Flore du Cameroun or Fl. Gabon accounts for this family. However, building on the 1963 work of J.K. Morton in FWTA 2, a series of publications by Pollard and Paton regarding Labiate species in Cameroon has helped delimit taxa, especially within the largest genus,

Plectranthus, which has been expanded to accommodate the genera *Solenostemon* and *Coleus*. Despite this, several genera are in dire need of revision and several potentially new species have identified in these, or poorly known names have been found, which may well prove to be threatened species in future: *Achyrospermum, Clerodendrum,* and *Vitex*.

Apart from the species treated as threatened below, the following species may prove to be threatened in future:

Achyrospermum erythrobotrys Perkins, *Achyrospermum schlechteri* Gürke and *Achyrospermum tisserantii* Letouzey are all poorly known taxa of Cameroon in a genus that needs taxonomic revision before species can be assessed for their IUCN status (treated here as DD).

Aeollanthus cucullatus Ryding (*A. repens* sensu Morton) of submontane (1200–2000 m) Nigeria and Cameroon is here treated as Near Threatened.

Clerodendrum buettneri Guerke and *Clerodendrum inaequipetiolatum* Good have both been treated as Near Threatened (Pollard in Cheek *et al.* 2004: 304) and are here maintained as that.

Pycnostachyus pallide-caerulea Perkins may be threatened, but the genus requires taxonomic revision before this can be established, and is here treated as DD.

Solenostemon sp. nov.? (Cable 216, 227, Cheek 5563 and Williams 52) cited as a possible Red data species in Cable & Cheek 1998: 64, is now published as *Plectranthus cataractarum,* a threatened species (see below).

Plectranthus sp. nov. [Ifyenge zvu (Kom) *fide* Peter Yama], *Pollard* 267, 2 Dec. 1998, Laikom Ridge, Ijim and also *Pollard* 385) was reported as a Solenostemoid *Plectranthus* which appeared to be new to science and a potential threatened species in Cheek *et al.* 2000: 138. Subsequently Pollard, following further research, identified it as *Plectranthus decumbens* Hook.f. This species is restricted to the Cameroon Highlands (Mt Cameroon and the Bamenda Highlands) and so may feature as threatened in future editions of this Red Data book. For the moment this Cameroon endemic is treated as Near Threatened following Pollard in Cheek *et al.* 2004: 306.

Plectranthus sp. nov. (*Etuge* 4192 and *Cheek* 11337 both from Nyale, Bakossi) is probably new to science and likely to prove threatened once published (Pollard in Cheek *et al.* 2004: 307).

Aeollanthus trifidus Ryding (assessed by Benedict Pollard updated by Martin Cheek)
VU B1ab(i,ii,iii,iv,v) + B2ab(i,ii,iii,iv,v)

Range: Nigeria (Mambilla plateau (2 loc.)); Cameroon (W: Bamboutos Mts. NW: Bali Ngemba FR; Jakiri, Bamenda, Bui; Mbiami (1 coll. each)).
The holotype (*Saxer* 64) was collected at Dschang, Bamboutos Mts, W Region, Cameroon on 29 June 1955. Two further specimens were collected on 10 Aug. 1973 across the border in Nigeria: *Chapman* 103 from Ngurogi, and *Medler* 941, near Chappal Wadi, probably on the road from Takum. J.F.F.E. de Wilde collected a specimen from Jakiri, Bamenda, NW Region, Cameroon, which remained unidentified until 2004, and represented the most recent collection until those of *Etuge* 4736 from Mbiami and *Pollard* 699 from the larger granite inselberg at Mantum by the Bali Ngemba FR, both in November 2000.
The assessment of this species by Pollard in Harvey *et al.* (2004: 101–102) is maintained here since only six locations are known (above; AOO 24 km^2 with 4 km^2 cells) with threats as below. EOO is calculated as 6372 km^2.
Subshrub to 30–50 cm; stem indumentum to 1.5 mm; leaves shortly petiolate, broadly elliptic, to 45–65 × 22–36 mm; apex subacute; base attenuate; margin crenate; veins prominent below; inflorescence paniculate, lax, conspicuously bracteate; bracts elliptic, to 4–6.5 × 2–3.5 mm; fruiting calyx circumscissile; corolla white; lower lip tinged purple distally; upper lip white, speckled purple; tube much-broadened towards the throat, 5.5–8 mm; upper lip 4-lobed; lower lip trifid.
Delimitation of this species was assisted by Ryding, O. (1986). The genus *Aeollanthus* s. lat. (Labiatae). Acta Univ. Ups. Symb. Bot. Ups. 26: 1. VI + 152 pp. Uppsala.
Habitat: grassland on rocks or shallow soil, subject to burning during the dry season (Dec.–Mar.); 1500–2200 m alt.
Threats: although this taxon is known mainly from rocky areas unsuitable for cultivation, management practices for grazing may threaten the species by increasing fire and trampling frequency.
Management suggestions: introduce this species into cultivation and distribute living specimens to local and regional botanic gardens for *ex-situ* conservation. Explain the importance of this rare species to local leaders and incorporate this species into land management plans. Collect basic populational data as a baseline for monitoring.

Clerodendrum anomalum Letouzey (assessed by Benedict Pollard)
EN B2ab(iii)

Range: Cameroon (SW: Kupe-Bakossi (1 coll.). Littoral: Ngambé (1 coll.)) and Gabon (3 coll. at 3 loc.).
Scandent shrub or climber; stem solid; leaves ovate, 12–14(–15) × 6–7(–8) cm, margin slightly undulate, long-acuminate,

cuneate; inflorescences axillary pseudoracemes of opposite fascicles, 5–25 cm; flowers strongly zygomorphic, green, lower lip white; fruit globose, a strongly-crested pyrene.
Described in 1974, based on *Leeuwenberg* 9540 (type), collected from near the edge of the eastern boundary of Kupe-Bakossi in 1972. *Letouzey* 11085, 100 km E of Douala is the only other Cameroonian subpopulation apart from that at Kupe Village. Three Gabonese collections were made by *Le Testu* (6096, 7594 & 8162), between 1926 and 1930.
The assessment of this species by Pollard in Cheek *et al.* (2004: 161) is maintained here since only five locations are known (above; AOO 20 km^2 with 4 km^2 cells) with threats as below. EOO is calculated as 96925 km^2.
Habitat: degraded lowland to submontane forest formations, overgrown edges to forest paths; c. 300–1000 m alt.
Threats: degradation of forest may actually favour this species, but wholesale clearance for urbanisation or plantation establishment is a serious threat at both the two Cameroonian locations.
Management suggestions: attempts should be made to refind this species, and to introduce it into cultivation. Basic populational data should be recorded as a baseline for monitoring.

Plectranthus cataractarum B.J.Pollard (syn. *Solenostemon sp. nov.?* of Cable & Cheek 1998: 64) (assessed by Benedict Pollard updated by Martin Cheek)
EN B1ab(iii)
Range: Equatorial Guinea (Bioko: Ureka-Moca (3 coll.)) and Cameroon (SW: Bakossi (3 coll.); Mt Cameroon (1 coll.); Etinde (2 coll.)).
This species was previously assessed in Pollard, B.J. & Paton, A.J. (2001) A new rheophytic species of *Plectranthus* L'Her. (*Labiatae*) from the Gulf of Guinea. Kew Bull. 56(4): 975–982 which was followed by Pollard in Cheek *et al.* 2004: 161.
The details presented in those publications (VU A2c, B2ab(ii,iv,v), C2a(i), D2) are set aside here, since bthe newly calculated EOO of 2017 km^2, together with the threats stated below and only three locations, allow EN status. Due to its extreme habitat specificity and the likely extinction of the Njonji subpopulation on the lower slopes of Mt Etinde due to clearance of forest for plantation agriculture here, it is maintained as Vulnerable.
Decumbent or rarely erect rheophytic herb to 0.6 m; leaves rhombic to obtrullate, c. 20–45 × 5–20 cm, attenuate, with basal half of margin entire, distally bluntly serrate; inflorescence terminal, to 35 × [illegible] cm, of 5–14, 6–8-flowered verticillasters; calyx 4–6 mm; pedicels 8–10(–14) mm; corolla mauve or rarely white.
Habitat: spray zone of waterfalls on wet rocks in or around fast flowing water within lowland to submontane forest, rarely non-epilithic on edge of watercourses; 300–1450 m alt.
Threats: see above; in addition, changes in run-off regime following deforestation within the stream catchments may result in habitat changes which are unfavourable to this species, such as increased silting along watercourses.
Management suggestions: protection of the Bakossi sites would help to preserve this species. Confirmation of its continued existence at Bioko and on Mt Etinde or otherwise should be made. Basic populational data should be collected as a baseline for monitoring.

Plectranthus dissitiflorus (Gürke) J.K.Morton (assessed by Martin Cheek)
CR B2ab(iii)
Range: Cameroon (SW: Mt. Cameroon (one coll.).
Known from a single specimen, the type (*Preuss* 1055). FWTA 2: 460 lists the collecting locality as "Buea to Bimbia", i.e. anywhere between 800 m alt. on the east flank of the main massif (Buea) and sea-level (Bimbia). However, this is contradicted by the protologue which reads "in Buschwald zwischen Buea und Mimbia, 940 m. 9 Oct. 1891". This narrows down the search area considerably.
The specimens at Kew are duplicates of the holotype which was presumably destroyed at Berlin.
This species is a herb 1 m tall, the stems glandular-pubescent. The leaves are as long as the petioles, 4–6 cm long, thinly pubescent on the nerves below, membranous and unusually in the genus, trullate and deeply crenate-sinuate. The flowers are large (2 cm long) and blue or purple-blue, the calyx 8 mm with the lower teeth lonest, acuminate. Related to the "Coleus" of commerce, and possibly with horticultural potential.
This species was first assessed, as CR A1c, by me in Cable & Cheek (1998: liii–liv). It is reassessed here in accordance to IUCN (2001) and so is updated as Critically Endangered since only a single location is known (above; AOO 4 km^2 with 4 km^2 cells) with threats as below. Benedict Pollard has recently identified what may be a second specimen-location for this species. If he confirms this, this species may be re-assessed as Endangered, especially if it is refound at Buea.
Habitat: submontane forest margin; 940 m alt.
Threats: forest clearance for firewood, timber, agriculture and urbanisation. This species may already be extinct on Mt Cameroon
Management suggestions: the vicinity of Buea and Mimbia should be searched for this species in October (when in flower) so as to try to refind this plant. This might be a good project for students at the University of Buea. If the search is successful, local communities and land-managers should be apprised and assisted to protect the plants, perhaps supported by a poster campaign. Basic populational data should be collected to form a baseline for monitoring. Seed should be

collected and banked to safeguard the species. Limbe BG could help to conserve the species by propagating it for introduction to safe sites and by educating the public about its importance.

Plectranthus insignis Hook.f. [Mbum (Kilum)] (assessed by Martin Cheek)
VU A2c + B1ab(iii)

Range: Cameroon endemic (SW: Mt Cameroon (several coll.); Rumpi Hills; Mwanenguba at Nsoung; Lebialem Highlands. NW: Mt Oku (7 coll.), Bafut Ngemba (1 coll.)).

Here *Plectranthus insignis* is assessed as Vulnerable since it is estimated that its population has declined by more than 30% in 30 years (three generations) due to destruction of its forest habitat: see threats below.

In addition its EOO supports this. EOO is calculated as 8505 km^2. This is a mass flowering species believed to reproduce only ever c. 9 years, at some locations in synchrony with one or more species of mass-flowering Acanthaceae.

When in flower this species appears common and conspicuous, but otherwise it is not easily observed and recorded e.g. seven collections were made at Kilum-Ijim in 1996, but despite survey teams being present there on other years, no other specimens are known from that location.

Large woody undershrub to 6 m tall., gregarious, leaves narrowly obovate-elliptic, 15–25 × 6–10 cm, acuminate, base cuneate into the petiole, veins pubescent, margin serrate, panicle terminal, lax and spreading; calyx 2 cm, corolla 2 cm, bright yellow with a red splash.

Habitat: submontane and montane forest; 1350–2450 m alt.

Threats: clearance of forest for firewood, timber and agricultural land. One area in the Bamenda Highlands saw 50% of surviving forest cleared 1988–2003 (Baena in Cheek *et al.*, 2010: back cover).

Management suggestions: the logical focus for conservation efforts for this species should be Mt Cameroon and Mt Oku since both locations have good pupulations of the species and have a degree of protection. Basic populational data should be collected as a baseline for monitoring. With good management, this species, together with other mass-flowering synchronous species in the Acanthaceae, could provide an income to local communities in the vicinity through sensible marketing of tourism opportunities.

Plectranthus punctatus L'Hér. subsp. ***lanatus*** J.K.Morton [Ijim (Kom)] M.Cheek
VU B1ab(iii)+D2

Range: Cameroon endemic (SW/W: Lebialem/Bamboutos Mts (5 coll.). NW: Bamenda Highlands (several coll.).

This subspecies was first collected by Maitland, probably either on Laikom Ridge or at 'Mbesa Swamp' (*Maitland* 1724, "Basenako-Lakom, on plateau in grassland, June 1931, 1800m"). In the Bamenda Highlands it is also known from Ndu and from the Bambili Lakes as well as Kilum-Ijim.

This plant is best distinguished from other members of the genus by the purple-blotched, white hairy stems and small stature. Its habitat is also distinctive.

This species is one of the Ijim grasses of the Kom people. Those who pick this plant will only return to Ijim as a corpse, it is believed.

Herb c. 30 cm tall. Stems ascending, rounded and sub-succulent when alive, nodes swollen, internodes 1.5–3(–5) cm long, whitish green, blotched purple, clothed in long white hairs c. 2 mm long. Older stem bases sometimes straggling, prostrate, rooting at nodes. Leaves of the midstem sessile, sublanceolate, c. 6 × 2 cm, apex rounded, margin serrate-crenate, with c. 15 teeth per side, white hairy above and below. Inflorescence 6–15 cm long, verticils 5–10, internodes c. 1.5 cm long, flowers subsessile. Flower with corolla c. 12 mm long, pale blue, prominently speckled with purple.

The largest population of this species seen was at Afua swamp at Kilum-Ijim in November 1999 where it lines the border of the swamp for about 100m or more. In total, five sites for the species are known at Kilum-Ijim, the number of individuals of which varies from 5–10 to 100–200.

The assessment VU D2, based on a low number of locations, yet without marked threats, was originally made in Cheek *et al.* (2000: 62–63) and is maintained here, with the addition of B1 status based on the newly calculated EOO which would permit EN status were the number of locations not seven (above). No data changes are available apart from an extra record from Lebialem (Harvey *et al.* 2010: 125). EOO is calculated as 2467 km^2.

Habitat: damp grassland at the edge of swamps, or banks; 1800–2600 m alt.

Threats: swamp drainage or development; shading-out by growth of grasses; possibly trampling by cattle.

Management suggestions: more research is needed on the management regime needed for this plant. However plants seen in long grass at the Mbesa swamp in 1998 seemed less healthy and were far fewer than those seen in close cropped grass at Afua swamp. Basic populational data are needed as a baseline for monitoring. If firmer threats can be demonstrated for this species it might be reassessed as Endangered.

Pogostemon micangensis G. Taylor (assessed by Martin Cheek)
CR B1ab(iii)+D

Range: Angola (Gangvellas, Cuiriri, River Micango) and Cameroon (C: Mefou proposed NP).

Here *Pogostemon mikangensis* is assessed as Critically Endangered on the basis that it is known from only an estimated 12 individuals seen by field botanists and with threats as below. And since there are threats(below) and the EOO is calculated as 1.03 km^2.

Prostrate annual herb, mint-scented when crushed; glabrescent, stem apex and inflorescence with sparse white hairs; stem square, internodes to 7 × 0.2 cm, nodes rooting; stem ascending slightly at apex; leaves opposite; leaf-blade elliptic, to 4.5 × 2 cm, acute, base unequally obtuse, crenate-serrate, teeth 1 × 3 mm, lateral nerves c. 7 pairs; petiole to 2 cm; spike terminal, dense, 3 × 1 cm; calyx bell-shaped, 3 mm, teeth 1 mm, white hairy; corolla pale purple, exserted 1 mm; stamens exserted 1.5 mm, purple.

The Mefou specimens were confirmed in their identification by Dr A. Paton at Kew; data on the Angolan material comes from the species protologue. In view of the geographical disjunction perhaps the identity of the Mefou material should be revisited. The tribe was analysed by Press (1982) in Bull. Brit. Mus. Nat. Hist. 10(1): 71–74 but species limits were not covered. Ours appears to be the first record of the genus in W-C Africa.

Habitat: river marshes (Angola); drying sandy river beds; 600–710 m alt. (Cameroon).

Threats: sand extraction for construction of buildings and infrastructure.

Management suggestions: gazettement of Mefou proposed NP would help secure resources to protect this species. A survey along river beds in Mefou and similar areas would yield a better conservation assessment, a baseline for future monitoring, and an evaluation of management needs.

Stachys pseudohumifusa Sebsebe subsp. ***saxeri*** Y.B.Harv. (assessed by Martin Cheek)

EN B1ab(iii)

Range: Nigeria (Mambilla Plateau, Gembu (1 coll.)) and adjoining Cameroon (NW (Bamenda Highlands): Ijim at Ndu; Laikom "Lakoni"; Mbesa swamp; Afua swamp (several coll.). W: Bamboutos Mts, Dschang (1 coll.)).

This subspecies was only recognised in 1996 (Harvey, Y. B (1996) The *Stachys aculeolata/aethiopica* complex in tropical Africa. *Kew Bull.* 51(3): 433–454). It was then known from only four localities (see above). We have located three further sites for this taxon in the Kilum-Ijim area, all at Ijim: the Afua Swamp (*Cheek* 9842), the Mbesa Swamp (*Cheek* 9811, *Pollard* 349) and at a swampy place in a valley between the Ardo of Ijim's compound and the Fon of Kom's Palace where about a dozen plants were seen (Cheek *pers obs.* Dec. 1999). The habitat requirements of this plant seem to be highly specific and so the number of localities that can support the species is likely to remain small.

The assessment of this species in Cheek *et al.* 2000: 63 as VU D2 is changed here to EN since the newly calculated EOO is 3109 km^2 and there are only three broad locations with threats as above.

Erect, weak-stemmed herb to 1 m tall, bright green. Stem pronouncedly square, finely prickly and sticking to clothing. Leaves erect, oblong lanceolate, up to 22 mm wide, margin crenate, revolute. Inflorescence terminal, verticils 4–6-flowered, 3–28 mm apart. Calyx 4–4.5 mm long. Corolla bright white, corolla 10–13 mm long; tube 6–7.5 mm long; lower lip 5–6 mm long; upper lip (2.5–)3–4 mm long. Nutlets c. 1.5 mm long. Description after Harvey 1996.

Stachys aculeolata var. *aculeolata* is the only other *Stachys* in our area. It differs principally in its leaves being patent, not erect, and in them being ovate, i.e. broader in proportion to their length. This last taxon in our area also differs in being usually prostrate, with grey-green leaves, and is found in woodland edges, not usually in wet spots.

Habitat: swamp grassland, often at the interface with montane forest or scrub; 1800–2460 m alt.

Threats: unknown but possibly trampling from cattle.

Management suggestions: more data is needed on the numbers of individuals at each site, and on the type and levels of regeneration. This species is conspicuous when in flower in October-December when baseline surveys should be conducted to aid future monitoring.

Vitex lehmbachii Gürke (assessed by Benedict Pollard)

EN A2c + B1+2ab(iii)

Range: Cameroon (SW: Buea (4 coll.); Mt Etinde (1 coll.); Barike-Manya, Kumba (1 coll.), Kupe-Bakossi (3 coll.). NW: Bamenda Highlands (1 coll.)).

The assessment of this species by Pollard in Cheek *et al.* (2004: 161) is maintained here, with the addition of criterion B1 and 2 status since only five locations are known (above; AOO 20 km^2 with 4 km^2 cells) with threats as below, and since EOO is calculated as 4363 km^2.

First collected at Buea, *Lehmbach* 11, in April 1897, this species has been collected at ± regular intervals since, but often in areas which have undergone considerable decline in habitat quality.

Tree to 12 m; branches glabrous; leaves compound, opposite; middle leaflets 10–25 × 4–10 cm, with 4–8 pairs of lateral nerves, obovate, abruptly acuminate, cuneate; petiole drying black; inflorescence of long-pedunculate axillary cymes; peduncle 10–20 cm; calyx 3–5 mm; corolla less than 1 cm, white with pink lower lip; fruits 1–1.5 cm.

Habitat: mid-elevation evergreen forest; 150–1330 m alt.

Threats: only the two records from Muambong and Ngomboku, in Bakossiland, are likely to have been unaffected by logging activities, and clearance of forest for agriculture. At all the other known locations, much clearance has occurred in the distant and recent past, and is likely to continue in the future. Pollard estimated that the population size will have reduced by ≥ 50% over the last three generations, which he estimate to be between c. 50 and 100 years.

Management suggestions: the two Bakossi sites should be revisited and seed collected, if available, to help introduce this species into cultivation. Searches around Buea could also be instigated, for the same purpose. Basic populational data should be collected as a baseline for monitoring.

Vitex yaundensis Gürke (assessed by Benedict Pollard)

CR A4c, B1ab(i,ii,iii,iv,v), 2ab(i,ii,iii,iv,v)

Range: Cameroon (SW: Mungo R FR (1 coll.). C: Yaoundé (1 coll.)).

The assessment of this species by Pollard in Cheek *et al.* (2004: 161– 162) is maintained here since no new data are available and because only two locations are known (above; AOO 8 km^2 with 4 km^2 cells) with threats as below.

This tree was only known from the type collection until *Cheek* 10139, from the Mungo R FR, in 1999. It was first collected in jungle by the station at Cameroon's capital city, Yaoundé: *Zenker* 1412, fl., June 1897, at 800 m altitude.

Tree 10–15 m tall; bole pale grey-brown; outer slash wet dark brown; inner pale yellow; leaves compound, opposite, 5-foliolate, with extremely long petioles to 30 cm; the secondary nerves conspicuously prominent beneath, close together and almost parallel; petiolules 1–2 cm; inflorescence long-pedunculate axillary cymes; bracts linear; corolla tube scarcely exceeding the calyx; fruits c. 2 cm long.

Habitat: low- to mid-elevation evergreen forest; 150–800 m alt.

Threats: there can be little doubt that in the intervening years the Yaoundé location will have suffered enormously from the ever-growing population size, and resulting expansion of urbanisation. At the Mungo R location, we know of illegal logging activity and clearance of habitat for agricultural purposes, which suggests that this taxon could be on the verge of extinction. Although known from two locations, they are severely fragmented, and indeed *V. yaundensis* could now be said to likely occur at just one location (Bakossiland).

Management suggestions: visits to the two known locations should be instigated to assess the chances of survival of this species. Inventory work should be continued in other uncollected areas of suitable habitat, between and around the two locations. If found, seed of this species should be distributed for germination at suitable sites, both *in situ* and *ex situ*.

LAURACEAE

assessed by Martin Cheek

This pantropical family of about 2500 species placed in c. 50 genera are named for the European species *Laurus nobilis* from which wreaths were made and awarded to victors in competitions in classical times, giving rise to the words Baccalaureat and Bachelor (of Science) for academic awards. Nowadays this species is most valued for its spicy leaves, used to impart their flavour in cooking (bay leaves). In Cameroon the introduced S American species *Persea americana*, the pear or poire is the best known member of this family, however there are numerous indigenous species also, mainly in the evergreen lowland forests.

In Cameroon most species of this family are understorey or canopy tees, with a spicy scent, but no coloured exudate, when wounded. The leaves are alternate, simple, entire and lacking stipules, while the flowers are usually borne in panicles, have parts in multiples of three, and are remarkable for their stamens, positioned in more than one ring, the fertile stamens usually alternating with fleshy staminodes as large as themselves, and with the anthers releasing their pollen by four separate flaps. The fruits are usually single–seeded berries resembling a miniature pear (avocado).

Although there is a Flore du Cameroun account for Lauraceae (Volume 18, by Fouilloy in 1974), it is difficult to use in identifying species so that the naming of most material collected is very uncertain. Until better guides to the identification of this family in Cameroon are available, and principally this concerns the species of the genus *Beilschmiedia*, any conservation assessments would be unreliable. An Africa wide revision of the genus would help redress this. Most of the species of this genus treated in Flore du Cameroun are listed with only a single specimen and if assessed would be Critically Endangered on the data available. These and others which could rate as threatened are listed below, but are not formally assessed. For the moment they should be considered DD. The Flore du Cameroun account accepts 44 species in four genera, *Beilschmiedia*, *Ocotea*, *Hypodaphnis* and *Cassytha,* the last three with only a single species in Cameroon. *Cassytha* at first sight does not resemble a plant at all; being a specialised parasite which festoons host vegetation with its spaghetti-like yellow stems which lack leaves and any green tissue. However, the flowers and fruits place it in *Lauraceae.*

Beilschmiedia acuta Kosterm.(Cameroon & CAR)
Beilschmiedia anacardioides (Engl. & K.Krause) Robyns & Wilczek (Cameroon and Gabon)
Beilschmiedia barensis (Engl. & K.Krause) Robyns & Wilczek (Cameroon & Gabon)
Beilschmiedia cinnamomea (Stapf) Robyns & Wilczek (Cameroon & Gabon)
Beilschmiedia congestiflora (Engl. & K.Krause) Robyns & Wilczek (Cameroon)
Beilschmiedia crassipes (Engl. & K.Krause) Robyns & Wilczek (Cameroon)
Beilschmiedia cuspidata (K.Krause) Robyns & Wilczek (Cameroon)
Beilschmiedia dinklagei (Engl.) Robyns & Wilczek (Cameroon and Gabon)
Beilschmiedia diversifolia Pierre ex Robyns & Wilczek (Cameroon & Gabon)
Beilschmiedia fruticosa Engl.(Cameroon & Gabon)

Beilschmiedia fulva Robyns & Wilczek (Cameroon, Gabon, Congo (Kinshasa))
Beilschmiedia gaboonensis (Meissn.) Benth. & Hook.f. (Cameroon & Gabon)
Beilschmiedia grandibracteata Robyns & Wilczek (Cameroon & Gabon)
Beilschmiedia grandifolia (Stapf) Robyns & Wilczek (Cameroon & Gabon)
Beilschmiedia hutchinsoniana Robyns & Wilczek (Nigeria & Cameroon)
Beilschmiedia jabasssensis (Engl. & K.Krausse) Robyns & Wilczek (Cameroon)
Beilschmiedia klainei Robyns & Wilczek (Cameroon & Gabon)
Beilschmiedia kostermansiana Robyns & Wilczek (Cameroon)
Beilschmiedia lancilimba Kosterm. (Cameroon)
Beilschmiedia letouzeyi Robyns & Wilczek (Cameroon & CAR)
Beilschmiedia louisii Robyns &Wilczek (Cameroon, CAR & Congo (Kinshasa))
Beilschmiedia membranifolia Kosterm. (Cameroon)
Beilschmiedia myricaefolia (Moore) Robyns & Wilczek (Nigeria & Cameroon)
Beilschmiedia ndongensis (Engl. & K.Krausse) Robyns & Wilczek (Cameroon)
Beilschmiedia nitida Engl. (Cameroon)
Beilschmiedia papyracea (Stapf) Robyns & Wilczek (Cameroon & Gabon)
Beilschmiedia pierreana Robyns & Wilczek (Cameroon & Gabon)
Beilschmieda preussii Engl. (Cameroon). Treated as CRA1c + 2c in Cable & Cheek 1998: liv
Beilschmiedia preussioides Fouilloy & N.Hallé (Cameroon & Gabon)
Beilschmiedia robynsiana Kosterm. (Cameroon)
Beilschmiedia sessilifolia (Stapf) Engl. ex Fouilloy (Cameroon)
Beilschmiedia staudtii Engl. (Cameroon & Gabon)
Beilschmiedia wilczekii Fouilloy (Cameroon)
Beilschmiedia zahnii (K.Krause) Robyns & Wilczek (Cameroon)

LECYTHIDACEAE - NAPOLEONOIDEAE OR NAPOLEONACEAE

assessed by Martin Cheek & Iain Darbyshire

Napoleonea and *Crateranthus* are easily recognised by their large distinctive flowers, 2–6 cm across.

The flowers consist [illegible]lly, ribbed petaloid structures, the innermo[illegible] outermost slightly reflexed, the middle often terminating in numerous, free, finger-like extensions, all variously coloured in rings of purple, yellow, brown or green, on white. The leaves are alternate, pinnately veined, and simple, lacking stipules. *Napoleonea* are recognised even when sterile by verticillate branching from the main axis, like *Diospyros*, and by gland pairs at the base of the leaf-blade. Cameroonian *Crateranthus* have winged stems and are likely to be confused only with *Oubanguia alata* (Scytopetalaceae) which also has wings on the stem although not so wide as those of *Crateranthus*. *Crateranthus* have similar flowers to *Napoleonea* but have dry, dehiscent fruits with winged seeds rather than leathery-fleshy indehiscent ones as in *Napoleonea*.

While *Napoleonea* was revised in 1971 by Liben (Bull. Jard. Bot. Nalt. Belg. 41: 363–382), *Crateranthus* lacks such an aid to species delimitation.

Both genera have been treated either as an endemic African subfamily of the pantropical Lecythidaceae, or as a distinct family. No Flore du Cameroun treatment exists for this group.

Napoleonaea talbotii Bak.f. of Nigeria and Cameroon, and *N. gabonensis* Liben of Cameroon and Gabon appear to be NT, but further study might reveal that they merit VU status.

Crateranthus talbotii Baker f.
VU A2c

Range: SE Nigeria (Oban; Stubb's Creek FR; Calabar–Mamfe Rd) and Cameroon.(SW: Korup (several coll.); Kumba–Mamfe Rd; and (possibly this species, sterile specimens only): Bakossi Mts; Banyang Mbo; Mone FR)
The assessment of Darbyshire & Cheek in Cheek *et al.* (2004: 162) is maintained here: VU B1+2ac is given in www.redlist.org, (IUCN 2003) which cites WCMC (1997) as its source, together with recent (1996) information from Klaus Schmitt on its status in Nigeria: "occurring in an area extending from south-east Nigeria into Cameroon. The largest, if not the only, remaining population occurs in the Oban Division of the Cross River NP in Nigeria and the contiguous Korup NP in Cameroon".
Darbyshire in Cheek *et al.* (2004: 162) considered that the species extends into the Rumpi Hills (*Mambo & Thomas* 6, 1986!) and into the Bakossi Mts (5 coll. at 3 sites), so that 13 sites were accepted (Darbyshire *obs.*). Accordingly, the taxon was considered no longer threatened under criterion B and so was reassessed under criterion A, it being suspected that over 30% of its habitat has been lost in the last 100 years, largely in Nigeria, from the statement of Schmitt.
The basic facts have been little altered since 2004 apart from the realisation that within current IUCN guidelines, the Bakossi locations should not be taken necessarily as three, but probably as just one. Moreover since a new species has

been found to the E and only sterile, not fertile material has not been seen from the Bakossi, Banyang Mbo and Mone areas, the identification at those locations is no longer as certain as it was.

Should the material at Bakossi etc prove to be specifically separable, the threat rating of *C. talbotii* is likely to be raised.

Tree to 20 m, 30 cm dbh; bark red-brown, smooth; leafy branchlets foliaceously winged, to 1 cm width; leaves oblong-elliptic to oblanceolate, 26–33.5 × 8–11 cm, grey-green above, acuminate, base unequally cordate, lateral nerves 10–14 pairs; petiole 0.5 cm, black; flowers axillary, subsessile; bracts leathery 1 cm; calyx 3-lobed, lobes obovate, 2.5 cm; corolla 3.5 cm, pink to magenta, margin pale.

Habitat: lowland freshwater swamp forest (where it is gregarious, *pers. obs.*) and (Bakossi etc. only) submontane forest near streams; 200 m (Bakossi etc 1000 m) alt.

Threats: forests outside protected areas have largely been logged and cleared for commercial crops and subsistence farming (www.redlist.org (IUCN 2003)).

Management suggestions: continued or improved protection of the Oban, Korup and Bakossi NPs is advised. Revisiting the Bakossi, Banyang Mbo, Mone locations in the flowering season, possibly February, might resolve whether plants at these sites are conspecific or not with *C. talbotii.* Within Korup the species is locally common and apparently secure in swamp forest and no conservation action beyond monitoring is needed.

Napoleonaea egertonii Baker f.

VU B2ab(iii)

Range: Nigeria (Cross River State, Oban (3 coll.)) and Cameroon (SW: Atolo to Mamfe (1 coll.); Kupe-Bakossi (3 coll., 3 sites), Nta Ali FR (1 coll.); Takamanda FR (1 coll.), Korup N.P. (1 coll.); Lebialem-Bechati (1 coll.)).

This striking medium sized (c. 40 cm dbh)forest tree , bearing large (to 70 × 30 cm) spiny fruits on the trunk, was only known from Oban Nigeria, prior to the plant inventory work in western Cameroon beginning in the 1980s. Discoveries of this species at Korup is important as it is relatively well protected at that site; however, it is not common at Korup, only 1–2 trees having been found (Cheek, *pers. obs.*), and its abundance at Takamanda is unknown. At Kupe Village and the adjacent Manehas FR, the species is again uncommon, one plant being found at each location. However, several specimens were observed within close proximity to Nyandong in W Bakossi. The assessment made here first featured in Cheek *et al.* (2004: 162) and is maintained here since only the Lebialem site has been added since. The assessment is supported on the basis of 12 spots (AOO 48 km² with 4 km² cells) at seven locations and threats as below.

Liben (1971 loc. cit.) united *N. letestui* Pellegr. of the Chaillu in Gabon with this species, understandably since both have panicles and massive spiny fruits, unusual in the genus. However he only saw the type specimen of *N. egertonii* and did not have access to the abundant material that has since been accumulated. Here *N. letestui* is resurrected since its fruit, although large, are of different shape and smaller size (broader than long, 14–16 × 12–13 cm). In addition, the absence of material of these conspicuous taxa from either S Region Cameroon or northern Gabon, such as the Cristal Mts, seems peculiar if the two are indeed conspecific. More comparative investigation is needed of this topic, and should Liben be vindicated and the range of *N. egertonii* follow his view, its rating is likely to drop to NT.

Tree to 20 m; leaves oblong-elliptic, 18–36.5 × 8–13 cm, apex shortly acuminate, base acute, lateral nerves 9–11 pairs; petiole to 1.2 cm; panicles cauliflorous, 12 × 12 cm, peduncles brown-puberulent; pedicel 0.3–1 cm; calyx 2.4 cm diam., 5-lobed, densely warted and puberulous; corolla c. 4.5 cm diam., white with purple veins, outer corona segments 0.45 cm, puberulent, inner corona glabrous; fruit to 30 × 30 cm, echinate, spines to 1.5 cm; seeds to 5 cm.

Habitat: low- to mid-elevation evergreen forest, often occurring on rocky slopes; 250–1000 m alt.

Threats: the Nigerian sites are likely to have been either lost or under severe threat from widespread logging of lowland forest here. This is the case at all other known sites, however threats in Gabon are unknown. The two sites at Mt Kupe are below the 1000 m alt. lower limit of effective forest protection and thus vulnerable to agricultural encroachment. At Nyandong, several trees were recorded close to the village and adjacent to tracks; these are highly vulnerable to future expansion of the village and road improvement.

Management suggestions: a survey of the number of trees of this species at Korup should be carried out to determine its abundance at this site, as it is the best protected, thus offering the best opportunity for conservation of this species. Informing local communities, most notably at Nyandong where this species appears most common and in closest proximity to human settlement, of the scarcity of this species may help to promote community-led conservation, particularly as it is such a striking and easily recognisable taxon. Enquiries are needed to determine whether or not the large and numerous seeds have any local uses which might encourage local protection. A comparative investigation with material of *N. letestui* in Gabon is needed to establish whether they are the same species or not.

LEGUMINOSAE (FABACEAE OF SOME AUTHORS)

This is one of the world's largest families of flowering plants, and one of great importance to man, containing many agricultural species, such as, in Cameroon, groundnut (*Arachis*) and local beans *Phaseolus*). Other species have been grown as ground-cover under tree plantations as for specialist products such as cigarette paper (*Crotalaria*). Yet more species are high value timber trees (e.g. *Pterocarpus, Guibourtia* and *Copaifera*). Numberous forest species have fruits gathered from the wild and eaten e.g. *Tetraberlinia tetraptera, Angylocalyx*) while others are planted on small-holder farms to enrich fertility using their nitrogen-fixing powers derived from their root-nodules (*Tephrosia vogelii, Cassia*). Despite the great utility of this family, most species have not been evaluated for their value to man and many of these, including some of those below, have high potential, yet are in danger of global extinction. For example, many species of the genus *Trifolium* have extremely high value in enriching pastures for livestock around the world and are commercially bred for this purpose. Yet Cameroon has a species, *T. gillettianum* known globally from only a single collection at a single site in Adamawa. This species could easily be lost for eternity by development or modification of its site by those who have no idea of the importance of the plant species present.

Three subfamilies are recognised:

Leguminosae subfamily Caesalpinoideae (Caesalpinacees of French botanists). Generally these have compound leaves lacking a terminal leaflet, and flowers which lack the pattern seen in better-known papilionoideae of outer standard, inner wings and well-defined keel which encloses the stamens. The herbaceous genera *Cassia, Chamaecrista* and *Senna* have pea-type flowers, but the standard is enclosed by the wings and the keel petals are not appressed to each other. Most Cameroonian species however, are lowland forest trees of the Amherstieae-Detarieae tribes, which show great variation in flower structure, some taxa e.g. *Talbotiella*, having flowers which, apart from the ovary, are actinomorphic, when zygomorphic flowers are the norm in this subfamily.

Leguminosae subfamily Mimosoideae (Mimosacees of French botanists). Herbs and trees (*Albizia, Parkia*), less usually lianas (*Entada*), the flowers are reduced in size, symmetrical with usually only the stamens forming the attractive structures. Grouped in spherical heads, essentially compound flowers, the best known examples of this subfamily are *Acacia* and *Mimosa*. The leaves are often bipinnate, lacking a terminal leaflet but the axes often bear cylinfrical gland structures.

Leguminosae subfamily Papilionoideae (Papilionacees of French botanists).

Herbs and shrubs of grassland, trees and lianas of forest, this subfamily contains the true peas and beans. Generally the leaves are once-pinnately compound (unifoliolate in *Baphia*) with a terminal leaflet. The standard encloses all other petals in the bud: the two wing petals and two keel petals. The keel petals are appressed to each other at their edges forming a cavity for the stamens in between.

Becuase of their economic importance, the Leguminosae are relatively well-served by plant taxonomists and for many genera, reliable, recent flora or monographic treatments are available to delimit taxa and to identify them. This is true, for example, for most of the papilionoid genera below, including *Baphia, Crotalaria, Dolichos, Eriosema* and *Trifolium*. However, for *Dalbergia, Leptoderris, Millettia, Rhynchosia,* to mention the smaller generic examples, there is no recent revision and the most recent available reference is E. G. Baker's (1920–1929, 3 vols) Leguminosae of Tropical Africa. It is to be hoped that this deficiency will be rectified in the near future. Fortunately Klitgaard has plans to co-ordinate revision of *Pterocarpus* and for Caesalpinoideae, Breteler and Wieringa at WAG, and Mackinder and van der Burgt at Kew, have made good progress recently delimiting and revising genera in the Amherstieae-Detarieae.

A useful work for deciding on whether, in the absence of a recent revision, published species names are likely to be acceptable or not, is Legumes of Africa, A Check-list (Lock 1989). Filling a large gap in our knowledge of the Leguminosae of Rio Muni, but also addressing Bioko, is Velayos *et al.* 2010, Leguminosae, Vol 5 of Flore de Guinea Ecuatorial.

Perhaps the greatest need for better conservation management of Leguminosae in Cameroon is the production of a Flore du Cameroun account so that all species present can be determined, and identified and assessed for their conservation needs. Of the three subfamilies, only one, the Caesalpinoideae, has a Flore du Cameroun treatment, (Aubréville, Flore du Cameroun 9 (1970)) and that is both out of date and out of print.

LEGUMINOSAE SUBFAMILY CAESALPINIOIDEAE

assessed by Barbara Mackinder

An initial list of caesalpinioid legume species recorded from Cameroon was generated from the International Legume Database and Information Service (ILDIS) and then modified to reflect recent taxonomic changes in several genera: *Anthonotha* (Breteler 2010); *Berlinia* (Mackinder & Harris 2006; Mackinder & van der Burgt 2009); *Englerodendron* (Breteler 2010; van der Burgt 2007); *Isomacrolobium*

(Breteler 2008) and *Talbotiella* (Mackinder *et al* 2010). Species known to be widespread were deleted from the list. All other species were evaluated for conservation status and those which qualified for an IUCN category of threat of Vulnerable, Endangered or Critically Endangered are included here. An earlier version of an unpublished red-data checklist for Cameroon prepared by Jean-Michel Onana (Head of the National Herbarium of Cameroon) was kindly made available to us. Locality records were taken from several sources, notably the RBG Kew Cameroon database, from exsiccatae included in the publications listed above and also from Breteler (1999), Breteler & Miyono Nguema (2008); de la Estrella *et al* (2010), Mackinder & Pennington (*in press,* 2011) and Wieringa (1999). From Wageningen Agricultural University, Netherlands (WAG) where there is a long history of specialist knowledge in African Caesalpinioideae, Jan Wieringa (Scientific Curator of Collections) generously provided additional locality records for several species from the WAG specimen database which allowed us to exclude some species. Their exclusion helped keep the focus of this account on those species which are of real conservation concern.

We compared the results of our evaluations with previously published assessments where they existed. In some cases there had been no change in category of threat. In other cases, after considering new data and observations an adjustment to the category of threat was warranted and if that resulted in a category of Vulnerable, Endangered or Critically Endangered, the prior assessment is referred to and a justification given for the change in status.

Three species previously thought to occur in Cameroon and to qualify for a category of threat are not included because their presence in Cameroon could not be confirmed. They are *Brachystegia nigerica* Hoyle & A.P.D.Jones, *Didelotia unifoliolata* J.Léonard and *Microberlinia brazzavillensis* A.Chev. Six other species, confirmed as present in Cameroon have wider distributions than previously known and consequently are not currently of conservation concern. They are *Cynometra sanagaensis* Aubrév., *Didelotia idae* J.Léonard, *Gossweilerodendron balsamiferum* (Harms) J.Léonard, *Guibourtia ehie* (A.Chev.) J.Léonard, *Pellegriniodendron diphyllum* (Harms) J.Léonard and *Plagiosiphon longitubus* (Harms) J.Léonard. Another two species no longer qualify for a category of threat due to recent advances in taxonomic knowledge. *Crudia bibundina* Harms, was apparently a narrow endemic known only from 2 collections but has recently been placed in synonymy with the more widespread *Crudia ledermanii* Harms by Breteler (2008) whilst *Brachystegia zenkeri* Harms is considered to be conspecific with, and a synonym of, the more widespread species *B. laurentii* (De Wild.) Louis ex Hoyle (Sosef *et al.* 2006).

Nevertheless insufficient data and the lack of basic taxonomic knowledge in some genera, for instance *Dialium*, *Gilbertiodendron* and *Plagiosiphon* continues to hamper our efforts to establish their conservation needs. Regrettably, there were no cases where a species which had previously qualified for a category of threat could now be excluded because the recorded threat(s) to its survival had abated or ceased since it was last evaluated.

Afzelia Sm., a genus of c. 11 species of which seven are African and the rest are Asian (Mackinder, 2005). Typically *Afzelia* species have leaflets in two or more pairs, twisted petiolules and lack translucent gland dots. Flowers have long narrow hypanthia and one large petal. The woody oblong fruit contain seeds with brightly coloured arils at their base. Four species occur in Cameroon, and all except *A. bella* Harms are threatened due to exploitation by the international timber trade.

Afzelia africana Sm.

VU A2c

Range: widespread in tropical Africa. Cameroon (NW: Bamemda (Mufung). Adamawa: Mai Idoana. C: between Deng Deng and Yaoundé. E: Babua [although this is almost certainly CAR]).

Assessed as Vulnerable in 1998 by Oldfield *et al.* (1998), who reported that although the species is widespread, the wood is exploited commercially and the species has declined in population numbers. That assessment is accepted but modified here to reflect changes in IUCN's categories and criteria.

Afzelia africana is characterised by the combination of large leaflets, greater than 6 cm long, in 3– 5 pairs; large petal is c. 1.4 cm long, white with red markings; seeds have a reddish-orange cup-shaped aril (Aubréville, 1970). Cameroonian collections were made between 1914 and 1955.

Habitat: wooded grassland, deciduous woodland, fringing forest and the drier part of forest regions.

Threats: heavy exploitation by the international timber trade; forest clearance for agriculture.

Management suggestions: mapping of locations on to protected areas is needed to find a site where this species can be protected. Seed propagation to multiply the species for introduction to safe sites should be considered to protect the survival of this valuable asset for Cameroon.

Afzelia bipindensis Harms

Vu A2c

Range: widespread in tropical Africa including from eight of nine provinces of Gabon. Cameroon (SW: Mt Etinde. NW: Ntali. Littoral: Yabassi. (3 loc.)).

Assessed as Vulnerable in 1998 by Oldfield *et al.* (1998) who reported that although the species has a wide distribution in Africa it is exploited by the international timber trade. That assessment is accepted but modified here to reflect changes in IUCN's categories and criteria. *Afzelia bipindensis* is characterised by the combination of large leaflets, greater than 6 cm long, in 4–8 pairs; large petal is 3–6.5 cm long, white with reddish purple claw and white lamina; seeds have an orange bi-lobed aril (Aubréville 1970).
Habitat: lowland rainforest, up to 900 m (Flora of Tropical East Africa).
Threats: heavily exploited throughout its range by the international timber trade. In some areas there are reported to be few seed trees remaining (www.iucnredlist.org); forest clearance for agriculture.
Management suggestions: mapping of locations on to protected areas is needed to find a site where this species can be protected. Seed propagation to multiply the species for introduction to safe sites should be considered to protect the survival of this valuable asset for Cameroon.

Afzelia pachyloba Harms

VU A2c

Range: Nigeria, Cameroon (2 loc., Littoral: Loum FR. SW: Dikulu), Gabon, Congo (Kinshasa) and Angola.
Assessed as Vulnerable by Oldfield *et al.* (1998) because it is a forest species which is heavily exploited for its commercial timber, with relatively few seed trees remaining through its range. That assessment is accepted but modified here to reflect changes in IUCN's categories and criteria. *Afzelia pachyloba* is characterised by the combination of small leaflets, not exceeding 6 cm in length, in 5–10 pairs; large petal is 2.5–4 cm long, white with red markings; seeds have a small, yellow, slightly 4-lobed aril (FWTA).
Habitat: evergreen forest and mixed deciduous woodland; forest clearance for agriculture.
Threats: heavy exploitation by the international timber trade.
Management suggestions: mapping of locations on to protected areas is needed to find a site where this species can be protected. Seed propagation to multiply the species for introduction to safe sites should be considered to protect the survival of this valuable asset for Cameroon.

Anthonotha P.Beauv. has recently been revised by Breteler (2010) whose strict circumscription of the genus includes only one of the five sections formerly include in *Anthonotha*. Two species were transferred to *Englerodendron* by Breteler (2006), and the remaining species of the four newly excluded sections are placed in the genus *Isomacrolobium* by Breteler (2008). *Anthonotha* is distinguished from *Isomacrolobium* in having an adaxial petal always present (not present or absent as in *Isomacrolobium*), the other four petals rudimentary or absent (not similarly developed or strongly reduced or absent as in *Isomacrolobium*); lower surface of leaflets hidden by a dense, whitish, silverish, or golden-brown, often lustrous, usually persistent indumentum (not lower surface glabrous or sparsely hairy as in *Isomacrolobium*). Five species of *Anthonotha* occur in Cameroon of which one is threatened.

Anthonotha wijmacampensis Breteler

CR D

Range: Cameroon endemic (S: 30 km Ébolowa-Minkok Rd;. 22 km E of Djoum).
Assessed here as CR D since the species is known from only from 2 mature individuals (Breteler, 2010). First collected near Djoum in 1966 by Letouzey and then only once more since along the Ébolowa-Minkok Rd in 1975 by J.J. de Wilde.
Anthonotha wijmacampensis has 4–5-pairs of leaflets, dense brown long or short hairs on the lower leaflet surface, inflorescences appearing with the new leaves, bracteoles thin, glabrous and distinctly veined inside. Pods are suborbicular to elliptic, up to 9 cm × 6 cm, 1–2-seeded.
Habitat: semi-deciduous forest in S Cameroon. Alt. up to c. 600 m (Breteler, 2010).
Threats: only two individuals known at localities c. 180 km apart, either of which could be vulnerable to stochastic events. Semi-deciduous forest is particularly rich in timber species and so under great logging pressure.
Management suggestions: rediscovery of this species in the wild is advised. Protection of surviving trees should be arranged with local leaders and area managers. Seed collection would allow propagation and reintroduction to the wild as well as cultivation in Botanic Gardens such as Limbe, as an extra safeguard.

Aphanocalyx Oliv. is a tropical African genus of 14 tree or shrub species. The leaflets of *Aphanocalyx* are unusual in that the distal (upper half) is missing so what would ordinarily be the mid-vein instead forms the distal margin of the leaflet. Four species occur in Cameroon of which one is threatened.

Aphanocalyx hedinii (A.Chev.) Wieringa

EN B1+2ab(iii)

Range: Cameroon endemic (SW: Mt Cameroon. C: Eséka. S: Kribi).
The last confirmed collection was in 1953, 10 km S of Eséka. Assessed here as Endangered. There are three locations (above: EOO <5000 km^2; AOO 12 km^2 with 4 km^2 cells) and threats as below. Wieringa (1999) notes that *A. hedinii* is restricted to three areas of Cameroon, one of which is between Kribi and Bipindi, but that despite searching the Kribi area extensively, no further individuals were located.

If the identity of several recent collections from other localities Ma'an, Ebienemeyong and Akom II, collected and cited by Tchouto (2004) but not seen, are confirmed then it will be necessary to re-evaluate the status of threat. *Aphanocalyx hedinii* has 35–57 pairs of narrow leaflets, not exceeding 4 mm wide. Inflorescence is a simple raceme (Wieringa, 1999).
Habitat: lowland humid forest, 0–200 m (Wieringa, 1999).
Threats: two of the three confirmed localities from which *A. hediniii* has been recorded are heavily logged and cultivated. Wieringa (1999) suggests the species may even be extinct.
Management suggestions: rediscovery of this species in the wild is advised. Protection of surviving trees should be arranged with local leaders and area managers. Seed collection would allow propagation and reintroduction to the wild as well as cultivation in Botanic Gardens such as Limbe, as an extra safeguard

Berlinia Sol. ex Hook.f. as recently monographed by Mackinder & Pennington (in press 2011). The genus now includes 21 tree species of lowland forest, all of which were assessed for conservation status during the course of the monographic study. Eleven species were evaluated as having a category of threat of which two occur in Cameroon. Typically *Berlinia* species have leaflets in two or more pairs and lack gland dots. Flowers are showy with a deep tubular hypanthium and narrowly oblong white sepals, which reflex after anthesis. The adaxial petals are large, white with green centres and the lateral petals are small to large and white. Fruits are large, woody and compressed with a broad upper suture. Eight species occur in Cameroon of which two are threatened.

Berlinia immaculata Mackinder & Wieringa
VU B2ab(iii) + D2
Range: Cameroon (S: Mt Elephant (1 loc.)) and Gabon (5 loc.).
Assessed as Vulnerable by Mackinder & Wieringa (2007). The assessment is maintained here. There are 6 locations (above; AOO 20 km^2 with 4 km^2 cells) and threats as below. The localities span a north-south range of 600 km (Mackinder & Wieringa, 2007). The flowers of *B. immaculata* are not known but the fruit have distinctive strongly ridged surfaces, a character which is unique within the genus.
Habitat: lowland humid forest; 50–150 m
Threats: Mt. Elephant is the only Cameroonian locality. It is unprotected and threatened by forest clearance.
Management suggestions: this species has been collected over a north-south range spanning 600 km, five in Gabon and the most northerly in Cameroon at Mt. Elephant. and some intermediate localities from which no gatherings have yet been made, are of suitable habitat. If a concerted effort was made at some of the intermediate localities of suitable habitat, it is quite probable that more individulas would be located, after which a reassessment of the conservation status would be needed. Ideally Mt. Elephant should be protected as a National Park since it harbours many narrowly endemic species.

Berlinia korupensis Mackinder & Burgt
CR D
Range: Cameron Endemic (SW: Korup NP (1 loc.)).
Assessed as Critically Endangered by Mackinder & van der Burgt (2009). The assessment is maintained here. The species is known from 17 individuals at a single locality in a protected area. Precise mapping by van der Burgt of 13 individuals in a study plot indicates that *B. korupensis* grows on well-drained sandy soils and avoids areas immediately adjacent to streams and gullies which become periodically inundated (Mackinder & van der Burgt, 2009). *Berlinia korupensis* is the only Cameroonian species in the genus to have flowers with one large broad petal and four others all narrower and about two-thirds as long as the largest.
Habitat: humid primary forest on well-drained sandy soil; c. 100 m alt.
Threats: only known from a single location albeit in a protected area. Stochastic events could wipe out the habitat known to be occupied by this geographically very restricted species
Management suggestions: van der Burgt recorded detailed data on the range, density, growth rate and habitat preference of this species. In total 17 individuals have been located, 13 inside a study plot and four nearby. The species is absent from a plot of the same forest type adjacent to the study plot and was also absent from another plot within Korup NP (Kenfack *et al.*, 2006). It is advisable that education on the species, especially images, be made available to the Korup project to promote its conservation locally and to tourists.

Brachystegia Benth is a genus of c. 26 African species of humid forest, seasonally dry forest, wooded grassland and bushland (Mackinder, 2005). Typically *Brachystegia* species have sessile leaflets in two or more pairs, petiolules not twisted and translucent gland dots lacking. Flowers have valvate sepals and petals are absent. Fruits are woody, compressed with winged sutures. Seven species occur in Cameroon of which one is threatened.

Brachystegia kennedyi Hoyle
EN B2ab(iii)
Range: Nigeria (4 loc.) and Cameroon (SW: Mamfe (1 loc.)).

Previously assesssed as Vulnerable (IUCN 1998) with three Cameroonian localities reported (Ejagham; Mamfe; Tabo–Agborkem) but reassessed here as Endangered. Only one Cameroonian locality, Mamfe is confirmed (*Letouzey* 13722). Therefore the species appear to have only five locations (AOO 20 km^2 with 4 km^2 cells). Nigerian localities are Amahor FR, Mamu FR, Sapoba FR and a collection made by Jones in the Cross River State, the most recent of these was made in 1965 (Sapoba FR). *Brachystegia kennedyi* has 6–7 pairs of leaflets, the terminal and basal pairs smaller than the others, lower surfaces hairy.
Habitat: lowland humid primary and secondary forest.
Threats: continued large-scale logging and clearance for agriculture has reduced forest coverage in unprotected areas where *B. kennedyi* may have occurred. Nigeria, in particular has seen massive losses of surviving forest in recent decades (www.mongabay.com)
Management suggestions: the species is a timber tree of local and cultural importance (IUCN 1998). Only five localities are confirmed and there have been no collections made of this species for 45 years. The largest and most stable population is thought to be in Cross River NP in Nigeria (IUCN 1998). Rediscovery of this species in the wild is critical so that protection an be arranged with local leaders and area managers, and so that seed can be collected for propagation and reintroduction in to the wild.

Cryptosepalum Benth. is a genus of c. 11 species, all African of which seven are restricted to the Guineo-Congolian region (Mackinder, 2005). *Cryptosepalum* species can have one to many leaflets and typically, petaloid bracteoles, a short hypanthium, sepals minute or absent and a single petal. Fruits are small, compressed and woody. Four species occur in Cameroon of which one is threatened.

Cryptosepalum ambamense Letouzey (syn: *Cryptosepalum elegans* Letouzey)
CR B1 + B2ab(iii)

Range: Cameroon endemic(S: vicinity of Ambam (1 loc.)).
Assessed here as Critically Endangered since there is one location (above; AOO 4 km^2 with 4 km^2 cells) and threats as below. *Cryptosepalum amabamense* is a small tree 10 m tall, known from a single collection, *Letouzey* 15281 (P) made in 1979 at Nsebito, 90 km WNW Ambam but not collected since. *Cryptosepalum ambanense* is unifoliolate, the midvein is not centrally placed but divides the leaflet blade into two parts of which one is approximately twice the width of the other.
Habitat: lowland semi-deciduous forest
Threats: forest clearance for timber extraction followed by agriculture

Management suggestions: rediscovery of this species in the wild is advised. Protection of surviving trees should be arranged with local leaders and area managers. Seed collection would allow propagation and reintroduction to the wild as well as cultivation in Botanic Gardens such as Limbe, as an extra safeguard.

Daniellia Benn., recently revised (Estrella *et al* 2010), the genus comprises ten species of medium to large trees from tropical and subtropical Africa. *Daniellia* is characterised by the combination of flowers with 4 imbricate sepals, 5 petals, 10 stamens (all free or 9 shortly connate) and fruits that are 'hemi-legumes', i.e. those where the single seed, attached by a long funicle is dispersed with one valve of the fruit. Five species occur in Cameroon of which one is threatened.

Daniellia oblonga Oliv.
CR D

Range: Cameroon (SW: S Bakundu FR; Mabeta camp, Likomba. S: Bella, nr Kribi. C: Yaoundé. Unlocated: Njoke (5 loc.)), Equatorial Guinea (Rio Muni (1 loc.)) and Gabon (1 loc.).
Assessed by Oldfield *et al.* (1998) as Vulnerable having been recorded from populations in Cameroon and Gabon that were under threat from logging (www.redlist.org). Reassessed here as Critically Endangered, *D. oblonga* is now known to occur at only eight localities with only one mature individual recorded from each locality (Estrella *et al.*), therefore only eight individuals are known globally.
Daniellia oblonga is closely related to *D. ogea* which also occurs in Cameroon, but differs in several floral characters which include having a glabrous ovary (not densely villous to pubescent as in D. *ogea*), glabrous sepals with a ciliate margin and a tuft of hairs at the top (not densely pubescent as in *D. ogea*).
Habitat: in forest at altitudes over 50 m (Estrella *et al.*).
Threats: substantial or complete loss of forest has occurred at three of the six Cameroonian locations; at Likomba the forest has been converted into plantations of oil-palm, rubber and banana; Yaoundé has been expanding due to urbanisation; S. Bakundu FR has seen conversion to agriculture.
Management suggestions: rediscovery of this species in the wild is advised. Protection of surviving trees should be arranged with local leaders and area managers. Seed collection would allow propagation and reintroduction to the wild as well as cultivation in Botanic Gardens such as Limbe, as an extra safeguard.

Englerodendron Harms. Until recently, this genus was regarded as monotypic, with one species in Tanzania.

Distinctive for the actinomorphic, not zygomorphic flowers, the inner stamens staminodial and shorter than the outer stamens. In re-evaluating *Anthonotha* for a revision of that genus, Breteler (2006) transferred two anomalous actinomorphic species of the second genus to the first, both of which occur in Cameroon and are assessed here as threatened. Additionally, van der Burgt (2007) described a fourth species from Korup NP, also threatened. Three species occur in Cameroon and are all threatened.

Englerodendron conchyliophorum (Pellegr.) Breteler (syn. *Isomacrolobium conchyliophorum* (Pellegr.) Aubrev. & Pellegr.)
EN B2ab(iii)

Range: Nigeria (Oban FR); Cameroon (SW: Mt Cameroon, Onge); Gabon (Doudou NP; Loango NP; Micouma-Lastoursville).
A tree with large (1–4.5 cm long) stipules enveloping the leaf axil, here assessed as Endangered, since only known from five locations (see above), A00 of 20 km² using 4 km² cells, and the threats indicated below.
Habitat: lowland evergreen forest; sea-level–100 m alt.
Threats: while two of the localities in Gabon are protected in National Parks, those in Cameroon and Nigeria are not, and are in localities threatened with clearance for plantations.
Management suggestions: more data is needed on population density, range and regeneration levels at one or several of the known localities. It is advisable to concentrate efforts on protection on trees in National Parks (Gabon).

Englerodendron gabunense (J.Léonard) Breteler (syn: *Anthonotha gabunensis* J.Léonard, *Leonardendron gabonense* (J.Léonard) Aubrév).
VU B2ab(iii)

Range: Cameroon (Littoral: Douala-Edéa at Lake Tissongo. S: Campo-Ma'an, Bibabimuoto [Bibamvoto]) and Gabon (NW Ndjolé; 30 km E Lastoursville; 8 km NE Bambidie; Makandé; Roungass; Koulamoutou; Ipassa-Makoukou).
A tree with stipules 1–6 mm long, not enveloping the leaf axil, here assessed as Vulnerable, since known from 10 specimens equating to seven locations (see above) resulting in an AOO of 28 km² using 4 km² cells and threats as below.
Habitat: lowland evergreen forest; sea-level to 500 m alt.
Threats: areas of Campo-Ma'an are under threat from plantation expansion and logging.
Management suggestions: surveys are needed to garner data on range, density and regeneration levels of this species at one or more of its localities so that plans can be made if needed to protect it. Apparently it does not occur in a protected area.

Englerodendron korupense Burgt
CR D

Range: Cameroon (SW: Korup NP).
This tree grows to 36 m high, differing from *E. gabunense* in its longer petioles (4–13 mm, not 2–4 mm), non-stipule-like, ovate petiolate lower leaflets (not stipule-like, cordate, sessile), apart from floral characters (van der Burgt *et al.*, 2007).
Sixteen trees of *E. korupense* are known, of which 12 are mature (found with pods), hence it has been assessed as CR D since less than 50 mature individuals are known (van der Burgt *et al.* 2007). This assessment is maintained here. The 16 known trees occur in five clusters separated from each other by 280–1470 m, with up to nine trees in one cluster. Monotypic stands of caesalp species are a common phenomenon.
Detailed density data is available since nine of the 16 known trees occur within two plots within which all tree species have been identified. These two plots cover 155.75 Ha including 3181 trees of 750 cm diam. at breast height, of which six were *E. korupense.* Trees in the 10–50 cm dbh class were recorded only in subplots, amounting to 14 Ha, in which just three *E. korupense* occurred amongst 5755 trees (van der Burgt *et al.*, 2007). Seedling regeneration was noted but not quantified.
Growth data recorded by van der Burgt (2007) gives diameter increments of up to 3.3 mm per year (range -0.1–3.3 mm) for each of 5 trees measured in 1991 and 2005 the smaller trees having barely changed at all in this time.
Habitat: lowland evergreen forest c 100 m alt.
Threats: stochastic events such as an intense storm or lightening produced fire following a dry period, could wipe out the habitat known to be occupied by this geographically very restricted species.
Management suggestions: excellent data on tree density, growth rates, flowering and fruiting seasons and presence of regeneration was recorded by van der Burgt (2007) as cited above. Therefore relatively little further data gathering is needed. Existing data suggest that management intervention is not needed beyond publicising the conservation importance of this species further, such as through a conservation poster.

Eurypetalum Harms, a genus of three tropical African tree species of lowland humid evergreen forest (Mackinder, 2005). *Eurypetalum* species have 1–2 pairs of leaflets with translucent gland; bracteoles minute, sepals four, unequal, overlappings, one large petal, other four minute, stamens 10, fused at base. Fruit are compressed and woody. Two species occur in Cameroon of which one is threatened.

Eurypetalum unijugum Harms
VU B2ab(iii)

Range: Cameroon endemic (SW: Baduma (1 coll.); Lake Ejagham (1 coll.); S Bakundu (3 coll.). C: Monatéle (1 coll). S: Campo (1 coll.), Bipindi (3 colls). (6 loc.)).
Assessed as Vulnerable by Cheek in Cheek *et al.* (2004: 163) and maintained here. There are 6 locations (above; AOO 20 km^2 with 4 km^2 cells) and threats as below. Although widespread in Cameroon this species is not known in Nigeria, Equatorial Guinea or Gabon. The range data above are taken from material at K and Flore du Cameroun. *Eurypetalum unijugum* is the only species in the genus with a single pair of leaflets.
Habitat: lowland evergreen forest; 0–400 m alt.
Threats: the rarity of this species makes it vulnerable to forest clearance. At S Bakundu yam farming has threatened the forest (Cable & Cheek (1998: xxxvi) while at Campo, extensive illegal commercial logging was observed in Feb. 1998 (Cheek *pers. obs.*). Threats at the other sites are unknown.
Management suggestions: both the Baduma and Lake Ejagham collections record the species as locally abundant and at the second site profuse regeneration was recorded (*Letouzey* 14612 & 13529). This may be a gregarious, grove-forming species, which reduces the likelihood of extinction at any one site due to loss of a single individual. Surveys are advised at all known sites to assess the state and size of subpopulations and to investigate how well these sites are protected. This information could form the basis of a species management plan

Gilbertiodendron J.Léonard is a genus of c. 26 tropical African tree species of lowland humid forest and seasonally dry forest (Mackinder, 2005). The species are generally poorly known and a taxonomic revision of the genus is much needeed and is currently underway (Estrella *in prep.*). The Cameroonian species are also the subject of a taxonomic study by van der Burgt so the assessments here must be considered very preliminary and will need to be updated following the findings of both studies. *Gilbertiodendron* species typically have several pairs of leaflets; stipules may have reniform bases and may or may not be persistent; bracteoles 2, valvate, one large petal, others much smaller, linear, stamens 3. Fruits are woody, compressed and with 1–4 lateral nerves present on the valves. Eleven species occur in Cameroon of which three are threatened.

Gilbertiodendron pachyanthum (Harms) J.Léonard
EN B2 ab(iii)

Range: Cameroon endemic (S: Ebianemeyong; Kom; Lolodorf; Massif des Mamelles (each 1 coll.); Bipindi (2 coll.). (5 loc.)).
Assessed as Vulnerable by Oldfield *et al.* (1998) and reassessed here as Endangered since there are five locations (above; AOO 20 km^2 with 4 km^2 cells) and threats as below. *Gilbertiodendron pachyanthum* has 2–3 pairs of large leaflets up to 30 cm long. Stipules are lanceolate and without reniform bases; flowers white. Species is poorly known and not known in fruit.
Habitat: lowland forest.
Threats: forest clearance for agriculture; open-cast mining iron ore mining (Massif des Mammelles).
Management suggestions: Rediscovery of this species in the wild is advised. Protection of surviving trees should be arranged with local leaders and area managers. Seed collection would allow propagation and reintroduction to the wild as well as cultivation in Botanic Gardens such as Limbe, as an extra safeguard.

Gilbertiodendron quadrifolium (Harms) J.Léonard
CR B2ab(iii)

Range: Cameroon endemic (C: Yaoundé).
Known only from the type collection *Mildbraed* 7774, made in 1914, 35 km SSW Yaoundé (région de Mbalmayo). Assessed here as Critically Endangered since there is only one location(above; AOO 4 km^2 with 4 km^2 cells) and threats as below. The K duplicate bears an annotation made by Letouzey in 1952. He comments that *G. quadrifolium* is provisionally considered distinct from *G. dewevrei* (De Wild.) Léonard because the stipules are smaller (1.5 cm long, not 2–8), the large (adaxial) petal is a different colour (white to yellow, not red) and the secondary nerves of the leaflets are less numerous (4–13, not 14–22). Species not known in fruit. The taxon was not recorded in Flore du Cameroun (Aubréville 1970).
Habitat: lowland forest
Threats: forest at the single locality from which *G. quadrifolium* has been recorded has long been cleared and the species may be extinct.
Management suggestions: rediscovery of this species in the wild is advised. Protection of surviving trees should be arranged with local leaders and area managers. Seed collection would allow propagation and reintroduction to the wild as well as cultivation in Botanic Gardens such as Limbe, as an extra safeguard.

Gilbertiodendron zenkeri (Harms) J.Léonard
EN B2ab(iii)

Range: Cameroon endemic (S: Bipindi (5 colls); Eséka (1 coll.)).
The collection at Eséka was made by Leeuwenberg in 1965, those at Bipindi were made by Zenker between 1896–1913. Here assesssed as Endangered since only two locations are known (AOO 8 km^2 with 4 km^2 cells) and because threats are as given below. *Gilbertiodendron zenkeri* can be distinguished by the combination of 5–6 pairs of large leaflets, leaf rachis 20–30 cm long; stipules conspicuous,

persistant, 3–4 cm long and with reniform bases. The large (adaxial) petal is a white and fruit are unknown.
Habitat: lowland forest.
Threats: localities from which *G. zenkerii* has been recorded are heavily logged and cultivated.
Management suggestions: rediscovery of this species in the wild is advised. Protection of surviving trees should be arranged with local leaders and area managers. Seed collection would allow propagation and reintroduction to the wild as well as cultivation in Botanic Gardens such as Limbe, as an extra safeguard.

Gossweilerodendron Harms is a genus of two tropical African tree species, both of lowland evergreen humid forest (Mackinder, 2005). *Gossweilerodendron* species have alternate leaflets with translucent gland dots and twisted petiolues. Bracteoles are very small, sepals 4, overlapping, petals absent. Fruit is winged, the single seed is apical. Two species occur in Cameroon, one of which is threatened.

Gossweilerodendron joveri Normand ex Aubrév. (syn: *Prioria joveri* (Normand ex Aubrév.) Breteler)
VU B2ab(iii)
Range: Cameroon (S: W Ngoulemakong. SW: N Kumba, C: Yaoundé-Edéa, Yaoundé-Mbalmayo (each 1 coll.) Unknown locations: Nolbewoa; Kong (each 1 coll.)), Angola (Rio Muni: Sendye, Okuamkos (each 1 coll.)) and Gabon (Goualé (1 coll.); Oveng (3 coll.).
The range data above are taken from the revision of *Prioria* by Breteler (1999) where the species is treated as *Prioria joveri* (Normand ex Aubrév.) Breteler, but the occurrence of *Prioria* in Africa is not widely accepted. Here assessed as Vulnerable since eight locations are known (above; AOO 32 km^2 with 4 km^2 cells) and threats as below. Differs from the more common *G. balsamiferum* (Verm.) Harms in having 4–5 alternate leaflets (not 6–10), numerous secondary nerves visible only on the lower surface, (not fewer secondary nerves forming a fine reticulate network with tertiary nerves, nervation visible on both surfaces).
Habitat: lowland evergreen forest; 0–500m alt.
Threats: forest clearance due to agriculture and logging.
Management suggestions: a survey to rediscover and establish subpopulation sizes and demography would inform a management plan for the species and act as a baseline for future monitoring.

Hymenostegia J.Léonard. As currently delimited, a genus of 15 tropical African tree species although about half the species are misplaced in the genus (Mackinder *et al.*, 2010). Species included in the genus have leaflets in pairs, lack translucent gland dots, petiolules not twisted. Flowers have paired persistant petaloid bracteoles, petals (1–)2–3(–5), stamens 8–10 (-numerous). Fruits are compressed, woody and dehiscent and often single seeded.

Hymenostegia bakeriana Hutch. & Dalziel
VU D2
Range: Nigeria and Cameroon (SW: Korup NP (numerous coll.)).
Assessed by Oldfield *et al* as Vulnerable (1998) and maintained here here with the addition of category D2 to denote the few localities. Known from Oban, Nigeria the type locality and 4 localities in Cameroon all of which are in Korup NP. All localities are from still forested areas within protected areas although the four Cameroon localities are all close together and could be considered a single population. *H. bakeriana* has 8–10 pairs of glabrous similar-sized leaflets, c. 4–8 cm long, leaf rachis is terete (not flattened or channelled), bracteoles are narrow, about 1 cm long, petals yellowish, stamens reported to be numerous 16–26 (FWTA).
Habitat: lowland humid forest 50–400 m.
Threats: any destructive events such as storms, lightening or disease could destroy either the Korup NP population or the population in Oban reducing this already restricted species to a single locality.
Management suggestions: a survey to rediscover and establish subpopulation sizes and demography would inform a management plan for the species and act as a baseline for future monitoring.

Hymenostegia brachyura (Harms) J.Léonard
VU B2ab(ii,iii)
Range: Cameroon endemic ((8 colls, 7 loc.): S: Bidou I; Bipindi; Ébolowa (1 coll. each). Littoral: Ebo, Dicam, Ndokbaembi, (each 1 coll.). C: Makak (2 colls)).
All collections made between 1912 and 2007. Assessed here as Vulnerable since ther are seven locations (above; AOO 28 km^2 with 4 km^2 cells) and threats as below. None of the known localities fall within protected areas and several have already been logged. Hymenostegia brachyura has 2–4 pairs of leaflets, the upper pair(s) larger than the lower pair(s), lower surface pubescent, petal colour unknown, stamens 10.
Habitat: rocky sites in evergreen forest.
Threats: forest clearance for agriculture.
Management suggestions: only a single collection is recorded from all but one of the known localities. Further investigation of the localities where forest remains is needed to see if more individuals exist. If the rather specialised habitat exists at other localities within the extent of occurrence, then those localities should also be searched. Conservation efforts are perhaps best focussed on Ebo since this is a proposed National Park.

Julbernardia Pellegr. A tropical African tree genus of c. 11 species of humid or seasonally dry forest, woodland and thicket (Mackinder, 2005). *Julbernardia* species have leaflets in 1–8 pairs, some species have translucent gland dots, petiolules not twisted. Sepals 5, overlapping, one well-developed petal, other four minute, stamens 10. Fruit compressed, woody, lateral nerves not present on valves, upper and/or lower suture broadened. Three species occur in Cameroon of which one is threatened.

Julbernardia letouzeyi Villiers
CR B1+2ab(iii)

Range: Cameroon endemic (S: 18–23 km NW of Bipindi (8 coll., 1 loc.)).
Assessed here as Critically Endangered. Since there is one location (EOO and AOO 4 km^2 with 4 km^2 cells) and threats as below. *Julbernardia letouzeyi* is known from 8 collections, made by Letouzey (2 coll.) and Villiers (6 coll.) in an area of forest 18–23 km NW of Bipindi much of which has already been cleared so the species is possibly extinct. The species is unique in the genus in having only a single pair of leaflets.
Habitat: forest on hillside on sandy soil.
Threats: known from a single location, which if lost due to habitat destruction or stochastic events could lead to the extinction of this rare species of restricted distribution if it is not already lost. Logging followed by agriculture is a threat.
Management suggestions: a thorough search at and near the locality to establish the exact number of individuals is needed to discover if this species is extant before a plan can be devised. Protection of surviving trees should be arranged with local leaders and area managers. Seed collection would allow propagation and reintroduction to the wild as well as cultivation in Botanic Gardens such as Limbe, as an extra safeguard.

Leonardoxa Aubrév. The genus comprises a single species, *L. africana* (Baill.) Aubrév. within which four subspecies are recognised: subsp. *africana*; subsp. *gracilicaulis* McKey; subsp. *letouzeyi* McKey; and subsp. *rumpiensis* McKey (McKey, 2000). All four subspecies occur in Cameroon, two are endemic, subsp. *letouzeyi* McKey extends into SE Nigeria and subsp. *gracilicaulis* is also recorded from Equatorial Guinea and Gabon. Two subspecies are threatened.

Leonardoxa africana Aubrév. subsp. ***letouzeyi*** McKey
VU B1+ B2ab(iii)

Range: SE Nigeria (Oban, Cal[illegible] coll.)) and Cameroon (SW: Baro; Bombe; Bechati; Mamfe; Takamanda FR (each 1 coll.); Korup NP (3 coll.)).
Here assessed as Vulnerable since there are eight locations (above; AOO 32 km^2 with 4 km^2 cells) and threats are as below.
McKey (2000) distinguishes *Leonardoxa africana* subsp. *letouzeyi* using the following unique combination of characters; young twigs with swollen internodes (at least on plants >1.5 m tall), leaves often 4(3–5)-jugate; nectaries often absent at base.
Habitat: lowland rain forest up to 300 m.
Threats: at Bechati: clearance of forest for agriculture is occurring.
Management suggestions: this taxon is probably secure at Korup NP which seems a logical focus for conservation efforts. Basic populational data (density, range at Korup, regeneration rates) to inform future monitoring and to gauge the need for management intervention. Future surveys may discover further locations for this tree which may then need downrating to NT.

Leonardoxa africana Aubrév. subsp. ***rumpiensis*** McKey
EN B2ab(iii)

Range: Cameroon (SW: Rumpi Hills (6 coll., 5 loc.)).
This tree is here assesed as Endangered since there are 5 locations (above; AOO 20 km^2 with 4 km^2 cells) and threats as below.
McKey (2000) `distinguishes *Leonardoxa africana* subsp. *rumpiensis* using the following unique combination of characters; young twigs with swollen internodes (at least on plants >1.5 m tall), leaves usually 2(2–3)-jugate; nectaries at base of each proximal leaflet present on almost all plants, usually 2 in number.
Habitat: submontane rain forest in the Rumpi Hills (above 800 m).
Threats: slash and burn agriculture; logging.
Management suggestions: the creation of a Rumpi Hills NP would help secure protection of this and numerous other threatened species. A survey to recollect basic population data would form a baseline for future monitoring operations.

Loesenera Harms. A tropical African tree genus of four species of lowland forest (Mackinder, 2005). *Loesenera* species have leaflets in 1–5 pairs. One species is know from Cameroon and is threatened.

***Loesenera talbotii* Bakerf.**
VU A2c

Range: SE Nigeria: Oban and Calabar-Mamfe (1 coll. each); SW Cameroon: Yingui-Yabassi, WSW Mamfe (1 coll. each), Mt Kupe-Bakossi (numerous sites and colls). Assessed as Vulnerable by Cheek (2004) and maintained here. This gregarious tree is only known from four sites outside Kupe-

Bakossi, within which it is rather common across its altitudinal range. If this taxon were to be treated under Criterion B, interpretation of what defines a location is critical to its status as VU. It could be argued that the Kupe-Bakossi area houses as many as 10 sites or as few as five. This taxon is considered Vulnerable under Criterion A given at least an estimated 30% reduction in population due to loss of habitat, particularly in the lower part of its altitudinal range, in the next 100 years. *Loesenera talbotii* has 3–4 pairs of leaflets.
Habitat: lowland and submontane evergreen forest; 180–1000 m alt.
Threats: forest clearance for agriculture and wood, e.g. at Kupe village.
Management suggestions: research into the demography of this gregarious tree would inform development of a management plan. Since *L. talbotii* is so frequent in Kupe-Bakossi, it is advisable that conservation efforts are focussed here. This tree could form the basis of a public education campaign due to its conspicuousness in Bakossi.

Microberlinia A.Chev. A genus restricted to Cameroon and Gabon comprising two species of tree of lowland forest which grow gregariously (Mackinder, 2005). *Microberlinia* species have leaflets in pairs. Sepals 4, petals 5, adaxial petal much longer than others, or petals of similar length. Fruit woody, compressed, single lateral nerve on valve, upper suture broadened. Both species occur in Cameroon and are both threatened.

Microberlinia bisulcata A.Chev.
VU A4c
Range: Nigeria (1 loc) and Cameroon (SW: Korup NP; S Bakundu (both numerous coll.); Mt Cameroon at Mokoko; 20 km W Nguti; Akwaya-Mamfe; Banyang Mbo. Littoral: Mangombé; Yabassi (10 loc.)).
Assessed as Critically Endangered (CR A1c+2c) by Cheek & Cable in 2000 (www.redlist.org) and maintained as such in Cheek *et al.* (2004: 163). Reassessed here as Vulnerable. It is here estimated that more than 30% of the population reduction to habitat loss will have occurred over a 100 year span. Although records exist from 11 localities, only the Korup NP population is in a protected area and other localities have been subject to large scale reduction of habitat through clearance of forest for agriculture. The leaf rachis of *M. bisulcata* is slightly winged distinguishing it from the unwinged rachis of *M. brazzavillensis.*
Habitat: lowland humid forest, usually on sandy soils in flat areas.
Threats: decline of habitat following clearance of forest for agriculture.
Management suggestions: conservation efforts should be focused on the Korup NP since the subpopulation there is well monitored.

Plagiosiphon Harms. A tropical African genus of five species of lowland forest of the Guineo-Congolian region. The genus is poorly known and Cameroon is the centre of diversity of the genus. Species sometimes grow gregariously. *Plagiosiphon* species have leaflets in 1–12 pairs, the lower leaflet surface is somewhat bullate. All five species occur in Cameroon and one is rare and threatened, possibly extinct.

Plagiosiphon discifer Harms
Range: Cameroon endemic (S: Lolodorf (1 loc.)).
Assessed here as Critically Endangered since there is one locations (above; AOO 4 km^2 with 4 km^2 cells) and threats as below. This species is known only from the type collection made by Staudt in Lolodorf in 1896 and not collected since. *Plagiosiphon discifer* is the only species in the genus to have 1–2 pairs of leaflets; other species have many more pairs.
Habitat: not recorded
Threats: clearance of forest in the area for plantations; slash and burn agriculture.
Management suggestions: an intensive search of Lolodorf and surrounding areas is needed to discover if the species is still extant there. Rediscovery of this species in the wild is advised. Protection of surviving trees should be arranged with local leaders and area managers. Seed collection would allow propagation and reintroduction to the wild as well as cultivation in Botanic Gardens such as Limbe, as an extra safeguard.

Talbotiella Baker.f. In a recent revision, the genus has been expanded from three to eight species (Mackinder *et al.* 2010). Four of the additions are new species discovered in Cameroon, while one is transferred from *Hymenostegia*. Cameroon is the centre of diversity for *Talbotiella*, with six of the eight species, all of which are threatened, including five which are endemic. All of the species in Cameroon are trees of evergreen forest. They differ from *Hymenostegia* in the absence of petals, a hairy ovary and fully tectate pollen with narrow appertures (Mackinder *et al.* 2010).

Talbotiella batesii Baker.f.
CR D
Range: Cameroon (S: Bitye; 33 km S of Djoum), Gabon (Minkébé NP; 25 km NW Oveng; Ivindo NP; Mt Sassamongo) and Congo (Brazzaville) (Mt Naemba).
Assessed as Critically Endangered (CR D) by Mackinder *et al.* (2010) and maintained. The assessment is based on the

assumption that only a single tree was seen at each of the known collection localities.

First collected at Bitye, the type locality, in 1929 by the American missionary, Bates, *T. batesii* was refound by Letouzey in 1966 S of Djoum. Between 1987–2002, five more collections were made in Gabon and Congo (Brazzaville). *Talbotiella batesii* is distinguished by the combination of leaves with 9–13 pairs of narrow leaflets, 1–3.5 mm wide and up to 10 × longer than wide (Mackinder *et al.* 2010).

Habitat: rocky areas and river banks; 500–900 m alt.

Threats: logging is a concern in the vicinity of both the Cameroonian sites. However, two of the Gabon localities occur in National Parks where they are protected.

Management suggestions: both Cameroon localities should be revisited to see if the species can still be found there. The newly discovered Gabonese sites should also be revisited. When refound the area should be surveyed to assess numbers of individuals, regeneration levels so as to develop a management plan for the species.

Talbotiella bakossiensis Cheek

CR D

Range: Cameroon endemic (SW: Rumpi Hills; Bakossi Mts).

This submontane forest tree is known from only six collections at three broad localities: Dikome Balue, Nyale and Ngomboku. It was assessed as CR D by Cheek (in Mackinder, B.A., Weiringa, J.J. & van der Burgt, X.M. (2010 (2010 publ. Jan. 2011). A revision of the genus *Talbotiella* Baker.f. (Caesalpinioideae: Leguminosae). Kew Bull. 65(3): 404) on the basis that less than 50 mature individuals are known. That assessment is maintained here. *Talbotiella bakossiensis* is distinguished by adult leaves that are 7.5–13 cm long, with 21–26 pairs of leaflets (Mackinder *et al.* 2010).

Habitat: submontane forest; 850–1800 m alt.

Threats: the localities in Bakossi are outside the Bakossi NP and other protected areas. Currently the Rumpi Hills do not have National Park status. The Ngomboku locality faces agricultural encroachment from the E. No threats are known at other sites.

Management suggestions: ideally part of the population should be included in a National Park or protected by other means. Public education of the rarity of this tree is advisable. Future surveys in Bakossi and Rumpi Hills would give more information on regeneration levels and population density enabling the possibility of better protecting this species. It is likely that more than 50 individuals would so be found, necessitating a downgrading of the threat level to EN under criteria B.

Talbotiella breteleri (Aubrév.) Mackinder & Wieringa

CR D

Range: endemic to Cameroon (C: Yaoundé (Mt Fébé and N of N'Kkolbison).

This tree is unique in the genus in having only two pairs of leaflets per leaf. First collected at Mt Fébé, the type locality, on 21 Dec. 1961 by Frans Breteler, its forest habitat has since been cleared and replaced by an upmarket hotel of the same name (pers.obs.). The second and only other known locality, also discovered by Breteler, several years later, 22 Dec. 1964, at 7 km W of Yaoundé, was revisited in March 2007 by Wieringa who refound the species, possibly the same individual (albeit in very degraded habitat). In transferring this species from *Hymenostegia* , Mackinder *et al.* (2010) assessed *T. breteleri* as Critically Endangered (CR D) since less than 50 mature trees are known. This assessment is maintained here.

Habitat: forest on inselbergs; 900–950 m alt.

Threats: forest clearance of inselbergs for urban development (see above), agriculture and rock extraction at second locality.

Management suggestions: since the species appears to be extinct at its type locality at Mt Fébé, it is imperative that the site N of N'Kolbison is formally protected. Further surveys around Yaoundé are recommended to hopefully discover new sites for *T. breteleri*. Studies of natural regeneration and urgent seed banking are recommended for this species.

Talbotiella ebo Mackinder & Wieringa

CR D

Range: Cameroon (Littoral: Ebo; 8 km E of Yingui; Loum FR).

Only three sites are known for this tree, which differs from *T. batesii* in having larger leaves and leaflets, and in having a shorter and sparser indumentum.

Assessed as CR D by Mackinder *et al.* (2010) since less than 50 mature individuals are known. This assessment is maintained here.

Habitat: evergreen forest on granitic outcrops; c 900 m alt.

Threats: logging concessions have been granted in the area, and the status of the proposed Ebo NP is unresolved.

Management suggestions: since recent accurate locality data is available for *T. ebo* inside the proposed Ebo NP, this is the obvious starting point for protecting the species. Some evidence was found there for regeneration of the species (Mackinder *et al.* 2010). Further survey work there and the third locality in Loum FR may yet yield additional individuals of the species. Seed-banking should be investigated.

Talbotiella korupensis Mackinder & Wieringa

EN B2 ab(iii)+D

Range: Cameroon, endemic (SW: S Korup NP; Besingi).

The assessment of EN D by Mackinder *et al.* (2010) implies that between 51 and 250 mature individuals are known, and is maintained here. In addition, since only two locations are known (above; AOO 8 km^2 with 4 km^2 cells) and threats as below, it is also EN under Criterion B.

Talbotiella korupensis has leaves 10.5–19.5 cm long, 16–22 pairs of leaflets, each 22–35 × 5–9 mm long. It can be confused with *T. velutina* which grows at the same localities. Differences are given under the last species. A great deal of data on this species is available due to the work of van der Burgt on the 'P transect' plot in S Korup NP as cited in Mackinder *et al.* 2010. In a survey of 56 randomly located sub-plots of 0.25 ha each, in which 5755 trees of 10–50 cm dbh were registered, 44 were of *T. korupensis.* However, it is gregarious. In one sub-plot 9 individuals occurred among a total of 127 trees over 10 cm dbh. The species is reported to be abundant in many parts of S Korup forest but completely absent in other parts that otherwise have a similar composition. Average annual increment per year of trunk girth is reported as 0.9 mm per year, based on 30 trees measured at intervals of 16–17 years at 1.3 m above ground level (Mackinder *et al.*, 2010).

Habitat: lowland evergreen forest on or near inundated soils, 50–100 m alt.

Threats: while secure inside the Korup NP, those trees outside, e.g. near Besingi village, are under threat from slash and burn agriculture and small-scale logging.

Management suggestions: thanks to the work of van der Burgt, a great deal of data on the range, density, growth rate and habitat preference of this species is known. However, information on regeneration levels would also be useful in assessing needs for intervention of this species. It is advisable that education on the species, especially images, be made available to the Korup project to promote its conservation locally and to tourists.

Talbotiella velutina Burgt & Wieringa

CR D

Range: Cameroon endemic (SW: S Korup NP; Besingi).

The assessment of CR D for this species, based on only 39 mainly mature individuals, being known (Mackinder *et al.*, 2010) is maintained here. This species is reported as growing with *T. korupensis* and the two are reported as potentially difficult to distinguish unless in flower. However, *T. velutina* is a canopy tree 20–35 m high, of well-drained soils, trunk to 76 cm diam., fluted or lobed from the ground to the crown, without stem shooots. In contrast *T. korupensis* trees are understorey, 5–20 m high, of inundated soils, trunk 10–35 cm diam., often leaning, cylindrical, often with stem shoots (Mackinder *et al.*).

Habitat: lowland evergreen forest, well-drained sandy soils, 50–100 m alt.

Threats: while secure inside the Korup NP, those trees outside, e.g. near Besingi village, are under threat from slash and burn agriculture and small-scale logging.

Management suggestions: thanks to the work of van der Burgt, a great deal of data on the range, density, growth rate and habitat preference of this species is known. However, information on regeneration levels would also be useful in assessing needs for intervention of this species. It is advisable that education on the species, especially images, be made available to the Korup project to promote its conservation locally and to tourists.

Tetraberlinia (Harms) Hauman. A genus of 7 tree species restricted to the Guineo-Congolian region (Wieringa, 1999). Leaflets opposite, in 1–15 pairs (in Cameroon). Petals are partly or completely yellow. Pods compressed, 1 or 2 lateral nerves on the valves. Three species occur in Cameroon of which one is threatened.

Tetraberlinia korupensis Wieringa

EN B2 ab(iii,iv)

Range: Cameroon endemic (SW: Korup NP (10 coll.); vicinity of Mundemba (3 coll.) (Wieringa, 1999) (14 coll., 2 loc.)).

Here assessed as Endangered since there are two locations (above; AOO 8 km^2 with 4 km^2 cells) and threats as below. There is a report of a seedling (MO data, specimen not seen) collected below a mature tree of this species at a third locality (Manja) but the habitat at the putative third locality is of secondary heavily degraded forest unlike the habitat of the other two localities and so the Manja locality is not included in the assessment here.

Tetraberlinia korupensis is a large forest tree, up to 55 m tall and typically has leaves with 10–15 pairs of leaflets, largest leaflet midway along leaf, c. 2–4.5 × 0.5–1 cm. Pods up to 1 cm long, usually 2 lateral nerves on the valves present, 2 seeded. *Tetraberlinia korupensis* has a restricted range but within it is known to grow gregariously and regenerate abundantly (Wieringa, 1999).

Habitat: primary and secondary lowland humid rainforest on sandy soils, 50–170 m.

Threats: while secure inside the Korup NP, those trees outside, e.g. vicinity of Mundemba are under threat from slash and burn agriculture and small-scale logging.

Management suggestions: the majority of the collections are from a population in Korup NP which is thriving. It would be useful to publicise the conservation importance of this species in that area. However, the population near Mundemba, outside Korup NP is not protected. A concerted effort to search at Manja to confirm or refute the existence of this species there is advised.

REFERENCES

Aubréville, A. (1970). Légumineuses - Caesalpinioidées. In: A. Aubréville (ed.) *Flore du Cameroun*. pp. 242–254. Paris: Musée Naturale d 'Histoire.

Breteler, F. J. & N.S. Miyono Nguema. (2008). Revision of the African species of *Crudia* (Leguminosae, Caesalpinioideae). *Systematics and Geography of Plants* 78(1): 81–110.

Breteler, F. J. (1999). A revision of *Prioria*, including *Gossweilerodendron*, *Kingiodendron*, *Oxystigma*, and *Pterygopodium* (Leguminosae - Caesalpinioideae - Detarieae) with emphasis on Africa. *Wageningen Agricultural University Papers* 99(3): 1–61.

Breteler, F. J. (2006). Novitates Gabonenses 56. Two *Anthonotha* species from Gabon transferred to *Englerodendron* (Fabaceae, Caesalpinioideae). *Adansonia* series 3, 28(1): 105–111.

Breteler, F. J. (2008). *Anthonotha* and *Isomacrolobium* (Leguminosae, Caesalpiniodeae): two distinct genera. *Systematics and Geography of Plants* 78(2): 137–144.

Breteler, F. J. (2010). Revision of the African genus *Anthonotha* (Leguminosae, Caesalpinioideae). *Plant Ecology and Evolution* 143(1): 70–99.

Cable, S. & M. Cheek. (eds.) (1998). *The Plants of Mount Cameroon, A Conservation Checklist.* Royal Botanic Gardens, Kew.

Cheek, M., B.J. Pollard, I. Darbyshire, J-M. Onana & C. Wild. (eds.) (2004). *The Plants of Kupe, Mwanenguba and the Bakossi Mountains, Cameroon: A Conservation Checklist.* Royal Botanic Gardens, Kew.

Estrella, M. de la, C. Acedo, B. Mackinder & M. Velayos (2010). Taxonomic Revision of *Daniellia* (Leguminosae: Caesalpinioideae). *Systematic Botany* 35(2): 296–324.

Kenfack, D., D.W. Thomas, G.B. Chuyong & R. Condit. (2006). Rarity and abundance in a diverse African forest. *Biodiversity and Conservation* 15(13): 4097–4117.

Lewis, G. P. (2005). Cassieae. In: G. P. Lewis, B.D. Schrire, B.A. Mackinder & J. M. Lock (eds.) *Legumes of the World.* Royal Botanic Gardens, Kew.

Lock, J.M. (1989). *Legumes of Africa: a check-list.* Royal Botanic Gardens, Kew.

Mackinder, B.A. Leguminosae. (2004). In: Cheek, M., Pollard, B.J., Darbyshire, I., Onana, J-M. and C. Wild. (eds.). The plants of Kupe, Mwanenguba and the Bakossi Mountains, Cameroon: a conservation checklist: 311–325. Royal Botanic Gardens, Kew.

Mackinder, B. A. (2005). Detarieae. In: G.P. Lewis, B.D. Schrire, B.A. Mackinder & J. M. Lock (eds.) *Legumes of the World.* Royal Botanic Gardens, Kew.

Mackinder, B. A. & X. M. van der Burgt, X. M. (2009). *Berlinia korupensis* (Leguminosae – Caesalpinioideae), a new tree species from Cameroon. *Kew Bull.* 64(1): 129–134.

Mackinder, B. A. & J. J. Wieringa. (2007). Novitates Gabonenses 58: Two New Species of *Berlinia* (Leguminosae – Caesalpinioideae: Detarieae). *Kew Bull.* 62: 159–164.

Mackinder, B. A., J. J. Wieringa, I. Lunenburg & H. Banks. (2010). Clarifying the generic limits of *Talbotiella* and *Hymenostegia* (Detarieae: Caesalpinioideae: Leguminosae). In: S. A. Ghazanfar, P. P. Lowry & B. Sonké (eds.). *XVIIIth AETFAT Congress* : 43–56. Yaoundé, Cameroon: Royal Botanic Gardens, Kew.

Mackinder, B., J. J. Wieringa & X. M. van der Burgt. (2010). A revision of the genus *Talbotiella* (Caesalpinioideae: Leguminosae). *Kew Bull.* 65: 401–420.

Mackinder, B. & R. T. Pennington. (*in press*, 2011). A monograph of *Berlinia. Systematic Botany Monographs* 91.

McKey, D. *Leonardoxa africana* (Leguminosae: Caesalpinioideae): a complex of mostly allopatric subspecies (2001). *Adansonia*, sér. 3, 22(1): 71–109.

Oldfield, S., C. Lusty & A. MacKinven. (1998). *The World List of Threatened Trees*. Cambridge: World Conservation Press.

Sosef, M. S. M., J. J. Wieringa, C. C. H. Jongkind, G. Achoundong, Y. Azizet Issembe, D. Bedigian, R. G. van den Berg, F. J. Breteler, M. Cheek, J. Degreef, R. Faden, P. Goldblatt, L. J. G. van der Maesen, L. Ngok Banak, R. Niangadouma, T. Nzabi, B. Nziengui, Z. S. Rogers, T. Stuart, J. L. C. H. van Valkenburg, G. Walters & J. J. F. E. de Wilde (2006). Check-list des plantes vasculaires du Gabon. Scripta Botanica 35. 438 pp.

van der Burgt, X. M., M. B. Eyakwe & D. M. Newbery. (2007). *Englerodendron korupense* (Leguminosae – Caesalpinioideae), a new tree species from Korup National Park, Cameroon. *Adansonia* 29(1): 59–65.

Wieringa, J. J. (1999). *Monopetalanthus* exit. A systematic study of *Aphanocalyx*, *Bikinia*, *Icuria*, *Michelsonia* and *Tetraberlinia*. *Wageningen Agricultural University Papers* 99(4): 1–320.

LEGUMINOSAE SUBFAMILY MIMOSOIDEAE

assessed by Martin Cheek

Newtonia duncanthomasii Mackinder & Cheek (Kew Bull. 58: 447 (2003) is restricted to the Bakossi Mts, Rumpi Hills, Mt Juahan of Korup but despite its low AOO, EOO and number of locations, faces no known threats, so was assessed as NT in the protologue. Since no changes have since occurred, this assessment is maintained here.

Albizia eriorhachis Harms (syn: *Cathormium eriorhachis* (Harms) Dandy)

EN B2ab(iii)

Range: Cameroon (Adamawa: Tibati; Carnot near Meïganga), CAR (Golo to Mansala) and Sudan (Equatoria: Raga to Said Bundas)

Slender tree c. 9 m, crown spreading, drooping; young branchlets densely short-rusty tomentose; leaf-rhachises sparsely tomentose, 5–8 cm; pinnae 8–12 pairs, leaflets 15–35 pairs, lanceolate or oblong, to 1 cm long; flowers greenish on solitary or clustered brown-tomentose peduncles to 2.5 cm long. Pod 7.5–12.5 cm, thick, crimson-brown, rounded at both ends, transversely impressed (Andrews (1952), The Plants of A-E Sudan 2: 155).

This elusive species is accepted by Lock (1989 op. cit.). Published by Harms (Bot. Jahrb. Syst. 53: 456 (1915)) on the basis of the two records from Adamawa in Cameroon made in 1909 and 1914. It has not been seen in Cameroon since. This tree has not been recorded globally since before 1950, we believe.

Here *Albizia eriorhachis* is assessed as Endangered since only four locations are known (AOO 16 km² with 4 km² cells, and threats as below). EOO is calculated as 165139 km².

Habitat: subsahelian Guinean woodland – wooded grassland (deduced).

Threats: lopping of leafy branches for fodder in the dry season; wood for firewood and construction.

Management suggestions: this species should be searched for in the Adamawa Highlands, initially near Tibati and Meïganga. Care should be taken to distinguish it from other species of *Albizia* (see description above). The herbarium at Garoua should be consulted – it may be that there are additional, more recent site records held there that have been overlooked. Similarly, the Paris herbarium should also be consulted for specimens. If found, any new records should help planning of survey work. Should the tree be found, the usual populational data should be gathered to inform future monitoring and to gauge the need for management intervention. Local communities should be informed of the importance of protecting this tree. This species is a prime candidate for seed-banking if it can be refound.

Calpocalyx was revised by Villiers in Bull. Mus. Nat. Hist. Nat., B Adansonia sér. 4, 6 (3): 299 (1985). However this volume was mislaid during the course of compilation of this book, so we were not able to access it and instead we had to depend on other, incomplete, sources. Access to this volume was obtained again, but too late to map the several additional species that were then discovered to probably merit threatened status: *Calpocalyx brevifolius, C. cauliflorus, C. heitzii, C. klainei* and *C. ngouniensis*. For the moment they can be ascribed as NT. This is a genus of trees with bipinnate leaves distinct in having just one pair of pinnae, but each with numerous leaflets. Ants are often associated with these trees.

Calpocalyx atlanticus Villiers

EN B2ab(iii)

Range: Cameroon endemic (SW: 13 km SSW Kumba at Kédongue. S: Nyong).

Here assessed as Endangered since only two locations are known (AOO 8 km² with 4 km² cells) and threats as below.

Tree 7–10 m, bark greenish black; leafy branches shortly pubescent; leaflets 6–8 pairs, each elliptic-oblanceolate, to 23 × 9.5 cm, acuminate-mucronate, petiolule pubescent, 3–7 mm; rachis 31.5–40 cm, glands on upper surface; petiole 5.5–10 cm, apex with elliptic gland; panicle terminal, 60 cm; flowers unknown; pod dark green, curved, dehiscent, 20 × 4.5–5.5 cm, puberulent. (Villiers 1985 op.cit. – see note above).

Habitat: lowland evergreen forest (deduced).

Threats: the site near Kumba corresponds with the S Bakundu FR which has been afflicted by Nigerian yam farmers and informal logging. These threats are ongoing. Threats at the Nyong site are unknown.

Management suggestions: rediscovery of this species should be attempted. If refound, the usual basic populational data should be obtained to assist future monitoring and gauge needs for intervention. Local communities should be advised and assisted to identify and protect it. Ex situ conservation at Limbe BG of live plants is advisable, also seed-banking.

Calpocalyx winkleri (Harms) Harms [Opaghegui (Beni – *Kennedy* 600); Tegai (Gola, Unwin 68)].

EN B2ab(iii)

Range: Sierra Leone (Gola forest), Nigeria (Sapoba FR) and Cameroon (SW: Mt Cameroon at Likomba, Tiko; Lake Ejagham; and ? Nguti-Baro).

A forest tree to 20 m unusual in that the flowers are borne both among the leaves and on the trunk right down to the ground, and that the branchlets are hollow, inhabited by ants.

Leaflets 4–5 pairs, oblong, 5–12 × 2.5–5 cm, acumen short, base unequal, petiole 5–7 cm; petiolules 4–5 mm; inflorescences congested, pedunculate panicles; bracts not conspicuous in bud, shorter than flowers.

First published in 1907, no reliably identified specimen has been collected since 1950. The Nguti-Baro record (Missouri data) has not been confirmed by a legume specialist and although it is mapped, it should be treated with caution. The Gola record is from a redetermination by Lock in 1986 of an old collection misidentified as *Xylia*, so is absent as a country record in FWTA and lacking altogether in Hawthorne and Jongkind's book.

Here *Calpocalyx winkleri* is assessed as Endangered since only five, possiby six, locations are known (AOO 24 or 28km² with 4 km² cells) and with threats as recorded below. EOO is calculated as 211606 km². In fact, the species may well have been lost from the historic sites in the Mt Cameroon foothills due to clearance of forest for plantations of oil palm, rubber and bananas (pers. obs. 1984–2008) and also at Sapoba FR which has been highly reduced and impoverished by tree-felling in recent decades.

Habitat: lowland evergreen forest, at least sometimes on fertile soils.

Threats: conversion of forest to agriculture after tree-felling, see also above.

Management suggestions: if the Nguti-Baro record can be confirmed, this should be the focus of rediscovery sites together with nearby Lake Ejagham – these are the most recent site records. If refound basic populational data should be gathered to inform future monitoring and gauge management needs, and local communities should be advised and assisted to identify the tree.

Entada camerunensis Villiers

EN B2ab(iii)

Range: Cameroon endemic (Adamawa: W Kongolo at bank of Bayo R. E: Meïganga-Bétare Oya, N Touraké).

A woody liana reaching 15 cm diam. at the base, the stem successively bifurcates. The bipinnate leaves lack glands and have 2–4 pairs of pinnae, petiole grooved above, 5.5–7 cm, red, glabrous; rachis 3.5–10 cm; pinnae axis 4.5–10 cm pubescent along the ridge; stipels minute; leaflets 5–10 pairs per pinnae, obovate-oblong, largest at apex, where 16–25 × 7–11 mm, apex truncate or retuse, mucronate or not, base asymmetric, upper surface more or less pubescent, sessile or with a minute hairy petiolule; pod-loment oblong, 20–29 × 8 cm, glabrous; seeds elliptic 1.6–1.8 × 1 cm (Villiers (1982), Adansonia 4: 193–197 [illegible]).

Here *Entada camer*[illegible]ed since two or three location[illegible] with 4 km² cells) and with threats [illegible] below.

Habitat: woodland near forest or edge of gallery forest.

Threats: man-made fires from grassland extending into forest; forest clearance for agriculture and wood.

Management suggestions: rediscovery of this species should be attempted at the sites indicated above. If successful, basic populational data should be gathered to inform future monitoring and to gauge intervention needs. Local communities should be informed and advised on how to identify *Entada camerunensis*. The needs are possibly too large to survive seed-banking so ex situ cultivation should be considered if a botanic garden could be located in the Adamawa area.

The genus *Newtonia* was revised by Villiers in 1990 (Bull. Jard. Bot. Belg. 60 (1–2): 119–138), 14 species being accepted in Africa. Mainly trees, they are close to the well-known *Piptadeniastrum* in having flat straight pods that open at one edge, with winged, flat seeds and opposite leaflets. However, *Newtonia* have cylindric glands on the rachide and seed stalks attached at the end, not in the middle (Keay in FWTA 1(2): 485 (1958)).

Newtonia camerunensis Villiers

CR A2ac+4ac

Range: Cameroon endemic (NW: Bamenda Highlands. W: Bamboutos Mts).

Tree, trunk 50 cm diam. with flared flutes at the base. Leaves c. 25 cm long, petiole 1.2–1.5 cm long; rachis with 8–10 pairs of pinnae, a large gland between each pair; pinnae with 25 pairs of leaflets, margin ciliate; leaflets oblong, 15 × 3.5 mm, apex rounded, base rounded to slightly retuse, upper surface glossy with midrib prominent. Flowers unknown. Fruits elliptic-oblong, 19–30 × 1.8–2.3 cm, straight or gently curved, apex rounded-apiculate, base attenuate-stipitate, nervation oblique, not prominent in young pods. Seeds winged, elliptic to ovate, 4–8 × 1.5–2.2 cm.

This species is distinguished from *Newtonia buchananii* (Baker f.) Gilbert & Boutique by the higher altitudinal range (1600–1800 m, not 900–1500 m alt.), the presence of a prominent midrib, the larger leaflets and the more conspicuous rachis glands.

Newtonia camerunensis was first assessed as CR A1c in Cheek *et al.* (2000: 66), when it was known from only five collections made between 1932–1974 from locations where it is now thought to be absent so at that time was considered to be possibly extinct. The current assessment is maintained from that made by Pollard in Harvey *et al.* (2004: 65–66) and updated in Cheek *et al.* (2010: 98–99). EOO is calculated as 3093 km².

"Much excitem[illegible]ed when, in April–May 2002, we rediscovered the species at Bali Ngemba FR and at Laikom below the Ijim Ridge. In May 2002 detailed survey work at Laikom had resulted in two fruiting collections: *Ghogue*

1401 & *Pollard* 1097. Identifications were confirmed by B. Mackinder at Kew. At the Akwamofu sacred forest (Laikom) a large fruiting tree was seen, to 30 m and dbh c. 2.5 m, and on the way down the path as far as the bridge across the stream 6 very large trees were seen, with dbh of up to 3 m or more. Many seedlings were found, including c. 20 within 5 m radius of the tree collected under *Pollard* 1097. The largest tree observed was being smothered and strangled by *Schefflera abyssinica* (Araliaceae), and the second largest tree with two juvenile *Schefflera* growing on the upper part of trunk, resulting in considerable die-back in the crown. The canopy in the Akwamofu sacred forest consists mainly of *Albizia gummifera* and *Newtonia*; the latter having a darker appearance when viewed from the ground, partly due to the darker green colour of the leaflets, but perhaps also the closer arrangement of the leaflets that form a denser covering, allowing less light through to the understorey. Three mature fruiting specimens were seen on the way to the waterfall, and at least four other fruiting trees seen on the way back across to the main path. Seedlings were seen to occur in huge numbers in the understorey, including c. 50 in a 5 m^2 patch, but only perhaps c. 30 plants with dbh between 1 and 5 cm were seen in the whole forest, suggesting poor recruitment to maturity, with even fewer specimens of dbh between 5 and 100 cm. Fruit is set abundantly. *Ghogue* 1401 was collected from a further subpopulation recorded in a forest pocket up along the Ijim ridge, which suggests that these fragments are important for the survival of this species. A further subpopulation has been located by Kenneth Tah at Finge (Tah pers. comm. 2002). A sterile sapling collected at Bali Ngemba FR during the 25 m × 25 m plot survey (*BAL*52 in April 2002), was not identified until 2004 at Kew" (Pollard in Harvey *et al.* 2004: 65–66).

Despite exhaustive searching, the flowers of this species remain unknown to science, probably appearing earlier in the season, around March.

Kenneth Tah was first to discover that the forests of Dom have the world's greatest known density of *Newtonia camerunensi,* with possibly a hundred or more stems. It is known from the forest at Kinjinjang Rock and Dom patches 2 (Sagnere)and 3 (Sousi). At all these locations it is frequent (Cheek *et al.* 2010: 98–99).

Habitat: submontane to montane forest with *Albizia gummifera, Carapa grandiflora, Syzygium staudtii, Prunus africana, Pterygota mildbraedii;* 1600–2030 m alt.

Threats: the sites at Kilum-Ijim fall outside the boundary of protection, except at Akwamofu where some protection is afforded by the Kom, who do not allow clearance of forest for traditional reasons. Unfortunately this is only enforced on one side of the path, and most of the specimens of *N. camerunensis* appear to occur on the non-protected side. It seems that the proliferation of *Schefflera* poses a threat to mature individuals, from which most of the seed-rain originates. It is likely that these substantial trees are still used as timber and it may well be exploited elsewhere in its known range. At Bali Ngemba FR the threat from forest clearance for timber, firewood and small-scale agriculture is increasing every year (Pollard pers. obs.).

Management suggestions: there is scope to investigate the biology of this species more closely, particularly at Laikom and Dom, to investigate: the dynamics of recruitment to maturity; ecological relationships with *Albizia gummifera*; the morphology of the inflorescences and flowers; identification of pollinators and seed-dispersal agents; cultivation regimes, the causes of mature specimen mortality, including the effects of parasitic and strangler plants. Such studies would greatly assist our conservation efforts.

In recent years, funded by RBG, Kew, seedlings of this species have been incorporated into the forest restoration plantings led by ANCO at Dom (Mzeka, in Cheek *et al.* 2010: 32–33). However, often wild collected seedlings are used since seed germination in nurseries is reported to have a very low success rate (Tah, in Cheek *et al.* 2010: 30).

Newtonia scandens Villiers

CR D

Range: Cameroon endemic (Littoral: Doula-Edéa FR, Lac Tissongo).

Liana to 20 m or more (all other *Newtonia* are trees); leaves unijugate (one pair of pinnae), glabrous, gland cylindric at apex of upper surface of the 2.5–14.5 cm petiole, pubescent; rachis resembling petiole; leaflets 3–6 pairs, opposite, sessile; rhombic-oblong, to 6.5 × 4 cm, apex truncate, base asymmetric, midrib running diagonally, glabrous, nerves prominent; flowers pedicellate, petals 3.5 mm, glabrous apart from the ciliate margins; pod oblong-ellipsoid, 13.5–7 × 2.2–3cm; seeds to 5 × 1.5 cm, winged.

Here assessed as Critically Endangered since less than 50 individuals have been recorded, although no threats are known at its single site (AOO 4 km² with 4 km² cells).

Habitat: lowland evergreen forest, probably on at least seasonally inundated soil.

Threats: none are known. Lac Tissongo is accessible only by crossing a large river by pirogue (pers. obs. Cheek 1984). It is currently remote from any threats.

Management suggestions: if possible, rediscovery should be aided by the original discoverer, Duncan Thomas. Basic populational data should be obtained to aid future monitoring and gauge intervention needs if any. Further characterisation of the stem at ground level might aid census work. The population at Marienburg, the nearest settlement, should be advised about this rare plant. Ex situ conservation in a seed bank and at Limbe BG are suggested.

LEGUMINOSAE SUBFAMILY PAPILIONOIDEAE

assessed by Martin Cheek

Angylocalyx talbotii. In our conservation checklist, The Plants of Kupe, Mwanenguba and the Bakossi Mts, Cameroon, (Cheek *et al.* 2004: 163–164), we assessed this taxon as VU A3c, B2ab(iii). At that time six localities were recognised and the species was confined to SW Region Cameroon and the type locality in adjoining SE Nigeria. However, since that time three records have been found on the Gabon database, Tchoutou (2004 loc.cit.) cites it for Campo-Ma'an and Velayos *et al.* 210 loc.cit. for Rio Muni. We have not been able to verify these records but for the moment downrate the assessment from VU to NT in the expectation that these additional records are correctly determined.

Dalbergia altissima Baker f. is a rare species listed for Cameroon by ILIDIS and Lock (1989) on the basis of Flore du bas- et du Moyen-Congo (1903–1912), Ann. Mus. Congo Bot. Ser. V, 1: 1–345. But no specimen has been seen for Cameroon at K or YA so we have not been able to verify this assertion. If correct it is probably a threatened taxon.

Several species of *Desmodium* have been described from Cameroon and appear to be only known from the type collection. Examples are *Desmodium ledermannii* and *D. mildbraedii*. However, they are not accepted by Lock (1989) and the genus badly needs revision so they are omitted here.

Tephrosia capitata Verdc. A rare shrub of the Cameroon Highlands has been found to be a synonym of the widespread Lower Guinea montane *Adenocarpus mannii* by Onana.

Baphia Lodd. Thankfully revised by Soladoye (Kew Bull. 40: 291–386 (1985)) who recognised 45 species. Evergreen forest shrubs and lianas, they are unusual in having a single leaflet. The flowers are usually white with yellow markings. Related and similar genera are *Leucomphalos, Baphiastrum* and *Airyantha* which can be distinguished by using Soladoye's key (loc. cit. 295). Specimen data for Soladoye's four new species was taken from his protologues (Kew Bull. 37: 295–302 (1982)). Several taxa considered as NT here might prove to be threatened if re-examined: *B. buettneri* subsp. *hylophila, B. cuspidata, B. leptostemma* subsp. *gracilipes* and *B. pilosa* subsp. *batangensis.*

Baphia breteleriana Soladoye
EN B2ab(iii)

Range: Cameroon (SW: "Kumba division". S: Lolodorf. C: 53 km SW Eséka at Songbong) and Gabon (Mayibout II; Cristal Mts at Asok).

Tree c. 6 m, glabrous; leaves leathery, glossy, obovate 7.5–14.5 × 3.2–7 cm, acumen less than 10 mm abrupt, obtuse, lateral nerves 5–6 pairs; petiole 4–10 mm, pulvina contiguous; inflorescence of short racemes in March and Aug.–Sept., brown pubescent, pedicels 5–20 mm; calyx spathe-like, 10–13 mm, standard white, with basal yellow blotch, suborbicular, 1–2 cm diam., ovary 7–9 mm with 4 longitudinal hair bands; pod 11.2 × 2.5 cm, in July.

Here *Baphia breteleriana* is assessed as Endangered since only five locations are known (see above; AOO 20 km² with 4 km² cells) and since threats are as below. EOO is calculated as 72643 km².

Habitat: lowland evergreen forest, usually near the coast.

Threats: slash and burn agriculture in S Region. Since the 1895 collection of Staudt listed as 'Kumba division' by Soladoye, large areas of forest habitat have been cleared for oil palm, rubber, cocoa and banana plantations.

Management suggestions: the best hope for refinding this species in Cameroon is probably in S Region where the most recent records derive. If refound, the usual populational data should be gathered to serve as a baseline for future monitoring and to gauge needs for management intervention. Leaders of local communities should be assisted to identify the species so as to ensure its survival.

Baphia dewildeana Soladoye
VU B2ab(iii)

Range: Nigeria (Cross River: Calabar by edge of Naifor fence; Apabuyo beach; Obubra Dist., Okumerate; Anambra State: Awka Dist, Mamu FR) and Cameroon (C: Goura, 40 km SE of Bafia. SW: Mamfe).

On currently available data *Baphia dewildeana* is assessed as Vulnerable since six sites (above; AOO 24 km² with 4 km² cells) are known with threats as below. If the suspected widespread loss of habitat at most of the Nigerian sites is confirmed it is likely that the species will be reassessed as EN or CR. EOO is calculated as 32889 km².

This is a small tree to 8 m, stems glabrous, rarely with brown tomentum. The leaf-blade is oblong to ovate, 4–10 × 1.6–3.6 cm, acuminate, the base rounded, lateral nerves 5–8 pairs; petiole 4–10 mm, glabrous, pulvini not contiguous. Flowers solitary or in pairs in axils, bracts caducous, pedicel 10–15 mm, glabrous; bracteoles paired, below calyx, ovate, cupped, 2.6–3 × 1–2 mm, margin ciliate; calyx spathe-like, 12–17 mm; standard white with basal orange blotch, suborbicular, 14–20 mm; ovary glabrous.

Habitat: lowland evergreen forest.

Threats: massive forest losses due to logging for timber followed by agriculture, have removed most of this habitat from Nigeria.

Management suggestions: rediscovery of this species in the Bafia area of Cameroon, where forest has less drastic losses than Nigeria, is most likely to be successful. If rediscovered, basic populational data should be gathered as a baseline for future monitoring and to assess needs for management intervention. Leaders of local communities should be assisted to identify and protect *B. dewildeana.*

Baphia latiloi Soladoye (syn. *Baphia* sp. A of FWTA 1: 513 (1958))
EN B1+B2ab(iii)
Range: SE Nigeria (Calabar-Mamfe, mile 66; Ikom Distr., Bendeghe-Ayuk village) and Cameroon (SW: Abonando; Mamfe).
A shrub or tree to 9 m, young stems red villous hairy. Leaves oblong or oblong-obovate 5–10.5 × 1.5–4 cm, gradually acuminate, lateral nerves 8–12 pairs, lower surface sparsely brown appressed hairy; petiole 5–10 mm, grooved, glabrescent, pulvini non-contiguous; stipules linear-oblong, subpersistent, striate, hirsute. Flowers 1–2 in leaf axils, pedicels 7–15 mm sparsely hairy; bracteoles semi-circular 1.5–2 × 0.5 mm inserted directly below the calyx, ciliate; calyx spathe-like 1.2–1.4 cm; standard suborbicular, 12–15 mm, white with basal yellow spot; ovary with long brown hairs on dorsal and sometimes also ventral edges.
Since the five collections equate to four locations (AOO 16 km² with 4 km² cells) and threats are as below, *Baphia latiloi* is here assessed as Endangered. EOO is calculated as 1813 km².
Habitat: lowland evergreen forest.
Threats: Nigeria has lost most of its original forest and that which survives is under great pressure for logging and agriculture, including National Parks such as those in Cross River State. The two sites in Cameroon are under much less pressure it is believed.
Management suggestions: since the Nigerian specimens have the most detailed collection localities it is logical to search first at these sites if forest habitat survives there. However, Cameroon probably offers the best hope for long term conservation since so much more forest habitat is intact here. Basis populational data (density, range, regeneration levels and threats) should be gathered as a baseline for future monitoring and so as to assess the need for management intervention.

Baphia leptobotrys Harms subsp. ***silvatica*** (Harms) Soladoye
VU 2ab(iii)
Range: Cameroon (S: Djoum. E: Boumba and Ngoko at Malandi) and Gabon (seven sites).
Small tree, sometimes lianescent, glabrous; leaves leathery, broadly ovate to oblong-elliptic, 8–23.8 × 3.5–9.5 cm, acumen to pulvini not contigurous; raceme 4–16 cm, bracteoles 3–6.5 × 1.5–2 mm; calyx spathe-like 9–14.5 mm; standard suborbicular 14–18 mm diam.
This taxon, sometimes maintained as a distinct species, is here assessed as Vulnerable since nine sites are known (AOO 36 km² with 4 km² cells) and since threats are as below.
Habitat: lowland evergreen to semi-deciduous forest.
Threats: the area from which the type collection was made (Boumba and Ngoko) is currently subjected to logging. Threats at the Gabon sites are known.
Management suggestions: the Gabon sites should be matched against the newly created National Parks to determine if any are included within. If confirmed, Park managers should be provided with photographic guides to identify this taxon. The usual basic populational data should be gathered (density, range, regeneration levels, threats) as a baseline for future monitoring and for gauging future management needs.

Baphia mambillensis Soladoye
EN B2ab(iii)
Range: Nigeria (NE: Gongola State, Sardauna Div., Mambilla, R Anterre; Dujire; R Nwum FR; Inkiri; Donga R below Inkiri; Akwaijatan forest; Anambra State, Ihiala Dist., Ihudim village) and Cameroon (NW: Bamenda Highlands, Fang, 30 km NE Wum. Adamawa: Banyo, Koti to Mayo Darlé).
The nine collections listed above of *Baphia mambillensis* equate to five locations, so here it is assessed as Endangered. (AOO 20 km² with 4 km² cells) since threats are as below. EOO is calculated as 14388 km².
The name derives from the Mambilla plateau whence five collections have been made (all by J. D. Chapman) – the greatest concentration within its range.
Since this appears to be the only submontane *Baphia* in the Cameroon Highlands, it should be easy to identify in the field.
Shrub or tree to 25 m, stems glabrescent. Leaves oblong or obovate-oblong, 4.5–15 × 2.6–5.5 cm, acumen to 2.2 cm, base usually asymmetric and rounded, lateral nerves 5–8 pairs, midrib hairy below; petiole 5–15 mm, grooved, pulvini not contiguous; stipules lanceolate-linear 7–10 mm, caducous. Flowers 1(–2) per axil, pedicel 8–26 mm yellow-tomentose when young, bracteoles semicircular, 2.5–4.2 mm, margin ciliate, opposite, inserted directly below calyx; calyx spathe-like 13–17 mm; standard suborbicular, 15–22 mm, white with basal orange blotch; ovary densely yellow-brown tomentose. Pods oblong, 10–12.5 × 2–2.5 cm, seeds 1–2.
Habitat: submontane forest, often along rivers; 800–1800 m alt.
Threats: the loss of submontane forest since the 1970s, which is ongoing, has been documented by J. D and H. D. Chapman (2001) The Forest of Taraba and Adamawa States,

Nigeria. Timber, firewood extraction, followed by agriculture, are the explanation.

Management suggestions: investigation is needed to discover whether any portion of the Mambilla submontane forest containing this species is protected in reality. If this is not the case it should be redressed. Local community leaders should be informed of the importance of this species and given advice on its identification and protection. The usual populational data should be gathered (density, range, regeneration and threat levels) as a baseline for future monitoring and to understand the need for management intervention.

Baphia obanensis Bak.f.

EN B2ab(iii)

Range: SE Nigeria (Oban), Cameroon (Littoral: Loum-Yabassi) and Gabon (Lopé Reserve; Ekonodo village)

A small tree to 10 m tall, stems red-brown tomentose; leaves papery elliptic to oblong-elliptic, 7.4–16.2 × 4.1–7.8 cm, acumen spatulate, base rounded, lateral nerves 5–7; petiole 5–10 mm, pulvini contiguous; racemes or fascicles of up to 35 flowers, pedicels 6–14 mm, hairy; bracteoles opposite, under calyx, ovate to suborbicular, cupped, 0.8–1.2 × 2.2–2.5 mm,calyx spathe-like, 7–10 mm, standard suborbicular 10–11 mm diam; ovary glabrous or with some hairs on the dorsal margin.

Here *Baphia obanensis* is assessed as Endangered since only four sites are known (AOO 16 km^2 with 4 km^2 cells) with threats as below. EOO is calculated as 82758 km^2.

Habitat: lowland evergreen forest.

Threats: threats at the sites in Gabon are not known, but forest in Nigeria has been under great pressure from logging for timber followed by agriculture, and although a National Park exists near the town of Oban, the site for this species must be considered likely to face these threats.

Management suggestions: attempts should be made to rediscover *B. obanensis* in the Loum-Yabassi area. Basic populational data (density, range, regeneration levels and threats) should be collected to inform future monitoring and management needs.

Crotalaria bamendae Hepper

VU D2

Range: Cameroon (NW: Bamenda Highlands (four pre-1996 coll.)), Nigeria (Mambilla Plateau (one coll.)) and Angola (two coll.).

This species was published by Hepper (1956) on the basis of two specimens from the Bamenda Highlands, *Egbuta* in FHI 3763 (farmland, Bamenda, Dec. 1951) and *Tamajong* in FHI 23479 (grassland at roadside, Kumbo, Oct. 1947, type of the species). A few years later (FWTA 2: 549 (1958)), he had added a third collection, *Latilo & Daramola* in FHI 34369, from Gembu in the Mambilla Plateau. Polhill (1982), in revising the genus for Africa and Madagascar, extended the range to Angola. The photograph of *Newton* 76 (Angola, Huila, Humpata, Feb. 1883) shows a specimen that differs from the Cameroonian material in having much larger and more broadly elliptic leaflets (2.9 × 1.4 cm) with inflorescences partly concealed by leaves and borne on short spur shoots. Perhaps the Angolan plants merit subspecific distinction. Polhill regards *Crotalaria bamendae* as an evolutionarily isolated species in section *Glaucae*.

More recent collections of *Crotalaria bamendae* are *Jacques-Félix* 8933 from Tchabal Mbabo and *Meurillon* in CNAD 125 from Lake Bambuluwe (November 1965). The only recent collection record known is *Munyenyembe* 882 (Kilum, Shambai, 2500 m alt.) November 1996. *Crotalaria bamendae* is evidently rare in its range. However, the fact that only this single collection was made during a month of collecting in its habitat, at the peak of its flowering season in 1996, by a large expedition of botanists from Kew and Cameroon suggests that even within its habitat it is rare and infrequent and so Vulnerable to threats. Recent surveys elsewhere in the Bamenda Highlands where it might reasonably be expected to occur, have failed to find new records for it (Harvey *et al.* 2004, Cheek *et al.* 2010). EOO is calculated as 286888 km^2.

This assessment, originally made in Cheek *et al.* (2000: 63-64), is maintained here after slight modification since no new data are available.

Subshrub 30–60(–90) cm tall. Stems densely pubescent. Leaves shortly petiolate, 3-foliolate, uppermost often reduced, rarely 1-foliolate; leaflets narrowly elliptic-oblong to elliptic-obovate, mostly 1.5–3.5(–4.3) × 0.4–1.5 cm, sparingly pilose above, silky-hairy beneath. Stipules lanceolate, 3–6 mm long. Racemes subsessile, dense, many-flowered, shortly cylindrical, c. 2.5 × 2 cm; bracts linear 4–5 mm long; pedicels 2–5 mm. Calyx 4–5 mm long; densely pilose, lobes triangular-lanceolate. Standard circular, yellow, with fine darker lines, hairy along midvein outside; wings longer than keel; keel abruptly rounded, shortly beaked, 4.5–5 mm long. Pod c. 5 mm long, pilose, glabrescent, 2–4-seeded. Seeds 1.5 mm long, yellow, glossy. Description after Polhill, Crotalaria of Africa, Balkema, 1982: 128–129.

Habitat: montane grassland, sometimes wet; 1800–2500 m alt.

Threats: unknown, but fire and/or grazing may effect this species adversely.

Management suggestions: the range and frequency of this species at Kilum-Ijim (Mt Oku) and factors influencing recruitment and survival need to be established.

Crotalaria ledermannii Baker f.

VU D2

Range: Nigeria (Mambilla Plateau (1 coll.)) and Cameroon (SW: Mwanenguba Mts (1 coll.). NW: Bamenda Highlands (7 coll.)).
This is a rare species, having been gathered only once in the course of our three inventories at Mt Oku and Ijim, once at Bali Ngemba FR, Bamenda and Mwanenguba Mts, and twice at Bafut Ngemba. The collection from the Mambilla Plateau is taken from a *Eucalyptus* plantation, so *Crotalaria ledermannii* may be able to tolerate some disturbance to its habitat. This is an annual or short–lived perennial species, so a deleterious change in the habitat or poor seed set in one year could drastically reduce the population within 12 months. This herb was assessed as above in Cheek *et al.* (2000: 64). No new data have come to light since then, apart from the recent collections at Bali Ngemba FR and Mwanenguba Mts, so the original conservation assessment was maintained in Cheek *et al.* (2004: 164) and Harvey *et al.* (2004: 66), and is maintained here. EOO is calculated as 8385 km².
Annual or short-lived perennial herb 20–70 cm tall, with several erect stems arising from near the base, branches appressed puberulous, the hairs dense above, glabrescent below. Leaves 3-foliolate; leaflets oblanceolate, 7–20 × 1–5 mm, apex rounded or truncate, apiculate, appressed puberulous beneath; petiole 4–13 mm long. Inflorescences 1.5–33 cm long, comprising a terminal head and a few laxly inserted flowers below; bracts linear-subulate, 1–1.5 mm long; pedicel 3–6 mm long. Calyx 2.5–3.5 mm long, densely brownish puberulous; upper lobes triangular, sometimes slightly acuminate, 1–1.5 times as long as the tube. Standard obovate to oblong-obovate, yellow, veined red, brownish puberulous outside, wings a little shorter than keel; keel angular, 5.5–6.5 mm long. Pod shortly stipitate, ovoid globose to ovoid-ellipsoid, 4–5 × 3–3.5 mm, appressed pubescent, 2–seeded. Seeds 2 mm long, with prominent aril. Description after Polhill (1982 loc. cit.)
Habitat: grassland and forest edge; 1200–2200 m alt.
Threats: unknown but possibly conversion of land to cultivation and grazing, trampling or fires.
Management suggestions: the natural habitat of *Crotalaria ledermannii* is not well characterised. More study is needed to rectify this. A survey to find the range of this species and levels of regeneration is advisable.

Crotalaria mentiens Polhill

EN B1+B2ab(iii)

Range: Cameroon endemic (NW: Bamenda Highlands (two coll.)).
For information on this species we depend entirely on the protologue (Polhill 1976). No specimens of *C. mentiens* are present at Kew, nor were any gathered during our several inventories in the Bamenda Highlands (Cheek *et al.* 2000; Harvey *et al.* 2004; Cheek *et al.* 2010). The type specimen (*Brouwers* 2, BR, Jakiri) was collected on the southeastern boundary of our Kilum-Ijim checklist area (Cheek *et al.* 2000: 141) and is the only specimen of the species cited by Polhill. However, the accompanying map shows two points, so it is assumed that there are two collections. Polhill states "Known only from the Bamenda area of W. Cameroon and deceptively similar to *C. ledermannii* in the same area". The collector Brouwers is not listed by Letouzey (1968) and so we surmise that he or she was a Belgian who collected a few specimens in the Bamenda Highlands in the late 1960s or early 1970s. This species is given an EN rating partly because of the fact that only two specimens are thought to exist, that it has not been re-collected during our inventories, and because it occurs in the most densely populated part of our area. Assuming that the two locations equate to two different locations (AOO 8 km² with 4 km² cells and threats as below) *Crotalaria mentiens* was assessed as EN B1+2c in Cheek *et al.* (2000: 64–65). Since no new data are available, that assessment is maintained here but updated to follow IUCN (2001) and the EOO data newly calculated as 109 km². It might be reasonable to assess this species under criterion D in which case CR status would apply.
Herb 20–40 cm tall, similar to but differing from *C. ledermannii* as follows: young branches strigulose, not puberulous, bracts 2–3 mm long, as long as pedicel, not 1–1.5 mm long, much shorter than the pedicel; calyx thinly, not densely puberulous; lobes 2–3 times longer than the tube, not 1–1.5 times longer; standard glabrous outside, not puberulous (Polhill loc. cit. 1982).
Habitat: unknown, but probably as *C. ledermannii.*
Threats: unknown, but probably as *C. ledermannii.*
Management suggestions: an attempt should be made to rediscover this species in the wild and to assess whether, as is suspected, it is threatened. The Brussels herbarium at Meise should be checked for the specimen cited, and any others of this species so that more detailed data on the locality and habitat might be obtained. This would assist in rediscovering *C. mentiens.*

Dalbergia ealaensis De Wild.

VU B2ab(iii)

Range: Cameroon (E: Bertoua-Batouri 6 km), Gabon (2 locs.), Congo (Kinshasa) (Eala; Bomaneh; Yangambi:Res. Fl. Isalowe) and Angola (site not located).
A large liana to tens of metres high, the stem reaching 25 cm diam., glabrous. Leaflets 8–15(–23), alternate, elliptic-oblong, 4–8 × 1.5–3.5 cm, apex and base rounded or retuse, yellowish-green below; petiole 2.5–5 cm. Inflorescences often on leafless stems, 30 cm, flowers white, pink or yellow; standard 9.5–13 mm; calyx and receptacle violet, ridged margins cilitate, otherwise glabrous, lobes 2.5–3.5 mm ovate, stamens in 2 groups of 5; pods 7–16 × 2–4 cm, strongly reticulate; seeds 1(–2) (Cronquist, Fl. Congo Belge 6: 60

(1954)). Several specimens from CAR, Cameroon and Gabon resemble this taxon but are distinct (sp. aff. *ealensis*). Here *Dalbergia ealaensis* is assessed as Vulnerable since seven sites are known (see above; AOO 28 km^2 with 4 km^2 cells) and threats as below. EOO is calculated as 398861 km^2.

Habitat: inundated evergreen forest and river edge; 400–650 m alt.

Threats: forest areas along large rivers such as the Congo are often cleared for settlement and agriculture. Slash and burn agriculture.

Management suggestions: the Bertoua site should be revisited so as to refind this plant and basic populational data (density, range, regeneration and threat levels) gathered as a baseline for monitoring and to inform management needs in future. Relevant local authorities should be assisted to identify, protect and understand the rarity of this species.

Dalbergia oligophylla Baker ex Hutch. & Dalziel

EN A3c

Range: Equatorial Guinea (Bioko (2 coll.)); Nigeria (Obudu Plateau (2 coll.) and Cameroon (SW: Mt Cameroon (4 coll.); Mt Kupe (6 coll.); Mwanenguba (1 coll.). NW: Bafut Ngemba (1 coll.); Bali-Ngemba (1 coll.)).

Known only from six sites, this shrub-liana dries a distinctive black, helping to distinguish it from similar species from Sierra Leone and Gabon, with which it has been confused. This is a highland species. Our map shows several records from lowland (c. 300 m alt.) Gabon. One of these *Louis* 748, examined at Kew, has leaflets twice as broad as those of the Cameroon Highland specimens while also lacking the conspicuous lateral nerves and golden-brown lower surface. It seems likely that the Gabonese records represent a different taxon.

This species was overlooked as threatened in Cable & Cheek 1998. It was assessed as EN A2c , A3c in Harvey *et al.* (2004: 66) and in Cheek *et al.* (2004: 164) on the basis that more than 50% of the population of this long-lived species has been lost in the last 100 years (see threats below). These assessments are maintained here since no new data are available, apart from the discovery of historic slight range-extensions into similarly threatened sites where it may already be lost. EOO is calculated as 28271 km^2.

Habitat: submontane and montane forest edge; (900–)1500–2000 m alt.

Threats: forest clearance for agriculture and wood: the subpopulations at Bafut Ngemba, Obudu Plateau and Mwanenguba have quite possibly been lost due to this already. The Bamenda Highlands may once have been the main range of this species (judging by records at Bali Ngemba and Bafut Ngemba) but forest loss here has been as high as 25% between 1987–1995 in one area studied (Moat in Cheek *et al.* 2000: back cover). In another, in the Dom area, 50% of forest cover has been lost in 15 years to 2003 (Baena in Cheek *et al.* 2010: back cover). Overall, over 50% habitat loss is postulated over the last 100 years, and at least 50% of that which remains could be lost in the next century.

Management suggestions: unless Bali Ngemba FR is protected from incursion more vigorously, the only hope for the survival of this species is the summit area of Mt Kupe and the upper tree line of Mt Cameroon both of which are reasonably secure from threat.

Dolichos reptans Verdc.

EN B2ab(iii)

Range: Nigeria (Adamawa Div: Vogel Peak, Jobdi to Wum Hills) and Cameroon (Adamawa (near Tibati): 80 km NW at Hosséré Ngo; 60 km NNE at Sabal Maba near Minim).

A far-creeping herb with 4- or 5-angled tough stems; leaves leathery, leaflets, subequal, each elliptic, entire or 3-lobed, with leathery elliptic stipels and foliaceous ovate 8 mm stipules. Inflorescence 4–6-flowered, flowers pink, keel white below.

Here *Dolichos reptans* is assessed as Endangered since only three sites are known (AOO 12 km^2 with 4 km^2 cells) with threats as below. EOO is calculated as 7030 km^2.

Habitat: submontane grassland and wooded grassland; 1580–1650 m

Threats: overgrazing by cattle; altered fire-regimes.

Management suggestions: this wild relative of the agricultural crop plant grows in isolated sites. More data is needed to understand the precise ecological niche and lifeform of this species – e.g. is there an underground rootstock or is *D. reptans* an annual? Field data on density, range, regeneration levels and threats of this species are needed to inform understanding of management needs, and to give a baseline for future monitoring.

Eriosema a genus of capitate, yellow-flowered shrubs, and subshrubs of grassland and woodland with distinctive oblong pods. Revised for W-C Africa by Jacques-Félix (Adansonia ser.2, 11(1): 141–169 (1971)).

Apart from those species listed here, further research on an African-wide basis may show that the following Cameroonian species are also threatened: *E. ramosum* Bak.f. (type from Angola), *E. chrysadenium* Taub., E. *gracilliumum* Bak.f. (type from Angola), *E. bauchiense* Hutch. & Dalz. (type from Nigeria), *E. shirense* Bak.f. (type from Malawi), *E. lebrunii* Stan. & de Craene (type from Congo (Kinshasa), *E. verdickii* De Wild. (ditto), *E. schoutedanium* Stan. & de Craene (ditto), *E. monticola* Taub. (type from Togo).

Eriosema adamouense Jac.-Fél.

CR B2ab(iii)

Range: Cameroon endemic (Adamawa: Hosséré Sillé).

Described by Jacques-Félix (loc. cit. 158–160) this is a 1.5–2 m subshrub, the stems ribbed from decurrent leaf-bases, with white indumentum on the ribs. The subsessile trifoliolate leaves have triangular 9 × 3 mm glandular stipules, petiolules 2–3 mm, leaflets rhombic-elliptic, 10 × 1.5 cm, lateral nerves c. 20 pairs, upper surface with fine appressed hairs, lower surface with a dense golden tomentum on the nerves, elsewhere white. The racemes are 12–15 × 4 cm on a 3–4 cm peduncle, bracts 10 × 5 mm, flowers 22 mm, yellow, pubescent and glandular.
Here *E. adamouense* is assessed as Critically Endangered since only a single site (above; AOO 4 km^2 with 4 km^2 cells) is known, with threats as below.
Habitat: grassland, 1600 m alt.
Threats: overgrazing by cattle, increased incidence of fires beyond that which is natural.
Management suggestions: the type locality should be revisited to refind this species and discover the density, range, regeneration levels and threats to this species. Traditional leaders should be enabled to identify this species so as to be able to ensure its survival.

Eriosema raynaliorum Jac.-Fél.
VU D2
Range: Cameroon (Adamawa: 37 km E of Ngaoundéré; Dzerkoka).
A fire-adapted subshrub 5–7 cm tall arising from a large underground tuber. Stems 2–7 from each tuber, with short internodes. Leaves 1-foliolate, broadly elliptic, 4–5 × 2.5–3.5 cm glabrous-above, tomentose below, petiole 3 mm. Raceme 4 cm, bracts lanceolate 5 × 1 mm, flowers yellow, 10–11 mm in Dec. and Jan. (from original decription by Jacques-Félix loc. cit. 196–197).
Here *E. raynaliorum* is assessed as Vulnerable since although two sites are known (AOO 8 km^2 with 4 km^2 cells) immediate threats are unknown at present.
Habitat: rocky grassland; 1700 m alt.
Threats: none are known but the sites should be visited to evaluate if these exist. Elsewhere species with this lifeform can be threatened by wildlife predation on tubers, or the absence of fire required to remove competition from shrubs.
Management suggestions: the two known sites should be revisited to acquire data on population density, range, regeneration levels, threats and to inform local community leaders about the species. The leafy stage is not yet known. Seeds should be banked to protect against extinction. Further survey of the Adamawa Mts is likely to produce more locations.

Eriosema letouzeyi Jac.-Fél.
VU B2ab(iii)
Range: Cameroon (Adamawa: Dankali, Meïganga), Tchad (Békao) and CAR (pl. de Ungourras; Yaloké, Savane Mayaka; Bambara region; Sav. Des Moroubas; ibid Yanguya; Bozoum; Manova-Gounda-St. Floris NP).
Described by Jacques-Félix (loc. cit. 154–156) , this is an 0.8–1.2 m pubescent subshrub with subsessile, 3-foliolate leaves, the leaflets oblong or oblanceolate, 3.5 × 0.9 cm pubescent above, white-tomentose below. Inflorescences globose, 2 cm, bracts obovate 6 × 3 mm, pilose. Flowers 12 mm, yellow. Pods 12 × 7 mm.
Jacques-Félix (loc. cit. 154–156) described *E. letouzeyi*, here assessed as Vulnerable since seven sites are known only, one of which (the last listed above) was found since the description. The first CAR site listed is not mapped since it was not located. AOO of 28 km^2 is estimated using 4 km^2 cells, and threats are as below. EOO is calculated as 183105 km^2.
Habitat: wooded grassland e.g. with *Terminalia laxiflora* and *Piliostigma thonningii*, 500–600 m alt.
Threats: overgrazing and trampling by stock; over-frequent fires set by man.
Management suggestions: on refinding this species, the density, and regeneration levels and threats should be evaluated. Excavation of the rootstock if present would aid undersand the ecology of this species. Is it perennial or annual, fire-adapted or not? Further survey work is likely to discover more locations for this species. The last CAR record listed is inside a National Park, but this might be the best site to focus on protecting the species. The park managers should be provided with the means to identify and protect the species.

Indigofera dasycephala Bak.f.
VU B1+B2ab(iii)
Range: Nigeria (Mambilla, Bauchi-Gombe Rd; Sokoto; Vogel Peak) and Cameroon (N: Mango; Orèfén; Diatomey-Dourbey. Extreme N: Bourha, 65 km SSW Mokolo).
Perennial pubescent herb 15–20 cm tall, hairs appressed, grey, long; leaves with 5–6 pairs of ovate, mucronate leaflets 5–7 × 3–4 mm; flowers red, in sessile heads c. 8 mm diam. with conspicuous exserted linear bracts; calyx 5 mm long, with slender; standard as long as calyx (Baker in Leg. Trop. Africa 1:100 and JSTOR website).
Here *Indigofera dasycephala* is assessed as Vulnerable since only seven locations are known (see above, AOO 28 km^2 with 4 km^2 cells) and with threats as below. EOO is calculated as 5721 km^2.
Habitat: open, gravelly and rocky hill slopes; 700 m alt.
Threats: overgrazing and trampling by cattle and goats; gravel and rock extraction.
Management suggestions: the ecology and life-style of this species is not understood and needs research. Data on density, range, threats and regeneration levels is needed to monitor and to assess management needs in future. Local

authorities should be apprised of the existence of this plant and given the means to identify it.

Indigofera patula Baker subsp. *okuensis* Schrire & Onana

VU D2

Range: Cameroon endemic (NW: Mt Oku and the Ijim Ridge (2 coll.)).

This subspecies came to light in August 1998 when Onana was naming the Leguminosae specimens from the 1996 Kilum-Ijim (Mt Oku and Ijim Ridge) inventory at Kew under the guidance of specialists, including Schrire (Schrire & Onana, Kew Bull. 55(1): 223 (2000)). The first specimen collected of this species *Munyenyembe* 814, was made between Oku-Elak and transects KD/KC at c. 2000 m alt. on 30 October 1996 when both flowers and fruit were present. The second specimen *Etuge* 3522, was collected on the other side of Mt Oku at 2252 m on 20 November 1996, when flowers and fruits were also present. This plant was found on the road from Aboh to Gikwang, towards the Nyasoso forest. Although only two specimens are known, this taxon may be undercollected because its habitat can be fallow fields, which were not the main target of plant collecting! Several other taxa of *Indigofera* occur in the area. This taxon is distinct from these in that the lower surface of the leaflets, apart from appressed biramous hairs, bear black, glandular hairs at the margins and sometimes along the midrib. These black, glandular hairs are also present on the stem.

The assessment above is taken from Cheek *et al.* (2000: 65–66) and is maintained here. Evidently the species is very rare since only two specimens have come to light among the thousands collected at Mt Oku. If threats can be identified this species would qualify as EN under criterion B, and if an estimate of numbers of individuals were available it is likely that it would rate CR D.

Perennial herb or subshrub 0.3–1 m tall. Stems trailing or erect, reddish, with appressed biramous hairs and blackish gland-tipped hairs, viscid. Leaves 3–13-foliolate; leaf axils 3–39 mm long including a petiole of 0.5–10 mm; terminal leaflets 2–14(–19) × 1.2–7(–9) mm, slightly larger than the laterals, elliptic-oblong to lanceolate or oblanceolate, about 2–3 times as long as wide, apiculate, sparsely to densely strigose with appressed biramous hairs, margins and sometimes midribs beneath usually fringed with short, blackish gland-tipped hairs up to 0.3 mm long. Racemes (4–)5–10 cm long, not zigzag, equalling or up to 3 times the length of the subtending leaf, including a usually densely glandular peduncle 22–38 mm long, laxly to densely c. 15–50-flowered, pedicels 1.5–2.5 mm long. Calyx (2.5–)3–8 mm long. Corolla 5–8 mm long, pink to carmine-red. Pods 5–14 × 1.5–2 mm, straight, cylindrical, spreading, sparsely hyaline strigose with hairs spreading at the tips, reddish to dark brown, densely covered with erect gland-tipped hairs 0.1–0.5 mm long, style-base forming a short beak. Seeds 2–6, 1–1.5 × 1–1.2 mm, quadrangular-cylindrical to globose, yellow to brown, shiny, not pitted. Description after Schrire & Onana (2000).

Habitat: montane grassland & fallow areas; 2000–2300 m alt.

Threats: unknown.

Management suggestions: a field survey of *Indigofera* is recommended at Kilum-Ijim in order to facilitate identification of this narrowly endemic subspecies and to gather data on its threats, distribution and frequency. This taxon may have potential as a nitrogen-fixing, soil fertility-improving fallow-field species.

Leproderris is a genus of woody climbers with pinnately compound leaves with leaflets opposite, bearing stipels and 1-few-seeded indehiscent fruits. Closely related and similar to *Lonchocarpus ostryocarpus* and *Ostyroderris* in the tribe Dalbergieae, this genus is separated by having wing petals part adnate to the keel, flowers arranged in clusters on the raceme and fruits winged narrowly on one side only.

Leptoderris aurantiaca Dunn

VU B2ab(iii)

Range: W. Nigeria (Abeku; Ijebu Ode Prov.; near and in Omo FR), Cameroon (SW (near Mt Cameroon): Barombi Koto; Idenao; Onge FR. S: Kribi-Bipindi), Equatorial Guinea (Rio Muni: Bata-Pembe-Entuba) and Gabon (Libreville).

Here *Leptoderris aurantiaca* is assessed under criterion B as Vulnerable since eight sites are known (AOO 32 km² with 4 km² cells) with threats as below. However, at all these sites the species is under immediate threat if it survives at them (see below). In future it is likely that this species will be assessed as EN or CR under criterion A. *Leptoderris aurantiaca* is an example of a species which although relatively widespread and frequent, is nonetheless probably on the brink of extinction due to habitat loss. EOO is calculated as 147561 km².

A woody climber with stems covered in flaking scales. Leaves 7-foliolate, oblong to obovate-oblong, rounded at base, shortly acuminate, lateral nerves 14–16 pairs; stipels 7–10 mm; flowers white in large pilose panicles, pedicels 2 mm, bracts lanceolate 10–12 mm, appressed pilose; calyx 3 mm, 4-lobed; fruits flat, 6–9 × 2.5 cm, appressed pubescent, upper edge broadly winged.

Habitat: lowland evergreen forest near the coast.

Threats: Nigeria has seen huge losses of forest for timber and agriculture in recent decades, including that in Forest reserves so it cannot be certain that *L. aurantiaca* survives at any of the historic sites in that country. In Gabon the only known site was at Libreville but has probably now been lost due to urbanisation. In Cameroon, Kribi is vulnerable to urbanisation and an increasing footprint for extractive industries. At Mt Cameroon intact forest still exists at Onge FR but is under pressure for commercial logging while Idenao is a centre for oil palm production and Barombi Koto has lost most of its natural vegetation due to settlement.

Management suggestions: the best hope for refinding *Leptoderris aurantiaca* is Mt Cameroon which has had the greatest concentration, and the most recent, records. If refound, live material should be brought to Limbe BG for propagation and reintroduction to safe natural sites. Data on population density, range, regeneration levels and threats should be collected to give a baseline for monitoring and to inform needs for management intervention.

Leptoderris ledermannii Harms

EN B1+B2ab(iii)

Range: Cameroon endemic (NW (Bamenda Highlands): Muti Mt, Bamenda; Sabga Range; 10 miles W of Nkambe).

Woody climber to 10 m high, petioles and rachis with soft brown pilose hairs 3–4 mm long, leaflets 2–3 pairs, oblong-elliptic, c. 12 × 5 cm, lateral nerves 10–12 pairs, acumen abrupt, base rounded, stipels 3 mm; stipules persistent, conspicuous; inflorescence yellow-pilose, 30 cm, paniculate, bracts ovate 10–13 × 8 mm, flowers sessile, white, 6 mm (E.G. Baker, Leg. Trop. Afric. 2 (1929)).

Here *Leptoderris ledermannii* is assessed as Endangered since three sites (AOO 12 km² with 4 km² cells) are known, with threats as below. EOO is calculated as 3789 km². Our map includes a spot at Korup based on two records not verified by us and regarded as unlikely since disjunct from and of highly different habitat (low alt.) from the group of specimens that include the type. We are also doubtful, for the same reason, of specimens cited in Velayos *et al.* (2010: 125) from Rio Muni.

This is a genuinely extremely rare species. Despite fairly intensive searching at three sites: Mt Oku and the Ijim Ridge (Cheek *et al.* 2000); Bali Ngemba (Harvey *et al.* 2004) and Dom (Cheek *et al.* 2010), all within its range, the species was found at none of these.

Habitat: remnants of submontane forest; c. 1500 m alt.

Threats: forest clearance for wood and agricultural land.

Management suggestions: efforts to refind this species should focus on forest scraps at Sabga, where it was rediscovered and last seen, by Nigel Hepper in 1958. This site is readily accessible from Bamenda, on the ring road. Failing this the last site listed above should be targetted since the first requires research to understand which modern site equates with it. If refound, basic populational data (range, density, regeneration levels, threats) should be recorded to inform future managemnt intervention and monitoring. This species should be incorporated into forest restoration plantings in the Bamenda Highlands.

Leptoderris macrothyrsa Harms

EN B2ab(iii)

Range: Cameroon (C: Yaoundé) and Gabon (Mekambo-Madjingo).

A woody climber with glabrous stems and leaves, the leaflets in 3 pairs, elliptic, c. 10 × 5.5 cm, glaucous below, lateral nerves 7–8 pairs, subacuminate, stipels 3 mm, stipules not seen. The inflorescence axillary 8–15 cm, the branches densely covered in 4 mm flowers.

Here *L. macrothyrsa* is assessed as Endangered since two sites are known (AOO 8 km² with 4 km² cells) and threats are as below.

Habitat: forest, type unknown.

Threats: since the type collection was made in 1895, Yaoundé has increased to more than a million people and continues to expand, so urbanisation is a major threat, if the species has not already been lost if it is not extinct at Yaoundé already. Threats at the Gabon site are not known.

Management suggestions: efforts to refind this species could focus on the Yaoundé area. The usual subpopulational data (density, range, regeneration levels, threats) should be gathered if the plant is refound so as to inform future moniitoring and management needs.

Millettia is a large genus of many tens of species of shrub, tree and liana. The last account on an Africa wide basis was that of E.G. Baker in Leguminosae of Tropical Africa (3 vols, 1926, 1929, 1930, Ghent).

In the 1980's Tobias Fonki Mbenkum at the Univ. Reading worked towards the revision of this genus, but no formal account seems to have been published. Amongst the papilionoids this genus can be recognised by being woody, pinnately leaved, with dehiscent pods, by the large flowers (often 2 cm long) with uniform stamens united in a sheath.

Millettia conraui Harms

EN A3c

Range: Nigeria (Mambilla) and Cameroon (SW: Fontem-Bangwa. NW: Bamenda Highlands).

Small tree or shrub 3–4 m tall. Leaves with c. 6 pairs of glabrous leaflets, each to 5–10 × 3–4 cm, stipels 2 mm. Flowers in terminal panicles, flowers pale purple, calyx and standard yellow pubescent; calyx broadly cup-shaped, lobes orbicular.

Restricted to the Cameroon Highlands (and possibly outliers) especially the Bamiléké Plateau and Bamenda Highlands, being unknown from Kupe-Bakossi and further south. EOO is calculated as 16936 km². This section has been the greatest loss of the original forest habitat. About 96% has been lost (Cheek *et al.* 2000: 6) and this loss is ongoing, forest cover in the Dom area being reduced by c. 50% in 15 years 1988–2003 (Baena in Cheek *et al.* 2010: back cover). Here it is estimated that the duration of one generation is about 10 years for *M. conraui*, and that in the last 30 years over the whole of its range 50%, or more, of the population of this species has been lost and is ongoing. Therefore *M. conraui* is here assessed as Endangered under criterion A.

Habitat: forest edge of upper submontane and montane forest; 1500–2200 m alt.

Management suggestions: within its range *M. conraui* appears to be an infrequent species. Despite fairly intensive surveys and specimen-searches at Dom and Bali Ngemba (within its range) it was not found (Cheek *et al.* 2010 and Harvey *et al.* 2004 respectively). At Kilum-Ijim, three specimens are known on and near Mt Oku (Cheek *et al.* 2000: 143) therefore since this area receives protection from the local communities, this is probably the best site on which to focus conservation efforts. Further data on the density, range, regeneration levels and threats to this subpopulation would assist assessment of the needs for management intervention. Seed should be collected for and seedlings raised in cooporation in the forst restoration projects in the Bamenda Highlands.

Millettia coruscans Dunn

EN B1+B2ab(iii)

Range: Cameroon (S: Grand Batanga; Kribi-Ébolowa 10 km) and Equatorial Guinea (Rio Muni).

Liana to 8 m or more, smooth pale brown stem, twigs shortly velvety hairy. Leaflets 3 pairs, each elliptic, 6 × 4 cm, acumen 1 cm, rounded below, lateral nerves 8–10 pairs, drying chestnut brown above, velvety and shining silver-copper below. Panicle loose, 30 cm; calyx bracts pubescent, corolla white 1 cm (from *Bos* 5178, Kribi).

Here *Millettia coruscans* is assessed as Endangered since only three sites are known (AOO 12 km² with 4 km² cells) with threats as below. EOO is calculated as 1089 km².

First collected in Rio Muni by Mann in 1861, later by Dinklage at Batanga, a gap of over 70 years intervened until Bos rediscovered it (det. by Mbenkum) nearby the Dinklage site.

Habitat: lowland evergreen coastal forest.

Threats: while no threats are known at the Rio Muni sites, forest habitat in the Kribi area is threatened by urban expansion; infrastructure expansion for industry (transport, storage and export from ports) and tourist development along the coast.

Management suggestions: this species should be tracked down in the wild using data from the most recent collection, so that seed may be collected for banking and growing in a botanic garden for multiplication and reintroduction into the wild. Data on density, range, threats and natural regeneration should be gathered to inform future management needs and traditional elders provided with the means to identify this species so as to be enabled to protect it against extinction.

Millettia hypolampra Harms

VU B2ab(iii)

Range: Nigeria (Calabar; Oban), Cameroon (SW: Korup; Mamfe. Littoral: Douala-Edéa. S: Kribi), Equatorial Guinea (Rio Muni: Bata) and Gabon (Loango; Mts Doudou; Moanda).

A liana, distinctive for the silver lower surface of the leaflets due to appressed silky hairs; leaflets 3–4 pairs, oblong-oblanceolate, long acuminate 9–15 × 3-6 cm; panicle terminal, tomentose, pedicel 3 mm, calyx lobes keeled in bud; flowers white.

Here *M. hypolampra* is assessed as Vulnerable since nine sites (AOO 36 km² with 4 km² cells) are known, with threats as below. EOO is calculated as 248004 km².

Habitat: lowland evergreen forest

Threats: urban expansion and new infrastructure for extractive industries at Calabar and Kribi; forest clearance for logging, followed by agriculture at Oban and Mamfe.

Management suggestions: fortunately, at least three of the known sites are in or near protected areas, these being Korup, Douala-Edéa, Loango and Mts Doudou. The usual populational data should be gathered to provide data for future management interventions. Lianas can be very difficult to census but the distinctive silvery leaflets should assist detection of *M. hypolampra.*

Millettia macrophylla Benth.

VU A3c

Range: SE Nigeria, Equatorial Guinea (Bioko) and Cameroon (SW: Mt Cameroon along coast from Mabeta to Onge (numerous coll.); Ekondo Titi (several coll.); Besingi; near Korup; Mt Kupe (several coll.). S: Batanga; Djoum; Bipindi).

This is a sparingly branched tree or shrub 1.5–10 m tall with large pink-purple and white flowers.

The assessment VU A1c, B1, 2c was made by WCMC (1997) on the basis that it is threatened by heavy logging and clearance for agriculture (www.redlist.org (IUCN 2003)). In Cheek *et al.* (2004: 164) it was stated that "only eight collections are held at K from Cameroon and the species occurs only in threatened lowland forest, thus the assessment seems justified and is maintained here". We now have 51 specimen records for this taxon and field data suggests that it is a forest pioneer and benefits from disturbance in forest. However forest habitat within its range is under great threat, or has already been lost, in over 30% of its range and for this reason it is hesitatingly concluded that within the last 100 years, it has suffered a drop in population of this percentage and that this is ongoing, allowing the updated (IUCN 2001) assessment above. Further analysis may well downrate this assessment. EOO is calculated as 63638 km².

Habitat: lowland evergreen forest; c. 400 m alt.

Threats: see above.

Management suggestions: In Cheek *et al.* (2004: 164) it was stated that "the best hope for the survival of this species is in well-protected lowland evergreen forest reserves, such as Korup. Future plans to better protect the Mungo R FR should also help secure the survival of this tree". However, as a pioneer, this species while needing forest, also requires disturbance, so would become rarer if forest was completely protected!

Millettia pilosa Hutch & Dalz.

EN B2ab(iii)

Range: SE Nigeria (Calabar-Mamfe Rd; Umu Afia) and Cameroon (SW: Bimbia)

Climbing shrub to 3 m or more, young stems dark brown velvety, leaflets 3 pairs, elliptic, to 10 × 5 cm, acumen slender, abrupt, 2 cm, lateral nerves 5–7 pairs; panicle 20 cm at spikes 10 cm, flowers pale purple, 1 cm.

Here assessed as Endangered since less than five locations are known (AOO 40 km² using 4 km² cells) with threats as below. EOO is calculated as 13310 km². Only three sites are known beyond dispute: those along the Calabar-Mamfe Rd at miles 56–57, mile 66, and miles 72–73. A fourth Nigerian specimen, at Umu Afia (?Omauhia) has not been located with certainty. Seven species from Bimbia in the foothills of Mt Cameroon, are plot vouchers and their identity requires confirmation.

Habitat: lowland evergreen, coastal forest.

Threats: Nigeria has seen massive forest losses for timber and agricultural land in recent decades and the species may no longer survive there. In Cameroon the Bimbia-Bonadikombo forest has seen great losses since the inventory in which this species was found in 1992 and it may no longer survive there either. Recent searches for other rare species discovered at Bimbia at that time have proved fruitless.

Management suggestions: attempts to refind this species should focus on the Calabar-Mamfe Rd and on the Bimbia area. Protected areas exist in both areas although incursions and reductions of habitat quality and extent are believed to have taken place. Limbe BG could propagate this species for multiplication and reintroduction in safe sites if it is rediscovered. Data on natural regeneration levels, density and range would aid future monitoring and enable management needs to be assessed. Community leaders should be enabled to identify this species so as to protect it.

Ormocarpum klainei Tisserant (syn. *Ormocarpum* sp. A of FWTA 1: 577 (1958))

EN B2ab(iii)

Range: Cameroon (SW: Mt Cameroon-Buea. S: Campo-Ma'an), Gabon (Libreville-Sibange) and Congo (Brazzaville) (Chaillu-Placongo).

Small shrub, stems angled, black, with patent short, bristles, scabrid; leaves c. 15 cm, rachis and stems, leaflets 9–15, narrow elliptic to 6 × 1.8 cm acute, glabrous, thinly papery, crenate, petiolule 1 mm, stipels absent; racemes nearly as long as leaves, axillary, internally 5–8 mm, flowers 10 cm, standard 7 mm, blade orbicular, claw 2 mm. Lomentum with articles 1–3, 15 × 6 mm, striate, scabrid.

Here *Ormocarpum klainei* is assessed as Endangered since only four sites are known (AOO 16 km² with 4 km² cells) and since threats are as below. EOO is calculated as 84505 km². The Campo-Ma'an record is based on the evidence of Tchoutou (2004 op.cit.) and has not been verified by us. The species was first collected in Libreville but has not be recorded in Gabon since 1881 (Soyaux) despite many surveys by legume specialists from Wageningen in recent decades. The record from Mt Cameroon dates from Jan 1931. Despite numerous surveys in the 1990's, no new records come to light. This appears to be an extremely rare although relatively widespread species.

The species was originally assessed as CR A1c in Cable & Cheek (1998: lv–lvi), when only three specimen records were known, the latest being from 1931. The Maitland specimen from Mt Cameroon appears to have first been linked with the Gabonese material by Jan Gillett.

Habitat: lowland forest; 0–500 m alt.

Threats: all known sites are threatened or already lost for *Ormocarpum klainei*, except that at Campo, which remains to be confirmed. The Limbe-Buea route has seen great loss of habitat to plantations of tropical crops, while Libreville has expanded it footprint greatly since 1881. At Placongo much destruction of forest habitat by ill-managed logging operations is reported (van der Burgt pers. comm. 2009).

Management suggestions: the Campo record is from a National Park, and should be verified and an attempt made to refind this species at that site and also at Mt Cameroon, although it may not survive there. Probably the best chance of refinding plants is at Placongo since the record is very recent and locality data are good. The basic populational data should be gathered as a baseline for future montioring and assessment of management needs. Local leaders and Park Managers should be helped to identify and protect the plant against extinction. Seed banking and cultivation in Botanic Gardens are advised.

Ostryocarpus zenkerianus (Harms) Dunn

CR B2ab(iii)

Range: Cameroon ("Sanaga").

A glabrous shrub. Leaflets in 3–5 pairs, oblong to obovate-oblong, apex shortly acuminate, subcoriaceous, glabrous.

Panicles very ample, branches glabrous or puberulous (Baker 1929 loc. cit.: 546).

Here *O. zenkerianus* is assessed as Critically Endangered since it is known only from a single site, last seen 100 years ago. (AOO 4 km² using 4 km² cells, threats as below).

The site is given as "Sanaga" but since this runs many hundreds of kilometres through several habitats, the exact site is not known. We have mapped it at the mouth, but equally it could be that Zenker collected it inland near Yaoundé, where he made other collections. Evaluation of his number sequence might yield an itinerary that would provide a solution.

This species was proposed by Harms, later transferred to the current genus by Dunn (Kew Bull.: 362–364 (1911)) who revised it there. Baker also accepted this taxon (loc. cit.).

Habitat: lowland evergreen riverine forest (deduced)

Threats: logging, hydro-electric schemes.

Management suggestions: research on Zenker's itinerary is needed to pin down his site for this species so that rediscovery can be attempted. If found, local leaders should be informed and given a photographic guide so as to identify and protect it. Basic populational data should be sought and recorded (density, regeneration levels, threats, range) together with details of the ecology.

Platysepalum scaberulum Harms

CR B2ab(iii)

Range: Cameroon endemic (S: Ébolowa-Sangmélima).

This woody climber is distinguished by the leaflets shortly puberulous and subscabrid to the touch beneath, from the otherwise very similar *P. pulchrum*.

J. B. Gillett in his key and notes to the genus (Kew Bull. 14: 464–467 (1960)) remarks: "no specimen has been seen: it is difficult from the description only to assess the importance in the indumentum of the leaves which alone seems to separate this species from *P. pulchrum*."

Here *Platysepalum scaberulum* is assessed as Critically Endangered since only a single collection is known (AOO 4 km² with 4 km² cells) and threats are as below.

Habitat: lowland evergreen forest (deduced)

Threats: slash and burn agriculture.

Management suggestions: a survey to target refinding this species is needed. If refound the usual basic populational data should be gathered to aid future monitoring and assess needs for management intervention. Authorities in the Ébolowa-Sangmélima area should be assisted to identify and protect this plant.

Pterocarpus zenkeri Harms

EN B2ab(iii)

Range: Cameroon endemic (C: Yaoundé. S: Dikop near Eséka).

Baker (Leg. Trop. Afr., loc. cit.: 542) accepted the species, "closely allied to (*P. soyauxii*) but differ in having larger leaflets and pods and fewer lateral nerves to the leaflets."

This species was provisionally accepted by Lock (1989) pending a revision. The specimens cited above were identified or confirmed at Kew by Rojo in 1969.

A tree 18–20 m tall, 5 m diam. Bark greyish white, fairly smooth, slash exudes a reddish gum, the wood is red. The leaves are held in bunches (*Bounougou* in SSRFK 2459). Leaflets 9–13, oblong, acute or shortly acuminate, lateral nerves in 5–7 pairs (Baker loc. cit.).

Here *Pterocarpus zenkeri* is assessed as Endangered since only two sites are known (listed above AOO 8 km² with 4 km² cells) and threats are as below.

Habitat: lowland forest, precise habitat unknown.

Threats: the type location of this species is probably destroyed due to urbanisation (Yaoundé, collected 1895). Threats at the Eséka site are unknown but since this species resembles a valued and well-known timber tree (*P.soyauxii*), there is every chance that trees have been felled in error. It may be that *P. zenkeri* has equally, or more valuable, timber.

Management suggestions: the most recently recorded site, at Dikop (1958) should be revisited and an attempt be made to refind this species when in flower (May). The usual basic populational data should be gathered if successful. Specimens of *P. soyauxii* should be examined to check that examples of *P. zenkeri* have not been erroneously included due to misidentification. A revision of the genus is needed to reassess the status of *P. zenkeri*.

Rhynchosia ambacensis (Hiern) K. Schum subsp. ***cameroonensis*** Verdc.

EN B1+B2ab(iii)

Range: Cameroon endemic (N: Béré, 10 km S Poli. Extreme N: Liri near Mogodé, 31 km SW Mokolo).

Probably a climber, softly and shortly grey pubescent; leaves simple and infoliolate on the same stem, entire leaves broadly ovate, c. 8 × 9 cm, apex acute-submucronate, base shallowly cordate, both surfaces with appressed soft grey hairs; petiole 3.5 cm; racemes glandular, 10–15 cm terminating axillary shoots, c. 6 per stem; pedicels 3 mm; flowers golden yellow inside, outside paler, and striped purple; calyx tube 3 mm, the basal lobe linear 1.5 cm, exceeding the corolla.

Here assessed as Endangered since only two sites (above, AOO 8 km² with 4 km² cells) are known, threats being as below. EOO is calculated as 267 km².

Habitat: disturbed areas near agriculture in wooded grassland area; 650–1000 m alt.

Threats: increased agricultural intensity at the only two sites might destroy this species.

Management suggestions: the two original sites (above) should be revisited to discover the exact habit, ecology and threats to this species. Local leaders should be provided with the means to identify this plant (e.g. a photographic guide or poster) so as to protect it. Basic populational data should be gathered (density, regeneration levels and range) at its known sites to give a baseline for monitoring and assessment of management needs.

Rhynchosia ledermannii Harms.
CR B2ab(iii)

Range: Cameroon endemic (NW: Bamenda Higlands, "Babadju").
Probably a climber. Stem densely hirsute, leaflets oval or ovate or subrhombic, softly greyish tomentose, especially below, petioles short. Racemes on short peduncles, dense; standard yellow, 12 mm; calyx 10–11 mm. Pods very hairy. (Baker Legs. Trop. Afr. 464 & 474 (1929)). Overlooked in FWTA, probably since no specimen survives. Known only from the location cited above (*Ledermann* 1843); this species was accepted by Baker (loc. cit.) but the only specimen has since been lost (unicate at Berlin) so cannot be examined. The genus needs a modern revision.
Here *Rhynchosia ledermannii* is assessed as Critically Endangered since only a single site is known (AOO 4 km² with 4 km² cells) and the threats are as below.
Habitat: submontane forest edge (deduced).
Threats: the Bamenda Highalnds have been c. 96% denuded of the natural habitat believed needed for this species (Cheek *et al.* 2000: 6). Inventories recently over several seasons at four sites along this range Oku-Ijim, Bali Ngemba, Dom and Lebialem) have failed to rediscover it. Clearance of surviving forest is ongoing with 50% lost in one areea over 15 years (Baena in Cheek *et al.* 2010: back cover), mainly for firewood and fertile land for agriculture.
Management suggestions: rediscovery of this species is the first priority in conserving it, if it is not already extinct. If refound, muliplication should be used to provide plants for inclusion in the existing reforestation projects in the Bamenda Highlands. The usual populational data should be gathered for monitoring and management purposes.

The genus *Trifolium* 'Clover' is well-known in Europe and elsewhere in the N Temperate zone for its agricultural importance in enriching pastures, providing protein-rich fodder for livestock and being sown in fields for this purpose. Several species occur at high altitudes in Africa including three in the Cameroon Highlands all of which occur in the E African Mts. However, Cameroon hosts one very rare, species, unique to the Adamawa Highlands.

Trifolium gillettianum Jac.-Fél.
CR B2ab(iii)

Range: Cameroon endemic (Adamawa: Ngaoundéré à Bélel, 40 km).
Annual herb, glabrous, 20–30 cm. Stems erect, simple or slightly branched. Petioles to 6 cm; stipules to oblong, 2 cm, adnate to petioles for half their length, free part triangular-lanceolate, leaflets narrowly oblanceolate, to 1.7 × 0.5 cm, nerves numerous, parallel. Inflorescences globose, 2 cm diam., 10–20 flowered, pedicels 2–3.5 cm, sparsely pilose. Calyx tube 15–16-nerved, 6 mm long, much longer than tube, with ciliate margins. Corolla purple, 1 cm long (Zohary & Heller (1984). The Genus Trifolium: 140).
Here *T. gillettianum* is assessed as Critically Endangered since only a single site is known (AOO 4 km² with 4 km² cells, threats as below).
Named for Jan Gillett who identified the material as unknown to science.
Habitat. damp depression in grassland on lateritic plateau; 1400 m alt.
Threats: overgrazing, trampling by livestock.
Management suggestions: *Trifolium gillettianum* has not been seen since its discovery in Oct. 1967 and should be refound. Seed should be banked against its extinction in the wild, and data on density, range, regeneration and threats gathered to inform decisions on needs for future management intervention. Leaders of communities managing its habitat should be provided with the means to identify this species so as to protect against its extinction.

LINACEAE (HUGONIACEAE)

assessed by Martin Cheek

Although the genus *Hugonia* has been placed in Linaceae traditionally (Flore du Cameroon 14 by F. Badré), it differs so radically from the temperate herbs such as *Radiola* and *Linum* (occurring in the montane grassland of the Cameroon Highlands) that strong molecular support for recognition of a separate Hugoniaceae was not unexpected. Hugoniacae is pantropical with six genera and 50–55 species but with only *Hugonia* (c. 30 species) in Africa and Madagascar (Brummitt in Heywood *et al.* (2007), Flowering Plant Families of the World). *Hugonia*, in common with the other Old World genera, are usually lianas with distinctive opposite or subopposite coiled hooks on the climbing stems. The stems produce no distinct scent or coloured exudate. The leaves are alternate, simple papery, often toothed, pinnately nerved, short petioled with dentate to laciniate stipules. The panicles, often axillary, bear showy flowers 1–2 cm across, with five free sepals and petals, and 10–15 stamens, filaments united at the base; the ovary is superior, 3–5-loculed, producing a fleshy drupe.

In the absence of flowers or fruit, *Hugonia* are sometimes confused with the rarer *Ancistrocladus* which also are lianas with coiled hooks. However, the last genus has entire leathery leaves and hooks on only one side of the stem. No important uses are known for *Hugonia*.

Hugonia micans Engl. was assessed as VU A2c, B2ab(iii) in Cheek *et al.* 2004: 164–165) but since then the number of locations and locations in Gabon known to us has increased, taking the total to beyond the VU B threshold, (ten), hence we now downrate this species to NT status.

Hugonia spicata Oliv. var. *spicata* (Cameroon and Bioko) should it be assessed, would probably rate EN due to the low number of locations, but infra-specific taxa are only included in this book in certain circumstances (see introductory chapters).

Hugonia batesii De Wildem.

VU B2ab(iii)

Range: Cameroon (S: Bitye), and Gabon (Ikobey-Evouta, 12 km; Rabi; Reserve Mts Doudou; Mouila-Yeno) and Congo (Brazzaville) (N: Souanke-Garabinzam).

Here *Hugonia batesii* is assessed as Endangered since six locations are known (above; AOO 24 km² with 4 km² cells) with threats as below. EOO is calculated as 51530 km².

Liana, densely short hairy on stems, with paired woody tendril-hooks; leaves with petioles 5–10 mm, stipules 4 mm, laciniate, tomentose on outer surface; blade elliptic 4–15 × 1.7–7 cm acumen short, base rounded, margin dentate, lateral nerves c. 16 pairs, ascending, conspicuous, lower surface yellow tomentellous; midrib hairy or not above; inflorescence tomentellous, of few-flowered cymes arranged in racemes grouped in terminal panicles; bracts as stipules; bracteoles palmatifid, subulate; sepals tomentellous, ovate, 4–6 × 4 mm; petals white, glabrous, 10 × 4 mm.

Habitat: swampy areas 400–600 m alt.

Threats: logging of swamp forest and adjoining semi-deciduous forest areas.

Management suggestions: revisiting Bitye to rediscover this and the many other rare and threatened species discovered at this location by the missionary Bates. Basic populational data should be collected for the species as a baseline for monitoring. Propagation of this species to multiply it and augment the wild population is suggested if it is confirmed that the species is as rare as it appears to be. The densest population appears to be in Southern Gabon, with one site in a protected area. Global conservation efforts should logically be focussed there.

Hugonia gabunensis Engl.

EN B2ab(iii)

Range: Cameroon (C: Sanaga) and Gabon (Libreville, Sibange Farm; Lopé Reserve; Doussala, 20 km SSE; Gamba).

Here *Hugonia gabunensis* is assessed as Endangered since only five locations are known (above; AOO 20 km² with 4 km² cells) with threats as below. EOO is calculated as 69258 km².

Liana 8 m, stem appressed hairy, finely striate, woody hooks opposite; leaf petioles 4–9 mm, stipules laciniate, petiole, 8–9 mm; blade oblong-obovate to elliptic, 6–15 × 3–5 cm, acuminate, base cuneate-acute, secondary nerves 8–10 pairs, arced and ascending, pubescent below; tertiary nerves scalariform; cymes axillary, 1–3-flowered, peduncle 2.5 cm, pedicels 1–2 mm; bracts and bracteoles as stipules; sepals tomentellous, ovate-lanceolate, 7–10 × 3–4 mm; petals white, glabrous, obovate 8 × 1.5 mm.

Habitat: lowland evergreen forest.

Threats: urbanisation (the plants at Libreville sites are probably extinct).

Management suggestions: rediscovery of plants at Zenker's Sanaga locality should be attempted so that basic populational data can be collected as a baseline for monitoring and so that protection of such plants can be attempted. The location at Lopé NP may be the most logical site for conservation investment.

Hugonia macrophylla Oliv.

VU A2c, B2ab(iii)

Range: S Nigeria (Omo FR; Degema; Oban; Old Calabar), Cameroon (SW: Mt Cameroon, Mokoko FR (2 coll.); W Bakossi (1 coll.). S: Ébolowa-Minkok) and Gabon (Cristal Mts; Tchibanga; Bélinga; Ngounie, chantier Leroy).

This liana, with nine known locations (above; AOO 36 km² with 4 km² cells) with threats as below, was assessed as Vulnerable in Cheek *et al.* (2004: 164–165) and this is maintained here since no additional data are available. EOO is calculated as 39764 km².

Hugonia macrophylla has a patchy distribution. This is possibly due in part to under-collecting, but even if some more sites are discovered, it is likely that, overall, its area of occupancy will be reduced by over 30% in the next 50 years. Range data are taken from FWTA, Flore du Cameroun, Fl. Gabon and the Gabon database.

Woody climber to 10 m; stems orange-brown scurfy, climbing by opposite woody hooks, usually on short spur-shoots; leaves papery, alternate on main axis or clustered at ends of spur shoots, oblanceolate, c. 20 × 7 cm, obtuse, cuneate, entire or subserrate, lateral nerves c. 16 pairs, densely orange-brown pubescent; petiole 0–5 mm; stipules 1–2 cm, deeply pectinate-laciniate; inflorescences few-flowered, subsessile; flowers yellow, 3–4 cm, petals 5, free, obovate 2 cm; stamens 10; ovary 1.

Habitat: lowland evergreen forest; to c. 650 m alt.

Threats: clearance of forest for agriculture (Mt Cameroon, W Bakossi), and logging. Nigeria has seen massive losses of its original forest cover), urban expansion (Oban, Calabar) or a combination of these. Bélinga is scheduled as an open-cast iron ore mine. Mokoko FR is being explored for oil and has ben mooted for logging.

Management suggestions: a survey of known sites is needed to find where the taxon is extant, and to assess the best means to conserve it. Basic populational data is needed as a baseline for monitoring.

LOGANIACEAE INCLUDING GENTIANACEAE

assessed by Martin Cheek

The tubular flowers with regular symmetry united a group of genera with opposite, simple leaves which lack an interpetiolar stipule and coloured exudates. Molecular evidence however has drawn attention to the morphological differences between the elements of traditional Loganiaceae. *Mostuea* is now placed in Gelsimiaceae (with *Gelsimium*), whereas *Anthocleista* should now placed in a redelimited Gentianaceae (but is maintained here to be consistent with former treatments). *Strychnos,* however, the largest genus of traditional Loganiaceae, remains, after the family has been divided, in the now more narrowly defined Loganiaceae.

Loganiaceae, is now newly restricted to just 15 genera. Although no Flore du Cameroun account exists, a revision of *Strychnos* for Tropical Africa by Leeuwenberg (The Loganiaceae of Africa V111. *Strychnos* III. Meded. Landbou. Wageningen 69-1(1969) and Belmontia 10: 1–316) means that species are well delimited. *Strychnos* species have three main nerves at the base of their entire leaf blades, so resemble closely, when sterile, *Warneckea* in the Melastomataceae. While a few Cameroonian *Strychnos* are trees, most are lianas with tendrils. The flowers are small, numerous in axilliary cymes, 4 or 5- merous, the calyx is five-lobed, while the corolla, tubular at the base, is also lobed, with the five stamens inserted upon it. The superior ovary is usually two locular and berry-like, up to 15 cm diam. with 2-many flat seeds which are toxic and used to poison animals, such as rats (strychnine).

Anthocleista was revised by the same author (Leeuwenberg, A.J.M. (1961) The *Loganiaceae* of Africa: I. *Anthocleista.* Acta Bot. Neerl. 10: 1–53). This is a genus of pioneer trees, some restricted to wet areas, with a few climbing and epiphytic species. They share large opposite, leathery, glabrous, obovate leaves with pinnate nerves and have large white flowers with numerous free, overlapping petals at the apex of the corolla tube.

Anthocleista obanensis Wernham (Nigeria, Cameroon and Congo (Kinshasa)) is considered Near Threatened here but might be reassessed in future as Vulnerable.

Strychnos ternata Gilg ex Leeuwenb. of Cameroon and Gabon, and *S. tricalysioides* Hutch. & M.B.Moss (Nigeria, Cameroon, Equatorial Guinea (Bioko), Gabon and Congo (Brazzaville)) both with a total of 11 locations are here assessed as NT, as is *S. ngouniensis* Pellegr. (Cameroon, Gabon and both Congos) with 14 locations. In future editions they may prove to rate as threatened when better data are available.

Anthocleista microphylla Wernham

VU A3c, B2ab(iii)

Range: Ghana (Atewa Hills (1 coll.)), SE Nigeria (2 coll.), Equatorial Guinea (Bioko (2 coll.)), São Tomé & Príncipe (São Tomé (7 coll. at 4 sites). Príncipe (1 coll.)), Cameroon (SW: Kupe-Bakossi (6 coll. at 2 sites)).

This strangling shrub of lower submontane forest has its stronghold in São Tomé with four sites and many more collections, the most recent in 1993. Range data are partly taken from the revision by Leeuwenberg (1961 loc. cit.). Loss of its habitat of over 30% in the next 50 years is predicted. This species was assessed as Vulnerable in Cheek *et al.* (2004: 165) and this is maintained here since there are ten or less locations (above, AOO 40 km² with 4 km² cells) and threats as below. EOO is calculated as 335039 km².

Limited habitat disturbance appears beneficial for *A. microphylla*; several collections occur in areas of regenerating (secondary) forest.

Epiphytic shrub, becoming scandent and then a strangler tree; leaves narrowly elliptic to oblanceolate, c. 17 × 6.5 cm, acuminate, acute, nerves c. 5 pairs; petiole 2 cm; inflorescence c. 25-flowered; flowers white or bluish white, outer sepals c. 0.4 × 0.4 cm.

Habitat: disturbed submontane forest;, 800–1200 m alt.

Threats: forest clearance for agriculture and wood; it may already have been lost from both Nigerian sites (Oban and Obudu) since significant forest loss has occurred there in recent decades. Although it is fairly common at Kupe village, it occurs there at such low altitudes (c. 800 m) that it is vulnerable to agricultural expansion from the valley bottom.

Management suggestions: more data are needed on the survival of this species at its various sites and the extent to which it is already protected. São Tomé should be the priority for a survey.

Anthocleista scandens Hook.f.

VU A2c

Range: Equatorial Guinea (Bioko: Clarence Peak, Moca), São Tomé & Príncipe (São Tomé (5 sites)), SE Nigeria (Obudu (1 site)), Cameroon (SW: Lebialem-Fosimondi; Mt Kupe (9 coll. at 3 sites); Mt Etinde (3 coll.). NW: Gepka, Nkambe (1 coll.); Mt Oku (2 sites); Bafut Ngemba (2 coll.)).

The assessment here is maintained from those in Cheek *et al.* (2004: 165) and Harvey *et al.* (2010: 78–79). It is estimated here that, over the last 100 years, over 30% of the Cameroon Highland submontane forest habitat of this species, mostly in the Bamenda Highlands and Bamboutos, has been lost due to forest clearance (see threats below). EOO is calculated as 511119 km^2.

In flower this strangling epiphytic shrub or small tree is spectacular with its clusters of large bright white flowers.

Epiphytic climbing shrub 7–17 m; branchlets square; leaves oblong-elliptic, 6–20 × 2.5–11 cm; flowers white; corolla tube c. 3 cm long.

Habitat: submontane forest; 1200–2000 m alt.

Threats: forest clearance for agriculture and wood, particularly in the Bamenda Highlands, where forest loss has been running at 25% over eight years at one sample area (Moat in Cheek *et al.* 2000: back cover), and at 50% in 15 years in another (Baena in Cheek *et al.* 2010: back cover). Forest losses in recent decades suggest that *A. scandens* may no longer survive at Bafut Ngemba, Gepka or Obudu. Submontane/Montane forest at Bafut Ngemba for example has been completely replaced by *Eucalyptus (*pers. obs. Cheek 2004) and clearance is ongoing at Fosimondi (Tchiengué pers. comm.. 2005-6).

Management suggestions: conservation efforts are probably best focussed where sites are most concentrated, namely São Tomé (5 sites, but protection levels unknown) and Mt Kupe (3 sites, protection levels currently high). At Mt Oku this taxon occurs at the lower altitudinal boundary of protection and is likely to become extinct; one of the two sites there occurring outside this boundary. The population at Mt Etinde occurs at the peak and is inaccessible to all but trekkers. Unless the summit area is cleared for touristic purposes, *A. scandens* seems secure here.

Monitoring of this species is best performed when in flower when plants are easily found due to the conspicuous fallen flowers below them. This species should be introduced to cultivation since it is spectacular in flower.

Strychnos canthioides Leeuwenb.

EN B2ab(iii)

Range: Cameroon (S: Edéa- Kribi, km 58) and Gabon (Doudou Mts).

Here *Strychnos canthioides* is assessed as Endangered since known from two locations (AOO 8 km^2 with 4 km^2 cells) with threats as below.

Liana, leafy stems glabrous, terete, lacking lenticels, tendrils in 1 pair; leaves opposite, thinly leathery, narrowly ovate to narrowly elliptic, 3-11 × 0.8-3.7 cm, acuminate, base cuneate-rounded, lateral nerves arising near the base and shorter than the leaves, axillary or terminal, few-flowered, 1 × 3 cm; glabrous; flowers 4-merous, sepals suborbicular 1mm, rounded; corolla in bud 5–7 mm, white.

Habitat: evergreen lowland forest for agriculture

Management suggestions: rediscovery of the species at the type locality in Cameroon should be attempted so that basic populational data can be collected as a baseline for monitoring, and so that local community leaders can be alerted to protect the plants. If possible mature fruit should be collected and used to propagate the species for introduction to forest in protected areas within its range so as to augment wild populations, and for botanic gardens.

Strychnos chromatoxylon Leewenb.

EN B2ab(iii)

Range: Cameroon (E: Bertoua- Doumé, km 28, near Dimako) and CAR (SW: Boukoko, N of Bangui).

Here *Strychnos chromatoxylon* is assessed as Endangered since known from two locations (above; AOO 8 km^2 with 4 km^2 cells) with threats as below.

Large liana, trunk 3–5 cm diam. or more, spines absent, leafy stems glabrous, tetrete, lacking lenticels, tendrils in 1–3 pairs; leaves opposite, leathery or papery, elliptic, 5–15 × 2.5–6 cm, acuminate, base cuneate to rounded, glabrous on both sides, basal nerve pair not enlarged as usual in genus, secondary nerves 4–6 pairs; tertiary nerves reticulate, prominent below; inflorescence axillary, congested, minutely pubescent; flowers 5-merous, sepals subulate, glabrous; corollas glabrous; fruit large, depressed globose, 6 × 6 × 7 cm, subcordate at base; thick-walled.

Habitat: semi-deciduous forest; 400–650 m alt.

Threats: logging, followed by agriculture.

Management suggestions: rediscovery of the species in Cameroon should be attempted so that basic populational data can be collected as a baseline for monitoring, and so that local community leaders can be alerted to protect the plants. If possible mature fruit should be collected and used to propagate the species for introduction to forest in protected areas within its range so as to augment wild populations, and for botanic gardens.

Strychnos elaeocarpa Gilg ex Leeuwenb.

VU B1+2ab(iii)

Range: Cameroon endemic (SW: Kumba-Mamfe mile 15; ; Likomba-Tiko; Mt Cameroon (Mabeta-Moliwe and Onge). Littoral: Edéa-Kribi 58–65 km); Loum-Kumba at Mungo Bridge. S: Kribi at Kienké R, numerous coll.).

This species was assessed as Vulnerable in Cheek *et al.* (2004: 165–166) and this is maintained here since there are six locations (above, AOO 24 km^2 with 4 km^2 cells) and threats as below. EOO is calculated as 12582 km^2, allowing B1 status.

Tree 6 m; stems terete; leaves thickly leathery, glossy above, elliptic, 11.5 × 6 cm, acumen broad and short, 0.7 cm; cuneate, palmately 3-nerved, marginal nerve, tertiaries open-scalariform; petiole 0.7 cm; inflorescence axillary,

congested, 1–1.5 cm; flowers 4.5 mm, white; fruit obliquely ellipsoid c. 1.3 × 2 cm, dark brown, soft.
Habitat: lowland evergreen forest; 50–200 m alt.
Threats: clearance of forest for agriculture, both small-holder (Kumba-Mamfe) and large-scale plantation (Likomba-Tiko, where forest is now absent, and both Onge and Mabeta-Moliwe, scheduled for plantation expansion); also building sand extraction at Mungo Bridge. The status of the Kienké R location at Kribi is not known.
Management suggestions: more intensive surveys in the Mungo R and Bakossi FRs might locate subpopulations which, if those boundaries are respected, are reasonably secure. This aside, no known site for the species appears to be within an officially protected area.
Basic populational data should be collected as a baseline for future monitoring.

Strychnos gnetifolia Onochie & Hepper
VU B1+2ab(iii)
Range: Nigeria (Oban) and Cameroon (SW: Lake Barombi Koto; Baro-Korup. S: Kribi-Lolodorf, 20 km; Mt Elephant; Bipindi).
Here *Strychnos gnetifolia* is assessed as Vulnerable since six locations (above; AOO 24 km ^{2} with 4 km^2 cells) are known, with threats as below. EOO is calculated as 17542 km^2.
Tree 10–20 m, leafy stems glabrous, terete, tendrils not seen; leaves opposite, leathery, blades matt, elliptic or narrowly elliptic 6–13 × 2–6.5 cm, acuminate, base cuneate, decurrent, glabrous, basal nerve pair curving along margin, tertiary nerves obscure; inflorescence axillary, few-flowered, shorter that the leaves, few-flowered; peduncle 5–7 mm, branches hairy; flowers 5-merous, sepals ovate 1.5–2 mm; corollas 3.5–4.5 mm, glabrous; fruit unknown.
Habitat: Lowland evergreen forest.
Threats: logging followed by agriculture.
Management suggestions: rediscovery of the species in Cameroon should be attempted so that basic populational data can be collected as a baseline for monitoring, and local community leaders should be alerted to protect the plants.

Strychnos mimfiensis Gilg ex Leeuwenb.
VU B1+2ab(iii)
Range: Cameroon (Littoral: Douala-Edéa, 24 km; ; Kopongo- Masok 17 km; Edéa-Kribi km 58. S: Mboké, 60 km S Edéa; Bipindi; Campo-Ma'an (Tchouto 2004)) and Gabon (Ngounié, Mouila-Yeno)
Here *Strychnos mimfiensis* is assessed as Vulnerable since seven locations (above; A00 28 km^2 with 4 km^2 cells) are known, with threats as below. EOO is calculated as 11103 km^2.
Large liana, trunk 4–8 cm diam. Bark pale brown, with some lenticels; leafy stems glabrous, terete, lacking lenticels; leaves opposite, papery to thinly leathery, elliptic, 5–11 × 2–5 cm, acuminate or caudate, base cuneate to rounded, glabrous, basal nerves curving along the margin, compressed above; tertiary nerves reticulate, prominent; inflorescences axillary, congested, 1–1.5 cm, flowers 5-merous, sessile, sepals ovate 1.2–1.4 mm, glabrous; corollas 2.5–3 mm, obtuse, pale yellow; fruit ellipsoid 23 × 15–25 × 17 mm, 1-seeded.
Habitat: lowland evergreen forest; 0–100 m alt.
Threats: logging followed by agriculture
Management Suggestions: rediscovery of the species in Cameroon should be attempted so that basic populational data can be collected as a baseline for monitoring, and local community leaders should be alerted to protect the plants.

Strychnos staudtii Gilg
VU A3c
Range: Cameroon (SW: Mt Cameroon (7 coll. at 3 sites); Kumba area (3 coll.); Kupe-Bakossi (21 coll. at 8 sites) and Gabon (7 loc.).
This lowland to submontane tree is unusual in a predominantly liana genus. Although the most commonly collected and widespread *Strychnos* in Kupe-Bakossi, it is vulnerable there to forest clearance because trees at the lower part of its altitudinal range adjoin farmland. It is also very heavily threatened in the lowlands around Mt Cameroon (Cable & Cheek 1998: xxxiv–xxxvi), as it is in the Kumba area. Although threats in the Gabon area are unknown, even assuming it is secure here, an ongoing overall population reduction due to loss of habitat for *S. staudtii* of over 30% in the next 100 years can be confidently predicted, hence the assessment of Vulnerable, maintained from that in Cheek *et al.* (2004: 166). EOO is calculated as 170399 km^2.
Tree or shrub 2–20 m; stems whitish brown, terete; leaves thinly leathery, pale green above, elliptic, c. 12 × 5 cm, acumen to 1 cm, obtuse, strongly palmately 3-nerved, scalariform; petiole 0.3 cm; inflorescence axillary, subsessile, 1 cm; flowers white; 3 mm; fruit ellipsoid or globose, pale brown, 3.5 cm
Habitat: evergreen forest; 200–1200 m alt.
Threats: clearance of lowland forest for commercial plantation expansion (Mt Cameroon area) and small-holder agriculture (elsewhere). Threats in Gabon are not known.
Management suggestions: *Strychnos staudtii* is currently probably most secure in Gabon. There is a good possibility that the several sites in the Cristal Mts correspond with one or both National Parks there. This might be the best focus for ptotecting the species. It is fairly secure at Mt Kupe on the slopes above the level of cultivation; such is the density of individuals there that currently no action is needed to ensure its survival.

Strychnos zenkeri Bak.
VU B2ab(iii)

Range: Cameroon (S: Bipindi), Gabon (Waka; Mouila-Yeno; Rabi; Cristal Mts at Tchimbele; Rabi-Kounga) and Congo (Kinshasa) (Ndolere in Kabare; Kivu, Kauumu-Walikale km 48).

Here *Strychnos zenkeri* is assessed as Vulnerable since only seven locations are known (above; AOO 28 km^2 with 4km^2 cells) with threats as below. EOO is calculated as 1136070 km^2.

Small tree to 15 m; leafy stems often with two lines of pubescence, leaves opposite, leathery, glabrous, elliptic or narrow elliptic, 4–14 × 1.5–5 cm, acuminate, base cuneate, basal nerves with two pairs extending to apex, tertiary nerves reticulate, conspicuous above; inflorescences axillary, congested, several together, c 1 cm, pubescent, flowers 5-merous, sepals suborbicular, 2 mm, corolla in bud 6.5 mm; fruit globose, 2 cm, 2-seeded, wall 1 mm.

Leeuwenberg notes the similarity with *S. staudtii* and separates as follows:

Pistil glabrous; corolla with two rings of hair inside...*S. staudtii*

Pistil pilose; corolla with one ring of hair inside...*S. zenkeri*

Habitat: lowland and submontane forest; 0–1800 m alt.

Threats: forest clearance for oil exploration (Gabon), Slash and burn agriculture (Cameroon and Gabon)

Management Suggestions: attempts to rediscover this species at the only known locality in Cameroon, at Bipindi, should be attempted, even though it has not been seen there in 100 years. If this fails conservation efforts should focus on Gabon where records are more recent and numerous and where forest is better protected by a network of National Parks. Basic population data should be collected as a baseline for monitoring.

LORANTHACEAE

assessed by Martin Cheek

Easily recognised and well-known in Cameroon as 'parasites' these are shrubs that establish in crop and wild trees and drain their hosts through an aerial haustorium. Two related families occur, the Viscaceae, with inconspicuous flowers and few species, and the Loranthaceae with match-stick like large, often showy, flowers 1–10 cm long. Mainly pollinated by birds, the flowers of Loranthaceae often open explosively, and are followed by juicy berries containing seed in a very sticky pulp, distributed by birds. In Cameroon the damage done to cultivated trees exceeds the uses of the Loranthaceae species which are rather minor. The fruit gum is used locally as an adhesive, e.g. for catching animals; the haustorium and associated branches are used by artists for sculpture in the Bamenda Highlands (pers. obs.).

While some of the species are common – mostly those destructive to crops – others are very rare and localised and threatened with extinction.

Thanks to many years of research by Mme Balle, the taxa of Cameroon are well delimited. Her work culminated in the Flore du Cameroun 23 (1982), recognising 26 species in 6 genera. Subsequently Polhill and Wiens (1998) published the sumptuous 'Mistletoes of Africa' (RBG, Kew) revising both families and raising many of Balle's subspecies to specific rank. The taxa delimited by Polhill and Wiens are the basis for this account, which should be used by those seeking to identify species since it provides identification keys and is heavily illustrated with colour photographs.

The main source of specimen data is this account is derived from records cited in the Onana manuscript, derived mainly from Balle (1982) and identifications at YA by Polhill. In some cases these have been supplemented by records at K, but K has not been exhaustively screened for this family regarding this project. Descriptions here are abbreviated from Polhill & Wiens (1998).

Many of the rarer Loranthaceae occur high in the canopy of lowland evergreen forest and so are very difficult to census by conventional means unless a way should be found to perambulate in this almost unknown domain. When surveying for Loranthaceae it is important to record the identity of the host tree – making a herbarium specimen if there is any doubt – since there is no doubt that each species of parasite has a preferred range of host species. It can be conjectured that rare parasites might be specific to only a few host species.

Agelanthus dichrous (Danser) Polhill & Wiens (syn. *Tapinanthus heteromorphous* subsp. *dichrous* (Danser) Balle)

VU B2ab(iii)

Range: Cameroon (C: Yaoundé; Makak. S: Bitye. E: 20 km NW Deng-Deng; 60 km SSW Batouri; 20 km E Bertoua; Tongo) and CAR (Dzanga-Sangha Reserve).

Parasitic epiphytic shrub to 1 m, densely shortly puberulous; leaves subopposite, petiole 3–9 mm, blade oblong to lanceolate, 4–8 × 1.5–3.5 cm, obtuse, base cuneate to rounded, glabrous, 4–8 pairs of lateral nerves. Umbels 6–12-flowered in March-Sept.; peduncle-pedicels 1–2 mm, densely puberulous. Calyx saucer-shaped, 0.5 mm, toothed. Corolla 2.5–3 cm, yellow to orange, green & abruptly swollen at base, red-brown at apex, ellipsoid-pentagonal, shortly pubescent; upper dilated part of corolla-lobes lanceolate, 3 mm long.

Here *Agelanthus dichrous* is assessed as Vulnerable since eight locations are known (above; AOO 32 km² with 4 km² cells) and threats as below. EOO is calculated as 90528 km².
Habitat: evergreen riverine forest in semi-deciduous forest bell; 350–750 m alt.
Threats: the range of this species coincides with a very actively logged zone. In principle riverine forest is less affected than elsewhere since regulations protect watercourses but if logging regulations are not implemented, as can happen, this habitat can be destroyed.
Management suggestions: this species is probably secure in the Dzanga-Sangha Reserve which may be the most logical focus for any conservation efforts. However, this species may be undercollected. Further surveys may yield more locations.

Agelanthus glaucoviridis (Engl.) Polhill & Wiens (syn. *Tapinanthus dodonaeifolius* (DC.) Danser subsp. *glaucoviridis* (Engl.) Ball.).
EN B2ab(iii)
Range: Cameroon (S: Bipindi) and Gabon (Ayem, SW de Ndjolé; Rabi-Kounga; 20 km N Koumameyong).
Parasitic epiphytic shrub, glabrous; leaves (sub-)opposite, well-spaced, elliptic to elliptic-obovate, 6–9.5 × 3–4.5 cm apex rounded, base cuneate, nerves 3–4 pairs from base, obscure, ascending. Umbels 3–4-flowered, peduncle 2 mm, pedicels in sockets, calyx tubular, 6 mm, corolla slightly swollen at base, narrowly constricted above, grey-violet, 6.5–7.8 cm, apicel swelling slight, lobes erect, 13 mm.
Here *Agelanthus glaucoviridis* is assessed as Endangered since four locations are known (above; AOO 16 km² with 4 km² cells). EOO is calculated as 55128 km².
Habitat: evergreen forest; 50–100 m alt.
Threats: slash and burn agriculture; logging of host trees.
Management suggestions: ideally this species should be rediscovered at Bipindi, censused, and protected there. However, it has not been seen in Cameroon in 100 years so it is more logical to secure protection at one of the recently discovered sites in Gabon.

Globimetula oreophila (Engl.) Tiegh.
VU A2c
Range: Nigeria (Sardauna- Gangiriwal, Chappal Waddi; Obudu) and Cameroon (SW: Mt Cameroon; Bangem; Mt Kupe. W: Bamboutos Mts. NW: Bali Ngemba; Bafut Ngemba; Kilum-Ijim; Acha. Adamawa: Mt Nganha).
Epiphytic shrub, twigs compressed. Leaves (sub-)opposite, petiole 1–2 cm, elliptic to lanceolate, 8–13 × 2.5–6 cm, acuminate, base slightly cordate to broadly cuneate, lateral nerves 6–12, well-spaced. Umbels 9–21-flowered, in Nov-March, 1–4 in axils or at nodes below; peduncle 5–35 mm; pedicels 2–4 mm; calyx 1–2 mm. Corolla 2.5–4 cm, red or red-purple, top green or cream, c. 2.5–3 mm diam., slightly 5-lobed; basal swelling 5-angled.
Here *Globimetula oreophila* is assessed as Vulnerable since >30% population reduction is estimated in the last 100 years (due to loss of host trees – see the threats, below). This is ongoing. EOO is calculated as 87530 km².
Habitat: montane forest margins and bush (then on *Hypericum*); 1200–1700 m alt.
Threats: forest loss due to timber extraction and demand for agricultural land. In one of the Bamenda Highlands, already c. 96% deforested, c. 50% of the remaining forest was lost between 1988-2003 (Baena in Cheek *et al.* 2010: back cover).
Management suggestions: this is a common species within its habitat. 55 records are known (several from sites now deforested). So long as some forest sites are protected within its range, the species is likely to survive. In Cameroon, Kilum-Ijim and Mt Cameroon are the best contenders for protected forest sites where this species has a good chance of surviving.

Helixanthera periclymenoides (Engl. & K.Krause) Balle
CR B2ab(iii)
Range: Cameroon endemic (S: Bipindi, Ilot de Mbiave).
Considered probably extinct according to Polhill & Wiens (1988, loc. cit.). Not seen since the type collection *Zenker* 3406 in Feb. 1908, despite considerable search for the species around the locality in 1984 by Loranthaceae specialists.
Epiphytic shrub, stems soon thick, to 5–10 mm diam., terete. Leaves (sub-)opposite, sessile, ovate or elliptic-oblong, 8–15 × 3.5–8cm obtuse or rounded apex, lateral nerves few, obscure. Peduncles 1-several, axillary or at old nodes, 1–1.5 cm, umbel of 12–18 flowers spirally arranged on swollen head, pedicels 1–3.5 mm; corolla in bud tapering from base; petals red, 3.2–3.8 cm.
Here assessed as Critically Endangered and possibly extinct since only a single location is known (above; AOO 4 km² with 4 km² cells) and threats as below.
Habitat: lowland evergreen forest.
Threats: slash and burn agriculture
Management suggestions: before this species is declared extinct, further searches at the type locality should be conducted.

Phragmanthera kamerunensis (Engl.) Balle
VU B2ab(iii)
Range: Nigeria (Kontagora; Oban), Cameroon (SW: Mamfe, 45 km E; Mt Cameroon–Mokoko; Meanja) and Gabon (Nkoulounga; 10 km SW Ndjolé; Abanga; Mouila-Yeno; Rabi-Koumaga; Lastoursville, 28 km E; Lambaréné).
Here *Phragmanthera kamerunensis* is assessed as Vulnerable since 10 locations are known (above, AOO 40 km² with 4

km² cells) and threats as below. EOO is calculated as 355007 km².

Large epiphytic shrub, branches and leaves scurfy with scales and minute red dendritic hairs; leaves (sub-)opposite elliptic or oblong, 5–12 × 2–6 cm, apex rounded or obtuse, base cuneate and decurrent, lateral nerves 4–6 pairs, petiole 1–2 cm. Umbels 1-several in axil at nodes, flowers 2–4, peduncle 2–3 mm, pedicels 2–3 mm. Corolla 6.5–11.5 cm, yellow tinged orange or orange-red, scurfy with scales and stellate hair, apical swelling slight, ellipsoid, 5-angled, 1.2–1.5 cm, corolla lobes 2.2–2.5 cm.

Habitat: lowland evergreen forest; 0–500 m alt.

Threats: logging – many records seem to be from trees felled during logging operations in more or less intact forest. Frequently slash and burn agriculture follows such logging activity.

Management suggestions: known locations should be mapped onto boundaries of existing National Parks to see if any fall into protected areas which might then be the best focus for basic populational studies and conservation actions, if needed. If this species grows in the forest canopy, censusing individuals will be very dificult.

Phragmanthera longiflora (Balle) Polhill & Wiens (syn. *Phragmanthera rufescens* subsp. *longiflora* Balle)
EN B2ab(iii)

Range: Cameroon (S: Ébolowa–Yaoundé) and Gabon (Lopé Reserve; Tchibanga; Lastoursville; Nkoulounga).

Here *P. longiflora* is assessed as Endangered since there are five locations (above; AOO 20 km² with 4 km² cells) and threats as below. EOO is calculated as 113479 km².

Parasitic epiphytic shrub, stems, leaves and flowers tomentellous with brown 0.5 mm hairs, glabrescent; leaves (sub-)opposite, blade elliptic, 6–13 × 3–7.5 cm, apex rounded, base cuneate to rounded, lateral nerves obscure, 6–8 pairs. Umbels at nodes, 2–4 flowers, peduncle 2–6 mm; pedicels 1–3 mm. Corolla 5.5–8 cm, apical head ovoid-ellipsoid 3.5–4 × 2–3 mm, slightly ribbed; basal swelling hardly developed; lobes erect, 1.4–1.6 cm.

Habitat: lowland evergreen forests; c. 500 m alt.

Threats: logging followed by slash and burn agriculture, e.g. at Ébolowa-Yaoundé.

Management suggestions: the Lopé Reserve is now a National Park and is probably the best choice to focus conservation activities for this species. Ideally basic populational data should be collected to form a baseline for future monitoring although this will be difficult if this is a canopy epiphyte.

***Phragmanthera** raynaliana* (Balle) Polhill & Wiens (syn. *P. polycrypta* (F.Didrichsen) Balle subsp. *raynaliorum* Balle)
EN B1+2ab(iii)

Range: Nigeria (Gotel Mts) and Cameroon (Adamawa: Hosséré Sillé; Dzerkoka 90 km ESE Ngaoundéré; Meïganga).

Epiphytic shrub, branches pendulous, 1.5–2 m; stems and leaves tomentose to floccose with dendritic hairs to 1 mm. Leaves (sub-)opposite, oblong-lanceolate to ovate-elliptic 6–14 × 2.5–8 cm, obtuse, base round to shortly cordate, petiole 1–2 cm. Umbels crowded at old nodes, 2–4-flowered, peduncle 0–2 mm; pedicels 2–4 mm. Corolla 4–4.8 cm, dull golden yellow, apical head ovoid to obovoid 3 × 2.5 mm, 5-ribbed; basal swelling neglible; lobes erect 1.2 cm.

Here *P. raynaliana* is assessed as Endangered since four locations (above; AOO 16 km² with 4 km² cells) are known with threats as below. EOO is calculated as 3643 km².

Habitat: submontane forest and scrub; 1000–1900 m alt.

Threats: submontane forest has seen many losses in the Adamawa area due to clearance for timber and agriculture.

Management suggestions: *Phragmanthera raynaliana* should be searched for at its known locations in Adamawa so that basic populational data can be gathered to aid future monitoring and also assessments of need for interventions such as reintroduction programmes.

Phragmanthera talbotiorum (Sprague) Balle
EN B2ab(iii)

Range: SE Nigeria (Eket; Oban) and Cameroon (S: Kribi).

Here assessed as Endangered since only three locations are known (above; AOO 12 km² with 4 km² cells) and threats as below. EOO is calculated as 14029 km².

Epiphytic shrub, branches pendulous to 1 m; stems, leaf midribs and flowers dark purple tomentellous, early glabrescent, leaves (sub-)opposite oblong-lanceolate, 9–19 × 4.5–8.5 cm, acute to acuminate, base cordate, lateral nerves obscure 5–12 pairs, petiole 5–8 mm. Umbels at nodes, 2–4-flowered; peduncle 0–2.5 mm; pedicels 1–2 mm. Corolla 3.8–4.5 cm, apicel head 3 × 3 mm; basal swelling 5–8 × 3 mm; lobes erect, 10–11 mm.

Habitat: evergreen lowland forest near the coast; 0–200 m alt (estimated).

Threats: two of the three known locations derive from the collections of Talbot a century ago in Nigeria which has lost nearly 90% of its original forest. Kribi is a fast-developing international port but necessary infra-structure development is occurring at the expense of forest habitat.

Management sugestions: every effort should be made to refind *P. talbotiorum* at Kribi, the most recent record, and where forest habitat is least threatened in the range of the [illegible] If successful, basic populational data should be [illegible] possible ([illegible] introduction) to gauge management [illegible] monito[illegible]

Tapinanthus letouzeyi (Balle) Polhill & Wiens (syn. *T. globiferus* Tiegh. subsp. *letouzeyi* Balle)

EN A2c B1+2ab(iii)

Range: Cameroon endemic (NW: Bafut Ngemba FR (believed extinct there due to forest clearance); Lake Oku; Mt Neshele, 10 km ESE Bamenda; Santa; Kondo 40 km NW Bamenda).

Epiphytic shrub, stems c. 1 m long, glabrous; leaves (sub-)opposite, elliptic, 6–12 × 1.5–6 cm, acuminate, base cuneate to rounded, lateral nerves 4–8 pairs. Umbels at nodes May-August, 4–6-flowered; peduncle 3–5 mm; pedicels 0.5–1 mm. Corolla tube red, 2.7–3 cm, bud-heads white, turning black, ellipsoid, angled, 3.5–4 × 2–2.5 mm; basal swelling obovoid 5–6 mm; lobes 6–7 mm, reflexed.

Tapinanthus letouzeyi was assessed as VU A1c, D2 in Cheek *et al.* (2000: 66–67). Since it seems that only five locations are known (above; AOO 20 km² with 4 km² cells) and threats being as below, that 2000 assessment was clearly erroneous and the species is here reassessed as EN. EOO is calculated as 3807 km².

Habitat: forest edges and associated bush; 1300–2500 m alt.

Threats: clearance of forest for wood and for agricultural land. A large part of the range of this species has been shown to have lost c. 50% of its surviving forest in the 15 years between 1988 and 2003. Therefore, loss of 80% of habitat over 100 years (estimated duration of three generations) seems extremely plausible. Forest losses are ongoing.

Management suggestions: 99% of the original forest in Bafut Ngemba has been lost (see sources in Cheek *et al.* 2000). Therefore searches to refind this species should be concentrated at the other locations, especially at Lake Oku, which is believed to have intact forest due to staunchly held traditional beliefs of the Kilum-Oku people. Basic populational data should be gathered as a baseline for future monitoring and to assess needs for management intervention. Additional data on this rare species, its identification, its ecology and threats, can be found in Cheek *et al.* (2000: 66-67).

MALPIGHIACEAE

assessed by Martin Cheek

Lianas, less usually shrubs in Cameroon, with medifixed hairs, leaves alternate or opposite, base with flat, circular surface glands c. 1 mm diam., often extending to petioles, stipules present or absent; flowers actinomorphic or zygomorphic, sepals 5, with glands at base; petals 5, free, usually clawed and yellow, crumpled; stamens 10, free at base; ovary superior, 3-lobed, styles 3, free; fruit of two samaras, winged.

A pantropical family with 56 genera and over 800 species, mainly in the neotropics, poorly represented in Africa. Cameroon has 11 species in 5 genera (Badré in Flore du Cameroun 14 (1972)): *Acridocarpus* (leaves alternate), *Triaspis, Flabellariopsis, Flabellaria* and *Heteroptens.*

No major uses are known in Cameroon for members of this family, which is restricted in that country to the forest zone, where even there it is rarely encountered, comprising far less than 1% of all plants encountered.

Apart from the species treated below, *Triaspis emarginata* De Wildem. (Cameroon and Congo (Kinshasa)) may also in future be assessed as threatened, while for the present being treated as NT, since there are very few known locations; *Acridocarpus macrocalyx* Engler (Guinea (Conakry), Togo, Cameroon and Congo (Kinshasa)) and *Acridocarpus congolensis* Sprague (Cameroon and Congo (Kinshasa)) are also treated as NT for the same reason.

Acridocarpus camerunensis Niedenzu

VU B2ab(iii)

Range: Cameroon (S: Kribi; Campo-Ma'an) and Gabon (Lopé Reserve; Makande, 65 km SSW Booué; Lastoursville; Rabi, 50 km SSW Doussala; ibid 35 km NW).

Here *Acridocarpus cameronensis* is assessed as Vulnerable since eight locations are known (above; AOO 32 km² with 4 km² cells) and threats are as below. EOO is calculated as 119018 km²

Liana, stems shortly hairy; leaves alternate, petiole 2.5–8 mm, pubescent; stipules absent; blades elliptic to obovate, 9–18 × 2.5–7 cm, acuminate, obtuse to cordate at base, with 2–4 glands, entire, nerves pinnate, ascending, prominent below, finer nerves inconspicuous; racemes terminal and axillary, forming large panicles, pubescent, bracts 2–4 mm; pedicels 4–12 mm; calyx with 2–5 glands; petals glabrous, pink-orange; anthers with two horns at apex.

Distinguished from other species by the horned anthers, the acuminate leaves and 2–4 mm bracts.

Habitat: evergreen lowland forest.

Threats: urbanisation; infrastructure for extractive industry (Kribi).

Management suggestions: this species is probably secure in Cameroon in the Campo-Ma'an NP. Plants should be refound with the support of Tchouto (who records it there in 2004) and basic populational data recorded as a baseline for monitoring.

Acridocarpus staudtii (Engl.) Hutch & Dalz.

CR B2ab(iii)

Range: Cameroon endemic (SW: Kumba at Lake Barombi Mbo).

Here *Acridocarpus staudtii* is assessed as Critically Endangered since it is known from a single location (above; AOO 4 km² with 4km² cells) and with threats as below.

Shrub or climber, stems shortly hairy when young; leaves alternate, petiole pubescent, 0.7–1.5 cm; stipules not seen; blades obovate to oblong-obovate, 10–15 × 6–7 cm, shortly acuminate, base obtuse with 2–4 glands, secondary nerves ascending, prominent below; racemes pubescent, terminal, bracts triangular, 1–1.5 mm, pedicel 6–12 mm; calyx with 2–3 glands, sepals orbicular, 2 mm, pubescent; petals 10 mm; anthers acute at apex; fruit with two samaras, 4–5 cm.
First collected in 1896–1897 by Staudt, refound at Barombi Mbo by D. W. Thomas in 1987 and then by C. C. Davis in 1999.
Habitat: lowland evergreen forest.
Threats: clearance of habitat for timber, followed by agriculture.
Management suggestions: the plants found by Thomas and Davis should be refound and, if this succeeds, protected in concert with local land-owners or community representatives. It may be possible to augment the wild population at secure sites by propagating the species at Limbe BG. The lake acts as the reservoir for the large town of Kumba and has much natural vegetation surviving, but from time to time this is cut and small holder farming of both cash and food crops occurs.

MEDUSANDRACEAE

assessed by Martin Cheek

This family was discovered in in 1952 by Brenan, Director of Kew, from material collected by the Richards expedition in Cameroon in the 1960s in the S Bakundu Reserve near Mt Cameroon. The single species known to Brenan at that time, he named *Medusandra richarsdiana.* Subsequently he argued that the anomalous genus *Soyauxia* be included in his family, but later work has shown this to be erroneous. *Soyauxia* belongs in the Peridiscaceae. Letouzey described a second species, *Medusandra mpomiana,* also confined to Cameroon, but less threatened than the type species and here assessed as Near Threatened.

The two species of this family, restricted to Cameroon are medium-sized pioneer trees with alternate, simple, narrowly elliptic leaves, lacking stipules; the petioles are swollen at the apex. The flowers are held in dense catkin-like racemes, each flower is c. 3 mm long with five free sepals and petals, five free stamens alternating with five very long filamentous papillose-hairy staminodes; the superior ovary has three apical, separated styles and a single ovary with apical, pendulous ovules, producing a 3-valved capsule and a single large horny discus-like seed.

Medusandra richardsiana Brenan
VU A3c

Range: Cameroon endemic (SW Region).
This tree behaves as a pioneer (D.W.Thomas, *pers. comm.*) so some forest disturbance may favour it, although forest clearance is a threat. It is here estimated that over 30% of its habitat either has been or will be destroyed in the next 100 years. It is almost confined to SW Region at the locations indicated below. This assessment, made in Cheek *et al.* (2004: 166), is maintained here since no new data are available.
Tree to 18 m, puberulous; leaves oblong-elliptic, 10–30 × 5–14 cm, acutely acuminate, cuneate-rounded, entire, lateral nerves 8 pairs, petiole 3–9 cm, bipulvinate; inflorescence spikes axillary, pendulous, 3–15 cm, petals 2 mm, fruit hard, 3-valved, 1-seeded, 1.5 cm.
Habitat: lowland evergreen forest; 200–600 m alt.
Threats: clearance of forest for expansion of plantations lower slopes of Mt Cameroon, eastern slopes at Bimbia), increased smallholder agriculture and logging along the upgraded Kumba-Mamfe Rd, clearance of forest for yam farming in S Bakundu.
Management suggestions: increased and more effective policing of existing areas gazetted as protected, particularly the S Bakundu FR, would help secure the future of this species.

MELASTOMATACEAE

assessed by Martin Cheek

Melastomataceae have opposite, simple leaves, usually with 3-5 main basal nerves, showy flowers with five pink to purple-blue petals and 5–10 exserted stamens with appendages, dilated connectives, or both. The genera are easily divided into two groups which are so distinct that some have proposed recognising them as families – an idea rejected here since the two units share so many characters not seen elsewhere. The Melastomatoids (Melastomoideae) are herbs, predominantly in open habitats such as montane grassland (e.g. *Dissotis*) but with a few genera (*Amphiblemma, Calvoa*) occurring mainly in forest which have a disproportionately high number of rare, threatened species. The Memecyloids (Olisbeoideae) are woody, shrubs and small trees, exclusively in forest. Confusingly the main genus, *Memecylon*, generally lacks the prominent 3–5 basal nerves seen in the rest of the family (the nerves are usually almost invisible) and so is easily confused with *Eugenia* (Myrtaceae) but lacks the translucent gland dots of that taxon.

The species and genera of Melastomataceae in Cameroon are well-defined thanks to decades of research by Jacques-Félix culminating in his account of the family in Flore du Cameroun 24 (1983). He documented 20 genera and 93 species. Since that date new collections have produced additional species discovered by Stone (e.g. *Memecylon* in Kew Bull. 63: 227–241 (2008) and *Warneckea* in Kew Bull. 64: 307–312 (2009)). Stone has also elevated to species rank some of Jacques-Félix's varieties (op. cit.). He has further species in course of publication, some of which (*Memecylon alipes* and *Warneckea ngutiensis*) are included below.

In addition to the species assessed as threatened below, several others with either restricted distributions or only slightly more than ten locations are assessed here as NT, but, if better data were available in future on population reduction, or numbers of individuals, they might well rate as threatened in future: *Calvoa pulcherrima, Amphiblemma mildbraedii, Heterotis buettneriana, Memecylon arcuato-marginatum, M. calophyllum, M. candidum, M. macrodendron* and *M.nodosum.*

Memecylon dasyanthum Gilg & Ledermann ex Engl. was assessed as VU A2c in both Cheek *et al.* (2004: 167) and Harvey *et al.* (2004: 67), given a range of Mt Kupe (1 collection at one site) to Bamenda Highlands and Bamoun (numerous collections at several sites). However Doug Stone has since brought to light so many more locations for this species (both low and high altitudes) that we here downrate it to NT.

MELASTOMATOIDS (Melastomatoideae)

Herbs, prostrate or erect, usually in non-forest habitats with high light levels; forest taxa often climbers, rock plants, or epiphytes; leaves conspicuouly palmately nerved, often membranous or thinly papery, with sparse appressed simple hairs; flowers usually exceeding 1–2 cm diam, showy.

Amphiblemma amoenum Jacq.-Fél.

EN B1+B2ab(iii)

Range: Cameroon endemic (SW: Kumba to Mamfe, 4 localities in northern Rumpi Hills and western Bakossi Mts: Boubaji; Okoroba-Nguti; Nta Ali; 25 km W Mamfe).

This is a prostrate herb with broadly ovate, cordate, membranous leaves up to 8 × 9 cm and long-pedunculate terminal cyme of sparse pink, 1 cm wide flowers. The shortly cylindric calyces have long glandular hairs.

Previously assessed as Endangered in Cheek *et al.* (2004: 167), the assessment is maintained here since no new data are available excepting EOO data allowing the additional B1 status. EOO is calculated as 1256 km^2. Four locations (see range, above; AOO of 16 km^2 with 4 km^2 cells) with threats as below.

Known only from four *Letouzey* collections, this prostrate herb was first described in 1976. It is particularly distinct within *Amphiblemma.*

Habitat: lowland to submontane forest, sometimes on rocks near streams; 850 m alt.

Threats: clearance of forest for agriculture and wood: only one site (Nta Ali) is protected by a reserve, as far as is known.

Management suggestions: the known sites should be revisited and surveyed for the species to assess whether they are included in protected areas or not. Public education might help to reduce pressure on the surviving plants. Basic populational data is needed as a baseline for future monitoring.

Amphiblemma lanceatum Jacq.-Fél.

EN B2ab(iii)

Range: Cameroon endemic (C: Mbalmayo. E: Deng Deng; Bertoua; Abong-Mbang).

Known from four localities (AOO 16 km^2 using 4 km^2 cells) and with the threats listed below, *A. lanceatum* is here assessed as Endangered. EOO is calculated as 25891 km^2.

This is an erect herb to 1.5 m tall with glabrescent lanceolate leaves 12–18 × 6–9 cm. The 12 cm terminal inflorescence bears c. 50 pink flowers in 2-ranks, each c. 1 cm wide.

Habitat: marshy or riverine sites in semi-deciduous forest; c. 700 m alt.

Threats: semi-deciduous forest is heavily impacted by logging, followed by farming in the AOO of this species (pers.obs.MC). None of the sites are in a protected area, it is thought.

Management suggestions: one or more of the original sites should be revisited to find this species and ascertain density and regeneration levels, so as to determine needs for intervention. None of the sites appears to be in a National Park but the species may yet be found in the proposed Mefou NP.

Amphiblemma letouzeyi Jacq.-Fél.

CR B2ab(iii)

Range: Cameroon endemic (S: 18 km NW Bipindi at Nkoltsia)

Known only from three collections at a single site (AOO 4 km^2 using a 4k m^2 cell), this bizarre herb is here assessed as Critically Endangered in view of the threats it faces (below). An erect herb to 0.8 m tall, leaves sometimes in whorls of four, of highly remarkable appearance, being lanceolate, c. 7 × 2 cm with an auriculate and angular base from the angles of which radiate 8 bristles, each 3 cm long. The species is most distinctive.

Habitat: inselberg in evergreen forest.

Threats: quarrying for the rock that forms the habitat for this species is a risk given the demand to supply infrastructure for future development projects in the Kribi area e.g. rail-line ballast, construction hard-core.

Management suggestions: Nkoltsia should be visited to refind this species and density and regeneration levels evaluated to aid management planning. *Amphiblemma letouzeyi* might have a future as a pot plant, it is so unusual and distinctive.

Amphiblemma monticola Jacq.-Fél. (syn. *Amphiblemma* sp. 1 of Cheek *et al.* 2004: 333)
VU B1+B2ab(iii)

Range: Cameroon endemic (SW: Bakossi Mts (Nyangdong, Nyale; Basseng; W Banguem, Bimie Rock and Abwoh)
This prostrate, creeping herb of rock faces near streams is easily recognised being strongly anisophyllous, in each pair of the lanceolate leaves one is much longer than the other.
Formerly assessed as Vulnerable since six sites are known (AOO 24 km² with 4 km² cells) and threats as below (Cheek & Woodgyer (2006). Kew Bull. 61: 601–604). The assessment is maintained here since no new data are available, excepting EOO data allowing the additional B1 status. EOO is calculated as 374.56 km². However, the sites are so close to each other that in future they might be interpreted as fewer, perhaps a single location, which would result in an assessment of EN or CR.
Habitat: rock faces either sloping or near rapids, moderately shaded by forest; (460–)750–1500 m alt.
Threats: rock falls or forest clearance.
Management suggestions: public education about this rare species in Bakossi to which it appears unique. While not under imminent and major threat, *A. monticola* should be brought to the attention of local communities so that they can better protect it.

Amphiblemma soyauxii Cogn.
VU B1+B2ab(iii)

Range: Cameroon (S: Lolodorf, Mt Mili) and Gabon (Sibange, Libreville; Koulounga; Mbei; Kinguélé; Essia)
This is a 2 m subshrub often purple and with red scurfy hairs when young. The numerous cymes are axillary, each 4–6-fruited, bifurcate, contracted, the flowers pink c. 2 cm wide.
Known from 6 locations (see above, AOO 24 km² with 4 km² cells) *A. soyauxii* is here assessed as Vulnerable in view of the threats below. EOO is calculated as 17738 km².
Habitat: soil on rocks in shade of subshrubs; 700 m alt.
Threats: Libreville is an expanding city and urbanisation of habitat is a threat for the subpopulation there if it has not been lost already.
Management suggestions: Mili Mt should be revisited and regeneration, dens[illegible]ts assessed to determine the needs [illegible]

Antherotoma clandestina Jacq.-Fél.
EN B2ab(iii)

Range: Cameroon endemic (N: Mt Vokré. Adamawa: Nganha, Ngaoundéré).
Known only from the two sites above (AOO 8 km² with 4 km² cells) and probably threatened by cattle grazing, *A. clandestina* is here assessed as Endangered.
A 12 cm tall erect annual herb with elliptic leaves and 10 × 3 mm, florets c. 1 cm wide, pink with orange throat. Sepals terminating in 2 wings, intersepal lobes truncate, otherwise resembling the common *A. naudinii.*
Habitat: soil on basalt rocks, probably in full sun; 1500–1800 m alt.
Threats: cattle grazing and trampling.
Management suggestions: the two sites should be revisited in October when plants flower at the end of the wet season and when seed might be formed. Population density and regeneration levels can then be assessed and seed obtained for banking purposes. Threats to the species could be gauged thoroughly.

Calvoa calliantha Jacq.-Fél.
EN B2ab(iii)

Range: Cameroon endemic (S: Nkoltsia near Bipindi; Campo Ma'an fide Tchouto 2004).
Here assessed as Endangered since known from two sites (AOO 8 km² with 4 km² cells) and threats as below. An ascending herb to 60 cm tall, *C. calliantha* has glabrous, rounded, slender stems and ovate, narrowly cordate leaves c. 9 × 4.5 cm. The flowers are 2.5–3 cm wide.
Habitat: inselberg in evergreen forest.
Threats: quarrying for rock (as for *Amphiblemma letouzeyi*).
Management suggestions: Nkoltsia should be considered for permanent protection in conjunction with the local community. It harbours several extremely rare or point endemic species. Data collection for population size; density, regeneration and threats would aid planning management.

Calvoa stenophylla Jacq.-Fél.
CR B1ab(iii)

Range: Cameroon endemic (S: 10 km SSE Zingui near Kribi).
This 20 cm tall subshrub is restricted to the bed of the Minsomo R. The 4-angled stems bear strap-like leaves c. 8 × 1 cm. The terminal inflorescences comprise 1–8 flowers each 3 cm wide.
Calvoa stenophylla is here assessed as Critically Endangered since known from a single site (AOO 4 km² using 4k m² cells) with threats as below.
The record of this species, *P. Mambo* 163 from Mbu-Bakundu on the Missouri database needs further checking before it can be accepted since its habitat is given as farm fallow and secondary forest, quite unlike that of the geographically type specimen. It may be a misdetermination.
Habitat: river bed in evergreen forest belt.

Threats: hydroelectric schemes; siltation of water derived from logging.
Management suggestions: revising the only known site would allow a more accurate assessment of population size, density, regeneration and threats.

Cincinnobotrys letouzeyi Jacq.-Fél.
EN B2ab(iii)
Range: Cameroon endemic (SW: Mamfe, 45 km ENE at Numba; Bechati in Lebialem Highlands).
From a creeping rhizome like a string of beads, this species produces an erect single elliptic leaf blade c. 12 × 5 cm on an 8–12 cm long and an inflorescence almost as long of c. 12 pink flowers c. 2 cm wide.
Here *C. letouzeyi* is assessed as Endangered since only two locations are known (AOO 8 km² with 4 km² cells) with threats as below.
Habitat: on rock near spray from streams in evergreen forest belt.
Threats: forest destruction at the Bechati site.
Management suggestions: protection at either or both of the two sites is desirable. Data on population size, density and regeneration levels is desirable.

Dissotis bambutorum Engl. (syn. *Heterotis angolensis* (Cogniaux) Jacq.-Fél. var. *bambutorum* (Engl.) Jacq.-Fél.)
VU B1+B2ab(iii)
Range: Nigeria (Gangirwal; Gembu; Sardauna-Nguroje) and Cameroon (W: Bamboutos Mt. NW: Bali Ngemba; Bafut-Ngemba; Mba Kokeka; Kilum-Ijim; Ndop-Bambui).
A several stemmed subshrub 0.6–1.2 m tall, stems red, 4-angled, stellate-hairy, glandular near inflorescence; leaves sessile, elliptic, c. 5 × 2 cm, 7-nerved; panicles contracted, terminal, flowers c. 5 cm across, sepals red, 8 × 4 mm, glandular hairy, petals red-purple, anthers 10, equal.
Restricted to a short section of the Cameroon Highlands, this is a showy, but ecologically restricted species. Here assessed as Vulnerable since nine locations are known (AOO 36 km² with 4 km² cells and threats as below). EOO is calculated as 15032 km².
Habitat: stream edges in submontane/montane grassland, often near forest edge; 1400–2300 m alt.
Threats: trampling by cattle when drinking at streams; man-made fires.
Management suggestions: this species has great horticultural potential for its large showy flowers, so ex situ cultivation would be especially welcome. Several sites occur at Kilum-Ijim, protected by traditional communities. The village elders should be advised of the rarity of the species and assisted in identifying and protecting it.

Dissotis bamendae Brenan & Keay
VU B1+B2ab(iii)
Range: Nigeria (Mambilla Plateau, Nguroje FR (1 coll.)) and Cameroon (NW: Bamenda (2 coll.); Nchan (1 coll.); Bafut-Ngemba FR–Bali Ngemba FR (3 coll.); Bamboutos–Bamenda Highlands (Santa, 1 coll.). SW: Lebialem–Fosimondi).
This many-stemmed, shapely shrub was sunk into *D. princeps*, a widespread species (South Africa to Ethiopia) by Jacques-Felix, but since it is abundantly distinct (e.g. 3 leaves per node, not 2) we follow FWTA in maintaining it as different. Only 9 locations are known (AOO 36 km² with 4 km² cells, and threats as below). The assessment made by Cheek in Harvey *et al.* (2004: 66) is maintained here, since only one extra record (Lebialem) is added, excepting EOO data allowing the additional B1 status. EOO is calculated as 2609 km². Several of the locations are so close together that they might be better be taken as one.
Habitat: grassland, often near streams, possibly requiring proximity to forest; 1500–2200 m alt.
Threats: poorly known, but including conversion of natural habitat to farmland, whether for grazing or for cultivation of beans or potatoes. If plants are dependent upon proximity to forest, then ongoing loss of surviving forest in the Bamenda Highlands is a major threat and this species could be assessed as EN A3c.
Management suggestions: long term studies of exclusion plots to study the effect of removing fire and/or grazing from this species would help to understand management regimes for this and other ecologically similar species in the Bamenda Highlands, but would need a site where the species is frequent.

Dissotis longisetosa Gilg & Ledermann ex Engl.
VU B2ab(iii)
Range: Nigeria (Obudu (2 coll.); Mambilla Plateau (Ngurdje FR to Gangirwal (4 coll.)) and Cameroon (NW: Bafut-Ngemba FR (2 coll.); Bali Ngemba FR (2 coll.); Dom; Tabenken (1 coll. each). Adamawa: Ngaoundéré (4 coll.). SW/W: Lebialem/Bamboutos Mts (2 coll.)).
This striking long-setose *Dissotis* is not easily confused with anything else. *Lock* 84/56, from Togo, 740 m, referred to this taxon at K, appears to belong to another taxon, it has soft hairs and a subsessile inflorescence, so is not mapped here. Since there are ten locations this species is here maintained as VU (see range above; AOO is calculated as 40 km² with 4 km² cells, and threats are as below), following Cheek *et al.* (2010: 99–100). EOO is calculated as 76471 km².
Usually only 1–3 plants occur at a site: they are not spread uniformly through the grassland habitat. Plants are thought to have an underground rootstock. The absence of this species and so many other Cameroon Highland species of *Dissotis* from Mt Cameroon, is remarkable.

Habitat: heavily grazed and sometimes burnt submontane/montane grassland; 1600–2100 m.
Threats: conversion of montane grassland to tilled land. The rootstocks may be vulnerable to predation.
Management suggestions: ecologically this appears such an undemanding species that it may be expanding its range since on the whole, burnt, heavily grazed grassland appears to be increasing within its range! Known from only ten locations, the factors accounting for its rarity and a detailed study of its ecology are desirable. It may be that the species is especially vulnerable to cattle trampling. Further surveys may reveal more sites, so downgrading the threat level to this taxon.

Tristemma camerunense Jacq.-Fél.
EN B2ab(iii)
Range: Cameroon endemic (S: Kribi–Edéa, 60 km; É bolowa, 12km S at Nkoemvom).
This 1 m tall subshrub has strigose subangular stems and elliptic-lanceolate leaves c. 9 × 4cm. The terminal inflorescence is a head of c. 6 sessile flowers subtended by 2 pairs of leaves and a ring of leafy bracts nearly as long as the flowers. The fruit has 2 or 3 complete rings of appressed, long hairs.
Here *T. camerunense* is assessed as Endangered since two locations are known (AOO 8 km² with 4 km² cells) with threats as below.
Habitat: old lowland evergreen secondary forest.
Threats: slash and burn agriculture; transport infrastructure: the Kribi-Edéa Rd is the main access to a major Cameroon port, road widening might destroy this site.
Management suggestions: the Kribi-Edéa site should be visited to refind this species and chart the extent, density and levels of regeneration and threat to *T. camerunense*. Since *Tristemma* species generally produce fruit freely, each with numerous small seeds, seedbanking and ex situ conservation in gardens are well worth attempting.

MEMECYLONOIDS (Olisbeoideae)
Forest shrubs or trees with four genera in Cameroon, two of which, *Spathandra* and *Lijndenia* have only one or two species which are not considered threatened here. Leaves are usually leathery and glabrous; flowers less than 1 cm diam. *Warneckea* have prominently 3-nerved leaves while in *Memecylon* nervation is usually obscure and pinnate.

Memecylon alipes R.D.Stone ined.
EN B2ab(iii)
Range: Cameroon (S: Campo-Ma'an area, Efoulan) and Gabon (Woleu-Ntem, Inselberg Milobo).
Here assessed as Endangered since two localities (above; AOO 8 km² with 4 km² cells) with threats as below. This unpublished taxon is described by Stone (pers. comm. to Cheek 9th Dec 2009) as "very distinctive but easily placed in the *M. normandie* species-group sensu Jacq.-Fél.
Habitat: submontane evergreen forest; 630–960 m alt.
Threats: slash and burn agriculture; logging, and plantation agriculture.
Management suggestions: means to identify this taxon are required so that it can be assessed for population size, density, levels of regeneration and threat.

Memecylon amshoffiae Jacq.-Fél.
EN B1+B2ab(iii)
Range: Cameroon endemic (C: SW Eséka, on the Nyong R; 40 km NW Eséka on the Késlé R; 49 km Eséka on the Nyong near Songbong; Eséka-Lolodorf bridge).
Although *M. amshoffiae* has been listed as occurring in Gabon, such references probably refer to allied species that have been misidentified (Stone pers. comm.. to M. Cheek 9th Dec. 2009). In the opinion of Stone, *M. amshoffiae* must be regarded as an endemic of the Eséka region (C & S Regions, Cameroon), known only from riverine forest on the banks of the lower Nyong and Kéllé Rs (Stone pers. comm. Dec. 2009 to Cheek).
A 5–10 m shrub or small tree with green, globular fruits and a thick, 4-lobed calyx, stems rounded, petioles 3–5 mm long, blades oblong-elliptic c. 10 × 4 cm, cymes 2-3 cm long. *Memcylong amshoffiae* is here assessed as Endangered since there are four locations (AOO 16 km², with 4 km² cells) and threats as below. The four locations are so close to each other that they might better be taken as one, in which case CR status would be appropriate. EOO is calculated as 597 km².
Habitat: lowland evergreen riverine forest.
Threats: slash and burn agriculture; road widening.
Management suggestions: further data on population size, density, regeneration and threat levels for *M. amshoffiae* are desirable to assess management needs. None of the sites appear to be within existing protected areas.

Memeylon bakossiense R.D.Stone, Ghogue & Cheek (syn. *Memecylon* Sect. *Mouririoides* sp. nov. 2. Stone & Cheek in Cheek *et al.* (2004: 336, t.10A))
EN B1+B2ab(iii)
Range: Cameroon endemic (SW: western Bakossi Mts)
Evergreen shrub or small tree 1.5 to 4 m tall; young branchlets rounded to obscurely quadrangular; leaves thinly coriaceous, dark green and somewhat shining on the upper surface, paler below; surface texture minutely granular when dry; blades broadly elliptic, (7–)[illegible]–12.5(–14) cm long, (3–)4–5.7(–7) cm wide, cuneate at base, acuminate at apex, the acumen 1–1.5 cm long; mid-nerve finely impressed on the upper surface, prominent on the lower; transverse nerves much thinner than the mid-nerve, obscure or faintly

prominent on the upper surface, conspicuously prominent below, 6–10 pairs; inflorescences solitary or geminate in the axils 2.5–4 cm long, (7–)10–20(–30)-flowered; corolla exposed in bud; petals pale purple, 2.5 mm long, 2.5–3 mm wide, broadly triangular to rhomboid, apiculate, at anthesis ± concealed beneath the anthers; nthers ± the same color as the petals, 2 × 1 mm; connective conical, acute, slightly curved or saddle-shaped, 2 mm long, distinctly keeled on the dorsal side, gland absent; filaments 3 mm long.

Memecylon bakossiense is only known from four areas in the lowlands of the western Bakossi Mts, with most of the six collections being made along the Nyandong-Messaka footpath during a period of two weeks of general collecting by c. 15 plant collectors based at Nyangdong, when *Memecylon* was being specifically targeted. The known area of occupancy of the species is estimated as 16 km². Accordingly, *M. bakossiense* is here assessed as EN B2a,b(iii). The assessment of this species in Stone *et al.* (op.cit) 2008 is maintained and repeated here, no further data being available excepting EOO data allowing the additional B1 status. EOO is calculated as 55.44 km². However, the locations are so close together that they might be reassessed as one in which case CR status would be appropriate.

Habitat: lowland evergreen forest, usually (but not always) on shaded stream banks; 350–650 m alt.

Threats: the proposed construction of a vehicular road in place of the path may impact some of the sites directly, or open up areas of forest containing the species to clearance.

Management suggestions: this species is one of 16 that are believed to be endemic to the Bakossi Mts, although collecting in other areas may yet extend their ranges (Cheek *et al.* 2004: 78). The Bakossi people have agreed to protect forest in their mountains, and an area has recently been gazetted as National Park, but sadly this species is not known to occur in the park boundary. However it is possible that the species occurs to the north and south of Nyandong-Messaka, which have not yet been well explored botanically. To the east, around Kodmin (c. 1400 m alt.) the forest has been well-explored and the species has not been collected there, so appears to shun higher altitudes.

Memecylon breteleranum Jacq.-Fél.

CR B2ab(iii)

Range: Cameroon endemic (E: Ebaka, SW Ndemba, 40 km NW Bertoua).

Here *M. breteleranum* is assessed as Critically Endangered since known from a single site (AOO 4 km² with 4 km² cell size) and threats as below.

A small tree c. 8 m tall with ± rounded stems, petioles 5–8 mm and elliptic, leathery leaves 8–9 × 5–6 cm, acumen 5 mm long and 4–6 pairs of obscure nerves. The cymes are axillary in leafy nodes, 3–4 cm long, 1–2-branched, peduncles 1–2 cm, hypanthium 3 mm wide.

Habitat: probably in evergreen riverine forest in the semi-deciduous forest belt.

Threats: logging; slash and burn agriculture.

Management suggestions: refinding this species at its only known locality and assessing number of individuals, population density and range, regeneration and threat levels, is needed to develop a management plant. The site is not protected so far as is known.

Memecylon fugax R.D.Stone ined.

EN B2ab(iii)

Range: Cameroon endemic (S:Akom II at Nkol Ndangueng; Efoulan and Egongo hills).

This new species in the *M. normandii* complex (sensu Jacques-Félix, 1979) is known only from the Campo-Ma'an area of Cameroon's South Region. It gets its name from the young branchlets which are conspicuously 4-winged. This character is also found in *M. macrodendron* Gilg ex Engl. and *M. alipes* R.D.Stone, ined., but in *M. macrodendron* the leaves are a bit smaller and differently shaped, and in *M. alipes* the peduncles of the inflorescence are much elongated (Stone pers. comm. to Cheek, June 2010).

Memecylon fugax is known from only three collections at two closely neighbouring locations (Stone pers. comm. to Cheek, June 2010). The area of occupancy is thus calculated as 8 km², using a 4 km² grid cell size, threats as below. Accordingly the species is here assessed as EN B2a,b(iii), that is Critically Endangered.

Habitat: lowland to submontane evergreen forest; 700–880 m alt.

Threats: slash and burn agriculture. Forest coverage in Cameroon generally is declining slowly but steadily in recent years, largely due to clearance for timber and agriculture, and the site for this species does not benefit from any protection.

Management suggestions: since the two known sites for this species are not formally protected it, the elders of Akom II village and other communities nearby should be informed of the rarity of this species which is unique to their forest, on current evidence. The village elders should be shown examples of the species and advised of measures that might support its survival, such as maintaining an intact forest canopy in the area and avoiding cutting of the understorey.

Memecylon hyleastrum R.D.Stone & Ghogue (syn. *M. afzelii* G.Don var. *pedunculatum* Jacq.-Fél.)

CR B2ab(iii)

The assessment of this species in Stone *et al.*(op.cit.) 2008 is maintained and repeated here, no further data being available.

Range: Cameroon endemic (S: 15km S Ébolowa, c. 2 km SE N'kolandom village).

Shrub with elliptic-acuminate leaves 4.5–6.5 cm long; inflorescences 3–5-flowered and peduncles 5–15 mm long has primary axes each terminated by a single flower on 0–1.5 mm pedicels. Corolla exposed in bud; petals white, c. 1 mm long and 1.5 mm wide, broadly triangular, at anthesis ± concealed beneath the anthers. Anthers white, 1.25 × 0.75 mm; connective conical, acute, slightly curved or saddle-shaped, distinctly keeled on the dorsal side, gland absent. Distinctive in the white flowers with connectives which lack glands.

Memecylon hyleastrum is known from only two collections at a single site. The area of occupancy is thus calculated as 4 km², using a 4 km² grid cell size. Accordingly the species is here assessed as CR B2a,b(iii), that is Critically Endangered. This assessment of taken from Stone *et al.* (2008: 232 loc. cit.) is maintained here since no additional data are available.

Habitat: lowland to submontane evergreen forest; 700–750 m alt.

Threats: slash and burn agriculture. Forest coverage in Cameroon generally is declining slowly but steadily in recent years, largely due to clearance for timber and agriculture, and the site for this species does not benefit from any protection.

Management suggestions: since the only known site for this species is not formally protected it is imperative that the elders of N'kolandom village are informed of the rarity of this species which is unique to their forest, on current evidence. The village elders should be shown examples of the species and advised of measures that might support its survival, such as maintaining an intact forest canopy in the area and avoiding cutting of the understorey.

Memecylon korupense R.D.Stone ined.

CR B2ab(iii)

Range: Cameroon endemic (SW: Korup, Ikengue-Bajo villages).

Within the *M. normandii* complex (sensu Jacques-Félix, 1979) this species appears most closely related to *M. fugax* R.D.Stone ined., but the young branchlets of *M. korupense* are very slender and simply quadrangular whilst in *M. fugax* they are prominently winged. *Memecylon korupense* also superficially resembles *M. macrodendron* but in that species the young branches are again winged, and the leaves are a bit smaller and differently shaped (Stone pers. comm. to Cheek, June 2010).

Memecylon korupense is known from only one collection at a single site (Stone pers. comm. to Cheek, June 2010). The area of occupancy is thus calculated as 4 km², using a 4 km² grid cell size. Accordingly the species is here assessed as CR B2a,b(iii), that is Critically Endangered with threats as below.

Habitat: lowland evergreen forest; 150 m alt.

Threats: slash and burn agriculture. Generally forest in Korup NP is well protected, but near villages the usual threats apply.

Management suggestions: the elders of Ikengue and Bajo villages, and the park managers of Korup, should be informed of the rarity of this species which is unique to their forest, on current evidence. The village elders should be shown examples of the species and advised of measures that might support its survival, such as maintaining an intact forest canopy in the area and avoiding cutting of the understorey.

Memecylon kupeanum R.D.Stone, Ghogue & Cheek

(syn. *Memecylon* sect. *Mouririoidea* sp. nov. 1. Stone & Cheek in Cheek *et al.* (2004: 336)).

EN B1+2ab(iii)

Range: Cameroon endemic (SW: Mt Kupe).

Evergreen tree 2–3 m tall; young branchlets rounded; leaf blades broadly elliptic 13–16 × 5–7 cm, broadly cuneate to ± rounded at base, acumen 1–1.5 cm long, acute; mid-nerve finely impressed on the upper surface, prominent on the lower; transverse nerves much thinner than the mid-nerve, faintly prominent on the upper surface, conspicuously prominent below, c. 8–10 pairs; inflorescence c. 2 cm long, 5–10-flowered, geminate or solitary at the axils; flowers with hypantho-calyx in bud subglobose and almost completely covering the corolla, at anthesis broadly campanulate, 2.5–3 mm long, 4–5 mm wide; calyx lobes broadly rounded, scarious and tearing ± unevenly along their margins, the intervening sinuses c. 0.75 mm deep; petals greenish white, ± ovate, 2 mm long, 2 mm wide, apiculate; anthers 1.5 × 1 mm; connective ± conical and abruptly saddle-shaped by the blackish, punctiform dorsal gland situated c. 0.5 mm from the posterior end; filaments 2–2.5 mm long.

Memecylon kupeanum is known from three areas at Mt Kupe: to the south, above Kupe Village (one coll.), the west (Nyasoso, one coll.), and the north, at Etube-Tape to Ndum (two collections). The area of occupancy is thus calculated as 12 km², using a 4 km² grid cell size. The mountain is now fairly well botanised. Threats are listed below. This species is one of 26 that are believed to be endemic to Mt Kupe, although collecting in other areas may yet extend their ranges (Cheek *et al.* 2004: 78). Accordingly, *M. kupeanum* was assessed as EN B2a,b(iii) in Stone *et al.* (2008 loc. cit.). This assessment is maintained here, excepting EOO data allowing the additional B1 status. EOO is calculated as 12 km². However, it would also be valid to take Mt Kupe as a single location, in which case the species would be assessed as CR.

Habitat: submontane evergreen forest; 1100–1200 m alt.

Threats: threats to the species include forest clearance for small scale timber production, followed by clearance for small-holder agriculture, especially a concern at lower altitudes, such as 800–1100 m. However, the Bakossi people have agreed to protect forest on the mountain, which has great traditional importance, and it was in the process of being gazetted as an Integrated Ecological Reserve (Wild *et* *al.* in Cheek *et al.* 2004: 111–116). However the status of the gazettement is unknown.

Management suggestions: since the only known site for this species is not yet formally protected it is imperative that the elders of the villages concerned are informed of the rarity of this species which is unique to their forest, on current evidence. The village elders should be shown examples of the species and advised of measures that might support its survival, such as maintaining an intact forest canopy in the area and avoiding cutting of the understorey. Propagation of the species for multiplication and reintroduction in safe sites is advisable.

Memecylon mamfeanum (Jacq.-Fél.) R.D.Stone, Ghogue & Cheek (syn. *Memecylon afzelii* G.Don var. *mamfeanum* Jacq.-Fél.)

EN B1+B2ab(iii)

Range: Nigeria (Oban) and Cameroon (SW: 15 km SW Akwaya; 20 km W Nguti; S. Korup; Korup, Juahan Mt).

Shrub, young branchlets, inflorescence axes, and pedicels terete, reddish-green (often drying dark purplish green); leaf-blades oblanceolate or elliptic-acuminate, acumen long (± 2 cm); transverse nerves ± obscure; inframarginal nerves ± curvilinear; inflorescences very lax with branches often terminated by solitary flowers; pedicels 3–4 mm long; petals broadly triangular to rhombic, at anthesis ± concealed beneath the anthers; anther connectives 1 mm long, obtuse; dorsal gland on anther connectives reduced or absent. Concerning the treatment of this taxon in the Flore du Cameroun, it should be noted that the information provided in the key (Jacques-Félix 1983: 124) is not correct in stating that the cymes of *M. mamfeanum* can be up to 7.5 cm long; the maximum length is later reported as 4.5 cm (cf. description on p. 140 of that treatment), and this is consistent with all of the specimens that we have examined. The species is thought to be closely related to two other newly discovered, locally endemic taxa in Cameroon's SW Region, namely *Memecylon bakossiense* and *Memecylon rheophyticum* also threatened and included in this account. *Memecylon mamfeanum* is known from a total of seven collections at five sites, all between Akwaya and the coast, and all but two collections are in or very near to the relatively well collected Korup NP. The area of occupancy is thus calculated as 20 km², using the current IUCN preference of a 4 km² grid cell per site (Hilton-Taylor pers. comm.). Within its range, in terms of numbers of specimens collected, it is rare compared with some other members of the genus in Cameroon. The sites inside Korup are well protected, but those outside are vulnerable to forest degradation and clearance for timber and agriculture, especially that at Oban in Nigeria (Cheek pers. obs.). Accordingly the species was assessed as EN B2a,b(iii), that is endangered, in Stone *et al.* 2008 loc. cit.) and this assessment is maintained here, no further data being available excepting EOO data allowing the additional B1 status. EOO is calculated as 4799 km².

Habitat: lowland and submontane evergreen forest; 60–600(–900)m alt.

Threats: see above.

Management suggestions: four of the seven records of this species derive from the southern and middle parts of the Korup NP so the species is secure here. Data on regeneration and density levels are desirable to inform management decision. Means to identify the species should be made available to the Park Managers.

Memecylon oubanguianum Jacq.-Fél.

VU B2ab(iii)

Range: Cameroon (E: confluence of rivers Lom & Djerem, 185 km NE Yaoundé; Bertoua–Dimako. S: Bitye near Dja), Gabon (Cristal Mts, 20 km NW Asok; SW Lambarene), CAR (Boukoko) and Congo (Kinshasa) (Businga (Ubangi)).

Here *M. oubanguianum* is assessed as Vulnerable since there are seven sites (see above, AOO 28 km² with 4 km² cells) with threats as below. EOO is calculated as 442750 km².

A 3–4 m tall shrub-tree with 4 angled stems and leathery, elliptic leaves 7–9 × 2.5–3 cm, obscurely 6–8-nerved, with acuminate, long-cuneate apices. The cymes are c. 2.5 cm long with 0.3–0.8 cm peduncles, 10–20-flowers, the hypanthia 4–5 mm, with calyx truncate to sinuate.

Habitat: evergreen forest, both riverine and submontane; 15–650 m alt.

Threats: slash and burn agriculture and logging.

Management suggestions: surveys to assess population size, density, regeneration and threat levels are needed to develop a management plan. The Cristal Mts and Lambarene sites may fall within Gabonese NPs.

Memecylon rheophyticum R.D.Stone, Ghogue & Cheek

VU D2

Range: Cameroon endemic (SW: rapids in lower stretches of the Mana R and tributary).

Rheophytic evergreen shrub or small tree 0.8–2 m tall; branchlets ± obscurely quadrangular when young, rapidly becoming rounded with age; petioles 2–4 mm long; blades thinly coriaceous, narrowly elliptic to lanceolate or oblanceolate, 6–11 × 1.2–2.5 cm wide, attenuate at both ends, base ± confluent with the petiole, apex with a narrow, acute acumen 1.5–2 cm long; mid-nerve impressed on the upper surface, prominent on the lower; intramarginal and transverse nerves obscure to ± evident on both upper and lower surfaces, at least when dry; transverse veins 5–6 pairs; cymes axillary, rather slender and lax, 3–4 cm long, branched once or twice above the peduncle, several-flowered; peduncles quadrangular, slender, 5–16(–20) mm long; branches of the inflorescence about the same length as the peduncle or generally somewhat shorter; true flowering pedicels 1.5–2 mm long, very slender. Hypantho-calyx cupulo-patellate, 2.5–3 mm long, 3.5–4 mm wide, limb shallowly sinuate-dentate. Corolla well-exposed in bud, rounded, apiculate; petals reportedly blue-green or mauve. Petals 2–2.5 × 2.5–3 mm wide, broadly triangular to subrectancular, the apex rounded and apiculate; anther connective conical, ± curved or saddle-shaped, keeled on the dorsal side, gland absent; fruits elliptic, c. 13 × 7 mm wide, borne on stout pedicels c. 3 mm long.

Memecylon rheophyticum is only known from four collections at three sites near Mundemba. The extent of occurrence is 11 km². The area of occupancy is 12 km², using a 4 km² grid cell size. Given that the habitat of rocky river banks faces no known threats in this area, the taxon does not qualify for assessment under criterion B. There is no evidence of a reduction in range or population, so only criterion D2 can be invoked, under which *M. rheophyticum* is here assessed as Vulnerable.

The extremely narrow and long-acuminate leaves of *M. rheophyticum* immediately distinguish it from all other currently recognised *Memecylon* species in Cameroon. Also distinctive is its ecology which is evidently restricted to rocky stream banks. At the Mana R rapids site, several other narrowly endemic species of rheophytic shrubs are known in genera that otherwise occur in evergreen forest understorey habitats (Cheek pers. obs.).

Habitat: rocky banks of fast flowing, clear-water rivers in lowland evergreen forest; 50–100m alt.

Threats: none are known currently (see above).

Management suggestions: data on regeneration and density levels are desirable to inform future monitoring and management decisions. Means to identify the species should be made available to the Korup Park Managers, and those responsible for the Palm Oil plantation that borders on the river habitat of this species.

Memecylon simulans (Jacq.-Fél.) R.D.Stone & Ghogue (syn. *M. arcuatomarginatum* Gilg ex Engl. var. *simulans* Jacq.-Fél.)

EN B1+B2ab(iii)

Range: Cameroon endemic (S: Bipindi; Kribi–Lolodorf at Mt Calvary; 23 km NW Bipindi at Nkoltsia hill).

This shrub, young branchlets rounded, branchlets conspicuously thickened at the nodes, the nodes 3–4 times as thick as the internode above; leaves sub-bullate. lanceolate, c. 10 × 4 cm, dull on upper surface, with 3–7 pairs of lateral nerves, transverse and [illegible] nerves strongly impressed on the upper [illegible] on the lower, the intramarginal nerves [illegible] m long.

Memecylon simulans is known from only three sites, none of which are protected and so all of which are vulnerable to forest clearance and degradation that has slowly but steadily been effecting forests in Cameroon. The area of occupancy is calculated as 12 km², using a 4 km² grid cell size. Accordingly the species was assessed as EN B2a,b(iii), that is Critically Endangered, in Stone *et al.* (2008, loc. cit.). That assessment is maintained here, no further data being available excepting EOO data allowing the additional B1 status. EOO is calculated as 253 km².

Habitat: lowland and submontane evergreen forest on coastal hill tops; c. 300 m alt.

Threats: see above; slash and burn agriculture.

Management suggestions: since the only known sites for this species are not yet formally protected it is imperative that the elders of the villages concerned are informed of the rarity of this species which is unique to their forest, on current evidence. The village elders should be shown examples of the species and advised of measures that might support its survival, such as maintaining an intact forest canopy in the area and avoiding cutting of the understorey. Propagation of the species for multiplication and reintroduction in safe sites is advisable. Basic populational data should be collected to inform future monitoring and management decisions.

Warneckea austro-occidentalis R.D.Stone (syn. *Warneckea* cf. *memecyloides* sensu R.D.Stone & Cheek in Cheek *et al.* (2004: 337)).

EN B2ab(iii)

Range: SE Nigeria (Calabar R at Akamkpa Rubber Plantation) and Cameroon (SW: Korup NP (1 coll.); Kumba-Mamfe Rd at Ikiliwindi (several coll.); Bimbia (several coll.))

Trees 6–20 m tall; bark whitish, branchlets reddish brown, the youngest terete in cross section but somewhat compressed and grooved on each face just below the proximal node; leaf petioles 7–10 mm, quadrangular or narrowly 2-winged when young, becoming stout (up to 3 mm wide) and subligneous with age; blades subcoriaceous, shining on both upper and lower surfaces, broadly ovate to broadly obovate, 12–19 × 7.5–10 cm wide, base broadly cuneate to sometimes rounded (never cordate!), apex broadly short-acuminate and acute; midnerve and principal lateral nerves impressed on the upper surface, prominent on the lower, the lateral nerves diverging from the midnerve c. 5–10 mm above the base of the blade, curvilinear except near the leaf apex where forming weak arches between the junctions with the transverse veins; secondary lateral nerves 1 pair (rarely 2 pairs), much thinner than the principal laterals, intramarginal and weakly arched for their entire length; transverse veins 8–1[illegible] pairs, of about the same thickness as the secondary later[illegible]florescence of 5–10 long-pedunculate, many-flowered cymes fascicled at the thickened nodes of older branches (never in the leaf axils or at recently defoliated nodes); peduncles 1.7–3.2 cm long, compressed or sometimes quadrangular below the apex; secondary axes reduced (the inflorescence thus subumbellate); flowers often 20–30 per infl., borne on slender pedicels 5–7 mm long (becoming thicker and 8–11 mm long in fruit); hypanthocalyx campanulate, 3 × 3 mm, the lobes triangular-acute; corolla conical in bud; petals white, elliptic-unguiculate, c. 3 mm long and 1.5–2 mm wide; stamens with slender filaments c. 2 mm long; anthers 1.5 mm long, the connective blue, strongly incurved by the oblong gland occupying almost the entire length on the dorsal side; thecae yellow; fruits ellipsoid-ovate, 15–16 × 10–11(–14) mm, lacking a persistent calycinal crown.

Warneckea austro-occidentalis was assessed by Stone *et al.* (2009 loc. cit.) as endangered EN B2ab(iii) in view of the threats stated below, an AOO of 20 km^2 and only five historic sites being known (listed above), although it may survive at only three of them. This assessment is maintained here since no additional data are available. EOO is calculated as 8740 km^2.

Habitat: lowland evergreen forest; alt. 5–250 m.

Threats: of the five known sites of *Warneckea austro-occidenalis*, only one, inside Korup NP, can be considered secure.

Two sites are not far off the Kumba–Mamfe Rd, scheduled for upgrading for many years. When this happens anticipated development along its length will threaten the species. Nigeria has seen large forest losses in recent decades and the survival of original forest on a plantation is unlikely.

In recent years much forest at Bimbia has been cut. Efforts over several days in March 2008 to refind Bimbian endemics such as *Ancistrocladus grandiflorus* Cheek and *Cola cecidifolia* Cheek by David Alicha with the herbarium team from Limbe BG who had been involved in their discovery in the 1990s, failed to find them due to loss of habitat (Alicha pers. comm. 2008).

Management suggestions: the species can be considered secure at its Korup NP site. Data on regeneration and density levels are desirable to inform management decision. Means to identify the species should be made available to the Korup Park Managers, and also those managing the Bimbia community forest. Since the other sites for this species are not yet formally protected it is imperative that the elders of the villages concerned are informed of the rarity of this species. The elders should be shown examples of the species and advised of measures that might support its survival, such as maintaining an intact forest canopy in the area and avoiding cutting of the understorey.

Warneckea mangrovensis (Jacq.-Fél.) R.D.Stone & Ghogue (syn. *Warneckea fascicularis* (Planch. ex Benth.) Jacq.-Fél. var. *mangrovensis* Jacq.-Fél.; *Warneckea* sp., Jacq.-Fél.)

EN B1+B2ab(iii)

Range: Cameroon (S: Kribi area) and NW Gabon (Estuaire Province, rivière Malibé).

Warneckea mangrovensis is known from a total of four collections at three localities (two sites in southern Cameroon, one in northwestern Gabon). On 1 May 2003 the type region on the seashore N of Kribi was surveyed by Stone & Ghoghe, who found the area to be almost totally deforested. In more than a half day of searching, only one small tree of *W. mangrovensis* was found (on a forested rock outcrop at the rear of the beach). *Warneckea mangrovensis* was assessed as Endangered EN B2ab(iii) in Stone *et al.* (2009, loc. cit.), on the basis that two of the three sites (AOO 12 km^2) are on the coast near Kribi which is being developed for retirement and holiday accommodation, and natural forest habitat has and is being cleared there in that connection. A new port and industrial facilities connected with the oil and gas industry are also expected at Kribi. That assessment is maintained excepting that B1 status is now included since EOO is newly calculated as 655 km^2.

Habitat: lowland evergreen forest, along and near the immediate coast, sometimes near mangrove.

Threats: see above.

Management suggestions: while atlantic mangrove itself contains only c. 6 woody species which are all common and widespread, forest behind mangrove on the landward side can contain rare species which are threatened. Apart from *Warneckea mangrovensis*, other examples in Lower Guinea are *Ternstroemia africana* and *Oxystigma mannii*. Representative areas should be protected for long term survival.

Warneckea ngutiensis R.D.Stone ined.

CR B2ab(iii)

Range: Cameroon endemic (SW: Banyang Mbo forest near Nguti).

Here *W. ngutiensis* is assessed as Critically Endangered since known from a single collection (*Pollard* 552) equating to an AOO of 4 km^2 (4 km^2 cell size) with threats as below. The species lacks anther oil glands in its flowers (Stone pers. comm.. to Cheek, 2009).

Habitat: lowland evergreen forest.

Threats: formerly a Wildlife Sanctuary, Banyang Mbo has had it protected status reduced due to high hunting pressure. Consequently it is more vulnerable to logging pressures.

Management suggestions: those managing Banyang Mbo should be given the means to recognise and protect this species, and data should be acquired on numbers of individuals, density, regeneration levels of this *Warneckea*.

Warneckea wildeana Jacq.-Fél.

EN B2ab(iii)

Range: Cameroon (S: Kribi-Ébolowa, Km 15; Kribi-Edéa, E of Bipaga 2) and Gabon (Cristal Mts).
Listed as VU D2 by WCMC in 1998 and accepted as such by IUCN in 2002, this taxon is here re-assessed as Endangered since only five locations are known (AOO 20 km² with 4 km² cells and threats as below). This is a very distinctive species on account of its greatly thickened nodes on older branches (Stone pers. comm. to Cheek 9 Dec. 2009).
A small tree to 8 m it has rounded branches and papery, elliptic leaves 20 × 9cm with 1.5 cm acumen and 1cm petioles. The inflorescences appear as small cushions 3–4 cm wide on older, leafless stems. There are 16–20 flowers in each.
Habitat: lowland evergreen forest.
Threats: slash and burn agriculture; road widening for upgrading transport infrastructure to the part of Kribi; future mineral extraction in the Cristal Mts.
Management suggestions: inclusion of a site within a protected area. Creation of a protected micro-site in consultation with the elders of Bipaga 2. Data on numbers of individuals, population density, regeneration levels and threats would inform future monitoring and plans to aid the survival of *W. wildeana*.

MELIACEAE

assessed by Martin Cheek

A pantropical family of trees, less usually shrubs and very rarely climbers. Most species occur in evergreen forest and have alternate, prinately compound leaves which lack stipules. The stems lack exudate (except for some *Trichilia*) but have a distinctive sweet clear odour. The flowers are symmetrical, with 4–5 free petals and a characteristic tubular staminal tube surrounding a nectar disc and superior ovary.

There is no Flore du Cameroun account for Meliaceae, but some genera have recently been revised for Africa, namely *Trichilia* by J.J.F.E. de Wilde (Meded. Landbouwhogeschool Wageningen 68-2, 207 (1968) and *Heckeldora* (Blumea 52: 179–199 (2007)), while publication of *Carapa* by D. Kenfack is awaited. When the last is available several additional species will probably need to be added to this account. Revisions of *Turreanthus* and *Guarea* for Africa are in progress under the supervision of J.J.F.E. de Wilde.

Heckeldora, for decades regarded as a monotypic genus, has recently been revealed by J.J.F.E. de Wilde (2007, op. cit.) to contain six species. Under[illegible]s and small trees, the genus is distinguished [illegible] indehiscent, orange, unilocular fruit with parietal placentation. In Cameroon *H. ledermannii* is the only threatened species but *H. leptotricha* (Harms) J.J. de Wilde, restricted to S Region and adjoining areas, is near threatened and may be shown in future to merit Vulnerable status under criteria A.

Entandophragma angolense (Welw.) C.DC.
Entandophragma cylindricum (Sprague) Sprague
Entandophragma utile (Dowe & Sprague) Sprague
Guarea cedrata (A.Chev.) Pellegr.
Khaya ivorensis A.Chev.
Lovoa trichilioides Harms
VU A2cd

The above are all internationally traded timber species of the mahogany family which were listed as Vulnerable by Hawthorne (Hawthorne 1997 www.redlist.org (IUCN 2003)) using the 1994 criteria of IUCN. They all have a wide range in Africa and, were they reassessed in this book, without reference to their use as timber, they would probably be downlisted. Hawthorne variously cites over-exploitation, poor levels of regeneration, fire damage, and slow growth to support his assessments of these species.

The assessment of VU under criterion A refers to the fact that more than 30% of the population has been lost in the last 100 years, mainly due to replacement of forest by agriculture and farm fallow in countries such as Ghana, Nigeria, Ivory Coast and Guinea (Conakry) which represent major parts of the range for these species. On the other hand these species are all diligently raised from seed and planted in reserves or otherwise favoured by governmental forest departments in most countries within their range, so as to maintain national timber revenues, so their extinction is unlikely, although large trees are certainly becoming increasingly rare because of the high value of their timber.

Heckeldora ledermannii (Harms) J.J. de Wilde
EN B2ab(iii)

Range: Cameroon endemic (SW: Fosimondi and Bakossi Mts. C: Mefou proposed NP. Littoral: Mts Ekomane (Kongoa)).
A shrub or small tree up to 5 m high, 5–9-foliolate leaves, glabrous apart from the puberulous midrib, calyx <1 mm long, young inflorescences 12–30 cm long, fruits apiculate but not beaked, nor ribbed, nor moniliform.
Here *H. ledermannii* is assessed as Endangered since three locations are known (AOO 12 km² with 4 km² cells) with threats as below. The type locality, Kongoa Mts, have not been visited by botanists in many decades. A fourth possible location has been identified from a specimen at the proposed Mefou NP near Yaoundé, but better material is needed to confirm the identification.
Habitat: submontane evergreen forest; 900–1500 m alt.

Threats: forest clearance for agriculture and wood at lower altitudes only on Mt Kupe, but wholesale at the Fosimondi site where extremely threatened by farm expansion.
Management suggestions: *Heckeldora ledermannii* appears secure in Bakossi, especially at Mt Kupe where it occurs at high density (given the number of specimens made there), so long as submontane forest continues to be protected here, the species is likely to survive.

Pterorhachis zenkeri Harms
VU B2ab(iii)
Range: Cameroon (C: Yaoundé e.g. Mt Eloumden) and Gabon (six locations).
This very strange species was accorded its own genus before a second species was recognised in Gabon. A treelet 1–2 m tall with a single unbranched stem and leaves with winged rachises and stellate hairs, *P. zenkeri* is easily recognised. Here it is assessed as Vulnerable since seven locations are known, all but the type locality are in Gabon (specimens det. by de Wilde, not seen by authors). AOO is 28 km² with 4 km² cells with threats as below.
Postscript: Yvette Harvey has brought to my attention that the Gabonese specimens previously ascribed to this taxon have now been re-named on the Gabon database as *P. le-testui,* suggesting that *P. zenkeri* may be considered by the WAG school as endemic to Cameroon (which seems perfectly credible on geographic grounds). If this is confirmed it is likely that this species will be re-assessed as CR in future.
Habitat: submontane evergreen forest on inselbergs in semi-deciduous forest belt, c 900 m alt.
Threats: urban expansion of the city of Yaoundé threatens the survival of forest on the hills that surround it.
Management suggestions: protection of natural habitat on one or more of the Yaoundé hills is highly advisable since they harbour a set of unique plant species apart from this taxon.

Trichilia zewaldae J.J. de Wilde
EN B2ab(iii)
Range: Cameroon endemic (S: Bipindi; 10 km NW Eséka. C: Mefou proposed NP).
A small or medium-sized tree to 25 m tall with 4–6-jugate leaves 29–62 cm long, distal leaflets obovate, more or less glabrous. Easily recognised by the beaked, 3-locular fruits (J.J. de Wilde op.cit.).
Known from six specimens at two or three locations (AOO 8 km² with 4 km² cells) and threats as below, *T. zewaldae* is here assessed as Endangered.
Habitat: lowland evergreen forest; 200 m alt.
Threats: slash and burn agriculture in S Region is a low level threat.
Management suggestions: inclusion of a site in a protected area is advisable. The Mefou proposed NP offers this potential if gazetted. Data on population density, range, regeneration levels and threats need to be gathered to inform management of this species.

Turraea pellegriniana Keay
VU B2ab(iii)
Range: Nigeria (Jemaa; Bauchi Plateau; Ijua in Obudu division) and Cameroon (E: Bertoua; Bétaré Oya. Adamawa: Minim near Tibati).
Described by Keay in 1956 (Bull. Jard. Bot. Brux. 26) this is a shrub or small tree to 8 m high with fascicles of up to 20 flowers and elliptic leaves 5–15 × 1.3–6 cm long with broad apical lobes 1 mm long including the anthers. The 1.3–1.5 cm long petals are green. *Turraea pellegriniana* is here assessed as Vulnerable since nine locations are known (AOO 36 km² with 4 km² cells) and threats as below. The species occurs most densely in the Bauchi Plateau of Nigeria which, if taken as one location, would decrease the total to five and qualify for Endangered status.
Habitat: submontane deciduous woodland.
Threats: clearance for agriculture and habitation in Bauchi Plateau of Nigeria.
Management suggestions: the locations in Cameroon should be evaluated for population density, range, regeneration and threats to inform management planning. Consideration to inclusion of a site in a protected area is advised.

Turraeanthus mannii Baill
VU B2ab(iii)
Range: Nigeria (Calabar; Oban) and Cameroon (SW: Korup. Central: Yaoundé. S: Kribi–Campo area).
A glabrous forest shrub or treelet 1.5-3m tall with 3-foliolate leaves and flowers in axillary cymes to 2 cm long. The 0.8–1.4 cm long flowers have petals united into the tube, characteristing the genus. The fruits are indehiscent, glossy, fleshy, yellow and irregularly 3-lobed; 5–6 cm wide.
Turreanthus mannii is here assessed as Vulnerable since ten locations are known (AOO 40 km² with 4 km² cells) with threats as below. At one location, Korup, the species is fairly common over a large area. The records from Yaoundé and from the Kribi–Campo area require checking since they may be another species. They have not been seen for this project.
Habitat: lowland evergreen forest.
Threats: forest clearance for agriculture and urban expansion, particularly in Oban and Calabar where it may be extinct already. The Kribi area is also at risk due to planned infrastructure development associated with the port.
Management suggestions: this taxon appears relatively well protected in the

Korup NP where the population appears most dense. It may also occur in the Campo Ma'an NP. Further data on population density, regeneration and threats would be advantageous in managing this taxon.

MENISPERMACEAE

assessed by Martin Cheek (based in part on the assessments of Pollard in Cheek *et al.* 2004: 167–169)).

A family of stem-twisting lianas, rarely shrubs (e.g. *Penianthus* in Cameroon), they have simple, alternate leaves with stalks swollen at base and apex but lacking stipules. The flowers are usually numerous, minute, in axillary clusters, 3-merous, with 3 free carpels giving apocarpous fruit, each fruitlet with a single curved, moon-shaped seed.
There is no Flore du Cameroun account for the family, but fortunately in 1962 G. Troupin published his Monographie des *Menispermaceae* africaines. Acad. Roy. Sci. Outre-Mer, Classe Sci. Nat. Méd. Mémoires in 8º, N.S. 13/2, 312 pp. (IRSAC, Brussels). This has provided a solid basis for the taxonomy of the family, that is, the recognition of genera and species. Most genera are identified by obscure floral characters.

Albertisia capituliflora (Diels) Forman (syn. *Epinetrum capituliflorum* (Diels) Troupin op. cit: 44)
EN B2ab(iii)

Range: Cameroon (S: Grand Batanga, Bipindi and Campo-Ma'an) and Gabon (Plateaux Batéké NP).
A climber with long-hispid stems, petioles old, leaf veins and fruits. Leaf-blades are papery,oblong or obovate-elliptic 15–25 × 5.5–13cm acuminate, shallowly cordate; basal nerves 5. Male inflorescences are capitulate, 7 cm long, outer sepals hispid. The fruits 3–4.5 × 1.5–2.3 cm (Troupin op. cit.: 44)
Here *Albertisia capituliflora* is assessed as Endangered since only four sites are known (AOO 16 km² with 4 km² cells) and threats are as below. EOO was calculated as 28185 km².
The Gabon record, *Walters* 1245, was identified by Ortiz. It has not been seen by us and requires checking since it is so disjunct from other records.
Habitat: lowland evergreen forest.
Threats: the Grand Batanga area, immediately S of Kribi, faces development for port and industrial infrastructure which would require loss of forest habitat.
Management suggestions: the Campo Ma'an record of Tchoutou offers the best hope for preserving this species, if the identification is confirmed, since it is a National Park. Surveys there should gather data on numbers of individuals, population density and regeneration [illegible] to provide plans and an identification guide for the park managers to aid intervention if required.

Albertisia glabra (Troupin) Forman (syn. *Epinetrum glabra* Troup op. cit.: 32).
EN B2ab(iii)

Range: Cameroon (C: Yaoundé. S: Dipikar Isl.; Bipindi fide Tchouto (2004) and Akoakas Rock).
This climber has glabrous stems and leaves as the name suggests. The stems are longitudinally striated, the petioles 1–2.3 cm long, bearing elliptic blades 7–11 × 3.5–5.5 cm, apex rounded-apiculate, base rounded or cuneate and with 7–10 nerves on each side of the midrib. Male inflorescence 3-flowered, peduncle 1 mm; sepals 12, the exterior triangular, 1–2 mm long, the inner 3 elliptic 3–4 mm long; synandrium conical, 2.5–3 mm long (Troupin op. cit. 42)
Here *Albertisia glabra* is assessed as Endangered since only the four sites above are known (AOO 16 km² with 4 km² cells) with the threats given below. EOO was calculated as 12272 km²
Habitat: lowland evergreen forest.
Threats: Yaoundé, the capital city of Cameroon is expanding and if the type locality has not already been destroyed by building in the near century since it was discovered it is at risk.
Management suggestions: if the species cannot be refound in the Yaoundé area, the best hopes for protecting it might be at Dipikar Is. A survey to refind the plant and gauge population range, density and levels of regeneration is needed to understand needs for management intervention. Consideration should be given to including one of the sites in a formal protected area and/or working with a local community to protect the same.

Platytinospora bucholzii (Engl.) Diels var ***bucholzii***
EN B1+2ab(iii)

Range: Cameroon endemic (Littoral: Abo. S: chutes Edéa; 11 km S Kribi).
A climber with petioles 4–5 cm long, peltate, leaf-blade leathery, ovate to ovate-oblong 6–9 × 4.5–6 cm, acumen linear 6–8 mm long, base rounded, basal nerves five; male inflorescences racemose 4–6 cm, pedicels 1.5 mm, sepals 6 outermost ovate, 1 mm long, petals and stames 6. Drupes ovoid 2 × 1.5 cm (Troupin op. cit.: 215)
Here this variety is assessed as Endangered since three sites are known (AOO 12 km² with 4 km² cells). EOO was calculated as 1785 km².
Habitat: lowland evergreen forest.
Threats: the chutes (rapids) at Edéa are now a hydroelectric dam. The coast road S of Kribi has seen much development of beach-front houses. The location of Abo is uncertain.
Management suggestions: this taxon should be refound in the wild and data gathered on population size, density and regeneration levels to enable an assessment of need for management intervention.

Platytinospora buchholzii (Engl.) Diels var. ***macrophylla*** Diels (originally assessed by Benedict Pollard, updated by Martin Cheek)
EN B2ab(iii)

Range: Nigeria (1 coll.); Cameroon: (SW: Mt Kupe (2 coll.). Littoral: Douala-Edéa Reserve (1 coll.). S: Bipindi (1 coll.). C: Yankafor II near Bafia (1 coll.); Mefou proposed NP (2 coll.)).

Differing from var. *bucholzii* in the membranous (not leathery) leaves 10–13 × 5.5–7.5 cm (not 6–9 × 4.5–6 cm). Considered by Troupin to be possibly a shade form of the typical variety. (Troupin op. cit.).

Zenker 3014a was cited by Diels (in Engler, Pflanzenr. 46: 168 (1910) as the type, collected at Bipindi, whence *Zenker* 4008, st., 1911 was also collected. From 1969, three collections were made within 10 years of each other: *Letouzey* 9610, fr. 26 Nov., at Yankafor II, 25 km ENE from Bafia, the fruiting Nigerian collection of *Wit & Daramola* in FHI 64887, from the Ago-Owu FR, W State, 28 Dec. 1971, and *McKey* 260, st., 26 May 1979, fallow fields by the village of Cité–Lac Tissongo, Douala-Edéa Reserve. Two further collections were made from Bakossi, along Walter's trail on Mt. Kupe itself in October 1995 (*Cheek* 7535, st., 1000 m) and June 1996 (*Cable* 2898B, st., 1000 m). Additional material is desirable to enhance our knowledge of this monospecific genus.

This taxon was assessed as CR A2c + 3c by Pollard in Cheek *et al.* (2004: 167–168) but is reassessed here as Endangered since the evidence of 80% population reduction is ambiguous, but it is only known from five locations (above; AOO 20 km^2 with 4 km^2 cells) with threats as below. EOO was calculated as 69545 km^2

Habitat: lowland evergreen and semi-deciduous forest and submontane forest; 0–1000 m alt.

Threats: forest clearance, especially at Mefou proposed NP and at lower altitudes at Mt Kupe, for agriculture. Even if the trees are left standing, but the undergrowth is cleared, vines can all be cut at ground level, as in the Loum FR understorey.

Management suggestions: the location of the subpopulations noted above should be revisited, and an assessment made of the number of individuals and regeneration levels as a baseline for future monitoring.gazettement of the proposed Mefou NP would provide a secure station for this taxon in the wild.

Sarcolophium suberosum (Diels) Troupin
VU B2 ab(iii)

Range: Cameroon (S: Ébolowa-Sangmélima; Fenda) and Gabon (Libreville, Lambarene to Fougamou; 10 km SSW Makoukou; Sibang; Moula-Yeno 45 km, Lastoursville-Okond).

Climber, the stem with corky plates. Leaf-blade ovate 7–25 × 15 cm, long acuminate, deeply cordate, leathery, glabrous, 3–5 basal nerves. Male panicules 8–12 cm, puberulent: flowers 2 mm long. Drupes ovoid 1–1.5 cm long, endocarp covered with ridges and wings (Troupin op. cit: 175–177). Troupin notes that *S. suberosum* when sterile could be confused with *Rhigiocarya racemosa* which also has corky stems, but that the first differs in having narrower leaves which lack the black patch at the base of the blade of the second. Here *Sarcolophium suberosum* is assessed as Vulnerable since eight sites are known (AOO 32 km^2 with 4 km^2 cells) with threats as below. EOO was calculated as 133617 km^2.

Habitat: lowland evergreen forest

Threats: in the hundred years since the type collection was made at Libreville, that settlement has become the capital of Gabon and expanded greatly, and continues to expand, at the extent of natural forest habitat so that the species if not already lost at its type locality is highly threatened.

Management suggestions: every effort should be made to refind this plant at its historic sites and to evaluate the possibilities for protecting one or more sites. Data on numbers of surviving individuals, density and regeneration levels would be useful to decide on needs for management intervention.

Tiliacora lehmbachii Engl. (originally assessed by Benedict Pollard, updated by Martin Cheek)
EN B2ab(iii)

Range: Cameroon: (SW: Mt Cameroon (2 coll.), Mt Kupe (1 coll.)) and Congo (Kinshasa) (1 coll.).

Climber, stem to 2 cm diam., young stems striated, glabrous; leaf-blade ovate-elliptic 9–18 × 5–9.5 cm long acuminate, base obtuse, glabrous, glossy above, petiole 2–3 cm. Male inflorescence cauliflorous, fascicles 5–15 cm, of several racemes of cymes, pubescent, flowers 2–4 mm long. Fruit unknown. (Troupin op. cit.: 56).

Engler described *Tiliacora lehmbachii* in 1899 based on *Lehmbach* 905 (type), fl., Nov. 1897 collected at Buea, 1000 m, the same approximate location as the second known collection, *Maitland* 313, fr., Jan. 1929. A Congo (Kinshasa) collection, *Germain* 404, fl., 29 July 1943, was made in 'Forestier Central, Yangambi, île Tutuku' at 470 m. The only other collection is *Cheek* 7809, fl., 15 Nov. 1995 at 960 m near Kupe Village. The data for this assessment is taken from Pollard in Cheek *et al.* (2004: 168), no new data are available so it is maintained here, excepting the removal of the Criterion A statement since loss of >50% of the population in three generations is not clear cut at this juncture. Three locations are known (AOO 12 km^2 with 4 km^2 cells) with threats as below. EOO was calculated as 69432 km^2.

Habitat: lowland and submontane forest; 470–1000 m alt.

Threats: forest clearance, especially at lower altitudes, for agriculture. Even if the trees are left standing, but the undergrowth is cleared, vines can all be cut at ground level, as in the Loum FR understorey. Buea is a large University town and has seen urbanisation in recent decades which has degraded forest habitat nearby.
Management suggestions: the subpopulation near Kupe village should be revisited, and an assessment of the number of individuals and regeneration made. There has been much vegetation clearance around Buea, so Mt Kupe offers the best opportunity for refinding this species in Cameroon.

Tiliacora odorata Engl.
EN B2ab(iii)
Range: Cameroon (S: Bipindi) and Gabon (Mouila-Kembélé, *Le Testu* 5222).
Climber with stems striated, leaf-blades sublanceolate, 12–15 × 4.5–5.5 cm, apex acuminate, base attenuate, glossy above and below, midrib with 6–8 lateral nerves on each side. Male inflorescences on stems below the leaves, in false racemes 12–22 cm long. Sepals 9, c 1 mm long; petals 6.
Here *Tiliacora odorata* is assessed as Endangered since only two locations are known (AOO 8 km² with 4 km² cells) and threats as below.
Habitat: lowland evergreen forest.
Threats: slash and burn agriculture in the Bipindi area.
Management suggestions: the Bipindi area should be revisited to refind this species and assess population density, range, regeneration and threats. If the species cannot be formally protected then local communities should be tasked with this objective.

Triclisia lanceolata Troupin (originally assessed by Benedict Pollard, updated by Martin Cheek)
EN B2ab(iii)
Range: Cameroon (SW: Bakossi Mts (1 coll.). C: Mbalmayo (1 coll.) and Congo (Kinshasa) (3 coll.).
Climber, stems striate, ferruginous to glabrescent; leaf-blade elliptic-lanceolate to elliptic 12–18 × 4.5–7cm, acuminate, base rounded. Female inflorescence cauliflorous, cymes 1–2 cm long, densely pubescent, flowers 4–5 mm long. Drupes subovate 1.5 cm long, shortly pubescent.
Troupin described *Triclisia lanceolata* in 1949, from two Gillett specimens collected in the 'District du Bas-Congo, environs de Kisantu' citing *Gillett* 3395, from 1903 as the type and *Gillett* s.n., from 1909. A third Congolian collection, *Compère* 1239, was made on 21 Jan. 1960 (Leopoldsville Province, Territ Thysville). *W.J.J.O de Wilde* 1900 was the first Cameroon collection, made on 12 Feb. 1964, about 5 km S of Mbalmayo, S of Yaoundé, on the border of the Nyong R. It then remained uncollected until *Cheek* 9169, on 2 Nov. 1998 at Edib village in the heart of Bakossiland, Cameroon. This collection was of fallen fruits, which match well those on all the other specimens. A 1946 collection from SW Nigeria, *Jones* & *Onochie* in FHI 17554, during the vegetational reconnaissance to Omo and Shasha FRs, is annotated (at K) as *sp. nr. lanceolata*. It differs in having larger, membranaceous (not coriaceous) leaves, and is not included in this assessment. The ♂ flowers are unknown, a fact less surprising when one considers that these vines are probably undercollected. This could be due to the difficulty of collecting fertile specimens from high up in the canopy. Although widespread, it appears to be rare within its range.
The data for this assessment is taken from Pollard in Cheek *et al.* 2004: 168, no new data are available so it is maintained here, excepting the removal of the Criterion A statement since loss of >50% of the population in three generations is not clear cut at this juncture. Four locations are known (AOO 16 km² with 4 km² cells) with threats as below. EOO was calculated as 93508 km²
Habitat: lowland and submontane forest; 0–1500 m alt.
Threats: forest clearance, especially at lower altitudes, for agriculture. Even if the trees are left standing, but the undergrowth is cleared, vines can all be cut at ground level, as in the Loum FR understorey.
Management suggestions: the subpopulation near Edib village in Bakossi should be revisited, and an assessment of the number of individuals and regeneration made.

Triclisia macrophylla Oliv. (originally assessed by Benedict Pollard, updated by Martin Cheek)
EN B2ab(iii)
Range: Sierra Leone (1 coll.), Equatorial Guinea (Bioko (1 coll.)) and Cameroon (SW: Mt Cameroon (1 coll.); Mt Kupe (1 coll.); Bakossi Mts (1 coll.)).
Climber to 10–20 m, young stems red puberulent; leaf-blade ovate-oblong to broadly elliptic, 18–25 × 10–18cm, acumen short, base deeply cordate, petiole 5–10 cm. Male inflorescence of contracted fascicles 2 cm long, flowers with 2 bracteoles, sepals 9–12, 3–4 mm long, puberulent.
Daniel Oliver described *Triclisia macrophylla* in 1868, based on Mann's collection in Jan. 1860 from Bioko. It was subsequently collected on the African mainland at Mt Cameroon in Nov. 1891, *Jungner* 144 (UPS), and soon after by the Swedish botanist Afzelius in Sierra Leone (1896, *Afzelius* s.n., UPS). It remained uncollected for 100 years, before *Cable* 2739, fl., 30 May 1996, 1000 m alt., near Kupe Village. As part of a series of 25 m × 25 m plots conducted during botanical surveys in Cameroon, a final collection (st.) was made in the Bakossi Mts, in 1998, as *Plot Voucher* B96, from Kodmin at 1500 m. Although widespread, it is rare within its range.
The data for this assessment is taken from Pollard in Cheek *et al.* (2004: 168–169), no new data are available so it is maintained here, excepting the removal of the Criterion A statement since loss of >80% of the population in three

generations is not clear cut at this juncture, and downrating to EN status. Five locations are known (AOO 20 km² with 4 km² cells) with threats as below. EOO was calculated as 263307 km².
Habitat: lowland and submontane forest; 0–1500 m alt.
Threats: continued clearance of low altitude forest for plantations, agriculture and urban expansion. Troupin (1962: 93) describes its habitat as "fôrets denses humides sempervirentes de basse altitude", which would suggest below 1000 m. It is likely that all of these collections were from at or below 1000 m, and on Mt Kupe the level of protection is above this altitude. Some subpopulations have probably already disappeared.
Management suggestions: the subpopulations near Kupe Village and Kodmin should be revisited, and an assessment of the number of individuals made. Local communities should be involved in their protection.

MORACEAE

assessed by Martin Cheek

Best known for the genus *Ficus*, this family are characterised by white exudate when wounded and stipules which completely, or nearly so, embrace the circumference of the stem. Ranging from herbs (most *Dorstenia*) to large forest trees (*Ficus* and *Milicia*), all species are united in having minute flowers with bifurcate styles in, usually inconspicuous, compound inflorescences and also in having alternate simple leaves. Cornelius Berg of the Netherlands has revised all the African species of each genus and, with collaborators completed the Flore du Cameroun treatment in 1985. These assessments depend on his work.

Dorstenia is an amphi-atlantic genus of herbs, rarely forest shrubs. They were treated by Berg & Hijman in Flore du Cameroun 28 (1985) where full specimen data for Cameroon is cited. Berg & Hijman produced a revision of the genus in 1999 (Ilicifolia 2, Univ. Bergensis) recognising 105 species. Two new varieties and a range extension were added for Cameroon by Pollard *et al.* in Kew Bull. 58: 185–193 (2003).

Generally we do not assess infra-specific taxa in this book due to resource constraints, unless such taxa have already been assessed elsewhere. As a result numerous local varieties of *Dorstenia* have not been included in this treatment. Those which would probably merit threatened status are: *D. barteri* var. *paucinervis* (Hijman & Berg) Hijman, *D. poinsettifolia* var. *angularis* Hijman & Berg, *D. poinsettifolia* var. *angusta* (Engl.) Hijman & Berg, *D. poinsettifolia* var. *librevillensis* (De. Wild.) Hijman & Berg, and *D. poinsettifolia* var. *staudtii* (Engl.) Hijman & Berg.

Dorstenia angusticornis Engl.

VU B1+2ab(iii)

Range: Cameroon endemic (SW: S Bakundu FR; Limbe. Littoral: Loum–Yabassi 20 km; 50 km NE Edéa. C: Eséka, 10 km SSE. S: Bipindi).
Six locations are know for *D. angusticornis*, based on nine specimens equating to an AOO of of 36 km² with 4 km² cells. EOO was calculated as 12874 km². Given the threats adduced below it is here assessed as Vulnerable.
Dorstenia angusticornis is a 0.5–1 m tall branched shrub with twigs with retrorse uncinate hairs. The obovate or elliptic leaves are sometimes teethed. The inflorescence is strongly zygomorphic, each with two diametrically opposed strap-shaped appendages c. 2 cm long, fringe and bracts absent.
Habitat: lowland evergreen forest.
Threats: the S Bakundu FR between Kumba and Limbe which is home to four of the nine records of *D. angusticornis* has seen great loss to agricultural incursions, especially yam farming.
Management suggestions: since the greatest concentration of records for this species is the S Bakundu area, it is logical to look at protecting it here if it has survived agricultural incursions. However pressure in this area may be so high that the idea is impractical. Data on regeneration and density levels would inform needs for management intervention. Limbe BG would be a logical base for managing this species and devising propagation protocols and ex situ approaches if needed.

Dorstenia astyanactis Aké Assi

VU B2ab(iii)

Range: Guinea (Conakry) (Pic de Fon; Mts Bero; Mt Ziama), Ivory Coast (Mt Tonkui (several coll.) and Cameroon (SW: Bakossi Mts (Nyale Rock; Ngomboku); Mt Kupe (Nyasoso-Nature Trail). NW: Nkom-Wum FR, below Bu village).
Initially assessed as Endangered (EN B2ab(iii))on the basis of only five locations being known, with threats (Pollard *et al.* 2003 loc. cit: and Cheek *et al.* 2004:169). In 2006–2009 three new locations were found in Guinea (Conakry) in the Guinea Highlands taking the total to eight with an AOO of 32 km² (4 km² cell size). EOO was calculated as 287834 km². Threats are documented below. Therefore, the species is reassessed here as Vulnerable. 500 plants are estimated at Bero and 10–20 at Ziama (Haba & Haba pers. comm. 2010).
To Martin Etuge goes the distinction of discovering all four of the Cameroonian sites for this species, probably the most bizarre, and certainly the only epiphytic, *Dorstenia* in Africa. First discovered in Ivory Coast by Aké Assi where it is only known from Mt Tonkui. *Dorstenia astyanactis*, published in 1997, is there either extinct or on the verge of extinction due to forest loss (Laurent Gautier *pers. comm.*). In Cameroon it

is on the edge of extirpation by agricultural expansion at all its sites but Nyale. It is to be hoped that this expansion will cease.

Since the 2004 assessment no new data has come to light on the Cameroon populations. The Ivory Coast locality has been inaccessible due to civil war. Several plants have been recorded at each of the new sites in Guinea. Indeed at one of these, Pic de Fon, thousands of individuals over a large area have been recorded due to detailed surveys in connection with a proposed open cast iron ore mine (van der Burgt, Thomas Williams, Seydou Cisse pers. comm. 2010). Fortunately almost all this subpopulation occurs outside the proposed mining area and Rio Tinto, the company concerned, are planning to protect *Dorstenia astyanactis* at this site for its long term survival.

Epiphytic herb, stem pendent, ascending apically, suculent, c. 50–150 × 1 cm; leaves in spirals, 3–5 crowded at stem apices, elliptic, 5–14 × 3–9 cm, margin denticulate, glandular, puberulous; infloresences solitary; receptacle discoid, obovate, 2–3.5 x 1.5–2.5 cm; fringe c. 1 mm, with a single terminal appendage 2–3.5 cm, borne apically.

Habitat: submontane forest; 600–1000 m alt.

Threats: forest clearance for agriculture and wood in Cameroon.

Management suggestions: conservation education as to the significance of this species may help the survival of this species in Cameroon at Nyasoso, Nyale and at Bu. Collection of baseline populational data is needed for future monitoring. Propagation to multiply numbers of individuals for reintroduction at locations in Cameroon may be needed.

Dorstenia dorstenioides (Engl.) Hijman & C.C.Berg
EN B1+2ab(iii)

Range: Cameroon endemic (S: Kribi-Ébolowa 36 km; 58 km E Kribi, at Fenda; Assok, 60 km ESE Kribi; Ébolowa-Kribi 110 km).

Restricted to a small area between Kribi and Ébolowa where four sites are known (AOO 16 km² with 4 km² cells) with threats as below, *Dorstenia dorstenioides* is here assessed as Endangered. EOO was calculated as 45 km².

A branched, puberulous shrub to 1.5 m tall, the distichously arranged leaves are more or less oblong or elliptic 5.5–16 × 3–6.5 cm, abrubtly acuminate, with 6–10 pairs of lateral nerves and caducous 3–4 mm stipules. Inflorescences solitary, green, receptacle stellate, 0.8–1 cm diam., puberulous, flowering face suborbicular, flat, fringe ± absent; appendages 8–11, subtriangular to 3 mm long, obtuse, unequal.

Habitat: lowland evergreen forest; to 200 m alt.

Threats: as one of Cameroon's three main ports, Kribi is expanding its industrial and urban footprint as new development projects take place. New gas processing and mineral export facilities have been proposed. Upgrading and developing new transport links to the hinterland will threaten sites for *D. dorstenioides*.

Management suggestions: none of the known sites are in protected areas it is believed. This should be rectified. A programme is advisable to educate local communities within the AOO of the rarity of this species, how to recognise and protect it. Data on population density and regeneration levels on *D. dorstenioides* is needed to guage the need for management intervention

Dorstenia involuta Hijman & C.C. Berg
VU B1+2ab(iii)

Range: Cameroon (S: Kribi-Ébolowa-Edéa-Campo-Lolodorf area (17 specimens equating to c. 8 localities).

This species while restricted to the area indicated above, seems relatively common locally. Further survey work may remove it from the threatened list (using criterion B – heavily based on number of sites). However, its coastal forest habitat has major threats (see below). EOO was calculated as 6781 km². An AOO of 68 km² is estimated using 4 km² cells, therefore *Dorstenia involuta* is here assessed as Vulnerable.

This is a shrub 0.5–2 m tall, sometimes branched, with puberulous stems. The obovate, oblong or linear leaves are 6–20 × 2.5–6.5 cm, acuminate, with 7–14 pairs of looped lateral nerves. The pedicel gradually thickens into the top-shaped receptacle which is 0.5–1 cm diam. and with 6–8 radiating lobes which extend further as reflexed primary appendages 0.3–0.6 cm with involute margins.

Habitat: lowland evergreen coastal forest to 200 m alt.

Threats: forest along the Kribi-Campo coastal road is increasingly being cleared to make way for touristic developments while forest around Kribi itself is threatened by new industrial developments, urban expansion and slash and burn agriculture.

Management suggestions: the Kienké FR, site of one specimen, (km 13 Kribi-Ébolowa) could be considered for upgrading to National Park status if the species still exists there, and other threatened species also occur there. It is possible that the species also occurs in the Campo-Ma'an NP. Data on regeneration levels and population density would allow gauging of management intervention if needed. Educating local communities to recognise and protect this species is suggested.

Dorstenia poinsettiifolia *Engl.* var. ***achoundongiana*** *Cheek & Bygrave* (syn. *Dorstenia poinsettiifolia* Engl. var. nov. *sensu* Cheek in Cable & Cheek 1998: 87)
EN B2ab (iii)

Range: Cameroon endemic (SW: Mt Cameroon at Bimbia-Bondadikombo and at Etinde, Upper Boando).

Herb to 60 cm, stems fleshy, often decumbent with basal nodal rooting, ascending to erect, swollen at the nodes, hirsute and increasingly so towards apex, whitish hairs dimorphic: short patent hairs c. 0.2 mm long interspersed with patent-curved, c. 0.6 mm long hairs. Leaves in spirals; lamina elliptic-oblong, 11–18 × 3–8 cm, apex bluntly acuminate, base cuneate, truncate to subcordate, margin crenate, lateral veins c. 10 pairs, upper surface glabrous, dark green, underside pale green, sparsely hairy along the midrib and lateral veins with a mixture of uncinate hairs, c. 0.2 and 0.6 mm long; petiole 1–3 cm long, longest towards stem base; stipules narrowly triangular, c. 2 mm long, persistent. Inflorescences strongly zygomorphic, axillary, in fascicles of 1–4 developing in sequence (contracted cyme); peduncles 3–4 mm long, indumentum as for stem; receptacle narrowly elliptic, navicular, 15–22 × 5–9 mm, underside with two longitudinal ridges, sparsely puberulous, only hairy at the edges; upper surface with green and orange fringe, 1 mm broad, crenate, extended into two opposing subcylindrical primary dorsiventral appendages, each c. 5 mm long; flowering area deep purple, convex, subangular-oblong, 12–15 × 3–5 mm, pustulate.

Differing from other varieties of the species in the navicular receptacle, with two subcylindric 5 mm long primary appendages, one at each end.

Apparently restricted to two locations in lowland evergreen forest on the foothills southeast of Mt Cameroon. The forests around both of these locations are under enormous pressure from conversion of land to agriculture and urbanisation due to local population increase in nearby towns and villages.

Dorstenia poinsettiifolia Engl. var. *achoundongiana* Cheek & Bygrave was assessed as Endangered in Pollard *et al.* 2003 loc. cit. That assessment is maintained here since no new data are available, except that threats in the Bimbia area have increased in recent years and several rare forest species there are believed to have been extirpated.

Habitat: in deep shade, herb layer of relatively undisturbed lowland evergreen forest, 0–500 m altitude.

Threats: see above.

Management suggestions: surveys are needed at Bimbia and Upper Boando to determine if this variety survives there, and if so, to collect basic populational data as a baseline for future monitoring. Propagation and multiplication of plants at Limbe BG for reintroduction to the Bimbia Community Forest may be needed.

Dorstenia poinsettifolia Engl. var. ***etugeana*** B.J.Pollard

EN B1+2ab(iii)

Range: Cameroon (SW: Rumpi Hills (1 site); Bakossi Mts (Nyale Rock, Nyale-Kodmin and Enyandong)).

Four sites in two broad locations (above) are known for this very distinct variety of *Dorstenia*. AOO is 8 km² with 4 km² cells and threats are as below. EOO was calculated as 12 km². The assessment of Endangered was first published in Pollard *et al.* (2003), followed in Cheek *et al.* (2004: 169) and is maintained here since no additional data are available.

D. poinsettiifolia Engl. var. *etugeana* B.J.Pollard seems most likely to be confused with *D. poinsettiifolia* Engl. var. *librevillensis* (De Wild.) Hijman & C.C.Berg but differs in having only 1 basal primary appendage (not 2, basal and apical), the flowering face of the receptacle being narrowly ovate (not elliptic-rhombic) and the stipules 1.2–2 mm long (not 3–6 mm).

Herb 35–70 cm tall; stem monopodial, ± fleshy, rarely woody, often decumbent with basal nodal rooting, ascending to erect, swollen at the nodes; indumentum with short white hairs, 0.2–0.5 mm long; leaves in spirals; lamina obovate, elliptic or oblanceolate, usually widest in the apical third, often narrowed below the middle, 5–13 × 2.5–6.0 cm, apex acute or bluntly acuminate, base rounded to subcordate, ± decurrent on the petiole for 1–3 mm, margin entire to subdenticulate, subpandurate or bluntly dentate with 1–8 teeth; venation penninerved, brochidodromous, prominent beneath, impressed above, secondary veins in 9–13 subopposite or ± alternate pairs; petiole 15–20 mm; stipules caducous or persistent towards the stem apex, subulate, 1.2–2.0 mm long; inflorescences strongly zygomorphic, solitary, axillary, developing successively, 1–3 borne at any one time, green or purplish; peduncle 15–20 × 0.5–2.5 mm; receptacle discoid, strongly convex, lower surface densely and minutely puberulent; fringe 1.5–5.0 mm broad, reflexed; primary appendages 1, basal, 5.9–10.2 cm × 0.9–2.3 mm, often slightly inflated apically, secondary appendages both apical, 1–8, filiform, 0.4–1.1 cm long and lateral, 2–9, ± triangular, 0.1–0.2 cm long; flowering face narrowly ovate.

Habitat: lowland and submontane forest; 450–1000 m alt.

Threats: forest loss due to expansion of agriculture. There has been a decline in habitat quality due to an observed increase in production of cash crops such as cocoa by local villagers throughout the region.

Management suggestions: annual monitoring of sites to detect deleterious changes, and further surveys of unexplored areas to detect potential new sites for the taxon. Consultation with communities to explain the rarity of this plant.

Dorstenia prorepens Engl.

VU B1+2ab(iii)

Range: Nigeria: (Okomu FR; Gambari FR; Oban) and Cameroon: (SW: Mt Cameroon (at Etome, Kumba–S Bakundu and Buea); Mt Kupe (at Kupe village and Nyasoso). S: Lolodorf).

This species was previously assessed as VU in Cheek *et al.* (2004: 169). Since that date an additional location, Lolodorf, has come to light, and one, Bioko, has been removed as an error. Eight locations, are known. AOO is 32 km² with 4 km² cells and threats are as below. EOO was calculated as 11269 km².

Herb to 40 cm, stem creeping to ascending, often forming small mats on the ground; leaves elliptic to oblong or obovate, 2–9 × 1–5 cm, margin repand or with 1–3 blunt broad teeth, almost 3-lobed; petiole 0.5–1.5 cm; stipules persistent, 0.1–0.4 cm; inflorescence zygomorphic, usually solitary, purplish; receptacle naviculate, 1.2–2 × 0.2–0.5cm; fringe 0.5 mm; appendages numerous, black-purple, 0.1–0.2 cm.

Habitat: lowland and submontane forest; 400–1550 m alt.

Threats: forest loss due to agricultural expansion and wood excavation, particularly at Bambuko (Cable & Cheek 1998: xxxvi–xxxvii), and due to urban expansion at Kumba and at Buea.

Management suggestions: surveys of subpopulations at each known site to detect whether the taxon survives at these, and if so, to gather quantitative data on individuals, regeneration and threats. This data could inform a plan to guide management of the survival of this species. The densest and least pressured subpopulations of this species in Cameroon are at Etome and at Mt Kupe. These might be the best centres to focus on conservation of *D. prorepens*.

Ficus L., the figs, are most recently treeated by C.C. Berg and J.T. Wiebes (1992) in African Fig Trees and Fig Wasps, North-Holland. Berg draws upon his Flore du Cameroun 28 (1985) and Fl. Gabon 26 (1984) accounts, as well as those for other African floras. He recognises 105 species for Africa, of which 34 occur in Cameroon. Most of these are widespread and common. Only those listed below are rare, localised and threatened. *Ficus chlamydocarpa* subsp. *chlamydocarpa* was previously assessed as VU A2c; B2ab(iii) in Cheek *et al.* 2004: 169 then known only from 10 sites in Cameroon, but is now known to extend to Equatorial Guinea (Bioko and Annóbon) and São Tomé and Príncipe so is downrated to NT (Onana ms).

Ficus tremula subsp. *kimuenzensis* was assessed in Cheek *et al.* 2004: 170 as VU B2ab(iii) being then known from 7 localities in Nigeria, Cameroon, Congo (Kinshasa) and Angola. However, this omitted three localities mentioned in Fl. Gabon 26: 208 (1984) and one in Flore du Congo Belge 1: 132 (1948). Therefore 11 locations are now known and the assessment is downrated to NT.

Ficus abscondita C.C.Berg

VU B2ab(iii)

Range: Cameroon ([illegible] [illegible]-Loum; Loum-Solé. C: Ndikini[illegible] [illegible] and Congo (Brazzaville) (Bouba – [illegible] mapped and not that in Lekoumou region).

In addition to the five localities listed above, the Checklist of Gabonese Plants, p. 282 lists *Breteler* 9779, determined by Jongkind, from a different province to that in which Bélinga occurs. Bélinga is the only other site from which *F. abscondita* is known in Gabon. Further locality data on *Breteler* 9779 is not available to us so it is not mapped here, however we can take it that there are six sites for *F. abscondita*, equating to an AOO of 24 km² with 4 km² cells, and with the threats below, allowing the assessment here of Vulnerable. EOO was calculated as 100495 km².

Berg described this species in Adansonia 20: 264, fig.1 (1981) and notes that *F. asbscondita* is an epiphytic or epilithic, sometimes lianescent shrub up to 3m tall. The oblonceolate leaves are 28–40 × 8–15 cm with an abrut acumen and long free broad persistent stipules 2.5–6.5 cm long. Berg (op. cit.) states that the species is characterised by its relatively small figs hidden by the long stipules. The figs have distinctively swollen apices. Sterile specimens differ from the similar *F. preussii* in having hollow not solid twigs.

Habitat: lowland or lower submontane forest.

Threats: Bélinga in Gabon is about to be a Chinese open-cast iron mine. Forest in the Nkongsamba-Loum area has seen major habitat loss for both small-scale and plantation agriculture since the volcanic soils are fertile and major centres of population are near.

Management suggestions: the Loum-Solé-Nkongsamba-Ndikiniméki area with half the world's sites for *F. abscondita* is the logical area in which to concentrate conservation efforts. One of the sites is at Bakaka FR which also houses several other rare species, discovered by Dutch botanists in the 1970s. This forest reserve should be revisited, and if not too degraded considered for upgrading to National Park status if such species as *F. abscondita* can still be found there. Data on population density and regeneration levels of this species are needed to evaluate the need for management intervention.

Ficus jansii Boutique

EN B2ab(iii)

Range: Cameroon (W: Bangwa, 12 km NW Bangangté; Bafoussam-Bangangté at Bayangam), Gabon (Ogooué-Ivindo) and Congo (Kinshasa) (Bas-Congo: Gombé).

In addition to the four sites above, a fifth, represented by *Floret* 1544 exists (cited in The Checklist of Gabonese Plants) being the second site in Gabon. However, since the locality is not given it has not been included on our map. Given five localities (AOO 20 km² with 4 km² cells) and the threats adduced below, *F. jansii* is here assessed as Endangered under criterion B. EOO was calculated as 58863 km².

Ficus jansii is a tree up to 7 m tall, probably terrestrial. The leathery leaf blades have up to 14 vein pairs and are oblong to elliptic, 16–27 × 9–16 cm with a rounded apex and subacute base, Berg (1992: 61 and 112) keys it out on having basal bracts caducous and stipules to 1 cm long.

Habitat: lowland to submontane evergreen forest, to 1500 m alt.
Threats: both localities in Cameroon are in the centre of the Bamiléké area which owing to the agricultural industry of those people has seen almost total destruction of natural habitat, especially forest. The most recent collection, *De Wilde* 2595 dates from 1964.
Management suggestions: efforts should be made to refind this species in Bangangté Cameroon, targetting ceremonial or sacred forests, and trees surviving in hedgerows or on steep or rocky slopes since all other forest areas are thought to have been extirpated.

Ficus oresbia C.C.Berg
CR B2ab(iii)
Range: Cameroon endemic: (W: Massif Mbepit, 30 km SW of Foumban. Adamawa: ? Tchabal Mbabo).
This montane tree is only known with certainty from the type collection at Mbepit. It is believed that the Tchabal Mbabo collection has not yet been confirmed by Berg.
Distinct in the white puberulous, partly retrorse hairs of the thick stems. The obovate to oblanceolate leaves are 7–8 vein paired, 15–25 × 5–9 cm with subacuminate apex and obtuse to acute base. The figs are axillary, single or in pairs, 2–2.5 cm diam., puberulous.
Habitat: montane evergreen forest; 1900 m alt.
Threats: fires set in the dry season threaten forest in the Foumban area (Ndam pers. comm.)
Management suggestions: the identity of the Tchabal Mbabo specimen needs review by Berg. While it is to be hoped that other mountains in the Cameroon Highland chain will yield this species, effort should focus on refining it at Mbepit, the type locality. Data on ecology, population regeneration and density are needed to gauge the need for management intervention, which is likely.

Ficus subsagittifolia Mildbr. ex C.C.Berg
VU B2ab(iii)
Range: Cameroon (C: Nanga Eboko, 120 km NE Yaoundé. S: Kribi; Ébolowa-Yaoundé; Bipindi) and Gabon (Ndjolé; Bokoué R; Petit Okano; Libreville; Makokou). Known from nine localities (AOO 40 km² with 4 km² cells), *Ficus subsagittifolia* is here assessed as Vulnerable due to the threats adduced below. EOO was calculated as 188488 km². It is easily recognised due to the lanceolate leathery leaves c. 30 × 8 cm, constricted at the second quarter from the base, the base itself dilated and cordate.
Habitat: lowland evergreen forest.
Threats: the sites at both Kribi and Libreville are under great threat from urban expansion and development projects if they have not already been lost. Elsewhere slash and burn agriculture poses a threat.

Management suggestions: within Cameroon, the Kribi-Ébolowa area should be the focus for conservation of this species since it holds most of the known sites. It is believed that none are in areas protected for conservation purposes. This should be rectified if possible. Data is needed on population density and regeneration to gauge management intervention if needed.

MYRICACEAE

assessed by Martin Cheek

A widespread family of trees and shrubs, often aromatic, usually in temperate, open areas. *Myrica*, the type genus, is well-known in north temperate bogs. The species of *Morella* were formerly included in *Myrica*, until researchers in E Africa discovered that the tropical African genera are a separate genus.

There is only a single species of the family in Cameroon. This shares with the rest of the family the exstipulate, simple, alternate leaves which are scaly-glandular on the lower surface. The flowers lack sepals and petals and are held in axillary spikes, the ovary superior, with 2 styles and 1 ovule, producing a warty drupe.

Morella arborea (Hutch.) Cheek (syn. *Myrica arborea* Hutch. FWTA 1: 589 (1958))
VU A2c, 3c
Range: Equatorial Guinea (Bioko: Moka area (4 coll.)) and Cameroon (SW: Mt Cameroon (two sites, 6 coll.); Mwanenguba (6 coll.). W: Bamboutos-Djuttitsa; Massif Mbam. NW: Bamenda Highlands; Djottin; Ijim Ridge; Dom (1 coll.)).
Morella arborea is one of the few montane tree species restricted to the Cameroon Highlands. Within this area it has wide range, but is generally rather rare. It does not grow in dense forest, but outside it, usually in rocky grassland areas, appearing as stunted, gnarled, aged trees. The thick bark of the trunk suggests fire resistance. The aromatic, simple, oblong leaves are distinctive, they are oblong, c. 5–9 × 1.3–3 cm, rounded to a mucronate apex and with an unequal, truncate base, lateral nerves are 12–16 pairs, the margin is subserrate. This text is modified from that in Cheek *et al.* (2004: 170). EOO is 18490 km².
While secure on Mt Cameroon, the species is highly threatened in the rest of its range.
Habitat: submontane and montane forest edge, or isolated in grassland; 1300–2400 m alt.
Threats: clearance of trees for agriculture (e.g. by fire for both pastoral and tilled land) and wood, particularly firewood. The assessment that 30% of the habitat of *Morella arborea* will be lost over the next hundred years, mainly in

the Mwanenguba-Bamenda Highlands sector of the Highlands, is maintained here. In addition it is estimated that this percentage loss has already occurred over a similar duration in the same sector. In the Dom area of the Bamenda Highlands, 50% loss of habitat occurred in 15 years recently (Baena in Cheek *et al.* 2010: back cover).

Management suggestions: research is needed on the ecology and demography of this species.

MYRSINACEAE (INCLUDING MAESACEAE)

assessed by Martin Cheek

Woody plants of forest occurring throughout the tropics, the Myrsinaceae of Africa are distinguished by having black or brown dots or streaks within their leaves and petals, best seen when held up to the light. The leaves are alternate, simple, often toothed, pinnately nerved, lacking stipules. A red-brown, scurfy covering is often present on young growth. The flowers have five petals united into a short inconspicuous tube at the base together with the five, often sagittate, stamens. The superior, single-styled ovary develops into a juicy berry.

In Cameroon, the family is represented by *Rapanaea melanophloeos*, a widespread afromontane forest tree now transferred to *Myrsine*. *Embelia* are lianas. *Maesa* are trees and lianas often treated as a separate family, Maesaceae. The largest genus in tropical Africa is *Ardisia* with 40–50 species, of which half are found in Cameroon. Although there is no Flore du Cameroun treatment of Myrsinaceae, de Wit revised *Ardisia* (as *Afrardisia*) in 1958, recognising 16 species (Blumea, suppl. IV: 241–262). Taton in 1979 updated this, recognising many more species, for a total of 39 (Bull. Jard. Bot. Nat. Belg. 49: 81–120).

Recently, two new species have been described from Cameroon (Cheek, pp. 72–77 in Cheek *et al.* 2010) and another is in press (Cheek & Xanthos in Kew Bull. in press). Many of the species are easily confused with each other by the non-specialist, as occurs in so many genera with numerous species. In addition, further difficulties and confusion have arisen from the fact that many *Ardisia* in Cameroon were described from specimens subsequently destroyed in Berlin in 1943 making interpretation of the names concerned difficult. De Wit (op.cit.) lectotypified many of these names, often using material distant from the type locality, and sometimes morphologically discordant and questionable. Taton (1979 op.cit.) resolved some of these difficulties but others remain, so that species identifiation can be troublesome.

Apart from the 16 species assessed as threatened below, *Ardisia ototomoensis* might be regarded as threatened but is suspected to be the same as *A. dewitiana* (see under that species). *Embelia mildbraedii* Gilg & Schellenb. (*E. sp. aff. welwitschii* of FWTA) was listed as LR nt in Cable & Cheek (1998: lvii) but may merit reassessment as Vulnerable.

Ardisia alabastro-alata Taton

VU D2

Range: Cameroon endemic (SW: Mt Kupe).

Erect shrub, 1 m, branches horizontal, red-puberulous; leaves elliptic 11.5–17.5 × 4–6 cm acuminate, shortly cuneate, lateral nerves 13–15 pairs, reticulum raised below, punctae numerous, conspicuous; petiole 1 cm; flowers in subfascicles of 6, flower buds distinctly winged; calyx 1.8 mm, lobes ovate-triangular, 1.5 mm wide, on ciliate, punctate; corolla 5.7 mm white, tube 2 mm, lobes 2.5 mm wide.

Distinctive and named for the distinctly winged and white (alabaster coloured) flower buds. Described only from the type at Mt Kupe but subsequently material from Gabon has been identified as this species (Gabon database). Omitted from Cheek *et al.* 2004 by oversight.

Here *Ardisia alabastro-alata* is assessed as Vulnerable under Criterion D since although no direct threats are currently known, only one at possibly a few locations are known and numbers of individuals currently known is probably less than ten.

Habitat: submontane forest; 1950 m alt. (Mt Kupe).

Threats: no direct threats are known but the species is vulnerable to path clearance for trails to the summit of Mt Kupe for ecotourism.

Management suggestions: existing material of *Ardisia* from Mt Kupe should be reviewed in case some additional specimens have been overlooked; Gabonese material should be verified; surveys on Mt Kupe near the summit should target this species and, apart from collecting basic populational data for future monitoring, make an assessment of numbers of individuals. It is possible that under Criterion C this taxon would rate EN or CR.

Ardisia atrobullata Taton

EN B2ab(iii)

Range: Cameroon (Littoral: route Razel, Douala-Edéa) and Gabon (Cristal Mts at Mt Muélakéné).

Subshrub 0.4–0.5 m, young stems yellow tomentellous; leaves elliptic, 9–18 × 3.5–5 cm, acuminate, acute, below minutely lepidote, lateral nerves 10–14 pairs, reticulum prominent, punctae few but large and conspicuous; petiole 0.5–1 cm; flowers in fascicles of 7–10, bracts 1.2 mm, lepidote, ciliate; flower 5.5 mm long, calyx lobes ovate, 1.8 mm, acute, yellow-puberulent, lepidote; corolla pale purple,

5 mm; ovary longitudinally ribbed, densely lepidote; fruit globose, 6 mm.
Distinct in the few but large and conspicuous leaf-punctae, visible in relief on both surfaces, and in the yellow-puberulous and lepidote indumentum.
Here *Ardisia atrobullata* is assessed as Endangered since only two locations are known (AOO 8 km^2 with 4 km^2 cells) with threats as below.
Habitat: lowland evergreen forest.
Threats: logging for timber, followed by agriculture; urban expansion. All apply in the Douala-Edéa area. Razel was a timber extraction enterprise.
Management suggestions: efforts should be made to pinpoint the original location and to rediscover the species so that local authorities can be helped in protecting it and so that basic populational data can be gathered as a baseline for monitoring. The Douala-Edéa reserve may be the best hope for rediscovering and then protecting this species in Cameroon. Two National Parks exist in the Cristal Mts.

Ardisia bamendae Cheek ([Echia (Oku)] (syn. *Ardisia kivuensis* Taton, see Cheek *et al.* 2010: 76–77; *Afrardisia cymosa* (Baker) Mez, FWTA 2: 31 (1963).
EN A2b,c
Range: Nigeria (Chappal Wadi) and Cameroon (SW: Mwanenguba. NW: Mt Oku and Ijim Ridge (14 coll.); Tabemken).
Shrub, 0.3–2 m; leaves ovate-elliptic or oblong, 13–18 cm long; petioles pink; flowers purple or reddish, in small clusters along branches; fruits bright red.
Ardisia bamendae, previously known in Cameroon as *A. cymosa* or *A. kivuensis,* was assessed as Endangered in Cheek *et al.* 2010 since population reduction was estimated as >50% in the last 100 years due to loss of its montane forest habitat (see threats below). That assessment is maintained here.
Habitat: upper submontane and montane evergreen forests with *Schefflera abyssinica, S. mannii, Prunus africana, Bersama abyssinica, Syzygium staudtii, Nuxia congesta;* 1650-2620 m alt.
Threats: it has been calculated that 96.5% of the original forest of the Bamenda Highlands has been lost (Cheek in Cheek *et al.* 2000: 49). That which remains is also under great threat, for example, between 1988-2003 (15 years) about 50% of the Kejojang forest was lost (Baena, in Cheek *et al.* 2010: back cover). The losses of forest habitat of this species in that part of its range outside the Bamenda Highlands (e.g. the Bamboutous Mts) appear to have been even more extreme. These losses are ongoing.
Management suggestions: efforts should be made to pinpoint the original location and to rediscover the species so that local authorities can be helped in protecting it and so that basic populational data can be gathered as a baseline for monitoring. The Douala-Edéa reserve may be the best hope for rediscovering and then protecting this species in Cameroon. Two National Parks exist in the Cristal Mts.

Ardisia conraui Gilg
EN B2ab(iii)
Range: Cameroon (Littoral: Baré near Nkongsamba; SW: Bangwe) and Congo (Kinshasa) (Konduë on the Sankuru in Kasai at Lualaba).
Monopodial shrub; leaves oblong or oblong-lanceolate, acute or broadly & shortly acuminate margin entire or weakly undulate-cuneate; punctae translucent, dense, 400 plus per cm^2, petiole less than 15 mm; inflorescence fasciculate; pedicel 6–12 mm; calyx less than 2.5 mm, not mucronate, ovate-triangular.
Ardisia conraui was united by de Wit 1958 with two other species, *A. rosacea* and *A. ledermannii.* Each of the three were described from a single specimen, all destroyed at B in 1943. De Wit attributed several more recent specimens to the species, but Taton (1979) specifically excluded these, while following de Wit in uniting the three species. It seems very unlikely that *A. ledermannii* of lowland Kasai in Congo (Kinshasa) should be conspecific with a species otherwise known only from submontane forest in the Cameroon Highlands. De Wit's conclusions were based purely on his reading of the protologues, which by today's standards, lack important details. Nevertheless, despite these doubts, Taton (1979) is followed here for the moment. Accordingly, *A. conraui* is here assessed as Endangered since known from only three locations (AOO 12 km^2 with 4 km^2 cells) with threats as below.
Habitat: submontane and lowland forest; 340–c. 800m alt.
Threats: not seen in 100 years, or at least not identified. Both sites in Cameroon have seen forest destruction and the species may be lost at these type localities. Threats in Kasai are not known.
Management suggestions: rediscovery of this species at Bangwe, the type locality, should be attempted by searching forest in the area. If successful, local authorities should be consulted and basic populational data should be collected to aid future monitoring. Seed should be collected and plants raised for reintroduction to safe sites within the natural range.

Ardisia dewitiana Taton
EN B2ab(iii)
Range: Cameroon endemic (Littoral: Mt Nlonako. C: Yaoundé at Ototomo. S: Kribi–Lolodorf km 13; Nkoakam, 38 km WSW Ébolowa).
Subshrub, stem erect, 50 cm, glabrous; lateral stems absent or few, horizontal; leaf-blade elliptic-oblong, 7.5–16.5 × 2.7–5.5 cm, acuminate, cuneate, margin entire or undulate, lateral nerves 7–14 pairs, glabrous above, below with dense yellow scales; reticulum conspicuous; translucent punctae

sparse, inconspicuous; fascicles 3–4-flowered; bracts lanceolate, lepidote; pedicel 4–5 mm; calyx lobes lanceolate, <2 mm, acuminate; corolla white, lobes 4.5 mm; fruit globose, 8 mm.

Here *A. dewitiana* is assessed as Endangered since only four locations are known (AOO 16 km² with 4 km² cells) with threats as described below.

Habitat: submontane evergreen forest.

Threats: Nlonako has seen forest clearance to provide agricultural land to supply nearby Nkongsamba town; the same threats, with urbanisation, face the Yaoundé sites. Slash and burn agriculture are threats in S Region.

Management suggestions: delimitation with *A. ototomoensis* should be revisited since the last shares its only location with *A. dewitiana* and differs in only a single, minor, character. Rediscovery at one or more of the four locations should be attempted and baseline populational data collected for future monitoring with local representatives involved in protection. None of the locations are believed to be formally protected.

Ardisia dom Cheek

CR D

Range: Cameroon (NW: Dom, two sites).

This 1.5–2m tall understorey forest shrub is distinct in its 2 cm long peduncles, ridged stems, and leaves with conspicuous, large and dense black dots within the blade. More information is available within Cheek *et al.* (2010: 72–75), where *Ardisia dom* was described as new to science and first assessed.

Only two individuals, at different sites, are known for this species. One site is unprotected and vulnerable to clearance.

Habitat: submontane forest; 1600–1830 m.

Threats: on current data this species is so rare that even low-level human activity in the forest habitat, such as path cutting could easily threaten the survival of this species.

Management suggestions: seed of *Ardisia dom* should be collected from the mother plants, and seedlings raised for incorporation in the existing forest restoration programme at Dom. All forests at Dom should be resurveyed to target *Ardisia dom* and other of the rare species discovered during the completion of the Dom checklist, so that we have a more accurate idea of how many individuals there are, and what the regeneration levels. Such data is needed to inform development of a management plan. A poster campaign featuring this species might heighten awareness of its rarity and increase the chances of its survival.

Ardisia ebolowensis Taton

EN B2ab(iii)

Range: Cameroon endemic (S: Ebolowa; Nkoemvon; Campo-Ma'an).

Here *Ardisia ebolowensis* is assessed as Endangered since known from only three locations (above; AOO 12 km² with 4 km² cells) and with threats as below.

Shrub 50 cm tall, branched, young stems yellow-puberulent; leaves narrowly elliptic-oblong or lanceolate, 5–11 × 1.2–2.6 cm, acuminate, cuneate, lateral nerves 20-22 pairs, entire, lower surface sparsely lepidote, translucent punctae irregular in form, numerous (500 per cm²), yellow; fascicles 1–3-flowered; bracts narrow, triangular, to 3 mm; calyx lobes narrowly ovate 1.8 mm; corolla lobes 4–8 mm; fruit globose, 4–8 mm. Recognised by the small leaves with numerous lateral nerves and very dense, irregular translucent punctae.

Habitat: lowland evergreen forest.

Threats: slash and burn agriculture; logging followed by agriculture.

Management suggestions: fortunately Tchouto (2004) has recorded this species in the Campo-Ma'an NP which is therefore a logical focus for protecting it. Tchouto should be engaged to rediscover this plant, collect basic populational data and engage with park and traditional authorities to protect and monitor it.

Ardisia etindensis Taton

EN B2ab(iii)

Range: Cameroon endemic (SW: Mt Etinde at Boando; S: 50 km NW Eséka).

Originally assessed by me as CR A1c + 2c in Cable & Cheek (1998: lvi), on the basis of population reduction. This now seems inappropriate given the lack of data on the population at Eséka. Here the species is reassessed as Endangered since only two locations are known (AOO 8 km² with 4 km² cells) with threats as below.

Erect shrub 75 cm, unbranched, young growth red tomentose; leaves elliptic, 10.5–16 × 3.7–5.5 cm, acuminate, cuneate, margin entire or undulate, minutely brown lepidote below; lateral nerves 10–15 pairs, reticulum lax and difficult to see; punctae translucent numerous, inconspicuous, white; petiole 12–18 mm, fascicles 3–8-flowered; pedicels 6–8.5 mm, bracts triangular, 2–3.5 mm; flowers 7 mm long lacking punctae, calyx 2.8 mm, lobes narrowly triangular; corolla 6–7 mm; ovary glabrous; fruit globose, 6 mm.

Unusual in the absence of punctae from the flowers.

Habitat: lowland to submontane forest; c. 600 m alt.

Threats: clearance of forest habitat for plantation and small-holder agriculture (Boando at Mt Etinde).

Management suggestions: rediscovery of the species at Boando and Eséka, if it survives there; collection of basic populational data as a baseline for monitoring; engagement with local community elders to protect; improved species recognition in Cameroonian *Ardisia*.

Ardisia koupensis Taton

EN B2ab(iii)

Range: Cameroon endemic (SW: Mt Kupe (numerous coll.) and Littoral: Mt Nlonako).

Shrub 2 m, glabrous, branches horizontal; leaf-blades elliptic, 11–16 × 3.5–6 cm, acuminate, cuneate, obscurely crenulate, lateral nerves 11–13 pairs, reticulum conspicuous, punctae warty, sparse, but at margin numerous; petiole 7–12 mm; fascicles 6–7-flowered, bracts caducous, pedicels 14–19 mm; calyx 1.5–1.7 mm, lobes broadly ovate, 1.6–1.9 mm wide, apex rounded, punctae translucent; corolla 6.2–6.4 mm, translucent lines; ovary ovoid, glabrous; fruit ellipsoid 12 × 17 mm, apiculate, pedicel accrescent, 3 cm.

Unusual in the broadly ovate calyx lobes with rounded apices, the ellipsoid fruit on pedicels extending from 1.4–1.9 cm in flower to 3 cm in fruit.

Ardisia koupensis was assessed as Endangered in Cheek *et al.* (2004: 170) since two locations were known (AOO 12 km² with 4 km² cells but allowing two cells for the broad location at Mt Kupe) with threats as below. This assessment is maintained here since no additional data are available.

Habitat: submontane forest; 900–1200 m alt.

Threats: forest clearance for agriculture is ongoing a Mt Nlonako, excouraged by proximity to the large town of Nkongsamba and the Douala-Bafoussam Rd. Currently forest at this altitude in Mt Kupe is mostly secure except below 1000 m alt.

Management suggestions: assessment of species density, regeneration, and range at both locations will aid future monitoring; engagement with communities, assisted with a conservation poster, will aid local understanding of protection needs; improved species recognition tools to *Ardisia* species would aid this work.

Ardisia letouzeyi Taton

CR B2ab(iii)

Range: Cameroon endemic (S: Kribi-Ébolowa, km 81 at Mbanga).

Here *Ardisia letouzeyi* is assessed as Critically Endangered since known from a single location (AOO 4 km² with 4 km² cells) and since threatened by slash and burn agriculture.

Subshrub 60 cm; stems erect, unbranched, glabrous, longitudinally grooved when dried; leaf-blade narrowly elliptic, 11–20 × 2.3–2.8 cm, long acuminate, cuneate, margin entire, lower surface with orange scales, punctae sparse, inconspicuous, translucent-yellow; fascicles 2–3-flowered, subsessile, bracts narrowly triangular; pedicel 3–6 mm; calyx 1.5 mm, lobes ovate acuminate, punctate; corolla 5 mm; ovary ovoid, glabrous.

Unusual in the long slender leaves with only sparse, inconspicuous, translucid punctae.

Habitat: lowland evergreen forest.

Threats: as above.

Management suggestions: rediscovery of this species at its only known location is the first priority. Basic populational data should be recorded and the people of Mbanga informed and asked to protect their species. Monitoring of the species is recommended. If not refound at Mbanga, searches in intact forest between Ébolowa and Kribi are suggested. Improved identification aids are needed for *Ardisia* species.

Ardisia oligantha (Gilg & Schellenb.) Taton

CR B2ab(iii)

Range: Cameroon endemic (SW: Mt Cameroon).

Here *Ardisia oligantha* is assessed as Critically Endangered since known from a single location much threatened by habitat destruction (see below; AOO 4 km² with 4 km² cells). The assessment replaces that of CR A1c+2c in Cable & Cheek (1998: lvi–lvii) for the reasons explained under *A. etindensis* (above).

Probably a shrub, glabrous except for the puberulous young growth; leaves elliptic 8–14 × 3.5–5.5 cm, acuminate, base cuneate, membranous, margin crenulate-repand; lower surface sparsely lepidote with numerous dark punctae, petioles 1–1.5 cm. fascicles 10–12-flowered, flower buds ovoid, subtended by bracts; pedicels 6–8 mm; sepals free, broadly ovate, mucronate, punctate; petals elliptic 4–5 mm mucronate, punctate; anthers mucronate, punctate on back; ovary ovoid, lepidote on top.

The description is taken from de Wit (1958) who took it from the protologue, presumably, since he stated that he had "seen no specimen belonging to *A. oligantha*". This speaks for the distinct nature of this species since for other 'lost' German *Ardisia*, de Wit usually attributed new material to them. Taton (1979) also accepted and keyed this species. The exact location of the type on Mt Cameroon (*Weberbauer* 48) is not given in the references cited but would be useful in targetting efforts to refind this species.

Habitat: probably lowland at submontane evergreen forest.

Threats: urbanisation (Limbe), plantation agriculture (bananas, rubber, oil palm, tea, cacao) and timber extraction are all threats on and around Mt Cameroon.

Management suggestions: research in Germany, perhaps at Berlin, might reveal where Weberbauer collected on Mt Cameroon, so as to refind this plant, if it is not already extinct. If the search is successful the usual populational data should be collected as a baseline for monitoring. Local community leaders should be notified and engaged in recognising and protecting this plant. Seed should be collected for banking (if proved to be of the orthodox type) and sown at Limbe BG to multiply plants for reintroduction in safe sites, and for educational purposes.

Ardisia platyphylla Gilg & Schellenb.) Taton

EN B2ab(iii)

Range: Cameroon endemic (S: Bipindi; near Batanga at Nkolebunde).

Here *A. platyphylla* is assessed as Endangered since it is known from only two locations (above; AOO 8 km² with 4 km² cells) and with threats as below.
Shrub to 1 m, glabrous; leaves elliptic 12–20 × 6–9 cm acuminate or obtuse, base acute, lateral nerves c. 12 pairs., forming a looping marginal nerve, together with an irregular tertiary marginal vein; lower surface sparsely brown lepidote; punctate on both surfaces; margin more or less crenulate; petiole 5 mm; fascicles c. 5-flowered, pedicels 8 mm; sepals lanceolate 2.5 mm, highly acute, punctate; petals white, as sepals, 6 mm; anthers not punctate; ovary lepidote at apex.
Recognised by the broad leaves 2–2.3 times as long as broad, and the 2.5 mm slender and acute sepals.
Habitat: lowland evergreen forest.
Threats: slash and burn agriculture.
Management suggestions: rediscovery of the species at the known locations is advised, in consultation with local leaders and officials. Basic populational data as a baseline for future monitoring is needed. Local leaders should be involved in protecting the species.

Ardisia polyadenia Gilg
CR B2ab(iii)
Range: Cameroon endemic (NW/SW: 'Kebo', probably modern Batibo).
Probably a shrub, height unknown; young shoots rusty tomentellous; leaves elliptic 12–6 × 4–6 cm, acuminate-acute, base cuneate, glabrous, lower surface with numerous small prominent dots, closely reticulate, crenulate; petiole 2–2.5 cm; fascicles c. 10-flowered in June; pedicels c. 5 mm; calyx lobes triangular-ovate, acute, obscurely punctate; petals red, c. 5 mm long, with dark lines and dots.
Only known from the type specimen, *Conrau* 215, lost in Berlin 1943, so unseen by de Wit and Taton. Although accepted by both, de Wit commented that it appears close to *A. staudtii*, a common species, but that new material from the type locality is needed before a decision can be taken. Since this is still awaited, the species is here maintained. *Ardisia polyadenia* is here assessed as Critically Endangered since only a single location is known (AOO 4 km² with 4 km² cells) and threats as below.
Habitat: lowland evergreen forest
Threats: forest clearance for timber extraction and agriculture.
Management suggestions: research on Conrau's collection sites is needed to pin down exactly where his 'Kebo' is. A search for *Ardisia* in the location is then needed. If this succeeds, basic populational data should be gathered as a baseline for future monitoring and leaders of the community nearest the forest concerned should be involved in its protection.

Ardisia sadebeckiana Gilg
EN B2ab(iii)
Range: Cameroon (S: Grand Batanga) and Gabon (Mfoa; La Nkoulounga; Mbel, station CTFT; Libreville).
Herbaceous or low shrublet, rhizome creeping, long, apex of stem densely and shortly hairy, soon glabrous; leaves broadly ovate, 6–12(–23) × 5–8(–12) cm, acute, base rounded or cordate-auriculate, entire, glabrous, punctae black, dense on both sides, lateral nerves 6–8(–12) on each side of the midrib, tertiary nerves inconspicuous; fascicles c. 6-flowered; pedicels 8–10 mm; sepals ovate-triangular, 2 mm, punctate, petals c. 6 mm; anthers not punctate; ovary lepidote at apex.
One of only four herbaceous African Ardisia species. Among these distinct in the large size of the leaves, often with a cordate base, and in the swamp habitat.
Here *A. sadebeckiana* is assessed as Endangered since only five locations are known (AOO 20 km², with 4 km² cells) with threats as below.
Habitat: lowland evergreen swamp forest.
Threats: urbanisation (Libreville), commercial infrastructure including extractive industries; touristic development (Batanga).
Management suggestions: rediscovery in the Batanga-Kribi area should be attempted, failing which Gabon must be the best prospect for the survival of *A. sadebackiana*. If successful, the usual basic populational data should be gathered as a baseline for future monitoring and to assess needs for management intervention. Local community leaders should be informed and involved in protecting the species.

Ardisia schlechteri Gilg
CR B2ab(iii)
Range: Cameroon endemic (SW: Mt Cameroon at Bibundi).
Here assessed as CR since a single location (above; AOO 4 km² with 4 km² cells) is known, with threats as below. Assessed in Cable & Cheek (1998: lvii) as CR A1c+2c.
Subherbaceous shrublet ± 30 cm tall, stem rhizomatous; very densely tomentose when young, then glabrous; leaves oblong, obovate-elliptic or broadly-elliptic, 8–12 × 4.5–5 cm, shortly acuminate, base cuneate, 10–14 pairs lateral nerves; ± entire; densely, minutely punctate above and below; strongly reticulate; petiole 10–13 mm; fascicles 3–6-flowered, subsessile; pedicels 3–5 mm; sepals triangular-ovate, 2 mm; punctate; petals 4.5–5 mm, punctate.
One of only four herbaceous African *Ardisia* species. Differing from the other three in characters set out under *A. ebo* above.
Accepted by de Wit 1958 (op. cit.) and Taton (1979 op. cit.) although the type specimen, *Schlechter* 12417, was not seen by them, being destroyed at Berlin in 1943. Subsequently two modern specimens from Gabon have been determined as

A. schlechteri (Gabon database), but these are treated with caution here until verified. Should they be confirmed the assessment will be changed to EN.
Habitat: lowland evergreen forest.
Threats: Bibundi still exists as a small village on the coast. Most of the forest around it has been converted to large-scale plantations of tree-crops such as oil palm (per. obs.).
Management suggestions: searching forest scraps near Bibundi with community representatives to find *A. schlechteri* should be attempted with support from the Limbe BG. If successful, basic populational data should be collected as a baseline for future monitoring and seed collected for propagation at Limbe BG followed by reintroduction at safe sites. Investigation of the Gabonese specimens referred to above is needed.

Embelia mildbraedii Gilg & Schellenb. (*E. sp. aff. welwitschii* (Hiern) K. Schum., FWTA 2: 32). Ntoh (Oku).
VU A2c
Range: Cameroon endemic (SW: Mt Cameroon (six coll.); Mt Kupe. NW: Bamenda Highlands (Mt Oku and Ijim Ridge)).
This taxon was assessed as LR/nt in Cable & Cheek (1998: lvii) and as NT in Cheek *et al.* (2004: 350). Here it is reassessed as Vulnerable since it is estimated that >30% population reduction has occurred due to loss of montane forest habitat in the Bamenda Highlands. 50% was lost in 15 years recently in one study area (Baena and Cheek *et al.* 2010: back cover).
Liana, about 4 m tall; twigs dark brown, covered with white lenticels; leaves elliptic, 4–8 × 2.5–4.5 cm, apex acuminate, base obtuse-acute, margin serrulate, abaxial surface silvery green, secondary veins 12–18, venation brochidodromous, glabrous; petiole about 1 cm long; inflorescence racemose, axillary; peduncle about 2–2.5 cm long; pedicel 0.2–0.4 cm long; flowers small, pinkish; fruit spherical, red, 4 mm diam., stigma persistent.
Eight collections of *E. mildbraedii* were made in the 1996 and 1998 inventories at Mt Oku and the Ijim Ridge, showing it to be relatively abundant at Mt Oku and the Ijim Ridge which is the only secure site for the species in the Bamenda Highlands. Most of these collections were made above Oku-Elak in 1996. Although *E. mildbraedii* is highly threatened in the Bamenda Highlands by forest clearance, there is little pressure on its survival in the higher altitude forest on Mt Cameroon. This plant is used in beehive construction at Oku (*Munyenyembe* 830).
Habitat: montane forest; 2100–3200 m alt.
Threats: clearance of forest for agriculture.
Management suggestions: protection of surviving forest in the Cameroon Highlands.

MYRTACEAE

assessed by Martin Cheek

The Myrtaceae in Africa are recognised by being woody plants with opposite simple leaves punctuated with translucent gland dots which, when the leaf is torn, release aromatic oils into the air. The stems lack both coloured exudate and stipules. The lateral nerves usually unite near the margin into an intramarginal nerve. The flowers have numerous free stamens and an inferior ovary producing a many-seeded berry. The family is best known for the non-African *Eucalyptus*, cloves (*Syzygium aromaticum*), and Guava (*Psidium* spp.). The two African native genera are *Syzygium* and *Eugenia*, separated as follows:

> Flowers axillary; sepal lobes distinct; petals falling separately from each other............................. *Eugenia*
> Flowers in terminal cymes; calyx truncate; petals forming a cap and falling off as one unit.......*Syzygium*

Most species of *Syzygium* are widespread and fairly common. They are often medium sized forest or woodland/grassland trees. Of these only *S. staudtii,* a montane tree with white trunk, root buttresss and square stems, may rate as threatened in Cameroon and can be treated as NT.

The genus *Eugenia* has many forest shrub species which are rare and localised. Amshoff worked on the S American species in the 1940's and 1950's before starting work on Africa in the 1960's culminating in her Fl. Gabon account of 1974. However, her Flore du Cameroun account was sadly, never finished, and no synopsis or revision of the genus resulted. This leaves species delimitation and identification in Cameroon problematic and uncertain. Keay's 1954 FWTA treatment is useful but covers only SW and NW regions of Cameroon. Two published species which are both poorly known, and not represented by material available to us, are *E. ledermannii* Engl. & Brehm. listed as such in FWTA (it may be a synonym of *E. gilgii* from its range) *and E. poliensis* Aubrev. & Pellegr., known only from the type at Mt Poli). It is perfectly possible that further research will reveal this to be valid species which are threatened.

Eugenia fernandopoana Engl. & Brehmer was assessed by Cheek in Cheek *et al.* (2004: 171) as VU B2ab(iii). Here it is downrated to NT since five locations are now known in Gabon (Gabon database) taking the total to 11, beyond the threshold for VU. Should any of these sites be lost or populations drop by 30% or more, the species will be elevated to VU status once again.

Eugenia ancorifera Amshoff

CR B2ab(iii)

Range: Cameroon endemic (Littoral: Douala-Bafia at Kikot Bridge on Sanaga R, 70 km SSW of Bafia).

Described by Amshoff (Adansonia 14: 481-483 (1974)) this species is still known from a single specimen. The species is remarkable for the small leaves and gland-free stigma.

A many stemmed shrub to 2 m with pubescent branches. The glabrous, oblong-elliptic leaves are 2–4 × 0.8–1.8 cm with an acute apex and 10 pairs of lateral nerves. The petioles are 3 mm long, pubescent. The flowers are axillary, often in pairs, the pubescent pedicel c. 2 mm, sepals 1.5 mm pubescent, petals white, 3 mm. Here *E. ancorifera* is assessed as Critically Endangered since known from a single site (AOO 4 km^2, with 4 km^2 cells) and threats as below.

Habitat: river bank, alt. Unknown.

Threats: renewal of the bridge at the type locality, or river extractive industries focussed at the bridge and its road might threaten this species.

Management suggestions: a revision of African *Eugenia* might uncover more specimen-sites for this rare species. Meanwhile, it seems logical to attempt to rediscover this species by searching at the type locality, Kikot Bridge, also a site for *Polycarpaea rheophytica* (EN). If refound, a search should be made in the area for further plants, to estimate the range, density, regeneration levels and detailed threats that the species may face.

Eugenia bucholzii Engl.

CR B2 ab(iii)

Range: Cameroon endemic (Mungo R)

Not seen since the type gathering in the 1870's although two specimens from Nigeria have been considered as possibly conspecific: *Hepburn* 69 (Mada Hills) and *Magajie and Tuley* 218810A (Obudu Plateau) but are not mapped here since they do not fit the FWTA description.

A shrub with branchlets, petioles and flowers densely rusty-pilose. The leaves are ovate, 5–6 × 2.5–3 cm, long acuminate or long caudate, the lateral nerves 5–6 pairs, flowers solitary, fruit globose.

Here *Eugenia bucholzii* is assessed as Critically Endangered since only a single site is known (AOO 4 km^2 with 4 km^2 cells) and threats are as below. It is not known where along the Mungo R Bucholz collected his specimen. It is possible that his site has since been lost to Plantation or other development.

Habitat: lowland evergreen forest.

Threats: slash and burn agriculture, plantation agriculture, logging followed by agriculture, all affect the Mungo R forests although much remains that is intact.

Management suggestions: a revision of existing specimens of *Eugenia* might reveal additonal specimen sites for this rare taxon. Failing this, any plant survey teams in the Mungo R area should be briefed to look for this taxon. If found, the usual populational, regeneration and threat data should be collected and options for protection reviewed with local communities and officials.

Eugenia dusenii Engl.

VU D2

Range: Cameroon endemic (SW: Mana R, rapids E border of Korup NP).

This rehophytic shrub is known only from the rocky banks of rapids in the Mana R near Mundemba where in the wet season it is submerged and most survive the fast moving currents. Its area of occupancy is estimated as 4 km^2. EOO is calculated as 156 km^2. It is probably intermittent along a stretch of river of several kilometres. One event could conceivably wipe out the species although mercifully there is no immediate threat to it. For these reasons it is here assessed as Vulnerable under Criterion D.

An erect shrub 0.3–0.6 m tall with short internodes and erect, narrowly lanceolate leaves 2–4 × 0.2–0.4 cm. The flowers are subsessile and sparse.

Habitat: rocky river banks at rapids in evergreen forest; 100–200 m alt.

Threats: none are known.

Management suggestions: education of the public as to the interest and rarity of this species through the offices of the Korup NP is advisable. Mapping of this species by surveying the length of the Mana and its tributaries, recording population densities and regeneration levels, would assist needs for management intervention, if any.

Eugenia gilgii Engl. & Brehm.

CR A2C

Range: Cameroon (W: Bamboutos Mts (three coll.). NW: Bamenda Highlands (seven coll.). Adamawa: Ngaoundéré (one coll.) and Nigeria (Obudu; Mambilla Plateau & Vogel Massif (13 coll.).

Tree 3–10(–15) m tall, bole pale brown, spiny when juvenile, exudate absent. Young stems brown-puberulent, glabrescent. Leaves opposite, elliptic, c. 9 × 5 cm, apex subacuminate, rounded, base obtuse, secondary nerves 5–6 on each side of the raised midrib, margin revolute at leaf-base; petiole 0.5–1 cm long. Inflorescence fasciculate, 5–10-flowered, below the leaves; pedicels c. 7 mm long, with a minute pair of bracts in the upper half. Flowers 6–8 mm diam., petals pink or pinkish white. Fruit red, shortly ellipsoid, c. 1 × 0.7cm.

The original assessment of CR A1c in Cheek *et al.* (2000: 67–68) was maintained by Cheek in Harvey *et al.* (2004: 67–68) and is maintained here albeit modified from A1 to A2 since A1 erroneously indicated that forest loss had ceased, while in fact it is ongoing. Although new sites, such as at Dom, have come to light, the measured rates of forest loss have increased (see Baena in Cheek *et al.* 2010: back cover).

The assessment is based on the estimate (founded on GIS studies such as Baena op. cit.), that >80% of the forest habitat of this tree species has been lost in the last 100 years. *Eugenia gilgii* appears restricted to the few forest patches of the Bamboutos and Bamenda Highlands in Cameroon, and neighbouring areas in Nigeria, which are the most deforested sections of the Cameroon Highlands. EOO is calculated as 3960 km².

This species is easily recognised by its spiny trunk when young, and the leaf blade margins which are folded back on themselves near the petiole. The brown puberulent stems and pink flowers and also distinctive. Extensive detail on populations in the Mt Oku area is given in Cheek *et al.* (2000: 67–68).

Habitat: drier lower montane forest, often at edges; (1200–)1500–2000 m alt.

Threats: clearance of forest for wood and land for agriculture.

Management suggestions: continued and improved educational support for individuals and communities managing submontane forest in the Bamenda Highlands is advised. Reforestation schemes in the Bamenda Highlands should incorporate this species into their planting mix, as already happens with those of ANCO in the Dom area, sponsored by RBG, Kew.

Eugenia kalbreyeri Engl. (syn. *E. dawei* Hutch. & Dalz.)

VU B2ab(iii)

Range: Sierra Leone (4 locations); Guinea (Conakry), Liberia and Ivory Coast (one site each) and Cameroon (SW: Mt Cameroon).

First described and named for *Kalbreyer* 157 from Mapanja, 600 m alt, on Mt Cameroon, this 3 m shrub has oblong-lanceolate acuminate leaves that exceed 10 × 4 cm, but bear only 6 widely looping marginal nerves that do not form a definite marginal nerve and are minutely pubescent below. The pedicels are 4–10 mm and, with the ovary, densely pilose.

Here *E. kalbreyeri*, united by Keay (op. cit.) with *E. dawei* of Sierra Leone, is assessed as Vulnerable since only seven locations are known (AOO 28 km² with 4 km² cells) and since threats are as below. EOO is calculated as 392250 km². The species is discussed by Cheek in Cable & Cheek (1998: lvii).

Habitat: lowland evergreen forest; c. 600 m alt.

Threats: forest on Mt Cameroon, the type location, is severely threatened at lower altitudes and has seen great losses in the last century. It is possible that the species has already been lost from its only site in Cameroon. Most lowland forest in Sierra Leone has already been lost and that which remains is under pressure from agriculture. None of its sites are known to be in protected areas.

Management suggestions: a revision of African *Eugenia*, particularly those in Cameroon is needed and might reveal that *E. kalbreyeri* has a wider distribution in Cameroon than is currently indicated. The Mapanja area should be searched for this species but since it was not found in the Mt Cameroon surveys of the 1990's the prospects of success do not seem good and Sierra Leone, where it seems most common, probably offers the best chance of survival for it.

Eugenia kameruniana Engl. (syn. *E. hankeana* Winkler)

VU B2ab(iii)

Range: Cameroon endemic (SW: Mt Cameroon; Bipindi to Debundsha. C: 50 km from Yaoundé at Ottotomo Res.; Mt Eloumden at Yaoundé. S (according to Tchoutou (2004)): Ebianemeyong; Ma'an; Nyabéssan).

The original specimens being lost at Berlin, Amshoff was able to identify specimens of this species from the last two named sites near Yaoundé by reference to a drawing of the type at BM by Baker (Amshoff op. cit. (1974)). The exact locality of the type specimen, *Dusen* 9, is unknown.

Eugenia kameruniana is here assessed as vulnerable since six sites are known (AOO 24 km² with 4 km² cells) and threats as below. EOO is calculated as 29237 km². Earlier I assessed this species as CR A1c (Cheek in Cable & Cheek 1998: lvii–lviii) in ignorance of Amshoff's paper (op. cit.) and the discoveries of Tchoutou that were to follow in 2004.

Shrub usually <0.5 m tall, leaves elliptic, 12–16 × 5–8 cm, obtuse-acuminate, base attenuate; lateral nerves 8–10 pairs; petiole 3 mm, stout; flowers in axillary fascicles, pedicels 5 mm, glabrous, sepals 2.5 mm, petals pink, ovary glabrous; fruit glabose, red.

Habitat: evergreen forest; sea-level–800 m alt.

Threats: the coastal forest between Bibundi-Debundscha has mostly already been lost to oil palm plantations while Mt Eloumden, in the suburbs of Yaoundé is threatened by small-holder agricultural and urban expansion.

Management suggestions: a revision or Flora account for Cameroon is needed to help aid identification of this plant. The Ebianemeyong, Ma'an and Nyabéssan sites, recently discovered, offer the best hope for refinding and protecting *E. kameruniana.* Data on population density, threats and regeneration levels would inform plans to assist the survival of this species.

OCHNACEAE

assessed by Martin Cheek & Iain Darbyshire

A pantropical woody (except (*Sauvegesia*) family of shrubs and trees especially diverse in the understorey of evergreen forest in Cameroon. Easily recognised by the simple, alternate, pinnately-veined, stipulate leaves which are very acutely serrate, the teeth often terminating in fine bristles. Coloured exudate, scented indumentum, and glands are absent from leaves and stems. The flowers are fruits are equally characteristic. The five free petals are usually crumpled, yellow and stalked, subtended by five red oblong sepals which persist red in fruit. Numerous free stamens conceal the superior ovary. In fruit the five carpels develop separately as glossy black drupoids seated on a fleshy red receptacule (the torus).

Commercially only a single species has great importance in Cameroon: *Lophira alata* or Azobé which has silica-rich timber much valued for constructing jetties and marinas in Europe and other areas. This species is exceptional in Ochnaceae in Cameroon in being an emergent tree with winged, wind-dispersed fruit. *Lophira alata* is a very widespread species in W Africa so is not mapped here. Although common in Cameroon, it has been assessed by other researchers as VU on the basis of population reduction due to great loss of its forest habitat in the last 100 years in countries such as Ivory Coast, Ghana and Nigeria.

Most species of Tropical African forest Ochnaceae have been treated in the genus *Ouratea*. Recent Flora accounts perpetuate this tradition, even though Claude Farron's work in the 1960's made it clear that *Ouratea* is confined to the New World and that Tropical African material under this name should be treated as three separate genera; divisible with the leaves alone:

> Leaves with secondary nerves few, reaching the leaf margin and often exiting as bristles .... *Idertia*
> Leaves with secondary nerves numerous and parallel, or if few, curved, becoming parallel to the margin, not exiting the margin
> Secondary nerves >30 ....*Rhabdophyllum*
> Secondary nerves 10–25 ....*Campylospermim*

Refs: Candollea 23/2: 177-228 (1968) and Bot. Helvetica 95/1 (1985).

The genus *Rhabdophyllum* has recently been revised by Sosef who recognises 8 species, none of which appear to be threatened in Cameroon (Adansonia 30(1): 119–135 (2008)). *Idertia* has four species (Farron 1985 loc. cit.).

Campylospermum is a far larger genus. Although Farron did preliminary work towards a revision, a complete monograph is needed if *Campylospermum* species are still to be delimited and assessed for their conservation status. Below only the more distinct species are assessed until such a monograph is available. Pulcherie Bissiengou of LBV is working towards this at WAG for her PhD with Marc Sosef. Apart from those taxa below, *C. mannii* (Oliv.) Tiegh. and *C. zenkeri* (Engl. ex Tiegh.) Farron are both known from only slightly more than 10 locations and are here assessed as NT while accepting that when better population reduction data is available they are likely to be rated as either VU or EN. *C. monticola* (Gilg) Cheek is also almost certainly threatened.

Ochna is the second largest group of species of Ochnaceae in Cameroon. They are generally deciduous trees or shrubs of woodland, forest edge or savanna, distinguished from the '*Ouratea*' genera in having numerous, not only ten stamens, and filaments longer than the anthers, not vice versa. This genus also needs revision.
Testuluea is a monotypic genus sometimes placed in a separate family, Luxembourgiaceae since it differs in so many features from typical Ochnaceae.

Campylospermum dusenii (Gilg) Bis. & Sosef (syn. *Ouratea dusenii* Gilg)
EN B1+2ab(iii)

Range: Cameroon (SW: Mana R at border of Korup NP) and Gabon (Ogooué R near and downstream of Lope Res.).
Until recently this species was believed restricted to a short length of the Ndian or Mana R at the edge of the Korup NP in SW Cameroon. In transferring the species to *Campylospermum*, the authors identified new records from along the Ogooué R in Gabon. Since this species is known from two rivers, it is here assessed as Endangered based on there being two linear locations (above; AOO 2 km² with 1 km² cells since aquatic) and threats as below. EOO is calculated as 195 km².
Erect rheophytic shrub 20–80 cm tall, branches sparse, erect, forming a compact crown. Leaves clustered at stem apices, leathery, glossy dark green, glabrous, (3–)4.5–6.5 × (0.6–)1.3 cm, apex acute, mucronate, base narrowly acute, margin finely serrate, numerous lateral nerves slightly prominent below; petiole 2–3 mm long. Inflorescence a simple, unbranched raceme, 4–9 cm long. Sepals 5, triangular 8 × 5 mm, persistent, red in fruit; petals yellow, obovate, 8 × 4 mm. Fruits with calyx persistent, red, enclosing 1–2 glossy black fruits.
This treatment is taken mainly from a book in preparation on the Rheophytes of Africa.
Habitat: rocky river banks in lowland evergreen forest; c. 200 m alt.
Threats: unknown and perhaps absent in Gabon, but in Cameroon at the Mana, the E riverbank is part of a oil palm plantation and so vulnerable to connected management practices.
Management suggestions: a population census at the Mana R of this and the other rheophytic species present would provide baseline data for future monitoring. Publicity about the species for the local populace would help to promote an

understanding of the importance of protecting this important rheophytic species and its community.

Campylospermum letouzeyi Farron
VU A2c; B1+2ab(iii)
Range: Cameroon endemic (SW: W Bakossi (2 coll.); Mt Cameroon (4 coll.). Littoral: nr. Melong (2 coll.). W: Bangati (1 coll.)).
Shrub or small tree 2–4 m, stems slightly ridged, bearing long persistent broad, scale-leaves c. 8 mm long with several raised longitudinal nerves; leaves long oblong-elliptic, c. 35 × 8 cm, acumen 1 cm, base rounded-truncate, secondary nerves at 1 cm intervals, each with an intersecondary, margin densely and finely acute-serrate; petiole 5–10 mm; inflorescence 40 cm, with a single c. 7 cm branch; internodes 5–15 mm, bracts spreading, oblong, 6 mm; flowers 1–6 per node, 1.5 cm diam, yellow.
First described from lowland forest near Bouda, Melong, Littoral, in 1969, this taxon has been recorded at 8 further sites equating to six broad locations (above; AOO 24 km² with 4 km² cells, threats as below, hence the assessment as Vulnerable) with the botanical inventory work of the 1980s and 1990s identifying SW Region as the stronghold of this species. At Mt Cameroon, it has been collected from Munyenge, Mundongo, Onge and the S Bakundu FR, all in the foothills. In the Kupe-Bakossi area the taxon appears rare, being recorded only twice, from Kurume and Mekom in western Bakossi; however, the paucity of collections here may be in part due to the limited inventory work carried out at lower altitudes. EOO is calculated as 2926 km².
Habitat: lowland forest especially in secondary regrowth, appearing to favour only light shade; 0–300 m alt.
Threats: conversion of large areas of lowland forest in western Cameroon to permanent agricultural land, particularly intensive plantation agriculture, have severely reduced this species' habitat. Proposals for further conversion of forest to plantation around Mt Cameroon will result in further losses and this species is likely to become EN or even CR under criterion A.
Management suggestions: further botanical surveys of lowland forest, for example in S Bakossi and Kumba District may reveal further sites for this species; care should be taken to separate it from other large-leaved *Campylospermum* spp. More data on its ecology are required; if it does require early secondary regrowth, areas of such vegetation should be maintained through active management in lowland protected areas, such as the Mungo-Bakossi FRs.

Campylospermum umbricola (Tiegh.) Farron
EN B2ab(iii)
Range: Cameroon endemic (SW: Korup NP; Mt Cameroon-Mokoko FR. S: Mt Elephant; Bipindi).
Tree 7–15 m, glabrous; leaves oblanceolate 35 × 13 cm, apex rounded with abrupt 0.5 cm acumen, base obtuse-acute, lateral nerves numerous, blade drying brown below, acutely serrate; petiole 1 cm; inflorescence terminal, 40 cm, branches ascending, 4 or 5, each 20 cm, nodes 2–3-flowered; fruiting calyx 8 cm wide, bright red, fleshy, 5-lobed, expanding from the flower; mericarps red, 9 mm.
Here *C. umbricola* is assessed as Endangered since only four location are known (AOO 16 km² with 4 km² cells) and threats are as below. EOO is calculated as 10601 km².
Habitat: low altitude evergreen forest in hilly areas; 200–250 m alt.
Threats: Mt Elephant is threatened with iron-ore mining; Mokoko which has the greatest denisty (seven) records for the species, is being proposed for logging.
Management suggestions: the most recent records of this species, in the 1990's, are from Mokoko and Mt Elephant, both severely threatened (see above). Efforts should be made to protect the species at these sites – perhaps parts could be protected. Basic populational data would assist future monitoring.

Idertia axillaris (Oliv.) Farron (syn. *Ouratea axillaris* (Oliv.) Engl.)
VU B2ab(iii)
Range: Cameroon (SW: Mt Cameroon–S Bakundu FR; Nguti. C: Eséka, 6 km SW; Nyong, Kellé & Kribi FR) and Gabon (Kougouleu; Mviadi; Kongui R).
Shrub 1–4 m, scale leaves brown, persistent, 7 mm; leaves petiolate, 12–22 × 3.5–7.5 cm, leathery, glossy, acutely acuminate, acumen 1.5 cm, lateral nerves 10–12 pairs, margins serrate, with bristles 1–2 mm, petioles 5 mm, pulvinate; flowers axillary, in sessile pairs, 22 mm wide, petals yellow; sepals red, pedicel 8 mm.
Here *Idertia axillaris* is assessed as Vulnerable (seven locations: above; AOO 28 km² with 4 km² cells, threats as below). EOO is calculated as 49199 km².
Habitat: lowland evergreen forest.
Threats: S Bakundu FR has seen conversion of forest to agriculture, particularly yam cultivation. None of the known locations occur in National Parks in Cameroon.
Management suggestions: rediscovery of this species at its known locations and collection of basic populational data as a baseline for future monitoring is advised.

Lophira alata Banks ex Gaertn.f.
VU A1cd
Range: Guinea (Conakry) to Congo (Kinshasa).
This tree was assessed as above by the African Regional Workshop of 1997, as cited in IUCN (2003; www.redlist.org), on the basis of large scale destruction of wet evergreen forest throughout its range, over-exploitation as a timber source, slow rates of growth and poor

regeneration levels in less than optimum conditions. Its timber, Azobe, rich in silica, is resistant to marine borer and favoured for use in jetties. In Cameroon it is common, widespread and regenerates easily in many areas and does not appear threatened; however, this is evidently not the case in other parts of its range, so this assessment is retained here. No description is included since this is such a well-known species. Even when juvenile trees are easily recognised at speed, e.g. along the Douala–Yaoundé Rd, by the bright red young leaves clustered at the stem apex and contrasting with the dark green older leaves. Since the species is very widespread it is not mapped.
Habitat: lowland evergreen forest; 100–1000(–1300) m alt.
Threats: see above.
Management suggestions: improved protection and management (e.g. felling regimes) of existing forest reserves.

Ochna calodendron Gilg & Mildbr.
VU B2ab(iii)

Range: Nigeria (Gangoro FR) and Cameroon (E: Lom, 250 km NE Yaoundé; Lobéké; Deng-Deng; 10–15 km NW Ngola, Abong-Mbang. Littoral: Ebo Forest).
Calodendron means beautiful tree in Greek, describing the spectacular trunk.
Tree 4–18 m, deciduous, bole smooth beautifully and vividly mottled in brown, cream, green and grey; leaves narrowly elliptic-oblong to 15 × 4.5 cm, subacuminate, base acute to obtuse, margin finely serrate, lateral nerves c. 12 pairs, obscure, petiole 1 cm; stipules strap-like, brown, membranous, 2.5 cm; racemes 10–15 cm, several on stems below leaves., pedicels 2 cm, internodes 5 mm; flowers white, 1.5–2 cm wide, produced when stems are leafless.
Here *Ochna calodendron* is assessed as Vulnerable since known from seven locations (above), with AOO 28 km² with 4 km² cells and threats as below. EOO is calculated as 119012 km².
Habitat: lowland to submontane evergreen and deciduous forest; 800–1500 m alt. At Ebo this species was most frequent on dome-like igneous outcrops, the thin soils of which support a deciduous woodland amid the surrounding evergreen forest.
Threats: many of the locations for this species in E Region are subject to logging which can lead to a decline in habitat quality and encourage slash and burn agriculture.
Management suggestions: observations at Ebo (above) suggest this species may have specialised ecological requirements – these need confirmation at other locations to verify their general applicability to the species. Basic populational data should be gathered as a baseline for monitoring. At Ebo the tree has traditional significance for local communities and receives some protection for that reason. The Nigerian record notes that the leaves of the tree have medicinal uses. These observations may be useful in encouraging protection of the species.
Censusing individuals in the wild will be facilitated by the distinctive patterned trunks which resemble those of Guava (*Psidium*) tree, but are more vividly coloured.

Testuluea gabonensis Pellegr.
VU B2ab(iii)

Range: Cameroon (S: 30 km ESE Campo; Dipikar, Ma'an) and Gabon (seven localities).
Here *Testulea gabonensis* is assessed as Vulnerable, since nine locations are known (AOO 36 km² with 4 km cells) with threats as below. Much of the information presented here derives from Lee White's remarkable book, A Guide to the Vegetation of the Lopé Reserve Gabon, WCS, New York (1997).
Tree to 50 m, bark yellow-brown, flaking in irregular scales, buttresses narrow, round; leaves oblanceolate, 20 × 5 cm, clustered at stem apices, subsessile, subacuminate, lateral nerves c. 20 pairs; racemes with pink, 4-petalled flowers, the petals unequal, the longest c. 10 mm; fruits 4 cm, leathery, 2-valved, with numerous winged seeds.
Habitat: lowland evergreen forest.
Threats: there is a high demand for the wood in Libreville for furniture making and carpentry (White, loc.cit.).
Management suggestions: that there is a high demand for this species suggests that it is more frequent in Gabon than the limited number of herbarium records suggests. However, work is needed to determine whether this demand is causing a decline in numbers of wild individuals. Analysis of data from logging companies, verified by local populational studies, would help address this question. In Cameroon both records are in or near the Campo-Ma'an NP so have the prospect of protection. At least one Gabonese location, that at Lopé, is within a NP.

OLACACAE/OCTOKNEMATACEAE

assessed by Martin Cheek

A pantropical family of c. 200 species in 28 genera which molecular work has suggested may need to be rearranged. In Cameroon this family is mainly found in evergreen forest and species are often indicators of undisturbed forest.

The genus *Octoknema*, unusual in the Olacaceae in having stellate hairs and invaginated seeds, is in the process of being segregated as a monogeneric family. It has been revised by Gosline and Malécot (in press), with the recognition of many new and often narrowly endemic species in Cameroon, such as *Octoknema mokoko*, restricted to the Mokoko Reserve on the Western slopes of Mt Cameroon. Such species are not treated here, but when this revision is published they will augment the second edition of this Red Data book.

Olacaceae are treated by J. F. Villiers in Flore du Cameroun 15 (1973), who recognised 20 species in 10 genera: *Ximenia*, *Olax*, *Heisteria*, *Strombosia*, *Coula*, *Ptychopetalum*, *Diogoa*, *Strombosiopsis*, *Aptandra* and *Ongokea*.

In Cameroon, Olacaceae are mainly evergreen understorey shrubs (*Olax*, *Heisteria parvifolia*) or small to medium-sized trees lacking coloured exudate (*Heisteria* has white exudate however) or distinctive scent, glabrous (except *Coula* and *Octoknema*), but often with the leafy stems characteristically 2-winged or angled. The leaves are alternate, lacking stipules, simple, entire, pinnately-nerved, lacking gland dots, but sometimes with minute pustules on the lower surface. The flowers are held in axillary racemes, often contracted into fascicles. The pedicellate flowers have a 5-lobed or cup-like calyx and five white or yellow, free or basally united petals, often with the triangular apex reflexed. The five or ten stamens, often inserted on the petal base, have flattened filaments and surround a usually superior, single-styled ovary that gives rise to a 1-seeded fleshy-coated fruit, often with the calyx then expanded and green (*Heisteria*, *Olax*, *Aptandra*).

None of the species are commercially important timbers, but ecologically the family can be important in evergreen forest, comprising 5% or so of the woody plant individuals and their fruits are probably significant in the diet of both birds and primates. Famously, *Coula edulis* seeds are opened by some Chimpanzee groups in Cameroun (Ebo Forest area) using stone tools. Both *Coula* and *Heisteria* fruits are harvested locally in Cameroon and consumed on a small-scale.

None of the published Cameroonian species are known to be threatened. However *Strombosia* sp. 1 of Bali Ngemba, apparently restricted to the Cameroon highland submontane forest is threatened by forest loss, and when published will be assessed as such (Cheek in Harvey *et al.* (2004: 115), Cheek *et al.*(2004: 355) and Cheek *et al.* (2010: 137)).

OPILIACEAE

assessed by Martin Cheek

A small pantropical family of shrubs, trees and climbers of evergreen forest, unusual in that all species are believed to be root-parasites, tapping into the roots of other species and extracting nutrients. The species in Cameroon are usually glabrous, with yellowish-green leaves, and lacking stipules. They are most likely to be confused, if flowers are absent, with Olacaceae (but that family often have ridged or winged stems).

The alternate simple leaves are pinnately nerved, with an entire margin. The stems lack coloured exudates or a distinct scent. The inflorescences are axillary umbels, or more usually long, pendulous,racemes which in bud resemble long cones or catkins with scales. The calyx is highly reduced, but the petals are oblong-linear or triangular, three, four or five-merous with stamens opposing and equal in number to the petals, free or joined to them, and alternating with unusual gland/staminode structures, together surrounding the superior or epigynous single-styled ovary.

Flore du Cameroun 15 by J. F. Villiers in 1973 treats this family, recognising six species in two genera, *Opilia* and *Rhopolopilia* and is the main source of the data produced here. These taxa were revised by Hiepko, P in 1982 in the journal Willdenowia 12: 161–182 (A Revision of Opiliaceae II. *Opilia* Roxb.) and in Bot. Jahrb. Syst. 108: 271–291 (1987) (A Revision of Opiliaceae IV. *Rhopalopilia* Pierre and *Pentarhopalopilia* (Engl.) Hiepko) transferring some species to *Pentarhopalopilia*. More recently they have received attention from Nickrent.

No important uses are currently known for the Cameroonian species. In addition to the species treated as threatened below, *Rhopalopilia pallens* Pierre (Cameroon, Gabon and Congo (Kinshasa)) is here assessed as NT. It may yet be shown to be threatened.

Rhopalopilia altescandens Engl.

VU B2ab(iii)

Range: Cameroon (E: Moloundou; Deng Deng; Bertoua-Batouri 6 km), CAR (Boukoko (3 coll.)) and Congo (Kinshasa) (Station Ineac Boketa, Elegili R; Likimi; Ile Tuli en face isangi; Yangambi (7 coll.)).

Here *Rhopalopilia altescandens* is assessed as Vulnerable since only eight locations are known (above; AOO 32 km^2 with 4 km^2 cells) and threats are as below. EOO is calculated as 458314 km^2.

Liana, stem with helically twisted ridge, purple-brown, glabrous; leaves alternate, lacking stipules; petiole 1–2 mm, ridged; blade elliptic to narrowly lanceolate, 3–6.5 × 1.1–2.1 cm, apex acute-mucronate, base strongly cuneate; greenish yellow; nerves pinnate, 4–6 pairs, often red above, ascending, with the nervelets conspicuous on both surfaces; umbels axillary, peduncle 3–5 mm, pedicel-bearing apex globose, fleshy, pedicels 2 mm, flower 4-merous, 2.5 mm diam, petals triangular 1.25 mm.

Distinct from other species in Cameroon in being glabrous, with umbels.

Habitat: lowland evergreen or semi-deciduous forest (presumed).

Threats: logging followed by agriculture is prevalent in the Moloundou, Bertoua and CAR locations.

Management suggestions: refinding this species in the wild, so that it can be brought to the attention of local community leaders and government officials, and so that protection can be achieved against its extinction. Propagation to augment the wild population should be attempted if a basic populational survey shows it to be as rare as it appears to be. Research will be needed to study which host species are preferred, if any, since this could potentially restrict attempts to increase the population.

PIPERACEAE

assessed by Martin Cheek

A pantropical family, mainly herbaceous, but with some subshrubs. Best known for the Indian species *Piper nigrum*, cultivated widely for its hot spicy fruits (black pepper and white pepper), including in Cameroon by CDC (Cameroon Development Corporation). African *Piper* species are mostly subshrubs 1–2 m tall but one, *Piper guineensis* or 'bush pepé' widespread in African forest, is harvested from the wild for its fruits which are as effective as *P. nigrum*. None of the three species of *Piper* in Cameroon is threatened. The second genus in Africa, *Peperomia*, apart from differing from *Piper* in minute floral and stipule characters, are almost all epiphytic herbs of cloud forest and smaller than *Piper*, 1–25 cm tall. 17 species occur in Africa, most being in Cameroon, of which four appear to be threatened.

Piperaceae generally have succulent stems, alternate (rarely opposite or whorled) simple leaves without obvious stipules. Crushed, they often produce a peppery scent. The flowers are minute, grouped in drab green terminal spikes, believed to be wind pollinated.

There is no Flore du Cameroun account, but the species are well delimited due to the revision of African species (in German) by R. Dűll (Bot. Jahrb. 93: 56–129 (1973)).

Peperomia dusenii C.DC

CR B2ab(iii)

Range: Cameroon endemic (SW: Mt Cameroon?).

Epiphytic stoloniferous herb, erect shoots to 10 cm, stems and leaves densely pubescent, grey and spreading in young shoots; leaves opposite, obovate-cuneate, 6–11 × 4–8 mm, lower surface with distinct lateral nerves.

Known from a single specimen, *Dusen* 436, which we surmise to have originated from Mt Cameroon since many of Dusen's collections were made there and since the altitude '1450 m' is given.

Peperomia dusenii was accepted by both FWTA 1:83 and by Dűll in his revision of the African species (1973, loc. cit.). Since it has not been seen for a century despite numerous botanists visiting Mt Cameroon during this time, and despite intensive surveys in the 1990s, its survival must be in doubt. Yet, there is a real possibility that it survives yet. Forest at the 1450 m contour has seen losses above Buea due to urbanisation-related destruction in recent decades (Cheek pers.obs.).

Here, *Peperomia dusenii* is assessed as Critically Endangered since it is known from only a single location (AOO 4 km² with 4 km² cells) and threats as detailed above.

Habitat: submontane evergreen forest (deduced); 1450 m alt.

Threats: see above.

Management suggestions: research of Dusen's field notes or diaries, probably in Sweden if they survive, should help narrow the search options needed to refind this and other rare species that Dusen discovered in SW Region. If rediscovered, basic populational data should be gathered to inform future monitoring. The possibility of multiplying individuals ex situ, for reintroduction, may be needed.

Peperomia kamerunana C.DC.

EN B2ab(iii)

Range: Equatorial Guinea (Bioko: Moca Lago de Biaó) and Cameroon (SW: Mt Cameroon (Buea and Mann's Spring); Mt Kupe–Nyasoso; Mwanenguba. NW: Tinachong 30 km WNW Bamenda).

Epiphytic stoloniferous herb, erect shoot 7–15 cm, stems and leaves spreading brown pubescent; leaves opposite, elliptic-obovate, 1–2.5 × 1–1.8 cm, lateral nerves inconspicuous on lower surface.

Here *Peperomia kamerunana* is assessed as Endangered since only five locations (above, AOO 20 km² with 4 km² cells) are known with threats as below, maintaining the assessment level in Cheek *et al.* (2004: 171–172). It was assessed as LR nt in Cable & Cheek (1998: lviii). EOO is calculated as 9090 km².

Habitat: submontane & lower montane forest; 1300–2500 m alt.

Threats: submontane forest clearance due to urbanisation around Buea; due to agriculture at Mwanenguba and in the Bamenda Highlands where in one area c. 50% of surviving forest was cleared in 15 years 1988–2003 (Baena in Cheek *et al.* 2010: back cover). Threats at the Bioko location are not known. The Nyasoso and Mann's Spring locations are believed not to be threatened (Cheek, pers. obs.).

Management suggestions: the best hopes for the long-term survival of *P. kamerunana* are probably at Nyasoso and Mann's Spring, where pressures on forest survival are lowest. Surveys to measure basic populational data are needed as a baseline for future monitoring. Care must be taken not to confuse this with the other epiphytic species of the genus that are much more common in the Cameroon Highlands. Surveys to rediscover the species at other historic locations Buea, Mwanenguba and Tinachong are needed to prove whether or not it still survives at these. If still present, representations should be made to leaders of local communities to protect the species and its habitat.

Notes: further data on the threats and population of this taxon are given are given in the assessment in Cheek *et al.* (2004: 171-172).

Peperomia laeteviridis Engl.
EN B2ab(iii)
Range: Liberia (Webbo Distr., Toroke), Equatorial Guinea (Bioko: El Pico; Lago de Biaó; Pico Basile) and Cameroon (SW: Mt Cameroon Hut 1–Hut 2. NW: Bafut-Ngemba. S: Lolodorf).
Epiphytic stoloniferous herb, erect stems succulent, to 20 cm tall, glabrous; leaves opposite or in 3's, elliptic, 4–7 × 1.5–4.5 cm, rounded obtuse at apex, obtuse to acute at base, 3-nerved from base, petiole c. 1 cm.
The range of this species geographically seems unlikely (the disjunct location in Liberia) and altitudinally, since all are at 2000–3000 m alt. apart from the ± lowland locations at Lolodorf and Liberia. Perhaps the morphology needs re-examination. Here *P. laeteviridis* is assessed as Endangered since five locations (above, AOO 20 km² with 4 km² cells) are known, with threats as below. EOO is calculated as 314969 km².
Habitat: submontane & montane forest; (900–)2000–2957 m alt.
Threats: Liberia has seen large forest losses in the last 20 years; the status at Pico Basile, Bioko and at Lolodorf is unknown. Bafut-Ngemba has been entirely logged and replaced by *Eucalyptus*. Hut 1–Hut 2 above Buea is the main route to the summit and has seen wood cutting to provide fuel for campers.
Management suggestions: surveys are needed to refind this species in the wild, to assess more accurately levels of threat, to collect basic populational data as a baseline for future monitoring and to assess needs for intervention.

Peperomia thomeana C.DC. (syn. *P. vaccinifolia* DC)
VU A2c
Range: Nigeria (Manbilla, Ngel Nyaki), São Tomé & Príncipe (São Tomé: Angolares; Lagoa and Angra S. João; Juliana de Sousa), Equatorial Guinea (Bioko: Moka) and Cameroon (SW: Mt Cameroon (Musake, Mann's Spring and Mimbia); Bakossi Mts & Mt Kupe; Mamfe-hill 55 km SW. Littoral: Nlonako Mt. NW: Lake Aweng; Bafut-Ngemba; Tadu; Bambui; Acha Tugui; Mt Oku; Bamenda).
Epiphytic stoloniferous herb, glabrous, flowering from erect stems 4–6 cm; leaves alternate but at uppermost node ,subtending the inflorescence, they are opposite, obovate or elliptic to 3 × 2 cm, apex retuse; inflorescences 1–2 in uppermost inflorescences, to 6 cm.
Peperomia thomeana was assessed as Lower Risk/NT in Cheek *et al.* 2000: 68 (also in Cable & Cheek 1998: lviii) since it was considered secure and well protected at its then key sites at Mt Oku and in the Bamenda Highlands. However, since that time protection of forest at Mt Oku has declined and surviving forest in the Bamenda Highlands has seen rapid and recent losses of c. 50% between 1998 and 2003 (15 years) in the Dom area (Baena in Cheek *et al.* 2010: back cover). Given this, and since *Peperomia* plants live indefinitely, it is reasonable to estimate that, over 100 years more than 30% of the population of this species has been lost due to deforestation over its entire range, hence the species is here reassessed as Vulerable under Criterion A. EOO is calculated as 23637 km².
Habitat: submontane forest 800–2300 m alt.
Threats: forest clearance for timber and agriculture.
Management suggestions: the best hopes for the survival of this species are in parts of Mt Cameroon above the plantation line and away from habitation, such as Mann's Spring, in Kupe-Bakossi, and if protection of forest can be stabilised, Mt Oku (Kilum-Ijim), since, where this species occurs, it is usually fairly abundant, its long-term future seems safe so long as forest survives at these places. However, gathering of baseline populational data for future monitoring is advisable.

PODOSTEMACEAE

assessed by Martin Cheek

This is a pantropical family of herbs which only occur in or near waterfalls or rapids, appearing to need highly aerated water, or nearly constant water spray. They mainly appear to grow and spread when submerged in the wet season and only flower as the water level drops, exposing them, in the dry season. Since they resemble mosses or algae, they are easily passed over and ignored, even by experienced botanists. Because they are highly adapted to their environment, they have great interest from scientists of several kinds. For example, they stick themselves to rock by producing a sort of biological 'super-glue 'which might have industrial applications in future. Seed germinates to develop a thallus which adheres to the rock, and from which arise in many species stems bearing either scale-leaves or long, forking leaves, whether densely, or sparsely. In some species stems are highly reduced, and leaf- rosettes are produced directly from the thallus.

The flowers develop inside a bud-like structure known as a spathellum from which the single flower emerges with a single ovary bearing two (rarely three) styles and one or two stamens, the perianth being absent or inconspicuous.

Thanks to the superb work of Colette Cusset, the African species of Podostemaceae are well-worked out. She wrote Flore du Cameroun account (vol. 30) in 1987, recognising 32 species in 10 genera, since when six new species and one genus (see treatments below) have been added to the tally. Most of these species are threatened because often they are only known from a single set of waterfalls, or at least very few, and such locations are favoured sites for the numerous planned hydroelectric schemes that have been proposed in recent years. Cameroon has more species of this family than any other country in Africa and possibly in the Old World. The Lobé Falls, a well-known tourist spot, S of Kribi, with nine species, has the greatest diversity of Podostemaceae yet known in the world for any single site. Podostemaceae, as riverine species, under current IUCN guidelines, qualify for the AOO cell size of 1 km^2 instead of the usual 4 km^2 cell size recommended for terrestrial species.

In addition to the species treated as below, the following should be treated as NT, and might in future, with better data available, be assessed as threatened: *Ledermanniella bifurcata* (Engler) C.Cusset (Cameroon, Gabon and Congo (Brazzaville)) and *L. pusilla* (Warming) C.Cusset (Cameroon, Gabon and Congo(Kinshasa)).

Dicraeanthus zehnderi H.Hess

CR B2ab(iii)

Range: Cameroon endemic (Littoral: Chutes de Sanaga, Edéa).

Thallus foliaceous, deeply divided, the lobes 4–5 mm wide, sinuate margin, bearing branched stems 10–60 cm long. Leaves sparsely scattered, ribbon-like 1.5–3 cm long, 1 mm wide, highly divided in capillary segments. Spathellum ellipsoid-obovoid, 2–3 mm, in small clusters. Pedicel to 3 cm. Tepals filiform 0.2–0.5 mm. Androecium shorter than ovary, filaments united for less than half their length. Ovary oblong to 3.5 × 1 mm, gynophore 2–4 mm; stigmas conical, straight, united at base. Capsule oblong, 8-ridged, dehiscing by 1 suture, the 2 valves long attached at the optical pole.

Here *Dicreanthus zehnderi* is assessed as Critically Endangered since it is known from a single collection made many decades ago and the species has not been seen since (AOO 1 km^2 with 1 km^2 cells). The location is now a major hydro-electric dam and it is thought that the species might have become extinct.

Habitat: falls in large river in evergreen forest belt; near sea-level.

Threats: as above.

Management suggestions: surveys to rediscover the species, recording of basic populational data, and annual monitoring in the flowering season if the species survives.

Djinga felicis C.Cusset

CR B2ab(iii)

Range: Cameroon endemic (Adamawa: Mt Djinga).

Thallus foliaceous bearing long branched stems 10–12 cm, scale leaves absent; long leaves scattered, disappearing rapidly, 5–15 mm, linear, stipulate, once branched. Spathellum enclosed in two bracts, ovoid-ellipsoid, 2 mm, flower erect in bud. Pedicel 3–4 mm. Tepals 2, linear, 0.2 mm. Androecium of a single stamen, about as long as the ovary. Pollen in monads. Ovary globose, sessile, longitudinally ridged; stigmas linear, 0.8 mm, reflexed toward androecium. Capsule 8-ridged, ridges broad and shallow, dehiscing by 2 sutures.

Here *Djinga felicis* is assessed as Critically Endangered since one location (AOO 1 km^2 with 1 km^2 cell), believed threatened by a hydroelectric installation, is known. This genus has but a single species.

Threats: see above.

Habitat: waterfalls in streams; alt. 1000 m (estimated).

Management suggestions: surveys to rediscover the species, recording of basic populational data, and annual monitoring in the flowering season.

Ledermanniella. Flowers inverted in the spathellum, ovary and fruit ellipsoid, circular in transverse section, 8-ribbed when dry.

Subgenus *Phyllosma.* Thallus mostly ribbon-like, aerial stems erect, c. 1 cm tall, numerous, covered in scale-like leaves mostly about as long as wide.

Ledermanniella annithomae C.Cusset

EN B2ab(iii)

Range: Cameroon (S: 60 km E Campo at Chutes du Ntem or Menvé élé near Nyabéssan) and Gabon (Booué on Ogooué R).

Thallus unknown, floating stems unbranched, 5-25 cm long, internodes 0.25 cm, long-leaves distichous, alternate, 2–2.5 cm long, filamentous, 4–5 dichotomously branched; scale-leaves orbicular, 1 mm. Spathellum ellipsoid to obovoid, 2.5 × 1 mm. Pedicel 10 mm. Tepals 2, 1 mm long. Androecium as long or longer than the ovary, stamens 2; filaments 1.8–2 mm, united for most of their length; pollen in monads. Ovary fusiform, 2 mm; gynophore 0.8 mm; stigmas c. 1 mm. Capsule 8-ridged.

Here *Ledermanniella annithomae* is assessed as Endangered since it is known from two locations (AOO 2 km^2 with 1 km^2 cells) and threatened by proposed hydro-electric installations so is here assessed as Critically Endangered.

Habitat: rapids in river, in full sun; evergreen forest belt; low alt.

Threats: as above.

Management suggestions: surveys to rediscover the species, recording of basic populational data, and annual monitoring in the flowering season.

Ledermanniella bosii C.Cusset

EN B2ab(iii)

Range: Cameroon endemic (S: Lobé Falls S of Kribi; Ntem Falls, Bongola, 40 km ESE Campo).

Thallus bearing several naked dichotomous stems 5–7 cm; long leaves few, lacking stipules, once dichotomous branched or simple; scale-leaves ovate, 0.6 mm, entire or toothed. Spathellum solitary at branch ends, ovoid, 2 × 0.9 mm. Pedicel 5 mm after anthesis. Tepals 2, 0.5–0.6 mm. Androecium longer than ovary, c. 3 mm, filaments united for half their length or less. Ovary ellipsoid, 2.5 mm. Gynophore 0.2 mm; stigmas filiform 0.8 mm. Capsule 8-ridged dehiscing by 1 suture.

Here, *Ledermanniella bosii* is assessed as Endangered since two locations (AOO 2 km^2 with 1 km cells) are known, and since the Lobé Falls are threatened by touristic pressure and the Ntem falls are at risk of hydro-electric development.

A record of this species (Missouri Data, Thomas specimen) from the Mbié R near Nguti is not included here since it requires verification.

Habitat: rapids and falls in lowland evergreen forest belt, sea-level.

Threats: as above

Management suggestions: surveys to rediscover the species, recording of basic populational data, and annual monitoring in the flowering season.

Ledermanniella boumiensis C.Cusset

EN B2ab(iii)

Range: Cameroon (S: Ntem Falls, Bongola, 40 km ESE Campo, Dec.) and Gabon (Boumi Falls, Mbigou, Massif du Chaillu).

Thallus unknown, stems erect, robust, branched, densely covered in scale leaves especially at apex. Scale-leaves ovate or subrhombic, 0.8–0.9 × 1–1.2 mm. Spathellum ellipsoid 2–3 mm. Pedicel 5 mm. Tepals linear 0.4 mm. Androecium longer than ovary, filaments united for about half their length, pollen in diads. Ovary fusiform, 1.6–1.8 mm; gynophore 0.1 mm; stigmas 0.5 mm. Capsule 8-ridged, dehiscing by 1 suture.

Here *Ledermanniella boumiensis* is assessed as Endangered since a two locations are known (AOO 2 km^2 with 1 km cells) one of which is threatened by a proposed hydro-electric scheme.

Habitat: falls in large river in evergreen forest belt; near sea-level.

Threats: as above.

Management suggestions: surveys to rediscover the species, recording of basic populational data, and annual monitoring in the flowering season.

Ledermanniella kamerunensis (Engl.) C.Cusset

CR B2ab(iii)

Range: Cameroon endemic (S: rapids on the Campo R near Dipikar Is.).

Species poorly known, stems repeatedly branched, covered in scale-leaves, scale-leaves linear-oblong. Long leaves few, linear, shortly stipulate, 2–4-dichotomously branched; flowers unknown; capsule 8-ridged, valves dehiscing by 1 suture.

Here *L. kamerunensis* is assessed as Critically Endangered since it is known from a single collection made a century ago and the species has not been seen since (AOO 1 km^2 with 1 km^2 cells) and since there are threats from hydro-electric schemes.

Habitat: falls in large river in evergreen forest belt; near sea-level.

Threats: as above.

Management suggestions: surveys to rediscover the species, recording of basic populational data, and annual monitoring in the flowering season.

Subgenus *Ledermanniella*. Thallus mostly crustose, aerial stems often several cm tall when present, leaves much longer than broad, usually filiform and dichotomously branched.

Ledermanniella aloides (Engler) C.Cusset

EN B2a,b(iii)

Range: Sierra Leone (Bumbuna), Cameroon (Tschappe Pass near Tchabal Mbabo), CAR and Angola.

Thallus crustose, surface with leaf rosettes, lacking stems. Leaf rosettes sessile, terminating in a single flower enclosed at its base by 2–7 alternate distichous leaves. Leaves with blade subulate 0.1–0.8 mm long, entire base dilated, about as long as blade, the stipules 2, conspicuous, inserted at apex of leaf-base. Spathellum ovoid, 2.5 mm long. Pedicel 4–5 mm long. Tepals filiform, 0.1–0.2 mm long. Androecium as long or less than the ovary, filaments 3.5 mm long, united for less than half their length, anthers 0.8 mm, pollen in monads. Ovary ellipsoid, 1–1.2 mm long; gynophore 0.4–0.6 mm long; stigmas linear, 0.2 mm long. Capsule 8-ridged, dehiscing by 1 suture.

Ledermanniella aloides is here assessed as Endangered since it is only known from five locations (above; AOO 5 km^2 with 1 km^2 cells) one of which, Bumbuna in Sierra Leone, is in the process of becoming a hydroelectric dam. *Hawthorne* 206a–302, 303 & 305 collected in 2006, are vouchers from this site (Cheek & Ameka pers. obs. 2008). The other four sites are very widely spread, one in each country, from Sierra Leone to Angola (Cusset 1984 op. cit.), all of which were recorded between c. 1900–1950. EOO is calculated as 3486340 km^2.

Notes: Cusset (1984) considered *L. pellucida* Engler (Cusset) to be probably a synonym of *L. aloides*, but later, on finding original material of the first taxon, maintained it as a distinct species, transferring it to *Macropodiella* (Cusset 1987).

It is likely that *L. aloides* is under-recorded because lacking conspicuous leaves, it is particularly inconspicuous and is only exposed above the water for a short period in the late dry season.

Habitat: deeper strata in rapids in evergreen forest belt; c. 800 m alt. At Bumbuna it occurs with *L. ledermannii* in the shallow strata of the rapids. The two species occupy different 'niches' at the same site (observations by Cheek & Ameka of Hawthorne's photos).

Threats: Bumbuna hydroelectric scheme.

Management suggestions: surveys to rediscover the species, recording of basic populational data, and annual monitoring in the flowering season.

Ledermanniella batangensis (Engler) C.Cusset

CR B2 a,b(iii)

Range: Cameroon endemic (S: Lobé Falls).

Thallus crustose, surface with leaf rosettes and stems. Leaf rosettes sessile terminating in a single flower enclosed at its base by 2 opposite leaves. Leaves with base enlarged, capuliform, with two lateral stipules; blade linear, entire or dichotomous. Stems erect to c. 5 cm, naked, sometimes many-branched, each branch terminating in a group of flowers. Spathellum ovoid, 3 mm long. Pedicel 1.2 cm long. Tepals 2, minute. Androecium longer than ovary, filaments united for most of their length, anthers 2, pollen in diads. Ovary ellipsoid, 2 mm long, gynophore 0.5 mm long, stigmas linear 0.8–1 mm long. Capsule 8-ridged, dehiscing by 1 suture.

The species is known from a single location (AOO 1 km² with 1 km² cells) and threatened by unregulated tourist trampling so is here assessed as Critically Endangered.

Not seen since the single type collection was made c. 1908, despite being an easily accessible tourist location. Possibly extinct but more site-searching needed to confirm.

Habitat: rapids in evergreen forest belt, sea-level.

Threats: as above.

Management suggestions: surveys to rediscover the species, e.g. at Tchabal Mbabo, recording of basic populational data, and annual monitoring in the flowering season.

Ledermanniella keayi (G.Taylor) C.Cusset (syn. *Inversodicrea keayi* G.Taylor

EN B2ab(iii)

Range: Cameroon endemic (NW: Mt Oku area at Mboh; Sagbo near Ndop; Kumbo).

Thallus crustose, surface with dense erect stems 6–10 cm tall. Stems with main axis unbranched, lacking leaves for most of their length; apex with numerous leaf rosettes terminating in a spathellum clasped by 4–6 distichous, alternate scale-leaves. Leaves concave [illegible] oblong, 1.5–2 mm long, acute, base not dilated. Spathellum ellipsoid to ovoid, apiculate, c. 2.5 mm long, opening laterally. Pedicel 3 mm long. Tepals minute. Androecium as long as ovary, filaments 1 mm long, united for less than half their length; anthers 2, 0.75 mm long; pollen in diads. Ovary ellipsoid 1.5 mm long, gynophore 0.5 mm to absent; stigmas sausage-like, 0.5 mm long. Capsule 8-ridged, dehiscing by 2 valves.

Here assessed as Endangered since three locations are known (AOO 3 km² with 1 km² cells) and since plants are threatened by continued forest clearance which is likely to increase river siltation due to increased surface run-off. Formerly assessed as CR B1+2c in Cheek *et al.* (2000: 68–69), updated here following IUCN (2001) and IUCN guidelines of 2003.

Seen (*Cheek* 8546A, 1 November 1996) on boulders in a small stream, 2–3m wide, running through intensively cultivated land below forest, fed from a high, unnamed waterfall east of Mboh (above Jikijem on N side of range). This species is otherwise known only from two other collections, both also from the Mt Oku checklist area. These are *Keay* in FHI 28457 from Banso and *Adams* 11073 from Sagbo near Ndop. At Mboh, *L. keayi* was seen only on 3 or 4 boulders, but each had numerous plants. It was difficult to determine where one plant ended and another began. In November 1999, what was probably this species was observed in the stream running from Ajung cliff.

Habitat: basalt boulders in clear, turbulent running water in full sunlight in deforested highlands; 1500–2000 m alt.

Threats: surface run-off from land cleared of forest for agriculture; threatened also by streams being diverted or lowered by irrigation or coffee-processing demands which are fairly common in cultivated areas around mountain. Pollution at the site observed is unlikely as it is above any habitation. However, if cultivation upstream of the site is extended up to the edge of stream, erosion may introduce sufficient silt to eliminate the population.

Management suggestions: surveys to rediscover the species, recording of basic populational data, and annual monitoring in the flowering season. Alerting local farmers to the importance of this plant for conservation and advising them of what the threats are might help this species survive.

Ledermanniella letouzeyi C.Cusset

EN B1+B2ab(iii)

Range: Cameroon (SW: Rumpi Hills (1 coll.), Bakossi Mts (2 sites)).

Here maintained as Endangered following Cheek *et al.* (2004: 172), since three locations are known (AOO 3 km² with 1 km² cells) and since plants are threatened by continued forest clearance which is likely to increase river siltation due to increased surface run-off. EOO is calculated as 193.69 km².

Thallus crustose, stems 4–10 cm, erect, stout, rubbery, branches sparse or nil, with leaves conspicuous, scattered over upper part, distichous. Leaves dark green, parallel-nerved, strap-like, entire (type) or dichotomously branched (Bakossi), 2–4 cm (type) or to 15 cm (Bakossi) long, 0.5–1.5

cm wide, apices acute, base cuneate to a terete petiole; petiole base clasping the stem for half its circumference and sheathing it for c. 1 cm; stipules minute, at petiole apex, caducous. Flowers unrecorded. Fruits in clusters from the stem apices; capsules 8-ridged, dehiscing by a suture.

The flowering Bakossi population differs in having much longer and more branched leaves than the (fruiting) type from Rumpi Hills, but this may be entirely attributable to the differences in development. After fruiting the leaves drop from the stems. The thallus and leaves may be perennial.

Habitat: vertical, shaded rock faces in spray zone of waterfalls, often covering large areas, less usually on horizontal rocks in streams near falls in submontane forest; 750–1350 m alt.

Threats: stochastic changes due to the small actual area of occupancy (probably less than 0.2 Ha) and low number of sites (see above); future logging is likely to contaminate its aquatic environment due to surface run-off after logging.

Management suggestions: the Rumpi Hills site should be rediscovered and assessed. Efforts should be made to monitor all subpopulations for numbers of individuals annually. *Podostemaceae* are vulnerable to increased turbidity of water which reduces the ability of seedlings to establish.

Ledermanniella linearifolia Engl.

EN B1+B2a,b(iii)

Range: Cameroon endemic (Littoral: Nkam R, Bafang-Yabassi. S: Lobé Falls, 7 km S Kribi; Falls on Campo R at Dipikar).

Thallus crustose, short-stemmed leaf-rosettes scattered densely, leaves distichous. Stems erect 0.6–3 mm long, clad in leaves terminating in a single flower. Leaves linear c. 1.5 mm long, entire; base enlarged, with two stipules. Spathellum narrowly obovoid, long-stipitate, 7–8 x 1 mm, pedicel 1–1.2 cm. Tepals 2, filiform, 0.3–0.4 mm long. Androecium longer than ovary, filaments 3.5 mm long, united for ⅔–¾ their length; anthers 0.8–0.9 mm long; pollen in diads. Ovary ellipsoid, 1.2 mm long; gynophore 0.2–0.5 mm long; stigmas subulate, 0.5–0.6 mm long. Capsule 8-ridged, dehiscing by 2 sutures, 1 valve falling.

The species is known from a three locations (above: AOO 3 km² with 1 km² cells) and threatened by unregulated tourist trampling at the Lobé Falls site, so is here assessed as Endangered. EOO is calculated as 1477.16 km².

Habitat: rapids in evergreen forest belt, sea-level to 200 m alt..

Threats: as above.

Management suggestions: surveys to rediscover the species, recording of basic populational data, and annual monitoring in the flowering season.

Ledermanniella monandra C.Cusset

CR B2a,b(iii)

Range: Cameroon endemic (Adamawa: Mayo Béka near Dodéo, 60 km W Tignère).

Thallus crustose; surface with leafy stems up to 1 cm long bearing c. 3 branches at apex. Leaves ?distichous ± covering the stem, 1–2 mm long, the greater part sheathing the stem, blade c. 0.5 mm long, apex rounded; dilated base and with two short lateral stipules. Spathellum terminal, single, obovoid, 2 mm long equalling the apical leaves, opening apically but also with a short lateral slit. Pedicel 3–4 mm long after anthesis. Tepals 2, linear, c. 0.5 mm long. Androecium about as long as the ovary, filament c. 0.6 mm long, anther single, as long as filament, pollen in monads. Ovary ellipsoid, 0.8–1 x 0.4–0.5 mm, gynophore 0.2 mm, stigmas linear, 0.2 mm long. Capsule 8-ridged, dehiscing by 2 sutures.

Known only from a single collection (*Ledermann* 2872) made 100 years ago (1908/1909) at one site on the (Cusset 1987, loc. cit.). This site is believed unprotected. Habitat degradation has been extensive in these highlands (Cheek *et al.* 2000).

Ledermanniella monandra is here assessed as Critically Endangered since three locations are known (AOO 3 km² with 1 km² cells) and since plants are threatened by habitat degradation.

Habitat: to be researched. Probably rapids in wooded hills.

Threats: habitat degradation.

Management suggestions: it is recommended that an attempt be made to revisit this species at its type locality to see if it survives, to see if it occurs at other sites and to explore means of its long term protection. It reproduces in March. If found, basic populational data should be collected as a baseline for monitoring.

Ledermanniella musciformis (G.Taylor) C.Cusset

EN B2ab(iii)

Range: Cameroon (NW: Bamenda at Mba Kokeka N: Alantika Monts at Nakalba, 21 km WSW Tchamba).

Thallus broadly ribbon-like, with numerous short, simple or branched leafy stems 2–3 cm long. Leaves ± clothing the stem, densely at apex, sparsely at base, distichous, lanceolate-linear, of two types, 3–5 mm long and 1(–2.5) cm long, apex long, acute, base gradually dilating, sheathing; stipules absent. Spathellum single, terminal, ovoid, 2–3 mm long, apex beaked, opening by a lateral window. Pedicel 1 cm long after anthesis. Tepals 2, 0.6 mm long. Androecium shorter than ovary, filaments united at the base, anthers c. 0.7 mm long; pollen in monads. Ovary ellipsoid 1.75 mm long, gynophore 0–1 mm long; stigmas sausage-like 0.1 mm long. Capsule 8-ridged, dehiscing by 1 suture.

Ledermanniella musciformis is here assessed as Endangered since it is known from two locations (above: AOO 2 km² with 1 km² cells) and its habitat is thought to have been degraded due to deforestation leading to increased turbidity and siltation of streams in the Bamenda Highlands.

Habitat: waterfalls; c. 1200 m alt.

Threats: habitat degradation due to increased surface run-off because of deforestation.

Management suggestions: surveys to rediscover the species, recording of basic populational data, and annual monitoring in the flowering season. A potential record of this species from near Anajyua near Belo (*Cheek* 9920 see Cheek *et al.* 2000: 69) needs further investigation to confirm its identity.

Ledermanniella ntemensis Y.Kita, Koi, Rutish. & M.Kato

CR B2ab(iii)

Range: Cameroon (S: Campo, 40 km ESE at Bongola Falls on Ntem).

Unique in its genus in its three styles, this species was first collected by Letouzey, and recollected, about 30 years later by members of the group that subsequently published it in 2003 (Kita, Y., Koi, S., Rutishauser & Kato, M., A new species of *Ledermanniella* (*Podostemaceae*) from Cameroon, Acta Phytotax.Geobot. 59(3): 223–227). Since it is only known from a single location (AOO 1 km² with 1 km² cells, it is here assessed as Critically Endangered in view of the threats below.

Probably an annual herb, erect, 35 × 25 cm, thallus reduced to a disc, stems branching, sparingly clothed with scale-leaves ovate, 0.5–1 mm, long, entire; long-leaves filiform 5–12 mm, repeatedly dichotomously branched; spathellum 4–8 mm, dehiscing with triangular lobes; pedicel 15–25 mm; tepals 2, filiform, 1.5 mm; stamens 2, 5–7 mm, filaments united for half their length; anthers 1.5 mm with monad pollen; ovary narrowly ellipsoid, 2.5–3 × 0.5–0.8 mm, stigmas 3, 0.3–0.5 mm, gynophores 1–2 mm: Capsule dehiscing into 2 boat-shaped valves.

Habitat: waterfalls in lowland evergreen forest.

Threats: a proposed hydroelectric scheme on the Ntem.

Management suggestions: seed-banking; discussion with HE project managers should the scheme go ahead; further surveys to find additional locations, if they exist; surveys to record basic populational data as a baseline for monitoring: publicity regarding the existence and rarity of this plant in the Campo area.

Ledermanniella onanae Cheek

EN B2ab(iii)

Range: Cameroon [illegible]: N of Kodmin; near Muambong).

This rheophyte resembling a *Lycopodium*, is known from only two sites hence has been assessed Endangered (above: AOO 2km² with 1 km² cells; threats as below). The assessment above was published in Cheek (2003), A new species of *Ledermanniella* (*Podostemaceae*) from western Cameroon. Kew Bull. 58(3): 733–737. This was maintained in Cheek *et al.* 2004: 172) and is maintained here, no further new data being available. However, if it is confirmed that the two locations are on the same stream, in future it would be justified to assess the species as CR.

Rheophyte; thallus less than 1 × 1 cm; aerial stems 5–7 cm, completely covered in stiff distichous leaves, each laterally flattened, c. 1 mm apart, in side view suboblong to ligulate, 5–8 × 1–1.5 mm, apex acute, base clasping the stem, forming a sheath; spathellae single, axillary, subtended by a curved bract about as long as the leaf; spathellum 4 mm; pedicels 7–8 mm; androecia 2; ovary laterally-flattened, 2 mm; styles 2.

Habitat: perennial waterfalls and rapids in submontane forest; 1000–1200 m alt.

Threats: stochastic changes due to the small actual area of occupancy (probably less than 0.2 ha) and low number of sites (see above); future logging is likely to contaminate its aquatic environment due to surface run-off after logging.

Management suggestions: efforts should be made to monitor the subpopulations for numbers of individuals annually.

Ledermanniella pollardiana Cheek & Ameka

CR B2ab(iii)

Range: Cameroon endemic (NW: near Bali).

Discovered by Benedict Pollard while bathing, this species was published in 2008 (Nordic. J. Bot. 26(3–4): 214–217) and the assessment published there is maintained in view of the single location is known (above: AOO 1 km² with 1 km² cells and threats as low. In fact the species was only seen over an area of 5 m × 5 m).

Probably perennial, thallus ribbon-like, 1.25–1.5 mm wide, forking repeatedly, forming a crust-like matrix or several m². Stems dense, usually 2–3 mm apart, erect, rubbery, terete, c. 2.5(–4) cm long, internodes 4–6 mm long, spur branches alternate, c. 2 mm long, bearing a dense sessile cluster of 3–12 spathellae and associated leaf rosettes. Leaf rosettes produced from both stems and thallus, rosettes with a single, opposite pair of leaves, each leaf pair subtending a single spathellum. Leaf-blades filiform, terete, (0.1–)2–2.5(–4) cm long, 0.1–0.4 mm diam. at the base, gradually tapering to the apex; variable in length and extent of branching, the shortest leaves arising from immature spathellae on actively growing thalli; longest leaves usually dichotomously branched, usually once or twice, but up to four times, usually in the distal half, rarely in the distal two-thirds; some long leaves simple; leaf base with short sheath to 0.2 mm long, limb oblong, c. 0.5 × 0.3 mm, not cupped, stipular lobes apical, 0.1–0.3 mm. Inflorescence single-flowered, sessile. Spathellum at anthesis cylindrical to narrowly obovoid, c. 3 mm long, opening irregularly at the tip. Pedicel 3.75–4 mm long at anthesis, extending up to 5.7 mm in fruit. Tepals filiform, 0.5–0.7 mm long. Androecium of a single stamen, about as long as ovary, filament 1.2–1.8 mm long; anther rounded-oblong. Ovary ellipsoid, 1.5–1.75 × 0.5–0.75 mm long; styles 2, erect, linear, 0.8 mm long, diverging from each other at c. 20°. Fruit an ellipsoid capsule 1.65 mm with 8 ribs running its entire length; valves 2, each valve with 3 ribs and 2 rib-like sutures; seeds ellipsoid, 0.125 mm

Habitat: perennial waterfall, in full sun.

Threats: increased siltation due to deforestation in the surrounding area of the Bamenda Highlands.

Management suggestions: the caretakers of this location for this species should be informed about the rarity of this species and asked to preserve it. Meanwhile further searches in the surrounding area are advisable to attempt to discover additional sites. Basic populational data should be collected as a baseline for monitoring.

Ledermanniella prasina J.Schenk & D.W.Thomas
VU D2

Range: Cameroon endemic (SW: Mundemba, 14 km NE, at Mana R falls).

Described in 2004 (Schenk & Thomas, Novon 14(2): 227–232, A new species of *Ledermanniella* (*Podostemaceae*) from Cameroon) this species is here assessed as Vulnerable since only a single location is known (AOO 1 km^2 with 1 km^2 cells) although threats are unknown.

Semi-aquatic herb from the spray-zone of a waterfall: thallus disc-shaped, plants 9–13 cm tall, stems dichotomously branched; leaves to 4 cm, terete, dichotomously branched; spathellae in sessile clusters of 1–6 terminal above the stem bifurcations, subtended by numerous bracts or 'intrapetiolar scales'; spathellae ovoid, then clavate, 3–8 × 1–2.2 mm, dehiscence irregular, pedicel to 13 mm, tepals 0.4–0.7 mm; stamens single, filament 1.1–1.3 mm at spathellum dehiscence, extending to 5 mm later; pollen white, in monads; gynophores 0.1 mm, ovary ellipsoid, 1.5 × 0.8 mm, styles 2, linear, 0.6–0.8 mm, stigmas bilobed; capsule 2.2–2.7 × 1 mm, bivalved, one valve persisting, the other caducous; seeds bright green, 25–37 per capsule.

Remarkable for not growing submerged, but in the spray zone of waterfalls, as does also *L. letouzeyi* (Cheek *et al.* 2004: 172); also in the erect habit and the bifid stigmas. The actual area of occupying is recorded as being 200 m^2, being a shelf above the high water level of a plunge pool at the base of a waterfall

Habitat: waterfall in evergreen forest zone; 300 m alt.

Threats: not known

Management suggestions: the population of the villages of Miangwe II and Meta, between which the location occurs, should be alerted to the importance of this species and asked to assist with its preservation.

Ledermanniella raynaliorum C.Cusset
EN B2ab(iii)

Range: Nigeria (Adamawa Highlands, 1 coll.) and Cameroon (N: Mt Poli, Massif de Vokré).

This species is very similar to *L. musciformis*, separated by the androecium being longer than ovary and the leaves being uniformly short in length. In addition the pedicel is 2(–2.5) cm long and the gynophore 0.7 mm long, far exceeding those of *L. musciformis*. However, it is possible that additional material might unite these two species. The single collection of *L. raynaliorum* occurs within the altitudinal and geographical range of *L. musciformis*, and they are fertile in the same month.

Ledermanniella raynaliorum is here assessed as Endangered, EN B2a,b(iii) being known from only two locations(above: AOO 2 km^2 with 1 km^2 cells) in the Cameroon Highlands, where habitat degradation is the same as discussed for *L. musciformis* above.

Habitat: waterfalls in the Cameroon Highlands, c. 1150 m alt.

Threats: habitat degradation due to increased surface run off because of deforestation.

Management suggestions: surveys to rediscover the species, recording of basic populational data, and annual monitoring in the flowering season.

Ledermanniella sanagaensis Cusset
VU D2

Range: Cameroon endemic (C: Sanaga at Nachtigal).

Thallus crustose, surface with leaf rosettes, lacking stems. Leaf rosettes sessile, terminating in 1(–2) flowers, enclosed at its base by 6–7 distichous leaves. Leaves linear, 2–4 × 0.15 mm, apex acute, base broad, with two stipules. Spathellum ovoid, c. 1.5 mm long. Pedicel 3.5 mm long before anthesis. Tepals 0.4 mm long. Androecium with filaments c. 0.7 mm long, partly united, shorter than ovary before anthesis. Anthers c. 1 mm long; pollen in monads. Ovary ellipsoid, 2 mm long; gynophore 0.4 mm long; stigmas linear, 0.2 mm long. Capsule 8-ridged dehiscing by 2 sutures, 1 valve falling.

Assessed here as Vulnerable since no threats are known at its single site. Were this to change, *L. sanagaensis* would be re-assessed as Critically Endangered under Criterion B, all else being equal.

Habitat: rapids near semi-deciduous forest-savanna junction; c. 800 m alt.

Threats: none are known.

Management suggestions: surveys to rediscover the species, recording of basic populational data, and annual monitoring in the flowering season.

Ledermanniella. thalloidea (Engler) C.Cusset
VU B1+B2ab(iii)

Range: Cameroon endemic (SW: Bakossi at Nyandong. Littoral: Nkongsamba; Chutes Edéa, Sanaga R; 10 km N Edéa. C: Nachtigal, Sanaga. S: Bipindi).

Thallus crustose, surface with leaf rosettes and sometimes also with long stems. Leaf rosettes sessile, terminating in a single flower enclosed at its base by 2–4 distichous leaves. Leaves with base enlarged, cup-like, membranous; blade linear, 1–8 mm long, simple or dichotomous; stipules united around the leaf-base so that the blade is peltate. Stems sometimes present, straight (branched?) carrying rosettes, either singly or in clusters of 2–3. Spathellum ellipsoid to obovoid, 5–9 x 1.5 mm. Pedicel c. 1 mm long after anthesis. Tepals 2, 0.5 mm long. Androecium longer than the ovary, filaments 2 mm, united for part of their length; pollen in diads. Ovary ellipsoid, 2 mm long; gynophore 0.5 mm; stigmas c. 1 mm. Capsule 8-ridged, dehiscing by 1 suture.

Here *L. thalloidea* is maintained as Vulnerable following Cheek *et al.* (2004: 172), since it is known from six locations (above; AOO 6 km² with 1 km² cells) and since there are threats from hydro-electric schemes. It was first discovered at Ndoungué near Nkongsamba. EOO is calculated as 4314 km².

Habitat: falls in large rivers in evergreen forest belt; near sea-level to 500 m alt..

Threats: a probable future decline in habitat quality is predicted due to contamination of its aquatic environment from surface run-off after logging.

Management suggestions: a survey of the known sites for this species would provide data on whether the species survives at these, and could obtain quantitative subpopulations data needed for a management plan and annual monitoring in the flowering season. Further surveys in the Bakossi-Rumpi Hills area are likely to yield new sites for the species. Unless in flower, the species is easily overlooked.

Ledermanniella variabilis (G.Taylor) C.Cusset
EN B1+2ab(iii).

Range: Cameroon (S: Lobé Waterfall (Bos 3594). SW: Munaya R near Mamfe (Keay FHI 28688) fide Cusset (1987: 88); Mawne R near Mamfe (Cheek)).

Thallus form uncertain but possibly ribbon-like, c. 5 mm wide with short stems inserted at the margins in two ranks; stems 0.5–0.8 cm long; usually concealed by leaves, but longer stems more sparsely covered. Leaves distichous, terete, linear, entire, or (distal leaves) 1(–2)-dichotomous, apices rounded, base sheathing, slightly dilated, with or without two lateral, supraxillary stipules. Spathellae terminal on the stems, 1–2(–7), lacking bracts, ellipsoid or subovoid, narrowly stipitate, apex umbonate, to 3.25 mm long, dehiscing irregularly at the apex. Pedicel c. 4 mm long at anthesis. Tepals 2, c. 0.5 mm long. Androecium as long as the ovary, stamens 2(–3–4), filaments 1–3.5 mm long, united for 0–1 mm. Pollen in diads. Ovary ellipsoid, 1.5 mm long, gynophore c. 0.3 mm; stigmas variable, club-shaped to palm-shaped, apices entire or 2–3-lobed, c. 0.2 mm long, free. Capsule 8-ridged, dehiscing in 2 valves.

Known only from the type location from Mamfe district, river Munaya, at the old Ikom-Mamfe Rd ferry in flower 20 Jan. 1951 (*Keay* FHI 28688), until 1987 when Cusset identified *Bos* 3594 as a second specimen, from Lobé Falls, Kribi, a site threatened by trampling from tourism. Accordingly this species is here assessed as Endangered since there are only two locations are known (above: AOO 2 km² with 1 km² cells) with threats as below. EOO is calculated as 4653 km².

Closely similar to *L. bifurcata* and *L. pusilla* but unusual in the club- and palmately shaped stamens, and the largely free filaments.

Habitat: rapids in evergreen forest; alt. c. 500 m.

Threats: habitat degradation due to increased surface run off because of deforestation.

Management suggestions: surveys to rediscover the species, recording of basic populational data, and annual monitoring in the flowering season.

Macropodiella pellucida (Engl.) C.Cusset
EN B1+B2ab(iii)

Range: Cameroon endemic (SW: Mana at Mundemba; Rumpi Hills; Jide at Muambong. L: Baré near Nkongsamba).

This species is maintained here as Endangered following the assessment in Cheek *et al.* (2004: 172–173), since there are only four locations are known (above: AOO 4 km² with 1 km² cells) with threats as below. EOO is calculated as 994 km². This species is very easily overlooked since the leaves are highly reduced and stems are absent. Unless the species is in flower it looks like an algae.

Rheophyte; thallus 3–5 cm diam., enveloping rock surface; stems absent; leaves bract-like, two subtending each spathellum, ovate, c. 1 mm; spathellae scattered, 4–5 mm, flower inverted in bud; pedicel 1–1.5 cm; androecia 2; ovary laterally-flattened, 8-ridged; styles 2.

Habitat: perennial waterfalls and rapids in submontane forest; 1000–1200 m alt.

Threats: stochastic changes due to the small actual area of occupancy (probably less than 0.2 ha) and low number of sites (see above); future logging is likely to contaminate its aquatic environment due to surface run-off after logging.

Management suggestions: monitoring the subpopulations for numbers of individuals annually is advised.

Saxicolella flabellata (G.Taylor) C.Cusset (syn. *Pohliella flabellata* G.Taylor)

EN B2ab(iii)

Range: Nigeria (Ikom, Afi R FR) and Cameroon (Mana R rapids near Korup-Mundemba fl. Dec.).

Thallus foliaceous carrying highly divided long stems, to 20 cm; leaves to 3 × 2 cm, highly divided, ultimate segments capillary. Spathellum sessile on thallus or grouped in leaf axils or grooved at stem apices. Pedicel short. Tepals 2 subulate, 0.3 mm. Androecium with a single stamen, filament 4.5 mm. Ovary ellipsoid to fusiform 3.5 × 1 mm, gynophore 0.2 mm; stigmas 2, filiform to 2 mm. Capsule 8-ridged, 2-valved.

Here *Saxicolella flabellata* is assessed as Endangered since only two locations are known (AOO 2 km² with 1 km² cells) and since the Afi location is suspected to be threatened by surface run-off leading to increased silt levels originating in clearance of forest for timber and agriculture. Nigeria has lost nearly 90% of its original forest cover and losses are ongoing (mongabay.com).

Habitat: rapids in lowland evergreen forest.

Threats: see above.

Management suggestions: surveys to rediscover the species at Mana R, recording of basic populational data, and annual monitoring in the flowering season.

Saxicolella laciniata (Engler) C.Cusset

EN B1+B2ab(iii)

Range: Cameroon endemic (SW: Mamfe, 20 km W at Tabo on Bawan R; Mawne R FR, Dec. Littoral: Bakaka FR near Nkongsamba).

Thallus cylindric, thread-like, attatching to subtrate at nodes only, where densely leafy little-branched stems to 10 cm long or more arise. Leaves 10 × 7–8 mm deeply divided and lobed irregularly. Spathellum ellipsoid, stipulate, 5 × 1 mm, grouped at apex. Pedicel 4–5 mm. Tepals linear 0.3–0.4 mm. Androecium with a single stamen, filament 3–4 mm. Ovary fusiform, gynophore 0.2 mm, stigmas linear to 1.5 mm. Capsule 8-ridged, dehiscing by 1 suture.

Here *Saxicolella laciniata* is assessed as Endangered since three locations are known (AOO 3 km² with 1 km² cells) and since plants at Mawne are threatened by future logging of the forest reserve which is likely to increase river siltation due to increased surface run-off (pers. obs. 2009). EOO is calculated as 3006 km².

Habitat: rapids in lowland evergreen forest.

Threats: see above.

Management suggestions: surveys to rediscover the species at Mana R, recording of basic populational data, and annual monitoring in the flowering season.

Saxicolella marginalis (G.Taylor) Cheek (syn. *Butumia marginalis* G.Taylor)

EN B2ab(iii)

Range: Nigeria (Obudu area, Butum Falls) and Cameroon (NW: Fundong).

Rheophyte, probably annual, thallus dark green and inconspicuous when alive, white when dried, adhering firmly to the rock, c. 0.75 cm wide, 5–6 cm long, dichotomously branching like a liverwort. Margin of thallus bearing at intervals sessile tufts of linear leaves 2–3 mm long, surrounding a single subsessile flower. Flower about as long as leaves, with sessile, olive green ovary bearing two purple stigmas. Stamen single, about as long as ovary. Fruit 6-ribbed. Flowering and fruiting in late November. Illustrated in FWTA 1: 125 (1954).

This very rare rheophyte was known only from one collection at the Butum falls, Obudu Plateau in adjoining Nigeria before it was newly recorded by us from Cameroon (Cheek *et al.* 1997; ibid 2000: 69, 153). In Cameroon it is known only from the waterfall at Fundong near the Touristic Hotel (*Cheek* 8740, 22 November 1996). What may prove to be a second site for this species in Cameroon was seen at Ajung in November 1999 (pers. obs.) but a specimen is needed to confirm the identification. All streams and rivers between Fundong and Belo were examined for Podostemaceae at road crossing points, but only this site was discovered to have Podostemaceae. The waterfalls of the Bamenda Highlands have been fairly well investigated for Podostemaceae thanks to the work of Ledermann and Keay. *Saxicolella marginalis* is locally common. At least several hundred plants were seen here in 1996, and probably many more were present. Observation of this species is difficult in its habitat due to spray, with turbulent water and slippery boulders. Monitoring at this site in December 1998 and November 1999 suggested that this species was present in approximately similar numbers to 1996. The species was assessed as CR B1+2c by me in Cheek *et al.* (2000: 69) from whence this treatment is modified. It is here re-assessed as Endangered since two locations are known (AOO 2 km² with 1 km² cells) and threats are as below.

Habitat: on exposed basalt boulders below a c. 30 m high waterfall; c. 1400 m alt. (Fundong)

Threats: probably requiring clean, well-oxygenated water as usual in this family. Silt in water known to be inimical to survival of Podostemaceae. Therefore this population is possibly Endangered at this site from pollution from laundry operations at the town of Fundong just upstream: much debris was seen at side of this pool.

Management suggestions: monitoring of the population is suggested from year to year to establish if there is any diminution of numbers of individuals. Cultivation of Podostemaceae known to be extremely difficult and so *ex situ* conservation of live plants is not an option although seed-banking should be feasible.

Saxicolella nana Engler

CR B2ab(iii)

Range: Cameroon endemic (C: Mbalmayo, Nyong R).

Thallus foliaceous, deeply divided, lacking stems but bearing rosettes of distichous linear leaves, 1.5-3 mm, irregular, divided into 2–4 segments. Spathellum terminal with flower erect in bud, sessile or with peduncle 2 mm. Pedicel 3–4 mm. Tepals filiform 0.4 mm. Androecium with 1 stamen, filament 2–3 mm. Pollen in diads. Ovary fusiform 1.5 × 0.6 mm, gynophore 0.5 mm, stigmas linear to 0.5 mm. Capsule 8-ridged, opening by a single suture.

Here *Saxicolella nana* is assessed as Critically Endangered, being known from a single location (AOO 1 km² with 1 km² cells) and not seen in a century, possibly extinct due to siltation resulting from surface run-off after conversion of forest to timber and agriculture within the Nyong catchment area.

Habitat: rapids of large river in semi-deciduous forest belt.

Threats: as above.

Management suggestions: surveys to rediscover the species in the Nyong R at Mbalmayo, recording of basic populational data, and annual monitoring in the flowering season.

Stonesia ghoguei Pfeifer & Rutis.

VU D2

Range: Cameroon endemic (Adamawa: Ngaoundéré, Tello falls).

Described in 2009 (Pfeifer,, E., Grob, V., Thiv, M., & Rutishauser, R., *Stonesia ghoguei*, Peculiar morphology of a new Cameroonian species (Podostemaceae Podostemoideae), Novon 19: 102–116), this species is maintained as Vulnerable under criterion D since although no threats are known, only a single site is known. Within the genus *Stonesia* this species is anomalous in occurring outside Upper Guinea, having pollen in monads (not diads), and one stamen and two tepals (not two stamens and three tepals).

Thallus ribbon-like, 5 mm wide; stems to 10 cm unbranched, producing leaves on one side only; long-leaves repeatedly forked, to 8.5 cm, linear, slightly flattened when young; spathellae in Feb. and March, in dense lines on thallus and/or stem base, ovoid, papillate, pedicel c. 2 mm, elongating to 4 mm, tepals 2, stamen single, pollen in monads, ovary ellipsoid, 1–1.5 mm, stigmas 2, filamentous; fruit with 12–16 ribs.

The Tello falls area were collected by several expatriate botanists in the 1960s and 1970s, so it is great credit to Jean-Paul Ghogue that he was the first to discover this plant in the early 21st Century.

Habitat: waterfalls, [illegible] m.

Threats: none are recorded.

Management suggestions: revisiting the only known locality is advised so that basic populational data can be recorded as a baseline for monitoring, and so that local land managers can be requested to protect the species, perhaps assisted by an educational poster on the species. Evaluation of threats is needed.

Winklerella dichotoma Engl.

CR B2ab(iii)

Range: Cameroon endemic (Littoral: Chutes Edéa, Sanaga R).

Basal thallus ribbon-like, 4–5 mm wide, the margins bearing short stems 1–3 cm long; scale-leaves absent; long-leaves linear, 1–2-dichotomous-branched, 5–6 mm, exstipulate, scattered. Spathellum with flower inverted in bud, obovoid-stipitate, c. 5 mm. Pedicel 1 cm. Tepals filiform 0.1–0.2 mm. Androecium about as long as ovary, stamens 2, filaments united for most of their length. Pollen in monads. Ovary broadly ellipsoid, 1.5 mm, laterally flattened, apex retuse, gynophore 0.5–1 mm; stigmas linear 0.7 mm. Capsule resembling ovary but with two longitudinal ridges on each side.

Here *Winklerella dichotoma* is assessed as Critically Endangered since a single location is known (AOO 1 km² with 1 km² cells) at which a hydro-electric plant has since been developed.

Habitat: rapids on large river in evergreen forest belt; low altitude.

Threats: as above.

Management suggestions: surveys to rediscover the species at Edéa if it survives there, recording of basic populational data, and annual monitoring in the flowering season.

Zehnderia microgyna C.Cusset

CR B2ab(iii)

Range: Cameroon endemic (Littoral: Chutes Edéa, Sanaga R).

Thallus present, bearing simple or branched stems 1–3 cm, leaves stipulate, mainly distichous, linear, unbranched, 2–6 mm, scattered along the shorter stems. Spathellum ovoid 1.5 × 0.8 mm, flower inverted in bud. Pedicel 1–1.5 cm. Tepals 2, filiform, 0.2 mm. Androecium much longer than the ovary, united for half their length or less. Pollen in monads. Ovary spherical 0.6 mm diam., gynophore 7–8 mm. Stigmas 2, linear, 0.7–0.8 mm. Capsule 8-ridged, opening by 2 sutures.

Zehnderia is remarkable for the small, globose, 8-ridged capsule and the long gynophore. The single species is known from a single location (AOO 1 km² with 1 km² cells) and threatened by a hydro-electric installation so is here assessed as Critically Endangered.

Habitat: rapids in river, in full sun; evergreen forest belt; low alt.

Threats: as above.

Management suggestions: surveys to rediscover the species if it survives, recording of basic populational data, annual monitoring in January (flowering season) and discussion with HE plant managers.

REFERENCE

Cheek, M., Satabie, B. & Onana, J.-M. (1997). Interim report on botanical survey and inventory for Kilum and Ijim Mountain Forest Projects by the National Herbarium Cameroon and RBG Kew, Oct./Nov. 96. Cyclostyled. 16pp.

POLYGALACEAE

assessed by Martin Cheek

A cosmopolitan family of about 1000 species best known for the largest genus *Polygala*, with 500 species which is globally widespread in open, sunny habitats, especially in Europe and the Mediterranean. The family have about 10 species in Cameroon. They are characterised by their legume-like flowers. Two of the five sepals are petaloid resembling the 'wings' of a papilionoid flower, while of the three true petals, one is cupped and contains the stamens, as in the 'keel' of the papilionoid flower. The leaves are alternate, simple, lacking stipules.

In Cameroon, *Carpolobia* is a small tree valued for its dense wood used for mattock handles and mortars. Its fleshy fruits are edible. *Securidaca* are lianas with winged mericarps, and *Atroxima* are also climbers but with juicy, 3-lobed fruits. The only threatened species in Cameroon appear to be *Polygala*, perennial or annual grassland herbs with unbranched stems 10–200 cm tall, and slender leaves. The genus was recently revised synoptically by Jorge Paiva: Fontqueria 50: 1–346 (1998). No Flore du Cameroun treatment exists for the family.

Apart from the species below, *P. atacorensis* Jacq.-Fél., widespread but rare in subsahelian grassland, appears rare and may be threatened if investigated further. In addition, among the genus *Heterosamara*, close relatives of *Polygala* but restricted to forest, the Cameroonian species *H. cabrae* (Chodat) Paiva rates as Near Threatened being known from only slightly more than 10 locations in Cameroon, Gabon, Congo (Kinshasa) and Angola (Cabinda).

Polygala tenuicaulis Hook. f. subsp. ***tenuicaulis***
VU D2

Range: Cameroon (SW: Mt Cameroon).

This subspecies seems to be restricted to Mt Cameroon although I stated in Cable & Cheek (1998: lviii–lix) that it overlapped in the Bamboutos and Bamenda Highlands with *P. tenuicaulis* subsp. *tayloriana* Paiva. The latter generally has larger flowers, c. 8 mm long compared with those of subsp. *tenuicaulis* on Mt Cameroon (5 or 6 mm long). The first collections of subsp. *tenuicaulis* are those of Mann on Mt Cameroon near Mann's Spring (*Mann* 1282, Dec. 1861 and *Mann* 1982, Nov. 1862) and subsequent collections on the mountain have been made in the same area, and not from above Buea, so its range may be limited.

On Mt Cameroon this plant can be locally common and gregarious, forming little patches of about 30 cm × 30 cm. The plants have two or three slender, unbranched stems c. 30 cm tall, of a purplish green hue, and purple flowers. Anthocyanin-free mutants are seen intermixed on occasion (pers. obs. 1992, 1993). These have yellow flowers and yellowish green stems. It is not clear whether this species is an annual, or perennates from an underground rootstock.

The assessment above is maintained from Cable & Cheek (1998: lviii–lix).

Habitat: montane grassland often on lava flows, within a few hundred metres of the forest boundary, but not at the forest edge; 2100–2400 m alt.

Threats: this subspecies may require fire for regeneration, but no threats are known since there is no grazing of livestock in its habitat and little interference from man.

Management suggestions: a survey of the range of this subspecies on the mountain is advisable to gather baseline data for future monitoring. Rootstock presence should be investigated since this has implications for the fire ecology of the taxon.

Polygala tenuicaulis Hook.f. subsp. ***tayloriana*** J.Paiva
VU A2c

Range: Nigeria (Mambilla Plateau, (two coll.) and Vogel Peak (two coll.) and Cameroon (W: Bamboutos Mts (two coll.). NW (Bamenda Highlands): Mt Oku (numerous coll.); Bambili; Bambui; Dom; Bafut-Ngemba; Bali Ngemba, Lac Bambuluwe). Adamawa: Mt Nganha, 60 km E Ngaoundéré (two coll.)).

Erect annual herb 30–45 cm tall. Stem unbranched in basal half, c. 2 mm diam., internodes 0.5–1 cm long, wiry, glabrous. Leaves alternate, sessile, linear-lanceolate, 25(–35) × 1–2 mm, acute, margin revolute, glabrous. Inflorescence a cluster of (1–)2–5(–10) terminal one-sided racemes, each bearing c. 20 flowers; bracts linear, 1.75 mm long; pedicels 1–2 mm long. Flowers pale pink, obovate in side profile, c. 7 mm long.

This taxon is only likely to be confused in the field with the type subspecies, unknown in the Bamenda Highlands. The second is much more slender and shorter in habit, with smaller, purple flowers.

This plant seems relatively common at Kilum-Ijim, in view of the fact that seven collections were made in 1996 and

1998. However, it tends to be highly localised there. It appears to be an annual that can compete with tall-growing grasses. Usually it is gregarious, 6–20 plants being seen at one site. Above Oku-Elak in fallow fields, some plants of this species persisted at one place, suggesting that it just might survive as a weed if its natural habitat is destroyed, but since this is a single observation, too much weight should not be given to it.
This species was assessed as LR/nt in Cheek *et al.* (2000: 69–70) on the basis that there were no threats, but it is here rated as Vulnerable since it is estimated that 30% of its population has been reduced due to the ongoing threats described below which increasingly have been realised to be deleterious to montane grassland herbs.
Habitat: rocky grassland; 1900–2400 m alt.
Threats: trampling by cattle, and frequent fires set by graziers, in montane grassland
Management suggestions: a monitoring experiment to exclude cattle and effects of frequent fires from areas of grassland might help resolve the extent to which this threat damages populations of different species of grassland herbs. Collection of baseline population data would aid future monitoring of populations.

RHIZOPHORACEAE

assessed by Martin Cheek

A pantropical family of trees, rarely shrubs, with opposite simple leaves, interpetiolar stipules and an absence of coloured exudates or scent when wounded, consequently they are often confused with Rubiaceae but differ in the superior ovary and the often toothed leaf margins.

Poga and *Anisophyllea*, once included e.g. in FWTA, are now placed in Anisophylleaceae, (Heywood *et al.* op. cit. 2007). There is no Flore du Cameroun account for the family.

Rhizophora is restricted to mangrove. Four species of this genus occur in Cameroon where they form 90% of the trees in that habitat. These species form the bulk of mangrove throughout the coasts of the Atlantic and are economically important in stabilising coastlines, providing a nursery to the young of commercial fish species, as well as prawns, and providing wood for specialised purposes.

Cassipourea was revised by Alston in 1925 in Kew Bull. 1925: 241–276, but many specimens have been collected since then. While several new species have since been published, there is a measure of taxonomic difficulty in identifying Cameroonian species of this genus. A new revision of the genus for Africa and a Flore du Cameroun account would address this problem.

In addition to the species treated below, the following species have been considered of importance for conservation and with further research might very well be shown to merit threatened status: *Cassipourea adamauensis* Jacq.-Fél. (Adamawa, *Jacques-Félix* 4047). The Kew copy of the protologue has been annotated, presumably by Alston, to suggest that this is a synonym of *C. ruwenzoriensis*. Otherwise it would rate CR. The following species are apparently endemic to Cameroon and for the moment are treated as NT:
Cassipourea dinklagei (Engl.) Alston
Cassipoura kamerunensis (Engl.) Alston
Cassipourea zenkeri Alston

Anopyxis klaineana (Pierre) Engl. was designated by W. Hawthorne (1997) at www.redlist.org (2001) as VU A1cd on the basis of it being a timber tree and in regeneration being poor. However, this appears unmerited in light of the wide range of this taxon and in its remaining widespread in the Lower Guinea and Congolian forest where exploitation of this species for timber is very limited; it was therefore reassessed as NT in Cheek *et al.* (2004: 366) and this is maintained here..

Cassipourea acuminata Liben (Iain Darbyshire, updated by Martin Cheek)
EN B2ab(iii)
Range: Cameroon (SW: Menyum (1 coll.)), Gabon (Miledi (1 coll.)) and Congo (Kinshasa) (environs of Mobanga, Bas-Katanga, (1 coll.); Yangambi (5 coll.)).
This species, described in 1986 by Liben (Liben, L. (1986). Deux *Cassipourea* (*Rhizophoraceae*) nouveaux d'Afrique centrale. Bull. Jard. Bot. Nat. Belg. 56(1): 139–144.) was previously known from 3 disjunct locations in Gabon and Congo (Kinshasa). The collections from Yangambi in the forests of central Congo (Kinshasa), seemingly the most significant site for this taxon, were all made in 1939. The collection from Menyum in W Bakossi is the first for Cameroon and greatly extends the species' extent of occurrence. However, it appears to be absent from large areas of apparently suitable forest stands within its range. Since four locations are known (above; AOO 16 km² with 4 km² cells, and threats as below), it was assessed by Darbyshire in Cheek *et al.* (2004: 173) as Endangered and this is maintained here, no additional data being available. EOO is calculated as 1202830 km².
Tree 4(–15) m, to 60 cm diam.; leaves opposite, elliptic, 7–10 × 3–4 cm, acumen pronounced, oblong-triangular, to 1 cm long, base cuneate, margins serrate, lateral nerves 7–10 pairs; petiole c. 5 mm; stipules ovate, 4.5 × 2 mm; fascicles

axillary, 2–3-flowered; pedicels c. 3 mm, pubescent; calyx lobes c. 3 mm; petals to 5 mm.
Habitat: lowland to mid-altitude rainforest, particularly along watercourses, including periodically flooded forest; 450–1000 m alt.
Threats: low level anthropogenic disturbance at the Cameroon site may threaten this taxon here; stochastic events such as severe river flooding at the Congo (Kinshasa) sites may result in loss of local populations.
Management suggestions: rediscovery of the populations at all the listed locations is important, as the majority refer to old collections. Future surveys of *Cassipourea* taxa in the Lower Guinea & Congolian forests may reveal further populations of this species; care should be taken to separate it from similar species such as *C. malosana*. Basic populational data should be recorded to assist mionitoring.

Cassipourea alternifolia Breteler
CR D
Range: Cameroon endemic (S: Campo-Ma'an area, Bibabimwoto)
Breteler described this remarkable species in 2007 (a new species of *Cassipourea* Aublet (Rhizophoraceae) from Cameroon, Kew Bull. 62: 609–612)
Here *Cassipourea alternifolia* is assessed as Critically Endangered since only a single collection with two duplicates is known, and presumably a single individual, *Tchouto et al.* 316, collected in flower in August 2000.
Understorey shrub to 40 cm, stems appressed pubescent; leaves alternate; stipules caducous, 4–6 × 1–1.5 mm; petioles 3–6 mm; blade obovate, 10–13 × 3–5 cm, acumen 1–2 cm, base cuneate, margin more or less entire, appressed hairy on lower surface, lateral nerves 8–9 pairs; fascicles 1–2-flowered, axillary, pedicels 1–2 mm, flowers white, oblanceolate, 5 mm, deeply laciniate, stamens 15–20, united in bundles of 3–6, glabrous; ovary 3-locular; fruit globose.
Differing from all other species of the genus in having alternate leaves. Unusual additionally in having stamens in bundles and the small habit.
Habitat: lowland evergreen forest; 40 m alt.
Threats: none are known
Management suggestions: Peguy Tchouto should be engaged to refind this species and to conduct a survey to collect baseline populational data on the species for future monitoring. Campo-Ma'an staff should be provided with the means to identify and so protect this species.

Cassipourea korupensis Kenfack & Sainge
CR D
Range: Cameroon endemic (SW: Korup NP).
Discovered during the course of research on the Smithsonian 50 Ha plot in Korup, this species was published in 2006 by Kenfack *et al.* (A new species of *Cassipourea* (Rhizophoraceae) from Western Cameroon, Novon 16(1): 61–64). There it was assessed as VU D1 provisionally, citing 47 individuals within the plot, 37 with diameters less than 5 cm (Thomas *et al.* op. cit. 2003). Only a single specimen (*Kenfack* 1186) with three duplicates, was cited. While no threats are known at the single location, the species is reassessed here since a maximum of 10 mature individuals are currently known. Should more locations and individuals be found as hoped the assessment will be revised downwards.
Tree 10 m, 30 cm diam. at 1.5 m from ground, stems appressed puberulous when young; stipules present at the terminal two nodes; triangular 8–11 × 4–5 mm; petiole 5–11 mm; blade elliptic 7–16 × 3.5–8.5 cm, acuminate, base cuneate-decurrent, lateral nerves 4–6 pairs sometimes uniting near margin, margin entire, leathery, glabrous; fascicles in terminal axils, 1–4-flowered, pedicel 2–2.5 cm: calyx cupular, 11 × 12–15 mm,deeply 5 lobed, petals unknown; stamens c. 60, about as long as calyx cup, filaments united at base in a short tube, anthers oblong, hairy, with apical connective; ovary 3–4 locular; fruit oblate, glabrous, 1 cm.
Habitat: lowland evergreen forest
Threats: none are known.
Management suggestions: excellent baseline populational data for this species already exists thanks to the Smithsonian 50 Ha plot in the Korup. Moreover, monitoring will be carried out automatically as a result of the re-mensuration programme. Since most of the individuals (37) are far smaller than the mature specimens, regeneration appears to be ongoing, as stated by Kenfack *et al.(*op. cit.). It is to be hoped that further surveys will uncover new locations for this species. Publicity, perhaps by a poster campaign, is advised to promote the importance of the species.

ROSACEAE
assessed by Martin Cheek

A large, mainly temperate family, mostly of herbs, with a few species extending into tropical African mountainous areas.

Formerly included in FWTA with the Chrysobalanaceae for no good reason, Rosaceae generally have alternate, stipulate leaves, actinomorphic flowers often with an epicalyx, superior ovary, free sepals, petals and stamens. In Cameroon best known for the widespread African *Prunus africana* native, but much planted in the Bamenda Highalnds for its medicinally valuable bark. All Cameroonian species of the family are montane, being found in scrub or forest: grassland transition. Flore du Cameroun 20 (Letouzey (1978)).

Alchemilla fischeri Engl. subsp. ***camerunensis*** Letouzey [Mbaklum (Oku)]

CR B2ab(iii)

Range: Cameroon endemic (NW: summit of Mt Oku).

Robust, carpet-forming, erect subshrub; main stems c. 10–15(–25) cm high, 6–8 mm diam., stolons over 50 cm long; leaves up to 8 × 10 cm, 7–9-lobed, median lobe 25–31-toothed, bidentate, thickly silvery hairy. Petiole up to 12 cm long. Stipules oblanceolate, apex acute, 6-toothed. Inflorescences paniculate cymes, borne on stolons, c. 20 cm tall; peduncle c. 8 cm long; bracts resembling stipules. Flowers green, 2.5 mm long. Description after Letouzey (Flore du Cameroun 20, Rosaceae (1978)).

This attractive, silvery-leaved herb is relatively abundant in the summit area of Mt Oku (four collections were made in the 1996 expedition), but is unknown elsewhere. Letouzey (1978), in his protologue for the taxon, reports in detail on its habitat and relationships. Pollard (pers. comm.) confirms Letouzey's (1978) observations that *A. fischeri* subsp. *camerunensis* forms a continuous silvery carpet near the summit of Mt Oku and I was able to see this myself in 2000, so many square metres being covered that it was concluded that the species might be detectable with satellite imagery! Reproduction by stolons in this taxon may be more important than seed. The other subspecies occurs many thousands of kilometres to the E on high mountains E of the Congo basin, such as Mt Kenya.

This assessment is mainly derived from Cheek *et al.* (2000: 70) where it was assessed as CR B1+2c, updated here to CR B2ab(iii) since there is only a single site (AOO is estimated as 4 km² using 4 km² cells) with threats as below.

Habitat: rocky grassland and scrub; 2800–3000 m alt.

Threats: man-made fires and grazing by cattle, goats and sheep and by insect larvae (Pollard pers. comm.).

Management suggestions: monitoring of the population is needed to determine the extent to which this taxon is declining.

Prunus africana (Hook f.) Kalkman

NT

Range: montane Africa (to E and South Africa) and Madagascar.

Evergreen tree 20(–30) m high, 0.4(–1) m diam. Breast height, sometimes bearing buttresses; bark brown, slash pink, soon oxidizing to brown. Leaves alternate, elliptic, 6–15 × 3–6 cm, acute or acuminate, base rounded, margin serrate; petiole red, glandular at apex, 1–2 cm long; stipules soon falling. Inflorescence axillary racemes c. 10 cm long, several inserted at the base of young leafy shoots. Flowers white, c. 5 mm diam. Drupes 1-seeded, red, c. 1 cm diam.

This species is one of about ten Pan-African montane tree species (including e.g. *Agarista salicifolia, Ilex mitis* and *Nuxia congesta)* and is not remotely in danger of extinction, so long as some montane forest survives somewhere within its enormous range. Locally it can be very common. However, on Mt Oku and Ijim Ridge, as with other areas within the range of this species, many trees have died as a result of girdling caused by bark removal. The bark from the trees was transported to the Plantecam factory at Mutengene where it was extracted to produce a powder for export to France. However, mature trees still survive in the wild. A great number of individuals of *Prunus africana* have been planted as the boundary of the protected area on the Kilum side. They are now 6–8 m high and appear to be growing well. A great deal of attention, and funding has been paid by international conservation organisations to investigate and address this harvest and, perhaps for this reason, the species has received a 'threatened' conservation rating of VU A1cd (www.redlist.org). In the opinion of this author, it merits Near Threatened.

This assessment is modified here from Cheek *et al.* (2000: 70–71).

Habitat: montane forest, usually at about 1800–2200 m alt.

Threats: harvesting of bark for the European medicinal market.

Management suggestions: none are needed. Large numbers of individuals in the Bamenda Highlands have planted this tree in large numbers, seeking to benefit from sale of the bark in future years (Assessment of *Prunus africana* (Pygeum) in Bui Division - North West Province - Report to the Ministry of Forest and Wildlife, N.W. Province, 2008). *Prunus africana* is almost certainly the most commonly planted native tree species in the Cameroon Highlands.

RUBIACEAE

assessed by Martin Cheek

In the forests of Cameroon, Rubiaceae are supreme in terms of species diversity, far surpassing legumes, orchids and euphorbs. They are often known as the coffee family since the world's highest value internationally traded commodity, *Coffea arabica* belongs in the Rubiaceae. *Coffea arabica* is native to the forests of Ethiopia, but related species are also important or of interest to the coffee trade. *Coffea canephora*, 'robusta coffee' and *Coffea liberica*, producing more bitter coffees used in 'instant coffee' products are both native in the forests of Cameroon. These wild stocks provide an important genetic pool for coffee breeders seeking to develop new commercial strains which might have improved traits, such as disease resistance, higher yields, and greater tolerance to low rainfall for example. Cameroon rivals Tanzania in numbers of endemic coffee species and in coffee species-diversity. Most famously, *Coffea charrieriana* of Bakossi in SW Region, is Africa's only naturally caffeine-free species, making it of great interest to breeders. However it is believed only to have been known as a single bush in the wild. The efforts we have made to refind it have failed so far: it may be extinct in the wild.

Quinine, the original anti-malarial drug, derives from the bark of *Cinchona*, a S American genus of Rubiaceae. Several timber trees *Nauclea diderrichii* and *Hallea stipulosa* of Cameroon also belong to this family.

In forest habitats, many herbs, climbers, and especially small trees and shrubs, are Rubiaceae. Many species are restricted in their geographical range and so feature among the threatened species below. New species are constantly being discovered, as are new locations for existing species.

There is no Flore du Cameroon account for Rubiaceae. FWTA 2 (1963) and especially, the Fl. Gabon accounts by N. Hallé are useful. Several genera have African-wide revisions, but these are often out-of-date due to new discoveries. However, Prof. Bonaventure Sonké of Yaoundé is a particularly prolific researcher on this family in Central Africa and thanks to him there has been taxonomic progress in genera such as *Oxyanthus*, *Rothmannia* and *Sherbournia*. At his *alma mater* in Brussels, Prof. Robbrecht has led his students to produce valuable revisions of such genera as *Tricalysia* (Robbrecht himself), *Ixora* (DeBlock), and *Pauridiantha* (Ntoré). Let us hope that this excellent work continues with new Robbrecht recruits such as Lachenaud.

Rubiaceae in Cameroon are easily recognised by the absence of coloured exudates, or essential oils when wounded, by their opposite, simple, entire leaves which have petioles connected on each side of the stem by a common, "interpetiolar" stipule, and by the flowers with a tubular (usually radially symmetrical) corolla and an inferior ovary.

Calycosiphonia macrochlamys (K.Schum.) Robbr. was assessed as VU B2ab(iii) in Cheek *et al.* (2004: 174) but is reassessed here as NT since 11 locations are now known.

Anthospermum are a genus of c. 40 species of ericoid shrubs often at high altitude, widespread in Tropical Africa and further E, last revised by Puff in 1986. Only one species occurs W of the Congo basin.

Anthospermum asperuloides Hook.f. (syn. *A. cameroonense* Hutch. & Dalziel)
EN B2ab(iii)

Range: Equatorial Guinea (Bioko: Pic Clarence [Pico Basile]) and Cameroon (SW: Mt Cameroon. W: Bamboutos Mts–Dschang. NW: Mt Oku).

"8 ft high growing on the very top of Peak Clarence" is how the first collection of this species was described (*Mann* 593, Dec. 1860, Bioko). More recent collections exist from Bioko, apparently from the same locality. At Mt Cameroon only two sites are known: Hut 2 above Buea and Mann's Spring. In the Bamboutos Mts there is but a single collection (*Sanford* 5590: "not common") and likewise for relatively well surveyed Mt Oku (*Munyenyembe* 835) where it was found on a cliff, suggesting that some factors prevent its survival in grassland: perhaps the fire and grazing prevalent there.

Assessed in Cheek *et al.* (2000: 71) as LR/nt, that is lower risk, we re-evaluate data presented there to reassess the species here as Endangered under IUCN (2001) since only four locations are known globally (using 4 km² cells AOO is 20 km²) and there are indications of threats (see below).

Shrub or subshrub 0.2–1 m tall. Stems c. 5 mm diam. at the base, branches sparse, persistently leafy, sinuous. Leaves subsessile, leathery, opposite or in whorls of three; blades 5–10(–15) × 0.8–1.5(–2) mm oblanceolate to linear-lanceolate, apex subspinulose, and base slightly attenuate, margin flat or revolute, inconspicuously white-hairy; stipules sheathing, cup-like, 1–1.5 mm long, with (1–)3–5(–6) gland-tipped setae 1.4–2.4 mm long, the lateral ones shortest. Inflorescences inconspicuous, of axillary, subsessile clusters of flowers. Flowers subsessile. Corolla yellowish (rarely purplish) green, hairy or glabrous, tube in bisexual flowers narrowly funnel-shaped, 0.5–0.7 mm long, lobes lanceolate, recurved, 1.4–2 × 0.6–0.7 mm long. Stamens 4, filaments 1–1.4 mm long; anthers 0.9–1.7 mm long. Fruits reddish brown, breaking into two fruitlets when ripe. Fruitlets oblong, 1.9–2.7 × 1–1.4 mm, outer side convex, with white hairs; calyx lobes minute or absent in fruit.

Habitat: upper montane grassland and scrub; 2400–3000 m alt.

Threats: grazing, fire and cattle-trampling, known in this habitat, probably have adverse effects on this species.

Management suggestions: a survey of this species on Mt Oku and/or Mt Cameroon is needed. The information required includes numbers of plants, their distribution, regeneration levels and threats.

Argocoffeopsis fosimondi Tchiengué & Cheek
CR B2ab(iii)

Range: Cameroon endemic (SW: Cameroon Highlands, known only from Fosimondi–Lebialem Highlands near W Bamboutos Mts).

Assessed in Harvey *et al.* (2010: 80) as Critically Endangered according to the criteria of IUCN (2001) since it is only known from a single location (area of occupancy 4 km^2 using 4 km^2 cells) and since this location is under severe threat of forest clearance for cash and food crops (Tchiengué pers. obs. 2006).

Shrub, rarely small tree, 2–3.5 m tall, glabrous. Leaves narrowly elliptic, 4–12 × 1.5–4 cm, acuminate, base cuneate, domatia absent; petiole 2–12 mm long; stipule shortly sheathing, sheath 2–3 mm long, becoming chaffy, limb broadly triangular, 2–3 × 5.5 mm, apex mucronate, outer surface conspicuously scattered with minute, yellow, rod-shaped raphides, each c. 0.1 mm long. Flowers in loosely branched terminal inflorescences of 8–30 flowers, often with accessory lateral branches from the axils of the first pair of leaves, overall 5–8.5 × 2.8–5.8 cm. Flowers 5(–6)-merous; calyx lobes triangular, c. 0.5 × 0.5 mm; corolla white, slightly curved in bud; tube 6.4–10 mm long; lobes ligulate-oblong, lacking appendages; stamens included or excluded; style exserted by 2–5 mm. Fruit a black drupe held obliquely on white peduncles, ovoid or spherical, 6–9 mm long; pyrenes usually single, with pre-formed germination slit.

Habitat: submontane (cloud) forest with *Santiria trimera* (Burseraceae), *Cola verticillata* (Sterculiaceae), *Macaranga occidentalis* (Euphorbiaceae), *Chassalia laikomensis*, *Pauridiantha paucinervis* (both Rubiaceae), *Carapa procera* (Meliaceae):1330—1380 m alt.

Threats: see above.

Management suggestions: it is to be hoped that the people of Fosimondi, led by their elders and Fon, will seek to protect this beautiful shrub from extinction. Educational material on the importance of this material will be passed to them by the authors via ERUDEF.

Argocoffeopsis kupensis Cheek & Sonké **sp. nov.** ined.
EN B2ab(iii)

Range: Cameroon endemic (SW: Mt Kupe)

Shrub or small tree (2–)3–8 m tall, glabrous; main axis brown. Stem terete, 2.5–4 mm diam., drying matt black in first and second internodes, in the third to fourth internodes the epidermis becoming white, spongy, subglossy and exfoliating as in a *Chazaliella*; internodes 5.2–6.8 cm long, apical bud with gum covering; buds supra-axillary, covered in gum, inserted 3–4.5 mm above the petiole bases. Leaf-blades opposite, equal, elliptic-oblong, rarely ovate-lanceolate, 13–18 × 6.2–7.3 cm, acumen tapering, slender, acute, 1 cm long, 3–4 mm wide at base, base obtuse to acute, slightly to conspicuously asymmetric, midrib impressed above, lateral nerves 8–10 on each side of the midrib and uniting near the margin to form a weak, looping, connecting nerve, tertiary and quaternary nerves sparse, reticulate, not conspicuous, domatia absent; margin entire; drying very dark green above, pale green below, with midrib nearly black; petiole 0.7–1.7 cm long, drying black. Stipule with basal cylindrical sheath 2–4 [illegible] mm, truncate, distal limb absent to narrowly trian[illegible] mm long with central ridge absent. [illegible] branches, inserted 2–4 mm above the leaf axil; fasciculate, peduncles 1-flowered, 2–4 per axil at both axils of node; 2–4 axils fertile per stem. Calyculi 3(–4), the uppermost slightly larger, each calyculus cylindrical-campanulate, more or less 4-lobed (the uppermost: 2 foliar lobes and 2 stipular lobes) or 2-lobed (the lower calyculi, with two foliar lobes only), ± sessile, glabrous. Flowers hermaphrodite, homostylous (5–or) 6–merous, sessile; calyx limb barely detectable. Corolla glabrous, white; lobes contorted to the left in bud; corolla tube cylindrical, widening slightly in the apical 2 mm, (9–)10–12 mm long; corolla lobes oblong-elliptic, 12–17 × 3–4 mm. Anthers completely exserted. Style filiform, 9.8–1.15 cm stigma arms 2, exserted. Fruits black, at length ripening orange, ellipsoid, 22–27 × 14–17 mm including a 2–3 mm broad, 3–5 mm long rostrum with apex truncate. Seeds 1–2, plano-convex and elliptic-orbicular in outline 10–13 × 8 × 2–3 mm

Argocoffeopsis kupensis is assessed as CR B2ab(iii), Critically Endangered, since there is one location with two sites (area of occupancy 8 km^2 using 4 km^2 cells for each of the two sites known at this location) with threats as below. This assessment is maintained here since there are no new data.

Habitat: submontane evergreen forest 1000–1600(–2000) m alt.

Threats: under threat of forest clearance for coffee (the type collection was made at a coffee plantation inside the forest) suggesting that the population of the species is at risk.

Management suggestions: further searching is needed to find more locations for this plant. It is to be hoped that the people of Mt Kupe, once they are informed of the rarity of this species, will seek to protect this beautiful shrub from extinction.

Argocoffeopsis pulchella (K.Schum.) Robbr.
EN A2c

Range: SE Nigeria (Eket and Oban), Cameroon (fide Robbrecht 1981) and Gabon (6 collections at 4 locations).

A liana to 24 m with white, sweetly fragrant 6-lobed flowers 1.5 cm across, this species was first described as *Coffea pulchella* K.Schum. based on a specimen from Sibange Farm, Gabon (*Soyaux* 19, 23 Aug. 1879). Robbrecht (Bull. Jard. Bot. Natl. Belg. 51(3-4): 364 (1981)) last revised the genus and mentioned Cameroon as part of the range of this species without citing a specimen.

Habitat: lowland evergreen forest; sea level–650 m alt.

Threats: the type locality is now a suburb of Libreville (pers. obs. Cheek 2007) and forest at another site, at Cap Esterias is also being cleared (Wilks pers. comm. to Cheek). Other locations in Gabon are more secure, and some are quite recent, but the species has not been seen in Nigeria since 19[illegible] (both Talbot collections) and since that time forest losses have been severe there. We estimate that more than 50% of the seven locations for *A. pulchella* have been lost and that such losses are continuing hence it is here assessed as Endangered under Criterion A2c.

Management suggestions: rediscovery of this species at its (probably) more secure locations in Gabon (e.g. *Breteler & J.J.F.E. de Wilde* 780, Nov. 1978, Lastoursville–Moanda Rd, 35 km) and their inclusion in a protected area, appears to be the best hope for protecting this species. It is to be hoped that survey work will extend the known range of this liana.

Argocoffeopsis spathulata A.P.Davis & Sonké
VU B1ab(i, ii, iii)

Range: Cameroon endemic (S: 21 collections from 19 locations).
This small single-stemmed forest treelet is 0.5–2.5 m tall. Its affinites are with the Calycosiphonioid group. The axillary 5-petalled white flowers are about 7 mm wide. The species is both distinctive, and named for, the spathulate leaf apices (Davis & Sonké Blumea 53(3): 528-531 (2008), protologue). Davis & Sonké give an EOO of 8105 km², an AOO of 2930 km² based on 19 cells of 12.4 km² using the Willis *et al.* Methodology, 2003: op. cit. This assessment is maintained here since no additional data are available.
Habitat: lowland evergreen forest; 500-600m alt.
Threats: Davis & Sonké (op. cit. 2008) cite a continuing decline inferred or projected for extent of occurrence, area of occupancy and quality of habitat. Forest clearance for agriculture and development is the main cause of habitat loss.
Management suggestions: *Argocoffeopsis spathulata* appears not to be included in any protected area. It is advisable that this be rectified.

Argostemma are a genus of small evergreen forest herbs which are only in growth during the wet season. Most species occur in SE Asia with only two accepted in Africa. However, observations by Letouzey have called into question the delimitation of the African species and until this is resolved their conservation status cannot be assessed.

Aulacocalyx was revised in 1997 by Figueiredo (Kew Bull. 52: 637-658) who recognised eight species, one of which, *A. talbotii* (Wernham) Keay should be rated as near threatened. In the last ten years the excellent Cameroonian botanist Sonké has discovered two more species both of which are localised and threatened. This genus, with *Rothmannia* and *Randia,* is recognised even when sterile, by commonly having three leaves at a node.

Aulacocalyx camerooniana Sonké & Dawson
CR D

Range: Cameroon, endemic (S: (1 location near Akom II)).
A 2–3 m forest shrub unusual in *Aulacocalyx* in being conspicuously anisophyllous: leaves in each pair are unequal, one being twice the area of the other. This species is known only from forest at Engon, 5 km S of Akom II where it is known from two collections (*Sonké & Beina* 3348, 3350,), fl. fr. 6 March 2004 (Sonké *et al.* Kew Bull. 60: 301–304 (2005)). The conservation assessment made by Sonké (EN D, loc. cit.) is raised here since only two individuals are known, fitting the CR threshold.
Habitat: evergreen forest; 620–800 m alt.
Threats: *Aulacocalyx camerooniana* is highly localised, as it is only known from the western part of Campo Ma'an NP, from two individuals several km apart. Despite the fact that the species is highly conspicuous in flower, extensive exploration in the area has failed to locate any other plants of this distinctive species. *Aulacocalyx camerooniana* is particularly endangered by the possibility of stochastic events. It has a limited range and a very small number known individuals. Unless more plants and new locations are discovered for this species it may prove to be one of the rarest and most threatened plants in Cameroon. (Sonké *et al* 2005: 304).
Management suggestions: it is fortunate that this species grows within a National Park where it has a measure of protection. Informing the population of Akom II, possibly via a poster dedicated to this species, may assist its protection further, and possibly bring to light new locations. Further survey work in Camp Ma'an might also find new sites. Sonké *et* al (2005: 304) suggest that mature fruits be obtained to initiate *ex situ* conservation: this is an excellent suggestion.

Aulacocalyx mapiana Sonké & Bridson
EN B2ab(iii)

Range: Cameroon (SW: Ebamut in Banyang Mbo; Rumpi Hills).
A glabrous shrub or small tree to 15 m tall, this species is unusual in the genus in having 4-lobed calyxes but a 7-lobed corolla and 7 stamens. It is further distinguished from the related *A. talbotii* in the 10–17 mm long pedicels and the corolla tube 5 × longer than the lobes. (Sonké & Bridson, Syst. Geogr. Pl. 71: 17–23 (2001). Currently only five collections are known from two locations equating to an AOO of 8 km² (4 km² cell), with threats as below, hence Endangered under criterion B.
Habitat: submontane evergreen forest; 900–1500 m alt.
Threats: a new road (c. 2005) from Bangem to Nguti is likely to have allowed expansion of Ebamut village and consequent loss of forest habitat to more numerous cocoa and coffee farms. The former Banyang Mbo sanctuary has had its protected status revised downwards in recent years and has lost the NGO resources allocated to conservation at the turn of the millenium.
Management suggestions: educating communities at its two locations about the variety of this species and renting micro-sites would further the chance of this species surviving.

Might one of its two known locations be formally protected? Further survey work may uncover new locations for this tree. It may well occur in the newly created but largely unsurveyed Bakossi NP for example, since this is adjacent to Ebamut.

Belonophora was revised by Cheek & Dawson in 2000 (Kew Bull. 55: 63-80) but no IUCN conservation assessments were made there. Globally, the genus is restricted to Tropical Africa. Five species are known, all of which occur in SW Cameroon. In addition to the two species assessed as threatened below, *Belonophora wernhamii* Hutch. & Dalziel may also be rated as threatened in future. Known from 13 collections from Oban to Kribi (Cheek & Dawson 2000: 72–73) it may yet be assessed under criterion B when better habitat loss data are available.

Belonophora ongensis S.E.Dawson & Cheek

CR A1c + 2c

Range: Cameroon (SW: known only from foothills NW of Mt Cameroon (5 coll.), the disjunct northern part of Korup NP (1 coll.), Kupe-Bakossi (1 coll.) and Ebo (1 coll.)).

Earlier assessments of this species (Cheek & Dawson 2000, listed as Cheek & Cable 1999 in IUCN 2003, www.redlist.org and Cheek in Cheek *et al.* 2004: 173) are maintained here. This monopodial treelet is often very rare at its known locations, e.g. the site in the Kupe-Bakossi area, in the Mungo R FR, just north of Bulutu on the Ebonji-Ngusi Rd ascending the ridge to the east (*Cheek* 10193, fr., 30 Nov. 1999). At this site two plants only were found, growing with *Hylodendron, Guarea cedrata, Guibourtia demeuseii, Pachystela sp.* and *Garcinia mannii*. The forest at this site was threatened by upgrading of the Ebonji-Ngusi Rd – then in progress under EU sponsorship (*pers. obs.* Cheek 1999).

Habitat: lowland evergreen forest; 200–500 m alt.

Threats: forest clearance for agriculture. In the western foothills of Mt Cameroon, the type locality, *Belonophora ongensis* is threatened by expansion of plantations.

Management suggestions: the site at Bulutu should be revisited. If this forest has not been cleared, as is feared, a survey of the subpopulation of *Belonophora ongensis* should be conducted as a baseline for future monitoring and an education programme initiated concerning the species. Conservation efforts for the species should be concentrated at the Onge Forest (Mt Cameroon) since the densest population is found here.

Belonophora talbotii (Wernham) Keay

CR D

Range: Nigeria (Oban, 1 coll.) and Cameroon (SW: Korup, Mokoko–Mt Cameroon; Banyang Mbo; Mone FR)

Discovered by Talbot in Oban in 1912, this 1–3 m forest treelet is the rarest species in its genus. At most of its sites, only one or two individuals are known. *Belonophora talbotii* is easily recognised by its very large 2.4–4.5 × 12–12.5 cm stipule which is broadly ovate and acuminate. It is here assessed as Critically Endangered since less than 50 individuals are known. It is estimated that only 35 plants have been enumerated. Most of these (c. 30) are at the Mone FR in SW Region (Cheek pers.obs. Dec. 2009). However the identification of the specimens from Mone, and that from Banyang Mbo, have yet to be formally confirmed.

Habitat: lowland evergreen forest on sand in areas with 3–5 m rainfall per annum; sea-level to 400 m alt.

Threats: logging, followed by agriculture. Only one of the five known locations is protected in a National Park (Korup) most others are production forest reserves where farming also occurs.

Management suggestions: upgrading the Mone FR to National Park status would protect the largest global population of this species.

Bertiera was revised for Africa by Nguembou in 2008 (Phylogénie, distribution, écologie et revision taxonomique du genre *Bertiera (Rubiaceae)* en Afrique. Thèse de Doctorat, Université Libre de Bruxelles, Lab. de Bot. Syst. et de Phytosoc. 187 p.). Nguembou recognised 47 species for Africa, Madagascar and the Mascareignes. The greatest global species diversity per degree square occurs in the forest of SW and S Regions of Cameroon with 20 and 21 species respectively (Nguembou 2008: 49).

Bertiera heterophylla Nguembou & Sonké

EN B2 ab (iii)

Range: Cameroon: endemic (S: 4 locations in vicinity of Bipindi & Ébolowa).

Discovered in 2006 (Nguembou & Sonké, Syst. Geogr. Pl. 76: 212) this is a very remarkable species of *Bertiera* having oblanceolate leaves on the main axis but elliptic ones on the lateral branches. A 1 m shrub, it has sessile, axillary pseudocapitate inflorescences bearing white flowers with corolla tubes c. 1 cm long (Nguembou op. cit. 122–123). *Bertiera heterophylla* is here assessed as EN B2 ab (iii), that is Endangered, since known from only four locations with an AOO of 16 km² (using 4 km² cells) and with threats indicated below.

Habitat: lowland-submontane evergreen forest; to 1200 m alt.

Threats: clearance of forest for timber and agriculture.

Management suggestions: incorporation of one or more of the locations of this species in a protected area is advised.

Bertiera rosseeliana Sonké, Esono & Nguembou

VU B2 ab (iii)

Range: Cameroon (S: 18 coll. at c. 4 loc.) and Equatorial Guinea (Rio Muni: 5 coll. at 2 loc.).

A 2–6m forest shrub described by Sonké *et al* in 2005 (Adansonia sér. 3, 27(2): 309-315) fide Nguembou (op. cit. 164-165). This species has terminal pseudocapitulate inflorescences 2–3.5 cm diam. The flowers are white, with a corolla tube of c. 2 cm and much resembles the better known *B. naucleoides*. However, *B. rosseeliana* has ovoid fruits with a caducous calyx, a persistent disc and a 5 mm pedicel (versus globose, persistent calyx, invisible disc and a 15 mm pedicel (Nguembou op. cit: 81).

In the protologue the species was assessed as "EN B1", but we here reassess it on the basis of the data in Nguembou 2008 as Vulnerable, or VU B2 ab (iii) since 6 locations can be recognised with an AOO of 24 km² (4 km² cells) and due to the threats below.

Habitat: submontane evergreen forest; 500–1000 m alt.

Threats: clearance of forest for logging and agriculture.

Management suggestions: inclusion in a protected area of at least part of the core area (e.g. forest near Akom II, Bindem, Bifa) for this species is advisable.

Calochone acuminata Keay

VU A3c

Range: Cameroon (SW: Mt Etinde; Rumpi Hills; Kupe-Bakossi.); Gabon (Cristal Mts (7 coll.)) and Angola (Cabinda (1 coll.)).

This spectacular species was first collected in 1951 in the Mungo valley between the Rumpi Hills and Bakossi Mts and published as a new genus by Keay (Bull. Jard. Bot. État Bruxelles 28: 30 (1958)). A further six collections were made over the next 45 years in western Cameroon, all along the roads that mark the western and eastern boundaries of the Kupe-Bakossi area, apart from one collection in the Rumpi Hills (*Letouzey* 14542) and one at Mt Etinde (*Watts* 567). Seven collections are cited in Fl. Gabon (17: 216 (1970)), all from the Cristal Mts. Between 1995–1999 11 collections were made in the Mt Kupe-Bakossi Mts area where it appears relatively frequent and is used for treating headaches. In conservation terms, Mt Kupe-Bakossi Mts appears to be the main stronghold for the species. When in flower (Jan. and Feb.) the plants are so spectacular that they are easily detected, so long as they have not reached the forest canopy. The assessment above was made in Cheek *et al.* 2004: 173 and is maintained here. No new data are available. It was postulated that over 30% of the habitat of this species will be lost over the next 50 years, mainly in SW Region Cameroon.

Shrub or climber to 10 m; stems brown pubescent; leaves papery, obovate or elliptic, 20(–30) × 11(–17) cm, acuminate, truncate to subcordate, pubescent, nerves 10, petiole 2–3 cm; stipules triangular, 15 mm; terminal panicle 10–20-flowered; corolla orange-pink, tube funnel-shaped, 5 cm, lobes 3 cm.

Habitat: submontane evergreen forest, (250–)800–1000(–1250) m alt.

Threats: forest clearance for agriculture has been observed in the Loum FR, one of the locations for this species (Cheek, Onana pers. obs.).

Management suggestions: monitoring of the population of *Calochone acuminata* is recommended. The survival of this species might be enhanced by targeted conservation education in Bakossi.

Reference: Keay, R.W.J. (1958) *Randia* and *Gardenia* in West Africa. Bull. Jard. Bot. Nat. Belg. 28: 15–72.

Chassalia is a genus of shrubs and climbers similar to *Psychotria*, but differing in their usually slightly S-shaped carolla tube; fleshy, brightly coloured inflorescence axes, chaffy stipules and endocarps with preformed germination slits. The genus has not been revised for Africa and while species in the Flora Zambesiaca and Flora Tropical East Africa areas are well resolved, those of Guineo-Congolian Africa need taxonomic revision. When this is done it is likely that numerous taxa, additional to those treated below, will require red-listing, and several of those treated here will be shown not to be as rare and geographically restricted as they appear at present. Several specimens at Kew are annotated 'Revision of Chassalia in C. Africa, J. Leon 2000' but so far, no publications appear to have resulted.

Chassalia laikomensis Cheek

CR A2c

Range: Nigeria (Mambilla Plateau (1 coll.)); Cameroon (SW: Mwanenguba (1 coll.); Lebialem-Fosimondi. NW: Bamenda Highlands (several locs: Bali Ngemba, Ijim and Dom)).

The assessment above, equating to a loss of >80% of the population in 100 years due to forest habitat loss, was made in Cheek & Csiba (2000), listed as having been assessed by Cheek *et al.* (2000: 71–72) in IUCN (2003) (www.iucnredlist.org). That assessment was maintained in Cheek in Harvey *et al.* (2004: 68), Harvey *et al.* (2010: 80–81), Cheek *et al.* (2010: 101–102) and is maintained here also, despite the addition of new locations (above). Several records from Bakossi of a similar plant may represent a separate taxon, but if shown to be conspecific the threat level of the taxon will drop.

This is a forest understorey shrub, probably very long-lived.

Shrub, rarely small tree, 2–3.5 m tall, glabrous. Leaves narrowly elliptic, 4–12 × 1.5–4 cm, acuminate, base cuneate, domatia absent; petiole 2–12 mm long; stipule shortly sheathing, sheath 2–3 mm long, becoming chaffy, limb broadly triangular, 2–3 × 5.5 mm, apex mucronate, outer surface conspicuously scattered with minute, yellow, rod-shaped raphides, each c. 0.1 mm long. Flowers in loosely branched terminal inflorescences of 8–30 flowers, often with accessory lateral branches from the axils of the first pair of leaves, overall 5–8.5 × 2.8–5.8 cm. Flowers 5(–6)-merous; calyx lobes triangular, c. 0.5 × 0.5 mm; corolla white, slightly curved in bud; tube 6.4–10 mm long; lobes ligulate-oblong, lacking appendages; stamens included or excluded; style exserted by 2–5 mm. Fruit a black drupe held obliquely on white peduncles, ovoid or spherical, 6–9 mm long; pyrenes usually single, with pre-formed germination slit.

Habitat: montane evergreen forest; 1650–2000(–2400) m alt.

Threats: about 95% of the original forest cover of the Bamenda Highlands has been lost to e.g. agriculture (Cheek *et al.* 2000: 6; Cheek & Csiba 2000) and there have been similar losses at Mambilla and Mwanenguba. In the Dom area 50% of forest cover was lost in the 15 years 1988–2003 (Baena in Cheek *et. al.* 2010: back cover).

Management suggestions: more information is needed on the numbers of individuals at the known sites and levels of regeneration. Enforcement of existing protected area boundaries would help protect a significant portion of the surviving population.

Chassalia manningii O.Lachenaud ined.
EN B2ab(iii)

Range: Nigeria (Benin Province, Usonigbe FR, 3 coll.), Cameroon (SW: S Bakundu FR, 2 coll.. C: Mefou proposed NP, 1 coll.).

Here assessed as EN since three locations (above, AOO 12 km^2 with 4 km^2 cells, threats as below). This assessment appears also in Cheek *et al.* 2011.

Stout, unbranched shrub, 45–115 cm; stems puberulous; stipules asymmetric, broad ovate, to 12 × 10 mm, indumentum as blade; leaves clustered in funnel at apex, oblanceolate, to 35 × 13 cm, acumen 1 cm, base cuneate, abruptly obtuse, lateral nerves c. 12 pairs forming a strong, looping submarginal nerve in addition to the weaker marginal nerve, upper surface glabrous, lower surface with patent pale brown hairs, dense on the nerves; petiole 1–2.5 cm, indumentum as blade; infructescence erect, fleshy, pink, 7 × 5 cm, lateral branches to 7 cm, stout; fruits black, bilobed, to 6 × 8 mm; endocarp bony, brown, strongly keeled, PGS elliptic.

Habitat: lowland evergreen forest; sea-level–700 m alt.

Threats: nearly 90% of the original forest of Nigeria has been lost to timber extraction and agriculture, including in Forest Reserves; the species may no longer survive in Nigeria. These same threats also occur in S Bakundu.

Management suggestions: Mefou proposed NP should be upgraded to full National Park status to protect this and the other threatened species that it contains. A census of basic populational data of this species at Mefou would provide a baseline for future monitoring and gauge needs for management interventions.

Chassalia petitiana F.Piesschaert
VU B2ab(iii)

Range: Equatorial Guinea (Bioko (1 site)), Cameroon (SW: Mt Kupe (2 loc.)), Gabon (Cristal Mts (2 loc.)) and Congo (Kinshasa) (Kivu (2 loc.)).

This species was originally assessed as VU in Cheek *et al.* (2004: 174) since seven locations are known (AOO of 28 km^2 with 4 km^2 cells), threats as below. That assessment is maintained here. No new data are available.

Known from only seven locations, this rare epiphytic shrub has a highly disjunct distribution. Although given a name as early as 1922 (*Psychotria epiphytica* Mildbr., based on the Bioko material), it was not legitimised as a species, or fully described, until many decades had elapsed (Piesschaert *et al.* 1999, from which most of the range data above are taken). This was the first published fully epiphytic species of African Rubiaceae.

Epiphytic shrub, glabrous, c. 30 cm; bark corky, flaking; leaves leathery, drying pale grey-green, elliptic, to 8 × 3 cm, broadly acute, cuneate, lateral nerves 4–6 pairs; petiole 1–1.5 cm; stipule sheathing, 6 mm; panicle 6 cm; peduncle 1.5 cm; flowers not seen; 20-fruited; fruits ellipsoid, 4 mm, dark purple; calyx 4-lobed

Habitat: submontane forest; 750–1200 m alt.

Threats: forest clearance for agriculture and wood at Mt Kupe, especially at the lowest altitudes.

Management suggestions: more data is needed on the survival of this species at its known sites, and on the extent of the subpopulations. At Mt Kupe the survival of this species might be enhanced by targeted conservation education.

Reference: Piesschaert, F., Jansen, S., Huysmans, S., Smets, E. & Robbrecht, E. (1999). *Chassalia petitiana* (*Rubiaceae-Psychotrieae*), an overlooked epiphytic species hidden in the African canopy. Syst. Bot. 24(3): 315–322.

Chazaliella was revised by Verdcourt (Kew Bull. 31: 785–818 (1977)). He recognised 24 species (six of which are poorly known) and numerous subspecies and varieties. Owing to time and space constraints, as usual, infra-specific taxa are not assessed here otherwise many would be rated as threatened. *Chazaliella coffeosperm,* a borderline VU/NT species, may appear in a second edition.

Chazaliella letouzeyi Robbr.
VU B2 ab (iii)

Range: Cameroon (S: Ébolowa–Ambam; Mvam-Oveng; Mvangan (3 locs)), Gabon (NW of Booué; NE of Oveng; Macocou-Kambouia (3 locs)) and Congo (Brazzaville) (Ogooué-Leketi proposed NP; Moutienne (2 locs)).
A single-stemmed forest treelet with dramatically spatulate, black-drying, leathery, subsessile leaves 20–32 × 8–16 cm, clustered at the stem apex. This appears to be the only litter-gathering monopodial species in the genus. It is further atypical of the genus inhaving apparently axillary infructescences.
It is known from eight collections equating to 6 locations in N–C Gabon and S Cameroon (Robbrecht, Adansonia 4 sér., 11: 341–349 (1989) and two more locations recently discovered in Congo. Since only 8 locations are known, using a 4 km² cell an AOO of 24 km² is indicated. Threats as below. Here assessed as Vulnerable (VU B2 ab (iii)).
Habitat: lowland evergreen forest, alt. range not indicated.
Threats: threat levels appear to be low since part of the Cameroon population is protected by the Campo-Ma'an NP, however, slash and burn agriculture remains a threat.
Management suggestions: the managers of Camp Ma'an NP need the means to identify this species so that it can be rediscovered, marked, and monitored and included in management plans.

Chazaliella obanensis (Wernham) Petit & Verdc.
VU A3c

Range: Nigeria (1 coll.) and Cameroon (SW: lower slopes of Mt Cameroon; Mt Kupe).
This shrub has the largest leaves known in its genus apart from the bizarre funnel-leaved *C. letouzeyi*. First published as a *Psychotria* in 1913, it was only known from a single specimen, from Oban, Nigeria, until the 1990s when botanical surveys on the lower slopes of Mt Cameroon found it to be fairly common in the Onge forest and also present at Bimbia; here it is highly threatened by expansion of plantations.
Chazaliella obanensis is also fairly common on the lower slopes of the western side of Mt Kupe, indeed it is the commonest member of the genus in Bakossi, with over 20 specimens known. However most of these specimens occur below 1000 m alt. and so are particularly vulnerable to expansion from agriculture. It is estimated here that 30–50% of the global habitat of this species is likely to be lost in the next ten years. The assessment here is maintained from that in Cheek *et al.* (2004: 174–175) since no new data area available.
Shrub 0.6–2 m, branching, glabrous; stem bark sloughing; leaves oblanceolate, to 25 × 10 cm, acumen 0.5 cm, lateral nerves 10 pairs, puberulous below, domatia absent; petiole to 3.5 cm; panicle sessile, loosely capitate, 2–3 cm.
Habitat: lowland evergreen rainforest; rainfall exceeding 3 m p.a.
Threats: clearance of forest for expansion of agriculture is imminent at all known locations.
Management suggestions: efforts should be made to protect a portion of this species' habitat so that it does not become extinct as a result of the apparently unrelenting expansion of agriculture in the lowlands of SW Cameroon.

Chazaliella obovoidea Verdc. subsp. ***villosistipula*** Verdc.
VU B2ab(iii)

Range: Cameroon (S: Bipindi; Lolodorf; Kribi-Lolodorf 40 km; NE Sangmelima. C: Mefou proposed NP. E: Kadei R (left bank); Batouri-Djampiel; Yaoundé-Deng Deng) and Congo (Brazzaville) (Odzala NP).
Here assessed as Vulnerable since nine locations (above, AOO 36 km² with 4 km² cells, threats as below). This assessment appears also in Cheek *et al.* 2011. It is perfectly possible that discoveries of additional locations will reduce the threat rating of this taxon.
Shrub, 1.5 m; bark sloughing, dark brown, minutely puberulent; stems with a ridge on each side; stipules broadly triangular, 2 mm, hairy; leaf-blade elliptic, 4–7.5 × 1.8–3.2 cm, acumen long, pointed, 1 cm, base decurrent, lateral nerves 8–9 pairs, forming a looping inframarginal nerve in the distal half, midrib orange-brown, prominent below; petiole 6–8 mm; inflorescence terminal, single, peduncle patent, 1 cm, puberulent, flowers few, subcapitate; pedicel 2 mm; fruit ellipsoid, stipitate, 9 × 6 mm.
Unusual in the genus is being so small, shrubby and deciduous. The stipules are not always villose!
Habitat: semi-deciduous and evergreen forest; 350–700 m alt.
Threats: clearance of forest especially semi-deciduous, for logging, followed by agriculture: cocoa, pineapples, food-crops.
Management suggestions: formal ratification of the Mefou proposed NP would provide the first protected area for this shrub. Within the Park it is rather common and easily viewed, even characteristic of semi-deciduous forest remnants. However, globally it currently appears threatened on current data. Survey of basic population parameters would serve as a baseline for future monitoring.

Coffea is a genus of 103 species restricted to Africa and Madagascar, best known for producing coffee from several of the native African species, mainly *C. canephora* and *C.arabica*. Coffee is the most valuable internationally traded commodity in the world. (Davis *et al.*, Bot. J. Linn. Soc. 152: 465–512 (2006)).

In continental Africa, Cameroon rivals Tanzania in terms of highest total diversity of *Coffea* (12 wild species) and numbers of endemics (six species).

Evergreen shrubs, rarely trees, of evergreen forest, *Coffea* species are can usually be recognised by their congested axillary inflorescences, short corolla tubes which lack emerging hairs from the apex, lack of calyx lobes and a highly reduced calyx limb and gum-covered apical bud.

Coffea was last revised for Tropical Africa in 1947 by Chevalier. Since that time the genus has been revised only on a regional basis. That relevant to Cameroon is the PhD Thesis from Katholicke Univ. Leuven by Stoffelen in 1998 entitled *Coffea* and *Psilanthus* in Tropical Africa. Between 1997–2008 all of Cameroon's six endemic coffee species were published. There is every possibility that more remain to be found. Davis *et al.* (op. cit. 2006) have assigned IUCN codes (but without explanatory text) to each species of *Coffea* and this is the starting point for the assessments below. In 2010 Davis *et al.* plan to produce full conservation assessments so those presented here may be further modified.

Coffea bakossii Cheek & Bridson
EN B2ab(iii)
Range: Cameroon (SW: Mt Kupe (1 site), Bakossi Mts (2 loc.)).
This rare tree was assessed in the species protologue (Cheek *et al.* 2002) as VU, but is reassessed here as Endangered since it is only known from three locations at such low altitudes that it is vulnerable to agricultural pressure. The other endemic *Coffea*, *C. montekupensis* Stoffelen, with 30 collections, is much commoner than *C. bakossii*, for which only four are know.
The forgoing assessment was made in Cheek *et al.* (2004: 175) and is maintained here. It was erroneously attributed to Hilton-Taylor & Pollock by Davis *et al.*(2006: 479 op. cit.) who accepted the assessment. A new location (see accompanying map) discovered by Sene at Mone FR near Mamfe was been identified there in December 2006 (Report to WCS-Cameroon by Cheek, Jan. 2009) but identification of the specimen needs to be confirmed. If positive the number of locations will rise to four and AOO to 16 km² using 4 km cells but the assessment will remain unchanged.
Shrub or tree 5–20 m, glabrous; leaves elliptic to obovate, 20–32 × 7.5–17 cm, acumen 1.5 cm, base rounded or subcordate, secondary nerves 8–12, domatia over axillary web, pit-like; petiole 2–4 mm; stipules triangular-ovate, 10–17 × 7–10 mm, acumen 3–4 mm, midrib conspicuous, ridge-like; flowers 3–4 per axil, 7-merous; corolla white, tube 5–6 mm, lobes 15 × 5 mm; fruit ellipsoid, 16–20 × 10–12 mm; disc 3–5 mm, large and conspicuous
Habitat: evergreen lowland to submontane forest; 700–900 m alt.
Threats: forest clearance for small-holder agriculture, and wood.
Management suggestions: a targeted conservation education campaign, with a conservation poster, might well help the survival of this species in those parts of its range outside of the proposed protected areas in Bakossi.
Reference: Cheek, M., Csiba, L. & Bridson, D. (2002) A new species of *Coffea* (*Rubiaceae*) from western Cameroon. Kew Bull. 57(3): 675–680.

Coffea charrieriana Stoff.& F.Anthony
CR B1ab(iii) + B2 ab(iii) + D
Range: Cameroon (SW: Bakossi FR).
Here assessed as Critically Endangered since known from only a single collection in an area where habitat destruction is occurring (pers.obs. Cheek 2006). AOO of 4 km² is estimated using 4 km cells. Davis *et al.* (op. cit.) accurately assign CR B1ab(iii) to this taxon since its EOO is minimal (a single bush). Since only a single individual is known in the wild, this species also merits CR D status.
Coffea charrieriana is the first record of a caffeine-free *coffea* species for C Africa. It differs from other C African species in its small leaves, (4–)5–7(–8) × 2.2–3.5 cm, corolla tube (1 mm broad), corolla lobes (linear, (5–)6–8 × 2–3 mm) and fruits (9–10 × 7 mm). It is a pubescent shrub 1–1.5 m high (Stoffelen *et al.*, Bot. J. Linn. Soc. 158: 62–72 (2008)).
Habitat:lowland evergreen (with some semi-deciduous elements) forest; 150 m alt.
Threats: at the only site, 2 km E of the Mungo Bridge in the Bakossi FR the forest is in many cases cleared in patches or else the understorey is for cultivation of crops.
We have searched but not found this species at this site in 2006. It is possible that the species is extinct in the wild, but more searching is needed before this judgement can reasonably be made.
Management suggestions: further searching at the type locality is needed to rediscover this species and to record its extent, density and generation so that it can be managed for long term survival. Clearly, as a caffeine-free coffee this species has potential commercial value.
If the species is not found in the wild, the possibility of it being erroneously recorded from this site will be raised. Molecular phylogenetic analysis might help confirm or deny its geographic placement by reference to sister species. It is fortunate that the species is already maintained in cultivation but is is advisable, if not already done, that live material be transferred from Ivory Coast to Cameroon if a secure site can be found – perhaps at Limbe BG.

Coffea fotsoana Stoff. & Sonké

CR B2ab(iii)

Range: Cameroon endemic (C: 40 km NW Yaoundé at Akoas Mbam Minkom massif).

We agree with the rating of CR B2 ab(iii) given by Davis *et al.* (2006 op. cit.: 485) since only a single location is known (therefore AOO 4 km^2 using 4 km^2 cells) with threats as below. In describing the species, Sonké & Stoffelen (Adansonia sér.3, 26:153–160 (2004)) give an assessment of CR B2 but this is not complete IUCN level code.

A glabrous 1.5 m tall shrub, its authors distinguish it from *C. mayumbensis* in having fruit with a crater-forming, dilated disc, flowers with corolla tube only 3–5 × 1.5–2mm and lobes 7 × 2.5–3 mm.

Habitat: submontane evergreen forest above 800 m alt.

Threats: Sonké in Sonké & Stoffelen (op.cit.: 158) states that the species is very localised. Although conspicuous in flower and fruit, he saw it at only a single site where it is extremely vulnerable to forest clearance for agriculture. He gives its area as less than 10 km^2 and states that less than 250 individuals are known.

Management suggestions: Sonké (op. cit.) suggests ex situ cultivation and seedbanking to help conserve the species. We would suggest, additionally, that since WCS Cameroon are believed to co-manage Mbam Minkom with Min. Environment and Forest as a protected area, that the farmers clearing the forests are educated using a poster campaign as to the importance of protecting this rare species.

Coffea leonimontana Stoffelen

CR B2ab(iii)

Range: Cameroon endemic (Littoral: W of km 58 on road Douala–Loum, S of Kompina).

Olivier Maurin in Davis *et al.* 2006: 487 (op. cit.) reattributes two specimens from Mt Etinde (*Faucher* 5, *Mbatchou* 399) which Stoffelen had placed in this species in his protologue (Belgian J.Bot.129: 72 (1997)) to a different species, *C. brevipes*. This seems logical based on the differences in altitude and vegetation types between Mt Etinde and the location of the type and now only remaining specimen of *C. leonimontana*. In view of this single location (AOO 4 km^2 using 4 km^2 cells), with the threats below, we here assess this species as Critically Endangered in place of the EN B1ab(iii) rating given in Davis *et al* 2006: 487 op.cit.

While Stoffelen mentions statements by Anthony (regarding a coffee from Koto that he treated as *C. liberica* as possibly this taxon, it is clear that no other specimens have been seen by Stoffelen.

Coffea leonimontana is unique to the genus *Coffea* in its spathulate leaf apex (Stoffelen op. cit.)

Habitat: lowland evergreen or semi-deciduous.

Threats: forest remnants in the area in which *Coffea leonimontana* was first collected in the 1970s are degraded. The species has not been seen since then, despite a mission to Cameroon to study *Coffea* species in the wild by Davis and Maurin, with Sonké in the mid 2000s. Forest degradation appears to be ongoing in this general area, in which no protected currently exists (Onana and Cheek pers.obs.).

Management suggestions: a mission specifically to rediscover *Coffea leonimontana* at its only known location S of Kompina is needed. If found, data on extent, density and regeneration of the population would help develop a management plan for the species, and local authorities and communities should be alerted to this species and be involved in its protection.

Coffea magnistipula P.Stoff. & Robbr.

EN B2ab(iii)

Range: Cameroon (S: forêt de Mamba près de Boga) and Gabon (Cristal Mts; Massif du Chaillu; Mouyanama)

Here *Coffea magnistipula* is assessed as Endangered since known from only four locations with threats as below. At the Cristal Mts three collections are known (Nkan, Ababga Chantier CETA and W of Balakala R) so if each is attributed a cell, a total AOO of 24 km^2 can be calculated using 4 km^2 cells. Assessed as NT by Davis *et al* 2006: 490 (op. cit.).

Coffea magnistipula is monocaulous shrub 0.8–1.25(–3) m tall, having an erect unbranched stem with a funnel of leaves at the apex. The obovate leaves are 23–48 × 7.5–13.5 cm, acuminate, the base rounded to cordate, the petiole glabrous, at most 5 mm long. The stipules are ± triangular, 10.25 × 15 mm acuminate, with fimbriate margins. The fruit is ellipsoid 18–33 × 14–24 mm, longitudinally 10-ribbed (Stoffelen & Robbrecht, Taxon 46: 39 (1997)).

Habitat: submontane evergreen or lowland forest, 400–800 m alt.

Threats: slash and burn agriculture; in the Cristal Mts future mining for precious minerals (pers. obs.).

Management suggestions: work is needed to determine whether any of the known sites fall within a National Park. Cristal Mts has two NPs. An attempt should be made to refind this species and to gather data on extent, density and regeneration in the subpopulation concerned to form a basis for managing the species. Sonké and Nguembou spent two weeks in Jan. 2005 in the area of S Cameroon where *C. magnistipula* was found, but did not find it. The same area was visited in Sept. 2005 but again they did not find it. It appears, at the very least, rare in S Cameroon (Sonké *et al.* Bot. J. Linn. Soc. 151: 429 (2006)).

Coffea mapiana Sonké, Nguembou and A.P.Davis

EN B1ab(iii) + B2ab(iii)

Range: Cameroon endemic (S: Mvini; Bibondi; W of Ngoyang).

Sonké *et al.* 2006 op. cit.: 425–430, in describing *C. mapiana*, assessed its conservation status as EN B1ab(iii) +

B2ab(iii), calculating an EOO of 378 km² and an AOO of 356 km² based on the three localities indicated, which being less than 5 km away from human habitation one considered at risk. This assessment was maintained in Davis *et al* (2006: 491, op.cit.) and is also maintained here. We have no new data to add.

Thanks to the industry of Sonké and Nguembou this species is known in some detail in all its stages – seven collections are known.

Coffea mapiana resembles *C. magnistipula* (see above) in appearance, but has petioles 15–40 mm long (not 1–2(–5) with a cuneate leaf base and acumen 9–20 mm long (not cordate or rounded, acumen 1–1.5 mm long).

Habitat: lowland evergreen forest; 100–650 m alt.

Threats: see above.

Management suggestions: if *C. mapiana* occurs outside the boundaries of the Campo-Ma'an NP as it appears to, consideration should be given to adding one or several of its sites as satellite areas to the NP if the boundaries cannot be altered. Information on *C. mapiana* should be passed to the local communities and authorities near which it occurs, to reduce the risk of the species being inadvertently destroyed. Sonké and Nguembou may have additional data on this species and are well placed to be involved in efforts to refind and protect it.

Coffea montekupensis Stoff.

VU B1ab(iii) + B2ab(iii)

Range: Cameroon endemic (SW: Mts Kupe & Bakossi; Lebialem Highlands, Fosimondi).

Assessed by Davis *et al* (2006: 492 op.cit.) as VU B1ab(iii). We accepted this assessment in Harvey *et al.* (2010: 81) and maintain this assessment here even though in Cheek *et al* (2004: 371) we assessed the species at NT.

Coffea montekupensis is readily recognised by its pink flowers. African *Coffea* species usually have white flowers. Within its extent of occurrence it is often rather frequent, reflected in the 39 specimens cited in Cheek *et al* 2004: 371, and the seven communities within the forests of which it was found. We have not measured its EOO accurately but believe it to be less than the 20000 km² required to trigger B1. In 2005/06 Tchiengué discovered this species as Fosimondi – a significant range extension. It is likely that it once extended between Fosimondi and Bakossi and may yet be found in this area where suitable habitat survives.

Habitat: submontane evergreen forest; 700–1500 m alt.

Threats: in Bakossi it is threatened by agriculture extending uphill from the fertile valley bottoms. At Fosimondi very heavy forest losses have been observed by Tchiengué in successive years because of clearance for agriculture.

Management suggestions: the newly gazetted Bakossi NP probably contains this species in a similar density to elsewhere in Bakossi and is likely to provide a long term future for the species. Meanwhile forest on the Mt Kupe, where it is best studied and documented, appear secure above 800–1000 m altitude and local communities should be supported to maintain this protection.

Craterispermum was revised by Verdcourt in 1974 (Kew Bull. 28: 434). 16 species of tree and shrub are recognised in Africa and its islands, of which only one appears threatened in Cameroon. They have supra-axillary inflorescences and obscure leaf venation.

Craterispermum aristatum Wernham

EN B2ab(iii)

Range: SE Nigeria (Oban) and Cameroon (SW: Korup; Mundemba, Mt Cameroon at Mokoko and Onge. S: Mémel II; Campo-Ma'an).

This glabrous tree, to 8 m tall is distinct in having sessile, axillary inflorescences with 8 mm long-aristate bracts. The oblong-elliptic leaves are 15–20 × 6–9 cm with 7 pairs of lateral nerves.

First discovered in Oban c.1912 by Talbot, the species appears not to have been seen in Nigeria since then and may have been lost due to forest destruction so may now be endemic to Cameroon! There are records from S Korup NP where it can be considered secure but the tree in Mundemba town must be at risk if it still survives. There is uncertainty as to the future of both Mokoko and Onge forest of Mt. Cameroon where it was found in the surveys of the 1990s. While Tchouto indicates the species in Campo Ma'an, no specimen has been seen to support this assertion. However, the recent record from nearby Mémel II by Sonké makes this eminently feasible.

Craterispermum aristatum is here assessed as Vulnerable since seven locations are known (AOO 28 km² at 4 km² cells) with the threats given above. Unless new sites are found it is likely that this will rise to Endangered when reassessed if it is confirmed, as likely, that is either at risk or already lost at four of its seven locations.

Habitate: lowland evergreen forest.

Threats: see above.

Management suggestions: surveys are needed to rediscover *C. aristatum* prioritising Korup and Campo Ma'an since they are protected. Data is needed on extent, density and regeneration so that management plans can be developed for the area managers, whom should be provided with the means to identify this tree.

Cuviera is placed in the Vangueirieae and shows the globular or cylindrical stylar head of that group. The species of the genus are trees or shrubs, sometimes spiny, often with swollen branches and axillary cymose inflorescences. The

24 species of *Cuviera* DC are restricted to Tropical Africa. All of the threatened species of Cameroon occur in *Cuviera* subg. *Globulostylis* (Onana, Kew Bull. 63: 401–403 (2008)).

This subgenus is distinguished by having 1–5-flowered cymes and styles with hairy swellings. Subgenus *Cuviera* has 8–20- flowered cymes and styles which lack swellings (Onana op. cit). Onana's work has clarified the taxonomy considerably but a revision of the genus is still desirable.

Cuviera cuvieroides (Wernham) Onana

EN B2ab(iii)

Range: Cameroon (S: 32 km ENE Djoum, near Meyas Mela; Bitye 27 km S Djolimpou. E: Dja, 12 km E of Somalomo) and Gabon (E of Lopé-Okando FR).

Five locations (AOO 20 km² with 4 km² cells) with the threats below enable *C. cuvierioides* to be assessed here as Endangered.

Characterised by narrowly oblong calyx lobes, 2-flowered cymes, pedicels 1.5–2.5 cm long.

Habitat: lowland evergreen to semi-deciduous forest.

Threats: logging in the Bitye-Dja-Djoum area.

Management suggestions: the best hope for the survival of this species is the Dja NP. Here an assessment of the extent, density and regeneration of *C. cuvierioides* is needed (if it can be refound there) to develop a plan for its management.

Cuviera leniochlamys K. Schum

EN B2ab(iii)

Range: Cameroon endemic (SW: S Korup NP; Mt Cameroon at Mokoko FR; Etinde. S: Mt Elephant at Kribi; Ngovayang; Bipindi)

Five locations enable *C. leniochlamys* to be assessed here as Endangered with the threats below. The Korup specimen is incomplete so the identification is not certain. This species appears patchy in its distribution. For example, at Mt Cameroon there are 11 records from within the Mokoko FR but none for neighbouring Onge forest, nor for Bambuko, S Bakundu, Bimbia-Bonadikombo. However, there is a record at Mt Etinde (Batoke). Similarly, there are three specimens for different locations near Ngovayang. For this reason, the usual cell size of 4 km² for each location seems too low. Estimating a 16 km² cell to be adequate to cover the locations at Mokoko and extending this to the four other locations, gives an AOO of 80 km².

A glabrous shrub 0.5–8 m tall with elliptic leaves 25 × 10 cm, with 5 pairs of lateral nerves and a 2 cm acumen. Distinguished by the 2–5-flowered cymes, lanceolate calyx lobes, pubescent corolla tube with lobes lacking tails (Onana op. cit.).

Habitat: lowland evergreen forest; 0–750 m alt.

Threats: Mokoko is under threat of plantation expansion, while Mt Elephant is a proposed iron ore mine. No threats are known at Ngovayang and Bipindi apart from slash and burn agriculture.

Management suggestions: since more than 50% of the known records of *C. leniochlamys* derive from Mokoko, and numerous other rare species occur there, some endemic to that location, it should be protected from clearing and given National Park status.

Cuviera talbotii (Wernham) Verdc.

VU B2ab(iii)

Range: SE Nigeria (Oban (2 coll.)) and Cameroon (SW: Korup (1 coll.); Mt Kupe (13 coll. at 4 sites)).

This shrub or small tree appears to be narrowly distributed (e.g. absent from Mt Cameroon) and rare, except at Mt Kupe where it is abundant at several sites on the lower slopes. The assessment above, first made in Cheek *et al.* (2004: 175) is maintained here since no further data are available.

Shrub or small tree to 5 m, glabrous; stems with ants; leaves elliptic, 15–23 × 7–10 cm, shortly acuminate, obtuse, nerves 5–7; petiole 1.5 cm; stipules sheathing, 5 mm, mucron awn-like; flowers 5–6, axillary; peduncle 1 cm; bracts inconspicuous; calyx leafy; corolla tube c. 4 mm, lobes c. 2 mm; fruit ellipsoid, 3.5 × 3 cm, brown, fleshy, pyrenes 5.

Habitat: submontane evergreen forest; 840–1300 m alt.

Threats: forest clearance (at lower altitudes) for wood and agricultural land.

Management suggestions: a taxonomic reassessment of this species with *C. leiochlamys* is needed. Mt Kupe, with two-thirds of the known sites for *C. talbotii*, is the key area for the conservation of this species. So long as the boundary of the proposed Mt Kupe protected area is respected, this species seems secure, although some pressure can be expected in the lower part of the altitudinal range.

Cuviera wernhamii Cheek (syn. *C. minor* (Wernh.) Verdc. non Wright)

EN B2ab(iii)

Range: Nigeria (Oban) and Cameroon (SW: Mt Cameroon (Onge and Etinde Forests); Mt Kupe).

Here *C. wernhamii* is asessed as Endangered since only five gatherings from four locations are known (AOO 20 km² at 4 km² cells), all of which are threatened (see below). It is likely that this species will rate Critically Endangered in future. A glabrous tree 2.5–9m tall, it resembles *C. leniochlamys* and *C. talbotii* to which it is related, but can be distinguished by the ovate, not lanceolate calyx lobes, and the glabrous, not pubescent corolla (Onana op. cit.).

Habitat: lowland, rarely submontane, evergreen forest; sea-level–450(–1000) m alt.

Threats: 36% of forest cover was lost 1990–2005 in Nigeria, while Onge forest is unprotected and in danger of conversion to plantations, as is lower altitude forest near Mt Etinde. The site at Kupe, at 1000 m alt., being above the line of cultivation, is relatively secure.

Management suggestions: Onge, with the highest concentration of records of *C. wernhamii* should be considered for protection, perhaps as a National Park.

Ecpoma. A genus of six unbranched subshrubs confined to the Gulf of Guinea, in the *Sabicea* alliance and sometimes sunk in that genus on molecular grounds. Resembling *Stipularia* but lacking tubular bracts and calyx, but rather with leafy calyx lobes, and differing also in the bilocular ovary with broad, cordate placental masses. See further discussion under *Sabicea.* Treated by N. Hallé (Adansonia 3: 173 (1963) and in Fl. Gabon 12: 220–226 (1966)) but a complete revision is still desirable.

Ecpoma apocynaceum K. Schum (syn. *Sabicea apocynaceum* Auctt.)

EN A2c + B2ab(iii)

Range: Cameroon endemic (S: Lolodorf; Campo-Ma'an).

Assessed here as Endangered under criterion A and B since only two locations are known (AOO 8 km² at 4 km² cell size) with threats as below. The leaves are narrowly elliptic, c. 30 × 6 cm, acute to cuneate at base and apex with 15–20 lateral nerves on each side of the midrib and 3 cm petioles. The 3 × 0.5 cm stipules are narrowly triangular. The 4 cm long white, tubular flowers are held in dense cauliflorous cushions. N. Hallé (loc.cit. 1966: 222) remarked that this species is very close to if not identical with, *E. geanthum* (Hiern) N. Hallé.

Habitat: lowland evergreen forest.

Threats: Onana (pers.obs.) remarks that about 50% of the area of occurrence of this species is degraded and points out that since the gathering of the type specimen (*Staudt* 208, Lolodorf) a century elapsed before Tchoutou (2004) reported it from Campo Ma'an.

Management suggestions: Tchoutou should be engaged to rediscover the site in Campo-Ma'an and make a survey to determine the extent, density and regeneration levels of this plant so as to develop a management plan. A revision of the genus is needed to confirm the distinctness of this taxon.

Ecpoma gigantostipulum (K.Schum.) N.Hallé (syn. *Sabicea gigantostipula* K.Schum.)

EN B2a,(iii)

Range: Nigeria (Oban), Cameroon (SW: Mt Cameroon at Bakingili and Mokoko), Equatorial Guinea (Bioko: Mt Balen; El Pico) and São Tomé & Príncipe (Príncipe).

Here assessed as Endangered since only five locations (AOO 20 km² at 4 km² cells) exist with threats given below.

Distinguished by the very broad (3–4 cm wide) stipules which embrace the two inflorescences one of which is axillary to a petiole, the other to a stipule.

Habitat: lowland to submontane evergreen forest.

Threats: forest at Bakingili is under threat from plantation expansion, as is that at Mokoko. 89% of Nigerian forest is reported destroyed of which 36% between 1990–2005.

Management suggestions: a revision of the genus is needed. Verification of the presence of this species in Gabon, noted in Davis & Figueirdo (2007: 165) but absent in N. Hallé (2006 op.cit.) is required. Surveys to refind this species at Mt Cameroon and to determine basic populational information so as to devise a protection plan, are suggested.

Gaertnera was last reviewed in Tropical Africa (11 species) by Petit (Bull. Jard. Bot. Brux 29: 37–53 (1959)) in treating the species of Congo (Kinshasa). Small soft-wooded trees, they are unique in the Rubiaceae in having superior (not inferior) ovaries and fruits. They can be recognised when sterile since their stipules have exceptionally long sheaths. A modern revision of the genus is needed.

Gaertnera fissistipula K.Schum.

VU B2ab(iii)

Range: Cameroon endemic (SW: Korup; Mundemba; Mt Cameroon at Onge & Mokoko. Littoral: Masok. S: Bipindi; Eséka-Songbong; Kribi).

Here *G. fissistipula* is assessed as Vulnerable since eight locations are known with the threats given below. Seven collections were recorded at the Mokoko FR but otherwise only a single specimen is known at a site. The AOO is estimated as 40 km², using 4 km² cells for all locations, but 16 km² for Mokoko (to reflect the larger area occupied there).

This is a 1–3 m shrub or tree, drying black, with minute white patent hairs on stems and inflorescences. The 15 × 4 cm oblanceolate leaves have an abrupt 2–3 cm long acumen and sheathing stipules. The 5 cm inflorescence doubles in length when in fruit.

Habitat: lowland evergreen forest.

Threats: the site at Korup is on the edge of a Palmoil plantation and that in Mundemba near a Catholic mission. At Mt Cameroon Mokoko and Onge are threatened with plantation expansion and forest in the Kribi area is at risk from multiple development projects.

Managements suggestions: Mokoko FR should be elevated to National Park status since it has not only the densest known global population of this threatened species but is the only known site for several other species of plant.

Gardenia epiphytica Jongkind

EN B2ab(iii)

Range: Cameroon (SW: Korup NP, transect P), Gabon (near Libreville; Santa Clara–Cap Esterias; Shell oil exploitation Rabi next to OMS camp).
Here *Gardenia epiphytica* is assessed as Endangered since only four locations (AOO 16 km² at 4 km² cells) are known and proximity of one these to a major city (Libreville) and their inclusion in an oil exploitation area (Rabi) is a threat (see note below).
Unique in the genus is being epiphytic, this species can form a shrub metres wide inside the canopy of large forest trees. The scented flowers occur in pairs at branch ends, the green corolla is up to 14 cm long with creamy white lobes c. 4 × 1.5 cm. Closely related to *Gardenia imperialis*, *G. epiphytica* lacks the pouched leaf bases and dilated corolla mouth of this species (Jongkind, Syst. Geogr. Pl. 75: 259–265 (2006)).
Habitat: within the crown of large forest trees in lowland evergreen forest.
Threats: trees in or near industrial or urban developments in developing countries are at risk from being utilised for construction by small enterprises and by being 'in the way' of new vehicle tracks, equipment stores etc. (see also above).
Management suggestions: since it has so much horticultural potential *G. epiphytica* should be introduced to cultivation in order to help protect it. It is likely that is is more common than the limited number of collections suggest, because plants are so inaccessible in the crowns of tall trees. Clearly, a skilled tree climber will be needed if this species is to be rediscovered and conducting a census of the size of a subpopulation, its density and regeneration levels, will be highly difficult. The Korup NP offers great security for this species since its forests are so well protected.

Hallea stipulosa (DC.) Leroy
VU A1cd

Range: Gambia to Congo (Kinshasa).
In view of the very wide range and numerous locations from which this species has been recorded, it is not mapped here.
This widespread timber tree was given the assessment above by WCMC (1997) cited in IUCN (2003; www.redlist.org), on the basis that in many places it suffers from over exploitation, e.g. clear cutting. This has not been observed by us in Cameroon, but evidently occurs elsewhere. The rating VU A1cd indicates that 30% or more population reduction has occurred in the last 100 years, reflecting the widespread loss of its forest habitat in W Africa in this time, as has occurred for many other timber trees with this range. However, the species is unlikely to become extinct since forestry departments and local communities protect young trees until they are large enough to fell, and even plant young trees to ensure future income.
Tree 35 m; leaves obovate, 12–45 × 8.5–26 cm, rounded, acute, nerves to 20; petiole to 5 cm; stipule oblong to suborbicular, 4–8 × 2.5–5 cm, pubescent; inflorescence of 3–10 terminal capitula, each 5–10 mm diam.
Habitat: lowland evergreen forest, often in swamps where it may be dominant.
Threats: as above.
Management suggestions: none.

Hymenocoleus was proposed by Robbrecht in 1975 (Bull. Jard. Bot. Nat. Belg 45: 273–300) for a group of species formerly included other genera, e.g. *Geophila*. True *Geophila* species differ by having ovate, cordate leaves among other characters. *Hymenocoleus* was revised by Robbrecht in 1977 (Bull. Jard. Bot. Nat. Belg. 47: 3–29), where 12 species are recognised. The species are generally prostrate or ascending forest floor herbs with bright orange fruits surrounded by a persistent cupular involucre. When assessing species distributions, both Robbrecht publications cited must be consulted since the second does not repeat those contained in the first.

Apart from those species assessed here, several others in Cameroon appear to very rare, probably rating NT status on available data, e.g. *H. nervopilosus* and *H. libericus* (14 locations globally each), *H. rotundifolius* (20 locations), H. *subipecacuanha* (11 locations), *H. scaphus* (18 locations), *H. globulifer* (12 locations).

Hymenocoleus axillaris Robbr.
VU B2ab(iii)

Range: Liberia (Sinoe country, Kulo), Cameroon (C: 8 km E of Awaé, road to Ayos. S: Méyos Méla), Gabon (Cristal Mts; Ovala) and Congo (Brazzaville) (Mayombe, Les Saras) – Robbrecht op.cit: 18.
Forest shrub to 60 cm tall, unique in the genus in having 6–7 pairs of axillary inflorescences – in other species the inflorescence is terminal. Here *H. axillaris* is assessed as Endangered since six locations are known (AOO 24 km² using 4 km² cells) with threats as below. Only a single collection is known at each location.
Habitat: lowland evergreen forest.
Threats: slash and burn agriculture.
Management suggestions: surveys to rediscover this species are advisable. At least one site should be protected, perhaps as a 'micro-site' with management by a local community if inclusion in a formal protected area is not possible: none of the sites appear to fall in a National Park. Data on extent, density and regeneration at a site would assist management planning.

Hymenocoleus barbatus Robbr.
VU B2ab(iii)

Range: Cameroon (Littoral: Masok), Gabon (Abanga; Cristal Mts), Congo (Brazzaville) (Mayombe, Dimonika-Makuba and Ngonyo), Angola (Cabinda: M'Bulu Hills) and Congo (Kinshasa) (Central Forest Distr: Bambesu).
A forest creeping herb unique in having tomentose involucral bracts, *H. barbatus* is here assessed as Endangered since seven locations are known (AOO 28 km² with 4 km² cells) and threats as below.
Habitat: lowland to submontane evergreen forest.
Threats: slash and burn agriculture.
Management suggestions: the known sites should be plotted onto a map of protected areas to identify which best merits targeting for rediscovery, survey of local extent, density and regeneration so as to develop a protection plan for the area managers.

Hymenocoleus glaber Robbr.
VU B2ab(iii)

Range: Cameroon (SW: Limbe; Kupe-Bakossi. Littoral: Masok. C: Boga. S: Campo Ma'an, fide Tchouto 2004).
Described in 1977, this species was first collected at Limbe (*Schlechter* 12366). The two other localities recorded by its author, Robbrecht, were 8 km west of Masok (*Leeuwenberg* 5416) and Mambe Forest near Boga (*Letouzey* 12279). A further collection was made near Limbe in 1993 (*Cable* 260 cited in Cable & Cheek 1998: 108). Numerous extra collections were made at three new locations in Kupe-Bakossi in 1995–present. This species is assessed as Vulnerable, being known only from six locations. Each colony appears to occupy only 1–2 m^2. This assessment, made in Cheek *et al.* 2004: 175 is maintained here, no new data being available apart from the addition of the Tchouto record.
Habitat: lowland evergreen and submontane forest; 750–1400 m alt.
Threats: clearance of forest for agricultural expansion and wood, particularly in the lower part of its range in the Mt Cameroon area (Cable & Cheek 1998) and also below 1000 m in Kupe-Bakossi.
Management suggestions: since Kupe-Bakossi has the densest number of sites and collections for the species, it seems to offer the best possibility of survival for the species. It seems likely that several areas in which the species occurs will be protected in future, though not those at lower altitudes.
Reference: Robbrecht, E. (1977) The tropical African genus *Hymenocoleus* (*Rubiaceae: Psychotrieae*): additions. Bull. Jard. Bot. Nat. Belg. 47(1): 3–29.

Hymenocoleus nervopilosus Robbr. var. *orientalis* Robbr.
VU B2ab(iii)

Range: Cameroon (S: Ébolowa-Kribi at Kienké R; 4–5 km E Kribi. C: Mefou proposed NP) and Gabon (Cristal Mts at Nkam; Makokou; Libreville; Mouila).
Here assessed as Vulnerable since six locations are known (above, AOO 24 km^2 with 4 km^2 cells, threats as below). This assessment appears also in Cheek *et al.* 2011.
Gregarious herb, with rooting horizontal stems, ascending to 15–20 cm at apex; stems long-hairy; stipules green, triangular, 4 mm, with apical bristle 5 mm, long hairy; leaf-blade drying green with white nerves above, oblanceolate, 7 × 2.5 cm, obtuse-acute, abruptly rounded at base, lateral nerves c. 12 pairs, tertiary nerves scalariform, raised, bristled; inflorescence terminal, sessile, c. 1 cm; calyx lobes erect, green, narrowly triangular, 4–5 mm, long brown hairy; corolla white, tube 3–4 mm, lobes 1.5 mm.
Habitat: lowland evergreen forest; 600–700 m alt.
Threats: clearance of forest for timber and/or agriculture.
Management suggestions: several records exist from Mefou proposed NP so this is a logical centre to conserve the plant if only governmental approval for the National Park is achieved. Baseline population data is needed against which to monitor in future and to gauge need for management intervention.

Hymenodictyon is an Old World genus of 22 species recently revised by Razafimandimbison and B. Bremer (Bot. J. Linn. Soc.152: 331–386 (2006)). Five species occur in Africa of which all but one occur in Cameroon. Only one of these is threatened, although the newly described *H. epiphyticum* Razafim & B.Bremer is known from only 12 locations and must rate as nearly threatened. The *Hymenodictyon* species are deciduous shrubs or trees with dehiscent woody fruits releasing numerous dry winged seeds as in several other genera in subgenus *Cinchonoideae*. *Hymenodictyon* is usually distinguished by a pair of coloured, often leaf-like bracts below the inflorescence that persists in fruit.

Hymenodictyon pachyantha K.Krause (syn. *H. gobiense* Aubrév. & Pellegr.)
EN B2ab(iii).

Range: Ivory Coast (NE Oumé), Nigeria (Benin; Iyamoyong FR in Ogoja; Orem at Calabar; Nwum R near Mambilla), Cameroon (E: Yokadouma–Deng Deng, 210 km NE Yâoundé; Yokadouma–Moloundou km 22; 65 km SSW Yokadouma).
Here assessed as Endangered since it is estimated that more than 50% population reduction has occurred in the last 100 years due to habitat loss principally in Nigeria and Ivory Coast (see below). Only nine locations have been recorded (see above).
A tree up to 31 m tall with medium-sized buttresses and a pyramidal crown. The leaves are clustered at the apices of

the branches, broadly obovate, 14–31 × 5–11 cm, acute to rounded with attenuate base, petioles 3.5–5 cm, stipules oblong, 6–8 mm; inflorescences 10–32 cm long, branched, calyx lobes 4.5–7 mm long.

Hymenodictyon pachyantha is the giant of the genus in Africa and not likely to be confused with any other species in Cameroon which are usually shrubs or small trees, with smaller parts than this species. The authors of the recent revision of the genus mistakenly attribute two collections by Farquhar from Benin in Nigeria to Benin the country.

Habitat: semi-deciduous lowland forest.

Threats: while not a known timber species, *H. pachyantha* grows in areas where, at least in Cameroon, logging impacts are high. In the course of logging, many non-timber species are impacted, posing a risk to such species as *H. pachyantha.* However, logging impacts in Nigeria and Ivory Coast have been higher still and often succeeded by agriculture. In those countries there is every possibility that the species is extinct.

Management suggestions: the best hope for the global survival of *H. pachyantha* is the Yokadouma area of Cameroon where the most recent (latest collection 1973) and most numerous records derive. Botanists in this area working in the several national parks there e.g. Boumba-Bek, should attempt to rediscover this species and, if successful, attempt to survey the extent, density and regeneration of the population, and to inform park managers as to how to identify and protect *H. pachyantha.*

Ixora was revised for Africa by De Block (Op. Bot. Belg. 9 (1998)) who recognised 37 species. One or two additional species have since been found in Cameroon and await description. The genus is easily recognised by the terminal inflorescences, 4-merous flowers with exserted, splayed stigmas, and by the articulated petioles.

Ixora batesii Wernham

CR B2 a,b(iii) or, possibly, EX

Range: Cameroon.endemic (C: Mt Fébé and Bitye (6 coll. total)).

This forest shrub possibly sometimes climbing, reaches up to 5 m tall. The elliptic leaves are 11–19.5 × 4–5.5 cm, the corolla tube white, 5–13 mm long. It is close to the common and widespread *I. guineensis* but differs in the ± glabrous ovary and calyx, the widely triangular calyx lobes and the smaller ratio of corolla tube: lobe length which is only 1 or 2:1. (De Block op. cit. 86). De Block examined 6 collections (op. cit. 86) but only cited 4 (op. cit.) all being those of Bates from Bitye collected c. 1912. She cites Mt Fébé as a site, so it is likely that either 1 or 2 of the uncited specimens is from this location. This species may be extinct since Bitye is believed to be in the semi-deciduous zone around Yaoundé which has been heavily logged and farmed, while Mt Fébé, one of the hills on which Yaoundé is built, is one of the two top hotels of the city and has had its forest cleared. Searches for other endemic forest taxa at Mt Fébé in 2007, notably the tree *Talbotiella breteleri*, failed to discover any taxa of it for this reason. Since the *Talbotiella* was later found at a new site, on another hill further outside Yaoundé there is hope that this *Ixora* might also be found again. For the present this species is assessed as Critically Endangered since it may survive only at one site Bitye, its other, Mt Fébé, being destroyed for an hotel, an AOO of 4 km² can be calculated using a cell of this size and threats as have already been detailed.

Habitat: evergreen forest; c. 700 m alt.

Threats: see above.

Management suggestions: *Ixora batesii* should be searched for at Bitye and at other forest sites around Yaoundé where these survive, so that it can be protected for the future, if it has not been lost altogether.

Ixora delicatula Keay

VU B2 ab (iii)

Range: Nigeria (Lagos: Ajilil and Degema Distr., 1 coll. each) and Cameroon (S: Campo-Ma'an (2 coll.). SW: S Bakundu FR; Korup; W of Nguti, Mt Cameroon–Etinde (1 coll. each)).

This 2 m shrub is unique in having pedunculate inflorescences subtended by a type of bract otherwise only seen in taxa with sessile inflorescences (De Block op. cit. 96: 168). It is rather widespread in forest near the coast in the Lower Guinea domain, but always infrequent. At Mt Cameroon about 9000 specimens have been made but only one of them, *Faucher* 9, is *I. delicatula.* In all, seven locations are known (AOO 28 km² with a 4 km² cell) but the species may have been lost from two or three of these. For the present it is here assessed as Vulnerable (see threats below) but this may need to rise to Endangered if the site losses are confirmed.

Shrub 2 m, stems minutely and sparingly puberulous; leaves elliptic, oblong or lanceolate, 6–16 × 2–5.8 cm, acuminate, cuneate to obtuse at base, lateral nerves 10–16 pairs; inflorescence to 3 cm, primary branches to 1.2 cm; calyx with short, broad lobes, corolla tube white, 10–11 mm, lobes 9–11 mm.

Habitat: lowland evergreen forest; sea level–800 m alt.

Threats: S Bakundu FR has seen large incursions for agriculture in recent decades despite its importance as a site for rare species. Lagos has lost almost all its natural forest habitat. Degema District lies in the Delta region and is believed to have suffered from oil industry activity, while habitat on the lower slopes of Etinde is under pressure from agricultural expansion.

Management suggestions: this species is most secure at Campo-Ma'an and Korup National Parks. The Park

Managers should be given the means to identify and so seek out and protect *Ixora delicatula*.

Ixora foliosa Hiern

VU A2c + 3c

Range: Nigeria (Chappal Waddi and Chappal Hendu (not shown on map)) and Cameroon (SW: Mt Cameroon (numerous coll.); Mt Kupe; Mwanenguba. NW: Bamenda Highlands (numerous collections from many sites)).

About half the range within which this characteristic tree of wet montane forest once occurred was in the Bamenda Highlands. These are now destitute of natural forest except for a very few exceptions. It is estimated that over 30% of the population of this tree has been lost over the last century due to habitat destruction and that over 30% of that remaining will be lost in the next century. The assessment of this taxon made in Cheek *et al.* (2004: 175–176) is maintained here. No new data are available.

Habitat: montane forest, mainly above 2000 m alt..

Threats: forest clearance for agriculture and wood, especially in the Bamenda Highlands, once probably the main area for this species. In one study area of these highlands, 25% of forest was lost between 1987-1995 (Moat in Cheek *et al.* 2000: back cover) and in another 50% in 15 years (Baena in Cheek *et al.* 2010: back cover). The absence of this species at high altitude at some locations where forest survives such as at Dom and the Lebialem Highalnds, was unexpected (Cheek *et al.* and Harvey *et al.* 2010 2010).

Management suggestions: implementation and policing of protected area boundaries would ensure the survival of this species in most of its range.

Ixora inundata Hiern

EN B2 ab (iii)

Range: Cameroon (SW: Mana Falls, Korup, 3 coll.) and Gabon (Cristal Mts at Kinguélé Falls on Mbe R (5 coll.); Ndjolé area (3 coll.); Ofooué R, Ilembari Falls (1 coll.)).

This shrub is a rheophyte, restricted to rapids and waterfalls. Submerged in the wet season, it resists being washed away by the currents by having long, narrow, glabrous streamlined leaves 6.5–17.5 × 1–2.2 cm. The inflorescences are sessile and few-flowered, the flowers are subsessile with white or pink corolla tubes 13–27 mm long.

Ixora inundata appears to be restricted to the four locations listed above, a possible fifth location is indicated by *N. Hallé* 836 (Wébé R near Ngongolana) but has not been traced on maps and may equate with one of the three Gabon locations. Using a 1 km² cell, an AOO of 4 [illegible] using current IUCN conventions for [illegible] this is a gross over-estimate since this [illegible] rapids in rivers less than 100 m wide, and the rapids do not extend for the 10 km that would be needed to make a 1 km² cell!

Ixora inundata is here assessed as Endangered (see above) due to the data provided here.

Habitat: rapids in evergreen forest with 3–5 m rainfall per annum; c. 100–200 m alt.

Threats: during the IUCN aquatic plants workshop for C Africa in Yaoundé June 2008, numerous sites for proposed new hydroelectric plants were raised which typically, are placed at rapids, thus threatening rheophytes. This species was not found at the Kinguélé falls/hydroelectric dam during a search for rheophytes in 2007 (Cheek pers. obs.) while c. 5 specimens are known from this site, none has been seen since 1986 and it is possible that the population has declined strongly due to habitat disruption caused by the placing of the dam and subsequent change in the flow patterns of the falls.

Management suggestions: a more detailed search as Kinguélé should be made to establish if the species survives there, if so, monitoring of the population to detect trends is suggested. Are new hydroelectric dams planned at the other Gabonese sites? The sole known site in Cameroon is on the edge of a National Park and well protected, but monitoring of the population is still advisable.

Ixora synactica De Block

EN B2 a,b (iii)

Range: Cameroon endemic (S: 25 km NNE Bipindi, 5 km W of Atog-Boga at Bog Hill; Mémel 11 03N, 10 25E; Zingui).

This unbranched forest shrub bears a funnel of oblanceolate, basally cordate leaves at the apex of the 0.5–1.5 m tall stem. This habit is unique in the genus. Minerals leached from litter gathered in the funnel are filtered out and absorbed by adventitious roots on the aerial stem below, it can be supposed.

Ixora synactica is here assessed as Endangered since effectively only two locations are known (Zingu and Mémel II–Bog Hill), making it susceptible to stochastic events. Described (De Block op.cit. 156) from a single Letouzey specimen, the species was refound by Sonké and Beina (*Sonké* 3260) in March 2004 at Mémel II nearby, and Tchouto records it from Zingui.

Habitat: probably lowland evergreen forest; altitude unknown.

Threats: stochastic events, such as tree falls or slash and burn agriculture could wipe out one of the two known locations.

Managements suggestions: surveys are needed to refind this species in the wild. If successful the assistance of local leaders and area managers should be sought to protect the plants, and basic populational data collected to enable future monitoring and necessary assessments to be made.

Keetia was reinstated by Bridson (Kew. Bull. 41: 965–994 (1986)) after the fragmentation of *Canthium* sens. lat. *Canthium* in the strict sense does not now occur in W-C Africa. *Keetia* is distinguished from *Psydrax* (the other split from former *Canthium* occurring in W-C Africa) in being climbers or scandent shrubs, often hairy, with woody pyrenes having a lid-like area (not usually trees, usually glabrous, pyrenes cartilaginous, lacking lids).

Bridson (op.cit. 985–993) effectively gives a synoptic revision of the W-C African species but stresses that a more detailed revision is needed, when several new taxa are likely to emerge. *Keeria purseglovei* Bridson and *K. venoisissima* Hutch. & Dalz. (Bridson) both probably merit threatened status in addition to the species below.

Keetia bakossiorum Cheek

CR D

Range: Cameroon: Bakossi Mts, Kodmin area (3 coll.).

Only three collections are known of this striking lianescent shrub. All appear to be within an area of c. 1 km^2 at or near which a new reservoir has been proposed, which may threaten the species. This assessment was first published in Cheek *et al.* 2004: 176 and maintained in Cheek *et al.* Kew Bull. 61: 592 (2007). Since no new data are available, that assessment is maintained here.

Shrub or climber 6 m; stems rounded, with ants, pubescent; leaves thinly papery, elliptic, 18 × 10 cm, acuminate, rounded to cordate, thinly scurfy-pilose below, margin ciliate, nerves 8–10; petiole 5 mm; stipule 18 × 5 mm; fruiting peduncle c. 5 mm; partial-peduncle 1.5 cm; fruit 2.8 × 2 cm.

Habitat: submontane evergreen forest; 1400–1500 m alt.

Threats: see above; slash and burn agriculture.

Management suggestions: if reservoir construction goes ahead, a survey to refind the plants is advised. If any are in danger of destruction, their relocation to another location should be investigated.

Leptactina Hook.f. consists of c. 25 species in Tropical Africa. Although thankfully, treated in Fl. Gabon 17: 70–85 (1970) by N. Hallé, a revision is still needed. One or two new species have been described recently (e.g. Adansonia 28 (2006)). *Leptactina surongaensis* De Wild. of Cameroon, CAR and Congo (Brazzaville) may also merit threatened status when species delimitation is clearer.

Leptactina latifolia K.Schum.

EN B2ab(iii)

Range: Cameroon (S: Lolodorf; Edéa-Kribi km 58. Littoral: Douala 20 miles on road to Edéa) and Gabon (Libreville; 18 km E of Libreville; 6 km SW Méla at Mt Muélakéne).

Recorded on Tropicos from Mundemba (*Nemba* 729) but not named by a specialist so requires verification before it can be included.

A woody climber or shrub to 4 m tall with terminal clusters of strongly gardenia-scented white flowers each c. 15 cm long, with 5 strap-like petals 6 cm long. The leaves are elliptic, c. 12 × 6 cm. Distinguished from other *Leptactina* in the broadly ovate, pointed glabrous stipules 2–4 × 1–2.6 cm, the leaves glabrous on both sides and the included stigma.

Here *L. latifolia* is assessed as Endangered since only five locations at the most survive for it (AOO 20 km^2 with 4 km^2 cells) and threats are as below:

Habitat: lowland evergreen forest.

Threats: despite the eight records from Libreville a century ago (Klaine and Pierre) we consider it extinct there since in that time the now capital city of Gabon has expanded greatly (MC pers.obs.) and despite significant national collection effort by Dutch, French, Gabonese and American botanists, this spectacular species has not been seen at Libreville in recent years. According to Onana (pers. obs) the subpopulation at Lolodorf is perhaps extinct (habitat destruction).

Management suggestions: *Leptactina latifolia* would be a desirable garden plant, with its large, numerous, highly scented flowers and compact scent. Introduction to horticulture would help its survival. Surveys are needed to rediscover it and introduce it to the horticultural world. Surveys of the Douala-Edéa reserve might well reveal this species, which could then be included in plans for protection of the species.

Mitriostigma Hook.f. is a genus of seven localised species restricted to forest on the coast of Africa. Until 2009 only one species, *M. barteri* was recognised from W African coast but now (Cheek & Bridson in press) it is clear that material in Cameroon is a separate species, *M. cameroonense.* True *M. barteri* is confined to Bioko. At the same time two unbranched, monocaulous species have come to light, *M. bakweri* at Mt Cameroon (Cheek & Bridson in press) and *M. monocaule* S of Kribi at Mt Elephant (Sonké & Dessein op. cit. below).

Mitriostigma is related to *Oxyanthus* but distinguished by the short (1–2 cm long) , coloured tubular flowers with flanged styler heads. In *Oxyanthus* the corolla tube is white, 15 cm long or more and the style head cylindrical. The genera also differ in anther appendage and seed surfaces.

Mitriostigma bakweri Cheek & Bridson *ined.*

CR B2ab(iii)

Range: Cameroon endemic (SW: Mt Cameroon, Ongé forest).

Only four records of *M. bakweri* exist from a single forest in the western foothills of Mt Cameroon with the threats indicated below, accordingly it is here assessed as Critically Endangered. An AOO of 4 km² is estimated (4 km² cells) and an EOO of 2 km² can be measured from the specimen grid references. An unbranched, densely puberlent unbranched, erect shrublet c. 30 cm tall with a terminal funnel of litter-gathering leaves with tubular red, purple-spotted flowers, this is a distinctive species.

Habitat: lowland evergreen forest.

Threats: Ongé forest is unprotected and threatened by conversion to plantations.

Management suggestions: it is imperative, if this species is to survive, that the Ongé forest is protected.

Mitriostigma cameroonense Cheek & Bridson *ined.*

EN A4bc

Range: Cameroon (SW: Mt Cameroon, Bimbia-Bonadikombo; Mt Etinde; Bakossi-Bolo, Korup; Mbongé. Littoral Region: Ebo, Yabassi).

A 0.3–1.5 m tall glabrous shrub with long lateral branches, red tubular flowers c. 1 cm long followed by bright orange cylindrical persistent fruits c. 2 cm long. Formerly confused with *M. barteri* Hook.f. (a species restricted to Bioko) and assessed as Endangered (EN A3c) in Cheek *et al.* 2004: 176 under that name. That assessment is essentially maintained here although altered to EN A4 to reflect that heavy habitat loss noted in that reference has already begun and is ongoing. In 2008 it emerged that much of the forest at Bimbia-Bonadikombo surveyed in the early 1990s (where this species was found there in relatively high density) has been lost. The same prospects await the sites at the base of Mt Etinde in the vicinity of the villages of Etome, Batoke and Upper Boando, which host a second concentration of the species. Fortunately the species is probably secure in the Korup NP and in the proposed Ebo NP although at each of these sites only a single collection is known.

Habitat: undisturbed lowland evergreen forest; 0–450 m alt.

Threats: clearance for agricultural expansion, particularly plantations.

Management suggestions: it is proposed that the sites identified above are surveyed to assess the size of their subpopulations of this species, and that measures be taken to protect at least the most significant sites.

Mitriostigma monocaule Sonké & Dessein

CR 2ab(iii)

Range: Cameroon endemic (S: Mt Elephant, S of Kribi)

Here *M. monocaule* is assessed as Critically Endangered since known from a single location (AOO 4 km² using 4 km² cells) threatened by iron ore extraction. Described in 2009 (Nord. J. Bot. 27: 305–312), the species was not found anywhere other than the southern edge of the Mts despite its authors making several lengthy visits there over recent years and despite the species being highly conspicuous in flower and fruit (Sonké & Dessein op. cit. 309).

With a single erect unbranched stem 20–30 cm tall, *M. monocaule* is unlikely to be confused with any other species of the genus apart from *M. bakweri* (which see). The second differs in having leaves clustered at the stemp apex (not scattered over the length of the stem) and with densely, persistent hairy stems, and in having non-attenuate based, hairy (not glabrous) leaves.

Habitat: lowland evergreen forest (altitude not given).

Threats: Mt Elephant is mooted as an iron-ore mine.

Management suggestions: data on the exact location, extent, density and regeneration of this species is desirable and its protection. Development at this site at Mt Elephant should not be permitted unless other sites can be discovered for the species.

Morinda is a genus of small pithy, sappy trees and climbers easily recognised by their fleshy compound fruits, composed of the products of a small head of flowers. The stipules are usually triangular.

Morinda mefou Cheek

VU B2ab(iii)

Range: Cameroon (S: Station du Cacaoyer de N'Koemvone; Bipindi. C: Mefou proposed NP), Gabon (Route Owendo–Ikoi Komo; Moanda-Franceville; Région des Abeilles) and Congo (Brazzaville) (Environs de Brazzaville).

Morinda mefou was formally named Cheek *et al.* 2011, where it was assessed as Vulnerable VU B2ab(iii) since it is only known from six locations (see above) equating to an area of occupancy of 24 km² using the recommended 4 km² cells, and since the species is threatened by forest clearance for agriculture, for example at Mefou itself.

Liana, scrambling on trees to 6 m tall; internodes (3–)4–6.5(–8) cm long, 3–4(–5) mm diam., glabrous or sparsely to densely hairy; stipule sheath (1–)2–3 × 4–6(–10)mm the basal half swollen, thickened, the apical half submembranous with conspicuous white rod-like raphides visible at the surface; leaves opposite, equal; blades elliptic or oblong-elliptic, rarely ovate (5.7–)7–10(–11.2) × (2.8–) 3.2–5.8(–6) cm, acumen 0.9–1.2 cm long, gradually tapering, base rounded or truncate, rarely obtuse, quaternary nerves reticulate, raised on the lower surface, indumentum dense on the lower surface midrib, remainder sparsely hairy, upper surface glabrous or moderately hairy, lateral nerves on each side of the midrib 5–9, parting at 45° from the midrib, forming a looping marginal nerve; petiole 5–8(–9) mm ×

0.75–1 mm, glabrous or hairy; inflorescence and flowers unknown; infructescence a syncarp comprising 3–4 united fruits, subglobose-lobed, ripening yellow or orange, hard, 2.2–3.1 cm diam., subglabrous to long hairy; individual fruits subovoid, the free part volcano shaped, c. 15 × 15 mm; calyx throat aperture c. 4 mm; disc persisting, annular, flat, 3 mm diam.; seed 10 × 8 × 5 mm, U-shaped, folded transversely.

Habitat: lowland evergreen semi-deciduous to submontane forest; either forest edge or secondary; 350–600 m alt.

Threats: clearance of forest for timber and/or agriculture.

Management suggestions: it is to be hoped that this and other Red Data species at Mefou proposed NP will be incorporated into a future management plan for the Park and protected for the future. The first step, in this instance, is to refind the original plant, survey for additional individuals in the area, establish actual and possible threats to those individuals so that they may be addressed, evaluate actual regeneration levels in the wild, begin seed collection for propagation in a nursery and future planting out in suitable safe sites to bolster the wild population.

Mussaenda is a genus of bristly hairy canopy climbers or scandent shrubs, very rarely epiphytic. They are recognised by the feature of one of the five calyx lobes usually expanding to 50–100 times the size of the others and being coloured in contrast to the corolla, presumably so as to better attract pollinators. The fruits are juicy berries with numerous small seeds on two intrusive placentre. The genus has not been revised since Wernham in J. Bot. 51 (1913) and a modern update is highly desirable.

Mussaenda conopharyngifolia Stapf
VU B2ab(iii)

Range: Liberia (Sinoe Basin), Ivory Coast (location unknown), Cameroon (SW: Edjagham FR. Littoral Region: Dibombari. S: Kribi, 10 km, near airfield) and Equatorial Guinea (Rio Muni: 4 km N of Mitong).

Hepper in FWTA 2:164 (1963) distinguishes *M. conopharyngifolia* from congeners in having leaves glabrous on the upper surface (not pubescent) with the corolla and calyx pilose outside. Onana (pers.obs.) notes that confirmation of the identity of some of the Cameroonian material is needed, referring to the Edjagham specimen. The Mitong specimen named thus is discordant since the floral indumentum is appressed, not patent and the upper leaf suface is hairy. The Kribi specimen is in fruit only so identification cannot be confirmed. This is not a satisfactory circumstance.

It would be preferable if there was less doubt on the delimitation of this taxa. However, on balance it seems better to assess it as is, despite the shortfall, accepting that in future it might be revealed that the Cameroonian records represent a species separate from that of Liberia–Ivory Coast. With six locations, *M conopharyngifolia* is here assessed as Vulnerable (AOO 24 km² with 4 km² cells and threats below).

Habitat: lowland evergreen forest.

Threats: most sites are not known to be threatened, however, the site at Kribi airport may be under threat from development.

Management suggestions: a satisfactory dilineation of this taxan is need before efforts are made to seek its protection.

Mussaenda epiphytica Cheek (syn. *Mussaenda* sp. nov. in Cheek *et al.* 2004: 376)
VU B2ab(iii)

Range: Cameroon endemic (SW: Kupe-Bakossi, 10 coll.)

Although the Bakossi NP was created in 2007, all known sites of this Bakossi endemic fall outside its boundary. While ten specimens are known, they equate to only six locations separated by more than 1 km (AOO 24 km² with 4 km² cells). Accordingly this species was assessed in the protologue as Vulnerable VU B2ab(iii).

Epiphytic shrub with stems to 60 cm, glabrous; leaves elliptic, dull white below, c. 8 × 5 cm; flowers several, bract-like calyx lobe yellow or white; corolla yellow, tube 5cm, lobes c. 1cm; fruit subellipsoid, with persistent linear sepals 1.5 cm.

Habitat: submontane forest; 760–1500 m alt.

Threats: three of the ten specimens known are from altitudes of below 1000 m and so are especially vulnerable to ongoing forest clearance for agriculture from inhabited areas below, for example at Nyasoso.

Management suggestions: involvement of community leaders and public education to explain the importance of this species and forest conservation in Bakossi are advised.

Nauclea diderrichii (De Wild. & T.Durand) Merrill
VU A1cd

Range: Sierra Leone to Congo (Kinshasa).

In view of the very wide range and numerous locations from which this species has been recorded, it is not mapped here.

This widespread timber tree was given the assessment above by an African Regional Workshop in 1997 (cited in IUCN 2003; www.redlist.org), on the basis that it is heavily exploited for its timber used in general construction work. Regeneration is good in large canopy gaps but the species is out-competed by other pioneers after clear felling. The rating VU A1cd indicates that 30% or more population reduction has occurred in the last 100 years, reflecting the widespread loss of its forest habitat in W Africa in this time, as has occurred for many other timber trees with this range. However, the species is unlikely to become extinct since forestry departments and local communities protect young trees until they are large enough to fell, and even plant young trees to ensure future income.

Tree c. 30 m; stems with ants; leaves elliptic or oblong, 7–15 × 4–9.5 cm, larger on young, sterile shoots, base rounded, nerves 6–8; petiole 8–13 mm; stipule ovate to obovate, 10–25 × 4–12 mm; inflorescence spherical, 3 cm diam.

Habitat: lowland evergreen forest; to 800 m alt.

Threats: see above.

Management suggestions: good management of lowland production forest reserves should ensure the survival of this species.

Oligocodon cunliffeae (Wernham) Keay

EN B2ab(iii)

Range: Nigeria (Calabar-Akamkpa; Eket District; Lagos-Ikeja, below Igboedun un Ogunk.) and Cameroon (Littoral: Douala-Edéa reserve. SW: Meme R, Kumba).

This gardenia relative has its own genus, delineated by the truncate calyx: corolla tube abruptly expanded at apex, glabrous apart from a narrow ring of hairs inside near the base and a 1-celled ovary. This is a glabrous climber, drying black, reaching 5 m high or more, with rough, splitting bark and obovate or elliptic leaves borne on 1–3 cm long spur shoots, terminating in fragrant flowers with 5–6 cm long, bell-shaped, pale green corolla tubes and white rounded lobes with purple splashes followed by 3 cm long, orange, egg-shaped fruits. *Oligocodon cunliffeae* is here assessed as Endangered since only five locations are known (AOO 20 km² at 4 km² cells) with threats as below.

Habitat: lowland evergreen forest, usually near the edge of a river or lake.

Threats: several collections issue from Lagos where it is probably extinct due to the expansion of this vast city. Elsewhere slash and burn agriculture poses a threat.

Management suggestions: the Meme R FR and Douala-Edéa FR offer the possibility of rediscovery this monotypic genus. Surveys are needed to refind it and establish basic populational data to develop a management plan. The protected status of these reserves may need re-assessment.

Oxyanthus was revised for C Africa by Sonké (Op. Bot. Belg. 8 (1999) who accepted 19 species and possible new species A–F, of which species D was later elaborated as *O. okuensis* (see below). A genus of forest shrubs, *Oxyanthus* are recognised by the short (1–2 cm long) but densely branched, axillary inflorescences held on the upper side of the lateral branches. The corolla tubes are long (10–20 cm) usually white and opening in the late afternoon. Leaves are generally glabrous and leathery, with broad, ovate-lanceolate leathery entire stipules. In addition to the species selected as threatened below, *O. setosus* may feature in future editions when better habitat loss data are available.

Oxyanthus brevicaulis Krause

VU B2 ab (iii)

Range: Cameroon (SW: Korup. S: Ébolowa-Zingui), Gabon (Bélinga and Minkebe), Congo (Brazzaville) (Chaillu) and Congo (Kinshasa) (5 collections at 4 locations in Forestier Central).

This species is unique in the genus due to its monopodial habit (only rarely with weak lateral branches). The leaves are oblanceolate (17–)21–39 × (3–)5–10 cm (Sonké 1999: 55). It seems likely that *Oxyanthus brevicaule* may be adapted as a litter-gatherer. This species is genuinely rare since in several intensive surveyed areas of Cameroon with the habitat required (e.g. Mts Cameroon, Mt Kupe) it has not been collected. Each of its collection sites in Lower Guinea coincides with a refuge area. The 10 locations equate to an AOO of 40 km².

Habitat: lowland evergreen forest; altitudinal range unknown.

Threats: one of its locations, at Bélinga, Gabon, is scheduled for destruction due to an iron mine.

Management suggestions: both of the sites in Cameroon appear to lie in National Parks but this needs to be confirmed and the park authorities notified and provided with the information needed to identify and protect this species, possible in the form of an educational poster.

Oxyanthus montanus Sonké

VU B2ab(iii)

Range: Equatorial Guinea (Bioko (2 coll. at 2 locations)) and Cameroon (SW: Mt Cameroon (6 coll. at 3 or more locations); Bakossi Mts (1 coll.)).

This taxon was first assessed as above in Cheek *et al.* 2004: 176. That assessment is maintained here. No further data are available. The species was first formally published in 1994, but was initially collected at Mt Cameroon in the 1930s, near Buea at Musake camp; a second location there, given as 'Nyanga camp', is now lost. In all, six specimens are known from Mt Cameroon, which appears to be the stronghold for the species. There it appears confined to the belt of forest between Etinde and Buea; this belt is steadily being reduced by agriculture along its lower boundary and when forest cover change data are available this species is likely to be reassessed as EN or CR. A total of six locations are known for the species (AOO 24 km² at 4 km² cells).

Shrub 5–9 m, glabrous; leaves elliptic, 10–14 × 3–6 cm, acuminate, cuneate, lateral nerves 16–21 pairs, with domatia; petiole 1 cm; stipule triangular, 8 mm; inflorescence 5–8-flowered; bracts persistent; pedicel 5 mm; corolla tube 11 cm, lobes 1.5–1.8 cm; fruit ellipsoid, 3–4 × 1.5 cm.

Habitat: submontane evergreen forest; 1400–1600 m alt.

Threats: forest clearance for agriculture, particularly at Mt Cameroon (see above).

Management suggestions: a survey is advised to rediscover the species at its known sites, to quantify the subpopulations and assess regeneration. The forested eastern slope of Mt Cameroon should be given priority.
Reference: Sonké, B. (1999) *Oxyanthus* (*Rubiaceae*) en Afrique Centrale. Opera Botanica Belgica 8: 106pp.

Oxyanthus okuensis Cheek & Sonké
CR A2b,c
Range: Cameroon endemic (NW: Mt Oku and the Ijim ridge; Dom).
Dom has the world's largest known population of this magnificent forest understorey shrub. At least 20 plants were counted at Dom in Sept. 2006, and there are probably many more than that present. Given that 50% of forest in this part of the Bamenda Highlands was lost 1988–2003 (Baena, in Cheek *et al.* 2010: back cover), it is likely that more than 80% of the habitat of this species has been destroyed over three generations.
This species was first collected, in fruit, by Duncan Thomas from forest near Lake Oku in Feb. 1985 (*Thomas* 4377) and mistakenly identified as *Oxyanthus formosus* (Thomas in McLeod 1986: 62). Thought to be restricted to the forest immediately around Lake Oku (where seven plants are known, Kemei *pers. comm.*), until c. four sterile treelets were seen in the reconnaissance to the Ajung Cliff (*Cheek* 10103) led by DeMarco in November 1999. It is notable that surveys in other areas of forest on Mt Oku and the Ijim ridge have not located this species. However, according to Kemei (*pers. comm.*), there is a third site (a single plant) for this species near Ntum at Ijim. A full description and notes on this species can be found in Cheek & Sonké (A new species of *Oxyanthus* (Rubiaceae-Gardiniinae) from Western Cameroon. Kew Bull. 55(4): 889–893 (2000)). The assessment above is taken from the Dom conservation checklist (Cheek *et al.* 2010: 102) and updates that in the Kilum-Ijim conservation checklist (Cheek *et al.* 2000: 72–73).
Oxyanthus okuensis is probably most closely related to *Oxyanthus montanus* Sonké which is endemic to Bioko, Mt Cameroon and the Bakossi Mts.
Shrub or small tree 3–8 m tall. Leaf blade elliptic or elliptic oblong, 9–13 × 2.5–6 cm, apex acuminate, nerves 8–9 on each side of the midrib, domatia hairy; petiole 5–8 mm long. Stipule oblong in the lower 2/3, triangular in upper 1/3, 12–19 × 3–9 mm, apex aristate, 2–4 mm long. Inflorescences c. 3 per branch, alternate on consecutive nodes, erect, 20–50-flowered, condensed panicles c. 2–3.5 cm diam. Flowers white, calyx cup-shaped, lobes 1.5 mm long. Corolla tube 3.3 cm long, lobes 5, lanceolate, 7.5 × 2 mm. Fruit with pedicel 12–14 mm long; fruit body ellipsoid, 35–45 × 13–16 mm, including an apical rostrum. Seeds c. 10 per fruit, faceted, c. 6–8 mm diam.
Habitat: understorey of submontane and montane evergreen forest; 1590–2200 m.
Threats: possibly cut for firewood; forest clearance for agriculture. At Dom it appears secure. Both the Lake Oku and Ajung cliff sites are inside the Kilum-Ijim boundary, and so are protected, so long as the boundaries are respected. However, the forest at Ajung was only recently included inside the boundary and there is still evidence that the more accessible part of this forest was still being cleared, possibly as late as early 1999. The forest at Ntum is not protected.
Management suggestions: more information is needed on the number of individuals of *Oxyanthus okuensis* present at the two known locations, and upon levels of regeneration. The plant at the Ntum site should be vouchered. Fertile material is still needed from the Ajung site in order to confirm the specific identity. This species should be looked for in other areas of forest in the Bamenda Highlands. Confusion with other species of *Oxyanthus* is unlikely since this is the only known member of the genus at high altitude in the Bamenda Highlands.

Oxyanthus oliganthus K.Schum.
VU B2 ab (iii)
Range: Cameroon endemic (SW: (Korup 1 coll.). C: (1 coll., 1 site). S: (14 coll., 7 sites).
A glabrous forest shrub with elliptic leaves 12–19 × 4–8 cm, lateral branches with a single inflorescence bearing subglobose fruits 2.5–3 × 1.5–1.8 cm, wrinkled when dry, with a rounded apex. It is here assessed as Vulnerable (see above) since there are 9 locations (AOO 36 km² with 4 km² cells) with threats as below.
Habitat: lowland evergreen forest c. 250 m alt.
Threats: forest clearance followed by agriculture in the main centre for distribution of this species, around Bipindi, is a concern.
Management suggestions: this species is already placed in both Campo-Ma'an (Tchoutou 2004) and Korup NPs. The former is most important for the viability of this species since it is much more frequent in S Region than elsewhere.

Pauridiantha was revised by Salvator Ntore (Opera Botanica Belgica 15 (2008)), recognising 38 species. Confined to Tropical Africa and Madagascar, they are mainly shrubs or small trees with one liana species, and one species a tree to 15(–32) m tall (Ntore op. cit.). Characteristically, species are densely and softly pubescent, with elongated free, narrowly triangular stipules, numerous short axillary inflorescences with small (corolla tube c. 1.5–4 mm long) dull flowers followed by small globose multi-seeded berries. The genus is concentrated in Lower Guinea where 50% of the species are endemic (Ntore op. cit.). Formerly (Cheek *et al* 2004: 177) *Pauridiantha venusta* N. Hallé was assessed as VU but

Ntore (op. cit.) has synonymised it with *P. talbotii,* now a widespread, non-threatened species. *Pauridiantha efferata* N. Hallé just falls outside the scope of criterion B, having 11 locations. When better habitat loss data are available it may qualify as threatened under criterion A. *Pauridiantha divaricata* (K. Schum.) Bremk. while endemic to Cameroon and previously assessed as VU B2ab(iii), now has 14 locations and is re-assessed here as NT.

Pauridiantha multiflora K. Schum.
VU B2ab(iii)

Range: Cameroon endemic (SW: Rumpi Hills, Butu-Lokando; Kumba-Mamfe, Mbu village; Toko village. Littoral & S: Edéa-Kribi; Mbouma; Batanga; Eséka).

Here assessed as Vulnerable since seven locations are known (AOO 28 km² using 4 km² cells) with threats given below. Location data from Ntore (op. cit. p.133). The seven locations are in 3 clusters: Rumpi Hills area; W. of Yaoundé near Eséka; Kribi area.

Pauridiantha multiflora is a 5–8 m shrub with glabrous or weakly pubescent, quadrangular interoides. The leaves are elliptic-oblong, 5–12 × 4–6 cm, acuminate. Inflorescences are to 20-flowered, 2 per axil, to 1.5 cm long.

Habitat: lowland evergreen to submontane forest; 100–800 m alt.

Threats: all sites are threatened with slash and burn agriculture which is often accelerated by logging activities. Habitat at Batanga is threatened by development activities.

Management suggestions: the Rumpi Hills area, being considered for a future National Park, offers a possibility for preserving this species since several sites are known there. Data on frequency, range and regeneration are needed to gauge levels of management intervention.

Pauridiantha siderophila N.Hallé
VU B2ab(iii)

Range: Cameroon (S: Lélé-Mbalam. Littoral: Massif de Nkohom 40 km S Ndikiniméki), Gabon (25 km SW Mitzic; Bélinga; 37 km SE de Makoukou) and Congo (Brazzaville) (Chaillu Massif).

Here *P. siderophila* is assessed as Vulnerable since 6 locations are known (above), AOO 24 km² with 4 km² cells, and threats as below.

This is a shrub 1–4 m tall with stems square, densely pubescent, and leaves narrowly obovate, 5–12.5 × 2–4 cm with 5–8 pairs of secondary nerves crypt domatia and petioles 4.5–11 mm long.

Habitat: evergreen forest 800–1000 m alt.

Threats: of the 12 [illegible] known, se[illegible] Bélinga in Gabon, im[illegible] [illegible]ened by a [illegible] cast iron ore mine. Elsewhere slash and burn agriculture also presents a risk, albeit lower.

Management suggestions: arranging the Bélinga mine to avoid destroying this species is advised. Elsewhere a poster campaign might inform local communities of the rarity of this species and secure their support for its preservation. Data are needed on population density, range and regeneration levels to decide on the need for management intervention.

Pausinystalia was revised by Stoffelen & Robbrecht (Bot. J. Linn. Soc. 120: 287–326 (1996)). They recognised five species all of which are trees with 2 cm long dehiscent capsules releasing numerous winged seeds. Related genera are *Hymenodictyon* and *Corynanthe. Pausinystalia* differs in having the corolla tube differentiated into an apical bladder and a narrow basal cylindric part, and with lobe appendages linear. The bark of one species, *P. johimbe,* is a well known male aphrodisiac.

Pausinystalia brachythyrsum (K.Schum.) W.Brandt
CR B2ab(iii) + D+ possibly EX

Range: Cameroon endemic (S: Bipindi).

Stoffelen and Robbrecht were resolute in maintaining this taxon as a good species. We here assess it as Critically Endangered since it is known from a single location (AOO 4 km² using 4 km² cells) with threats as below. Since our only evidence is that there was only a single tree known, the species can also be assessed under Criterion D. Distinguished by the small (c. 3 × 2 cm) inflorescence with persistent stipules at its base and corolla appendages <3 mm. Listed as extinct by IUCN in 2006, but this must be uncertain, while possible.

Habitat: lowland evergreen forest (assumed).

Threats: slash and burn agriculture; urban expansion; felling for construction timber.

Management suggestions: a survey to refind this species is needed. If located, trees should be marked and signposted for protection against random felling so that seed for propagation can be collected in season, allowing the reintroduction of what appears to be an incredibly rare species since it has not been seen in a hundred years.

Postscript: the above was written after overlooking that Charlotte Taylor (MO) has identified *Etuge* 505 (March 1987) from Ikiliwindi in Bakossi as this species (Tropicos). This is potentially excellent news, and after verification, all efforts to refind this tree should be redirected accordingly.

Pausinystalia talbotii Wernham
[illegible]

[illegible] (Benin Prov., Iyekusulu Distr.; Oni; Oyo in Ondo Prov.; Ekenwan; Calabar, Dukwe; Oban) and Cameroon (SW: S Korup; Kumba-Mamfe Rd at Bolo; 15-35 km NE Limbe; Kumba-Baduma mile 15. S: Bipindi; Lake Tissongo)).

Eleven locations (above) being recorded for *P. talbotii* we cannot assess it as threatened under criterion B. However, since six of these are in Nigeria and it has not been recorded there in 40 years, and since 36% of the remaining Nigerian forest was lost in the 15 years between 1990 and 2005 (mongabay.com), we hereby estimate that more than 30% of population loss has occurred in one generation of this tree species, allowing an assessment of VU A2b,c.

This is a 10–12 m tree distinguished by having axillary inflorescences and hairy domatia on elliptic or ovate leaf blades.

Habitat: lowland evergreen forest.

Threats: as for *P. brachythyrsum.* The Lake Tissongo, and especially the Korup NP sites must be considered relatively secure but all those in Nigeria are at risk (if still in existence) with the remainder in Cameroon. Trees close to roads are especially vulnerable to cutting for their wood or clearing for the land they occupy so that crops may be grown.

Management suggestions: it would seem sensible to focus efforts on protecting this species at Korup NP where lowland forest is secure. Accordingly surveys to gain populational data such as extent, density and regeneration of the subpopulation so as to assess management needs should be executed here. It would be wise to draw upon data from the Univ. Bern and Smithsonian large forest plots to this effect.

Pavetta is a genus of c. 400 species of shrubs (less usually trees and subshrubs) mainly occurring in the understorey of evergreen forest (less usually savanna) in the tropics of the Old World, with the main centre of diversity in Africa. All species have bacterial nodules in the leaf-blades, visible as raised black lumps on the lower surface but often more easily seen as translucent or black areas when the leaves are held up to the light. The stipules are sheathing and awned. The inflorescences are terminal on distinctive lateral shoots ("floriferous twiglets") which have leaves only at the stem apex. Flowers are generally 4-merous with exserted anthers and inconspicuous stigmatic lobes. The genus was last revised by Bremekamp between 1934-1939. It is fortunate that the Cameroonian species of subgenus *Baconia* were revised by Manning in 1996 (Ann. Miss. Bot. Gard. 83: 87-150 (1996)) based on his own extensive fieldwork with Thomas in 1986. Manning's work delimiting taxa provides a sound basis for assessing the conservation status of these taxa. Sadly, such a work is not available for subgenus *Pavetta* so except in a few cases, that subgenus is not assessed here.

Subgenus *Baconia* is characterised by flowers with corolla lobes as long or longer than the tube, bearded at the throat. The generic characteristics referred to above of distinctive inflorescence shoots and bacterial nodules are often either poorly developed or difficult to view in this subgenus.

Pavetta baconiella Bremekamp
VU D2

Range: Cameroon endemic (C/E: 'Monjala to Mole'. SW: Mana R, Korup).

The location of the only original specimen, now destroyed, was given as 'between Monjala and Mole' (*Mildbraed* 8364). Manning was not able to find these places but Onana believes they may occur between Nanga Eboko and Deng Deng (new observation) and so we have mapped it here. Manning remarks that this species has the smallest leaf-size of all Cameroonian *Baconia* and presumably this was a factor in his deciding that *Thomas* 2212 from the Mana R, represents a second specimen of this species.

Pavetta baconiella, represented by the Mana R material, is a rheophytic shrub 0.8 m tall, subglabrous, the leaves are elliptic, 2–10 × 0.5–3 cm with 5–12 lateral nerves on each side of the midrib. It is here assessed as Vulnerable under criterion D2 since, while there is no evidence of direct threats (see below), only two locations are known. If further data such as on population size, is obtained, or if the situation and status of the type locality is resolved it is possible that *P. baconiella* will need to be reassessed as EN or CR.

Habitat: rapids over rock in lowland evergreen forest.

Threats: the only known location, the Mana R rapids are protected by inclusion in the Korup NP. Since the type locality is not known precisely, threats there are unknown.

Management suggestions: researching Mildbraed's itinerary might reveal the type locality with precision so that it can be rediscovered and so that the population, if still present, can be assessed. A survey of the species at the Mana R rapids is needed to assess numbers of individuals and regeneration.

Pavetta bidentata Hiern var. ***sessilifolia*** S.D.Manning
EN B2ab(iii)

Range: Cameroon endemic (C: Mt Kala and Mt Kombeng).

This understory forest shrub is recognised by its sessile leaves with petioles <1 mm long, and cordate to rounded, sometimes pouched leaf-bases. It may merit elevation to specific rank. It is here assessed as Endangered since only two locations are known (AOO 8 km^2) and forest at Mt Kala is threatened by slash and burn farming (Sonké pers. comm. to Cheek).

Habitat: submontane evergreen forest in Yaoundé area.

Threats: see above.

Management suggestions: the two localities named should be searched to attempt to refind this plant and assess the number of individuals that survive. Other hill forests in the area may also house this taxon although several, such as Mt

Fébé, have been deforested due to development. Protection of hill forest habitat in the Yaoundé area might help to secure the future of this plant.

Pavetta brachycalyx Hiern

EN B2ab(iii)

Range: Cameroon endemic (SW: Mt Cameroon (9 coll. from 2 sites); Kupe-Bakossi (3 coll. from 2 sites); Lebialem-Fosimondi (1 site)).

Once thought restricted to Mt Cameroon, the range of this nondescript species was extended to Bakossi (Cheek *et al.* 2004: 177 where this assessment, maintained here, was made), in recent years a further site, at Fosimondi has been found. Five locations (AOO 20 km² with 4 km² cells) and threats as below justify EN status.

Shrub 3–6 m, glabrous; floriferous twigs 20 cm; leaves papery, drying green, elliptic to 20 × 9 cm, acumen 0.5–1 cm, acute, lateral nerves 9 pairs, tertiary nerves inconspicuous, domatia small pit-like, bacterial nodules rod-like, 2–3 mm, sparse; petioles 3 cm; panicles 10 cm across; calyx lobes 0.5 mm; corolla white, tube 6mm, lobe 6 mm.

Habitat: lowland and submontane forest; 300–1500 m alt.

Threats: clearance of forest for agriculture especially in the lower part of the altitudinal range. Its recent discovery at Fosimondi reveals that its range extends to areas where forest survival is highly threatened.

Management suggestions: this species is reasonably secure where it occurs above c. 1000 m alt. in both the Bakossi and Etinde parts of its range. Elsewhere it is threatened. Populations outside of protected areas could be assisted in their survival by a poster campaign.

Pavetta brachysiphon Bremekamp

VU D2

Range: Cameroon (E: near Deng Deng, near confluence of Lom and Djérem rivers).

Shrub with elliptic-oblong leaves 3–12 × 1–3.4 cm, minutely puberulent or papillose on upper surface, with domatia along the lateral nerves; the flowers are small for the genus (corolla tube 1–2 mm long, lobes 2–4 mm) in very congested globular inflorescences only 1–3 cm across.

Pavetta brachysiphon has not been seen since the type collection was made in March 1914 (*Mildbraed* 8536). However, since no threats are known at its locality, and no data on population size have been reported, it is here assessed as Vulnerable under criterion D2 (less than 5 localities known).

Habitat: 'near a forest-savanna boundary' (Manning loc. cit: 106); 750 m alt.

Threats: unknown.

Management suggestions: attempts should be made to rediscover the site of this species and establish the size of, and threats to, its population, so that a management plan can be made.

Pavetta cellulosa Bremekamp

EN B2ab(iii)

Range: Cameroon (S: Bitye, Dja R), Gabon (1 site) and Congo (Kinshasa) (3 sites).

Forest shrub or small tree, stem to 5 cm diam. or more; leaves elliptic to oblong, obovate or ovate, 5–19 × 1.5–6.5 cm. The species is unusual in having septate anthers. When sterile the partly deciduous stipules which leave behind bases of the stipule sheaths are distinctive.

Since 5 locations are known (AOO 20 km² at 4 km² cells) and the site at Bitye is considered threatened by Onana (pers. obs.) we here assess the species as Endangered.

Habitat: lowland evergreen forest.

Threats: slash and burn agriculture.

Management suggestions: attempts should be made to refind this taxon at its only known site in Cameroon so as to assess its status and develop a management plan for its survival.

Pavetta grossissima S.D.Manning

EN B2ab(iii)

Range: Cameroon (SW: in or near Korup NP).

Shrub to 1 m tall, pubescent to subglabrous; leaves oblong-elliptic at obovate, 4–22 × 2–9 cm, glabrous; infructescences globose, 0.1–2 cm across, peduncle 5 mm or absent; distinguished by the orange fruits (otherwise unknown in the subgenus except in *P. gabonica*) and in the most coarse fine nerves of the subgenus. *Pavetta grossissima* is known from nine specimens at three locations all within 10–15 km of each other. An AOO of 12 km² is estimated. The species is here assessed as Endangered under criterion B in view of the threats below.

Habitat: lowland evergreen forest; sea-level–120 m alt.

Threats: slash and burn agriculture and small-scale logging outside the Korup NP (van der Burgt pers. comm.)

Management suggestions: five of the nine specimens derive from inside Korup NP where they are fairly secure. A survey to establish the extent of the population there is advisable. Observations should be made on regeneration levels. Sites outside the NP should be revised and assessed to see if the species a) survives there, and b) can be protected there, perhaps supported by public education.

Pavetta hookeriana Hiern var. ***hookeriana***

VU A2bc

Range: Cameroon (SW: Mt Cameroon (numerous coll.);Mwanenguba (c. 3 coll.). NW: Bamboutos/Bamenda Highlands (numerous coll. at several sites: Lebialem, Bali Ngemba, Kilum-Ijim and Dom)).

It is estimated that over 30% of the habitat of this woody forest understorey species has been lost in the last century.
This is the highest altitude *Pavetta* known W of the Congo basin. Generally a good indicator of montane forest at 2000 m, its occurrence of altitudes of 1500–1700 m at Lebialem,Bali Ngemba and Dom is puzzling.
The assessment above follows that in Cheek *et al.* (2010: 103), Harvey *et al.* (2010: 82), which are updated from those in Cheek *et al.* (2004: 177) and Cheek in Harvey *et al.* (2004: 68).
Shrub 2–3 m, subglabrous; floriferous twigs 15 cm; leaves papery, elliptic to 13 × 6 cm, acumen to 1 cm, cuneate, lateral nerves 10 pairs, domatia arched, hairy, nodules not seen, tertiary venation inconspicuous; petiole 2 cm; inflorescence to 10cm across; flowers to 100; calyx lobes rotund, 2 mm; corolla white; tube 2–5 mm; lobes 4–8 mm.
Habitat: montane forest; (1500–)1900–2000(–2400) m alt.
Threats: secure from threat at Mt Cameroon, *P. hookeriana* is threatened by forest clearance for agriculture and wood throughout the extensive Bamenda Highlands, probably once the main area for the species. Study of one large area in the highlands between 1987–1995 showed that 25% of the surviving forest was lost (Moat in Cheek *et al.* 2000: back cover) and elsewhere 50% was lost 1988–2003 (Baena in Cheek *et al.* 2010: back cover).
Management suggestions: improved policing of existing forest reserve boundaries could prevent extinction of this species in the Bamenda Highlands, where its survival is precarious, except at Kilum-Ijum. At Mwanenguba and Bamboutos Mts (presence inferred) it may not survive for much longer. The species is most secure at Mt Cameroon, where the narrowly endemic variety *pubinervia* also occurs.

Pavetta kribiensis S.D.Manning
EN B2a b(iii)

Range: Cameroon endemic (S: Kribi to Campo).
A glabrous forest shrub c. 1 m tall with elliptic, oblong or obovate leaves 2.5–21.5 × 1–7.5 cm with a conspicuous grid-like pattern of fine nerves below and non-fimbriate bracts which largely conceal the inflorescence axes. *Pavetta kribiensis* is known from only five gatherings all made in the 1970s, three of which are along the Kribi-Lolodorf Rd (*Bos* 4137, 4638, 6202), one 45 km SSE of Kribi and the last 25 km ENE of Campo. Since there are only five locations (AOO estimated as 20 km² using 4 km² cells) and due to the threats below, *Pavetta kribiensis* is here assessed as Endangered under criterion B.
Habitat: lowland evergreen forest sometimes along streams.
Threats: coastal forest around Kribi has been and will be much affected by expansion for beach-side accommodation and by large-scale international industrial developments such as the Chad-Kribi pipeline, proposed mineral ore ports, and a proposed gas storage depot.
Management suggestions: there is a good prospect that one of the sites for this species (25 km ENE Campo) occurs inside the Campo-Ma'an NP. This should be verified and the numbers of individuals present and regeneration rates assessed so that a management plan can be drawn up.

Pavetta kupensis S.D.Manning
CR B2ab(iii)

Range: Cameroon endemic (SW: Mt Kupe (6 coll.) and western Bakossi (1 coll.)).
This shrub, which bears some resemblance to *P. owariensis* var. *satabiei*, is more or less confined to the summit of Mt Kupe, apart from an anomalous specimen in western Bakossi which needs investigation. The assessment of Critically Endangered is based on the taxon occurring at a single location (AOO 20 km² at 4 km² cells) with threats as below. This follows Cheek *et al.* 2004: 177, maintained here since no further data are available.
Shrub or tree 3–8 m, subglabrous; floriferous twigs c. 20 cm; leaves leathery, oblanceolate, 15 × 5 cm, acumen 0.5–1 cm, cuneate, midrib thick, orange below, lateral nerves 7–8 pairs, brochidodromous, tertiary venation inconspicuous, nodules few, domatia crypts; petiole 2 cm; panicles 2–9 cm across; flowers 25–75; calyx lobes 0.1–1 mm; corolla greenish-white, tube 6–8 mm, lobes 8–12 mm.
Habitat: montane forest; c. 2000 m alt. (*Thomas* 5197 at c. 400 m alt.).
Threats: the subpopulations on the summits of Mt Kupe are vulnerable to stochastic events, such as landslides.
Management suggestions: the geographically and altitudinally anomalous *Thomas* 5197 (MO) should be studied to confirm or reject is placement in this taxon; the latter is most likely. A study to quantify the demography of the population at Mt Kupe would help develop a plan to manage the species.

Pavetta laxa S.D. Manning
CR B2a b(iii)

Range: Cameroon (S: near Mékomo, 8 km SW of the Dja and Lobo river confluence).
This 5 m forest shrub has not been seen since the only known collection (*Letouzey* 4581, March 1962). The elliptic to obovate leaves are 6–15.5 × 4.5–6.5 cm, the upper surface minutely papillate, nodules are absent or very rare. *Pavetta laxa* is named for the distinctive lax inflorescences which are c. 35-flowered, 3–4 cm across and sessile. The flowers are white, very small (corolla tube & lobes c. 2–3 mm long) except for the large (1–1.5 × 1 mm) rotund calyx lobes. It is here assessed as Critically Endangered since only a single location is known (AOO estimated as 4 km²) with threats discussed below.
Habitat: lowland evergreen forest (probably).
Threats: slash and burn farming.

Management suggestions: the original site should be revisited to attempt to rediscover *P. laxa.* If successful, numbers of individuals should be inventoried, mapped, regeneration levels and threats assessed in detail so that a management plan can be made to protect the species.

Pavetta longistyla S.D.Manning

CR B2ab(iii)

Range: Cameroon (W: Dschang-Santchou (*Meurillon* 645, March 1967)).

Known only from a single specimen with imprecise location data: "Route des Mbos". Since Meurillon was based at Univ. Dschang, this probably refers to the road from that town to the Plaine des Mbos (main settlement: Santchou). Immediately on leaving Dschang this road descends steeply through forest and it is on this forested scarp that the specimen was probably taken – an altitude of 1100 m is given.

Pavetta longistyla lacks bracteoles and has style-heads long-exserted (20–30 mm) both features of subgenus *Pavetta.* However, the bearded corolla throat agrees with subgenus *Baconia* (Manning op. cit.:127).

Since *P. longistyla* is only known from a single location (AOO 4 km^2) it is here assessed as Critically Endangered due to the threats below.

Habitat: forest beside a stream; alt. 1100 m.

Threats: personal observation of the forest at the site concerned in the early 2000s showed that much loss of habitat had occurred due to conversion to farms.

Management suggestions: attempts should be made to refind this taxon at its only known site so as to assess its status and develop a mangement plan for its survival.

Pavetta mpomii S.D.Manning

EN B2ab(iii)

Range: Cameroon (S: Kribi-Campo (Bipindi; Mt Elephant; Lolodorf; Nkolémenlong; 10km ESE Campo)

A rainforest shrub 2 m tall, stems pubescent; leaves obovate to elliptic, 3–22 × 1–9 cm puberulent, tertiary nerves inconspicuous, floriferous twiglets 5.5–19.5 cm long, corolla white, tube 5–8 mm, lobes 8-13 mm; calyx lobes ovate 1.2–3 × 1–2 mm. *Pavetta mpomii* is here assessed as Endangered since five locations are known (AOO 20 km^2 with 4 km^2 cells) and due to the threats below.

Habitat: lowland evergreen forest; sea level to 200 m alt.

Threats: Mt Elephant is reportedly threatened by open cast iron ore mining while forest habitat in the neighbourhoood of the towns Bipindi and Lolodorf is threatened by small-scale farming.

Managment suggestions: the last two sites listed under 'range' above are believed to be in the Campo-Ma'an NP. This is probably the best prospect for the survival of this species. Efforts should be made to rediscover the species at these sites, assess its status and develop a plant to manage its survival.

Pavetta muiriana S.D.Manning

EN B2ab(iii)

Range: Cameroon (SW: Bakossi Mts (4 coll. at 2 sites) and Barombi Mbo (1 coll. at 1 site)).

This very distinctive shrub, known from only three locations, is almost entirely restricted to the Bakossi Mts. This assessment is taken from Cheek *et al.* (2004: 178), which is maintained here since no further data are available.

Shrub 1–2 m, densely grey appressed puberulent; floriferous twigs c. 7 cm; leaves papery, narrowly elliptic, to 17 × 6 cm, acumen 1 cm, acute, lateral nerves c. 14 pairs, tertiary nerves linking, quaternary nerves plain, baterial nodules and domatia not seen; petiole 2 cm; panicles to 4 cm across, to 20-flowered, densely grey puberulent; calyx lobes 0.8 mm; corolla white, drying black.

Habitat: lowland and submontane forest; 400–1780 m alt.

Threats: clearance of forest for agriculture, wood and public events. Aaron Davis (*pers. comm.* 2003) reports that much, perhaps most, of the crater forest of Barombi Mbo has been cleared.

Management suggestions: Kodmin, in the heart of the Bakossi Mts, with three collections, has the densest subpopulation of *Pavetta muiriana* and is the logical site from which to conduct a demographic survey that will also quantify individuals of this taxon, to provide data for a management plan.

Pavetta namatae S.D.Manning

CR B2ab(iii)

Range: Cameroon (C: near Makak).

A 6 m shrub with obovate-elliptic leaves 4–21 × 1.5–8 cm and 10–15 lateral nerves on each side of the midrib, tertiary nerves forming a rectangular grid with the secondary nerves but not prominent below. Inflorescences large, 6–11 cm across, bracts fimbriate, corolla white, tube 5–7 mm long, lobes 6–7 mm long. The location of the type specimen *Gocker* 13, is not known (Manning op. cit.: 132). Since only a single location (AOO 4 km^2) is known for the species, 7 km ESE Makak *(Manning* 2097**)** and since this part of C Region has seen intense logging of its semi-deciduous forest, followed by farming of *Cocoa* and other crops, habitat deterioration has occurred. *Pavetta namatae* is here assessed as Critically Endangered under criterion B. It is notable that both gatherings have few (one or two duplicates) suggesting that only single individuals were seen.

Habitat: unknown

Threats: see above.

Management suggestions: research on Gocker's itinerary might lead to rediscovery of the type locality. If found it should be visited and numbers of individuals, regeneration and threats assessed so as to plan the protection of this species.

Pavetta owariensis P.Beauv. var. ***satabiei*** S.D.Manning
EN B2ab(iii)

Range: Cameroon (SW: Rumpi Hills (1 coll. at 1 site); Bakossi Mts and Mt Kupe (5 coll. at 3 sites)).

Known from only four locations, this shrub differs from other varieties in the absence of domatia, the large (10 mm diam.) fruit with persistent calyces and tall (to 5 m) stature. This assessment is taken from Cheek *et al.* 2004: 178, which is maintained here since no further data are available.

Shrub 3–5 m, minutely papillate-puberulent, glabrescent; floriferous twigs 30 cm; leaves papery, drying pale brown below, elliptic-oblong, 12–20 × 5–8 cm, acumen 1 cm, cuneate, lateral nerves 8 pairs, tertiary nerves few, sparse, domatia and bacterial nodules not seen; petiole 3 cm; panicle 10 cm across, c. 50-flowered; calyx lobes to 1.5 mm; corolla white, not seen; fruit with persistent calyx, to 10 mm diam.

Habitat: submontane forest; 800–1400 m alt.

Threats: forest clearance for agriculture and wood, especially at the lower part of its altitudinal range at the densely populated Nyasoso.

Management suggestions: this taxon should prove reasonably easy to refind, given the number of recent collections. A survey to do this should also quantify the subpopulations and assess regeneration levels and demography. A poster campaign would aid public education on the importance of protecting this shrub.

Pavetta robusta Bremekamp
EN B2ab(iii)

Range: Cameroon (S: Bitye near Dja) and Gabon (Poungui).

This species is a 5 m forest shrub with large oblong to obovate leaves 5–28 × 2–13 cm, the inflorescences are also large and robust being pyramidal, 4–14 cm across with 75–400 fragrant flowers. The white corolla has a 5–9 mm long tube and lobes the same length.

Since two locations (AOO 8 km² using 4 km² cells) and habitat threats are known, *P. robusta* is here assessed as Endangered. It has not been seen since 1924.

Habitat: lowland evergreen forest (probably).

Threats: slash and burn farming.

Management suggestions: the sites at Poungui (*Le Testu* 6420, 1924) and at Bitye (*Bates* 1046) should be checked to discover if this species survives and if so, the number of individuals, levels of regeneration and threat present to protect this species properly.

Pavetta rubentifolia S.D.Manning
CR B2ab(iii)

Range: Cameroon (SW: Bakossi Mts (1 coll. at 1 site)).

Known only from the type collection (*Thomas & McLeod* 5343, west of Bangem, 1986). This assessment is taken from Cheek *et al.* 2004: 178, which is maintained here since no further data are available.

Shrub 1 m, glabrous; floriferous twigs 14 cm; leaves red, elliptic-oblong, 13–16 × 3.5–4.5 cm, acumen to 1.5 cm, cuneate, lateral nerves 8–13 pairs, domatia and nodules absent; petiole 1 cm; panicles 0.5–1 cm wide, subglabrous to puberulent; flowers 20–35; calyx tube to 1 mm, lobes valvate, to 0.5 mm, glabrous; corolla white, tube to 4 mm, lobes 3–5 mm, pubescent near throat

Habitat: submontane forest; 800–1600 m alt.

Threats: forest clearance for agriculture and wood.

Management suggestions: a survey should be made to refind this distinctive species (leaves maroon when live), quantify the subpopulation and develop a plan to protect it.

Pavetta tenuissima S.D.Manning
EN B2ab(iii)

Range: Cameroon (SW: S Bakundu. S: near Eséka, Ma'an and Ambam).

Endemic to Cameroon and known from only five locations (AOO 20 km² at 4 km² calls) with threats as below, *P. tenuissima* is here assessed as Endangered.

This rainforest shrub can reach 5 m tall, it is puberulent on most parts. The leaves are oblong-elliptic, 14–31 × 6–11 cm, distinctive in the extremely fine reticulation of the higher order nerves for which the species is named.

Habitat: lowland evergreen forest.

Threats: none of the sites of *P. tenuissima* appear to be in a protected area. The S Bakundu FR has suffered decline in habitat quality due especially to farming over many years. Although most sites occur in S, none of them occur inside the Campo-Ma'an NP, sadly.

Management suggestions: revisiting the known sites for this taxon should enable its rediscovery and an assessment of population size, regeneration levels and threats to inform a management plan.

Pavetta urophylla Bremekamp subsp. ***bosii*** S.D.Manning
CR B2ab(iii)

Range: Cameroon (S: top of Calvary Mt, 28 km ENE of Kribi, Lolodorf Rd (*Bos* 6611).

Named for the Dutch botanist Bos, this forest shrub has brown flowers with nodules more conspicuous below than above, which place it with *P. urophylla* subsp. *urophylla* of Congo (Kinshasa). However, the *Bos* collection lacks the septate anthers, truncate calyx lobes and long leaf acumina of the taxon so may merit species rank. This subspecies is here assessed as Critically Endangered since known from only one location (AOO 4 km²) with threats as below.

Habitat: lowland or submontane forest, probably.

Threats: slash and burn agriculture.

Management suggestions: revisiting the known sites for this taxon should enable its rediscovery and an assessment of population size, regeneration levels and threats to inform a management plan.

Pentas is a genus of about 40 species in Tropical Africa and Madagascar, last revised by Verdcourt (Bull. Jard. Bot. Brux. 23: 237 (1953)) with subsequent additions and amendments. These are generally 60 cm tall herbs, sometimes from rootstocks, in open grassland or at the forest edge, with terminal white to purple flowers. Apart from *P.ledermannii*, treated here, *Pentas nervosa* Hepper, described from Vogel Peak in former N British Cameroons (now Nigeria) may yet be found in Cameroon and could be assessed as either EN or CR.

Pentas ledermannii Krause emend Verdc. (syn. *P. pubiflora* subsp. *bamendensis* Verdc. in part)

VU B2ab(iii)

Range: SE Nigeria (Obudu (1 coll.)) and Cameroon (SW: Mt Kupe-Mwanenguba (2 sites). W: Bamboutos (1 coll.):). NW: Mt Oku area (3 coll.); Bali Ngemba FR (2 coll.); Santa, Lake Aweng (1 coll. each)).

Two taxa have been confused under the name *P. pubiflora* subsp. *bamendensis*, and probably also under *P. ledermannii* K.Krause before being emended by Verdcourt in Kew Bull. 31: 184–186 (1976).

They can be separated as follows (Verdcourt op. cit.)

	Pentas pubiflora var.	*Pentas ledermannii* emend.
Geography	Cameroon Highlands & E Africa	Cameroon Highlands
Stem indumentum	dries pale yellow	dries rich rusty red brown
Calyx lobes	markedly unequal	equal

This taxon was first mentioned as being of possible conservation concern in Cheek *et al.* (2000: 73). The assessment here is repeated from that given in Cheek *et al.* (2004: 178) and in Harvey *et al.* (2004: 69). It is maintained here since no further data are available. In future it is likely that this taxon will be assessed under criterion A. Eight locations are known at present. At Mt Kupe it is known from 4 collections on the grassy summits (*Cable* 108 (1992); *Cheek* 7573 (1995), S[illegible] also known from Mwanenguba ([illegible]). Mt Kupe is the southernmost poi[illegible]: it does not appear to extend to Mt Cameroon. Mt Oku and the Ijim Ridge appear to be the northern extreme of its range. It is absent from some montane areas within its range which have been studied recently in some detail, such as Dom and Lebialem Highlands (Cheek *et al.* 2010 and Harvey *et al.* 2010 respectively).

Weak shrub 0.6–1 m, densely pubescent; leaves elliptic to 10 × 5 cm, subacuminate, acute, lateral nerves 10 pairs; petiole 1 cm; stipules divided into 7–10 aristae, 4 mm; corymbs terminal, 5 cm wide, 10–20-flowered; peduncle 3 cm; calyx lobes 3 mm; corolla white; tube 5 mm; lobes 5.

Habitat: montane forest-grassland edge; 1000–2060 m alt.

Threats: forest clearance is a major threat in the Bamenda Highlands, with 25% loss in one area between 1987 and 1995 (Moat in Cheek *et al.* 2000: back cover) and 50% in 15 years in another (Baena in Cheek *et al.* 2010: back cover). This rate of loss probably continues and probably also extends to the 1800–2000 m altitudinal range in Obudu, Bamboutos and Mwanenguba. Frequent human-set fires in grasslands in these same areas probably also adversely effect the grassland-forest interface as a habitat for this taxon, although occasional natural fires may aid its regeneration.

Management suggestions: research to resolve these uncertainties would aid in management planning for the conservation of this and other species in this habitat in the Cameroon uplands. For the moment, only the summit area of Mt Kupe lacks the two threats outlined above, and so is alone in offering a secure base for *Pentas ledermannii*.

Poecilocalyx setiflorus (R.Good) Bremek.

VU B2ab(iii)

Range: Cameroon (C: Mefou proposed NP; 35km SE Mbalmayo and 28km S of Atozok village on Soa R), Gabon (Etéké, distr. Mimongo; Bélinga; Mouila; Moukalaba-Doudou NP; Rabi-Kounga; Massif du Chaillu) and Angola (Cabinda).

Here assessed as Vulnerable since nine locations are known (AOO 36 km^2 with 4 km^2 cells) and threats as below. This assessment appears also in Cheek *et al.* 2011.

Shrub, 1–4 m; hairs dense, spreading, 1 mm, dull white on stem, midrib, petiole, stipule and fruit; stipule lanceolate, 10 × 4 mm, with a tooth 1–3 mm on each side, or entire, subsessile, axillary, 1–2 per axil; leaf blade bullate, obovate or oblong-elliptic, 8–13 × 3–4.5 cm, acumen 1 cm, with mucro 1–2 mm, base abruptly rounded, lateral nerves 10–12 pairs, domatia absent, tertiary nerves scalariform; petiole 5 mm; fruit globose, 4 mm, with 5–7 erect green calyx lobes 4 × 1 mm.

Distinguished from the similar *P. schumannii* in the stipule being entire and acute, not multifid, secondary nerves 7–10 pairs not 11–12 pairs.

Habitat: evergreen riverine forest; 600–700 m alt.

Threats: clearance of forest for agriculture and timber.

Management suggestions: this is a relatively common understorey shrub in *Gilbertiodendron dewevrei* forest in Mefou proposed NP, but globally it is rare. Formal ratification of the NP would help to protect this species from extinction. So long as its habitat is protected at Mefou it will be secure. Public education at Mefou for visitors might help understanding of rare plant species in Cameroon.

Pseudosabicea batesii (Wernham) N.Hallé
VU B2ab(iii)

Range: Cameroon (SW: Mokoko (7. coll.), Nyandong, Menyum. S: Bipindi) and Gabon (Abanga, Moumba, Libreville, Kinguele, Mfoa (1 coll. each)).

Known from only nine locations (above, AOO 36 km^2 with 4 km^2 cells) with threats as below.an assessment of VU is warranted. This assessment is taken from Cheek *et al.* (2004: 179), which is maintained here since no further data are available. This climber appears thinly scattered throughout its range, being absent from some very well-collected areas, such as Douala-Edéa and Mt Cameroon (apart from Mokoko, where unusually it appears common).

Climber, appressed white arachnoid hairy; leaves appearing alternate, whitish-brown tomentose below, thinly leathery, c. 22 × 11 cm, subacuminate, truncate-highly oblique, nerves 14; petiole 3 cm; flowers numerous in sessile capitula 1 cm diam.

Habitat: lowland evergreen forest; up to 1000 m alt.

Threats: forest clearance for agriculture and wood (Mokoko FR) or due to urbanisation (Libreville).

Management suggestions: this bizarre *Rubiaceae* (appearing to have alternate leaves) is, by the number of collections, particularly common at Mokoko FR in the western foothills of Mt Cameroon. This would therefore be the logical site to concentrate on the protection of this species, were it not for the threat of its deforestation for plantation expansion (Cable & Cheek 1998: xxxv). Ideally, a survey of all the known sites to quantify subpopulations, regeneration and vulnerability, is suggested.

Pseudosabicea medusula (K.Schum.) N.Hallé
VU A3c

Range: Cameroon (SW: Mokoko (3 coll.), Mamfe (2 coll.), Baduma. Littoral: Douala; Lake Tissongo. S: Kribi-Lolodorf; Bipindi; Ébolowa-Lolodorf; Ébolowa-Nkondo (2 coll.). E: Dja) and Gabon (Minkebe).

Known from 12 locations widely scattered in the evergreen coastal forest belt of Cameroon and Gabon (above, AOO 48 km^2 with 4 km^2 cells) with threats as below, an assessment of VU is warranted since it is here estimated that population reduction of 30% has occurred or is likely. Five of the 18 records of this species are in the Mokoko Forest reserve which has been proposed for logging, and is about to be explored by drilling for oil production. Several other locations are unlikely to suport the species now, being Cocoa plantations (Kribi-Lolodorf) or roadside areas likely to be cleared for farms (Baduma). The assessment under criterion B in Cheek *et al.* (2004: 179), has been modified here since more than 10 locations are now known. The preponderance of collections are from the south-eastern part of the range.

Creeping herb, densely brown pubescent; leaves appearing alternate, brownish-white below, elliptic, c. 10 × 5 cm, acuminate, unequally cordate, nerves c. 10; petiole 2 cm; flowers c. 5, in sessile capitula, 1.5 cm diam.

Habitat: lowland evergreen forest; 0–200 m alt.

Threats: forest clearance for agriculture or wood (e.g. Mokoko; Cable & Cheek 1998: xxxv).

Management suggestions: this creeper seems most common (6 coll.) in Littoral/S (Douala-Edéa-Ébolowa-Lolodorf-Bipindi) and conservation efforts are presumably best focussed here.

Pseudosabicea pedicellata (Wernham) N.Hallé
VU B2ab(iii)

Range: SE Nigeria (Afi R, Orem, Okwangwo, Boshi, Oban) and Cameroon (SW: Korup-Fabe Rd, Kumba-Mamfe Rd (1 coll. each)).

A scandent shrub apparently more common in SE Nigeria than adjoining Cameroon, this species grows in a particularly threatened habitat. Only seven locations are known (above, AOO 28 km^2 with 4 km^2 cells) with threats as below, warranting VU status. This assessment is taken from Cheek *et al.* (2004: 179), which is maintained here since no further data are available.

Scandent shrub, 3 m, densely grey silky hairy; leaves brownish-white below, ovate-oblong, 13 × 7 cm, acuminate, truncate, nerves 14; petiole 1 cm; flowers c. 30 in capitula; peduncles 5 cm; pedicels 1.5 cm; corolla tube green, 7 mm.

Habitat: lowland evergreen forest; c. 400 m alt.

Threats: clearance of forest for agriculture and wood; this has been particularly prevalent in Nigeria in recent decades, although several collection sites of this species are from protected areas. However, those in Cameroon appear unprotected.

Management suggestions: the status of this plant in Nigeria needs surveying, in particular since historic collections are concentrated there. Quantitative evaluation of the size of subpopulations, regeneration and threats would help in the formulation of a management plan.

Pseudosabicea sthenula N.Hallé
VU B2ab(iii)

Range: Cameroon (C: Mefou Proposed NP) and Gabon (Akoga; 37 km SW Makokou; Mimongo-Mbigou; Lopé NP; Boyan-Oyan; Koumameyong; Minkébé; St. Germain).

Here *Pseudosabicea sthenula* is assessed as Vulnerable since 10 locations (above, AOO 40 km^2 with 4 km^2 cells) are known with threats as below. This assessment appears also in Cheek *et al.* 2011.

Prostrate, rooting herb, internodes 5–8 cm, long red patent hairy; leaves apparently alternate (opposing leaf highly reduced); stipule triangule, 8–9 – 3mm; petiole 3cm, hairs as stem; leaf-blade elliptic, drying dark brown above, with large inconspicuous simple hairs, lower surface white with dark brown veinlets with brown long simple hairs, lateral nerves c. 12 pairs; inflorescences axillary, globose dense, 1 cm diam.

Habitat: lowland evergreen forest; 710m alt.

Threats: road clearance and widening (Mefou) since the site is along an overgrown sideroad which if cleared might damage the population.

Management suggestions: official recognition of Mefou as a NP would safeguard Cameroon's only known population of this species. At Mefou the species is locally abundant and so long as it is protected and monitored, if likely to persist.

Psychotria has more African species than any other genus in *Rubiaceae.* With the exception of sect. *Involucratae* Petit & Verdc., which still needs attention, the genus was revised by Petit (Bull. Jard. Bot. Brux 34: 1–229 (1964)) with those species with bacterial nodules, revised in 1966 (Bull. Jard. Bot. Brux 36: 65–190). In the intervening decades very little has been published on the genus in the Gulf of Guinea area apart from a number of papers revising some species groups and/or describing new species, most of which are threatened: Cheek, M. & Csiba, L. (2002). A revision of the *Psychotria chalconeura* complex (*Rubiaceae*) in Guineo-Congolian Africa. Kew Bull. 57: 375–387; Cheek, M. & Bridson, D. (2002). Two new species of *Psychotria* (Rubiaceae) from western Cameroon. Kew Bull. 57: 389–395; Cheek, M. & Sonké, B. (2005). Two further new species of *Psychotria* (*Rubiaceae*) from western Cameroon. Kew Bull. 60: 293–300; Cheek, M., Horwath, A., Haynes, D. (2008). *Psychotria kupensis* (*Rubiaceae*) a new dwarf, litter-gathering species from western Cameroon. Kew Bull. 63: 243–246; Cheek, M., Corcoran, M. & Horwath, A. (2008/09). Four new submontane species of *Psychotria* (*Rubiaceae*) with bacterial nodules from western Cameroon. Kew Bull. 63: 405–418; Séné, O. & Cheek, M. (2010). *Psychotria monensis* sp. nov. (Rubiaceae) a tree species from Cameroon. Nordic J. Bot. 28(2): 130–133.

Several more papers are in press. Lachenaud has been active in the taxonomy of this genus in Africa for some years and publications from him are expected with further new species, new taxon delimitations and with range extensions of known taxa which will necessitate modification of many of the assessments given below. The genus remains incompletely sampled throughout the Lower Guinea region. Many specimens from the region remain unidentified.

Psychotria are recognised by a terminal inflorescence, 4-or 5-merous flowers, with usually included stamens; berries with two coffee-like seeds often with longitudinal invaginations leading to ridges in the dried fruit, lack of a pre-formed germination slit in the endocarp and stipules which are often bifid.

Allied bombing of the Berlin herbarium in 1943 destroyed most of its specimens including the types of many *Psychotria* species that had been described by German authors from specimens made by German collectors in Kamerun. Petit listed several of these as incompletely known due to complete loss of material (1964, 1966 op. cit.) since their descriptions are not precise enough to attribute them with certainty. Some are probably synonyms of accepted species but others may be narrowly endemic species now extinct or awaiting rediscovery. For Cameroon these are (Petit 1964):

Psychotria aemulans *K.Schum.* Grand Batanga 1890 *Dinklage* 845

Psychotria alluviorum *K.Krause* Elabi near Ilendi *Ledermann* 660

Psychotria bangweana *K.Schum.* Bangwe *Conrau* 202

Psychotria barensis *K.Krause* Bare *Ledermann* 5397

Psychotria batangana *K.Schum.* Batanga *Dinklage* 1068 + 1317

Psychotria ceratalabastrum *K.Schum.* Mt Cameroon, Victoria-Bimbia, *Preuss* 1044; Kebo, *Conrau* 211.

Psychotria dimorphophylla *K.Schum.* Batanga, Lobe, *Dinklage* 1242

Psychotria dodoensis *K.Krause* Djairo Ptuju to Dodo *Ledermann* 5397

Psychotria erythropus *K.Schum.* Buea, Mt Cameroon *Preuss* 1044.

Psychotria hypsophila *K.Krause* W Buea *Preuss* 1071.

Psychotria ilendensis *K.Krause* Ilende *Ledermann* 600; Elabi *Ledermann* 612.

Psychotria leucocentron *K.Schum.* Near Edéa on Sanaga R *Preuss* 1358.

Psychotria nebulosa *K.Krause* Ndonge near Nlonako *Ledermann* 6276.

Psychotria oligocarpa *K.Schum.* Batanga, *Dinklage* 1408.

Psychotria perbrevis *K.Schum.* Mungo, *Bucholz.* s.n.

Psychotria trichanthera *K.Schum.* Bipindi near Lokundje and Macao, *Zenker* 2136.

Tchouto (2004) has applied some of these names to new material and Lachenaud has also applied some to specimens, but these need to be justified in print and neotypified if they are to be accepted.

Psychotria articulata (Hiern) Petit
VU B2ab(iii)

Range: Nigeria (Lagos; Epe; Shasha FR; Okomu FR; Sapoba; Ibeji Okojo NA FR; Aliya Village, Bouche Distr.) and Cameroon (SW: Batanga. Littoral: Lake Tissongo).
Shrub 5 m or tree to 10 m tall, young stems pubescent; leaves purple, pubescent, obovate or elliptic 6–18 × 2.5–9 cm, lateral nerves 10–16 on each side of the midrib; petiole 0.5–3.5 cm long; stipule obovate-oblong, acuminate, entire, 8–17 mm long, sparsely pubescent. Panicles 10–20 cm long, sparsely puberulent to glabrous, peducle 4–10 cm long. Flowers white, subsessile, 5-merous, corolla tube 3–4 mm long.
Psychotria articulata is here assessed as Vulnerable: nine locations are known (above; AOO 36 km² using 4 km² cells) with threats as below.
Habitat: lowland evergreen swamp forest.
Threats: seven of the nine locations are in Nigeria which has seen immense loss of forest over the last century and it is very likely that the species no longer survives at some of these locations, such as Lagos.
Management suggestions: one of the most secure locations for *P. articulata* is that at the remote Lake Tissongo in the Douala-Edéa forest reserve of Cameroon. Here a survey to determine the extent of the population should be made, regeneration levels evaluated and a management plan made to protect this species.

Psychotria babatwoensis Cheek (syn. *Psychotria* sp. A of Bali Ngemba Checklist, Cheek in Harvey *et al.* (2004: 123)).
EN B2a,b(iii)
Range: Cameroon endemic (SW: Lebialem Highlands at Fosimondi. NW: Bali Ngemba FR and Baba 2 community forest).
Shrub 1–2 m tall, glabrous. Leaf-blade elliptic, 7.1–15.4 × 2.3–8.3 cm, secondary nerves (8–)9–12 on each side of the midrib, bacterial nodules dull black, visible on lower surface only, increasingly conspicuous as the leaves age, being often undetectable in the distal leaf pair; nodules irregularly elliptic, c. 0.2–0.5 mm long, almost entirely in archipelago-like clusters 3–8(–20) mm wide, 2–7 clusters per cm², or forming bands of chains, especially along the nerves; the elliptic nodules sometimes uniting, forming lines 3–8 mm long. Inflorescence erect, paniculate, 2–8 × 2–6 cm, 20–70-flowered, peduncle 1–4 cm long, terete, flowers white, 3–4 mm long, 4-merous. Infructescence pendulous, 2-winged. Fruit red when mature, ellipsoid 1.3 × 0.9 cm.
Psychotria babatwoensis is restricted to two, almost adjacent, small patches of submontane forest in the Bamenda Highlands of Cameroon and a single collection in the Lebialem Highlands. The first three collections were made in the Bali Ngemba FR, SW of Bali. This forest is about 10 km² in extent and the understorey, the habitat of this species, is being steadily cleared for the planting of crops such as cocoyams (*Colocasia esculenta*). At this location the species was fairly rare (Cheek pers. obs.). Four other collections were made from the very much smaller community forest of the village of Baba 2, perhaps 2 km² in extent, probably the stronghold of the species, where it is relatively common. The species has newly been found at Lebialem (Cheek in Harvey *et al.* 2010: 139), but has not been found in any of the few other surviving submontane forest fragments in the Bamenda Highlands, such as those of Dom or Mbiame, but is to be hoped that in future it will be located elsewhere. Since the area of occupancy of this species at the two Bamenda Highland locations can be no more than 12 km² and there is habitat loss due to past, ongoing, and expected future clearance at Bali Ngemba FR, *P. babatwoensis* was assessed when first published (Cheek et al. op. cit. 2009) as Endangered. This assessment is maintained here.
Habitat. "Submontane forest with *Pterygota mildbraedii*" (Cheek in Harvey et al. 2004: 17); 1310–1750 m alt. *Psychotria babatwoensis* is one of the species characterising the above vegetation type, accompanying species being *Alangium chinense, Kigelia africana, Carapa grandiflora, Turreanthus africanus, Strombosia scheffleri, Euclinia longiflora* and *Chlorophytum sparsiflorum.*
Threats: see above.
Management suggestions: conservation efforts should be concentrated with the traditional authorities of the communities at the only known locations for this species, Fosimondi, Baba 2 and those around Bali Ngemba FR since these hold the future of the species in their hands. These authorities should be provided with the information needed to identify and protect this species. Data on regeneration levels and population size is needed to develop a plan to manage the species.

Psychotria bakossiensis Cheek & Sonké (syn. *Psychotria sp A aff. gabonica* Hiern, in Cheek *et al.* 2004: 384).
EN B2a,b(iii)
Range: Cameroon endemic (SW: Mt Kupe & the Bakossi Mts).
Shrub 2–5 m tall, fertile stems pale brownish white, distal-most internode drying black, glabrous. Leaf-blade lacking bacterial nodules, obovate or elliptic, 11–16 × 4.2–7.2 cm, acumen 0.3–0.5 cm long, secondary nerves (6–)7–8(–9) on each side of the midrib. Stipules sheathing, soon detaching, and then often persisting for 2–3 nodes below the stem apex, 6–10 mm long, sheath cylindrical, 2–3.5 mm long, 6.5 mm wide, limb triangular, slightly reflexed, c. 5 mm wide, apex entire. Inflorescence sessile, capitate, c. 1 × 1 cm, densely c. 8–9-flowered; involucral bracts in two sheathing pairs, leafy, leathery. Flowers with corolla white, 6mm long, hermaphrodite, (5–)6-merous. Calyx lobes 6, narrowly triangular, 2.5 × 0.8 mm. Fruit bright glossy red when ripe,

drying black, with irregular pale brown corky blotches, ovoid, 20–25 x 12–13(–15) mm including the calyx, c. 10-ridged, the ridges longitudinal.

Of the 12 specimens known, two were located at Kodmin in the Bakossi Mts, 10 arise from Mt Kupe, seven of which were collected above Nyasoso on the west side of the mountain, the remainder on the south side, above Kupe Village. While forest at Mt Kupe seems fairly secure at the altitudes at which the species occurs, that at Kodmin is uncertain, since there are plans for a small reservoir, construction of which could cause a decline in forest habitat. For this reason, and the fact that only three locations are known for the species, *P. bakossiensis* was assessed as EN B2ab(iii), when first published in Cheek & Sonké 2005. That assessment is maintained here since no new data are available.

The preponderance of specimens at Nyasoso may be partly a reflection of the high collecting intensity at that locality, but since the other two locations mentioned have also been fairly intensively collected, it may also reflect that the species is genuinely most common at Nyasoso. Several other Bakossi localities at which plant survey teams have been based e.g. should this be Tape Etube, Ngomboku and Enyangdong, have not yielded any specimens at all, suggesting that it is patchy in distribution.

Habitat: submontane evergreen forest; 1100–1550(–1650) m alt.

Threats: see above.

Management suggestions: conservation efforts should be at Kupe village and Nyasoso where the species is concentrated. The village elders should be given the means to identify and manage this species. Further data on regeneration levels are required. A protected area on Mt Kupe has been proposed but not enacted.

Psychotria bimbiensis Bridson & Cheek (syn. *Psychotria sp. nov. 1 aff. dorotheae* Wernham; *Psychotria sp. nov.* 1 sensu Cheek *et al.*)

CR A4b,c

Range: Cameroon endemic (SW: foothills of Mt Cameroon and Banyang Mbo Wildlife Sanctuary).

Psychotria bimbiensis is distinctive amongst African species in the genus in being a treelet with enormous leaves, possibly the largest in the genus in Africa. On the evidence of the stipule shape (usually ovate-triangular with an entire apex), the orange-tinged (if not exactly yellow) flowers, and the seeds lacking conspicuous ribs, this species falls in sect. *Flaviflorae* Petit (Petit 1964: 41). This species is a small tree or large shrub 2–5 m tall, the leaves are 29–43 × 12–27 cm, apex rounded or shortly acuminate, base obtuse to acute, margin very shallowly sinuate, nerves 9–11 on each side of the midrib, leathery, drying greyish blue, glabrous apart from the puberulent midrib on the lower surface; petioles triangular in section, 7.5–10.5 × 0.2–0.4 cm, glabrous; stipules ovate-triangular to obovate, 1.5 × 0.9 cm, apex rounded or acuminate, not bifurcate, glabrous, caducous.

Psychotria bimbiensis had been allocated the IUCN rating CR A1c+2c (Cable & Cheek 1998: lx as *Psychotria sp. nov. 1 aff. dorotheae*), meaning Critically Endangered, on the basis of the threat to the foothills of Mt. Cameroon to which, on the basis of available evidence, it appears restricted, apart from the location at the Banyang Mbo Wildlife Sanctuary near Nguti. That assessment was maintained when the species was formally described (Cheek & Bridson 2002 op. cit.) and is modified here to CR A4c, since the current position is that the losses of habitat in the last 10 years and those in the next 90, are likely to amount to more than 80%. While no remote sensing data are available, recent observations at Bonadikombe-Bimbia (Mabeta-Moliwe) forest in which this species seems concentrated report very heavy losses of habitat to agriculture (David Alicha, Elias Ndive pers. comm. 2008). Almost all of this forest stands on land reserved for plantation expansion it is believed, although a more immediate threat is forest clearance from small-holder agriculturalists from the nearby town of Limbe (Cable & Cheek 1998: xxxvi).

Habitat: disturbed evergreen forest on old volcanic soils with (Mt. Cameroon) *Lophira (Ochnaceae), Desbordesia (Irvingiaceae)* and *Strombosia (Olacaceae);* 80–100 m alt.

Threats: see above.

Management suggestions: efforts in 2009 to refind this species in forest at Bimbia where specimen evidence suggest it is most frequent have not yet succeeded. More effort is needed by botanists at Limbe BG to work with the managers of the Bimbia community forest to find and protect this rare species, if it has not already been lost due to forest clearance in the area in recent years. If live seeds could be obtained, plants could be multiplied at the nursery of Limbe BG for re-introduction at Bimbia and also maintained in cultivation and distributed as a safeguard against further losses in the wild.

Psychotria camerunensis Petit

VU B2ab(iii)

Range: Cameroon endemic (SW: Mt Cameroon–S Bakundu, Mt Etinde and Mokoko; Bakossi. C: N'kolbison, Yaoundé, Minka on Nyong R. S: Bipindi).

This bacterial-noduled taxon was described by Petit in 1966 from eight specimens centred around three towns in the evergreen forest area of Cameroon: Kumba, Yaoundé and Bipindi. More specimens have since come to light, e.g. from surveys at Mt Cameroon. Several other specimens have also been attributed to this taxon since, but erroneously so since they lack the diagnostic densely brown pubescent stems and obovate leaves that dry brown. Forest clearance around Yaoundé and Tombel has been extensive and is continuing. This species was assessed in Cheek *et al.* (2004: 179) as Vulnerable, since there were less than 10 locations, and an area of occupancy less than 2000 km^2, with threats. This assessment for *P. camerunensis* is maintained here since although numbers of locations has risen (see range above), these still total only eight (see map).

Habitat: evergreen forest understorey; 150–1700 m alt.

Threats: clearance of forest for urban expansion (Yaoundé area) and cultivation of crops (Tombel and Yaoundé).

Management suggestions: an attempt should be made to rediscover this species in the wild in the Bakossi area since it has not been seen since 1986 despite intensive collection there since 1995. It should also be looked for in the Bipindi area, and threats there assessed, since currently these are unknown.

Psychotria darwiniana Cheek (*Psychotria* sp. aff. *alatipes* Wernham, Cheek *et al.* 2004: 382).
VU B2ab(iii)

Range: Cameroon endemic (SW: Mt Kupe; Bakossi Mts. Littoral: Ekomane)

Shrub 1–2.5 m tall, glabrous. Stems terete, fistular, 3–4 mm diam. Leaf-blade elliptic, 6.5–19.8 × 3.2–10.1 cm, acumen (0.3–)1–1.4(–1.5) cm long, base acute-cuneate, often asymmetric and slightly decurrent, secondary nerves 11–13 on each side of the midrib, bacterial nodules black, conspicuous on lower leaf surface only, 3–4-lobed, ink stain-like, longest diam. 1–3.4 mm. Inflorescence paniculate, erect 10.5–21.4 × 3–5 cm, c. 100-flowered, peduncle 3-winged, branches in whorls of 3, flowers white, 4-merous, 1.6–2.2 mm long. Fruit red, globose, 3–6 mm diam.

Psychotria darwiniana is currently known from 33 specimens and appears to be relatively common in some parts of its range on Mt Kupe being collected at almost all of the localities botanised within its altitudinal range. It is also known from the Bakossi Mts, near the villages of Edib, Kodmin and Muambong and one collection is known from Mt Ekomane, just to the N of the Bakossi Mts. It may in future be found at the Rumpi Hills to the W, and the Nyale Plateau to the N. The probable maximum area of occupancy of *P. darwiniana* is about 800 km^2 (see under *P. ngollengollei*). Threats are as below. Accordingly when first described (Cheek *et al.* op. cit. 2009) the species was assessed as Vulnerable (above) and this assessment is maintained here since no additional data are available.

Habitat: lower submontane evergreen forest; 870–1300(–1550) m.

Threats: loss of its lower submontane forest habitat is ongoing at the lower edge of the altitudinal range (Cheek pers. obs.).

Management suggestions: Mt Kupe has been proposed as an Integrated Ecological Reserve, Cameroon's highest ranking category of protection. However it is not known when this proposal will come into effect. Further data on regeneration levels for this species are desirable. Involvement of village communities around Mt Kupe in protecting this and the many other rare species in its forests is advisable. Production of a poster illustrating this species and explaining its importance for conservation is suggested.

Psychotria densinervia (K.Krause) Verdc.
EN B2ab(iii)

Range: Cameroon endemic (S: Bipindi–Mimfia. SW: Bakossi (possibly also at Kagwene)).

This 6 m tree is remarkable for the pendulous, cord-like peduncles several metres long which bear fist-sized, globular inflorescences. Only the much commoner *P. camptopus* is likely to be confused with it (differences are that in *P. densinervia* lateral nerve pairs are 12–15, involucral bracts are pubescent). First described from a single specimen from the NW slopes of the Mimfia Mts of Bipindi (*Zenker* 4683) it was not recollected until the 1950s, in Mungo Ndaw, western Bakossi (*Dundas* in FHI 15324). A further specimen was found in 1996 near Kupe village at the base of Kupe Rock (*Ryan* 225). Lowland forest in these areas is generally under threat of clearance although the site at Kupe Rock seems reasonably secure at present. In summary, only three locations are known, at each of which no more than a single tree was recorded. The above assessment was made in Cheek *et al.* (2004: 179–180) and is maintained here. However the species may also occur at Kagwene since there are field identifications of specimens from that location. If confirmed the conservation rating will be unchanged since total locations will rise to four.

Habitat: lowland evergreen forest; up to 1000 m alt.

Threats: clearance for agriculture.

Management suggestions: it is recommended that attempts be made to rediscover this rare species at Bakossi and to educate local villages as to the importance of its conservation. See also the comments made for *P. minimicalyx*. It is assumed that neither of the Bakossi sites can be formally protected due to their low altitude. Specimens from Kagwene should be prioritised for final identification, if confirmed, this area offers good prospects for the species since it has some level of protection.

Psychotria elephantina Lachenaud & Cheek ined.
EN B2ab(iii)

Range: Cameroon endemic (SW: Mt Cameroon western foothills at Onge FR; Mokoko FR; Bonjare; Boa).

Shrub, probably sparsely branched or unbranched, 1–2.5 m tall, glabrous. Leaves elliptic or oblong-elliptic, 22.4–36 × 12.2–19 cm, acumen 0.5–1 cm long, lateral nerves 17–20 pairs. Stipules not sheathing, lacking appendages or ridges, oblong-elliptic or ovate, 3.5–6 × c. 2.5 cm, apex entire, not or very weakly bifid. Inflorescence capitate, terminal, c. 3.5

× 4.5 cm, 60–90-flowered; peduncle c. 2 cm long, terete. Involucral bracts two pairs, free, enclosing the inflorescence; outer bracts broadly ovate-triangular, c. 1 × 2 cm, apex rounded, inner bracts similar, c. 1.5 × 1.5 cm. Base of involucral chamber flat, reticulate, elliptic, c. 2.5 × 1 cm, being c. 20 evenly scattered partial-peduncles. Partial-peduncles c. 5 mm long, erect, 3–5-flowered, bracts inconspicuous. Flowers 5-merous. Corolla tube spindle-shaped just before anthesis, c. 7 mm long, 2 mm wide at midpoint, c. 1 mm wide near base and apex; apex 5-grooved, corresponding with the lobes; corolla lobes triangular, c. 1 × 2 mm, apex hooded, with a downward pointing apical protrusion c. 3 mm long, weakly keeled, appendage absent. Fruits fleshy, white; oblong when dried and flattened, c. 1.7 × 1.1 cm.

Psychotria elephantina is assessed in the journal in which it is to be published as Endangered (EN B2ab(iii)), since it is known from four locations with an area of occupancy of only 16 km² with 4 km² cells. In addition, its natural habitat, the forest of the western foothills of Mt Cameroon, is under great pressure for clearance in order to expand commercial plantations of oil palm and other tropical crops. Indeed at some of its few known locations it may already be extinct.

Habitat: lowland evergreen forest, rainfall 3–4 m p.a.

Threats: see above

Management suggestions: consideration should be given to raising the status of either or both of Onge and Mokoko from FR to National Park status given their importance in hosting so many unique plant species among which is *P. elephantina*. More data is needed on the numbers of individuals, density and regeneration levels of this species to inform management planning.

Psychotria geophylax Cheek & Sonké (syn. *Psychotria sp. B aff. gabonica* Hiern, in Cheek *et al.* 2004: 384).

VU B2a,b(iii)

Range: Cameroon endemic (SW: Etinde at Mt Cameroon; Mt Kupe & the Bakossi Mts).

Shrub 2–5(–6) m tall, stems and leaves glabrous. Leaf-blade lacking bacterial nodules, obovate or elliptic, 17.5–31 × 8–12.5 cm, acumen 0.3–0.6 cm, base acute-decurrent, secondary nerves 12–14 on each side of the midrib. Stipules ovate or elliptic, free, green when live, 18–35 × 15–20 mm on stem below inflorescence, to c. 13 mm long on subsidiary stems, persisting for two nodes below the inflorescence, apex slightly acuminate or acute, entire or bifid by up to 4 mm, outer surface glabrous. Inflorescence sessile, densely capitate, 3–4.5 cm diam., c. 500-flowered, only 2–3 of which are at anthesis at one time (Cheek pers. obs.), peduncle and rachis highly reduced, inconspicuous. Flowers hermaphrodite, 5-merous. Calyx matt brown, subcylindrical, c. 13 × 3 mm long, lobes oblong-ligulate, 6 × 1.5 mm. Fruit leathery, dull orange, ovoid, 17–25 × 7–10mm including the foliaceous calyx.

This taxon was assessed as VU B2a,b(iii), using the criteria of IUCN (2001) when first published (Cheek & Sonké (2005)). Although 23 specimens were recorded of *P. geophylax*, it is only known from eight localities, at several of which (e.g. Bakole Bakossi and Mt Etinde) its forest habitat is believed to be threatened with clearance for agriculture. This assessment is maintained here since no further data are available.

Local populations of this species appear to vary greatly in the numbers of individuals. In comparing two sites which have received broadly similar levels of collecting activity, Mt Etinde and Nyasoso, it is noteworthy that whereas only two specimens are recorded at the first, eleven are known at the second. Within the Kupe-Bakossi area, the stronghold of this species, the distribution of *P. geophylax* is extremely patchy. Although it is evidently common above Nyasoso, it has not been recorded at all at other sites in the area, such as Kodmin, Ngomboku and Muambong, which have also been subjected to fairly intensive botanical survey by the same set of collectors. Since the species appears to be fertile most of the year (it has been collected in every month bar April, September and December, usually with fruit), this cannot be attributed to collections at these sites being at times inappropriate to record presence of the species, but rather to real absences. Consequently it is possible to be fairly precise about the distribution of the species: it is known only from the western and southern slopes of Mt Kupe, where it is common, and from the western foothills of the Bakossi Mts where it is rarer, and from Mt Etinde at Mt Cameroon, where it is scarcer still.

Habitat: lowland to submontane evergreen forest; (300–)700–1500(–1600) m alt.

Threats: see above. The habitat at the lower end of the altitudinal range is particularly threatened by conversion to small-scale farms.

Management suggestions: conservation efforts should be focussed at Mt Kupe, where individuals of this species are concentrated. So long as forest here is secure, so will be *Psychotria geophylax.*

Psychotria humilis Hiern var. ***humilis***

VU B2ab(iii)

Range: Nigeria (Benin, Okomu FR; Usonigbe FR new Sapoba; Oban), Cameroon (Mt Cameroon at Onge and S Bakundu) and Gabon (Corisco Bay).

A tiny, decumbent subshrub 5–15 cm tall, pubescent to glabrous. The leaves are ovate to obovate, 2–8 × 1–3.5 cm, acute-acuminate those of the type with white, crinkled hairs on the upper surface. The bacterial nodules are maculate to elongated, numerous. The panicle is only c. 1 cm long,

glabrous, slightly winged, with 4-merous flowers and appendaged petal lobes.
Petit treated this taxon as one of three varieties but the case could be made to give each of these species rank.
Six locations (AOO 24 km² at 4 km² cells) are known so this taxon is here assessed as Vulnerable in view of the threats listed below.
Habitat: lowland evergreen forest.
Threats: forest habitat in Nigeria has seen massive losses, 89% of the original being destroyed of which c. 36% in the 15 years 1990–2005 (Mongabay.com). While losses in Cameroon have not been so extreme, S Bakundu has also seen great forest loss and the future of the Onge forest is under threat from plantation expansion.
Management suggestions: further work to evaluate the extent and regeneration of the population around Mt Cameroon is desirable for forming a management plan and to allocate resources for protection. If loss of the species at any of the Nigerian locations is confirmed, EN status will be merited.

Psychotria kupensis Cheek (syn. *Psychotria* sp. aff. *foliosa* Hiern (Cheek *et al.* 2004: 384)
EN B2ab(iii)
Range: Cameroon endemic (SW: Mt Kupe).
Erect, monopodial, litter-gathering shrublet 0.3–0.5 m tall. Stem grey-black, 5–6 mm diam., matt, glabrous, leafy portion usually densely clothed in a mass of branched adventitious roots. Leaves equal at each node, held at c. 45° from the horizontal, in a funnel-like terminal rosette of c. 12 leaves, each oblanceolate to subspathulate, 17.5–31 × 5.1–9.5 cm, obtuse to acuminate, acumen 0.7–1.2 cm long, base abruptly obtuse to acute, lateral nerves 15–23 on each side of the midrib, lower surface with midrib and secondary nerves black, raised, densely pubescent, hairs patent, hollow, broad-based, c. 0.15 mm long. Petiole 1–2 cm long. Inflorescence capitate; peduncle terminal, longitudinally ridged, erect, robust 7 × 0.2 cm, very densely grey pubescent with patent hairs, as the lower surface of the midrib; capitulum globose, 1.5 cm diam., flowers 100–150, arranged in fascicles of 6–15 on 8–12 branches. Fruits glossy, smooth, fleshy, orange or red when mature, ovoid, 0.6–1.2 × 0.6–0.9 cm.
The area of occupancy was estimated to be less than 10 km², with only two main locations and threats as below. Accordingly *Psychotria kupensis* was assessed as Endangered when first published (Cheek *et al.* op. cit. 2008). This assessment is maintained here, no further data being available.
Habitat: submontane evergreen forest, rainfall 3–4 m per annum; 750–950 m alt.
Threats: although much of Mt Kupe is due to be designated as an Integrated Ecological Reserve, the highest level of protected area in Cameroon (Wild *et al.* pp. 111–116 in Cheek *et. al.* 2004), the two sites for this species, at Ngomboku and Kupe village, are at relatively low altitudes, vulnerable to ongoing forest clearance for small-scale agriculture, and not likely to be completely included in the proposed Integrated Ecological Reserve.
Management suggestions: conservation efforts should be focussed with the elders of Kupe village where individuals of this species appear to be concentrated. The elders should be provided with the means to identify this species and to protect it for long term survival.

Psychotria lanceifolia K.Schum.
VU B2ab(iii)
Range: Cameroon endemic (SW: Mundemba; Korup; S Bakundu; Barombi Mbo; Bakossi; Mamfe. S: Bipindi-Kribi-Ébolowa; Memel 2; Efoulan. E: Somalomo, Dja)
This taxon was described from several Zenker collections (6 at K) from Bipindi in S Region, made between 1904 and 1912 and otherwise unlocated. It was not re-recorded until Bos (WAG) made at least two collections in 1968 and 1969 at km 19 and 20 from Kribi on the Bipindi Rd. The two specimens listed from two sites in Bakossi (*Manning* 481, *Etuge* 414, MO) have not been seen by us. This is a very distinct, even ornamental, species "small tree about 2.5 m high, slender main stem and gracefully arching, very dark green branches, leaves thinly leathery, smooth, glossy" (*Bos* 3061).
In Cheek *et al.* (2004: 180), *P. lanceifolia* was assessed as VU D2, Vulnerable, on the basis that it was known only from three locations, occupying only a few m^2 at each location, and with the threats given. Since that date further survey work, especially by Sonké and Davis, has extended the range of the species considerably (see above) and now nine locations are known (AOO 36 km² at 4 km² cell size) with threats as below allowing the species to be assessed as Vulnerable under criterion B. In future it is likely that the taxon will be assessed under criterion A.
Habitat: lowland evergreen forest, usually stream-sides; 0–1600 m alt.
Threats: clearance of forest for agriculture (e.g. Mundemba), development and infrastructure projects (e.g. Kribi area).
Management suggestions: the identity of the MO specimens cited needs checking, although this is a very distinct species, not likely to be confused with any other. The collection sites should be resurveyed in order to (a) determine whether the taxon is still extant, (b) decide whether it is indeed rheophytic (restricted to water courses), (c) determine the extent to which it is threatened by human or other factors, and (d) levels of regeneration and numbers of individuals.

Psychotria letouzeyi Petit
VU B2ab(iii)

Range: Cameroon (C: 15 km W of N'kolbison. S: Mvilé (Ngovayang); Bibondi. E: Medjo, 25 km SSW Messaména) and Gabon (Bélinga projected Iron Mine; Mouila-Yeno; 30 km NE Makoukou; NE Mekambo).
Known from the eight locations recorded above (AOO 32 km^2 at 4 km^2 cells) with the threats given below, *P. letouzeyi* (Petit 1966: 163–164) is here assessed as Vulnerable.
A well-branched puberulent shrub c.1 m tall with elliptic, c.10 × 4 cm leaves conspicuously nodulated, the nodules elongated or lobed. The 1–2 cm peduncle terminates in a whorl of 3–4 branches. Remarkably, in the fruit the calyx tube is a c 2 mm long, crowned by 4 longer lobes. In flower the calyx equals the corolla in length.
Habitat: submontane forest; 600–1000 m alt.
Threats: Bélinga is threatened by an iron mine, while hills around Yaoundé are threatened by development and farming.
Management suggestions: not seen since the 1980s, *P. letouzeyi* should be searched for at its former sites and, if found, the extent, density and regeneration levels assessed to aid conservation of a plan to manage its long term survival.

Psychotria microthyrsa E. Petit
CR B2ab(iii)
Range: Cameroon endemic (E: Ngoko R).
Known from a single specimen (AOO 4 km^2 at 4 km^2 cell size), *P. microthyrsa* is here assessed as Critically Endangered in view of the threats given below.
An ascending forest shrublet 30 cm tall, the white stems are completely covered by a layer of short dense dark brown hairs. Internodes are c. 5 cm, leaves elliptic, c. 8 × 4 cm on 1.5–3 cm long petioles with entire ovate-trianglular 2–4 mm long stipules. The 1 cm panicles are more or less winged, glabrous with 4-merous flowers. The lower surface of the leaves is spotted with bacterial nodules. Placed by Petit (1966: 178) next to *P.humilis*.
Habitat: mixed semi-deciduous and Dja congolian forest at c. 400 m alt. (Onana pers. obs.).
Threats: this corner of Cameroon, with neighbouring Congo (Brazzaville), has seen a vast amount of logging activity in recent decades.
Management suggestions: every effort should be made by botanists active in this area to rediscover this species before it is lost. Data on distribution, density and regeneration of the population will enable development of a plan to protect this species.

Psychotria minimicalyx K.Schum.
CR A3c
Range: Cameroon (SW: [illegible] Kumba-Mamfe areas).
Described in 1899 on the [illegible] single specimen (*Staudt* 606) from Kumba, it was rediscovered nearby at Banga, S Bakunda FR in 1948 (*Brenan* 9269). Subsequently it has been found at two sites along the Mungo valley running from Kumba towards Mamfe (*Thomas & Nemba* 5185, Konye, 1985 and *Mambo & Thomas* 64, Mbu, 10 km W of Wone, 1986). It has been confused with *P. bifaria* on account of its similar leaf-shape and size but is distinguished by the terete, evenly densely pubescent stem (not 4-angular, with 2 lines of hairs).
This shrublet was assessed under criterion A as Critically Endangered due to the likelihood of 80% of its habitat being destroyed in the next 30 years (Cheek *et al.* 2004: 180). That assessment is maintained here since no new data are available apart from a field identification by Séné from the Mone forest reserve in Dec. 2008 that extends the range further N than previously known. A generation duration of ten years was estimated but may well be an underestimate since forest shrublets can be very long-lived.
Habitat: lowland evergreen forest; c. 300 m alt.
Threats: clearance of its habitat for agriculture, such as cocoa plantations or food crops, is likely along the Mungo valley owing to the prevalence of fertile soils.
Management suggestions: the lowland forest habitat of *P. minimicalyx* does not lend itself to formal protection in a gazetted area because it grows in fertile soils close to centres of population and lines of communication. Assuming that a new location cannot be found in one of the areas already mooted for protection, it is recommended that an attempt be made to rediscover plants at one of the known locations and to educate the population as to the desirability of protecting it, possibly by using posters or rental agreements.

Psychotria moliwensis Bridson & Cheek (syn. *Psychotria sp. nov. 2 aff. bidentata* (Thunb. ex Roem & K.Schult.) Hiern sensu Cheek *et al.* 1992, Cable & Cheek 1998: 118).
CR A4c
Range: Cameroon endemic (SW: Mt Cameroon at Bimbia-Bonadikombo and Kumba).
Shrub 2 m tall, stems and leaves glabrous, leaves broadly elliptic 19–27 × 5.7–12.5 cm, acumen 1–1.25 cm long, base acute-decurrent, lateral nerves ± 12 main pairs; petioles 3.5–5 cm long, puberulent above. Stipules caducous, ovate-elliptic 1.8–2.5 × 0.9–1.3 cm, deeply bilobed, the lobes subacuminate, ± 1.2 cm long, hairs dense, red-brown, up to 0.1 cm long, conspicuous at the upper margins of the lobes and in a triangular area 0.1 × 0.7 cm at the base of the outer stipule surface, the remainder of the outer surface glabrous. Inflorescences terminal, congested, with numerous flowers forming a dense mass 1.5–1.8 × 1.5–2.0 cm, not fully exserted from the stipules; peduncles stout, up to 2.2 cm [illegible] 0.15 cm wide, sparsely puberulent, primary [illegible] branches 0.4 cm long;
Psychotria moliwensis, as with *P. bimbiensis,* had been allocated the IUCN rating CR A1c+2c (Cable & Cheek 1998: lx as *Psychotria sp. nov. 2 aff. bidentata*), meaning Critically

Endangered, on the basis of the threat to the foothills of Mt. Cameroon to which, on the basis of available evidence, it appears restricted, specifically to the Bimbia-Bonadikombo area. That assessment was maintained when the species was formally described (Cheek & Bridson 2002 op.cit.) and is modified here to CR A4c, since the current position is that the losses of habitat in the last 10 years and those in the next 90, are likely to amount to more than 80%. While no recent remote sensing data are available, recent observations at Bonadikombo-Bimbia (Mabeta-Moliwe) forest the only known location for this species report very heavy losses of habitat to agriculture (David Alicha, Elias Ndive pers. comm. 2008). Almost all of this forest stands on land reserved for plantation expansion it is believed, although a more immediate threat is forest clearance from small-holder agriculturalists from the nearby town of Limbe (Cable & Cheek 1998: xxxvi). The one collection that derives from Limbe was made in 1899, since when that town has expanded greatly, no more collections have been made and it is now considered extinct at that location. A record from Kumba, even more heavily impacted by man, has recently come to light based on a det. by Lachenaud.

Habitat: evergreen forest on old volcanic soils with *Lophira (Ochnaceae), Desbordesia (Irvingiaceae)* and *Strombosia (Olacaceae)*; sea-level–100 m alt.

Threats: the same observations apply to *P. moliwensis* as have already been noted (above) for *P. bimbiensis* excepting that *P. moliwensis* is so far solely known from the Bonadikombo-Bimbia (Mabeta-Moliwe) forest and thus more vulnerable to extinction than *P. bimbiensis*.

Management suggestions: efforts in 2009 to refind this species in forest at Bimbia have not succeeded. More effort is needed by botanists at Limbe Botanic Garden (LBG) to work with the managers of the Bimbia community forest to find and protect this rare species, if it has not already been lost due to forest clearance in the area in recent years. If live seeds could be obtained, plants could be multiplied at the nursery of LBG for re-introduction at Bimbia and also maintained in cultivation and distributed as a safeguard against further losses in the wild.

Psychotria monensis Cheek & Séné

CR D

Range: Cameroon endemic (SW: Mone FR near Mamfe).

Tree 6–9 m tall, trunk cylindrical, c. 15 cm diam. 1.5 m from ground, stems and leaf blades glabrous. Leaf blades leathery oblong or elliptic, 32.5–37 × 14.5–18 cm, nerves 10–11 on each side of the midrib, domatia absent; petioles 3–4 cm, white puberulent; stipules ovate to lanceolate, c. 28 × 15 mm, leathery, apex rounded, not bifurcate, exterior margin minutely hairy, otherwise glabrous. Inflorescence erect, c. 9 × 10 cm, in side view with four well-marked lateral lobes, 500–600 flowered, compact, the flowers held side by side, close together, forming a concave surface; inflorescence axes minutely white pubescent, peduncle c. 2 cm long, stout, terete; rachis with c. 4 nodes. Flowers hermaphrodite, 5–6-merous. Calyx-hypanthium white puberulent, lobes absent or minute. Corolla whitish green at anthesis; tube cylindrical, 5–7 × 3 mm, lobes reflexed, 3–3.5 × 2–3 mm, apices slightly hooded. Infructescence spreading, 100–200-fruited, c. 15 × 20–30 cm; peduncle c. 5.5 cm long, stout. Fruit leathery, red, ellipsoid, 15 × 13 mm, smooth, when live.

In its large leaves and tree habit this species is close to *P. bimbiensis*, differing in its shorter petioles, larger habit, and white puberulent, not glabrous, petiole, inflorescence branches and calyx-hypanthium.

Psychotria monensis, was assessed as CR D, that is, Critically Endangered, since only two individuals are known, both of which are in a production logging reserve and further, one of which is within a farm inside the forest reserve and so could be felled at any time.

Habitat: lowland evergreen forest; 100–200 m alt.

Threats: see above.

Management suggestions: it is to be hoped that more individuals and more sites will be found for this species and also that the Mone FR will be upgraded to full National Park status so as to protect better this species and the many others of conservation importance that occur there.

Psychotria moseskemei Cheek (syn. *Psychotria chalconeura sensu* Petit *pro parte* op. cit.: 60–62 (1964); FWTA ed. 2: 202 (1963), *quoad Maitland* 1744; *Psychotria* sp. nov.? in Cheek *et al.* 2000: 73, 160)

CR A4c

Range: Cameroon Highlands of E Nigeria (Mambilla Division, Chappal Waddi (Gangirwal); Gashaka, Chappal Hendu) and Cameroon (NW: Bamenda Highlands, Dom; Kilum-Ijim (Mt Oku & Ijim Ridge: Boyo, Laikom head of valley; Chubuh; Zitum Rd; Bui; Elak, Upkim forest). W: Bamboutos, Bangou, 10 km W of Bangwa).

Shrub, rarely a small tree, 2–4(–5) m tall, glabrous. Leaf-blades elliptic, 6–14 × 3.5–6 cm, acumen 0.2–0.4 cm, base cuneate, slightly asymmetric, midrib sunken above, prominent below, midrib and secondary nerves pale red both alive and dried, domatia conspicuous, situated on junction of midrib and secondary nerves, circular or transversely elliptic, 0.2–0.5 mm long, glabrous; secondary nerves 8–14 on each side of the midrib.

In its natural habitat *Psychotria moseskemei* is easily confused, when sterile, with *Chassalia laikomensis* (Cheek *pers. obs.*). The two are similar in stature, leaf shape and size, and stipule shape and size. They can grow side by side. *Psychotria moseskemei* can be distinguished by the lack of persistent chaffy stipule bases (the entire stipule falls off leaving a clean scar), the dull red (not green) nerves of the lower leaf surface and the domatia, which are glabrous, circular lacunae c. 0.5 mm in the axils between the midrib and secondary nerves. Domatia are absent in *Chassalia laikomensis*.

This species first came to light as possibly new to science and threatened in Cheek *et al.* 2000: 73 where notes are given but no formal assessment was made. In Cheek & Csiba 2002, *P. moseskemei* was formally published and assessed as CR A1 using the criteria proposed by IUCN (2001) on the basis of a high degree of past habitat loss. The justification being that it had been estimated that as much as 96.5% of the original forest habitat of the Bamenda Highlands had been lost (Cheek *et al.* 2000: 6) and in one study, Moat in Cheek *et al.* (2000: back cover), it had been shown that 25–30% of forest habitat had been lost in the Kilum-Ijim area in the eight years Jan. 1987–Jan. 1995. The earlier assessment, in hindsight, is inappropriate since criterion A1 currently relates to cases where population size reduction (for which habitat loss is a proxy in this case) is reversible and understood and ceased. In fact, habitat loss is ongoing, so the assessment of CR A4c is more appropriate and is made here, it being estimated that more than 80% of the habitat of *P. moseskemei* has been lost over a hundred year time frame and that this loss is ongoing (understorey forest shrubs such as this species can live for more than 40 years, it is believed). A recent study by Baena (in Cheek *et al.* 2010: back cover) of forest loss in the Dom area where *P.moseskemei* was found recently, shows that even higher losses have occurred in more recent years than in the earlier Moat study.

Habitat: understory of submontane and lower montane evergreen forest with *Tabernaemontana* cf. *ventricosa* Hochst. ex A.DC., *Chassalia laikomensis* Cheek, *Psychotria peduncularis* (Salisb.) Steyerm., *Croton macrostachyus* Hochst. ex Delile, *Albizia gummifera* (J.F.Gmel.) C.A.Sm., *Bridelia speciosa* Müll.Arg., *Xymalos monospora* (Har.) Baill. ex Warb., *Coffea liberica* Bull. ex Hiern and *Keetia venosa* (Oliv.) Bridson; 1700–2060 m alt.

Threats: see above.

Management suggestions: currently none of the six known locations for *P.moseskemei* are formally protected at governmental level, but several are in or near areas protected by local communities, such as those at Dom, at Elak (Upkim forest) and near Laikom. More effort is needed to support these communities in protecting this species, for example in providing better identification aids, joint surveys to establish locations and numbers of individuals, and levels of regeneration, so as to help those communities develop a management plan for this Critically Endangered species.

Psychotria ngollengollei Cheek (syn. *Psychotria* sp. aff. *camerunensis* Petit in Cheek *et al.* 2004: 383).
VU B2ab(iii)

Range: Cameroon endemic (SW: Bakossi Mts including Mt Kupe)

Shrub 1–1.5m tall, glabrous. Stems sometimes, when fallen or leaning, rooting adventitiously, producing new vertical shoots and so reproducing apomictically. Stems otherwise erect, solid, terete, but often finely longitudinally ridged, 1–2(–2.5) mm diam., lenticels absent, glabrous. Leaf-blades thickly papery, upper surface drying black, lower surface dull orange-green or brownish green, bacterial nodules conspicuous on lower surface, more or less circular, cone-like, black, raised, 0.15–0.3 mm diam., the centre crater-like, density, blade elliptic, rarely slightly ovate, oblong, or oblanceolate, 5.9–14.7 × 1.4–5.9 cm, domatia absent; secondary nerves orange, (6–)7–12(–13) on each side of the midrib. Inflorescence paniculate, 50–70 × 28–40 mm, flowers 5-merous, white, c. 3.5 mm long. Fruit red when mature, ellipsoid, 0.8–1 × 0.7 cm.

The 15 specimens known occur at seven locations. The extent of submontane forest believed to survive in Kupe-Bakossi, the distribution of the species, is about 800 km², equating to the maximum likely EOO of *Psychotria ngollengollei*. Accordingly the species was assessed as Vulnerable when originally described in Cheek *et al.* op. cit. 2009. That assessment is maintained here, no further data being available.

Habitat: Evergreen submontane forest; (550–)850–1500(–1700) m alt.

Threats: loss of habitat is ongoing at the lower edge of its altitudinal range, i.e. c. 800–900 m alt.(Cheek pers. obs.).

Management suggestions: Mt Kupe has been proposed as an Integrated Ecological Reserve, Cameroon's highest ranking category of protection. However it is not known when this proposal will come into effect. Further data on regeneration levels for this species are desirable. Involvement of village communities around Mt Kupe in protecting this and the many other rare species in its forests is advisable. Production of a poster illustrating this species and explaining its importance for conservation is suggested.

Psychotria njumei Cheek (syn. *Psychotria* sp. A aff. *leptophylla* Hiern (Cheek *et al.* 2004: 385))
CR B1a,b(iii)

Range: Cameroon endemic (SW: Mt Kupe).

This is a 1–5 m tall shrub with cylindric-hollow, dark-brown puberulent stems, 3–4 mm diam.

The leaves are elliptic 14–23 × 6–11 cm, with 13–16 lateral nerves on each side of the midrib and bacterial nodules 0.4–0.7 mm diam. The flowers are 5-merous, producing red globose fruits c. 5 mm diam.

The five known specimens occur at one of two locations, both within a c. 10 km² area near and just to the north of Kupe village, which is sited on the southern slopes of Mt

Kupe. Lachenaud (pers. comm. to Cheek 2007) reports that he has seen specimens from S Cameroon and Gabon that are very close and probably not specifically distinct from *P. njumei*. If confirmed this is likely to reduce the conservation rating for this species.

Psychotria njumei was assessed as Critically Endangered when originally described (Cheek et al. op. cit. 2009) on the basis of the information above. This assessment is maintained here since no new data are available.

Habitat: submontane evergreen forest; 750–1200 m alt.

Threats: these include clearance for small-scale agriculture (Cheek pers. obs.).

Management suggestions: it is to be hoped that the species will be found or confirmed at other locations, so decreasing its conservation threat. Mt Kupe has been proposed as an Integrated Ecological Reserve, Cameroon's highest ranking category of protection. However it is not known when this proposal will come into effect. Further data on regeneration levels for this species is desirable. Involvement of the Kupe village community in protecting this and the many other rare species in its surrounding forests is advisable. Production of a poster illustrating this species and explaining its importance for conservation is suggested.

Psychotria piolampra K.Schum.

VU B2ab(iii)

Range: SE Nigeria (Obubra FR; lyamogong) and Cameroon (SW: Mamfe; Boa at Mt Cameroon; Mokoko; Kumba. Littoral: Dschang-Mbo Plains).

First discovered in Kumba in the late 19th Century, *P. piolampra* is a glabrous shrub 20–60 cm tall. The leaves are elliptic to obovate 11–21 × 4–11 cm, acuminate, cuneate; lateral nerves 7–9 on each side of the midrib, lacking domatia. Stipules triangular-ovate, 4 mm long, entire, glabrous. Panicles 1.5–3 cm long. Flowers 1 mm long, white with orange calyx.

Although *Psychotria piolampra* occurs at Mt Cameroon, it was overlooked in Cable & Cheek (1998) and is assessed here, for the first time, as Vulnerable. Six locations (AOO 24 km^2 at 4 km^2 cells) are known (above). See threats below.

Habitat: lowland evergreen forest.

Threats: since the type specimen was gathered at Kumba in the 19th Century, that settlement has expanded into a vast town and so much forest habitat in the area has been lost – the species may no longer survive there. The records from Mokoko and Boa date from the 1990s, but both locations are threatened with plantation expansion. Forest in Nigeria has seen massive losses in recent years. With better forest loss data this species may be reassessed as EN or CR.

Management suggestions: the Mt Cameroon area has half the known sites for *P. piolampra* including all the most recent records. It is logical to seek to focus protection of this species here. Further data on population distribution and regeneration is needed to develop a management plan to this end.

Psychotria podocarpa Petit

VU B2ab(iii)

Range: Nigeria (1 coll.) and Cameroon (SW: Mt Cameroon at Njonji, Bakingili, Bomana Bakweri; Bakossi Mts: at Ngomboku, Nyale, Konye, Enyandong; NNW Nguti. NW: Nkom-Wum, Bu)

This shrub is highly attractive in fruit and has great potential as a pot plant. The profuse glossy black, pendulous fruits are held on contrasting bright red, carrot-shaped pedicels. When first described in 1964 by Petit, it was known from only a single Nigerian collection but in the early 1990s it was found at Bomana Bakweri and Njonji, both at the foot of Mt Cameroon, and since then sites have been found at Bakossi, namely Ngomboku, Nyale and Konye. The first two of these seem secure sites for this species (*pers. obs.*). A total of six locations are known; where site observations have been made, it is known to occupy only 1–2 m^2. Threats are given below. The preceeding assessment of Vulnerable was made in Cheek *et al.* (2004: 180–181) and is maintained here. Although numbers of specimens have risen to 15 and number of locations to nine, this is still within the threshold of the earlier rating. It is likely that in future this species will be reassessed under criterion A since forest habitat at so many of its sites is under threat.

Shrub 0.3–4 m, glabrescent, resembling *P. latistipula*, but fruiting pedicel 1.2 cm, fleshy, bright red; fruits black, bracteoles ovate, foliaceous, 7 mm.

Habitat: lowland evergreen forest; 300–800 m alt.

Threats: clearance of forest for agriculture especially in the Mt Cameroon area, where the planned expansion of plantations is likely to destroy the subpopulations listed above.

Management suggestions: a survey of the subpopulations present at each of the sites is desirable. At the Njonji site, hundreds of individuals were recorded in a small area, but this has not been noted at other sites.

Psychotria rubripilis K.Schum.

VU B2ab(iii)

Range: Cameroon (SW: Mt Cameroon, 30 km WNW Muyuka. E: Zilly, 4 km S Nguélémendouka. S: Bipindi; Yaoundé-Bitye), Gabon (Lebamba) and Congo (Kinshasha) (Dist. Du Bas-Congo: Kipako, Wulu; Mputu; Sanda).

With eight locations (AOO 32 km^2 at 4 km^2 cell size) and the threats identified beow, we here assess *P. rubripilis* as Vulnerable.

While widely distributed this species is rare within its range. Notably it appears not to have been seen since 1976, despite much collection activity in the last two decades.

Distinctive among the bacteria-noduled species (densely black dotted on the leaves) in the moderately dense long red hairs on all surfaces. The leaves are oblanceolate, c. 10 × 4cm, the base rounded with a 0.5–1 cm petiole. The c. 1 cm panicle is dense, subcapitate, sessile, with 5–10 5-merous flowers, the calyx lobes are 2 mm long, becoming leafy in the fruit. A shrub 0.2–1 m tall.

Habitat: submontane (extending to lowland?) evergreen forest.

Threats: its site on Mt Cameroon is threatened by agriculture while no data is available on its other locations excepting Bitye, threatened by logging.

Management suggestions: efforts should be made to refind this distinctive species at one or several of its known sites, to evaluate the extent, density and degeneration of the sub populations and so to develop a plan to protect *P. rubripilis.*

Psychotria subpunctata Hiern (syn. *Psychotria* sp. A aff. *calva* Hiern in Cheek *et al.* 2004: 383)

VU B2ab(iii)

Range: Equatorial Guinea (Bioko: Clarence Peak. Rio Muni/Gabon border: Kongui R), Cameroon (SW: Mt Cameroon–Upper Boando; Mt Kupe. NW: Bali Ngemba. S: Longji R), Gabon (Mt Méla; 30 km SW Makoukou) and Congo (Kinshasa) (Central Forest Region: Yabahondo, terr. Isangi). Note that the Bioko and Congo (Kinshasa) locations are unmapped.

One of only five W African species currently known with inconspicuous linear bacterial nodules held along the nerves, *P.subpunctata* is distinguished from the similar *P. calvoa* by the 2-ridged stems and papery leaves (Petit 1966).

A glabrous shrub up to 4 m tall with chalky grey stems and oblong-elliptic leaves c. 20 × 9 cm, lateral nerves 10–12 pairs, petiole to 3 cm, stipule ovate-aristate, bifid to 8 mm. The 4 cm panicles develop red, subglabrous, 6 mm fruit.

With nine locations (AOO 32 km² at 4 km² cell size) and the threats below, we have assessed *P. subpunctata* as Vulnerable. While it appears rare, with only a single collection at each location, this is not so at Mt Kupe, where 15 specimens have been made in the 700–1400 m alt. band, (here entered as one location but requiring a recalculation of the AOO). These were cited at *Psychotria* sp. A aff. *calva* in Cheek *et al.* (2004: 383). While strictly submontane in most of its range, collections from Longji R and Yabahondo appear to be from lowland areas. Further investigation may show that two species are involved.

Habitat: submontane evergreen forest 700–1700 m (but see above).

Threats: submontane forest is threatened with clearance for agriculture in the Cameroon Highlands. Plantation expansion near Upper Boando is a concern, while in the Bamenda Highlands around Bali Ngemba, more than 95% of the original forest habitat has been lost. At Mt Kupe submontane forest is fairly secure, but its lower reaches are being nibbled by agriculture. Threat data for the remaining locations are absent.

Management suggestions: Mt Kupe, with more than 2/3rds of all known collections, is the logical focus for conservation efforts for this species. Formal protection of its habitat would be advantageous and studies of regeneration levels would also be useful to inform a management plan.

Psychotria talbotii Wernh.

EN B2ab(iii)

Range: Nigeria (Ondo, Akure at Idanre; Calabar, Kwa falls; Akampa; Oban) and Cameroon (SW: Rio del Rey; Ndian-Mana R).

Apparently restricted to rapids, this rheophytic shrub may only occur at three locations since the Kwa Falls and Akampa are near Oban as Rio del Rey is to the Mana Rapids. If five locations are accepted, (AOO 20 km²) *P. talbotii* can be here assessed as Endangered given the threats raised below.

One of five W African species with inconspicuous linear bacterial nodules held along the nerves, related to *P. calva* and *P. subpunctata.* Distinguished by the slender leaves c. 8–10 × 1 cm, the longitudinally wrinkled stems, the long corolla tube. A shrub 0.6–2 m tall, with dense, 1–2 cm long panicles on 3 cm long peduncles.

Habitat: rocks in river rapids in lowland evrgreen forest.

Threats: 36% of natural forest in Nigeria was lost between 1990–2005 although that at the Mana Rapids in Cameroon are in the well-protected Korup NP. Increasingly rapids and waterfalls are being degraded as habitats as hydro-electric schemes are implemented although whether any are proposed for sites of *P. talbotii* is unknown.

Management suggestions: population studies are needed at the Kwa falls and Mana rapids to assess the extent, density and regeneration of their populations. It is unlikely that direct management will be needed other than to educate those managing these sites as to the rarity of the species they contain and perhaps to promote this to tourists attracted by the falls.

Psydrax bridsoniana Cheek & Sonké

EN B2ab(iii)

Range: Cameroon (SW: Jide valley).

Known from only two locations (above; AOO 8 km² at 4 km² cells), this rare tree is in great danger of extinction from agricultural expansion. The assessment above is taken from the protologue.

Tree 12–20 m; stems above nodes dilated, ant-inhabited, glabrous; leaves coriaceous, glossy, oblong, c. 25 × 15 cm, shortly acuminate, truncate-subcordate, nerves 9–10; petiole 1.5 cm; stipule ovate-triangular, 18 mm, midrib conspicuous, often reflexed; inflorescence bracts and bracteoles caducous;

fruit numerous, axillary; peduncle 6 cm, endocarp 1.25 cm, slightly curved
Habitat: lowland to submontane evergreen forest; 800–1100 m alt.
Threats: clearance for agriculture and wood. Both sites, near Kupe village and Nyasoso, are on the edge of cultivated land. Further clearance for agriculture could result in extinction of this species.
Management suggestions: a survey to rediscover the species should be conducted urgently. The Bakossi public should be informed of the uniqueness of these rare trees, and to help ensure their survival. New protected areas in Bakossi are likely to help secure the future of this tree species.

Rothmannia ebamutensis Sonké
EN B2ab(iii)
Range: Cameroon (SW: known only from the Bakossi Mts including the montane part of the Banyang Mbo Wildlife Sanctuary).
This shrub or small tree was discovered to be a new species by Dr. Bonaventure Sonké, one of Cameroon's leading botanists (Sonké 2001), on the basis of eight specimens collected by him from at least six plants in the neighbourhood of the village of Ebamut. All these specimens derive from the mountainous eastern part of the Banyang Mbo Wildlife Sanctuary, to the north of Ebamut. In September 2001, while determining specimens at the Kew Herbarium of the genera *Oxyanthus*, *Rothmannia* and *Aulacocalyx* for the checklist of Mt Kupe and the Bakossi Mts, Dr Sonké discovered a further specimen of this species, *Biye* 50, collected in 1998 from the heart of the Bakossi Mts, near Kodmin.
Rothmannia ebamutensis is assessed here as Endangered due to its estimated area of occupancy of 8 km^2, and its being known from only two sites, in an area where there is decline in habitat quality, albeit at a low level.
Tree 10–15 m; stems glossy, pale brown; leaves coriaceous, obovate or elliptic, c. 20 × 10 cm, obtuse or shortly acuminate, c. 8 pairs nerves, glabrous; petiole 1 cm; flowers single, yellow-white; pedicel 2.5 cm; calyx tube 1.5 cm, wrinkled limb 0.7 cm, teeth 0.5 cm; corolla with basal tube 12 × 1 cm, upper tube 8 × 4.5 cm, lobes broadly ovate, 3.5 × 3.5 cm, outer surface densely pale brown-pubescent.
Habitat: submontane forest; 1100–1500 m alt.
Threats: forest clearance even on a local scale, of a fraction of a hectare, if coinciding with one of the sites of this species, could reduce or destroy a subpopulation. Placement of a proposed new reservoir near Kodmin should be done with caution!
Management suggestions: Dr. Sonké's material has already been used by the management team of Banyang Mbo to promote public awareness and pride in this very local and spectacular species. This could be emulated in the Kodmin area. Survey teams in explored parts of Bakossi should look out for new sites for this plant.
Reference: Sonké, B. (2000) Une nouvelle espèce de *Rothmannia* (Rubiaceae, Gardenieae) de Banyang Mbo, Cameroon. Syst. Geogr. Pl. 70: 149–153.

Rutidea nigerica Bridson
VU B2ab(iii)
Range: Benin, Nigeria (Lagos; Egbada; Benin-Iyek; Akpaka; Ijebu; Idranre) and Cameroon (SW: Nyasoso; Baduma; Kumba).
This liana is known from only ten locations (AOO 40 km^2 at 4 km^2 cells), at some of which (Lagos, Kumba) it may very well already be extinct due to the threats discussed below.
Climber, hispid, with brown hairs; leaves papery, often sub-bullate, lanceolate or elliptic, to 17 × 7 cm, base cordate' nerves c. 7, domatia absent; petiole to 12 mm; stipule awn to 14 mm; inflorescence with central inflorescence branch 1.5 times or more as long as lateral braches; corolla white, tube c. 15 mm; fruit orange, 6 mm.
Habitat: lowland forest; 250–800 m alt.
Threats: forest clearance for agriculture and wood; it is very likely that its habitat has been lost at Lagos and Kumba where there has been extensive forest clearance in recent decades' as there has been in much of Nigeria.
Management suggestions: surveys should be made to attempt to rediscover this species at its known sites, and to evaluate the size of subpopulations, regeneration, local threats and possibilities for conservation. When this data is available it may well be possible and necessary to re-evaluation this taxon under criterion A, at a higher level of threat.

Sabicea is a genus of about 150 species. It is centred in Africa but with 40–50 species in Tropical America and four in Madagascar. The species are recognised by being climbers (rarely erect or creeping herbs), the leaves are softly even silkily, hairy below, and the flowers are axillary, in compact inflorescences with the white corolla tubes including the stamens, often with species-specific bracts of various shapes and numbers. Fruits are variedly coloured multi-seeded berries. *Sabicea* was last monographed in 1914, by Wernham (British Museum (Nat. Hist.)) consequently a new revision is desirable to re-assess species limits. In the 1960s, two segregate genera, *Ecpoma* (unbranched shrubs with large stipules) and *Pseudosabicea* (strongly anisophyllous, densely white-woolly hairy, mainly ground creepers) were recognised by N. Hallé in Fl. Gabon and Adansonia 3:127 (1963). Recent molecular work has shown these to be embedded within *Sabicea* as a whole and they, with *Stipularia* were formally sunk back into *Sabicea*.

However, since the morphological justifications for these sinkings were not made, for the present we maintain the segregate genera

While there is no recent work specifically on *Sabicea* of Cameroon, N. Hallé, in his excellent Fl. Gabon treatment, did refer to some species in Cameroon and so provides a useful frame of reference.

Apart from the species listed here, several other Cameroonian species accepted by Wernham may merit inclusion as threatened in future since they have not been seen since the types were collected. They are not included here since we have seen no original material and N. Hallé (1966 op.cit.) did not refer to them although this does not invalidate them. Examples are: *S. brachiata* Wernham (Tibati, 2900 ft, *Ledermann* 2450); *S. gracilis* and *S. geophiloides* of Abonando (*Rudatis* 44).

Sabicea bigerrica N.Hallé.

EN B2ab(iii)

Range: Cameroon (S: Mémel II) and Gabon (Makokou and 20 km SE Makokou).

Described by N. Hallé in 1966 (Fl. Gabon 12, Rubiaceae pt I: 192) from the Makokou specimens, *S. bigerrica* was identified from Cameroon for the first time in 2004 by Sonké from his own collection (YA, K). Here it is assessed as Endangered since there are only three locations (AOO 12 km² using 4 km² calls) with threats as below.

N. Hallé (op.cit.) placed *S. bigerrica* in his group 4, characterised by multiflowered umbellate-spherical inflorescences subtended by pseudoverticillate bracts. The slender pedicel, ± pendant inflorescence with slender, long outcurving sepals help characterise *S. bigerrica*.

Habitat: lowland evergreen forest (presumed).

Threats: slash and burn agriculture, proximity to large town (Makokou) hence urban expansion.

Management suggestions: a revision of *Sabicea* is needed to delimit species with more confidence. Field visits are needed to acquire subpopulational data to manage this species and determine whether any of its locations can be included under the management of a National Park.

Sabicea cameroonensis Wernham

VU B2ab(iii)

Range: Cameroon (E: Lomié; Batouri to Kenzou; Deng Deng; Echambot), CAR (omitted from map because doubtful, 25 km E Bayanga) and Angola (Cabinda: Belize, Luali R).

A climber with elliptic leaves 6–12 × 4–8 cm, distinguished in having the lower surface densely grey-felted, petioles c. 2 cm long, and calyx lobes persistent in the fruit, lanceolate-oblong, c. 4 mm long. Fruits are obovoid, 13–14 x 10–12 mm, silver, following whitish green flowers.

Assessed here as Vulnerable since only six locations are known, AOO is 24 km² (with 4 km² cells) and given the threats below.

Habitat: semi-deciduous forest, probably: c. 750 m alt.

Threats: logging and slash and burn agriculture are thought to pose a risk.

Management suggestions: a revision of *Sabicea* is needed to delimit species with more confidence. Field visits are needed to acquire subpopulational data to manage this species and determine whether any of its locations can be included under the management of a National Park.

Sabicea cruciata Wernh.

EN B2ab(iii)

Range: Cameroon (E: Lomié) and Equatorial Guinea (Rio Muni: Bebai).

Here *S. cruciata* is assessed as Endangered since only two locations are known (AOO 8 km² at 4 km² cells) with threats as below.

Sabicea cruciata is accepted as a species by N. Hallé (1966: 164 op. cit.) who keys it in group 4 with *S. bigerrica* (above), distinguishing it by the leaves, including the midrib, being glabrous above, the calyx lobes being 4–6 mm long, the fruits sessile and the peduncle less than 3 cm.

Only a single duplicate is known to survive, that at Kew. That this species has not been seen for nearly 100 years is an indication of its rarity.

Habitat: lowland forest (assumed).

Threats: slash and burn agriculture following logging is a risk in the Lomié region.

Management suggestions: a revision of *Sabicea* is needed to delimit species with more confidence. Field visits are needed to acquire subpopulational data to manage this species and determine whether any of its locations can be included under the management of a National Park.

Sabicea laxa Wernh.

EN B2ab(iii)

Range: Cameroon (S: Bipindi; Lolodorf).

Here assessed as Endangered since known from only two locations (AOO 8 km² at 4 km² cells) with threats as below. Not seen the original collections made a century ago by Zenker and Staudt, despite intensive targeted collections of Rubiaceae in the same area by Sonké and his team.

"Distinguished readily by the lax inflorescence, the long slender silky corolla and short calyx-teeth and also by the elongated petioles and secondary veins more or less distant" (Wernham 1914: 33 op. cit.). *Sabicea laxa* was accepted by N. Hallé (1966: 165 op. cit.) and placed in group V.

Habitat: lowland evergreen forest, probably.

Threats: slash and burn agriculture.

Management suggestions: a revision of *Sabicea* is needed to delimit species with more confidence. Field visits are needed to acquire subpopulational data to manage this species and determine whether any of its locations can be included under the management of a National Park.

Sabicea leucocarpa (K.Krause) Mildbraed
CR B2ab(iii)

Range: Cameroon endemic (S: E of Ébolowa at Ekouk).

Unusually in the genus, *S. leucocarpa* is a creeping herb. Moreover the inflorescence lacks a peduncle, branches and bracts and is reduced to a few-flowered fascicle bearing white flowers. In fact, it was first described as a *Geophila* and only transferred to *Sabicea* after Wernham's monograph was published. Not collected since 1911 and with only a single specimen known (duplicate HBG), *S. leucocarpa* is here assessed as Critically Endangered AOO 4 km² with 4 km² cells and threats as below.

Habitat: lowland evergreen forest (assumed).

Threats: slash and burn agriculture.

Management suggestions: a revision of S*abicea* is needed to delimit species with more confidence. Field visits are needed to acquire subpopulational data to manage this species and determine whether its location can be included under the management of a National Park, and to try to find new locations.

Sabicea najatrix N.Hallé
EN B2ab(iii)

Range: Cameroon (S: Nkolendom I) and Gabon (Libreville-Sibanga; Cristal Mts;Tchibanga).

Here *S. najatrix* is assessed as Endangered since only four locations (above) are known giving an AOO of 16 km² using 4 km² cells and the location at Libreville-Sibanga is now largely occupied by ongoing urban expansion. Described by N. Hallé (op. cit. 1966: 182) this is a distinctive species with pilose hairs. The flowers are borne on the leafless stem, at first concealed in a pair of nearly glabrous heart-shaped bracts 2–3 × 2–3 cm which when open are united in a cup at the base and bear 5 or 6 tubular white flowers.

Habitat: lowland evergreen forest.

Threats: see above.

Management suggestions: a revision of *Sabicea* is needed to delimit species with more confidence. Field visits are needed to acquire subpopulational data to manage this species and determine whether any of its locations can be included under the management of a National Park.

Note: Hallé (loc.cit.) cites 'fôret du Mayombe' for a Thollon specimen but since Tchibanga falls in this generalised area it has not been separately mapped.

Sabicea rufa Wernh.
VU B2ab(iii)

Range: Cameroon (S. Region: Bipindi; Zingui, 40 km ESE Kribi; Lobé R, SE of Grand Batanga; Bwambe Catholic Mission, 6 km S Kribi) and Gabon (Libreville-Sibanga; 70 km SSW Moanda).

Sabicea rufa is here assessed as Vulnerable since only six locations are known (AOO 24 km² with 4 km² cells) and threats are as below. Since the Sibanga area is now a suburb of Libreville, it is likely extinct here. If confirmed and known locations drop to five, *S. rufa* will qualify for EN status. Distinguished from neighbouring species by the calyces extending past the involucral bracts, peduncles c. 3 cm long, axillary and all parts carrying patent, 2 mm long red hairs.

Habitat: lowland evergreen forest.

Threats: urbanisation is the cause of the probable loss of the Sibanga location. Three of the six locations are near Kribi, an expanding town where several major infrastructure projects are being considered and will damage natural forest habitat if they go ahead.

Management suggestions: a revision of *Sabicea* is needed to delimit species with more confidence. Field visits are needed to acquire subpopulational data to manage this species and determine whether any of its locations can be included under the management of a National Park.

Sabicea schaefferi Wernh.
EN B2ab(iii)

Range: Cameroon (Littoral Region: Baré, near Nkongsamba. SW: Mamfe- identification doubtful) & Equatorial Guinea (Bioko: Moca; Lago de Biaó).

Here *S. schaefferi* is assessed as Endangered since only four locations are known (AOO 16 km² at 4 km² cells) with the threats given below. One of these locations is based on a very doubtfully identified specimen only. N. Hallé indicated acceptance of this taxon (op. cit. 1966: 164).

This climber has 5–9 cm long, axillary peduncles supporting subglobose, 3 cm diam. capitula dominated by leafy, elliptic calyx lobes c. 6 × 3 mm. However, the protologue, based on the lost Baré specimen gives the calyx as "limbo fere ad basin in laciniis lineari-oblongis 8 mm longes acutis ciliatis divisio" fitting the Mamfe, not the Biokan material. It may be that the Biokan material represents a separate species from that in Cameroon.

Habitat: lowland evergreen forest (Cameroon); submontane to 1200 m alt. in Bioko.

Threats: slash and burn agriculture and urban expansion (Cameroon).

Management suggestions: species delimitation needs resolution before an effort is made to rediscover this species in the wild at Baré and Mamfe. It appears genuinely rare since it was not found in the survey of the adjoining Kupe, Mwanenguba, Bakossi Mts, Cameroon (Cheek *et al.* 2004). If refound, the usual populational data should be gathered and the possibility of including locations in a protected area examined.

Sabicea trigemina K.Schum.

CR B2ab(iii)

Range: Cameroon (S: Bipindi).

N. Hallé indicates that while some material attributed to this species is *S. duparquetiana*, the type of *S. trigemina* K.Schum. is a taxon he has not seen in Gabon. Elsewhere he refers to it in such a way that it is implied that he accepts it as a good species (N. Hallé op. cit. 1966: 163, 173).

The type specimen at Kew shows a species with a very shorly puberulent stem bearing leaves glabrous apart from the secondary nerves below and inflorescences with peduncles only 2–3 mm long, a navicular involucre hosting broad, scale-like bracts subtending centrally constricted calyx tubes.

Habitat: lowland forest presumably.

Threats: slash and burn agriculture.

Management suggestions: a revision of *Sabicea* is needed to delimit species with more confidence. Field visits are needed to acquire populational data to manage this species and determine whether it survives and can be protected.

Sabicea xanthotricha Wernh.

VU B2ab(iii)

Range: SE Nigeria (Oban) and Cameroon (SW: Mt Cameroon at Etinde and Mokoko; Bakossi Mts (Kodmin); Fosimondi. Littoral: Ebo).

Previously *S. xanthotricha* was assessed as EN B2ab(iii) since only four locations were known (Cheek *et al.* 2004: 181–182). New locations have been discovered at Ebo and Fosimondi, raising the total to six locations, hence AOO 24 km² at 4 km² cells. Accordingly we downrate the threat assessment to VU. In future it is likely to revert to EN or even CR since so many of its locations are threatened with habitat destruction and none are yet formally protected.

The species is easily identifiable since it is a pithy shrub 2–4 m tall, in a genus of lianas.

Habitat: lowland and submontane evergreen forest; 0–1400 m alt.

Threats: forest clearance for agriculture and wood, particularly at Oban and at Mokoko FR.

Management suggestions: a survey to rediscover plants of this taxon at the known locations and gather the usual data on each of the subpopulations is recommended. The location at Ebo might offer the best prospect for conservation being placed in a .proposed National Park

Sericanthe was erected by Robbrecht (Bull. Jard. Bot. Nat. Belg. 48: 3–78 (1978)), essentially [illegible] the name *Neorosea* proposed by N. Hallé (Fl. Gabon 1970). The 15 species, all African, were formerly included in the large genus *Tricalysia* (except those described since *Sericanthe* was erected) and naturally show many similarities with that genus. *Sericanthe* species, however, have basifixed stamens with a broad flat connective and bacterial nodules in the leaves, showing as black vein knots or dashes. More conspicuously, most *Sericanthe* species have their flower buds covered in a hairy, cap-like calyx which splits as the bud opens. However this also occurs in *Tricalysia*, although rarely, in e.g. *Tricalysia jasminiflora.*

Sericanthe auriculata (Keay) Robbr.

VU B2ab(iii)

Range: Nigeria (Calabar) and Cameroon (SW: S Korup; Mundemba; Kumba; Mt Cameroon, Mokoko, Onge-Liwenyi and Bambuko FR; 20 km WNW Nguti; Akwaya-Mamfe. S. Region: Bipindi; Mamelles, 40 km S Kribi; Campo-Ma'an).

Here *S. auriculata* is assessed as Vulnerable since 10 locations are known (AOO 40 km² with 4 km² cells) with the threats below. Mt Cameroon is taken as one extended location: eight collections are known in three continuous forests of the Western foothills. Therefore an AOO of 126 km² is estimated, allowing 90 km² for the subpopulation at Mt Cameroon and 36 km² for the remaining nine locations using 4 km² cells.

This taxon was initially rejected from *Sericanthe* by Robbrecht, until 1981 (Bull. Jard. Bot. Nat. Belg. 51: 172). It is recognisable in the genus by its auriculate leaves.

Habitat: lowland evergreen forest.

Threats: several of the known locations have seen habitat destruction or exist with this threat (Calabar, Mundemba, Kumba, Mt Cameroon Western foothills, Mamelles) either due to plantation expansion, slash and burn agriculture, mining, or high pressures for timber (in Nigeria).

Management suggestions: Korup and Campo-Ma'an, being NPs, may represent the best hope for long term protection of this species. Surveys to establish extent, density and regeneration levels of subpopulations would help assess priorities of sites for protection and help assess management interventions, if needed.

Sericanthe jacfelicis (Keay) Robbr.

VU B2ab(iii)

Range: Cameroon (C: Ndikiniméki; 8 km SSE of Matonb, Kombeng Hill), Equatorial Guinea (Bioko: no locality data), Gabon (M'passa; Cristal Mts) and Congo (Brazzaville) (Bouyala-Kimboto, along Djoueli R).

Here assessed as Vulnerable since there are only six locations (AOO 24 km² with 4 km² cells) and the threats are as below.

Sericanthe jacfelicis is a glabrous shrub or small tree (trunk c. 20 cm diam.) with 16-20 cm long blades, hairy only on the nerves, 1-3 cm long petioles, and unusually, lacking bacterial nodules and domatia, the calyx open in bud, truncate.

Habitat: lowland to submontane evergreen forest.

Threats: slash and burn agriculture,.
Management suggestions: surveys to establish extent, density and regeneration levels of subpopulations would help assess priorities of locations for protection and help assess management interventions, if needed.

Sericanthe raynaliorum (N.Hallé) Robbr.
EN B2ab(iii)

Range: Cameroon endemic (Adamawa: Ngaoundal; 47 km E of Ngaoundéré, Tello R; 36 km E of Ngaoundéré, Sadolkoulay; Meïganga-Yarban).
The only upland species of the genus W of the Congo basin, and known only from the Ngaoundéré area, *S. raynaliorum* is here assessed as Endangered since only four locations are known (AOO 16 km² with 4 km² cells), with the threats below.
A shrub with leaves clustered at stem apices and flowers on the naked stems below. The leaves are 8–18 × 4–9 cm, lacking obvious bacterial nodules, on slightly hairy, stout 3–7 mm wide, scaley twigs.
Habitat: submontane gallery forest; c 1200 m alt.
Threats: slash and burn agriculture.
Management suggestions: none of the four known localities occurs within a protected area so far as is known since no National Parks have yet been gazetted in the Adamawa area although this upland area contains numerous endemic species. It is advised that this is rectified. In the meantime the best hope for this species is that local communities, having been alerted to its importance, are involved in protecting it. Surveys are needed to determine that it still exists in the wild (it was last seen in the 1970s) and to determine the full range, population density and levels of regeneration for the species.

Sericanthe testui (N.Hallé) Robbrecht
VU B2ab(iii)

Range: Cameroon (S: Bitye on Dja R; Evélessi), Gabon (Oyem; Bélinga; Moucombo; Lebagny) and Congo (Kinshasa) (Kasai: Kahemba, Kwango).
This is a 3–7 m shrub with puberulent stems and obovate-acuminate leaves 7–12 x 2.7–5.5 cm. It is distinguished from related species by its leaves, which dry chestnut brown, and the short spreading hairs on the style, and the glabrous ovary. Seven locations are known (above) giving an AOO of 28 km² using 4 km² cells, allowing the assessment made here of Vulnerable. Threats are indicated below.
Habitat: evergreen forest, sometimes in galleries; c. 800 m alt.
Threats: Bélinga habitat is threatened by a large open-cast iron mine. Sites in Cameroon are threatened by logging and slash and burn agriculture.
Management suggestions: none of the localities of *S. testui* are known to be within a protected area which should be rectified. Gabon may be the best option for protecting the species since it has the highest density of protected areas and of localities for this species. Further survey work might produce new sites for this species and would allow gathering data needed to assess the needs for management intervention, if any (see above under *S. raynaliorum).*
Re-evaluation of the specimens from Kasai and the subpopulation it represents is needed to confirm that it is conspecific despite the large range disjunction.

Stelechantha is a genus of four species of shrub or small tree restricted to W-C Africa. All species are rare. Three of the species occur in Cameroon, of which two are endemic. In addition to the species assessed below, *S. makana* N.Hallé of Makak, Cameroon and *S. cauliflora* (R.Good) Bremek. of Cameroon, Gabon and Angola (Cabinda) are probably also threatened. The genus is characterised by having inflorescences of blue flowers borne on stems below the leaves, succeeded by many-seeded berries. The valvate, heterostylous, flowers with falsely septate ovary, and the opresence of raphides indicates a relationship with *Pauridiantha* and related genera.

Stelechantha arcuata S.E.Dawson
CR A2c

Range: Cameroon (SW: Mt Cameroon at Bimbia-Bonadikombo and at Mokoko, Kupe-Bakossi in western Bakossi and at Kupe Village. S: N'Kolandom (1 coll. each).
This blue-flowered, cauliflorous treelet, even within its range, seems very rare given the few collections known despite intensive collecting over recent years in several of the forests where it occurs, e.g. at Mt Cameroon. The assessment above is listed in IUCN (2003; www.redlist.org) under the citation Dawson (2003); the species was published with an assessment in 2002 (Dawson 2002). The assessment signifies that an 80% or more population reduction has is estimated to have occurred due to habitat loss. In the last few years the Bimbia site has almost certainly been lost to this species for that reason, the Bakossi site is close to a village and at risk of being cleared for agriculture, while Mokoko FR is likely to be logged imminently and is at risk from oil exploration (late 2010). No new locations have come to light for the species in recent years.
Shrub 2.5 m, glabrous; leaves coriaceous, narrowly obovate, 12–22 × 3.5–7.5 cm, shortly acuminate, cuneate, nerves 7; petiole 1–2.5 cm; stipule lanceolate, 0.4–1.4 × 0.2–0.4 cm, arching away from the stem; cauliflorous; flowers blue, 1–16, single or in subsessile clusters; calyx tube c. 2 × 2 mm, slightly lobed; corolla tube 8 mm, lobes 2 mm; fruit globose, 6 mm.
Habitat: lowland evergreen forest; 200–950 m alt.

Threats: forest clearance for agriculture, especially plantations in the Mt Cameroon area (Cable & Cheek 1998: xxxv) and in Bakossi, small-holder agriculture. See above also.

Management suggestions: individual trees of this taxon merit special protection given their rarity. A survey to relocate plants at their known sites should assess what measures are most suitable at each site. At Kupe village for example, renting the forest area concerned together with a poster campaign might be the best means of supporting the survival of the subpopulation there.

Reference: Dawson, S.E. (2002) A new species of *Stelechantha* Bremek. (*Rubiaceae: Urophylleae*) from Cameroon. Kew Bull. 57(2): 397–402.

Tarenna are a pantropical genus of c. 180 species of which 54 occur in Madagascar, the Mascarenes and Africa. Forest shrubs and climbers they are easily recognised by their glossy black stipules. Often mistaken for *Pavetta*, they are 5-merous not 4-merous and lack bacterial nodules. The African species were revised by De Greef (Op. Bot. Belg. 14 (2006)). There are no species endemic to Cameroon and only one that is threatened:

Tarenna baconioides Wernh.

EN A2

Range: SE Nigeria (Calabar/ Cross River State area: Afi FR; Oban; Akampa Distr.: Awi-Nsau; Iyila-Ibere), Cameroon (SW: Mt Cameroon–Onge R, Bakossi-Baduma) and Gabon (Bélinga; Libreville; Lastoursville).

This liana was assessed in Cheek *et al.* (2004: 182) as VU B2ab(iii), but is reassessed under criterion A as Endangered here since it is considered that it has been lost due to habitat destruction at >50% of its sites (Libreville and those in Nigeria).

Climber, glabrous; leaves elliptic, 9 × 5 cm, acuminate, unequally rounded-obtuse, nerves 4–5, subinvolute, pit-domatia; flowers 15–20; peduncle 2 cm; pedicel 2 mm; calyx appressed-hairy, lobes 1 cm; fruit ovoid, 1.8 × 1.1 cm.

In Fl. Gabon a var. *nephrosperma* was erected for specimens with elongated fruits and seeds, (not globose) and so Cameroonian material was assessed as var. *baconoides.* However De Greef (loc. cit.) states that intermediates between the states occur and treats the species without recognising varieties. However, the absence of the species from S Region Cameroon and from the Cristal Mts suggests that it may be worth looking again to see if there are any corresponding morph[illegible].

Habitat: lowland ev[illegible]

Threats: forest clearance for agriculture and wood, in Cameroon particularly at Onge (see Cable & Cheek 1998: xxxv). Forest clearance in Nigeria has been extensive in recent decades: see elsewhere in this volume.

Management suggestions: quantitative data on the subpopulations, including regeneration levels and local threats, are needed to inform a management plan for the taxon, and should be obtained by visits to the known sites.

Tricalysia has been revised for Africa in four stages by Robbrecht, recognising 99 species:

1. Bull. Jard. Bot. Nat. Belg. 49: 239–360 (1979)
2. ibid. 52: 311–339 (1982)
3. ibid. 53: 299–320 (1983)
4. ibid. 57: 39–208 (1987)

Several additional species have since been described from Cameroon. This genus is commonly represented in African forest and woodland. Many species were first described in the genus *Coffea* and they are often known locally or confused with this genus in the wild, having white flowers in axillary inflorescences, a shrub-small tree habit and berried fruit. However, they are easily distinguished by having a conspicuous, often lobed or toothed calyx, and what resembles a series of three, cupped calyces (in fact bracts) below the flower and fruit. Most species have puberulent stems and aristate stipules. They lack the gummed buds usual in Cameroon *Coffea* species.

Tricalysia achoundongiana Robbr., Sonké & Kenfack

EN B2ab(iii)

Range: Cameroon endemic (SW: Korup NP; Mundemba town (3 coll.)).

Named for the great Cameroonian botanist Dr Achoundong, the first General-Secretary of the African Botanical Congress (AETFAT) from W and C Africa and specialist in Violaceae, this is a puberulous shrub up to 10 m tall. It differs from other members of the *T. atherura* group in having sessile, 5–6 merous flowers and cuneate leaves (Sonké *et al.* (2002). Adansonia sér. 3, 24: 173–177). So far it is known from only three collections, two in the S of the Korup NP and one in nearby Mundemba town.

Habitat: lowland evergreen forest with 3–5 m rainfall per annum; sea-level–100 m alt.

Threats: while the sites of *T. achoundongiana* in the Korup NP are fairly secure, that in Mundemba town (*Thomas and Mambo* 4239) is not, and is either highly threatened or may have been lost already due to urban infilling and expansion. Sonké *et al.* 2002 loc. cit. assessed this species as "VU B" but here it is re-assessed as EN B2 a, b(iii), that is Endangered, since the three collections indicate three locations for an AOO of 12 km². If numbers of individuals were known it is likely that the species would be assessed as CR under criterion C. The great rarity of this species, even within its small range, is indicated by the fact that although Duncan Thomas collected thousands of specimens in

southern Korup and Mundemba, only a single Thomas specimen of this taxon exists.
Management suggestions: unless further survey work shows this species to be common in Korup it is advisable that the known individuals are refound and marked for protection against the possibility that they might be cut during path or transect-making operations. It is also advisable that a poster be produced to enable Park Managers to recognise and identify this threatened species. An effort should be made to refind and protect the plant in Mundemba town, if it survives there.

Tricalysia amplexicaulis Robbr.
VU A2c,d
Range: Cameroon endemic (SW: S Bakundu. C: Mt Kala; N'kolbison; Mbam-Minkoom; Makak; Mefou proposed NP. S: just E of Kribi, Kienke R; Campo-Ma'an).
A forest shrub or treelet, 1–2.5 m tall, the stems are erect, unbranched densely puberulent when young, bearing pairs of sessile, oblanceolate cordate-based leaves 20–35 × 6–11.5 cm, making it very distinctive within the genus (Robbrecht op.cit. 4: 157). The main distribution of *T. amplexicaulis* is Yaoundé to Kribi and Campo. More than 10 locations are known. At the Mefou proposed NP it occurs in undisturbed evergreen forest areas, but only rarely and this forest is under severe threat from agricultural clearance (Cheek pers. obs.). At S Bakundu large areas have suffered from agricultural operations, while at Mt Kala slash and burn agriculture is also a threat. It is here estimated that over 30% of habitat at its known locations has been lost in the last 3 generations/100 years (understorey shrubs in forest are believed to be long-lived) and losses are ongoing so *T. amplexicaulis* is here assessed as VU A2c,d. This assessment appears also in Cheek *et al.* 2011.
Habitat: lowland evergreen forest; sea-level to c. 700 m alt.
Threats: see above. At Mefou proposed NP several plants were found by us in a forest patch destined for a new chimpanzee enclosure. Chimpanzees are very destructive of forest habitat when in enclosures, so this posed a threat to these rare plants. Fortunately it was possible to move the plants to a safe location nearby.
Management suggestions: marking, publicising, protecting and monitoring clusters of plants in protected areas such as at Mefou proposed NP is advisable to avoid accidental destruction of this often very local species.
Note: Robbrecht (loc.cit.) indicated that four specimens are close to *T. amplexicaulis* but differ in having smaller, shortly petiolate leaves. These need more research to elucidate their taxonomic status.

Tricalysia atherura N.Hallé
EN A3b,c
Range: Cameroon (SW: Kupe village; Etinde. Littoral: Loum. C: Mt Kala; Mt Fébé. S: Ngoasé; Nteigne) and Gabon (Bélinga, Mt Babiel).
This species was earlier assessed as VU B2ab(iii) in Cheek *et al.* (2004: 182) since known from nine locations, with a low AOO and documented threats. Here the assessment is changed to Endangered using criterion A since > 50% habitat loss over a 100 year span at known locations for this taxon has either occurred or is anticipated. This understorey shrub, although first described correctly from Mt Babiel, in Gabon, earlier received a name and description based on material from Loum in Cameroon. The locations above are taken from Robbrecht (1987: 76), apart from those of Etinde (Cable & Cheek 1998: 123) and Kupe village (recorded in Cheek *et al.* 2004: 393).
Shrub 1–4(–8) m, glabrous; leaves elliptic, c.15 × 6 cm, acumen apex rounded, base acute, nerves 4–5; petiole 6 mm; flowers 1 per axil, bracteolar cup sheathing pedicel, truncate 2 mm; calyx subcylindrical, truncate, 7mm; corolla tube 20–34 mm, lobes 7–8, 15 mm; fruit orange or red, juicy, ellipsoid, fusiform or globular-rostrate to 4 × 2 cm; pedicel accrescent.
Habitat: lowland forest often near the boundary with semi-deciduous forest (*pers. obs.*); 200–1250 m alt.
Threats: clearance or part-clearance of forest for agriculture; much of the understorey of the Loum FR. has been cleared in recent decades and replaced by crops; other parts have been cleared of natural forest and replaced by cocoa (*pers. obs.*); Mt Kala has had its forest damaged in recent years and Mt Fébé has been cleared of forest and had prestige buildings erected upon it; the lower slopes of Mt Etinde (where one plant was recorded at 400–600 m alt.) is earmarked for plantation expansion. The Bélinga location is set to become an open-cast iron ore mine, Chinese owned.
Management suggestions: a survey to refind this species at each of its sites, gather quantitative data on subpopulation size, recruitment and protection levels would inform development of a management plan for the species. In the absence of this, conservation efforts should be concentrated around Kupe village, which adjoins the Loum area. At the former, 15 collections were made in 1995–2002 (compared with one at Etinde, despite similar sampling effort) suggesting Kupe village to have a very high density of the species. The proposed Mt Kupe protected area should help secure the future of the species.

Tricalysia fangana (N.Hallé) Robbr.
EN B2ab(iii)
Range: Cameroon: (E: Essélégué, 92 km from Bertoua) and Gabon (Libreville and Sibang; Komo R, Mbel; Makoukou).
With minute white axillary 6-merous flowers (corolla tube 3–5 mm long, lobes 3–4 mm) this 1–3(–9) m forest shrub has 4–8 mm petioles and leaf-blades 5–15 × 2.5–7 cm. It is

densely golden puberulent (Robbrecht op. cit. 4: 126–128). Known from only five locations (AOO 20 km² using 4 km² cells) it is almost certainly lost from two of these (Libreville and Sibang) due to urban expansion (Cheek pers. obs.) so the correct number of extant locations is likely to be three and it is here assessed as Endangered under criterion B.
Habitat: lowland evergreen forest; alt. range not known.
Threats: urban expansion has been a major threat at two locations. Specific threats at other locations are not known but slash and burn agriculture is likely to pose a risk.
Managements suggestions: attempts should be made to refind the plant at the location near Bertoua, establish regeneration levels, density and range of this subpopulation and educate local communities as to how to identify and protect this rare plant which currently appears not to fall within a protected area within Cameroon.

Tricalysia ferorum Robbr.
EN B2ab(iii)

Range: Cameroon endemic (SW: S Korup. Littoral: 50 km NW Eséka. C: 15 km Eséka. S: 10 km NE Bipindi; 4 km E of Kribi, Kienké R).
Robbrecht (op. cit. 3: 318-320) considered this species to be in an isolated position. The fruits have an unusually large disc which occupies most of its width. The 5-merous sessile, white flowers have corolla tube and lobes both 3–4.5mm long, hairy stamens and style. The five known locations equate to an AOO of 20 km² (4 km² cells) and the location at Kribi must be considered at risk due to the intensity of major commercial infrastructure being developed in and around the town. The Korup specimen was identified by Bridson.
Habitat: lowland evergreen forest on sand.
Threats: see above.
Management suggestions: consideration to inclusion of one or several of the Eséka-Bipindi locations in a protected area is advisable once they have been refound. Means should be provided to managers to identify, mark and monitor the plants.

Tricalysia lejolyana Sonké & Cheek
CR D

Range: Cameroon endemic (SW: E of Nguti, Banyang Mbo (3 coll.)).
A 4–8 m tall treelet or shrub, *Tricalysia lejolyana* has 3–3.5 cm wide white, perfumed flowers with 7–8 petal lobes and a split calyx. This and the very large (35–40 mm diam.) orange fruit separate it from other species of the *T. atherura* group which are characterised by large leaves that lack domatia, and the large flowers and fruit (Sonké *et al.* (2002). Kew Bull. 57: 681-686). Discovered by Sonké in Banyang Mbo from two locations separated by several kilometres, the species was first assessed as “EN D” but since it is only known from two individuals despite much searching (Sonké *et al.* op.cit. 682) it is here re-assessed as Critically Endangered (<50 individuals).
Habitat: lowland evergreen forest; 200–300 m alt.
Threats: logging, followed by agriculture (M. Cheek pers.obs.).
Management suggestsions: since Banyang Mbo, despite having its protected status recently downgraded, is the only known home of this and several other rare species of plant, consideration should be given to ensuring that the plants in its forest are protected for the future.

Tricalysia obstetrix N. Hallé
VU B2ab(iii)

Range: Cameroon (S: 2 locations), Gabon (3 locations) and Congo (Brazzaville) (1 location).
First discovered in Gabon, this 2–7 m tall small forest tree is unusual in having young stems hairy only at the nodes and in its glabrous leaves. The flowers are 6-merous. It falls in sect. *Tricalysia* (Robbrecht op.cit 4: 143–144). The species is here assessed as Vulnerable since known from only six locations (AOO 24 km² with 4 km² cell) and threatened (see below).
Habitat: submontane forest; 800–900 m alt. (*Hallé & Thomas* 182).
Threats: the type locality, Bélinga in Gabon is being developed as an open-cast iron mine. Should the plant be lost at this location the species should be re-assessed as EN.
Management suggestions: a visit to rediscover the plant at the Akonétyé location so that it can be protected and monitored, is advisable. Means should be provided to Park Managers to identify the species.

Tricalysia pangolina N.Hallé
VU B2ab(iii)

Range: Equatorial Guinea (Bioko (1 location)), Cameroon (SW: Korup, Nta Ali. C/E: Djolimpoum), CAR (Baboundji (1 location)), Gabon (Bélinga; Mbigou; Nzoumou) and Congo (Brazzaville) (Moukoundou).
Within the *T. atherura* group (see notes under *T. lejolyana), T. pangolina*, a 1–5 m tall puberulent forest treelet, is unusual in sometimes having, swollen, ant-inhabited internodes. The 5-merous flowers are held on unusually long peduncle-pedicels (up to c. 1.5 cm) and appear fasciculate. The ellipsoid fruits (c. 3 × 2 cm) have a persistent toothed-tubular calyx (Robbrecht op. cit. 4: 77–78). With eight locations, leading to an AOO of 32 km² at 4 km² cells and with the threats below, we here assess *T. pangolina* as Vulnerable.
Habitat: submontane evergreen forest; 700–900 m alt. (Robbr. loc. cit.)
Threats: Bélinga is threatened by open-cast iron-ore mining. There is no data about threats at other locations available to

us currently. Nta Ali and Korup are regarded as protected and secure.
Management suggestions: the means should be provided to identify, mark, protect and monitor individuals at the last two named locations.

Tricalysia sylvae Robbr.

VU B2ab(iii)

Range: Cameroon (Littoral: Douala; Edéa. S: Kribi area; Bipindi; Bella from 12 coll.) and Gabon (Abanga; NE Oveng; Lope NP, 1 coll. each).
A forest shrub or shrublet (0.2–)1–2.5 m tall, densely patent-hairy, with 5–6 merous flowers in condensed, sessile inflorescences. The leaves have only small, few domatia and the subulate calyx lobes are as long as the tube (Robbrecht op. cit. 4: 156). Most collections of this species are from Bipindi (Zenker) and Kribi (Bos). Forest at these locations especially at Kribi, has seen losses due to urban expansion and international development projects. When forest loss data in these areas is available this species could rate as EN or CR under criterion A. For the present it is assessed under criterion B as Vulnerable since seven global locations are known, or nine if the two potential records from Gabon referred to above are verified and geographically distinct. With 4 km² cells a maximum AOO of 36 km² can be calculated. Threats are indicated above.
Habitat: lowland evergreen rainforest; alt. range unknown.
Threats: see above. Kribi is one of only three major ports in Cameroon. New port facilities and hinterland infrastructures are scheduled for installation in connection with planned mineral extraction and energy projects.
Management suggestions: forest habitat protection as Kribi and Bipindi are advisable since not all threatened species occurring at these locations extend to the Campo-Ma'an NP.

Tricalysia talbotii (Wernham) Keay

EN A3b,c

Range: SE Nigeria (Oban; Awi-Akampka; Boli-Bateriko, Iyamoyong FR) and Cameroon (Littoral: Eboné-Yabassi. SW: Inokum-Mbenyan; Ndikiniméki; Minso; western Bakossi (Kumba-Mamfe, 2 locations)).
Known from only the ten locations listed above. All but the last two are taken from Robbrecht (1979: 353).
This species was earlier assessed as VU B2ab(iii) in Cheek *et al.* (2004: 182–183) since known from nine locations, with a low AOO and documented threats. Here the assessment is changed to Endangered using criterion A since >50% population loss over a 100 year span at known locations for this taxon has either occurred or is anticipated due to reduction in habitat. Destruction has occurred and is believed ongoing at Eboné-Yabassi, Kumba-Mamfe Rd and all Nigerian locations.
Tree 5 m, densely pale brown patent-pubescent; leaves oblong-elliptic, 15 × 4.5 cm, acuminate, acute, nerves 6–8; petiole 1 cm; inflorescence 3–10-flowered, pubescent, bracteoles opposite, free; calyx 5-lobed; corolla tube 4 mm; fruit globose, 8 mm white to purple.
Habitat: lowland forest; c. 250–350 m alt.
Threats: clearance of forest for agriculture and wood.
Management suggestions: a survey to rediscover this taxon, quantify subpopulations, demography, regeneration and threats, is advisable. Given far-reaching forest loss in Nigeria and anticipated increased forest loss along the upgraded Kumba-Mamfe Rd, the prospects for this species do not seem good.

Tricalysia vadensis Robbr.

EN B2ab(iii)

Range: Cameroon (S: 60 km W of Eséka; Kribi-Lolodorf 10 km).
This 2 m shrub of the *T. atherura* group (see under *T. lejolyana)* has subsessile inflorescences and flowers, with white 4-merous corollas 5 mm long. It has a dirty-green densely hairy stems and calyx teeth shorter than the tube (Robbrecht op. cit. 4: 78–79).
Here *T. vadensis* is assessed as Endangered since only two locations are known (AOO of 8 km² with 4 km² cells) with threats as below.
Habitat: lowland forest; c. 200 m alt.
Threats: clearance of forest for agriculture and wood.
Management suggestions: a survey to rediscover this taxon, quantify subpopulations, demography, regeneration and threats, is advisable.

Trichostachys is an African genus of about eight species restricted to evergreen forests in W and Congolian Africa but centred in Cameroon and Gabon. They are single-stemmed subshrubs, often creeping, pubescent and blue-fruited. The inflorescence is terminal in a dense pedunculate spike. Close to *Psychotria*, they differ in endocarp characters. The genus needs revision but species delimitation is not a challenge. In addition to the three taxa listed below, *T. lehmbachii* K. Schum. may also be included here in future. Its description needs analysis to assess its status since the only known specimen, which was from Etinde, Mt Cameroon, was destroyed in Berlin.

Trichostachys interrupta K.Schum.

VU B2ab(iii)

Range: SE Nigeria (2 coll.), Cameroon (SW: Korup NP (1 coll.); Lake Barombi Mbo (1 coll.); Bakossi Mts (2 coll. at 2 sites)) and CAR (1 site).
Published in 1903 on the basis of *Preuss* 466 from Barombi near Kumba, two additional localities were later found in Nigeria (Oban, *Talbot* 1045 and Kwa Falls, *Brenan* 9238).

Rediscovered in Cameroon in 1983 (*Thomas* 2291, S Korup NP), the only other locations for this sub-shrub are in the Bakossi Mts below Nyale Rock (*Etuge* 4176, Feb. 1998) and between Nyale and Kodmin (*Etuge* 4461, Nov. 1998). The foregoing assessment was made in Cheek *et al.* (2004: 183) and is maintained here. One additional record from CAR has since been discovered, seven locations in total now being known, (above; AOO 28 km² at 4 km² cells) and threats as below. The species is distinguished by the interrupted inflorescence spike.

Rhizomatous subshrub; aerial stems erect, monopodial, to 30 cm, densely puberulent, internodes 2 cm; leaves at 3–4 nodes, drying black above, obovate, to 12 × 6 cm, obtuse or subacuminate, cuneate, lateral nerves 8 pairs, dirt-gathering; petiole 7 mm; stipule ovate, 5 × 4 mm, black, glabrescent; spike 4.5 cm, erect; peduncle 0.5–1 cm, appressed-hairy; rachis interrupted; flowers white, 1 mm; fruit 2 mm, pink, hairy.

Habitat: understorey of lowland evergreen forest; 400–1000 m alt.

Threats: clearance of forest for agriculture has occurred extensively in Nigeria, and at Barombi Mbo, where the species may no longer occur.

Management suggestions: the Nyale-Kodmin area in the Bakossi Mts of Cameroon has least pressure of all known sites and offers the best hope for the survival of this herb. Surveys to determine whether the species survives at the other sites are suggested. If so, an assessment of the best means of protecting them, if feasible, should be made. Reinvestigation of the geographically disjunct CAR material is advised to confirm that it is conspecific.

Trichostachys petiolata Hiern (syn. *Trichostachys* sp. 1 Cheek in Cheek *et al.* 2004: 395)

EN B2ab(iii)

Range: Cameroon endemic (Littoral: 'Cameroons R' = Wouri R (type). SW: Bakossi-Nyale; Bechati-Lebialem).

First collected in Jan. 1861 by the Kew botanist Gustav Mann. *T. petiolata* was not identified again until 2010, as this book was being assembled, when *van der Burgt* 885 from forest near Bechati, was identified as this species.

Assessed by Cheek in Harvey *et al.* (2010: 83–84) as Endangered since only three locations are known (AOO 12 km² using 4 km² cells) with threats as given below, this status is maintained here since no further new data are available.

Subshrub resembling *T. interrupta*, but leaves obovate, held to ground, 8–10 × 5 cm, apex and base rounded, lateral nerves 10 pairs; petiole 1 cm; stipule triangular; spike 7 cm, not interrupted.

Habitat: lowland evergreen forest: 400–650 m alt.

Threats: forest clearance followed by agriculture e.g. at Bechati (van der Burgt pers. comm.)

Management instructions: none of the sites occurs in a formally protected area. The exact location of the type collection is unknown, while that in Bakossi, at Nyale is outside the new NP. Conservation efforts should be focussed at the only two known sites where neighbouring communities should be consulted and advised on the rarity of this species and their help sought in protecting it.

Trichostachys zenkeri De Wild.

EN B2ab(iii)

Range: Cameroon (S: Bipindi) and Equatorial Guinea (Rio Muni: Mitong).

Here assessed as Endangered since only two locations are known (AOO 8 km² using 4 km² cells) and threats are as below. *Trichostachys zenkeri* might also be assessed under criterion D since less than 20 plants have probably been recorded by way of herbarium specimens. The apparent rediscovery by Sonké at Mitong of this species a century after its last sighting and first discovery by Zenker is a triumph of hope over despair. Resembling *T. aurea* this species differs in having very large leaves c. 30 × 10–15 cm, which are glabrous below and in the erect, not pendant peduncles. Further,the stipules are persistent, triangular and large, c. 1 × 1 cm.

Habitat: lowland evergreen forest.

Threats: slash and burn agriculture; tree felling for timber.

Management suggestions: rediscovery of this species should be attempted at Bipindi and populational data gathered to assess management needs, if any. Options for protection should then be assessed.

Vangueriella Verdc. was segregated from *Vangueriopsis* Robyns on the basis of much smaller flowers, calyx lobes patent, corolla lobes subequal to the tube and style and stamens long-exserted. Eighteen Tropical African species of shrub and tree, sometimes scandent and/or spiny are recognised (Verdcourt (1987). Kew. Bull. 42: 189–196). Species are concentrated in Lower Guinea and five species occur in Cameroon, one of which is very rare. Two other Cameroonian taxa are range-restricted but both have more than ten locations so do not qualify under criterion B although, with better habitat loss data, might be assessed under criterion A. These are *V. laxiflora* and *V. chlorantha.*

Vangueriella zenkeri Verdc. (syn *Plectronia macrocarpa* K.Schum.)

EN B2ab(iii)

Range: Cameroon endemic (S: Mt Elephant & Bipindi)

"Seemingly a rare and elusive species" (Verdcourt op. cit.)

Shrub 0.5 m tall, patent dark brown hairy. Leaves elliptic, 10 × 6 cm, lateral nerves five pairs, petiole c. 3 mm long. Inflorescence 5 cm long, flowers 4 mm diam. Corolla white

with pink lobes. Here *V. zenkeri* is assessed as Endangered since it is known from only two locations (AOO 8 km^2 using 4 km^2 cells), with the threats below.
The first collection of this species (*Zenker* 1114) was made in the early 20th century. Unusually for a Zenker collection, only two duplicates are known, suggesting that only a single shrub was seen by him. *Bos* 5654A made in 1969 from Mt Elephant, within 50 km of Bipindi (Verdcourt), exists as only a single duplicate, both collections were made on 18 Nov. suggesting a narrow flowering window.
Habitat: lowland evergreen forest, near a stream; 90 m alt.
Threats: Mt. Elephant is believed to be proposed as an open cast iron ore mine.
Management suggestions: effort to refind this species at either of its two localities should take place on or near 18 Nov. when flowering is most likely and plants will be most identifiable. Data are needed on density, range and regeneration levels of this species. Neither known localities are in a protected area. If there is no prospect of this being rectified it is advisable that the species, if rediscovered, is taken into cultivation. Local communities should be apprised of the importance of this species using illustrated material.

REFERENCES

Cheek, M. & Csiba, L. (2000).A new species and new combination in *Chassalia (Rubiaceae)* of western Cameroon. Kew Bull. 55: 883–888.

Davis, A. P. & Figueiredo, E. (2007). A checklist of the Rubiaceae of Bioko and Annobón (Equatorial Guinea, Gulf of Guinea). Systematics and Biodiversity 5(2): 159–186.

Willis, F., Moat, J. & Paton, A. (2003). Defining a Role for herbarium data in Red List assessments: a case study of *Plectranthus* from eastern and southern tropical Africa. Biodiversity and Conservation 12(7): 1537–1552.

RUTACEAE

assessed by Martin Cheek

A cosmopolitan family, in Cameroon all species are shrubs or trees, mainly of evergreen forest, easily recognised by their translucent, gland spotted, digitately compound leaves, often with 3, rarely with 1, leaflet(s), or with pinnately compound leaves (e.g. *Zanthoxylum*). The leaves lack stipules, and are alternate, (rarely subopposite). Flowers are usually 2-5 mm wide, symmetric radially, with 4–5 free sepals, petals and stamens, and a superior pistil.

The Rutaceae are best known for the genus C*itrus* (orange, lemon, lime, grapefruit, tangerine, etc) which derives from Asia-Australia. The Cameroonian species were treated by Letouzey in Flore du Cameroun 1 (1960). He recognised 26 species. Forest genus *Fagara* has since been transferred to *Zanthoxylum*, while a revision of the African Toddalieae by Mziray (Symbol. Bot. Upsal., 30 (1) (1992)) has resulted in the Cameroonian species of the genera *Oricia, Diphasia, Araliopsis Orciopsis* and *Teclea* being placed in *Vepris*.

Apart from the species treated as threatened below, several others are known from only slightly more than ten locations so are here assessed as NT: *Vepris gabonensis* (Pierre) Mziray, *V. glaberrima* (Engl.) D.J.Harris, *V. grandifolia* (Engl.) Mziray, *Zanthoxylum buesgenii* Engl., and *Zanthoxylum dinklagei* (Engl.) P.G.Waterman. Several species are so poorly known that their taxonomic status is not clear – all reference specimens appear to have been lost, therefore their conservation status is also uncertain: *Teclea campestris* Engl., *Diphasia mildbraedii* Engl. and *Teclea ebolowensis* Engl.

Several taxa not included in Flore du Cameroun have come to light, both those new to science and those that appear to have been overlooked. Most of these, when published, will be assessed as threatened. An example is *Vepris* sp. B of Bali
Ngemba (Cheek in Harvey *et al.* 2004: 124).

Vepris lecomteana (Pierre) Cheek & T.Heller (syn. *Oricia lecomteana* Pierre) assessed as VU B2ab(iii) in Cheek *et al.* (2004: 183), was reassessed as NT in Cheek *et al.* 2009 (Kew Bull. 64: 509–512) when formally transferred to *Vepris,* since 13 locations are now known.

Afraegle asso Engl. [Kpakpo (Lissongo fide Letouzey 1963)]

EN B2ab(iii)

Range: Cameroon (E: Moloundou; Lobéké FR) and CAR (Mbaiki and Boukoko).
Large tree reaching 30 m, spiny; petiole 2.5–5.5 cm; petiolules 0.2–2 cm; leaflets three, ovate or ovate-elliptic, 8–4 × 3.5–1.5 cm apex obtuse-rounded, base acute to rounded, margin crenulate, lateral nerves 6–8 pairs; axillary panicles 4–6 cm, pedicels puberulent, tetramerous, stamens 12; young fruits pubescent.
This is a poorly known species, attached with doubt to this genus. Better material is required to clarify its delimitation and placement (Letouzey 1963, loc.cit.)
Here *Afraegle asso* is assessed as Endangered since only three locations (see above; AOO 12 km^2 with 4 km^2 cells) are known and threats are as below.
Habitat: lowland mixed semi-deciduous and Dja congolian forest (Onana pers. obs.).
Threats: logging, followed by slash and burn agriculture.

Management suggestions: two of the known locations are at possible secure areas (Lobéké has been proposed as a National Park and Boukoko is a research station). These should be the best choices for conservation action for *Afraegle asso*, beginning with efforts to refind the species, to collect basic population data as a baseline for monitoring and to gauge intervention needs with the Park/Station managers.

Balsamocitrus camerunensis Letouzey
VU B2ab(iii)

Range: Cameroon (E: Bertoua–Koubité; SSE Batouri; Moundia; Lobéké NP), CAR (Boukoko) and Congo (Kinshasa) (Lékoli R).

Tree or shrub to 5 m, spiny, glabrous; leaves (1–)3-foliolate, petiole to 8 cm, petiolules 0–1 cm; leaflets ovate-elliptic or obovate, central largest, to 20 × 11 cm subacuminate, base cuneate, margin crenulate, lateral nerves 6–9 pairs; panicles axillary, c. 2 cm; pedicel 3–4 mm; flowers 4 merous, stamens 8–10, ovary 7–8 locular; fruit globose, 9–10 cm, in April.

Here *Balsamocitrus camerunensis* is assessed as Vulnerable since only six locations (see above; AOO 24 km² with 4 km² cells) are known and threats are as below.

Habitat: unknown, probably semi-deciduous forest.

Threats: logging, followed by slash and burn agriculture.

Management suggestions: two of the six known locations are at possible secure areas (Lobéké has been proposed as a National Park and Boukoko is a research station). These should be the best choices for conservation action for *Balsamocitrus camerunensis*, beginning with efforts to refind the species, to collect basic population data as a baseline for monitoring and to gauge intervention needs with the Park/Station managers.

Vepris oubanguiensis (Aubrév. & Pellegr.) Onana comb. nov. (Basionym: *Teclea oubanguiensis* Aubrév. & Pellegr. Not. Syst. 49: 60 (1950); Fl. For. Soudano-Guinéenne: 362, pl.75, 2 (1950). Type: CAR, 60 km N of Bambari, *Tisserant* 1737 (P)).
EN B2ab(iii)

Range: Cameroon (Adamawa: Ngaoundéré) and CAR (60 km N Bambari).

Teclea oubanguiensis is here formally transferred from *Teclea* to *Vepris* by J.-M. Onana. It is here assessed as Endangered since two locations (above, AOO 8 km² with 4 km² cells) are known and threats are as below.

Shrub 3–4 m, stems grey with pale lenticels; leaves 1–3-foliolate, petiole to 6 cm; leaflets lanceolate, 9–20 × 3–6cm, acuminte-mucronate, [illegible] glands apparent on both surfaces; racemes [illegible] axillary and terminal, 8–12 cm, pubescent, bracteoles [illegible]ate; flowers sessile, 4-merous, c. 3 mm; fruit ellipsoid, 8 × 6 mm, non warty.

Habitat: not known, possibly evergreen forest along water courses.

Threats: this species has not been recorded for fifty years and is probably extremely rare. Conversion of forest to agriculture, grazing of livestock and management of habitat to favour this practice, are possible threats.

Management suggestions: efforts should be made to refind this species and to enumerate threats, regeneration levels, number of individuals and population density to assist future monitoring and assess needs for management interventions. Local communitites should be engaged in protection if shrubs are found within their areas of influence.

Vepris soyauxii (Engl.) Mziray (syn. *Araliopsis soyauxii* Engl.)
VU B2ab(iii)

Range: SE Nigeria (Calabar; Umon Ndealichi FR), Cameroon (SW: SW Korup NP; Mt Cameroon – S Bakundu FR; Bimbia-Bonadikombe. Littoral: Yingui-Yabassi; 5 km E Mbanga) and Gabon (Sibang, Libreville; Doudou Mts).

Here *Vepris soyauxii* is assessed as Vulnerable since seven locations (above; AOO 28 km² with 4 km² cells) are known, with threats as below.

Tree 25–30 m, trunk slightly fluted with cork outgrowths, otherwise smooth, bark ash-grey, outer slash yellow-orange with red specks; leaves digitately 5–7-foliolate, petiole 30 cm; petiolule 1–3 cm, grooved above; leaflets obovate to oblanceolate or oblong, to 40 × 14 cm, acumen to 2 cm, lateral nerves 10–20 pairs, straight; tertiary nerves parallel to secondaries, numerous; panicles subglobose, 15 cm in fruit, puberulent; fruit globose, 1 cm, 4-seeded.

Habitat: lowland evergreen forest in coastal districts.

Threats: harvesting of bark for medicinal purposes in SW Region Cameroon (pers. obs.), urbanisation (Calabar, Bimbia & Libreville); logging followed by agriculture.

Management suggestions: the best hopes for conserving this tree are in National Parks such as Korup and Doudou. Painting trunks with noxious paint could relieve pressure from debarking. Data on tree density, regeneration levels and population data is probably available in Korup from the two large plot programmes there. This will serves as a baseline for monitoring activities in future.

Vepris trifoliolata (Engl.) Mziray (syn. *Oricia trifoliolata* (Engl.) Verdoorn)
EN B2ab(iii)

Range: Cameroon endemic: (SW: Korup; Mt Cameroon; Mt Kupe and the Bakossi Mts).

Tree, branches pubescent; leaves 3-foliolate, petiole 3–16 cm, pubescent; petiolules 2–3 cm, blade narrow oblong to broadly ovate, [illegible] × 4–12 cm, acumen short, lateral nerves 10–12 pairs, glabrous; panicles puberulent, pedicels tomentose, 3 mm; flowers 4-merous, petals 3.5 mm, puberulent.

This species, under its synonym, was assessed as VU B1 + 2c by WCMC (1997) and is listed thus in IUCN (2003; www.redlist.org). The assessment was based on data from Peguy provided in 1997 and uncritically maintained in Cheek *et al.* (2004: 183). However this is reassessed as EN since there are only five locations (above; AOO 20 km² with 4 km² cells), with threats as below.
Two collections were made at Limbe before 1931 (Flore du Cameroun 30 (1963)) but the taxon was not refound in the Mt Cameroon area in the intensive surveys of the 1990s (Cable & Cheek 1998).
Habitat: lowland and submontane evergreen forest; 0–1600 m alt.
Threats: habitat decline through logging and the expansion of areas under cultivation (Peguy data). Lower altitude forest is under severe threat in much of SW Region, especially around Mt Cameroon, where it has not already been destroyed.
Management suggestions: a revision of central African *Vepris* species would enable development of a more effective identification guide than is now available. Confusion with other species of *Vepris* is possible currently. Mt Kupe and the Bakossi Mts may now be the stronghold for this species, although Korup, as lowland forest NP, is even more secure. Data on numbers of individuals per site and demography should be gathered to develop a management plan for the species.

SAPINDACEAE

assessed by Martin Cheek

A pantropical family which are common in tropical African forest. The Flore du Cameroun account for Sapindaceae (Fouilloy & N. Hallé 1973, Flore du Cameroun 16) recognises 21 genera and 81 species for Cameroon. Most Cameroonian species are understorey evergreen forest trees, but many are canopy trees; rarely are species climbers (e.g. *Paullinia*) or herbaceous (*Cardiospermum*). With the exception of the genera already mentioned, and *Allophylus*, most genera have alternate, pinnately compound leaves, lacking stipules, distinguished from other woody families with such leaves by the highly reduced, vestigial terminal leaflet. However, in *Pancovia* and *Placodiscus* this is not present. The flowers are usually slightly zygomorphic, small, and numerous in the inflorescences. Usually the petals are free, with an adaxial flap. Very often the carpels of the ovary are incompletely joined, lending to strongly 3–5-lobed fruits, or even (*Deinbollia*) apocarpous fruits, with only a single carpel usually developing to maturity and the other carpels, while present, remaining small. Generally speaking, Sapindaceae are not very useful to man except a few species cultivated for their fleshy seed stalks, such as akee, litchi, rambutan. Since the Flore du Cameroun account, several papers describing extra species, and one new genus have been published. Even today new taxa remain to be published in Cameroon.

Allophylus is a large pantropical genus of trees and shrubs distinguished from the other Sapindaceae by having three leaflets. Famously, the numerous species were reduced to one, *A. cobbe* L. by a Dutch researcher in the 1960's, but most subsequent workers have not taken up his idea. Fouilloy & Hallé (1973 op. cit.) recognised 20 species in Cameroon, since when one, *A. ujori* Cheek, has been added and another, *A. conraui* Gilg, redefined.

Deinbollia laurentii De Wilde. var. *laurentii* is known from one location, Lobeké in Cameroon but with c. 10 locations in CAR and, mainly Congo (Kinshasa), it is likely to be NT.

Deinbollia maxima Gilg was previously assessed in Cheek *et al* (2004: 184) as VU B2ab(iii). It is here downrated to NT on discovery of numerous additional records (34 on Gabon database) from Gabon, taking the locality number beyond the threshold for threatened categories.

Placodiscus opacus Radlk. was assessed in Cheek *et al.* (2004: 184) as VU B2ab(iii) but is reassessed here as NT. Formerly known from seven locations in Cameroon, CAR, Equatorial Guinea (Rio Muni) and Gabon, numerous additional locations in Gabon have come to light through the Gabon database where a total of 24 records are cited!

Allophylus bullatus Radlk.

VU A2c

Range: SE Nigeria (Obudu and Mambilla) and Cameroon (SW–NW: Mt Cameroon to Bamenda Highlands).
A 15–18 m tree unusual in the genus in occurring at high altitude and in having leaves that dry brown, with white-tufted domatia occurring along the secondary nerves of the leaflets. The bullate leaves of the epithet are mainly seen only on Mt Cameroon, they type locality.
This understorey tree of upper submontane to montane forest, while secure on Mt Cameroon and Mt Kupe, has lost large tracts of its habitat in recent decades in the Bamenda Highlands. Over 30% of its overall habitat is estimated to have been lost in the last 100 years. This species was first assessed in Cheek *et al.* (2004: 183–184) and that assessment is maintained here since the additional records of the species from Dom and Lebialem (Cheek *et al.* 2010: 104, Harvey *et al.* 2010: 84) while providing evidence that a higher proportion of the range is under threat than was previously known, are believed not to take the species across the threshold to EN status. In Cheek *et al.* 2004: 183–184 the

range was erroneously stated to extend to Príncipe & São Tomé.

Habitat: upper submontane and montane forest; 1600–2400 m alt.

Threats: clearance of forest for agriculture and wood, particularly in the Bamenda Highlands of Cameroon, once probably the main area for *A. bullatus*. It has been estimated that as much as 96.5% of the original forest has been lost (Cheek *et al.* 2000: 49–50). Study of one area here (Moat in Cheek *et al.* 2000: back cover) showed that c. 25% of surviving forest was lost between 1987–1995 and in another (Baena in Cheek *et al.* 2010: back cover) c. 50% was lost in the five years 1998–2003.

Management suggestions: improved policing of the existing protected areas might secure the future of this species. Support from local communities in protecting this species and its habitat should be sought and assisted.

Allophylus conraui Gilg

EN B2a,b(iii)

Range: Cameroon (SW: Bakossi Mts; Banyang Mbo; Mone FR; "Kebo", Lebialem-Bechati)

The assessment made here is maintained from that in Cheek M. & Etuge, M. (2009). *Allophylus conraui* (*Sapindaceae*) reassessed and *Allophylus ujori* described from western Cameroon. Kew Bull. 64(3): 495–502. Since that publication only one other location, at Mone FR has come to light. Here it is locally common at several spots.

An area of occupancy of 20 km² is estimated for the species, based on a grid cell size of 4 km² and five locations being known. Threats are given below. Accordingly *A. conraui* is here maintained as Endangered, EN B2a,b(iii).

False records: Harris (2002: 189) records *Allophylus conraui* from the CAR, but indicates some doubt.as to the delimitation of the taxon. Subsequent observations (Harris pers. comm.. to Cheek 2008) of the specimens on which the record is based, *Harris* 3308 and 4815 (E) indicates that these represent another taxon. The CAR taxon has hairs only 0.3–0.4 mm (not 3 mm) long, inflorescences twice as long as broad (not c. four times as long as broad); infructescences accrescent, 11.5 cm long (not non-accrescent, 2–4 cm long). Jongkind (2006: 379) cites *A. conraui* as occurring in Gabon, based on *Reitsma* 2996 and *Wilks* 2721. Observations of the first (the second specimen was not available), loaned from WAG to K, shows it to be another taxon, differing in the hairs <0.5 mm long (not 3 mm long); leaflets papery, marginal teeth inconspicuous,1–2 per side (not membranous, marginal [illegible]).

Habitat: lowland evergreen [illegible] alt.

Threats: the site in the [illegible] Mts between Banyemem and Ayong is vulnerable to logging as is that to the N, in the Banyang Mbo Reserve which has had its protected status downrated. At Bechati in the Lebialem Highlands, the plant is under threat of forest clearance for agriculture, while there is a possibility that Mone FR will be legally logged. In short, it is secure at none of its locations.

Management suggestions: any moves to log the Mone FR, the site of the largest and most important subpopulation of *A. conraui* should be rejected since this is likely to destroy the habitat of this rare species. Mone Forest is the logical choice for the focus of conservation efforts for this species.

Allophylus hamatus Hauman

VU B2ab(iii)

Range: Cameroon (C: Nanga Eboko. E: Bertoua; Garoua Sambé; Mt Songogo; Batouri), CAR (1 loc.) and Congo (Kinshasa) (Eala; Ikela; Dundusana; Yangambi)

Extremely distinctive in the genus in being both a liana and in being spiny, *A. hamatus* is here assessed as Vulnerable since only ten locations (AOO 40 km² with 4 km² cells) are known, with threats as below.

Habitat: lowland inundated (Congo (Kinshasa)) or semi-evergreen Congolian forest, often along rivers (Cameroon); low-mid alt.

Threats: logging of forest followed by agriculture is a threat within the Cameroonian range of this species.

Management suggestions: taxonomic investigation is needed to ensure that the Cameroonian material, disjunct from the Congolian, is conspecific, inclusion of one site within a National Park should be investigated and if impossible, communities proximal to one or more sites should be engaged and educated regarding its protection. Data on density and regeneration levels are needed.

Allophylus megaphyllus Hutch. & Dalz.

EN B2ab(iii)

Range: SE Nigeria (Oban) and Cameroon (SW: SW Korup; Mt Cameroon at Onge; S Bakundu).

This shrub is one of the few in the genus with unifoliolate leaves and unbranched inflorescences. The oblong-oblanceolate leaves are 18–45 × 7–15 cm, much larger than in most other species. With five locations (above) and an AOO of 20 km² (using 4 km² cells) and threats as below, *A. megaphyllus* is here assessed as EN.

Habitat: lowland evergreen forest.

Threats: tree felling followed by agriculture especially yam farming, has been long standing in S Bakundu; Onge is under threat of logging, while Nigeria has seen very heavy forest losses in recent decades, including in SE Nigeria.

Management suggestions: the logical focus for protecting *A. megaphyllus* should be the Korup NP since this is the most secure, possibly the only secure site, at which the species occurs. More data on population density and regeneration levels are needed.

Allophylus oyemensis Pellegr.

EN B2ab(iii)

Range: Cameroon (C: Mfou-Akonolinga near Yaoundé) and Gabon (Acam; Oyem; Moyen Ogooue)

A shrub with stems and nerves of the underside of the blades inverted with 0.5 mm long hairs, lacking domatia, secondary nerves terminating in mucronate teeth; leaves 11–15 × 5–6 cm; inflorescence dense, 15–20 cm long.

Here *A. oyemensis* is assessed as EN since three or four locations (AOO 12 km² with 4 km² cells) exist, with threats as below.

Habitat: lowland forest, probably evergreen.

Threats: in Gabon unknown. At Yaoundé, capital city of Cameroon, urban expansion and conversion of forest to agriculture after logging (pers. obs.).

Management suggestions: rediscovering of this taxon at its Yaoundé site should be sought and the means to protect it in Cameroon. Education and community engagement should be attempted if formed protection is problematic. One specimen locality, Acam, was not located so is not mapped.

Allophylus ujori Cheek

EN B2ab(iii)

Range: Nigeria (Mambilla-Gembu, Mayo Naga-Njawe (1 collection each)) and Cameroon (NW: Bali Ngemba FR (3 coll.); Dom (2 coll.)).

This 6 m forest understorey tree is unusual in its genus for being spiny. Published in 1908, it is only known from four locations. There may be a fifth, Kebo, but this has not been located, and may be a Germanic version of the modern Batibo with the usual Bantu place name prefix Ba- omitted. The assessment above is updated from that in Cheek *et al.* (2010: 104–105) and Cheek in Harvey *et al.* (2004: 69–70 [as *A. conraui*]). There are four locations (AOO 16 km² with 4 km² cells) exist, with threats as below.

Habitat: evergreen forest; 1400–1600 m alt.

Threats: forest clearance for agriculture (e.g. Bali Ngemba FR where farming has increased inside the forest year by year) and in the Mambilla plateau area of Nigeria (H. Chapman pers. comm.). At Dom, its forest habitat at one site, near Kinjanjang Rock, remains outside the protected areas.

Management suggestions: the status of forest at the two Nigerian sites needs to be ascertained. This would best be done by reference to Chapman and Olsen, who are conducting surveys in the area. In Cameroon, survey work should focus on Bali Ngemba and Dom where several individuals of *A. ujori* have been seen recently (pers. obs.). Seed of this species should be collected and raised for incorporation in forest restoration plantings such as that at Dom.

A definitive enumeration of the subpopulation, and assessment of regeneration and demography, would provide the data needed to formulate a management plan for this species.

Note: this species was mistakenly named as *Allophylus conraui* Gilg ex Radlk. in the Bali Ngemba checklist.

Allophylus zenkeri Radlk.

VU B2ab(iii)

Range: Cameroon (SW: Mt Cameroon–Bomana. S: Bipindi. Littoral: Douala?), Gabon (Libreville and Mondah; Lope) and Congo-Brazzaville (La Patte d'Oie).

A 2–3 m shrub or tree to 5–10 m tall, stems minutely hirsute, leaves 12–18 × 5–6 cm, midrib velvety on both surfaces, otherwise glabrous, with minute domatia. The inflorescences are as long as the leaves.

Here *A. zenkeri* is assessed as Vulnerable since there are potentially seven locations (AOO 28 km² with 4 km² cells) and threats as below. However, while five of the locations derive from Fouilloy and Hallé, two others are less certain and confirmation of identification is needed.

Habitat: lowland evergreen forest.

Threats: since Douala is Cameroon's largest city, urban expansion at this site must be a threat, as at Libreville in Gabon, if the species has not already been lost at these locations.

Management suggestions: options for protecting plants at those sites outside cities, such as at Bipindi, should be investigated if the species can be refound at one of these. Population density and regeneration levels should be recorded if possible, to gauge managemnt needs.

Aporrhiza are branched shrubs or small trees, rarely climbers, with petiolate leaves often hairy on both surfaces, which lack or have only a short acumen and have an entire margin. The 2-valved fruits have an arillate seed in each of the 2 locules.

Aporrhiza multijuga Gilg

CR B2ab(iii)

Range: Cameroon (S: Lolodorf, type, destroyed. ? SW: Korup).

A climbing shrub with 10–11 pairs of leaflets in each 60–100 cm leaf, the leaflets are 12–30 × 6–8 cm with 15–18 pairs of nerves and inflorescences 15–25 cm long. Here *A. multijuga* is assessed as CR since 1 location (AOO 4 km² with 4 km² cells) is known with threats as below. The type collection (*Staudt* 9) was made at the end of the 19th century. A second specimen, *Thomas* 3354 from Korup has been identified by Gereau (MO data) as this taxon, but consists of leaves only and so must be treated with caution. If verified with fertile material this will be excellent news for the future of this species.

Habitat: lowland evergreen forest.

Threats: slash and burn agriculture at Lolodorf.
Management suggestions: the assistance of Thomas is needed to attempt to rediscover his 3354 at Korup and verify its identification. If successful a census of the population is advisable and data on regeneration levels should also be obtained. Otherwise efforts should focus on refinding the species at Lolodorf.

Aporrhiza tessmannii Radlk.
EN B2ab(iii)

Range: Equatorial Guinea (Rio Muni: Nkolentangan), Cameroon (S: Ébolowa; Ébolowa–Sangmélima) and Gabon (Nyanga).
This species has not been seen for about a hundred years. A shrub with 30–40 cm long glabrous, glossy, 4–7-jugate leathery leaves; the inflorescences are larger than the leaves and densely flowered. Here *A.. tessmannii* is assessed as Endangered since only three locations are known (AOO 12 km² with 4 km² cells) with threats as below.
Habitat: lowland evergreen forest.
Threats: unknown in Rio Muni and Gabon, but in Cameroon slash and burn agriculture is a concern around the large towns of Ébolowa and Sangmélima.
Management suggestions: effort should be made to refind this species in the wild and to protect it through the support of local communities since none of its sites seem to be in protected areas. Basic populational data should be gathered to aid management.

Deinbollia is an African-Malagasy genus of c. 40 species of shrub or small tree with entire, regular paripinnate leaflets and thyrse-like inflorescences. It is easily distinguished by the fleshy indehiscent fruits often with more than one locule developing apocarpously.

Deinbollia angustifolia D.W.Thomas
VU D2

Range: Cameroon endemic (SW: Korup NP).
Published in 1986 (Ann. Missouri Bot. Gard. 73(1): 291).
A rheophytic shrub known only from the banks of the Mana R and tributaries within a small area of a few km². Essentially this equates to a single location since a single disastrous event could destroy the species.
This is a slender shrub c. 1 m tall, the leaves have a single pair of slender leaflets 10–30 × 0.8–2.2 cm, petioles 0.6–1.5 cm long. The panicles are often terminal, to 12 cm long bearing white flowers 1–2 mm wide, followed by yellow, 1.5 cm fruit.
Here *D. angustifolia* is assessed as Vulnerable under Criterion D2 since no obvious immediate threats are known. The rocky river banks of its habitat are not under any agricultural or development threat, so while the AOO may be only 4 km² at one site, with possibly only on 200 individuals, in the absence of such threats only criterion D is applicable.
Habitat: rocky river banks of the Mana in evergreen forest; 50–100 m alt.
Threats: see above. None are immediate, but in future hydroelectric projects might threaten the species. The river marks the boundary of the National Park and much of one side is occupied by Palm Oil plantations which might in future threaten the plants on one bank, conceivably.
Management suggestions: basic populational data should be gathered as a baseline for future monitoring by the Korup authorities.

Deinbollia dasybotrys Radlk.
EN B2ab(iii)

Range: Cameroon (S: Fenda, Kribi) and Gabon (Estuaire: La Mvoum. Woleu-Ntem: Lalara-Makoukou).
A single-stemmed shrub with 4–5 jugate leaves, petiole 10–12 cm leaflets 18–20 × 6–8 cm and very short inflorescences (2–6 cm) with glabrous flowers buds.
Here *D. dasybotrys* is assessed as Endangered since only 3 localities are known (AOO 12 km² with 4 km² cells) and threats are as below.
Habitat: lowland evergreen forest.
Threats: unknown in Gabon, but numerous in the Kribi area: logging, slash and burn agriculture; road widening and development in connection with extractive industries, development of new ports, urban expansion, tourist infrastructure.
Management suggestions: investigation is needed to discover if any of the sites fall within Gabon's numerous new National Parks, so that their management authorities can be informed. In Cameroon, Fenda should be visisted and an attempt made to refind the sites so as to inform traditional authorities and invoke their assistance in protecting this species.

Deinbollia insignis Hook.f.
VU B2ab(iii)

Range: Nigeria (Sapoba FR (4 coll.)), Equatorial Guinea (Bioko (2 coll.)) and Cameroon (SW: Bambuko FR (2 coll.); Mt Kupe (2 coll., two sites)).
This is a large treelet, with leaves to 1 m long. Previously assessed in Cheek *et al.* (2004: 184) as VU, that assessment is raised here to EN despite no new data being available, since only five locations (AOO 20 km² with 4 km² cells) are known, with threats as below. Despite survey work in the past five years, no new records of the species have been made, and every one of its known sites has suffered from ongoing habitat degradation. It may even be extinct in Nigeria and Bioko. It is suspected that, when better data are available, this species may prove to be Critically Endangered.

Habitat: lowland forest; 400 m alt.
Threats: extremely vulnerable due to clearance of lowland forest for agriculture. It may well be extinct on Bioko due to extensive forest clearance there for cocoa plantations in the late 19th and 20th centuries. It may also be extinct in Nigeria due to extensive forest loss there in the late 20th century. Forest loss at Bambuko is documented in Cable & Cheek (1998: xxxiv–xxxv). It is notable that the species was not found elsewhere around Mt Cameroon during the intensive surveys of the early 1990s. At Mt Kupe it is vulnerable due to its low altitude, placing it outside of the proposed new protected area.
Management suggestions: surveys should be made to rediscover this species at its known sites. It seems possible that it may only survive at Mt Kupe (where the most recent collections are recorded). It is therefore advised that efforts to protect the tree might be centred here. Individuals need to be rediscovered, demographic data obtained and protection secured with the help of local communities.

Deinbollia macrantha Radlk.
CR B2ab(iii)
Range: Cameroon endemic (SW: Kumba).
Small branched tree with leaves 1 m long, leaflets oblong, 20–22 × 6–7 cm, acumen 1.5 cm long, petiolule 0–2 mm long. Inflorescences 40 cm long, flower buds glabrous, 6–7 mm diam.
Here *Deinbollia macrantha* is assessed as Critically Endangered since a single location (AOO 4 km² with 4 km² cells) is known with threats as below. Not seen since the type collection 100 years ago.
Habitat: lowland evergreen forest.
Threats: urbanisation and agriculture. Kumba, the only known location, has expanded greatly in the century since this species was last seen.
Management suggestions: this species should be searched for in intact forest habitat near Kumba, such as Barombi Mbo, which should be considered for National Park status in view of the concentration of rare species of plant there. Basic populational data on the species should be acquired to aid management for its survival, if it is not already extinct.

Deinbollia macroura Radlk.
EN B2ab(iii)
Range: Cameroon (S: Ébolowa to Campo at Kribi; Campo-Ma'an) and Gabon (Lopé).
A branched shrub. The 3-jugate leaves have membranous, oblong, 21 × 6 cm leaflets, the acumen 2 cm long, petioles 8–11 cm long, inflorescences 2–6 cm long.
Here *D. macroura* is assessed as Endangered since three locations are known (above, AOO 12 km² with 4 km² cells), with threats as below.
Habitat: lowland evergreen forest.
Threats: none are known, or likely at two of the three locations, but at the type locality, Kribi, there has been habitat loss which is likely to continue due to new infrastructure projects. Urban expansion and touristic infrastructure are also threats to the forest habitat there.
Management suggestions: if specimen records for Lopé and Campo-Ma'an are verified, these, as National Parks, are ideal locations to focus conservation efforts. The species should be refound and census data taken of regeneration levels, range and density, so as to gauge need for managemenet interventions.

Deinbollia mezilii D.W.Thomas & D.J.Harris
EN B2ab(iii)
Range: Cameroon endemic (S: Campo-Ma'an NP and Dipikar ls.; 20 km E Campo at Nkoelon).
Also reported from Campo-Ma'an by Tchoutou (2004).
An unbranched treelet to 1.5 m with leaves clustered in a terminal, litter-gathering rosette, petiole 1 cm, swollen at base and apex, bearing a single pair of oblanceolate leaflets, inflorescence terminal, 5–6 cm.
Published in 2000 (Kew Bull. 54: 954), *D. mezilii* is here assessed as Endangered since there are two locations (above, AOO 8 km² with 4 km² cells) with threats as below.
Habitat: lowland evergreen forest.
Threats: slash and burn agriculture for *Manihot* cultivation at the site near Nkoelon village. The other site(s) is/are not reported near villages and are within the Campo-Ma'an NP.
Management suggestions: meeting with the elders of Nkoelon to explain the importance of *D. mezilii* and provide the means to identify it (e.g. a poster) is advisable, similarly for the managers of the National Park. Generally forest at the southern end of Campo-Ma'an, in the Dipikar area, is thought to be intact and relatively secure.

Deinbollia oreophila Cheek
VU A2c
Range: Nigeria (Obudu plateau) and Cameroon (SW–NW: Cameroon Highlands from Mt Cameroon to Bali Ngemba FR).
The assessment above, which appeared with the original publication of this species is maintained here (Cheek, M. & Etuge, M. (2009). A new submontane species of *Deinbollia* (*Sapindaceae*) from Western Cameroon and adjoining Nigeria. Kew Bull. 64(3): 503–508). *Deinbollia oreophila* was rated as VU A2c, i.e. Vulnerable, given the continued loss of its submontane forest habitat, rated at 30% over its whole range over the last 100 years. In the northern part of its range, recent loss of its surviving forest habitat varies from 25% in 8 years to 50% in 15 years.
Habitat: understorey of submontane evergreen forest; (880–)1000–2050 m.

Threats: forest clearance for agriculture, especially at Mt Cameroon-Bambuko, Obudu, Lebialem-Fosimondi, Bali Ngemba, where it is active and ongoing.
Management suggestions: the logical focus for conservation of this species is Mt Kupe, where it is especially common and relatively well protected. Although reasonable data on population density is available, information is still lacking on regeneration levels and growth rates.

Deinbollia pycnophylla Radlk.
EN B2ab(iii)
Range: Cameroon (SW: Korup; Mt Cameroon-Batoke. S: Bipindi; Campo-Ma'an, Tchoutou 2004) and Gabon (Libreville).
A treelet of 3–5 m, usually unbranched, with a cluster of terminal leaves with 16–19 pairs of leaflets, the leaflets 11–22 × 2.5–5 cm, lateral nerves 14–20 pairs, arching and uniting near the margin. Inflorescence terminal, 40 cm, bearing c. 20 branches.
Here *D. pycnophylla* is assessed as Endangered since there are four extant localities known (AOO 16 km² with 4 km² cells) and threats as below.
Four Zenker collections are known from Bipindi. The Libreville record is the type of *D. klainei*, given as a synonym of *D. pycnophylla* by Fouilloy & Hallé (1973, op. cit.).
Habitat: lowland evergreen coastal forest.
Threats: urbanisation at the Libreville site is a major threat to this species, if it still survives there, which is unlikely. At Bipindi and Campo-Ma'an, slash and burn agriculture are lower-ranking threats.
Management suggestions: the Korup site, if verified with a specimen record, is the logical best focus for conservation efforts since it is the best protected. Data on population density and regeneration levels should be obtained (if not already available from large plot programmes at Korup) so as to inform management of this species.

Deinbollia saligna Keay
VU D2
Range SE Nigeria (Kwa falls) and Cameroon (SW: Ndian R).
A shrub or small tree 0.2–1(–2.5) m tall, sparingly to much branched, c. 1 cm diam. at base. Leaves in terminal clusters, with 10–14 leaflets on rachis 6.5–15 cm, petiole 3–9 cm. Leaflets very narrowly lanceolate to strap-shaped, glabrous, leathery, (4.5–)8–12.5 × 0.7–[illegible] apex long acuminate–rounded, base cuneate, ner[illegible] the leaves, c. 10 × 2 cm, bearing flowers [illegible] short stemmed clusters along the length of the axis; flowers 3 × 3 mm, white. Fruits 1-seeded, orange, obliquely ovoid, c. 14 × 12 mm. seeds with green embryo. Published in Bull. Jard. Bot. Brux. 26: 193 (1956).
Here *D. saligna* is assessed as Vulnerable under Criterion D2 since no obvious immediate threats are known. The rocky river banks of its habitat are not under any agricultural or development threat, so while the AOO may be only 4 km² at one site, with possibly only 50 individuals, in the absence of such threats only criterion D is applicable.
Habitat: "Seasonally inundated rocks by Ndian waterfall" (*Thomas* 4268); "waterfall, on the rocks" (*Ujor* in FHI 31776); "Crevices in fully exposed rocks at top of falls" (*D.H.Maggs* 153); "Banks of the Ndian R." (*J.Smith* /80/36). rocky river banks of the Mana in evergreen forest; 50–100 m alt.
Threats: none are immediate, but in future hydroelectric projects might threaten the species. The river marks the boundary of the National Park and much of one side is occupied by palm oil plantations which might in future threaten the plants on one bank, conceivably.
Management suggestions: basic populational data should be gathered as a baseline for future monitoring by the Korup authorities.

Deinbollia unijuga D.W.Thomas
EN B2ab(iii)
Range: Cameroon endemic (SW: Ekondo Titi to Mundemba; Ekumbako; Mundemba).
A forest shrub (–3) m tall, usually unbranched, leaves clustered at stem apex, glabrous; petioles 1–4 cm, leaflets 1-pair, oblanceolate, 17–50 × 5–14 cm, acuminate, base unequal, cuneate lateral nerves 12–18 pairs, looped close to margin, prominent on both surfaces. Inflorescences lateral, borne among the leaves, panicles to 6 cm.
Described by Thomas (Ann. Missouri Bot.Gard. 73: 219–221 (1986)) and known only from specimens collected by him, this is a very rare species, even within its small range.
Habitat: lowland evergreen forest on sand; 50–100 m alt.
Threats: urban expansion at Mundemba; slash and burn agriculture at the other two locations.
Management suggestions: although close, none of the three closely placed locations are within the protection of the Korup NP. Therefore conservation of this species is dependent upon the interest and support of the communities nearest each site, whose attention should be drawn as the rarity of this species and provided with the means to protect it. Further data on population density, range and regeneration would help understand the ecology and management needs of this species.

Leconiodiscus punctatus J.B.Hall
EN B2ab(iii)

Range: Ghana (a few forests around Atewa/Kibi: Bunsu, Oboden – Hawthorne & Jongkind: 752) and Cameroon (SW: Mt Cameroon–Bambuko FR near Kuke Bova).
Published in Bull. Jard. Bot. Natl. Belg. 50(1–2): 262 (1980).
This species is reported to be endemic to a few forests around Atewa/Kibi in Ghana, and to Cameroon at Bambuko. Here it is assessed as EN since four locations are known (AOO 16 km² with 4 km cells) and threats as below.
At Bambuko, Keay reported it as abundant in one part of the reserve.
A tree with spreading crown, to 18 m high, numerous erect shoots from the base, branching low, branches arched, irregular in section. The 20–30 cm long leaf has 4(–8) pairs of glabrous leaflets, each with 14–19 pairs of nerves, translucid dots; the glandular hairs of the inflorescence and fruit give a red juice.
Habitat: lowland evergreen forest.
Threats: Bambuko is reported not to be part of the new Mt Cameroon NP. Increasing large areas of Bambuko, being with fertile volcanic soils, are cleared to grow fruit and vegetables for the Douala and Gabonese markets. Threats in Ghana are unknown.
Management suggestions: efforts to refind this tree at Bambuko should be made before it is extinct in Cameroon. Cultivation at Limbe BG is a possibility that should be supported. This species may already be protected in Ghana.

Namatea simplicifolia D.W.Thomas & D.J.Harris
EN B2ab(iii)
Range: Nigeria (Oban) and Cameroon (SW: Korup NP; Mt Cameroon–Onge).
Published in Kew Bull. 54: 952 (2000), this is a much branched tree up to 6 m tall with glabrescent branches and has, very unusually in the Sapindaceae, simple leaves. The leaves are oblanceolate, 8–43 × 3–13 cm, acumen 1–3 cm, base cordate-auriculate, lateral nerves 11–14 pairs, with pit domatia glabrous or with scattered hairs; inflorescence spike-like, 2–10(–28)cm.
Namatea is a monotypic genus, for the moment. Here it is assessed as Endangered since only, three locations are known (AOO 12 km² with 4 km² cells) and threats are as below.
Habitat: lowland evergreen forest.
Threats: Nigeria has seen great losses of forest in recent years and Oban is also thought to have suffered degradation. Onge is under threat of logging.
Managements suggestions: Korup is the logical focus for conservation efforts for this species, being a National Park. Data on range, density and regeneration levels there should be obtained so as to enable future monitoring and intervention if needed. If refound at Oban and Onge locations, local communities should be empowered to recognise and protect the plants.

Pancovia polyantha Engler
CR B2ab(iii)
Range: Cameroon endemic (SW: Kumba).
Included by Fouilloy & Hallé (op. cit: 124) as an unsatisfactorily known species, *P. polyantha* is included here with caution.
Probably a shrub, it has 3–4 jugate leaves with leaflets 14–20 × 4.5–8 cm, petiolules 1 cm long. The inflorescence is 15–30 cm long, flower buds 4 × 3 mm, the calyx tomentellous.
Not seen in the wild since the type collection, *Staudt* 906, in March 1897. Here *P. polyantha* is assessed as CR since only a single location is known (AOO 4 km² with 4 km² cells) with threats as below.
Habitat: lowland evergreen forest.
Threats: Kumba is a major town. Urbanisation and small-scale agriculture threaten forest habitat nearby. The original habitat of the species may have been lost already and the species may be extinct.
Management suggestions: searches for this species around Kumba are needed in March. Research for Staudt archives in Germany might give clues as to the exact type location and habitat. If refound, basic populational data should be gathered and authorities/land owners enabled to identify to protect the species.

Placodiscus angustifolius Radlk.
EN B2ab(iii)
Range: Cameroon (Mimfia–Bipindi), CAR (1 loc.), Gabon (Lopé) and Congo (Kinshasa) (Kasai).
See Fouilloy & Hallé (op. cit. 132) where this species is described as an 8–15 m tree with angular branches. The 7–12-jugate leaves bear linear leaflets 11–22(–30) × 2–4(–6) cm with 10–15 pairs of nerves, 8 mm petiolules. The 50–70 cm rachis has a 5 cm peduncle. Inflorescences are 40–50 cm with 3–5-flowered cymules.
Here *P. angustifolius* is assessed as EN since five locations (above; AOO 20 km² with 4 km² cells) are known, with threats as below.
Habitat: lowland evergreen forest.
Threats: while Lopé is a National Park other locations are not known to be protected. That of the type location in Cameroon near Bipindi faces slash and burn agriculture. Threats at the other locations are not known.
Management suggestions: at Lopé basic populational data should be gathered so that a baseline for monitoring exists and so that intervention needs can be assessed. At other locations, outside of protected areas, such as Bipindi, local authorities should be enabled to recognise and protect this plant.

Placodiscus caudatus Pierre ex Radlk.

EN B2ab(iii)

Range: Cameroon (SW: Korup (1 coll.); Bakossi Mts–Konye (confirmation of identity needed), CAR (1 coll.) and Gabon (Libreville and at Rabi-Kounga).

Placodiscus caudatus was assessed as EN in Cheek *et al.* (2004: 184). That assessment is maintained here since only one extra location has since come to light (Rabi-Kounga). The assessment is merited since the species is known from five locations (AOO 20 km² with 4 km² cells) with threats as below.

This is a 1 m tall cauliflorous treelet with 7–14 cm long flower spikes; the leaves are 3–4-jugate with 8–9 cm petioles and a 25 cm rachis, leaflets 12–15 × 7–8cm, 8–10 pairs of nerves.

Habitat: lowland evergreen forest; c. 300 m alt.

Threats: forest clearance for logging, followed by agriculture and urbanisation (Libreville location).

Management suggestions: a revision of the species of *Placodiscus*, many of which are poorly known, would result in a better understanding of their geographic ranges and identification. From the available evidence Gabon, with two known locations, most warrants concentration of resources in conserving this poorly known species. A survey is advised to rediscover the plant there, and gather demographic data to aid formulate a management plan. Currently, the species is only known to be protected at its Korup site.

Placodiscus turbinatus Radlk.

EN B2ab(iii)

Range: Nigeria (Ikom-Obudu; Oban FR at Orem; Iyamoyong FR, Old Calabar R) and Cameroon (SW: S Bakundu FR).

A tree 5–12 m tall with fruits borne on the trunk. The leaves are 30–40 cm long, with 4 pairs of leaflets, each lanceolate, acumen 10–15 mm long.

Here *Placodiscus turbinatus* is assessed as EN since five locations are known (AOO 20 km² with 4 km² cells) and threats as below.

Habitat: lowland evergreen forest

Threats: logging of forest followed by agriculture, at all locations. Most of the Nigerian locations may already be lost due to these threats. Almost all surviving Nigerian forest has been lost in recent decades. At S Bakundu, yam farming has degraded large areas of the forest.

Management suggestions: efforts should be made to refind this tree in Cameroon at S Bakundu, initially, if it is not already extinct. Basic [illegible] to aid future interven[illegible] for reintroduction prog[illegible] authorities should be empowered to identify and protect this tree.

REFERENCES

Harris, D.J. (2002). Vascular Plants of Dzangha-Sangha Reserve, Central African Republic. National Botanic Garden of Belgium, Scripta Botanica Belgica 23.

Hawthorne, W. & Jongkind, C. (2006). Woody Plants of Western African Forests, RBG, Kew.

Jongkind, C. in Sosef M.S.M. *et al.* (2006). Checklist of Gabonese Vascular Plants. National Botanic Garden of Belgium, Scripta Botanica Belgica 35

SAPOTACEAE

assessed by Martin Cheek

A pantropical family of trees and shrubs, Sapotaceae are characterised by their white, rubbery exudate when wounded- the source of the original chewing gum is a S American species. *Terminalia* branching is often seen in this family, as well as a tendency to a short brown indumentum of biramous hairs and clustering of leaves at the stem apex. The leaves are alternate, entire and simple, with pinnate nerves, and with or without a pair of stipules, persistent or not. The flowers are often held in dense axillary fascicles, they are regular, white, with free sepal lobes, sometimes in more than one whorl. The petals frequently with a pair of petaloid appendages, are often united in a tube at the base with the stamens attached to the inner part. The superior ovary has several to many locules, but a single style, and produces a fleshy, one to many-seeded berry or 1-seeded capsule with distinctive hard glossy, laterally flattened seeds which have a large, long, rough hilum (seed attachment area) along one edge.

Moabi (*Baillonella toxisperma*) and Douka (*Tieghemella africana*) are well-known Cameroonian timber trees both previously assessed as threatened, while *Synsepalum dulcificium*, the 'miracle berry' is of great interest to biochemistry researchers due to its fruit, containing FDC a protein sweetener with application to the food industry since it alters the taste bud receptors themselves.

Although Flore du Cameroun 2 by Aubréville in 1964 treated Sapotaceae, generic limits were altered by Pennington's global generic revision (The Genera of Sapotaceae, T.D. Pennington, New York Botanic Garden and RBG, Kew, 1991) and more recently data from molecular sources has resulted in upheaval and uncertainty of generic limits in the family. For example, suggesting that *Gambeya*, previously sunk into *Chrysophyllum*, should be resurrected in part (Mackinder pers. comm., Ewango pers. comm.). Revisions are thankfully in progress for several genera, *Manilkara* (Plana), *Synsepalum* (Ewango), *Englerophytum* (Gautier), and *Omphalocarpum* (Harvey), while *Mimusops* still requires a champion. Consequently, until these are completed and published, this treatment is far from perfect and highly incomplete, and future editions will probably see an increase in the number of threatened species. Already, new material from Cameroon has resulted in several newly published endemic species (see *Gluema korupensis, Lecomtedoxa plumose* and *Manilkara lososiana* below, all discovered in Korup, and yet more remain undescribed, at least three further new species in Korup, for example, also *Pradosia* sp. nov. (fide Onana ms (SW: Banyang Mbo. C: 30 km SW Ndikiniméki. S: 5 km SE Sangmélima). These are also likely to feature in a second edition of this book.

Additional possibly threatened species of Cameroonian Sapotaceae are:

In the genus *Manilkara*;
M. dukensis (Engl. & K.Krause) H.J.Lam (N: Douka near Garoua (*Ledermann* 3638). This is an old name of uncertain attribution
M. fouilloyana Aubrev & Pellegr. (Cameroon, Gabon, CAR, Congo (Kinshasa)) here assessed as Near Threatened since there are more than 10 locations known.
M. ilendensis (Engl.) H.J.Lam (S: Campo (*Ledermann* 627)). This is an old name of uncertain attribution.
M. kribiensis (Engl.) H.J. Lam (Cameroon). Four locations and seven collections are known to Onana but it is not clear if it is a good species or not (Ewango pers. comm.).
M. mabokeensis Aubrév. (Cameroon and CAR). Possibly only a variant of *M. obovata* (Ewango pers. comm.).
M. pellegriniana Tisser. & Sillans (Cameroon and CAR). Possibly not a good species (Ewango pers. Comm..)

In the genus *Mimusops* which requires revision before conservation assessments can be made, the following are old names of uncertain attribution:
M. dodoensis Engl. (Adamawa: at Dodéo (*Ledermann* 2976, 2891)).
M. ebolowensis Engl & K.Krause (S: Ébolowa-Sangmélima (*Ledermann* 5562)).
M. mildbraedii Engl & K.Krause (S: Campo (*Schorrkoff* 42)).
M. ngembe Engl & K.Krause (S: Campo (*Schorkoff* 42)).

Pouteria hexastemon Baehni (SW: Limbe (*Ludwigs* 302)).

Baillonella toxisperma Pierre
VU A1cd

Range: E Nigeria, Cameroon, Gabon, W Congo (Kinshasa) and Angola: Cabinda.

This species was assessed as Vulnerable by L.White (1997) and is listed so by IUCN (2003; www.redlist.org), on the basis that the species is overexploited for its timber and is therefore declining in large parts of its range, that maturation rates are slow (90–100 years), and that regeneration is limited by occurring only under a closed canopy. *B. toxisperma* is recorded as the second most important exported timber species in Gabon. We maintain this assessment here. This species is not mapped since so may records are known. EOO is calculated as 211981 km^2 based on an incomplete data set.

Tree to 12 m tall, c. 80 cm dbh; bole nearly black, ridged, hard; slash yellow, white exudate; leaves clustered at ends of branches; stipules lanceolate, blades 20–30 × 6–10 cm, rounded with acuminate apex, cuneate at base, young leaves with chestnut pubescence, subglabrous when mature although hairs persistent on midrib; inflorescence of dense flowering fascicles at the branch tips; pedicels 2–3 cm, pubescent; calyx c. 1 cm long, with 8 lobes, 4 inner and 4 outer, pubescent on exterior surface, corolla with 8 lobes, each with 2 dorsal appendages longer than the lobes (5.5 mm), tube 2.5 mm long, lobes c. 4 mm long; fruits large, spherical, c. 6.5 cm diam., grey-green; 1–2-seeded in a yellowish white pulp; seeds ellipsoid, c. 4.2 × 2.5 × 2 cm, ventral scar nearly the entire length of the convex shaped ventral face. (Description from Harvey pp. 400 in Cheek *et al.* 2004.).

Habitat: closed canopy lowland primary rainforest and old secondary forest, 0–500 m alt.

Threats: see above.

Management suggestions: minimum exploitable diameters have been set in several countries (White, www.redlist.org (IUCN 2003)); these should be enforced where possible. Better protection and prevention of illegal logging in protected areas may ensure the future survival of this species.

Englerophytum oubanguiense (Aubrév. & Pellegr.) Aubrév. & Pellgr.
EN B2ab(iii)

Range: Nigeria (Sapoba), Cameroon (S: Dja at Bitye. C: Nanga Eboko between Koumbou and Azonkout) and CAR (N. Bambari; Yalinga).

Shrub or small tree of river edges and marshes, leaves alternate, simple, stipulate, petiole 1.5 cm; blade obovate-oblong, 12–18 × 3–6 cm, apex obtuse or emarginate, base attenuate, lower surface red tomentose, lateral nerves numerous, slender; flower fascicles on old wood, pedicels 1–4 cm, red tomentose; sepals 5, 4 mm; corolla 5–6 mm, glabrous, lobes as long as tube; stamens 5, united in a tube exserted from throat; fruit 3 cm, red tomentose; seed 2.5 × 1.5 × 1 cm.

Here *Englerophytum oubanguiense* is assessed as Endangered since only five locations are known (above; AOO 20 km^2 with 4 km^2 cells) and with threats as stated. EOO is calculated as 379027 km^2.

Most of the data in this treatment is taken from the protologue in Bull. Soc. Bot. France 104: 277 (1957) and from Flore du Cameroun 2. A field identification of this species from Mt Cameroon, at Onge, needs confirmation before it can be accepted (*Ndam* 673).

Habitat: river edge and swamp and swamp in inland forest habitats.

Threats: forest clearance for timber and agriculture (Nigeria). Threats unknown in CAR and Cameroon.

Management suggestions: rediscovery at the two Cameroon sites should be attempted and basic population data recorded so that a baseline for monitoring is available. Involvement of local communities in protecting the species is advised.

Gluema korupensis Burgt

EN D

Range: Nigeria (Stubbs Creek) and Cameroon (SW: Korup NP (2 sites); Makeke, 2 km E of Korup).

This species was assessed in the protologue (van der Burgt & Newbery in Kew Bull. 61: 79–84 (2006)) as Endangered since only 176 individuals are known. That assessment is maintained here and the data for this treatment is taken from that reference. EOO is calculated as 2050 km².

Tree to 42 m, buttresses absent broadest trunks fluted, exudate white, indumentum red, on young stems, leaves, flowers and fruit; stipules absent, leaves in clusters of 6–9 at stem apices, petioles 0.5–1.2 cm, hairy; blade leathery, dull, glabrescent, narrowly obovate, 17–54 × 6–16 cm, apex acuminate, acumen 9–46 mm, lateral nerves 22–29 pairs; fascicles 1–7–flowered, densely covering the stem for 20 cm blow its apex pedicels 15-20 mm, calyx of five unequal sepals 8-9 mm, corolla at 1.5 mm, lobes 5, divided into 3 subequal units 7–9 × 2–3 mm, stamens 5, filaments 4 mm anthers 5 mm, staminodes opposite stamens, triangular,6–7 mm; ovary 5-locular; fruit globose, 9–13 cm, verrucate-subspinose, projections 1–2.5 cm, 1-seeded, remaining locules empty; seeds 75–82 × 3 ×2 cm.

Habitat: lowland evergreen forest on sand; 100 m alt.

Threats: clearance of forest for timber and agriculture (Stubbs Creek and Makeke locations).

Management suggestions: the Univ.Bern-Newbery plot site within Korup NP is the logical focus for future conservation efforts since it is protected, has > 90% of the known individuals of the species and since basic populational data has already been assembled, useful as a baseline for future monitoring. The importance of conserving the species should be publicised by Park Management for the benefit of staff and visitors, perhaps via a poster.

Inhambanella guereensis (Aubrév.& Pellegr.) T.D.Penn.

(syn. *Kantou gueerensis* Aubrév.& Pellegr.)

VU B2ab(iii)

Range: Liberia (Wanau), Ivory Coast (Oumé; Guijio; Tai) and Cameroon (S: Kribi-Ébolowa at Akok; Ébolowa-Mvangane 45 km, near Nkolénieng; Mvangane, 25 km SE at Ekowong, 40 km SSW de Sangmélima).

Here *Inhambanella gureensis* is assessed as Vulnerable since six locations are known (above; AOO 24 km^2 with 4 km^2 cells) and threats as below. EOO is calculated as 151869 km².

First described from Ivory Coast this species was still unknown in Cameroon when Flore du Cameroun was published in 1964, being revealed only by three collections by Letouzey in 1968–1970, which are cited above. The description here is based mainly from data in FWTA 2: 23 (1963) by Heine, (as *Kantou gueerensis*).

Tree 40 m, white exudate, leaves crowded at the top of branches, alternate; stipules minute; petiole 4–5 cm; blades oblong-lanceolate, 15–25 × 6–7 cm, lateral nerves 10–15 pairs, glabrescent: flowers fasciculate, pedicels 1 cm, pubescent, bracts ovate, 5–6 mm; sepals 6mm pubescent; corolla 7–8 mm, lobes 5, each deeply 3-lobed, tube 1.5 mm; ovary 5-locular, pubescent; fruit globose, red, 1-seeded: seed oblong, 5 cm, compressed

Habitat: evergreen lowland forest.

Threats: logging, sometimes followed by agriculture, is prevalent in the known Cameroon locations; in Ivory Coast, massive losses of forest cover have occurred due to logging followed by cocoa plantations and food crops.

Management suggestions: re-discovery of the species at the Cameroon site should be followed by collection of basic populational data as a baseline for monitoring. Local people should be supported to protect individuals of this species in the wild.

Lecomtedoxa klaineana (Pierre ex Engl.) Dubard

VU B2a(iii)

Range: Cameroon (SW: Korup NP numerous coll.; Mt Cameroon at Muyuka (1 coll.)) and Gabon (Libreville-Cap Esterias numerous coll.; Rabi; Setté Cama; NW Doussala, 40 km; Nyomié; Malibé).

Large gregarious tree to 40 m with triangular buttresses 4 m high, 2 m wide, with white exudate, glabrous, leaves alternate, petiole 2 cm, stipules not noted; blade oblong, to 13.5 × 5.5 cm, apex shortly acuminate or rounded, base cuneate, lateral nerves 12–8 pairs, with parallel intermediaries, all uniting in a looping marginal nerve; flowers with corolla 5 mm; lobes lanceolate, 3.5 mm; tube 1.5 mm; each lobe with two appendages resembling the lobe; stamens 5, filaments short, alternating with pubescent staminodes; fruit pyriform, curved, 5 × 2.5 × 2 cm, apiculate, leathery splitting down one side to expel a single seed ; seed flat, 3.2 × 1.8 × 1 cm, yellow-brown.

Here *Lecomtedoxa Klaineana* is assessed as Vulnerable since six broad locations, occupying an estimated 12 × 4 km^2 cells

(AOO 48 km^2) are known, with threats as below. EOO is calculated as 70210 km^2.
Habitat: lowland evergreen forest, often in sandy soils.
Threats: habitat clearance and degradation due to urban expansion (Libreville, and Cap Esterias and Muyuka)
Management suggestions: although this species is likely to be lost at the locations mentioned above under threats, elsewhere it is secure and, at Korup, relatively frequent. Since it occurs in the Smithsonian 50 Ha plot, detailed populational data already exists as a basis for monitoring which will take place as part of the usual Smithsonian procedure (see Thomas *et al.* (2003). Tree species of Southwestern Cameroon CTFS Washington, detailing 303 stems > 1 cm diam. in the 50 Ha plot). It is likely that further locations for the species will be found in Cameroon and Gabon. Publicising the rarity of this species and the need to protect it is recommended. Care should be taken not to confuse it with *L. plumosa* Burgt.

Lecomtedoxa plumosa Burgt
EN D
Range: Cameroon endemic (SW: Korup NP numerous coll.).
The assessment of this species as Endangered based on the number of trees existing being estimated as 51–100 (although only 28 have actually been seen) is maintained here following that in the protologue (van der Burgt, Kew Bull. 64: 313–317 (2009). *Lecomtedoxa plumosa* (Sapotaceae), a new tree species from Korup NP, Cameroon). This paper should be consulted for full details on the species. EOO is calculated as 13.72 km².
Gregarious tree 18–36 m, buttresses absent, bole c 15 m, fluted; exudate white, short red hairs on all young grass; leaves alternate, stipules absent, petioles 1–3.5 cm, pulvinate; blade obovate, 7–22 × 3–8 cm, lateral nerves 8–12 pairs, minutely reticulate below, midrib hairy below; fascilles 1–15-flowered on leafy stems, pedicels 9–13 mm, calyx 4–5 mm, lobes ovate; corolla tube 1 mm, lobes 5, divided into 3, each c. 4 × 1.5 mm, stamens 5, opposite the lobes, staminodes 5 alternating with the stamens, each divided into numerous laciniae, ovary 5-locular; fruit dehiscent, 1-seeded, rubbery, ovoid, beaked, 48 × 28 × 18 mm; seed 30–40 × 15–20 × 11–13 mm.
Habitat: lowland evergreen forest on sand
Threats: none are known.
Management suggestions: basic populational data sufficient as a baseline for future monitoring has already been collected (see protologue). Publicity for the conservation of this species for park visitors and staff is suggested, perhaps using a poster.

Manilkara lososiana Kenfack & Ewango (syn. *Manilkara* sp. ENG 2 in Thomas *et al.* 2003 op. cit.).
CR D
Range: Cameroon endemic (SW: Korup NP).
This species was uncovered and published through the work of the Smithsonian 50 Ha plot project in Korup NP (Kenfack *et al., Manilkara lososiana*, a new species of Sapotaceae from Cameroon. Kew Bull. 59: 609–612 (2005)). The assessment there of Critically Endangered is maintained here since only two adults and 15 saplings are known.
Tree to 35 m, bole lacking buttresses or flutes, grey-brown; white exudate, leaves clustered at stem apex in contracted spirals, stipules triangular, persistent, 4–7 × 2 mm, petiole 5–15 mm, pulvinate at base, tomentose; leaves leathery, obovate-elliptic, 4.5–20 × 1.5–7.5 cm, acumen 0.2–0.5 cm, lower surface with red tomentose scales, lateral nerves 12–20 pairs, visible, looping at margin, intersecondaries 1–3 between secondary pairs; fascicles axillary, 4–8-flowered, pedicels 2–3 cm, shortly red hairy; calyx of two whorls of 3 lobes, outer 4–6 × 3 mm, inner to 6 × 1 mm; corolla tube 1 mm, lobes 6, each 3-lobed, 4 mm; stamens 6; staminodes 6, alternating with stamens, lacinate, slightly shorter; fruit oblong, apiculate, 1–2.5 × 0.9–1.5 cm, probably 1-seeded.
Habitat: lowland evergreen forest; 180 m alt.
Threats: none are known.
Management suggestions: detailed populational data has already been recorded thanks to the 50 Ha Smithsonian plot at Korup (Thomas *et al.* op. cit. 2003). This is more than adequate for monitoring which in any case will occur as part of the mensuration programme there. If new sites/ locations are not discovered, consideration should be given to augmenting the wild population since it is so small and localised.

Manilkara zenkeri Aubrév. & Pellegr.
EN B2ab (iii)
Range: Cameroon (SW: Bakossi at Menyum village. S: Bipindi), Gabon (Minkebe; La Nké) and Congo (Kinshasa, Orientale Prov.).
Here *Manilkara zenkeri* is assessed as Endangered since five locations are known (above; AOO 20 km^2 with 4 km^2 cells) and with threats as below. EOO is calculated as 531038 km².
Tree 6–22 m, bark smooth, scaly, white exudate, stem and leaves glabrous; leaves alternate, stipules not noted, petiole 1.5–2 cm; blade oblanceolate, 12–22 × 5–8 cm, acumen short, obtuse and emarginate, base cuneate lateral nerves c. 15 pairs, discernable, uniting in a marginal nerve, finer nerves not conspicuous; fascicles on old wood, pedicles short; sepals 5–6 mm tomentose, united in a tube at base; corolla lobes five oblong, each 4–5 mm, reflexed, margin involute, each with two petaloid appendages similar to the lobe but flax and bifid; stamens and staminodes each five, equal in length, united for half their length in a tube.
Recognised by the notched obtuse leaf acumen and the inconspicuous nervelets. Known in Flore du Cameroun only from the type at Bipindi until Gautier identified *Doumenge*

526 as this species from Bakossi. The records from Gabon were each identified by their collectors, none of which was a Sapotaceae specialist (Wilks and Hallé), and likewise of Congo (Kinshasa) (*Gereau* 4861)

Habitat: lowland evergreen forest

Threats: slash and burn agriculture is thought to b a risk at the Cameroon locations, but is probably not a high level threat.

Management suggestions: attempts should be made to refind the species at the two Cameroon locations and collects basic populational data as a baseline for future monitoring. A revision of *Manilkara* is needed to clarify identifications and so the range of this species.

Neolemonniera batesii (Engl.) Heine

ENB2ab (iii)

Range: Ivory Coast (Tabou), Cameroon (Littoral: Yabassi, 15 km E. S: Bipindi, 20 km NNE) and Gabon (Cristal Mts at Kinguélé and at Mfoa)

First collected in Gabon at Mfoa by Bates for whom the species was named. Its range was extended to Cameroon by Letouzey. Heine in FWTA 2 (1963) did not accept Aubréville's record of the species from Tabou in Ivory Coast, not having seen the specimen at P, so this should be verified, although it should be noted that Aubréville was noted for his work on Sapotaceae.

Here *Neolemonniera batesii* is assessed as Endangered since four locations are known (above; AOO 16 km^2 with 4 km^2 cells) and with threats as below. EOO is calculated as 444447 km^2.

Tree, glabrous, leaves subverticillate with long internodes: petioles 2-4 cm, blade oblong to 35 × 8 cm, acuminate, base cuneate, lateral nerves leathery 12–18 pairs, not very prominent, united in an indistinct marginal nerve, inter-secondary nerves parallel, finer nerves numerous, resembling fingerprints; fascicles terminal and axillary; pedicles 2–4 cm; sepals 5, ovate, 8 × 3.5–4 mm; corolla lobes oblong, 10 mm, appendages subulate, 1 mm; tube 2 mm; stamens 5 (outer whorl); staminodes 5 (inner whorl) elliptic with apex aristate. The fingerprint-like finer nervation characterises the genus.

Habitat: submontane evergreen forest, sometimes near rivers.

Threats: slash and burn agriculture

Management suggestions: refinding this species in Yabassi and Bipindi should be attempted, and basic populational data collected as a baseline for monitoring. Investigation of the Tabou specimen at P is needed to confirm its identification

Neolemmoniera ogouensis (Pierre ex Dubard) Heine

EN B2ab (iii)

Range: Cameroon (S: Campo, NE d'Ipono) and Gabon (Ogooué at Samkita).

This, the type species of the genus, is based on the Ogooué specimen made by Thollon. Aubréville, in Fl. Gabon 1 in 1961 remarks that the species is close to the threatened *Neolemmoniera clitandrifolia* of W Africa. Here *N. ogouensis* is assessed as Endangered since only two locations are known (above; AOO 8 km^2 with 4 km^2 cells) and threats are as below.

Tree, glabrous stems and leaves, with leaves subverticillate, stipulate, petioles 1–1.5 cm; blades elliptic 5–7 × 2–2.5 cm, acuminate, base cuneate-decurrent, lateral nerves 8–10 pairs, finer nerves conspicuous, finger print-like; fascicles axillary, pedicels 9–10 mm, sepals 5, oblong, 3–5 mm, sometimes slightly pubescent, corolla 3.5–4 mm, tube 1–1.5 mm, lobes 2.5 mm, 3-lobed; stamens 5, alternating with 5 petaloid ovate, aristate, pubescent staminodes; ovary 5-locular.

Habitat: lowland evergreen forest, along rivers sometimes.

Threats: slash and burn agriculture

Management suggestions: investigation is needed as to whether the Cameroon location falls within the Campo-Ma'an NP, or not. If so, this is the logical focus for conservation efforts. Efforts are needed to refind the species at either of its two known locations and to characterise its microhabitat; is it indeed river specific? Basic populational data should be collected as a baseline for monitoring and publicity used to promote its conservation especially with local community leaders that might support its protection.

Synsepalum batesii (A.Chev.) Aubrév. & Pellegr.

CR D

Range: Cameroon endemic (S: Dja at Bitye).

Named for the American missionary-botanist Bates who collected the type, this species has not been seen since, although this may change soon, now that Ewango has begun revising *Synsepalum*.

Here *Synsepalum batesii* is assessed as Critically Endangered since only a single location is known (AOO 4 km^2 with 4 km^2 cells) and with threats as below.

Probably a small-medium-sized tree with white exudate and alternate leaves. Stipules not noted, petiole 1.5–3 cm, blade leathery, oblong-lanceolate, 15–18 × 6–7 cm, obtuse, base cuneate, appressed pubescent below, secondary nerves 12–15 pairs, raised below, extending to the margin, tertiary nerves scalariform; fascicles axillary on stems below the leaves, flowers subsessile, calyx appressed hairy, shortly obtusely lobed; corolla 7 mm, tube 3 mm, lobes undivided, 3.5–4 mm, stamens opposite lobes and about as long, alternating with shorter, rhombic-aristate staminodes in the same whorl.

Aubréville in his key in Flore du Cameroun 2: 89 gives the range as S Nigeria, Cameroon but there appears to be no evidence for its occurrence in Nigeria.

Habitat: lowland evergreen forest along rivers within a semi-deciduous forest area (surmised)

Threats: the Dja area sees logging for timber species in the course of which this species might be inadvertently destroyed.

Management suggestions: Ewango should be encouraged to complete his revision of *Synsepalum*. This species might be found if surveys are sent to the type locality at Bitye whence so many other rare species were first discussed by Bates. If successful, local communities should be alerted to the importance of the species and involved in its protection. Basic populational data should be collected as a baseline for monitoring and consideration given to augmenting the wild population by raising seed and planting it within the range.

Synsepalum brenanii (Heine) T.D.Penn. (syn. *Vincentella brenanii* Heine)

CR B1+2ab(iii)

Range: Cameroon endemic (SW. Banga, S. Bakundu FR).

Described as a Vincentella by Heine from Brenan's S Bakundu collection in 1948, a second collection from the same broad location was made in 1956. A record from the Onge area needs confirmation although it has been mapped.

Tree 4–5 m, white exudate, leaves obovate-lanceolate, 10–25 × 4.5–10 cm, red or dark brown when drying, cuneate from the upper third to the base and there mostly slightly and abruptly rotundate, with about 10–13 pairs of lateral nerves; stipules 3–5 mm; flowers fasciculate, 3–6 together in the axils below the leaves; pedicels rusty tomentose, 2–4 m; sepals oblong-lanceolate, 2 mm; petals 3.5 mm; staminodes 5.5 mm very conspicuous; ovary pilose (description from FWTA 2: 23 by Heine (1963)).

I first assessed this species as CR A1c + 2c in Cable & Cheek (1998: lxi) but this is here updated to follow changes in the IUCN 2001 and in the guidelines of 2003. In addition criterion B can be invoked since essentially only a single location is known unless the Onge record can be confirmed. AOO is here estimated as 8 km^2 with 4 km^2 cells, threats are as below. EOO is calculated as 21.62 km^2.

Brenan, for whom the species is named, gives the following notes which may help to relocate it, if it survives: "Frequent in undergrowth of high forest. Small tree to about 15' high. Trunk to c. 3" diam., cauliflorous and flowers on twigs up to those of second year. Young shoots brown-lanate, older purplish brown. Leaves subcoriaceous, deep green and glossy above, lateral nerve slightly impressed, pale green beneath. Petals and calyx brownish. Flowers dirty greenish-cream" (Brenan 9273, March 1948). It was recollected in the same area in March 1956 (Binuyo & Daramola in FHI 35589) where it was described as a "shrub to 6 ft high".

There is considerable interest in the allied species S. dulcificium `the miraculous berry' which has the property to depress appetite and make sour and salty subjects taste sweet, due to the glycoprotein miraculin. It is not known whether the fruits of S. brenanii have the same property.

Habitat: lowland forest

Threats: the S Bakundu FR has suffered from illegal timber extraction in recent decades. Nigerian yam farmers are reported to be cultivating the southern part.

Management suggestions: if this species can be relocated, the forestry department authorities at Kumba, who manage this forest, together with the local population, should be notified and involved in any protection measures taken.

Synsepalum letouzei Aubrév.

EN B2ab(iii)

Range: Cameroon (E: banks of the Dja R, between Meu and Edjumé; banks of Kadei near Batouri. C: Akonolinga, 20 km, WSW. S: Lobo near Bengbis) and Congo (Kinshasa)? (Ituri FR).

Ewango (pers.comm.) confirms that this is a good species but states that the Ituri record needs confirmation so this species may prove to be a Cameroonian endemic. All records cited are taken from Flore du Cameroun 2, no more records appear to have been made in the last 50 years.

Here Synsepalum letouzei is assessed as Endangered since four locations are known (above; AOO 12 km^2 with 4 km^2 cells) and since threats are as below. EOO is calculated as 111794 km^2.

Shrub or small tree 8–10 m low-branching, bark reddish, wood red; white exudate, leaves in clusters, alternate; stipules linear caducous; petiole 6–8 mm; blade oblanceolate to 30 × 12 cm, acumen short, apex rounded, base cuneate terminating in two auricles, lateral nerves 12–15 pairs, lacking a marginal nerve, weakly scalariform, lower surface grey-silver; fascicles axillary on stems among and below at the leaves, flowers sessile, calyx brown pubescent 4mm, very shortly lobed; corolla 7 mm, tube 3 mm; stamens 5, opposite and as long as petals or lobes, staminodes rhombic, nearly as long as stamens, dentate, with two basal subulate appendages; ovary 5-locular, pubescent.

Habitat: riverine and seasonally flooded forest with Manilkara multinervis, Uapaca hudelotii and Arthrosamanea altissima; c. 600 m alt.

Threats: the locations for this species are in an area where large-scale logging, sometimes followed by agriculture, occurs.

Management suggestions: re-evaluation of the Ituri specimen is needed to confirm its suspected misidentification. Re-discovery of the species at the Cameroon site should be followed by collection of basic populational data as a baseline for monitoring. Local people should be supported to protect individuals of this species in the wild.

Synsepalum zenkeri Aubr. & Pellegr.

ENB2ab(iii)

Range: Cameroon endemic (S: Mbiave; Campo-Ma'an fide Tchouto 2004).

Here Synsepalum zenkeri is assessed as Endangered since there are only two locations (above; AOO 8 km^2 with 4 km^2 cells)and threats as below.

Probably a shrub or small tree; buds velvety; stipules not noted; petiole 2.5–4 cm, leaves alternate, oblong, 14–20 × 5–7 cm, attenuate-acuminate, base cuneate; lateral nerves 12 pairs parallel, well marked and uniting in a marginal nerve below, appressed hairy but soon glabrous; flowers axillary, pedicel 1 mm; sepals 5-lobed 2 mm; corolla with lobes oblong 3 mm, longer than the 2 mm tube, 5 mm; stamens and staminodes shorter than corolla.

The Tchouto record has not been seen but is accepted here pending confirmation. This species is maintained by Ewango (pers. comm.) for the moment although it is close to S. bequaertii

Habitat: lowland evergreen forest (deduced)

Threats: slash and burn agriculture.

Management suggestions: it is advisable that Ewango be supported to view the Tchouto specimens to confirm the Campo-Ma'an site, hopefully. If confirmed this, as a protected area, would be a logical location to focus conservation efforts. Basic populational data should be collected for monitoring in future. If the species is as rare as suspected, augmentation of the wild population may be advisable.

Tieghemella africana Pierre

EN A1cd

Range: Cameroon, Gabon and Congo (Brazzaville).

This species was assessed as above and is listed as threatened by the African Regional Workshop (Conservation & Sustainable Management of Trees, Zimbabwe) 1998. Tieghemella africana. In: IUCN 2010. IUCN Red List of Threatened Species. Version 2010.4. Sierra Leone is erroneously cited there as within the distribution of the species. This assessment is maintained here since the species is known to be highly valued by loggers for its valuable timber, for example in Congo (Brazzaville). However, data on habitat loss or population reduction that would support the assessment above have not been seen. The statement that in much of West Africa "habitat degradation is a problem" see threats below, does not relate to this species because it does not occur in W Africa. Losses of forest in Gabon, for example, are extrem[illegible]. Because this is rather a frequent species it is not ma[illegible]

Tree c. 40 m, tru[illegible] diam., white exudate; buttresses absent; slash p[illegible]; leaves clustered in tight spirals at branch ends, glabrous; terminal buds sticky, stipules absent; petiole 1.5–3.5 cm; blade obovate, 8–16 × 4–7 cm, obtuse or subacuminate at apex, base cuneate-decurrent, secondary nerves 16–18 pairs; fascicles with 2–3 flowers numerous among the axils of the leaves; pedicels 10–15 mm; flower parts in 8s; sepals ovate, 6 mm; in two whorls; corolla tube 1.5 mm, lobes reduced to filaments but each with two lateral petaloid appendages, elliptic 4 mm, stamens opposite petals, alternating with oblanceolate-oblong petaloid staminodes; fruit orange-brown, globose 8 cm; seeds several, laterally flattened, 5–7 × 2.5–3.5 × 2–2.5 cm.

Aubréville (Flore du Cameroun 2: 44) describes this species as one of the largest canopy emergents of Gabon, the trunk below the branches alone attaining 20–30 m.

Habitat: lowland evergreen forest.

Threats: "No direct information is available on population declines but levels of exploitation are moderate. In much of West Africa habitat degradation is also a problem". (IUCN 2001 op. cit.)

Management suggestions: more data should be acquired on this species's habitat/population decline in relation to its actual area of occupation.

Zeyherella letestui Aubrév. & Pellegr.

VU B2ab (iii)

Range: Cameroon (SW: Korup, NP. S: Campo, 25 km N at Lolabé) and Gabon (Kinguele in Cristal Mts; Lastoursville; Doussala 15 km SSW; Lastoursville c. 30 km at Bambidie).

Here Zeyherella letestui is assessed as Vulnerable since six locations are known (above; AOO 24 km^2 with 4 km^2 cells) and threats as below. EOO is calculated as 137654 km^2.

Tree, white exudate, young stems brown tomentose; leaves alternate, stipules persistent, linear 1 cm, tomentose; leaves obovate-oblong, leathery 9–20 × 2.5–5.5 cm, apex emarginate, base attenuate, lateral nerves c. 40 pairs with intersecondaries brown tomentose below; fascicles on old wood, pedicels 8 mm, sepals 5 mm, brown tomentose, 5-lobed; corolla tube 2.5 mm, lobes 4.5 mm, filaments exceeding lobes appendages absent; ovary 5-locular.

Habitat: lowland evergreen forest

Threats: slash and burn agriculture (Gabon and Lolabé), coastal developments for housing (Lolabé).

Management suggestions: since this species is recorded in the 50 ha Smithsonian plot in Korup (Thomas et al. op. cit. 2003), excellent populational data is already available and a sound baseline for monitoring not only exist, but will be carried out as part of the Smithsonian program and the plants protected for the future. Conservation at other sites should be promoted, perhaps using a conservation poster about the species.

SCROPHULARIACEAE (LINDERNIACEAE, OROBANCHACEAE, PLANTAGINACEAE)

assessed by Martin Cheek

This large, worldwide family of mainly herbs of open sunny habitats are found from sea-level to 3000 m alt in Cameroon. The leaves are usually opposite, lacking stipules, bearing single axillary flowers in creeping species, or held in terminal spikes in those with erect stems. The tubular corolla is zygomorphic with an upper lip of one or two lobes, lower lip of three, usually with 2–4 stamens inserted on the inside and a single style with two flat apical stigma lobes. The ovary and fruit are the most reliable features for supporting this family from the similar looking Labiatae (which have four nutlets, and leaves and stems with scented essential oils) and Acanthaceae (fruit splitting into two equal valves, usually then exposing hooks, and 2–4 flat seeds). Scrophulariaceae have a superior ovary partly or wholly separated into two internal locules or cavities containing numerous dust-like seeds developed on a large central globose placenta, the seeds being released by slits and pores in the ovary wall. *Lindernia* and *Torenia* are often confused with *Lobelia* (Campanulaceae) but the last has an inferior ovary.

The family lacks economic uses of note in Cameroon, although some European species are important medicinally (*Digitalis*) and others as garden plants (*Verbascum*, *Mimulus*). One genus *Striga,* has several species that reduce crop yields of cereals such as *Sorghum* by parasitising the roots.

An indication of the diversity of this family in W-C Africa are the 31 genera recognised in 1963 by Hepper in FWTA 2. Most of these species occur as seasonal herbs in Guinean savanna or wooded grassland, flowering as the rains and some in seasonal pools. Others are hemi-parasites of the roots of other plants. Many species are fairly widespread in this habitat across W Africa in the Sudanian zone. There is no Flore du Cameroun account, creating a major obstacle to the identification and assessment of threatened species of this genus in all but NW and SW Regions (covered by FWTA). Fortunately, for one tribe of the family (covering the genera *Craterostigma*, *Crepidorhopalon*, *Hartiella*, *Chamaegigas* and *Lindernia*, there is an African-wide revision due to the doctoral studies of Eberhard Fischer, Systematic det africanischen Lindernieae (Scrophulariaceae) Tropische and subtropische Planzenwelt 81 (1992).

The Scrophulariacae have "been affected more than most other families by the application of cladistics in the last decade." stated Brummitt in Heywood *et al.* 2007, Flowering Plant Families of the World, RBG, Kew, 2007- whose approach is followed here- in turn Brummitt largely follows Fischer, Scrophulariaceae, pp.33–432. In Kaderit, J.W. (ed.), Families and Genera of Vascular Plants. V11. Berlin, Springer-Verlag 2004, in essentially, maintaining the status quo and finding the arguments used by others for redefining the Plantaginaceae and Phrymaceae, while erecting or resurrecting the Calcolariaceae, Linderniaceae and Stilbaceae, unconvincing.

Craterostigma sp. nov? (*Pollard* 273, 2nd Dec. 1998, Laikom Ridge, Ijim) was referred to as being a potential threatened taxon in Cheek *et al.* (2000: 73) but has since been identified by Fischer as *Lindernia abyssinica* which is not of conservation concern.

Lindernia yaundensis (S.Moore) Fischer (syn. *Ilysanthes yaundensis* S.Moore)

EN B2ab(iii)

Range: Cameroon endemic (C: Yaoundé at Mt Mbankolo; rocher de Bamélap (Nanga- Eboko). E: 4-5 km SW de Chouam (40 km de Messaména). S: Mékas, 72 km SE Akonolinga).

Here *Lindernia yaundensis* is assessed as Endangered since four locations are known (above; AOO 16 km^2 with 4 km^2 cells), and threats as below.

Annual, erect herb with dense basal rosette of slender oblanceolate-oblong leaves c. 10 × 2 mm, apex rounded-obtuse, inflorescence 10–14 cm tall, several arising from centre of rosette, internodes regular,< 1 cm long, bearing opposite bracts 2 mm long, resembling reduced leaves, and often a single flower from each of the uppermost five nodes; pedicel 6–7 mm; calyx 2–3 mm, corolla 3–4 mm, blue-purple in Oct-April. Data derived from Fischer (loc. cit. 285–289)

Habitat: humus rich crevices in granite rock outcrops; c. 700 m alt.

Threats: quarrying for construction material (pers. obs. 2006).

Management suggestions: efforts should are made to rediscover this species in one of its known locations and to confirm its habitat and any threats. Basic populational data should be recorded as a baseline for monitoring. If it is confirmed that this species is specific to rock outcrops, further consideration should be given to formal protection of inselburgs around Yaoundé since these also host so many other rare and threatened species.

Rhabdotosperma (*Verbascum*). Two montane forest-grassland species occur in the Cameroon Highlands. The generic name has altered several times in recent decades. In FWTA 2: 355 (1963) they appear as *Celsia*, formerly distinguished from *Verbascum* by having four, not five stamens, but later this feature was set aside and the names *Verbascum densifolia* (Hook.f.) Huber-Mor. and *V. ledermannii* (Schlecht.) Huber-Mor. were adopted, and are followed at Kew. More recently these species have been transferred to *Rhabdotosperma*, supported by the Scrophulariaceae authority Eberhard Fischer and which I took up in an earlier work (Cheek *et al.* 2004) and so maintain here.

In Tropical Africa this genus occurs mainly in high altitude open habitats. To the W of the Cameroon Highlands, only one species, *Verbascum sundingii* Lobin & Porembski (syn. *V. capitis-viridis* Huber-Mor.) occurs, on the Cape Verde Islands. To the E the nearest species occur on the Albertine Rift (*V. brevipedicellatum* and *V. sinaiticum* Benth.) extending into E Africa.

The two Cameroonian species are separated by Hepper in FWTA 2: 355 as follows:

> Leaves tomentose beneath, often very densely so, lanceolate, closely serrulate, 2–7 cm long, 7–20 mm broad; stems densely tomentose; bracts ovate-lanceolate, nearly as long as pedicels in flower; pedicels 1 cm long in flower, 1.5 cm long in fruit, glandular; flowers about 2 cm diam.; fruits 6–8 mm long ....**1. *densifolia***
>
> Leaves sparsely tomentose on the nerves beneath, oblong-lanceolate, crenate, up to 16 cm long and 6 cm broad; stems sparsely tomentose; bracts ovate-lanceolate, lower ones cordate, shorter than the pedicels; pedicels 1–2 cm long in flower, 1.5–3 cm long in fruit, glandular; flowers about 2.5 cm diam.; fruits 8–10 mm. long ....**2. *ledermannii***

Rhabdotosperma densifolia (Hook.f.) Hartl (syn. *Celsia densifolia* Hook.f., *Verbascum densifolia* (Hook.f.) Huber-Mar.)

VU A2c

Range: Equatorial Guinea (Bioko: Pico Basile (3 coll.)) and Cameroon (SW: Mt Cameroon (numerous coll.); Mwanenguba (1 coll.). NW: Bamenda Highlands (several loc.)).

This tall, yellow-flowered herb is spectacular in flower. It partly overlaps in range (in the Bamenda Highlands) with *R. ledermannii*, which is also restricted to the Cameroon mountain line, but extends further N, to Nigeria and is less common and more threatened.

Here *R. densifolia* is assessed as Vulnerable since it is estimated that more than 30% of its population has been reduced, mainly in the Bamenda Highlands, in the last 10 years, due to clearance of montane and submontane forest, which is continuing.

Robust erect herb to 90 cm; stems pithy, woody at base, tomentose; leaves tomentose beneath, lanceolate, closely serrate, 2–7 × 0.7–2 cm; inflorescence a terminal raceme; pedicels 1.5 cm in fruit; flowers yellow, c. 2 cm diam.; fruits 6–8 mm.

Habitat: grassland near edge of forest; c. 2000 m alt.

Threats: loss of montane forest due to agriculture is thought to be the main concern for this species. 25% of forest cover was lost in one sample area of the Bamenda Highlands between 1987–1995 (Moat in Cheek *et al.* 2000: back cover). This equates to about a 30% loss in area of occupancy for this species over 10 years. In another area, Dom, 50% of surviving forest was lost over the 15 years 1988–2003 (Baena in Cheek *et al.* 2010: back cover).

Management suggestions: conservation efforts should be concentrated in the Bamenda Highlands where the species seems to be most abundant (six collection sites). The ecological relationship of this species with forest edge needs further confirmation. Research is also needed on its demography, (for example is it a biennial or perennial?) and requirements for establishment in the wild. The demography data may alter the conservation assessment in future, since under criterion A, population reduction is linked to duration of the species generation

Rhabdotosperma ledermannii (Murb.) Hartl. (syn. *Celsia ledermannii* Schltr. ex Murb., *Verbascum ledermannii* (Schlecht.) Huber-Mor.)

VU A2c

Range: Nigeria (Chappal Wadi) and Cameroon (SW: Mwanenguba. NW (Bamenda Highlands): Ndu, Bambili Lakes, Bamenda–Mba Kokeka, Bambui; Laikom).

This 1 m or more tall, yellow-flowered herb is difficult to overlook. It partly overlaps in range (in the Bamenda Highlands) with *R. densifolia*, which is also restricted to the Cameroon mountain line, but extends further S and is more common and less threatened. Only eight locations are known (above). Here *R. ledermannii* is maintained (following Cheek *et al.* 2004: 185) as Vulnerable since it is estimated that more than 30% of its population has been reduced, mainly in the Bamenda Highlands, in the last 10 years, due to clearance of montane and submontane forest, which is continuing.

Stout erect herb 1.2–1.6 m; stems sparsely tomentose, woody at base; leaves oblong-lanceolate, sparsely tomentose on

nerves beneath, up to 16 cm; pedicels 1.5–3 cm in fruit; flowers yellow, c. 2.5 cm diam.; fruits 8–10 mm.
Habitat: grassland near edge of forest; c. 2000 m alt.
Threats: loss of montane forest due to agriculture is thought to be the main concern for this species. 25% of forest cover was lost in a sample area of the Bamenda Highlands between 1987–1995 (Moat in Cheek *et al.* (2000: back cover)). This equates to about a 30% loss in area of occupancy for this species over 10 years.
Management suggestions: conservation efforts should be concentrated in the Bamenda Highlands where the species seems to be most abundant (six collection sites). The ecological relationship of *R. ledermannii* with forest edge needs further confirmation. Research is also needed on its demography, (for example is it a biennial or perennial?) and requirements for establishment in the wild.

Torenia silvicola A.Raynal
VU B2 ab (iii)

Range: Cameroon (E: Yokadouma, 30 km E at Ngola; Ngola, 10-15 km NW) and Gabon (Lastoursville, Lméno; Billagone; Makokou, 15 km SW; ibid 37 km SE at Mboundou).
Here *Torenia silvicola* is assessed as Vulnerable since six locations (above; AOO 24 km^2 with 4 km^2 cells) are known, with threats as below.
Ascending annual herb, probably mat-forming, stems at 7.5–20 cm long, internodes rooting, 3 cm long, leaves opposite, petioles 1–1.5 cm, blades broadly ovate, 2–2.8 × 1.25–2 cm obtuse- acute, base obtuse to cuneate, margin serrate, ten per side, flowers single in axils, pedicel 1–1.5 cm, calyx c. 10 mm with leafy wings, lobes 2–3 mm; corolla exserted 15 mm, lower lip white with 2 yellow bars upper lip elliptic, pale orange, hooded; lower lip with three orbicular lobes 5 mm; fruit 10 mm; flowering in April in Cameroon.
Habitat: half-aquatic in muddy or sandy seasonally wet open places in evergreen or semi-deciduous forest areas or low altitude, e.g in *Raphia* swamps and stream edges.
Threats: logging in adjoining habitat (e.g. Yokodouma area) is likely to cause increased surface run-off leading to siltation of the habitat of this species.
Management suggestions: rediscovery of this species in the Yokadouma area is advised. If successful, basic populational data should be collected so as to allow future monitoring. Annual species such as this can show large fluctuations from year to year. The stated threat requires verification.

Veronica mannii Hook. f.
EN B2ab(iii)

Range: Equatorial Guinea (Bioko (four coll.) and Cameroon (SW: Mt. Cameroon (24 pre-1988 collections). NW: Mt Oku (two recent coll.).
Formerly considered endemic to Mt Cameroon and to Bioko, the range of *V. mannii* was extended to the Bamenda Highlands by the discoveries of two of our colleagues on the 1996 expedition (*Munyenyembe* 820 and *Zapfack* 1133, both found above Oku-Elak on 31 Oct. 1996).
First collected in December 1860: "on the very top of Clarence Peak, Fernando Po" (*Mann* 604), this species has often been collected from Mt Cameroon. However, nearly half of those collections were made by two people in two days collecting which has inflated considerably the number of specimens! On Mt Cameroon, most of the collections have been made above Buea, between Hut 2 and Hut 3, but Morton records it from a location 3 miles from Mann's Spring. Descriptions of flower colour vary from "vivid royal blue" to "violet" and it has generally been collected in flower in the dry season from December to April.
First assessed as LR/nt (Cheek in Cable & Cheek (1998: lxii) and Cheek *et al.* (2000: 73–74)), it is here reassessed as Endangered since three locations (above; AOO 12 km^2 with 4 km^2 cells) are known with threats as below.
Multistemmed perennial herb 20(–40) cm tall. Stems erect, woody at base, 3–4 mm diam., glabrous. Leaves opposite, subsessile, ovate or lanceolate, 1.5(–3.5) × 0.6–0.7(–1.3) cm, apex obtuse, margin finely serrate. Flowers solitary in axils of uppermost 3–5 cm of stem, purple-blue, c. 1 cm diam.
Veronica abyssinica, the only other species of the genus to occur in the Bamenda Highlands, is not likely to be confused with *V. mannii*. It has prostrate, herbaceous (not erect and woody) stems, and is widespread in Africa.
Habitat: montane grassland; (2700–)3000–4000 m alt.
Threats: grazing pressure and/or fires set by graziers is thought to be deleterious for this species at Mt Oku and to account for the scarcity of the species there, compared with Mt Cameroon, where grazing of animals is not practiced, and number of specimens known is far higher. The extent of threats on the summit of Bioko are not yet known.
Management suggestions: more information is needed on the phenology, regeneration levels, locations and numbers of individuals of *V. mannii* at all its sites, and on the question of the threats posed by fire and grazing to this species.

SCYTOPETALACEAE

assessed by Martin Cheek

A family of lowland, rarely submontane, evergreen forest, shrubs or small trees, confined to W-C Tropical Africa apart from *Asteranthos* in S America. Flore du Cameroun 20 (Letouzey 1978) includes this family, delimiting five genera and 14 species for Cameroon. The leaves are simple, alternate, lacking stipules, on stems that are sometimes ridged, lacking exudate or scent. The flowers have a cupular calyx and an unusual fleshy corolla that in bud is completely unlobed, forming petals or anthesis by splitting into lobes and revealing a mass of free anthers concealing the superior or half-superior single-styled ovary.

One new species was described in 2004 (see *Rhaptopetalum geophylax* below). *Rhaptopetalum* sp. nov. was assessed in Cable & Cheek (1998: lxii) as CR A1c+2c but remains unpublished. It comprises the Cameroonian material attributed to *R. pachyphyllum* in Flore du Cameroun. Most threatened species of Scytopetalaceae belong to the genus *Rhaptopetalum,* characterised by an articulated pedicel and ruminate endosperm, easily seen if the seed is cut in half.

Pierrina zenkeri Engler
VU B2ab(iii)

Range: Cameroon endemic (SW: Mt Cameroon at Muyuka. Littoral: Makak, Edéa. S: Kribi area; Bipindi area; Campo, 60 km E at Nyabéssan).

Six locations (above) are currently known for *Pierrina zenkeri*, here assessed as Vulnerable (AOO 24 km² with 4 km² cells, threats as below). EOO is calculated as 21443 km². Four or five specimens are known from each of three of the locations: Muyuka, Kribi and Bipindi, suggesting that the species can be locally common. In future this species might well rate EN A3c. *Louis* 217 is attributed to this species on the Gabon database, listing the locality 'Estuaire' but with no other data. More investigation of this specimen is needed before it can be accepted.

Shrub 0.5–4 m, glabrous; stems 4-angled; leaf-blade oblong-elliptic 8–20 × 4–7 cm, acumen 1–2 cm, base obtuse-elliptic, lateral nerves 7–9 pairs, uniting near the biserrate margin; petiole to 5 mm; flowers on stems below the leaves, pedicel 3–5 mm; calyx 8 mm wide, corolla purple, opening in 4 reflexing lobes each 1 cm; capsule 2-valved, leathery, long and beaked, 5–7 × 1.5–2 cm, seeds flat, to 2 cm long.

Habitat: lowland evergreen forest.

Threats: Muyuka, Kribi and Edéa locations all have their forest habitat threatened by urban expansion. In addition at Kribi, placement of new infrastructure for extractive industries poses a threat.

Management suggestions: confirmation is needed that the site near Campo is included in the National Park; if confirmed this would be the logical focus of conservation actions since protected. However, efforts should also be made to refind this species at Muyuka, Bipindi and Kribi where populations are probably densest, and if plants survive their importance should be raised with local leaders. Basic populational data should be gathered as a baseline for future monitoring and to assess needs for management intervention.

Rhaptopetalum breteleri R.Letouzey
CR B2ab(iii)

Range: Cameroon endemic (E: Zilly, near Nguélémendouka, 50 km NW Abong Mbang).

Known only from *Breteler* 2754 at the location cited. Here assessed as Critically Endangered (AOO 4 km² with 4 km² cells) and threats as below.

Several specimens from Mt Cameroon, collected in 1997 had been identified as this taxon (*Nning* 222, *Tchoutou* 1643 etc) but these are regarded as uncertain here since they were collected in a different habitat, far distant from the type locality, and were not named by specialists.

This is a glabrous shrub of 6 m, the stems terct, leaves oblong-elliptic, 15–22 × 7–10 cm, acumen short, lateral nerves 6–10 pairs, base truncate, glands present on lower surface; calyx cupular 4-lobed, 4–6 × 8–10 mm, lobes elliptic; ovary superior, hemispherical, 3 mm.

Habitat: lowland swamp forest.

Threats: slash and burn agriculture; forest clearance for timber.

Management suggestions: a survey is needed to rediscover this species. Basic populational data should be gathered to provide a baseline for future monitoring and to gauge needs for intervention. The leaders of Nguélémendouka should be advised on how to identify and protect this plant.

Rhaptopetalum depressum R.Letouzey
CR B2ab(iii)

Range: Cameroon endemic (S: Lolodorf, 20 km SE at Mébandé).

Here *R. depressum* is assessed as Critically Endangered since only a single site is known (AOO 4 km² with 4 km² cells) with threats as below. When first published, *Leeuwenberg* 9298 from Mt Kupe was included in this taxon but has since been shown to be a separate species, *R. geophylax* (q.v.).

Shrub or small tree, glabrous, leaves oblong-obovate 20–28 × 8–11 cm, acumen obtuse and emarginate, base obtuse, lateral nerves 8–12 pairs, distinct glands on lower surface; flowers articulated directly below the calyx; calyx bowl-shaped, 3–4 × 6–8 mm, truncate, entire or with slits, corolla bud obtuse, 6 mm; ovary semi-inferior.

Habitat: evergreen lower submontane forest; 950 m.

Threats: slash and burn agriculture.

Management suggestions: every effort should be made to refind this incompletely known species at its only site, to gather basic populational data and, if it appears as rare as existing data suggests, propagation material should be gathered so that it can be multiplied for reintroduction. The community of Mébandé should be informed of the rarity of this tree and involved in its protection.

Rhaptopetalum geophylax Cheek & Gosline
EN B1+2ab(iii)

Range: Cameroon endemic (SW: Bakossi Mts; Rumpi Hills; Lebialem Highlands).

Here *Rhaptopetalum geophylax* is assessed as Endangered since three locations are known (above). At the type locality,

Kupe-Bakossi, it occurs over large areas at a density of c. 1 tree per km travelled, and was not considered threatened (Cheek *et al.* 2004) leading to a rating of NT. The AOO at his location estimated as 50 km². At the Rumpi Hills and at Lebialem where newly discovered it is known from a single collection, each taken as 4 km², giving a total AOO of 58 km². EOO is calculated as 3038 km².

Letouzey included a fruiting specimen of this species from Mt Kupe in his *R. depressum* but further collections in Kupe-Bakossi showed it to be distinct. It was described in Kew Bull. 57: 662 (2002).

Tree 6–10 m, stems terete, glabrous; leaves obovate-oblong (14–)21–28 × 11–15 cm; acumen 1 cm, base unequally rounded; inflorescence ramiflorous, pedicel 3–6 mm, articulated; flowers pink, buds 9–15 mm, ovary superior; fruit orange, ovoid, 27–38 mm, receptacle covered in sinuous corky ridges.

Habitat: submontane evergreen forest; 900–1530 m alt.

Threats: once thought restricted to the Kupe-Bakossi area where it faces few threats, its discovery at Lebialem shows that its natural range actually extends N, along the Cameroon Highlands. But, since forest continues to be lost here, and the Lebialem trees are in an area being cleared for farms we now know it to be highly threatened in the northern half of its range.

Management suggestions: public education through a poster campaign is advised, especially at Lebialem where threats are greatest. Surveys are needed in the northern part of the range to see if trees survive between Lebialem and Bakossi, also in Rumpi to gather basic populational data.

Rhaptopetalum sessilifolium Engler

EN B1+2ab(iii)

Range: Cameroon endemic (S (Bipindi area): colline Nkol Tsia; 20 km at NE Bipindi pres Gouap; Mbiave; Mimfia)

Immediately recognised by the deeply cordate leaf bases, the leaf base clasping the stem. A shrub or small tree with strongly angled stems. The oblong-lyrate leaves are 14–32 × 6–10 cm with 8–12 pairs of arced ascending nerves and no glands on the lower surface of the leaf.

Here *R. sessifolium* is assessed as Endangered since four sites (above; AOO 16 km² with 4 km² cells) and threats as below are known. The sites are so close that they might be treated as one location, qualifying for CR status. EOO is calculated as 168 km².

Habitat: submontane/lowland evergreen forest.

Threats: slash and burn agriculture.

Management suggestions: this species should be refound and local authorities enabled to identify and understand its conservation importance. Basic populational data should be gathered to inform management interventions and monitoring. None of the sites is in a protected area it is believed.

SIMAROUBACEAE

assessed by Martin Cheek

Quassia sanguinea Cheek & Jongkind (syn: *Hannoa ferruginea* Engl.)

VU A2c B1ab(iii)

Range: Nigeria (Mt Koloishie at Obudu Plateau) and Cameroon: (SW: Mt Kupe and Bakossi Mts; Nyale Mts at Ebamut; Lebialem Highlands. NW: Bali Ngemba. W: Bangwa).

Restricted to the submontane forests of the Cameroon Highlands, from Mt Kupe (not Mt Cameroon as erroneously stated in Cheek *et al.* (2004: 185) and Cheek in Harvey *et al.* (2010: 85–86) from whence this assessment is maintained) in the S, to Bali Ngemba in the Bamenda Highlands in the N, with an extension into Nigeria at Obudu, this small tree, often only 3-4 m tall, is distinctive in its red leaf-axes. Previously named as *Hannoa ferruginea*, justification for the name change and additional data on the species is given in: Cheek, M. & Jongkind, C. (2008). Two new names in West-Central African *Quassia* (Simaroubaceae). Kew Bull. 63: 247–250.

The original VU A2c assessment is now supported under criterion B with a newly available EOO calculation of 8239 km².

Tree (1.5–)2–4(–6) m, glabrous; leaves alternate, lacking stipules, pinnately compound, 3(4)-jugate, leaflets c. 13 × 5 cm, subacuminate; rachis and midrib violet; inflorescence a terminal panicle, c. 15 cm, flowers white c. 4 mm, petals free; fruits apocarpous, mericarps fleshy, white, ellipsoid c. 6 cm.

Habitat: submontane forest; 800–1750 m alt.

Threats: forest clearance for wood, followed by agriculture, particularly in the northern three-quarters of its range, the Bamboutos Mts and the Bamenda Highlands. In the latter, a remote sensing study over 15 years (1988–2003) by Baena (in Cheek *et al.* 2010: back cover) of one area showed c. 50% loss of forest. By extrapolation, it is here estimated that well over 30% of its overall population has been lost due to habitat destruction over the last three generations, or sixty years (estimating one generation at twenty years). Most of the range of this species, e.g. the Bamiléké Plateau, has been almost totally deforested for agriculture. These losses are ongoing in e.g. the Lebialem Highlands where its habitat is being cleared for farming (Harvey *et al.* 2010: 31). With better data on habitat loss over time over the whole range of this species it is likely that the assessment would be raised to EN.

Management suggestions: although forest loss in W and NW Regions has seriously reduced the population of *Q. sanguinea* in those areas, it seems relatively secure at Mt

Kupe-Bakossi Mts in SW Region for the moment. So long as these areas remain protected, no further action is needed to ensure the survival of the species. However, data on generation duration and other aspects of demography, together with data on densities, are desirable.

STERCULIACEAE

assessed by Martin Cheek

Traditionally, and in all Flora accounts for West and Central Africa, genera placed in Sterculiaceae (in the traditional sense) have only one or two whorls of anthers (usually 5–10 in number) and also often have staminodes, as distinct from traditional Tiliaceae, which have numerous free stamens, lacking staminodes. However, recent phylogenetic molecular analysis of core Malvales has shown that one unit of the first family is much more closely related to the major unit of traditional Tiliaceae than to anything else. Within core Malvales (Malvaceae, Bombacaceae, Tiliaceae and Sterculiaceae) it is now clear that the existing families cannot be recognised as traditionally defined. Instead, one of two approaches must be taken:

1. recognise ten families in place of the former four.
2. recognise one 'super-Malvaceae' with ten subfamilies.

I favour the former (Cheek in Heywood *et al.* 2007, Flowering Plant families of the World, RBG, Kew). Thus the threatened Sterculiaceae in Cameroon should be divided into three families, Byttneriaceae, Sterculiaceae and Dombeyaceae. However, for the sake of consistency with earlier literature they are all grouped here together in broad, traditional Sterculiaceae. The delimitation of those families is as follows:

Byttneriaceae (or Malvaceae-Byttneroideae) best known for cocoa, *Theobroma cacao*, a native of S America, this diverse segregate family of Sterculiaceae are characterised by having small flowers, in inconspicuous axillary inflorescences, cupped petals or petal bases; the stamens united at the base into a very short ring in which staminodes alternate with stamens; the stamen ring surrounds an ovary in which all the carpels are united.

Principal genera in Cameroon, with 20–25 species, are understorey shrubs of evergreen forest: *Leptonychia* (see e.g. Cheek *et al.* in press (Three new or resurrected species of *Leptonychia* (Sterculiaceae-Byttneriaceae-Malvaceae) from West-Central Africa. Kew Bull.) and *Scaphopetalum*.

Dombeyaceae (or Malvaceae-Dombeyoideae) are also referred to as Pentapetaceae. Best known for the genus *Dombeya* in Cameroon, this family have showy flowers with large white or pink flat, petals which persist, drying brown, long after flowering. The stamens are united in a short tube at the base, the staminodes far exceed the stamens in length; carpels are united. Principal genus in Cameroon, although with only 3–4 species: *Dombeya*.

Sterculiaceae sens. stricto (Malvaceae-Sterculioideae).
A pantropical family of c. 15 genera including some timber trees, but best known for the African genus *Cola* which I am in the process of revising (e.g. Cheek, M. (2002). Three new species of *Cola* (Sterculiaceae) from western Cameroon. Kew Bull. 57(2): 403–415). *Cola* has more species than any other in the family, and includes *Cola nitida* and *acuminata* which produce cola nuts, culturally important in W Africa and which awaken the brain, from whence Cola drinks are derived. Flowers lack petals, having a single petaloid calyx whorl, usually with five lobes; the male flowers have filaments united in an androphore on which sits a head of anthers; female flowers have an ovary in which the carpels are separate, producing several mericarps in fruit. Principal genera in Cameroon: *Cola, Sterculia* and *Pterygota.*

Despite the efforts of Prof. Nkongmeneck in the 1980s, no Flore du Cameroun account yet exists for Sterculiaceae sens. lat. and there is thus difficulty delimiting species in several of the larger genera in which threatened species occur. Furthermore, there are no recent African-wide revisions for any genus excepting *Dombeya* (Seyani, A Taxonomic Study of *Dombeya* Cav. (Sterculiaceae) in Africa, DPhil thesis, University of Oxford and in Opera Botanica Belgica, 1982). In future therefore, many more threatened species are likely to be revealed in this family.

<u>Poorly known species</u>
In addition to the species treated as threatened below, several other species may also merit threatened status in future when revisions of their genera have been completed. For the moment they must be treated as poorly known since their type specimens were destroyed at Berlin. These are:

Cola bipindensis Engl. & K.Krause (Bipindi, *Zenker* 3873, 4479)
Cola edeensis Engl. & K.Krause (Edéa, *Buesgen* 501)
Cola fibrillosa Engl. & K.Krause (several specimens in S Region have been attributed to this taxon)
Cola lomensis Engl. & K.Krause (Ndonge near Mt Nlonako, *Ledermann* 6234)
Cola nana Engl. & K.Krause (Fenda, 48 km from Kribi, *Mildbraed* 5898)
Cola ndongensis Engl. & K.Krause (Ndonge near Mt Nlonako, *Ledermann* 6231)
Cola subglaucescens Engl. & K.Krause (Kribi, *Mildbraed* 6027)

Cola sulcata Engl. (Bipindi, *Zenker* 3724)
Pterygota adolfi-friderici Engl. & K.Krause (Moloundou, *Mildbraed* 3969)
Leptonychia lokundjensis Engl. & K.Krause (Bipindi, *Zenker* 3215, 3619)
Scaphopetalum acuminatum Engl. & K.Krause (Kribi, *Mildbraed* 5889; Bipindi, *Zenker* 2582 and Gabon)
Scaphopetalum brunneo-purpureum Engl. & K.Krause (Kribi, *Mildbraed* 5917)
Scaphopetalum pallidinerve Engl. & K.Krause (Lom, *Mildbraed* 6403, 6474).

Near Threatened species

Several species, while not treated here as threatened, should be regarded as Near Threatened since they are either geographically restricted, or known from only 11-20 locations, and might, with better habitat loss data, be shown in future to be threatened:
Cola anomala K.Schum
Cola argentea Mast.
Cola flaviflora Engl. & K.Krause.

Cola cecidiifolia Cheek (syn. *Cola* sp. nov. aff. *flavo-velutina* K.Schum., sensu Cheek in Cable & Cheek, The Plants of Mt. Cameroon, lxiii & 134 (1998)).
CR A2c

Range: Cameroon endemic (SW: Mt Cameroon at Mokoko R FR, Mundongo also Bonjare village; Bonadikombo-Bimbia; S Bakundu FR).

Here *Cola cecidifolia* is maintained as Critically Endangered, following Cheek in Cable & Cheek 1998: lxii (CR A1c + 2c as *Cola* sp. nov. *aff. flavo-velutina*), since it is estimated that more than 80% of the population has been lost/will be lost over three generations/ 100 years, due to habitat loss at all known locations. EOO is calculated as 2035 km^2.

Much branched shrub 1.5–2 m, internodes 4–6 cm, young stems dark brown puberlent with simple hairs; petiole 7–15 mm pulvini hairy; stipules long-persistent, needle-like, 8–11 mm, white, glabrous; blade elliptic 13–19 × 4.5–7 cm, acumen ligulate 2–3.5 cm, base acute, lateral nerves 6–8 pairs, yellow; matt and with hemispherical galls on each surface; inflorescence 1–5-flowered peduncle 1–2 mm; fruits one per stem, among leaves, fruitlets 1–2 red or orange, globose 1.7–3.2 cm, round glossy; seeds 1–3 per fruit.

In the infertile state, *Cola cecidiifolia* closely resembles the much more common *C. flavo-velutina* K.Schum. in leaf-blade shape, venation and colour, and in the very similar, persistent, awl-shaped, off-white stipules. However, *C. flavo-velutina* differs in being 4–5 m tall (not 2 m), in having scurfy orange-stellate hairs on the stems and petioles (not with a dark brown puberulent indumentum of simple hairs), in having distinctly long or short petioles (not with the petioles more or less constant in length), and in the fruits borne on the trunk, or on stems below the leaves, sessile, with 5 densely pubescent, narrowly ovoid, rostrate fruitlets 5–7 cm long, (not borne amongst the leaves, long-pedunculate, with 1–2 glabrous spherical fruitlets 1.7–2.3 cm long).

This species is named for the distinctively galled leaves. It appears to be endemic to the foothill forests north and south of the main massif of Mt. Cameroon. Within its range it appears to be rare and easily overlooked. Despite an intensive botanical inventory in the Bonadikombo-Bimbia forest by many botanists over several months in 1992 (Cheek, 1992 op. cit.), this species was entirely overlooked until the type specimen was collected there the following year by Lyn Hunt of the Oxford University Expedition to Cameroon (Baker *et al.*, 1995).

First collected in the western foothills in March 1993 (*Watts* 670 from Mundongo "Near MC 03"), the only other collection was that from near Bonjare village, made in May 1994 (*Ndam* 1160 "Plot M 11-2") until the discovery of the Hunt specimen at Bimbia, followed by the record from S Bakundu. This shrub, about 2 m tall, is known only from fruiting collections.

Habitat: lowland evergreen forest;

Threats: all of the known sites for this species are believed to be on land threatened by clearance for plantations.

Management suggestions: improved policing of existing protected areas would help to secure the survival of this species. A survey of the population of this species is advisable as baseline for monitoring, as are measures to protect what remains from further destruction. Given its rarity and geography it is suggested that seed be collected for multiplication of plants at Limbe BG, to be used for conservation education of the public and reintroduction to secure sites in the wild to maintain the wild population.

Cola hypochrysea K.Schum.
EN B2ab(iii)

Range: Nigeria (Eket) and Cameroon (S: Kribi-Grand Batanga; Eséka at Badjob; Bidou I; Campo-Ma'an (Tchoutou, 2004)).

Tree 25 m tall, leaf-blade, thickly leathery, entire ovate to suborbicular, 7.5–40 × 5–30 cm, apex obtuse to rounded, base obtuse to subcordate, lateral nerves 4–5 pairs, lower surface covered densely in golden scales; panicles axillary, covered in golden scales; flowers subsessile, 5 mm long, anthers in two ranks; fruiting carpels 4–5, 3–7 × 2–4 cm, sessile, shortly and abruptly beaked, longitudinally ridged.

Here *Cola hypochrysea* is assessed as Endangered: five locations are known (AOO 20 km² with 4 km² cells) and threats are as below. EOO is calculated as 25538 km^2.

Habitat: lowland forest, often in swampy areas.

Threats: both Eket and Kribi are oil export ports with associated infrastructure; slash and burn agriculture; timber extraction.
Management suggestions: Campo-Ma'an NP would be the logical locations to focus conservation efforts if the species can be verified and rediscovered there. Basic populational data is needed to form a baseline for monitoring in future, and to gauge needs for management intervention, if any.

Cola letouzeyana Nkong.
VU B2ab(iii)
Range: Cameroon endemic (SW: Mt Cameroon, Bonja-Onge. C: Mefou proposed NP; Mbam-Minkom; Boga, 30 km N Eséka. S: Kribi-Lolodorf; Ébolowa-Ambam. Littoral: Hikoa Mahouda; Ebo).
Here *Cola letouzeyana* is maintained as Vulnerable following Cheek in Cheek *et al.* 2011, since 8 locations are known (above, AOO 36 km^2 with 4 km^2 cells) and threats are as below. EOO is calculated as 36916 km^2.
Shrub or tree, 2–3 m; stems black, corky-ridged; leaves alternate, simple; leaf-blades oblanceolate-oblong, to 17 × 6 cm, acumen 1.25 cm, base shortly cordate, lateral nerves 9–11 pairs, white, uniting, forming a looping marginal nerve, puberulent, older leaves densely galled with dense brown tomentose domes 2 mm wide arising from both surfaces of the gall; petioles 2–4 mm, hairs as stem; stipules persistent, filiform, 7 mm; flowers single on leafy stems, pale yellow, axillary; pedicel filiform, 2 cm; perianth 4 mm diam.
This species was described by Nkongmeneck in Adansonia (Sér. 2) 7: 337 (1986).
Habitat: lowland evergreen forest including *Gilbertiodendron dewevrei* dominated; 200–700 m alt.
Threats: forest clearance for agriculture, both plantation (Mt Cameroon) and slash and burn (S and C Regions).
Management suggestions: formal recognition of the Mefou NP would protect a major centre for this species since it is relatively frequent in *Gilbertiodendron* forest here. Considerable data on the population at Mefou already exists and could be a basis for future monitoring. Studies on rates of regeneration would be valuable. A record of this species in Bakossi requires confirmation.

Cola megalophylla Brenan & Keay.
EN B2ab(iii)
Range: Nigeria (Okomu FR and Oban) and Cameroon (SW: Korup NP; Mundemba; Kumba-Mulonge).
Unbranched tree to 36 m, bole straight, unbranched, leaves compound, digitate, thickly leathery, to 85 cm, leaflets 5–7(–9) deeply lobed, apices obtuse; lower surface densely stellate-puberulous; petioles 8–15 mm diam., stipules caducous; flowers in axillary clusters among the leaves; male calyx 16–22 mm long, lobed to middle, lobes erect, lacking a crisped margin, anthers grouped in bundles within the capitate head; fruiting carpels boat-shaped, not dehiscing.
Here *Cola megalophylla* is assessed as Endangered since there are five locations (AOO 20 km^2 with 4 km^2 cells) and threats are known as below. EOO is calculated as 16571 km^2.
Habitat: lowland evergreen forest.
Threats: forest in Nigeria, including in forest reserves, has seen immense losses in recent years. The Okumu and Oban forest habitats may no longer survive. In Cameroon, urbanisation is a threat at Mundemba.
Management suggestions: Korup NP is the logical focus for conservation efforts for this species since forest habitat is well protected there. Basic populational data is needed to form a baseline for future monitoring and to gauge needs for management intervention.

Cola metallica Cheek (syn. *Cola* sp. nov.1 aff. *philipii-jonesii* sensu Cheek in Cable & Cheek, The Plants of Mt. Cameroon, lxiii & 134 (1998)).
CR A2c
Range: Cameroon endemic (SW: Mt Cameroon at Batoke, Upper Boando, Ekombe- Mofako; Bakossi; F.R; Banyang Mbo)
This shrub was assessed as CR A1c + 2c by Cheek in Cable & Cheek (1998: lxiii) and in Cheek (2002 op. cit.). That assessment was maintained in modified form in Cheek *et al.* (2004: 186) since use of A1 using IUCN (2003) guidelines now implies that causes of population reduction have ceased. This assessment is maintained here on the basis that more than 80% of the population of this species has and is being lost over 3 generations or 100 years due to habitat destruction. EOO is calculated as 3925 km^2.
Branched shrub 0.5–1.5 m, stems nearly black, with white lenticels; internodes 0.5–1 cm, densely black puberulent with simple hairs; petioles 5–9 mm, not pulvinate, indumentums as stems; stipules caducous; leaf-blade simple, alternate, oblanceolate, 13–22 × 4–7 cm, subspathulate 1.25–2.5 cm, lateral nerves 7 pairs, uniting in a marginal nerve 1 cm from the base, drying metallic black or grey below, glossy above.
Extremely rare since only known from five specimens among the c. 20,000 that have been collected within the area of its range. Mainly it occurs around Mt Cameroon where it is only known from the Etinde coastal strip, the western foothills (Mokoko forest) and S Bakundu.
This shrub is distinct in its shortly petioled, homophyllous leaves which dry a distinctive metallic grey. *Kwangue* 97 (June 1992, Batoke, Mile 9, c. 200 m alt.) is the first record of this species, followed by *Kwangue* 132 (Nov. 1992, between Upper Boando and Etome, c. 700 m alt.). The third record, *Watts* 1169 is from Ekombe-Mofako (May 1994).
Habitat: lowland evergreen forest, c. 50–700 m alt.

Threats: all known sites for this species are believed to be on land threatened by clearance for plantations, other forms of agriculture, or logging.

Management suggestions: improved policing of existing protected areas would help to secure the survival of this species. A survey of the population of this species is advisable as baseline for monitoring, as are measures to protect what remains from further destruction. Given its rarity and geography it is suggested that seed be collected for multiplication of plants at Limbe BG, to be used for conservation education of the public and reintroduction to secure sites in the wild to maintain the wild population.

Cola nigerica Brenan & Keay

CR A2c

Range: Nigeria (Shasha FR; Omo FR; Oban) and Cameroon (SW: Mt Cameroon at Onge, Bambuko FR, Bimbia forest).

Shrub or tree to 8 m, leaves alternate, unequal in size, petioles 2.5–8cm, upper pulvinus glabrous; stipules lanceolate, caducous; leaves elliptic or elongate-elliptic, to 27 × 12.5 cm; flowers white, in cushions on the stem, mainly below the leaves, calyx broadly campanulate, 9–10 mm, stellate outside; fruit red, carpels subglobular, 4–5, tomentose drying matt, pedicels hairy.

Here *Cola nigerica* is maintained as Critically Endangered, following but updating Cheek in Cable & Cheek (1998: lxii–lxiii) (CR A1c+ 2c), since it is estimated that more than 80% of the population has been lost in the last three generations or 100 years, and since this loss is ongoing due to partial or more or less complete loss of its lowland evergreen habitat at all known localities (cited above). However several additional records at YA have been identified as this species and if verified will extend the range of the species considerably, necessitating a re-assessment. These collections are: *Raynal* 9787 (Ambam), *Letouzey* 13275 (Bafang) and *Letouzey* 15134 (Korup). EOO is calculated as 47353 km^2

This tree is one of the many simple-leaved, heterophyllous, cauliflorous, species of the genus.

Cola nigerica is extremely rare in each of the five disjunct areas in its range. For example in the 1992 survey of Mabeta-Moliwe, while 22 specimens of *Cola flavo-velutina* were recorded, only a single specimen of *Cola nigerica* was found (*Watts* 244 at TD 6250 m, 24th April 1992). The only other specimens known on or around the mountain are Keay in FHI 37412 (Bambuko FR), and Tchouto 694 (Onge Forest).

Habitat: lowland evergreen forest.

Threats: all the areas in which *Cola nigerica* occurs(ed) have been cleared or are under threat of forest clearance and cultivation for timber, small-holder agriculture, oil exploration, rubber, oil palm, cocoa or banana plantation.

Management suggestions: marking and protecting individual trees might be a useful stopgap before making detailed population surveys which might form the basis of future monitoring. Seed should be collected of any trees that are found to survive for propagation at Limbe BG to safeguard the species, educate the public and for augumenting surviving wild populations in secure areas within the range of the species.

Cola praeacuta Brenan & Keay

CR A2c

Range: Cameroon endemic (SW: Korup; Mosongosele; Mt Cameroon at Onge, - Mokoko (10 coll.), Bambuko FR and Bakossi–Mungo R FR).

Monopodial tree 8–10 m, stems dull white, leaves alternate, unequal in size, petioles 0.7–4 cm, minutely black puberulent; stipules conspicuous at stem apex, lanceolate, 6 × 2 mm, not persistent; blades elliptic or obovate-elliptic, to 15.5 × 6 cm, matt below, acute-caudate, rounded to obtuse; flower fascicles among the leaves or below, pedicels to 13 mm; calyx puberulent outside.

Here *Cola praeacuta* is maintained as Critically Endangered, following Cheek in Cable & Cheek (1998: lxiii) (CR A1c + 2c) and Cheek *et al.* (2004: 186), since it is estimated that more than 80% of the population has been lost/will be lost over three generations/ 100 years, due to habitat loss at all known locations except Korup (see above). Several additional records from S Region (*Tchoutou* 2004) have not been included since they require verification. No specimens were cited so they may only be sight records which are unreliable in this group of species.

At Mt Cameroon, the main location for *Cola praeacuta,* it is only known from about 10 collections in the northern and western foothills, and in the coastal strip. It is unknown from Mabeta-Moliwe. It seems relatively common in the western foothills and this area seems most suited for developing a plan for the conservation of this species if difficulties of habitat destruction can be resolved. EOO is calculated as 4716 km^2.

Habitat: lowland evergreen forest, often on poor sandy soils; 0–400 m alt.

Threats: forest clearance for logging followed by plantation and/or slash and burn agriculture. Except at Korup this is a severe and ongoing problem at all known locations and the species is thought to have been probably lost at Bambuko and severely reduced at Onge/Mokoko where there are plans to increase logging activities. Oil exploration drilling is now being planned in Mokoko and may pose a further threat.

Management suggestions: improved policing of existing protected areas and reinforcement of the level of official protected status, would help secure the future of the species. The main site for this species is the Onge forest in the western foothills of Mt Cameroon. It is advised that resources for the conservation of this species be focussed here. Marking and protecting individual trees might be a useful stopgap before making detailed population surveys which might form the basis of future monitoring. Seed should be collected of any trees that are found to survive for propagation at Limbe BG to safeguard the species, educate the public and for augumenting surviving wild populations in secure areas within the range of the species.

Cola suboppositifolia Cheek (syn. *Cola* sp. nov. 2 aff. *philipi-jonesii* Brenan & Keay, Cheek in Cable & Cheek, The Plants of Mt. Cameroon, lxiii: 134 (1998)).
EN B1+B2ab (iii)

Range: Cameroon endemic (SW: Korup (3 coll.); Banyang Mbo (1 coll.); Mt Cameroon at Onge (5 coll.); Mokoko (1 coll.)).

Assessed as CRA1c+ 2C (as *C. sp. nov. 2 aff. philippi-jonesii* by Cheek in Cable & Cheek, 1998: lxiii), reassessed as VUA2c in Cheek (op. cit. 2002) after the discovery of the species at Korup and Banyang Mbo, it is here reassessed as Endangered in view of the threats described (above; with three locations, AOO 36 km^2 since nine sites and 4 km^2 cells are being used). EOO is calculated as 3794 km^2.

Shrub or small tree 2–6 m, leafy stems greyish white, internodes 3–5.5 cm, glabrous; leaves opposite, or subopposite, or alternate; petiole 7–10 mm, pulvini absent; stipules caducous, narrowly triangular 1.5–2.5 mm; blades simple, ovate or elliptic, 12.5–18.5 × 5–7.5 cm, acumen 1–1.5 cm, base abruptly attenuate secondary nerves 5 pairs, glossy; fascicles on perennial burrs, borne below the leaves pedicels 2 cm; flowers 1–1.8 cm diam., calyx pale-yellow, 5-lobed; fruit with two yellow pendant fruitlets, cylindrical 7–8 × 1.8–2.5 cm rostrate and stipulate, outer surface reticulate, seeds 4–5 per fruitlet.

Wheatley 792 (March 1993) is the earliest record of this species ("Onge, Likengi, 3.5 hr S on Wonge R"). Later that year it was collected nearby (*Tchouto* 658, Sept. 1993) and then near Mount Thump (*Thomas* 9869) and at the Bomana-Koto junction (*Ndam* 797). The only record from the Mokoko forest is adjacent to the Onge area: *Ndam* 1240 (May 1994, "Boa, transect to plot 14).

Cola suboppositifolia is a shrub or small tree 2–8 m tall, it takes its name from a feature unusual in *Cola*: the leaf insertion is subopposite, rather than being evenly alternate or whorled. The species is also unusual in that there are usually just two carpels present in the flowers. I know of no other species in the genus that has either of these two characters.

Cola suboppositifolia closely resembles *C. philipi-jonesii* of adjoining Nigeria in leaf shape, venation, the constantly short petioles and the caducous stipules. However, the two differ in that *C. suboppositifolia* has leaves subopposite and unevenly alternate on one stem, not evenly alternate; leaves 12.5–18.5(–23) × 5–7.5(–9) cm, not 5–10.8 × 2–4.5 cm; inflorescences cauliflorous or ramiflorous, fasciculate from burrs, not amongst the leaves, single-flowered.

Habitat: lowland evergreen forest.

Threats: clearance for timber and oil exploration, and agriculture, particularly in the western foothills of Mt Cameroon for plantations of bananas, oil palm, rubber and cocoa, and also small-holder agriculture.

Management suggestions: a survey of the population of this species is advisable to form a baseline for monitoring, as are measures to protect what remains from further destruction. The best hope for the survival of this species is the Korup NP which is the most secure location

Dombeya ledermannii Engl.
CR A2b,c

Range: Nigeria (Mambilla Plateau (4 coll.), Jos Plateau (2 coll.)) and Cameroon (NW (Bamenda Highlands): Kilum-Ijim; Bali Ngemba; Dom (7 coll.)).

This species is easily censused since, in April, trees are clearly visible from a distance on account of their white flowers. Flowering can occur as early in the dry season as November (pers. obs.). The habitat of this species is highly threatened in all its known localities. Seyani in his revision of the genus in continental Africa (Opera Botanica Belgica 1982), reports that *D. ledermannii* is characteristic of forest edges and the early stages of forest succession, but that on exposed rocky slopes it is a normal component of stunted, more open forest. In the Bamboutos Mts, it is propagated by cuttings and planted to form hedges (*Letouzey* 201, cited by Seyani 1982 op. cit). EOO is calculated as 5552 km^2.

This tree has previously been assessed as Vulnerable in Cheek *et al.* (2000: 74), Cheek, in Harvey *et al.* (2004: 70) and in Cheek *et al.* (2010: 105) since it was estimated that more than 80% population reduction of this tree species has occurred in the last 100 years due to habitat loss (see below). This assessment is maintained here, no new data being available.

Tree 3.5–15 m tall, usually more than 7 m tall. Leaf-blade suborbicular to ovate, slightly 5-lobed, 11–19 × 5.3–15 cm, shortly acuminate, cordate, margin slightly denticulate; petiole 4–7 cm long. Inflorescence 6.8–12 cm long, axillary cyme; peduncle 3–7 cm long, 3–4-branched; pedicels 1–2 cm long. Epicalyx bracts 3, 0.2–0.35 cm, obtuse. Calyx lobes lanceolate, 0.5 cm long. Petals 5, white, 0.8–1.3 cm long; staminodes c. 1 cm long. Fruit capsule depressed globose, 0.3–0.5 cm diam.

Habitat: submontane woodland and forest:grassland transition; (700–)1220–1980 m alt.

Threats: clearance of submontane forest and woodland for agricultural land, over-exploitation for bast fibre. Studies in the Dom general area of the Bamenda Highlands showed that 50% of the surviving forest was lost between the 15 years 1988–2003 (Baena in Cheek *et al.* 2010: back cover).

Nigeria has seen loss of over 90% of its original forest cover according to Mongabay.com

Management suggestions: more information is needed on the population structure, extent, distribution and threats to *D. ledermannii* within the Bamenda Highlands which appears to be the centre of distribution of the species. This would form a baseline for monitoring in future. The practice of propagating the species as in the Bamboutos Mts (see above) should be encouraged and supported elsewhere within the species range.

Leptonychia kamerunensis Engl. & K.Krause (syn. *Leptonychia montana* Engl. & K.Krause; *Leptonychia* sp. 1 of Kupe-Bakossi checklist (Cheek *et al.* 2004: 411); *Leptonychia* sp. 1 of Bali Ngemba checklist (Cheek in Harvey *et al.* 2004: 129)).

EN B1+B2a, b(iii)

Range: Cameroon endemic (SW: Mt Kupe; Lebialem Highlands. NW: Bali Ngemba FR).

This treatment is taken from Cheek *et al.* in press (Three new or resurrected species of *Leptonychia* (Sterculiaceae-Byttneriaceae-Malvaceae) from West-Central Africa, Kew Bull.) in which the species is resurrected.

Tree (2–)3.5–9 m, puberulent; leaves obovate or narrowly elliptic, c. 21 × 7 cm, acumen c. 3 cm, base obtuse, entire, lateral nerves c. 8 pairs, white on upper surface, scalariform; petiole 1.5 cm; inflorescence axillary, rachis 1 cm; flowers pale green, 1.5 cm; fruit globose c. 5 cm.

Leptonychia kamerunensis is the only member of the genus known to occur above 1000 m altitude W of the Albertine Rift, that marks the eastern boundary of Congo (Kinshasa). It is further distinguished by having by far the largest fruits of any member of the genus. Most species have fruits only 1–1.5 cm diam. when mature. The local name at Fosimondi is Keliteh (Banwa language). The fruits can be stewed like those of okra (*Abelmoschus esculentus)* according to Nwopiemb and Tanjong of Fosimondi (*Tchiengué* 2208).

Leptonychia kamerunensis probably once occurred more or less continuously in the forests of the Cameroon Highlands from Mt Kupe in the S to the Bamenda Highlands in the N. Today it still has this range, but in the now largely deforested Bamenda Highlands it is restricted to the forest relic which is the Bali Ngemba FR. *Leptonychia kamerunensis* appears to be largely absent from the area in between. However it does survive just to the W of the Bamboutos Mts at Fosimondi in the Lebialem Highlands. It is now only known at four sites: at Mt Kupe (above Kupe village and above Nyasoso), at Fosimondi, and at the Bali Ngemba FR. At the latter two it is under immediate and ongoing threat from clearance of the understorey for the cultivation of crops such as *Colocasia.* At Fosimondi, in addition, the forest canopy itself is also being progressively cleared downslope from the village of Fosimondi (c. 1900 m alt.), fide Tchiengué (pers. comm. 2006). The area of occupancy is estimated as being 50 km² (2007). At the site where it was first discovered, Mt Ekomane, SW of the the Bamboutos Mts, not far distant from Fosimondi between Mt Kupe and Bali Ngemba, forest clearance has been so extensive that the species is believed no longer to exist there.

Leptonychia kamerunensis is here maintained as EN B2ab (iii) following Cheek *et al.* (op. cit. in press) since there are only three locations (AOO 12 km² with 4 km² cells) and threats are known as described above and below. EOO is calculated as 1018 km². The species, while not common, is not infrequent, at either Bali Ngemba (six specimens) or Mt Kupe (eight collections). At Bali Ngemba it is listed as one of 17 taxa that characterise the submontane forest since collections occur there throughout the altitudinal range 1300 to 1900–2000 m alt. (Cheek in Harvey *et al.* 2004: 16). B1 status has been added here since the newly calculated EOO permits this.

Habitat: submontane forest with *Santiria trimera* (Burseraceae), *Cola verticillata* (Sterculiaceae), *Macaranga occidentalis* (Euphorbiaceae); 1190–2000 m alt.

Threats: clearance of surviving forest for timber, followed by agriculture especially at Fosimondi and Bali Ngemba (see above).

Management suggestions: at Fosimondi and Bali Ngemba the assistance of local communities should be sought and supported to protect this species and it should be incorporated into reforestation plantings. The logical focal point for conserving the species is at Mt Kupe where threats are lowest and the species has been most frequently recorded. Data on regeneration levels and growth rates are needed, as well as educational material to support its protection and management.

Leptonychia moyesiae Cheek ined. (syn: *Leptonychia* sp. nr. *echinocarpa* pro parte quoad Olurunfemi in FHI 30641 (Keay 1958: 316); *Leptonychia* sp. aff. *echinocarpa* K.Schum., Cable & Cheek (1998: 134); *Leptonychia multiflora* K.Schum. excl. Khayota 579, Cable & Cheek (1998: 134); *Leptonychia* sp. aff. *multiflora* K.Schum. excl. Wheatley 823, Cable & Cheek (1998: 134)).

VU B1a,b(iii)

Range: Equatorial Guinea (Bioko (5 coll.)); Nigeria (Oban area (3 coll.)) and Cameroon (SW: Korup (20 coll.); Mt cameroon at Onge-Mokoko, Njonji, Bakingili and Bimbia).

Leptonychia moyesiae has an extent of occurrence of 14914 km² and is known from nine locations, at several of which its lowland evergreen forest habitat is known to be threatened by clearance for agriculture especially in the foothills of Mt Cameroon e.g. at Bimbia, Njonji and Mokoko. For these reasons it is here maintained, following Cheek *et al.* in press, op. cit. as VU B1a,b(iii) i.e. Vulnerable. Fragmentation is set to continue as the coastal strip around Mt Cameroon continues to be developed for tourist hotels, and oil palm plantations. Fortunately the stronghold of the species, Korup, with 20 of the 38 known collections, appears secure, as this National Park is probably the best protected in Cameroon.

Evergreen shrub 1–3(–10) m; internodes 0.5–2 cm, drying smooth, black, shortly and densely pubescent with dull pale brown, simple and 5–7-armed stellate hairs; leaves distichous, drying dark-brown to black, very rarely green, narrowly elliptic 7.5–21 × 2.5–7.5 cm, acumen 1.7–4 cm long, base cuneate-acute, margin entire, midrib raised above, domatia absent between basal secondary nerves and midrib, but often present in axils of higher secondary nerves, elliptic, parallel to midrib c. 0.75 × 0.3 mm, crater-like; lateral nerves 7–8 on each side of the midrib, concolorous with the lamina, midrib and nerves glabrous, or on lower surface with sporadic stellate hairs c. 0.3 mm diam., lamina glabrous above, lower surface inconspicuously stellate-hairy, hairs 0.5–1 mm diam., 4–5-armed, arms extremely fine, white, straight, held parallel to the blade surface, hairs separated by 1–15 hair diameters; petiole canaliculate 0.5–1.1 cm long, lacking distinct pulvini, moderately densely hairy with appressed 5-armed stellate hairs, 0.1 mm diam.; stipules ligulate-falcate 2–4.7 mm, apex acute, moderately densely hairy, hairs appressed, robust, 0.25 mm long on outer surface, glabrous inside. Inflorescences axillary, 8–10 mm; flowers pale green, 4–6.5 mm; sepals 5, drying brown-black, ligulate-oblong, 4.7–5.7 mm, mucronate, outer surface sparsely 3–7-armed stellate hairy, hairs white; petals slightly concave, inversely subtriangular to transversely elliptic, 0.3–0.6 cm, apex truncate to concave-emarginate, staminal tube dilating towards apex slightly, 0.1–0.3 mm long; phalanges 5, each of two fertile stamens, lacking filamentous staminodes; small staminodes inserted on the rim of the staminal tube, between the phalanges, flattened; ovary ovoid-pyramidal 0.9–1.2 mm diam., densely pubescent; fruit globose before dehiscence, 1.2–1.5 cm diam., rostrum acute, c. 1 mm long; ripening green (field notes), drying glossy black, surface verrucate to sub-tuberculate, thickly scattered with 5–10-armed white stellate hairs c. 0.3 mm diam.; valves 3–4. Seeds 1–2, ellipsoid, 1–1.5 cm black, 60–100% covered by an orange fleshy layer.

According to Ferdinand Namata, in the Korup area the species is known as 'Bush Iodine' for its use as a wound disinfectant (*Cheek* 7232).

Habitat: Evergreen forest; 50–300(–350) m. alt.

Threats: clearance for timber and oil exploration, and agriculture, particularly in the western foothills of Mt Cameroon for plantations of bananas, oil palm, rubber and cocoa, and also small-holder agriculture.

Management suggestions: a survey of the population of this species is advisable to form a baseline for monitoring, as are measures to protect what remains from further destruction. The best hope for the survival of this species is the Korup NP which is the most secure location

Leptonychia subtomentosa K.Schum.

EN B1a,b(iii)

Range: Cameroon endemic (C: Yaoundé area at Mefou proposed NP; Ototomo Reservoir; Colline Kombeng, 8 km SSE, Matonb).

The ten specimens known for *Leptonychia subtomentosa* derive from four locations: the proposed Mefou NP (five coll.), Colline Kombeng, Ototomo Reservoir and an unknown location in Yaoundé (the type collection). So far as is known, these all fall in Yaoundé's SW quadrant with an extent of occurrence calculated here as 1158 km^2. Yaoundé is the capital city of Cameroon with a population of over a million people and steadily expanding. It is very likely that the original Zenker and Staudt collections from over a century ago were made on one of the city's now deforested hills. Even at Mefou proposed NP destruction of natural habitat continues, albeit at a smaller scale than in the wider area. Accordingly, *L. subtomentosa* is here maintained as EN B1a, b(iii) following the assessment in Cheek *et al.* op. cit. in press and in Cheek *et al.* 2011. *Leptonychia subtomentosa* is easily identified by the long length and persistence of the stem hairs. No other *Leptonychia* species of central Africa known has these features.

Shrub, 3 m; stems persistently softly tomentose with brown 1 mm hairs extending to petiole and nerves; leaves alternate, needle-like, 6–7 mm; leaf-blade elliptic-oblong, 12–18 × 6–9 cm, excluding the 2–4 cm slender acumen, base shallowly cordate to obtuse, weakly 3-nerved at base, lateral nerves extending 1/3 the length of the blade, lateral nerves 6–7 pairs, weakly uniting near the margin, tertiary nerves scalariform, domatia absent; petiole 4–8 mm; flowers resembling *L. lasiogyne*; fruits 1.5 cm, densely long brown tomentose.

Habitat: within semi-deciduous forested areas: submontane forest in hilly areas and also along rivers in *Gilbertiodendron dewevrei* gallery forest; 550–880 m alt.

Threats: see above.

Management suggestions: since the most recent records derive from the Mefou proposed NP, this is the logical place to focus conservation efforts. Basic populational data should be gathered to inform future monitoring and to assess needs for intervention.

THYMELAEACEAE

assessed by Martin Cheek

A worldwide family of alternate leaved, stipule-free shrubs, with some trees, rarely lianas with opposite leaves (*Craterosiphon*). In the absence of flowers, members of this family can readily be recognised by their especially silky fibrous bark. When the end of a cut stem is agitated by a thumb-nail, a brush of fine fibres results. No coloured exudates or scents are known in the family. The leaves are simple, pinnately nerved, entire, the flowers in the main subfamily Thymelaeoideae, are distinctive in having a hollowed-out receptacle making a deep tube, from the rim of which the sepals and petals arise, with the stamens inserted within the throat in one or two whorls. The superior ovary has 2–5, or more locules, with a single style and style-head which, produces a drupe or berry, less usually a capsule (eg. *Octolepis* or *Synandrodaphne*).

800 species in 45 genera are recognised by Heywood in Heywood *et al.* 2007 (op. cit.) The family is treated in Flore du Cameroon 5 by Aymonin who accepted 24 species in 6 genera: *Dicranolepis*, *Craterosiphon*, *Peddiea*, *Gnidia*, *Octolepis*, and *Synandrodaphne*. None are of major economic value. The first four genera are placed in subfamily Thymlaeoideae while *Octolepis* is placed in subfamily Gonystyloideae and *Synandrodaphne paradoxa*, with a single species, is accorded its own subfamily, Gilgodaphnoideae, restricted to Western Tropical Africa.

All the Cameroonian genera are found in forest, usually lowland and submontane evergreen forest, excepting *Gnidia*, with numerous species, which is associated with grassland habitats or forest edges.

Craterosiphon montanum Domke
EN B2ab(iii)

Range: Cameroon (Adamawa: Pass Tchappe, forest of Gendro (Tchabal Mbabo)) and Gabon (Lastoursville).

Here *Craterosiphon montanum* is assessed as Endangered since two locations are known (above; AOO 8 km^2 with 4 km^2 cells) with threats as below.

Liana, stem twining, glabrous, epidermis light red-brown with numerous pustular lenticels, leaves opposite, stipules absent; petioles 3–5 mm; blades ovate-oblong 4–6 × 3–4 cm, acumen triangular, 1–1.5 cm, lateral nerves c. 20 pairs uniting near the margin; flowers axillary, solitary, pedicels 1 mm bracteoles minute, 4–10; receptacle tube 12–15 mm, lobes 4–6 × 1–1.5 mm, stamens inserted in throat; petaloid glands not present in throat; fruit unknown

Habitat: submontane forest (deduced); 800–1420 m alt.

Threats: over 90% of submontane forest in that part of the Cameroon highlands whence this species was found has since been cleared for agriculture.

Management suggestions: rediscovery of this species in the Cameroon Highlands at and around Tchabal Mbabao should be attempted so as to record basic populational data for monitoring and to secure local support to protect the species, if it survives.

Craterosiphon pseudoscandens Domke
ENB2ab(iii)

Range: Togo (unlocalised, *Buttner* 388 and 686 cited by Domke), Cameroon (S: Bipindi) and Gabon (Libreville; Cristal Mts at Tchimbele; Moula).

Here *Craterosiphon pseudoscandens* is assessed as Endangered since five locations are known (above; AOO km^2 with 4 km^2 cells) and threats as below. EOO is calculated as 355126 km².

Liana, finely ridged grey-orange or red-grey, lenticels numerous; leaves opposite, petiole 3–5 mm; blade elliptic, 4–8 × 2.5–4 cm, acumen acute, lateral nerves c. 16 pairs; flowers axillary, solitary, pedicels unknown; receptacle tube 12–15 mm; lobes 2 × 1.5 mm; stamens inserted in throat; petaloid glands not present in throat; fruit 4 × 1.5–2 cm, spongy, seed single oblong-conic, 2–2.5 × 0.75 cm.

Habitat: lowland evergreen forest

Threats: slash and burn agriculture (Bipindi area).

Management suggestions: rediscovery of this species in the Bipindi area should be attempted, and the local people involved in its protection if it is refound, perhaps supported by a species-specific conservation poster. If re-discovered, basic populational data should be collected as a baseline for monitoring.

Dicranolepis polygaloides Gilg ex H.H.W.Pearson
VU B2ab(iii)

Range: Cameroon (SW: Mundame-Etam, Bakossi FR, Nyasoso, Bambuko FR. C: Essam (Nanga-Eboko). S: Bipindi).

Here the species is maintained as Vulnerable following Cheek in Cheek *et al.* (2004: 186) since six localities are known (above; AOO 12 km^2 with 4 km^2 cells) and with threats as below. EOO is calculated as 65747 km². This very distinctive shrub is the smallest by far of all Cameroonian *Dicranolepis*. A former record of this species from Gabon has since been redetermined as *D.bucholzii*.

Shrub, 0.5 m; leaves 3 × 1.5 cm; calyx tube 2.5 cm, sepals 0.7 cm, both densely white puberulent.

Habitat: lowland evergreen forest; to c. 900 m alt.

Threats: forest clearance due to logging, followed by agriculture. Although the species is centred in SW Region, all the known sites here have been affected by habitat destruction to some extent or another. Threats to the last two sites (outside of SW Region) are unknown.

Management suggestions: revisiting all known sites for the species is advised in order to attempt to rediscover it and to assess levels and nature of threats, if ongoing. Data on numbers of individuals per population and demography are also required to inform a management plan for the species.

Gnidia bambutana Gilg & Lederm ex Engl.

EN B2ab (iii)

Range: Cameroon (W: Bamboutos Mts, La Pastorale (2 coll.)) and Nigeria (Mambilla at Nguroj F.R. (5 coll.); Vogel Peak (1 coll))

This submontane-montane grassland species is restricted to the inland, drier parts of the Cameroon highlands. Here it is assessed as Endangered since three localities are known (AOO 12 km^2 with 4 km^2 cells) and with threats as below. EOO is calculated as 13488 km^2. The absence in the Bamenda Highlands is remarkable.

Twiggy shrub 1–2.5 m, glabrous; stems dark brown, matt, 4–5 mm diam; leaves alternate, petiole 1 mm, blade linear, 10–15 × 1–2 mm, acute, numerous; flower heads globose, 2 cm, flowers greenish-yellow, c. 100 per head, in bud pubescent, matchstick-like with a swollen head, interspersed with ovate bracts c. 5 × 4 mm: flowering late Nov. to end Jan.

Habitat: forest thicket along streams or in open grasslands; 1720–2100 m alt.

Threats: forest clearance for agriculture; grassland trampling due to intense cattle stocking. This section of the Cameroon highlands has seen 95% loss of its original forest.

Management suggestions: globally-speaking Nguroje FR in Mambilla would be the logical focus for conservation efforts since most recent records derive from there and there is an active conservation project in the area (Nigeria Montane Forests project). In a Cameroon context, the Bamboutos Mts should be visited to rediscover the species and to seek means to protect it with local support since there is no protected area in the vicinity. Basic populational data are needed and an evaluation of threats, as a baseline for monitoring.

Synandrodaphne paradoxa Gilg

VU B2ab(iii)

Range: Cameroon (E: Yokadouma-Lomié, 7 km; Moloundou-Yokadouma; 25 km SW Yokadouma), CAR and Congo (Kinshasa) (Oriental Prov. Ubangi-Uele Distr at Tupwo; Businga-Mobayi; Kisangani area; Bas-Uele; Bas-Katanga at PeneYumbi).

Here *Synandrodaphne paradoxa* is assessed as Vulnerable since ten locations are known (above; AOO 40 km^2 with 4 km^2 cells) and threats as below. EOO is calculated as 270181 km^2. The CAR locations remain unknown for the present.

Subshrub 1.25 m, stems ap[illegible] leaves alternate, petiole 3–5 mm, blade e[illegible] cm, acutely subacuminate, base acute-obtuse, lateral nerves c 8 pairs, uniting in a looping marginal nerve, membranous, appressed pubescent on both surfaces; flowers in terminal heads; involucral bracts ovate, 5–8 × 2.5–5 mm densely appressed hairy, ribbed; flowers sessile, recepatacular tube absent; sepals four, similar to bracts, the two inner smaller than the two outer, petals absent; filaments four, surrounding a tube; ovary 2 locular superior: fruit 2- valved.

This species is so unusual that it is placed in its own subfamily in Thymelaeaceae. Further surveys may well discover new locations, removing the species from the threatened category. However it appears genuinely rare and localised within much of its range. While fairly common in the Yokodouma area and also around Kisangani, it is entirely absent from the Dzanga-Sangha Reserve (Harris 2002, The Vascular Plants of the Dzanga-Sangha Reserve, Scripta Botanica Belgica 23) which has similar forest vegetation.

Habitat: mixed semi-deciduous-evergreen forest (with *Terminalia*) deduced; 500–600 m alt.

Threats: logging in semi-deciduous forest, sometimes followed by agriculture (Yokadouma area). Threats in CAR and Congo (Kinshasa) are less well understood.

Management suggestions: rediscovery of their plants in the Yokodouma-Moloundou area is advisable so that basic populational data can be collected as a baseline for monitoring. Boumba-Bek NP should be screened for presence of this species, since if present this would be a logical focus for national conservation efforts.

UMBELLIFERAE (APIACEAE)

assessed by Martin Cheek (advised by Iain Darbyshire)

The Umbelliferae are a cosmopolitan family, but predominate in open, sunny habitats in cooler climates. Most Cameroon species and all those that are threatened, occur in the Cameroon Highlands just above the forest in grassy habitats at the forest edge, making them vulnerable to forest loss. They are absent in trampled and grazed areas. *Samicula elata* occurs in submontane forest, *Hydrocotyle bonariensis* near coastal beaches and *Centella asiatica* is a weed, but these are exceptions.

Most species are erect herbs 0.3–2 m tall with compound leaves and pungent aroma when crushed. The family are easily recognised by, and named for, their umbellate inflorescences. The ovary is ribbed, bears two curved styles, and divides into two flat dry, often winged fruitlets.

Numerous vegetable and herb crops belong to this family, such as the carrot (*Daucus*), parsnip (*Pastinaca*), fennel (*Foeniculum*) to name only a few.

The taxa in Cameroon are well delimited due to Flore du Cameroun 10 by Jacques-Félix (1970). However, one or two apparently endemic species recognised by him have since been placed in synonymy under more widespread E African taxa, e.g. *Lefebvrea stuhlmannii* Engl. is now included under *L. abyssinica* A.Rich. Following the FTEA Umbelliferae become a synonym of the very widespread *Lefebvrea grantii.* Similarly *Pimpinella ledermannii* is no longer endemic to Cameroon, being found also in E Africa. However, the E and W populations are recognised as different subspecies (Townsend (1989), loc. cit.). In contrast, *Peucedanum winkleri* was regarded in Flore du Cameroun as a widespread tropical African species but the Cameroon Highlands element has since been shown to be specifically distinct (see under *Afroligusticum townsendii* below).

One additional new species, *Peucedanum kupense* I. Darbysh. & Cheek has since been described (Kew Bull. 59: 133–136 (2004)). *Peucedanum*, long known to be an artificial genus has been analysed on a molecular basis by Winter *et al.* in Taxon 57 (2): 347–364 (2008). Several of the Cameroonian taxa were shown to be better treated in the genus *Lefebvrea*, and *Peucedanum angustisectum* was transferred there, as *Lefebvrea angustisecta*. Winter *et al.* sank *Peucedanum camerunense* Jacque-Felix, and *Peucedanum kupense* into the latter, without morphological explanation. Since these two taxa appear perfectly distinct from each other and from *Lefebvrea angustisectum*, they are here maintained as separate species, transferred to *Lefebrvrea*, pending a revision.

Pimpinella cf. praeventa Norman
Numerous collections of what seems to be a new species closely related to *P. praeventa* (which occurs in the Jos Plateau area of Nigeria) were made during our inventories of 1996–1999 at Kilum-Ijim (see Cheek *et al.* 2000: 76). This material differs in being robust in habit, with densely pubescent rays. Another taxon, with sparsely, long-hairy rays occurs in the Tchabal Mbabo area, while the Jos material has glabrous rays and a gracile habit.

More work is still needed to delimit these taxa. If each is confirmed as separate, they are likely to be assessed as Red Data taxa in future.

Afroligusticum townsendii (Charp. & Fern. Casas) P.J.D.Winter (syn. *Peucedanum townsendii* Charp. & Fern. Casas; *Peucedanum winkleri* Wolff pro. parte)
EN B1+B2ab(iii)
Range: Equatorial Guinea (Bioko: Pico Basile, numerous coll.) and Cameroon (SW: Mt Cameroon, Wikile (several coll.). Bamboutos Mts. NW: Mt Oku (1 coll.)).
Herb to 1.5 m, highly branched, stems finely ridged when young; leaves bipinnate to –sect, to 35 × 30 cm, homomorphous, cauline leaves reduced, to 35 × 30 cm, pinnules ovate in outline, 3 cm, deeply pinnatifid, apex acute, account by Townsend (1989), and *Peucedanum zenkeri* has

base cuneate, margins serrate-mucronate, stem and leaves tomentose at nodes; inflorescence of dispersed umbels, peduncle hispid at apex; involucral bracts linear, 0.5–1 cm; rays 8–20, 5 cm, puberulent, umbels 10–20-flowered; pedicels 1 cm; flowers white, petals 2.5–3 mm. Fruit oblong, 10 × 7.5mm, wings exceeding fruit body in width and at apex.
Treated in FWTA ed 2 and Flore du Cameroun 10 (1970) as *Peucedanum winkleri,* occurring in both the Cameroon Highlands and in E Africa and therefore not previously regarded by us as threatened and so not included as a threatened taxon in Cable & Cheek 1998, nor in Cheek *et al.* 2000. The references below only came to light in 2010 thanks to a question raised by Jo Osborne.
Townsend in a precursor to his FTEA Umbelliferae (1989) states that three taxa have been included under the name *Peucedanum winkleri*, moreover that both syntypes from E and W Africa are lost and that there are difficulties applying the name; he lectotypifies on the Tanzanian material and states that the name should certainly not be used for the W African (and Ethiopian) material which appear to require new names (Kew Bull. 42: 593–596 (1987)).
Charpin and Fernandes Casas provide a new name for the Cameroon Highlands material, as *Peucedanum townsendii* (Anales Jard Bot Madrid 54(1): 276 (1996)).
Winter *et al.* review African *Peucedenaum* in Taxon (57(2): 359 (2008)) and split it up, moving this species to another genus, hence the current name.
Afroligusticum townsendii is here assessed as Endangered, since four locations are known (above, under range, giving an AOO of 16 km² with 4 km² cells, threats as below). EOO is calculated as 2814 km².
Habitat: montane forest: grassland transition 2000–2800 m alt.
Threats: at the Mt Cameroon location, no threats have been seen, while threats at Pico Basile are not known to us. However, at Bamboutos the forest element of the habitat has mostly been destroyed and that which remains is under high pressure from agriculture. At Mt Oku, goats and cattle are grazed in the habitat of this species, potentially threatening it by trampling and by the setting of more frequent fires by graziers to benefit their animals. Only a single specimen of the species was seen among many thousands recorded at this location.
Management suggestions: detailed data on regeneration levels, frequency and range of this species is needed. October and November, when the plant is conspicuous since in flower, is probably the best time to conduct this work. Direct observations of threats to this species are desirable.

Lefebvrea angustisecta Engl. (*Peucedanum angustisectum* (Engl.)Norman)
VU B2ab(iii)

Range: Nigeria (Mambilla Plateau, Chappal Waddi, 2 coll.) and Cameroon Highlands (SW: Mt Cameroon (12 coll.). W: Bamboutos Mts. NW: Bali Ngemba; Baba 2 (1 coll. each); Kilum-Ijim (2 coll.). Adamawa: Tchabal Mbabo (1 coll.)). Dubiously identified specimens from Congo (*Kassner*) and Kaduna, Nigeria (*Sharland*).

Annual herb up to 1.2 m tall, slightly branched, glabrous. Stems smooth or slightly striate. Leaves biternate to bipinnate, gradually reducing in length towards the apex of the stem, the largest 25 × 20 cm, 2–3-jugate, petiole 10 cm long, basal sheath c. 3 cm long, slightly canaliculate; lowermost pinnae with petiolules c. 2 cm long, blade often subpalmatisect; pinnules linear-lanceolate to narrowly lanceolate, the terminal the longest, up to 8 × 1.5 cm, coarsely and unequally acutely serrate to incised serrate. Umbels dispersed in a slender inflorescence; involucre absent; rays 6–8, 1.5–1.8 cm long; involucel reduced to a few short straight bracts c. 5 × 1.5 mm; pedicels subequal, 4 mm long. Fruits narrowly elliptic, 6–7 × 3–3.5 mm, 2.2 mm broad, margin straight, dorsal face with slightly raised ridges; commissural face with four bands, the inner broad, the outer pair submarginal and sometimes evanescent. Description after Jacques-Félix (1970: 92–93).

Lefebvrea angustisecta is known from the length of the Cameroon Highlands, but was not collected from the Bamenda Highlands until located at Ijim in November 1996 (*Cheek* 8698, grassland, 1900 m alt., path from Fon's Palace to Mbororo settlement). Described by Engler as *Lefebvrea angustisecta* in 1921, this species was transferred by Norman to *Peucedanum* in 1934 (Jacques-Félix 1970: 92). Winter *et al.* (2008) divided African species of this genus, placing this and other species back in *Lefebvrea* (see family introduction).

Previously assessed as LR/nt (Cheek *et al.* 2000 as *P. angustisecta*), *Lefebvrea angustisecta* is here assessed as Vulnerable, since six locations are known (above, under range, giving an AOO of 24 km² with 4 km² cells, threats as below). EOO is calculated as 20598 km².

Habitat: rocky montane grassland; forest:grassland transition; 1900–3200 m alt.

Threats: as for *Lefebvrea camerunensis* (q.v.).

Management suggestions: as for *Lefebvrea camerunensis* (q.v.).

Lefebvrea camerunensis (Jacq.–Fél.) Cheek & I. Darbysh. comb. nov. (Basionym: *Peucedanum camerunensis* Jacq.–Fél. Adansonia 10(1): 47 (1970)).
Typus: Cameroon, region Foumbam, Mt Nko Gam, *Koechlin* 7541 (holotypus P)
VU B1+B2ab(iii)

Range: Cameroon (SW: Mt Cameroon. W: Bamboutos Mts: Nkounden; Massif du Mbam; Massif du Nkogam; (1 coll. each); NW: Bamenda Highlands (Bambili; Kilum-Ijim (3 coll.)).

Annual (?) herb, 30–40 cm tall, erect, glabrous. Stems cylindrical, smooth, pale green, firm to subwoody. Basal leaves 18–20 cm long, petiole 5–6 cm, sheath 0.5 cm, rachis 2–(–3-)jugate, petiolules 2.5 cm long, the pinnae ternate, rarely pinnate, leaflets sessile, linear, 7–8 × 0.2–0.3 cm, variable in size and number of leaflets, from 3 to 1. Umbels usually terminal, on relatively numerous branches; involucre nil; rays 5–7, straight, 1.5 cm long; umbellules poorly developed; involucel of several filiform bracts 1–1.5 mm long; flowers 8–12; pedicels 1 cm long. Fruits elliptic, 5 × 3 mm, moderately flattened on the back, wing straight, 0.5–0.6 mm wide, commissural bands 4, the inner thick and conspicuous, the outer pair filiform, situated in the wing. Description after Jacques-Félix (1970: 93–94).

This species is easily recognised by the linear leaf divisions. It was described by Jacques-Félix in 1970, from collections at two locations straddling the Bamenda Highlands (Nkounden, Massif du Nkogam, 2400 m, Oct. 1964, *Koechlin* 7541 (type) and Bamboutos, 2300 m, (sterile, and therefore uncertainly identified), *Jacques-Félix* 5482). Since then, two further collections have been identified (Massif du Mbam, 2335 m alt., 35km NW Foumban, *Letouzey* 13099, November 1974 and Lake Bambili, Aug. 1951, *Ujor* in FHI 29968).

Three collections were made in the course of our 1996 inventory at Kilum-Ijim. They were at locations each c. 10 km or so from each other: cliff-face, Oku-Elak, *Munyenyembe* 819 (31 Oct. 1996); Ewook Etele Mbae, Oku-Kumbo, 10 km, *Zapfack* 1229 (6 November 1996) and on the southern (Ijim) side of Lake Oku, *Onana* 639 (26 November 1996). No collections were made on Laikom/Ijim ridge during the expedition there in Dec. 1998 probably because, although the habitat there might be suitable, it was then later in the year and the flowering season for that species (which seems to be restricted to Oct. and November) was past.

There is little information on the population density of this species available from the herbarium specimens seen. However, given that 2–3 duplicates of each of our specimens were made, it seems likely that plants of this species are fairly thinly scattered, although not completely isolated (e.g. by a kilometre or more) from each other at each location. Since many thousands of specimen records exist from Mt Cameroon in total, and only one represents this species, it appears genuinely rare there

Lefebvrea camerunensis appears to be an annual species (Jacques-Félix 1970), given the relatively diminutive size of the plant and absence of a substantial rootstock. If this fact is confirmed, it makes this species more vulnerable to adverse environmental effects than if it were a perennial, and makes seedling recruitment of utmost importance for this species.

This reassessment here made is Vulnerable, since six locations are recognised (above, under range, and AOO is 24 km² using 4 km² cells, threats as below). EOO is calculated as 8714 km².
This is modified from the EN B1+2c given (under the name of *Peucedanum camerunense*) in Cheek *et al.* (2000: 75–76). The need for the name change above is explained in the family introduction.
Habitat: rocky montane grassland or scrub; 2300–2800 m alt.
Threats: man-made grassland fires and grazing and trampling by domesticated animals are believed to affect the survival and establishment of this species and are a threat at all known locations apart from at Mt Cameroon where, unusually, no domestic stock are grazed in the montane grasslands.
Management suggestions: more detailed data on regeneration levels, frequency and range of this species is needed. October and November, when the plant is conspicuous since in flower, is probably the best time to conduct this work. Direct observations of threats to this species are desirable. The ecology of this species also needs clarification. Most Umbelliferae in the Cameroon Highlands occur near forest margins but currently there is no evidence that this is true in this case.

Lefebvrea kupense (I. Darbysh. & Cheek) Cheek & I. Darbysh. comb. nov. (Basionym: *Peucedanum kupense* I.Darbysh. & Cheek in Kew Bull. 59: 133 (2004). Type: Cameroon, SW Region, Mt Kupe summit, 1900 m alt., fl. fr. 31 Oct. 1995, *Cheek* 7586 (holotypus K; isotypi MO, P, SCA, WAG, YA))
VU D2
Range: Cameroon endemic (SW: Mt Kupe (2 coll.)).
Upright herb c. 2 m; stems finely striate, glabrous; stem leaves biternate, or with a terminal ternate pinna and two lateral pinnae of paired unequally sized leaflets, leaflets ovate to trullate, central leaflet of pinna 7–10 × 2–3 cm, acuminate, mucro to 0.7 mm long, base attenuate, asymmetric, margins deeply serrate-mucronate or biserrate, incisions 5–6.5 mm; inflorescence much-branched, of 4–10 umbels; primary peduncle 0.4–5 cm, subtended by a reduced trilobed-foliolate leaf; umbel diam. 3–3.5 cm in flower; 5–5.5 cm in fruit, 8–10-rayed; rays 1.5–2.8 cm, glabrous; involucre of 5–7 linear bracts 0.5–0.65 mm; flowers globose, 1.6 mm diam., yellow (not white as reported in protologue); mericarps elliptic, 4.7–6.3 × 3–3.6 mm, base and apex shallow-emarginate, body with 3 longitudinal ridges, wings not confluent at base or apex, 0.5–0.6 mm wide.
The area of occupancy of this taxon is less than 1–2 km², this being our estimate for the grassland areas comprising the two (of three) peaks of Mt Kupe from which it has been collected. Since these grassy areas are very precipitous below the peaks, we have not fully explored them and so can only roughly estimate the size of the populations of *P. kupense*. It is likely that the number of mature individuals is only a few hundred, but could fall in the range 50 to several thousand. We here assess this taxon as vulnerable due to the very restricted area of occupancy and number of locations such that it is vulnerable to stochastic change (see below). Subsequent attempts to relocate this species in 2003 to 2004 have been unsuccessful to date (M. Etuge, *pers. comm.*).
It is possible that this conservation assessment will be down-rated in future if the plant is found in other adjacent montane grassland areas, such as the poorly known Mont Nlonako to the NE. However, investigations of montane grassland in the Bakossi Mts, the Mwanenguba Mts and Mt Cameroon have so far not revealed populations of this plant. This assessment and the description are taken from Darbyshire & Cheek (Kew Bull. 59. 133–136 (2004)) and also appears in Cheek *et al.* (2004: 186–187). Since no new data area available, the assessment is unchanged here.
It is worth adding that no other related taxa occur on Mt Kupe, although we speculated on a hybrid origin for the species.
The need for the name change above is explained in the family introduction.
Habitat: montane grassland; 1900–2000 m alt.
Threats: stochastic change such as lightning-induced fires or landslides. The species is not endangered by man at present, being found far above the highest limits (c. 1100 m alt.) of agriculture on the mountain.
Management suggestions: rediscover and investigate more fully the populations of this taxon in the grassy areas on the summit of Mt Kupe, possibly using ropes and harnesses to access the steeper slopes. Surveys of other grassy areas on the mountain, such as Kupe Rock, might reveal further sub-populations. Ideally, surveys should be carried out annually over several years so that fluctuations can be recorded and their significance assessed. Surveys are best conducted between October and March, the dry season, when the plant is flowering and fruiting and so conspicuous and easily identifiable.

Pimpinella ledermannii H. Wolff. subsp. ***ledermannii***
VU B2ab(iii).
Range: Nigeria (Mt Vogel) and Cameroon (W: Bamboutos Mts, near Dschang. Adamawa:Tchabal Mbabo; Hosséré Sille; Meïganga; Hosséré Ngo, 80 km NW Tibati).
Erect herb, unbranched 0.3–1.2 m, stems terete, glabrous; cauline leaves oblong, c. 7 × 3 cm, apex acute, base cordate, margin dentate, surface densely puberulous: head 4–5 cm diam, slightly convex, axes densely puberulous, composed of 3–5 smaller capitula, each with a linear involucral bracts as long as the rays; flowers white, petals orbicular 1–2 mm.

Distinguished from the E African subspecies by glabrous styles, tomentellous rays and pedicels, umbel rays to 9 in number (Townsend 1989).

Easily identifiable in Cameroon due to the simple, oblong leaves.

This taxon is here assessed as Vulnerable since six locations (above, AOO 24 km² with 4 km² cells, threats as below) are known. The type location, Markt Singwa needs localisation (*Ledermann* 1620). EOO is calculated as 37952 km².

Habitat: high altitude grassland 1500–2500 m alt.

Threats: man-made grassland fires and grazing and trampling by domesticated animals are believed to affect the survival and establishment of species such as this and are thought to be a threat at all known locations.

Management suggestions: more detailed data on regeneration levels, frequency and range of this species is needed. October and November, when the plant is conspicuous since in flower, is probably the best time to conduct this work. Direct observations of threats to this species are desirable.

VIOLACEAE

assessed by Martin Cheek, Gaston Achoundong & Iain Darbyshire

This worldwide family are represented in Cameroon by four genera, *Viola*, *Hybanthus*, *Allexis* and *Rinorea*. They share alternate simple leaves and similar flower structure: zygomorphic flowers with sagittate stamens and an ovary with three parietal placentas. However, at first sight, *Viola,* a herb of temperate woodland and forest edge habitats, and *Rinorea*, shrubs and trees of the understory of lowland and submontane forest, appear to have little similarity. There is no Flore du Cameroon account, but Dr Gaston Achoundong former head of the Herbier National Camerounais, and General-Secretary of AETFAT, having researched the taxonomy of this family for many years, is working to that end. *Rinorea* is the most diverse genus of its family in Tropical Africa. Several *Rinorea* species are rare or range-restricted and consequently most of the threatened taxa below derive from this genus. *Rinorea* are easily recognised when in flower due to their small zygomorphic flowers, often densely arranged in cymes. In fruit the 3-valved capsules with parietal placentas aid identification to genus. When sterile *Rinorea* can be distinguished from other groups with alternate, simple leaves by the presence of a long, arched apical buds, unequal petiole lengths on each stem, the absence of stellate hairs [illegible] leaf-blade margin. Stipules are [illegible]. The threatened species below were all delimited by Dr Achoundong (Kew Bull. 53(4): 1009 (1998); ibid. 58(4): 960 (2004); ibid.60(4): 581–583 (2006)).

Undoubtedly additional species of *Allexis* and *Rinorea* in Cameroon will be found to be threatened as Dr Achoundong's work progresses. Those that appear currently to rate as Threatened/NT are: *Allexis obanensis* (Bak.f.) Melchior, *Rinorea albidiflora* Engler, *Rinorea caudate* (Oliv.) Kuntze, *Rinorea crassiflora* (Bak.f.) De Wild., *Rinorea ebolowensis* M.Brandt, *Rinorea exappendiculata* Engl., *Rinorea keayi* Brenan, *Rinorea ledermannii* Engl., *Rinorea leiophylla* M.Brandt, *Rinorea letouzeyi* Achoundong, *Rinorea microglossa* Engl., *Rinorea ovata* Chipp, *Rinorea sinuate* Chipp, *Rinorea umbricola* Engl., *Rinorea verrucosa* Chipp,.*Rinorea woermanniana* (Buettner) Engl. and *Rinorea zenkeri* Engl.

Allexis zygomorpha Achoundong & Onana
VU B2ab(iii)

Range: Cameroon endemic (S: Edéa-Kribi, 27 km; Kribi-Lolodorf, 28 km; Akok, near Campo; Calvary Mt; Kribi, 2 km N of Longji; Bipaga area).

Here *Allexis zygomorpha* is assessed as Vulnerable, since there are six locations (above; 24 km^2 with 4 km^2 cells, threats as below). EOO is calcutated as 233 km^2.

Shrub or small tree, to 10 m high. Stems dull grey, smooth, several together from the base; twigs hollow, dirty white, young parts white, glabrous. Stipules triangular, 10 × 1 cm, caducous. Leaves with petiole red when young, 9–10 cm long, glabrous; blade smooth, glossy bright to dark green above, shiny pale green below, broadly obovate, 20–90 × 4–15 cm; shortly acuminate, base acute, margin remotely serrulate; midrib and lateral nerves prominent beneath. Flowers fasciculate from nodes on the old wood, below the leaves and on upper stems within the crown; pedicels glabrous, greenish white, 5–6 mm long. Calyx greenish white, red in very young bud; sepals circular, unequal, 2–4 × 2–2.5 mm. Corolla yellow in bud; petals yellow outside, violet inside, the two upper petals united, coriaceous, cap-shaped, 4 × 2 mm, concave. Anthers yellow, 3.5 × 1.5 mm, subsessile, the uppermost anthers hairy outside; connective appendages white, ovate, closely cochlear-imbricate around the ovary, 1.5 × 1 mm; thecal appendage lacking. Ovary black, conical, 2 × 1 mm; style elongate, 2 mm long, rounded at top. Young fruit shiny white, very densely red-dotted becoming dull pale greenish white with pale dull-red mottling when mature; capsule oblong, coriaceous, reticulate, 4 × 2.5 cm. Seeds flat, obovate, pink, caruncGulate, 1.5 × 1.5 cm.

This species was described in Kew Bull. 53(4): 1009–1010 (1998), as the fourth species of the genus. It is close to *Rinorea* in its zygomorphic flowers, the imbricate anther appendages and elongated petioles, but was placed in *Allexis* due to its monocauly, cauliflory and the fusiform shape of the capsule (Achoundong & Onana loc. cit.). All data in this treatment derives from that publication.

Habitat: lowland evergreen forest.

Management suggestions: if the Akok locality falls in the Campo Ma'an NP, this would be the logical location for focussing conservation efforts since it is a formally protected area. Basic populational data should be collected as a baseline for monitoring and the park management provided with the means to identify the species, and encouraged to incorporate it in management plans; community leaders in Akok, and at other communities near locations for this species, should also be assisted to this end.

Rinorea dewitii Achoundong

VU B2a,b(iii)

Range: Cameroon endemic (SW: Mt Cameroon at Mokoko and Dakingili. Littoral. Nkongsamba, Loum area; Solé area, Loum-Yabassi Rd. S: Njabilobé, Akom II area, Ébolowa-Kribi Rd, Kienké riparian forest; Bidou II area, 18 km W of Kribi, Kribi–Ebolowa Rd).

Rinorea dewitii was assessed in the protologue (Achoundong & Cheek, op. cit. (2006)), on the basis of the data above, as Vulnerable, since there are six locations (above; AOO 24 km^2 with 4 km^2 cells, threats as below), this is maintained here in the absence of any additional data apart from the EOO which is calculated as 18649 km^2.

Shrub up to 5 m high; leaves subcoriaceous, elliptic, ovate or oblong, 12–20(–25) × 4–6.5(–7) cm; acuminate, base cuneate to obtuse, margin serrulate, lateral nerves 7–9 on each side of the midrib, prominent beneath; petiole up to 2 cm long; stipules triangular, 8 × 1.5 mm; inflorescence terminal or subterminal, thyrsiform, curved, not straight, 1.5–2.5 cm. long, glabrous; bracts triangular, 2.5 × 2 mm, mucronate, longitudinal nerves 5–7, prominent, margin ciliate; lateral branches 3–5-flowered; flowers zygomorphic, 6 × 3 mm; sepals 5, unequal, green, concave, ovate to orbicular, apex rounded, 3.5 × 4 mm, longitudinal nerves prominent, often branched at apex; petals 5, white, obovate to oblong, unequal, 4–5.2 × 2 mm, longitudinally nerved; fruit 3-angled, 2 × 2.5 cm, green when young, white at maturity. Seeds six, creamy white, unequally 4-angled.

This species is strictly endemic to the Cameroon littoral plain. Its distribution extends in the west to the Mokoko R FR immediately W of Mt Cameroon. There large populations of individuals of up to 4 m tall have been found. On the contrary, in the Nkongsamba-Loum-Yabassi area, where the species is most abundant, it occurs as a very short shrub no higher than 1 m. Far to the S, a small disjunct population is found E of Kribi, where it also occurs as a small shrub c. 2 m tall.

Habitat: lowland evergreen forest, generally near rivers, up to 250 m alt.

Threats: slash and burn agriculture; infrastructure for extractive industries (Kribi area).

Threats: Mokoko is scheduled for logging, and the Bakingili and Loum areas are also threatened by forest clearance for agriculture.

Management suggestions: basic populational data should be collected as a baseline for future monitoring. Currently none of the locations is protected so local community leaders should be consulted

Rinorea fausteana Achoundong

EN B1ab(iii)

Range: Cameroon endemic (SW: Banyang Mbo Wildlife Sanctuary (1 coll.); Rumpi Hills (2 coll.); Bakossi (3 coll. at 3 locs, 1 further unlocated specimen).

This species has an extent of occurrence calculated as 479 km^2 and occurs at five locations (above; AOO 20 km^2 with 4 km^2 cells, threats as below) with a continuing decline being anticipated in the quality of habitat in this area, and was thus assessed as Endangered by Cheek and Darbyshire in Cheek *et al.* (2004: 187), maintained here in the absence of any additional data apart from the EOO cited above..

Despite being a submontane species, it is not known from Mt Kupe; it seems that the Jide valley forms a barrier to its expansion eastwards from the Bakossi Mts. It is also not known from Mt Cameroon; here a band of semi-deciduous lowland forest may form a barrier to expansion southwards.

Shrub to 2.5 m; stems glabrescent; leaves subcoriaceous, (oblong-)elliptic, 13.5–19 × 5.5–9 cm, acumen to 1.8 cm, base cuneate to obtuse, margin crenulate, lateral nerves 9–10 pairs, glabrous, lower surface glandular; petiole 1–2 cm; inflorescence terminal, thyrsiform, to 5 cm, finely puberulent; sepals triangular, 1.5 mm, petals ovate-lanceolate, 4 mm, yellow; fruit minutely tuberculate.

The description is based on the protologue by Achoundong & Cheek (Kew Bull. 58: 958 (2003)).

Habitat: dense submontane evergreen forest;1100–1500 m alt.

Threats: in Bakossi its habitat is threatened by continued sporadic illegal logging operations.

Management suggestions: recent formal protection of part of the Bakossi Mts Forests as the Bakossi NP may well help preserve this species. Basic populational data should be collected to form a baseline for future monitoring.

Rinorea simoneae Achoundong

EN B1, B2 a,b (iii)

Range: Cameroon (S: Bella, 2 locs).

This species is only known from the two localities at Bella village, E of Edéa, the type locality of the widespread and well known *Afzelia bella* Harms. With 13 species per degree square, Bella is amongst the richest areas for *Rinorea*

diversity in Cameroon. *Rinorea simoneae*, although an evergreen forest species, is not known from the intensively collected Campo Ma'an area (Tchouto 2004). It is not known from the intensively collected Edéa area either. Neither is it known from Bipindi, close by to Bella, which has been well-collected by such botanists as Zenker, Letouzey and Villiers. Therefore its narrow distribution is unlikely to be an artefact of undercollection. The two locations near Bella village are at the borders of the Moha and Bella rivers, at both of which only 15–20 individuals are known, occupying about 40 m^2. Illegal logging and shifting cultivation are very common in the Bella area which is unprotected. On the basis of the above data, gathered by Dr Achoundong during seven site visits, *Rinorea simoneae* was assessed as Endangered in the protologue (Achoundong & Cheek, op. cit.(2006)). This assessment is maintained here although equally it appears that the taxon could be assessed as CR D since there is a single broad location, and AOO and EOO, with 4 km^2 cells, is 4 km^2.
Shrub 1.5–2.5(–3) m high, stems and leaves glabrous; leaves coriaceous, elliptic to slightly ovate, 17–22(–26) × 4–6 (–8) cm, acuminate, base cuneate, margin sightly serrate to subentire, lateral nerves 6–8 on each side of the midrib; petiole 2–3.5(–5) cm long; stipules triangular, 7–14 mm long; inflorescences terminal, thyrsiform, 2–3(–4) × 0.8–1.5 cm; bracts triangular, 1–1.5 × 1.5–2 mm, apiculate, longitudinal nerves 5–7, prominent; lateral branches 5–10-flowered; flowers zygomorphic, 5–6 × 3–4 mm; sepals 5, green to creamy yellow when mature, ovate 1.5–2.7 × 1–2 mm, margin ciliate, longitudinal nerves prominent, often branched at apex; petals 5, creamy yellow, ovate to elliptic, 4–6 × 2–3 mm, apex rounded, inner surface hairy near the apex, 4–5-nerved, margin ciliate. Lower petals broader than the upper and lateral petals, slightly constricted towards the apex. fruit tuberculate, ovoid 3–3.5 × 2.2–3 cm, not strongly 3-lobed as in other *Rinorea*. Seeds six, white, acutely angled, dorsal face convex, ventral face flat c. 1.7 × 1 cm
Habitat: dense evergreen humid forest, altitude 300 m.
Threats: illegal logging and shifting cultivation are very common in the Bella area which is unprotected.
Management suggestions: discussion with village elders of Bella to explain the rarity of this species and to enable them to identify the plant is suggested. Otherwise the few known individuals might be unknowingly cleared for cultivation. Basic populational data should be collected for future monitoring.

Rinorea thomasii Achoundong
VU A3c
Range: Cameroon endemic (SW: Mt Cameroon (8 coll. at 3 locs); Korup NP (6 coll.); Bakossi Mts (1 coll.)).
This species was assessed by Cheek and Darbyshire in Cheek *et al.* (2004: 187) as Vulnerable on account of a suspected decline of the population due to reduction in the extent of occurrence of 30%, and also a decline in its quality of habitat, over the next three generations (estimated as c. 45 years). This is maintained here, however, in future this rating is likely to increase to EN under criterion B. EOO is calculated as 7323 km^2.
Shrub 3 m; stems glabrous; leaves subcoriaceous, (ovate-)elliptic, 11–21 × 5–8.5 cm, acumen to 2.5 cm, base cuneate to obtuse, margin subentire, lateral nerves 7–9 pairs, glabrous, eglandular; petiole 1.5–2.5 cm; panicles terminal, to 5 cm, glabrous, lateral branches 2–3-flowered; sepals triangular, 1.5–3 cm, petals ovate-elliptic, c. 4 mm, white; fruit fusiform, conspicuously nerved. This description is taken from the protologue (Achoundong & Cheek, Kew Bull. 58: 960 (2003)).
Habitat: lowland evergreen forest, generally near rivers; 0–800 m alt.
Threats: of the three areas in which the taxon is known to occur, it seems secure only at the Korup NP, where it is afforded a high level of protection. The surviving forest in the Jide valley of the Bakossi Mts has mostly already been extirpated by agriculture owing to the high fertility of the soil; this is likely to continue and the species is already rare here, as evidenced by the single collection, made from the bottom of a steep-sided gorge. Its habitat at Mt Cameroon is vulnerable to clearance for timber extraction and plantation agriculture; already much of the lower altitude forest has been converted to rubber, banana and oil palm plantations.
Management suggestions: census of the populations at the most protected areas mentioned above should provide better data on which to formulate a conservation strategy and will also form a baseline for monitoring.

VITACEAE

assessed by Martin Cheek

A cosmopolitan family best known for *Vitis vinifera*, the grape vine, one of the most important species for sustaining mankind. Most species of Vitaceae, like the grape vine, are climbers. Species of Vitaceae are distinguished in Cameroon from those of other mainly herbaceous climbing families with alternate simple leaves and tendrils, (such as the cucumbers (Cucurbitaceae) and passion-flowers (Passifloraceae)) by having the tendril inserted at nodes on the opposite side from the leaf (not axillary or subaxillary). Species of savanna habitats may be shrubs which lack tendrils and have underground tubers but are easily recognised as Vitaceae due to the usually digitately compound or lobed leaves, the minute, radially symmetrical 4 or 5-merous flowers with petals and stamens free, stamens opposite the petals; ovary superior, producing a 1–4-seeded berry.

Flore du Cameroun 13 (1972) by Descoings is the most authoritative work on the Vitaceae of Cameroon. He recognises 48 species in 4 genera in his excellent study. No Syst. 16: 124 (1960) in which new combinations are made, Natur. Monsp. 18: 277 (1967) and Bull. Soc. Bot. Fr. 114: 353 (1966) in which new W African taxa are published.

Vitaceae are mainly absent from undisturbed lowland evergreen forest, being much more common in Chablis, secondary forest, forest edges and other habitats, such as semi-deciduous forest and woodland or savanna, where light-levels are higher.

Forest disturbance favours Vitaceae, but if disturbance is maintained by, e.g. cultivation of land, then the rarer Vitaceae will be lost. Savanna species of Vitaceae are not well-known ecologically, but are believed to be threatened by grazing pressures, grassland fires set by man, and conversion to agriculture.

It is notable that six of the nine species below are from drier open areas in Adamawa, N and Extreme N regions of Cameroon. Only one threatened species is known from the Cameroon Highlands. In our our first six Conservation Checklists, all covering different parts of the Cameroon Highlands, no Vitaceae species was assessed as threatened.

Ampelocissus macrocirrha Gilg & Brandt
EN B2ab(iii)

Range: Liberia (Mt Nimba), Equatorial Guinea (Bioko: Malabo-Riaba; Luba) and Cameroon (SW: Kumba; Nyandong. S: Bipinde).
Known only from Cameroon & Bioko until discovered at Mt Nimba by J.G. Adam in 1971. Both FWTA 1: 682 and Enum. Pl. Afr. Trop. 2: 191 state that *A. macrocirrha* is doubtfully distinct from *A. abyssinica* (Hochst. ex A.Rich.) Planc., of which *A. cavicaulis* (Baker) Planch. of Flore du Cameroun 13: 14 is now considered synonymous. However, Flore du Cameroun omits *A. macrocirrha*; it is therefore maintained as a distinct taxan pending further analysis (Darbyshire pp. 422 in Cheek *et al.* 2004). A record from Rio Muni tentatively identified as this species has since been shown to be an *Adenia.*
Robust climber; stems glabrous; leaves ovate, to 22 × 20.5cm, apex subacuminate, base deeply cordate, margin finely dentate, glabrous except principal veins puberulent beneath, petiole to 14 cm, glabrescent; tendrils borne on the peduncle, bifid; inflorescence of compound cymes, to 5 × 5 cm, dense; primary peduncle to 13 cm, sparsely pubescent; bracts triangular, 1 mm; flowers c. 2 mm long; calyx cupular; corolla buds rounded at apex, pedicels 1 mm, glabrescent.
new species have been published for Cameroon since this work, which is the basis for the treatment below. Descoing's 1972 work was based several earlier papers, such as in Not. Here *A. macrocirrha* is assessed as Endangered since five locations are known (AOO 20 km² with 4 km² cells) with threats as below. EOO is calculated as 214025 km².
Habitat: lowland & submontane forest; often disturbed; 300–1100 m alt.
Threats: iron ore extraction (Nimba); forest clearance (Kumba).
Management suggestions: complete forest protection is unlikely to favour this species. Like many forest Vitaceae some disturbance is favourable to their increase. However, total or close to total, forest destruction will remove them. The sites least pressured for this species, where are the best chances of it surviving, are probably Nyandong and Bipinde. Local communities should be consulted about its protection and involved in surveying for basic populational data which will be a baseline for future monitoring. Propagation for reinforcing population numbers is likely to be easily achievable by cuttings and seed.

Cyphostemma cameroonense Desc.
EN B2ab(iii)

Range: Cameroon endemic (C: N'kolbison; Mefou proposed NP; Nanga Eboko. E: Bertoua).
Climber, stem cylindric 2.5–3.5 mm, hairs sparse, long, patent, glandular and simple, flowers on stems and leaves, internodes 8–15 cm; stipules broadly falcate, 7–8 × 3–4 mm; leaves compound, digitate, 5-foliolate; petiole 7–10 cm; petiolules 0.8–2.2 cm; leaflets obovate-elliptic, 8–15 × 3.5–5 cm, acumen abrupt, base acute, margin crenate-denticulate; inflorescence 8–12 cm, cymes subumbellate, few-flowered, peduncle 4–6 cm; berries ovoid, 7–10 × 5–6 mm, pubescent.
Here *Cyphostemma cameroonense* is assessed as Endangered since there are only four locations (above, AOO 16 km² with 4 km² cells) and threats as below. EOO is calculated as 11168 km².
Habitat: semi-deciduous forest; c. 700 m alt.
Threats: the habitat of this species has been logged in most parts of its range, often followed by cultivation of cocoa or manioc (pers. obs. Cheek & Onana).
Managment suggestions: since the most recent record derives from the Mefou proposed NP, this is the logical place to focus conservation efforts. Basic populational data should be gathered to inform future monitoring and to assess needs for intervention. Propagation and reintroduction may be needed.

Cyphostemma cuneatum (Gilg & Brandt) Desc.
EN B2ab(iii)

Range: Cameroon endemic (N: Garoua, Boglé collines; Guider).

Climber, stem cylindric 2.5–3.5 mm, hairs dense, glandular hairs sparse; internodes 6–12 cm; stipules oblong-linear 7–9 × 1.5–3 mm; leaves compound, digitate, 5-foliolate; petiole 2–5 cm; petiolules 2–5 mm; leaflets oblong-lanceolate 8–15 × 2.5–4.6 cm, apex obtuse, base decunent, margin serrate; inflorescence 20 cm, cymes subumbellate, peduncle 11 cm; berries 2 × 1 cm.

Here *Cyphostemma cuneatum* is assessed as Endangered since two locations are known (above, AOO 8 km² with 4 km² cells) and threats as below. The two locations are so close that they might be taken as one in which case CR status is needed.

Habitat: wooded grassland (deduced); alt. unknown.

Threats: little is known of the locations but grazing by livestock and pasture management regimes are thought to be unfavourable to Vitaceae.

Management suggestions: the location(s) for this species should be rediscovered being gathered to aid future monitoring and management needs. Leaders of local communities nearby should be informed and consulted to help secure protection for the plants.

Cyphostemma letouzeyanum Desc.

EN B1+2ab(iii)

Range: Cameroon endemic (Adamawa: Ngaoundéré; 45 km NW Bagodo; Meïganga-Yarbang).

Climber, stems cylindric, 2.5–3 mm, densely pubescent, glandular; internodes 6–12 cm; stipules oblong-obovate, apex falcate, 8–10 × 2–3mm; leaves compound, digitate, 3-foliate, petiole 1–2 cm; petiolules 0.3–1 cm; leaflets, obovate to elliptic, 8–13 × 3.5–6 cm, apex rounded-subobtuse, base cuneate; basal half with margin entire; upper half irregular toothed, teeth denticulate; inflorescence 8–20 cm, peduncle 5–12 cm; berries ovoid 8–10 × 4–5 mm, berries dense.

Here *Cyphostemma letouzeyanum* is assessed as Endangered since only three locations are known (AOO 12 km² with 4 km² cells) with threats as below. EOO is calculated as 1954 km².

Habitat: savanna; c. 800 m alt (deduced).

Threats: clearance of savanna for cultivation.

Management suggestions: the location(s) for this species should be rediscovered being gathered to aid future monitoring and management needs. Leaders of local communities nearby should be informed and consulted to help secure protection for the plants.

Cyphostemma leucotrichum (Gilg & Brandt) Desc.

EN B1+2ab(iii)

Range: Cameroon endemic (Adamawa: Ngaoundéré, 8 km S. N: Godé, Hoséré Face (18 km NW Poli)).

Erect herb, 1 m, lacking tendrils; from underground tuber; stems ± cylindric 5–7 mm diam., stems, petioles, inflorescence densely pilose with a mixture of grey simple and glandular hairs, internodes 3–10 cm; stipules ovate-deltoid 20–25 × 10–15 mm, acuminate, margin ciliate; petiole 2–6 mm; leaves digitately compound, 5-foliolate, leaflets narrowly elliptic, 10–17 × 2–4 cm, acute, cuneate, serrate, lower surface with sessile red gland; inflorescence 8–15 cm, umbels fairly dense flowered. Berries ovoid, 8–10 × 5–6 mm, densely pubescent-glandular.

Here *Cyphostemma leucotrichum* is assessed as Endangered since only two locations are known above; (AOO 8 km² with 4 km² cells) and since threats are as below, seven records are known, implying that this species is not uncommon at its locations.

Habita: savanna; 800–1200 m alt.

Threats: clearance of savanna for cultivation

Management suggestions: the location(s) for this species should be rediscovered being gathered to aid future monitoring and management needs. Leaders of local communities nearby should be informed and consulted to help secure protection for the plants.

Cyphostemma ornatum (Hutch & Dalz.) Desc.

EN B2ab(iii)

Range: Ivory Coast (Azaguié) and Cameroon (E: Moloundou).

Last collected in Cameroon in 1910-1911 by Mildbrand (4007), this is Cameroon's rarest Vitaceae. Here *Cyphostemma ornatum* is assessed as Endangered since two locations are known (AOO 8 km² with 4 km² cells) and threats as below.

Large climber, stems 5–7 mm diam., deeply grooved, glabrescent; stipules oblong to ovate-deltoid, 10–20 × 7–10 mm, densely puberulent or glabrous; leaves digitately compound 5-foliolate, petiole 10–17 mm; petiolules 3–5 cm; leaflets ovate-elliptic, obovate to suborbicular 8–13 × 5–9 cm, obtuse to rounded apex, base cuneate, margins revolute, with shallow, wide-spread teeth terminating a secured hook, upper surface glabrous, glossy, nerves 4–5 pairs; inflorescence 25–30 cm long, multi-flowered, peduncle 10–15 cm. Since the species has not been recorded in Cameroon in 100 years despite surveys of forest species around Moloundou and since Ivory Coast has lost most of its original forest, *Cyphostemma ornatum* may be extinct.

Habitat: lowland semi-deciduous forest (deduced).

Threats: logging of forest habitat at both known locations.

Management suggestions: botanical surveys in the Moloundou area should target Vitaceae species in the hopes of rediscovering this plant. If found, basic populational data should be gathered as a baseline for future monitoring. It is likely that propagation by seed or cutting for introduction to safe sites by proficient horticulturalists will be needed as a management intervention.

Cyphostemma ouakense Desc.

EN B2ab(iii)
Range: Cameroon (Adamawa: Gandoua; Bam Yanga, Ngao, 45 km NW Bagodo) and CAR (Bambari Region, Moroubas; 15 km NW Moroubas; Region de la Ouaka, Deyade river).
foliolate, leaflets lanceolate or narrowly elliptic 20–40 × 6–12 cm acute to obtuse, base decurrent, sessile or petiolules to 15 mm, margin regularly acute-serrate, teeth denticulate; inflorescence 8–10 cm, highly branched, peduncle 0–5 cm; berries ovoid, 8–9 mm, glabrous.
Here *Cyphostemma ouakense* is assessed as Endangered since only five locations are known (AOO 20 km² with 4 km² cells) and threats as below. EOO is calculated as 58255 km².
Habitat: savanna (wooded grassland); c. 800 m alt.
Threats: clearance of savanna for cultivation.
Management suggestions: the location(s) for this species should be rediscovered being gathered to aid future monitoring and management needs. Leaders of local communities nearby should be informed and consulted to help secure protection for the plants.

Cyphostemma rupicolum (Gilg & Brandt) Desc.
EN B1+2ab(iii)
Range: Cameroon (Extreme N: col de Moutourova, 45 km SSW Maroua; Maroua to Guider; Laf, 40 km SSW Maroua. N: Garoua) and Chad (cited by Descoings, (1972) without locality data).
Erect herb 0.5–1 m, in clumps, lacking tendrils, stem thick, fleshy 5–13 mm diam., striate, hairs scurfy, white, crisped; internodes 6–15 cm; stipules linear-oblong 15–30 × 2–7 mm; leaves digitately 3–5-foliolate, petioles fleshy, 1–8 × 0.4–0.5 cm, leaflets elliptic 20–30 × 5–10 cm, apex and base long acute, margin biserrate-dentate, tooth apex denticulate, inflorescence 20–35 cm, lax, peduncle 8–15 cm; berry ovoid 10–12 mm, glabrous.
Here *Cyphostemma rupicolum* is assesses as Endangered since only five locations (above, AOO 20 km² with 4 km² cells) are known, with threats as below. EOO is calculated as 132 km².
Habitat: sandy savanna on gneiss, granite or quartz; alt. unknown.
Threats: grazing pressure and pasture management.
Management suggestions: the location(s) for this species should be rediscovered being gathered to aid future monitoring and management needs. Leaders of local communities nearby should be informed and consulted to help secure protection for the plants.

Cyphostemma tisserantii Desc.
VU B2ab(iii)
Range: Cameroon (Adamawa: Tékel to Béka Goto, 45 km NNW Bagodo; 40 km NNW Bagodo), Chad (Region de Iro, Bolbitdgia) and CAR(Krebedje; Yalinga-Wadda; N Moroubas; 60 km N Bambari; Bamingui-Bangoran).

Erect herb 1 m; stem stout, grooved, lacking tendrils, 4–6 mm diam. red pubescent; internodes 6–12 cm; stipules ovate acuminate, 10–14 × 4–6mm; leaves (sub)sessile, digitately 3-

Erect herb 0.5–1 m from woody rootstock, lacking tendrils, stem cylindric, 3–4 mm diam., densely long glandular hairy, mixed with sparse, sessile glands, internodes 5–8 cm; stipules ovate-triangular, 1.5–3 × 0.6–1.2 cm, acuminate, cordate; leaves digitately 5-foliolate, petiole 2.5–6 cm; leaflets sessile, narrowly elliptic-linear 12–20 × 2–5 cm, acute, cuneate; serrate, teeth denticulate; inflorescence 4–6 cm, dense, pilose-glandular; peduncle 1.5–3 cm; berry ovoid 6–7 × 4 mm, with short-stalked glandular hairs.
Here *Cyphostemma tisserantii* is assessed as Vulnerable since seven locations (above, AOO 28 km² with 4 km² cells) and threats as below are known. EOO is calculated as 301396 km².
Habitat: wooded savanna.
Threats: clearance of habitat for cultivation.
Management suggestions: the location(s) for this species should be rediscovered being gathered to aid future monitoring and management needs. Leaders of local communities nearby should be informed and consulted to help secure protection for the plants.

VOCHYSIACEAE

assessed by Martin Cheek

A mainly S American (six genera) family of trees which lack coloured exudates, glands or scent, but have opposite, simple, entire leaves each with a pair of stipules. In flower they are even more distinctive, the calyx usually spurred; the zygomorphic corolla is showy, of 3–5 petals, with a single style and a single, exerted stamen; their inferior fruits have the calyx lobes unequally expanded to form wings to aid dispersal of the fruit (Africa). Only two genera, *Erismadelphus* and *Korupodendron* , occur in Africa, being found mainly in the countries of Lower Guinea. There is no Flore du Cameroun account. No major uses are known for these rare trees.

Erismadelphus was described by Mildbraed as a single species in 1913. Keay and Stafleu revised the genus, recognising two species, one with two varieties, in 1953, (*Erimadelphus*, Acta Botanica Neerlandica 1: 594–599). A second genus for Africa was then added: Litt, A. & Cheek, M. (2002), *Korupodendron songweanum*, a new genus and species of Vochysiaceae from West-Central Africa. Brittonia, 54: 13–17. Much very useful further data including range extensions then appeared: Senterre, B. & Obiang, D. (2005), Nouvelles découvertes à propos des Vochysiaceae africaines: *Erismadelphus* Mildbr. et *Korupodendron* Litt & Cheek. Taxonomania 17: 3–18

All African Vochysiaceae belong in the tribe Erismeae Dumort. on the basis of their inferior unilocular ovary and winged fruit. *Korupodendron* is similar to *Erismadelphus* in having simple hairs, five petals, and five wings on the fruit (three large, two much smaller, corresponding to the proportions of the five sepals). Like *Erismadelphus* it has simple hairs; the Erismeae have been described as having stellate hairs e.g. by Kawasaki in 1998, but this is only true of the neotropical *Erisma* Rudge.

Key to genera of Erismeae:

1. One petal. Four-winged fruit. S America ....*Erisma*
1. Five petals. Five-winged fruit. W-C Africa 2
2. Three sepals conspicuous and petaloid, two sepals inconspicuous ....*Korupodendron*
2. Five sepals subequal in size, none petaloid *Erismadelphus*

Erismadelphus sessilis Keay & Stafl. [Bobowajoubé (Bibaye)]

VU B2ab(iii)

Range: Cameroon (E: Abong Mbang at Djolempoum, 4–8 km ENE; Nkoum (Bertoua)), Equatorial Guinea (Rio Muni), Gabon (Mbé NP, Cristal Mts; 25 km NNE Koumamayang; Kembélé) and Congo (Kinshasa) (3 coll.).

Here *Erismadelphus sessilis* is assessed as Vulnerable since up to 9 locations are known (above; AOO 36 km^2 with 4 km^2 cells) with threats as below.

Described in 1953 by Keay and Stafleu from a single specimen (Kembélé, *Le Testu* 5559), new records, mainly sterile specimens, have been added by Senterre and Obiang (2005 loc.cit.) and by the staff of WAG (Gabon database). These are accepted here although unseen and cited above since this is a vegetatively distinct species unusual in the family in the small, sessile, hairy leaves.

The description derives from that of Keay and Stafleu (loc.cit.)

Tree to 30 m, diam. 40 cm, with four small buttresses; young stems tomentellous; stipules subovate, thickened, 0.5–1 mm at the base of the petioles, connected by a ridge across the stem, leaves sessile, oblong- elliptic or elliptic, 6–8 × 2.5–3.5 cm, obtuse, base cuneate, lateral nerves 6–9 on each side, hairy below, lamina otherwise puberulent-glabrescent below, finer nerves prominent; inflorescence tomentellous when young, bracts reniform, cordate, 6 × 2.5 mm; petals 5 mm; fruit unknown.

Habitat: semi-deciduous and submontane evergreen forest (probably).

Threats: logging followed by agriculture (Cameroon).

Management suggestions: rediscovery of this species at its two Cameroon locations should be followed by collection of basic populational data to support monitoring. Consideration should be given to protecting one of the sites in a National park, or failing this, alerting the local leader(s) of nearby community(ies) to the rarity of this tree so that it might be protected.

Korupodendron songweanum Litt & Cheek [Kosowe-ke obon (Durop)]

EN B2ab(iii)

Range: Cameroon (SW: Korup NP), Equatorial Guinea (Rio Muni: Mt Alen area) and Gabon (Cristal Mts; Mondah FR, NW of Libreville; Lake Ezanga, SW Lambarene).

When first described (Litt & Cheek 2002, loc. cit.) this species was known only from the slopes of Mt Juahan in Korup NP it was assessed as Critically Endangered. Subsequently, it was found a little to the S, still within Korup. Within a few years, botanists in Gabon realised that several previously unidentified sterile specimens from their surveys were the same tree. Senterre and Obiang (2005 loc.cit.) cited many records from the Cristal Mts of Gabon and Rio Muni where it appears rather common, occupying a large area. Two other isolated and rather surprising records appear on the Gabon database, one at Mondah FR, the other at Lake Ezanga. Both are near sea-level and not associated with hilly areas.

Here *Korupodendron songweanum* is re-assessed as Endangered since four locations appear to be known, one of which, the Cristal Mts, has an area of occupation of about 1000 km^2 (above; AOO 1012 km^2 with 4 km^2 cells excepting the Cristal Mts location) with threats as below.

Tree c. 35 m tall, 45 cm diam. at breast height, with 7–8 short buttresses, bark white. Leaves opposite, simple; petiole 1 cm; blade elliptic to ovate, entire, glabrous, 8–12.5 × 4–6 cm, base acute or obtuse, apex acuminate; the major lateral veins forming a looping marginal vein 3 mm from the margin except the basalmost lateral vein which terminates at the margin; major lateral veins 3–4(–6) pairs, 1–2.5 cm apart, the basal pair inserted near the petiole and decurrent, the upper lateral nerves nearly patent; raceme with rachis slender, 3–5 cm long, mostly glabrous or with very sparse simple golden hairs but pubescent at nodes; bracts hemi-orbicular, c. 1 mm long, deciduous, leaving raised scar, opposite scars connected across the rachis by a slight raised ridge; flowers zygomorphic, sweetly scented (at 10 AM), opposite; pedicel 1 cm, slender, or flowers sessile; calyx white, sepals five, fused basally to the inferior ovary, glabrous except for a somewhat sparse fringe of simple golden hairs along the edge; three larger sepals elliptic, the dorsal (uppermost) larger sepal with a red-speckled, pale yellow band along the midrib, sepals 7–10 × 2.5–3 mm, two smaller sepals deltoid-elliptic, green at the base, 1 mm long; spur of fourth sepal cylindric, straight, white, 2–4 × 1–1.5 mm, tip rounded, greenish or brownish; corolla white, petals five, free, glabrous, clawed, elliptic, similar in shape to the large sepals but slightly narrower, 9.5 × 1.5–2.5 mm, apex subacuminate; stamen single, filament straight or slightly curved, 2–3 mm, glabrous; anther sagittate, 2 mm; ovary inferior, unilocular, glabrous; ovule one, inserted laterally; style straight or slightly curved, slightly tapered, 3 × 0.5 mm; stigma lateral, oval, 0.5 mm, wet; fruit indehiscent, winged, the body turbinate, c. 1.5 × 1.5 cm, the five wings formed by the accrescent growth of the sepals, three wings large, two small, reflecting the proportions of the sepals; larger wings oblanceolate, 4.5–5 × c. 1.7 cm.

Habitat: lowland evergreen and submontane forest; 0–500 m alt.

Threats: forest clearance for timber and firewood (Mondah FR fide Chris Wilks).

Management suggestions: re-examination of the Mondah and Lake Ezanga specimens is needed to confirm their identity. In Cameroon, mapping of the species at the only known location, Korup, is suggested, followed by collection of basic populational data as a baseline for monitoring.

MONOCOTYLEDONAE

ALISMATACEAE

assessed by Martin Cheek

This family of aquatic plants have six genera and seven species in Cameroon according to J.-J. Symoens in Flore du Cameroun 26: 3-26 (1984). All are erect herbs which root at the bottom of shallow water, often with hastate leaf-blades. They are unique among other mainly aquatic families in Cameroon in that the 3-merous petaloid flowers, usually in interrupted panicles, have numerous free carpels.

Limnophtyon fluitans Graebner

EN B2ab(iii)

Range: SE Nigeria and Cameroon (SW: Korup NP; Nguti. S: Campo. Littoral: Edéa).

Limnophyton fluitans is here assessed as Endangered since is known from only five localities (above; AOO 20 km² with 4 km² cells), with threats as below. In each location it appears rare. It has not been observed in Nigeria, nor in Edéa since 1908 and is possibly extinct at those localities. At Nguti and Campo the species is known from single collections, but at Korup from c. 4 collections in at least 3 places.

Limnophyton fluitans is the only rheophyte in a genus of 5 species restricted to Australia and the old world.

Submerged, bottom-rooting, acaulous herb; white exudate produced from cut surfaces; rhizome pointing down stream, about 1 cm diam., covered in long-persistent leaves; leaves 20–30 per rhizome, dark bronze-green, translucent when young, linear-ligulate, 60–80 × 4–5 cm, basal 20 cm gradually narrowed into a winged petiole c. 1 cm wide, longitudinal nerves 3, apex acute; inflorescence erect, 30–40 cm long, rachis 2–3 cm long, with 3 nodes, bracts opposite, patent, narrowly triangular, up to 17 × 2 mm, each subtending 2–3 flowers; pedicels c. 10 mm long; flowers with 3 white tepals c. 4 mm long; fruit globose.

Habitat: forest-shaded, perennial clear, fast flowing streams at least 20 cm deep, on stones and sand, with *Anubias barteri* and *Crinum natans* (pers. obs. 1996, 1998).

Threats: this species is considered vulnerable to logging since run-off from from exposed soil can be expected to introduce deleterious turbidity into its stream habitat. Extensive logging is known to have occurred in the last decade at Oban, C[illegible].

Management su[illegible] for conservation e[illegible] since there are more [illegible] location than any other. Here it appears unthreatened, but basic populational data should be collected as a baseline for future monitoring. Elsewhere public education and community involvement are advised to reduce threats to this species.

ANTHERICACEAE

assessed by Martin Cheek

Treated in FWTA as Liliaceae, having superior ovaries with 6 stamens and 6 perianth lobes, Anthericaceae are represented in Cameroon by a single genus, *Chlorophytum*. The genus *Anthericum*, formerly accepted e.g. in FWTA, then distinguished by the entire, not deeply lobed fruit, was sunk into the former genus by Kativu and Nordal (Nord. J. Bot. 13(1): 59–65 (1993)) apart from two non-Cameroonian species. The rainforest species of Guineo-Congolian Africa were revised by Poulsen and Nordal (Bot. J. Linn. Soc. 148: 1–20 (2005)). These authors recognised 15 taxa in eight species only one of which is both range-restricted and occurs in Cameroon (*C. petrophilum*). However, several more species occur outside the rainforest in open habitats. Several of these are rare and threatened and are treated below.

While no Flore du Cameroun account nor modern revision exists for the non-forest species of W Africa or Cameroon, fortunately these species are represented by photos at Kew of the original German collections now thought lost. The species are easily distinguished from each other.

Chlorophytum, well-known in Europe as the 'spider-plant', pot-plant of bathrooms, are easily recognised when dug-up. Their thick unbranched roots terminate in distinctive fleshy tubers. All Cameroon species lack an aerial stem and have a loose rosette of often strap-like leaves. The tubers of some species are eaten by humans in times of drought in some parts of Africa.

Chlorophytum altum Engl. & K.Krause (syn. *C. garuense* Engl. & K.Krause)

EN B2ab(iii)

Range: Ghana (20 km SE Bossomo, Upper Volta; White Volta R) and Cameroon (N: Mayo Mali to Mayo Douka; Garoua; Ssagdje-Alhadjori)

Perennial, tuberous rooted herb; aerial stem absent, leaves 2–6, erect, strap-like, 25–50 × 1–3 cm apex long-tapering, pseudo-petiole absent; panicle 40–60(–90) × 15–20 cm, branches long and sparse, to 15 cm, interrupted, internodes 5–10 mm; flowers white, 4–5 mm in May-July.

[illegible]ation given above for Cameroon is not mapped [illegible]ot located. Here assessed as Endangered since four or five locations are known (above; AOO 20 km² with 4 km² cells) and with threats as below.

Last seen in Cameroon on 5 July 1909 at Garoua (*Ledermann* 4496).
Habitat: swampy grassland (Ghana specimen data).
Threats: cattle trampling at waterside drinking holes; drainage of swampy areas.
Management suggestions: rediscovery of this plant, and a detailed assessment of its threats, regeneration, range and density in Cameroon is needed so as to develop a plan for its survival, if it still exists.

Chlorophytum benuense Engl. & K.Krause
EN B2ab(iii)

Range: Cameroon endemic: (N: Garoua area, Bengi; Bossoum-Buar).
Terrestrial herb, 30 cm tall; leaves c. 10 per rosette, erect, strap-like, 20–28 × 0.4–1.1 cm, tapering to an acute apex, pseudo-petiole constriction c. 6 cm from the base; base sheathing shortly, white, c. 9 mm wide; inflorescence c. 8 × 4cm, c. 3-branched from base, branches stout, internodes 3–4 mm, bracts 12–17 mm, boat-shaped, subtending white flowers 5 mm in June.
Last collected in June 1914 (*Mildbraed* 9730).
Here *C. benuense* is assessed here as Endangered since only two locations (AOO 8 km² with 4 km² cells) are known, with threats as below.
The second location may cross the boundary into CAR.
Habitat: 300–900 m alt., probably in wooded grassland, possibly in swampy conditions.
Threats: cattle trampling, cultivation of habitat for food crops; artificial burning of grassland and woodland.
Management suggestions: rediscovery of this species, not seen in 100 years is advisable so that its threats, range, density and regeneration can be surveyed. Its conservation needs can then be assessed and addressed.

Chlorophytum camporum Engl. & K.Krause
CR B2ab(iii)

Range: Cameroon endemic (Littoral: Baré near Nkongsamba) known only from the type (*Ledermann* 1401)
Perennial tuberous herb, leaves c. 15 per rosette, erect, strap-like, 18–25 × 0.7–0.9 cm, apex tapering to a point, base lacking pseudo-petiole; inflorescence spike-like, erect, 25–38 × 2–3 cm, peduncle 15 cm; bracts subulate, 7 mm, subtending 2–3 flowers, pedicels 2–3 mm; flowers white, 5–6 mm, November.
Here *C. camporum* is assessed as Critically Endangered since only a single location (above; AOO 4 km² with 4 km² cells) is known, and since threats are as above.
Habitat: submontane, presumably forest or grassland; 800 m alt.
Threats: the proximity of Baré to Nkongsamba is likely to drive cultivation of food crops for export to that market.
Management suggestions: several mysteries remain to be solved regarding this species. Its location at Baré probably places it within forest, yet it was not treated by Poulsen & Nordal (2005). Evaluation of Ledermann's itinerary by Hepper is needed to confirm that it does derive from near Nkongsamba and not from another Baré further North. Poulsen and Nordal should also be consulted to confirm the status of the species, before attempts are made to refind it at Baré and establish whether or not it survives, and, if so, its threats, habitat, range and regeneration levels. This is probably a truly rare species since it was not picked up in the intensive plant survey of Mt Kupe, despite Baré being on its E flank (Cheek *et al.* 2004).

Chlorophytum caudatibracteatum Engl. & K.Krause
CR B2ab(iii)

Range: Cameroon endemic (N: Garoua).
Here *C. caudatibracteatum* is assessed as Critically Endangered since only a single location (above; AOO 4 km² with 4 km² cells) is known, and since threats are as above.
Terrestrial herb, 30 cm tall; leaves 3–5 per rosette, erect, 20–25 × 0.8–1.5 cm, acute, base lacking pseudopetiole; panicle slender 45 × 2–3 cm, peduncle 38–40 cm, internodes 3–6, 2.5 cm, bracts 2.5 cm, branches 1–6-flowered.
Habitat: wooded grassland; 300 m alt.
Threats: artificial burning of habitat, and clearing for cultivation; cattle trampling.
Management suggestions: rediscovery of this species in the wild is a priority since it has not been seen in 100 years. Little or nothing is known about its range, density, and regeneration levels so, if refound a survey to collect this data is advisable. An assessment of needs can then be made.

Chlorophytum petrophilum K.Krause
EN B2ab(iii)

Range: Nigeria (Oban FR), Equatorial Guinea (Bioko) and Cameroon (SW: Mt Cameroon above both Batoke and Njonji. S: Bipindi, type collection).
Formerly assessed as CR A1c+2c by Cheek in Cable & Cheek (1998: lxiv) before Poulsen and Nordal's revision revealed the Nigerian material. The species is here reassessed as Endangered since only four locations are known (above; AOO 16 km² with 4 km² cells) are known, and since threats are as below. *Cheek* 10604 from Banyang Mbo has a field determination as this species but needs verification before accepted as a new location.
Slender-stemmed herb 10 cm tall, supported by stilt-roots, internodes to 4 cm on horizontal stems, at length erect when internodes short; leaf-blades elliptic or ovate, 4.5–8 × 1.5–3.5 cm, acute-acuminate with mucron, base rounded, lower surface with longitudinal vein stripes; pseudopetiole strongly defined; inflorescence single, unbranched, filiform, c. 30 cm, internodes c. 4 cm, bracts 7 mm, subtending 1–2 flowers,

petals white 6 mm; inflorescence often at length prostrate and viviparous.
Habitat: lowland evergreen forest; 300–500 m alt.
Threats: conversion of forest to oil-palm and/or other plantations (Mt Cameroon); slash and burn agriculture (Bipindi).
Management suggestions: site protection with support of local communities; possibly multiplication at Limbe BG for public education and reintroduction into the wild at safe sites. Mt Cameroon with the densest concentration of records (three) for the species seems the logical place to focus conservation efforts despite the threats referred to.

Chlorophytum staudtii Nordal (syn. *Anthericum zenkeri* Engl.)
EN B2ab(iii)
Range: Nigeria (Vogel Peak) and Cameroon (C: Yaoundé area (N'kolbison)).
Erect herb 30–50 cm high; rhizome stout, truncated c. 4 cm; leaves 4–6 per rosette, flat, 17–50 × 0.2–0.4 cm, long-acute, base of larger leaves with pseudopetiole constriction between c. 7–14 cm from the base, basal 7 cm dilating and sheathing the culm; inflorescence usually unbranched, 22–35 × 2 cm, rachis 3–7 cm, internodes 4–20 mm bracts 1 mm, flowers single, pedicel 6 mm, flower 15 mm diam., white in Nov., fruit globose, papery 3-lobed, 5 mm.
Zenker & Staudt 268 from Yaoundé in the 1890's is the oldest record seen of this species, the most recent being the de Wilde's in Nov. 1964 from N'kolbison. The Vogel peak collection is more robust and differs in some characters, so may need confirmation as to its identity.
Here *C. staudtii* is assessed as Endangered since only two sites are known (AOO 8 km² with 4 km² cells) and since threats are as below.
Habitat: inselbergs, in humus in open among rocks; 600–1540 m alt.
Threats: rock extraction for building aggregate.
Management suggestions: inselbergs in the Yaoundé area should be visited in November to refind this species and to assess its range, regeneration, density and threats in detail so that the need for management can be assessed. Re-evaluation of the Nigerian material is needed to see if it is conspecific or not.

ARACEAE

assessed by Martin Cheek

Easily recognised when reproductive since the 'flowers' are inflorescences arranged in a distinctive way: the reduced, sessile flowers are placed on an erect fleshy cylindrical axis, the spadix, of which the lower portion is concealed by the basal part of a single large, cloth-like bract, the spathe, which then either partly conceals the rest of the spadix on one side, or is reflexed. Economically important in Cameroon for the 'coco-yam' *Colocasia esculenta* and *Xanthosoma sagittifolium*, both are non-African species cultivated in the evergreen forest belt of Cameroon for their starchy tubers. The aerial roots of some wild forest species of *Cercestis* and the stems of *Culcasia* are highly favoured for tying and known as 'bush rope'. *Anchomanes* is important medicinally. About 50 species and 13 genera of Araceae occur in Cameroon.

Mme Ntépé-Nyamè wrote the highly esteemed Flore du Cameroun 31 (1988) which is followed in this treatment. No further revisions have appeared since then apart from the 2003 revision of African *Amorphophallus* by Ittenbach (Englera 25: 3–263). Ittenbach has changed the delimitation of several species which were once thought to be widespread are now known to be more restricted in range. The threatened species recognised here fall into two genera:

Amorphophallus, recognised by having a single leaf with a cylindrical petiole arising from an underground disc-like tuber, the leaf blade being highly divided. Species are distinguished from the similar *Anchomanes* by their lack of petiole prickles and lack of a horizontal rhizome. They occur in both forest and grassland.

Culcasia is the most species-rich genus in Cameroon. They are confined to forest and mainly occur as epiphytes on tree trunks 1–3 m from the ground (hemi-epiphytes), they have long stems and adventitious roots. They are easily overlooked due to their dull inflorescences but their vivid orange-red berries attract attention. Felling and burning of forest destroys plants of both species. *Amorphophallus* may benefit from low-levels of disturbance but most species are forest-dependent.

Several species of *Culcasia*, identified by Ntépé-Nyamè and later workers, await description and are likely to have threatened status. Data on generation duration should be available from botanic gardens in Europe. In identifying *Amorphophallus*, flowers are essential to get to species. Finding one in the wild is always an event since they are so striking, and most species in Cameroon are rare and threatened. Not included below is *Amorphophallus staudtii* which has 10 or 11 locations, so is borderline Vulnerable/NT.

In contrast, *Culcasia* are common in undisturbed forest, but most species are known from more than 10 locations apart from those listed below. Species can often be identified to species by leaf characteristics with the aid of Ntépé-Nyamè's excellent Flore du Cameroun treatment. *Nepthytis swainei* records from Cameroon are now considered to be variants of

N. gravenreuthii, a common species and not as rare as the former.

Amorphophallus calabaricus N.E.Brown subsp. ***calabaricus***

VU B2ab(iii)

Range: Nigeria (Oyo Prov., Ibadan Distr., Ibanda SFR & IITA compound; Old Calabar R.; Akumi (Okumi); Benin Prov., Sapoba; Okomu FR) and Cameroon (C: Mbikai, 18 km NW Yaoundé, Massif de Mbam Minkom. SW: Mt Kupe).

Here *A. calabaricus* subsp. *calabaricus* is assessed as Vulnerable since eight locations (above; AOO 32 km^2 with 4 km^2 cells) with threats as below are known.

The second subspecies, subsp. *mayoi* Ittenb. occurs around Lake Victoria in E Africa. FWTA and Flore du Cameroun treated both taxa as one undivided species. Data is mainly derived from Ittenbach (2003: 122–124).

Tuber depressed globose, 1.5–4 × 3–8 cm; petiole 0.6–1.2 m, appearing after the inflorescence; blade 0.8–1 m diam.; inflorescence in Oct., peduncle 0.8–1.5 m; spathe erect, 10–25 cm long, a conic tube 5–9 × 3–4 cm outside green and triangular, open limb erect, 5–16 cm, purple-brown, without a constriction; spadix erect, sessile, 15–50 cm, at least 1.5 times as long as the spathe.

Habitat: lowland to submontane evergreen forest up to 1000 m alt.

Threats: forest clearance in Nigeria for agricultural land and timber. If loss of the species in most of the Nigerian sites is confirmed as being credible, this taxon will qualify under Criterion A2c as Endangered. Fortunately both Cameroon sites appear reasonably secure.

Management suggestions: low-level and temporary forest disturbance may favour reproduction by creating light gaps which seem to trigger flowering (pers. obs. Cheek on Mt Kupe). Basic populational data should be gathered at both Cameroon sites locations to inform future managetment, and area managers including traditional leaders, should be made aware of this species.

Amorphophallus mildbraedii K.Krause

CR B2ab(iii)

Range: Cameroon endemic (Adamawa (?): between 'Babua' [=Gbaboua?] und Budaje).

While not included in the Flore du Cameroun account, this species was accepted by Ittenbach 2003: 201–205. Known only from the type collection (above; AOO 4 km^2 with 4 km^2 cells; threats as below), therefore it is here assessed as Critically Endangered.

"Certainly a good species and well delimitable by the spherical spathe tube and the slitted and circular open limb area." "The leaves and the habitat of *A. mildbraedii* are very similar to those of *A. dracontioides*; it is therefore possible that unknown plants exist in herbaria." (Ittenbach 2003 loc. cit.).

Unexpanded leaf distinctly above ground during flowering in early March; petiole 7–8 cm; blade 40 cm diam., terminal leaflets lanceolate, 10–15 × 1–2 cm; inflorescence 12–13 cm high; peduncle 5–7 cm; spathe c. 5 cm high, with a compressed globose tube 3 × 4–6 cm; limb 3–4 cm wide sticking out horizontally, round and irregularly incised by up to 1 cm. Spadix sessile, 5.5 cm, erect.

Habitat: probably in forest or woodland; 1000 m alt.

Threats: clearance of habitat for timber and agricultual land.

Management suggestions: efforts should be made to retrace the itinerary of Mildbraed in 1914, to revisit Gbaboua and so to refind this long-lost species. It is not clear if Mildbraed's Babua equates with Baboua in CAR or Gbaboua nearby in Cameroon (spellings as in the Road Map of Cameroon, Macmillan). Gbaboua is also spelled Baboua, to add to the confusion. "Budaja", also referred to, may be Badjer, in Cameroon, c. 80 km W of Gbaboua. Searches should be made in early March so as to find the species in flower and confirm its identity. If found, basic populational data should be gathered to inform future monitoring and local authoritites informed.

Amorphophallus preussii (Engl.) N.E.Brown

EN B2ab(iii)

Range: Cameroon endemic (SW: Mt Cameroon at Buea and at Etinde; Mt Kupe; ; Bakossi-Banguem. Littoral: Mt Nlonako. W: 10 km W Bangwa).

Here *A. preussii* is assessed as Endangered since five locations (above; AOO 20 km^2 with 4 km^2 cells) are known, with threats as below. Data are mainly from Ittenbach 2003: 213–218. This species was previously assessed as EN A1c+2c by Cheek (Cable & Cheek 1998: lxiv–lxv) and as VU B2ab(iii) by Darbyshire (Cheek *et al.* 2004: 187–188). Both should be consulted for additional information.

Tuber 4–6 × 3.5–10 cm; petiole 35–70 cm; blade 40–80 cm diam., with 24–40 cm long segments; inflorescence emerging Dec.–Feb., before the leaf, peduncle 12–60 cm; spathe 8–15 cm long, erect, tube cylindric–conic, green, purple to off-white, limb broadly elliptic to triangular; spadix sessile, 6–8 cm, shorter than the spathe.

Habitat: submontane evergreen forest; often on shaded rock-faces; 800–1600 m alt.

Threats: forest clearance for timber and agricultural land, and, at Buea, urbanisation. Nlonako, Buea and Bangwa locations may already by lost to this species. It has not been seen at Buea, the type locality, since 1892.

Management suggestions: the Kupe-Bakossi and Etinde locations offer best hope for the long-term survival of *A. preussii*, being so far, fairly intact and with local communities interested in forest conservation. Here basic populational data should be collected, addressing density and range of individuals together with threats and regeneration levels. This will help inform future monitoring and also management intervention needs.

Amorphophallus zenkeri (Engl.) N.E.Brown subsp. ***zenkeri***

EN B2ab(iii)

Range: Nigeria (Oban) and Cameroon (SW: Mt Cameroon at Bimbia. C: Yaoundé-N'Kolbison; Zamakoé near Yaoundé. S: 38 km WSW Ébolowa at Nkoakam).

Here *A. zenkeri* subsp. *zenkeri* is assessed as Endangered since five locations are known (above; AOO 20 km² with 4 km² cells) with threats as below. Data here are mainly derived from Ittenbach loc. cit. 2003: 242–244. The last two locations are derived from specimens cited in Flore du Cameroun only.

Subspecies *mannii* is restricted to Bioko and has formerly been treated as a distinct species. It appears to merit specific recognition but is poorly known and may be extinct in view of extensive conversion of lowland forest to cocoa plantations.

Tuber flat-globose, 1.5–4 × 2–6 cm; petiole 0.3–1.3 m; blade 0.4–1 m diam.; inflorescence emerging in Oct.–Jan., after the leaf and withering with the leaf; peduncle below ground for a third of its length, 3–8 cm; spathe 8–24 × 11–26 cm, divided by a weak constriction into a globose to ovoid tube and an open limb; the open limb is tilted erect, horizontal, or hanging down, purple; spadix sessile, 25–60 cm, at least twice as long as the spathe.

Habitat: evergreen lowland and submontane forst; sea-level–800 m alt.

Threats: forest in Nigeria has declined by nearly 90% from the original cover (mongabay.com) while the Bimbia-Bonadikombo forest of the 1990's has been reduced to a much smaller community forest, due to plantation and agricultural pressures, so the species may not survive there. Natural habitat near Yaoundé is under pressure from urbanisation. Threats at the Ébolowa site are not known but slash and burn agriculture in that area is usual.

Management suggestions: searches to refind this plant in the Yaoundé area, where records are densest, should be encouraged, and also at the other sites. If refound in flower, plants should be marked and revisited when fruiting for seed-collection. Seeds should be banked and some sown to produce new plants for re-introduction to safe sites, such as Bimbia community forest, if population levels are as low as suspected. Basic populational data should be gathered to aid future monitoring. If found to be extinct in the wild there is every chance of re-introduction from cultivated material in Europe.

Culcasia bosii Ntépé Nyamè

VU B2ab(iii)

Range: Cameroon (SW: Mt Cameroon-Bimbia. S: Massif des Mamelles 50 km S Kribi; 4 km N Bipindi; 45 km SE Kribi at Elom; Ile Dipikar) and Gabon (Massif du Chaillu).

Here *Calcasia bosii* is assessed as Vulnerable since six locations (above; AOO 24 km² with 4 km² cells) are known, with threats as below.

Hemi-epiphyte, climbing, internodes not warty; petiole 2.5–5 cm, deeply canaliculate, basal 2/3 sheathed, the sheath terminating in a rounded auriele; blade lanceolate, asymmetric, 4.5–10 × 1.4 cm, mucronate; nerves 4–6 pairs, united in a collecting nerve, in transmitted light with black dashes and dots; inflorescence peduncle filiform, curved at apex, 5.5–8 cm; spathe green 2 cm; spadix at length exserted; flowers white.

Habitat: lowland evergreen forest; sea-level-100m alt.

Threats: since this species was discovered at Bimbia in the 1990's, much of the original forest has been lost to agriculture, wood extraction and plantation expansion. Massif de Mamelles is an iron ore deposit threatened by extraction.

Management suggestions: the best prospect for rediscovery and long term preservation of this species is the Kribi-Campo area where records are concentrated. The Dipikar site is possibly the most secure in Cameroon. Surveys are needed to collect basic populational data as a baseline for future monitoring at one or more locations.

Culcasia obliquifolia Engl.

VU B2ab(iii)

Range: Cameroon (SW: Mt Cameroon at Bimbia and at S Bakundu FR; Mt Kupe. S: Mt Calvary at Bidou I, Kribi-Lolodorf km 28; ibid km 12; Kribi-Edéa at Chutes Ebéa on Lokoundjé; 40 km S Kribi), Gabon (Libreville; Cristal Mts and Massif du Chaillu) and Angola (Cabinda).

Here *Culcasia obliquifolia* is assessed as Vulnerable since ten locations are known (above; AOO 40 km² with 4 km² cells) with threats as below.

Hemi-epiphyte, climbing with dropper roots to 1 m, adventitious roots few, stems with some warts; petiole long, 11–17 cm, canaliculate, lower 2/3 sheathed; blade ovate-triangular 16.5–28.5 × 8.5–13 cm, often marbled yellowish and dark green, acuminate, base broadly rounded to subtruncate, abruptly acute to decurrent, lateral nerves 3–5 pairs, in transmitted light with dashes, lower surface with raised circles; inflorescences enveloped in large persistent bracts, 5–6 cm; peduncle 4–6 cm, pendant, spathe green, 2.5–5 cm.

Habitat: lowland evergreen forest; sea-level–600 m alt.

Threats: several locations may well have been lost already to this species due to habitat destruction for timber and agricultural land (all sites at Mt Cameroon) and at Libreville (urbanisation). Locations on the coast at or near Kribi are at risk from development of infrastructure connected to extractive industries, urban expansion and tourism.

Management suggestions: while two of the three locations in Gabon are probably reasonably secure, investigation is needed to find at least one secure site around Kribi where the densest records for the species occur. Populational data should be gathered as a baseline for monitoring. It seems likely that further surveys will uncover extra locations for this species, taking it past the threshold of VU, into NT status if it survives at old locations.

Culcasia sanagensis Ntépé-Nyamè

VU B2ab(iii)

Range: Cameroon endemic (SW: Mt Cameroon at Bimbia. C: N'kolbison; Nachtigal, 50 km N Yaoundé; Essam, 20 km E Nanga Eboko; Bissaga, 25 km W Nanga Eboko; Mt Kala; Évodoula-Ndom km3; N Bafia at 38 km S Ngoro).

Here *Culcasia sanagensis* is assessed as Vulnerable since seven locations (above; AOO 28 km with 4 km² cells) are known, with threats as below.

Hemi-epiphytic climber, stems densely warty and folded; petioles 7–13 cm, broadly sheathing in lower ¼ to ½, blade oblong-ovate, 16–24 × 7–14 cm acumen 0–1 cm, base acute-obtuse, abruptly decurrent for 1.5 cm, lateral nerves 5–8 pairs, tertiary nerves finely reticulate, prominent; in transmitted light with black dashes; inflorescences clustered, peduncles 2.5–6 cm; spathe green 2–3.5 cm, mucronate, spadix orange-yellow.

Habitat: semi-deciduous forest and galleries in grassland; 600–800 m alt.

Threats: logging followed by agriculture.

Management suggestions: re-investigation of the Bimbia specimen is needed since it is disjunct geographically. Mt Kala is part of a protected area and so may be the best focus for conservation efforts. Basic populational data should be collected for future monitoring.

Rhaphidophora pusilla N.E.Br. (assessed by Iain Darbyshire)

VU D2

Range: Cameroon (SW Province: Bakossi Mts (1 coll.)) and Gabon (Cristal Mts (1 coll.)).

Hemi-epiphytic climber, internodes long, finely striate when dry; petioles 9–11 cm, sheathed for 9/10 its length; blade angled at 60–120° from the petiole, narrowly elliptic, 13–17 × 3–4 cm; gradually long acuminate 2 cm, base acute, slightly decurrent; lateral nerves numerous, parallel, conspicuous 3 mm apart, 45° from the midrib; peduncle 3–3.5 cm, spathe cylindric, 1.5–2.5 cm rostrate-subulate.

This taxon was previously known only from the type collection made in 1862 in the Cristal Mts, Gabon, by Gustav Mann. It has not subsequently been collected here. The discovery of this taxon in the Bakossi Mts near Kodmin therefore greatly increased the known extent of occurrence of this taxon. It was, however, classified as vulnerable by Darbyshire in Cheek *et al.* 2004: 188 on the basis that it is known from only 2 locations and is therefore threatened by stochastic events, such as local fires or landslides, or future anthropogenic pressures. This assessment is maintained here since no additional data are available.

This taxon was, until recently, considered synonymous with *R. africana* N.E.Br., but has been resurrected by P.Boyce (*pers. comm.*). It is readily separated from the latter taxon, being a slender, not robust, climber and having solitary, not clustered, and erect, not pendant, inflorescences.

Habitat: submontane forest; c. 1500 m alt.

Threats: the locations of the two known collections remain largely unthreatened, although the Bakossi site near Kodmin was adjacent to the well-used track from Kodmin to Muahunzum (and eventually to Nyandong), thus it may be threatened by future increased use, and thus expansion, of this route.

Management suggestions: rediscovery of the two populations, and an assessment of their size and any local threats, is imperative. Further taxonomic work on the *Rhaphidophora* complex in west Africa may reveal further populations until now treated as *R. africana*.

ASPHODELACEAE assessed by Martin Cheek

Treated in Liliaceae in FWTA, but currently placed in Asphodelaceae with several other genera, none of which appear to occur in Cameroon. The family have not been covered for Flore du Cameroun.

Kniphofia reflexa is the only species of this temperate, horticulturally important genus known from West-Central Africa. The other 64 species of the genus are known from Arabia, through the mountains of E Africa, to South Africa and Madagascar. They generally occur along streams in mountains. Various taxa are cultivated in Europe as garden ornamentals known in the UK as "red-hot pokers".

Kniphofia reflexa Hutch. ex Codd.

EN B2ab(iii)+C1

Range: Cameroon endemic (NW (Mt Oku): summit area; Ijim ridge; Mbesa swamp).

Long known only from a single collection from a single locality above Laikom ('Lakom', April 1931, *Maitland* 1624) where it was then recorded as "scattered in considerable numbers in the grass on the plateau", *Kniphofia reflexa* was then lost to science for over 60 years until Martin Etuge's team rediscovered it in the Afua swamp while searching for material of *Eriocaulon bamendae* in November

1996 (Cheek *et al.* 1997; Cheek *et al.*: 2000: 170). Etuge observed about seven plants at Afua. Maisels confirmed plants at this site in 1997. However, when this site was revisited in November 1998, no plants were found and the species may now be extinct at Afua. The reason for the loss is not clear. In May 1997 Boo Maisels found a new site for *Kniphofia reflexa* at Kinkolong, near the summit of Mt Oku where several hundred plants were recorded, and two more locations have since been found. One, the "Mbesa Swamp" on Ijim Ridge, is probably the site at which Maitland collected his specimen in 1931. It has about 200 individuals. Mbi crater is the third site. Maisels estimates that this also has about 200 individuals (Maisels *et al.* 2000).

It is remarkable that such a conspicuous species had remained unobserved by scientists for so many decades. The assessment of this species in Cheek *et al.* (2000: 76–77) as EN C1 is maintained and amplified here, since about 700 individuals are estimated from three locations (above; AOO is 3 km² with 1 km² cells recommended for aquatic species).

Perennial herb from yellow-staining rootstock 3–4 cm thick; leaf-rosettes one to few from each rootstock, about 1 m wide, acaulescent; eaves long-linear, 14–16 mm wide, soft; inflorescence a stout spike 0.6–1.6 m tall (to 2 m tall in fruit) bearing in the upper part numerous tubular flowers; bracts oblong, boat-shaped, sharply reflexed, 5–6 mm long, 2–3 mm wide, margin finely papillose-fimbriate; flowers yellow, 8–8.5 mm long, tube 5–5.5 mm long, campanulate; lobes 3–3.5 mm long, the outer 3 mm broad at the base, the inner 2.5 mm, ovate-triangular, rounded; filaments 7–8 mm long; anthers c. 2.5 mm long. Style 5 mm long; fruiting in November; flowers in April? Description partly after Marais (1973).

Habitat: swamp or stream edges; 1750–2900 m.

Threats: trampling by cattle accessing watering places. Asonganyi (1995) remarks upon the Afua area as suffering from extensive recent habitat destruction and so we consider this extremely rare species highly Endangered.

Management suggestions: annual monitoring of the three sites for this species is needed. Numbers of individuals should be counted on these occasions, signs of damage recorded and regeneration noted. Afua should be revisited to confirm that this species has indeed disappeared. Similar sites to those already known should be searched to discover potential new locations. Some evidence of fire scorching on the Afua specimens was noted, so some burning at the correct season may be required in management of habitat for this species or important for promoting flowering. Propagation by seed and replanting in e.g. Afua Swamp might be considered.

BURMANNIACEAE-THISMIACEAE

assessed by Martin Cheek

The perennial forest floor herbs of this family lack green tissue and except when in flower or fruit, are underground and invisible to observers. The roots are believed to attract and digest fungal hyphae. Most species of *Afrothismia* are known only from one or two locations.

Cameroon is by far the most species-diverse country in Tropical Africa for Burmanniaceae in the broad sense, above all for *Afrothismia* with 8 of the 14 species known in the world, 7 of which are unique to Cameroon. Schlechter erected this genus in 1906 with two species from sites on the slopes of Mt Cameroon (Moliwe and "Neu Tegel" now Muea) from which they have now almost certainly been wiped out by destruction of their forest habitat for plantations of oil palm and bananas, as documented e.g. in Blumea 49: 451–446 (2004) by Franke *et al.* Schlechter also erected the genus *Oxygyne* based on a single specimen from Moliwe. This species is probably globally extinct. In the early 1990s a resurgence of interest began in African achlorophyllous mycotrophs ('saprophytes') among which Burmanniaceae are prominent (Cheek & Ndam 1996: 612–617 in The Biodiversity of African Plants, Kluwer). Then, in 1999–2003, Thassilo Franke of Munich began a field programme of research focussed on this group, working with Moses Sainge to make many new discoveries. He in turn attracted Stepan Imhof and then Vincent Merckx from the Netherlands and Belgium to join his doctoral research on "saprophytes", also known as achlorophyllous mycoheterotrophs. Scientifically speaking, SW Region Cameroon is probably the most important study site in the world, and certainly in Africa, for these plants and their fungi.

All four African genera of Burmanniaceae occur in Cameroon: *Burmannia* and *Gymnosiphon* (species awaiting revision so not assessed here); and two genera sometimes placed in Thismiaceae: *Afrothismia* and *Oxygyne*. Research led by Merckx (2008) supports recognising Thismiaceae as a separate family from Burmanniaceae.

Afrothismia foertheriana Th. Franke, Sainge & Agerer
CR B2ab(iii)+ D

Range: Cameroon (SW: western foothills of Mt Cameroon, Mokoko, Diongo Community Forest).

This bizarre species in a bizarre genus is only known from a single site (AOO of 4 km² with 4 km² cells) where it is recorded as flowering and fruiting on 7 Oct. 2002 (*Franke & Sainge* 02/30, Blumea 49: 451–456 (2004)). The assessment by Franke *et al.* loc. cit. is maintained here since no new data are available. Peak flowering was considered to be probably August–September. Merckx (2008) reports seeing the species in 2006 at two spots at Diongo so it may be that two locations can be accepted, in which case the species can be

accepted as CR under criterion D1: less than 50 individuals are known.
Habitat: lowland evergreen forest, 220 m alt.
Threats: "severely threatened by illegal logging and the extent of agricultural activities (mainly oil palm and cocoa cultivation)" Franke *et al.* loc. cit.
Management suggestions: the importance of this plant should be discussed with the Diongo elders and the possibility of renting the site of this *Afrothismia* explored with them, if not already done so.

Afrothismia hydra Sainge & Th. Franke
EN B2ab(iii)
Range: Nigeria (Akure FR) and Cameroon (SW: Korup; Mt Cameroon–Diongo; Mt Kupe).
One of the commonest *Afrothismia* species, but still known from only four locations. First proposed as a species based on the Akure material by Mildbraed and Keay as *Afrothismia minor* but not published until 2005 (Franke & Sainge, Nordic. J. Bot. 23: 299–303) as *A. hydra*. Four locations are mentioned there, but only three are given. The Mt Kupe site is given as new here, based on a drawing by S. Williams of *Cable* 3806 at Kew. *Afrothismia hydra* resembles *A. winkleri* but is about half the size, has a more acutely angular and papillate perianth tube, and tepals that tend to be held vertically, not horizontally.
Afrothismia hydra is here assessed as Endangered since known from four locations with the threats listed. Although IUCN guidance results in an AOO of 16 km^2, the reality is that at each location the species occupies only one or two m^2.
Habitat: lowland and submontane forest; 60–1040 m alt.
Threats: see under *A. foertheriana* Diongo. The status of the Akure site is unknown, but large losses of forest habitat are reported from Nigeria. The sites at Korup and Kupe are believed secure.
Management suggestions: see under *A. foertheriana* re the Diongo site. Pinpointing of other locations, community education, protection and monitoring is advisable.

Afrothismia korupensis Sainge & Th. Franke
CR B2ab(iii)+D
Range: Cameroon endemic (SW: Korup)
Discovered in July 2002 in the Smithsonian 50 ha plot near Chimpanzee camp in Korup NP when 12 plants were seen by Sainge. Twenty individuals were found in the same area the following June (Sainge & Franke 2005, Willdenowia 35: 287–291). This species is unusual in its erect, externally flanged perianth tube with a horizontally directed mouth and appendages as long to twice as long as the tube. *Afrothismia korupensis* is here assessed as Critically Endangered since less than 50 individuals are known from a single site (AOO of 4 km^2 with 4 km^2 cells) with threats as below.
Habitat: lowland evergreen forest, 5 m rainfall p.a.; 230 m alt.
Threats: the only site is placed within the 50 ha Smithsonian Plot. During enumeration phases this is heavily staffed and intensively visited, leading to a risk of inadvertent trampling of these small plants.
Management suggestions: the site should be signposted, marked off, and monitored at yearly intervals.

Afrothismia gesnerioides H.Maas
CR B2ab(iii)+D
Range: Cameroon endemic (S: Nyangong).
This species is only known from the collection by the mollusc specialist *de Winter* 91 from a transect at Nyangong. The flowers are similar to those of *A. pachyantha* but the tepal lobes are reflexed and the perianth tube more slender. Flowering 24 April 1996 (*Maas* 2003, Blumea 48: 475–478). *Afrothismia gesnerioides* is here assessed as Critically Endangered, less than 50 plants being known from a single site (AOO of 4 km^2 with 4 km^2 cells) with threats as below.
Habitat: evergreen forest; 600 m alt.
Threats: slash and burn agriculture.
Management suggestions: an effort should be made to refind the original collection site (details are given in the protologue) so as to monitor it and to protect it with signage. It is not clear whether it is in a protected area or not.

Afrothismia amietii Cheek
CR B2ab(iii)+C2+D
Range: Cameroon endemic (C: Mt Kala)
In recent years Mt Kala, the only known locality for this species, has been visited by many botanists in connection with the vegetation study of the habitat of a rare ground-dwelling bird, *Pithacartes*. Species of *Afrothismia* usually occur at known locations as a cluster of a few to dozens of individuals at a site of up to some tens of m^2 (Cheek pers. obs.). However, no more plants of *Afrothismia amietii* have been found so far since the type specimen was found in late April 1970. This extremely rare species was assessed in the protologue as being Critically Endangered (CR) according to the criteria B, C and D. This was because it is only known from a single individual (AOO of 4 km^2 with 4 km^2 cells). For these reasons it merited the assessment of CR B2a,b(iii). It also deserved Critically Endangered status under the population size criteria C and D. Under criterion C it was assessed as C2 (a) (i) and (ii). These assessments are maintained here since no further data are available.
Habitat: submontane evergreen forest; 850 m alt.
Threats: single poorly protected site where habitat degradation (illegal logging followed by agriculture, fide Onana *pers. comm.*) has occurred recently and is expected to continue.

Management suggestions: searching for the species at the type locality is advised in order to pinpoint and protect its site within the surviving forest. It is to be hoped that *Afrothismia amietii* still survives at Mt Kala, or might be found at other locations where its protection can be ensured.

Afrothismia pachyantha Schltr.

CR A2c; B1 + 2ab(iii); C2a(i,ii); D

Range: Cameroon endemic (SW: Mt Cameroon at Moliwe (almost certainly extinct there); Mt Kupe (1 site)).

This species was first assessed as Critically Endangered in 1998 (Cable & Cheek 1998: lxv (CR A1c + 2c, B1 + 2abcde, C2b, D)). Since then no new locations have been located, and further searching on Mt Cameroon has failed to rediscover the species there (Franke *et al.* in press), where it is almost certainly extinct. Changes to the IUCN criteria in recent years, (IUCN 2003), resulted in the modification of the assessment in Cheek *et al.* (2004: 188) to CR A2c; B1 + 2ab(iii); C2a(i,ii); D, which is maintained here with since no new data are available. Less than 50 individuals are known at the only surviving site (AOO of 4 km² with 4 km² cells), and numbers appear to have declined there between 1995 and 2007 (Cheek pers. obs.).

Habitat: lowland evergreen forest; c. 700 m alt.

Threats: forest clearance for agriculture, stochastic events.

Management suggestions: continued monitoring and rental of the single known site for this species is advised.

Afrothismia saingei T. Franke

CR B2ab(iii)+ D

Range: Cameroon (SW: Mt Kupe, Mbule).

Discovered by *Sainge* (1052) in Oct. 2002, this strange species is known from a single site (AOO of 4 km² with 4 km² cells), at which 3 plants have been seen 50 m apart from each other (Franke, 2004, Syst. Geogr. Pl. 74: 27–33). It resembles an *A. winkleri* but has strongly unequal tepal lobes, the shortest pair being 1.3–2.2 cm long, the longest 4.2–7.3 cm long. *Afrothismia saingei* is maintained here as Critically Endangered, supporting the assessment made in Cheek *et al.* 2004 since no new data are available.

Habitat: evergreen forest, 3–4 m rainfall p.a.; 970 m alt.

Threats: being close to Mbule village it is vulnerable to forest clearance.

Management suggestions: the elders of Mbule should be consulted, and the possibility of renting the site of this *Afrothismia* explored with them, if not already done so.

Afrothismia winkleri (Engl.) Schltr.

EN B2ab(iii)

Range: Cameroon (SW: Mt Cameroon foothills; Mt Kupe; Banyang Mbo).

Formerly believed known from several locations, with a range from Nigeria to Uganda. The Nigerian material is now known to be *A. hydra*. The Ugandan specimen has also been given separate taxonomic status. First discovered c. 1906 from two sites in the eastern foothills of Mt Cameroon at both of which it is now almost certainly extinct since much searching by specialists has not rediscovered it.

Winkler's "Neu Tegel" (now Muea at Buea mile 14) near Mt Cameroon was the type locality but is now a CDC Banana plantation (Bate Oben pers. comm. 2008). The second historic site, Schlechter's Moliwe is now under oil palm plantations and chop farms (Cheek pers. obs. 1991–1993). Since it has not been seen at these places in about 100 years, despite considerable searching in surviving habitat nearby by Franke *et al.* and at Moliwe, first by Cheek and later by four Oxford University undergraduates over many weeks in the 1993, there seems little hope of its rediscovery there. Searches by Franke *et al.* in the western foothills of Mt Cameroon aimed at finding this and other Schlechter saprophyte species (*Oxygyne triandra, Afrothismia pachyantha*) failed to do so, but did uncover two further new species at Diongo in the eastern foothills (see *A. foetheriana, A. hydra*).

The species was assessed as CR A1c + 2c in Cable & Cheek (1998: lxv), but since then the IUCN criteria have changed and new sites have been discovered at Mt Kupe (at Nyasoso and Kupe Village) and at Banyang Mbo, both in SW Region, Cameroon (AOO of 12 km² with 4 km² cells); accordingly a new assessment was made in Cheek *et al.* (2004: 189) and is maintained here.

Habitat: lowland evergreen forest; 0–1200 m alt.

Threats: forest clearance for timber and agriculture. The area of occupancy is less than 100 km^2 (at each site only 1–3 m^2 are occupied), severely fragmented and declining in quality.

Management suggestions: continued monitoring and rental of the site at Kupe village, Mt Kupe, and protection of the new site at Banyang Mbo.

Oxygyne triandra Schltr.

CR B2ab(iii)+D

Range: Cameroon endemic (SW: Mt Cameroon at Moliwe).

First collected near Moliwe in 1905 (*Schlechter* 15790) and known from that single site (AOO of 4 km² with 4 km² cells) but not rediscovered since, despite searches by the author in 1991–92 and later by Franke *et al.* The Oxford University expedition to Mt Cameroon (July–Sept. 1993) had the rediscovery of this species, together with *Afrothismia,* as a key objective, but did not meet with success.

This species was assessed as CR A1c+2c by Cheek in Cable & Cheek (1998: lxv), which is here updated.

Habitat: lowland evergreen forest.

Threats: forest clearance for agriculture and plantations.

Management suggestions: continued searches for this species on Mt Cameroon could be made, but the prospects of refinding this species must be very low after three other groups have failed in this task.

COLCHICACEAE

assessed by Martin Cheek

Included in Liliaceae in FWTA 3: 107 (1968), *Wurmbea* is the only threatened taxon of Colchicaceae currently known in Cameroon. Other genera in Cameroon are *Drimia, Gloriosa, Iphigenia, Dipcadi* and *Scilla.* All have perennial underground bulbs which survive the dry season, and leaves and flowers which appear annually. With the exception of *Gloriosa* which is cultivated and also occurs at forest edges, all are grassland taxa.

Wurmbea in Africa was revised by Nordenstam (Notes Roy. Bot. Gard. Edinb. 36(2): 211–233 (1978) recognising four subspecies within *W. tenuis* one of which is endemic to the Cameroon Highlands, the other three being found in E & S Africa.

Wurmbea tenuis (Hook.f.) Bak. subsp.***tenuis***

VU A2c

Range: Nigeria (Mambilla Plateau: Chappal Waddi), Equatorial Guinea (Bioko) and Cameroon (SW: Mt Cameroon. W: Bamboutos. NW: Bamenda Highlands. (numerous locs.)).

Cormous herb; corm ovoid, about 1 cm long, tunicate, bulb-like; stem with basal sheaths membranous, truncate, 1–3 cm; stem leaves 2–3, decreasing in size upwards; leaves 1 or 2 arising from the corm, linear, acute, 6–16 × 0.15–4 cm, glabrous; spike 2–6-flowered; perianth segments white and purple, fading in older flowers, 6 mm (from Pollard in Cheek *et al.* 2000: 170).

Here *W. tenuis* subsp. *tenuis* is assessed as Vulnerable since it is estimated that more than 30% of the population has been lost within 3 generations and that this loss continues. See threats below.

On Mt Cameroon it seems secure: "Very common in places and present from around 2000 m. Also seen SW flank at same altitude. Frequently forms large colonies. Bulbous herb to 5/6 cm. Usually with 2 leaves. Flowers white, each petal with a purple spot at base. Montane grassland. Mt Cameroon, E flank, above Ekona fl. 25/5/90" (*Upson* 100).

Habitat: montane grassland; 1800–2500 m alt.

Threats: grazing by cattle, including trampling and fire-setting by graziers; conversion of grassland to cultivated fields and fallow.

Management suggestions: since this species is almost invisible unless in flower it should be surveyed in late March to mid May. Exclusion plots with and without cattle, studied over years, would help accurately quantify rates of loss of this taxon from this source. Fortunately this bulb is locally common on Mt Cameroon, and secure (cattle are absent) even if it is under threat in the other parts of its range. It is possible that dry season fires are needed to trigger flowering but this needs confirmation.

COMMELINACEAE

assessed by Martin Cheek

A large pantropical family of succulent-stemmed herbs, Commelinaceae are easily recognised by their usually sprawling stems which are partly concealed by tubular leaf-bases (sheaths) from which the broad, often elliptic or ovate blades arise. The flowers, usually only 5–10 mm wide, are very short-lived, often for only 2–3 hours, and have 3 delicate, stalked petals. Different species open at different times of day and it is important to record this when studying them. In two of the largest genera in Cameroon, *Cyanotis* and *Commelina*, the flowers emerge from a terminal half-moon shaped spathe, also seen in *Polyspatha.* Recently a genus new to Africa was discovered in this family from Yaoundé-N'kolbison: *Tricarpelema africana* (see below). Further discoveries are to be expected once research for a Flore du Cameroun account begins, which Robert Faden, world authority on the family in Africa, will hopefully complete now that his FTEA treatment is finished.

Several species of *Commelina* are weeds of habitation and most genera prefer open habitats with full or near full sunlight to survive. *Cyanotis,* with its underground tubers, is most diverse in grasslands. However, most of the threatened species with narrow distributions, are restricted to forest. *Palisota* only occurs in forest and is atypical in being more robust in habit than all other African genera, often with a stout erect aerial stem to 2 m tall.

No important uses are known, although there are reports that the family are favoured by elephants being high in tusk-building materials, but this is not substantiated. Some species are cultivated as ornamental plants in glasshouses in temperate regions such as Europe.

Two poorly known species names have uncertain status. If it is resolved that they are real entities and not synonyms of more widespread species it is likely that they will be assessed as threatened in future: *Palisota gracilior* Mildbr. (*Mildbraed* 5425, Lomié) and *P. waibelii* Mildbr (*Waibel* 208 and 242, Yoko). However Robert Faden advises that they are probably synonyms of accepted species (pers. comm.. 2010). *Commelina cameroonensis* J.K.Morton, restricted to montane forest in the Cameroon Highlands, is likely to be included in future editions of this Red Data book.

Aneilema silvaticum Brenan

VU B2ab(iii)

Range: Nigeria, Cameroon (SW, NW and Littoral) and Congo (Kinshasa).

Known at Kew from three collections in Nigeria (*Meikle* 637, 1949; *J.D.Kennedy* 2674 & 1758, respectively, 1935 and 1931, all from Sapoba), three in Congo (Kinshasa) (*Lemaire* 379, Mobwassa; *Seret* 69, Uele; *Tilquin* 183, Kinshasa) and six in Cameroon (*Zenker* 1110, Bipindi, 1896 and *Mbatchou* 470, Mt Cameroon, 1992; also Lebialem-Bechati and Kupe-Bakossi in SW; Chantier Sanaga in Littoral).

Weak, erect herb 15–30 cm tall; leaves ovate-lanceolate c. 9 × 3 cm, acuminate, petiolate, margin ciliate; inflorescence terminal, dense, <2 × 2 cm; flowers 10–15, white.

This is a tiny mat-forming herb in swamp forest, very likely to have been overlooked. It was first assessed in Cheek *et al.* (2004: 189). That assessment is maintained here since although new sites have been added, the threshold of ten has not yet been exceeded. The eight sites equate to an AOO of 32 km^2 with 4 km^2 cells. Where the author has seen this species, it only occupied a site of about 4–5 m^2.

Habitat: lowland forest; 300–880 m alt.

Threats: clearance of lowland forest for timber and/or agriculture is known to be a cause for concern in Nigeria generally including Sapoba, but also at Mt Cameroon, Kinshasa and the Loum FR.

Management suggestions: better policing and higher protection of existing forest reserves would help secure the future of this species. Further intensive survey work is likely to yield additional sites leading to its transfer to NT status in future.

Commelina zenkeri C.B.Clarke

CR B2ab(iii)

Range: Cameroon endemic (C: Yaoundé).

Known only from *Zenker and Staudt* 432 made between 1890–1894 at "Kamerun, Yaúnde-Station, 800 m.s.m.". The species author, C. B. Clarke was the World Authority on this family, so although the genus needs revision, this is almost certainly a valid species, and the specimen still exists, at Kew.

Here *C. zenkeri* is assessed as Critically Endangered since only a single location is known (above; AOO 4 km^2 with 4 km^2 cells) and with threats as below.

Herb with 2–5 main stems from the roots, stems rooting at nodes, horizontal stems to 20 cm; erect stems 8–10 cm; leaf-blades with sheath 13–15 mm long, with scattered patent white hairs, pseudopetiole 1 mm; blades 2–3(–4) per erect stem, increasing in size towards apex., 2.7–5.0 × 1.8–2.4 cm, acute-acuminate, mucronate; base rounded, glabrous, lower surface scalariform nerves; spathe single, stipe 5 mm, blade semicircular, at 90° from vertical, 10 × 15 mm, glabrous apart from external midrib with scattered patent white hairs, nerves strongly scalariform; flowers exserted 5 mm, c. 10 mm diam.

Habitat: unknown, 800 m.

Threats: urbanisation. Yaoundé, capital city of Cameroon, is expanding.

Management suggestions: rediscovery of this species, not seen in 120 years, is needed. It is possible that this is an inselberg specialist as are other Yaoundé endemics. Dr. Faden of the Smithsonian Institute, Washington, USA should be asked for an opinion and to verify any potential specimens that might be discovered of this species since he is the world specialist.

Floscopa mannii C.B.Clarke

VU B2ab(iii)

Range: Nigeria (Oban), Cameroon (SW: Mt Cameroon; Kupe-Bakossi. E: 3 km NE Lomié), Gabon (2 loc incl. type: *Mann* 1867, Corisco Bay, 1862.) and Congo (Brazzaville) (1 loc.).

Herb; stems prostrate, ascending to 5 cm; leaves ovate, to c. 4 × 2 cm, apex acute, base cordate, sessile; sheath white villous; inflorescence 4–8 cm long, diffuse, with 5–6 slender parent branches, each c. 2 cm long, glandular-hairy; petals white.

In Kupe-Bakossi it is known only from a single site to the east of Ngomboku, in Mbumbe forest (*Cheek* 10349, 13 Dec. 1999). Here it was gregarious, and thriving in a swampy valley bottom (c. 700 m alt.) where the forest had been partly cleared apart from *Raphia*.

This species is probably often overlooked due to its small size and unspectacular nature. However, it is nevertheless rare. Previously assessed as EN in Cheek *et al.* (2004: 189), it is reassessed here as VU since several new locations have been discovered, now totalling seven (above; equating to an AOO of 28 km^2 with 4 km^2 cells) with threats as below.

Habitat: lowland swamp forest; c. 700 m alt.

Threats: clearance of lowland forest for timber and for agriculture.

Management suggestions: it is not clear that any of the sites for this species are currently gazetted as protected: this should be rectified if possible, and if governmental protection cannot be secured, efforts should be made to invoke protection from local communities. Ngomboku in Bakossi, where the species was seen recently at a site that is easily revisited, might be a suitable focus for conservation efforts of *F. mannii*.

Palisota brachythyrsa Mildbr.

VU B2ab(iii)

Range: Cameroon (S: Campo; E: Moloundou at Boumba-Ngoko), CAR (Dzangha-Dzangha NP), Congo (Brazzaville) (Odzala NP; Brazzaville; Lekoumou), Gabon (Bateké;

Mayombe) and Congo (Kinshasa) (Kikwit; Equateur-Wendji).
Herb 0.5–1.5 m tall stems and lower surface of leaves densely hairy, often red; leaves in verticils of 3–6; internodes 7–15 cm; leaf-blades elliptic or oblonceolate-elliptic, 12–16 × 4.2–5.8 cm, acumen 1.5–2 cm, base acute; pseudopetiole 0.3–1 cm; flower spike 1.5–3 cm, 4–12-flowered , terminal, held within terminal, funnel-like leaf rosette; fruits vivid red 1cm.
Here *P. brachythyrsa* is assessed as Vulnerable since 10 locations (above; AOO 40 km^2 with 4 km^2 cells) are known with threats as below. This species has been confused with *P. ambigua* which generally has much longer inflorescences, and *P. thollonii* which is a climbing species.
More research is needed to see if more than one taxon has been included under this name. For example the specimen cluster in E Region Cameroon and CAR occur in swamp forest and have short, pale indumentums, while those in the Chaillu of Congo (Brazzaville) and Gabon occur on hilltops and have long red hairs.
Habitat: evergreen forest both swamp and hill; 200–700 m alt.
Threats: clearance of forest for mineral extraction (Congo); logging followed by agriculture.
Management suggestions: this species should be mapped to determine its representation in National Parks which might be centres of conservation for the species. Basic populational data should be collected at one or more of these sites as a baseline for future monitoring. Discovery of further locations is likely to remove this species from the threatened list.

Palisota ebo Cheek ined.

EN B2ab(iii)

Range: Cameroon endemic (Littoral: Ebo proposed NP).
Perennial rosulate herb, lacking an aerial stem, leaves 4–5 per rosette, rhizome ascending, short and stout, densely clothed in petiole bases and hairs; leaves alternate in a basal rosette c. 30 cm wide; blades held ± horizontally, ovate-lanceolate, rarely elliptic or oblong, 12–15 × 5.5–7.5 cm, acumen 0.8–1.5 cm long with an acute apex, base rounded, rarely obtuse or subtruncate, major lateral nerves 3–4 on each side of the midrib, sunken in life, giving the blade longitudinal, curved corrugations; upper surface bicolored, a stripe 1–1.5 cm wide along the midrib yellow-green the rest of the blade dark green, very rarely plants lacking the stripe; hairs on upper surface persistent, bristle-like, 2.5–3.25 mm long, patent, white, margin with dense, long, appressed hairs; petioles ascending, canaliculate, fleshy, 6.5–10 cm long, 2.5 mm wide (dried), densely clothed with patent (pilose) pale brown hairs. Flowering shoots 1–2 per plant, axillary, horizontal 1.5–7cm long, concealed under the leaves and amongst leaf litter, rarely branched, not rooting, each shoot bearing 1–2 inflorescences; inflorescences terminal, scale-leaf-opposed, pedunculate, with a single c. 6-flowered cincinnus subtended by a terminal pair of bracts; peduncle 0.8–1.8 cm long, mostly sheathed with spirally inserted bracts; bracts 4–6 per peduncle oblong-triangular, amplexicaul 6–9 × 4 mm; terminal bracts pair 1–1.5 cm long, forming a cup c. 0.4 cm wide, indumentum as scale-leaves; flowers pale pink-white 12 mm wide, odourless; pedicel erect in flower, becoming spirally contorted shortly after anthesis and in fruit, c. 2.5 cm long, white, pilose; fruits ovoid berries, dull, yellow-brown, pilose, 2.5 × 1.5 cm, apex slightly rostrate, base rounded to truncate, 3-seeded (1 per locule), mesocarp white, juicy, several mm thick. Pedicel dilated in fruit and 3–4 coiled; seeds ellipsoid 5.5 × 4 × 2.5 mm ventrally flattened, grey, hard, glossy.
Palisota ebo is assessed in the protologue as Endangered: EN B1a, b(iii) since it is known from two locations with an estimated area of occupancy of 8 km^2 and a projected decline of extent, area and quality of habitat due to the threat of logging followed by slash and burn agriculture. At the type location, thousands of plants are present since it is the dominant forest floor species along a 2 km length of trail, where it occurs at intervals of 0.3–0.9 m (*Cheek* 13073). The assessment is maintained since no new information is available.
The data above is contracted from Cheek *et al.* (in press) in which the species is formally described.
Habitat: locally dominant understorey herb of undisturbed lowland-submontane evergreen forest on sand, terrain flat; 800 m alt.
Threats: see above.
Management suggestions: official National Park status for Ebo is recommended to secure conservation efforts here. Further work is needed to alert local communities as to the importance of this species. Consideration should be given to introducing the species into cultivation as a pot-plant. This might secure further support for conservation at Ebo of this and other rare plant species.

Palisota flagelliflora Faden

EN B2ab(iii)

Range: Cameroon endemic (Littoral: Bakaka FR on Loum-Nkongsamba Rd. S: Kribi-Lolodorf Rd, km 18–20).
Here *Palisota flagelliflora* is assessed as Endangered since only two locations are known (AOO 8 km^2 with 4 km^2 cells: above) and threats being as below.
Erect, rosette-forming herb, aerial stem absent or very short, leaves numerous; petiole 4–19 cm appressed pubescent; blade elliptic to obovate, 10–40 × 3.5–10 cm acute or acuminate, base cuneate, glabrescent above, white sericeous below; inflorescences short 1–10 cm long, erect, borne on long prostrate, whip-like horizontal shoots 25–100(–175) cm

long; flowers white 1.2–1.5 cm wide; fruits red, triangular ovoid-ellipsoid 1.4–2.3 cm long.

This is a most bizarre species, given the strange origin of its inflorescences on prostrate whip-like shoots among the leaf-litter. It might be confused with *P. bogneri* and *P. ebo* which also have inflorescences borne among the leaf-litter, but which differ in having much shorter, non-whip-like inflorescences and upper surfaces of the leaves pilose, not glabrescent.

Habitat: lowland evergreen forest.

Threats: logging followed by agriculture. Both locations are in areas where logging occurs.

Management suggestions: one of the two locations should be considered for formal protection. Bakaka FR certainly has a large number of other threatened species present which might justify this. Basic populational data should be collected as a baseline for future monitoring. Propagation of this species to strengthen wild populations should be considered.

Palisota preussiana K.Schum. ex C.B.Clarke

EN B2ab(iii)

Range: Equatorial Guinea (Bioko (1 coll.)), Nigeria (Mambilla Plateau) and Cameroon (SW: Mt Cameroon (at Buea, Bokwango, Likombe, Mt Etinde and Bakingili); Mt Kupe. S: S of Ébolowa).

Previously assessed as VU D2 in Cheek *et al.* (2004: 190), it is reassessed here as EN despite the addition of new locations, now totalling five (above; equating to an AOO of 20 km² with 4 km² cells) because threats are as below.

The Bioko record has erroneously been omitted from our map. The Mambilla record (*Chapman* 2820) has not been confirmed by a specialist. The record S of Ébolowa, is extracted from Tropicos, and although geographically and ecologically discordant, has been identified by Faden so must be accepted.

Herb c. 1 m; stem erect; leaves attenuate, not in pseudowhorls, ovate, c. 17 × 9 cm, acuminate, base rounded; petioles 3–4 cm; inflorescence solitary, terminal, dense, cylindrical, 7 × 3 cm; peduncle 4 cm; flowers mauve.

Distinguished from all other *Palisota* species with aerial stems in that the stems are slender (c. 6 mm diam. when dried) and that the leaves are not in pseudowhorls, but evenly scattered along the stem. *P. preussiana* is also unusual in the single terminal inflorescence. The short petioles usually recorded in this species are longer than usual (3–7 cm) in the Mt Kupe specimen.

Despite c. 9,000 specimen records being made in Bakossi (Cheek *et al.* 2004 [illegible] to be this species: [illegible] Kupe, at 2000 m alt. on 26 June 1996 when it was recorded as bearing immature fruit. Only two duplicates were collected, suggesting that only two individuals were seen. This is genuinely a rare species.

Habitat: submontane forest; 1000–1600(–2000) m alt.

Threats: clearance of forest for agriculture (especially on the slopes of Mt Cameroon, cocoyam farms).

Management suggestions: efforts to protect this species should be centred at the eastern slopes of Mt Cameroon where it appears to be most common, although most under pressure from forest clearance for agricultural land. A campaign to publicise the species and to multiply it, perhaps with propagation efforts focussed at Limbe BG, should be considered. Basic populational should be gathered so that a baseline exists against which future monitoring can take place. Collection of more material of the Mt Kupe subpopulation is advised to determine whether *Cable* 3383 is either aberrant, or represents a distinct taxonomic entity; verification of the Mambilla specimen is also needed.

Tricarpelema africana Faden

VU B2ab(iii)

Range: Cameron (S: Ébolowa at Meyos; Akoakas; Nkoemvom to Ambam; C: N'kolbison, Yaoundé at Minlova), Equatorial Guinea (Rio Muni: Centro Sur: Bicurga; Wele Nzas: Anisok, Nzamayong; Nchoho Biworo; Dimu) and Gabon (Mangom, Nzec; Mitzic; Wolev-Ntem, Bordamur concession; Oyem-Bitam, 35 km; ESE Medouneu).

Here *Tricarpelema africana* is assessed as Vulnerable since 10 broad locations are known (above; AOO 40 km² with 4 km² cells) and since threats are as below.

All eight species of *Tricarpelema* are Asian except *T. africana* which is so distinct in its vegetative morphology that it is placed in its own subgenus (Faden, Novon 17: 160–171 (2007) from which data here derives).

Decumbent succulent perennial herb rotting at lower nodes, roots fibrous, internodes stiff, almost wiry, covered by overlapping leaf-sheaths except the flowering shoots, 2.8–6 cm, sparsely glandular pubescent when young; leaf-blades distichous, lanceolate-oblong or linear-lanceolate, 4–11.5 × 1–1.5 cm, acuminate, base rounded, sessile, amplexicaul, both surfaces sparsely glandular-pubescent, margins scabrid; inflorescence terminal, lax, compound thyrse, of up to 7 branches, bracteole, peduncle 5.5–12 cm; flowers open in morning 7–9 mm wide, sepals 3, oblong, 2.5–4 × 1–2 mm; petals 3, free, not clawed, reflexed, pink, purple, blue to white, ovate, 3–6 × 1.2–4mm. Flowering Jan.–June and Oct.–Nov.

Habitat: forest fringes of inselbergs in forested areas; 500–950 m.

Threats: rock extraction for aggregate used in building any general infrastructure (e.g. Yaoundé, pers. obs. 2006).

Management suggestions: consideration should be given to formal protection perhaps as a National Park, of a good example of the inselbergs around Yaoundé since they have such a large number of threatened species and are under threat.

COSTACEAE

assessed by Martin Cheek

Costaceae were included in the Zingiberaceae in Flore du Cameroun and Fl. Gabon accounts by Koechlin (1964 and 1965), sharing many similarities. However, *Costus* in Cameroon always lack the strong ginger-smell that results when stems of Zingiberaceae are cut, and the leafy aerial stems almost always twist in a spiral. In the 45 years since Flore du Cameroun Zingiberaceae was published, many new specimens from previously unsurveyed areas have come to light. Paul and Hiltje Maas at Wageningen are currently revising the family for Africa, and undoubtedly additional species will come to light from their researches. It is also likely that species will be redelimited, and additional records will come to light for previously rare species. So the species assessed below as threatened will need re-assessing once the Maas's research is published.

Costaceae in Cameroon have comprised a single genus, *Costus* until recently, when the Maas's transferred *Costus englerianus*, unusual in lacking an aerial stem and being prostrate, to a separate genus, *Paracostus*. *Costus* species parallel *Aframomum* of Zingiberaceae in usually having leafy stems separate from the shorter flowering stems but the weedier species have flowers terminal on the leafy stems. Several species of *Costus* are epiphytic but those assessed below are all terrestrial forest understorey herbs. *Costus megalobractentus* K. Schum. (Grand Batanga *Braun* 1888); *Costus nematotrichus* K. Schum. (Grand Batanga, *Dinklage* 954); *Costus ubangiensis* Gagnep. (SE Cameroon, *Mildbraed* 4180 and CAR) are poorly known species which, if proven to represent distinct taxa, are likely to be assessed as threatened in future. *Costus dinklagei* K. Schum (SE Nigeria and Cameroon) and *C. letestui* Pellegr. (Cameroon and Gabon) both have slightly more than 10 locations, 11 and 13 respectively so are here assessed as Near Threatened.

Costus phaeotrichus Loes.

VU B2ab(iii)

Range: Cameroon (S: Bipindi; Nyabéssan; Kribi-Ébolowa. Littoral: Ebo (to be confirmed)) and Gabon (Nzogholoumou, Ogooué to Cameroon; Mbel; Abanga; Kinguélé).

Here *Costus phaeotrichus* is assessed as Vulnerable since seven locations (above; AOO 28 km² with 4 km² cells) are known with threats as below.

Terrestrial herb with pubescent rhizome and rhizome scales; leafy stems 30–60 cm, 3–7 mm diam., sheaths red-brown, pubescent, apex truncate to lacerate; leaves few or single, pseudopetiole 1 6 mm, brown pubescent; blade broadly elliptic 10–16 × 4–9 cm acuminate, long brown pubescent on both surfaces; ligule short, 1–1.5 mm; inflorescence borne on a separate spike, peduncle 1.5–2 cm; bracts ovate, pink, 3 cm long, each with two flowers; calyx membranous, 12 mm, 3-toothed; corolla white, 6 cm; lip 7 cm edge fimbriate, pink with yellow median band.

Habitat: understorey of evergreen forest.

Threats: forest clearance for timber and agriculture.

Management suggestions: rediscovery of this species in Cameroon is most likely in S Region where most locations are sited. Congo-Ma'an NP is a logical focus for conservation efforts since it is a protected area. Basic populational data should be collected as a baseline for future monitoring and to assess needs, if any, for management intervention.

Costus tappenbeckianus J.Braun & K.Schum.

VU B2ab(iii)

Range: Cameroon (S: Batanga; Campo Ma'an) Gabon (Mougouyanga; La Maboumi, Mayumbe-Bavili; Tsande, Lastoursville; Sibange) and Congo (Brazzaville) (Massif du Chaillu).

Here *Costus tappenbeckianus* is assessed as Vulnerable since seven locations (above; AOO 28 km² with 4 km² cells) are known with threats as below.

Terrestrial herb, rhizome stoloniferous, leafy stems 0.5 m; blade broadly oblanceolate to elliptic, 10 × 4.5 cm, acuminate, base attenuate then rounded at base, pseudopetiole several mm; upper surface pubescent along midrib, otherwise glabrous or sparsely hairy; lower surface red, with fine appressed pubescence, ligule obliquely truncate, 5 mm, with the sheath finely pubescent; inflorescence borne directly from rhizome, peduncle 1–3 cm, covered in sheaths with dentate margins; bracts ovate, red, thick, 1.5–2 cm, each with a single flower; calyx 8–10 mm, 3-toothed; corolla white, labellum 6 cm; pink with white splash.

Habitat: understorey of lowland evergreen forest.

Threats: slash and burn agriculture; mineral extraction; urbanisation (Sibange-Libreville).

Management suggestions: Campo-Ma'an seems the logical focus of conservation efforts in Cameroon, being a protected area. Basic populational data should be collected as a baseline for future monitoring and to gauge needs for management intervention, if any.

CYPERACEAE

assessed by Martin Cheek, Iain Darbyshire & Benedict Pollard

Within the monocotyledons, this is the most species-diverse family following the Orchids, and the economically vital grasses. Cyperaceae, or sedges, are mainly confined to grassland and woodland habitats, often those which are seasonally or permanently inundated. However, some species (*Scleria shielae*) are restricted to the bare rock of inselbergs, others occur only in trees (*Coleochloa domensis*), while *Cypers rheophyticus* only occurs at the edge of fast-flowing upland streams in forest. Generally Cyperaceae are absent from forest and cannot tolerate shade. The exceptions are the genera *Hypolytrum* and *Mapania* which are specific to this vegetation type. Most botanists, on viewing these in the wild for the first time, are unaware that they are sedges and conclude that they are one of several other monocot families! They have broad leaves, atypical of their family, as are the bambusoid forest grasses *Puellia* and *Guaduella* of theirs (see Gramineae). Disproportionately, *Hypolytrum* and *Mapania* species are of higher conservation importance since a larger fraction of their species are narrowly distributed, with few locations, compared with genera of grassland habitats.

Identifications of Cyperaceae in Tropical Africa are not easy. While Sheila Hooper produced an excellent account for FWTA 3 (1) (1968) and Haynes & Lye did the same for E Africa in 1982, no account exists for Cameroon, Gabon, nor for the Congo basin countries. Superb papers in preparation to fill these gaps were published by Jean Raynal of France, but he died tragically young in a car crash and so was not able to complete any revisions or flora accounts. More recently Kare Lye of Norway has expanded his interests in Cyperaceae from E Africa and delineated many new, rare species in Cameroon. Few genera have been revised across Tropical Africa in the last 100 years. Nelmes revised *Hypolytrum* in 1951 but this is now greatly in need of update. *Nemum* was revised recently in part by Larridon. But the larger genera, *Cyperus, Kyllinga, Pycreus, Bulbostylis* all await revision for Africa.

Only *Mapania* of moderate to larger genera, has been revised recently (Simpson, A Revision Of The Genus *Mapania*, RBG Kew, 1992).

Disputes in generic delimitation exist. Traditionally *Pycreus* and *Kyllinga* have been maintained as separated genera from *Cyperus* at Kew, and are generally easily separated on morphology. However, in most of our Conservation Checklists we have followed Lye, who has adduced credible evidence for uniting all three under *Cyperus*. Lye kindly identified most of the specimens for those works. However, in recent years, for personal reasons, his expertise has not been available, and so we are reverting to the traditional generic system.

Recognising Cyperaceae, and distinguishing them from grasses (Gramineae) and rushes (Juncaceae) is usually not difficult since their stems are 3-angled, and not round.

Bulbostylis schoenoides (Kunth) C.B.Cl. subsp. *erratica* (Hook.f.) Lye was assessed as LR nt (equating to NT) in Cable & Cheek (1998: lxvi–lxvii). Its range is Mt Cameroon (c. nine pre-1988 coll.), Bamenda Highlands, Chappal Waddi and a specimen from the Loma Mts (Sierra Leone). If reassessed it might rate threatened.

Carex preussii K.Schum. was assessed as LR/nt (equating to NT) in Cable & Cheek (1998: lxvii) and Cheek *et al.* (2000: 77). Its range is Mt Cameroon (two pre-1988 coll.), Bamenda Highlands, particularly Mt Oku (nine pre-1996 coll.) & Mambila Plateau (one coll.). If reassessed it might rate threatened..

Bulbostylis sp. nov. referred to in Cheek *et al.* (2000: 77) as a possible new species from Laikom Ridge (*Pollard* 293, Dec. 1998) subsequently proved to be the E African *B. pusilla* (A.Rich.) C.B.Clarke subsp. *yalingensis* (Cherm.) R.W.Haines and is believed LC.

Cyperus niveus Retz. var. nov. (*Cheek* 9810, *Pollard* 354, both November 1998) referred to in Cheek et al. 2000: 78 as possibly new, subsequently proved to be *Pycreus scaettae* Cherm. and is believed LC.

Bulbostylis densa (Wall.) Hand.-Mazz. var. ***cameroonensis*** Hooper

EN B2ab(iii)

Range: Cameroon endemic (SW: Mt Cameroon (1 site, 2 coll.); Mt Kupe (1 coll.). Littoral: Mt Mwanenguba [Massif du Manengouba] (1 coll.). NW: Bali Ngemba FR (1 coll); Dom (1 coll)).

Annual herb to ± 30 cm; stem deeply grooved, 0.2–0.4 mm thick; leaves canaliculate, grooved, 0.2–0.3 mm broad; inflorescence usually a compact umbel, somewhat contracted, with 3–8 shortly-pedicellate spikelets, each one 2–5 × 1.5–3 mm; glumes few, each standing out from its neighbour, dark brown with conspicuous pale green or grey midrib.

Care must be taken not to confuse this plant with the sympatric typical variety. Hooper in FWTA notes: "Differing from all the other species in Africa in having spikelets on short pedicels producing a compact inflorescence and glumes with a distinctly excurrent nerve".

Until the 1990s, this variety was known only from the type collection (*Mann* 1360b) on Mt Cameroon), believed to be from the Mann's Spring area. It was then rediscovered there in 1992 (*Thomas* 9407), (Cheek, in Cable & Cheek (1998:

lxvi, 153) where assessed as VU D2). Subsequently it has been recorded at other montane grassland sites. The taxon is here reassessed as Endangered since there are five currently known sites (AOO 20 km² with 4 km² cells, threats as below) which are highly isolated and limited in size, thus they are still vulnerable to local stochastic change, such as lava flow on Mt Cameroon or natural fires, and the threats reported below.

The assessment above is updated from that of VU B2ab(iii) by Darbyshire *et al.* in Cheek *et al.* (2004: 190), Harvey *et al.* (2004: 70–71) and Cheek *et al.* (2010: 105–106).

Habitat: upper submontane and montane grassland; 1800–3000 m alt.

Threats: trampling by cattle in the northerly three locations.

Management suggestions: more data on the size of each population are required to assess further its vulnerability to local stochastic events. Further botanical inventory work on montane grassland sites in e.g. the Bamenda Highlands, or Mt Nlonako in Littoral Region, may reveal further populations, in which case the conservation status would be downgraded.

Coleochloa domensis Muasya & D.A.Simpson [Njing-Njing (Noni)]

CR D

Range: Cameroon (NW: Dom, two sites).

Tropical Africa's only obligate epiphytic sedge was discovered by Prof. M. Muasya while identifying specimens of this family (collected two years previously) at Kew in late 2008. So far, it is unique to Dom.

This species was assessed in Cheek *et al.* (2010: 106) as Critically Endangered (CR)) since it is only known from two sites at one location (AOO 4 km² with 4 km² cells), one of which is under threat of forest clearance due to agricultural pressure. This assessment is maintained here.

Epiphytic herb, 30 cm, erect, tufted; culms c. 6-leaved; leaf base sheath 2–4 cm, apex long-hairy; blade 25 × 0.4 cm; inflorescence spike-like, 6–8 × 1 cm, brown; flowers 3–5-clustered, c. 7 mm long, erect, cluster stalks to 1 cm; styles long exserted by 3 mm, black, recurved, conspicuous.

Habitat: submontane forest; 1600–1830 m alt.

Threats: this species, currently, is known from only three trees. While efforts are made to protect forest by the Dom community, some tree-felling continues. Overall it is estimated that as much as 96.5% of the original montane (including submontane) forest of the Bamenda Highlands has been lost (Cheek in Harvey *et al.* 2004: 9). GIS-based studies within another area of the Bamenda Highlands show that the loss of forest cover continues, with 25–30% lost in the area around Mt Oku between 1987–1995 (Moat in Cheek *et al.* 2000: back cover). Surveys in other surviving submontane forest fragments of the Bamenda Highlands carried out by British and Cameroonian scientists in recent years have not yet revealed further sites for this species.

Management suggestions: the Apicultural and Nature Conservation Organisation (ANCO), a non-governmental forest conservation organisation based in Bamenda, has alerted the population of Dom, which manages the forest fragment inhabited by *Coleochloa domensis*, to the existence of this rare species and its conservation importance. This needs to be followed up. Consideration should be given to seedbanking the species for long term protection, and to multiplying the species in a nursery to strengthen the wild population.

Cyperus microcristatus Lye

CR B2ab(ii,iii)

Range: Cameroon endemic (SW: Mt Kupe (1 coll.)).

Perennial herb to 50 cm; rhizome short, slender, with many densely set stems; culms 30–50 cm × 0.5–1.0 mm; leaves 3–4 per culm; blades 5–12 cm × 1.5–2.5 mm; inflorescence a congested greyish to pale brown head, 10 × 10 mm, consisting of 2–3 crowded spikes; involucral bracts 3–4, foliaceous; central spike 6–8 × 4 mm, with numerous crowded spikelets.

Despite the extensive botanical inventory work on Mt Kupe, this taxon is currently known from only a single collection (*Patterson* 11) collected in 1995 near Kupe Village, and is thus likely to be very rare within its extremely limited distribution. The assessment by Darbyshire *et al.* in Cheek *et al.* (2004: 190) is maintained here. CR is warranted since there is a single currently known site (AOO 4 km² with 4 km² cells, threats as below).

This species most closely resembles the sympatric *C. densicaespitosus* Mattf. & Kük., differing in the more perennial habit, and the smaller spikelets and glumes which are more prominently winged.

Habitat: roadsides in agricultural area, 500 m alt.

Threats: intensification of small-scale agriculture on Mt Kupe below 1000 m alt. is resulting in an inferred decline in habitat quality within this taxon's distribution. The only known site is by a road, thus it is vulnerable to disturbance, as dirt roads are often annually "improved" following wet season damage.

Management suggestions: a survey of the previous collection site and other suitable locations on the lower slopes of Mt Kupe is imperative in order to rediscover and assess the population size of this taxon. The extent to which it can tolerate disturbance should be studied further, as in its one known location, it was found in an area of relatively high disturbance.

Cyperus rheophyticus Lye

EN B2ab(iii)

Range: Cameroon (SW: Mt Cameroon at Njonji; Mungo R W of Mbu (1 coll.); Bakossi Mts (4 coll.)).

Slender rheophytic herb; culms crowded, 5–30 cm × 0.3–0.4 mm; leaves from the lower 8 cm only; largest blades to 10 cm × 0.5–1.3 mm, flat; inflorescence a single, terminal, globose, whitish, congested anthela, c. 3 mm diam.; involucral bracts usually 3, foliaceous, the largest 2–6 cm × 1.0–1.4 mm; spikelets 1.5–1.7 × 0.7–0.9 mm, 1-flowered. On rocks and stones in or beside streams and rivers, usually submerged during rains.

This species was first collected from the Mungo R in 1986 (*Manning* 542), and has subsequently been collected along the River Jide at Muambong (*Onana* 585) and near Nyale (*Cheek* 10405), at a stream near Kodmin (*Etuge* 4063), and at Ndile waterfall, Baseng (*Cheek* 10405).

The assessment of VU B2ab(iii) by Darbyshire *et al.* in Cheek *et al.* (2004: 190–191) is uprated here to EN since the sites above equate to only four locations (AOO 4 km² with 1 km² cells since an aquatic species, threats as below).

Habitat: on rocks and stones in or beside streams and rivers, usually submerged during rains; 500–1350 m alt.

Threats: illegal logging activity in catchment areas in the Bakossi Mts is likely to result in increased run-off and thus higher river levels with large sediment loads which may wash away existing populations of this taxon or destroy its habitat through excessive silting. At Mt Cameroon forest clearance for expanding plantation agriculture has been mooted at the Njonji site and poses a similar threat.

Management suggestions: this species should be sought for in all future botanical work along rivers within its range. Monitoring of populations in catchments affected by land-use change, such as the Mungo R, should be carried out to assess this taxon's sensitivity to increased run-off or silting.

Cyperus tenuiculmis Boeck. subsp. ***mutica*** Lye ined.
CR B2ab(iii)

Range: Cameroon (SW: Nyasoso, Mt Kupe (1 coll.)).

This new subspecies, still undescribed, has been collected only once, as recently as 1998 (*Pollard* 113), when it was found in a ditch by the Tombel to Bangem Rd in Nyasoso town, where it was recorded as common. However, due to the lack of other records of this taxon from this area despite extensive botanical inventory work, and the significant threat to the known location (see below), it is considered Critically Endangered. The assessment by Darbyshire *et al.* in Cheek *et al.* (2004: 191) is maintained here. CR is warranted since there is a single currently known site (AOO 4 km² with 4 km² cells, threats as below).

Habitat: damp roadside ditch; [illegible]

Threats: the Tombel to Bangem [illegible] route and so is prone to degradation [illegible] of heavy rain. Annual repairs are therefore made, including redigging of the drainage ditches. The only known population of this taxon is therefore severely threatened, if not already lost.

Management suggestions: the site should be revisited in order to determine if the population is extant; care should be taken to distinguish it from other *Cyperus* spp.; publication of the formal description is therefore important.

Hypolytrum pseudomapanioides D.A.Simpson & Lye
EN B2ab(iii)

Range: Cameroon (SW: Bakossi Mts (2 coll.)).

Robust tussocky perennial; culm 1, erect, central, 60–70 cm × 2–3 mm; leaves mostly basal, 1 cauline, to 100 × 1.3–1.6 cm; inflorescence an irregular subspherical brownish white head, 2.5–3.5 × 3–3.5 cm, comprising numerous, crowded, fairly distinct spikes; involucral bracts usually 4, foliaceous, very unequal, the largest to 50 × 1.6 cm; spikes ovoid to lanceolate, 10–14 × 3–5 mm, pale brown

This taxon is known from only 2 collections made in 1998, one 0.5 km from Kodmin towards Ndip (*Cheek* 8919) and the second on the path from Kodmin to Muawhojom (*Ghogue* 33). It has not been recorded from similar habitat on the adjacent Mt Kupe or elsewhere in the Cameroonian highlands. It is considered Endangered due to the highly restricted area of occupancy (<500 km^2) and the perceived decline in habitat quality (see below). The assessment by Darbyshire *et al.* in Cheek *et al.* (2004: 191) is maintained here. EN is warranted since there are two currently known sites (AOO 8 km² with 4 km² cells, threats as below).

Recent molecular studies (Simpson *et al.* 2003) have cast doubt over the separation of the genus *Hypolytrum* from *Mapania*; future, more detailed, work may support their amalgamation into a single genus.

Habitat: montane forest undergrowth; 1470–1500 m alt.

Threats: localised encroachment of small-scale agriculture along the paths from which the 2 collections were made threaten these populations, though this is limited at present.

Management suggestions: further investigation of sedge taxa in the montane forest of the Bakossi Mts may reveal further populations which are less threatened by agricultural encroachment. Care should be taken to separate this species from the superficially similar *H.subcompositus* Lye & D.A.Simpson.

Reference: Simpson, D.A., Furness, C.A., Hodkinson, T.R., Muasaya, A.M. & Chase, M.W. (2003). Phylogenetic relationships in *Cyperaceae* subfamily *Mapanioideae* inferred from pollen and platid DNA sequence data. Amer. J. Bot. 90: 1071–1087.

Hypolytrum secans (K.Schum.) J.Raynal (syn. *H. rhizomatatum* Cherm.)
VU B2ab(iii)

Range: Cameroon (S: Lolodorf; Col. Ngoa 16 km SE de Djoum; Col. Ongongo, près Mbanga (Kribi-Ébolowa, km 81)) and Gabon (four loc.).
Here *Hypolytrum secans* is assessed as Vulnerable since seven locations (above; AOO 28 km² with 4 km² cells) are known, with threats as below.
Herb, leaves several, probably erect in life, strap-like, c. 118 x 24 cm, apex acute, basal sheath 15 cm, blade 3-nerved at base, margin toothed, plicate in cross-section; inflorescence globose, 1 cm, pale brown, with numerous sessile spikes, capitulum subtended by three unequal foliaceus triangular bracts, the largest 14 x 4 mm, axis 15 cm long, 1.5 mm wide.
Habitat: lowland and submontane evergreen forest; alt. to 950 m.
Threats: slash and burn agriculture.
Management suggestions: rediscovery of this poorly known species should be attempted. If successful, basic populational data should be collected as a baseline for future monitoring.

Hypolytrum subcompositus Lye & D.A.Simpson
CR B2ab(iii)

Range: Cameroon (SW: Kodmin, Bakossi (1 coll.)).
Robust, rhizomatous perennial; culms 1-several, erect, lateral, 40–50 cm × 1.8 mm; leaf-blade linear, flat, mid-green, coriaceous, 60–120 × 1.3–2 cm; inflorescence a congested terminal subhemispherical corymb with up to 20 spikes and a few major branches still visible, 1.5–3 × 2.5–3.5 cm, arising laterally from rhizome, capitate; flowers spicoid with 2 floral bracts and 2 stamens.
Known only from the type collection from Kodmin in the Bakossi Mts, collected in January 1998 (*Etuge* 4007), this taxon is clearly very rare within its range, as it has not subsequently been found in Bakossi despite extensive survey of its montane forest habitat. The assessment by Darbyshire *et al.* in Cheek *et al.* (2004: 191) is maintained here. CR is warranted since there is a single currently known site (AOO 4 km² with 4 km² cells, threats as below).
Habitat: montane forest undergrowth; 1500 m alt.
Threats: due to its highly restricted area of occupancy, this species is highly threatened by local disturbance, such as small-scale agricultural encroachment which is widespread in the Bakossi Mts
Management suggestions: rediscovery, and a subsequent census, of the known population is a priority; subsequent surveys of sedge taxa in the Bakossi Mts may reveal further sites.

Hypolytrum unispicatum Sosef & D.A.Simpson
CR B2ab(iii)

Range: Cameroon endemic (S: Campo Ma'an area at Mt Elephant).
Robust perennial with a thick horizontal rhizome at least 8 cm long, 1–2 cm diam. including old sheath bases; scale-leaves ovate-lanceolate 3–5 × 1–3 mm, acute, brown; leaves basal, linear 63–125 × 0.8–1.5 cm, apex acute, base merging into sheath gradually, 3-nerved, slightly plicate in cross-section, margin scabrid, pseudopetiole absent; sheath oblong, 6–12 × 1.2–1.8 cm, inflorescences of a single spike each borne on the c. 20 slender stalks arising from the rhizome behind the leaves, each 3–15 cm long, 0.5 mm wide, 3-angled; spikes themselves ellipsoid to globose, 3–5 × 1.5–4 mm, brown, of elliptic bracts 1.5–3 × 1–1.5 mm.
This is truly a bizarre species, with numerous thread-like stems bearing single terminal spikes arising from behind the leaves on the stout horizontal rhizome. It is known only from two records, "both made in ignorance of the other, at Mt Elephant, in 2001 and 2002 (Sosef & Simpson, Blumea 50: 523–525 (2005)).
Habitat: lowland evergreen forest; 200 m alt. on a steep slope.
Threats: Mt Elephant is believed to be an iron ore deposit – mining is a threat.
Management suggestions: data is needed above the range of this species, its population size, regeneration, and detailed threats. *Hypolytrum unispicatum* appears unique to Mt Elephant, along with such species as *Begonia montis-elephantis*. It should be constituted as a protected area for plants. Efforts should be made to involve local communitites in protecting these species, and to attempt to propagate these species in cultivation for reintroduction, if needed.

Mapania africana Boek. subsp. *africana*
VU B2ab(iii)

Range: Cameroon (S: Bipindi; Kribi, 10 km N; 2–10 km S; 34 km S; Batanga; 25 km E Batanga), Equatorial Guinea (Rio Muni: St John) and Angola (Cabinda, Magumbe, M'bulu Hills).
Robust herb with rhizome 1 cm diam, scales-leaves ovate to lanceolate, to 4.5–0.3 cm; leaves strap-like c. 100 × 2.3–4 cm, apex abruptly acuminate, base abruptly narrowed into a pseudo-petiole 5–19 cm; basal sheath 7–9 × 1–2 cm, inflorescences axillary, peduncles erect 11.5–42 × 0.2 cm, spikes (3–)5–12 in capitate head 0.9–1.5 cm wide, lacking foliose bracts.
Here assessed as Vulnerable since six locations are known (above; AOO 24 km² with 4 km² cells) with threats as below.
Habitat: lowland evergreen forest.
Threats: infrastructure for extractive industries; urbanisation; tourism developments: all in the Kribi area, the stronghold for this species; slash and burn agriculture in the rural areas around Kribi.
Management suggestions: rediscovery of this subspecies at the locations noted above should be essayed, and basic

populational data gathered as a baseline for future monitoring.

Mapania africana Boek. subsp. ***filipes*** (J.Raynal) D.A.Simpson

EN B2ab(iii)

Range: Cameroon (S: 10 km ESE Campo) and Gabon (Cristal Mts. Balakabu)

Robust herb, as *M. africana* subsp. *africana*, but peduncles more slender, 0.4–0.7 mm, curved-deflexed spikes per head 1(–2). Here assessed as Endangered since only two locations are known (above; AOO 8 km² with 4 km² cells) with threats as below.

Habitat: lowland evergreen forest.

Threats: slash & burn agriculture.

Management suggestions: potentially both known locations for this taxon are in protected areas: this should be investigated and confirmed before basic populational data is collected with involvement of local communities. Future monitoring should be collected against this baseline.

Mapania ferruginea Ridl.

VU B2ab(iii)

Range: São Tomé & Príncipe (São Tomé: Contador (1 coll.); Macambrá, Vanhulst (1 coll.); Casa del Pico (2 coll.); Callario (1 coll.); unlocated (1 coll.). Príncipe: W of Pico (1 coll.)) and Cameroon (SW: Bakossi (1 coll.); Mt Kupe (2 coll.)).

Moderately robust herb; culm solitary, erect, central, 35–45 cm × 2–3 mm, subtriquetrous, glabrous; leaves basal, to 75 cm; blade linear, 40–60 × 2.6–2.9 cm; pseudopetiole absent; inflorescence terminal, globose, 2–3.5 cm wide, mid reddish brown, consisting of numerous spikes.

The collections above Kupe Village on Mt Kupe (*Cable* 3773 and *Etuge* 2853) in July 1996 represented the first records of this species on continental Africa. It has subsequently been found once in the Bakossi Mts at Kodmin. It was previously known only from São Tomé & Príncipe where it was recorded as "a rare species…confined to fairly high altitudes" (Simpson 1992). The assessment by Darbyshire *et al.* in Cheek *et al.* (2004: 191–192) is maintained here. VU is warranted since there are seven currently known locations (AOO 28 km² with 4 km² cells, threats as below).

Care should be taken to separate this taxon from the similar *M. soyauxii* (Boeck.) H.Pfeiffer which is recorded at lower altitudes in Cameroon; notes on the separation of these taxa can be found in Simpson's *Revision of Mapania*. RBG, Kew (1992).

Habitat: montane forest [illegible]

Threats: the subpopulation at the lower altitudinal range on Mt Kupe (1100 m) is threatened by future encroachment of agriculture and resultant forest clearance. Montane forest on São Tomé is also threatened by small-scale human encroachment.

Management suggestions: improved protection of the montane forest habitat on Mt Kupe. A survey of the São Tomé populations, this species' stronghold, should be carried out and important areas protected where necessary.

Mapania raynaliana D.A.Simpson

EN B2ab(iii)

Range: Cameroon (S: Batanga; S of Zingui; Nkoolong) and Gabon (Cristal Mts).

Here *M. raynaliana* is here assessed as Endangered since four locations are known (above; AOO 16 km² with 4 km² cells) with threats as below.

Robust herb from rhizome 0.5 cm diam.; leaves all basal to 116 cm, blade 79–103 × 3.8–5.4 cm, apex acute, base gradually narrowed into a sheath 14 × 3 cm, pesudopetiole absent; inflorescences at rhizome apex, central, peduncle 53–76 × 0.4–0.6 cm, 3-angled, bracts 3–4, foliose, unequal, 2.5–46 × 1.1–5 cm; spikes very numerous forming a dense, globose head 3.5–5.5 cm wide.

Habitat: lowland to submontane forest.

Threats: slash and burn agriculture; touristic development (Cameroon).

Management suggestions: rediscovery of this species at its known localities followed by collection of basic populational data should be attempted to form a baseline for future monitoring.

Nemum megastachyum (Cherm.) J.Raynal

EN B2ab(iii)

Range: Cameroon (Adamawa: 15 km SSW of Meïganga at Dankali) and CAR (four locations).

Here assessed as Endangered since only five locations are known (above; AOO 20 km² with 4 km² cells) and threats as below.

Robust, densely tufted, strongly rooted glabrous herb 75 cm tall, leaves all basal, curving, c. 35 × 0.15 cm; peduncles terete, numerous, to 70 cm, terminated in an umbel of 4–5 black, subglobose heads each 11 mm diam., comprised of obovate black & brown scales 4 mm long, apex filiform; partial-peduncles subsessile (one per inflorescence) or up to 3.5 cm; bracts subtending umbel 1–2, filiform, unequal, to 2.5 cm. Elevated to species level by Raynal in Adansonia 13(2): 145–171 (1973).

Habitat: inselbergs and bare, seasonally wet rock surfaces.

Threats: trampling by cattle.

Management suggestions: attempts should be made to rediscover this species and to collect basic populational data as a baseline for future monitoring. Local community leaders should be apprised of the importance of this plant. Work is needed on the delimitation of this taxon.

Scleria afroreflexa Lye
EN B2ab(iii)
Range: Cameroon (SW: Bakossi Mts (1 coll.). NW: Boyo (1 coll.), Bali Ngemba FR (2 coll.)).
Delicate annual; culms 10–50 cm × 0.3–0.8 mm; leaves 2–4 per stem, but only 1–3 perfecting leaf-blades; sheaths densely covered by retrorse white hairs 0.2–0.4 mm; blades 2–9 cm × 0.8–1.8 mm; inflorescence 3–9 × 1–2 cm, appearing spike-like with sessile glomerules above, but in fact a narrow panicle with 1-several reflexed branches with 1–3 glomerules below; glomerules 4–5 × 3–6 mm, consisting of 2–10 spreading spikelets.
This conservation assessment was originally made by B.J.Pollard in Lye & Pollard (Nordic J. Bot. 23(4): 431 (2005)). The updated assessments by Darbyshire *et al.* in Cheek *et al.* (2004: 192) and in Harvey *et al.* (2004: 71) are maintained here. EN is warranted since there are three currently known locations (above, AOO 12 km^2 with 4 km^2 cells, threats as below). It was first collected in 1999 in two disjunct locations 170 km apart in western Cameroon: between Laikom and Fundong in Boyo Division, NW, and near Kodmin in the Bakossi Mts, SW. It was recorded as locally common at the latter site. Further subpopulations are likely to exist in locations that have not yet been included in botanical inventories, given the large areas of grassland at altitudes of around 1500m in the Bamenda Highlands, Bamboutos and, to a lesser extent, in SW (Pollard, *pers. obs* 2004.).
Habitat: montane grassland and grassland patches in submontane forest; 1450–1550 m alt.
Threats: trampling by cattle and other livestock, and burning of the montane grasslands in NW may result in subpopulations losses or may lead to long-term habitat changes which do not favour this taxon. The site near Kodmin appears relatively unthreatened though may be lost to forest encroachment if human disturbance remains low.
Management suggestions: studies of the subpopulations in the NW Highlands could be made to better understand the ecology of this taxon, including its tolerance of fire and human disturbance. This species should also be searched for in suitable habitat elsewhere in western Cameroon; discovery of further subpopulations would lead to a downgrading of its conservation status.

Scleria shielae J.Raynal
CR B2ab(iii)
Range: Cameroon endemic (C: Yaoundé, N'kolbison, Mt Minloa).
Here assessed as Critically Endangered since only a single location (above; AOO 4km^2 with 4km^2 cells) is known with threats as below.
Herb 70 cm tall, probably from a rhizome, culms slender 1 mm wide, each bearing 2–6 leaves, basal leaves abrect, cauline leaves narrowly strap-like, 10 × 0.2 cm, sheath c. 2.5 cm; panicle 15 cm, open, peduncle 9 cm, branches c. 5, spreading, each 3–4 cm long, interupted, spike clusters 3–5 per branch mid-brown, 1–3 spikes per cluster.
Habitat: inselbergs, bare rock in the carpet around *Microdracoides* in the semi-deciduous forest belt; 800 m.
Threats: extraction of granite for construction aggregate (pers. obs. 2006).
Management suggestions: rediscovery of this species, followed by a census of basic populational data, should be attempted to form a baseline for future monitoring. Protection of selected inselberg areas in and around greater Yaoundé is advisable given the large numbers of very rare species that they support.

DRACAENACEAE

assessed by Martin Cheek advised by Geoffrey Mwachala

The genus *Dracaena*, here classified under the Dracaenaceae, has also been placed under several other family names in recent years: Agavaceae (FWTA), Liliaceae, Asparagaceae, Ruscaceae and Convallariaceae. Current molecular phylogenetic studies show that the tree *Dracaena* species of the Canary islands and Socotra belong to a different lineage to those of subsaharan forest which are more closely related to *Sanseviera* – a genus of open habitats, the species of which lack aerial stems and have succulent leaves.

In Cameroon, *Dracaena* are strictly species of forest. Apart from a few species, such as *Dracaena phrynoides*, they generally have aerial, woody stems, unusual in the monocots. The woody species vary from branched 1 m forest undershrubs to massive canopy trees in submontane forest (*Dracaena arborea*). The leaves are leathery, flattened, with or without a basal constriction or petiole. Leaf insertion is usually alternate-spiralled. Flowers are produced in an interupted terminal spike, the petioles tube being white, long tubular, with six lobes, each with a stamen attached. The fruits are fleshy orange or red, globose or 3-lobed and seeded berries, often reduced to 1 or 2 lobes by abortion. Many species are cultivated as ornamental plants in gardens.

J.J. Bos of Wageningen began a programme to revise the Tropical African species of *Dracaena*. His brilliant *Dracaena* in W Africa (Agric. Univ. Wageningen Papers 84-1 (1984)) achieves this for the countries W of Cameroon. Sadly he did not live to complete a similar work for Central Africa. Geoffrey Mwachala of Nairobi has now taken on this role. Much data in this account is derived from Bos, 1984.

Dracaena viridiflora Engl. & K.Krause (syn. *D. mannii* sensu FWTA p.p.) was assessed in Cheek *et al.* (2004: 192) as VU B2ab(iii) on the basis that seven locations were known, principally in Cameroon. However, access to the Gabon database has revealed that four locations occur in that country, and since they derive from Bos determinations, cannot be doubted. Therefore this species is here downrated to NT.

Dracaena ledermannii Engl. & K. Krause, based on *Ledermann* 1483 (B) from Santchou is of uncertain status, the type having been mislaid or destroyed. It may represent a threatened species if it can ever be delineated confidently. Additional, new threatened species are in the process of being discovered and published. One such is *Dracaena mokoko* restricted to the Mokoko reserve of Mt Cameroon.

Dracaena bicolor Hook.f.
VU B2ab(iii)
Range: Nigeria (Oban; Calabar), Equatorial Guinea (Bioko (4 coll.)) and Cameroon (SW: Mt Cameroon at Bimbia; Kumba. S: Bipindi), Gabon (Makokou; Monts de Cristal) and Congo (Brazzaville) (Massif du Chaillu).
Single-stemmed shrub 1–2 m; stems erect, to more than 1 cm diam.; leaves spread along the stem, but more congested towards the apex, oblanceolate, 10–35(–55) cm long. 4-10 cm wide in the distal part, apex acuminate, with a mucro to 5 mm, the proximal part narrowed into a long, winged pseudo-petiole 1.5–10(–16) × 0.4–0.8 cm, the basal sheath 1–2 cm long, dilated and clasping the stem; ribs absent; inflorescence terminal, erect, subcapitate to thyrsoid, 4–12 cm; perianth tube white, c. 1.5 cm, lobes 1 cm; fruit globose c. 1 cm.
Here *Dracaena bicolor* is assessed as Vulnerable since eight locations are known (above; AOO 32 km² with 4 km² cells), with threats as below.
The species is based on material collected between 100 and 150 years ago in Nigeria and Bioko. It has not been seen since in Nigeria. During the interval, large tracts of forest habitat have been degraded and lost (see below). The record at Kumba is based on the supposition that *D. preussii* (for which this is the type location) is a synonym. This needs confirmation. The Bimbia record is based on a sterile specimen annotated as *D.* cf. *bicolor* by Bos: it needs more research also.
The records from Gabon have not been seen but can be accepted since they are determined by Bos.
Habitat: lowland evergreen forest.
Threats: forest clearance for timber plantation agriculture (bananas, oil palm and rubber) and small-holder agriculture: all occur in the Mt Cameroon area; urnaisation at calabar may have resulted in loss of the species there.
Management suggestions: further investigation of the supposed records in Cameroon (see above) is required to confirm the presence of the species. If refound, seed or cuttings should be made without compromising the wild plants, so that propagation for multiplication at Limbe BG for reintroduction to safe sites can take place. Basic populational data should be collected if possible, but this appears to be such a rare plant that densitites may be less than one individual per 10 km².

Dracaena bueana Engl.
DD
Range: Mt Cameroon (one pre-1988 coll., possibly extinct) and Ghana (one coll.).
This species commemorates the town of Buea, capital of the Bakweri, and was described from the conservatory in Berlin, based on a specimen (*Deistel* 461) collected at Buea in flower in Feb. 1900, reportedly growing in a hedge at c. 1000 m alt. *Dracaena bueana* has not been identified from the mountain since and its taxonomic status is not yet resolved. Bos, the authority on the genus, had no knowledge of this plant until 1984, when Leeuwenburg brought a specimen from Ghana, which at that time was judged to be be either *D. fragrans* or *D. arborea*: they cannot be distinguished when sterile. However, as the plant developed, the leaves became twice as long as those of *D. arborea*, so it was suspected of being a new species. Eventually it flowered and Bos identified it as Engler's *D. bueana.* Its reproductive characters are intermediate between the two tree species mentioned above, and it may be of hybrid origin, however, the plant is fertile, and produces fruits. Living seedlings were distributed by Bos to a number of Botanic Gardens, including Limbe and Ghana (Bos pers. comm.). Collections of fertile tree *Dracaena* are needed if we are to determine whether *Dracaena bueana* persists on Mt Cameroon, and whether also *D. arborea* does. *D. fragrans* (more often shrub-like than tree-like) is already known from the mountain, as is *D. mannii*, another arborescent species, though from only a single specimen (*Thomas* 4507). There is no doubt that the difficulty in making good herbarium specimens of tree *Dracaena* has deterred most plant collectors from collecting what is all to easily assumed to be " *Dracaena arborea"*. Only the gathering of further fertile specimens will enable us to sort out the tree *Dracaena* on the mountain and establish which are rare and in need of protection. In the Bakossi mountains, tree *Dracaena* are propagated and are planted in lines as barriers to livestock, thus indirectly conserving the species concerned (pers. obs.), a case of conservation through cultivation!
Habitat: "Buea, im hohen lichten Wald, im Buschwald und an freien Plätzen, ungefähr um 1000 m, wo sie oft vereinzelt emporragt " (*Deistel* 461, quoted by Engler in the species protologue in Engl. Jahrb. lix. Beibl. 131, 20 (1924).
Threats: unknown.
Management suggestions: further research into the survival and taxonomic delimitation of this species is needed, as

outlined above, before its conservation status can be assessed. This account derives from Cheek in Cable & Cheek (1998: lxvii-lxviii). No new data has been acquired since that time.

Dracaena kupensis Mwachala, Cheek, Eb.Fischer & Muasya (syn. *D.* cf. *phanerophlebia* Baker, Bos, Mwachala & Cheek (2004: 441).
EN B2a,b(iii)

Range: Cameroon endemic (SW: Etinde, Mt Cameroon; Kupe-Mwanenguba).

Dracaena kupensis is only known from c. 11 specimens at three broad localities, at Mt Etinde (three specimens), at Mt Kupe: just above Nyasoso (six coll.), Kupe village and in the adjoining Bakossi Mts at Kodmin. Its area of occupancy is estimated as 20 km² based on these four sites but including two cells for Nyasoso since it occurs along several paths there over a larger area than a single cell. Accordingly it is here maintained as Endangered, following broadly the assessment given in the protologue since although an extra location (Mt Etinde) has since come to light, the EN threshold has not been exceeded. Threats are given below.

Monocaulous shrublet 30–60 cm tall from a sparsely branching, horizontal wiry-woody underground rootstock, glabrous; aerial stem erect, very short, terete, 0–9 cm long, 0.4 cm diam., enclosed by sheathing leaf bases and prophylls, the base of longer stems often naked, marked only by girdling white leaf scars; leaves pseudopetiolate, ascending weakly, distributed along the short stem, 3–7, alternating with the prophylls, forming a loose, litter-gathering rosette, blades elliptic to elliptic-oblong, 20–28 × 7.4–12 cm, apex acuminate, or acute, acumen if present 1–3 × 1 cm, base continuous with the pseudopetiole, midrib inconspicuous above, highly conspicuous and prominent on the lower surface, secondary nerves not conspicuous, not raised or sunken; pseudopetiole winged, 16–21 cm long, midrib 1–2 mm wide, wings 1.5–6 mm wide, more or less of even width in each pseudopetiole, but dilating at the base to sheath the stem; prophylls sheathing the stem, papery, foliose, scarious or green, oblong-elliptic, 1.5–5.5 cm long, 1–1.5 cm wide with a hyaline margin c. 2 mm wide; flowers not known; infructescence, terminal, or sometimes arising directly from the stem on short leafless branches, racemose, erect; peduncle red or purple, terete, but grooved on drying, 6–8 cm long, 0.15 cm diam., 4-2-fruited, peduncle bracts 7–8, entirely sheathing the lower few cm of the peduncle, ovate-concave, 1–6 × 0.8–3 cm, apex triangular; bracts subtending the pedicels triangular, 2 mm long, apex long-acuminate, base c. 1 mm wide; pedicels grouped in fascicles, terete, 3–10 mm long, apex dilated, minutely lenticellate; calyx persistent, campanulate, 2 × 3.5 mm, apex 6 lobed, lobes truncate or acute, margins thickened. Fruits ripening glossy orange from green, fleshy when live, the pericarp 1–2 mm thick, indehiscent, 3-seeded, globose or obscurely beaked, to 1 cm in diam., with a short stipe; seeds dull white, hemispherical, 8 × 8 mm.

Habitat: forest undergrowth in both primary and secondary submontane forest, 800–1250 m.

Threats: at its lower altitudinal range its habitat is vulnerable to forest clearance for agricultural expansion.

Management suggestions: Nyasoso is a logical base for conservation of this species since the evidence is that it is most common in the forests above the village. Basic populational data should be collected as a baseline for future monitoring and to gauge whether conservation action is needed or not. There is potential for multiplying the species from seed for strengthening wild populations if this appears advisable. This species also has potential as a pot plant so propagation efforts in that direction could also help safeguard the species.

Dracaena goldieana Masters & Moore
EN B2ab(iii)

Range: Nigeria (Oban FR; Uwet; Orem) and Cameroon (SW: Korup NP), possibly Gabon.

Discovered by the Rev. Hugh Goldie in Calabar who sent plants to Edinburgh in 1870–1871 from whence it became widely cultivated for its spectacular variegated leaves.

Shrublet 30–60 cm, stem diam. to 1 cm, leaves along length of stem; leaf-blade ovate or elliptic-oblong, 18–27 × 4.5–6.5 cm, acumen with 3–5 mm long threat-like mucro, base rounded; false petiole well-defined, 3–7 cm, sheath 1.5–3 cm long, clasping stem for its circumference; upper surface of blade dark green with transverse white bands, midrib raised; flowering material not known.

Here assessed as Endangered since only four locations are known (above; AOO 16 km² with 4 km² cells) and threats being as below.

Habitat: lowland evergreen forest.

Threats: forest clearance for timber and agricultural land particularly in Nigeria.

Management suggestions: the status of the Gabon material requires resolution as to whether or not it is *D. goldieana.* The best focus for protecting this species is probably Korup NP which is secure. Here several plants were seen growing in one area in undisturbed forest by the author in 1986. The usual basic populational data should be collected as a basis for monitoring in future. This species should be considered for propagation and commercial use as an ornamental plant since is is so spectacular. A cultivated plant of this species was purported to come from the wild in Boka, western Bakossi. If confirmed, this will represent a new location, as will a record from Banyang Mbo research station. Consequently the threat rating for this species may need to be lowered in future.

Dracaena talbotii Randle
EN B2ab(iii)

Range: S.E. Nigeria (Oban;) and Cameroon (SW: Korup NP).

Probably an unbranched shrub 1–2 m; foliage congested at stem apex; leaves linear-oblanceolate, 28–80 × 2.5 cm, acute, with mucro to 9 mm, base c. 0.5 cm wide, sheath short, clasping the circumference of the stem, midrib visible above, parallel nerves dense, secondary nerves irregular, transverse; inflorescence terminal 10–15 cm, peduncle with leaves merging into bracts, reducing in size towards the inflorescence apex, each bract with an auxillary cluster of flowers; flowers white, perianth tube 13–20 mm, lobes 10 mm; fruits and seeds unknown.

Here *Dracaena talbotii* is assessed as Endangered since only two locations (above; AOO 8 km² with 4 km² cells) are known, with threats as below.

Habitat: lowland evergreen forest.

Threats: although National Park boundaries in the area around Oban, the type locality, are currently respected (Andrew Dunn, pers. comm. 2010), this has not always been so and forest losses have occurred overall in the last 100 years since the species was last seen in Nigeria.

Management suggestions: the two records from Korup (*Letouzey* 15141 and *Thomas* 5436) require verification of identification since it is not clear that they were identified by Bos. However since species described from Oban, Nigeria are frequently rediscovered in Korup, and since this species is readily identifiable from the leaf shape and dimensions, there is every possibility that the identification is correct. Basic populational data should be collected in Korup for future monitoring. Here the species is likely to be secure.

ERIOCAULACEAE

assessed by Martin Cheek

This is a pan-tropical family of monocot herbs with only four or five genera, most diverse in Brazil where several species are used in the cut-flower industry. This family requires open habitats, most usually perennial or seasonal pools, and sometimes streams. In the wet season a leaf rostee develops under shallow water, at least the species discussed here. At the end of the wet season, as the leaves are exposed, the characteristic inflorescences are produced, the dense heads of minute flowers being button-like, produced on long stalks.

A Flore du Cameroon account for the family has been written by Dr Sylvia Phillips who is monographing the family for Africa.

It is likely that new rare species of eriocaul will be discovered as botanical inventory continues in Cameroon.

Eriocaulon asteroides S.M.Phillips (assessed by Martin Cheek and Benedict Pollard)
VU D2

Range: Nigeria (Mambilla Plateau (Chappal Waddi (1 coll.) and Cameroon (NW: Bamenda Highlands; Mount Oku and the Ijim Ridge area; Bali Ngemba FR (6 coll.)).

This species was assessed as VU D2 in Cheek *et al.* (2000: 78–79), and maintained as such in an account updated by Pollard in Harvey *et al.* (2004: 71–72). This assessment is maintained here since no new data are available. A low number of individuals, from very few sites (see above) is the basis for this assessment.

This minute annual often co-occurs (in 3 of the 7 known sites) with the superficially similar *Eriocaulon parvulum*. The second species is not known to occur without the first. Indeed, both species were originally described (Phillips, S.M. Two new species of *Eriocaulon* from West Africa. Kew Bull. 53(4): 943–948 (1998)) from an unwittingly mixed collection of the two species (*Zapfack* 1204, November 1996) from the Kumbo-Oku Rd (Iwooketele Mbai). At the time of its description, *Eriocaulon asteroides* was only known from that site and from Chappal Wadi, the highest mountain in Nigeria, situated on the border with Cameroon. Subsequently, in December 1998, a large colony of the two species was found on a headland of the Laikom spur of Ijim Ridge. This site was studied in some detail (see Cheek *et al.* 2000: 18–19). 1–2 km along the path from this site to the Ardo of Ijim's compound another site for this species was found. The basalt pavement area here was wetter than on the headland, perhaps accounting for the absence of *E. parvulum*. Finally at about 1700 m alt., between Laikom and Fundong, below the Fulani settlement, another colony of *E. asteroides* was encountered by Tadjouteu of our party, in November 1999. Again, the species was associated with an outcrop of wet basalt pavement, and again *E. parvulum* was absent, although about ten of us spent an hour searching for it. Finally, in listing the *Eriocaulon* holdings of YA for the monographer of the African species of this genus, Dr Phillips, another mixed collection of the two species was encountered. This was collected in September 1975 (*De Wilde* 8633) from km 21 on the Bamenda-Jakiri Rd at the southern boundary of our area. This taxon was assessed in Cheek *et al.* (2000: 78–79). The only change in the assessment has been the addition of Bali Ngemba FR to the range.

Annual herb with leaf rosette c. 2–3 cm diam. Leaves linear–subulate, 0.8–1.5 cm long, c. 1 mm wide, acute. Scapes up to 10, 1–2.5 cm high, 0.2–0.3 mm thick, 3–4-ribbed. Capitulum 5–7 mm wide, few-flowered, star-like; involucral bracts radiating, much exceeding the floral disc, mostly in one series, membranous, narrowly lanceolate, long acuminate, 3.0–3.8 × 0.7–1.2 mm, whitish buff or flushed grey. Seeds ellipsoid, 0.5 × 0.3 mm, brown, smooth.

Habitat: basalt pavement, i.e. thin, peaty, seasonally waterlogged soil in the cracks between blocks of basalt; with *Utricularia scandens*, *Loudetia simplex* and *Scleria interrupta*; c. 1700–2500 m alt.

Threats: unknown. However, too much trampling by cattle might cause damage to these small annual *Eriocaulon* plants by dislodging from the basalt substrate the thin layer of peaty soil in which they grow. Conversely, lack of grazing or of intermittent grassland fires might permit the build up of enough soil on the pavement to allow a *Sporobolus*-based community to encroach upon the basalt pavement and smother or compete with the *Eriocaulon.*

Management suggestions:

1. A survey of basalt pavement should be made in the Kilum-Ijim area. When areas are located, these species should be searched for and vouchered if found. A rough estimate of the area of occupation and total number of plants should be made. This will allow more complete mapping of the species and a more comprehensive understanding of their population size.
2. Consideration should be given to using experimental means to examine the effects of the possible threats mentioned above on these two rare annual species. Several m^2 of one population could be fenced off and protected from fire and grazing. The effect of this could be monitored on an annual basis. Another area could be subjected to cattle or horse trampling to look at the effect of this on the soil that hosts these species. Results of this experimentation could then be used to guide management of the habitat of these species.

Eriocaulon bamendae S.M.Phillips
VU D2

Range: Nigeria (Mayo Daga (1 coll.)) and Cameroon (NW: Mt Oku; Ijim Ridge; Bamenda Highlands (four pre-1996 coll.)).

This species was assessed as VU D2 in Cheek *et al.* (2000: 79). This assessment is maintained here since no new data are available. A low number of individuals, from very few sites (see above) is the basis for this assessment.

This perennial herb was only recently described as new (Phillips (2000). Two more new species of *Eriocaulon* from West Africa. Kew Bull. 55: 195–202). Previously it had been treated as a south-central African species, *E. zambesiense* Ruhland. It was first collected in June 1931 at Laikom "in a pond" (*Maitland* 1400). The second locality that was discovered, given as Kumbo-Oku 5 km (*Hepper* 2021, 15 Feb. 1958) has not been refound by us. The site of the third collection (*Brunt* 1092, 11 April 1963), is given as "near Pinyin" (south of Bali Ngemba FR). The fourth (*Bauer* 35, 28 Feb. 1970) is Bambili Lakes. A fifth collection, from nearby Nigeria (*Hall* 1748) is slightly morphologically anomalous.

We first located this species in the field in November 1996 at the Afua Swamp (*Etuge* s.n.). Revisiting this site in December 1998, an assessment of the population size produced the figure of 2000–3000 plants. The Afua Swamp is thus one of the two most important localities for this species, in terms of population size. In December 1998, in travelling along the Laikom ridge from Laikom towards Mbesa, two more localities were discovered. The first of these, just east of the Ardo of Ijim's compound, resembles a seasonal pond, and may be Maitland's locality of 1931. Here only five plants were seen. Further east still, at the "Mbesa Swamp", 1000–2000 plants were found (*Cheek* 9819) between the tussocks in the bed of a seasonal stream and this is the second most important population of the species that we have located. In the meantime, on the Kilum side of Mount Oku, Maisels had located two more populations. The first of these was at the Kinkolong swamp near the summit of Mount Oku (*Maisels* 115, 26 June 1998), the second at Tadu stream (*Maisels* 146, 11 July 1998). From the first of these localities Maisels collected the seed which enabled Phillips to confirm her hypothesis of the specific distinctness of this species. The most recently discovered site for this species is at Mbingo. Here, at the swamp which must be crossed to reach "Back Valley", a dozen plants were seen on 10 November 1999 (Cheek *pers.obs.*). In summary, seven of the nine sites of *E. bamendae* are at Kilum-Ijim.

Eriocaulon bamendae flowers between April and July, i.e. at the end of the dry season and the beginning of the wet season, setting seed in the wettest months of the year July, and presumably August. At this time its leaf rosettes are up to 40 cm diam. By the end of the wet season and the earlier part of the dry season (Oct.–Dec.) there is no evidence of flowering or fruiting and the leaf rosette is only c. 10 cm diam.

Perennial, leaf rosette c. 10 cm diam. (when sterile in dry season) to 30–40 cm diam. (when fertile in wet season). Leaves strap-shaped, 10–22 cm long in fertile specimens, 6–16 mm wide, bright green, apex rounded. Scapes up to 15, 30 cm high, 1.3–1.5 mm diam., 5–7-ribbed. Capitulum 4–7 mm wide, globose, black and white, often viviparous; involucral bracts as wide as the capitulum, straw coloured, tinged grey, leathery, 1.7–2.3 mm long, oblong with broadly rounded tip, reflexing at maturity. Seeds ellipsoid, 0.7 mm long, pale yellow-brown.

Habitat: swampy grassland, often at the seasonally inundated margins of streams or ponds, often between tussocks; 2000–2900 m alt.

Threats: *Eriocaulon bamendae* is vulnerable above all, to changes of the watertable. Drainage of swamps or, conversely, flooding for use as reservoirs would threaten this species with extinction. The species is also vulnerable to trampling by cattle.

Management suggestions: a population census of the known sites should be completed, and the sites monitored for changes in numbers of individuals annually. Information on the demography of this species is deficient. Work is needed to discover the relative importance and levels of recruitment from division of clumps, seed and parthenogenesis from viviparous capitula.

Eriocaulon mamfeense Meikle

EN B2ab(iii)

Range: Cameroon endemic (SW: Mamfe Rock; Mone FR).

This species was published in Kew Bull. 22: 141 (1968) and for the next forty years was only known from the type locality at Mamfe Rock, a large granite inselberg across which the road N of Mamfe town traverses for a kilometre or so after crossing the Cross R. Here it is locally abundant in seasonally wet pockets (pers. obs. Dec. 2008) and many specimen records exist. In late 2008 a second location was found to the N of Mamfe, in the Mone (also written Mawne) FR in the same habitat. Since there are only two locations, and the plant is threatened by road traffic and trampling at the type locality, albeit currently a t a low level, it is here assessed as Endangered since two locations (above; AOO 8 km² with 4 km² cells) are known with the threats referred to below.

Annual herb, with a rosette of numerous flaccid, white hairy leaves, each 3–7 cm long, 3–4 mm wide; inflorescences numerous from the centre of the rosette, each with peduncles 5–35 cm long, erect, dull yellow-white, bearing at the apex a white, softly hairy, button-like head 5–8 mm diam.

This species is unlikely to be confused with any other, since at the type locality, no other eriocaul was seen. At Mone, a second, undescribed species of eriocaul was found, but it is very much smaller than the this species.

Habitat: seasonally wet pockets and seepages on granite inselbergs; 200 m alt.

Threats: the main and type location of *Eriocaulon mamfeense* has a road running through it (see above).

Management suggestions: a population census of the known sites should be completed, and the sites monitored for changes in numbers of individuals annually. Information on the demography of this species is deficient.

Eriocaulon parvulum S.M.Phillips

VU D2

Range: Cameroon (NW: endemic to Mt Oku and the Ijim Ridge).

This species was assessed as VU D2 in Cheek *et al.* 2000: 79–80. That assessment is maintained here since no new data are available. A low number of individuals, from very few sites (see above) is the basis for this assessment.

The discovery of *Eriocaulon parvulum* runs in tandem with that of *Eriocaulon asteroides*. Both are annual species restricted, so far as is known, to basalt pavement. However, *E. parvulum* is much rarer than *E. asteroides*, being known from only three, rather than six sites (see *E. asteroides*). Having studied three sites at which *E. asteroides* occurs, I have noted that *E. parvulum* occurs at only the driest and flattest of these and I conclude that, compared with *E. asteroides*, it may require slightly drier conditions.

The two species are distinguished by the fact that *E. parvulum* (the epithet means small) has slightly smaller capitula than *E. asteroides*, and by the lack of radiating, pointed bracts from the margin of the capitulum which give to the capitulum of *E. asteroides* the resemblance to a little star from which it takes its name.

Annual herb, leaf-rosette 1.5–3 cm diam. Leaves linear subulate, 0.8–1.5 cm long, c. 1 mm wide, acute. Scapes c. 7, 1.5–3 cm high, 0.5 cm thick, 4–5-ribbed. Capitulum subglobose, 4–4.5 mm wide, dirty white, more or less glabrous, bracts loose and untidy. Seeds ellipsoid, light brown, almost smooth.

Habitat: basalt pavement, i.e. thin, peaty, seasonally waterlogged soil in the cracks between blocks of basalt; with *Utricularia scandens*, *Loudetia simplex* and *Scleria interrupta*; c. 1700–2500 m alt.

Threats: unknown. However, too much trampling by cattle might cause damage to these small annual *Eriocaulon* plants by dislodging from the basalt substrate the thin layer of peaty soil in which they grow. Conversely, lack of grazing or of intermittent grassland fires might permit the build up of enough soil on the pavement to allow a *Sporobolus*-based community to encroach upon the basalt pavement and smother or compete with the *Eriocaulon.*

Management suggestions:

1. A survey of basalt pavement should be made in the Kilum-Ijim area. When areas are located, these species should be searched for and vouchered if found. A rough estimate of the area of occupation and total number of plants should be made. This will allow more complete mapping of the species and a more comprehensive understanding of their population size.
2. Consideration should be given to using experimental means to examine the effects of the possible threats mentioned above on these two rare annual species. Several m^2 of one population could be fenced off and protected from fire and grazing. The effect of this could be monitored on an annual basis. Another area could be subjected to cattle or horse trampling to look at the effect of this on the soil that hosts these species. Results of this experimentation could then be used to guide management of the habitat of these species.

GRAMINEAE (POACEAE)

assessed by Martin Cheek

The Gramineae, or grasses, are of great economic importance, providing the main starch staples for mankind. Maize, *Zea mays*, was domesticated in the New World, *Triticum aestivum* and *Triticum durum*, the wheats, in the near East, while rice, *Oryza sativa* derives from China. However, several other species have been domesticated in Africa and although partly displaced now by those above, are still important in many countries, including in Cameroon. Sorghum, *Sorghum bicolor* is still very important as a cereal in the north of Cameroon with many different cultivars, also known as 'millet', 'sorgho' and 'gros mil', while *Eleusine coracana* is cultivated around Garoua and the Mts Mandara (finger millet, dana, sat, sade or tchai gari) and *Oryza glaberrima*, the indigenous African rice, is cultivated in the same areas. Numerous other species are important as forage grasses for livestock, and meat production in Cameroon depends upon these. Other species are important in providing roofing materials while the native Bamboo species *Oxytenanthera abyssinica* and *Arundinaria (Sinarundinaria) alpina* are important in house construction for their woody stems.

Thanks to the excellent work of A.P.M. van der Zon (Graminées du Cameroun, volume II, flore, Wageningen Agric. Univ. Papers 92–1 (1992)) the species of Gramineae in Cameroon can be considered well-delimited despite the absence of a formal Flore du Cameroun account. However, for the purposes of this Red Data book, some difficulties are caused by the fact that only type specimens are cited for each species so that our mapping of the threatened species may not always be comprehensive.

The identification of grasses is a specialist subject even among botanists since this family have peculiarly unique and reduced flowers and species are separated from each other using flower and inflorescence structures given terms used only in this family. The aerial 'stems' are known as culms, leaves have a basal cylindrical portion, the sheath, and an upper, free flat blade. At their junction, a ligule, axillary to the blade, continues the line of the sheath, comprising a line of hairs, or a membrane of tissue. The ovary has 2 or 3 feathery styles (most grasses are believed wind-pollinated), while there are often 3-6 exserted stamens. The perianth is reduced to 2 or 3 small fleshy lodicules which can function to open the pair of bracts that protect each flower, the outer lemma and inner palea. Flowers are usually termed florets, and grouped together in spikelets subtended by another pair of bracts, the glumes (outer and inner) on an axis known as the rachilla. Genera and species of this family are mainly separated by differences in these structures. Van der Zon (1992) recognised 125 genera of Gramineae in Cameroon in the 557 pages of the work cited above. Although short descriptions of each species, mainly based on his work, are given in the treatments below, to reliably identify grass species in Cameroon, botanists will need to refer to van der Zon (1992) and to compare their voucher specimens with those in herbaria with authoritatively named specimens.

The threatened grass species of Cameroon can be divided into three main ecological groups, with different threats, as follows:

1. Lowland evergreen forest grasses. In this habitat grasses are sparse and not very species-diverse. They have broad leaf-blades looking at first sight unlike grasses of other habitats and resemble those of commelinaceae or zingiberaceae. Those which are threatened are the Bambusoids *Guaduella* and *Puelia*. Several are known from very few locations and threatened by logging followed by agriculture.
2. Montane grassland species. Above about 2000 m alt on Mt Cameroon and at varying altitudes on other points of the Cameroon Highlands, forest gives way to grasslands. Because of the height, conditions are cooler and many genera of the North temperate zone, genera well-represented in Europe, but otherwise not seen in the lowlands of Africa, abound. Examples are *Agrostis, Festuca* and *Deschampsia.* Those species restricted to the Cameroon Highlands are thought to be threatened by use of these grasslands for pasturing cattle. In their volumes on the Soils and Ecology of W. Cameroon, Hawkins and Brunt (1965) showed how species-rich grassland in the Cameroon Highlands has and is being changed to species-poor *Sporobolus africanus* grassland by a combination of frequent burning (to provide flushes of new growth for cattle food) and by cattle trampling. Thus in general, endemic montane grass species of Cameroon appear to be in the process of being lost. Detailed studies of individual species are needed. Fortunately the highest mountain in the Highlands, Mt Cameroon, does not suffer from cattle grazing and many rare grassland species appear to be secure there.
3. Grasslands of central and norther Cameroon. North of the southern forest belt, natural grasslands and wooded grassland (savanna) occur as rainfall levels decline. Most of these grasses are widespread and common, occurring throughout this range of habitats which are spread from Senegal in the west to Ethiopia in the east, in a broad belt south of the Sahara and north of the forest. However, some are restricted to a few small areas or locations, often in hilly or rocky areas such as in the Adamawa Region, e.g. *Digitaria adamouensis*. Such species are threatened by clearance of natural vegetation for agriculture and by overgrazing by pastoralists.

Apart from the species treated below, several others might be included in future editions of this book once species delimitation issues have been resolved: *Guaduella mildbraedii* Pilg. is known only from the type, at Campo, now lost. Van der Zon accepts it but at Kew it is treated as a synonym of *Guaduella marantifolia* Franch. which has more than 10 locations in Gabon; *Pennisetum felicianum* Asonganyi is regarded by van der Zon as probably a synonym of *P. monostigma* itself assessed below as NT. However the former, known only from the type at Tchabal Mbabo, if accepted, should be assessed as CR. *Hyparrhenia* sp. nov? of Cheek *et al.* (2000: 80) known only from NW Region in swamps at Ijim Ridge, will also rate CR if confirmed as new.

Species which are borderline threatened, and best rated as NT since known from more than 10 locations, are: *Guaduella marantifolia, G. densiflora* Pilg., *Loudetiopsis trigemina* (C.E.Hubb.) Conert, *Maltebrunia letestui* (Koechl.) Koechl. and *Pennisetum monostigma* Pilg.

In future some of these may be reassessed as threatened when better data are available on the threats and habitat loss that they face.

Agrostis mannii (Hook.f.) Stapf subsp. ***mannii***
EN B2ab(iii)

Range: Equatorial Guinea (Bioko (several coll.)) and Cameroon (SW: Mt Cameroon (many coll.). W: Bamboutos Mts (1 coll.). NW: Mt Oku (3 coll.))
The second subspecies, subsp. *aethiopica*, occurs in Ethiopia. This taxon was assessed by me (in Cable & Cheek 1998: lxviii and in Cheek *et al.* 2000: 80) as LR nt, being relatively abundant on Mt Cameroon with no known threats there. Subsequently it was found that it does face threats in W and NW Regions. Here it is reassessed as Endangered since four locations are known (AOO 16 km² with 4 km² cells) and threats as below.
Perennial, tufted herb 0.15–1 m high. Leaves variable, convoluted, up to 20 cm long; ligule hyaline, 2–3 mm long. Inflorescence paniculate, lax, open, c. 25 cm long, branches flexuose, filiform, to 4 cm long; spikelets at branch extremities. Spikelets 3.5–5.5 mm long, purple; rachillae sometimes elongated. Glumes longer than the lemma. Lemma 3–3.5 mm long, pilose, carrying 3 setae, two lateral and one median. Palaea as long as the lemma.
Description based on van der Zon (1992).
Habitat: montane grassland, sometimes at forest edge; 1850–3800 m alt.
Threats: grazing, trampling and fires due to domestic livestock pasturing at [illegible] Oku locations.
Management sugges[illegible] be made to refind this taxon at Mt Oku and monitor effects of livestock upon it, if it can be refound. Threats at Bioko need investigation. In the long-term this plant appears secure at Mt Cameroon (no grazing).

Andropogon pusillus Hook.f.
EN B2ab(iii)

Range: NE Nigeria (Chappel Waddi) and Cameroon (SW: Mt Cameroon near Mann's Spring. W: Bamboutos. Adamawa: Tchabal Mbabo).
This dwarf annual, sometimes forming mats on wet rocks, is extremely rare, and vulnerable to domestic livestock at all locations but Mt Cameroon. As an annual, any disruption to the culms, as the reproductive phase advances, may destroy production of seed for the next generation. Here *Andropogon pusillus* is assessed as Endangered since only four locations are known (above; AOO 16 km² with 4 km² cells, threats as above).
Annual herb 15–20 cm; leaf-blades 1–3 × 0.1–0.3 cm; inflorescence of 1–3 pairs of racemes, peduncle enveloped by a green spathe 5–7 cm long; racemes lax, 1–2.5 cm, partial-peduncles nodes and pedicels white pubescent; spikelets sessile, elliptic, 5–6 mm; lower glume bidentate, with a circular gland on each side of the midrib at midpoint; lemma with awn 5 cm long.
Habitat: montane grassland, often near forest edge, sometimes on wet rocks; 1900–2400 m alt.
Threats: see above.
Management suggestions: surveys are needed to discover the size of the four known subpopulations, their ranges and density and regeneration levels. This will aid future monitoring by providing a baseline and also help decide whether management intervention is needed or not. Mt Cameroon might provide the best focus for any conservation actions since its grassland habitats are least threatened of the known locations.

Ctenium ledermannii Pilg.
VU A3c

Range: Nigeria (Obudu Plateau and Mambitla - Chappel Waddi), Cameroon (NW: Bafut Ngemba; Mt Oku. W: Bamboutos Mts; Mbepit-Foumban. Adamawa: Ngaoundéré; Tchabal Mbabo) and CAR (Bouar) possibly also Burundi (Bujumbura; Kitega *Reekmans* 11006 and van der Ben 1932).
"Grass growing on very thin, almost peaty soil overlying a steep rocky slope. Forms small, very firmly rooted clumps, the leaves soft and not over a foot in height. The flowering stem is above 2 ft.high" "The inflorescence is different from any other Mambilla grass I've seen". (*J.D. Chapman* 3310, 8 Nov. 1973, Mambilla Plateau).
Here assessed as Vulnerable since the 15 locations for this species all occur in livestock grazing areas where trampling and fire have changed the composition of species in natural grassland (see Gramineae introduction). All *Ctenium* species

have a one-sided, curved spike feathered by long awns on one side. In fruit the spike coils gracefully. This species is distinguished from the more common *C. newtonii* by the paired, not single, spikes.
Habitat: montane grassland on thin soils on rock; 1500–2200 m alt.
Threats: see above.
Management suggestions: monitoring of this species in the wild is needed to confirm the threats (above) faced by this species. If it can be shown that this species is immune to livestock damage, perhaps being distasteful, it might yet be downrated to NT status. None of the known locations appear to be protected.

Deschampsia mildbraedii Pilg.
EN B2ab(iii)

Range: Nigeria (Chappal Waddi, 1 coll.) and Cameroon (SW: Mt Cameroon, many coll.). Van der Zon gives Bamenda in NW region as a location but no specimen is cited.
Locally abundant at Mt Cameroon where it can form waist-deep swards on volcanic ash deposits, and it also found on old lava flows. Here no herds of domestic animals threaten it, but, at Chappal Waddi, and at Bamenda, this is not the case. *Deschampsia* is a genus more usual in Europe than Africa. Here *D. mildbraedii* is assessed as Endangered since it is known at most from only three locations with threats at all but one. AOO is 12 km² with cells of 4 km².
Perennial tufted herb 0.6-1 m; leaf-blades convolute, 15–30 × 0.1 cm, ligule hyaline, acute, 6–10 mm; panicle 10–25 cm, open, branches undulate, 4–7 cm; spikelets 2–3-flowered; glume violet, edge orange, outer 4–6 mm; inner 5–7 mm; rachilla pilose; lemma 4.5 mm, 4-dentate with a long awn arising from the base of the dorsal face.
Habitat: montane grassland; 2000–4000 m alt.
Threats: domestic grazing and associated management regimes.
Management suggestions: surveys are needed to discover the size of the three known subpopulations, their ranges and density and regeneration levels. This will aid future monitoring by providing a baseline and also help decide whether management intervention is needed or not. Mt Cameroon might provide the best focus for any conservation actions since its grassland habitats are least threatened of the known locations.

Digitaria adamouensis van der Zon
CR B2ab(iii)

Range: Cameroon endemic (Adamawa: Tello, 43 km E of Ngaoundéré). Van der Zon (1992) implies the possibility of a second location in the region but no details are given.
Here *Digitaria adamouensis* is here assessed as Critically Endangered since only a single location is known, with threats from pastured animals and their managers (see introduction; AOO 4 km² with 4 km² cells).
Annual herb 0.8–1.1 m, culms branched at base; leaf-blade linear, 8–12 × 3–4 mm; sheath ± glabrous, auriculate; inflorescence a panicle of 6–8 racemes from a 3-angled 8.5–11 cm axis; racemes 4–7 cm, pedunculate, rachis scabrid; spikelets in pairs, pedicels scabrid; spikelets ellipsoid, c. 2 mm; outer glume reduced, ovate, 0.7 mm, acute, truncate; inner glume 7-nerved, margin pubescent, hyaline; lemma similar to but slightly longer than glumes.
Habitat: inundated grassland in the Adamawa Highlands.
Threats: conversion of natural habitat to cultivation or to grazing lands.
Management suggestions: rediscovery of this species in the wild is advised. If successful, populational data should be collected to aid future monitoring and gauge need for management interventions. It is advisable to discuss the prescence of this species and its protection with traditional and national authorities. Van der Zon should be contacted in case he can provide additional data on the species that he discovered, such as additional site or populational data to that already cited here. Seed-banking is advisable.
Digitaria is a large genus with several species important to man, both crops (*D. miliaceium*, millet) and weeds. It is possible that *D. adamouensis* might be of importance in developing new strains of plant of importance to man.

Eragrostis camerunensis Clayton
VU A3c

Range: Nigeria (Obudu Plateau and Mambilla Plateau) and Cameroon (Bamenda Highlands (numerous coll.).
Perennial tuft-forming herb 30–60 cm; leaves mainly basal, blades inrolled, 3–7 × 0.1–0.2 cm; linear 10 cm, with few branches; spikelets dark grey, along the central axis, dense, concealing axis 5–7 × 2 mm, shortly stalked; lemma 1.5 mm, obtuse.
Here *Eragrostis camerunensis* is assessed as Vulnerable on the basis that it is here estimated that more than 30% of its population has been lost in three generations (see threats below). Its range corresponds with the most densely populated and cultivated part of the Cameroon Highlands – the Bamiléké Plateau and the Bamenda Highlands. Almost all natural habit in these areas has been converted to cultivation (50% loss in 15 years 1988–2003 in one part of the Bamenda Highlands (Baena in Cheek *et al.* 2010: back cover).
Habitat: montane grassland above 1500–2000 m alt.
Threats: see above.
Management suggestions: more information is needed on the precise threats faced by this species. If it is found that it is surviving tolerably well, despite man's impact, perhaps in farm fallow in certain conditions, then re-assessment, possibly to NT, might be in order. Research is needed to

define the natural ecological niche of the species more precisely and to determine its sensitivity to fire and trampling. Since the species is known from more than 20 records, several recent, it see unlikely that its extinction is imminent. The absence of the species S and W of the Bamiléké Plateau is remarkable.

Eragrostis mokensis Pilg.

VU A3c

Range: Equatorial Guinea (Bioko: Moka), Nigeria (Obudu Plateau; Mambilla Plateau; Vogel Peak area) and Cameroon (NW: Bamenda Highlands. Adamawa: Adamawa Highlands).

Named for the type location in Bioko, this species is widespread in the Cameroon Highlands with 16 records, but absent from Cameroon South of the Bamiléké Plateau. The absence needs explanation.

Annual herb 30–40 cm, forming small, open tufts; leaf-blades 6 × 0.3–0.5 cm, long white hairy; sheath margin hairy especially near the ligute; ligule a line of hairs; panicle open and rigid, 5–12 cm; branches perpendicular; spikelets suborbicular, olive green, 2–5 × 2–2.5 mm, inflated, 5–10-flowered; stalks 1–2 mm perpendicular to axis; rachilla covered by the 1 mm lemmas.

Here *Ergrostis mokensis* is assessed as Vulnerable since it is suspected that more than 30% of the population has been lost due to habitat destruction. It is possible that more detailed study might reveal that this is an overestimate, in which case it should be reassessed as NT. Rates of habitat loss for part of its range are given by Baena in Cheek *et al.* (2010: back cover).

Habitat: submontane and montane grassland 1000–2200 m alt.

Threats: conversion of natural habitat to cultivation; overgrazing, overburning and trampling of natural grassland for cattle.

Management suggestions: more detailed research on the precise ecology, niche, threats, regeneration levels, population density and range of this species is needed to confirm or correct the assessment given here, and to form a baseline for future monitoring and basis for management intervention, if needed.

Eragrostis raynaliana Lebrun

EN B2ab(iii)

Range: Cameroon (W: Mt Nkogham near Foumban. C: near Yaoundé at Mt Mbankolo and also Minloua Hill, N'kolbison) and CAR (1 location).

Here *Eragrostic raynaliana*is assessed as Endangered since only three locations (above; AOO 12 km² with 4 km² cells) are known, with threats as below. The two records are Yaoundé are taken as one location. Van der Zon indicates a record in NW Region, but this may be an error for Mt Nkogham.

Perennial herb 30–90 cm; dense tufts; leaf-blade glossy, inrolled, 20–60 × 0.1 cm, sheath base slightly fibrous; panicle purple, narrow ellipsoid, 10–30 cm; branches to 8 cm, not verticillate; spikelets lax, 4 × 2.5–3 mm, 3–7-flowered, stalk 3–4 mm; rachilla conspicuous between florets; glumes glabrous, the lower two-thirds length of the lemma; lemma violet, 3-nerved, 2–2.5 mm, long white hairy; palea glabrous, 1.5 mm.

Habitat: seepages on rock surfaces in open, submontane habitats sometimes in forest areas; 800–1000 m alt.

Threats: rock extraction for construction of infrastructure (Mt Minlo).

Management suggestions: formal protection for one or several of the Yaoundé inselbergs on which this species occurs would help ensure protection for the type locality of *E. raynaliana* and the many other rare and threatened species at these sites. A survey of basic populational parameters, such as range, density and regeneration levels, would form a baseline for monitoring and help decide whether management intervention is needed.

Guaduella humilis Clayton

EN B2ab(iii)

Range: Nigeria (Oban area); Cameroon (SW: Mt Cameroon foothills at Onge; W of Mamfe).

Here *Guaduella humilis* is asessed as Endangered since only three locations are known (above; AOO 12 km² with 4 km² cells) and threats being as below.

Perennial herb with horizontal rhizome, bearing erect leafless culms (inflorescence) 15–26 cm and leafy culms to 50 cm; leaf-blade ovate-elliptic, 2–6 × 3–3.5 cm, acuminate, base asymmetric, glabrous above and below; false petiole short; raceme with several subsessile spikelets, spikelets linear-elliptic 2–3.5 cm, lemma oblong, 8 mm; palea longer than lemma. Raceme sometimes terminal on leafy stems.

Habitat: lowland evergreen forest; sea-level–400 m alt.

Threats: logging of forest for timber, followed by agriculture – believed imminent at the Onge location but also a likely risk at the other two locations. None of the sites are in a National Park. The absence of records from comparatively well-surveyed sites such as Korup NP, suggests that is is far from ubiquitous even in intact forest within its range.

Management suggestions: rediscovery of this species is a priority. Local community leaders should be apprised of its rarity and vulnerability if found in their areas. Basic populational data should be gathered to help form a baseline for future monitoring and to decide needs for management intervention. Cultivation at Limbe BG for multiplication and re-introduction may be advisable.

Helictotrichon mannii (Pilg.) C.E.Hubb.
EN B2ab(iii)
Range: Equatorial Guinea (Bioko) and Cameroon (SW: Mt Cameroon (6 coll.). NW: Mt Oku) also cited from W: Bamboutos Mts by van der Zon 1992, but no records known.
Here *Helictotrichon mannii* is assessed as Endangered since three locations are known (above; AOO 12 km² with 4 km² cells and threats as below). Assessed as LR nt by Cheek in Cable & Cheek (1998: lxviii), its range has since been shown to extend to Mt Oku where it is threatened, hence the change in the assessment.
Perennial herb 0.8–1 m, tuft-forming; leaf-blade 20–35 × 0.4–0.8 cm; ligule 1.5–3 mm, membranous, glabrous; panicle slender, 25–30 cm, branches scattered, basal 4–12 cm; spikelets laterally flattened, 12–14 mm; outer glume 4–5 mm; inner glume 5–7 mm; rachilla terminating in pubescent awn; florets 3–4, longer than glumes, lemma 8–11 mm, bearing an awn 11–15 mm arising 1/3 from the apex.
The second species of the genus in Cameroon, *H. elongatum* is widespread in Africa. It differs in having outer glumes 6–8 mm, and leaf-blades 2–5 mm wide.
Habitat: submontane and montane grassland; 800–2200 m alt.
Threats: montane grassland on Mt Cameroon seems secure, but that at Mt Oku is threatened by cattle, goat and sheep grazing due to frequently set fires and trampling. If presence in the Bamboutos Mts is confirmed, identical threats obtain there. Threats at Bioko are not known.
Management suggestions: the most secure site seems tobe that at Mt. Cameroon and appears to be the most single important site in terms of density of records for the species. Conservation efforts might be focussed here. Basic populational data should be gathered to inform future management interventions, if recorded, and as a baseline for future monitoring. More information is needed on the phenology and ecological niche of this species.

Hypseochloa cameroonensis C.E.Hubb.
VU D2
Range: Cameroon endemic (SW: Mt Cameroon (7 coll.)). Referred to as in Nigeria by van der Zon 1992, presumably in error.
Here maintained as Vulnerable following Cheek in Cable & Cheek (1998: lxviii–lxix)), since although no threats are known, it is known from only a single location.
Delicate annual herb 10–15 cm, forming small tufts; leaf-blades 3.5 cm, ligule hyaline; panicle delicate, 5–10 cm, lax trichotomous; spikelets on stalks 2–5 mm; spikelets 9?-flowered, laterally flattened, 3 mm; glumes equal 5-nerved, acuminate; lemma bifid 2.5 mm, with awn 4–6 mm from centre of back; callus hairs dense, 1 mm.
A genus of African mountains, near to *Agrostis*, with a single species in Cameroon.
First collected (*Mildbraed* 10881) in 1928 (overlooked by Mann, unusually), and known from a total of only seven collections, this is the rarest and the only incontestably strictly endemic grass on the mountain. It is a diminutive annual 15–18 cm tall. The other species of *Hypseochloa*, *H. matengoensis* C.E.Hubb., is known from a single collection in grassland at Songea, S Tanzania. It is curious that *Hypseochloa cameroonensis* was not found in the inventories of the 1990s.
Habitat: montane grassland, frequent between tufts of high grass, 2800 m (Mildbraed), frequent in grassland strewn with boulders. Mostly collected between Hut 2 and Hut 3 above Buea; (1000–)2000–4000 m alt.
Threats: none are known.
Management suggestions: basic populational data should be gathered to form a baseline for future monitoring. It is possible that, as an annual of the lusher relatively lower altitude grassland it needs open areas for establishment, and regular firing of grassland may be found to favour the survival of this species.

Loudetia furtiva Jacq.-Fél.
VU B2ab(iii)
Range: Cameroon (Adamawa: Meïganga-Yarbang; Meïganga-Ngaoundéré; Tibati, 40 km N at Maka; Yafounou-Ouda Ngaoui, 60 km ENE Meïganga; Col de Zabondo) and CAR (two loc.).
Here *Loudetia furtiva* is assessed as Vulnerable since known from seven locations (above; AOO 28 km² with 4 km² cells) and with threats as above.
Annual herb 30–80 cm, culms slender, nodes with patent hairs; leaf-blade 10 × 0.3–0.5 cm, sheath 2–3 cm, densely hairy ligule a line of hairs; panicle 12–20 cm, diffuse, golden; spikelets linear 8–10 mm; outer glume elliptic-oblong, 3 mm, truncate; inner 8–10 mm, with lines of hairs; 2-flowered, lower male, upper 4 mm with awn 5–6.5 mm, callus bidentate, densely hairy.
A genus usually found on poor or thin soils on rock. Nine other species occur in Cameroon.
Habitat: wooded grassland on rocky, lateritic soils, c. 600 m alt.
Threats: conversion of habitat to cultivated land; overgrazing and other pastoral impacts.
Management suggestions: most records are near Meïganga. Perhaps this area is the most logical as a focus for conservation efforts for the species. A survey is advisable to rediscover the plant and to collect basic populational data as a baseline for future monitoring and assessment of intervention needs.

Panicum acrotrichum Hook.f.

VU A3c

Range: Equatorial Guinea (Bioko (2 coll.), Nigeria (Obudu and Mambilla Plateau (1 coll. each)) and Cameroon (SW: Mt Cameroon (6+ coll.). W: Bamboutos Mts (1 coll.). NW: Bamenda Highlands (numerous coll.)).

Here *Panicum acrotrichum* is assessed as Vulnerable since it is estimated that more than 30% of its population has been lost in the last 50 years which may equate to the length of three generations. Losses are ongoing: see below.

Perennial herb 25–35 cm, base rampant, branching, culms ascending, rooting at base; leaf-blades lanceolate 4–9.5 × 0.1–2 cm, base amplexicaul, minute transverse nerves; ligule membranous; panicle 7–11 cm, branches sparse, ascending, to 5–7 cm; spikelets lanceolate 2.5 mm, with 0–5 hairs each 2–3 mm, usually on the inner glume; lower glume c. 1.25 mm, 1–3-nerved; upper glume 2.5 mm, pilose; lower lemma as the upper glume; palea reduced.

A genus with 26 other species in Cameroon. Several other species also occur in submontane forest and might be confused with this one, e.g. *P. hochstetteri*.

Habitat: submontane and montane forest; 1300–2450 m alt.

Threats: forest clearance for agriculture except on Mt Cameroon where seemingly secure. In the Bamenda Highlands, c. 50% of its forest habitat was lost in the Dom area between 1988–2003 (Baena in Cheek *et al.* 2010: back cover).

Management suggestions: conservation of surviving submontane and montane forest in the Cameroon Highlands is a priority since in areas such as the Bamenda Highlands it is disappearing fast (see above). Local communities and national authorities should be encouraged and supported to protect this habitat. Mt Cameroon, where this species has often been recorded, might be the best focus for conservation efforts. Basic populational data should be collected to aid future monitoring and gauge needs for management intervention.`

Pentaschistis mannii C.E.Hubb. Subsp. ***mannii***

VU D2

Range: Cameroon endemic (SW: Mt Cameroon).

Formerly treated as endemic of Mt Cameroon, a closely similar population is now known in Rwanda in E Africa which is regarded by some as the same taxon and by others as a separate subspecies, as here. No threats are known to the Mt Cameroon population, but because it is known from a single location, it is vulnerable to stochastic events, such as a volcanic eruption, hence the assessment here of Vulnerable under criterion D. [illegible] the upper altitudinal quarter of the mountain.

Perennial tufted herb forming dense mats; culms slender, with 2–3 nodes, 10–30 cm; leaf-blades rigid 2–10 × 0.3 cm apex spiny; sheath base broad, papery; ligule a line of hairs; panicles spike-like, 2–4 cm, branches short; spikelets 2-flowered, 6–6.5 mm, laterally flattened; glumes persistent 5.5–6.5 mm, purple, glossy, midrib scabrid, acute; lemma 2–3 mm, apex 2-awned, awns 2–4 mm, a thick awn 6–8 mm arising in the sinus, base twisted; palea 2–3 mm.

The genus mostly occurs in S Africa, with only a few species further north, on mountains.

Habitat: sparse grassland on scree, sometimes the dominant species; 3000–4000 m alt.

Threats: none are known (see above).

Management suggestions: baseline populational data is needed against which future monitoring can take place.

Puelia schumanniana Pilg.

EN B2ab(iii)

Range: Cameroon endemic (C: Nanga Eboko. S: Bipindi; Campo-Ma'an).

Likely to be found in Gabon in future.

Here *Puelia schumanniana* is assessed as Endangered since only three locations are known (above; AOO 12 km² with 4 km² cells) with threats as below.

Perennial herb with horizontal rhizome; leafy culms 1 m, 1-leaved; leafless culms (inflorescence) 0.2–0.4 m; leaf-blades oblong 26–32 × 6–8 cm, acuminate, false petiole 1.5–4 cm, base asymmetric; external ligule; raceme short, cylindric, dense, 3–5 × 2–3 cm, spikelets 7–22, narrowly ovoid c. 1 cm.

This is the only *Puelia* in Cameroon with inflorescences on leafless culms. It might easily be confused with species of *Guaduella* (the other Bambusoid with non-woody culms) except that *Puelia* all have an external ligule comprising a line of hairs which are not seen in *Guaduella*.

Habitat: lowland evergreen forest; 0–400 m alt.

Threats: logging, followed by slash and burn agriculture.

Management suggestions: efforts should concentrate on rediscovering this species and collecting basic populational data as a baseline for future monitoring and to gauge needs for management intervention. Local authorities should be consulted to alert them of the existence and rarity of this species, and engaged in its protection.

Sporobolus montanus (Hook.f.) Engl.

EN B2ab(iii)

Range: Nigeria (Chappal Wadi, 1 coll.) and Cameroon (SW: Mt Cameroon (5 coll.). NW: Bamenda Highlands (1 coll.)).

Here *Sporobolus montanus* is assessed as Endangered since only three locations are known (above; AOO 12 km² with 4 km² cells) with threats as below.

Initially discovered on Mt Cameroon by Gustav Mann, and only found at other locations more recently. Although most records (five) are from Mt Cameroon, this number is low compared with some other endemic grassland species, suggesting that is is infrequent.

Perennial herb 15–40 cm, forming dense tufts; sheath bases fibrous; leaf-blades involute, 2–7 × 0.2 cm; panicle ovoid to pyramidal, 5–15 cm; branches in 2–4 verticels, 2–3 cm; spikelets dark-green, 4 mm; outer glume 2.5–3 mm, acute; inner glume longer than the single floret.
Sometimes considered an upland form of *S. subulatus* Hackel. *Sporobolus* is a large tropical and subtropical genus with 17 species in Cameroon.
Habitat: montane grassland on shallow soils or on lava; 2000–4000 m alt.
Threats: none are known at Mt Cameroon but in the Bamenda Highlands and Chappal Wadi, overgrazing and habitat management regimes introduced to support domestic livestock are a threat to all rare grassland species.
Management suggestions: conservation efforts might best be focussed on Mt Cameroon where basic populational data should be collected as a baseline for future monitoring and to gauge needs for management intervention. At the Bamenda Highlands, and at Chappal Wadi, more effort is likely to be needed to rediscover this species, and given the threats raised above, the needs for management intervention are likely to be higher. Discussion with the Ardos in whose domain the species survives is advisable, so that they can influence its protection.

HYPOXIDACEAE

assessed by Martin Cheek

These are mostly yellow-flowered herbs of mainly montane or submontane grassland. They appear, often after fire, from an underground corm or rhizome, bearing a rosette of linear leaves (often with white hairs). They develop one to several inflorescences, each with 1–12 star-like flowers. There are usually 6 free, subequal tepals, 6 free stamens and a single style from the inferior ovary.

The Flore du Cameroun 30 volume includes a treatment by Nordal and Iversen, pp. 33–49 (1987), recognising 2 genera and 5 species, of which one is treated as threatened here. A second, *H. camerooniana* Baker may also prove threatened when more research is done. In Flore du Cameroun it is stated that this species probably occurs as far East as Ethiopia. If this is found not to be correct, but that it is localised in the Cameroon Highlands, it is likely to be assessed as VU in future.

According to Wiland-Szymańska and Nordal (FTEA Hypoxidaceae (2006)) "some authors have been splitters – and probably largely described clones (e.g. Nel, Engl. Jahrb. 51: 259–286, 301–340 (1914)) – others have been lumpers e.g. Nordal *et al.* (Nordica J. Bot. 5: 15–30 (1985))".

Hypoxis suffruticosa Nel. (syn. *H. urceolata* auct. non Bak., sensu FWTA 3: 172 (1968)).
VU B2ab(iii)
Range: NE Nigeria (Naraguta FR at Jos Plateau; Werran Plateau; Gangoro FR; Mambilla (Gangirwal)) and Cameroon (Adamawa: Mt Gotel-Banyo; 75 km W Tignère; Ngaoundéré, 39 km E; 18 km SSE Ngaoundéré).
Terrestrial herb with massive vertical rootstock (corm?) to 7 × 7 cm, orange in section, truncate at base, the side with stout white roots, the apex with fibrous leaf-bases, flowering after fire when leafless in the dry season, leaves following, c. 6–12 per rosette, erect, blade linear, 20–40 × 1 cm, apex tapering, margin and midrib white-pilose, basal sheath c. 5 × 2.5 cm; racemes white-pilose, 1–4 central, 15–25 cm, flowers 2–7, each 2 cm long, tepals 1 cm, subtended by linear bracts 1–2 cm.
The type specimen is from Tschappe-Pass W of Tignère in Cameroon (*Ledermann* 2705 A-B, photo K).
Here *Hypoxis suffruticosa* is assessed as Vulnerable since 10 or less locations are known (AOO 40 km² with 4 km² cells) and threats are as below.
This massive *Hypoxis* is sometimes confused with *Curcilago pilosa*, but *Curcilago* have the ovary far below the tepals, separated by a long stipe and on indehiscent fruit placed underground rather than a dehiscent capsule held above ground.
Habitat: submontane grassland, often wooded; 1450–2100 m alt.
Threats: trampling and grazing by livestock, artificial fires set by graziers.
Management suggestions: surveys are needed to refind this species and obtain basic populational data as a baseline for monitoring and future management planning. It is possible that further work might reveal new locations and show that threats are not severe, in which case its status might be reduced to NT. Surveys should be conducted after late dry season fires and after early rains (March–April) when plants are most easily found, being in flower. The greatest number and density of specimens is from Nigeria. Naraguta FR in Jos Plateau has several records and so should be considered as a focus for conservation efforts. In Cameroon both Banyo and Ngaoundéré areas should be considered.

JUNCACEAE

assessed by Martin Cheek

This worldwide family of monocot herbs resemble grasses (Gramineae) and sedges (Cyperaceae). Open, wet habitats are their domain, where they often occur with Cyperaceae. Classically, the two families are separated by the stems of Cyperaceae being 3-angled, while those of Juncaceae are round. However several Cameroonian Cyperaceae, such as

Schoenoplectus, also have round stems. Juncaceae differ from both Cyperaceae and Gramineae in their fruits which are 3-valved capsules with numerous seeds, rather than having 1-seeded nutlets. There are only five or six Juncaceae species in Cameroon, all in montane grassland or grassland-forest transition habitats. There is no Flore du Cameroun account.

Luzula mannii (Buchenau) Kirschner & Cheek subsp. ***mannii*** (syn. *Luzula campestris* (L.) DC var. *mannii* Buchenau)
VU D2

Range: Equatorial Guinea (Bioko (4 coll.)) and Cameroon (SW: Mt Cameroon (7 coll.)).

This montane grassland herb is restricted to Bioko and Mt Cameroon. The second subspecies occur in Uganda on Mt Elgon (Kirschner, J. & Cheek, M. (2000). New combinations in Tropical African *Luzula* sect. Luzula (Juncaceae). Kew. Bull. 55: 899–903).

Here *Luzula mannii* subsp. *mannii* is assessed as Vulnerable under Criterion D, since although no immediate threats are known, this species has only two locations, is relatively rare (a total of 11 collections are known) and is thus vulnerable to stochastic changes. EOO is calculated as 121.76 km^2. Should direct threat be discovered, reassessment as EN is likely.

Perennial 25–60 cm tall, rhizome horizontal or ascending, stolons short, ascending; stem rigid, erect, c. 1.7 mm in diam; basal leaves usually 7–9(–13) cm long, (4–)5–6(–8) mm wide, cauline leaves 2–3, usually 3.5–7.0 cm long and 2.5–4.5 mm wide, all obtuse, densely hairy, margins densely serrulate; lower bract herbaceous, erect, 2–6.5 cm long, usually not overtopping the inflorescence; inflorescence composed of 3–4 sessile and 2–7 pedunculate clusters, peduncles usually with secondary branches and 1–3 sessile clusters on each branch; peduncles straight, rigid, erect and patent (very rarely one of them nodding), 0.3–0.5 mm in diam., smooth, 2–3(–11) cm long; clusters echinate, globular, 0.6–0.9(–1.2) cm in diam., each usually with more than 13 flowers; capsule globular, rounded above with an indistinct mucro, capsule segments 1.8–2.0 mm long and c. 1.5 mm wide, dark brown; seeds ovoid-subglobular, 0.8–1.1 mm long, 0.6–0.8 mm wide, appendage almost invisible, less than 0.1 mm long, hyaline.

Habitat: montane grassland, usually close to the forest boundary, on young, volcanic soils; 2800–3000 m alt.

Threats: no immediate threats are known, but lava flows erupt from summit areas of Mt [illegible] 20 years, posing a potential threat.

Management suggestions: mapping the subspecies, and recording basic populational data at both locations is advised to form a baseline for future monitoring.

MARANTACEAE

assessed by Martin Cheek

Closely allied to the Zingiberaceae, Marantaceae lack the ginger scent of the crushed stems, and the ligule of the leaf-sheath of that family, having instead a waxy, thickened pseudopetiole, often jointed at base and apex with sheath and blade. Several species of this pantropical family of understorey forest herbs are cultivated in Europe as pot plants, but they are best known as wrapping leaves in Cameroon, being used to package forest products of spices etc. in village markets, and to contain food being cooked or steamed, such as agusi, miondo or baton de manioc, throughout the forest zone. In some ecological cases, such as forest colonising grassland in CAR, Congo (Brazzaville) and Gabon, species of this family dominate the ground-layer and are very important as food-plants for elephants, gorilla and chimpanzee, so are well-known to wildlife workers. *Thaumatococcus daniellii*, a widespread forest pioneer, has been of great interest to researchers in the food industry of Europe and USA for decades. Its triangular pink fruits, borne in leaf-litter, open to reveal seeds cupped in a transparent jelly which is extremely sweet, not due to sugar, but a protein, thaumatococcin which, weight for weight is thousands of times more sweet than any sugar. It is not stable at high temperatures.

Cameroonian Marantaceae were created in Flore du Cameroun 4 (1965) by Koechlin; most genera are terrestrial herbs, but *Hypselodelphys* has aerial stems and can form thickets, as can *Trachyphrynium* and, with stem prickles, *Haumannia danckelmaniana*, a hazard when forest walking.

Hypselodelphys zenkeriania (K.Schum) Milne-Redhead
VU B2ab(iii)

Range: Cameroon (S: Bipindi. Littoral: Plaine des Mbo. SW: Buea. E: Yokadouma; Bertoua), Gabon (Mbigou-Malinga; Parc Waka) and Congo (Brazzaville) (Chaillu Massif).

Climbing herb, lacking spines, branched stems; petiole thick, 2.5 cm; blade linear-oblong to lanceolate, 25 × 10 cm, acuminate, rounded at base, glabrous or with midrib initially pubescent below; inflorescence dense, of several racemes, base enclosed by bracts 6–10 cm; fertile bracts 3 cm; corolla 1.5–2 cm; fruit 3-lobed, 1.7 × 5 cm, muricate.

Distinctive in the inflorescences being numerous racemes in a bunch.

Here assessed as Vulnerable since eight locations are known (above; AOO 32 km² with 4 km² cells) and threats being as below.

Habitat: lowland evergreen forest; sea-level–800 m alt.

Threats: clearance of forest for timber followed by agriculture; mineral extraction.

Management suggestions: not a single protected location for this species exists currently in Cameroon. Efforts should be made to refind it and to work with local communities to protect at least one subpopulation. Data on density, frequency and regeneration levels should be gathered as a baseline for future monitoring. Generally *Hypselodelphys* species benefit from higher light levels in forest so this should be considered in developing a management regime.

Sarcophrynium villosum (Benth.) K.Schum.
VU B2ab(iii)

Range: Cameroon (SW: Bakossi at Edip; Ehumseh and Kodmin) and Gabon (Chaillu at Songuu; Rabi; Gamba; Sindara; Ekobakoba).

Here assessed as Vulnerable since six locations are known (above; AOO 24 km² with 4 km² cells) and threats as below.

Terrestrial herb 2 m; leaf sheaths densely long hairy; petioles thickened, 2 cm; blade oblong-lanceolate, 27 × 6 cm, shortly acuminate, base rounded, lower surface pubescent, in florescence paniculate, densely pubescent; bracts 2.3 cm, subtending a single 2-flowered cyme, sepals 3 mm; corolla lobes 5 mm.

Habitat: lowland and submontane evergreen forest.

Threats: clearance of forest for timber followed by agriculture.

Management suggestions: conservation efforts should be developed in Cameroon with the Bakossi people since the Bakossi Mts are the only known location in Cameroon for this species. The only shaggy hairy species in the genus, this species will therefore easily be targeted during surveys to establish basic population data as a basis for future monitoring.

ORCHIDACEAE

assessed by Benedict Pollard

Aërangis bouarensis Chiron
EN B2ab(iii)

Range: Cameroon (Adamawa: Adamawa (1 coll.)) and CAR (2 loc., 4 coll.).

This species is known from three locations with an AOO of 12 km^2 and an EOO of 58070 km^2. The only known Cameroon location is from Adamawa, *Leroux* s.n. The CAR specimens are Préfecture de Nana-Membéré, près de Bouar, not far from the village of Bayangadré, April 1996, *Chirio s.n.*; Bŭar, 8th June 1914 *Mildbraed 9640*, July 1914, *Mildbraed 9743*, Haute Kotto District: 25 km S of Yalinga, Yadelé, *sine date*, *Le Testu 3969*.

Habitat: forest; 800 m.

Threats: threats to the lowland forest in CAR are considerd to be ongoing and considerable.

Management suggestions: search for this species in Adamawa, and attempt to introduce it into cultivation.

Aërangis gracillima (Kraenzl.) Arends & Stewart
VU A2c; B2ab(iii)

Range: Cameroon (SW: Barombi Mbo, Bakossi Mts. C: Mbalmayo. S: Batanga; Ekouk (1 coll. each)) and Gabon (5 loc., 7 coll.).

Status: ↑ assessed as LR nt in Cable & Cheek (1998: lxx). This species is here considered threatened, and assessed as VU on the basis of it having undergone a suspected population size reduction of ≥30%, especially around the Kumba area locations. It also qualifies under Criterion B as it is known from ten locations, with an AOO of 40 km^2 and an EOO of 139810 km^2. Discovered at Barombi Mbo crater lake, 2nd September 1890, *Preuss* 459, holotype B†, it was also collected in the Bakossi mountains at Supe between Ekona and Nongomadiba, 40 km north of Kumba, 1976, *Letouzey* 14361; Mbalmayo, 24th March 1975, *Leroux* s.n.; Grand Batanga, 1895–1904, *Bates* s.n.; Ekouk, 24 km route Ébolowa-Sangmélima, 1975, *de Wilde W.J.J.O.* 8496A. It is also known from seven collections in five locations from Gabon.

Habitat: epiphyte in lowland and submontane forest, sometimes in semi-deciduous forest; 470–800 m.

Threats: the type location, Barombi Mbo, is a crater lake with very steep walls that are unsuitable for agriculture or timber extraction, which may afford a level of protection at that site. Given that this is quite a spectacular species (flowers 7–8 cm across, spur up to 20cm long), it is surprising that more collections have not been made. This could be an indication of it being rare throughout its range where lowland to submontane forest is generally under threat from logging, farming practices, establishment of crop plantations, population growth and human migration, leading to urban expansion.

Management suggestions: efforts to propagate this species through *in vitro* methods have already yielded about 30 plantlets (Droissart *et al.* (2006: 14)), and so this approach could perhaps form the basis of a methodology for continuing *ex situ* conservation of threatened orchids of Cameroon. There is a risk that this kind of approach will produce plants with a narrow genotype, and so efforts to utilise specimens from throughout the range of this species could be more beneficial. It would perhaps be more appropriate to highlight the conservation significance of the known locations for this species and to redouble efforts to protect areas containing large numbers of threatened plant taxa, to promote conservation of living plants *in situ.* This suggestion really applies to all the taxa in this family account.

Aërangis stelligera Summerh.
NT

Range: Cameroon (Adamawa: Banyo. C: Mefou. E: Bélé Salo-Niadaba; Badékok à Mpan (1 coll. each). S (Dja Reserve): Mbassakok, Bouamir (1 coll. each)), CAR (2 loc., 4 coll.) and Congo (Kinshasa) (13 loc.,14 coll.).
This species is treated as NT on account of the 22 known locations.

Ancistrorhynchus constrictus Szlach. & Olszewski
LC

This species was until recently considered to be known from only three locations and as such would have qualified as EN according to IUCN (2001), but Droissart *et al.* (in press) now treat this as a synonym of *A. capitatus* (Lindl.) Summerh., which is considered to be a more widespread species known from more than ten locations, and is here now considered to be of Least Concern.

Ancistrorhynchus crystalensis P.J.Cribb & Laan
EN B2ab(iii)

Range: Cameroon (1 coll.), São Tomé & Príncipe (Príncipe (2 loc., 5 coll.)), Equatorial Guinea (Rio Muni) (unknown number of locations or collections) and Gabon (1 loc., 2 coll.).
Known from at least five locations with a minimum AOO of 20 km^2 and an EOO of 42120 km^2. This species was described from Gabon and subsequently found in Equatorial Guinea (Rio Muni) and Príncipe see Stévart & Cribb (2004a). The number of locations and collections from Rio Muni are unknown to the author but treated here as occurring at the minimum of one location. Should further locations come to light then this species would be reassessed as Vulnerable, being known from more than five locations. The Rio Muni locations(s) cannot be shown on the distribution map at present. Kew Bull. 59(2): 77–86 details the collections from Príncipe. The only known Cameroon specimen was collected from Akom II, E of Kribi, *Stévart & Droissart* 2093 (Droissart *et al.* (2006: 16)).
Habitat: forest; 550–800 m.
Threats: threats to lowland forest at Akom II are inferred from its proximity to the port town of Kribi whence timnber logs are exported. Satellite imagery of the area shows considerable deg[illegible] forest vegetation, and in [illegible] species Cameroon [illegible] threat.
Management suggestions: introduce it into cultivation if it has not already been as part of an *ex situ* conservation programme. Promote the Akom II area as a potential protected forest on account of the numerous threatened orchids that have been collected there in recent years.

Angraecopsis cryptantha P.J.Cribb
VU D2

Range: Cameroon (SW: Mt Cameroon (1 coll.)).
Described in 1996, this species has an AOO of 4 km^2 and is still only known from the type collection, made from Likombe to Mann's Spring during a botanical inventory of Mt Cameroon: *Thomas* 9443, 11th October 1992. As there is no obvious current or likely future decline in quality of habitat, Criterion B is not applicable here. Droissart *et al.* (2006: 18) assess this species as CR.
Habitat: montane mist forest dominated by *Agarista salicifolia* G.Don (*Ericaceae*) and *Schefflera* spp. (*Araliaceae*) on the old lava flow, adjacent to grassland; 1900–2000 m.
Threats: the higher elevations on Mt Cameroon from which this species was collected are generally not under threat from the more common pressures (timber extraction, farming, urban expansion), but they could still be vulnerable to further stochastic events such as volcanic eruption, to which the vegetation on Mt Cameroon is susceptible. The creation of the Mt Cameroon NP affords this species greater protection from potential future anthropogenic pressures, which is most welcome.
Management suggestions: this species was discovered on the old lava flow which indicates that there must have been a parent population somewhere in the vicinity. This area should be more thoroughly searched and numbers of any resultant individuals recorded. A small percentage of individual plants could be introduced into cultivation, perhaps at Limbe BG.

Angraecopsis lisowskii Szlach. & Olszewski
EN B2ab(iii)

Range: Cameroon (NW: Kondo; Tinachong (1 coll. each)).
This recently described species is only known from two collections, with an AOO of 8 km^2 which merits an assignation of the EN category of threat. The collections were made on August 7th 1975 at Kondo, between Oshie and Bako, 40 km NW Bamenda, *Letouzey* 14187, and on August 10th 1975 at Tinachong, 30 km WNW Bamenda, *Letouzey* 14224.
Habitat: epiphyte in submontane forest; 1150–1800 m.
Threats: Cheek *et al.* (2000: 49–50) quantified the loss of montane forest (1500–2100 m) in the Bamenda Highlands at 96.5% of the original cover, and most of the threats that precipitated this catastrophic decline are still ongoing and relevant today.
Management suggestions: further inventories in NW Region should be initiated to search for this species,

especially in the few remaining submontane forest patches. It could then be introduced into cultivation.

Angraecopsis tridens (Lindl.) Schltr.
VU A2c; B1ab(iii)+2ab(iii)

Range: Equatorial Guinea (Bioko (2 loc., 2 coll.)) and Cameroon (SW: Mt Cameroon (1 loc., 4 coll.); Mt Etinde; Mt Mwanenguba; Banyang Mbo Wildlife Sanctuary (1 coll. each); Kodmin (2 coll.). W: Bamboutos Mts (1 coll.)).

Status: ↓ this species was assessed as EN A1c+2c in Cable & Cheek (1998: lxx), but is downgraded here to Vulnerable on the basis of the recent addition of locations in the Bakossi Mountains. As such the reduction in population size is inferred to lie closer to the ≥ 30% level than the ≥ 50% level on the basis of habitat decline on Mt Cameroon, the Bamboutos Mountains and on Mt Mwanenguba. It is also assessed under criterion B, and being known from eight locations has an AOO of 32 km^2 and an EOO of 3230 km^2. It also qualifies as VU under subcriterion B1, having an EOO of less than 20000 km^2. This species was discovered on Bioko, *Mann* 646, and has since been collected at various places along the western Cameroon mountain chain, including at Buea in 1891 by Preuss, in 1947, *Gregory* 508 and by Bill Sanford in the 1960s, who also collected it between Dschang and Fongo Tongo in 1968, and on Mwanenguba, *Sanford* 5482. Fieldwork undertaken as part of the Darwin Initiative Project Conservation of the Plant Diversity of western Cameroon resulted in three recent collections (see Cribb & Pollard 2004: 452).

Habitat: submontane forest and savanna; 1000–1700 m.

Threats: conversion of forest to small-scale agriculture, grazing lands or crop plantations on Bioko, Mt Cameroon, Mt Mwanenguba and in the Bamboutos Mountains. Three collections from the Bakossi Mountains during the inventory work of 1998 and 2001 were from areas now included in the National Park, which should afford this species a measure of protection not currently available in the rest of its range, except perhaps the Banyang Mbo Wildlife Sanctuary. Recent events at this latter location suggest that it may now also be under threat, however, after recent comments (Cheek *pers. comm.*, 2010) suggest that it may have had its level of protection officially downgraded. Researchers from the New York Zoo's World Conservation Society had been active there, but no longer maintain a presence, which could further diminish its viability as a recognised protected area, and potentially open it up to logging companies.

Management suggestions: this species should be introduced into cultivation, after searching in the Bakossi Mountains and Banyang Mbo, where it could be introduced into cultivation at the Nguti Field Station. Further investigation is required into clarification of the official status of Banyang Mbo and the state of (dis-)repair of the Field Station. Efforts to rehabilitate this important facility should be prioritised; it could serve as a regional hub for conservation activities and a focal point for access into other nearby forest areas such as Korup NP and Takamanda FR.

Angraecum angustum (Rolfe) Summerh.
EN B2ab(iii)

Range: Nigeria (Old Calabar (1 coll.)) and Cameroon (S: Ngovayang (5 coll.); Bidou III/Nkolembonda (3 coll.); Monts des Eléphants (5 coll.)).

The type specimen, *Holland* 27 was collected on 20th May 1897 from Old Calabar, South Nigeria and until recently this species was known only by that collection. The first record of this species from Cameroon was made during extensive orchid surveys made between 2004 and 2007 that yielded ten collections overall (Droissart *et al.* 2009: 120). These were from three locations in S Region, the Ngovayang forest near Akom II, Bidou III/Nkolembonda (Kribi-Ébolowa Rd), and the summit of the 'Monts des Eléphants'. As such this species is considered EN on the basis of its four known locations, with an AOO of 16 km^2 and an EOO of 16510 km^2.

Habitat: epiphyte in submontane forest; 450–1050 m.

Threats: the Akom II forest area is near the Kribi-Ébolowa Rd and as such is close to an arterial vehicular route which it is inferred represents a present and future threat to suitable habitat for this species by way of potential increases in traffic volume, levels of human migration and settlement, urban expansion and small- or even large-scale crop production. All submontane forest in Cameroon, outside of protected areas, must be considered to be under threat from logging activities. The population at Old Calabar is most probably extinct given the urban development that has occurred there in the 113 years since its dicovery.

Management suggestions: maintain the existing specimens in the Yaoundé shade houses, and develop propagation techniques in Cameroon, with a view to introducing plants into protected areas. Effort should be invested into looking at ways of protecting the forests at the two Cameroon locations as these now probably harbour the only known surviving populations of this species.

Angraecum egertonii Rendle
EN B2ab(iii)

Range: Nigeria (2 loc., 2 coll), Cameroon (SW: Banyang Mbo Wildlife Sanctuary (1 coll.). C: Mbam Minkoum Hills (1 coll.)) and Gabon (1 loc., 1 coll.).

This species is confirmed from five locations with an AOO of 20 km^2 and an EOO of 217750 km^2. It was described based on the collection made in 1911 in Nigeria, Oban, Talbot 889. and a second Nigeria collection was made from Sapele, Sapoba District, 14th June 1954, *Wright, J.O.* 157 with a specimen cultivated at Kew being made on 8th September 1954, perhaps based on Wright's earlier collection. A location is not included for that specimen. Droissart *et al.* (2006: 20) report it from the Banyang Mbo Wildlife Sanctuary, *Simo* 92 and mention two additional collections from Cameroon, but provide no additional details. A collection from the Mbam Minkoum Hills, near Yaoundé, *Droissart & Simo,* M. 214 was cited in Simo *et al.* (2009: 114). It is likely that the Banyang Mbo collection was made by Murielle Simo, who has been working with the BRLU Orchid research group and collected there, as did Placède Simo, whilst working with the Yaoundé and Kew teams in 2001. The only known Gabon collection is from Chute de la Mboumi at Mbigou, 18th November 1925, *Le Testu* 5759.

Habitat: epiphyte in lowland forest; 200–450 m.

Threats: although this area of SW Region is relatively undisturbed, there is still pressure on all lowland evergreen forest in the region in general.

Management suggestions: investigate the locations and number of collections from Nigeria and cameroon. Conduct further orchid surveys in the Banyang Mbo area and introduce this species into cultivation, if discovered.

Angraecum firthii Summerh.

VU B2ab(iii)

Range: Cameroon (E: Deng Deng; Kongolo (1 coll. each)), CAR (1 coll.), Uganda (1 coll.) and Kenya (2 loc., 2 coll.).

This species is known in Cameroon from only two 1914 collections: Deng Deng, *Mildbraed* 8889 and Kongolo, 2nd April 1914, *Mildbraed* 9005, and six locations overall. It could be surmised that its presence in eastern Cameroon is the westernmost limit of its geographcial range. Although it has a large EOO (255840 km^2), its AOO of 24 km^2 ensures that it qualifies as VU under Criterion B. It is assessed here for the first time.

Habitat: epiphyte in submontane forest; 700–800 m (1450–1600 m in E Africa).

Threats: it is suspected that habitat cover in E Region will have declined significantly in the century since this species was last collected in Cameroon, due to extensive logging activities in that part of the country.

Management suggestions: conduct additional surveys in E Region around the known locations in an attempt to introduce this species into cultivation and conduct a more formal assessment of the ongoing threats to its submontane forest habitat.

Angraecum pyriforme Summerh.

VU B2ab(iii)

Range: Ivory Coast [illegible] Nigeria (3 loc., 4 coll.) and Cameroon (SW: [illegible]

Despite its large EO[illegible] from only eight locations, with an AOO of 32 km[illegible] of 264160 km^2 and is maintained as VU as per the assessment of Pollard & Darbyshire (Cheek *et al.* 2004: 193). Described from *Talbot* 888 (type), Oban, Nigeria in 1911, two more Nigerian collections near Sapele, *Wright* 132 and 154, were made in July 1953 and 14th June 1954 respectively, and *Richards* 3476, October 1935 from the Shasha FR. Three Ivory Coast collections were made in the 1960s and one in the 1970s: Tai forest, 21st October 1961, *Aké Assi* 6028, and near Dahiri, 27th June 1966, *Aké Assi* 9002, then *Breteler* 5232, border of Bandama, 10th July 1968, and *van der Burg, W.J.* 430, Yapo, 24th June 1975 (this latter specimen *fide* Szlachetko 2008: 370)). The only other known collection is *Balding & Sivell* 81, made at Nyasoso on 28th August 1993. This represented a new record for Cameroon, this taxon not having been included in the Flore du Cameroun Orchidaceae accounts (Szlachetko & Olszewski 1998–2001).

Habitat: an epiphyte in forest; c. 1450 m.

Threats: clearance of forest for conversion to agricultural small-holdings or plantations, particularly threatening the Nigerian and Ivory Coast sites where clearance has been widespread in recent decades, although the Tai Forest is thought to remain intact.

Management suggestions: further orchid surveys are to be encouraged around Nyasoso to attempt to relocate this species and, if successful, introduce it to an orchid garden there.

Angraecum sanfordii P.J.Cribb & B.J.Pollard

EN B1ab(iii)+2ab(iii)

Range: Cameroon (SW: Mt Cameroon (1 coll.); Mt Kupe (2 loc., 3 coll.)).

This orchid is known from three locations, with an AOO of 12 km^2 and an EOO of 334 km^2. It qualifies as EN under subcriterion B1, having an EOO of less than 5000 km^2. This First collected by W.W.Sanford above Buea on Mt Cameroon in 1968, *Sanford* 550/65, this species was not formally described until 2002 (Cribb & Pollard 2002: 653) where its IUCN assessment was first detailed, and followed by Droissart *et al.* (2006: 22). It has since been collected at 1100–2000 m on Mt Kupe at two different locations: above Nyasoso and on the summit of Kupe Rock.

Habitat: an epiphyte in submontane and montane forest; 800–2000 m.

Threats: the Buea site was at 800 m where clearance of forest for plantation agriculture and timber has been widespread since Sanford's collection was made. This species is now most likely under considerable threat on Mt Cameroon. On Mt Kupe, much of the forest above 1000 m has some form of protected status, thus disturbance is limited, although the site at 1100 m above Nyasoso is threatened by encroachment of agriculture following the continued expansion of this town, and particularly from migration into this general part of SW Region from people of NW and W Regions.

Management suggestions: efforts to rediscover the population on the eastern slopes of Mt Cameroon should be made. A survey of its population on Mt Kupe, particularly at lower elevations, would allow a better assessment of its status on the mountain, seemingly the stronghold of this species. It could be introduced into cultivation at the orchid garden at Nyasoso.

Bulbophyllum alinae Szlach.

EN B1ab(iii)+2ab(iii)

Range: Cameroon (S: Mt Eléphant; Bipindi; Akom II (1 coll. each)).

This species was described by Szlachetko based only on the type collection from S Region: SE slopes of Mt Eléphant, SE of Kribi, 26th February 1969, *Bos* 6423. It has subsequently been twice collected: *Stévart & Droissart* 2023, Mémel II, N of Bipindi, and *Stévart & Droissart* 2125, Akom II (East of Kribi). This species is treated by Govaerts *et al.* (2011) as a synonym of *B. saltatorium* Lindl. var. *saltatorium*, but is treated under the name *B. alinae* by Droissart *et al.* (2006: 23), where an explanation of the differentiating characters of these putative two taxa are also provided. They note that there may be specimens of the type variety of *B. saltatorium* that do in fact belong to this species and therefore assessed it as DD. Here the data are deemed sufficient to enable assignation of a category of threat and this species merits EN on the basis of it being known from just three locations with an AOO of 12 km^2 coupled with an inferred decline in quality of habitat over its known range. With an EOO of 2470 km^2, it qualifies as EN under subcriterion B1, by having an EOO of less than 5000 km^2.

Habitat: epiphyte in lowland rainforest with Caesalpinioid trees and/or an understorey of *Cola* spp; 200–900 m.

Threats: lowland forest is inferred to be under threat throughout Cameroon from logging activities.

Management suggestions: additional botanical surveys should be conducted at the known locations and efforts made to introduce this species into cultivation, whereby a propagation protocol can be developed.

Bulbophyllum bifarium Hook.f.

VU A2c; B2ab(iii)

Range: Cameroon (SW: Mt Cameroon (Buea, Limbe, Likombe (1 coll. each)); Bakossi Mts (1 loc., 3 coll.); Mt Kupe (1 loc., 2 coll.). NW: Mfongu near Bagangu (1 coll.). Littoral: Douala to Bimbia (1 coll.); Nkokom Massif near Ndom (1 coll.). W: Bana-Batcha near Fibé (1 coll.)) and Kenya (1 loc., 2 coll.).

This species was assessed as VU B2ab(iii) in Pollard & Darbyshire (Cheek *et al.* 2004: 194) and that treatment is maintained here on the basis of there being ten known locations with an AOO of 40 km^2 and an EOO of 600880 km^2 on account of the outlying collection from Uganda. The two Mt Kupe records, however, are here interpreted to represent one location, as opposed to the two detailed by Pollard & Darbyshire (op. cit.). Although this species is likely to occur elsewhere in Cameroon, which if discovered, would mean it no longer qualified under Criterion B, being known from more than ten locations, it would still qualify under Criterion A due to a suspected population size reduction of ≥ 30% across its range. This species was first described from a collection by Gustav Mann on Mt Cameroon, *Mann* 2121. *Aké Assi* 5606, the sole Ivory Coast specimen cited in Summerhayes (1968: 239) is now referred to *B. bidenticulatum* J.J.Verm. subsp. *bidenticulatum* (Vermeulen 1987: 167 and also according to a note on the herbarium sheet held at K). On Mt Kupe it was first recorded in April 1899 at Nyasoso, *Schlechter* 12896, and relocated there almost 90 years later, *Thomas* 5062, Nov. 1985, but was not recorded there during the intensive surveys of the 1990s. The Bakossi Mts appear to be a stronghold for this species, with three collections made in the Kodmin area in 1998. Govaerts *et al.* (2011) list this species as occurring in Angola, but the supporting evidence for this has yet to be located by the author, and so that country is not included in this assessment.

Habitat: epiphyte in lowland(?), submontane and montane forest; (0–)800–1800 m.

Threats: the Littoral, W and NW Region locations, as well as the Mt Cameroon subpopulations are inferred to be under considerable threat from human activities. The lower altitude sites for this taxon, particularly at Limbe, on the Douala to Bimbia Rd and at Nyasoso on Mt Kupe, where it was recorded at c. 800 m, are threatened by forest clearance for plantation agriculture or small-holder farming at the latter; the forest at these sites is likely to be already lost.

Management suggestions: conservation of this taxon should focus upon the Kodmin area of the Bakossi Mts where it appears to be most gregarious and where threats to its habitat are currently limited. A survey of its abundance should be carried out in November, its known flowering period there. It could be introduced into cultivation at the orchid garden at Nyasoso. This species could be used as an example to test the estimation of suspected population size reduction of ≥ 30%, provided here. This could be done by revisiting all the known locations for this species in an attempt to assess how threatened the forest at these locations is, and to help quantify the reduction of habitat area. It may be that the ≥ 30% figure could be revised upwards towards the ≥ 50% limit, as most of the known locations are under some form of threat, and as such this species may merit assignation of the EN category of threat under Criterion A. This could help us formulate a more refined methodology for calculating suspected population size reductions for future IUCN Red Data assessments.

Bulbophyllum blepharochilum Garay
VU D2
Range: Cameroon (1 loc. unknown, 1 coll.).
This species has an AOO of 4 km^2 and is known from only one collection that was introduced into cultivation at the University of Vienna BG: 27th May 1988, Sieder 098B07-1 (AMES holotype). It is not known whence it came or when.
Threats: it can be inferred to be known from only one location, although it cannot be said with any certainty that there are current or future threats ocurring there. It is therefore assessed as VU under Criterion D on the basis that an unpredictable stochastic event could in theory wipe out the only known population. It has not been possible to produce a distribution map for this species.
Management suggestions: the location of the original collection needs to be revealed to enable us to assess this species more fully.

Bulbophyllum calvum Summerh.
EN B2ab(iii)
Range: NE Nigeria (1 coll.) and Cameroon (SW: Mt Cameroon (1 coll.); Mukete (2 loc., 2 coll.). NW: Ijim Ridge (1 coll.); Dom (1 coll. unconfirmed)).
This species is treated as EN on the basis of its five known locations with an AOO of 20 km^2 and an EOO of 15720 km^2. This species was only known from the type collection from Nigeria on the Mambila plateau, Masaimari, in Kumi, 6th November 1961, *King* 178, until three records from SW Region, Cameroon were made in the 1980s and 1990s: Fabe-Mundemba Rd, camp Mukete, 1987, *Manning* 100; 10 km W Banga, near Mukete plantation, 1985, *Thomas* 4866; Mt Cameroon, Mapanja, 23rd September 1992, *Tekwe* 200. A record from NW Region was made at Ijim, 1998, *DeMarco* 4. A specimen collected from Dom, September 2006, *Cheek* 13616 (Cheek *et al.* 2010: 152) is a tentative identification and is not included in this assessment. Should it be confirmed as belonging to this species, then *B. calvum* would most likely merit a lowering of its threat status to VU on the basis of being known from more than five locations.
Habitat: epiphyte in forest; 640–2000 m.
Threats: vegetation clearance has been extensive on the lower slopes of Mt Cameroon (Mapanja) and in NW Region near the Ijim location.
Management suggestions: the DeMarco collection has not been located in the Kew herbarium and needs to be found to confirm the determination. The Ijim Ridge should be revisited to search for this species, as should the Mukete locations.

Bulbophyllum capituliforme [illegible]
EN B2ab(iii)
Range: Cameroon (S: Mindourou (1 coll.)), Gabon (2 loc., 11 coll., many cult.), Congo (Brazzaville) (1 coll.) and Congo (Kinshasa) (1 coll.).
This orchid is known from five locations with an AOO of 20 km^2 and an EOO of 85390 km^2. It was described in 1906 from material apparently originating from Gabon, and cultivated at Glasnevin BG, Ireland, from which several herbarium specimens were made and all mounted on a single sheet at K. Subsequently it has been recorded from Congo (Brazzaville) (*fide* Fl. Gabon 36: 188), Congo (Kinshasa): Central Forest, Bokatola – Bikoro, *Lebrun* 1403 and Gabon: Bélinga, *Hallé* 2843. A spirit collection at K of *Stévart* 14, 60km from Mindourou, 24th March 2001, represents the first country record for Cameroon.
Habitat: epiphyte in dense lowland and submontane forest; (?)400–740 m.
Threats: the Cameroon specimen was collected in a licenced timber extraction area, which suggests that pressure on the forests in this region are present, ongoing and are likely to continue to cause a decline in habitat quality.
Management suggestions: revisit Mindourou and search for this species and monitor the levels of historical and ongoing logging, perhaps with a view to identifying undisturbed areas that are not part of any felling licence. These could form part of a safe haven for this and other rare orchid species in the region.
Note: Flore du Cameroun (2001) and Fl. Gabon (2004) both refer to this species erroneously as *B. capituliforme*.

Bulbophyllum comatum Lindl. var. ***comatum***
EN B2ab(iii)
Range: Nigeria (Cross River State (1 coll.)), Equatorial Guinea (Bioko (2 loc., 2 coll.)) and Cameroon (SW: Mungo-Ndor (1 coll.). Littoral: Loum FR (1 coll.)).
This variety was assessed as EN B2ab(iii) by Pollard & Darbyshire (Cheek *et al.* 2004: 194) but for some reason has not been included on www.iucn.redlist.org, accessed December 6th 2010. This species is maintained as EN on the basis of it being known from just five locations, including the recent addition of a Cameroonian location: Banyang Mbo, giving it an AOO of 20 km^2 and an EOO of 12850 km^2. This variety, first collected on Bioko, *Mann* 642, is restricted to the western Cameroon uplands. It is seemingly absent from large areas of suitable altitudinal range and habitat within its extent of occurrence, for example on Mt Cameroon. *Zapfack* 944 was made in July 1996 from the Loum FR. The type variety is separated from the Guineo-Congolian var. *inflatum* (Rolfe) J.J.Verm. (which has a known distribution of Sierra Leone, Liberia, Ivory Coast, Gabon) by the denser and longer hairs on the ovary and by the more uniformly long-white-haired abaxial side of the sepals (Vermeulen 1987: 86). A recent collection from the Banyang Mbo Wildlife Sanctuary, *Simo* 126, belongs to this species.

Habitat: epiphyte in submontane forest; (300–)600–1400 m.
Threats: forest clearance is widespread at lower elevation sites throughout this taxon's range, for example on Bioko where much of the forest below 1000 m has been converted into crop plantations. In the Loum FR, small-holder farms continue to encroach into the forest despite its protected status. Although the canopy trees remain intact in this reserve most of the understorey has been given over to cultivated crops, such as bananas and cocoa, and as such the potentially available surface area of phorophyte has been massively reduced.
Management suggestions: enforcement of the Loum FR boundary and restoration of the understorey flora may help to protect this species there. Further research into the typical phorophyte species and their habits (e.g. tree or shrub, or both) for this orchid would be most welcome, so that a more precise understanding of its habitat requirements can be gained. Often epiphytic orchids can be found growing on coffee or cocoa plants in understorey crop regimes (Pollard, *pers. obs.*), so the disturbance at Loum may accidentally have assisted this species. It should also be brought into cultivation at an orchid garden as part of an *ex situ* conservation programme. The Mungo-Ndor collection forms part of the Banyang-Mbo Wildlife Sanctuary, which should afford it a measure of protection there, although concerns over the status of that location have been detailed earlier in this account.

Bulbophyllum coriscense Rchb.f.
VU B1ab(iii)+2ab(iii)
Range: Cameroon (S: Bifa (Kribi-Ébolowa Rd); Ebounja I (Kribi-Campo Rd) (1 coll. each)). Equatorial Guinea (Rio Muni (2 loc., 2 coll.)) and Gabon (3 loc., 4 coll.).
This orchid is known from seven locations with an AOO of 28 km^2 and an EOO of 16300 km^2, qualifying as VU under subcriteria B1 and B2. It was known only from the type specimen, *Mann* 1883 (holotype K!), collected in 1862 from Corisco Bay, Gabon until the work of Tariq Stévart's group revealed an additional five locations (Droissart *et al.* 2006: 122–123), two in Cameroon, two in Equatorial Guinea (Cataratas and Esamalan) and one near Tchimbélé and two specimens from Andocka-Foula in Gabon.
Habitat: lowland forest; 50–600 m.
Threats: Vermeulen (1987: 66) posited that this species might perhaps be extinct, but fortunately Stévart's recent work has conclusively shown that concern to be unfounded. Threats to lowland rainforest throughout its range exist from logging activities.
Management suggestions: this species is now established in cultivation at the Yaoundé shadehouse, where a viable population for a propagation programme should be developed, to help with establishment of this species within protected areas.

Bulbophyllum fayi J.J.Verm.
EN B2ab(iii)
Range: Cameroon (S: Sangha R, Mbassakok, Chutes Mbolo, Dja Reserve (1 coll. each)), Gabon (1 coll., cult. only) and Congo (Brazzaville) (1 coll., unmapped).
Known from just five locations, *B. fayi* has an AOO of 20 km^2 and an EOO of 127370 km^2. This species was described quite recently (Vermeulen, 1992: 137) based on a single collection from the Sangha (Sanaga) R made on 18th September 1988 (*Harris* 1144). Subsequent Cameroonian collections have arisen from Mbassakok, 21st December 2001, *Stévart & Pial* 311 and Chutes Mbolo, Dja Reserve, *Stévart & Pial* 684. A first country record for Congo (Brazzaville) was made by Isobyl la Croix on 13th August 1992 under *la Croix* 1118 but details of the precise location are currently unknown and this record has not been included on the distribution map. The Gabon record is not known from herbarium material, but is represented by a cultivated specimen in the garden of J.P.Biteau in Libreville (Droissart *et al.* 2006: 25).
Habitat: lowland, semi-deciduous and riverine forest; 350–515 m.
Threats: there will be a measure of protection afforded this species by its presence in the Dja Reserve, but actual or potential threats from logging to lowland forest across its range continue.
Management suggestions: it is here assumed that specimens are currently in cultivation in the Yaoundé shadehouse programme and as such there might exist the beginnings of a potential propagation programme. This needs to be investigated.

Bulbophyllum gravidum Lindl.
LC
Status: ↓ assessed as VU D2 in Cable & Cheek (1998: lxx); we do not follow Vermeulen's treatment of this taxon (1997: 42) at varietal rank under *B. cochleatum* Lindl., and follow Cribb & Pollard's (Cheek *et al.* 2004: 456) IUCN assessment of Least Concern, due to its broad geographical distribution within which it is relatvely common.
Range: Bioko, Cameroon, Congo (Kinshasa), Tanzania, Zambia, Malawi.

Bulbophyllum jaapii Szlach. & Olszweski
VU D2
Range: Cameroon (SW: Mt Kupe (1 coll.)).
This species was discovered in November 1985 at the summit of Mt Kupe by D.W.Thomas & H.L.MacLeod, *Thomas* 5049, where it was recorded as a 'common branch epiphyte' in forest and scrub. However, intensive botanical surveys on this mountain during the 1990s, including the summit forests and scrub, have revealed no further

collections, indicating that this species is possibly extremely limited in its range, with an AOO of 4 km^2. On Mt Kupe, it is perhaps restricted to the east-facing slopes near the summit, which were less well-collected during the 1990s. Here maintain the VU D2 assessment of Pollard & Darbyshire (Cheek *et al.* 2004: 194–195). Droissart *et al.* (2006: 25) assessed this species as DD.
Habitat: epiphyte of montane forest and scrub; 1800 m.
Threats: this species is found well beyond the highest altitude of current destructive anthropogenic activity on the mountain and is within the zone of protection, thus its habitat receives little disturbance. However, its extremely limited known range renders it vulnerable to stochastic events, particularly localised fire or landslides, which could cause small-scale loss of forest cover and thus loss of habitat. Such an event could threaten this species.
Management suggestions: the summit of Mt Kupe should be visited in November (its recorded flowering time) to assess the only known location for this taxon. Information on numbers of individuals, and estimates of population size may then allow Criteria A, C & D1 to be used in future. Prepare a management plan, to include both in situ and ex situ conservation measures.

Bulbophyllum kupense P.J.Cribb & B.J.Pollard
Status: ↓ previously assessed in Pollard & Darbyshire (Cheek *et al.* 2004: 195) as CR B1ab(iii)+2ab(iii), this species is now treated as a synonym of *B. teretifolium* Schltr. (see Droissart *et al.*, in press).

Bulbophyllum modicum Summerh.
Status: ↓ the assessment in Cable & Cheek (1998: lxxi) of EN A1c+2c is no longer valid, as this species is treated in synonymy under the widespread *B. josephi* (Kuntze) Summerh. var. *mahonii* (Rolfe) J.J.Verm. by Vermeulen (1987: 70).

Bulbophyllum nigericum Summerh.
VU A2c; B2ab(iii)
Range: Nigeria (2 loc., 5 coll.), Cameroon (SW: Mt Kupe (1 coll.); Bakossi Mts, Enyandong (1 coll.). NW: Bali Ngemba (1 coll.); Lebialem (1 coll.)).
This species is known from six locations with an AOO of 24 km^2 and an EOO of 60120 km^2. This assessment follows on from that of Cheek & Pollard (Harvey *et al.* 2010: 87–88). It was described from *King* 124 collected in October 1958 from Plateau Region, Nigeria, whence King made three additional collections. These latter specimens were collected from three different localities [illegible] somewhat odd specimen numbers [illegible] 124β [illegible] and were all mounted on the same herbarium sheet at K. They are considered here as one location. A further record, from the Mambilla Plateau in November 1993, *Spurrier* 18, remains unconfirmed, the specimen being labelled '*Bulbophyllum nigericum ?*'. The specimen cited in FWTA, *Aké Assi* 9218, from Mont Momi, Ivory Coast is now referred to *B. bidenticulatum* J.J.Verm. subsp. *bidenticulatum* (Vermeulen 1987: 167) although Szlachetko (2008: 152) has not followed Vermeulen's treatment and includes this as a record for Ivory Coast. It was first collected in Cameroon on the southern side of Mt Kupe, *Letouzey* 408, but has not since been recorded there. Two additional specimens are recorded from western Cameroon by Vermeulen (1987: 92), but no site locations are listed. Recent intensive surveys in Cameroon have revealed only two additional sites, at Enyandong in the Bakossi Mts (*Salazar* 6322, 28th October 2001) and at Bali Ngemba FR (*Plot voucher* BAL 25, April 2002) and a single specimen, *Simo, P.* 229, 17th April 2004 from Lebialem.
Habitat: an epilith or epiphyte in submontane and montane forest; c. 800–2050 m.
Threats: the Nigerian sites are threatened by continued extensive clearance of forest up to high elevations; one or more of these subpopulations are likely to have been lost. The plant at Enyandong was found growing in largely-cleared forest, on a tree in the village. It is therefore likely to occur in the surrounding forest, some of which is being cleared for small-holder farming, thus threatening this subpopulation. In all, a loss of over 30% of the population is estimated over the past three generations.
Management suggestions: as this species is found within the village of Enyandong, on a tree in front of the house of the Chief of the village, this is an ideal location for promoting community-based conservation. Local residents here could be encouraged to search for this species in the surrounding forest, perhaps using a species conservation poster as an aid to identification, and to promote protection of any locations where it is found.

Bulbophyllum pandanetorum Summerh.
EN B2ab(iii)
Range: Cameroon (SW: Mt Kupe (1 coll.)) and Gabon (Ngounyé (2 loc., 2 coll.)).
It has been recorded from three sites with an AOO of 12 km^2 and an EOO of 17750 km^2. This taxon was previously known from two collections made in September 1925 in the Ngounyé R area of Gabon, the type from near Kembélé, *LeTestu* 5527, and the second from along the River Dévèla, *LeTestu* 5547. Both were recorded growing on *Pandanus* trees, hence the choice of specific epithet. The discovery of this species growing in submontane forest on Mt Kupe in June 1996, *Zapfack* 607 greatly extends the extent of occurrence of this species, but it is clearly scarce, not having been recollected in Gabon and not having been found elsewhere in western or southern Cameroon, despite extensive botanical survey work there.

Habitat: epiphyte in lowland or submontane forest; c. 200–950 m.
Threats: the location above Nyasoso on Mt Kupe is below the lower altitudinal limit of effective protection on the mountain and thus is subject to habitat loss through encroachment of smallholder farms; threats to the Gabon sites are unknown.
Management suggestions: attempts should be made to rediscover this species in its previous collection localities, and to bring this species into cultivation for *ex situ* conservation. An investigation into the extent and quality of habitat at the Gabon locations needs to be researched to help inform conservation efforts there.

Bulbophyllum porphyrostachys Summerh.
VU A2c
Range: S Nigeria (Okumu FR; Sapoba; Uoonigbe FR; Akampke Rubber Estate; Calabar (total of 5 loc., 7 coll.)), Cameroon (SW: Mt Cameroon (4 loc., 4 coll.)) and Congo (Brazzaville) (1 loc., 1 coll.).
Status: ↑ *B. porphyrostachys* was assessed as LR nt in Cable & Cheek (1998: lxxi), where they state that 'it is debatable whether this species qualifies as sufficiently rare for this list in view of its widespread distribution'. It is upgraded here to VU using Criterion A on the basis that is known from eleven locations, with an AOO of 44 km^2, which is just outside the threshold of ten locations or fewer to qualify as VU under Criterion B2, but it is here suspected to have suffered a reduction in population size of > 30% across its known range. It has an EOO of 234810 km^2. Three older Cameroon collections have been made: 8th June 1937, on the 1922 Lava flow, *Rosevear* 60/37; Mt Cameroon, 24th November 1950, *Keay & Brenan in FHI* 28133; Ekondo Titi to Mundemba Rd, between River Lor and PALMOL plantation, Ikassa, 24th January 1985, *Thomas* 4396. The most recent Cameroon collection was from Memel II, N Bipindi, March 2004, *Stévart & Droissart* 2021. Isobyl la Croix discovered this species in Congo (Brazzaville) on 8th December 1990, *la Croix* 1126.
Habitat: epiphyte (also found as a terrestrial on the 1922 Mt Cameroon lava flow) in lowland forest or in oil palm plantation; 0–150 m.
Threats: vegetation clearance of lowland forest on and around Mt Cameroon has been extensive and is ongoing, and it is susepcted that lowland forest in Nigeria, whence this speces has been collected most often, will have been under even more pressure due to the very large human population size and rate of growth there. The presence of this species near an oil palm plantation near Ikassa, suggests that suitable habitat for this species has already been highly modified, although it may be adaptable to such change. Thomas's collection does record it as being from felled forest which provides us with clear historical evidence of a decline in habitat quality.
Management suggestions: an attempt to rediscover this species should be followed by an evaluation of its population in terms of size and regeneration levels. An exhaustive survey of the 1922 lava flow in 1995 did not record this species there.

Bulbophyllum resupinatum Ridl. var. ***filiforme*** (Kraenzl.) J.J.Verm. (syn. *Bulbophyllum filiforme* Kraenzl.)
LC
Status: ↓ this taxon was assessed as CR A1c+2c in Cable & Cheek (1998: lxx) under the name *Bulbophyllum filiforme* Kraenzl., but we follow Vermeulen's treatment of this taxon (1987: 120) at varietal rank here, and Cribb & Pollard's (Cheek *et al.* 2004: 458) IUCN assessment of Least Concern, due to its wide geographical distribution across Guineo-Congolian Africa and it being known from more than ten locations.
Range: Sierra Leone, Liberia, Ivory Coast, Cameroon, Gabon, Congo (Brazzaville) and Congo (Kinshasa).

Bulbophyllum scaberulum (Rolfe) Bolus var. ***fuerstenbergianum*** (De Wild.) J.J.Verm.
VU A2c
Range: Nigeria (Ogoja Region: Ikwette (1 loc., 2 coll.), Mambilla Plateau (1 coll.)), Equatorial Guinea (Bioko (1 coll.)), Cameroon (SW: Ekondo Titi; Mukete; Bomana Bakweri to Koto; Mt Kupe; Bakossi Mts; Banyang Mbo Wildlife Sanctuary. NW: Ijim Ridge, unlocated. S: Mbouma (1 coll. each)), Congo (Kinshasa) (Haut Katanga (1 coll.)).
This variety is known from 13 locations from disjunct populations in the western Cameroon uplands and in the Congo Basin, with an AOO of 52 km^2. It is here suspected that the population has declined by ≥30% over the last three generations which qualifies as VU under Criterion A. It was described from Bioko based on *Fuerstenberg* s.n., but has not since been re-recorded there. It was first noted as occurring in Cameroon by Vermeulen (1987: 118), but no site details were recorded. Three specimens were made in the between 1985 and 1993: 24th January 1985, *Thomas* 4795, Ekondo Titi; 30th October 1985, *Thomas* 4868, Mukete; 15th October 1993, *Tchoutou* 784. It has since been collected during the Kew/Herbier National du Cameroun-Earthwatch inventories once in NW Region, along the Ijim Ridge, *DeMarco* 19, flowering, December 1999, once at Kodmin in the Bakossi Mts, *Pollard* 252, November 1998, and once near Kupe Village on Mt Kupe, *Gosline* 253, December 1999. *Simo, P.* 49 was collected on 30th October 2001 at Ebamut in the Banyang Mbo Wildlife Sanctuary

Stévart has also recorded it once from Mbouma in the Dja reserve.

Habitat: an epiphyte in lowland to montane forest; c. 500–2000 m.

Threats: the Nigerian populations are threatened by continued, widespread forest clearance for agriculture. The two Duncan Thomas collections make reference to the fact that these two localities were in areas of mature forest being felled and cleared. The site near Kupe Village, along the trail to Madam Kupe area, at 800 m elevation, is below the lower limit of effective protection on Mt Kupe and thus threatened by potential future agricultural encroachment. Although the Dja Reserve affords this species a level of protection, it was collected from a cocoa plantation, which suggests that this reserve is also used by small-scale agricultural farmholders.

Management suggestions: in light of the significant threat to several of the known locations of this species, it should be a high priority to bring this taxon into cultivation and to begin an *ex situ* conservation programme, with eventual reintroduction into protected areas.

Bulbophyllum teretifolium Schltr. (syn. *Bulbophyllum kupense* P.J.Cribb & B.J.Pollard)

VU A2c; B1ab(iii)+2ab(iii)

Range: Cameroon (SW: Bibundi, Rumpi Hills, unlocated (1 coll. each), Nkud (2 coll.), Nyasoso (2 coll.), Banyang Mbo (Bejange) (2 coll.); Enyandong (1 coll.). C: Mbam-Minkom (3 loc., 5 coll.)).

Status: ↓ Droissart *et al.* (in press) reduce *Bulbophyllum kupense* P.J.Cribb & B.J.Pollard to synonymy under *B. teretifolium*, a view concurred with by this author! Notes are provided below by way of explanation. *B. teretifolium* is now known from ten locations with an AOO of 40 km^2 and an EOO of 17580 km^2 so it qualifies as VU under Criteria B1 and B2. Threats to vegetation at Nyasoso, Mbam-Minkom, Bangem and Bibundi also qualify this taxon as VU under Criterion A with a suspected population reduction of $\geq 30\%$ over the last three generations. It was described by Schlechter, based only on his type collection, *Schlechter* 12362, deposited at the Berlin herbarium where it was destroyed during World War II. It was recorded as common at Bibundi by Schlechter in April 1899 (Vermeulen 1987: 160), but has not since been collected there. The material from near Bangem, northern Bakossi, was collected and determined by D.W.Thomas in January 1986, *Thomas* 5281 & 5950, and deposited at the Missouri herbarium (MO). These two specimens refer to the same plant which was found growing on a [illegible] taken into cultivation in [illegible] flowering in [illegible] specimen collected by the present author, which was designated as the holotype of *B. kupense*, was made from a mango tree at Nyasoso, where Droissart and Stévart visited in 2009 as part of their investigation into the problem of *B. kupense*/*B. teretifolium*. Droissart *et al.* (in press) neotypify *B. teretifolium* and detail a number of further locations from which they have collected material recently. Droissart *et al.* (2006: 27) cite *Simo, P.* 71 from Enyandong in Bakossiland. Cribb & Pollard (2004: 139) state '*B. kupense* closely resembles the description in the protologue of *B. teretifolium*, but differs in having dorsiventrally compressed leaves (not terete), a pendent habit (not erect to erect-patent), and is known from fertile material (not sterile). It is possible that Schlechter erroneously described *B. teretifolium* as erect to erect-patent based on the herbarium sheet, but as he had observed it as being common, his interpretation of the habit was likely based on personal familiarity with its mode of growth. This difference, combined with that of leaf morphology, leads us to the conclusion that *B. kupense* represents an hitherto undescribed species'. It is unclear how this species came to be described predominantly on the basis of leaf morphology as the leaves in the holotype are clearly terete. The only argument for retaining *B. kupense* could be that the flowers of *B. teretifolium* were undescribed by Schlechter and as such we cannot be sure that the recent material belongs to that species. The persuading factor to treat it all as one species, however, is that the floral morphology of all of the recent material is consistent and as such seems to represent a single entity that shares the highly unusual leaf morphology with the description of Schlechter's species. Should any terete-leaved material of *Bulbophyllum* come to light that exhibits different floral morphology to that currently known then there would be a need to revisit this issue.

Habitat: an epiphyte in forest; 0–1200 m.

Threats: the Bibundi population has almost certainly been lost and attendant pressures at Mbam-Minkom, Nyasoso and Bangem are likely to have negatively impacted on this species' distribution.

Management suggestions: this species of orchid is now in cultivation at the Yaoundé shadehouse and this could be used as the basis for a propagation programme and reintroductions into the wild. As it is so easily recognisable when sterile it could be used as a target species when conducting botanical inventories for conservation management in other areas within its known range.

Bulbophyllum vanum J.J.Verm.

EN B2ab(iii)

Range: Cameroon (SW: Nguti (4 coll.)), Equatorial Guinea (Rio Muni (1 coll.)), Gabon (2 loc., 2 coll.) and Congo (Kinshasa) (1 loc., 2 coll., doubtful identifications).

Known from five locations, this species is assessed as EN with an AOO of 20 km^2 and an EOO of 64540 km^2. It was described by Vermeulen (1984) based on a single Gabonian collection, *Breteler & de Wilde* 642 and two sterile collections from Congo (Kinshasa), for which the identity is somewhat uncertain. A second Gabon collection, Ogooué-Ivindo, Bélinga, lower slopes of Babiel Nord, was made on 15th September 1978, *Hallé* 3244. All four Cameroon collections were made from a single location, Nguti, in an area suffering from timber exploitation and a single specimen was made from Bikurga inselberg in Rio Muni. These recent five specimens were all made during Stévart's orchid surveys in atlantic central Africa. An additional sterile specimen from S Region, *Droissart* 380, may belong to this species but cannot yet be confirmed. If it does belong to *B. vanum* then this species would be downgraded to VU on the basis of it then being known from six locations.

Habitat: epiphyte in rainforest rich in *Caesalpinioidae*, or epilithic on rock outcrops in Rio Muni; 150–900 m.

Threats: forest disturbance in the Nguti region is ongoing for small scale agriculture although the nearby Banyang Mbo Wildlife Sanctuary may currently afford a measure of protection for this species in Cameroon, if it is found to occur there in the wild.

Management suggestions: collection efforts should be made at the Congo (Kinshasa) site to attempt to assess its presence/absence at this location. This species could be introduced into cultivation at Nguti where an orchid shadehouse was in existence in 2001 (*pers. obs.*). All four specimens from Nguti are now in cultivation at the Yaoundé shadehouse, which could be used to distribute specimens to leading herbaria to increase the availability of material of this species for researchers. The living plants could also form the basis of a propagation programme for reintroduction into the wild, with Banyang Mbo being a possible location. Further information on the status of this location is urgently required.

Chamaeangis gabonensis Summerh.
EN B2ab(iii)

Range: Cameroon (S: Dja Reserve (1 loc., 2 coll.)) and Gabon (4 loc., 5 coll.).

This species was discovered at Djoun R Basin, banks of Liboumba R at Kemboma, October 1932, *Le Testu* 8952 (holotype) and considered endemic to Gabon until recent collections made by Stévart & Pial in 2002–3 from Cameroon, by the Mbolo falls, N of Djoun near the Dja Reserve in S Region. It is now known from five locations with an AOO of 20 km^2 and an EOO of 61040 km^2. This treatment follows Droissart et al. (2009: 294), which differs in its citation of specimens when compared to the Plants of Gabon database.

Habitat: Droissart et al. (2009: 294) do not provide information regarding the typical habitat of this species, but four of the seven known collections were made in close proximity to water: 'Mbolo falls, beach near falls on way to fisherman camp', *Stévart & Pial* 705, 3rd December 2002 and *Stévart & Pial* 968, 22nd December 2003; 'Estuaire, Tanne from Landing stage', *Louis* 1714; 'Estuaire, near Moka, Libreville – Cap Esterias', *Louis* 2648 which suggests that this might be a riverine epiphyte or a species requiring high levels of atmospheric moisture; altitude apparently not recorded.

Threats: two of the Gabon collections are from near the capital city, Libreville, and so a threat to the natural vegetation at these locations is inferred from the likelihood of urban expansion in the past, present and future.

Management suggestions: further investigation is needed into the habitat preferences of this species. Efforts could be made to relocate the Cameroon populations and introduce it into cultivation. Perhaps an area of the Dja Reserve could be designated to serve as an orchid sanctuary.

Chamaeangis lanceolata Summerh.
EN A2c, B2ab(iii)

Range: Nigeria (Sapoba (3 coll.), *sine loc.* (1 coll.)) and Cameroon (SW: Banyang Mbo (1 loc., 2 coll.)).

This species is known from three locations with an AOO of 12 km^2 although the system used to provide an EOO as laid out by Willis *et al.* (2003) does not allow for calculation of an EOO when only two locations are known. This species is now considered endemic to the western Cameroon Upland chain (see Droissart *et al.* 2009: 291). It was previously also reported by Govaerts *et al.* (2011) as occurring in Sierra Leone, Liberia, Ivory Coast, Gabon and Congo (Brazzaville), but these country records are now known to be specimens belonging to *C. lecomtei* (Finet) Schltr. Three of the four Nigerian collections were made from Benin Province, Sapoba, Jamieson R: *J.D.Kennedy* 2731 and 2732 in 1935, and November 1949, *Meikle* 607 (holotype). It is highly unlikely that this location can currently afford any realistic level of protection for this species, and it is suspected that there has been a population decline of ≥ 50% over the last three generations, and as such this species can also be considered to qualify as EN under Criterion A. The other Nigerian collection, *Ross* 186, made on 15th November 1933, does not state a location, and as such is not included in the distribution map. The two Cameroon collections were made on the way to the camp of the Banyang Mbo Wildlife Sanctuary in SW Region.

Habitat: Droissart *et al.* (2009) do not provide habitat information, but we can infer from the collection localities that it favours lowland evergreen forest; c. 50–100 m.

Threats: the status of conservation activities in Banyang Mbo is currently uncertain, but definitely diminishing (Cheek *pers. comm.* 2010), and one could infer that this state of affairs may in the long term represent a threat to this subpopulation. The Nigerian localities are very close to the large town of Sapele that lies within an area that has witnessed increased pressure from population growth over the last 70 years and beyond.

Management suggestions: investigate funding possibilities to strengthen the usage and utility of the Banyang Mbo Wildlife Sanctuary Field Station. This could then become more of a hub of conservation activities in the area, and in particular for orchids which have historically been a conservation focus there, with the construction and maintenance of small orchid shadehouses.

Chamaeangis letouzeyi Szlach. & Olszewski (syn. *Chamaeangis pauciflora* Pérez-Vera)

VU A2c; B2ab(iii)

Range: Guinea (Conakry) (Nzéré-Koré (1 coll.)), Sierra Leone (Bonabu (1 coll.)), Liberia (Nimba Mts (1 coll.); Ganta (2 coll.)), Ivory Coast (Zagné Forest (1 coll.); Douane to Mau (1 coll.)) Equatorial Guinea (Bioko (3 loc., 3 coll)) and Cameroon (NW: Tinachong (1 coll.)).

This species is known from ten locations with an AOO of 40 km^2 and an EOO of 585740 km^2. It is also apparently rare within its range, with only small numbers of collections being made from each country within its known geographical range. All of the collections were made between 1933 and 1975, with no more recent specimens coming to our attention. The only Cameroonian specimen was made on 10th August 1975 from Tinachong, 30 km WNW of Bamenda and 20 km NNE of Batibo, *Letouzey* 14218, with no more recent records appearing despite the intensive botanical surveys of the last 20 years. These facts suggest that it qualifies as threatened under Criterion A too, with an inferred population size reduction of ≥ 30%, particularly in Upper Guinea where forest clearance has been much more widespread than in parts of Lower Guinea.

Habitat: epiphyte in submontane and montane forest; 1800 m.

Threats: the type specimen was collected in the Bamenda Highlands, Cameroon which has suffered a very heavy decline in habitat quality through extreme levels (96.5%) of deforestation (Cheek *et al.* 2000: 49–50), as has much of the Upper Guinea phytochorion over the last 75 years, during which time the remaining specimens originated from that region. It is here inferred that there has been a continuing decline in habitat quality throughout its range, due to a number of socioeconomic factors.

Management suggestions: revisit the type locality, Tinachong, 30 km WNW of Bamenda and 20 km NNE of Batiba to search for this species and assess the extent of remaining suitable habitat. If this species can be relocated it could be introduced into cultivation and reintroduced into suitable submontane forest fragments in NW Region.

Chamaeangis spiralis Stév[illegible]n. *Chamaeangis sp. 1 sensu* Szlach. & Olszewski, Flore du Cameroun 36: 752 (2001); *Chamaeangis sp. nov. sensu* Segerbäck, Orchids of Nigeria: 82–83 (1983).

EN B2ab(iii)

Range: Nigeria (Ikom; Supaba (1 coll. each)) and Cameroon (SW: Banyang Mbo Wildlife Sanctuary (5 coll.), *sine loc.* (1 coll.). S: Bifa (1 coll.)).

Status: ↑ informally assessed by Droissart *et al.* (2009: 289) who state that 'additional information is needed to clarify its conservation status and entire distribution'. This species is currently known from four locations with an AOO of 16 km^2 and an EOO of 66920 km^2. Described only in 2009, it was first collected in Nigeria as *Segerbäck* 1169, and then by *Sanford* 369/66. *Rose* 46 from Cameroon was a cultivated specimen, and represented the only specimen until recent collections (2004–2007) by Droissart and others, mainly from the Banyang Mbo Wildlife Sanctuary, and also a solitary collection from S Region, SE of Bifa (village between Kribi and Ébolowa), path between Niété R and Nyamefoo R.

Habitat: epiphyte in lowland primary forest and secondary forest, 100–220 m.

Threats: although five of the collections from Cameroon were made in the Banyang Mbo Wildlife Sanctuary, recent information (Cheek 2010, *pers. comm.*) suggests that use of the field station there may no longer be funded and conservation efforts in the region are diminishing from their levels of ten years ago. It is reasonable to infer that if this state of affairs continues, then it constitutes a real threat to the Nguti subpopulation.

Management suggestions: efforts to revitalise the Banyang Mbo Wildlife Sanctuary field station should be made a priority for NGOs operating in the region. This could then once again serve as a focus for visiting researchers, field staff and conservation practitioners. It could serve as a focal point for orchid conservation through construction and maintenance of an orchid shadehouse, perhaps as part of developing a shadehouse network in Anglophone Cameroon to complement the ones elsewhere in the country.

Cheirostylis divina (Guinea) Summerh. var. ***ochyrae*** Szlach. & Olszewski

EN B2ab(iii)

Range: Cameroon (SW: Mt Cameroon (1 coll.). C: Zoabissim (1 coll.); Chaine de Nkohom (1 coll.)).

This recently described variety is known from just three locations in Cameroon with an AOO of 12 km^2 and an EOO of 12320 km^2. A single specimen from SW Region was collected from forest on the southern slope of Mt Cameroon above Batoke, 29th December 1983, *Thomas* 2780. Central Region has yielded the other two collections, from 'colline Zoabissim', 4 km from the village of Ekékam, Yaoundé, 9th December 1978, *Dang 752* and from 'chaîne Nkohom', 42 km SSW Ndikiniméki, 1985, *Nkongmeneck* 565 (holotype).

Habitat: terrestrial herb in submontane rainforest and semi-deciduous forest; 500–950 m.

Threats: the Mt Cameroon site is likely to be threatened by the various attendant pressures to the forest on that mountain at lower altitudes. The Ekékam location is very close to the capital city, Yaoundé and the main Douala-Yaoundé Rd, and as such is inferred to have been exposed to significant and increasing pressures on the forest from the typical threats posed by urban proximity and expansion. Ndikiniméki is a small town or village situated on a secondary road, and as such will likely be under some form of threat from farming activities that presumably continue along the road towards the location from which this species was collected. Inspection of satellite imagery, however, reveals a considerable area of extant, undisturbed forest to the south and west of this settlement whence the type specimen was collected, which may well represent the best chance for survival for this species within its currently known range.
Management suggestions: revisit Batoke on Mt Cameroon and assess the surrounding extent of suitable forest and investigate any evidence of decline in habitat quality. Conduct further surveys in the forest west of Ndikiniméki to better understand the orchid flora there, and to try to relocate this species there.

Corymborkis minima P.J.Cribb
VU B2ab(iii)
Range. Cameroon (SW: Korup NP (2 loc., 4 coll.)). S: Longii to Songkwari (1 coll.). C: Mefou (2 coll.)), Equatorial Guinea (Rio Muni (1 loc. presumed)) and Congo (Brazzaville) (1 coll.).
C. minima is presumed to be known from six locations with an AOO of 24 km^2 and an EOO of 95340 km^2. It was originally thought to be endemic to Cameroon, but according to Govaerts *et al.* (2011) and Droissart *et al.* (2006: 29) it is also known from Equatorial Guinea (Rio Muni) and a very recent collection has been made from Congo (Brazzaville). The author has not yet been able to locate specimens from Rio Muni, and so for the purpose of this assessment, a presumed occurrence at one location in Rio Muni is included, but remains doubtful and needs to be verified. The first country record for Congo (Brazzaville) was made between Moussaou and Dziba Dziba, 25th May 2010, *Kami E.* 4989. In Cameroon, it was first collected on 7th January 1970 in S Region between Longii and Songkwari, *Bos* 6069. Four other Cameroon collections from the Korup NP were all made by Duncan Thomas: 10th December 1983, *Thomas* 2672; 5th December 1984, *Thomas* 4102; 27th March 1985, *Thomas* 4574; near Mundemba, Ekundu-Kundu footpath on the east side of the River Mana, 2.5 km north of the PAMOL plantation, 26th March 1994, *Thomas* 10008B (holotype). It has also recently been found in the Mefou proposed NP, S Region, 21st October 2002, *Cheek in Plot Voucher MF* 259, and then again on 29th March 2004, *Onana* 2893 during botanical inventories there.
Habitat: forest floor herb on shady soil in lowland evergreen rainforest; 50–150(–700) m.
Threats: this species clearly prefers lowland rainforest habitat, which is the vegetation type most under threat in Cameroon. The Korup locations should prove to be well protected on account of them occurring within the National Park, although the type collection was made very close to an existing Oil Palm plantation which demonstrates that even in this part of Cameroon threats exist to old growth forest, in which this species is known to occur. The 700 m altitudinal record is from one of the Mefou collections and represents quite a change in altitudinal range. This could provide hope that *C. minima* is known from other locations within its known altitudinal range where it is yet to be discovered.
Management suggestions: relocate the Korup populations and introduce this species into cultivation. Continue to promote the importance of threatened plant species known from Mefou and its potential conservation value if it was to be gazetted as a National Park. Further investigate the Rio Muni specimens to determine the accuracy of the distribution for this species.

Diaphananthe bueae (Schltr.) Schltr.
VU A2c; B2ab(iii)
Range: Cameroon (SW: Buea (2 coll.), Banyang Mbo Wildlife Sanctuary (1 doubtful coll.); Lebialem. NW: Tadu-Kumbo; Nkambe (1 coll. each); Lake Oku, Bali Ngemba; Mujung; Dom (5 loc., 1 coll. each, four of them unconfirmed). C: Mbam Minkom (1 coll.), unlocated (1 or more coll.)), Uganda (1 coll).
This species was first assessed by Cable & Cheek (1998: lxxi) and then Cheek *et al.* (2000: 81) as EN A1c+2c and treated as a Cameroon endemic. Harvey *et al.* (2004: 72) and Cheek *et al.* (2010: 107) did not provide a Red Data Taxon treatment despite its presence in those checklists, but Cheek & Pollard (Harvey *et al.* 2010: 88–89) included an assessment, downgraded it from EN to VU assessed as B2ab(iii) with seven potential locations and an AOO of 28 km^2, and under Criterion A as A2c since they estimated that more than 30% of its habitat has been lost in the last thirty years. Here it is no longer considered to be a Cameroon endemic and now potentially present at 12 locations, but half of these are currently doubtful on the basis of unconfirmed identifications and as such this assessment considers it to be confirmed as present in six locations with an AOO of 24 km^2 and an EOO of 324300 km^2, and maintains the assessment as VU A2c; B2ab(iii). A probable record of this species from Ivory Coast, *Pérez-Vera* 725 *pro parte*, was previously always rejected in the series of Kew-National Herbarium of Cameroon Conservation Checklists, on the basis of it being only known from fruiting material, but Szlachetko (2008: 346) has apparently seen a flowering duplicate and does accept this species for that country, but we currently maintain

our doubt and do not include it here. A spirit collection from Uganda, *Meyer* 129, is accepted here. Stévart's group also have a live specimen in their orchid shadehouse network which suggests an additional, as yet unspecified, location. As the specific epithet suggests, *D. bueae* was first collected at Buea on Mt Cameroon (*Deistel* s.n., type, collected 24th July 1905, "auf der altern Rinde hoher Baumen in gesellschaft anderer Orchideen. Walden in d. Ungebung Buea"). It was rediscovered there over 40 years later, *Gregory* 153, and more recently it has been found in the Bamenda Highlands, at Nkambe, *Letouzey* 8889 and Tadu-Kumbo, *Mbenkum* 354. Recent work has rather unsatisfactorily led to a number of doubtful collections. Cribb *et al.* (2000: 188) list a DeMarco collection from the north side of Lake Oku, 22nd May 2000, but with the proviso that this was a provisional determination. Surveys at the nearby Ijim Ridge in 1996,1998, 1999 and 2002 did not yield further material. A sterile collection, *Simo* 228, made on 17th April 2004 from Lebialem has been provisionally identified as this species (Roberts *et al.*, 2010: 157). The collections from Banyang Mbo, 26th October 2001, *Simo, P.* 36; Bali Ngemba, 15th April 2004, *Simo, P.* 202 and Bamenda Highlands, Mujung, 15th April 2004, *Simo, P.* 270 are all field determinations and as such cannot yet be considered reliable determinations. Despite relatively high sampling levels at Mt Kupe and the Bakossi Mountains, the species was not found there (Cheek *et al.* 2004).

Habitat: epiphyte in submontane forest; 780–1830 m.

Threats: most of the known locations are believed to be under pressure from forest clearance for agricultural practices, firewood collection and at Buea particularly from urban expansion, and this species is considered to have experienced a suspected population reduction of $\geq$ 30% across its known range in the last 30 years to merit assignation of VU under Criterion A. Even if this species is confirmed from all of the possible locations from which it has been provisionally identified, many of these are in NW Region which has suffered an enormous decline in habitat quality and as such this species would still be considered threatened under Criterion A.

Management suggestions: further research is required into the identifications for all of the doubtful collections. It is unsatisfactory that so many provisional determinations have been made and this is to be avoided in future botanical inventories wherever possible. All of the locations for which the identifications are doubtful could be revisited, specimens collected and grown on. This should help to better understand the distribution of this species. It is likely that our knowledge of the distribution of this species has suffered from the problems posed by CITES regulations as discussed in the introduction to this account, and this species could therefore be used as an example to highlight this hindrance to effective conservation action.

Diaphananthe ceriflora J.B.Petersen

EN B2ab(iii)

Range: Cameroon (E: Ngom (1 coll.)) and Gabon (2 loc., 2 coll.).

This species was not included in the Flore du Cameroun account (Szlachetko & Olszewski 2001b), and until recently (Droissart *et al.* 2006: 31) was considered to be endemic to Cameroon and known only from the type specimen, 1949, Menyaye in the Nyong [Ngom] FR, cult. *Birket-Smith* s.n. in the Copenhagen BG. Two recent collections, *Biteau* 26 and *Stévart* 891 were made in Gabon, meaning it is currently known from three locations with an AOO of 12 km^2 and an EOO of 13990 km^2.

Habitat: epiphyte in savanna; 200–600 m.

Threats: it is inferred that all lowland forest in the known altitudinal range of this species is or will be under threat from logging and other destructive activities.

Management suggestions: revisit Ngom to search for this species and introduce it into cultivation.

Diaphananthe garayana Szlach. & Olszewski

EN B2ab(iii)

Range: Cameroon (SW: Banyang Mbo Wildlife Sanctuary (1 coll.). S: Bipindi to Déhané (1 coll.).

This Cameroon endemic is known from only two locations with an AOO of 8 km^2 and as such has not yet been given an EOO. It was described on the basis of the type collection made on the road from Bipindi to Déhané, June 1918, *Annet* 1405, with a second specimen, *Simo* 79, Banyang Mbo being collected in 2001.

Habitat: epiphyte in lowland forest; 50–250 m.

Threats: the fact that the collection loaclity is given as being along a road, as so many orchids are, tells us that the type locality for this species lies in close proximity to human activities and settlements. The threats to forest habitat from these are only likely to increase as the general trend in population growth is currently upwards. The Banyang Mbo Wildlife Sanctuary may or may not afford this species a sustainable level of protection.

Management suggestions: revisit the Bipindi-Déhané Rd and look out for this species. If located, a survey should be undertaken to increase our knowledge of its habitat and ecology. Useful information to record would be phenology, numbers of individual plants, habitat, typical species growing in the vicinity, phorophyte species. If a significant population was to be discovered then a small percentage of live individuals could be collected and introduced into cultivation for propagation, perhaps this latter activity would be well-suited for introduction at Banyang Mbo.

Diaphananthe sanfordiana Szlach. & Olszewski

EN A2c; B2ab(iii)

Range: Cameroon (C: Yaoundé, Mt Fébé (1 coll.)) and Gabon (Oyem (1 coll.)).
D. sanfordiana is known from two locations with an AOO of 8 km^2 and as such has not yet been given an EOO (see Willis *et al.* 2003), but is likely to be extinct at the Yaoundé location, and is suspected to have suffered ≥ 50% reduction in population size, qualifying it as EN under Criteria A and B. This species was first collected on Mt Fébé, on the outskirts of the Capital City of Cameroon, Yaoundé, Central Region on 1st December 1961, *Breteler* 2147 (holotype) and has not since been recollected in Cameroon. *Stévart* 26 between Oyem and Mitzik in Gabon represents the only other known collection for this species.
Habitat: epiphyte on shady rocks in forest; 1000 m.
Threats: Mt Fébé near Yaoundé has experienced considerable disturbance over the last 50 years and it is almost certain that this species does not survive at that location. Threats to the Gabon location under not currently well known to the author.
Management suggestions: revisit submontane forest patches surviving near or around Yaoundé and search for this species. Introduce it into cultivation at the Yaoundé shadehouse network if possible.

Diaphananthe sarcorhynchoides J.B.Hall
EN B2ab(iii)
Range: Ivory Coast (3 loc., 5 coll.), Ghana (1 coll.), Cameroon (S: Dja Reserve: Mbassakok de Mama (1 coll.)).
D. sarcorhynchoides is known from five locations with an AOO of 20 km^2. The only known Cameroon collection is from S Province, *Stévart, Chamba & Pial* 84, Mbassakok de Mama, 6th March 2001.
Habitat: epilith on inselbergs or epiphyte in forest; 670 m.
Threats: current threats to the Ivory Coast and Ghana subpopulations are not currently well-known, and the Cameroon location should have an element of protection from its origin within the Dja Reserve. Lowland forest across its range is, however, under threat and evidence of farming activities within the Dja Reserve have been recorded and reported on elsewhere in this chapter. It is therefore inferred to be under threat.
Management suggestions: investigate the current threats at its locations in the Upper Guinea phytochorion. Introduce it into cultivation at the Yaoundé shadehouse network.

Disperis kamerunensis Schltr.
EN A2c; B2ab(iii)
Range: Cameroon (SW: Mt Cameroon (2 loc., 2 coll.)).
Known from two locations only, with an AOO of 8 km^2. This herb was assessed as EN B1+2a by Cable & Cheek (1998: lxxii) and followed by Droissart *et al.* (2006: 33). *D. kamerunensis* is reassessed here according to all Criteria, and also to reflect the correct formatting of the Criteria and Subcriteria according to IUCN (2001). It was first collected near Buea, *Preuss* 609, 28th September 1891, 1100–1200 m, and later recollected in the forest at Musake Camp, in 1930, *Maitland* 106, 1800m. It is suspected that there has been a population reduction of ≥ 50% over the last three generations, meriting assignation under Criterion A as well as Criterion B.
Habitat: terrestrial herb in montane and submontane forest; 1100–1800 m.
Threats: clearance of forest for agriculture, firewood, construction materials.
Management suggestions: not seen since 1930, an attempt to rediscover this species should made. If successful, it could be followed by an evaluation of its population in terms of numbers of individuals, regeneration levels, and threats. The identification of the precise whereabouts of 'Musake Camp' would allow us to conduct a search for this species there and perhaps gain a better idea of the natural former range of this species.

Disperis mildbraedii Schltr. ex Summerh.
VU B2ab(iii)
Range: Nigeria (Obudu Plateau; Mambilla Plateau (1 coll. each), Equatorial Guinea (Bioko (1 coll.)) and Cameroon (NW: Bafut Ngemba (1 coll.). SW: Mt Kupe (2 coll.)).
This species was assessed by Pollard & Darbyshire (Cheek *et al.* 2004: 197) as VU B2ab(iii) on the basis of it being known from six locations with an AOO of 24 km^2 and an EOO of 45260 km^2, which is followed here. *D. mildbraedii* was described from *Mildbraed* 6312, collected at Santa Isabel, Bioko, 16th August 1911, 1100–1400 m. Two Nigerian collections were made in 1973: *Hall* 2950, 18th June, north slopes of Babanke, Obudu Plateau and *Chapman* 85, 16th August, 1800m, Mambilla Plateau. *Ujor in FHI* 29965 was made on 15th August 1951 in the Bamenda Highlands: Banja in Bafut Ngemba FR to Lake Bambili. *Cable* 117, 11th September 1992 and *Cable* 3378, 26th June 1996, both from Mt Kupe, are the only other known collections.
Habitat: forest, sometimes on wet rocks; c. 1100–2050 m.
Threats: continued clearance of forest on Bioko, in the Bamenda Highlands and in neighbouring parts of Nigeria all represent significant threats.
Management suggestions: attempts should be made to relocate this species and introduce it into cultivation. Mt Kupe would be a sensible place to start and a good opportunity to revitalise the orchid shadehouse at Nyasoso, whence this species is known.

Disperis nitida Summerh.
VU B1ab(iii)+2ab(iii)
Range: Cameroon (SW: Mt Cameroon (1 coll.); Mwanenguba (1 coll.). NW: Mt Oku (2 loc., 3 coll.); Mt Neshele (1 coll.); Lake Bambuluwe (1 coll.)).

Status: ↓ assessed as EN B1+2b in Cribb *et al.* (2000: 81) and Droissart *et al.* (2006: 34) and as EN B1ab(iii)+2ab(iii) in Pollard & Darbyshire (Cheek *et al.* 2004: 197), it is here re-assessed as VU on the basis of it being known from six locations with an AOO of 24 km^2 and an EOO of 2820 km^2. *Watts* 456, 5th October 1992, Buea was omitted from Cable & Cheek (1998: 174), Cheek *et al.* (2000: 81) and Pollard & Darbyshire (Cheek *et al.* 2004: 197), it representing the first record of this species from Mt Cameroon. *Munyenyembe* 793, 29th October 1996, Elak was omitted from the Mt Oku checklist (Cribb *et al.* 2000: 189). These additional Oku collections, which include: above Lake Oku, Bamenda, 4th September 1952, *Savory* UCI 451 (holotype); KDH, Kilum mountain, October 31st 1996, *Zapfack* 1152 and 1st November 1996, *Zapfack* 1156, can be considered to be from two locations. As such *D. nitida* is now considered to occur at six locations. Other collections are from the Bamenda Highlands: Lake Bambuluwe, 3rd September 1952, *Savory* UCI 475 (K), Mt Neshele (=Mba Mfeu), 10 km ESE Bamenda, 12th August 1975, *Letouzey* 14259 and SW Region: Mt Mwanenguba, 15 km NW Nkongsamba, 1976, *Letouzey* 14404.
Habitat: submontane to montane forest on lower branches or leaning trunks of trees in densely canopied areas, rarely terrestrial; 1800–2800 m.
Threats: continued clearance of remaining montane forest at Mts Oku and Mwanenguba and in the Bamenda Highlands outside of protected areas, together with possible future encroachment into protected areas, severely threaten this species.
Management suggestions: see Cheek *et al.* (2000: 81); management is best focused upon the Kilum-Ijim population which was the best protected area within this species' known range, and at which the species appears from our surveys to be at its most numerous. The precise status of the Kilum forest protected area is not currently known, and it is recommended that a field survey there should be conducted as soon as possible to establish the level of protection afforded to that forest, and in an attempt to introduce this species into cultivation.

Disperis parvifolia Schltr.
EN B2ab(iii)

Range: Cameroon (NW: Ijim Ridge (1 coll.)), Tanzania (2 loc., 2 coll.) and Malawi: (Nyika (1 loc, 1 coll.)).
Assessed as EN B1+2a-e in Cribb *et al.* (2000: 82), but not included on the IUCN Red List (accessed 15th November 2008) or in Droissart *et al.* (2006). This diminutive terrestrial orchid is only known from a single collection west of the Congo basin, at Ijim between Mbesa and Nchang, *Maitland* s.n., Basenako to Nchang, 5000', 1951. Despite its unusual distribution and very large EOO (162060 km^2), it is known from only four locations with an AOO of 16 km^2, and has not been collected in Cameroon for 80 years. It no longer qualifies as EN under Criterion B1, and is only accepted under Criterion B2. Isobyl la Croix recorded it for the first time from Malawi on 9th January 1983, and it is also known from two locations in Tanzania, Mbozi, *Leedal* 3675 and North of Lake Nyasa, February 1912, *Stolz* 1125.
Habitat: grassy banks near river crossing; c. 1660 m.
Threats: unknown, but potentially trampling by cattle.
Management suggestions: the reduced leaves suggest that *D. parvifolia* may be a mycoheterotroph, in which case flowering may occur infrequently and so render it less likely to be visible above ground, and therefore infrequently collected. A targeted collecting mission should be designed for this species, to try to relocate and protect it.

Eggelingia gabonensis P.J.Cribb & Laan (syn. *Tridactyle gabonensis* (P.J.Cribb & Laan) R.Rice)
EN B2ab(iii)

Range: Cameroon (S: Djoum (1 coll.); Akom II (1 coll.)) and Gabon (1 coll.).
This species was described in 1989, has not been recorded since the 1980s, and is known from only three locations with an AOO of 12 km^2 and an EOO of 32010 km^2. Two of the specimens were cultivated and derived from living collections made at two locations. It was first collected 18 km east of Djoum, 1980, *Beentje* s.n., fl. May 1983, cult. WAG in *van der Laan* 399, and subsequently by *de Wilde, Arends et al.* s.n. in Gabon, Cristal Mts, 10 km from Tchimbélé on the Kinguélé Rd, fl. 7th March 1983, cult. WAG in *van der Laan* 577. *Stévart & Droissart* 2186 was discovered at Akom II (east of Kribi), and despite being sterile, is determined as this species according to Droissart *et al.* (2009: 35) on account of its vegetative characteristics and distribution.
Habitat: epiphyte in lowland and submontane forest, 570–1050 m.
Threats: all lowland forest in S Region of Cameroon is inferred to be under actual or future threat from timber extraction. Djoum is a town at a crossroads, and 18 km east of this town, where *E. gabonensis* was collected is likely to lie close to the main road that runs in that direction. This area is close to the Dja Reserve, but not within its borders.
Management suggestions: revisit the known localities for this species and mount a search to relocate it for introduction into cultivation in Cameroon. Confirmation of the identification of the Akom II specimen would be welcome.

Eulophia thollonii Szlach. & Olszewski
This species was described recently but is treated by Govaerts *et al.* (2011) as a synonym of the common afrotropical species *E. milnei* Rchb.f.

Gastrodia africana Kraenzl.

CR A1c+2c

Range: Cameroon (SW: Mt Cameroon (1 coll.) and ?Mt Kala (1 doubtful coll.)).

Assessed in Cable & Cheek (1998: lxxii) and Droissart *et al.* (2006: 35) as CR A1c+2c, which is maintained here. Apparently overlooked by Summerhayes (1968) in his preparation of the orchid account for the FWTA, this appears to be the westernmost and only native African species of a genus of about 35 species of mycotrophic, sometimes edible and medicinal orchids, mostly found in SE Asia and tropical Australia. Collected by Dusen at Mt Cameroon between 'Love and Ndiva' in April 1900, *Dusen* 397 (holotype), which are not referred to in Letouzey's gazetteer of 1968. However, if the letters v in these words are converted to b (they can be interchanged in German) then the localities are rendered Lobe and Ndibe, recognisable as a major plantation area immediately to the north of Mt Cameroon. The type specimen survives at the Swedish Museum of Natural History (S). Sanford's unpublished manuscript for the Flore du Cameroun refers to this species a collection from Mt Kala, 20 km W Yaoundé, near Nkométou, 27th October 1968, *Letouzey* 9509, but with a great deal of doubt and discussion.

Habitat: terrestrial herb in submontane forest with *Allanblackia*, *Carapa* and *Cola* spp., 1120 m.

Threats: the likely location of the type collection is close to a known major crop plantation area. Habitat destruction of lowland and lower submontane forest all around Mt Cameroon is and has been considerable in extent and will almost certainly have negatively affected this species. The recent creation of the Mt Cameroon NP is most welcome but is not necessarily going to afford much protection for this species, as the protected area is concentrated more on higher elevations. The Mt Kala location is so close to the capital city Yaoundé that it is inferred this potential subpopulation will also have suffered commensurately.

Management suggestions: it is thought likely that Lobe and Ndibe are indeed the locations referred to by Dusen, although that is not certain. These places need to be revisited urgently in an attempt to relocate this species. It would be a useful exercise to map suitable habitats within its range to look at targeting other areas containing similar forest habitat in attempt to try to discover additional populations of this remarkable species.

Genyorchis macrantha Summerh.

EN B2ab(iii)

Range: Cameroon (SW: Mt Cameroon (2 loc., 2 coll.). NW: Ijim Ridge, Mbesa (1 coll.)).

Status: ↑ assessed as VU D2 in Cable & Cheek (1998: lxxii), Cribb *et al.* (2000: 82) and Droissart *et al.* (2006: 36) it is now considered to be EN, as it is known from three locations with an AOO of 12 km^2 and an EOO of 700 km^2, and threats are considered to be prevalent within its known range. First discovered on 2nd April 1948, *Brenan* 9570 at Mann's Spring on Mt Cameroon, it was not rediscovered until 4th March 1988 on the NE slope of that mountain: NE slopes, radio station road, 1500–2200 m, *Nemba* 938. Summerhayes (1957) states that 'it has the largest flowers so far recorded in the genus'. The fact that it has so rarely been collected, despite its relatively large flowers, suggests that it might be extremely scarce. A specimen believed to be this species was collected in forest near Mbesa in March 2000, *DeMarco* 56. This represents the first and only record of *Genyorchis macrantha* away from Mt Cameroon.

Habitat: epiphytic on tree trunks at forest margins, c. 1500–2260 m alt.

Threats: forest habitat at high altitude on Mt Cameroon remains relatively intact and the 2010 creation of the 58,000 ha Mt Cameroon NP should afford a high level of protection to higher submontane and montane plant species. Forest clearance in NW Region, however, has in places reached 96.5% of the original cover, and the Mbesa location must be considered threatened, especially since the winding down of conservation programmes run by BirdLife International in the area during the late 1990s and early 2000s.

Management suggestions: further surveys on Mt Cameroon should be instigated to revisit the type location, as well as the Mbesa forest, Ijim Ridge to try to relocate the probable population there. The Ijim specimen still remains to be confirmed, and an investigation into its identification must surely form a primary focus of research into the distribution and conservation of this rare species.

Genyorchis micropetala (Lindl.) Schltr.

EN B1ab(iii)+2ab(iii)

Range: Equatorial Guinea (Bioko (3 loc., 3 coll.)), Cameroon (SW: Banyang Mbo; Mt Kupe (1 coll. each)) and Congo (Kinshasa) (1 doubtful coll., excluded).

Status: ↑ known from five confirmed locations, with an AOO of 20 km^2 and an EOO of 4960 km^2, this species is here considered EN according to Criterion B. It was described from Bioko, based on *Mann* 144, December 1860, 1330 m, with two further Bioko collections: *Melville* 437A, 2nd September 1959, 1400 m, from Moka, and *Sanford* 4267, 4th January 1967, 1650m, Pico Boca. A fruiting collection from Congo (Kinshasa), *Burn* s.n., June 1948 is a doubtful record, and determined as *G. ? micropetala*, hence is not included here in the assignation of a category of threat and does not feature in the distribution map. Since the assessment by Pollard & Darbyshire (Cheek *et al.* 2004: 197), it has become clear that two of the Cameroon collections, *Sanford* 5256 & 5274, both made on 27th October 1968, Batouri district, near Dimako, have been referred to *G. sanfordii* Szlach. & Olszewski by Droissart whose redeterminations are marked on the K specimens. The two Cameroon collections are

Letouzey 477, November 1970, from Mt Kupe, and Banyang Mbo, *Simo, P.* 116.

Habitat: submontane forest, or tree in open grassland; c. 1330–1650 m.

Threats: small-scale forest clearance of montane forest on Bioko and in Cameroon may threaten the subpopulations. Though this is currently limited within this species' altitudinal range, expansion of agriculture into higher altitude forest is likely to occur in the near future due to growing demands for additional agricultural land, thus threatening unprotected sites.

Management suggestions: attempt to relocate this species at its Bioko and Cameroon locations, and introduce it into cultivation should be made. Further investigation of the Congo (Kinshasa) material is advised in order to establish its identity.

Genyorchis platybulbon Schltr.

VU B2ab(iii)

Range: Cameroon (SW: Mt Cameroon (Limbe (Victoria); Moliwe (2 loc., 2 coll.)). Littoral: Douala-Edéa Reserve (1 coll.). S: Nkoltsia (1 loc., 3 coll.); Akoakas (2 coll.); Djam-Oveng (1 coll.); Akom II (1 coll.)) and Gabon (1 coll.).

Status: ↓ assessed by Cable & Cheek (1998: lxxiii) as CR A1c+2c on the basis of Summerhayes' (1968: 242) listing this species as being known from only two collections. This species is here downgraded to VU on the basis of a number of additional specimen locations arising in recent years. It is now known from seven locations with an AOO of 28 km^2 and an EOO of 39740 km^2. The first specimen was from Moliwe, *Stammler* s.n., 1900 (holotype B+, drawing K!), the second from Victoria (now Limbe), *unknown collector in FHI* 11155. A third collection from Douala-Edéa Reserve near Lombe Camp, more or less at sea level (*Thomas* 311, February 1978) awaits confirmation. Three additional specimens in the Paris herbarium were collected in 1974 from 'Colline de Nkoltsia', 18 km west of Bipindi, *Villiers J.F.* 811, 22nd March; *Villiers J.F.* 824, 23rd March; *Villiers J.F.* 876, 24th April, and were all determined by Sanford as this species. Droissart *et al.* (2006: 34) list an additional four recent Cameroon collections from S Region, from Akoakas, *Parmentier & Kouob* 1939 & 2127; Djam-Oveng, *Stévart & Droissart* 2034; Akom II, *Stévart & Droissart* 2128. Szlachetko *et al.* (2004 : 209–212) cite *Hallé* 4636b, 21st January 1968, Chutes Kinguélé, Monts de Cristal, as belonging to this species, representing a new record for Gabon and a range extension.

Habitat: epiphyte in lowland to submontane evergreen forest, inselbergs; [illegible]

Threats: forest clearance for agriculture, particularly plantations, and urban expansion.

Management suggestions: if this species is already in cultivation in the orchid shadehouses, then those live specimens could be used as the basis for a programme of reintroduction into locations on Mt Cameroon where it may now be lost.

Genyorchis sanfordii Szlach. & Olszewski

EN B2ab(iii)

Range: Cameroon (E: Dimako (1 loc., 2 coll.); Mindourou (1 loc., 9 coll.)), Equatorial Guinea (Rio Muni: Akoak Benguan (1 loc., 2 coll.), Gabon (1 coll.); Congo (Brazzaville) (*fide* Fl. Gabon) (unconfirmed).

This species is known from four (or perhaps five) locations with an AOO of 16 (or 20) km^2 and and an EOO of 57900 km^2, and is thus considered as EN. It was recently described and based on *Sanford* 5256, 27th October 1968, in a forest exploitation area near Dimako, Haut-Nyong [Ngom], Cameroon. *Sanford* 5257 also belongs to this species (according to an annotation on the K specimen). Nine further collections were made by Stévart's group from the same location in Mindourou near Lomié between 2001 and 2007 (see Droissart *et al.* 2009: 125). The Gabon specimen is *Louis* 320, from forest eploited to S of Ekouk. The Congo specimen has not been confirmed by the author and is not included in this assessment. Droissart *et al.* (2009: 125) confirm the first two records from Equatorial Guinea (Rio Muni).

Habitat: epiphyte in lowland forest; 100–600–750 m.

Threats: all of the Cameroon specimens have been collected at or near logging concessions and must therefore be considered at risk.

Management suggestions: this species is already in cultivation with Stévart's group. Propagation of those specimens could then lead to reintroductions into the wild at suitable locations that have a measure of protection available.

Habenaria alinae Szlach.

CR B1ab(iii)+2ab(iii)

Range: Cameroon endemic (NW: Menda Nkwe (1 coll.)).

This species is known from just one location, with an EOO of less than 100 km^2 and an AOO of 4 km^2 which merits assignation of CR for this species. The holotype was collected from Menda Nkwe, 5 km SE Bamenda, August 1975, *Letouzey* 14298.

Habitat: terrestrial herb, on mossy rocks in submontane forest; 1700–1800 m.

Threats: reduction in forest cover has been extensive in the Bamenda Highlands, NW Region, in some places reaching levels as high as 96.5%.

Management suggestions: revisit the type location in an effort to relocate this species and assess if it can be introduced into cultivation, and perhaps any existing populations be protected in situ.

Note: Szlachetko (1998: 116) omitted the last digit of the type specimen of Habenaria alinae meaning that they designated Letouzey 1429 instead of Letouzey 14298 as the holotype. The error is corrected here following on from Jean-Michel Onana's original identification of this error in his manuscript for this publication.

Habenaria batesii la Croix (syn. *Podandriella batesii* (la Croix) Szlach. & Olszewski)
EN B2ab(iii)

Range: Cameroon endemic (SW: Bakossi Mts (1 coll.). S: Efoulan; Akom II, Ebianemeyong (1 coll. each).

This species is known from four locations with an AOO of 16 km^2 and an EOO of 12020 km^2 and therefore merits the EN category of threat. It was described in 1993 from a single collection from Grand Batanga, Efoulan, near Ébolowa in S Region, Cameroon, October 1895, *Bates* 453. It has not been recorded at this site since, and remained the only record of this species until *Pollard* 731, collected at Enyandong in Bakossiland, in October 2001. This taxon was originally published by Isobyl la Croix (1993: 371) but proved to be a later homonym of *Habenaria praetermissa* Seidenf., and this was corrected by la Croix (1996: 364). *Podandriella batesii* (la Croix) Szlach. & Olszewski is considered synonymous by Govaerts *et al.* (2011). Peguy Tchoutou (Mbatchou)'s PhD thesis lists two collections from Ebianemeyong and Akom II.

Habitat: a terrestrial herb of dense lowland and submontane forest; c. 200–900 m.

Threats: Efoulan is located on the main route from Lolodorf to Ébolowa, two sizable towns in S Region, and is therefore likely to have experienced significant habitat disturbance, due to both population and agricultural expansion; this site may well now be lost as the species was last recorded here over a century ago. The Bakossi site is threatened by encroachment of small-holder farming into the submontane forest habitat, although it may receive some longer term protection by the creation in 2007 of the Bakossi NP.

Management suggestions: attempts to rediscover the species at its known locations should be made, especially in Bakossiland, which probably represents its best long term chance of survival.

Habenaria bosseriana Szlach. & Olszewski
CR B1ab(iii)+2ab(iii)

Range: Cameroon (W: (1 coll.)).

Known only from one location with an EOO of less than 100 km^2 and an AOO of 4 km^2. *H. bosseriana* is considered to deserve the CR category of threat. It is based solely on the type specimen, collected near Lake Monoun, NNW Foumbot, 11th July 1972, *Leeuwenberg* 10199.

Habitat: terrestrial herb in savanna; 1200 m.

Threats: although savanna vegetation is probably less threatened than forest or submontane forest in W and NW Regions, satellite imagery of the area around Foumbot and Lake Monoun clearly show high levels of anthropogenic disturbance to the vegetation. It is here inferred that high human population densities will have resulted in conversion of land into cultivation, which would pose a serious threat to this species.

Management suggestions: revisit Lake Monoun in an effort to relocate this species.

Habenaria egregia Summerh.
EN B2ab(iii)

Range: Cameroon (Adamawa: Ndene at, Meïganga (1 coll.)), Ethiopia (1 loc., 2 coll.) and Kenya (1 loc., 1 coll.).

This magnificent species has an unusual Afromontane distribution, being known from three locations with an AOO of 12 km^2 and an EOO of 1221240 km^2 having been recorded from two E African countries, Ethiopia, Assossa, Mt Inzi, 27th June 2000, *Hermann* 124 and 125 (see Demissew *et al.* 2004: 90), and Kenya, Kakamega, July 1947, *Barney* 5194. It was previously considered to be a Cameroon endemic, collected in Adamawa at Ndene, on the way from Ngaoundéré to Meïganga, July 1939, *Jacques-Félix* 4379.

Habitat: submontane savanna, open grassland, meadow in shade, on rocky outcrop; 1600m.

Threats: the Cameroon collection is along a main road near urban development which suggests that there may exist some form of threat from the populations of those places.

Management suggestions: links with conservation organisations in E Africa could be developed to highlight the rarity of this species. The Cameroon location could be revisited and searched to better understand the habitat requirements for this species and assess its suitability for a propagation programme.

Habenaria maitlandii Summerh.
EN B2ab(iii)

Range: Nigeria (1 coll.), Cameroon (NW: Ijim Ridge (1 coll.)) and Gabon (1 coll., doubtful).

Status: ↓ Cheek *et al.* (2000: 83) assessed this species as CR A1c, B1+2c, when it was considered to be known only from the type collection, Maitland 1386, at Nchang, Ijim Ridge, Cameroon. That assessment was followed by Droissart *et al.* (2006: 39). It is here downgraded to EN on the basis of two further collections coming to light, one from Nigeria: Oyo, Ibadan, 9th November 1961, *King* 144 (K spirit collection only) and Gabon: Estuaire, Pointe Denis, Pongara, Ovin, 20th December 1999, *Simons, E.* 447. It could then be considered to be known from three locations with an AOO of 12 km^2 and an EOO of 253170 km^2. The Gabon collection seems somewhat incongruous having been collected near to sea level, whereas the type collection was from about 2000 m above sea level. As such the Gabon location has not been included in the distribution map.

Habitat: on mountain slope among rocky grassland; 1750–1860 m.
Threats: grazing from cattle and horses, or even the change of use of land from grazing land to a more settled form of agriculture.
Management suggestions: Nchang should be revisited and this species found, so that it may be protected or propagated or both. Investigation into the authenticity of the determinations of the other two specimens identified as representing this species should be made, especially the Gabon collection.

Habenaria microceras Hook.f.
VU B1ab(iii)+2ab(iii)
Range: Equatorial Guinea (Bioko (1 coll.)) and Cameroon (SW: Mt Cameroon (2 loc., 6 coll). NW: Mt Neshele (1 coll.); Lake Bambuluwe (1 coll.); Mt Oku (1 loc., 3 coll.); Laikom (1 coll.)).
Status: ↑ assessed as LR nt in Cable & Cheek (1998: lxxiii), Cheek *et al.* (2000: 83) and Droissart *et al.* (2006: 40), it is here considered threatened and upgraded to VU, being known from seven locations with an AOO of 28 km^2 and an EOO of 3450 km^2. The three early Mt Cameroon collections were November 1862, *Mann* 2116 (holotype); December 1886, *Johnston* 31 & 32, and Buea, 24th September 1891, *Preuss* 967. Rediscovered in October 1992 on Mt Cameroon, when two collections were made on 8th October 1992, near Bokwango Hut 3, *Thomas* 9381 and *Cheek* 3641, the latter at a study of the forest-grassland ecotone at about the same altitude. The NW Region collections are from Mts Neshele (= Mba Mfeu), 10 km ESE Bamenda, 12th August 1975, *Letouzey* 14271; Lake Bambuluwe, 1947 – 1952, *Savory* UCI 462, and four collections from Mt Oku and the Ijim Ridge: Mt Oku, summit grassland, 29th October 1996, *Munyenyembe* 775; *Zapfack* 1069 & 1139; 11th December 1998, foot of Akwamofu Sacred Forest, Laikom, *Etuge* 4598. The Bioko collection was made on 1st September 1959, Moka, *Melville* 417.
Habitat: forest and montane grassland, sometimes at the ecotone with forest (rarely epiphytic); (1400–)2000–3100 m.
Threats: although threats at this higher altitudinal range are fewer than at lower altitudes, there is evidence of continuing forest clearance in the Bamenda Highlands, and this species has been recorded as an epiphyte. Fire, grazing, trampling, migration of settlers from other areas and conversion of grassland to agricultural production are all potential threats at higher altitudes.
Management suggestions: a survey to estimate the threats to the various populations would certainly be worthwhile. Efforts to conserve the Mt Oku and Mt Cameroon locations may well be achievable given the levels of protection already afforded to those locations.

Habenaria nigrescens Summerh. (syn. *Roeperocharis nigrescens* (Summerh.) Szlach. & Olszewski)
VU A2c; B2ab(iii)
Range: Nigeria (Obudu Plateau (1 coll.)) and Cameroon (Littoral: Mwanenguba (1 coll.). NW: Bafut Ngemba FR (1 loc., 4 coll.); Bamenda Nkwe-Bafut Ngemba footpath (1 coll.); Mt Neshele, 10 km ESE Bamenda (1 coll.); Dom (1 coll.). Adamawa: Banyo, Mayo Tankou (1 coll.)).
It was assessed as VU A2c; B2ab(iii) by Pollard & Darbyshire (Cheek *et al.* 2004: 198) and Pollard (Cheek *et al.* 2010: 107) which is followed here, on the basis of seven locations with an AOO of 28 km^2 and an EOO of 28890 km^2, but one of the locations (Ngaoundéré) is considered dubious, and so is excluded, but the recent Dom collection is included (see the reasoning for this below). The previous stronghold of this highly localised taxon was the Bafut Ngemba FR in NW Province, Cameroon, from where the type specimen was collected, *Daramola in FHI* 41568, 3rd July 1959, whence four other collections were made between 1959 and 1975: *Brunt* 774; *Daramola in FHI* 41777; *Lowe* 72 and *FHI* 50718. *Daramola in FHI* 41568 was made along the Bamenda-Nkwe to Bafut Ngemba footpath. Other collections from NW Region are Mt Neshele, 10km ESE Bamenda, 12th August 1975, *Letouzey* 14257 and the very recent collection from Dom, 26th September 2006, *Cheek* 13592. Outside of the Bamenda Highlands, it is known from Banyo, vallée du Mayo Tankou, 1955, *Saxer* 340, a single collection on 21st June 1973 from the Obudu Plateau of Nigeria, *Hall* 2952, and from a 1971 collection from Mwanenguba, Littoral Region, 27th September 1971, *Leeuwenberg* 8473. An unconfirmed and somewhat disjunct record, *Meurillon in CNAD* 844, 5th August 1967, made near Ngaoundéré, Adamawa Province, is recorded at a considerably lower altitude (1100 m) and in much drier climatic conditions to the other known collections, the specimen being labelled as '*Habenaria* sp. aff. *nigrescens*' and likely refers to a separate entity. It is therefore excluded from this assessment.
Habitat: montane grassland, sometimes on wet rocky slopes; c. 1700–2300 m.
Threats: a significant reduction in both the quality and area of natural habitat has been recorded in the Bafut Ngemba FR (Pollard, *pers. obs.* 2002), and it is likely that this will have had deleterious effects on all wild plant taxa, irrespective of their habitat preferences. Similarly, high levels of pressures from fire, grazing, trampling, migration of settlers from other areas and conversion of grassland to agricultural production are all potential threats at higher altitudes. These human-led activities on the natural vegetation are continuing in the montane sites at Obudu, Nigeria, and Mwanenguba, Cameroon.
Management suggestions: a survey of this taxon should be made at the Bafut Ngemba FR and protective measures

should be put in place here if it remains extant there. Other locations should be resurveyed in an effort to introduce this species into cultivation.

Habenaria obovata Summerh.

EN B1ab(iii)+2ab(iii)

Range: Cameroon (SW: Mt Cameroon (4 loc., 8 coll.). NW: Mt Oku (1 coll.)).

Status: ↑ assessed as VU D2 by Cheek *et al.* (2000: 84), that category of threat is upgraded here to EN under Criterion B on account of it being known from five locations with an AOO of 20 km^2 and an EOO of 1460 km^2 and there being increased levels of threat to habitat at Mt Oku in recent years. The protologue of *Habenaria obovata* (Summerhayes 1932) stated: 'a remarkable species without any near relative known to me', yet curiously it was omitted from the Flore du Cameroon treatment (Szlachetko & Olszewski, 1998) that covers the terrestrial orchid genera. Until 1992, *Habenaria obovata* was known from six historical collections: December 1886, *Johnston* 29 (holotype), March 1896, *Mary Kingsley* s.n., 24th September 1891, *Preuss* 980, December 1929, *Maitland* 804, September 1954, *Boughey* s.n. who states that it was collected in grassland at 2820 m, above Mann's Spring, noted as being 'reasonably common in grassland, seen down to around 2400m', and 21st May 1956, *Cole* 526. Recent collections from Mt Cameroon include 9th October 1992, *Sidwell* 39, and also between Huts 2 and 3 above Buea, 2nd November 1995, *Banks* 104. Dr Zapfack Louis appears to be the first person to have collected this species away from Mt Cameroon, 31st October, 1996, summit of Mt Oku, *Zapfack* 1140.

Habitat: terrestrial herb in montane grassland; 2150–3050 m.

Threats: fire, grazing and trampling are potential threats to this species in its habitat at the summit of Mt Oku particularly now that the BirdLife International project at Kilum has now wound down and entry to the forests and grasslands of Mt Oku are now possibly unfettered. The 2010 gazettement of the Mt Cameroon NP is a welcome benefit for this species, whose primary stronghold is in the upper elevational grasslands there. The level of threat to those Mt Cameroon locations should now have reduced as a result, although habitat there is still vulnerable to large-scale stochastic events such as a volcanic eruption.

Management suggestions: a survey to estimate the size of the population on the summit of Mt Oku would be useful. An assessment of the significance of the threats mentioned, and annual monitoring of the population is advisable.

Habenaria phantasma la Croix (syn. *Podandriella phantasma* (la Croix) Szlach. & Olszewski in Flore du Cameroun 34: 198 (1998))

EN B2ab(iii)

Range: Cameroon (S: 1 loc., 3 coll.) and Congo (Brazzaville) (Foungouti (1 coll.)).

H. phantasma is known from only two locations with an AOO of 8 km^2. It was until May 2007 known by just the type specimen, from Congo (Brazzaville), Niari Region, Col de Bamba (Mayombe), 9th December 1990, *la Croix* 1090. Three additional collections were made from Bidou III/Nkolembonda (Kribi-Ébolowa Road): *Droissart & Simo* 560, 561, 568 (Droissart *et al.* 2009: 125).

Habitat: terrestrial herb in evergreen forest; 500 m

Threats: presumably the location along the Kribi-Ébolowa Rd is subject to increasing threat from urbanisation and migration and settlement along this arterial route.

Management suggestions: revisit the Cameroon location and introduce this species into cultivation.

Habenaria sanfordiana Szlach. & Olszewski (syn. *Ceratopetalorchis sanfordiana* (Szlach. & Olszewski) Szlach., Górniak & Tukałło Richardiana 3(4): 162 (2003); Szlachetko (2008: 52).

EN B2ab(iii)

Range: Cameroon (Adamawa: 1 coll.), Ivory Coast (1 coll.) and Congo (Kinshasa) *fide* Govaerts *et al.* (2011) (doubtful loc.).

Known only from two confirmed locations with an AOO of 8 km^2, this species was assessed as CR by Droissart *et al.* (2006: 41) but is here considered as EN according to Criterion B. It is known from the type specimen, Adamawa: Ngaou, route Ngaoundéré-Meïganga, July 1939, *Jacques-Félix 4004*, and Ivory Coast, Parc National de la Maraoué, 11th August 1975, *Pérez-Vera* 801 *pro parte*. The Congo (Kinshasa) specimen could not be located by the author and is considered doubtful.

Habitat: terrestrial herb, in swampy forest, montane grassland; altitude not recorded.

Threats: the Ngaoundéré-Meïganga Rd is a major thoroughfare and so it is here inferred that there has been a decline in suitable habitat quality at this location most probably through conversion of suitable habitat into agricultural land.

Management suggestions: revisit the type locality to assess the current threats to the typical habitat and levels of agricultural land conversion.

Habenaria thomana Rchb.f.

VU B1ab(iii)+2ab(iii)

Range: Equatorial Guinea (Bioko (1 coll.)), São Tomé & Príncipe (São Tomé: 4 loc., 4 coll.) and Cameroon (SW: Kumba (1 coll.); Bwassa (1 coll.); Mt Kupe (2 loc., 7 coll.)).

H. thomana is known from nine locations with an AOO of 36 km^2 and an EOO of 7440 km^2, and is assessed as VU. Pollard & Darbyshire (Cheek *et al.* 2004: 198) assessed this species as VU B2ab(iii), but it also here qualifies under Criterion B1 with an EOO of less than 20000 km^2. It was described based on *Mann* s.n., 1861, on São Tomé, one of four collections from that island, with two made in the late nineteenth century, between Lagoa Amelia and San Pedro, September 1885, *Moller* 95 and Agua Grande, March 1888, *Quintas* 96, then on 14th August 1997 at Calvario, *Stévart* 39. In 1972 it was collected for the first time in Cameroon, along the Victoria (Kumba-Douala) Rd, south of Bombe, *Leeuwenberg* 10617. It is now also known from above Bwassa village on Mt Cameroon, and from seven collections made in the 1990s from two locations on Mt Kupe, and from Bioko (see Cribb & Pollard: 463).

Habitat: a terrestrial herb of lowland to submontane forest, rarely in secondary forest; 50–1330 m.

Threats: clearance of forest throughout its range, especially below 1000 m.

Management suggestions: management should focus upon its two strongholds of São Tomé and Mt Kupe. A survey should be carried out at the former in order to assess its current status, as three of the four known collections were made in the nineteenth century. At the latter, material from the more-threatened lower altitude sites should be considered for cultivation and propagation at the former CRES headquarters' living orchid collection at Nyasoso on Mt Kupe, which is probably in need of funding for refurbishment.

Halleorchis aspidogynoides Szlach. & Olszewski
EN B2ab(iii)

Range: Cameroon (S: Mébandé (1 coll.)) and Gabon (Belinga (1 coll.)).

This species is known from only two locations with an AOO of 8 km^2 and as such qualifies as EN under Criterion B. In Cameroon it is known from SW of Mébandé, 20 km SE Lolodorf, 22nd January 1973, *Letouzey* 12750. The type collection was made in Gabon at Belinga, *Hallé* 1251 and represented a genus new to science.

Habitat: terrestrial or on wet rocks in lowland and submontane forest; 850–950 m.

Threats: submontane forest on the Kribi to Yaoundé Rd must be considered threatened in the long term as this route is likely to increase in its popularity, with visitors increasingly wishing to travel directly from the capital city towards the beaches and increasing numbers of holiday resorts on the coast at Kribi. It is suspected that this will increase the levels of migration of people towards this arterial route with a probable increase in consequent small-scale agriculture and subsistence farming.

Management suggestions: revisit the Cameroon locality in an attempt to rediscover this species and introduce it into cultivation.

Holothrix tridentata (Hook.f.) Rchb.f.
VU B2ab(iii)

Range: Cameroon (SW: Mt Cameroon (2 loc., 5 coll.)), Sudan (Jebel Marra (1 coll.)) and Ethiopia (4 loc., 5 coll.).

Status: ↑ this species was not included in the Mt Cameroon Red Data account (Cable & Cheek, 1998: lxxiii), but is here assessed as VU. It is known from seven locations with an AOO of 28 km^2 and an EOO of 1920240 km^2. The Cameroon collections are: Mt Cameroon, December 1886, *Johnston* s.n; November 1862, *Mann* 2128; Buea 1891, *Preuss* 1036, above Mann's Spring, 3rd October 1992, *Thomas* 9269, 1500 m west of Mann's Spring, 4th October 1992, *Tchouto* 336. The Sudan collection was made on 8th September 1964 in upland meadow, *Wickens* 2450. Five collections have also been made from four locations in Ethiopia.

Habitat: steep slopes, grassland with scattered thickets; 2100–3350 m.

Threats: the higher elevation grassland on Mt Cameroon does not currently suffer from any obvious threat, but the subpopulations there could be threatened by stochastic events, such as fire or volcanic eruptions. The potential threats to habitat at the Sudan and Ethiopia locations are not presently well known to the author, but it is inferred that changes in land use for agricultural production presents a potential threat in Ethiopia.

Management suggestions: revisit Mann's Spring and locate this species with a view to surveying numbers of individuals, ecological associations and suitability for introduction into cultivation. Investigate more fully the level of threat at the East African locations.

Liparis goodyeroides Schltr.
VU B2ab(iii)

Range: Liberia (1 coll.), Cameroon (SW: Mt Cameroon (1 coll.); Takamanda (1 coll.). W: Mbaw Plain (1 coll.)), São Tomé & Príncipe (São Tomé (1 coll.). Príncipe (2 loc., 4 coll.)).

Status: ↓ Cable & Cheek (1998: lxxiii – lxxiv) assessed this species as CR A1c+2c, but this species is here downgraded to VU due to country records for Liberia, São Tomé and Príncipe being reported by Droissart *et al.* (in press), and it overall being recorded from seven locations with an AOO of 28 km^2 and an EOO of 762290 km^2. The Nigerian record in Cable & Cheek (1998: lxxiv) was apparently erroneous, based on *Barter* 2029, which was actually collected from Príncipe Island. *Brunt* 479, collected south of Ngu at the Mbaw Plain, was designated as the holotype of Szlachetko's species, *Liparis* [illegible] Szlach. published in 1993, but [illegible] it to synonymy under *L. goo*[illegible] collections were from Moliwe, [illegible], *Stammler* s.n., Takamanda F.R., 30th April 1987, *Thomas* 7351, and the Plaine des Mbos (=Mbaw), 5 km south of Ngu, 22nd May 1962, *Brunt* 479. The Liberian collection was made from the Nimba Mts, 10th May 1965, *Yallah* 98. The Príncipe collections are *Barter* 2029; 6th October 1997, *Stévart* 225 and 1st November 1998, *Stévart* 435. The São Tomé specimen(s) have not yet been seen by the author, but the presence of this species there is confirmed by Stévart & de Oliveira (2000: 163) and Govaerts *et al.* (2011).

Habitat: terrestrial in lowland evergreen forest, usually on foothills; 300–550 m.

Threats: forest clearance for agriculture (particularly plantations) and firewood in the lowlands of SW Region is considerable, but the addition of other locations for this species now merits consideration of this species under Criterion B and a lowering of the category of threat.

Management suggestions: although this species was not rediscovered in the botanical inventory of Mabeta-Moliwe in 1992, nor in 1993, it may not yet be extinct on Mt Cameroon, and efforts should be made to rediscover it in Mabeta-Moliwe.

Liparis gracilenta Dandy

EN B2ab(iii)

Range: Cameroon (SW: Korup (1 coll.). S: Akom II (1 coll.)) and São Tomé & Príncipe (São Tomé (1 coll.). Príncipe (1 coll.)).

L. gracilenta is known with certainty from four locations with an AOO of 16 km^2 and an EOO of 108180 km^2, although further locations on São Tomé and Príncipe may come to light. The type collection was made from Príncipe Island, *sine loc.*, *Quintas* 9, where it occurs uniquely in the forests covering the summit. Stévart & de Oliveira (2000: 164) confirm its presence in montane forests and old secondary forest on São Tomé, but the numbers of locations are not precisely known to the author on either island. Two Cameroon collections are known, on a steep hillside south of Esukutang village, 25th May 1988, *Thomas* 7893 and Akom II, east of Kribi, *Stévart & Droissart* 2115.

Habitat: lowland to submontane evergreen forest; 300–500(–1050) m.

Threats: forest clearance in the lowlands of Cameroon must pose a threat at Akom II, although the Korup NP should afford this species some protection.

Management suggestions: revisit Esukutang village in Korup to try to relocate this species at that location and attempt to introduce this species into cultivation. Investigation is required into the confirmed numbers of collections from São Tomé and Príncipe, which may be more than cited here.

Liparis kamerunensis Schltr.

CR B1ab(iii)+2ab(iii)

Range: Cameroon (SW: Mt Cameroon (2 coll.)).

Status: ↑ *L. kamerunensis* is known from only one location, with an EOO of less than 100 km^2 and an AOO of 4km^2. Cable & Cheek (1998: 176) listed this as a synonym of *L. deistelii* Schltr., but Govaerts *et al.* (2011) treat it as an accepted species, which we follow here. The type collection was made on Mt Cameroon, November 1862, *Mann* 2129 and a second collection was made on 11th October 1992 in grassland and mist forest, Likombe to Mann's Spring, *Thomas* 9448.

Habitat: upper submontane evergreen mist forest, grassland; 1850–2300 m.

Threats: although the upper elevations of Mt Cameroon should now be protected by the recent gazettement of the National Park, forest clearance and other detrimental human activities could still pose a threat to this species in the long term. Until it becomes clear how effective the protected area is proving, this species will still be considered to be under threat, according to the precautionary principle.

Management suggestions: revisit the known locations in an attempt to relocate this species and assess its suitability for cultivation and propagation. If relocated, then perhaps its location could be fenced off and given more robust protection. Searches should be made at its preferred habitat at other locations around Mt Cameroon and perhaps other mountains in the western Cameroon upland chain, such as Mt Mwanenguba, Mt Kupe, Mt Oku and the Bamboutos Mts.

Liparis letouzeyana Szlach. & Olszewski

CR B1ab(iii)+2ab(iii)

Range: Cameroon (NW: Bambuluwe (1 coll.)).

L. letouzeyana is known from just one location, with an EOO of less than 100 km^2 and an AOO of 4 km^2, represented by only the type specimen, near Lake Bambuluwe, 10 km SSE Bamenda, 13th August 1972, *Letouzey* 11623.

Habitat: montane grassland; 2000 m.

Threats: a significant reduction in both the quality and area of natural habitat has been recorded in the Bafut Ngemba FR which corresponds to the Bambuluwe locality, and it is likely that this will have had deleterious effects on all wild plant taxa, irrespective of their habitat preferences. Much of this land has been turned over to *Eucalyptus* monoculture plantation and cropping and facilities for tourists or other visitors were seen to be under construction (Pollard pers. obs. 2002). Grassland habitat here is also subject to grazing and fire.

Management suggestions: a survey of the only known location should be instigated as well as efforts to understand the habitat management regime in this area and understand whether it is dominated by fire, grazing, forest management techniques or tourism and what the inter-relaitonships of these are.

Liparis suborbicularis Summerh.

VU A2c; B2ab(iii)

Range: Nigeria (Oban; Mambilla Plateau; Chappal Waddi (1 coll. each)) and Cameroon (SW: Korup (1 coll.). NW: Oku (2 coll.); Bambili (2 coll.); Bambui (1 coll.); Nkambe to Binka (1 coll.). S: Lolodorf (1 coll)).

Known from nine locations with an AOO of 36 km^2 and an EOO of 64260 km^2, *L. suborbicularis* is considered VU according to Criterion B, and is also here assessed as VU under Criterion A, with a suspected population reducton of ≥ 30%. The type collection was made in Nigeria, Oban, *Talbot* 871, with two additional Nigerian collections, 3rd June 1972, Chappal Waddi, *Chapman* 2866 and 29th June 1972, Mambilla Plateau, *Chapman in FHI* 46249. NW Region collections comprise: 3 miles NE of Bambili, 26th June 1970, *Bauer* 87; Bambui, 1976, *Botte* 511; Nkambe to Binka, *Savory* 380; Ewoof Etele, Oku-Mbae, 6th November 1996, *Etuge* 3344; Kilum mountain forest, KC, 30th October 1996, *Zapfack* 1155. The SW Region collection is from the south of Korup NP, 6th July 1983, *Thomas* 2254. Two specimens at K previously identified as belonging to this species were referred to *L. deistelii* by Stévart, on 17th May 2003 and were included under that name by Pollard & Darbyshire (Cheek *et al.* 2004: 464). The checklist of Gabonese Vascular Plants (Sosef *et al.* 2006: 312) cites four specimens identified by Stévart under the name *L. pentagonalis* Szlach., which was considered by Govaerts (2011) and Droissart *et al.* (2006: 45) to be a synonym of *L. suborbicularis*. Further recent examination of these Gabon collections by Droissart *et al.* (in press) has led to the conclusion that the two species are in fact distinct.

Habitat: epiphyte in forest; 50–2350 m.

Threats: the stronghold of this species is in NW Region, Cameroon and the adjoining high altitude regions of Nigeria. The reduction in forest cover in these areas is well documented (e.g. Cheek *et al.* 2000: 49–50) and as such this species is assessed to qualify as VU under Criterion A.

Management suggestions: revisit the various Cameroon locations in order to conduct surveys to investigate numbers of individuals, typical phorophyte species and other characteristic species of its forest habitat with a view to identifying areas of suitable forest habitat within its extent of occurrence that could be surveyed for this species, to identify key sites for its conservation. Areas that are already within protected areas could be targeted as the best options for *in situ* conservation.

Manniella cypripedioides Salazar, T.Franke, Zapfack & Beenken

EN B1ab(iii)+2ab(iii)

Range: Equatorial Guinea (Bioko (1 coll.)) and Cameroon (SW: Banyang Mbo (1 coll.); Mt Cameroon-Mt Etinde (1 coll.); Bakossi Mts (2 loc., 2 coll.)).

This taxon is currently known from only five locations with an AOO of 20 km^2 and an EOO of 1750 km^2, all within the western Cameroon uplands chain. It was first described by Salazar *et al.* (2002), and given a conservation rating of DD (data deficient), which was followed by Droissart *et al.* (2006: 46), on the basis that the status of the first Bakossi site, track to Mwanzoum, 8th December 1999, *Muasya* 2021, Kodmin and the Bioko population could not apparently be assessed. However, this assessment was modified to EN by Pollard & Darbyshire (Cheek *et al.* 2004: 198–199) under Criterion B2, and included an additional collection made in November 2001 at Bime Rock, Mwendolengo, Bakossi, 12th November 2001, *Etuge* 4515r. The assessment is updated to include Criterion B1 on account of its EOO being less than 5000 km^2. A series of Martin Etuge's collection numbers were accidentally repeated and so the suffix 'r' has been added to distinguish between these concurrent sequences of numbers. Other Cameroon collections are SW Region: eastern slope, just below the pass between Mount Etinde and Mount Cameroon, c. 6 km west of the village of Ekonjo, *Franke* 53, 14th October 2001; Nden, Banyang Mbo, 28th October 2010, *Salazar* 6323 (holotype) from an area which is currently a protected wildlife sanctuary. The Equatorial Guinea collection is Bioko, trail from Rinché to Caldua San Caulas, where forest at c. 1300 m alt. remains largely intact, 11th January 1967, *Sanford* 4376, recorded as 'frequent in dense shade', thus this species may be gregarious within its specialised habitat.

Habitat: a terrestrial herb along streams in primary submontane forest; 950–1350 m.

Threats: the proposal to expand agricultural plantations to the 1000 m contour around Mt Etinde-Mt Cameroon may well adversely affect the subpopulation there. A threat to the Kodmin population may exist from a proposed reservoir scheme has been proposed near the village. Its habitat is believed to be largely unthreatened at some of the other locations.

Management suggestions: attempts should be made to discover this species within its preferred habitat in other parts of the western Cameroon uplands (its recorded flowering period is October to December).

Microcoelia sanfordii L.Jonss. (syn. *Encheiridion sanfordii* (L.Jonss.) Senghas, Schlechter Orchideen 1(16–18): 1066 (1986)).

EN B1ab(iii)+2ab(iii)

Range: Cameroon (Adamawa (2 loc., 2 coll.)).

This species is considered EN, being known from two locations with an AOO of 8 km^2 and an EOO of less than 5000 km^2. The two collections are both from Adamawa Region: River Mba, near Lokoti, 1968, *Sanford* 6189; Gangi, près Meïganga, on the way to Djohong, 1968, *Sanford* 6223.

Habitat: epiphyte in dry or riverine forest and savanna; 800–1000 m.
Threats: the four place names mentioned in the specimen data (Lokoli, Gangi, Meïganga, Djohong) are all along the main N1 and/or D22 Rds in Adamawa Region and so are likely to have been subject to historic disturbance which is inferred only to have increased over the last 40 years with the increasing numbers of vehicles, transport and travel in Cameroon.
Management suggestions: revisit this stretch of road and search again for this species. If it is located, record additional information on typical habitat requirements, phorophyte species, and if a sufficient number of specimens are found, collect live material to introduce some into cultivation.

Nephrangis bertauxiana Szlach. & Olszewski

CR A2c
Range: Cameroon (S: Bindem (1 coll.)) and Gabon (1 coll.).
N. bertauxiana is known from just two locations with an AOO of 8 km^2 and is assessed as CR. This species was described based on a single collection from Gabon, collected from near the aerodrome, Bélinga, 28th October 1964, *Hallé* 2847. The second ever collection was collected from Cameroon, S Region, Bindem (Messama Rd), 15th August 2006, *Droissart & Simo* 258 (reported in Droissart *et al.* 2009: 126).
Habitat: epiphyte in lowland forest; 600 m.
Threats: both of the known locations are outside of protected areas and by an aerodrome and a road, which we can infer means there is considerable human activity there. As such, we suspect there may be associated threats to this species.
Management suggestions: revisit the two known locations for this species, study the actual or potential threats and introduce it into cultivation if it is relocated.

Orestias micrantha Summerh.

EN B2ab(iii)
Range: Cameroon (S: Bipindi; Ekouk (1 coll. each)) São Tomé & Príncipe (São Tomé (1 coll.)) and Equatorial Guinea (Rio Muni (1 loc., 3 coll.)).
O. micrantha is known from four locations with an AOO of 16 km^2 and an EOO of less than 39820 km^2. This species was originally collected in Camerooon, Ekouk, 20 km east of Ébolowa, 1911, *Mildbraed* 5733; Bipindi, 1897, *Zenker* 1380. It was collected from São Tomé, *Stévart* 84, and Stévart & de Oliveira (2001: 176) state that it is very frequent there, although no additional specimens have been seen by the author. Droissart *et al.* (2009: 126) report three collections from Monte Alen NP, made in 2002: *Senterre, Obiang & Esono* 756; *Stevart, Ndong Bokung & Ndong Maye* 1417, and *Parmentier & Esono* 2826.
Habitat: terrestrial herb in lowland and submontane forest; 200–1200 m.
Threats: both of the Cameroon locations are urban areas, with Bipindi sited on a main road and Ébolowa centred at the crux of a number of major thoroughfares. Satellite imagery of both towns shows considerable disturbance to natural habitat in the surrounding region, especially around Ébolowa, with clear evidence of habitat fragmentation visible. The São Tomé and Monte Alen locations therefore provide the best locations for the future of this species, on account of the large numbers occurring at the former and the latter being a designated protected area. In a Cameroonian context it can be considered particularly threatened.
Management suggestions: revisit the two known Cameroon locations for this species and attempt to introduce it into cultivation. Ensure the São Tomé populations' habitat is subject to some form of protection.

Ossiculum aurantiacum P.J.Cribb & Laan

CR A2c
Range: Cameroon (SW: Mungo R FR; Banyang Mbo Wildlife Sanctuary (1 coll. each)).
Ossiculum was assessed as CR B2ab(iii) in Cribb & Pollard (Cheek *et al.* 2004: 199) but this assessment is modified here and considered under Criterion A with a suspected population reduction of ≥ 80%, on the basis of disturbance at the type locality. The type collection was made 13 km from Kumba on the road to Loum, Mungo R FR, 16th December 1980, *Beentje* 1460A, and it has recently been collected in Banyang Mbo (*Stévart* pers. comm. 2010) which represents a second known location for this species. This represents a most exciting discovery and welcome news as it was otherwise considered to be on the edge of extinction. This monospecific genus has now apparently only been seen once since it was first discovered in 1980, despite several thousand botanical collections being made within 50 miles of the type location. BRLU's excellent orchid research programme has yielded upwards of 7000 living specimens and a huge amount of additional herbarium material of many orchid taxa, amongst which exists a living specimen of this rare genus, from the Banyang Mbo Wildlife Sanctuary.
Habitat: an epiphyte in lowland evergreen rainforest; 200 m.
Threats: recent road construction along the border of the Mungo R FR, aided by European Union funding, provides greater accessibility into the species' type location and may result in intensification of forest clearance outside of the reserve boundary and encroachment of small-scale cash-crop agriculture into the reserve, if not carefully managed. The Mungo R FR is understood to be a government forestry reserve, similar to the nearby Loum FR. The understorey at the latter has been largely removed, and it is posited that similar problems will have occurred at the former. Threats at the new location are currently unknown, although the

stability of Banyang Mbo as a Sanctuary appears to be under threat and its status may have even been officially downgraded (Cheek *pers. comm.* 2010).

Management suggestions: urgent action is required to address the issues of timber extraction from the Mungo R area, and to protect the existing forest reserves from illegal encroachment of agricultural activity. Intensive orchid surveying should be carried out for the CRES orchid programme at Mungo and other nearby lowland areas, such as the adjacent Loum FR. A better understanding of the typical phorophyte(s) would be most useful. Collaboration between the BRLU/Yaoundé orchid programme and the K/YA/SCA plant conservation programmes would be a welcome development. *Ossiculum aurantiacum* could represent an iconic flagship taxon to illustrate the benefits of taxonomy, specimen collecting, the value of herbaria and living collections working in tandem to help understand where priorities for conservation lie. Areas that conservation ought to focus more on could be interpreted as being of importance on account of both their high levels of species diversity as well as their high degree of endemism and phylogenetic diversity.

Platycoryne alinae Szlach.
CR B1ab(iii)+2ab(iii)

Range: Cameroon (Adamawa: Mt Nganha (1 coll.)).

This species is only known from one location, with an AOO of 4 km^2 and an EOO of less than 100 km^2. The type specimen was collected from Adamawa: Mt Nganha, 50 km East of Ngaoundéré, June 1977, *Fotius* 2510, and has not been collected since.

Habitat: terrestrial herb in shrubby savanna; 1650 m.

Threats: Amiet (2004) details threats to biodiveristy on Mt Nganha existing from smallholder farming activities and subsistence wood extraction, from which it is inferred this species is or will also be susceptible. He also records the presence of an endemic species of frog, *Astylosternus nghananus* Amiet, 1978 from this location, and assesses it as Critically Endangered.

Management suggestions: revisit this location to survey the population and gain an idea of numbers of individuals and habitat preferences, and conduct a mapping exercise to pinpoint surviving plants and investigate the possibilities of protecting some of these *in situ*. It would also be worthwhile to survey other areas nearby with suitable habitat and altitude. Attempt to introduce this species into cultivation for propagation and potential (re-)introduction at Mt Nganha and elsewhere.

Platycoryne megalorrhyncha Summerh.
VU B1ab(iii)+2ab(iii)

Range: Nigeria (2 loc., 2 coll.) and Cameroon (NW: Fougom; Bafut, Bambili; Wum (1 coll. each); Santa (4 coll.). W: Bamboutos Mts (2 loc., 4 coll.)).

Known from nine locations with an AOO of 36 km^2 and an EOO of 15190 km^2. This species occurs most frequently in NW Region in the Bamenda Highlands: Fougom, near Bamenda, April 1931, *Maitland* 1509 (holotype); Bafut, 22nd June 1951, *Ujor* in FHI 30450; Santa, 1st June 1955, *Saxer* 6 & 2nd June 1955, *Saxer* 28; Bambili, *Bauer* 88; Wum, 13th June 1959, *Daramola* in FHI 41084. Collections from West Region are: Djuttitsia (or Djittitsa), Mts Bamboutos, *Meurillon* CNAD 365, 5th June 1967, *Meurillon* CNAD 803 & 15th May 1968, *Meurillon* CNAD 1286. Droissart *et al.* (2006: 48) list two further specimens from Santa, *E.B.C. & J.A.* 2113, 2124. It is not clear to what or whom these initials refer, but they are included here as being valid records. The Nigerian collections are 17th June 1958, *Chapman J.* 11 and 21st June 1973, *Hall* 2960. Three additional collections were located in the on-line P herbarium including an additional specimen from the Bamboutos Mountains, *Jacques-Félix* 5494, apparently determined by Bill Sanford which is included in this assessment, plus two specimens from Congo (Kinshasa), which have not been reported elsewhere and are here excluded, treated as doubtful until the identifications can be corroborated.

Habitat: terrestrial herb in grassland, damp soil on rocks, savanna; 1050–2300 m.

Threats: forest clearance in the NW Region has reached levels as high as 96.5% loss of original cover and considerable clearance has also been effected in the Bamboutos Mountains. It could be argued that forest clearance favours establishment of savanna type vegetation that might suit this species. However, high population levels in both NW and W Regions present large increases in the area of land turned over to small-scale agriculture and are likely to present continued threats to even these savanna habitats.

Management suggestions: revisit the known locations for this species and, if it can be found, introduce it into cultivation. Select areas that already have a level of protection within its known range and extend surveys therein, to select sites for possible reintroductions in the future.

Plectrelminthus caudatus (Lindl.) Summerh. var. *trilobatus* Szlach. & Olszewski
EN B2ab(iii)

Range. Cameroon (Littoral: Douala-Nkongsamba (1 coll.)) and CAR (2 loc., 2 coll.).

Known from three locations with an AOO of 12 km^2 and an EOO of 89480 km^2. This recently described variety is known from just one location in Cameroon, Maléké, along the Douala-Nkongsamba Rd, 18th August 1959, *Mpom 342*,

and from two locations in the CAR, where it was collected 22nd July 1924, Baidou, *Tisserant 1577* and at Bania, 6th August 1949, *Sillans* 1560.
Habitat: epiphyte in lowland and submontane forest; 50 m.
Threats: the road from Douala to Nkongsamba is a main arterial road providing access to numerous towns and metropoli throughout Littoral, SW, W and NW Regions. As such, large numbers of street-side vendors and other small businesses have sprung up to serve the increasing population and high volume of through traffic. This has led to incremental creeping development all along the route over time, providing accommodation and workspace for the enterprise of economic migrants. There is also considerable acreage given over to monoculture cropping, particularly bananas, rubber and oil palm, which have devastated the original vegetation. As such we can infer that Maléké will have suffered considerable degradation of the habitat in the 50 years since the only known Cameroon specimen was collected.
Management suggestions: a dedicated orchid survey of habitat along this route may well be of benefit in helping us to understand the current orchid flora in such disturbed areas as well as to locate patches of original habitat that may be worthy of conservation management.

Polystachya albescens Ridl. subsp. ***angustifolia*** (Summerh.) Summerh.
EN A1c+2c; B1ab(iii)+2ab(iii)
Range: Cameroon (SW: Mt Cameroon (3 loc., 6 coll.). W: Bamboutos Mountains (1 coll)).
Known from four locations with an AOO of 20 km^2 and an EOO of 180 km^2, and a suspected population reduction of 50% across its range. Discovered by Miss Gregory, *Gregory* 165 (holotype), on 13th July 1947 at Buea: 'upper part of golf course, tree, 20 feet from ground, very dense growth reminiscent of mistletoe. Some pendulous stems three feet long'. The six collections from which this epiphyte is known, are all from near Buea, 23rd April 1948, *Gregory* 284; July 1948, *Keay* in FHI 22411 and 4th July 1949, *Ejiofor* in FHI 25340). The last two specimens were cultivated in 1951 from material obtained from Miss Gregory's original collections and so do not represent additional locations. The only recent collection was made at Bokwango, 21st July 1992, *Tekwe* 107. The only location not on Mt Cameroon is from the Bamboutos Mountains, from Dschang to Fongo Tongo and border at Mt Bana, 26th November 1968, *Sanford* 5649.
Habitat: epiphyte in submontane forest; c. 900–1000 m.
Threats: forest clearance for agriculture, firewood, housing.
Management suggestions: this species merits effort to rediscover it and to assess population size and threats. Charles Tekwe's assistance would be welcomed. From the collection data available, it is reasonable to surmise that this species may be centred around the large town of Buea, and so threatened by urban expansion. Relocation of plants would be desirable followed by efforts to introduce it into cultivation, particularly at Limbe BG.

Polystachya albescens Ridl. subsp. ***manengouba*** W.Sanford
CR A2c; B1ab(iii)+2ab(iii)
Range: Cameroon (Littoral: Mwanenguba (1 coll.)).
Known from only one location with an AOO of 4 km^2 and an EOO of less than 100 km^2. The type collection, Sanford 5557 was made from Mwanenguba (see Darbyshire (Cheek *et al.* 2004: 205), for notes on the orthography of this mountain's name) near Nkongsamba, 18th November 1968, but this subspecies has never been recollected there despite botanical inventory work there in both the 1970s by Leeuwenberg and in the 1990s and early 2000s by Onana, Cheek, Pollard and others. *P. albescens* Ridl. sensu lato comprises seven subspecies, ranging from Nigeria to Tanzania and south to Zimbabwe (see Stévart & Nguema 2004: 230). The similar *P. albescens* subsp. *albescens* is recorded at significantly lower altitudes than subsp. *manengouba*. Whilst subsp. *manengouba* is the scarcest subspecies, subsp. *angustifolia* (Summerh.) Summerh. is also restricted to Cameroon, being known from only six collections on Mt Cameroon and a single outlier in the Bamboutos Mountains.
Habitat: montane woodland; 1980–2100 m.
Threats: high levels of habitat disturbance and montane woodland clearance have occurred in recent decades at Mwanenguba.
Management suggestions: conduct surveys in an attempt to rediscover this subspecies and, if successful, introduce it into cultivation and create a propagation plan. Restoration of forest habitat on Mwanenguba is highly desirable and could include re-introduction of threatened orchids.

Polystachya albescens Ridl. subsp. ***polyphylla*** (Summerh.) Stévart
VU A2c
Range: Cameroon (C: Yaoundé (2 loc., 7 coll., plus 8 unlocated coll.)), Equatorial Guinea (Rio Muni (4 loc., 13 coll.)) and Gabon (4 loc., 4 coll.).
The collections and locations detailed above are taken from Stévart & Nguema (2004: 229–230), except for eight additional Cameroon collections reported in Droissart *et al.* (2006: 50) for which we do not currently have locations. It is inferred that the additional locations will result in greater than ten locations being known for this subspecies, although that is not yet confirmed, which means it does not qualify as VU under B2ab(iii), but the close proximity to Yaoundé of many collections for this subspecies suggests that overall we can infer a level of threat and a decline in quality of habitat.

It is likely that threats at those locations will qualify this species as VU A2c based on a suspected reduction in population size of ≥ 30%. It was assessed as VU by Droissart *et al.* (2006: 50). A distribution map for this species has not been prepared as there is considerable uncertainty on the exact locations for this subspecies.

Habitat: terrestrial on inselbergs in semi-humid grassland or dry grassland with *Afrotrilepis pilosa*; 700–950 m.

Threats: the typical inselberg habitat is not generally thought to be under threat across its range. However, the very close proximity of species to Yaoundé is here inferred to represent a level of threat to this taxon overall.

Management suggestions: additional details are required on the status of the vegetation on N'kolbison, near Yaoundé in order to better assess the level of current and likely future threat to this species. Surveys of habitat type, extent and disturbance would be most useful in understanding the status of that subpopulation. Investigate the origin of the unlocated collections.

Polystachya anthoceros la Croix & P.J.Cribb
EN B1ab(iii)+2ab(iii)

Range: Nigeria (1 loc., 1 coll.) and Cameroon (NW: Bali Ngemba FR; Baba II (2 loc., 3 coll.)).

Polystachya anthoceros is known from three locations with an AOO of 12 km^2 and an EOO of 310 km^2. It was assessed by Pollard in Harvey *et al.* (2004: 72) as EN B2ab(ii,iii), but this assessment is modified here to include B1 and to remove subcriteria (ii). It was described in 1996, known only from *Spurrier* N17, Nigeria, Mambilla Plateau, Ngel Nyaki Forest, 31st October 1993, fl. in cultivation, August 1994 (holotype). In October 2001, two collections were made within a week of each other, 6th October, *Etuge* 4302r from the Baba II forest, near Bali Ngemba FR, and 11th October, Pollard 696 from inside the Bali Ngemba FR. Another unconfirmed Bali Ngemba collection, Simo 158 was made in April 2004.

Habitat: epiphyte in montane forest or forest patches in grassland; 1500–2100 m.

Threats: conversion of forest to small-scale agriculture is ongoing, and timber extraction was observed in the Bali Ngemba FR during fieldwork in 2001. This was often being used to build furniture for sale in local markets, apparently to raise income for schooling of the local villagers' children.

Management suggestions: identification of Simo 158 needs to be updated, and it is quite unsatisfactory to have doubt hovering over the names of such collections. This species should be introduced into cultivation, perhaps at the nearby Savanna BG. It would be a most welcome project to investigate and [illegible] sustainable income from the Forest Reserve for local villagers that does not adversely affect the biodiversity dwelling therein.

Polystachya batkoi Szlach. & Olszewski
VU B2ab(iii)

Range: Cameroon (S: Bidjouka (1 coll.), Bifa (2 coll.)), Gabon (5 loc., 6 coll.) and Congo (Kinshasa) (1 doubtful coll. excluded).

This recently described species is known from only five locations with an AOO of 20 km^2 and an EOO of 31150 km^2. The type was collected on 20th February 1968, River Balakabo, Cristal Mountains, Gabon, *Hallé* 5416. It was first reported from Cameroon by Droissart *et al.* (2009: 126–127), where three recent collections are cited: Bidjouka (Ngovayang forest area), 17th June 2006, *Droissart* 159; Bifa (Kribi-Ébolowa Rd), 14th April 2007, *Droissart & Simo, M.* 391; *ibid.*, 13th September 2007, *Droissart, Stévart & Simo, M. (Yaoundé shadehouse)* 727. The record from Congo (Kinshasa) was made by Szlachetko *et al.* (2004) but without a specimen citation and so that country record is excluded from the distribution map.

Habitat: epiphyte in submontane primary and secondary forest; 100–600 m.

Threats: conversion of forest to small-scale agriculture presents a threat. The collection of this species from secondary forest implies anthropogenic factors have been at work within its known range.

Management suggestions: develop a propagation programme based on the existence of this species in cultivation at the Yaoundé shadehouse network.

Polystachya bicalcarata Kraenzl.
VU A2c+3c; B2ab(iii)

Range: Nigeria (1 loc. unknown, 1 coll. cult.), Equatorial Guinea (Bioko (1 coll.)), Cameroon (SW: Buea (1 loc., 3 coll.); Mt Cameroon (1 coll.); Mt Etinde (1 coll.); Mt Kupe (1 loc., 2 coll.); Muambong (1 coll.); Lebialem (1 coll.). W: Bamboutos Mts (1 coll.). NW: Mt Oku (1 coll.)) and Gabon (1 coll.).

Assessed as EN A1c+2c in both Cable & Cheek (1998: lxxiv) and Cheek *et al.* (2000: 84), downgraded to VU A2c+3c; B2ab(iii) in Pollard & Darbyshire (Cheek *et al.* 2004: 199), followed by Cheek & Pollard (Harvey *et al.* 2010: 89) and maintained here. This taxon is known from eleven locations, with an AOO of 44 km^2 and an EOO of 329640 km^2, but with all but one of the locations having yielded just a single specimen each. Although this species is now known from eleven locations and thus does not qualify as VU under B2a if it had been known from ≤ 10 locations, it does qualify under B2a as it is considered to be 'severely fragmented', according to the guidelines laid out in IUCN (2001: 14). This refers to "the situation in which increased extinction risk to the taxon results from the fact that most of its individuals are found in small and relatively isolated subpopulations", which appears to be the case for this species. Specimens include SW: Buea, *Deistel 62c*;

Maitland 730; 27th August 1946, *Dundas* 15303; Mt Etinde, 25th October 1992, *Thomas 9178*; Muambong, 22nd January 1998, *Mackinder 185*; Nyasoso, 28th August 1993, *Balding & Sivell 84*; 5th June 2007, *Russell* 81, Lebialem, 17th April 2004, *Simo, P.* 226; W: Bamboutos Mts, May 1940, *Jacques-Félix* 5439; NW: Kilum, 4th November 1996, *Zapfack 1187* (K). The Nigerian collection has no known location as it was taken from a cultivated specimen at Ibadan, April 1967, *Sanford* 488. We are making an assumption that this represents a valid location, different from the others. The Gabon collection is from the southern ascent of Mt Igoumbi, 16th April 2005, *van Valkenburg* 3074.

Habitat: an epiphyte of submontane and montane forest, also recorded growing on *Cola* sp., in coffee plantation; 950–2100 m.

Threats: forest clearance for agriculture and firewood, particularly in the Bamboutos Mts and above Buea on Mt Cameroon.

Management suggestions: surveys of this taxon at each of the collection localities should be made in order to assess its abundance. Continued protection of the submontane forest on Mt Kupe should ensure this species' future survival.

Polystachya bipoda Stévart

EN B1ab(iii)+2ab(iii)

Range: Cameroon (S: Akom II (1 loc., 2 coll.)), Equatorial Guinea (Rio Muni (1 loc., 4 coll.)) and Gabon (1 coll.).

This species is known from three locations with an AOO of 12 km^2 and an EOO of 200 km^2. In Cameroon, it has been collected twice, from Akom II, east of Kribi, *Stévart & Droissart* 2062 & 2103. All the specimens have been collected since 2000, and are cited in Stévart & Nguema (2004: 218) and Droissart *et al.* (2006: 52).

Habitat: an epiphyte of submontane forest, and inselbergs; 680–1110 m.

Threats: it is inferred that the Akom II location is under potential threat from logging activities as it is currently not under any form of formal protection and is in an area that has suffered considerable habitat degradation historically.

Management suggestions: introduce it into cultivation if it has not already been.

Polystachya camaridioides Summerh.

EN B2ab(iii)

Range: Nigeria (1 loc., 2 coll.), Equatorial Guinea (Bioko (unknown location, 1 coll. (?)) and Cameroon (SW: Lake Barombi Mbo (1 coll.); Banyang Mbo (1 coll.)).

This species is known from four locations with an AOO of 16 km^2 and an EOO of 29490 km^2. In Cameroon, it has been collected from two locations, firstly from farmland and secondary scrub near Lake Barombi Mbo, 15th May 1985, *Thomas* 4811 and more recently from Akom II, east of Kribi, *Stévart, Droissart & Simo* 2166, cited in Droissart *et al.* (2006: 52). The Nigerian collections are *Wright* 101 & Calabar, *Wright* 3776. The Equatorial Guinea record is included in Droissart *et al.* (2006: 52) but has not been referred to a specimen and as such the location shown in the distribution map for this species is perhaps not at exactly the right location on Bioko. It is also unclear whether there is more than one location from that country.

Habitat: an epiphyte of submontane forest; 50–200 m.

Threats: it is inferred that the Akom II location is under potential threat from logging activities as it is currently not under any form of formal protection, but it is close to the major port of Kribi in an area. The Barombi Mbo location already has experienced severe disturbance from encroachment of farming activities. The Calabar location is likely lost due to urban expansion since Wright's time.

Management suggestions: introduce it into cultivation, and utilise the Banyang Mbo records as additional weight to maintain the research station there and bolster efforts to safeguard that important sanctuary. Investigate the Equatorial Guinea collection(s) to determine its distribution there.

Polystachya carnosa P.J.Cribb & Podz.

EN B1ab(iii)+2ab(iii)

Range: Equatorial Guinea (Bioko (2 loc., 2 coll.)) and Cameroon (C: Mbam Minkoum Hills (1 coll.)).

This species is known from three locations with an AOO of 12 km^2 and an EOO of 1940 km^2. It was described from Bioko in 1979, based on two collections made by Bill Sanford in the 1960s. It was considered to be a Bioko endemic until it recently found in Cameroon, where it has been collected only from the Mbam Minkoum Hills, represented by *Droissart, Stévart & Simo 363*. Simo *et al.* (2009: 115) indicate that it has been recorded there from lowland and submontane forest and inselberg habitats. This suggests that there has been more than one collection there, although those putative additional data are not available to the present author, and the area is here treated as representing one location.

Habitat: epiphyte in lowland and submontane forest, inselbergs; 600–1300 m.

Threats: detailed information of the threats occurring at the Mbam Minkoum Hills is provided by Simo *et al.* (2009: 121) and summarised here. The main threat is loss and framentation of habitat due to encroachment by the local human population for agriculture, logging and hunting. A considerable proportion of the lowland forest has been converted to cocoa and plantain cultivation. Timber exploitation during the dry season is a significant local economic activity that is particularly destructive to these forests.

Management suggestions: develop a conservation strategy for the Mbam Minkoum Hills, perhaps incorporating the creation of an orchid shadehouse there, particularly given the difficulty of cultivation of this species at the Yaoundé shadehouse.

Polystachya cooperi Summerh.

EN B2ab(iii)

Range: Nigeria (Mambilla Plateau (1 coll.); Obudu Plateau (2 loc., 2 coll.)) and Cameroon (SW: Mwanenguba (1 coll.); Mt Kupe (1 coll.)).

This species is known from five locations with an AOO of 20 km^2 and an EOO of 24340 km^2. It was first collected in November 1961 on the Obudu Plateau, *Cooper* 2A, and was subsequently cultivated at Kew. Further collections were made at the Obudu Cattle Ranch on 1st August 1973, *Lowe* 2662 and on the Mambilla Plateau at Ngel Nyaki FR on 31st July 1976, *Chapman* 207, where it was recorded as a 'common epiphytic orchid' at 5000ft. It was first collected in Cameroon on Mt Mwanenguba on 16th November 1968, *Sanford* 5464, but has not since been rediscovered at that site. The only recent collection is from Mt Kupe on 5th August 1993, *Balding & Sivell* 43.

Habitat: epiphyte in submontane and montane forest and woodland; 1400–1700 m.

Threats: extensive and continued forest clearance at high altitudes in E Nigeria and at Mwanenguba threaten any remaining populations at these sites. The submontane location on Mt Kupe, however, has experienced little disturbance to date.

Management suggestions: as Mt Kupe appears to offer the best chance for survival of this taxon, attempts to rediscover and census the population here should be made. Continued protection of the forest above 1000 m alt. here should ensure this species' survival, though it is clearly rare, having been collected only once during the extensive surveys on this mountain.

Polystachya crassifolia Schltr.

EN B2ab(iii)

Range: Ivory Coast (1 coll.) and Cameroon (SW: Moliwe (1 coll.)).

This species is known from only two locations with an AOO of 8 km^2. The EOO has not been calculable using the methodology employed by Willis *et al.* (2003) for a species known from only two locations. It was collected from Moliwe, 12th February 1909, *Schlechter* 12841. In Ivory Coast it is known from one collection, according to Pérez-Vera (2003: 466), although it is not included in Szlachetko (2008).

Habitat: forest; 800–900 m.

Threats: the Moliwe location is under threat from logging activities and establishment of plantation monocultures.

Management suggestions: maintain awareness of this species in conservation planning for the lowland areas surrounding Mt Cameroon.

Polystachya farinosa Kraenzl.

EN B2ab(iii)

Range: Cameroon (SW: Buea; Mwanenguba (1 coll. each)) and São Tomé & Príncipe (São Tomé (1 coll.)).

This species is known from only three locations, with an AOO of 12 km^2 and an EOO of 1010 km^2. It is known by two collections from Cameroon, the first from above Buea on Mt Cameroon, collected October 1891, *Preuss* 1064, and the second from Mwanenguba, collected on 16th November 1968, *Sanford* 5463. It has also been recorded from São Tomé, though data are sparse on its range there. *P. farinosa* was considered in FWTA to be conspecific with *P. bifida* Lindl., but was reinstated as a separate entity by Szlachetko & Olszewski (Flore du Cameroun 35: 536 (2001)). The latter species is more widely distributed in the Lower Guinea and Congolian phytochorion.

Habitat: an epiphyte of dense montane forest; 1000–1700 m.

Threats: continued clearance of montane forest at Mwanenguba threatens this population which may no longer be extant, as this species was not rediscovered there during botanical survey work in the 1970s and 1990s.

Management suggestions: more data on the São Tomé population are required. Attempts to rediscover this species at both Mwanenguba and Mt Cameroon should be encouraged.

Polystachya geniculata Summerh.

EN B2ab(iii)

Range: Cameroon (SW: Mamfe (1 loc., 4 coll.); Mwanenguba (1 coll.)).

P. geniculata is known from two locations with an AOO of 8 km^2. The EOO has not been calculable using the methodology employed by Willis *et al.* (2003) as it is known from only two locations. First described from material collected in May 1947 growing on rock-slabs near Mamfe, *Gregory* 124, subsequent material from Banyang-Mamfe, *Eyeku in FHI* 22304, Mamfe, 9th September 1948, *Gregory* 323 and *Swarbrick* 2051 are likely to refer to the same location, as these specimens were again collected from 'outcrops of granite' and 'rock-slabs with running water in the wet season' respectively. The single collection from Mwanenguba was made on 5th June 1948, *Gregory* 302, where this species was found in marshy grassland.

Habitat: an epilith on seasonally-wet rocks in submontane forest, or terrestrial in marshy grassland; 100–1850 m.

Threats: the Mwanenguba population is threatened by habitat deterioration from farming at high elevations. The sites near Mamfe, although not thought to be directly threatened by current human activity, may deteriorate

through, for example, changes in run-off regime following human disturbance in the proximity of these rocky sites.
Management suggestions: attempt to relocate the area in which the collections were made near Mamfe; efforts should then be made to relocate this species there as it is a more likely location for its future survival than the heavily disturbed Mwanenguba Massif.

Polystachya kornasiana Szlach. & Olszewski
EN B2ab(iii)
Range: Cameroon (SW: Banyang Mbo (1 coll.)) and Congo (Brazzaville) (1 coll.).
This species is known from only two locations, with an AOO of 8 km^2, but for which the EOO has not been calculable using the methodology employed by Willis *et al.* (2003). In Cameroon it is known from Banyang Mbo, *Simo* 24. Szlachetko & Olszewski (2001: 510) described this species, based upon a specimen from Congo (Brazzaville), Zanaga, 9th December 1971, *Sita* 3213 (holotype).
Habitat: epiphyte in submontane forest; 1200 m.
Threats: the Banyang Mbo wildlife sanctuary should afford this species a measure of official protection in Cameroon, but recent events have suggested that its status as a sanctuary has been downgraded and that it could even be subject to the possibility of logging operations in the future.
Management suggestions: further investigation into the status of Banyang Mbo is urgently required. Introduce this species into cultivation.

Polystachya kubalae Szlach. & Olszewski
EN B2ab(iii)
Range: Cameroon (C: Mbam-Minkoum (1 coll.)) and Gabon (1 coll.).
This species is known from only two locations, with an AOO of 8 km^2, but for which the EOO has not been calculable. In Cameroon it is known from Mbam-Minkoum (region NW of Yaoundé), *Droissart* 6. The type collection was made at Piti Masango, Gabon, April 1926, *Le Testu* 5924A.
Habitat: epiphyte in submontane forest; 1200 m.
Threats: the only known collection from Cameroon is near to the capital city Yaoundé, and so it is inferred here to be under threat from urban expansion and/or all the other attendant stresses to the natural environment of high population density.
Management suggestions: further investigation into the status of Banyang Mbo is urgently required. Introduce this species into cultivation.

Polystachya kupensis P.J.Cribb & B.J.Pollard
CR B2ab(iii)
Range: Cameroon (SW: Mt Kupe (1 coll.)).
The following conservation assessment of CR is taken largely from Cribb & Pollard (2002: 636–638). This taxon is known from only the type specimen collected above Kupe Village on the path to Kupe Rock, made in May 1996, *Cable* 2521. It has not been rediscovered on the mountain despite attempts from Orchidaceae specialists during several months of collecting there, nor has it been found in the now well-collected Bakossi Mountains, to the west. It is therefore likely to be very rare within its extremely limited range.
Habitat: an epiphyte in submontane forest, growing on *Coffea* sp.; 1050 m.
Threats: the only known locality for this species is in an area of forest being encroached upon by plots of *Coffea canephora* Pierre ex A.Froehner, and may thus be lost in the near future. However, the type specimen was collected from a *Coffea* tree (possibly a wild species) so this taxon may be able to withstand such a change in habitat in the short term. However, well-tended farms are 'cleaned' of epiphytes, thus *P. kupensis* is unlikely to survive in the long-term in coffee plantations.
Management suggestions: revisiting the type location and rediscovering this species is a priority. Once this has been achieved, further assessment of its ecological requirements, including its potential for survival in plantation agriculture, can be assessed. Ultimately, enforced forest protection or cultivation followed by propagation may be the only means of conserving this taxon.

Polystachya lejolyana Stévart
EN B2ab(iii)
Range: Cameroon (S: Akom II (1 loc., 5 coll.)) and Gabon (1 loc., 2 coll.).
This species is known from only two locations with an AOO of 8 km^2, but for which the EOO has not been calculable. It was first collected from Mount Mbilan, Gabon, 7th April 2002, *Stévart & Kombila* 1289, and a second collection there on 18th August 2002, *Stévart* 1800. Five recent collections have been made from Akom II (Kribi-Ébolowa Rd) in Cameroon and are cited by Droissart *et al.* (2009: 127). These are all now in cultivation as part of the Yaoundé shadehouse programme.
Habitat: epiphyte in lowland forest; 600–650 m.
Threats: satellite imagery of lowland forest along the Kribi-Ébolowa Rd shows clear evidence of forest clearance all along the route as well as considerable areas of more-targeted clearance in the general area. The close proximity of this area to Kribi, one of Cameroon's largest ports, means that the costs of transporting timber to sea for export are likely to be much reduced for these forests when compared to more distant parts of Cameroon, such as E Region. It is inferred, therefore, that these forests are highly likely to be under serious levels of threat in the future as logging activities spread inland from the coast.
Management suggestions: use the specimens in cultivation as an *ex situ* conservation collection and gene bank, and look

to establish a propagation programme in Yaoundé. Use threatened species such as this as examples of the biodiversity significance of the lowland forests in this region in efforts to promote protection of large forest blocks.

Polystachya letouzeyana Szlach. & Olszewski
EN B2ab(iii)

Range: Cameroon (S: Ngwon Hill (1 coll.)) and Gabon (4 loc., 4 coll.).

P. letouzeyana is known from five locations with an AOO of 20 km^2 and an EOO of 106340 km^2. It was collected in Cameroon from Ngwon hill, 38 km east of Kribi, 18th April 1968, *Letouzey* 9372 (holotype). The Gabon collections are: western flanks of Monts Doudou, 6th December 1984, *Arends* 657; west bank Ovindo R opposite Mayibout Island, 12 km from Bélinga, October 1987, *Louis* 2441; between Rabi and Echira, 27th March 1990, *Breteler* 9567; forestry concession Bordamur, c. 60 km NE of Mitzik, 10th February 2003, *Sosef* 1982.

Habitat: epiphyte in submontane forest on inselbergs and rockfaces; 100–260 m.

Threats: satellite imagery reveal considerable forest clearance in the Kribi region, and it can only be expected that this will accelerate in the region so close to Kribi, a major international port. Low level clearance of forest also results from farming activities and extraction of fuelwood by locals.

Management suggestions: utilise the threatened status of this species as a conservation tool for highlighting the threats to biodiversity in the lowland forests of S Region, Cameroon.

Polystachya moniquetiana Stévart & Geerinck
EN B2ab(iii)

Range: Cameroon (S: Akom II (1 coll.)), São Tomé & Príncipe (São Tomé (1 coll.)), Equatorial Guinea (Rio Muni (2 loc., 2 coll.)) and Gabon (1 coll.).

This species is known from five locations with an AOO of 20 km^2 and an EOO of 24070 km^2. The single Cameroon collection was made from Akom II (Kribi-Ébolowa Rd), 13th September 2004, *Stévart, Nguembou & Djuikouo (Yaoundé shadehouse)* 106 and published in Droissart *et al.* (2009: 127). The holotype was collected from Equatorial Guinea (Rio Muni) at Engong, Monte Alén National Park, 11th September 2001, *Ndong Bokung S. & Stévart* 418, with a second country record being collected between Engong and Lago, 19th September 2001, *Ndong, Bokung & Stévart* 425.

Habitat: epiphyte in lowland forest; 650–850 m.

Threats: satellite imagery reveals considerable forest clearance in the Kr[illegible] only be exp[illegible] this will accelerat[illegible] close [illegible] major internationa[illegible] export o[illegible] a major source of foreign income for Cameroon.

Management suggestions: utilise the threatened status of this species as a conservation tool for highlighting the threats to biodiversity in the lowland forests of S Region, Cameroon.

Polystachya odorata Lindl. subsp. ***trilepidis*** (Summerh.) Stévart (syn. *Polystachya odorata* Lindl. var. *trilepidis* Summerh.)
VU D2

Range: Nigeria (1 loc., 4 coll.) and Cameroon (S: Nkoltsia (1 loc., 3 coll.)).

This subspecies is known with certainty from only two inselberg locations with an AOO of 8 km^2 but for which the EOO has not been calculable using the system used by Willis *et al.* (2003). Four collections were made on Carter's Peak, Ondo Province, Nigeria between 1943 and 1970 and three Cameroon collections from Nkoltsia, near Gouap, 18 km NE of Bipindi, 23rd March 1974, *Villiers* 823; 25th March 1974, *Villiers* 853; June 1974, *Villiers* 935. *Jones* 20729, Nigeria, also states that it is 'common on the rounded granite hills'. This suggests that it is a species that occupies highly specialised inselberg habitats.

Habitat: epiphyte on inselbergs in submontane forest, with *Afrotrilepis pilosa* (Cyperaceae). Stévart & Nguema (2004: 227) and Droissart *et al.* (2006: 58) detail its adaptations to seasonal fire, so it could be described as an epilithic pyrophyte; 350 m.

Threats: the summits of inselbergs tend generally not to be subject to typical threats from human acitivites on account of their inaccessibility and the unsuitable nature of the substrate for agricultural practices, but they certainly could be vulnerable to random stochastic events, such as rockfalls and landslides.

Management suggestions: introduce this subspecies into cultivation and revisit the known localities to make an assessment of any threats.

Polystachya pyramidalis Lindl.
NT

Range: SE Nigeria (1 coll.), Cameroon (S: Bipindi; Ebounja, Bifa (1 coll. each); Akom II (4 coll.)) Equatorial Guinea (Rio Muni (2 loc., 2 coll.)) and Gabon (7 loc., 14 coll.).

This species is treated as NT on the basis of it being known from 13 locations with some of the specimens thought to occur in areas not currently under imminent threat, for example in Gabon. Specimens are cited in Droissart *et al.* (2009: 128) and on the Plants of Gabon database.

Polystachya retusiloba Summerh.
[illegible]B2ab(iii)

[illegible]meroon (W & Littoral: 2 loc., 2 coll., one doubtful and excluded), Congo (Kinshasa) (2 loc., 2 coll.) and Rwanda (1 coll.).

This species is known from five locations with an AOO of 20 km^2 and an EOO of 375410 km^2. The Cameroon locations are West Region: Bakong, 10 km SW Baganté, 1964, *de Wilde W.C.S.* 2414; Littoral: Nyombé (cult. Hort. Bot. P). This latter specimen appears to occur at much lower altitude than the other known specimens and so the location is treated here with some doubt and excluded from the distribution map. Fischer *et al.* (2010: 350) suggest that the Cameroon record is unlikely to belong to this species which they consider to be an Albertine Rift high altitude endemic.

Habitat: epiphyte in submontane and montane forest, rocky gallery forest; (250–)1500–2300 m.

Threats: much of the vegetation in W Region is subject to heavy human pressures for agricultural land and satellite imagery of Bakong and Baganté show that this species is known from an area with forest to the SW but ever depleted forest cover to the NE. It is here inferred that these pressures are ongoing and are likely to continue to reduce forest habitat in a NE to SW direction, as has happened across the region historically.

Management suggestions: revisit Bakong and Baganté and look for this species. A full assessment of the decline in habitat quality in this area would be most useful. Perhaps a desktop GIS study could be commissioned to analyse the rate of advance of forest clearance in the aforementioned NE to SW plane. Further scrutiny of the Cameroon specimens is desirable to check their identity.

Polystachya stodolnyi Szlach. & Olszewski

EN B1ab(iii)+2ab(iii)

Range: Cameroon (S: Lolodorf (1 coll.)), Equatorial Guinea (Rio Muni (*fide* Droissart *et al.* 2006: 59, numbers of locations and collections unknown)).

P. stodolnyi is known from at least two locations with an estimated AOO of 8 km^2 and a probable EOO of less than 5000 km^2. It is here treated as EN on the assumption that the unlocated Equatorial Guinea specimens do not constitute more than four locations. Known in Cameroon from the type collection only, from S Region, Mt Finde, Lolodorf, 1916–1920, *Annet* 307. The Equatorial Guinea location(s) is(are) not included in the distribution map.

Habitat: epiphyte in submontane forest; 850 m.

Threats: lowland forest throughout this area of Cameroon is under potential and actual threat from logging concessions and as such a present and future threat is inferred upon this species.

Management suggestions: investigate the locations of the specimens from Equatorial Guinea. Include this species in a dedicated orchid survey of targeted areas within S Region to better understand species distributions and highlight the importance of this region in terms of biodiversity conservation. Introduce this species into cultivation as part of an *ex situ* conservation programme.

Polystachya superposita Rchb.f.

EN A1c+2c; B2ab(iii)

Range: Cameroon (SW: Mt Cameroon (2 loc., 3 coll.); Banyang Mbo (1 coll. unconfirmed). NW: (2 loc., 2 coll.)).

Cable & Cheek (1998: lxxiv) and Cheek *et al.* (2000: 85) assessed this species as Endangered using Criterion A: EN A1c+2c. Here it is reassessed using Criterion B on the basis that we do not hold precise enough information about the Mt Cameroon locations to be able to estimate a reduction in population size of ≥ 70%, but we can infer a continuing decline in area, extent and quality of habitat, and it being known for certain from five locations with an AOO of 20 km^2 and an EOO of 70010 km^2. *Polystachya superposita* was first discovered in November 1862 at 5,000 ft. on Mt Cameroon, *Mann* 2125 (holotype), where it was also collected near Buea under *Deistel* s.n. and as *Preuss* 1072. In the NW Region, it is known from Kilum, Elak, 9th June 1996, *Zapfack* 824 and the forest of the Kwifon, 3rd November 1996, *Zapfack* 1169. A recent collection from Ebamut in the Banyang Mbo Wildlife Sanctuary, 30th October 2001, *Simo* 51, needs to be confirmed. The specimen previously referred to this species, collected from Bioko, near Baca, 4th January 1967, *Sanford* 4281, is considered by Droissart *et al.* (in press) to belong to *P. fusiformis* (Thouars) Lindl. Two collections from the Plants of Gabon website identified as *P. superposita* do not belong to this species. *Reitsma* 1078 should now be referred to *P. calluniflora*, and *Reitsma* 1954 belongs to *P. seticaulis* (Stévart 2011, *pers. comm.*), having been recently inspected at the WAG herbarium. *P. superposita* is now considered to be endemic to Cameroon.

Habitat: epiphyte in submontane and montane forest, epiphytic; c. 900–2500 m.

Threats: the Mt Cameroon locations near Buea will have been subjected to tremendous pressures from urban expansion and small-scale agriculture since the 19th century collections and these populations can be deemed to have suffered as a result. The Kilum forest, once a beacon of hope in the NW Region in conservation terms is now by no means safe from disturbance, as the BirdLife International Projects no longer operate there and there have been stories of incursions into the forest by local farmers and others (Cheek, *pers. comm.* 2010). It seems as though the forest boundary may no longer being respected.

Management suggestions: further investigation into the two Gabon collections is required to establish their identity. The Banyang Mbo Wildlife Sanctuary may represent the location currently affording the most protection and as such this species should feature strongly in any orchid collecting there in future. Surveys of the Kilum forests and the Kilum-Ijim forest boundary are urgently required.

Polystachya victoriae Kraenzl.
NT

Range: Guinea (*fide* Govaerts *et al.* (2011)), Ivory Coast (1 coll.), Cameroon (SW, Littoral, C & S: 5 loc., 6 coll.), Equatorial Guinea (Rio Muni: 1 coll.) and Gabon (15 loc., 19 coll.).
Status: ↓ Cable & Cheek (1998: lxxv) assessed this species as Critically Endangered, CR A1c+2c on the basis of it being known from only eight collections. The Checklist of Gabonese Vascular Plants (Sosef *et al.* 2006) and the Plants of Gabon database list numerous additional locations and collections and as such it is now considered to be Near Threatened.

Rangaeris longicaudata (Rolfe) Summerh. (syn. *Mystacidium longicaudatum* Rolfe)
VU B2ab(iii)

Range: Ivory Coast (3 loc. (1 unknown), 3 coll.), SE Nigeria (3 loc., 3 coll.), Cameroon (SW: Banga (1 coll.)) and Gabon (1 coll.).
Known from eight locations with an AOO of 32 km^2 and an EOO of 571560 km^2. The Cameroon location was made 10 km west of Banga, near Mukete Plantation, 30th October 1985, *Thomas* 4907. Szlachetko (2008) cites three specimens from Ivory Coast, all collected in the 1970s, two of which are also cited by Cribb & Pérez-Vera (1975: 211). One of the Ivory Coast collections is without a locality and so is not included in the distribution map.
Habitat: epiphyte in lowland evergreen forest; 50 m.
Threats: the collection near Kumba is likely to be facing threat from urban expansion and small-scale agriculture around this large town. Thomas's collection notes state that it was collected from 'mature forest on heavy clay being felled and cleared', giving clear evidence of typical threats there.
Management suggestions: revisit the Banga area and search for this species, to introduce it into cultivation.

Rhipidoglossum obanense (Rendle) Summerh.
EN B2ab(iii)

Range: Nigeria (1 loc., 2 coll.) and Cameroon (SW: Ekondo Titi (1 coll.). S: Akoakas (2 coll.)).
Status: ↑ not included by Cable & Cheek (1998), but it is reassessed and included here on the basis of it being known from only four locations with an AOO of 16 km^2 and an EOO of 8270 km^2. The specimens include Nigeria, Oban, 1911, *Talbot* 896; Oban FR, *Markham in FHI* 50084, and Cameroon, Mbongé, [illegible] 19th December 1958, *Wright* 58/19. Dro[illegible]) also cite *Onana* 387 & 389 from Ako[illegible]olowa.
Habitat: epiphyte in lowland evergreen forest.
Threats: lowland forest in the region of Mt Cameroon (the Ekondo Titi location) has reduced in area considerably over the last 100 years to make way for extensive stands of exotic monoculture crops, such as rubber and oil palm. The Ébolowa subpopulation is likely to have suffered from urban expansion in recent years.
Management suggestions: initiate further surveys to try to recollect this species and introduce it into cultivation, perhaps at Limbe BG.

Rhipidoglossum ochyrae Szlach. & Olszewski
CR B1ab(iii)+2ab(iii)

Range. Cameroon (SW: Mamfe (1 coll.)).
Known only from the type specimen with an EOO <100 km^2 and an AOO of 4 km^2, SW Region, between Akwaya and Mamfe, 15 km NE Mamfe, 1976, *Letouzey* 14154.
Habitat: epiphyte in lowland evergreen forest; 150 m.
Threats: although this area of SW Region is relatively undisturbed, there is still pressure on all lowland evergreen forest in the region in general.
Management suggestions: conduct further orchid surveys in the Mamfé area and introduce this species into cultivation, if discovered.

Rhipidoglossum polydactylum (Kraenzl.) Garay (syn: *Diaphananthe polydactyla* (Kraenzl.) Summerh.)
VU B1ab(iii)+2ab(iii)

Range: Cameroon endemic (SW Province: Mwanenguba (2 coll.); Banguem; Ehumseh-Mejelet; Enyandong; Kodmin; Nyasoso (1 coll. each). NW Province: Tatum (1 coll.); Bafut Ngemba FR (1 obs.); Baba II (1 coll.)).
This taxon is known from nine confirmed locations with an AOO of 36 km^2 and an EOO of 3480 km^2. Pollard & Darbyshire (Cheek *et al.* 2004: 197) assessed this species as VU B2ab(iii) which is expanded here to take into consideration its EOO which also qualifies this species as VU under subcriterion B1. It was described from a single collection from Tatum (Kufum) 30 km ENE of Mt Oku in the Bamenda Highlands, 17th October 1909, *Ledermann* 5716A (B†). It has since been collected twice again in that region, Baba II, 9th October 2001, *Onana* 1868 and 2002. It was recorded at Mwanenguba in November 1968 (*Sanford* 5478 & 5486). Inventory work in the Kupe-Bakossi region in the 1980s and 1990s revealed five further locations for this taxon (see Cribb & Pollard 2004: 461): Banguem, Ehumseh-Mejelet, Enyandong, Kodmin and on Mt Kupe, above Nyasoso. Although considered rare within its range, this is probably a function of collection intensity, as it has appeared in recent surveys of suitable habitat, for example at Bafut Ngemba FR (Pollard, *pers. obs.*).
Habitat: an epiphyte of submontane to montane forest and woodland; c. 1400–2000 m.
Threats: clearance of montane forest is widespread in the Bamenda Highlands, and at Banguem and Mwanenguba to the south. Botanical inventory work in the 1990s at

Mwanenguba failed to provide new records, indicating that this population may no longer be extant.
Management suggestions: as Kupe-Bakossi now appears to be the stronghold of this taxon, a survey of the epiphytic orchids of the submontane and montane zone should be carried out to better assess the abundance of this, and other rare taxa. Maintenance of the protection of the montane forests in this area should ensure the survival of *R. polydactylum*.

Stolzia grandiflora (*Rolfe) Summerh. subsp. *lejolyana Stévart, Droissart & M.Simo
EN B2ab(i,ii,iii,iv)
Range: Cameroon (SW: Enyandong (1 coll.). C: (1 loc., 3 coll.)) and Equatorial Guinea (Rio Muni (1 loc., 4 coll.). Bioko (1 coll.)).
Previously assessed as EN by Droissart *et al.* (2009: 28), their assessment of level of threat is maintained here, but with a note regarding a probable but unconfirmed additional location added (see note below) which results in a modification to the Criteria and subcriteria applicable in this assessment. Known from four locations with an AOO of 16 km^2 and an EOO of 28360 km^2. The collection from Enyandong means that the EOO now exceeds the 5000 km^2 threshold for EN under Criterion B1, which is excluded from this treatment which is therefore an update of the assessment of EN B1ab(i,ii,iii,iv)+2ab(i,ii,iii,iv) provided by Droissart *et al.* (2009: 28). The type specimen was made from Bioko, Gran Caldeira de Luba. 'A montante do campamento e ao longo do rio na direcçao do Monte Pisarro', 27th Feb 1990, *Carvalho* 4265. The three additional Cameroon collections from Central Region are from Mbam-Minkoum, north of Yaoundé (see Droissart et al. (2009: 28)). Cribb & Pollard (2004: 470) include a putative new species of *Stolzia*, known from a single specimen, *Pollard* 906, collected at 1760 m in November 2001 from Bime Rock, an inselberg near Enyandong, Bakossi Mts, Cameroon. I consider the short description of that specimen as having a papillose labellum as being somewhat doubtful, as no evidence of this structure is present on the fruiting K specimen, on which the short description therein was based. Droissart *et al.* (2009: 27) refer to this specimen, and also comment on it as being only in a fruiting state and they do not consider it further. The reason *Pollard* 906 was proposed as a new species was on account of the inflorescence (interpreted from the infructescence) being 2-flowered, whereas all other known *Stolzia* species treated by Cribb (1978) were unifloral. Droissart *et al.* (2009: 27) confirm that all central African bifoliate *Stolzia* taxa bear 1-flowered inflorescences except for this newly described subspecies. The Bakossi Mts specimen accords well with the description and illustration for *S. grandiflora* subsp. *lejolyana* with regard to its habit, leaf and pseudobulb measurements and bract and inflorescence structure and measurements, except for the papillose labellum.
Habitat: submontane forest on inselbergs; 1100–1200(–1800) m.
Threats: its proximity to Yaoundé is enough to warrant an inferred or suspected threat on account of likely population growth, urban and agricultural expansion. The Massif of Mbam-Minkoum, despite the presence of many threatened plant species, does not yet have a dedicated conservation programme.
Management suggestions: Bime rock should be revisited in October and November to search for this diminutive *Stolzia* and assess this as an additional location which would lie within the Bakossi Mts NP. There is a clear need to understand the origin of the description of a papillose labellum, which if its presence is confirmed, could yet represent another undescribed taxon, or perhaps a variety of this subspecies. Focus efforts on instigation of a multilateral conservation programme on hills surrounding Yaoundé, especially Mbam-Minkoum.

Stolzia repens (Rolfe) Summerh. var. ***cleistogama*** Stévart, Droissart & M.Simo
EN B1ab(i,ii,iii,iv)+2ab(i,ii,iii,iv)
Range: Cameroon (C: Mbam-Minkoum (2 coll.); Mefou (1 coll.). S: Akom II (2 coll.)).
Known from three locations with an AOO of 12 km^2 and an EOO of 3040 km^2, it was previously assessed by Droissart *et al.* (2009: 33), and their assessment is maintained here. Mefou, 22 miles Yaoundé to Akonolinga Rd, 19th October 1968, *Sanford* 5204; Mbam-Minkoum, region to NW of Yaoundé, village of Nyemeyong, summit of a hill NW of camp, 13th May 2006, *Droissart* 68 (holotype); Nyemeyong, 14th April 2008, *Droissart, Stévart & Simo (Ombrière de Yaoundé)* 1002; S Region: Akom II, 24th April 2007, *Droissart & Simo* 427; ibid., 24th May 2008, *Simo, Sonké & Taedoumg* 51.
Habitat: epiphyte in submontane forest or on inselbergs with *Microdracoides* (Cyperaceae); 1000–1150 m.
Threats: its proximity to Yaoundé is enough to warrant an inferred or suspected threat on account of likely population growth, urban and agricultural expansion. The Massif of Mbam-Minkoum, despite the presence of many threatened plant species, does not yet have a dedicated conservation programme.
Management suggestions: continue with shade house cultivation and commence efforts to propagate this taxon. Focus efforts on instigation of a multilateral conservation programme on hills surrounding Yaoundé, especially Mbam-Minkoum.

Tridactyle anthomaniaca (Rchb.f.) Summerh. subsp. ***nana*** P.J.Cribb & Stévart (Kew Bull. 59: 195–205 (2004), but incorrectly published as *anthomoniaca*).
EN B2ab(iii)

Range: Cameroon (SW: Banyang Mbo. S: Akom II (1 coll. each.)) and Equatorial Guinea (Rio Muni (2 loc., 3 coll.)).

Known from four locations with an AOO of 12 km^2 and an EOO of 10900 km^2. Three collections from Equatorial Guinea are cited in Stévart & Cribb (2004: 205). Droissart *et al.* (2006: 62) cite the two Cameroon collections: Akom II, east of Kribi, *Stévart & Droissart* 2063; Banyang-Mbo Wildlife Sanctuary, *Simo* 25.

Habitat: epiphyte in lowland and submontane forest; 200(–1000) m.

Threats:threats to lowland forest in/near to Akom II and even Banyang Mbo are inferred on the basis of forest clearance that has already happened in or near to these locations and so this species is considered threatened in that regard.

Management suggestions: introduce this subspecies into cultivation in the Yaoundé shadehouse programme. Revisit Banyang Mbo and look for this taxon to include in a refurbished orchid garden at the Nguti Field Station there.

Tridactyle lagosensis (Rolfe) Schltr. (syn. *Angraecum lagosensis* Rolfe)
VU B2ab(iii)

Range. SE Nigeria (3 loc., 3 coll.), Cameroon (SW: Tissongo; Tabo (1 coll. each). C: Nyong (1 coll.)), Equatorial Guinea (Rio Muni (unknown number of loc., coll.)) and Gabon (2 loc., 5 coll.).

Known from at least nine locations with an AOO of 36 km^2 and an EOO of 240270 km^2, this species is here assessed as VU on the although the number of locations from Equatorial Guinea is not known and its presence in that country is not illustrated on the distribution map. Nigeria, *Barter* s.n. (without locality and so not included in distribution map); March 1891, *Moloney* 3/91; Eket, 1912, *Talbot* 3298. Cameroon: Tissongo, 1977, *Thomas* 96; Tabo and River Akoumayip, 20 km west of Mamfe, 2nd June 1975, *Letouzey* 13703, C: borders of Nyong, near Nkolnlong III, 20 km SW of Ngoumou, 55 km SW of Yaoundé, *Letouzey* 11515.

Habitat: epiphyte in lowland and submontane forest; 150–700 m.

Threats: its typical habitat is under threat across its range.

Management suggestions: investigate the precise number of locations from Rio Muni. Introduce this species into cultivation.

Tridactyle muriculata (Rolfe) Schltr.
EN B1ab(iii)+2ab(iii)

Range. SE Nigeria (1 loc., 1 coll.) and Cameroon (S: Komba Tida Rd (1 coll.)).

Known from only two locations with an AOO of 8 km^2 and an EOO of less than 5000 km^2. It was newly reported for Cameroon by Droissart *et al.* (2009: 129–130). It was known only by the type, Nigeria, Obam, 1911, *Talbot* 904 until a single collection from the Komba Tida Rd, 4th July 2005, *Droissart, Stévart, Nguembou & Djuikouo (Yaoundé shadehouse)* 303.

Habitat: epiphyte in lowland and submontane forest; 150–700 m.

Threats: its typical habitat is under threat at both locations.

Management suggestions: use the cultivated specimens as a source for introductions of live plants into protected areas at its known locations.

Vanilla africana Lindl. subsp. ***cucullata*** (Kraenzl. ex Braun & K.Shum.) Szlach. & Olszewski
EN B2ab(iii)

Range: Cameroon (S: Kribi; Bipindi (1 coll. each). E: Dja Reserve (1 coll.)) and Gabon (2 loc., 4 coll.).

Known from five locations with an AOO of 20 km^2 and an EOO of 557840 km^2. Cameroon specimens include Kribi, 1887–1888, *Braun* 4; Bipindi, *Zenker* 525; Dja Reserve, 10th July 2002, *Stévart* 448. Five Gabon specimens are listed on the Plants of Gabon database (accessed 20th December 2010) but the Libreville collection is not included in the assessment or the distribution map as it was collected from Sibang Arboretum and it is unclear if it was a cultivated or wild specimen.

Habitat: epiphyte in lowland forest; 400–515 m.

Threats: lowland forest across the region is under threat from actual or potential logging. Kribi is an ever-expanding port town where the subpopulation is most probably now lost. The Dja Reserve ought to provide a measure of protection for this species.

Management suggestions: search the known locations and introduce this species into cultivation.

Vanilla ochyrae Szlach. & Olszewski
EN B2ab(iii)

Range. Cameroon (C: Méven Nanga Eboko (1coll.). E: Bertoua (1coll.); Djomedja (3 coll.)).

Known from three collections in Cameroon with an AOO of 12 km^2 and an EOO of 17070 km^2. It was assessed as VU by Droissart *et al.* (2006: 64). Specimens are C Region: Méven Nanga Eboko, 1960, *Endenguele* 113. East Region: Bertoua, July 1939, *Jacques-Félix* 4727 (holotype), Djomedjo, *Stévart & Pial* 345, 367 and 527.

Habitat: epiphyte in semi-deciduous marshy submontane forest; 600–700 m.

Threats: lowland and submontane forest in the areas around Bertoua are subject to forest clearance by logging.

Management suggestions: revisit the known locations for this species and introduce it into cultivation.

Vanilla seretii De Wild.
VU B2ab(iii)

Range: Cameroon (S: Dja Reserve, Piste de Bouamir (1 coll.)), CAR (1 coll.) and Congo (Kinshasa) (5 loc., 6 coll.). This species is known from seven locations with an AOO of 28 km^2 and an EOO of 904730 km^2. It was only known from CAR and Congo (Kinshasa) until a recent Cameroon collection, from Piste de Bouamir, Réserve du Dja, entre Etou et le petit rocher, 19th June 2002, *Stévart & Pial* 382, cited in Stévart (2003: 128). The precise location for the CAR specimen is not known to the author and it has not been included in the distribution map.

Habitat: liana in forest; 620 m.

Threats: the Cameroon location should have an element of protection from its origin within the Dja Reserve. Lowland forest across its range is, however, under threat and evidence of farming activities within the Dja Reserve have been recorded and reported on elsewhere in this chapter. It is therefore inferred to be under threat.

Management suggestions: investigate the current threats at its Cameroon location. Introduce it into cultivation at the Yaoundé shadehouse network.

BIBLIOGRAPHY

Amiet, J.-L. (2004). *Astylosternus nganhanus*. In: IUCN 2010. IUCN Red List of Threatened Species. Version 2010.4. <www.iucnredlist.org>. Downloaded on 18 December 2010.

BirdLife International (2011). Important Bird Areas factsheet: Omo FR. Downloaded from http://www.birdlife.org on 05 January 2011.

Cable, S. & Cheek, M. (1998). The Plants of Mount Cameroon. A Conservation Checklist. Royal Botanic Gardens, Kew. lxxix + 198pp.

Cheek, M. & Pollard, B.J. (2010). *Orchidaceae*. In: Cheek, M. Red Data Taxa of Lebialem Highlands. In: Harvey, Y.B., Tchiengué & Cheek, M. The Plants of Lebialem Highlands, Cameroon. A Conservation Checklist. Royal Botanic Gardens, Kew: 88–89.

Cheek, M., Onana, J.-M. & Pollard, B.J. (2000). The Plants of Mount Oku and the Ijim Ridge, Cameroon. A Conservation Checklist. Royal Botanic Gardens, Kew. iv + 211 pp., 30 figs.

Cheek, M., Darbyshire, I. & Pollard, B.J. (2004). Taxon Red Data Treatments. *In*: Cheek, M., Pollard, B.J., Darbyshire, I., Onana, J.-M. & Wild, C. The Plants of Kupe, Mwanenguba and the Bakossi Mountains, Cameroon. A Conservation Checklist. Royal Botanic Gardens, Kew: 138–201.

Cheek, M., Pollard, B.J., Darbyshire, I., Onana, J.-M. & Wild, C. (2004). The Plants of Kupe, Mwanenguba and the Bakossi Mountains, Cameroon. A Conservation Checklist. Royal Botanic Gardens, Kew. iv + 508 pp, 22 figs, 16 plates.

Cheek, M., Harvey, Y.B. & Onana, J.-M. (2010). The Plants of Dom, Bamenda Highlands, Cameroon. A Conservation Checklist. Royal Botanic Gardens, Kew. iv + 162pp.

Cribb, P.J. (1978). A revision of *Stolzia* (*Orchidaceae*). Kew Bull. 33(1): 79–89.

Cribb, P.J. (1996). New species and records of *Orchidaceae* from West Africa. Kew Bull. 51(2): 353–363.

Cribb, P.J. (1998). *Orchidaceae*. In: Cable, S. & Cheek, M. The Plants of Mount Cameroon. A Conservation Checklist. Royal Botanic Gardens, Kew: 169–179.

Cribb, P.J. & Pérez-Vera, F. (1975). A contribution to the study of the *Orchidaceae* of the Côte d'Ivoire. Adansonia, sér. 2, 15(2): 199–214.

Cribb, P.J. & Pollard, B.J. (2002). New orchid discoveries in western Cameroon. Kew Bull. 57: 653–659.

Cribb, P.J. & Pollard, B.J. (2004a). *Bulbophyllum kupense* P.J.Cribb & B.J.Pollard, an unusual new orchid from western Cameroon. Kew Bull. 59: 137–139.

Cribb, P.J. & Pollard, B.J. (2004b). *Orchidaceae*. In: Cheek, M., Pollard, B.J., Darbyshire, I., Onana, J.-M. & Wild, C. The Plants of Kupe, Mwanenguba and the Bakossi Mountains, Cameroon. A Conservation Checklist: 451–471.

Cribb, P.J., Van der Laan, F.M. & Arends, J.C. (1989). Two New Species of *Orchidaceae* from West Africa. Kew Bull. 44(3): 479–483.

Cribb, P.J., Zapfack, L. & DeMarco, J. (2000). *Orchidaceae*. In: Cheek, M., Onana, J.-M. & Pollard, B.J. The Plants of Mount Oku and the Ijim Ridge, Cameroon. A Conservation Checklist. Royal Botanic Gardens, Kew: 185–191.

Darbyshire, I. (2004). Read this first! In: Cheek, M., Pollard, B.J., Darbyshire, I., Onana, J.-M. & Wild, C. The Plants of Kupe, Mwanenguba and the Bakossi Mountains, Cameroon. A Conservation Checklist: 202–206.

Demissew, S., Cribb, P.J. & Rasmussen, F. (2004). Field Guide to Ethiopian Orchids. Royal Botanic Gardens, Kew. 304pp.

Droissart, V., Simo, M., Sonké, B., Cawoy, V. & Stévart, T. (2009). Le genre *Stolzia* (Orchidaceae) en Afrique centrale avec deux nouveaux taxons. Adansonia sér. 3. 31(1): 25–40.

Droissart, V., Sonké, B. & Stévart, T. (2006). Les *Orchidaceae* endémiques d'Afrique centrale atlantique présentes au Cameroun. Syst. Geogr. Pl. 76: 3–84.

Droissart, V., Sonké, B., Nguembou, C.K., Djuikouo, M.-N.K., Parmentier, I. & Stévart, T. (2009). Synopsis of the Genus *Chamaeangis* (*Orchidaceae*),

Including Two New Taxa. Systematic Botany 34(2): 285–296.

Droissart, V., Sonké, B., Simo, M. & Stévart, T. (2009). New Orchid Records from Atlantic Central Africa. Edinburgh Journal of Botany 66(1): 115–132.

Droissart, V., Sonké, B., Lowry, P.P., Pollard, B.J. & Stévart, T. (in press). Notes taxonomiques et chorologiques sur les Orchidaceae endémiques d'Afrique centrale atlantique. Adansonia.

Fischer, E., Killmann, D., Delepierre, G. & Lebel, J.-P. (2010). The Orchids of Rwanda. Koblenz Geographical Colloquia Series. Biogeographical Monographs 2. Koblenz, Germany. 439pp.

Govaerts, R., Pfahl, J., Campacci, M.A., Holland Baptista, D., Tigges, H., Shaw, J., Cribb, P.J., George, A., Kreuz, K. & Wood, J. (2011). World Checklist of Orchidaceae. The Board of Trustees of the Royal Botanic Gardens, Kew. Published on the Internet; http://www.kew.org/wcsp/ accessed 15 February 2011; 16:30 GMT.

Hall, J.B. (1974). African Orchids XXXIV. Kew Bull. 29(2): 427–429.

Harvey, Y.B., Pollard, B.J., Darbyshire, I., Onana, J.-M. & Cheek, M. (2004). The Plants of Bali Ngemba Forest Reserve, Cameroon. A Conservation Checklist. Royal Botanic Gardens, Kew. iv + 154pp.

IUCN (2001). IUCN Red List Categories: Version 3.1. Prepared by the IUCN Species Survival Commission. IUCN, Gland, Switzerland and Cambridge, U.K.

La Croix, I. (1993). Two new species of *Habenaria* sect. *Podandria* (*Orchidaceae*) from West Africa. Kew Bull. 48(2): 369–373.

La Croix, I. (1996). A new name in *Habenaria*. Kew Bull. 51: 364.

Larsen, T.B. (2005). Hazards of butterfly collecting: some sticks have fallen over the road. Omo Forest, Nigeria, June 1967. The Entomologist's Record and Journal of Variation. 117(4): 171–173.

Podzorski, A.C. & Cribb, P.J. (1979). A Revision of *Polystachya* sect. *Cultriformes* (*Orchidaceae*). Kew Bull. 34(1): 147–186.

Pollard, B.J. (2004). *Orchidaceae*. In: Cheek, M. & Harvey, Y.B. Red Data Taxa of Bali Ngemba Forest Reserve. In: Harvey, Y.B., Pollard, B.J., Darbyshire, I., Onana, J.-M. & Cheek, M. (2004). The Plants of Bali Ngemba Forest Reserve, Cameroon. A Conservation Checklist. Royal Botanic Gardens, Kew: 72.

Pollard, B.J. & Darbyshire, I. (2004). *Orchidaceae*. In: Cheek, M., Darbyshire, I. & Pollard, B.J. Taxon Red Data Treatments. In: Cheek, M., Pollard, B.J., Darbyshire, I., Onana, J.-M. & Wild, C. (2004). The Plants of Kupe, Mwanenguba and the Bakossi Mountains, Cameroon. A Conservation Checklist. Royal Botanic Gardens, Kew: 193–201.

Pollard, B.J., Roberts, D.L. & Cribb, P.J. (2004). *Orchidaceae*. In: Harvey, Y.B., Pollard, B.J., Darbyshire, I., Onana, J.-M. & Cheek, M. The Plants of Bali Ngemba Forest Reserve, Cameroon. A Conservation Checklist. Royal Botanic Gardens, Kew: 141–150.

Roberts, D.L. & Solow, A. (2008). The effect of the Convention on International Trade in Endangered Species on scientific collections. Proceedings of the Royal Society B. 275: 987–989.

Roberts, D.L., Drinkell, K. & Pollard, B.J. (2010). *Orchidaceae*. In: Harvey, Y.B., Tchiengué, B. & Cheek, M. The Plants of Lebialem Highlands, Cameroon, a Conservation Checklist, Royal Botanic Gardens, Kew: 155–159.

Simo, M., Droissart, V., Sonké, B. & Stévart, T. (2009). The Orchid Flora of the Mbam Minkom Hills (Yaoundé, Cameroon). Belg. J. Bot. 142(2): 111–123.

Sonwa, D.J., Nkongmeneck, B.A., Weise, S.F., Tchatat, M., Adesina, A.A. & Janssens., M.J.J. (2007). Diversity of plants in cocoa agroforests in the humid forest zone of Southern Cameroon. Biodiversity and Conservation. 16(8): 2237–2244.

Stewart, J. (1979). A Revision of the African Species of *Aerangis* (*Orchidaceae*). Kew Bull. 34(2): 239–319.

Stévart, T. (2003). Étude taxonomique, écologique et phytogéographique des Orchidaceae en Afrique Centrale Atlantique. Thèse de Doctorat, Université Libre de Bruxelles, Belgique. 225pp.

Stévart, T. & Cribb, P.J. (2004a). New Species and Records of *Orchidaceae* from São Tomé and Príncipe. Kew Bull. 59: 77–86.

Stévart, T. & Cribb, P.J. (2004b). Five new taxa of *Tridactyle* (*Orchidaceae*) from West Central Africa. Kew Bull. 59: 195–205.

Stévart, T. & de Oliveira, F. (2000). Guide des Orchidees de São Tomé et Príncipe. Libreville, ECOFAC (ed.), cellule de coordination. 258pp.

Stévart & Nguema, N. (2004). Trois espèces et trois combinations nouvelles de *Polystachya* (*Orchidaceae*) du Cameroun, de Guinée Équatoriale et du Gabon. Adansonia, sér. 3. 26(2): 217–233.

Summerhayes, V.S. (1932). African Orchids III. Bull. Misc. Inform., Kew: 188–193.

Summerhayes, V.S. (1954). African Orchids XXIV. Kew Bull. 12(1): 107–126.

Summerhayes, V.S. (1968). *Orchidaceae*, in Hepper, F.N. (ed.). Flora of West Tropical Africa (2nd ed.). 3(1): 180–276.

Szlachetko, D.L. & Olszewski, T.S. (1998). *Orchidaceae* vol. 1. *In*: Satabié, B. & Morat, P. (eds.). Flore du Cameroun: 34. Ministère de la Recherche Scientifique et Technique (MINREST), Yaoundé, Cameroun: 1 – 321.

Szlachetko, D.L. & Olszewski, T.S. (2001a). *Orchidaceae* vol. 2. *In*: Achoundong, G. & Morat, P. (eds.). Flore du Cameroun: 35. Ministère de la Recherche Scientifique et Technique (MINREST), Yaoundé, Cameroun: 322 – 665.

Szlachetko, D.L. & Olszewski, T.S. (2001b). *Orchidaceae* vol. 3. *In*: Achoundong, G. & Morat, P. (eds.). Flore du Cameroun: 36. Ministère de la Recherche Scientifique et Technique (MINREST), Yaoundé, Cameroun: 666 – 948.

Szlachetko, D.L., Sawicka, M. & Kras-Kapińska, M. (2004). *Orchidaceae* I. Flore du Gabon 36. Muséum National D'Histoire Naturelle, Paris. 231 pp.

Szlachetko, D.L. (2008). *Orchidaceae* of Ivory Coast. Monografías del Jardín Botánico Madrid. 396pp.

Van der Laan, F.M. & Cribb, P.J. (1986). *Ossiculum* (*Orchidaceae*), a new genus from Cameroon. *Kew Bulletin*, 41, 823–826.

Vermeulen, J.J. (1984). Five new species and one new variety of continental African *Bulbophyllum* (*Orchidaceae*). Bull. Jard. Bot. Nat. Belg. 54: 137–150.

Vermeulen, J.J. (1987). A taxonomic revision of the Continental African *Bulbophyllinae*. Orchid Monographs 2. E.J.Brill, Leiden, The Netherlands. 300pp.

Vermeulen, J.J. (1992). Additions to the taxonomic revision of the continental African Bulbophyllinae (Orchidaceae). Kew Bull. 47(1): 137–140.

Willis, F., Moat, J. & Paton, A.J. (2003). Defining a role for herbarium data in Red List assessments: a case study of *Plectranthus* from eastern and southern tropical Africa. Biodiversity and Conservation 12: 1537–1552.

Zapfack, L., Engwald, S., Sonké, B., Achoundong, G. & Madong, B.A. (2002) The impact of land use conversion on plant biodiversity in the forest zone of Cameroon. Biodiversity and Conservation. 11: 2047–2061.

PALMAE (ARECACEAE)

assessed by Martin Cheek

Palms need no definition or explanation. Few people in Cameroon are not acquainted with this family. Throughout the forest belt, oil palm can be found wherever people have had settlements, while in the north borassus palms with their fan-like leaves are prominent in savanna. In wet areas, such as swamps, raphia palms are common. Spiny, climbing palms, rattans, are often to be found in forest.

African palm diversity with perhaps only 50 species, is much lower than in S America and SE Asia. However, within Africa, Cameroon probably has higher palm diversity than any other country. For example, 18 of the 22 known species of African rattans occur in Cameroon. Many of these species are useful to man. *Phoenix reclinata*, *Elaeis* and some *Raphia* species are tapped extensively for 'palm wine' in forest areas, while raphia leaves are very important for house construction, especially roofing, but also furniture; the introduced coconut has edible, oil-rich seeds; rattans are extensively harvested for cane production, used in basket-making and for furniture, and above all, *Elaeis guineensis* the oil palm a native African species, is immensely important locally in producing cooking oil and commercially on large plantations for palm oil and palm kernel oil which has a huge international market. Generally it is the more common species that are used, and all the groups mentioned above, apart from the rattans, have long been planted by people in Cameroon, so are not threatened.

There is no Flore du Cameroun account for the family, but for many genera, identification is not difficult since there is only a single species native in Cameroon (e.g. *Borassus, Elacis, Podococcus, Nipa, Phoenix*). However, for Africa's largest genus, *Raphia*, a modern revision is needed to build upon that by Otedoh which dates from 1962 (J. Nigerian last. Ford Oil Palm Research 6 (22): 145–189). Several of the species accepted by Otedoh are contenders for conservation status. *R. mambillensis* Ocedoh was proposed as LR/cd in Cheek *et al.* (2000: 85–86), but with better data might prove to be threatened. *Raphia regalis* was assessed as VU B2ab(iii) by Darbyshire in Cheek *et al.* (2004: 201) but is reassessed here as NT since more than 10 locations are now known though it remains very rare. *Raphia africana* Otedoh was described from one collection Nigeria and has been tentatively identified from Cameroon (Cheek *et al.* 2004: 472). If verified this should be assessed as EN B2ab(iii).

Fortunately, for the four genera that comprise the rattans, a recent "Field Guide to the Rattans of Africa" is now available (Sunderland (2007), RBG Kew) which is the main basis for the species treatments below.

In identifying rattan genera, features of the climbing stem are used, in particular the surface of the stem, which is made mainly by the leaf-sheaths, from which the compound leaf-blades arise. This sheath extends as a short, often papery, cylinder, the 'ocrea', beyond and above the point at which the leaf-blade arises. The shape and features of the ocrea are often useful for identification. Below the insertion of the leaf-blade the rhachis in some species appears to continue down the leaf-sheath for a few cm as a ridge, the 'knee' which ends abruptly.

A great deal of field research on rattans was conducted by Sunderland at the turn of the Millenium. He was based at Limbe, SW Region, over ten years. It is likely that much basic populational data on the species treated below has already been recorded by him. Therefore, he should be consulted before any plans are made to conduct field surveys on rattans, or to put in place any conservation actions.

Hyphaene guineensis Schumach. a rare W. African fan palm of coastal beaches recently rediscovered in Gabon by Van Valkenberg (Van Valkenberg & Dransfield in Palms 48 (1) (2004) may well be found in Cameroon and is probably threatened.

Eremospatha – climbing spiny rattans; leaf sheath and stem lacking spines or prickles.

Eremospatha barendii Sunderland

CR B1+2ab(iii) + D

Range: Cameroon endemic (S: Ebom near Lolodorf).

Here assessed as Critically Endangered since only one individual was known from a timber concession that has since been logged (information from Sunderland 2007 who assesses the species as Endangered). AOO is 4 km² with 4 km² cells.

Climber, multi-stemmed, to 25–30 m, leaf-sheaths about 2.5 cm diam., lacking prickles, knee present; ocrea striate, longitudinally splitting into a v-shape or tattering; leaves to 1.2 m, leaflets c. 20 pairs, linear-lanceolate, lowermost leaflets smaller than the rest and curving over the stem.

Habitat: lowland evergreen forest.

Threats: logging (see above). No direct uses are recorded.

Management suggestions: efforts should be made to refind this species at its only known locality and to obtain seed for ex situ propagation and reintroduction. A census should be made of the numbers of individuals present, regeneration levels and apparent threats. This will form a baseline for future monitoring. Although only one individual was recorded in the original discovery, others may have been overlooked. If none are found at the original locality the search should be widened.

Eremospatha quinquecostulata Becc.

VU B2ab(iii)

Range: Nigeria (Calabar/Oban) and Cameroon (SW: Takamanda; Rumpi Hills. S: Campo-Ma'an NP. E: 40 km S Messaména).

The Nigerian locations have not been mapped.

Here *Eremospatha quinquecostulata* is assessed as Vulnerable since six locations are known (AOO 24 km² with 4 km² cells) and threats are as below. It was assessed as Vulnerable by Sunderland (2007).

Climber, multi-stemmed, 10–15 m; leaf sheath striate, lacking prickles, 5–10 mm diam., ocrea c. 1.5 cm, truncate, entire; knee inconspicuous; leaves 60–80 cm, leaflets clustered in pairs or fours, narrowly oblanceolate, veins conspicuous.

Habitat: lowland evergreen forest in hilly areas.

Threats: clearance of forest for agriculture (Nigeria) – not used by man as a cane in Cameroon since other species have better properties.

Management suggestions: since Takamanda and Campo-Ma'an are National Parks, these may be the best locations to focus on protecting this species. Data on regeneration levels, density and range would help form a baseline for future monitoring and would help assess needs for management intervention.

Eremospatha tessmanniana Becc.

EN B2ab(iii)

Range: Cameroon (SW: Takamanda. S: Ébolowa) and Equatorial Guinea (Rio Muni).

The Rio Muni location has not been mapped.

Here *Eremospatha tessmanniana* is assessed as Endangered since only three locations are known (above; AOO 12 km² with 4 km² cells) and threats are as below. Vulnerable in Sunderland (2007).

Climber, multi-stemmemd, 60–80 cm; leaf-sheaths 12–15 mm diam., lacking prickles, striate; ocrea 1.5 cm long, truncate; knee absent; adult leaves to 80 cm, bluish-green, narrowly oblong, lower most leaflets lax, smallest, apex rounded, irregular, margins with forward pointing black spines near leaflet apex, reflexed spines in basal half.

Habitat: lowland evergreen forest in hilly areas.

Threats: slash and burn agriculture; logging in the Ébolowa area. No uses for the species have been recorded (Sunderland (2007)).

Management suggestions: the best focus for conservation activities is probably Takamanda since newly gazetted as a National Park, and also since recently recorded here. Basic populational data should be recorded, but may already be available from Sunderland's research at the turn of the Millenium (see family introduction).

Laccosperma. Leaf sheath spiny, spines long, slender, firmly attached; ocrea triangular [illegible]ic, drying brown and papery.

Laccosperma korupensis Sunderland

EN B2ab(iii)

Range: Cameroon endemic (SW: Korup NP; Mt Cameroon foothills at Onge-Kotto II – Idenao. S: Kribi at Bidou I).
Here assessed as Endangered since only known from five sites at three locations (above; AOO 20 km² with 4 km² cells) and with threats as below. Vulnerable in Sunderland (2007).
Climber, multi-stemmed, 10 m; leaf-sheath 15 mm diam. spiny, ocrea 7–10 cm long, gradually tapering to the apex, often tattered; leaves to 1 m, petiolate, leaflets peridant, narrowly long-elliptic, acuminate, margins with small reflexed spines.
Habitat: lowland evergreen forest; c. 200 m alt.
Threats: logging, followed by agriculture is a major risk at all known sites in the Mt Cameroon location where most of the records for this species derive.
Management suggestions: protection efforts are needed around Mt Cameroon since the foothill forest areas are under severe pressure of being logged. The most secure areas for the species is probably at Korup NP. Basic populational data is needed as a baseline for future monitoring and to assess needs for management intervention.

Oncocalamus. Leaf sheath spiny, spines irregularly spaced, short, triangular, easily detatching; ocrea tubular, not brown and dry but green.

Oncocalamus tuleyi Sunderland
VU B2ab(iii)
Range: Nigeria (near Cross River NP, 50 km S Obubra; Ojo Rd; Calabar-Ikot) and Cameroon (SW: Mt Cameroon at Onge R. S. Bakundu FR, Buea-Kumba mile 48; Rumpi Hills; Takamanda; Mamfe, 15 km S; Kumba-Nguti at Wone; Korup NP).
Here *Oncocalamus tuleyi* is assessed as Vulnerable since eight locations are known (above; AOO 32 km² with 4 km² cells) and threats as below. Previously (Sunderland *et al.* 2007 op. cit.) assessed as NT based on eleven locations being then accepted (revised here under new IUCN guidelines), but treated as Vulnerable by Sunderland 2007 op. cit.
Climber, multi-stemmed, to 30 m; leaf-sheaths 25–45 mm diam., patchily spiny; ocrea cylindric, c. 2.5 cm long, extending furthest on side opposite to the leaf blade; knee detectable; leaves sessile, 1.2–2 m; leaflets narrowly long-elliptic, arching; pendulous, long-acute, margins with prickles.
Habitat: lowland evergreen forest in slightly hilly areas; 200–400 m alt.
Threats: not used as a commercial rattan but leaf-sheaths used as a chew stick and stem epidermis for tying yams. Highly threatened by logging followed by agriculture in the foothills around Mt Cameroon (Onge and S Bakundu areas) where it has been most densely recorded.
Management suggestions: conservation efforts are needed around Mt Cameroon. The most secure areas for the species are those at Korup and Takamanda, both National Parks. Basic populational data is needed as a baseline for future monitoring and to assess needs for management intervention.

TRIURIDACEAE

assessed by Martin Cheek

All species of this family lack chlorophyll and derive the nutrition from assocations with underground fungi, being known as achlorophyllous heteromycotrophs, or more traditionally, as saprophytes.

Among the saprophytes this family is unique in having many separate carpels in each female flower, resembling those of Alismataceae or Ranunculaceae. The family is represented in Cameroon otherwise by *Sciaphila ledermannii*, which differes in having both male and female flowers on each spike (not on separate spikes) and glossy red-purple spikes (not dull white).

In conducting an inventory of the plant species of Mt Kupe in SW Region, Cameroon in 1995, two small populations of a new species of Triuridaceae were discovered by us. The species was so distinct that it merited a new genus and tribe. The bilaterally symmetrical, 2-seeded carpels and sessile flowers were then unknown and unique in the family.

Kupea martinetugei Cheek & S.Williams
CR C1
Range: Cameroon, SW Region (Mt Kupe (2 sites); Mt Cameroon, western foothills near Mokoko at Diongo)
This species was assessed as CR C1 in Cheek *et al.* (2003, 2004: 201), that is, as Critically Endangered, because less than 250 individuals are known (in fact only c. 70 flowering spikes were then known in total, at only two sites at one location at Mt Kupe), these being estimated to be likely to decline by more than 25% in the next generation (estimated at c. 50 years). In the last few years numbers have declined at the first discovered site from 20 spikes to only one spike over a ten year period, possibly due to visitors compacting the soil adversely.
In 2008 a new location, at Diongo, Mt Cameroon was published, evidenced by a photo record (Figure 10.1 p.158 in Merckx (2008) Myco-heterophy in Dioscoreales). Rumour of a third location in S Region has not yet been substantiated. The previous assessment is maintained here since there is no evidence yet that the 250 individual threshold has yet been passed.
Herb 3–7 cm tall, lacking chlorophyll, yellow or dull white, translucent, glabrous. Stems usually unbranched, erect or decumbent at the base, 0.7–1.2 mm diam., scales sparse.

Inflorescence a spike, 25–70-flowered, alternate at intervals of 1–4 mm along axis, lowest flower c. 1.2 cm above ground-level; bracts and pedicels absent. Flower tepals 4, fleshy, the lower slightly larger and overlapping the valvate remainder in bud, appendages absent. Male flowers bilaterally symmetrical, c. 5 mm high, 3–3.5 mm wide, with upper three tepals patent, elliptic, 1.25–1.5 × 0.9 mm, apex acute to rounded, usually slightly hooded. Lower tepal much larger, oblong, 1.5 × 2.2 mm, carinate, apex irregularly truncate, with c. 10 shallow lacinia, not hooded. Stamens 4, opposite tepals, staminodes absent, with filaments largely adnate to the tepals, swollen, obclavate, c. 0.6 × 0.4 mm, anther transversely elliptic, 0.5 mm wide, cells 4, collateral, connective appendages absent. Gynoecium rudiments entirely absent. Female flowers radially symmetrical, more or less globular, with tepals concealed by carpels. Tepals slightly fleshy, patent, ovate-elliptic, apex acute to obtuse, 0.8–1 × 0.5–0.6 mm. Gynoecium c. 2.5 mm, uppermost c. 1–9 flowers bun-shaped, wider than tall, lacking staminodes, carpels 25–60, transversely bilobed, 0.4–0.6 mm wide, c. 0.3 mm high and thick, older carpels with pericarp sometimes becoming abruptly black (when fertilized?); unilocular, ovules two; style filiform, terminal not excentric (except by abortion) but arising from between the two lobes, 0.6–0.8 mm long , apex rounded, glabrous.
Habitat: lowland to submontane evergreen forest; c. 700 m alt.
Threats: see above.
Management suggestions: continued monitoring of the known sites, both at Mt Kupe, is advised.

REFERENCE

Cheek, M., Willams, S.A. & Etuge, M. (2003) *Kupea martinetugei*, a new genus and species of *Triuridaceae* from western Cameroon. Kew Bull. 58(1): 225–228.

ZINGIBERACEAE

assessed by Martin Cheek

All members of this family have a strong spicy scent when cut, separating them at once from close relatives such as Costaceae (once included in Zingiberaceae) and Marantaceae, which are also herbaceous monocots with sheathing leaf-bases, numerous parallel nerves and delicate, orchid-like flowers. Best known for the several Asian species cultivated for spice (e.g. ginger, cardamom and turmeric), only *Aframomum* [illegible] or alligator pepper, is both cultivated as a spice (its [illegible]) and native to Africa. Flore du Cameroun 4 (1965) covers this family but is now outdated (see below). Two genera are commonly found in Cameroon's forest, these are:

1. *Aframomum* with large (3–8 cm wide) brightly coloured (pink, red, yellow or white), if short-lived flowers, opening one or two at a time from the top of a short peduncle separate from the 1–6 m tall leafy culm, the flowers producing large 5–8 cm narrowly ovoid to globular fruit which are usually red or orange and often have edible pulp. Several wild-gathered species are as dried fruits sold in markets in Cameroon as pepper, spice or as a sweet, depending on their flavour. This genus, restricted to Africa, has possibly a hundred or more species, many of which remain unknown to science. Still others are rare and localised, but have not been well characterised so are difficult to identify. What is needed is a taxonomic revision of the genus. David Harris at Edinburgh has been working towards this for more than 10 years and it is to be hoped that he will soon publish something that will address these taxonomic impediments. In the meantime it would be premature to assign IUCN ratings to most species of *Aframomum.*
2. *Renealmia* has fewer species than *Afromomum.* Their inflorescences are racemes or panicles with small inconspicuous flowers producing 4–5 mm diam., berry-like fruit. This genus is also in need of revision and until this is addressed, again, IUCN assessments seem inappropriate since they would lack firm foundations. In Cameroon most material of the genus is named *Renealmia africana* sensu lato, as a provisional measure until this taxonomic impediment can be lifted.

The following species currently known from Cameroon appear to rate either threatened or near-threatened status, judging by the paucity of the number of locations known to us. However, David Harris's revision is likely to contain new data on these that would affect any assessment made now:

Afromomum arundinaceum (Oliv. & Hanb.) K.Schum.)
A. kayserianum (K.Schum.) K.Schum.
A. longeligulatum Koechlin
A. pilosum (Oliv. & Hanb.) K.Schum.
A. pruinosum Gagnep.
A. pesudostipulare Loes. & Mildbr ex Koechlin
Renealmia cabrae De Wild. & Th. Dur.
R. congoensis Gagnep.
R. densispicata Koechlia
R. stenostachys K.Schum.

Aulotandra kamerunensis Loes.
EN B2ab(iii)
Range: Cameroon endemic (S: Bipindi; Ebianemeyong et Nyabéssan. C: Mefou proposed NP; Bodi, 20 km SW Eséka).

Here *Aulotandra kamerunensis* is assessed as Endangered since only four locations are known (above; AOO 16 km^2 with 4 km^2 cells) with threats as below.

Herb, 30 cm, glabrous; lacking aerial stem; leaves several, erect, from stout horizontal rhizome; sheath cylindric; erect 7–8 × 0.6 cm, papery; blade lanceolate, 20–26 × 6.5–8.8 cm, apex acute to acuminate, mucron 2 mm, base obtuse with midrib concealed by blade, secondary nerves parallel, arising at a steep angle from the midrib and discernible as ridges in the midrib for 1–2 cm before departing at an angle of c. 40°, slowly curving up to the apex and margin, inter-secondary nerves 3, weaker, parallel, with transverse nerves forming square or oblong compartments, each compartment with 7–15 black dots on the lower surface, in transmitted light, translucent; pseudopetiole 12–35 × 0.15–0.2 cm, finely ridged; inflorescence arising from rhizome, 2–3-flowered; peduncle erect, 5–6 cm, covered in sheathing papery oblong bracts 30–40 × 5–7 mm, internodes 1.5–2.7 cm; pedicels 2 cm, subtended by bracts; ovary 9 × 3 mm; calyx tube 9 × 4 mm; tepal tube 1.5 cm, outer tepal lobes 3, oblong-elliptic, 3.2 × 0.7 cm, inner tepals purple and white, forming a tube c. 7 × 2 cm.

First published in Bot. Jahrb. Syst. xliii: 389 (1909) based on *Zenker* 36908 from Bipindi collected in 1908, the species was not seen again until 1973, near Eséka (*Letouzey* 12504). It was then recorded from the Campo area by Tchoutou (2004 op. cit.) but we have not seen a specimen. The Mefou record, (*Tadjouteu* 630 4/2004), is only the fourth global location for this very rare species. After his specimen was identified at Kew by David Harris, Tadjouteu returned to Mefou to rediscover the plant but sadly it was not refound.

This is the only continental African species of the genus, the other five all occurring in Madagascar.

Habitat: evergreen *Gilbertiodendron dewevrei* forest; c. 650 m alt.

Threats: clearance of forest for cultivation.

Management suggestions: the best hope for the survival of this species is Mefou proposed NP, should it be gazetted. Tadjouteu should be engaged to attempt to refind this species there during April, when, in flower it is conspicuous. This species has potential as an ornamental plant and there is certainly potential to multiply the species in cultivation with a view to introducing it to the wild.

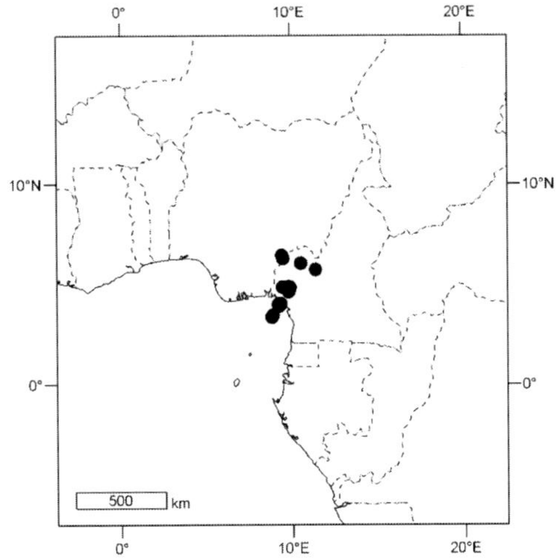

a. ***Acanthopale decempedalis*** VU (Acanthaceae)

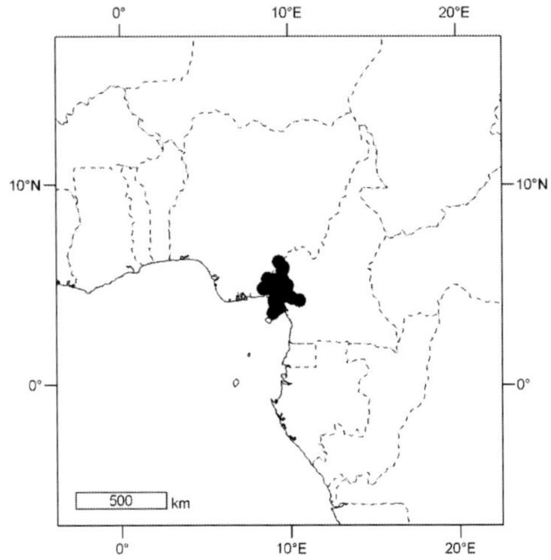

b. ***Afrofittonia silvestris*** VU (Acanthaceae)

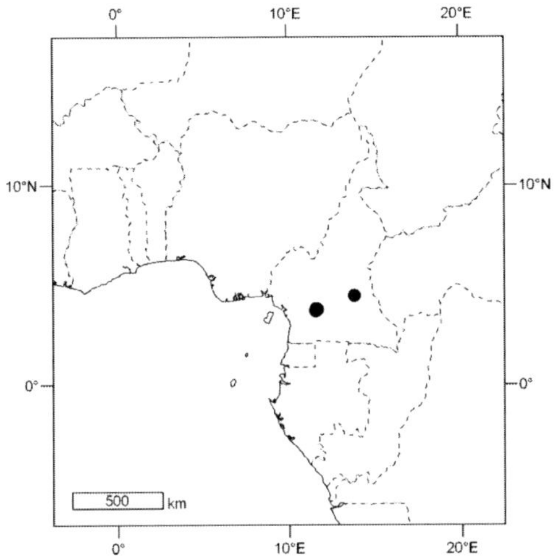

c. ***Anisotes zenkeri*** EN (Acanthaceae)

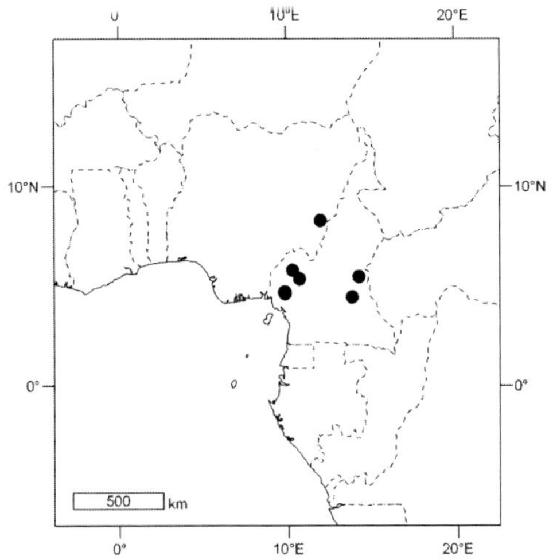

d. ***Asystasia glandulifera*** VU (Acanthaceae)

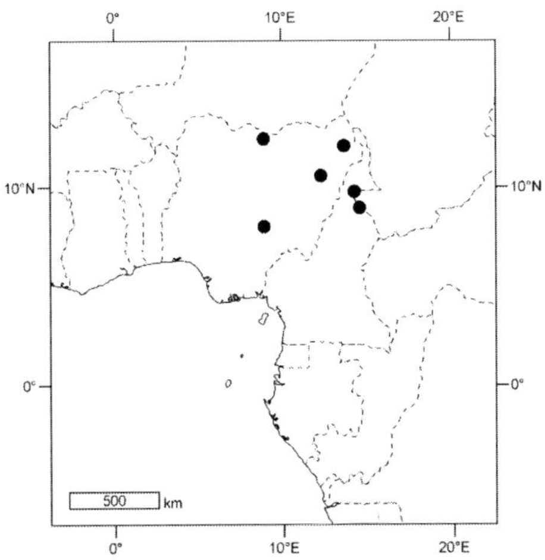

e. ***Barleria bornuensis*** VU (Acanthaceae)

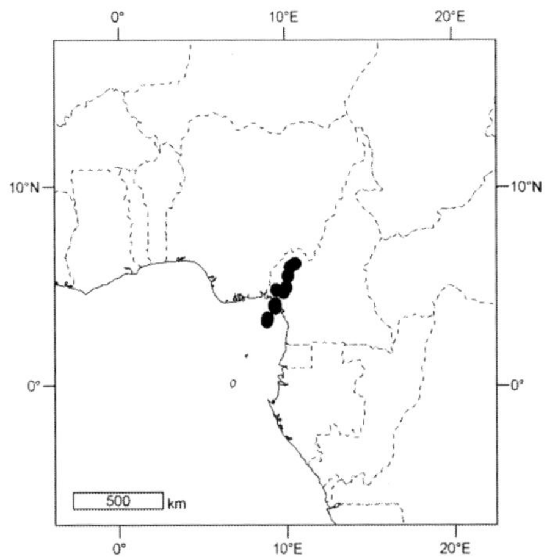

f. ***Brachystephanus giganteus*** VU (Acanthaceae)

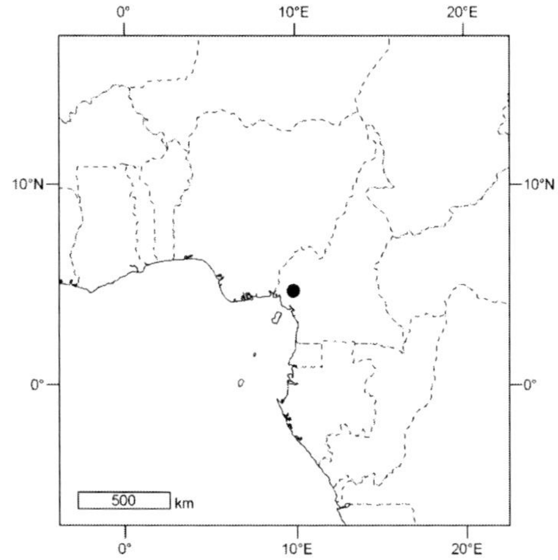

a. ***Brachystephanus kupeensis*** CR (Acanthaceae)

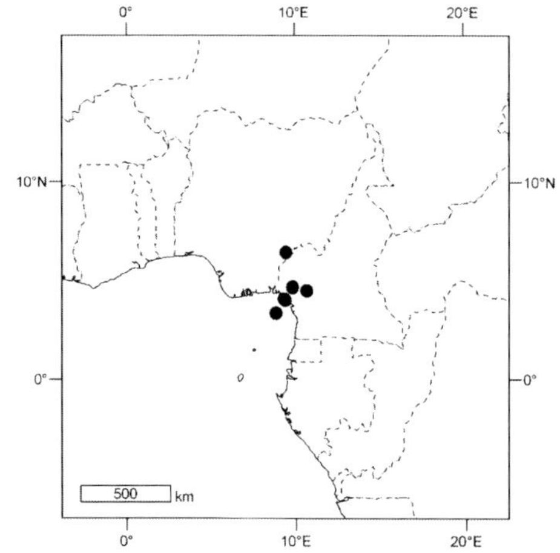

b. ***Brachystephanus longiflorus*** VU (Acanthaceae)

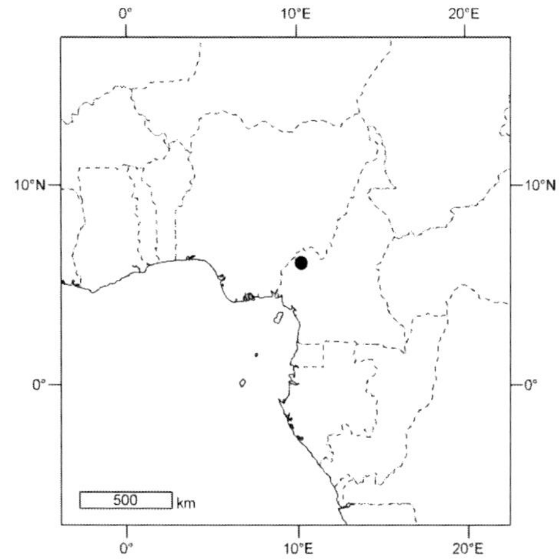

c. ***Brachystephanus oreacanthus*** EN (Acanthaceae)

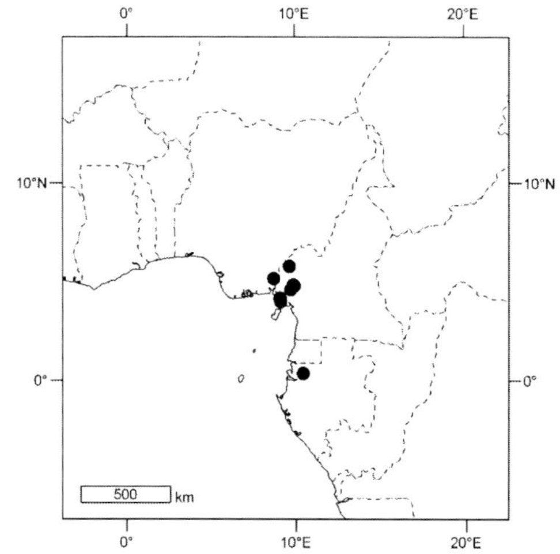

d. ***Brillantaisia lancifolia*** VU (Acanthaceae)

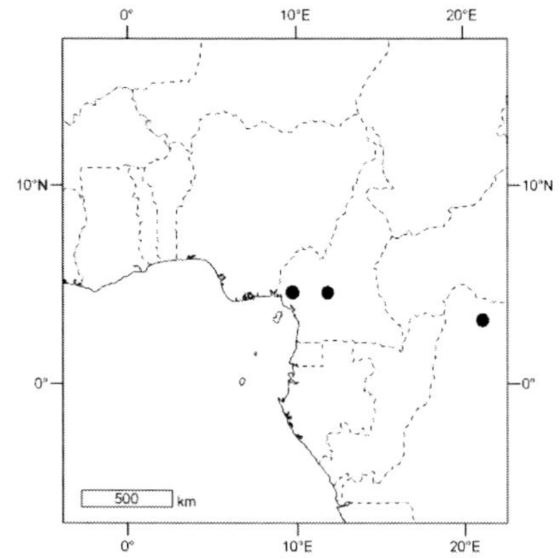

e. ***Chlamydocardia subrhomboidea*** EN (Acanthaceae)

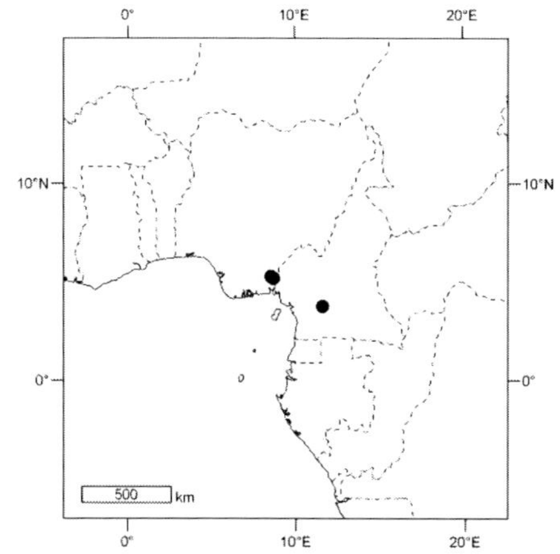

f. ***Crossandra obanensis*** EN (Acanthaceae)

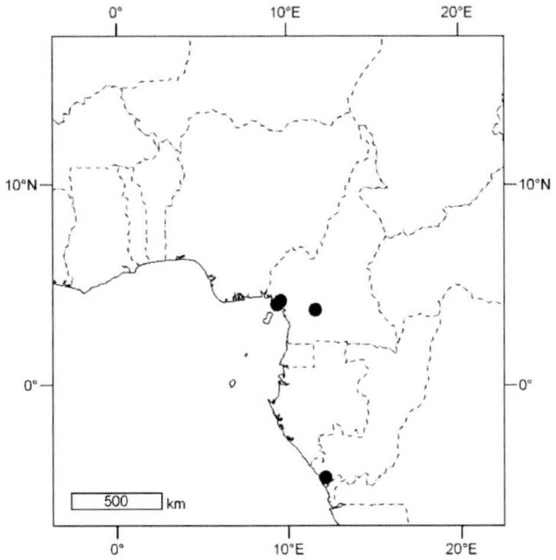

a. ***Dicliptera alternans*** VU (Acanthaceae)

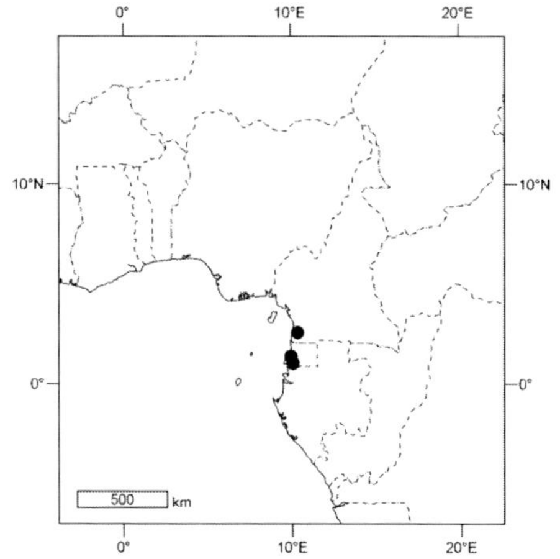

b. ***Dischistocalyx champluverianus*** EN (Acanthaceae)

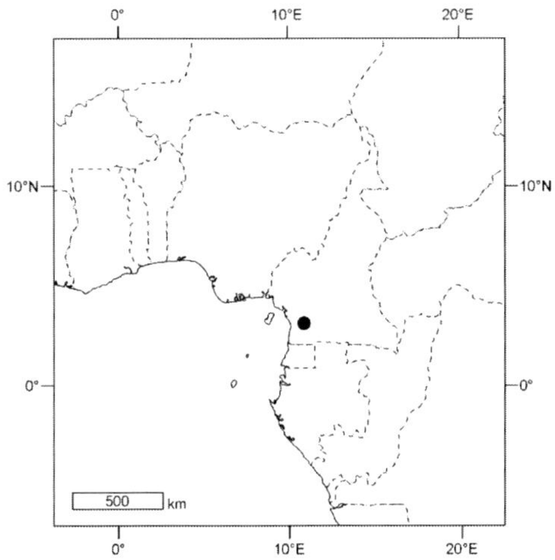

c. ***Dischistocalyx rivularis*** CR (Acanthaceae)

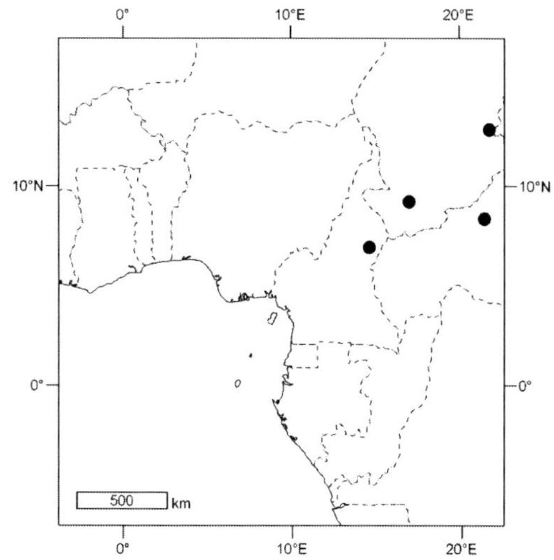

d. ***Hygrophila mediatrix*** EN (Acanthaceae)

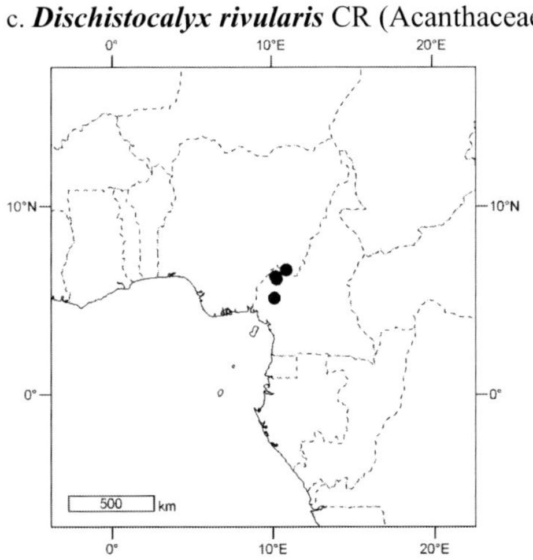

e. ***Isoglossa dispersa*** VU (Acanthaceae)

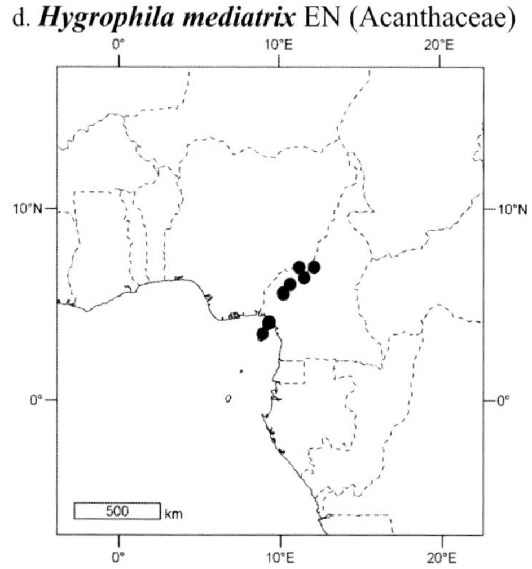

f. ***Isoglossa nervosa*** VU(Acanthaceae)

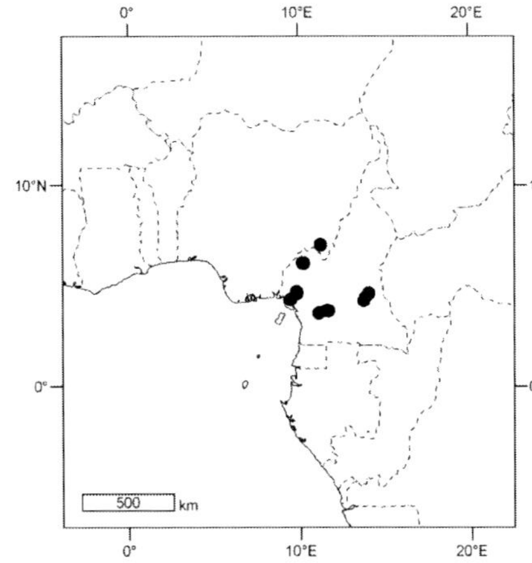

a. ***Justicia camerunensis*** VU(Acanthaceae)

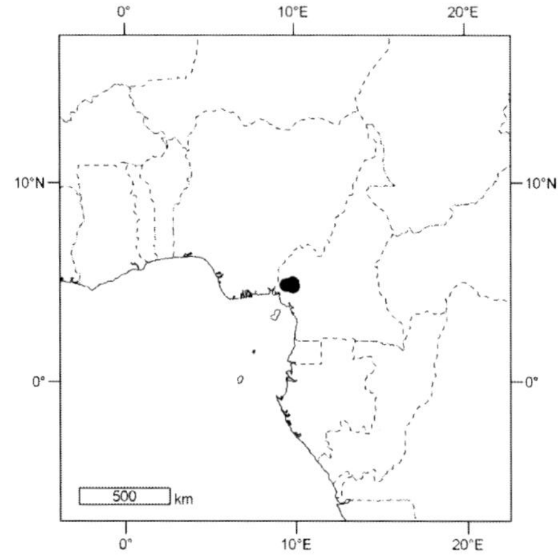

b. ***Justicia leucoxiphos*** EN (Acanthaceae)

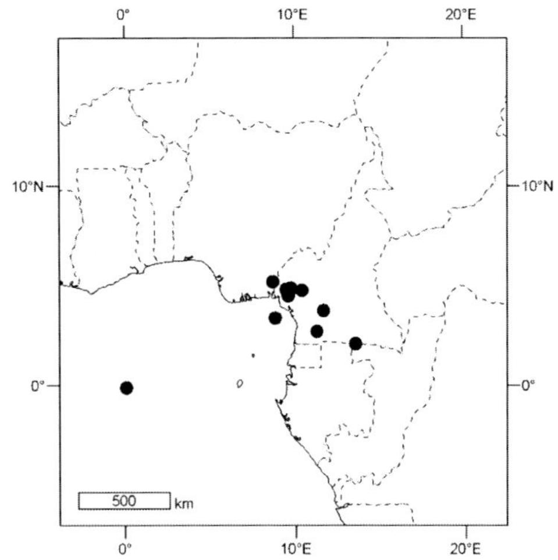

c. ***Justicia orbicularis*** VU (Acanthaceae)

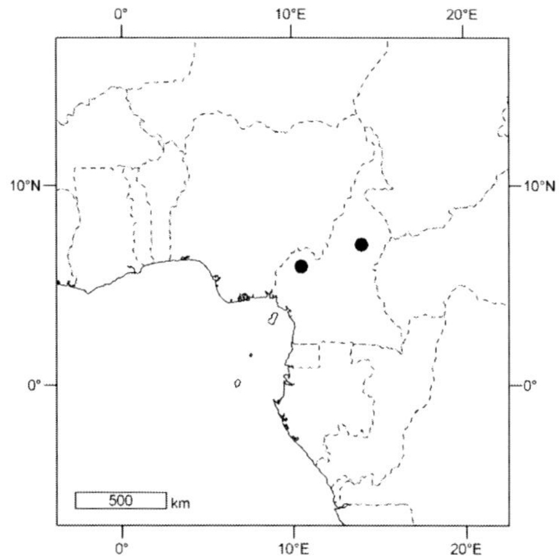

d. ***Justicia telloensis*** EN (Acanthaceae)

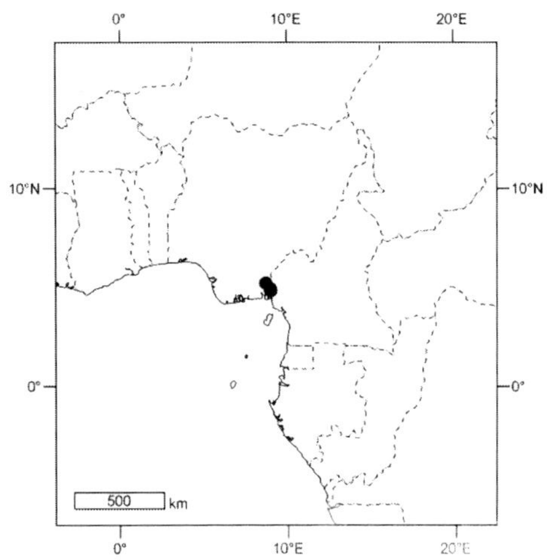

e. ***Justicia tenuipes*** EN (Acanthaceae)

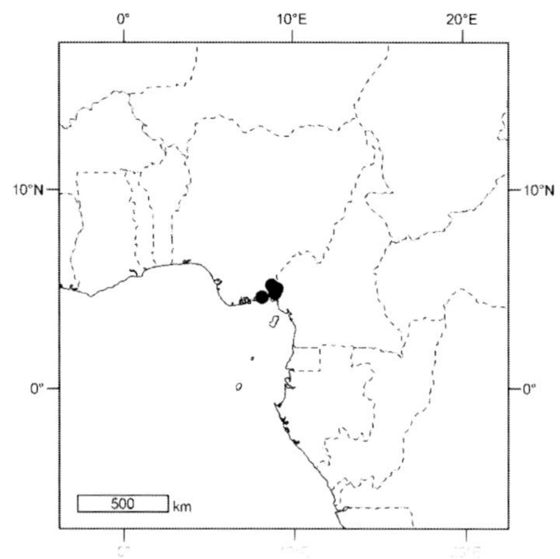

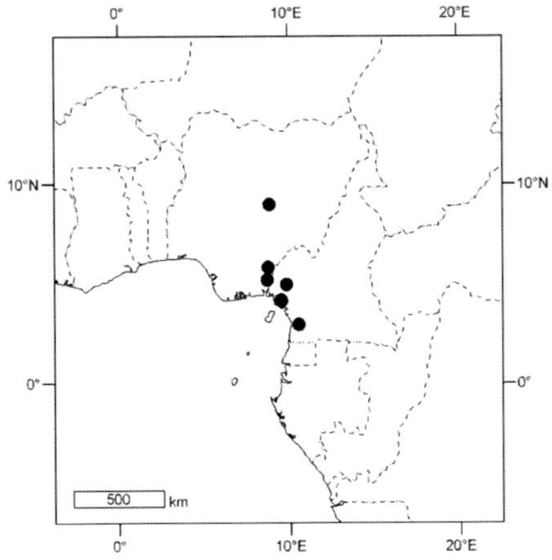

a. ***Pseuderanthemum dispersum*** VU (Acanthaceae)

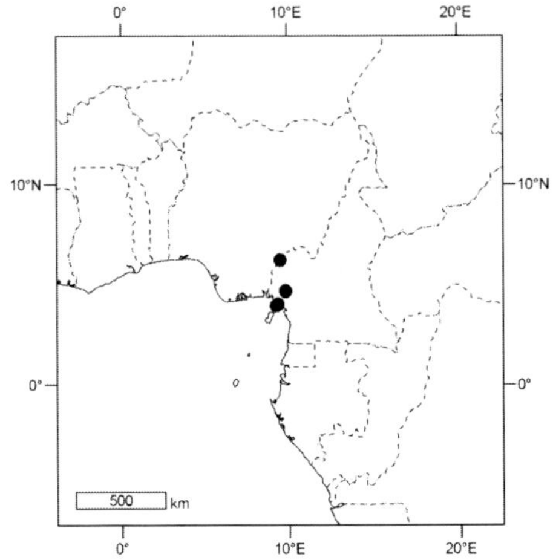

b. ***Sclerochiton preussii*** EN (Acanthaceae)

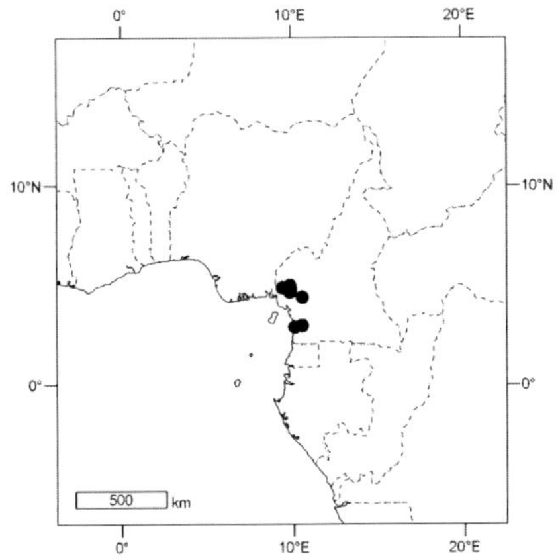

c. ***Staurogyne bicolour*** VU(Acanthaceae)

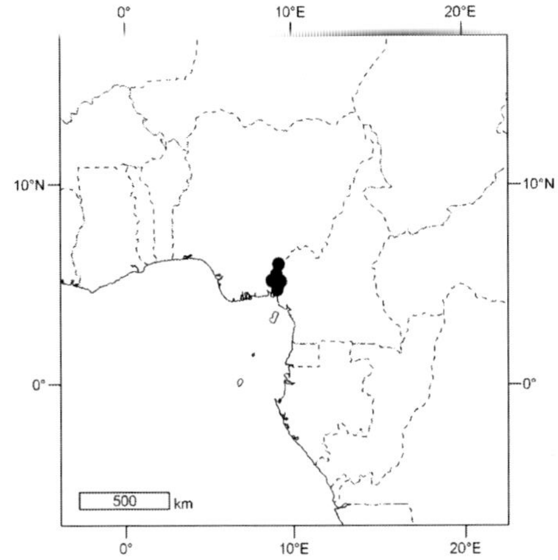

d. ***Staurogyne kamerunensis*** subsp. ***calabarensis*** VU (Acanthaceae)

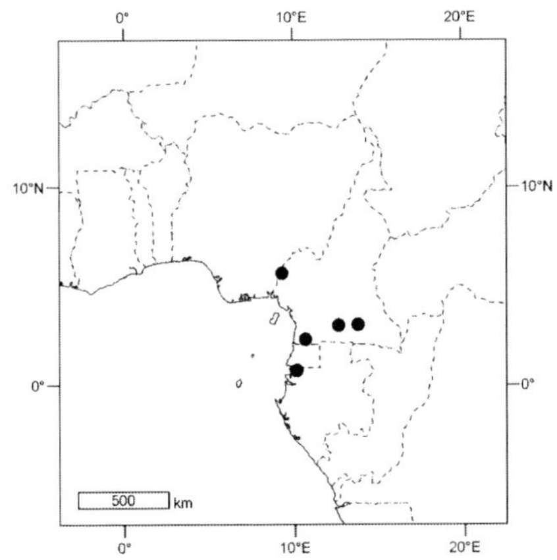

e. ***Staurogyne pseudocapitata*** EN (Acanthaceae)

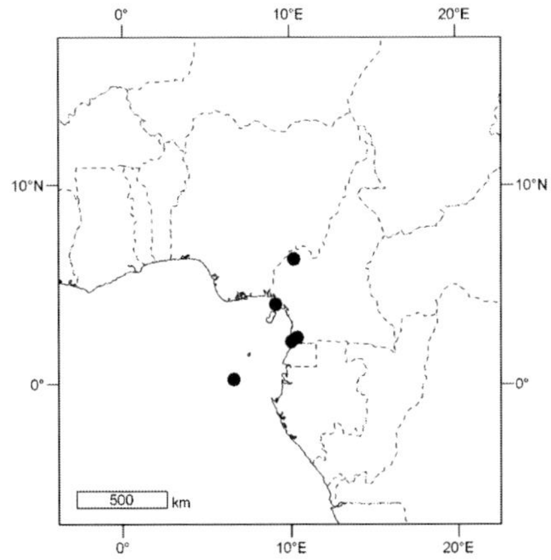

f. ***Stenandrium thomense*** EN (Acanthaceae)

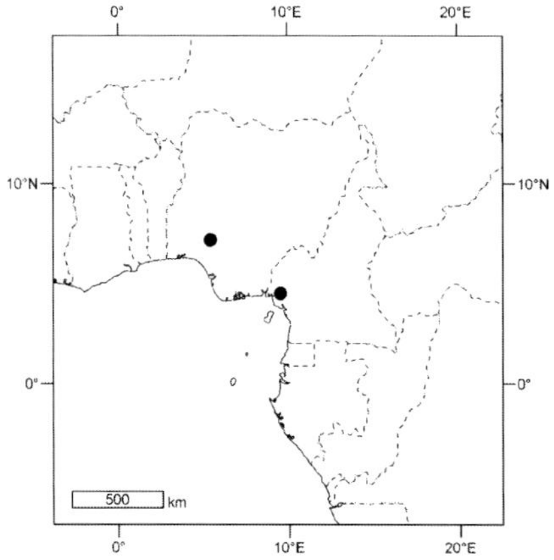

a. ***Thunbergia rufescens*** EN (Acanthaceae)

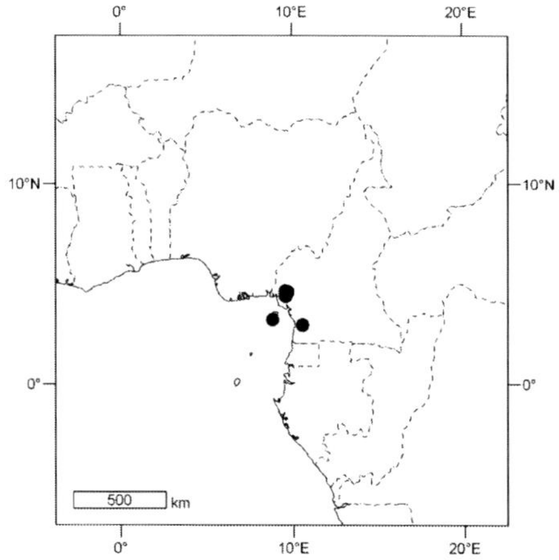

b. ***Whitfieldia preussii*** VU (Acanthaceae)

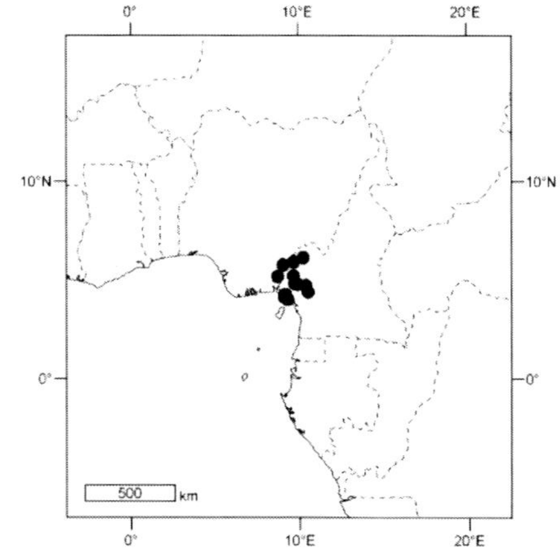

c. ***Achyranthes talbotii*** VU (Amaranthaceae)

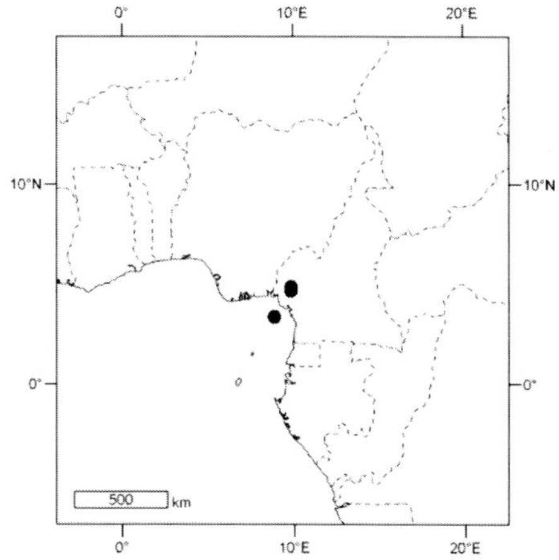

d. ***Cyathula fernando-poensis*** VU (Amaranthaceae)

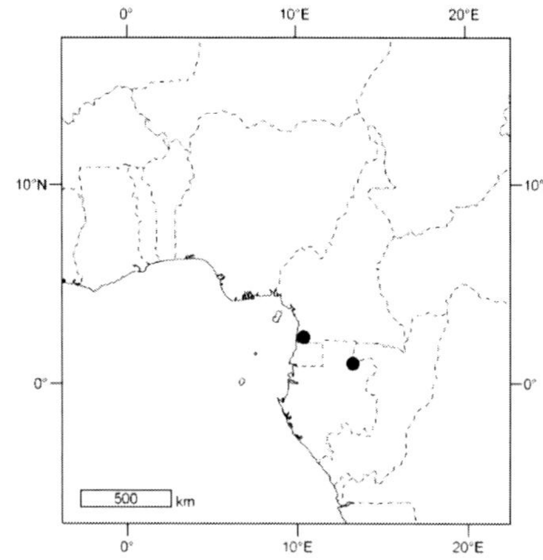

e. ***Trichoscypha hallei*** EN (Anacardiaceae)

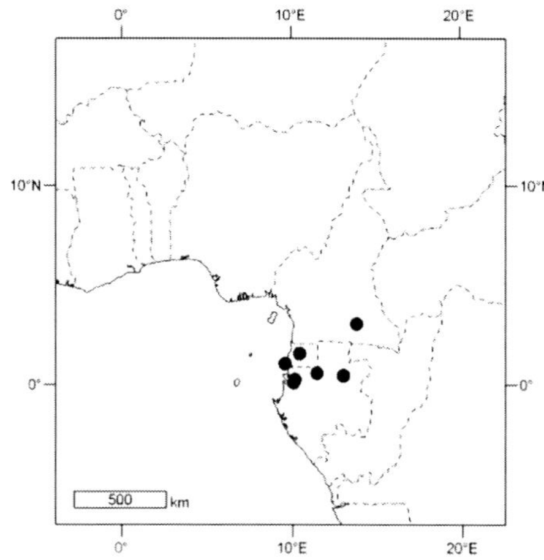

f. ***Trichoscypha engong*** VU (Anarcadiaceae)

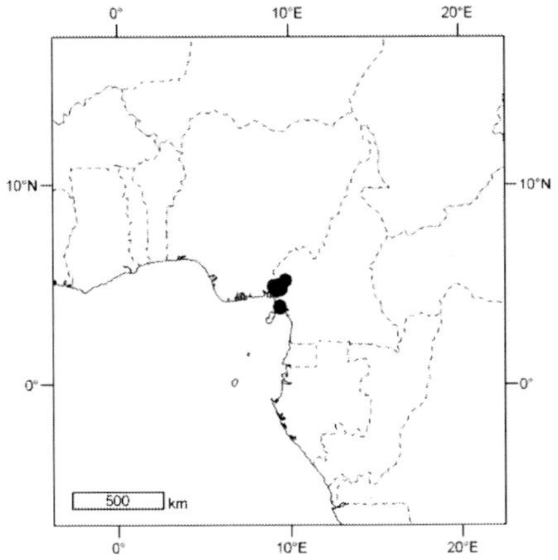

a. ***Ancistrocladus grandiflorus*** VU (Ancistrocladaceae

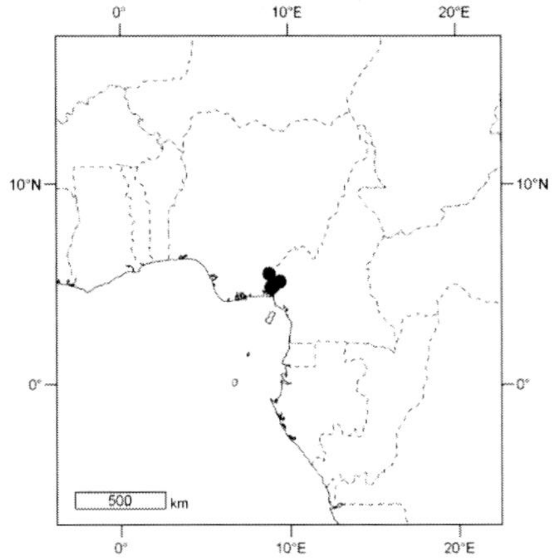

b. ***Ancistrocladus korupensis*** EN (Ancistrocladaceae)

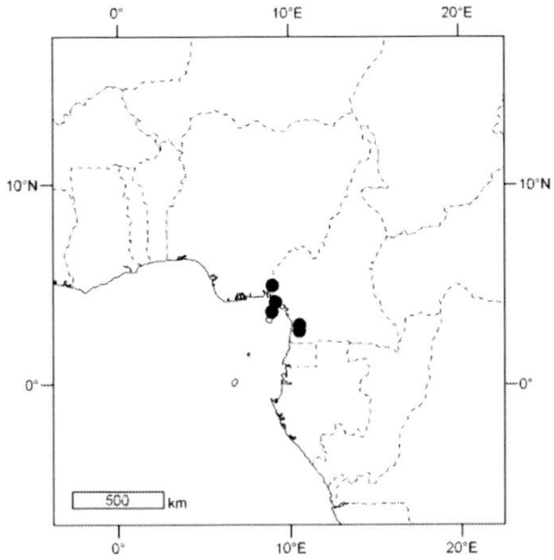

c. ***Boutiquea platypetala*** EN (Annonaceae)

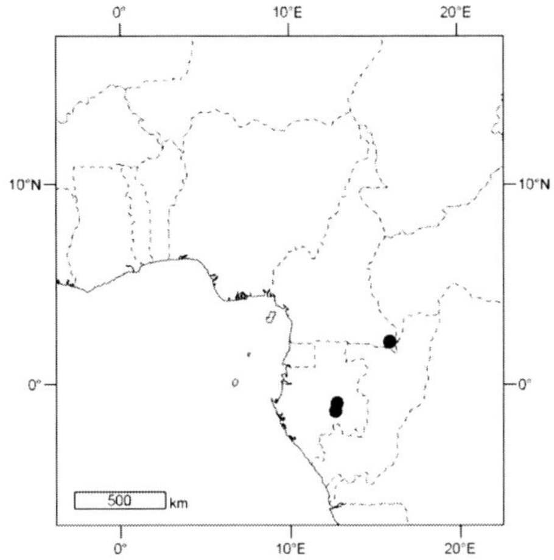

d. ***Isolona pilosa*** VU (Annonaceae)

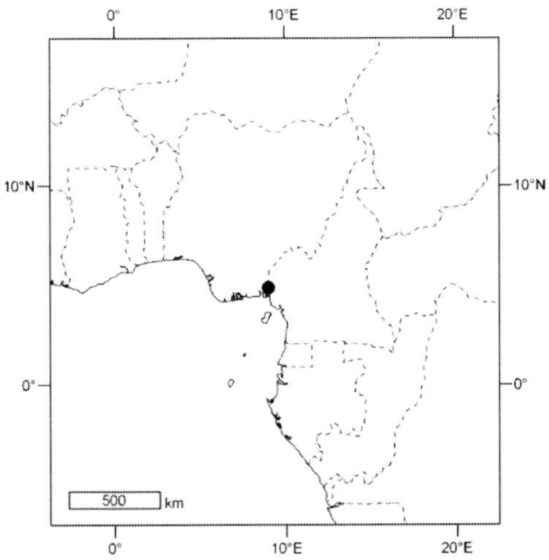

e. ***Isolona pleurocarpa*** VU (Annonaceae)

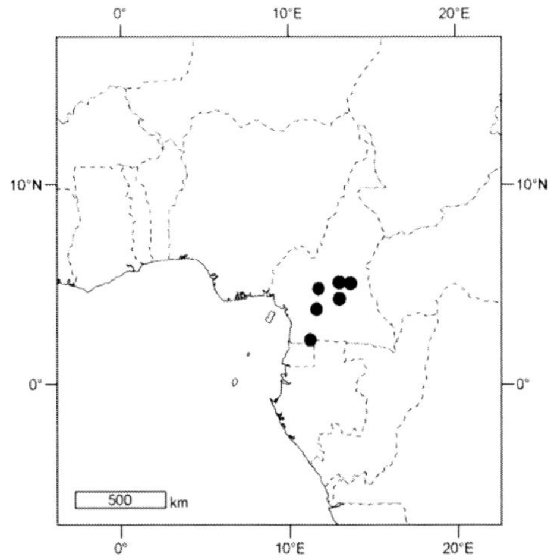

f. ***Monodora zenkeri*** VU (Annonaceae)

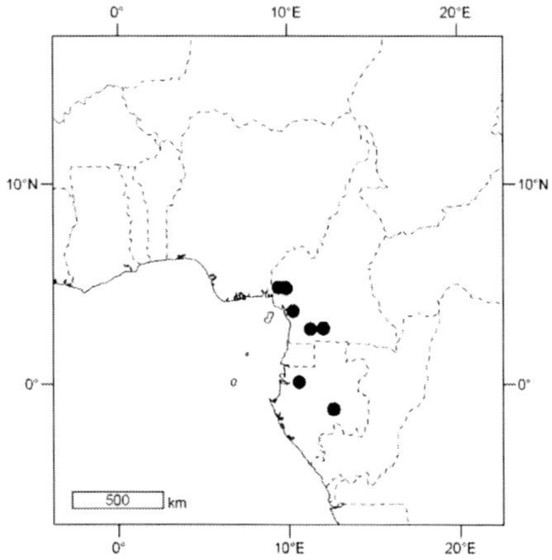

a. ***Piptostigma calophyllum*** VU (Annonaceae)

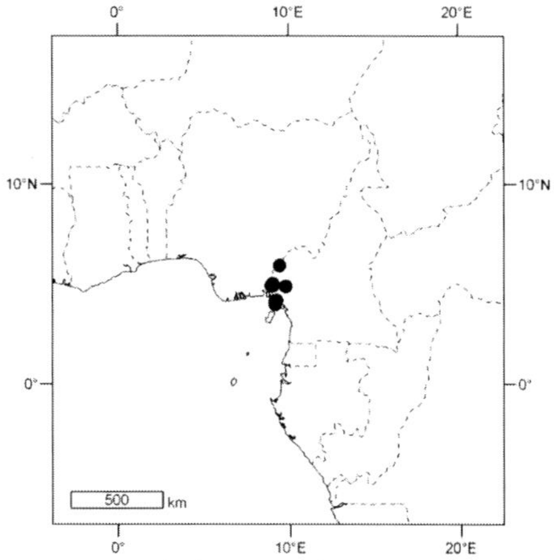

b. ***Uvariopsis korupensis*** EN (Annonaceae)

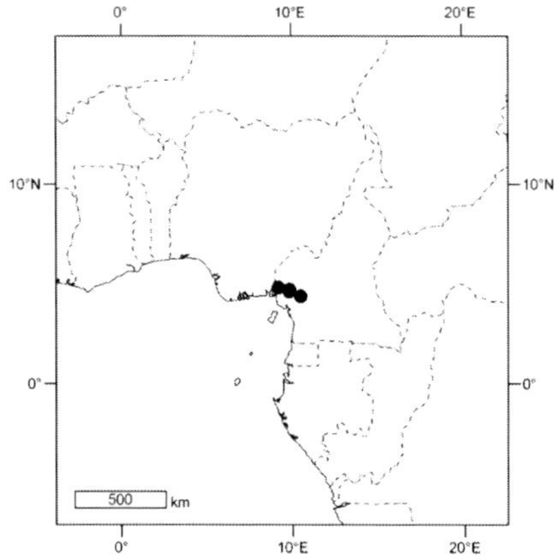

c. ***Uvariopsis submontana*** EN (Annonaceae)

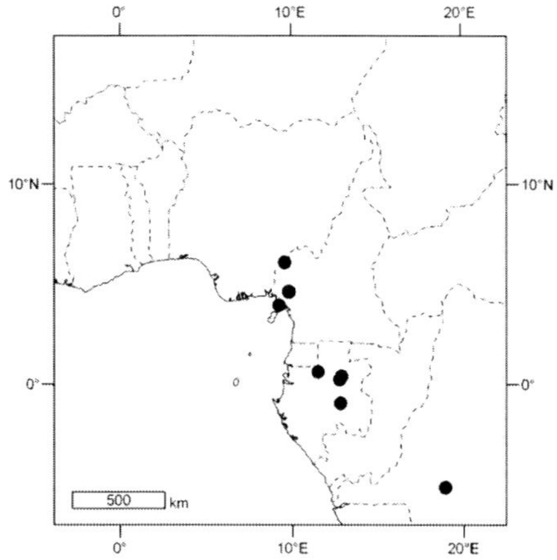

d. ***Uvariopsis vanderystii*** EN(Annonaceae)

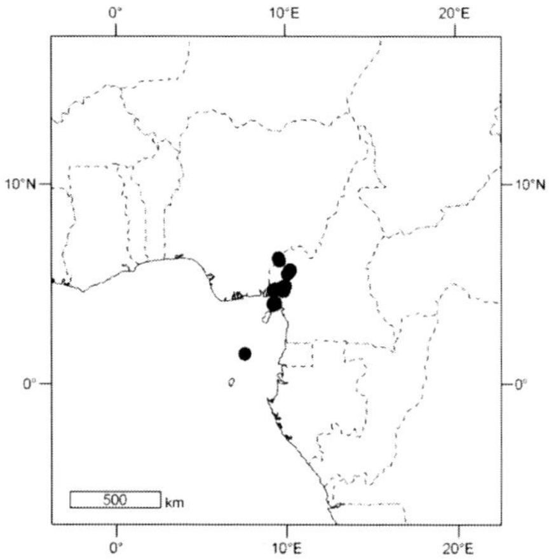

e. ***Xylopia africana*** VU (Annonaceae)

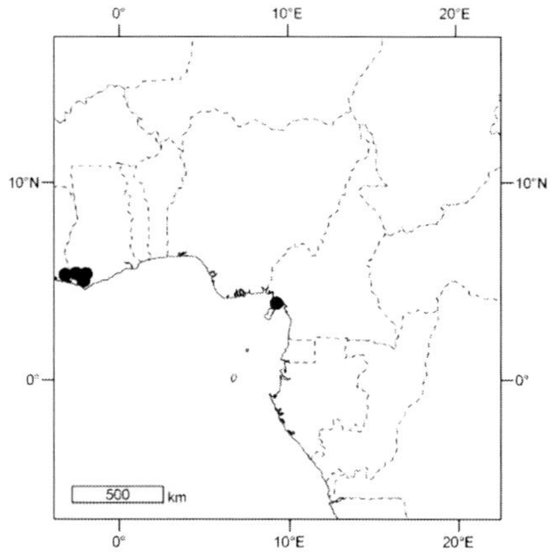

f. ***Alafia whytei*** VU (Apocynaceae)

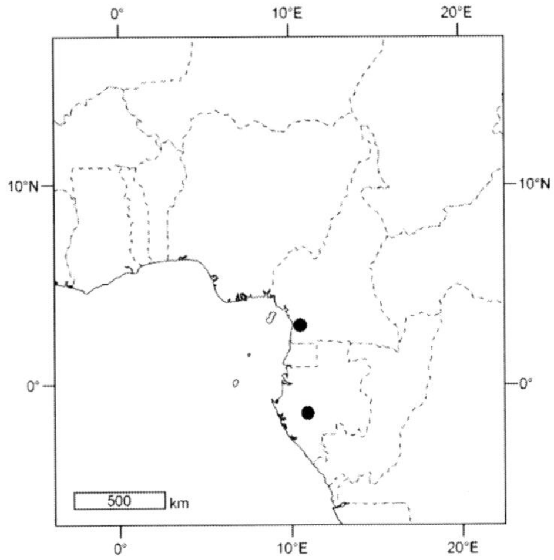

a. ***Baissea ochranta*** EN (Apocynaceae)

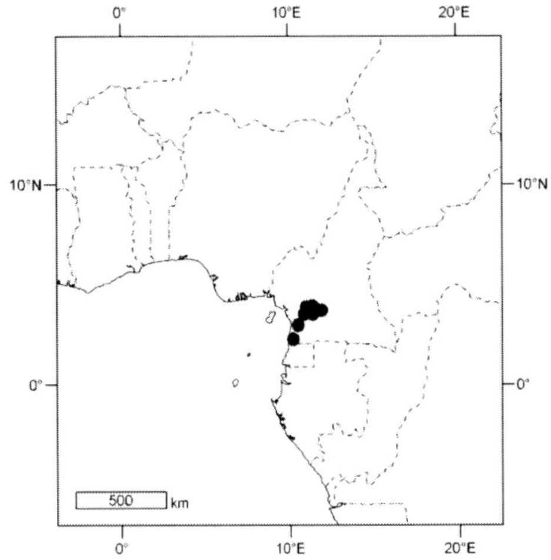

b. ***Callichilia monopodialis*** EN (Apocynaceae)

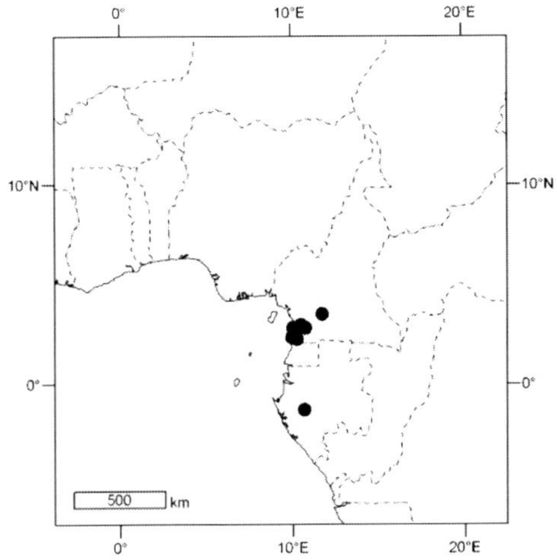

c. ***Dictyophleba setosa*** VU (Apocynaceae)

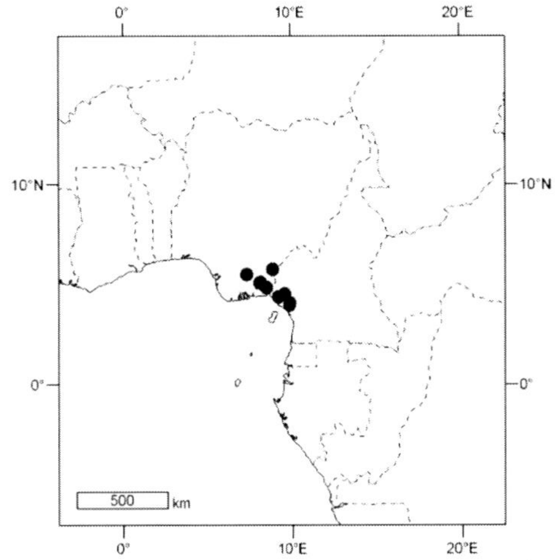

d. ***Isonema buchholzii*** VU (Apocynaceae)

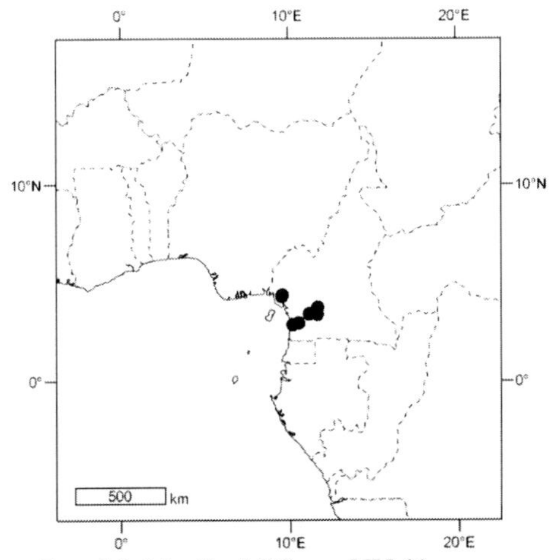

e. ***Landolphia flavidiflora*** VU (Apocynaceae)

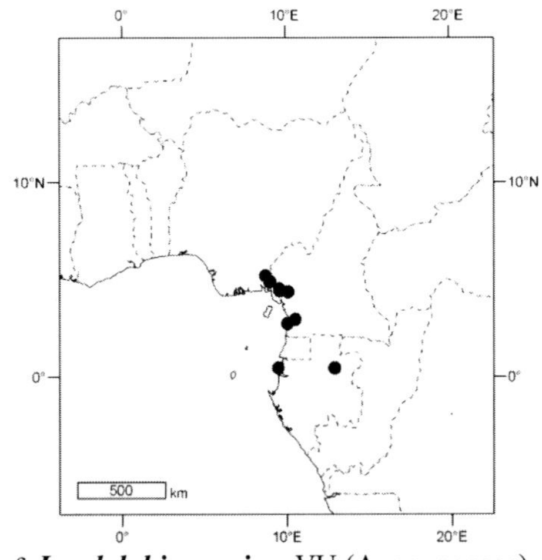

f. ***Landolphia maxima*** VU (Apocynaceae)

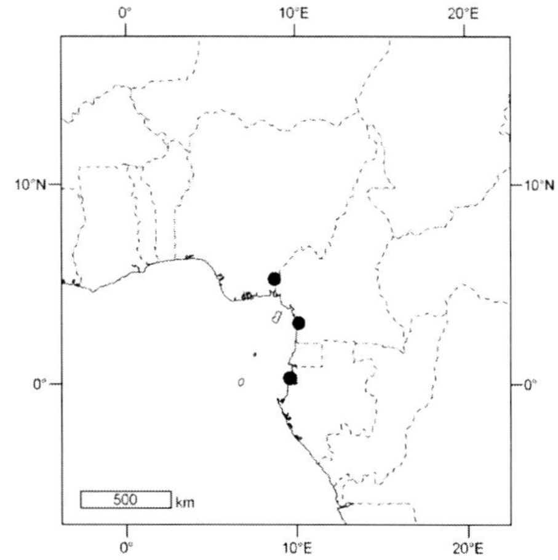

a. ***Landolphia uniflora*** VU (Apocynaceae)

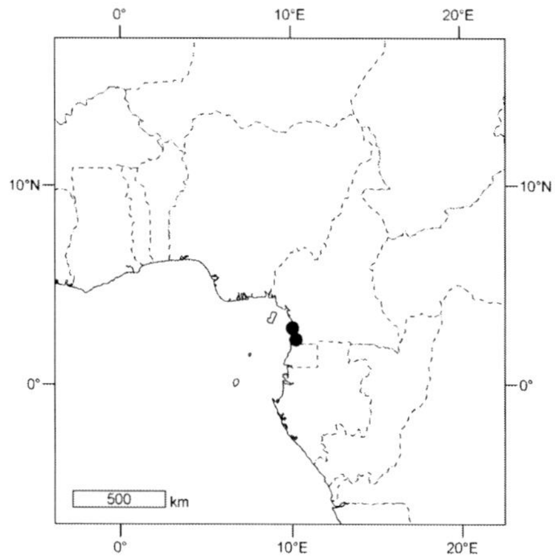

b. ***Malouetia barbata*** EN (Apocynaceae)

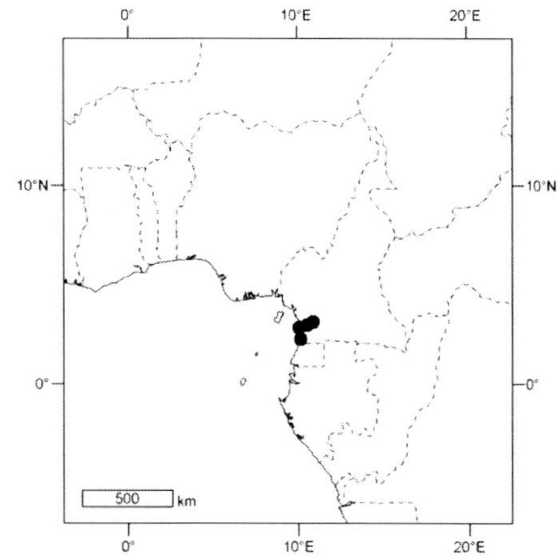

c. ***Petchia africana*** EN(Apocynaceae)

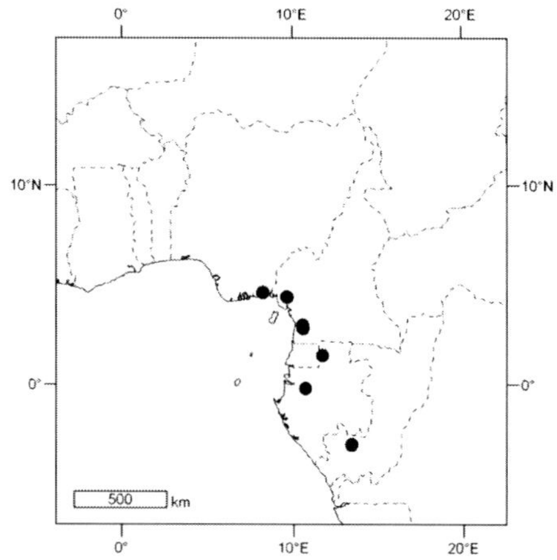

d. ***Pleioceras zenkeri*** VU (Apocynaceae)

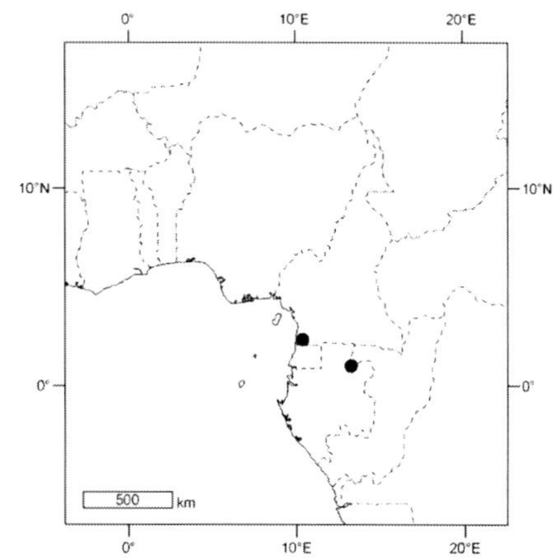

e. ***Tabernaemontana hallei*** VU (Apocynaceae)

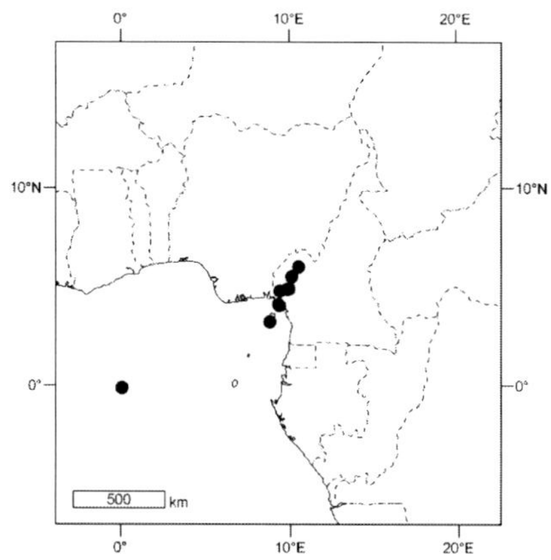

f. ***Schefflera hierniana*** VU (Araliaceae)

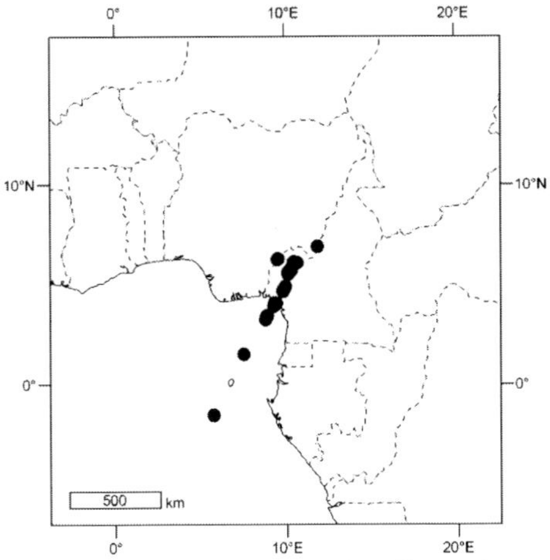

a. ***Schefflera mannii*** VU (Araliaceae)

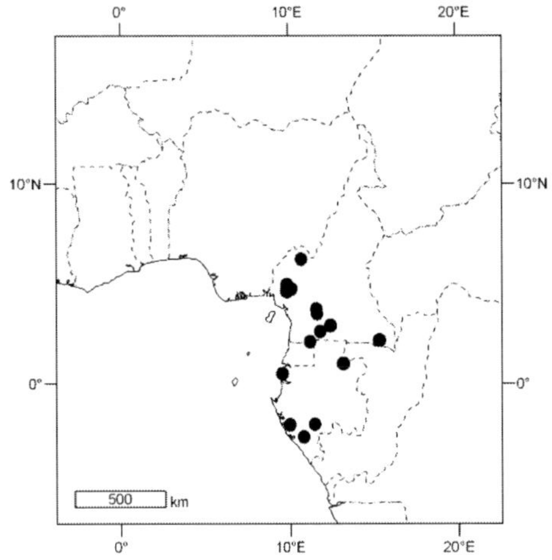

b. ***Pararistolochia ceropegioides*** VU (Aristolochiaceae)

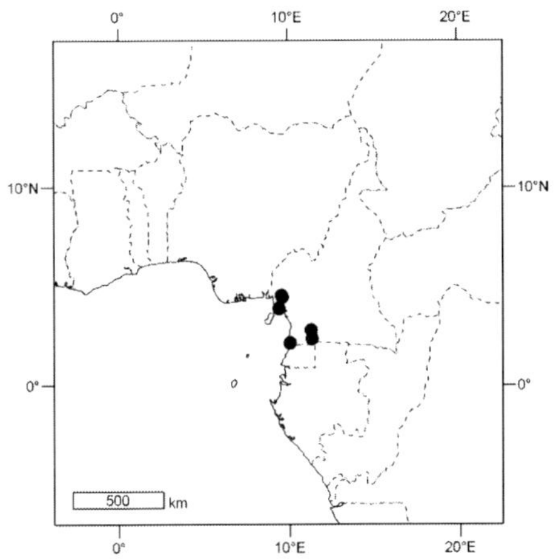

c. ***Pararistolochia preussii*** EN (Aristolochiaceae)

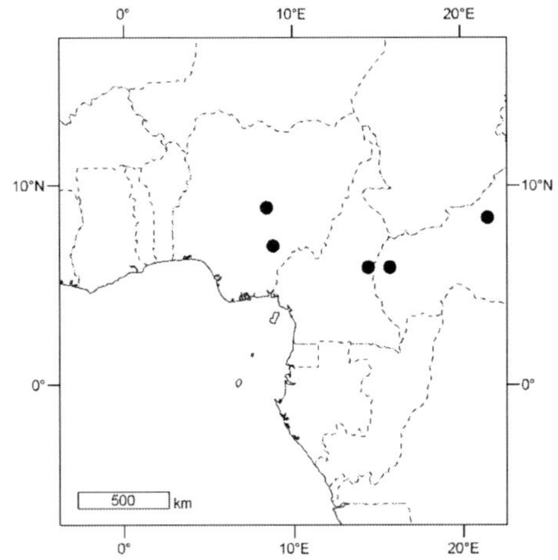

d. ***Asclepias kamerunensis*** **EN** (Asclepiadaceae)

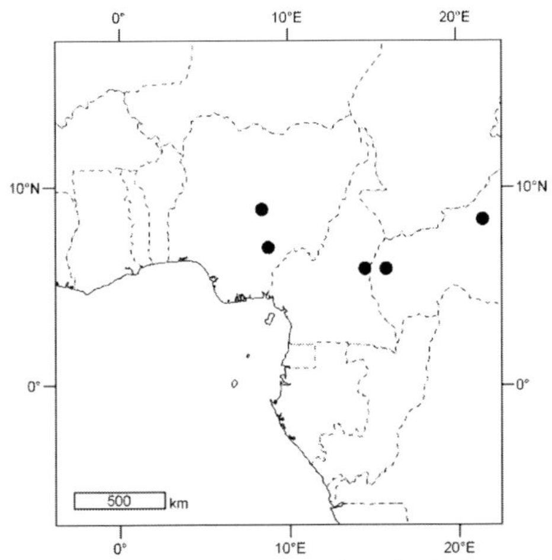

e. ***Brachystelma exile*** **EN** (Asclepiadaceae)

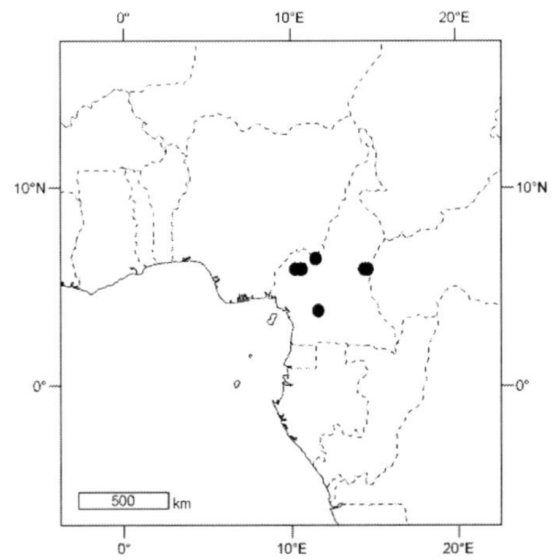

f. ***Brachystelma omissum*** **VU** (Asclepiadaceae)

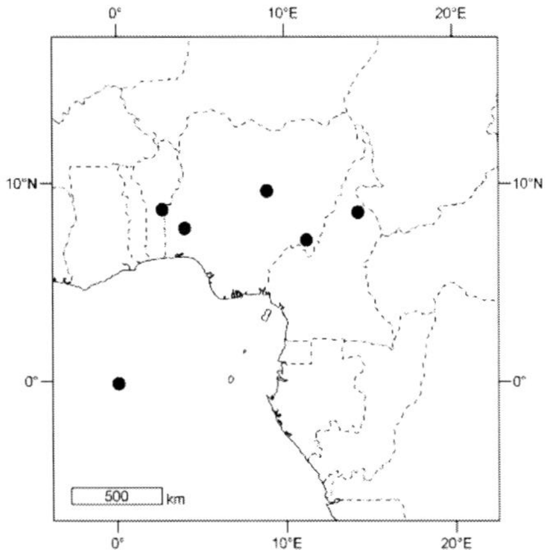

a. ***Ceropegia ledermannii* EN** (Asclepiadaceae)

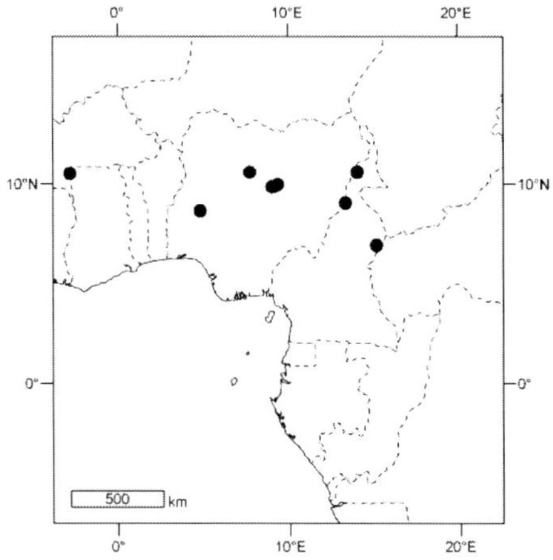

b. ***Ceropegia rhynchantha* VU** (Asclepiadaceae)

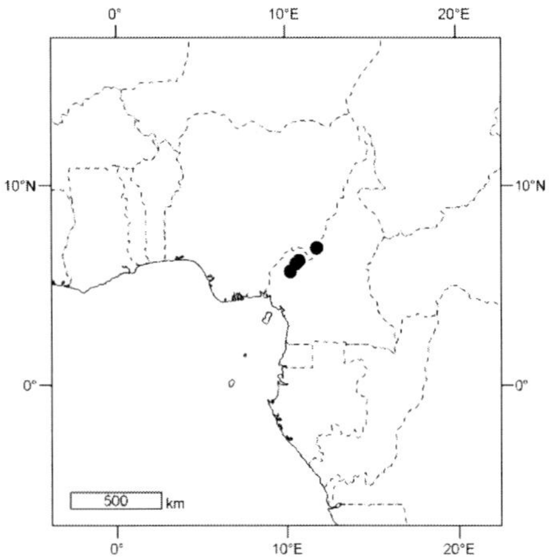

c. ***Epistemma decurrens* EN** (Asclepiadaceae)

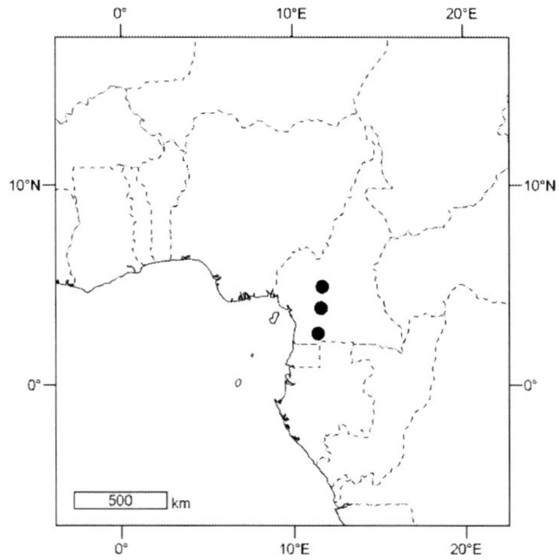

d. ***Epistemma rupestre* EN** (Asclepiadaceae)

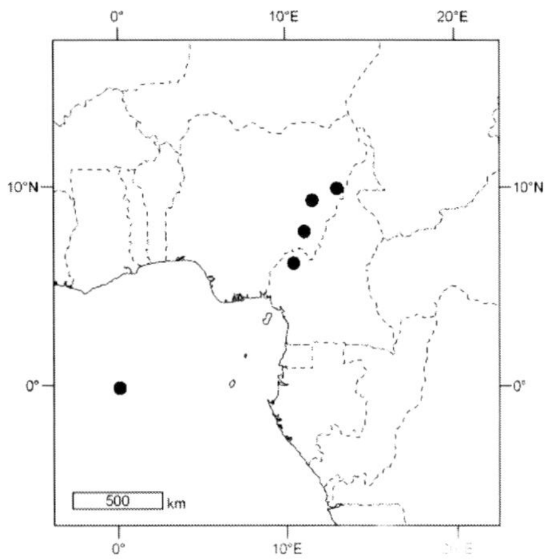

e. ***Maclaudia felixii* VU** (Asclepiadaceae)

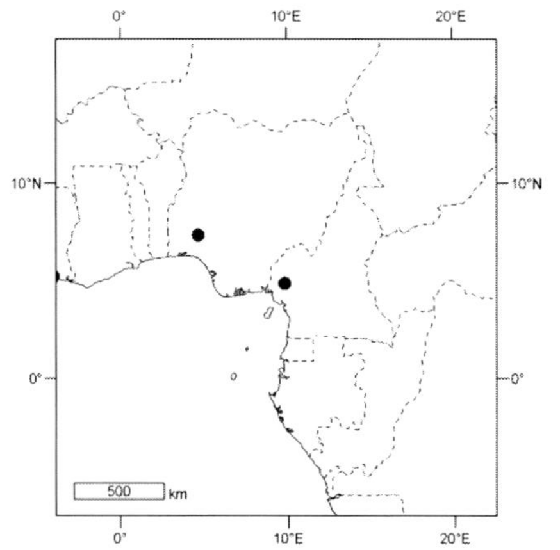

f. ***Marsdenia magniflora* VU** (Asclepiadaceae)

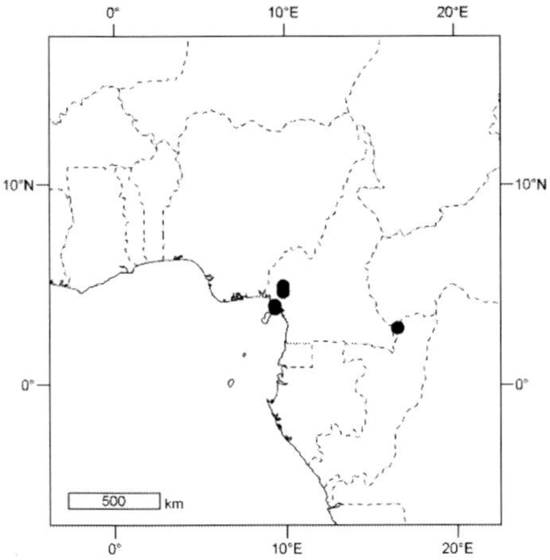

a. ***Neoschumannia kamerunensis*** CR (Asclepiadaceae)

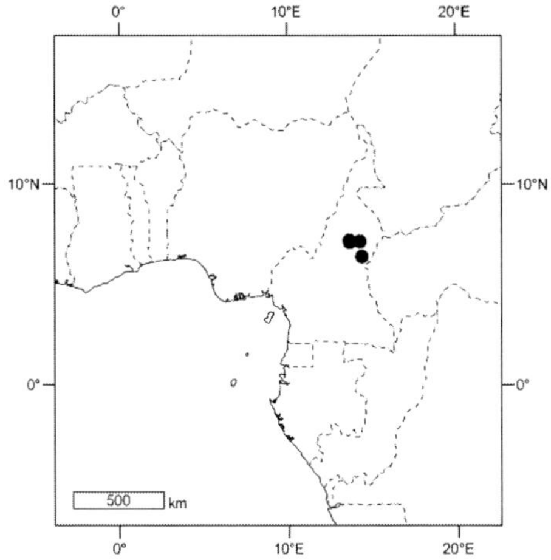

b. ***Pachycarpus medusonema*** **EN** (Asclepiadaceae)

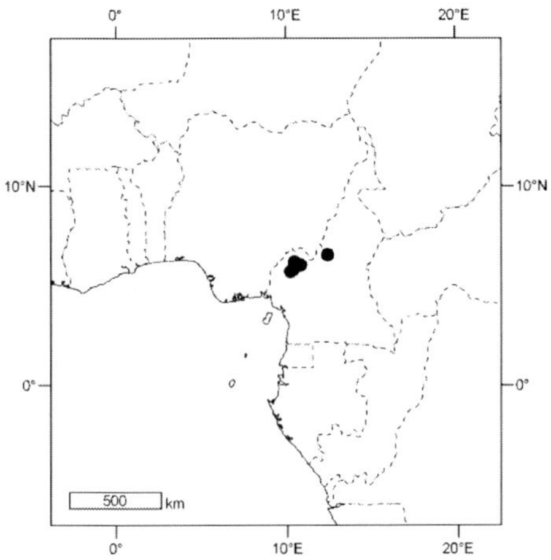

c. ***Pentarrhinum ledermannii*** VU (Asclepiadaceae)

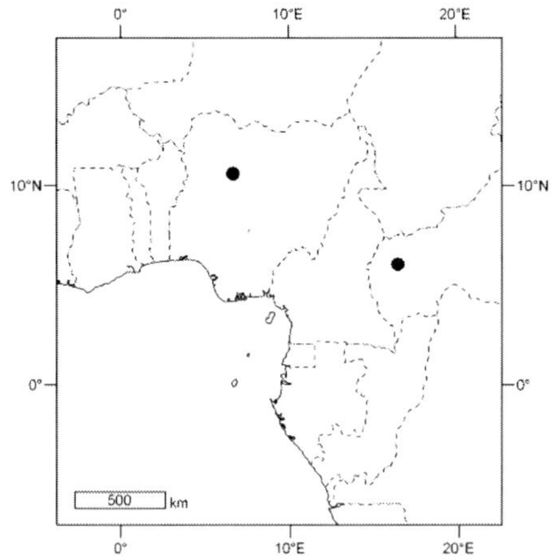

d. ***Raphionacme keayii*** Bullock **EN** (Asclepiadaceae)

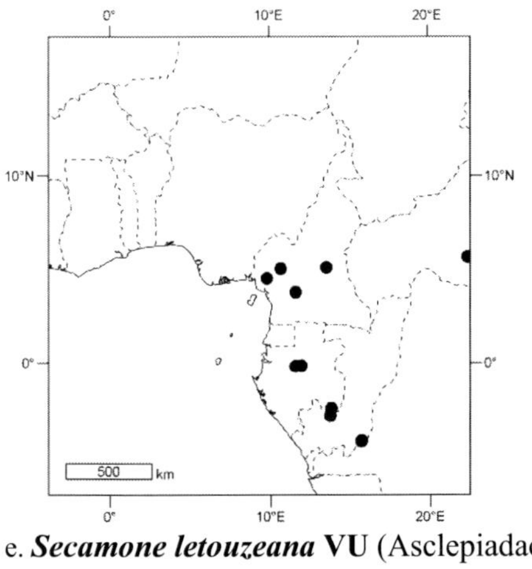

e. ***Secamone letouzeana*** **VU** (Asclepiadaceae)

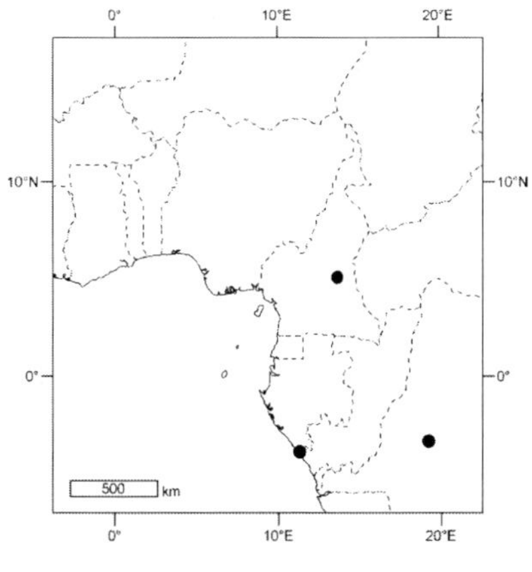

f. ***Chlamydophytum aphyllum*** EN (Balanophoraceae)

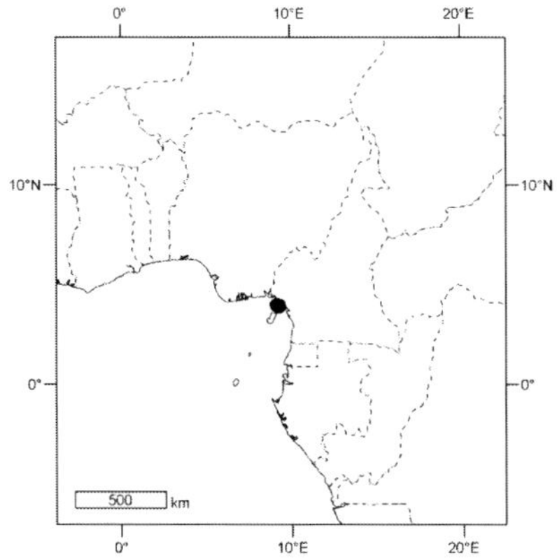

a. ***Impatiens etindensis*** EN (Balsaminaceae)

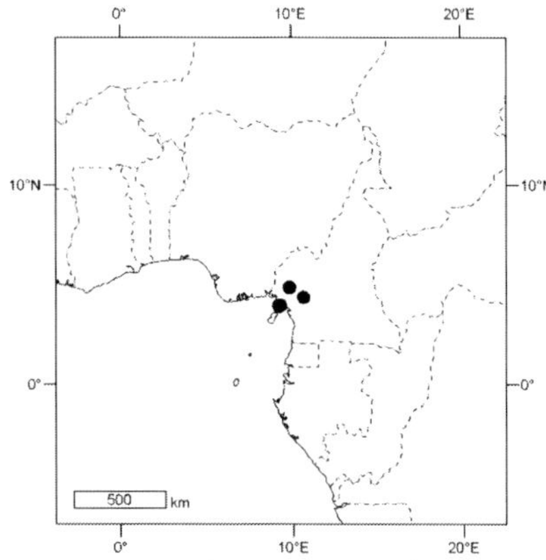

b. ***Impatiens frithii*** EN (Balsaminaceae)

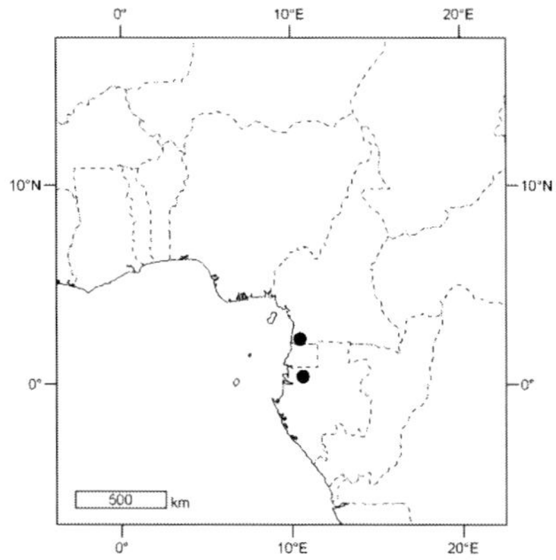

c. ***Impatiens gongolana*** EN (Balsaminaceae)

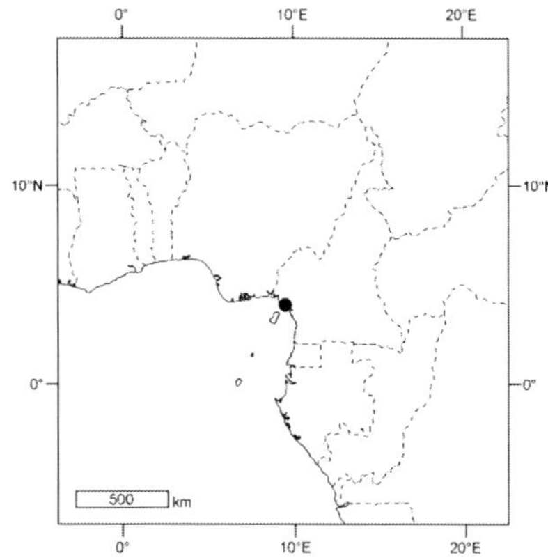

d. ***Impatiens grandisepala*** CR (Balsaminaceae)

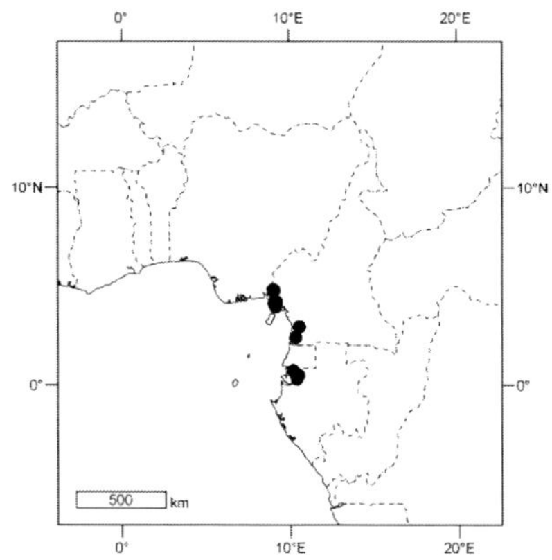

e. ***Impatiens hians*** *var.* ***bipindensis*** VU (Balsaminaceae)

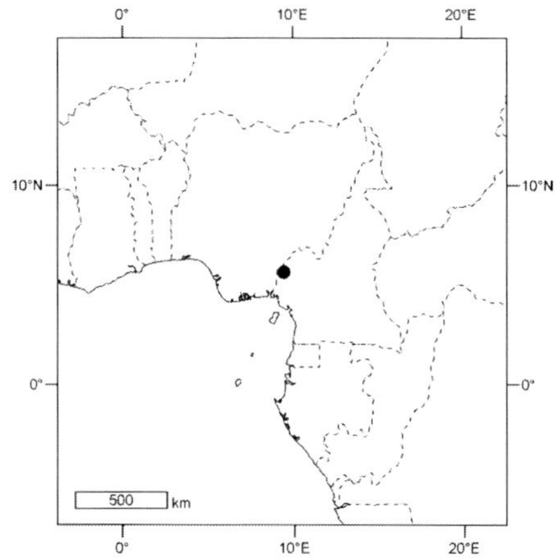

f. ***Impatiens kamerunensis*** subsp. ***obanensis*** CR (Balsaminaceae)

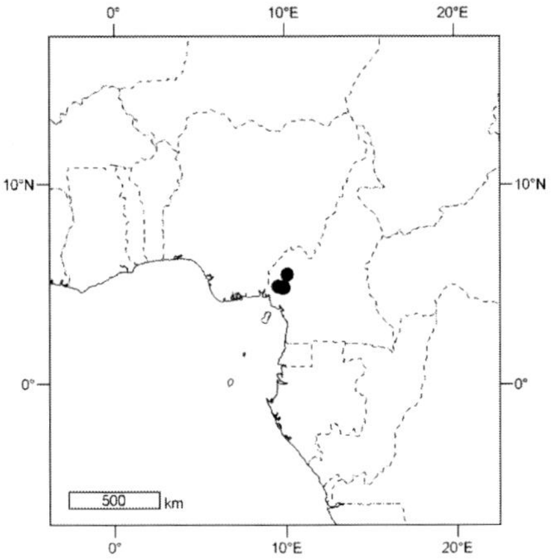

a. ***Impatiens letouzeyi*** EN (Balsaminaceae)

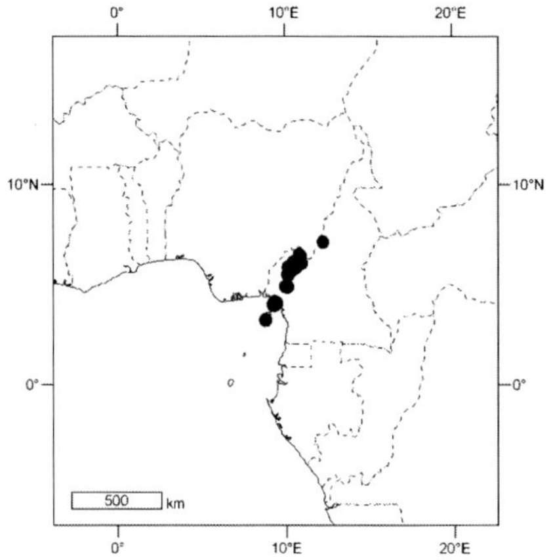

b. ***Impatiens sakeriana*** VU (Balsaminaceae)

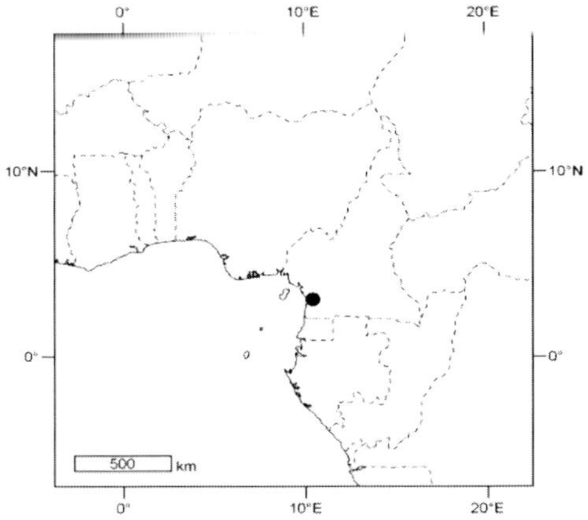

c. . ***Begonia minuta***CR (Begoniaceae)

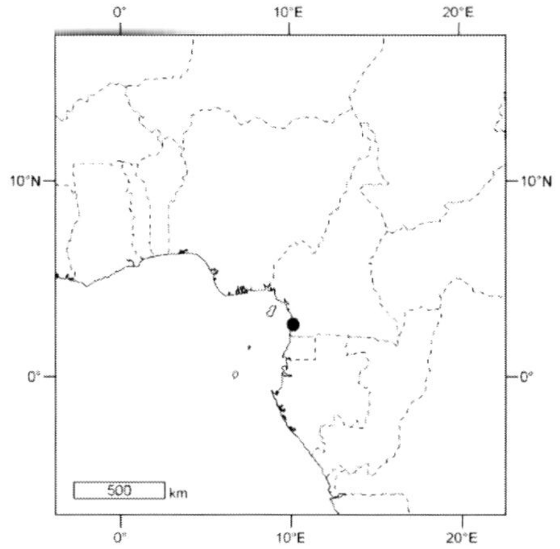

d. ***Begonia montis-elephantis*** CR (Begoniaceae)

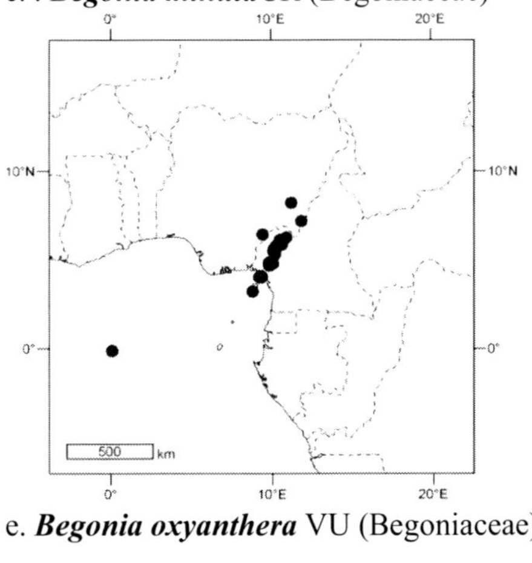

e. ***Begonia oxyanthera*** VU (Begoniaceae)

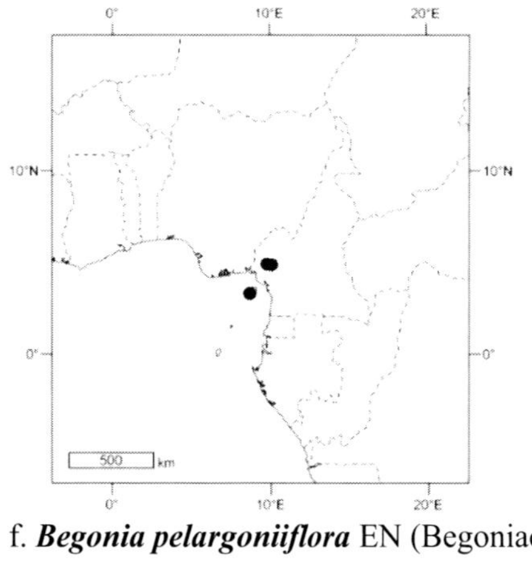

f. ***Begonia pelargoniiflora*** EN (Begoniaceae)

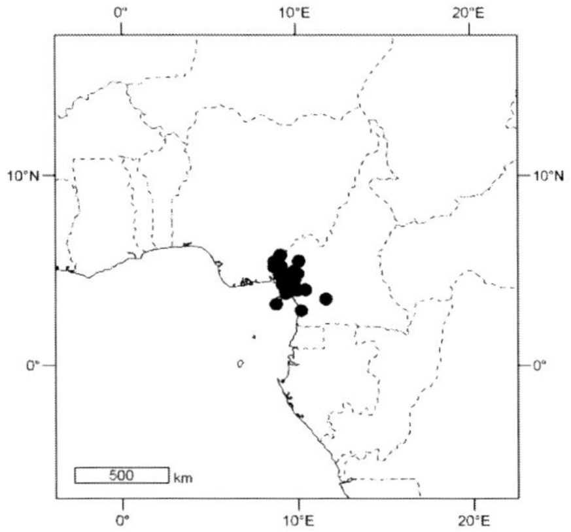

a. ***Begonia preussii*** VU (Begoniaceae)

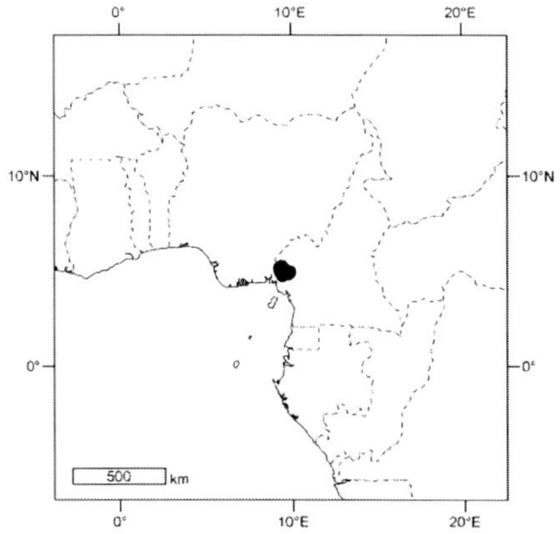

b. ***Begonia prismatocarpa*** subsp. ***delobata*** EN (Begoniaceae)

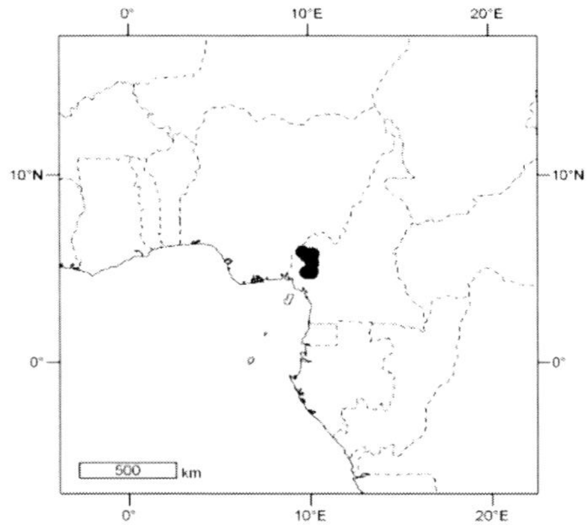

c. ***Begonia pseudoviola*** EN (Begoniaceae)

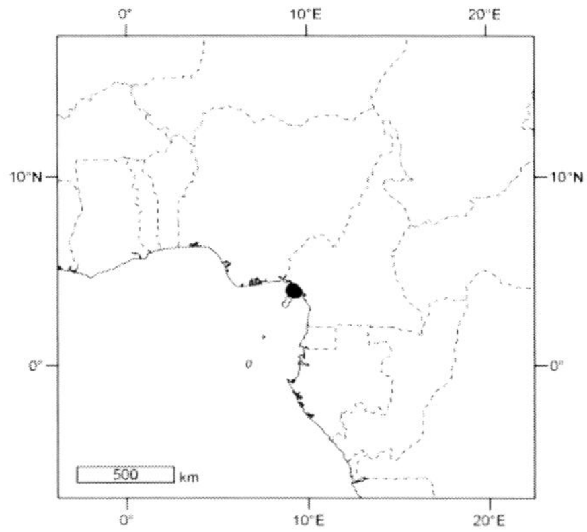

d. ***Begonia quadrialata*** subsp. ***dusenii*** CR (Begoniaceae)

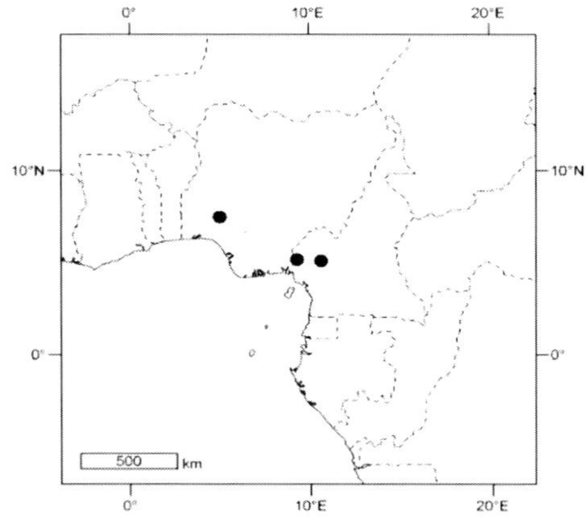

e. ***Begonia rubromarginata*** EN (Begoniaceae)

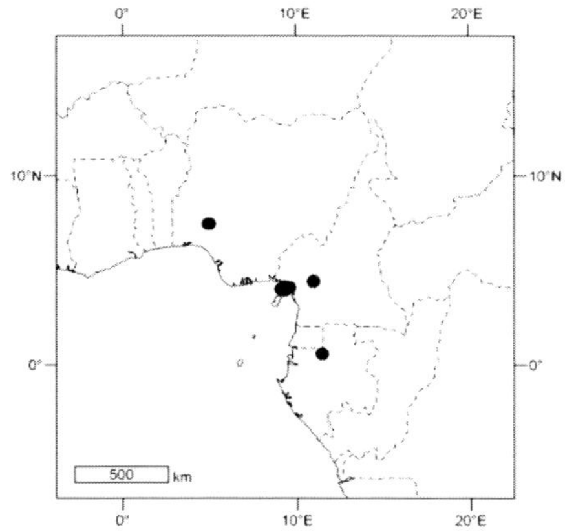

f. ***Begonia scapigera*** subsp. ***scapigera*** EN (Begoniaceae)

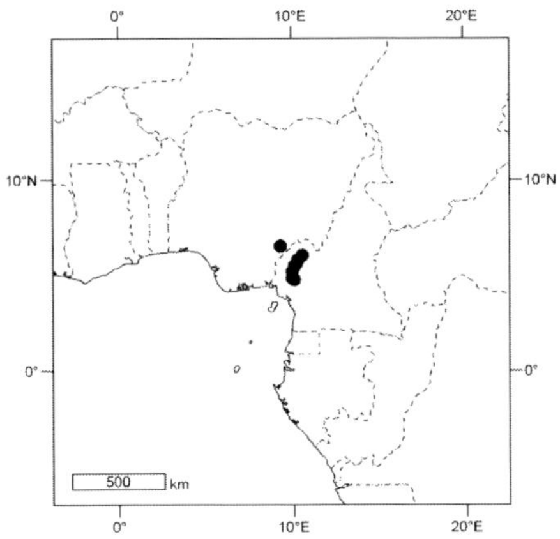

a. ***Begonia schaeferi*** VU (Begoniaceae)

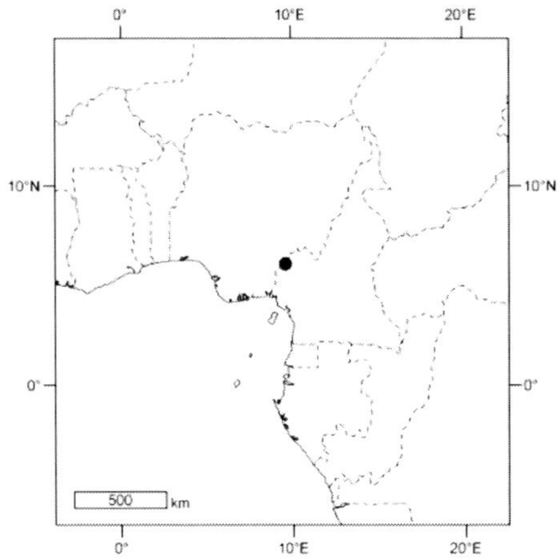

b. ***Begonia stellata*** CR (Begoniaceae)

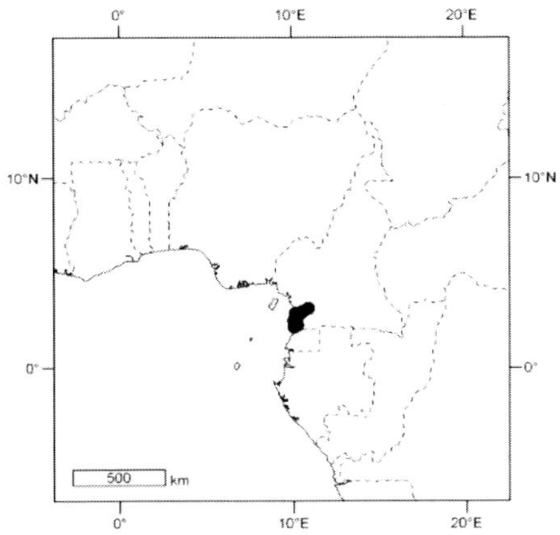

c. ***Begonia zenkeriana*** VU (Begoniaceae)

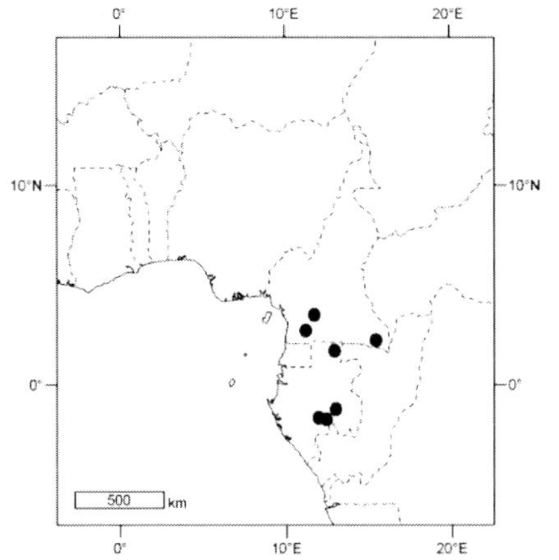

d. ***Fernandoa ferdinandi*** VU (Bignoniaceae)

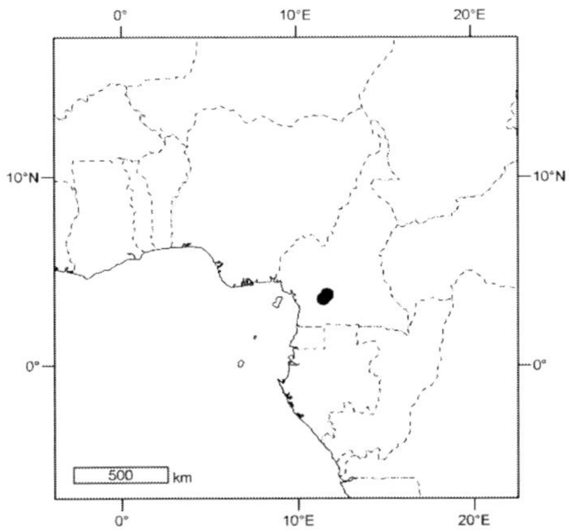

e. ***Stereospermum zenkeri*** CR (Bignoniaceae)

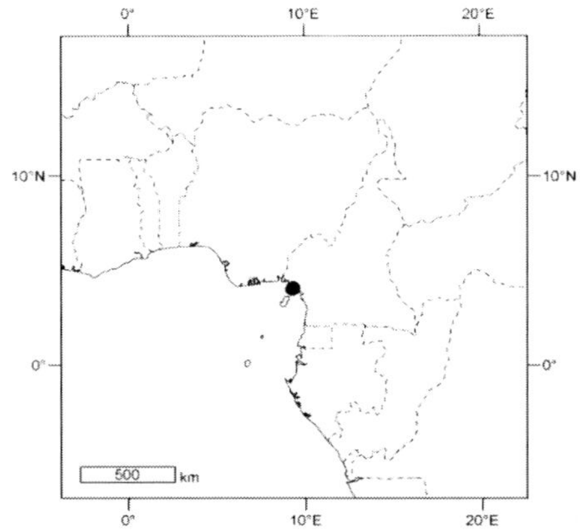

f. ***Myosotis cameroonensis*** VU (Boraginaceae)

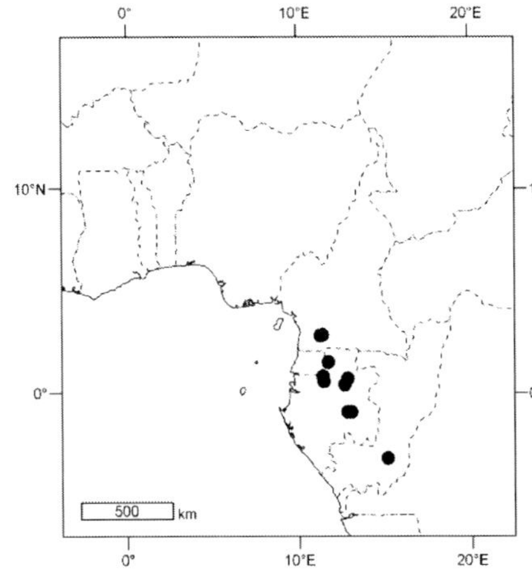

a. ***Dacryodes camerunensis*** VU (Burseraceae)

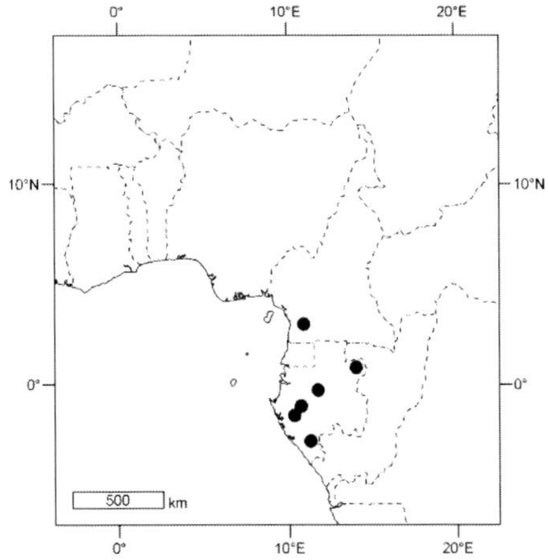

b. ***Dacryodes igaganga*** VU (Burseraceae)

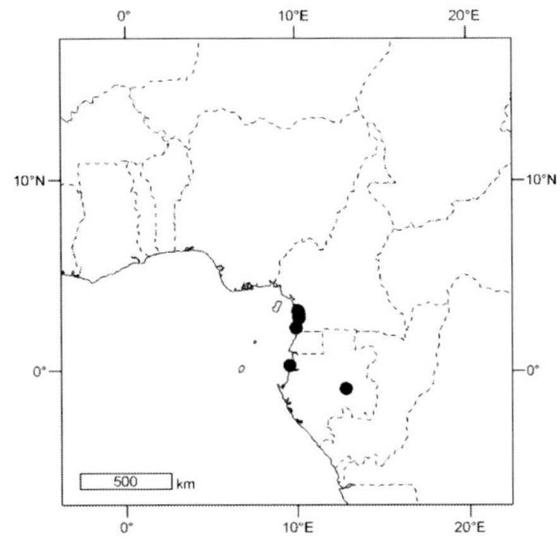

c. ***Dacryodes ledermannii*** EN (Burseraceae)

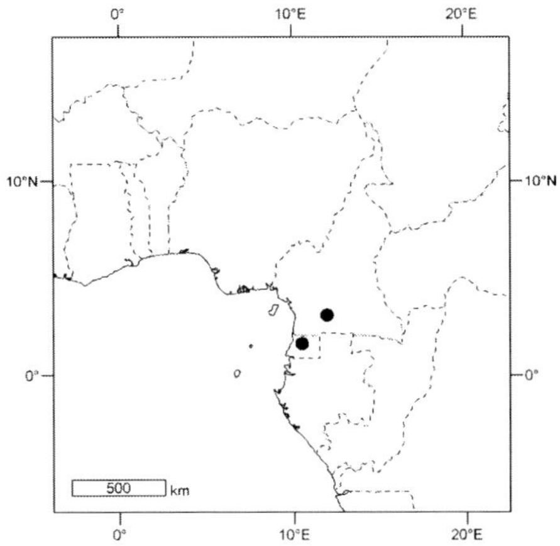

d. ***Dacryodes tessmannii*** CR (Burseraceae)

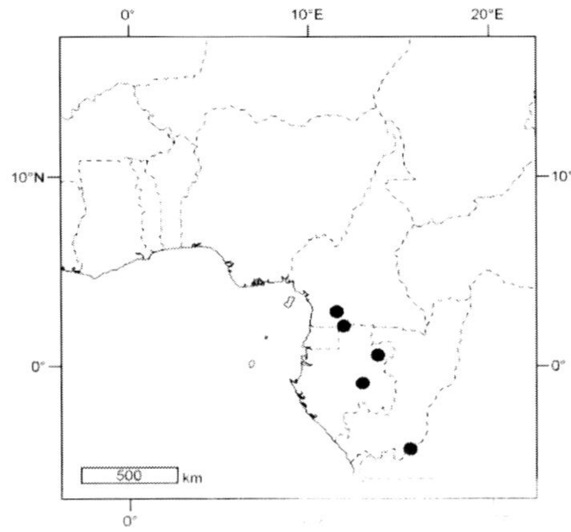

e. ***Dacryodes villier*** [illegible]

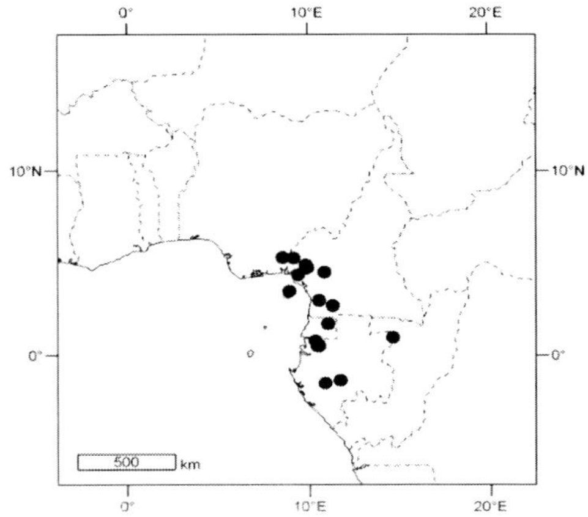

f. ***Dielsantha galeopsoides*** VU (Campanulaceae)

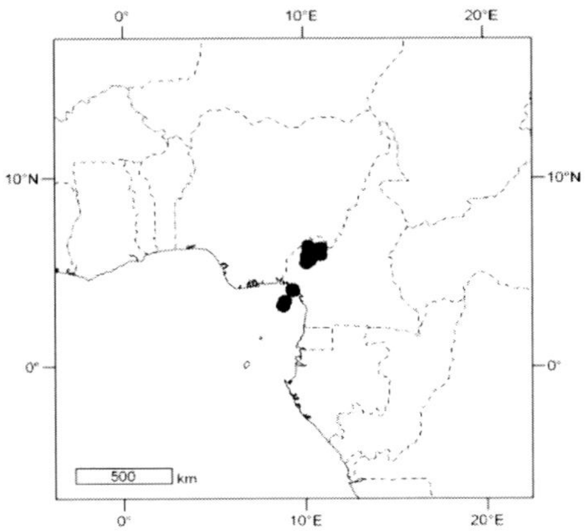

a. ***Lobelia columnaris*** VU (Campanulaceae)

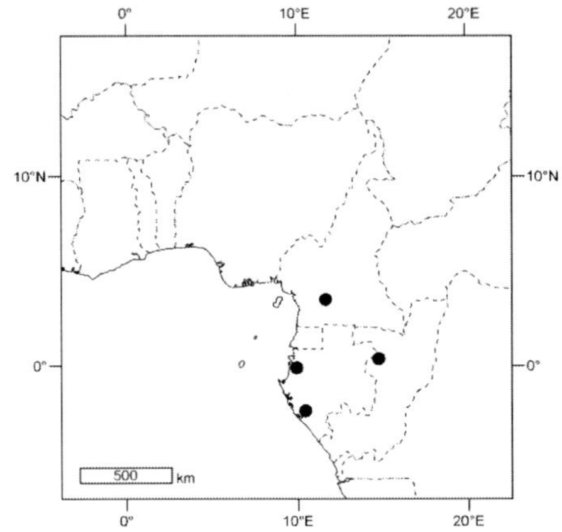

b. ***Lobelia gilletii*** EN (Campanulaceae)

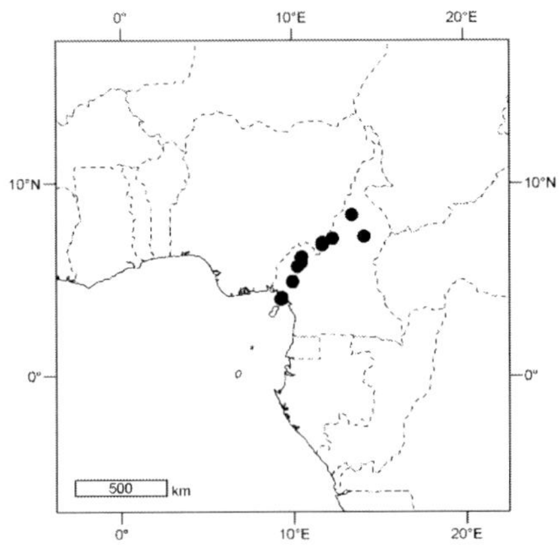

c. ***Wahlenbergia ramosissima*** subsp. ***ramosissima*** VU (Campanulaceae)

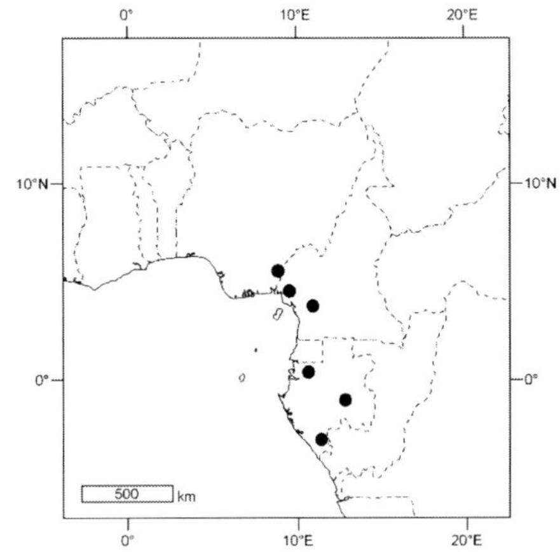

d. ***Ritchiea macrantha*** VU (Capparaceae)

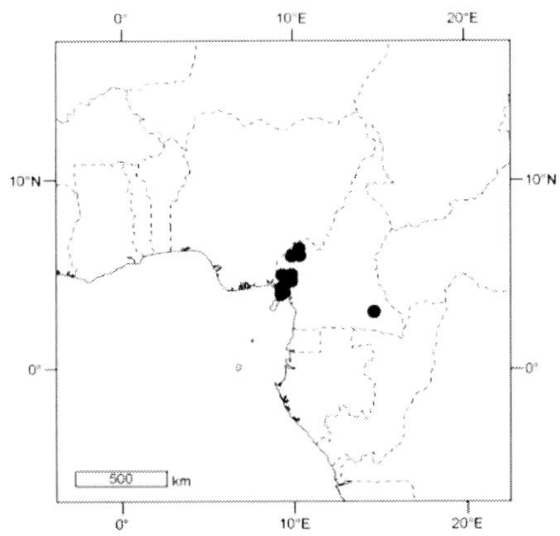

e. ***Cylicomorpha solmsii*** VU (Caricaceae)

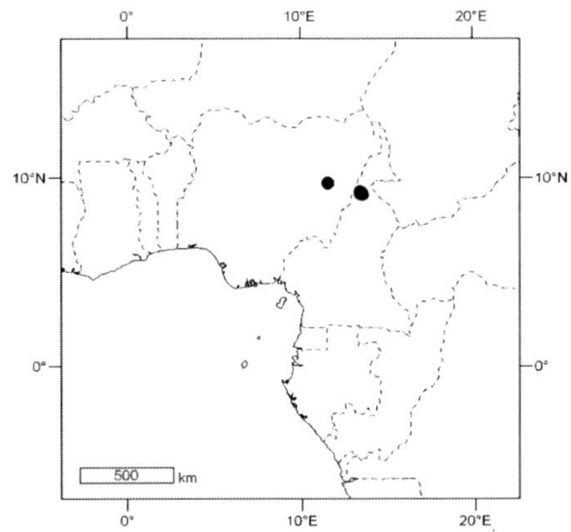

f. ***Polycarpaea garuensis*** EN (Caryophyllaceae)

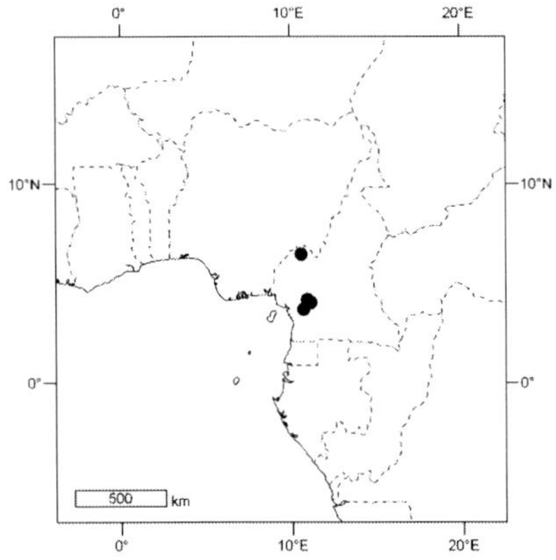

a. ***Polycarpaea rheophytica*** EN (Caryophyllaceae)

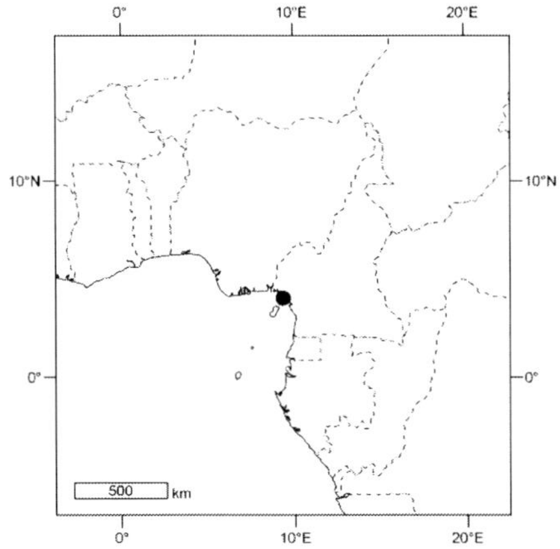

b. ***Silene biafrae*** VU (Caryophyllaceae)

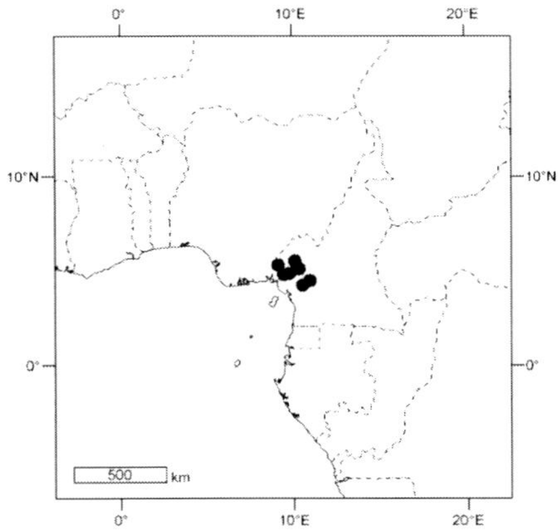

c. ***Myrianthus fosi*** VU (Cecropiaceae)

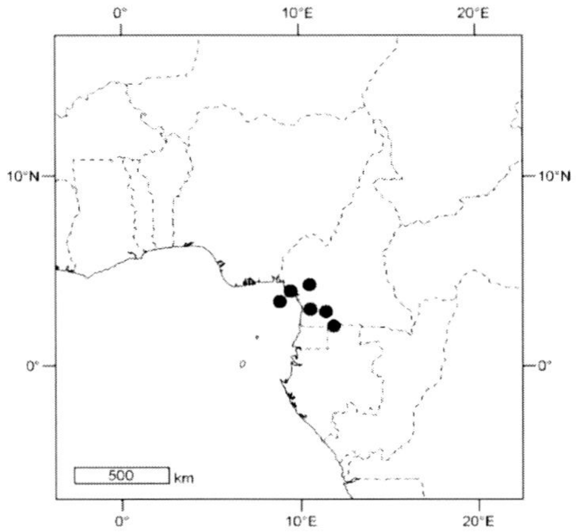

d. ***Campylostemon mitophorum*** VU (Celastraceae)

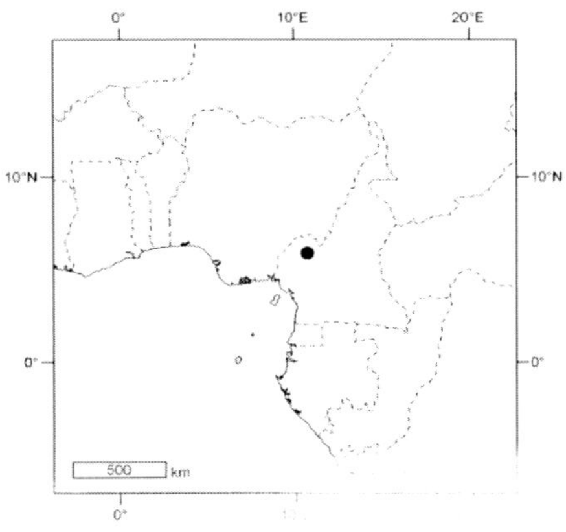

e. ***Elaeodendron kam*** [illegible] (Celastraceae)

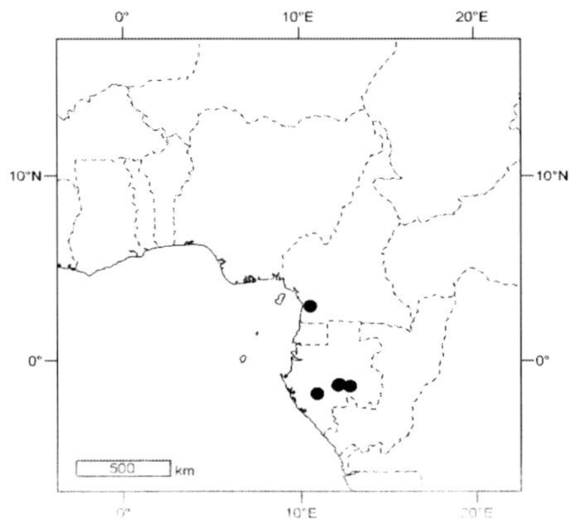

f. ***Loeseneriella camerunica*** EN (Celastraceae)

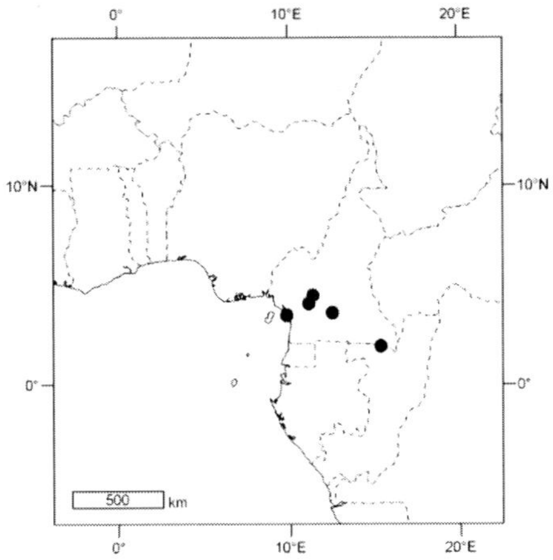

a. ***Pleurostylia serrulata*** EN (Celastraceae)

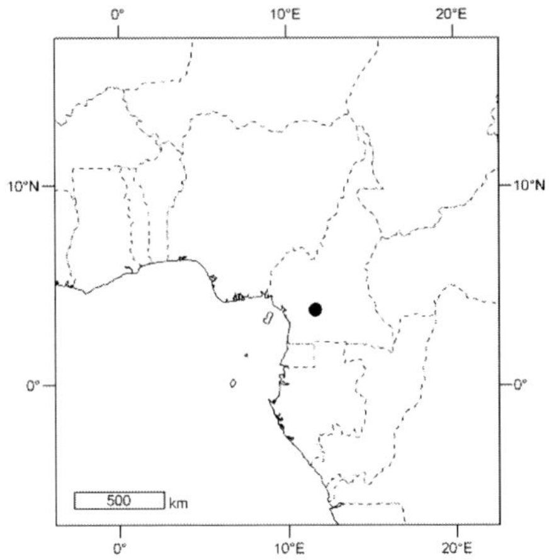

b. ***Pristimera biholongii*** CR (Celastraceae)

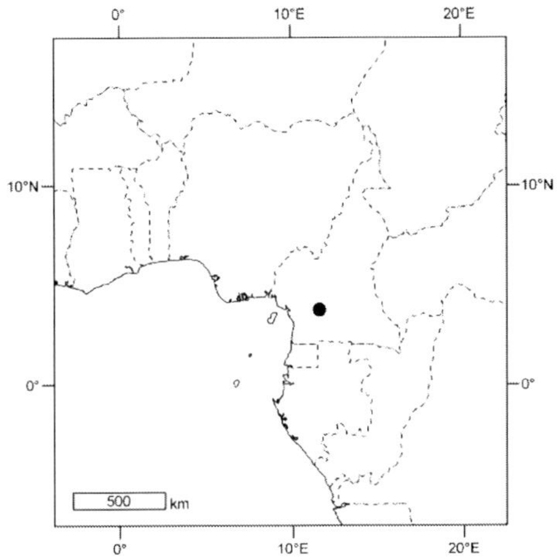

c. ***Pristimera breteleri*** CR(Celastraceae)

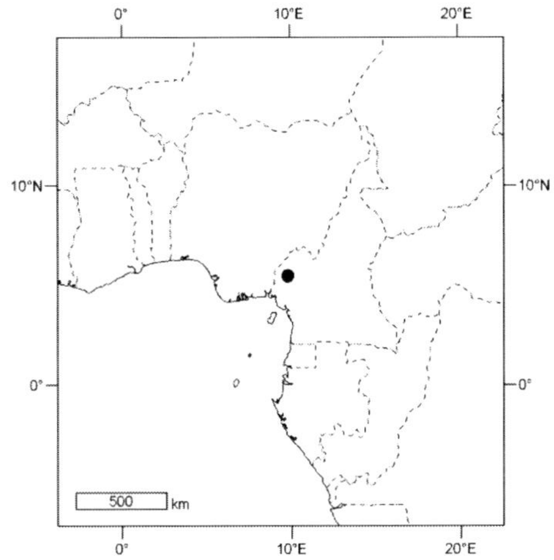

d. ***Salacia conraui*** CR (Celastraceae)

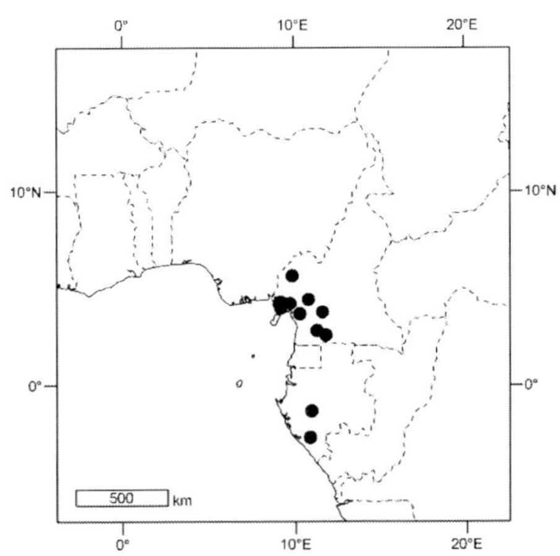

e. ***Salacia dimidia*** VU (Celastraceae)

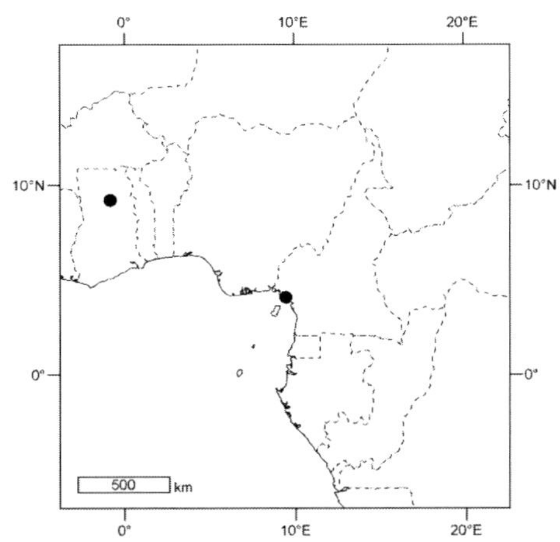

f. ***Salacia fimbrisepala*** EN (Celastraceae)

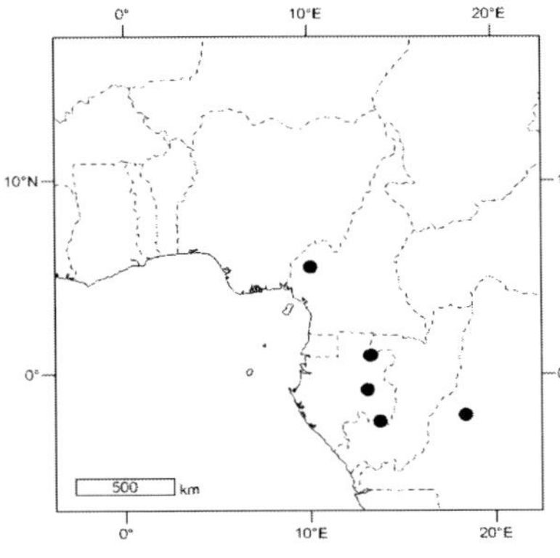

a. ***Salacia lebrunii*** VU (Celastraceae)

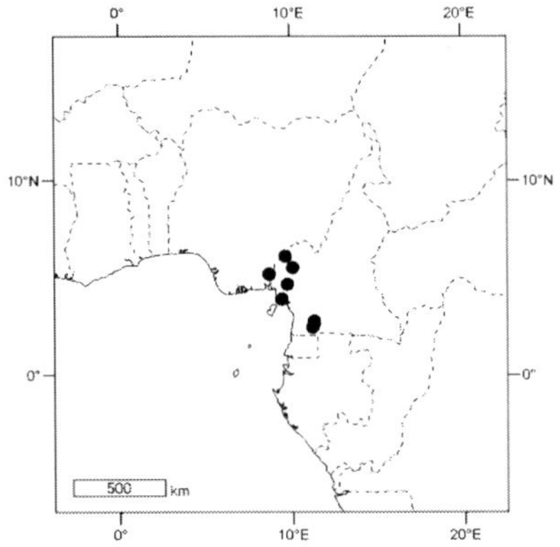

b. ***Salacia lehmbachii*** var. ***pes-ranulae*** VU (Celastraceae)

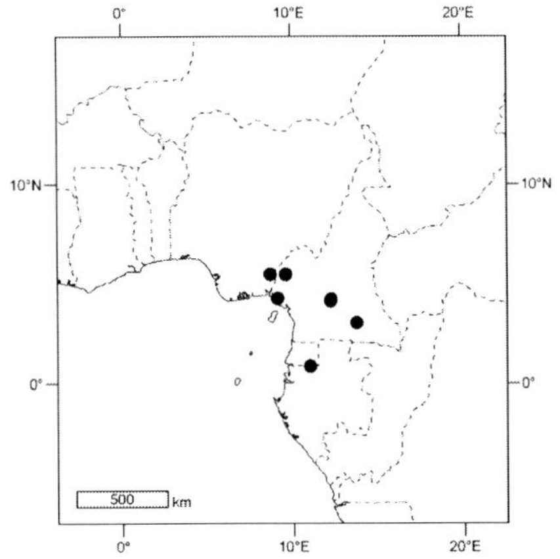

c. ***Salacia lenticellosa*** VU (Celastraceae)

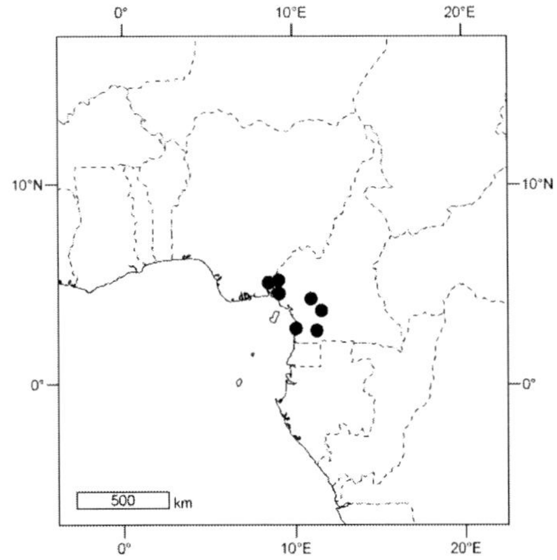

d. ***Salacia letouzeyana*** VU (Celastraceae)

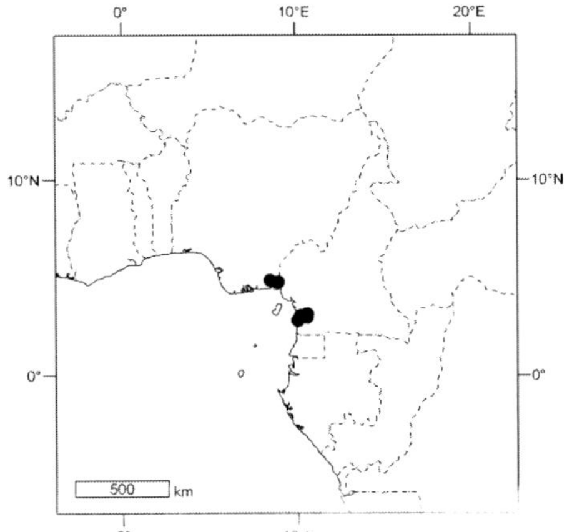

e. ***Salacia lucida*** VU (Celastraceae)

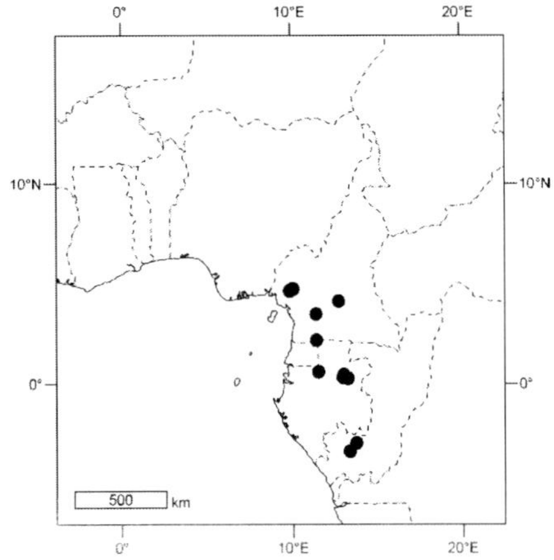

f. ***Salacia mamba*** VU (Celastraceae)

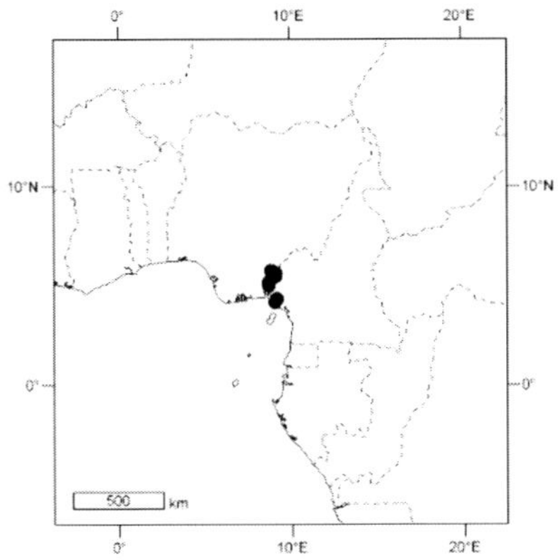

a. ***Salacia talbotii*** EN (Celastraceae)

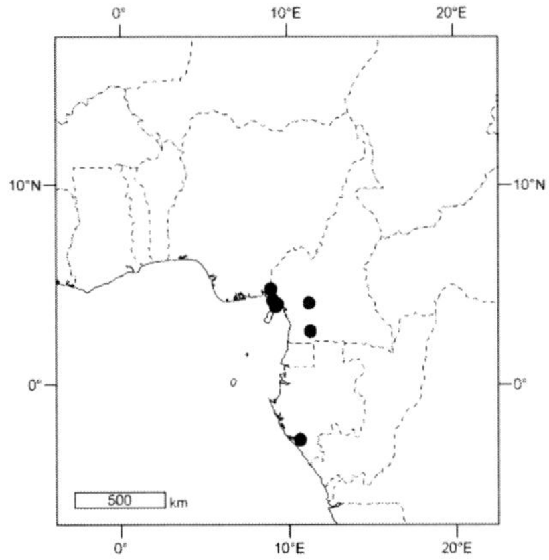

b. ***Salacia volubilis*** EN (Celastraceae)

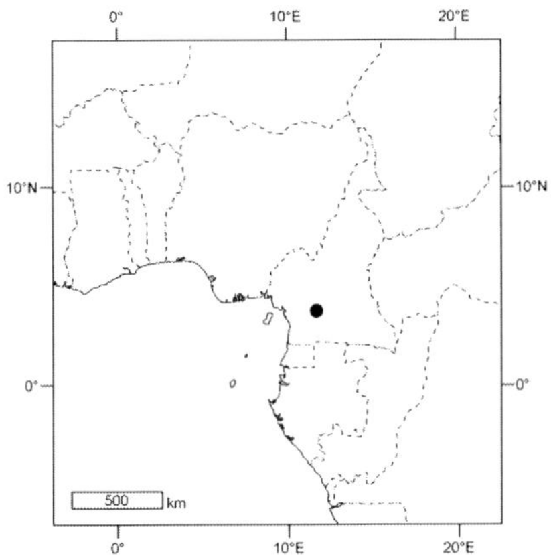

c. ***Simirestis staudtii*** EN (Celastraceae)

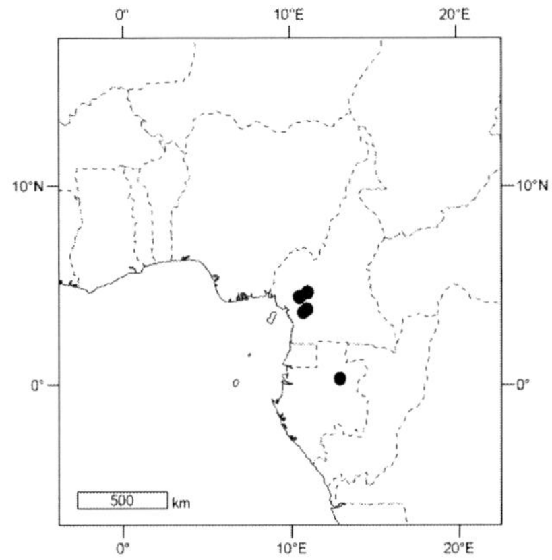

d. ***Thyrosalacia pararacemosa*** EN (Celastraceae)

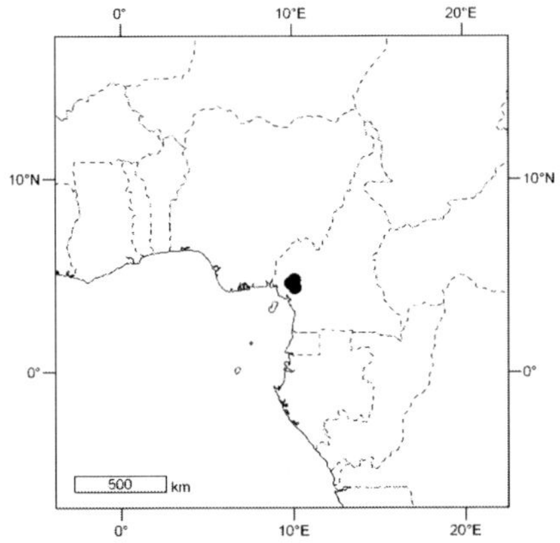

e. ***Thyrosalacia racemosa*** CR (Celastraceae)

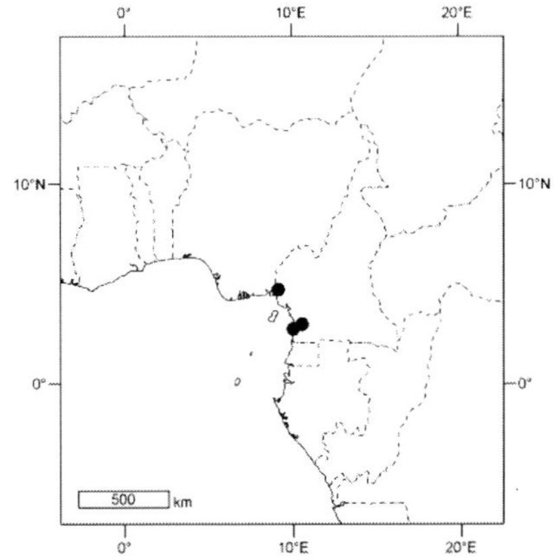

f. ***Dactyladenia cinerea*** EN (Chrysobalanaceae)

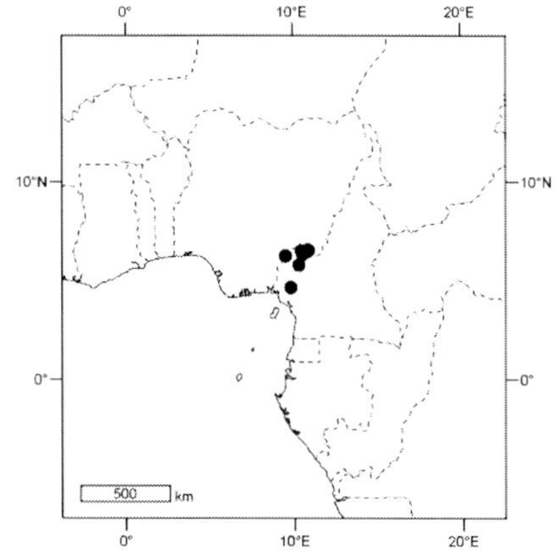

a. ***Dactyladenia johnstonei*** CR
(Chrysobalanaceae)

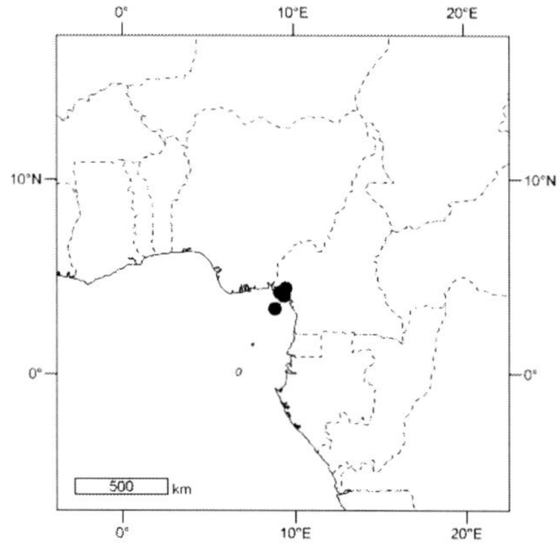

b. ***Dactyladenia mannii*** CR
(Chrysobalanaceae)

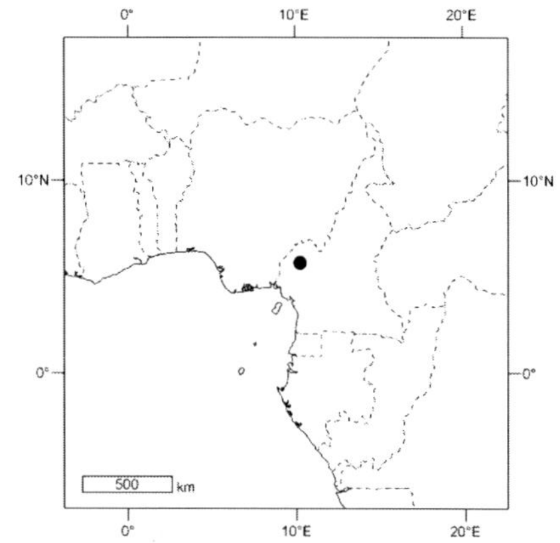

c. **Magnistipula butayei** subsp.
balingembaensis CR (Chrysobalanaceae)

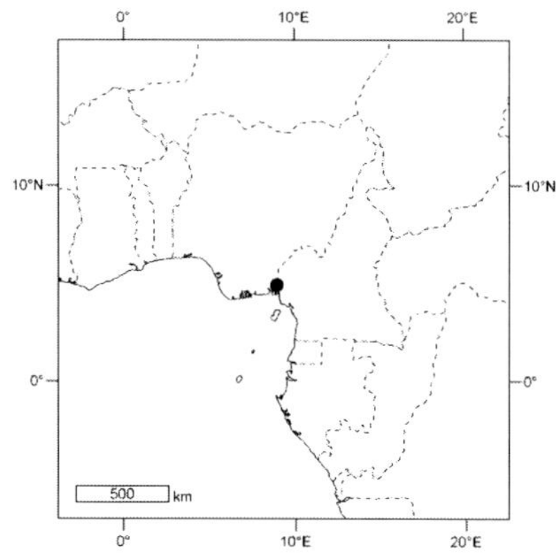

d. ***Magnistipula butayei*** . **subsp. *korupensis***
CR (Chrysobalanaceae)

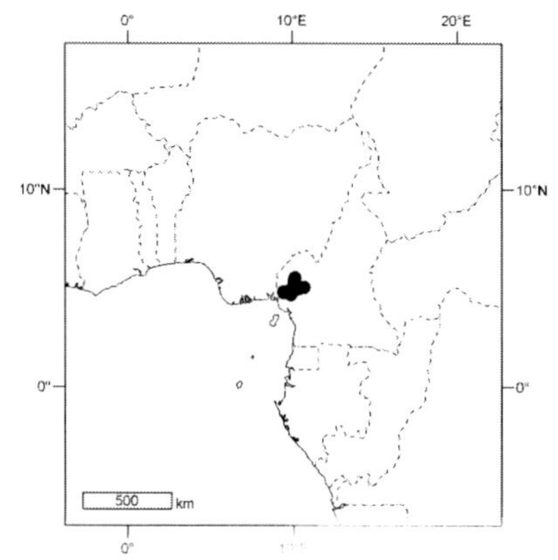

e. ***Magnistipula co***
(Chrysobalanaceae)

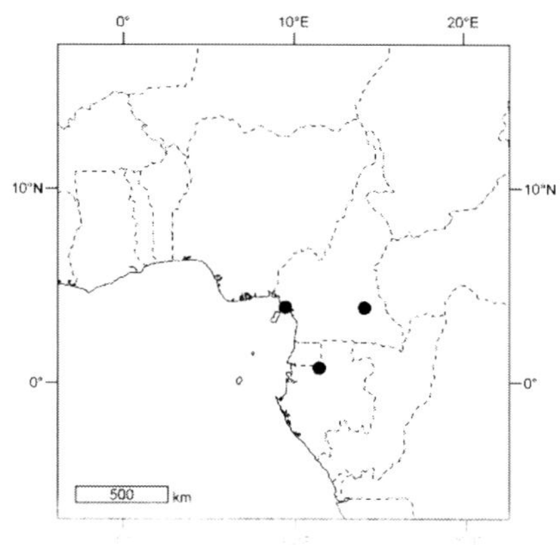

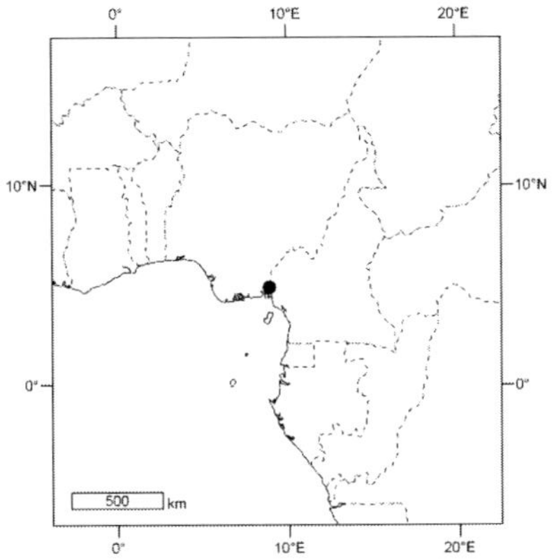

a. ***Magnistipula multinervia*** CR (Chrysobalanaceae)

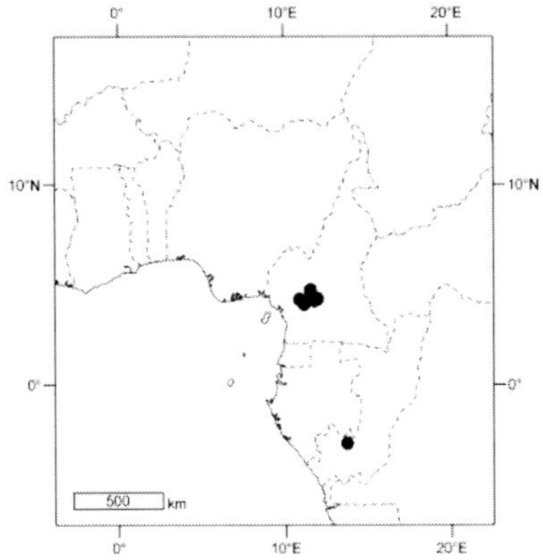

b. ***Maranthes sanagensis*** EN (Chrysobalanaceae)

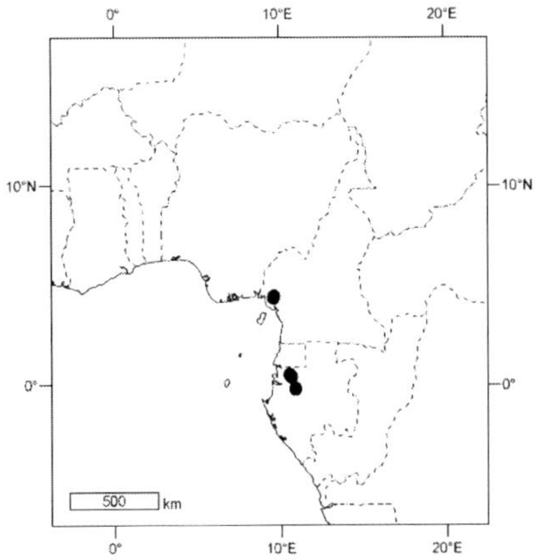

c. ***Combretum exellii*** EN (Combretaceae)

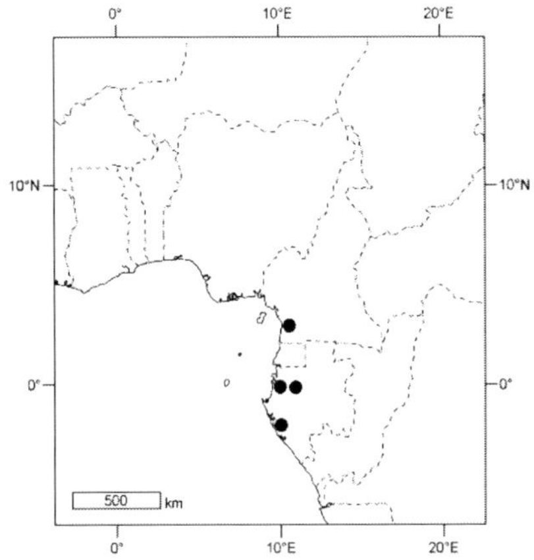

d. ***Combretum echirense*** EN (Combretaceae)

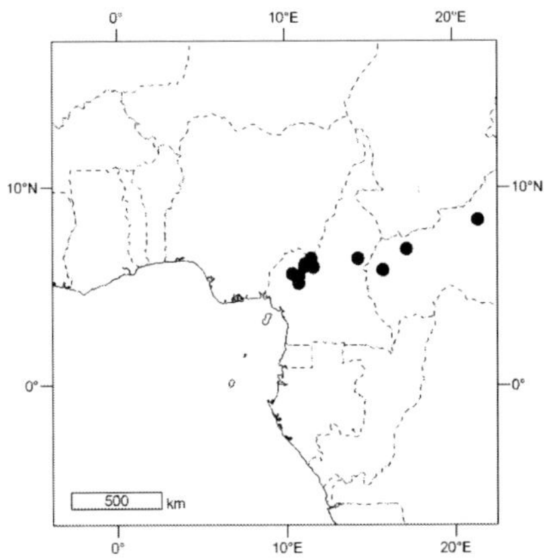

e. ***Aedesia spectabilis*** VU (Compositae)

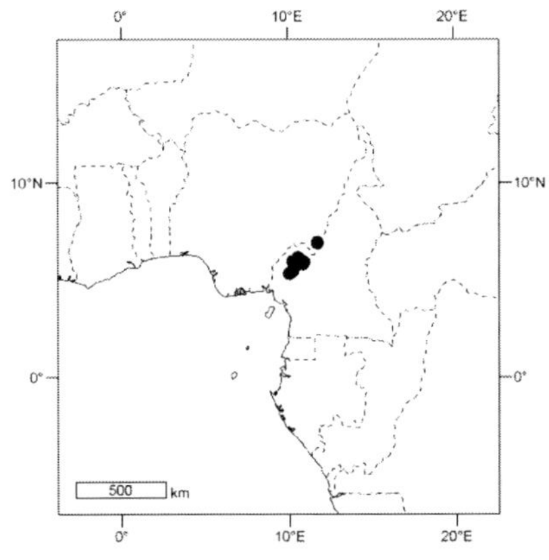

f. ***Bafutia tenuicaulis*** subsp. ***tenuicaulis*** VU (Compositae)

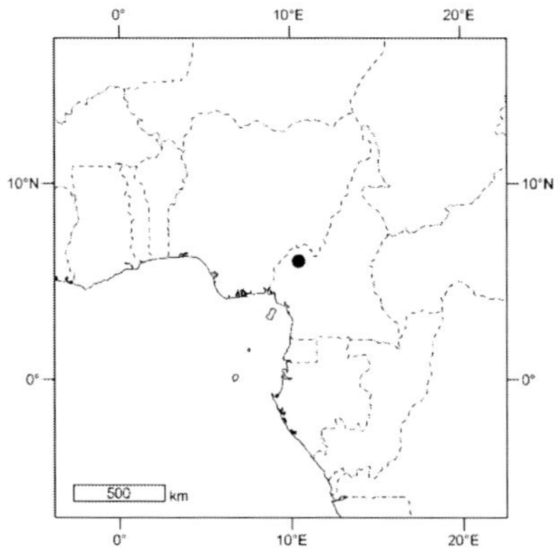

a. ***Bafutia tenuicaulis*** subsp. ***zapfackiana*** EN (Compositae)

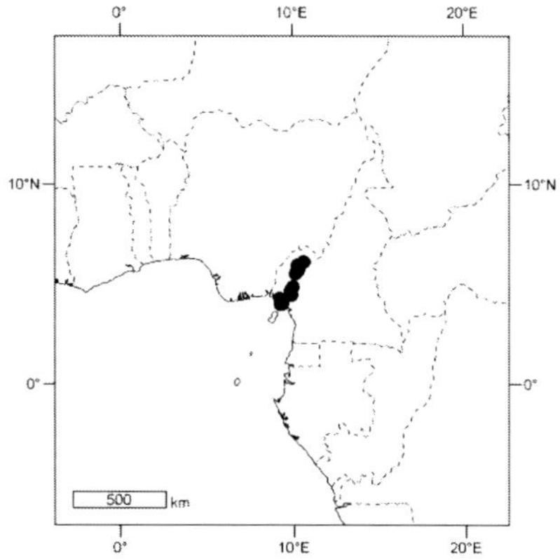

b. ***Bidens mannii*** VU (Compositae)

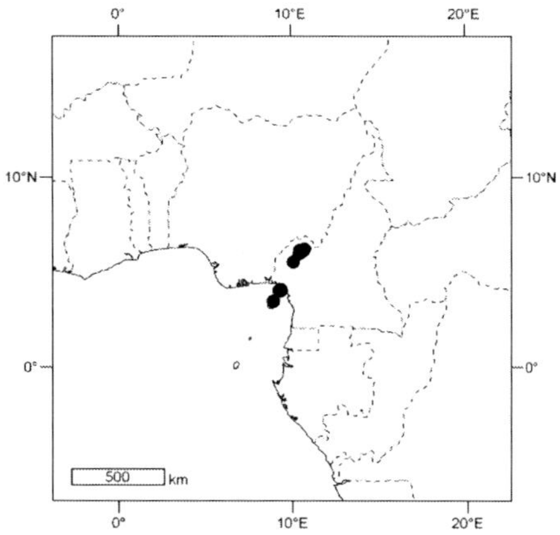

c. ***Crassocephalum bougheyanum*** VU (Compositae)

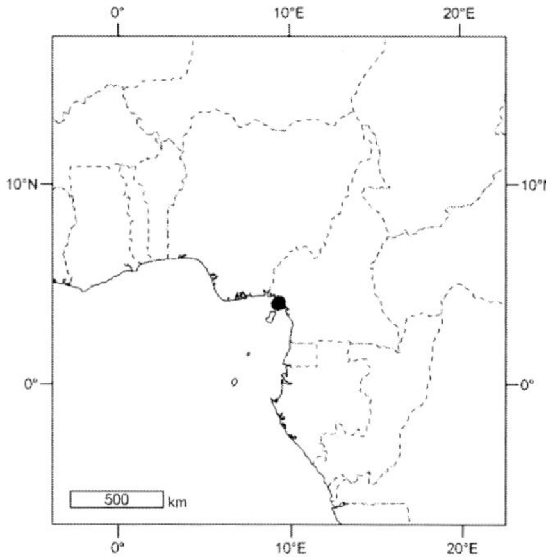

d. ***Helichrysum biafranum*** CR (Compositae)

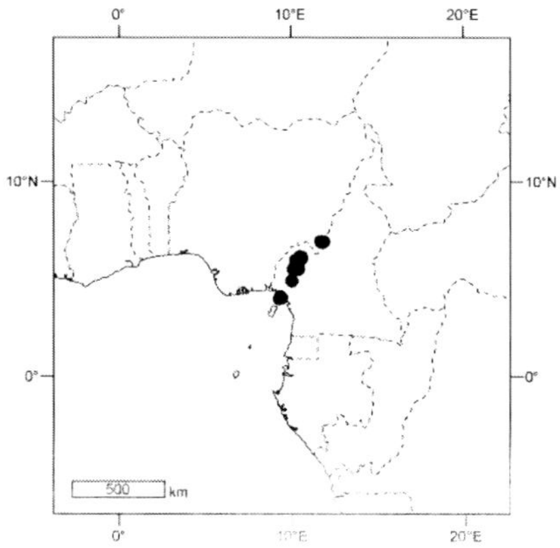

e. ***Helichrysum cameroonense*** EN (C[illegible]

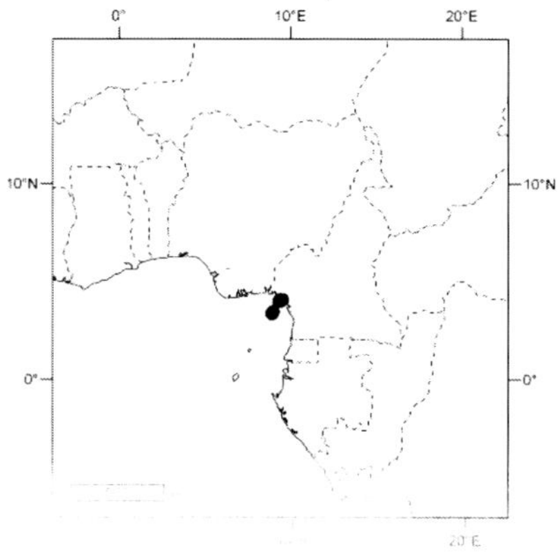

[illegible] EN (Compositae)

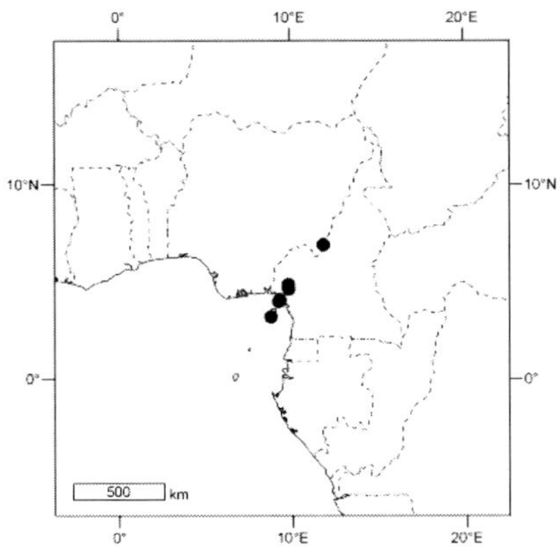

a. ***Mikaniopsis maitlandii*** EN (Compositae)

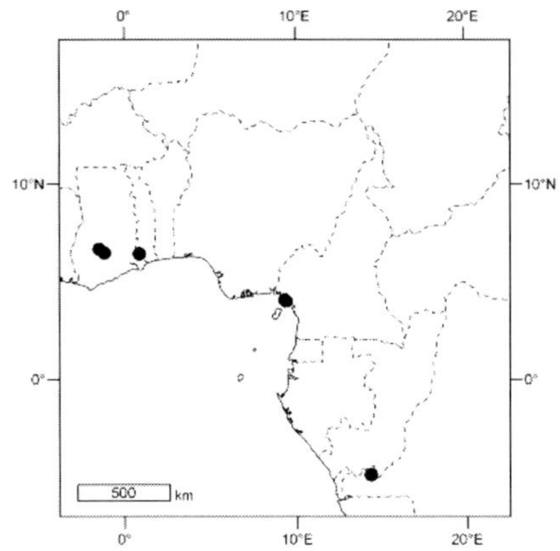

b. ***Mikaniopsis tedliei*** VU (Compositae)

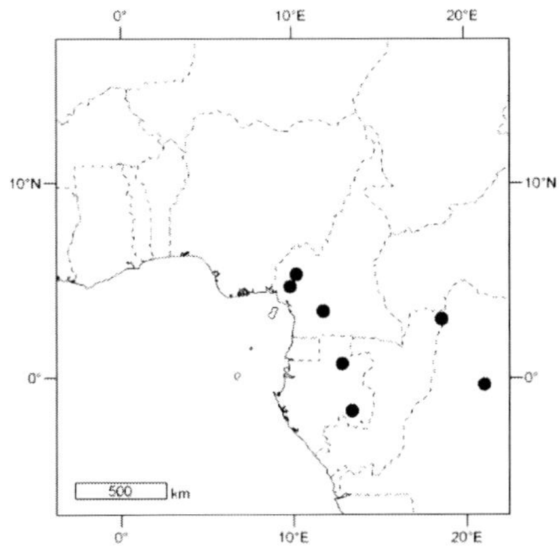

c. ***Mikaniopsis vitalba*** VU (Composiate)

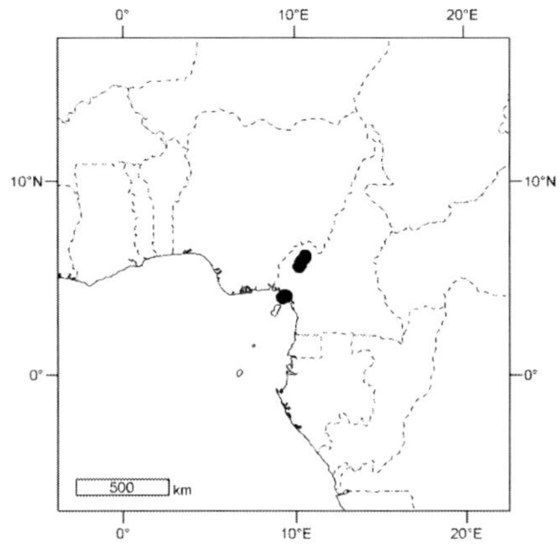

d. ***Senecio burtonii*** VU (Compositae)

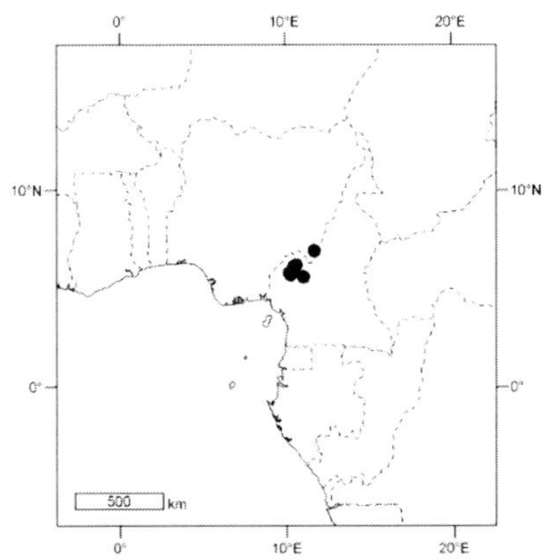

e. ***Vernonia bamendae*** VU (Compositae)

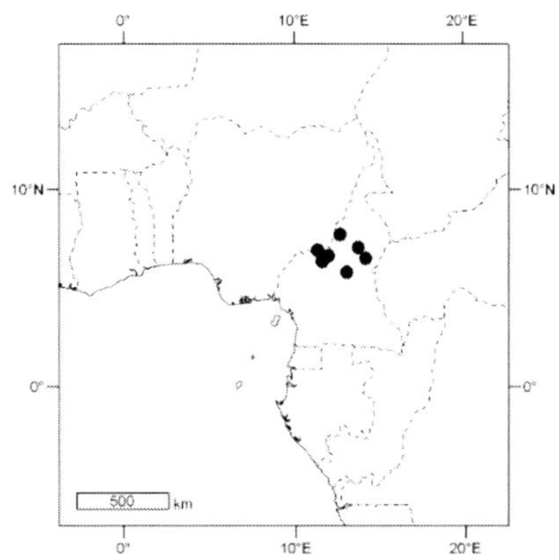

f. **Vernonia chapmanii** VU (Compositae)

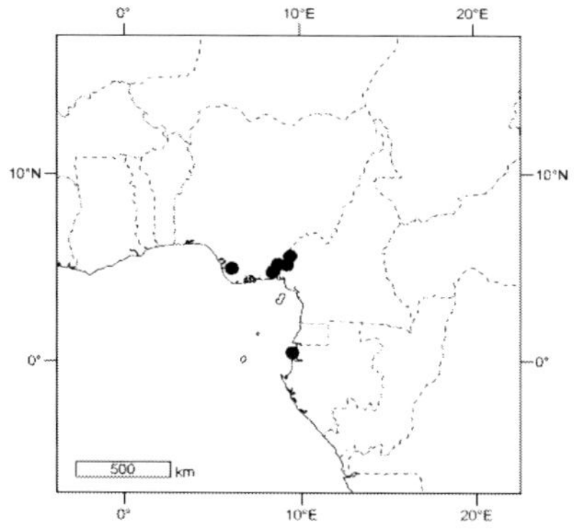

a. ***Cnestis macrantha*** EN (Connaraceae)

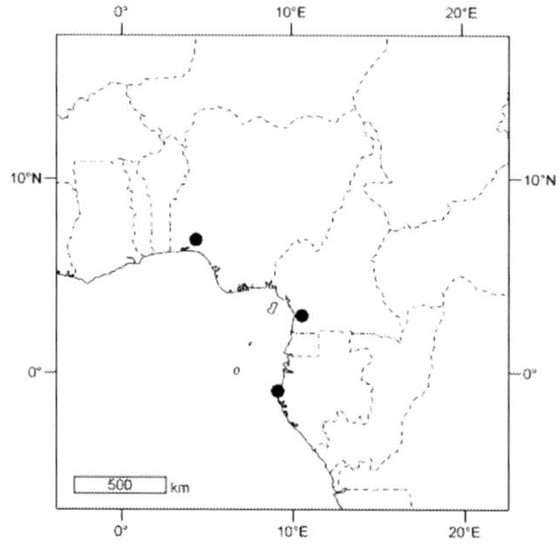

b. ***Cnestis macrophylla*** EN (Connaraceae)

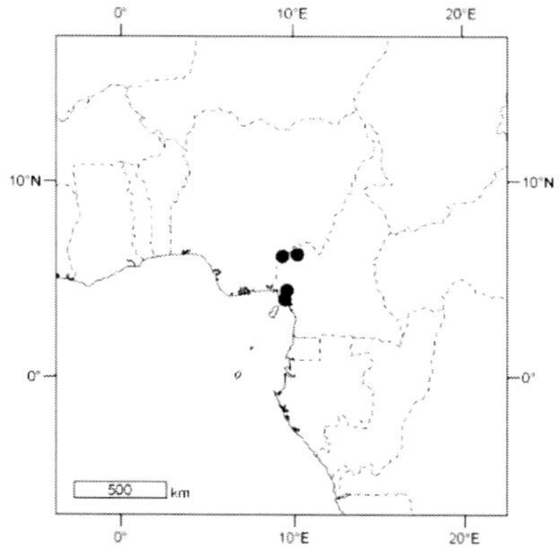

c. ***Jollydora glandulosa*** EN (Connaraceae)

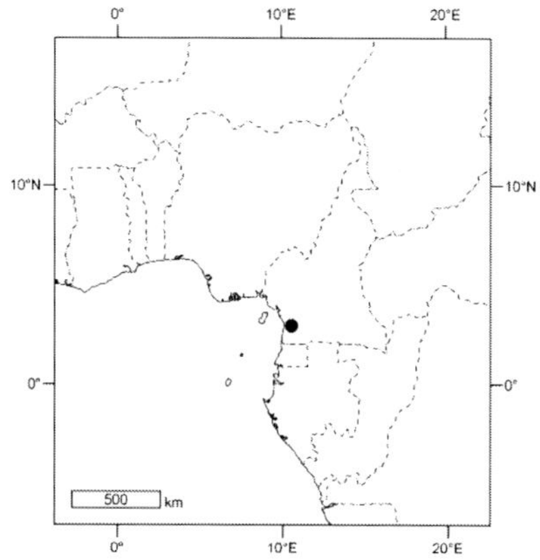

d. ***Calycobolus gilgianus*** CR (Convolvulaceae)

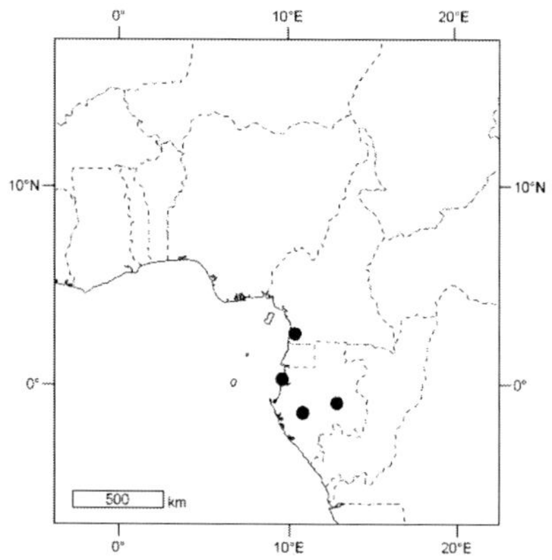

e. ***Neuropeltis laxiflora*** EN (Convolvulaceae)

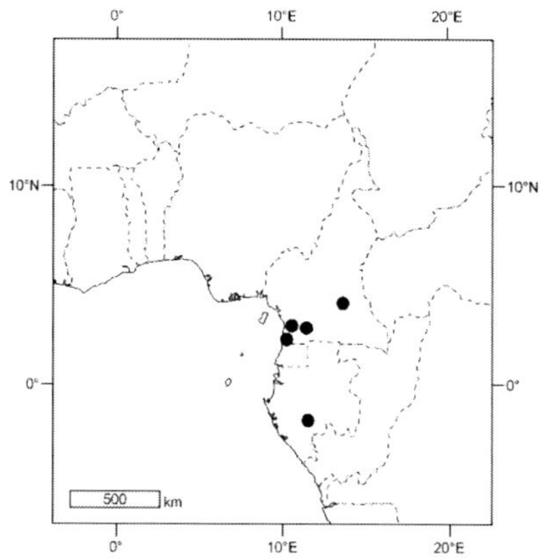

f. ***Neuropeltis pseudovelutina*** EN (Convolvulaceae)

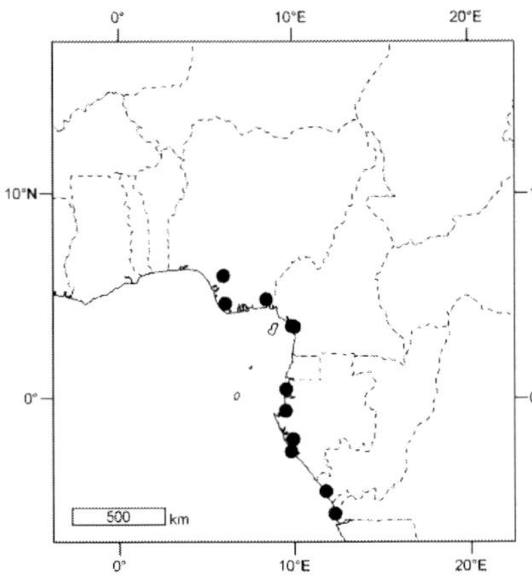

a. ***Ctenolophon englerianus*** VU (Ctenolophonaceae)

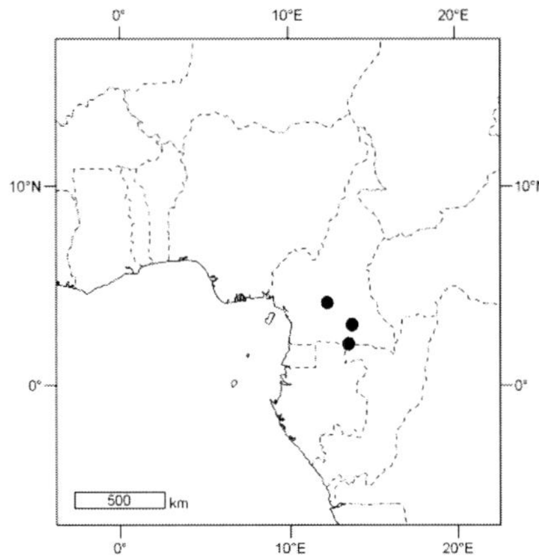

b. ***Cyclantheropsis occidentalis*** EN (Cucurbitaceae)

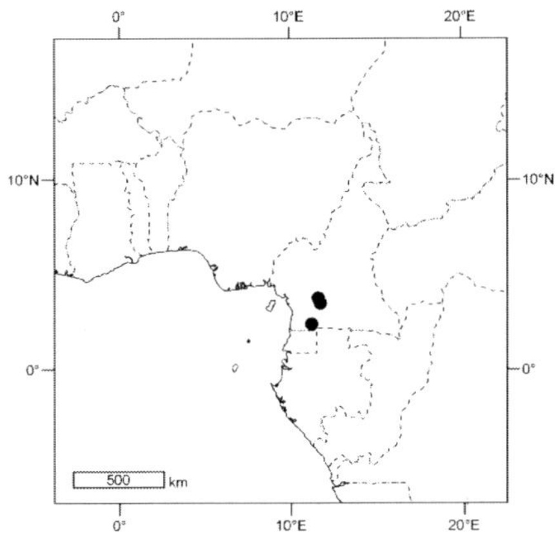

c. ***Momordica camerounensis*** EN (Cucurbitaceeae)

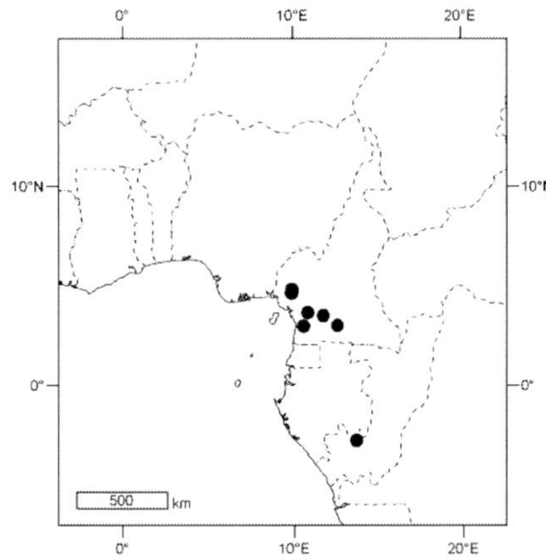

d. ***Momordica enneaphylla*** VU (Cucurbitaceae)

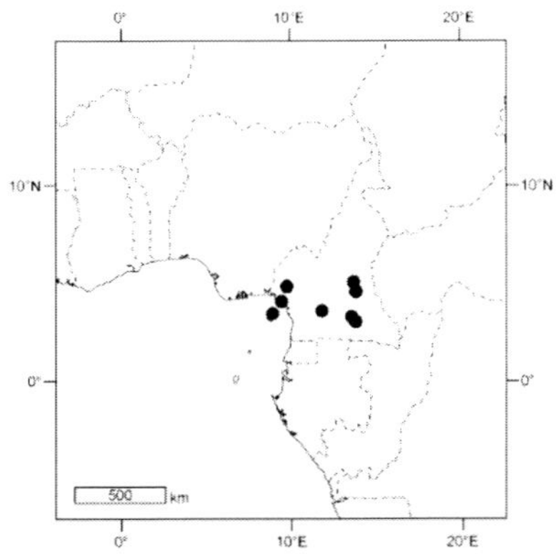

e. ***Momordica gilgiana*** VU (Cucurbitaceae)

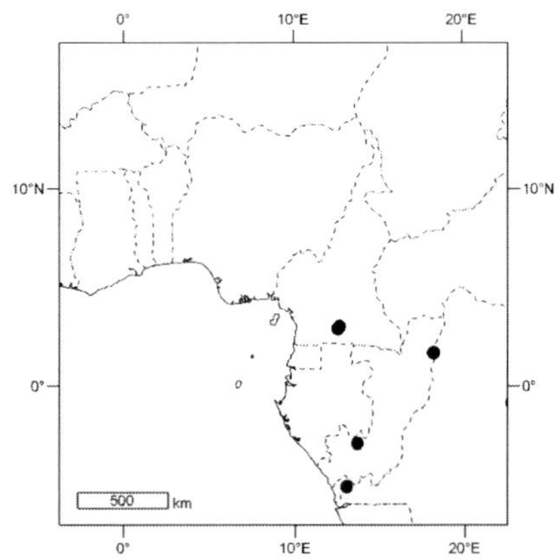

f. ***Momordica jeffreyana*** VU (Cucurbitaceae)

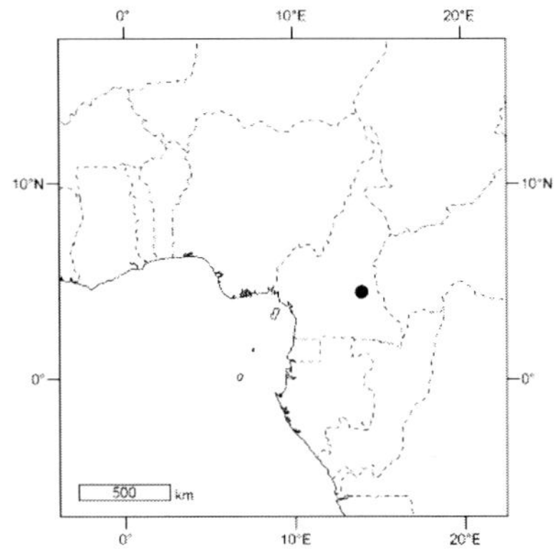

a. ***Momordica obtusisepala*** CR (Cucurbitaceae)

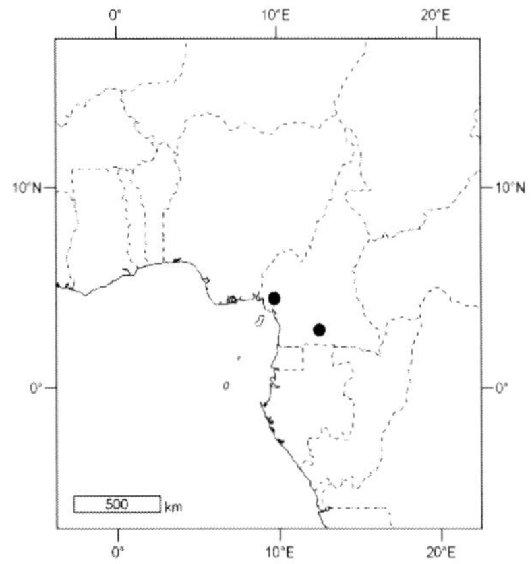

b. ***Telfairia batesii*** CR (Cucurbitaceae)

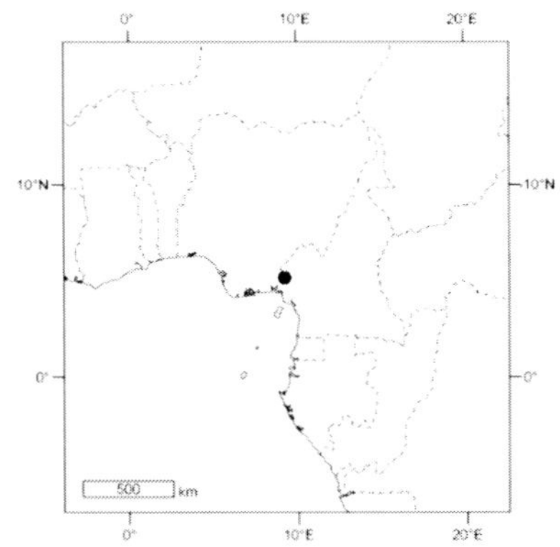

c. ***Dichapetalum korupinum*** CR (Dichapetalaceae)

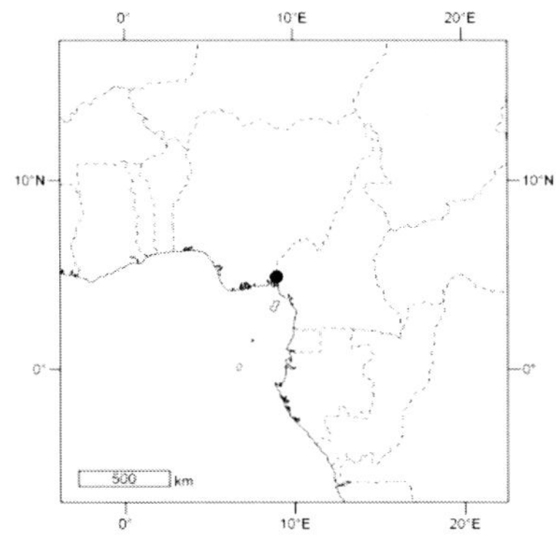

d. ***Dichapetalum letouzeyi*** CR (Dichapetalaceae)

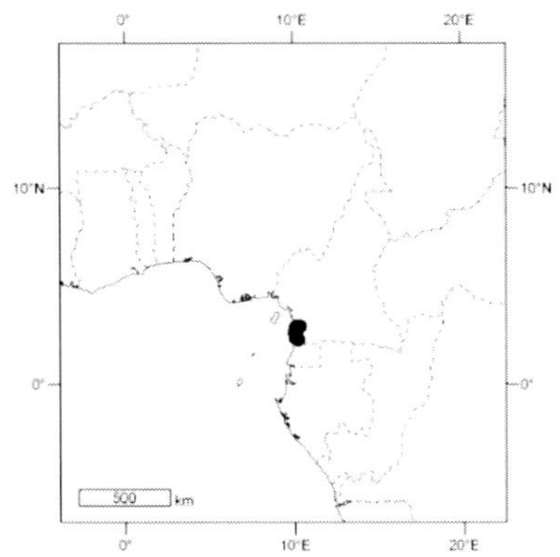

e. ***Dichapetalum oliganthum*** VU (Dichapetalaceae)

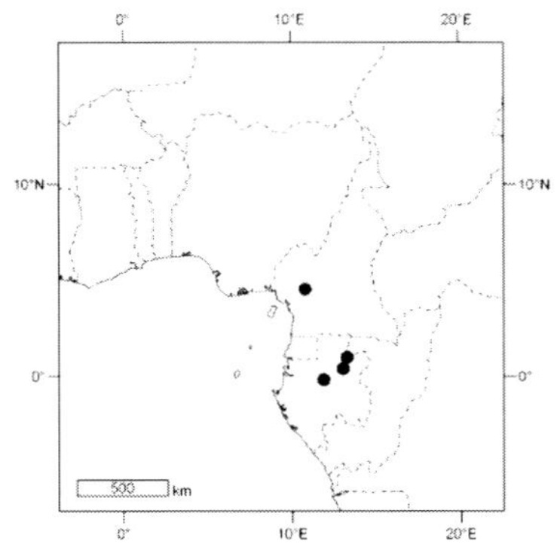

f. ***Dichapetalum potamophilum*** EN(Dichapetalaceae)

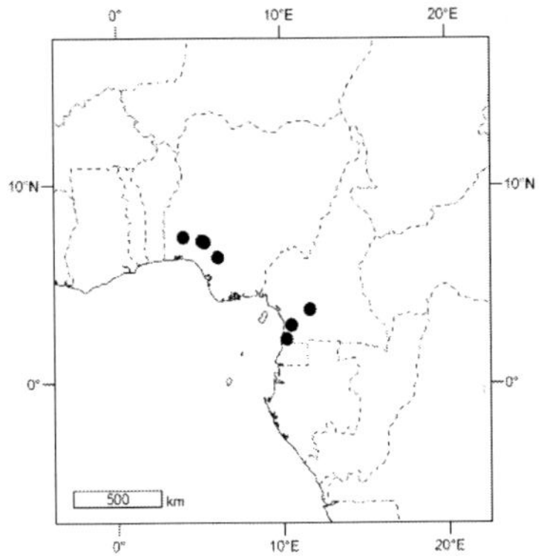

a. ***Dichapetalum reticulatum*** VU (Dichapetalaceae)

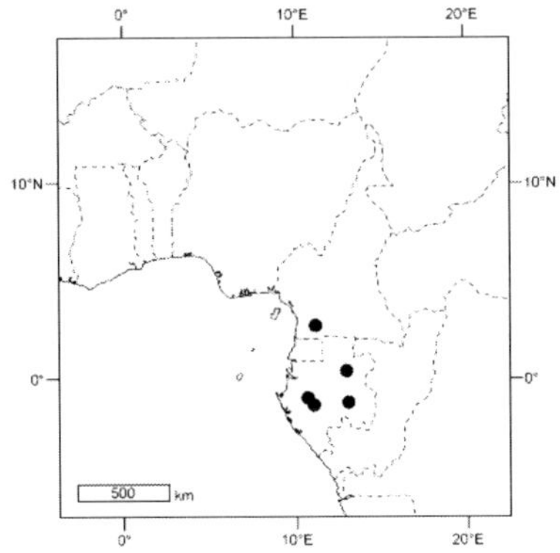

b. ***Dichapetalum witianum*** EN (Dichapetalaceae)

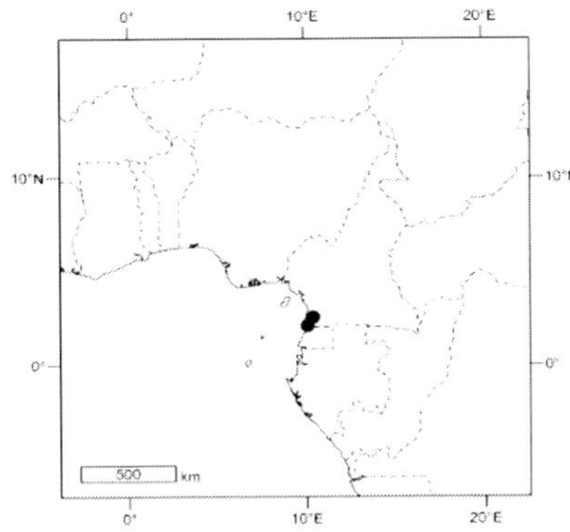

c. ***Tapura tchoutoi*** EN (Dichapetalaceae)

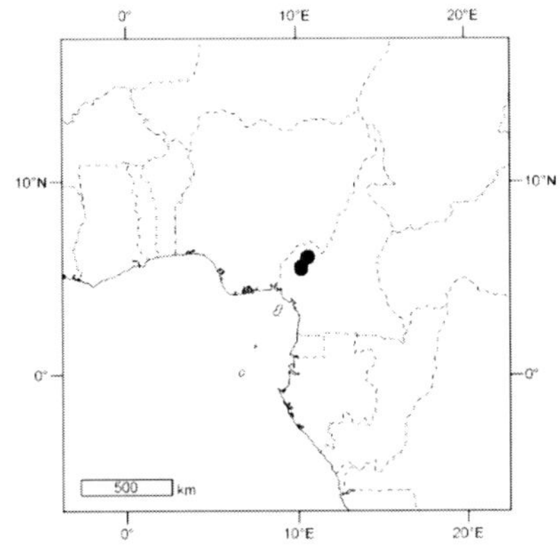

d. **Dipsacus narcisseanus** EN (Dipsacaceae)

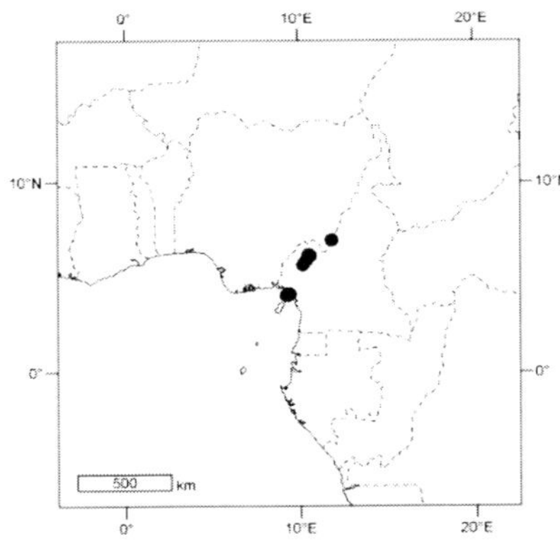

e. ***Succisa trichotocephala*** EN (Dipsacaceae)

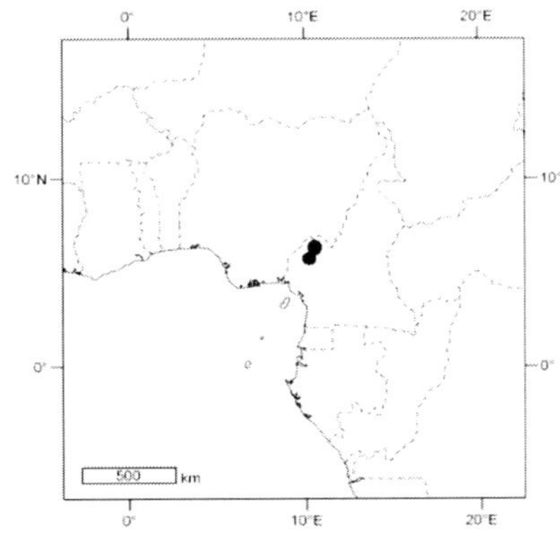

f. . ***Drosera pilosa*** VU (Droseraceae)

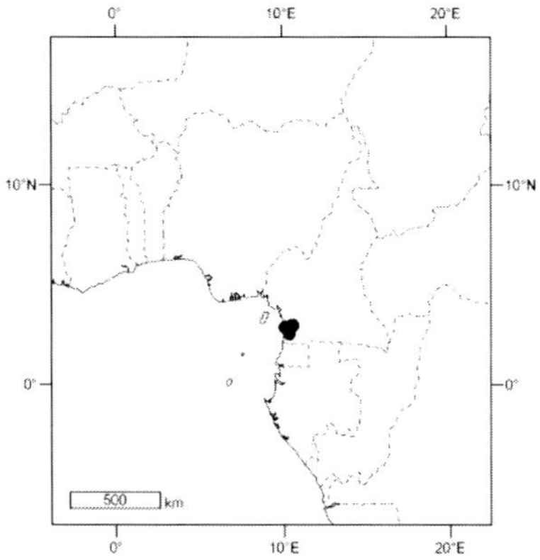

a. ***Diospyros alboflavescens*** EN (Ebenaceae)

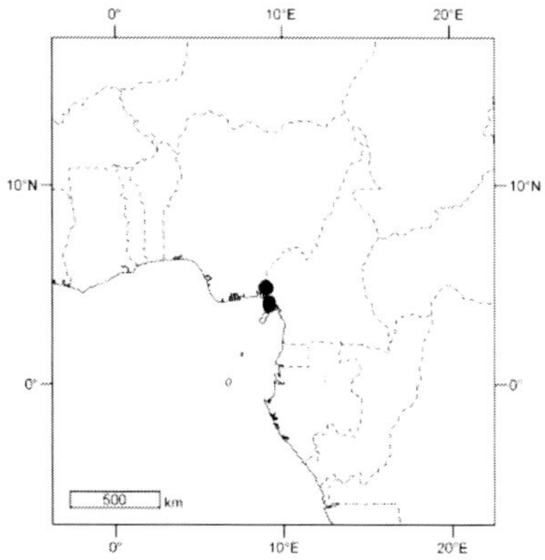

b. ***Diospyros korupensis*** VU (Ebenaceae)

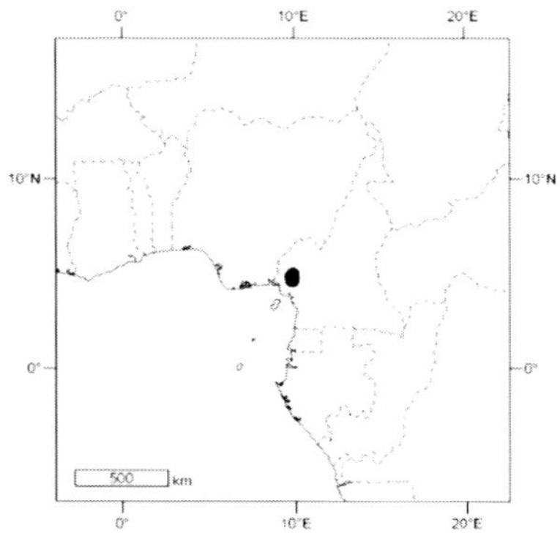

c. ***Diospyros kupensis*** VU (Ebenaceae)

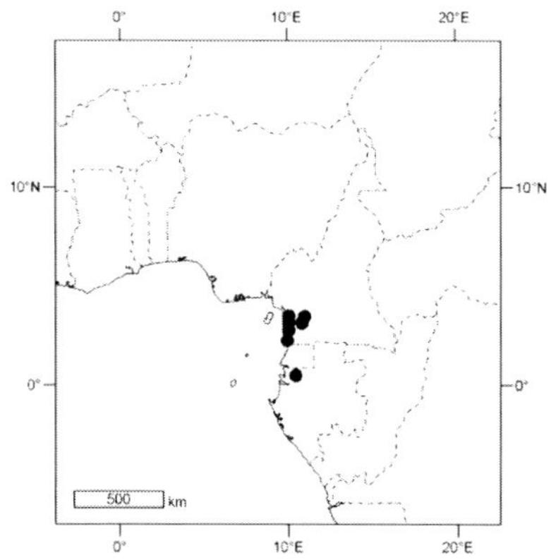

d. ***Diospyros longiflora*** VU (Ebenaceae)

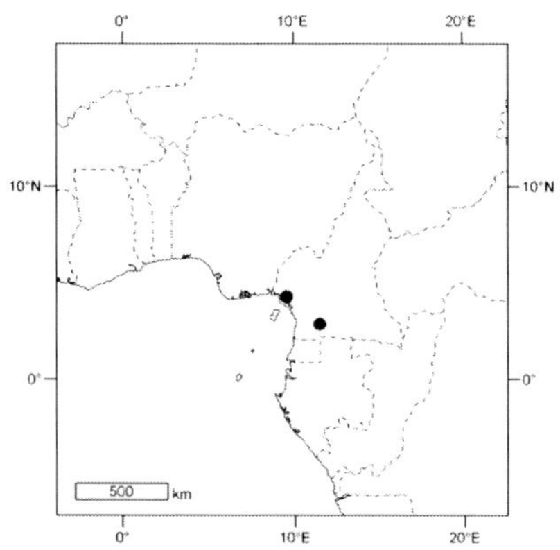

e. ***Diospyros onanae*** EN (Ebenaceae)

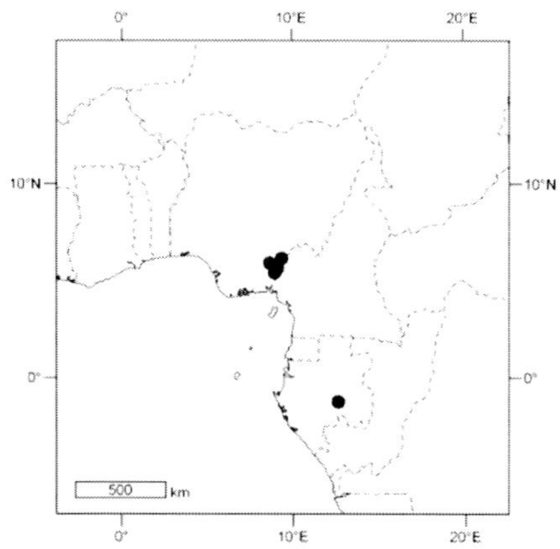

f. ***Diospyros platanoides*** EN (Ebenaceae)

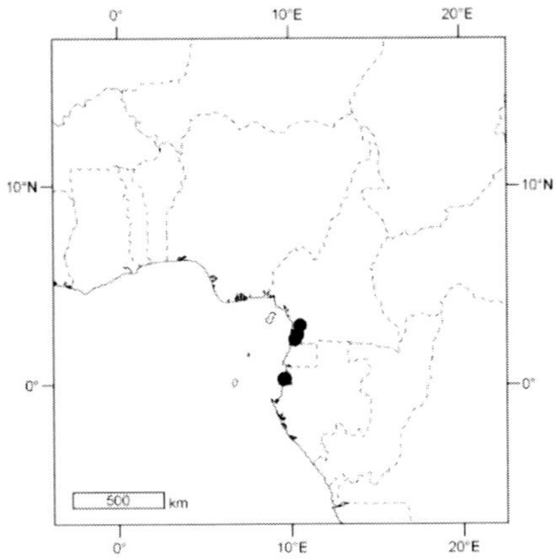

a. ***Diospyros soyauxii*** EN (Ebenaceae)

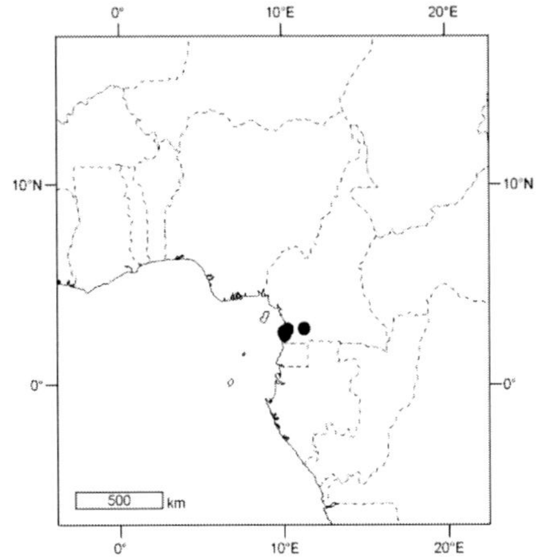

b. ***Afrotrewia kamerunica*** EN (Euphorbiaceae)

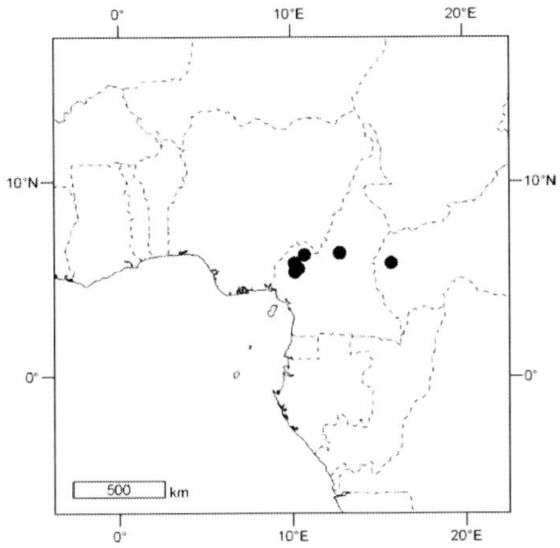

c. ***Antidesma pachybotryum*** VU (Euphorbiaceae)

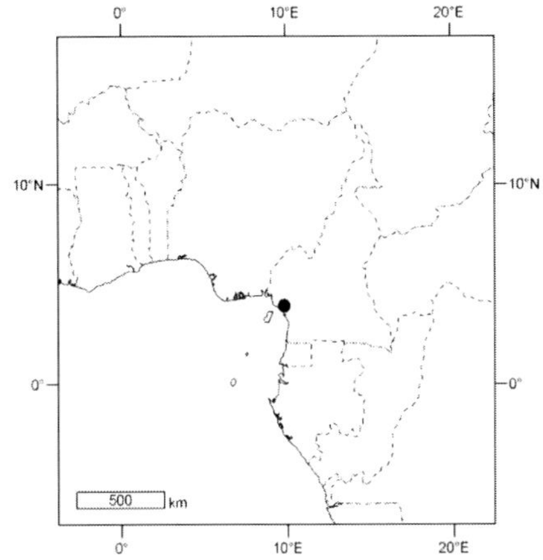

d. ***Cleistanthus camerunensis*** CR (Euphorbiaceae)

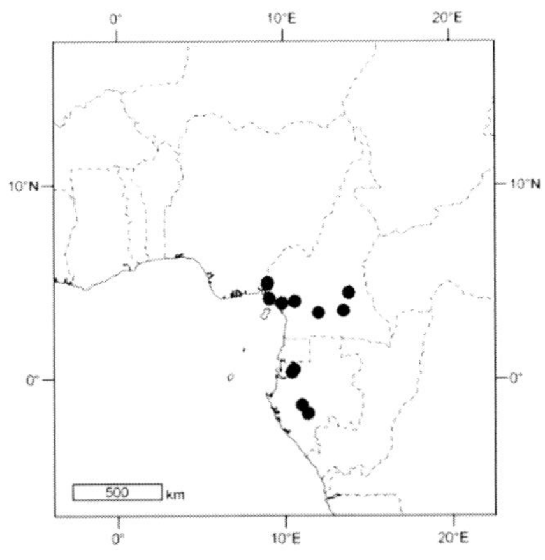

e. ***Cleistanthus letouzeyi*** VU (Euphorbiaceae)

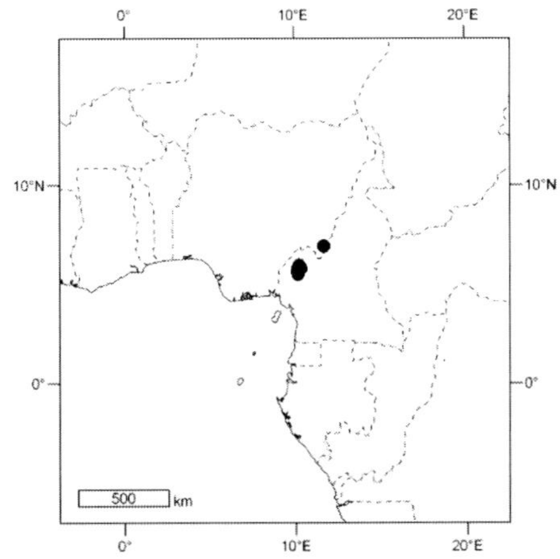

f. ***Clutia kamerunica*** EN (Euphorbiaceae)

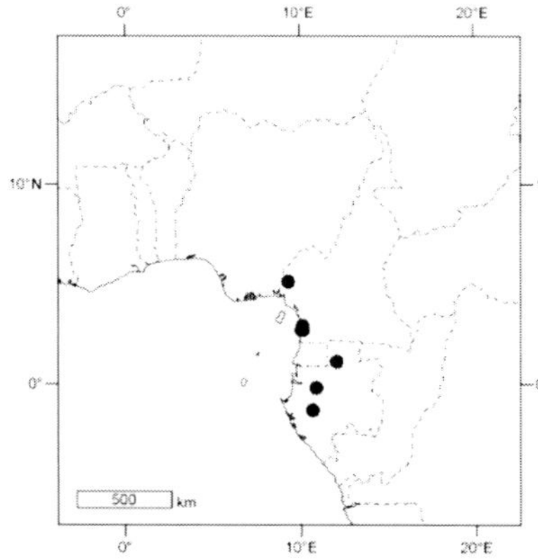

a. ***Conceveiba leptostachys*** VU (Euphorbiaceae)

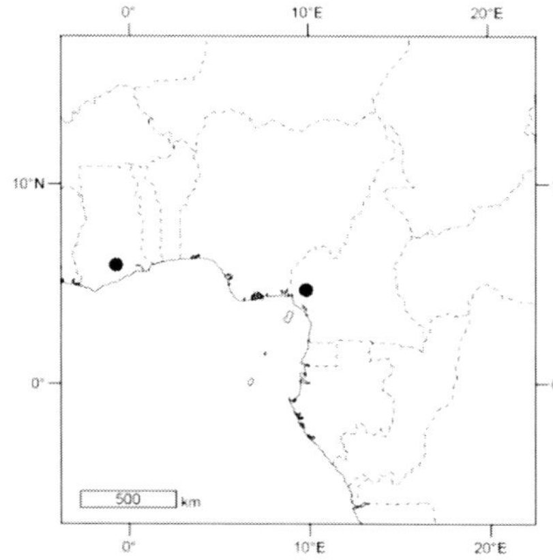

b. ***Croton aubrevillei*** VU (Euphorbiaceae)

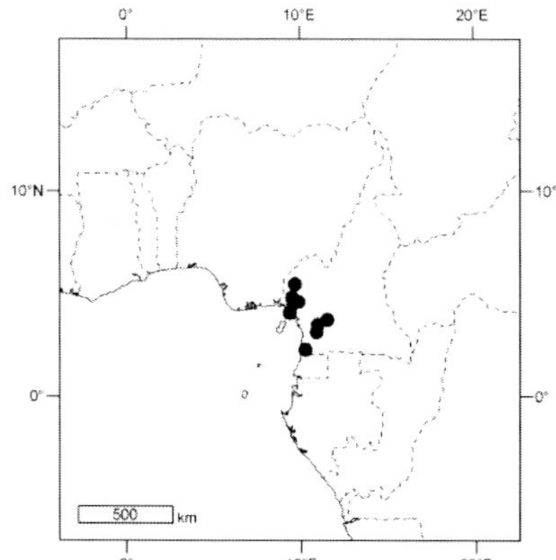

c. ***Crotonogyne impedita*** VU (Euphorbiaceae)

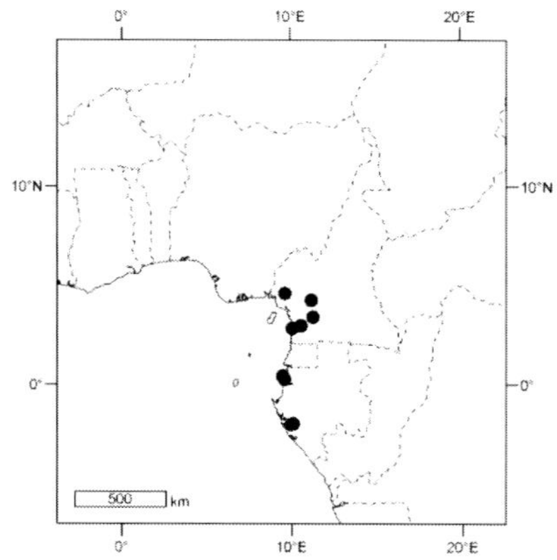

d. ***Crotonogyne zenkeri*** VU (Euphorbiaceae)

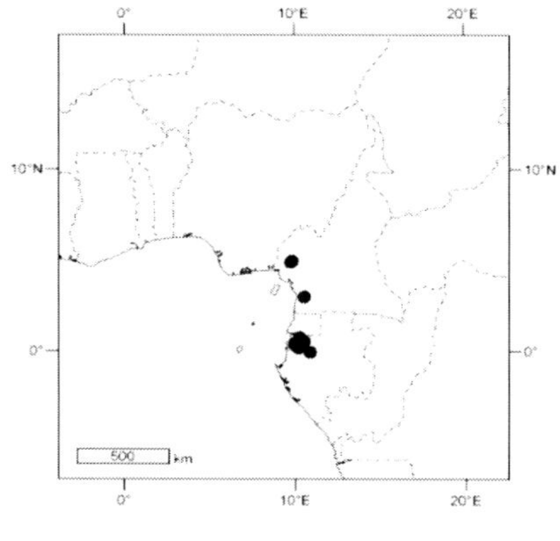

e. ***Drypetes magnistipula*** EN (Euphorbiaceae)

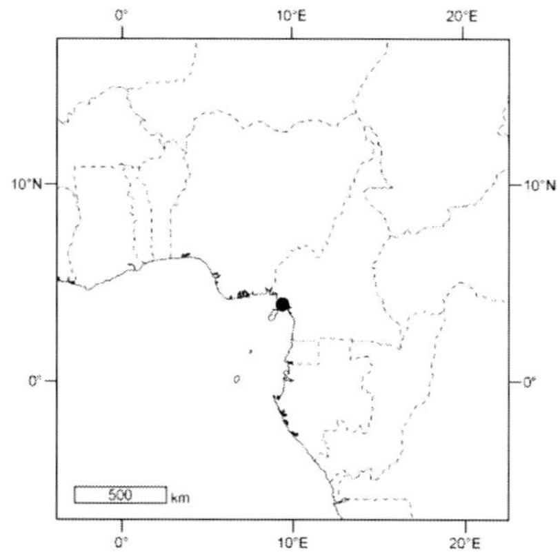

f. ***Drypetes moliwensis*** CR (Euphorbiaceae)

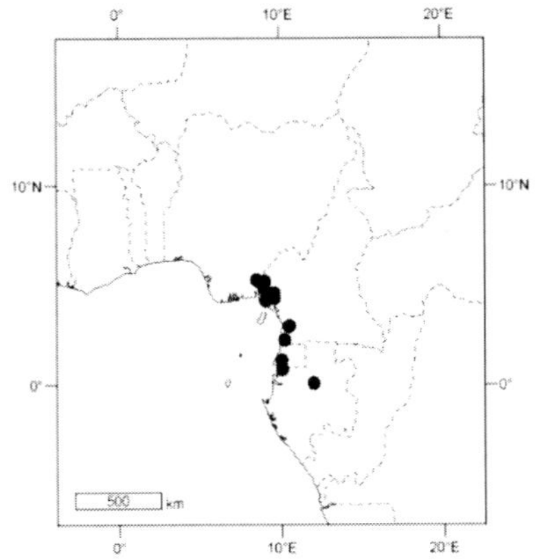

a. ***Drypetes preussii*** VU (Euphorbiaceae)

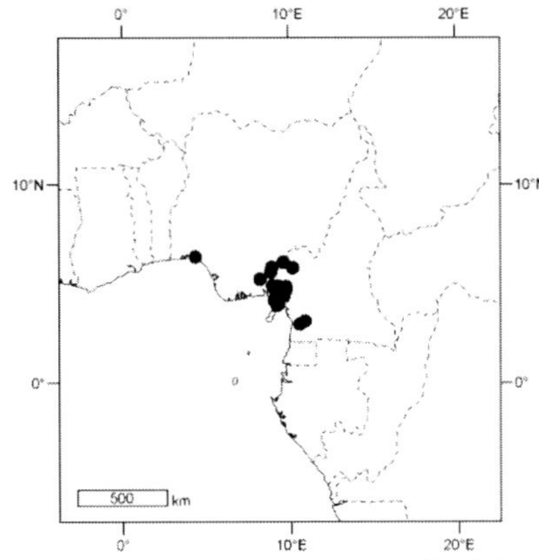

b. ***Drypetes staudtii*** VU (Euphorbiaceae)

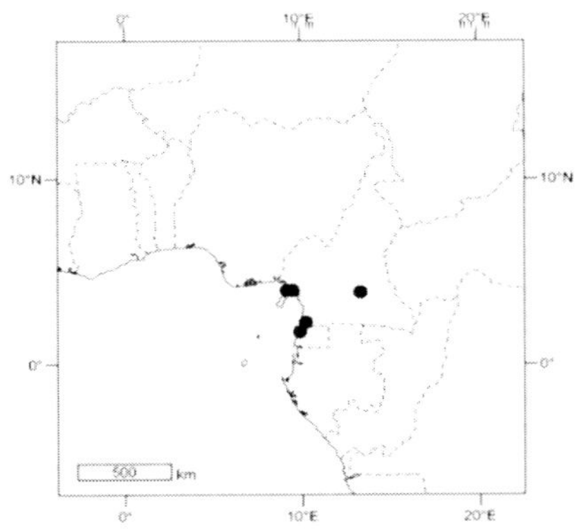

c. ***Drypetes tessmanniana*** EN (Euphorbiaceae)

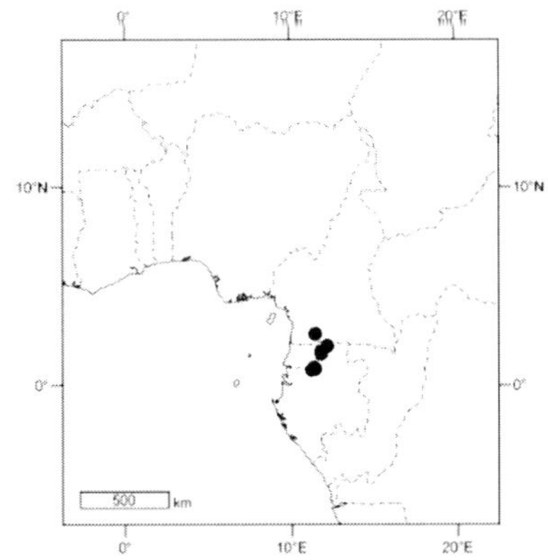

d. ***Euphorbia letestui*** EN (Euphorbiaceae)

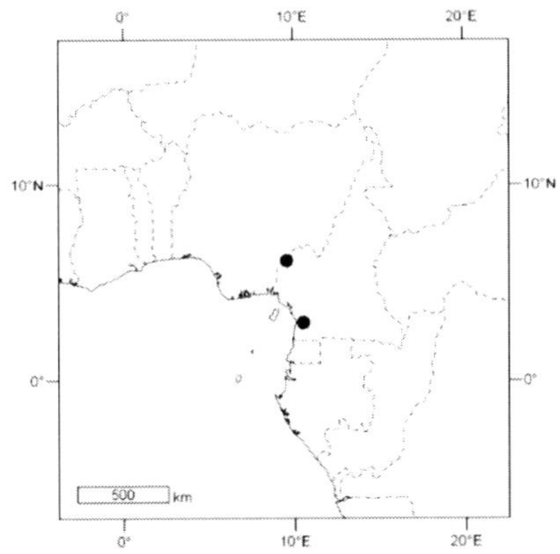

e. ***Grossera major*** EN (Euphorbiaceae)

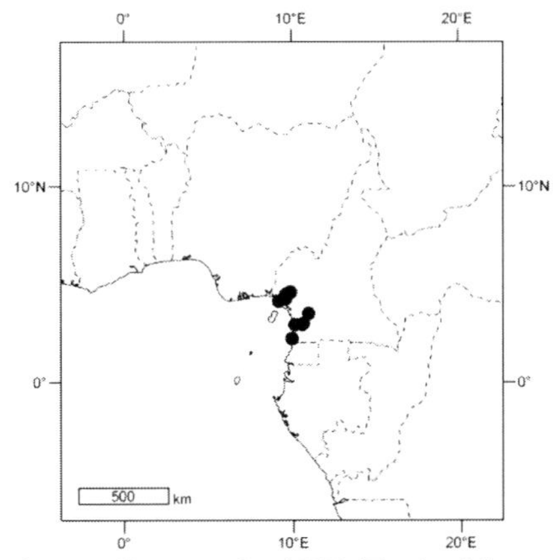

f. ***Hamilcoa zenkeri*** VU (Euphorbiaceae)

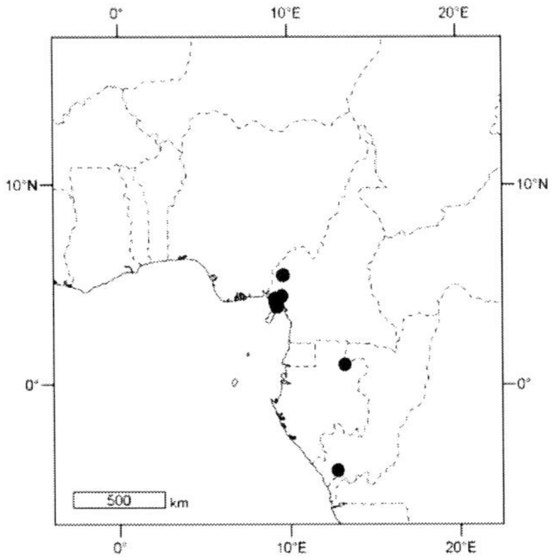

a. ***Leeuwenbergia africana*** EN (Euphorbiaceae)

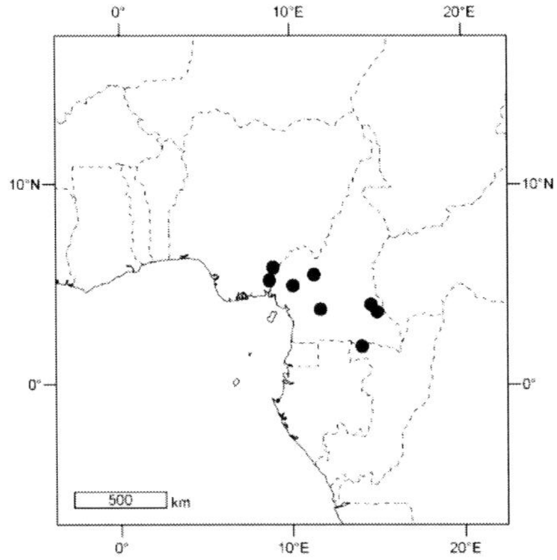

b. ***Macaranga paxii*** VU (Euphorbiaceae)

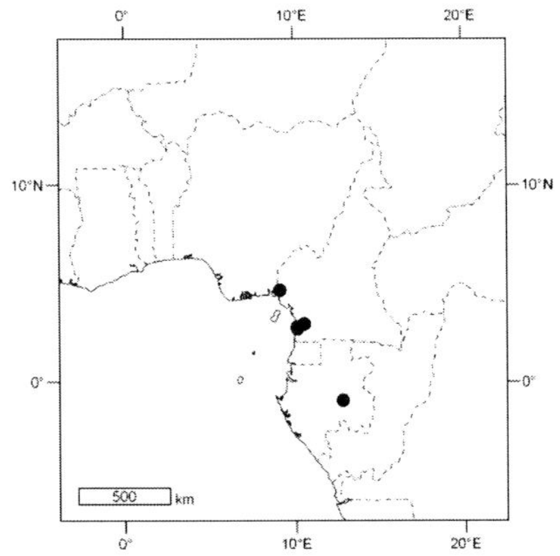

c. ***Necepsia afzelii*** subsp. ***zenkeri*** EN (Euphorbiaceae)

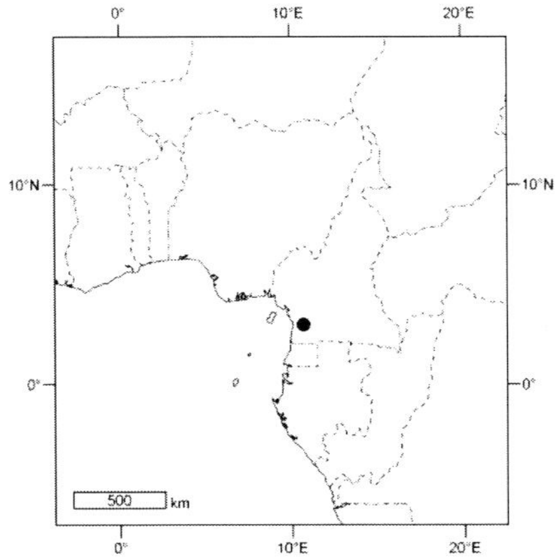

d. ***Phyllanthus aspersus*** CR (Euphorbiaceae)

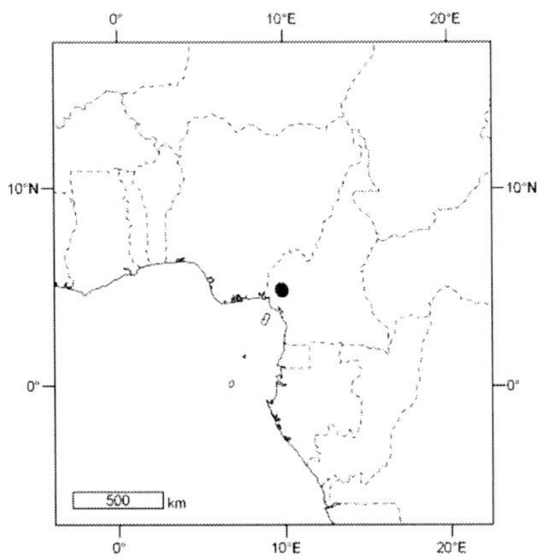

e. ***Phyllanthus caesiifolius*** CR (Euphorbiaceae)

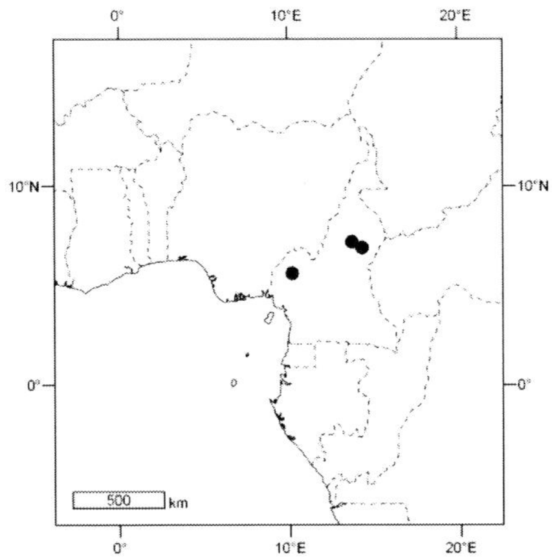

f. ***Phyllanthus caligatus*** EN (Euphorbiaceae)

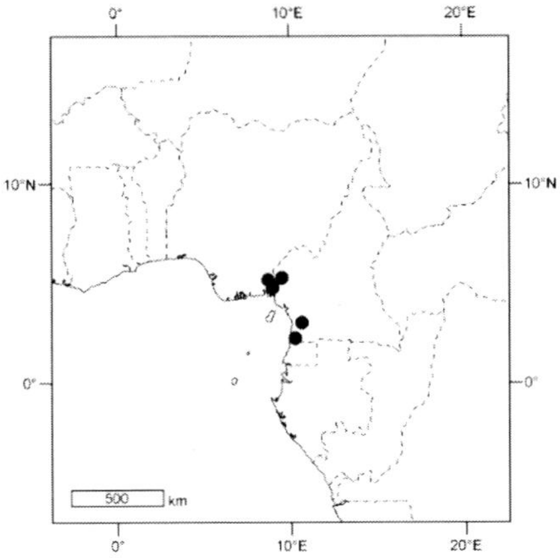

a. ***Phyllanthus dusenii*** EN (Euphorbiaceae)

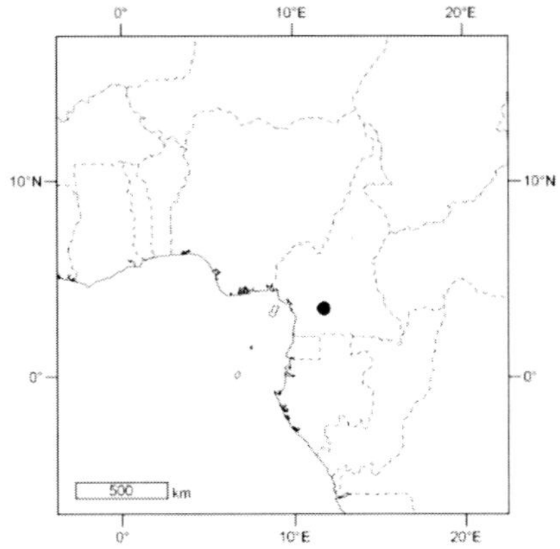

b. ***Phyllanthus kidna*** CR (Euphorbiaceae)

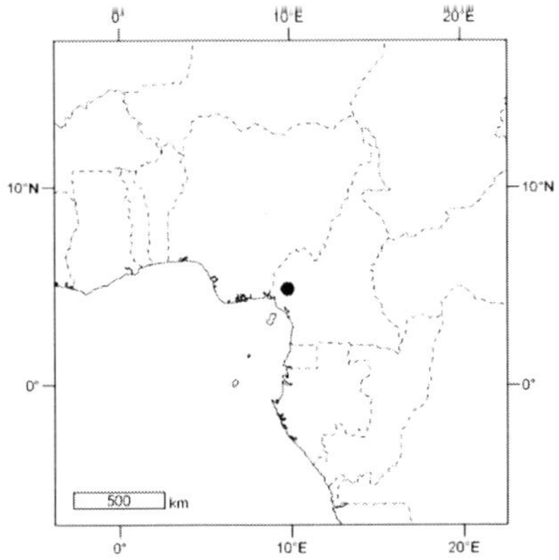

c. ***Phyllanthus nyale*** CR (Euphorbiaceae)

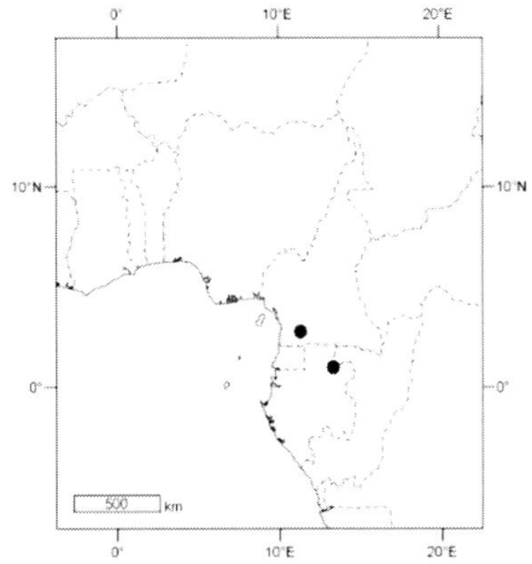

d. ***Phyllanthus raynalii*** EN (Euphorbiaceae)

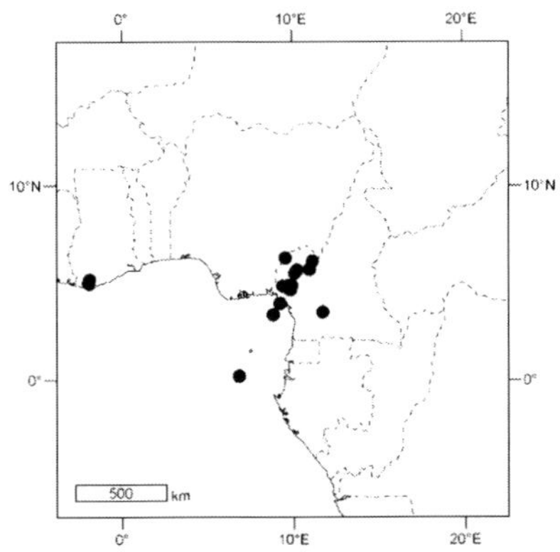

e. ***Pseudagrostistachys africana*** subsp. ***africana*** VU (Euphorbiaceae)

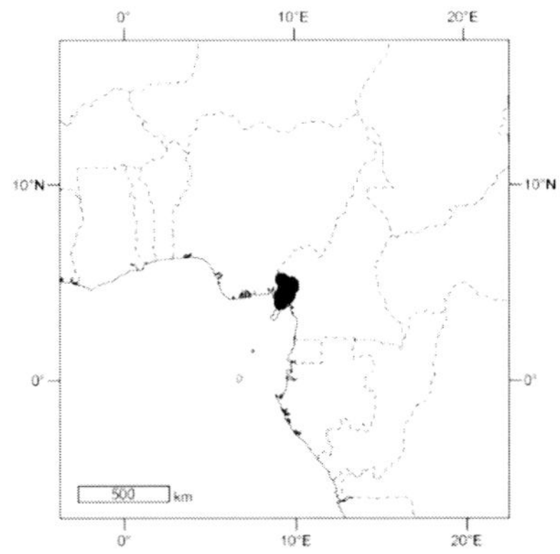

f. ***Thecacoris annobonae*** EN (Euphorbiaceae)

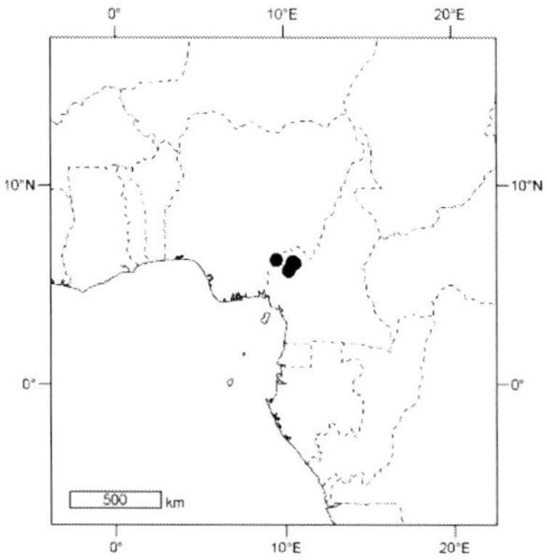

a. ***Dovyalis cameroonensis*** CR (Flacourtiaceae)

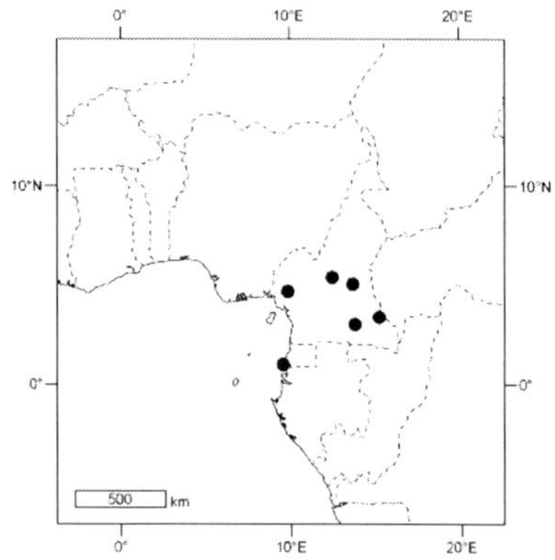

b. ***Homalium hypolasium*** VU (Flacourtiaceae)

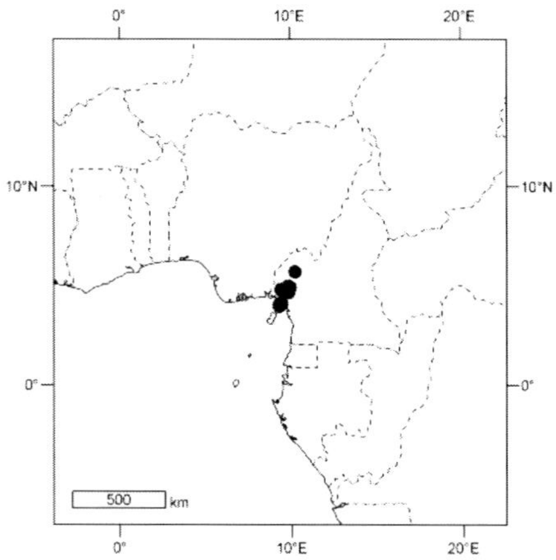

c. ***Oncoba lophocarpa*** VU (Flacourtiaceae)

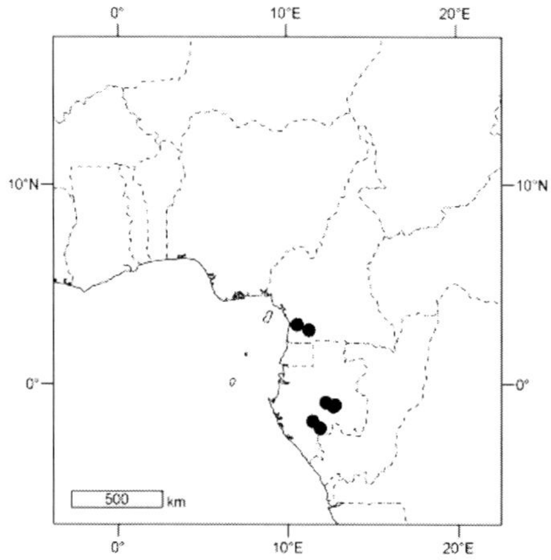

d. ***Trichostephanus acuminatus*** VU (Flacourtiaceae)

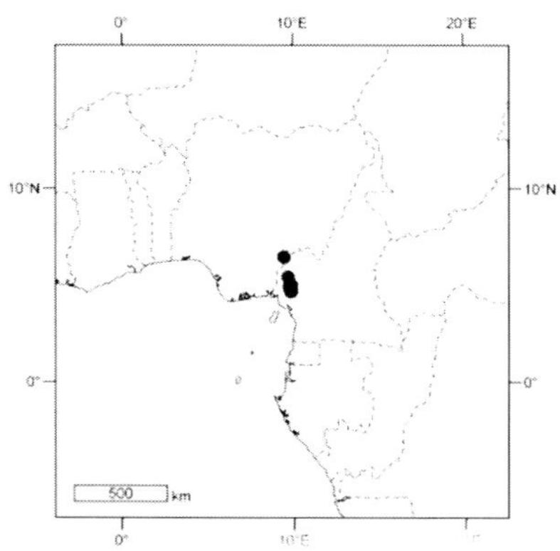

e. ***Nodonema linea*** [illegible]

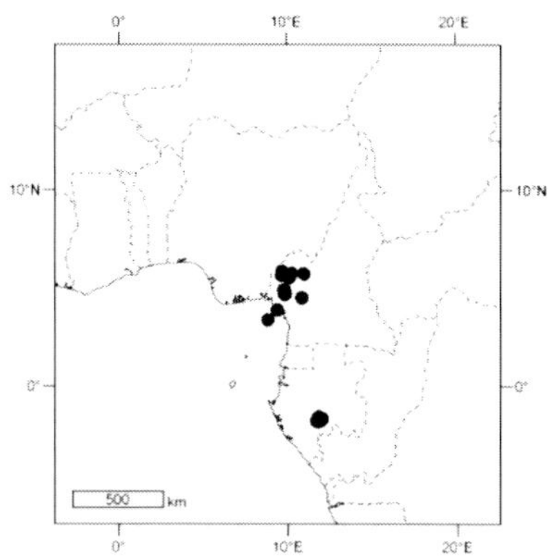

f. ***Allanblackia gabonensis*** VU (Guttiferae)

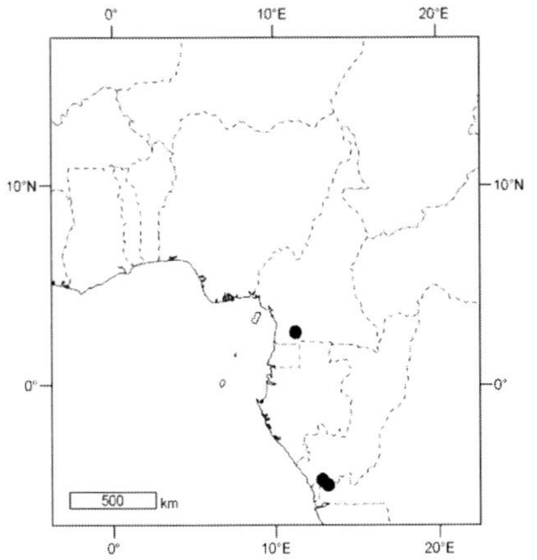

a. ***Allanblackia stanerana*** EN (Guttiferae)

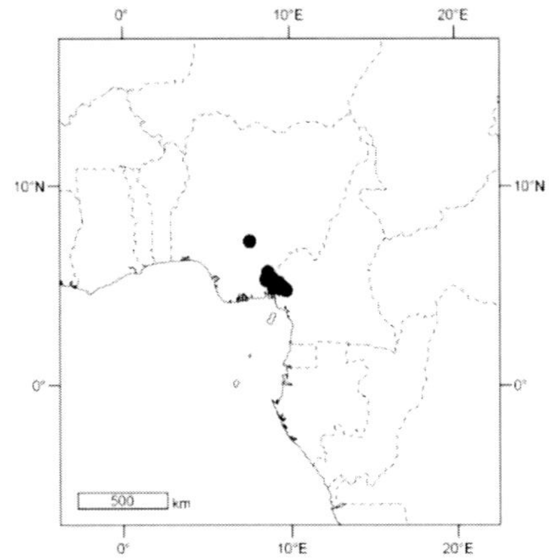

b. ***Garcinia brevipedicellata*** VU (Guttiferae)

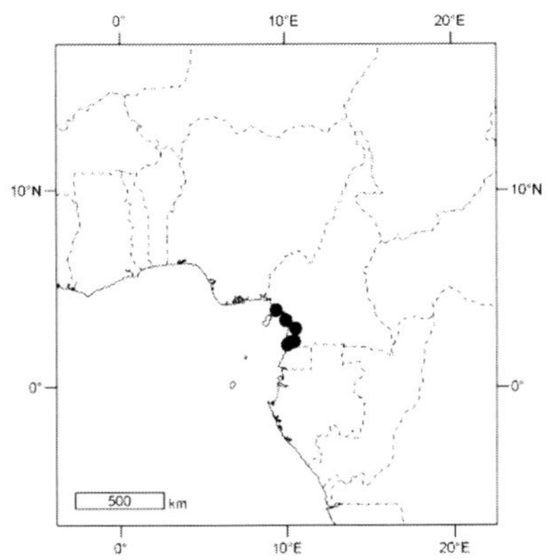

c. ***Garcinia densivenia*** EN (Guttiferae)

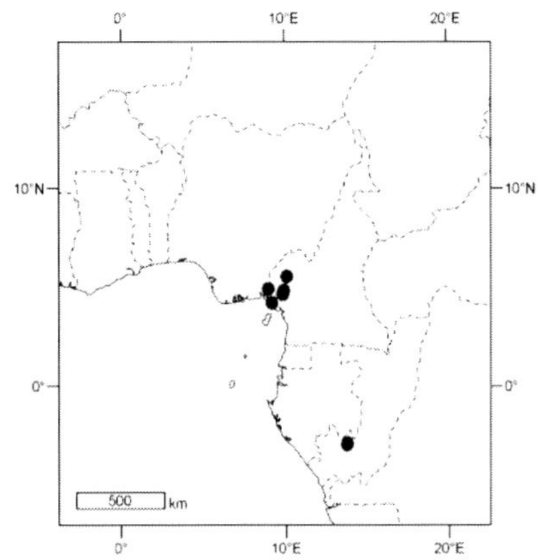

d. ***Garcinia kola*** VU (Guttiferae)

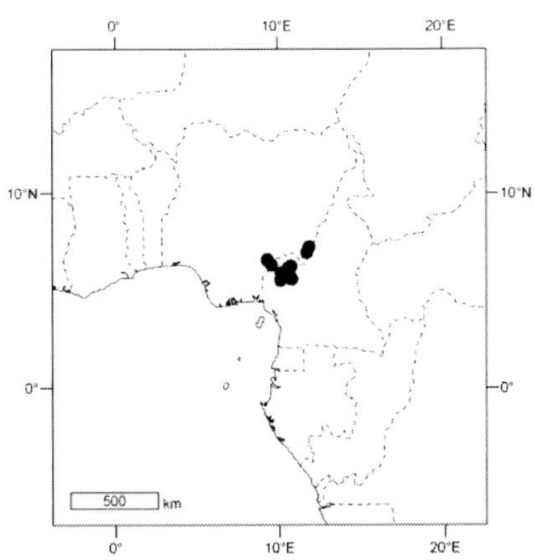

e. ***Psorospermum aurantiacum*** VU (Guttiferae)

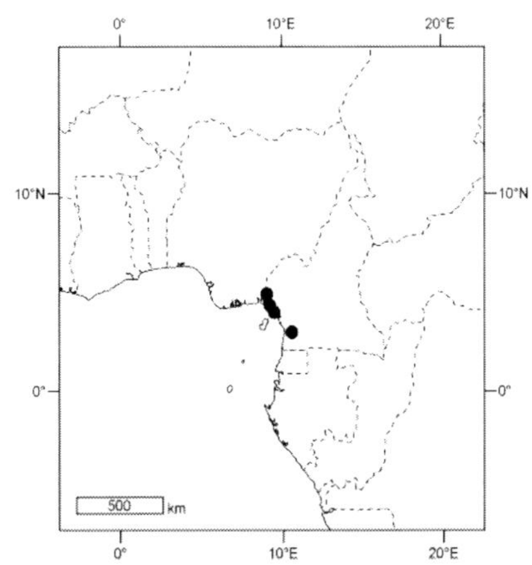

f. ***Hoplestigma pierreanum*** CR (Hoplestigmataceae)

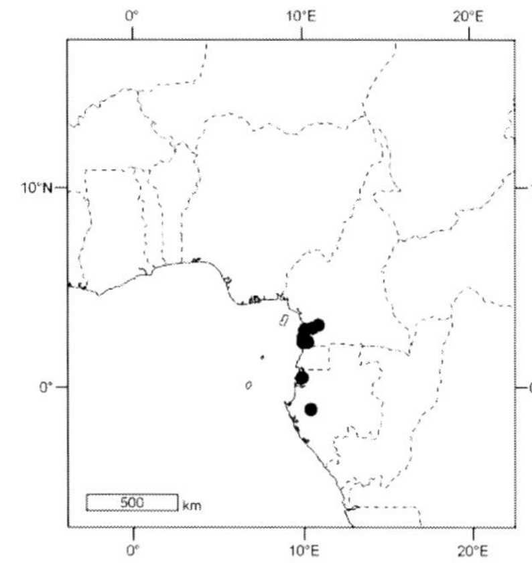

a. ***Afrostyrax macranthus*** VU (Huaceae)

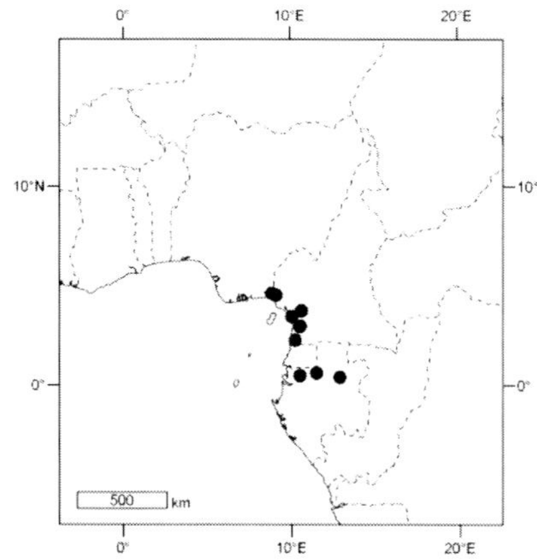

b. ***Alsodeiopsis rubra*** VU (Icacinaceae)

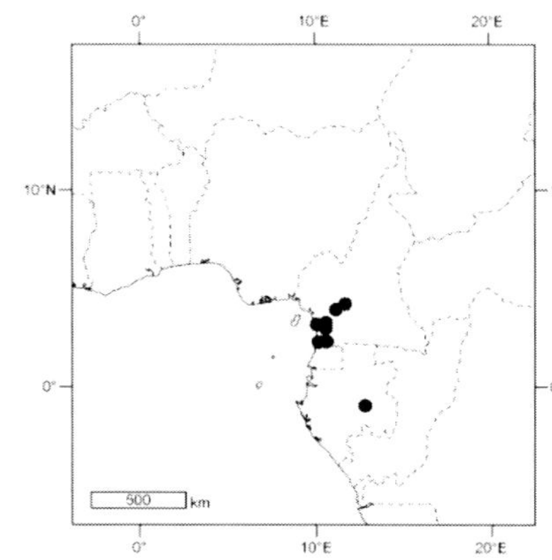

c. ***Alsodeiopsis zenkeri*** VU (Icacinaceae)

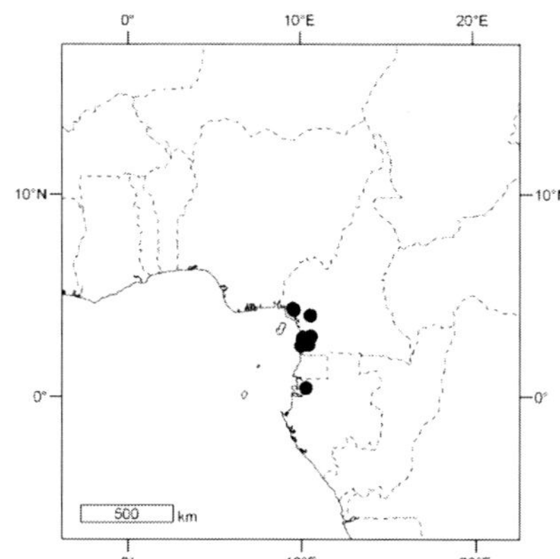

d. ***Leptaulus grandifolius*** VU (Icacinaceae)

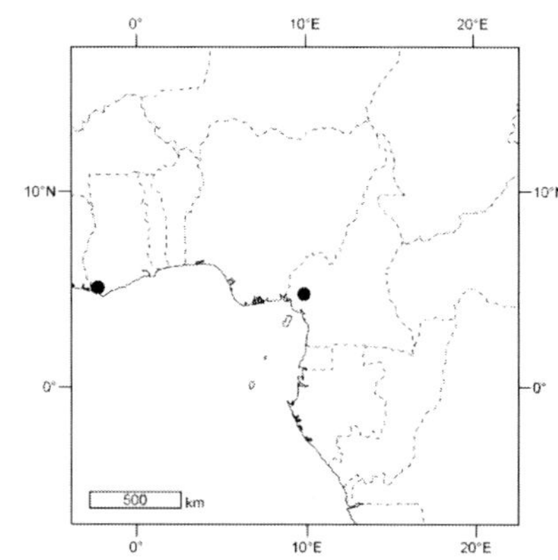

e. ***Pyrenacantha cordicula*** EN (Icacinaceae)

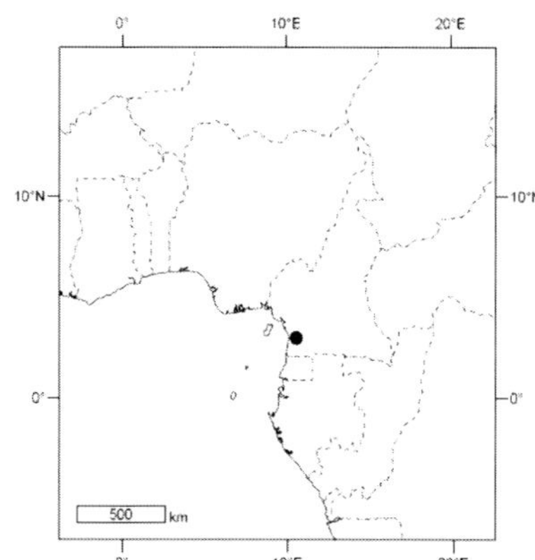

f. ***Pyrenacantha grandifolia*** CR (Icacinaceae)

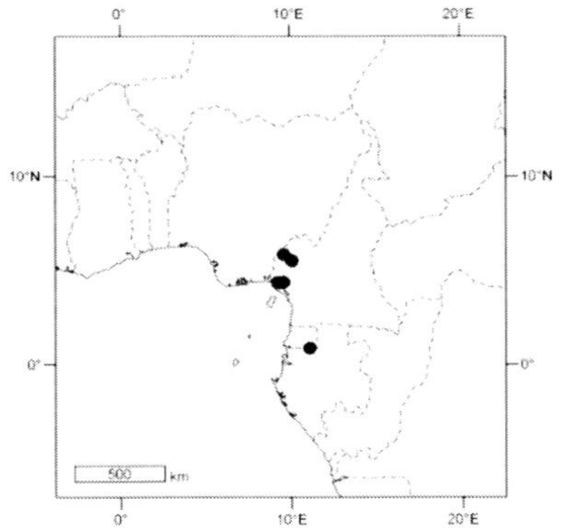

a. ***Pyrenacantha longirostrata*** EN (Icacinaceae)

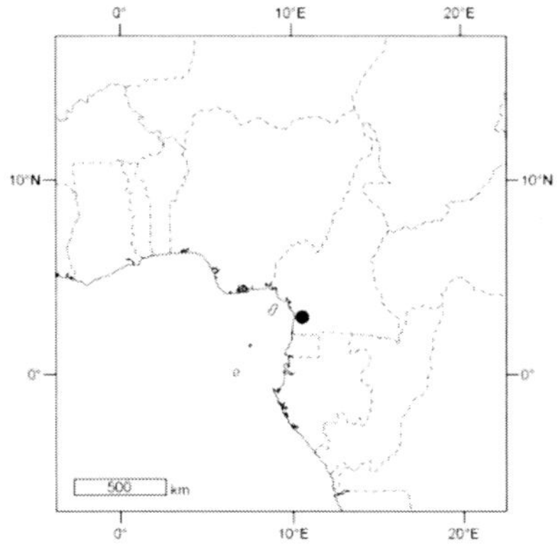

b. ***Rhaphiostylis elegans*** CR (Icacinaceae)

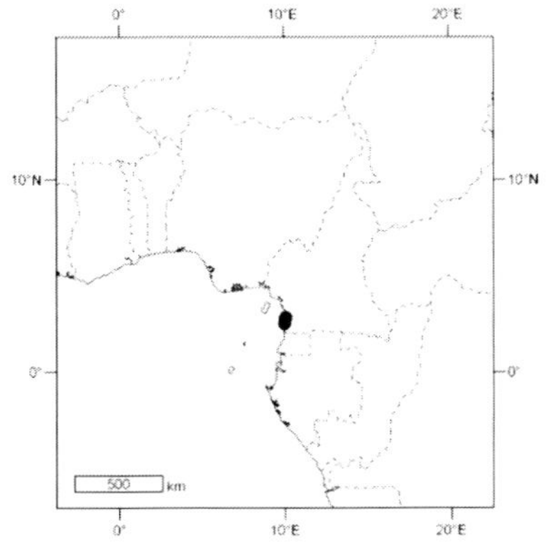

c. ***Rhaphiostylis ovatifolia*** EN (Icacinaceae)

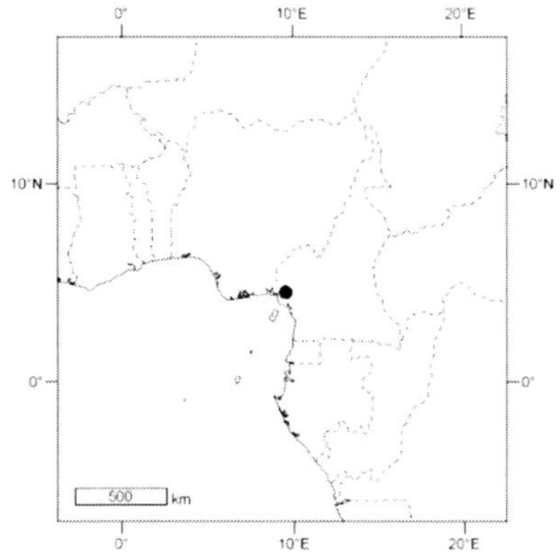

d. ***Rhaphiostylis poggei*** CR (Icacinaceae)

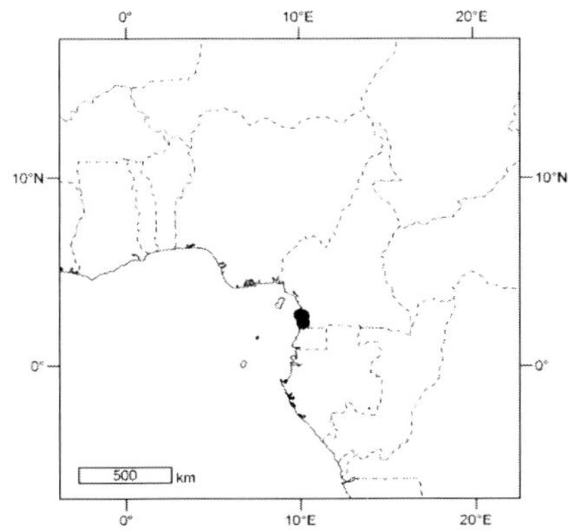

e. ***Rhaphiostylis subsessifolia*** EN (Icacinaceae)

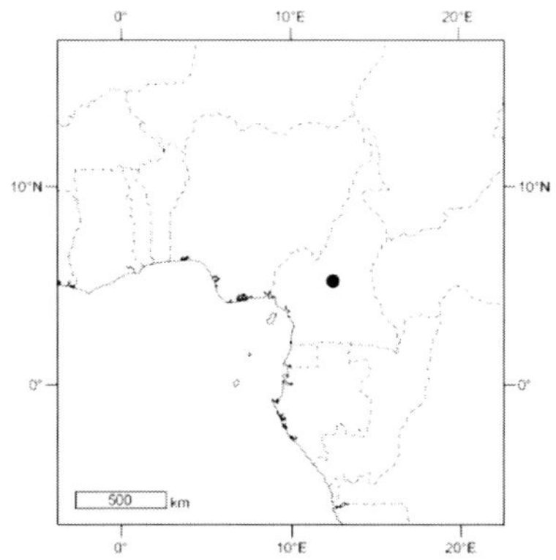

f. ***Stachyanthus cuneatus*** CR (Icacinaceae)

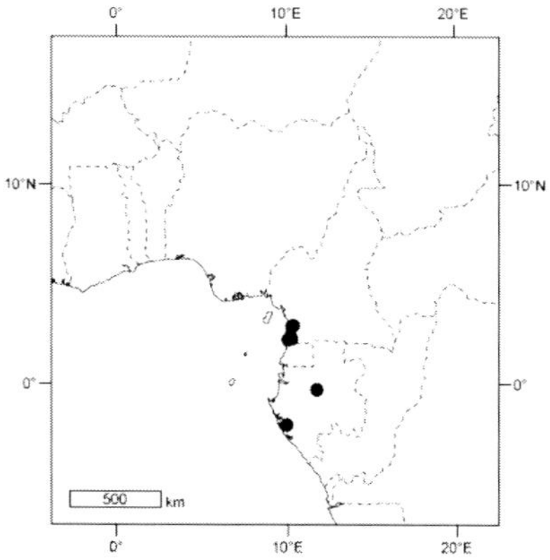

a. ***Ochthocosmus calothyrsus*** VU (Ixonanthaceae)

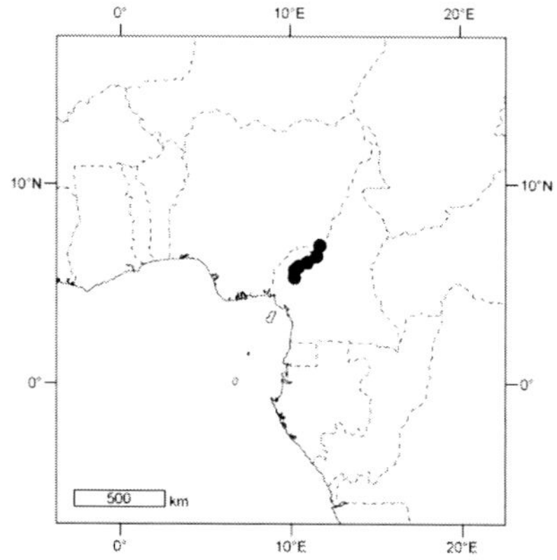

b. ***Aeollanthus trifidus*** VU (Labiatae)

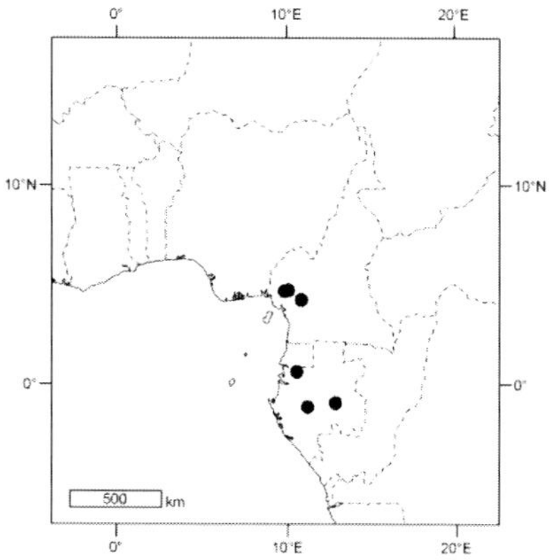

c. ***Clerodendrum anomalum*** EN (Labiatae)

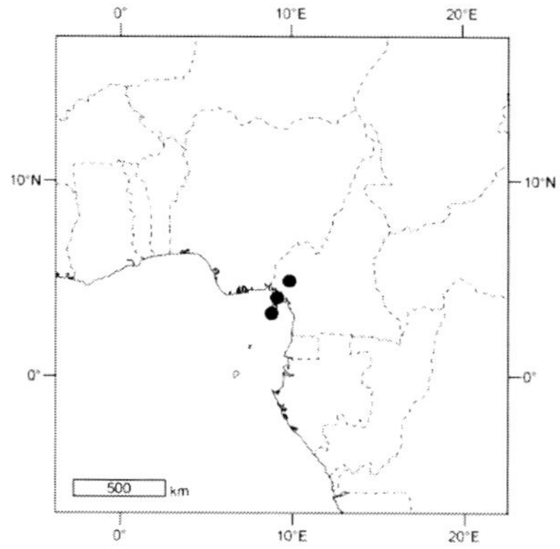

d. ***Plectranthus cataractarum*** EN (Labiatae)

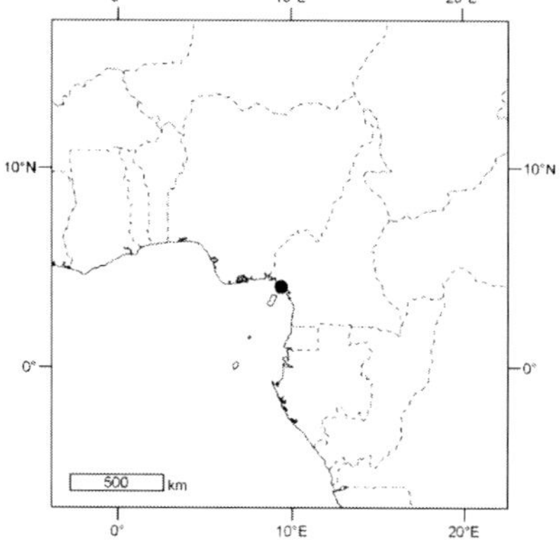

e. ***Plectranthus dissitiflorus*** CR (Labiatae)

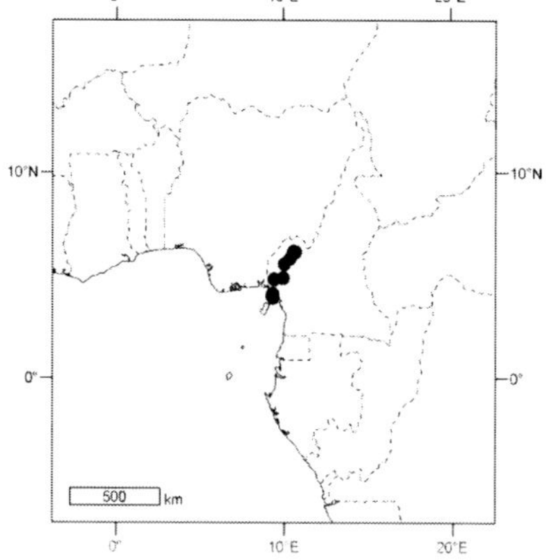

f. ***Plectranthus insignis*** VU (Labiatae)

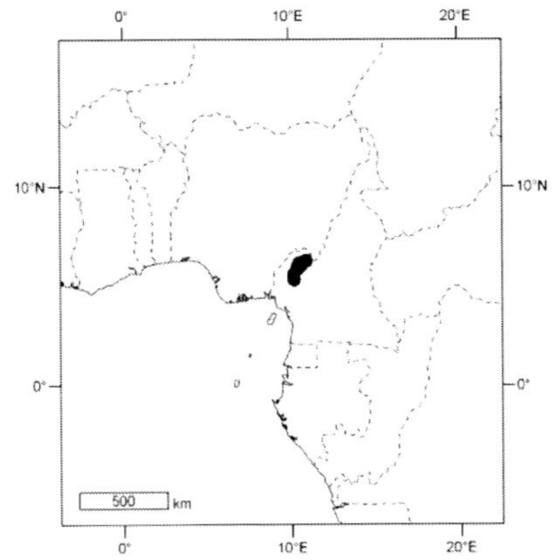

a. ***Plectranthus punctatus*** subsp. ***lanatus*** VU (Labiatae)

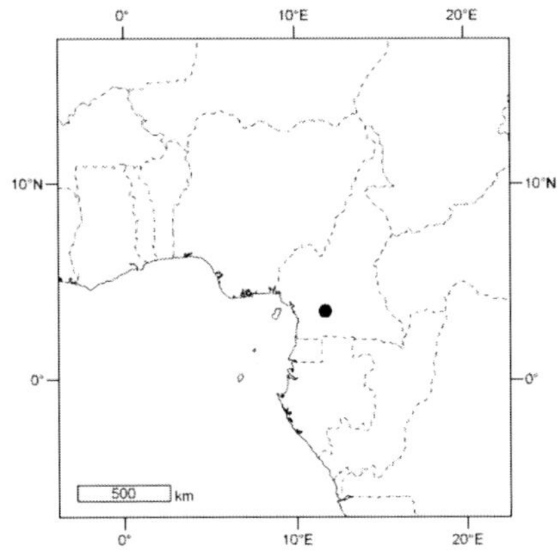

b. ***Pogostemon micangensis*** CR (Labiatae)

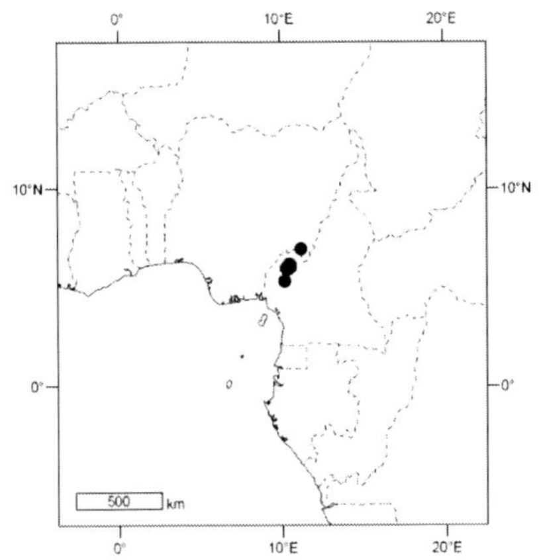

c. ***Stachys pseudohumifusa*** subsp. ***saxeri*** EN (Labiatae)

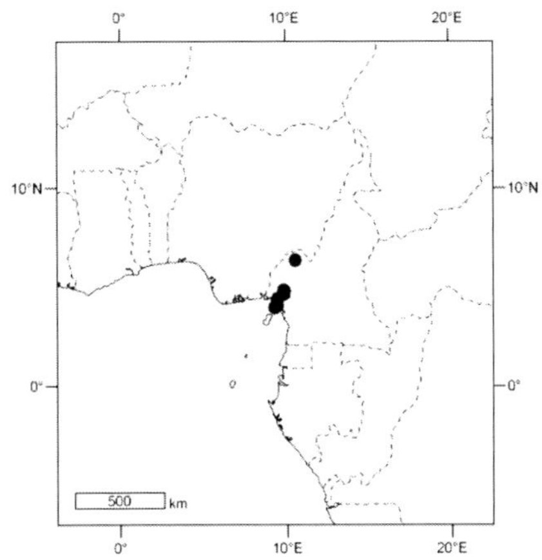

d. ***Vitex lehmbachii*** EN (Labiatae)

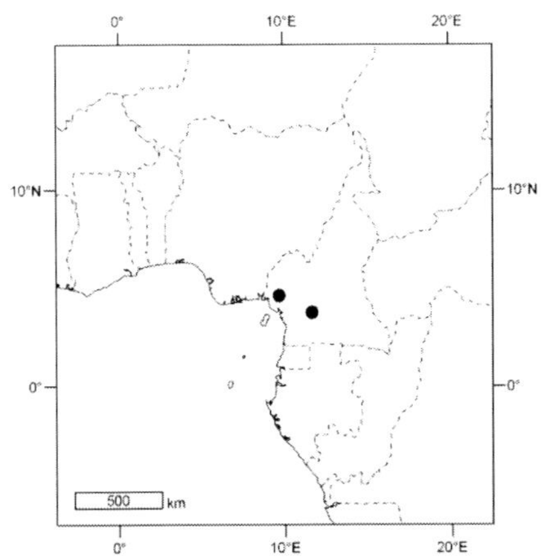

e. ***Vitex yaundensis*** CR (Labiatae)

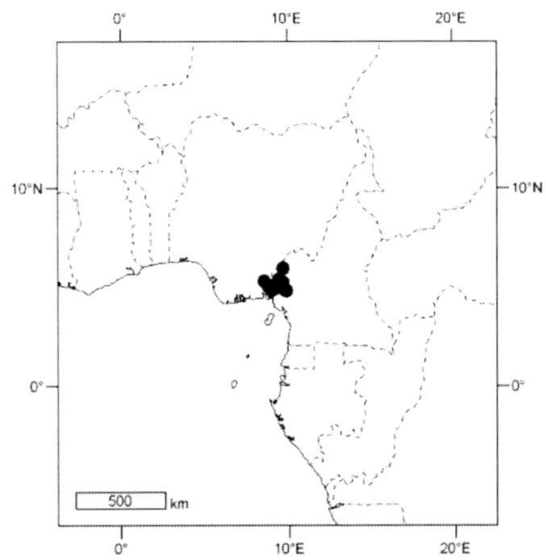

f. . ***Crateranthus talbotii*** VU (Lecythidaceae)

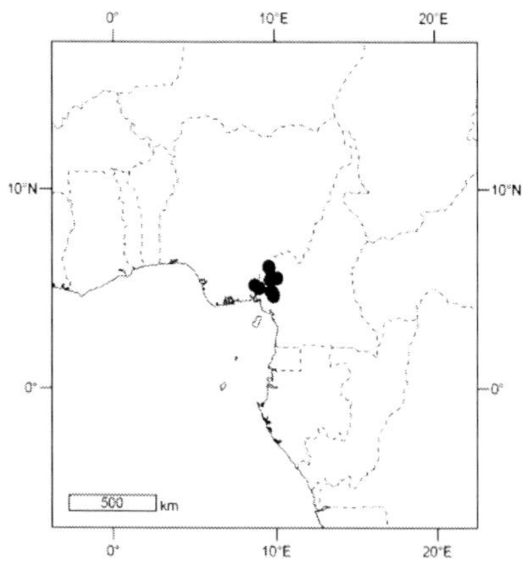

a. ***Napoleonaea egertonii*** VU (Lecythidaceae)

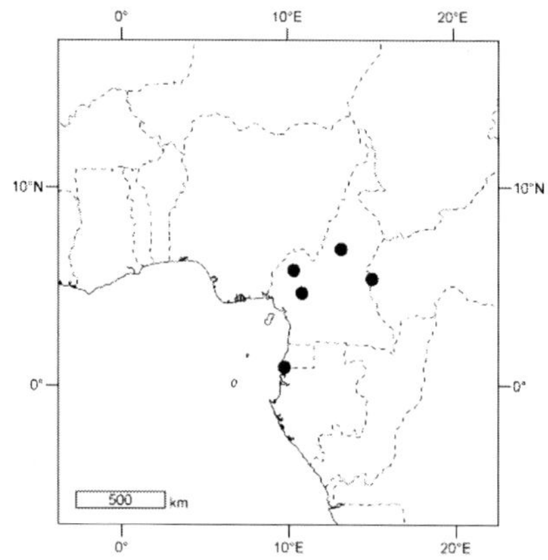

b. ***Afzelia africana*** VU (Leguminosae-Caesalpinioidae)

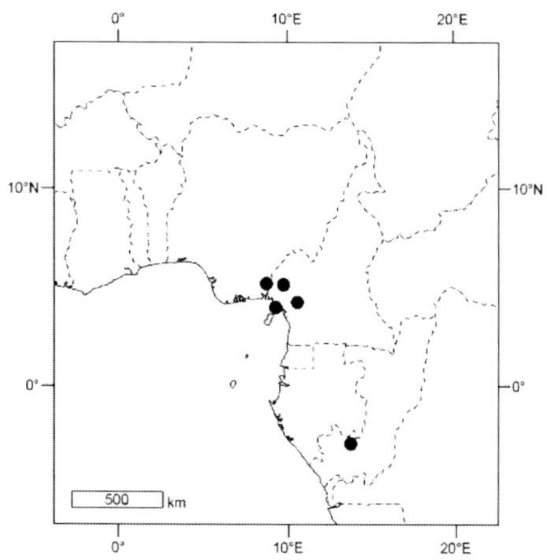

c. ***Afzelia bipindensis*** VU (Leguminosae-Caesalpinioidae)

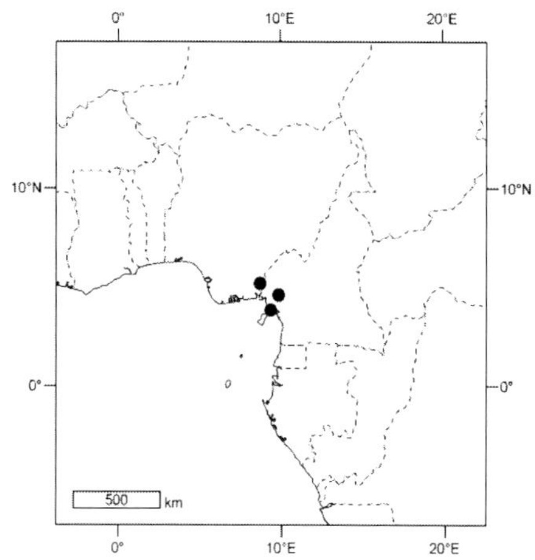

d. ***Afzelia pachyloba*** VU (Leguminosae-Caesalpinioidae)

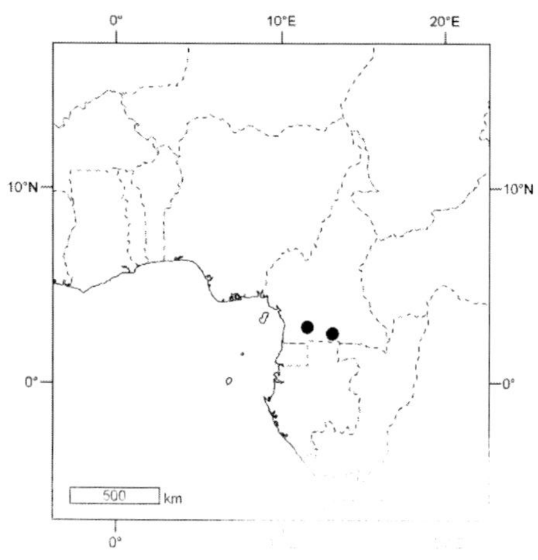

e. ***Anthonotha wij***[illegible] (Leguminosae- Caes[illegible]idae)

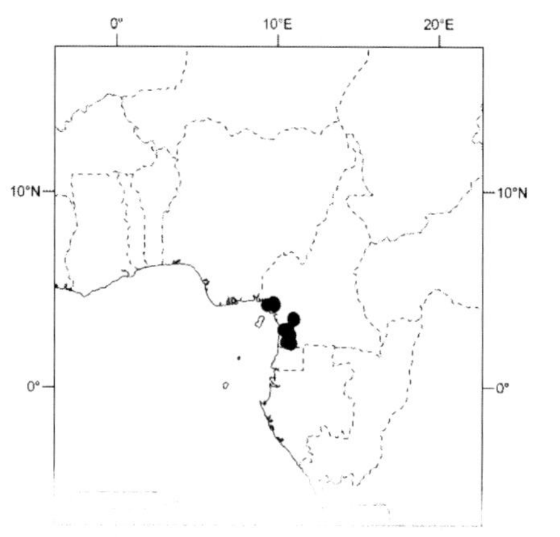

[illegible]sae-Caesalpinioidae)

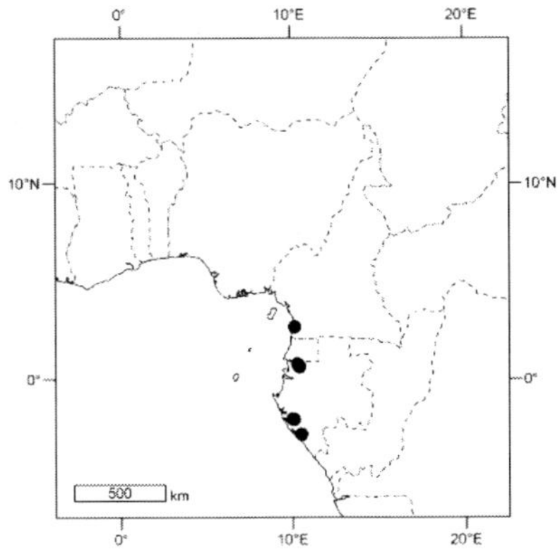

a. ***Berlinia immaculata*** VU (Leguminosae-Caesalpinioidae)

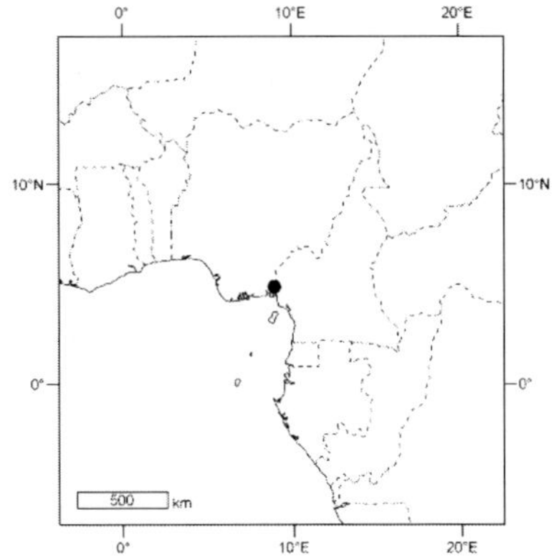

b. ***Berlinia korupensis*** CR (Leguminosae-Caesalpinioidae)

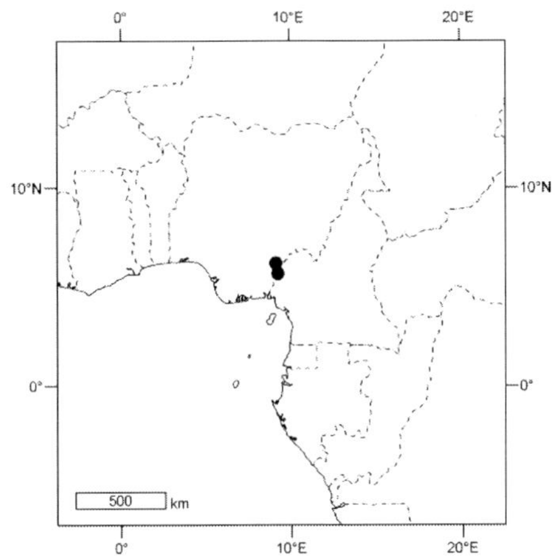

c. ***Brachystegia kennedyi*** EN (Leguminosae-Caesalpinioidae)

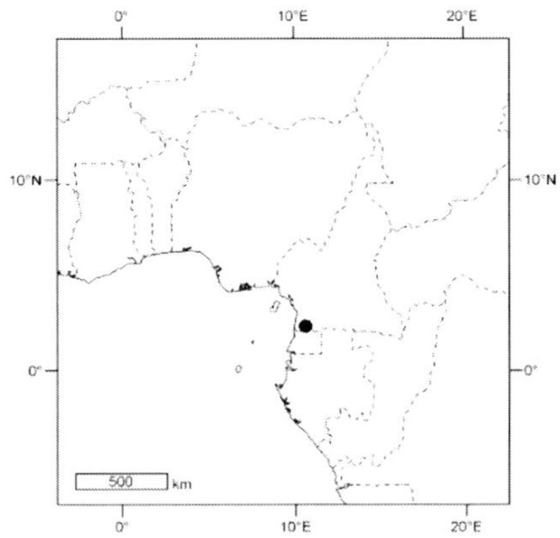

d. ***Cryptosepalum ambamense*** CR (Leguminosae- Caesalpinioidae)

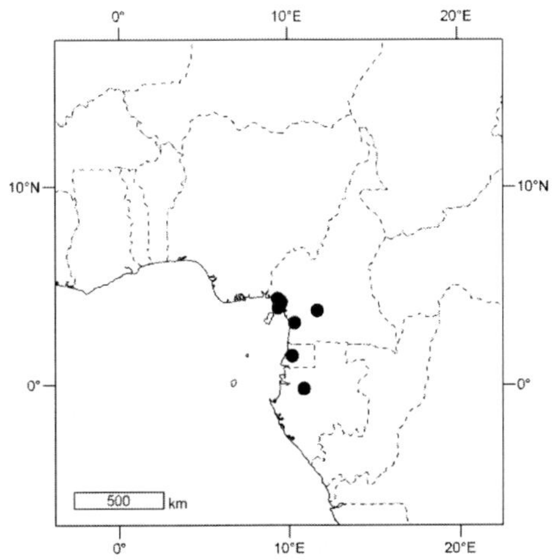

e. ***Daniellia oblonga*** CR (Leguminosae-Caesalpinioidae)

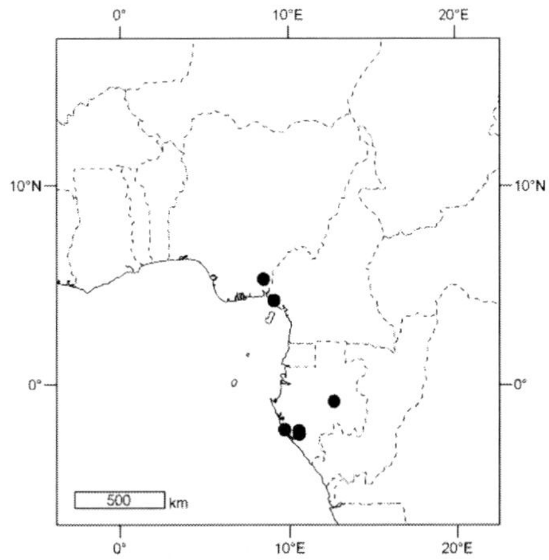

f. ***Englerodendron conchyliophorum*** EN (Leguminosae- Caesalpinioidae)

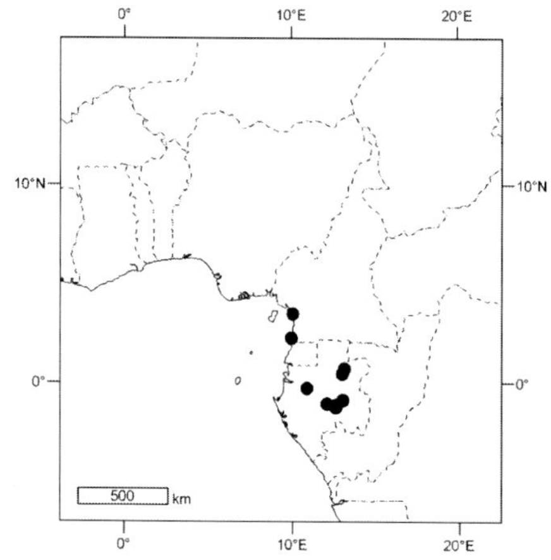

a. ***Englerodendron gabunense*** VU (Leguminosae- Caesalpinioidae)

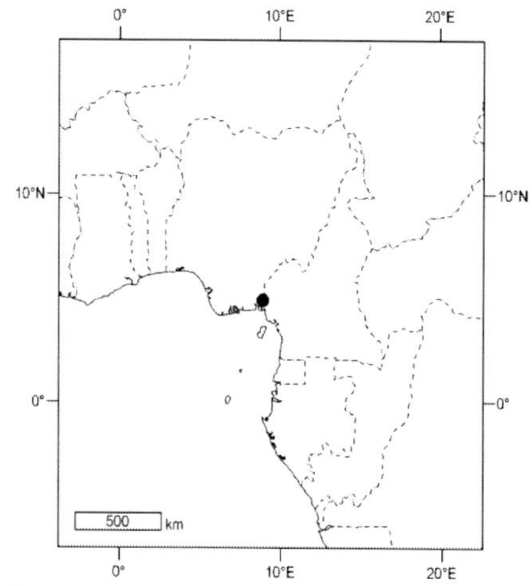

b. ***Englerodendron korupense*** CR (Leguminosae- Caesalpinioidae)

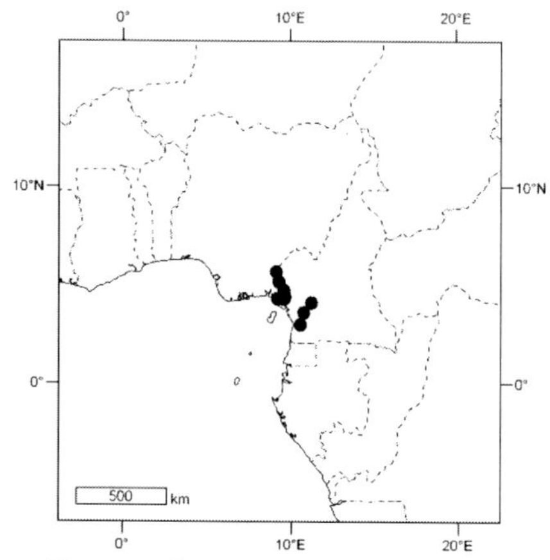

c. ***Eurypetalum unijugum*** VU (Leguminosae- Caesalpinioidae)

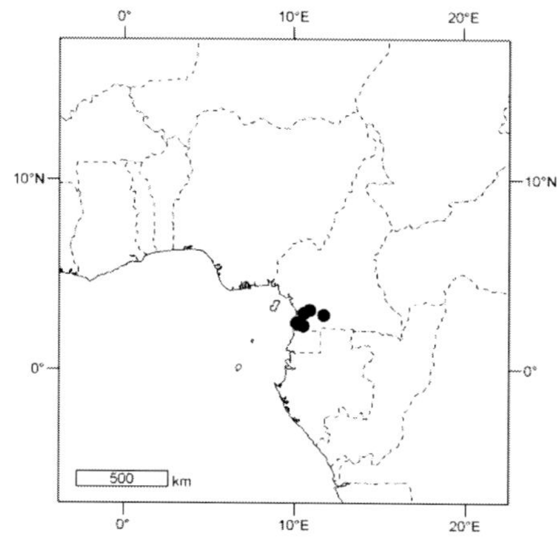

d. ***Gilbertiodendron pachyanthum*** EN (Leguminosae- Caesalpinioidae)

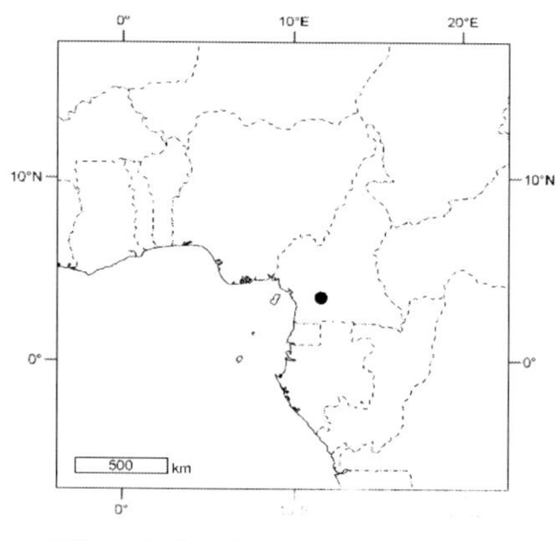

e. ***Gilbertiodendron*** [illegible] (Leguminosae- Caesa[illegible]

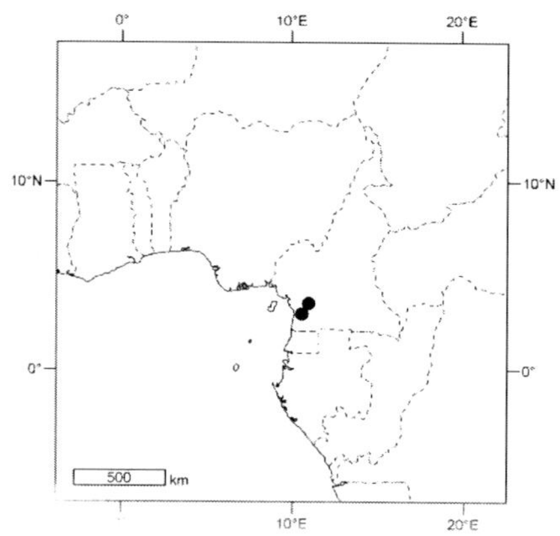

[illegible]***odendron zenkeri*** EN (Leguminosae- [illegible]oidae)

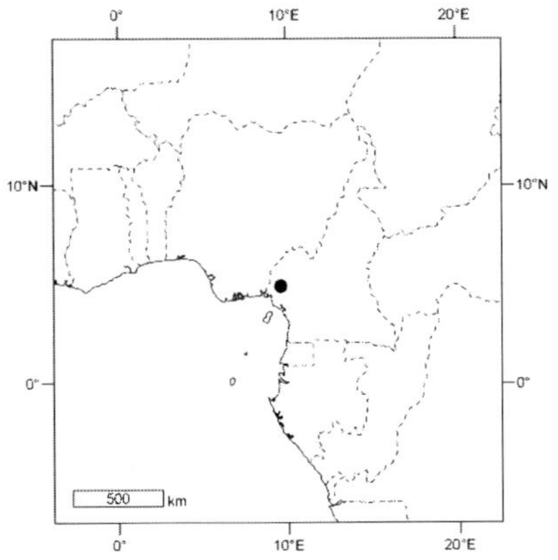

a. ***Gossweilerodendron joveri*** VU (Leguminosae- Caesalpinioidae)

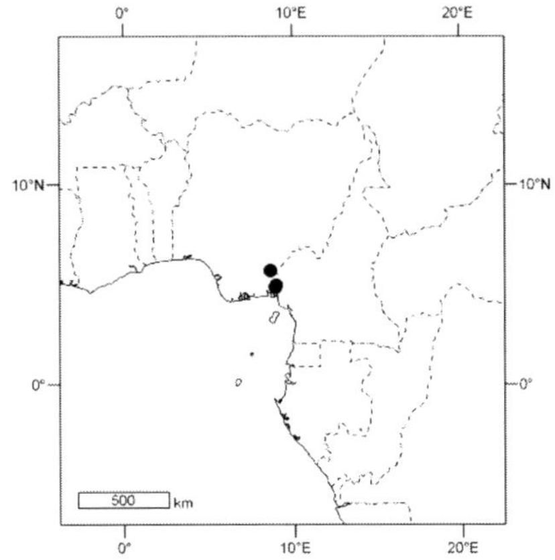

b. ***Hymenostegia bakeriana*** VU (Leguminosae- Caesalpinioidae)

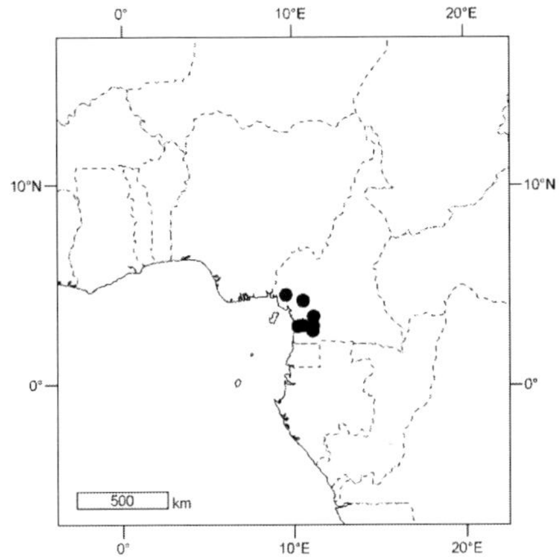

c. ***Hymenostegia brachyura*** VU (Leguminosae-Caesalpinoidae)

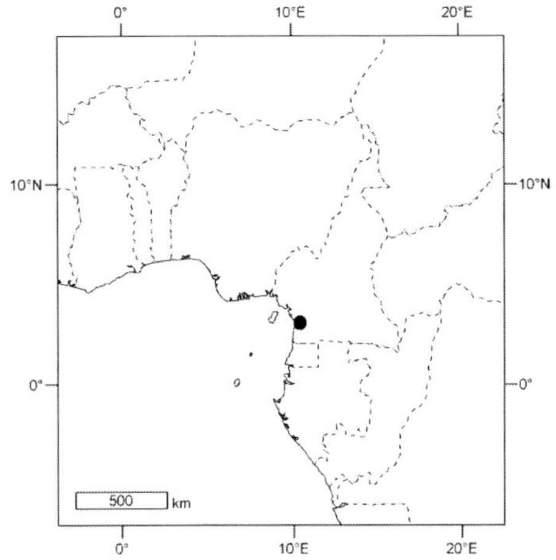

d. ***Julbernardia letouzeyi*** CR (Leguminosae-Caesalpinoidae)

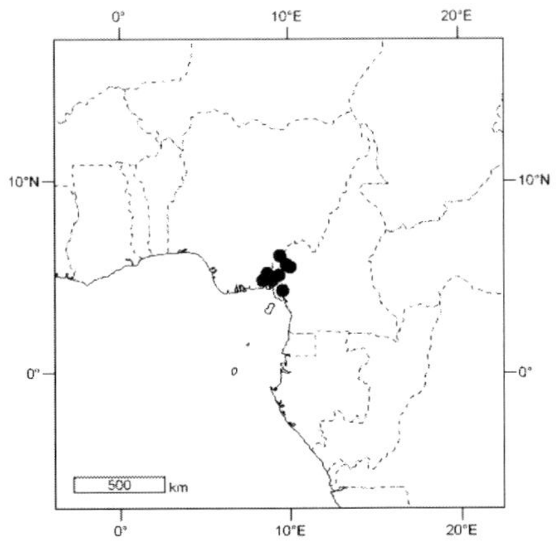

e. ***Leonardoxa africana*** subsp. ***letouzeyi*** EN (Leguminosae- Caesalpinoidae)

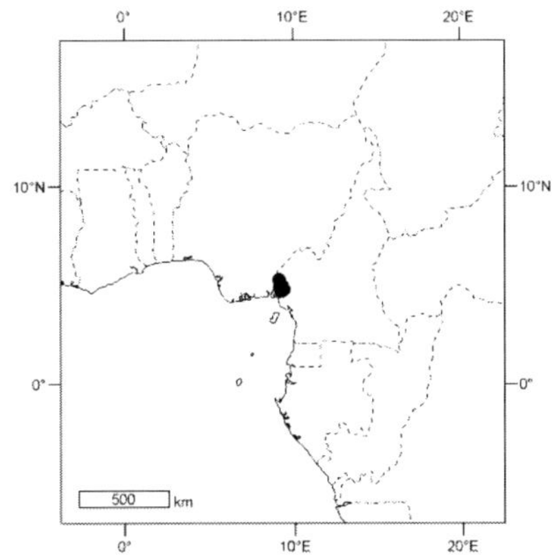

f. ***Leonardoxa africana*** subsp. ***rumpiensis*** EN (Leguminosae- Caesalpinoidae)

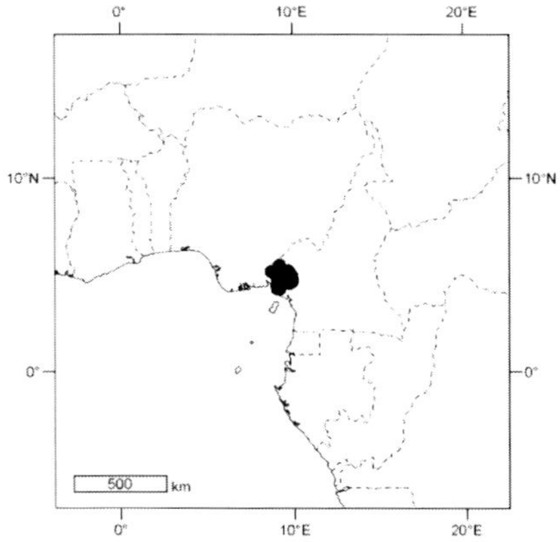

a. ***Loesenera talbotii*** VU (Leguminosae-Caesalpinoidae)

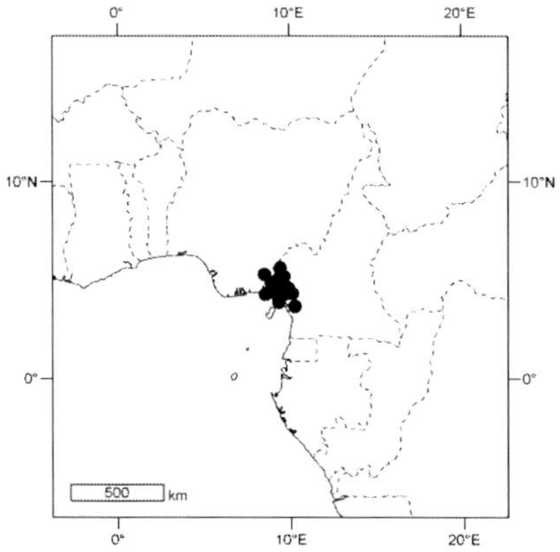

b. ***Microberlinia bisulcata*** VU(Leguminosae-Caesalpinoidae)

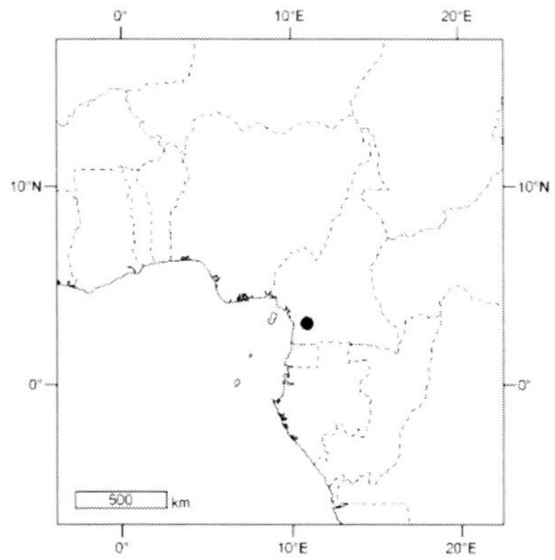

c. ***Plagiosiphon discifer*** (Leguminosae-Caesalpinioidae)

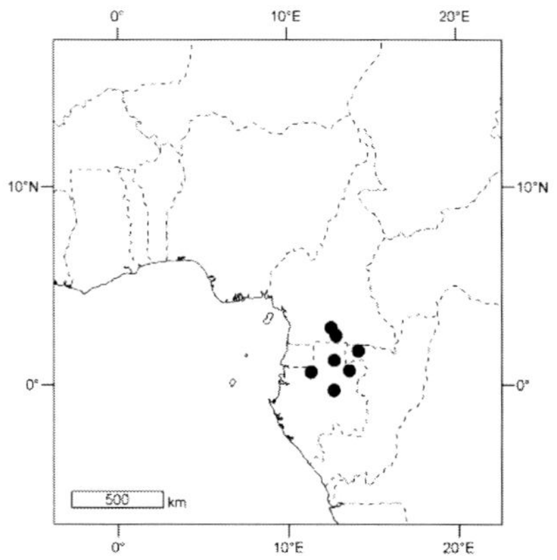

d. ***Talbotiella batesii*** CR (Leguminosae-Caesalpinioidae

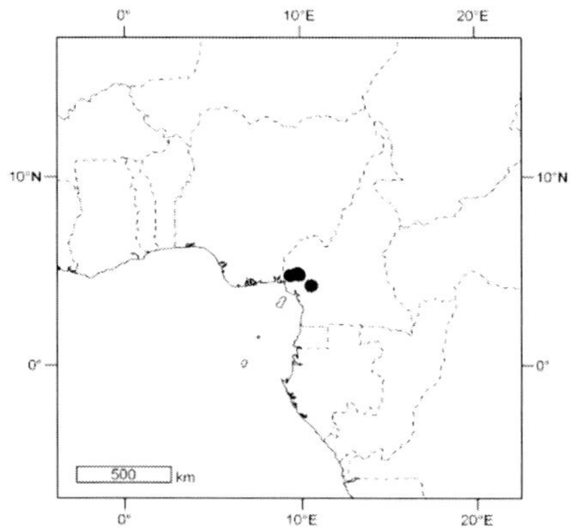

e. ***Talbotiella bakossiensis*** CR (Leguminosae-Caesalpinioidae)

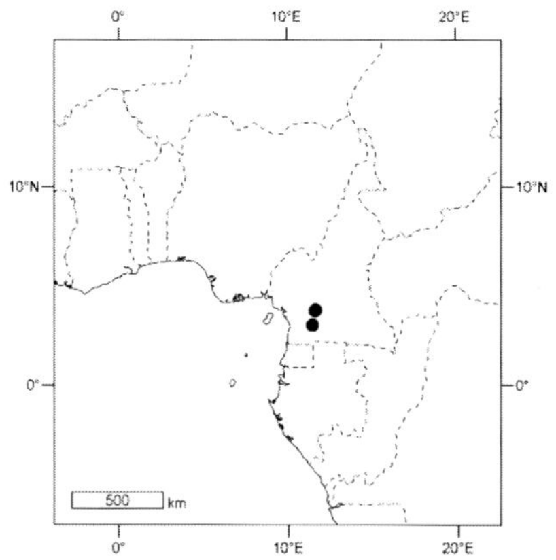

f. ***Talbotiella breteleri*** CR (Leguminosae-Caesalpinioidae)

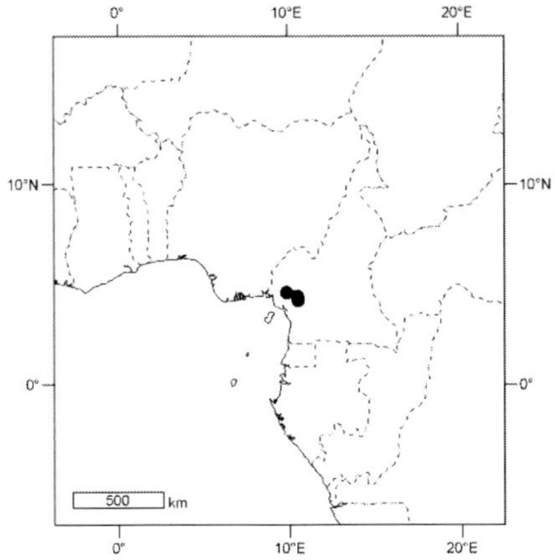

a. ***Talbotiella ebo*** CR (Leguminosae-Caesalpinioidae)

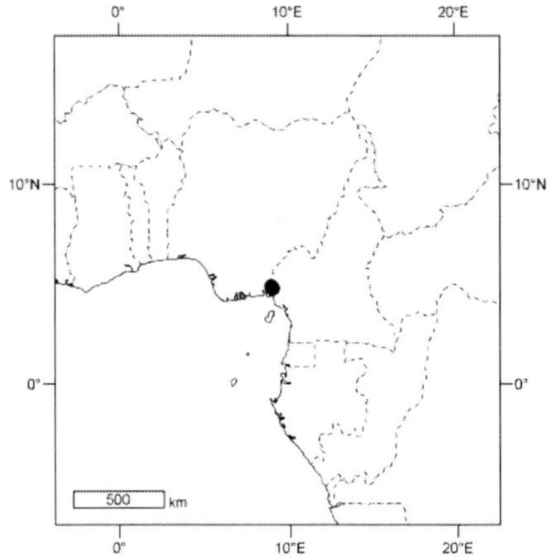

b. ***Talbotiella korupensis*** EN (Leguminosae-Caesalpinioidae)

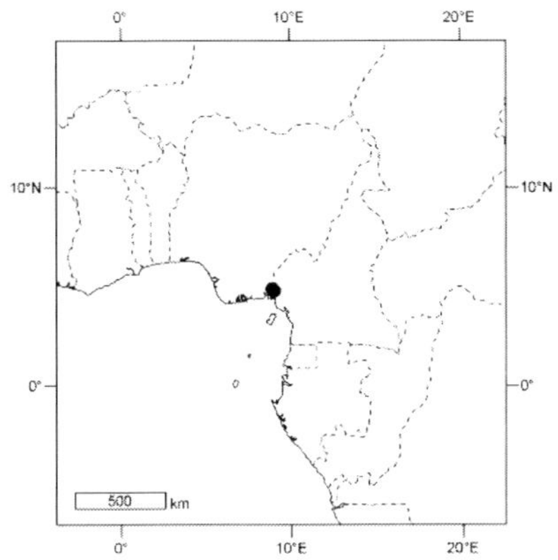

c. ***Talbotiella velutina*** CR (Leguminosae-Caesalpinioidae)

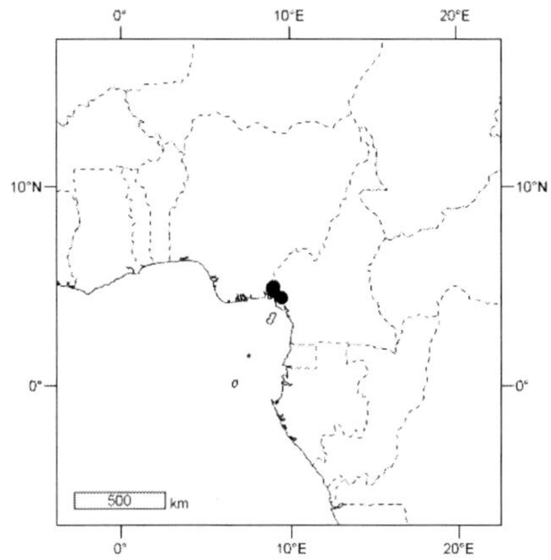

d. ***Tetraberlinia korupensis*** EN (Leguminosae- Caesalpinioidae)

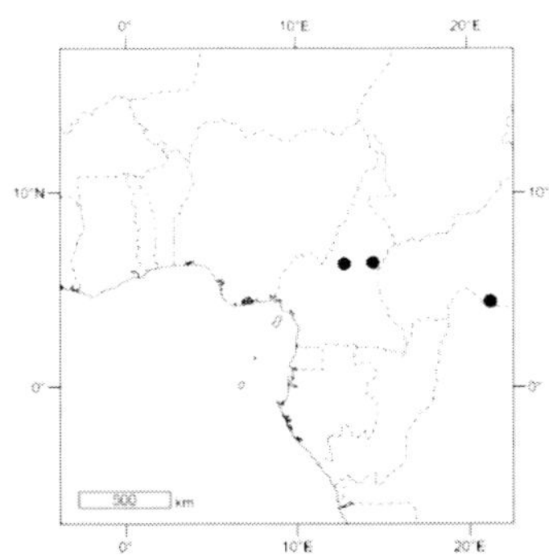

e. ***Albizia eriorhachis*** EN (Leguminosae-Mimosoideae)

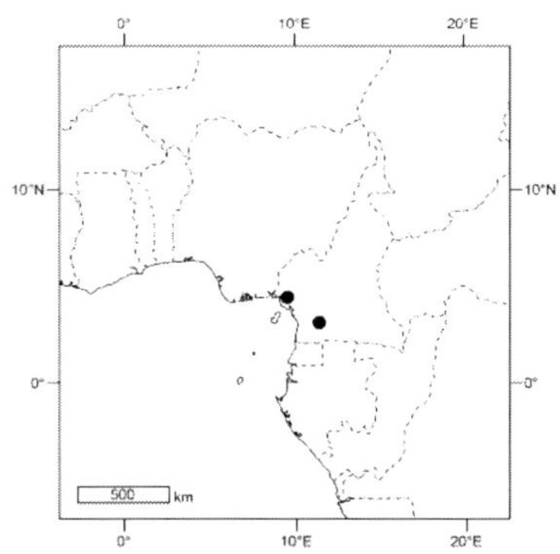

f. ***Calpocalyx atlanticus*** EN (Leguminosae-Mimosoideae)

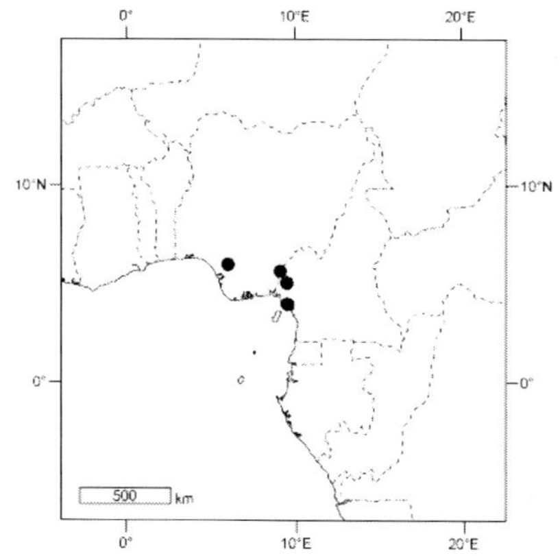

a. ***Calpocalyx winkleri*** EN (Leguminosae-Mimosoideae)

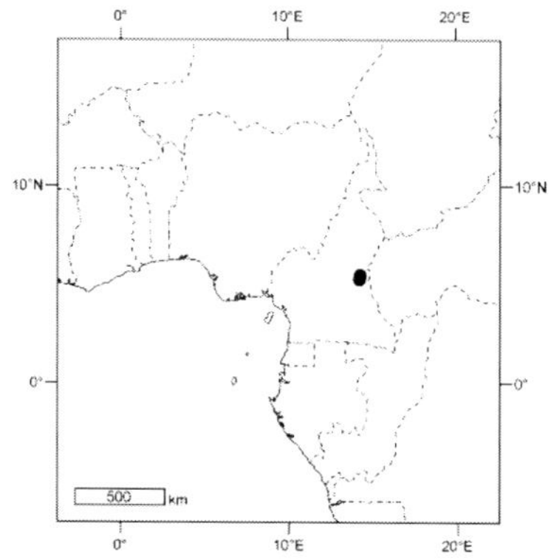

b. ***Entada camerunensis*** EN (Leguminosae-Mimosoideae)

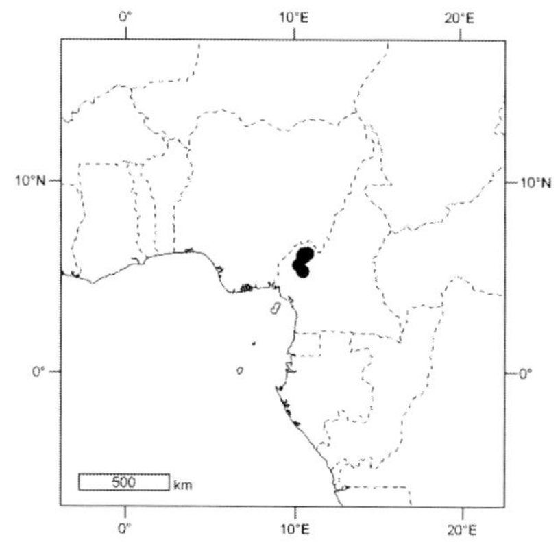

c. ***Newtonia camerunensis*** CR (Leguminosae-Mimosoideae)

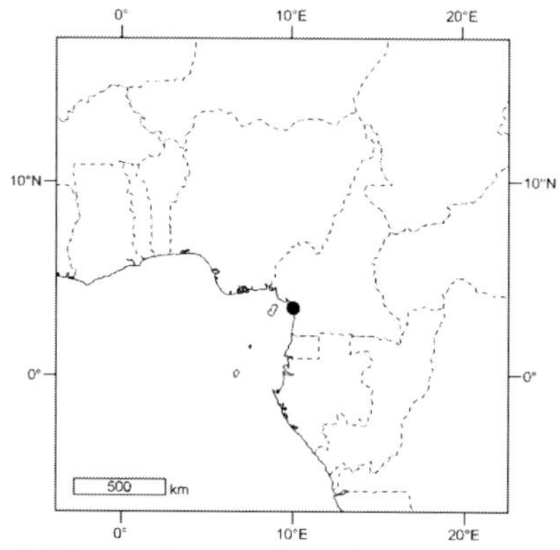

d. ***Newtonia scandens*** CR (Leguminosae-Mimosoideae)

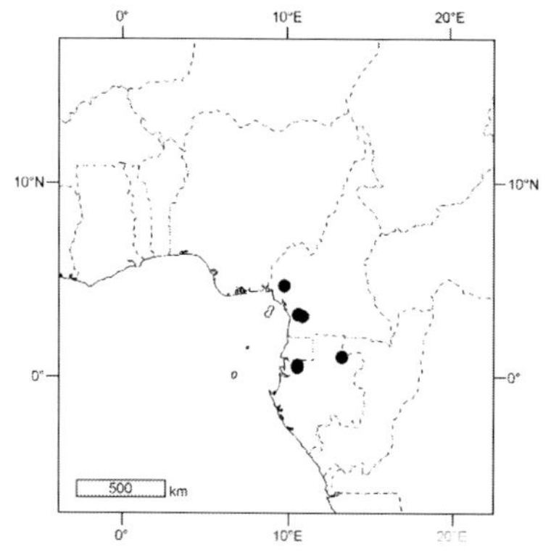

e. ***Baphia breteleriana*** EN (Leguminosae-Papilionoideae)

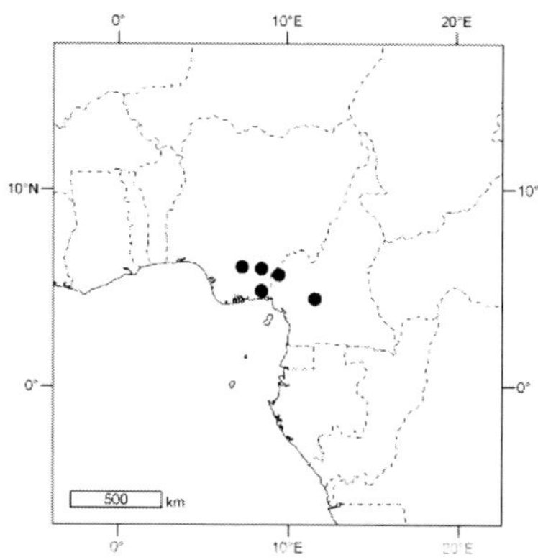

f. ***Baphia dewildeana*** VU (Leguminosae-Papilionoideae)

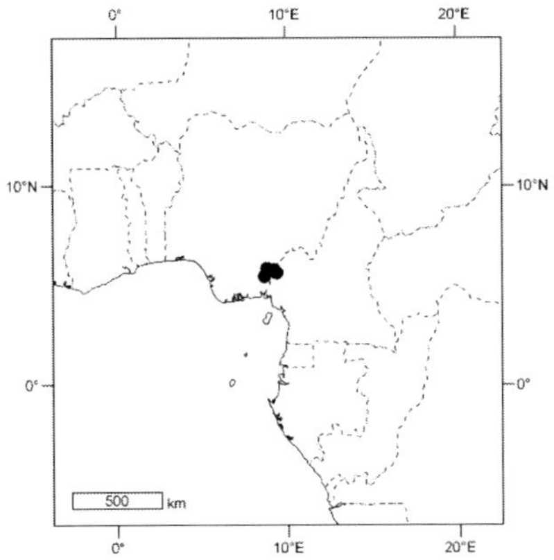

a. ***Baphia latiloi*** EN (Leguminosae-Papilionoideae)

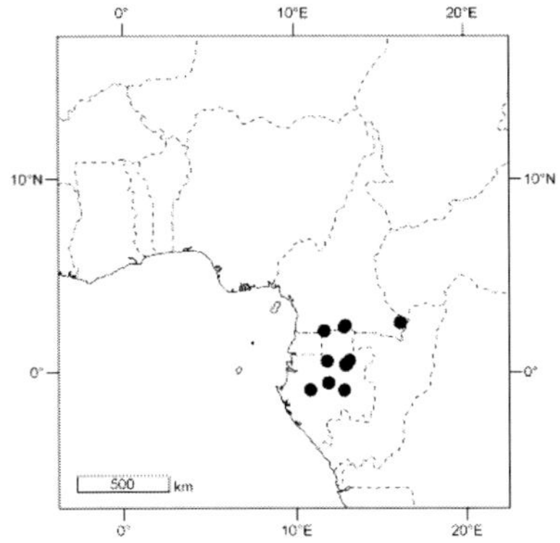

b. ***Baphia leptobotrys*** subsp. ***silvatica*** VU (Leguminosae- Papilionoideae)

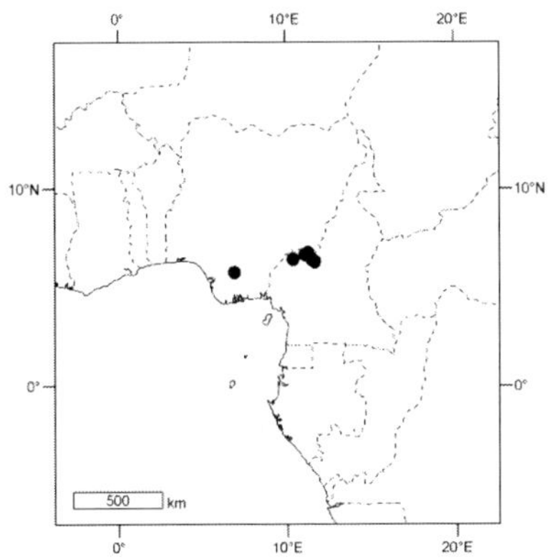

c. ***Baphia mambillensis*** EN (Leguminosae-Papilionoideae)

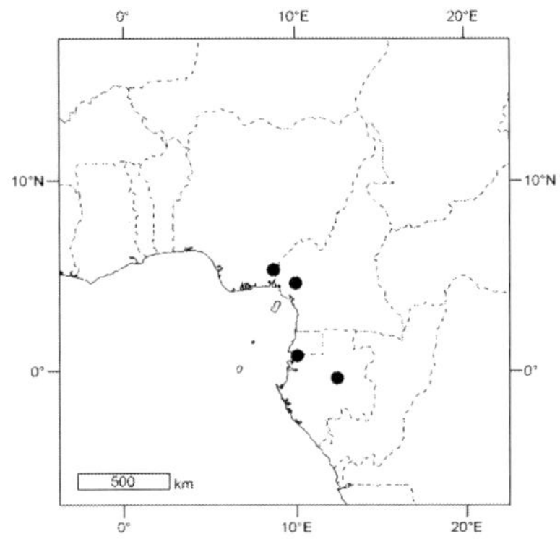

d. ***Baphia obanensis*** EN (Leguminosae-Papilionoideae)

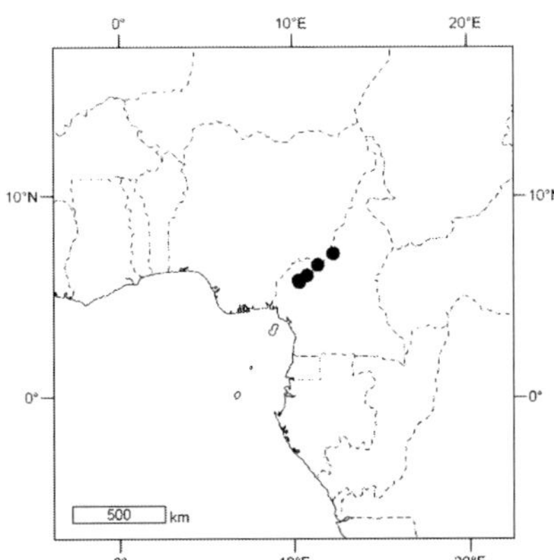

e. ***Crotalaria bamendae*** VU (Leguminosae-Papilionoideae)

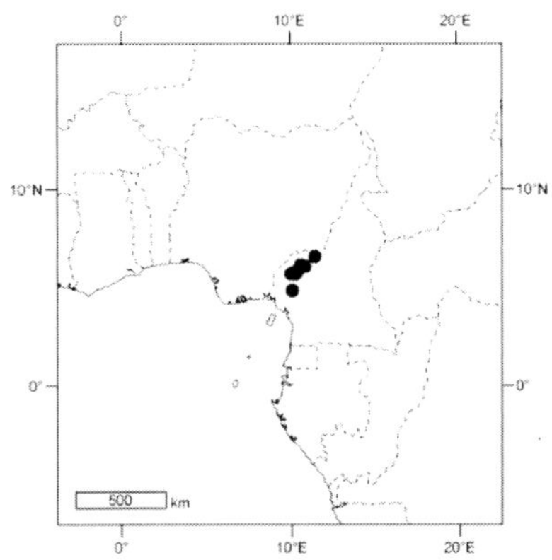

f. ***Crotalaria ledermannii*** VU (Leguminosae-Papilionoideae)

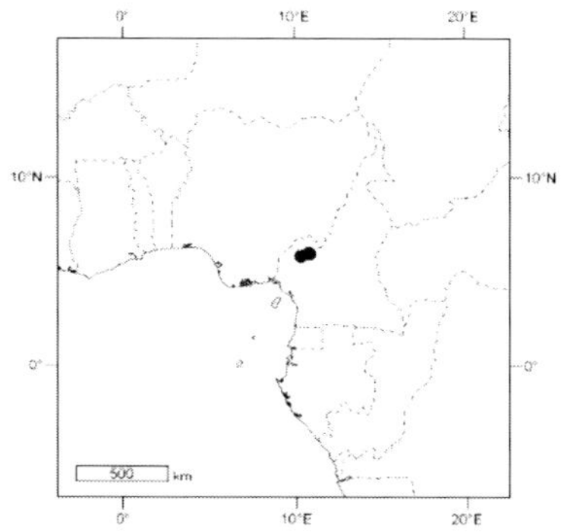

a. ***Crotalaria mentiens*** EN (Leguminosae-Papilionoideae)

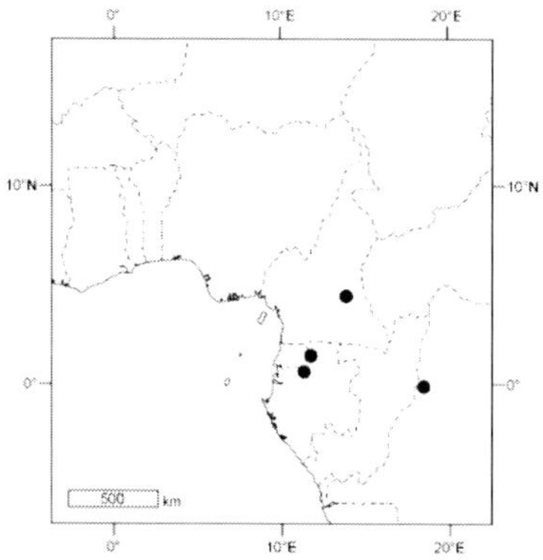

b. ***Dalbergia ealaensis*** VU (Leguminosae-Papilionoideae)

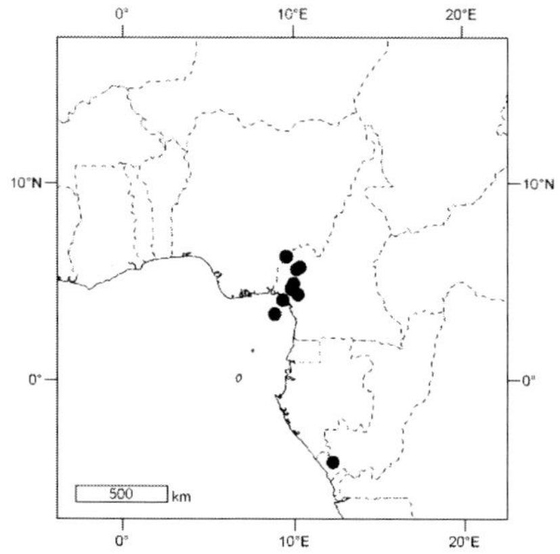

c. ***Dalbergia oligophylla*** EN (Leguminosae-Papilionoideae)

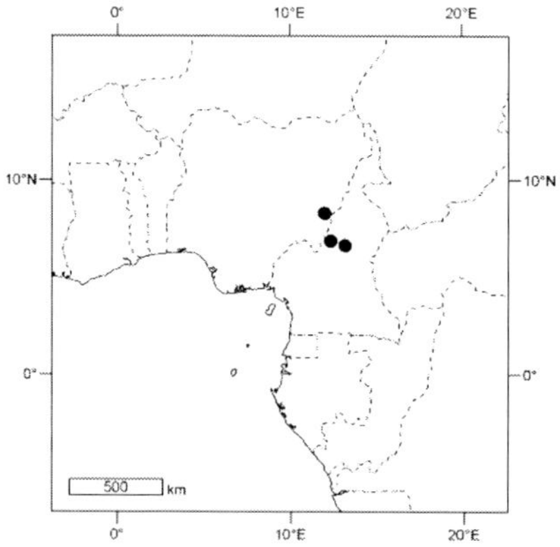

d. ***Dolichos reptans*** EN (Leguminosae-Papilionoideae)

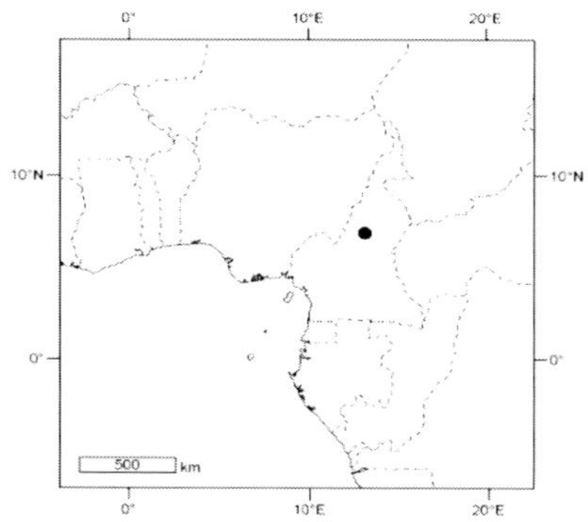

e. ***Eriosema adamouense*** CR (Leguminosae-Papilionoideae)

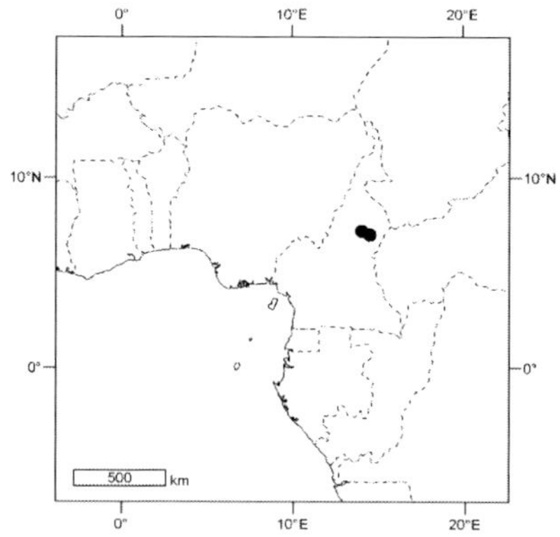

f. ***Eriosema raynaliorum*** VU (Leguminosae-Papilionoideae)

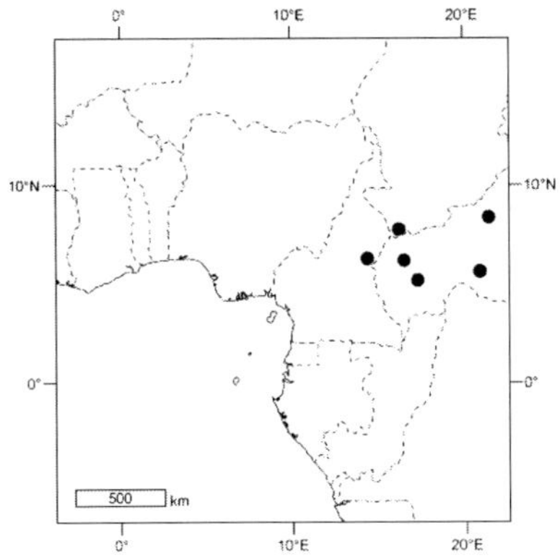

a. ***Eriosema letouzeyi*** VU (Leguminosae-Papilionoideae)

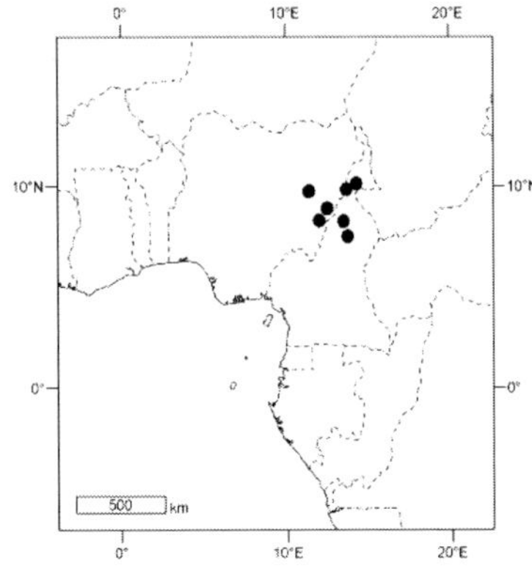

b. ***Indigofera dasycephala*** VU (Leguminosae-Papilionoideae)

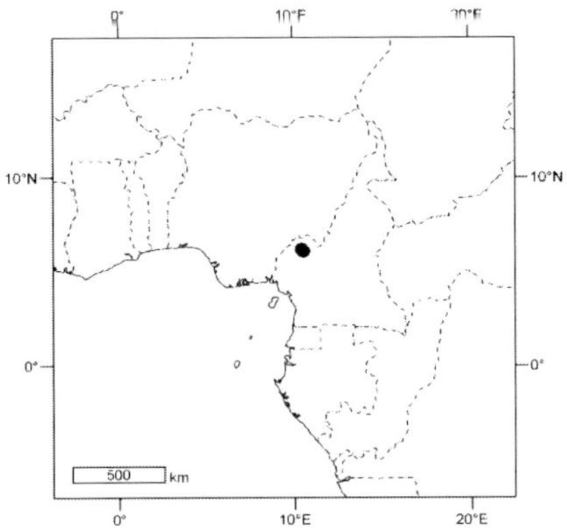

c. ***Indigofera patula*** subsp. ***okuensis*** VU (Leguminosae- Papilionoideae)

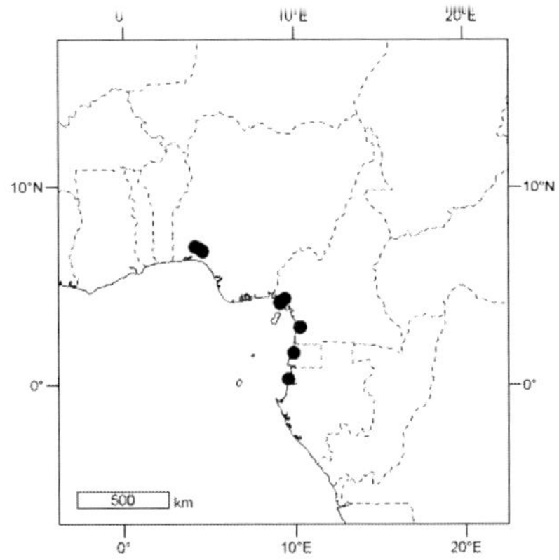

d. ***Leptoderris aurantiaca*** VU (Leguminosae-Papilionoideae)

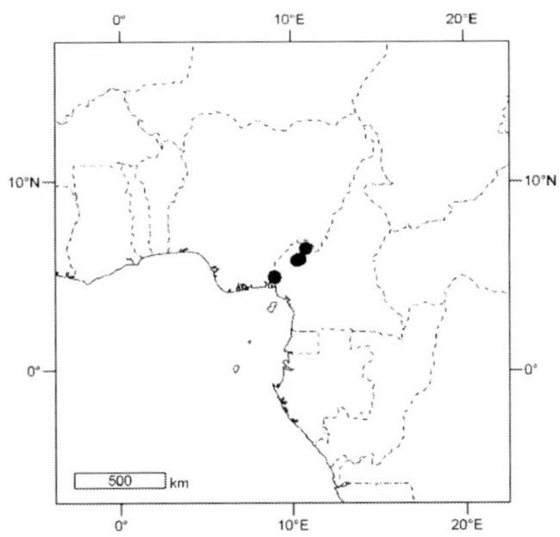

e. ***Leptoderris ledermannii*** EN (Leguminosae-Papilionoideae)

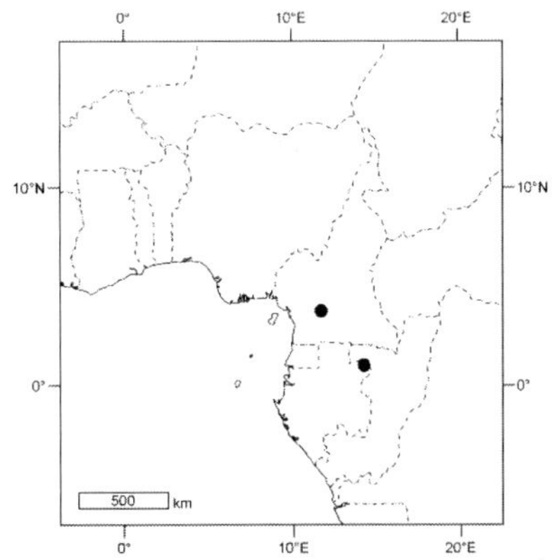

f. ***Leptoderris macrothyrsa*** EN (Leguminosae- Papilionoideae)

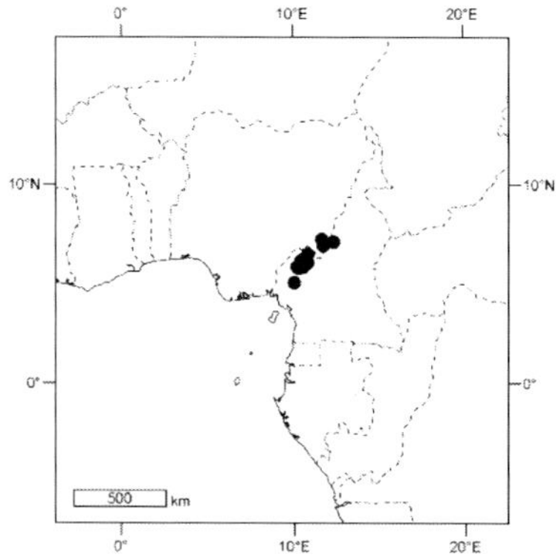

a. ***Millettia conraui*** EN (Leguminosae-Papilionoideae)

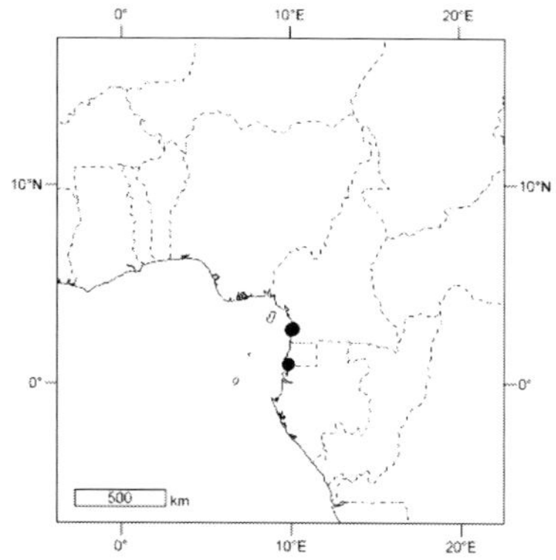

b. ***Millettia coruscans*** EN (Leguminosae-Papilionoideae)

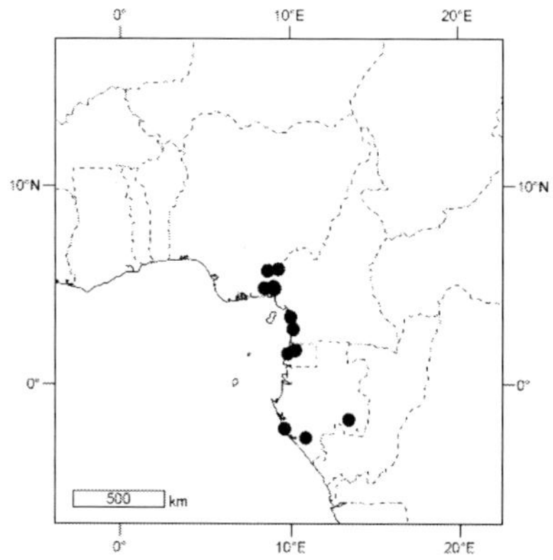

c. ***Millettia hypolampra*** VU(Leguminosae-Papilionoideae)

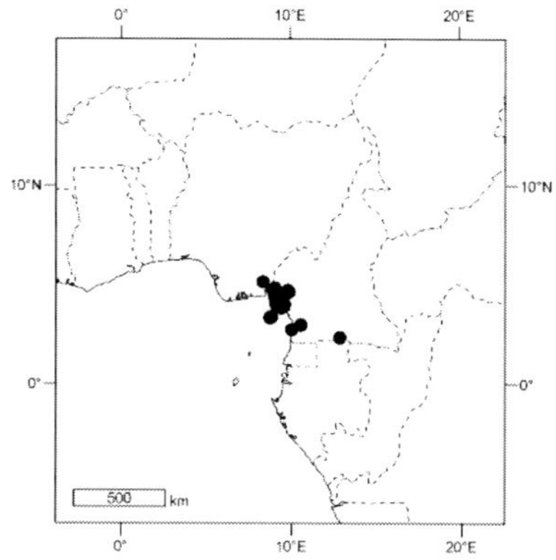

d. ***Millettia macrophylla*** VU (Leguminosae-Papilionoideae)

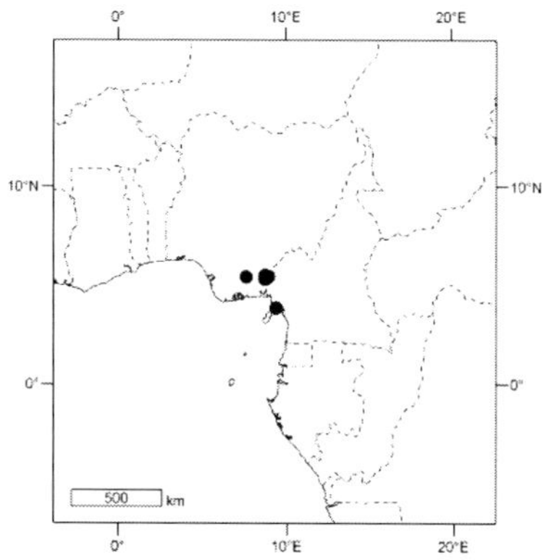

e. ***Millettia pilosa*** EN (Leguminosae-Papilionoideae)

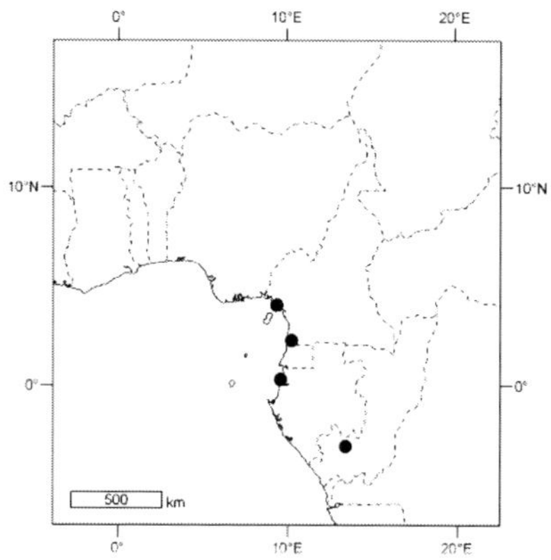

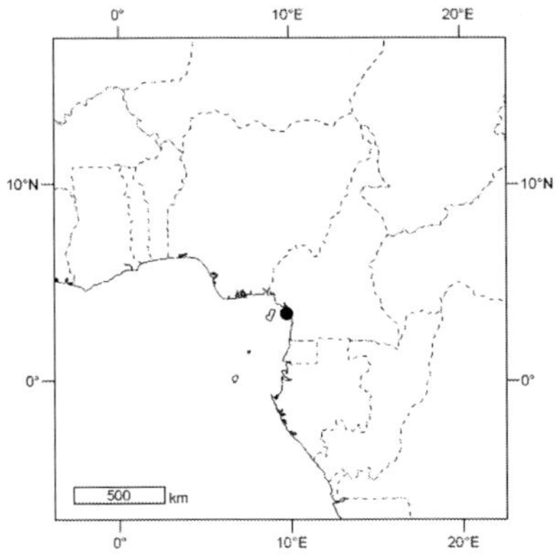

a. ***Ostryocarpus zenkerianus*** CR (Leguminosae-Papilionoideae)

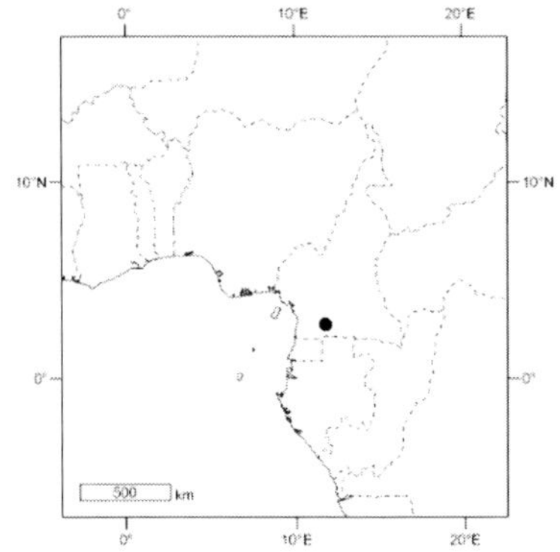

b. ***Platysepalum scaberulum*** CR (Leguminosae- Papilionoideae)

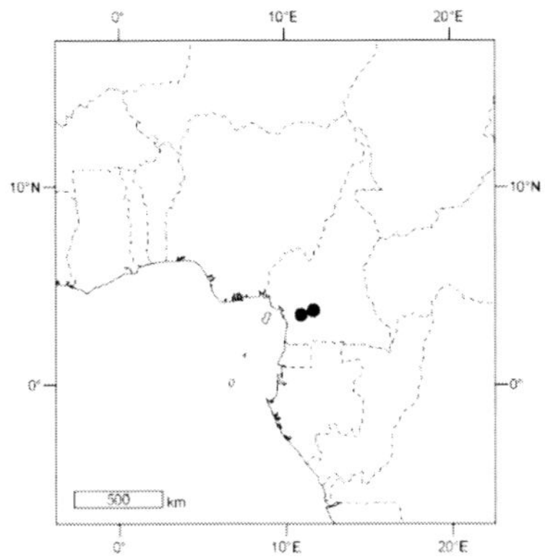

c. ***Pterocarpus zenkeri*** EN (Leguminosae-Papilionoideae)

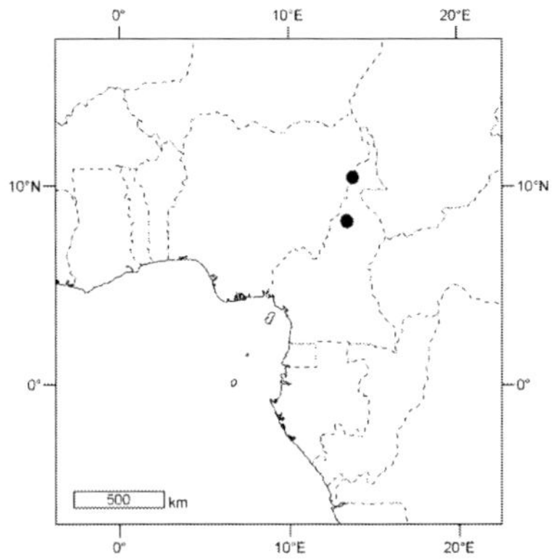

d. ***Rhynchosia ambacensis*** subsp. ***cameroonensis*** EN (Leguminosae-Papilionoideae)

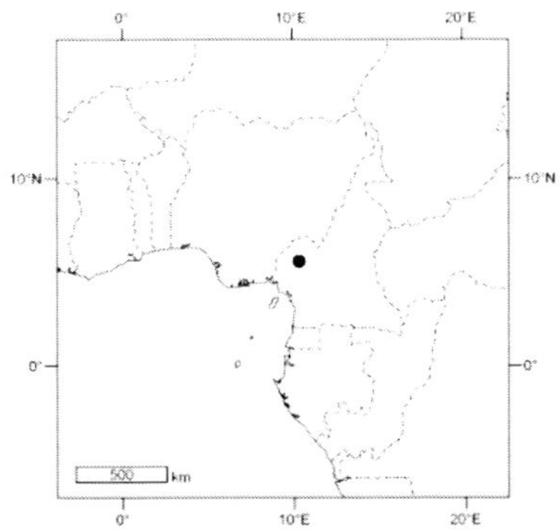

e. ***Rhynchosia ledermannii*** CR (Leguminosae-Papilionoideae)

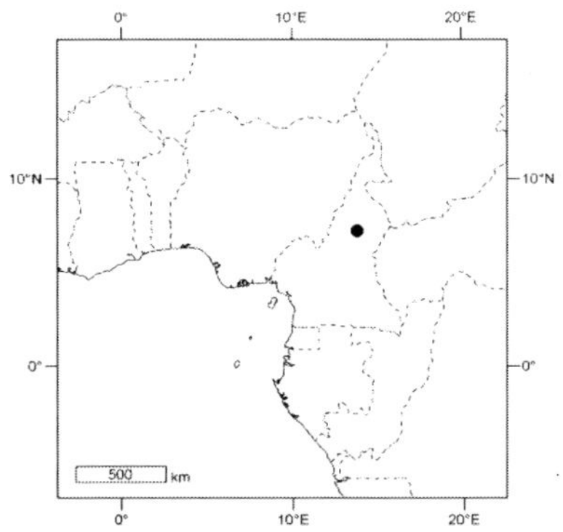

f. ***Trifolium gillettianum*** CR (Leguminosae-Papilionoideae)

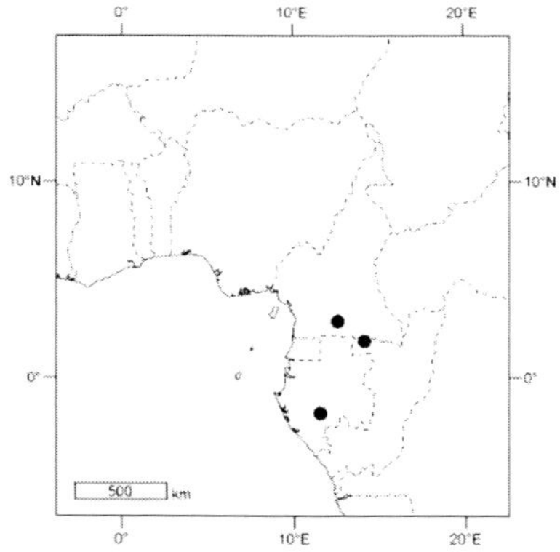

a. ***Hugonia batesii*** VU (Linaceae)

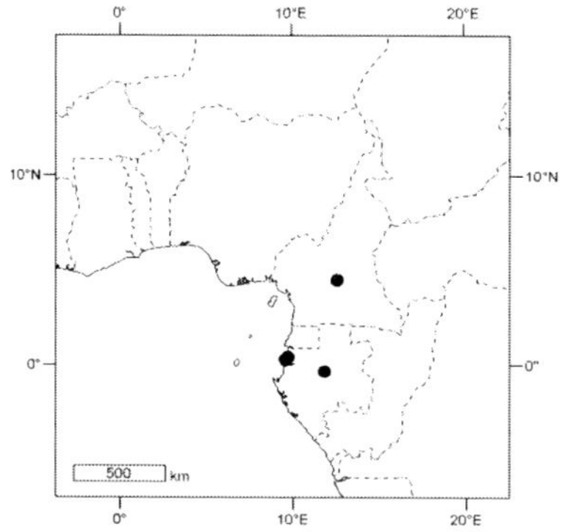

b. ***Hugonia gabunensis*** EN (Linaceae)

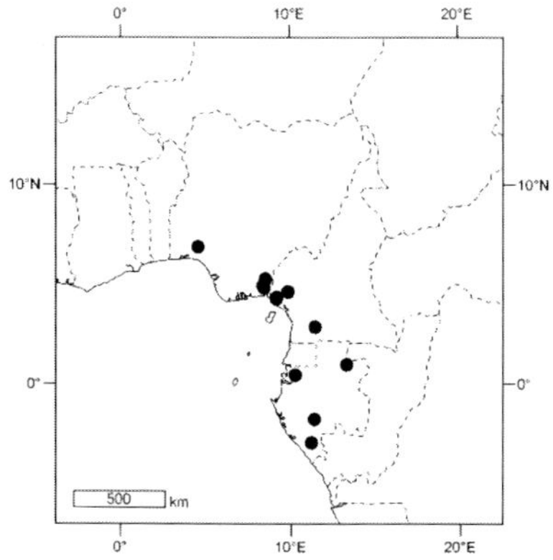

c. ***Hugonia macrophylla*** VU (Linaceae)

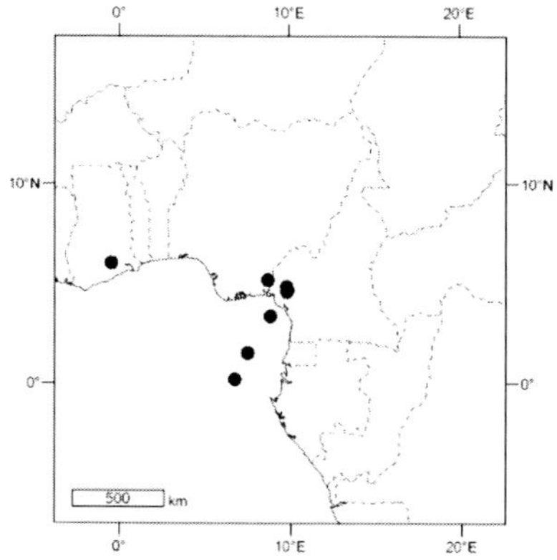

d. ***Anthocleista microphylla*** VU (Loganiaceae)

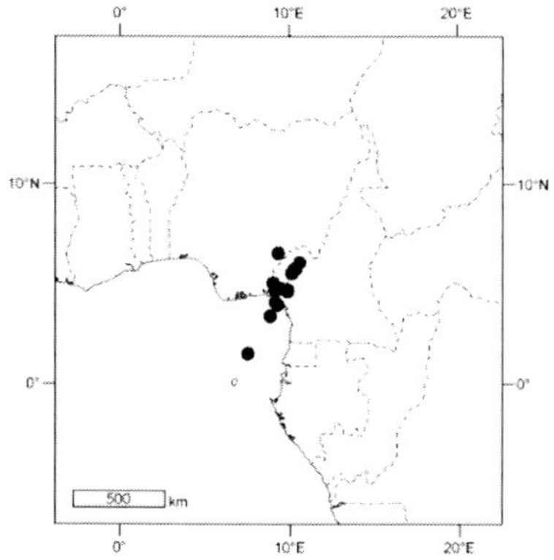

e. ***Anthocleista scandens*** VU (Loganiaceae)

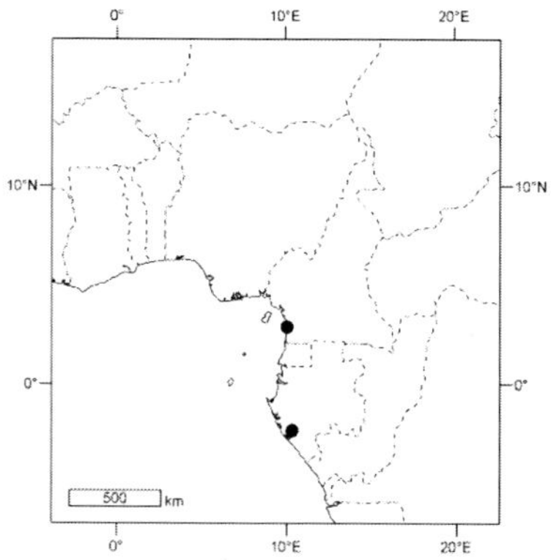

f. ***Strychnos canthioides*** EN (Loganiaceae)

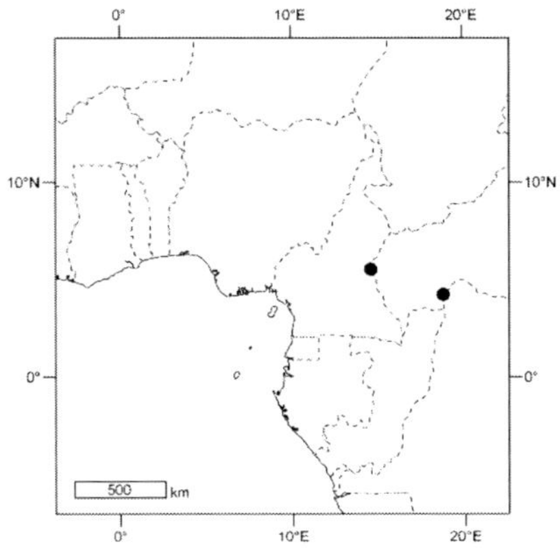

a. ***Strychnos chromatoxylon*** EN (Loganiaceae)

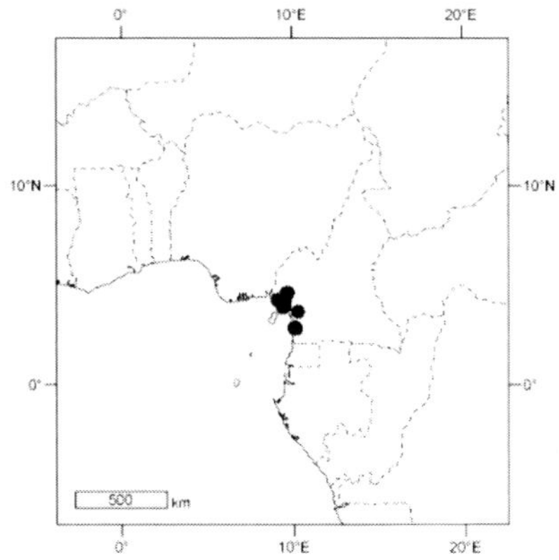

b. ***Strychnos elaeocarpa*** VU (Loganiaceae)

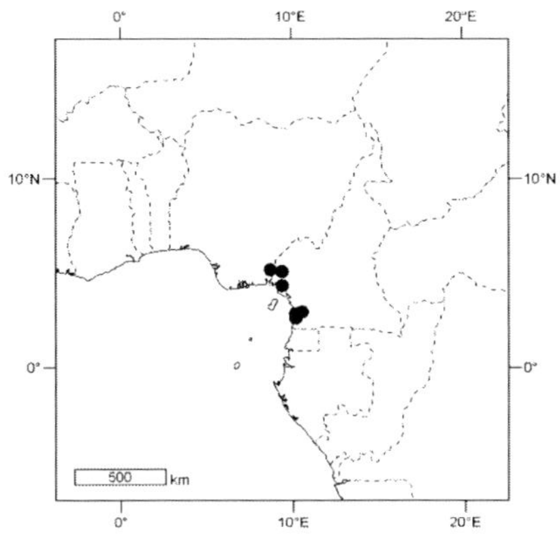

c. ***Strychnos gnetifolia*** VU (Loganiaceae)

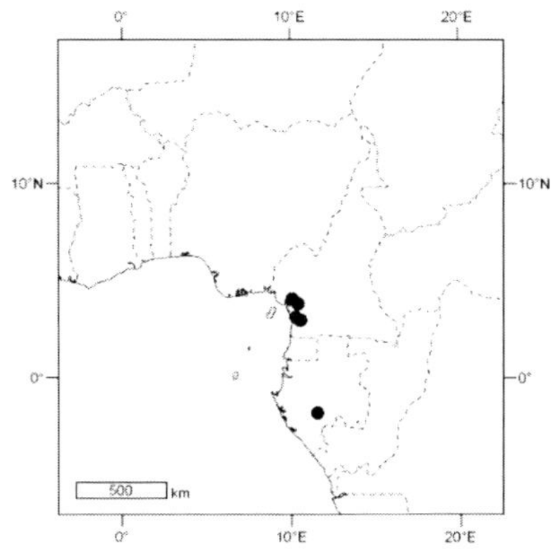

d. ***Strychnos mimfiensis*** VU (Loganiaceae)

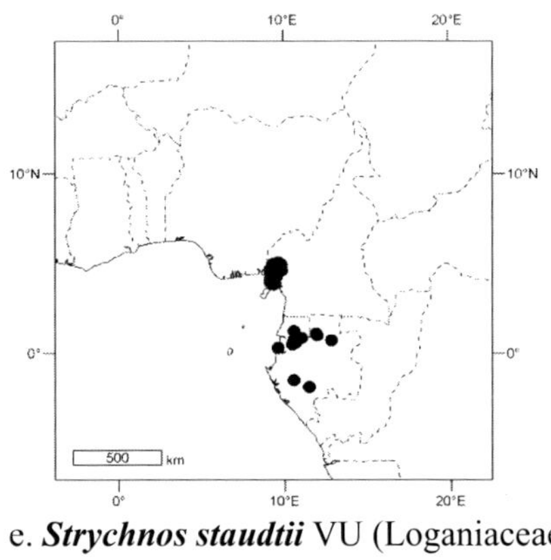

e. ***Strychnos staudtii*** VU (Loganiaceae)

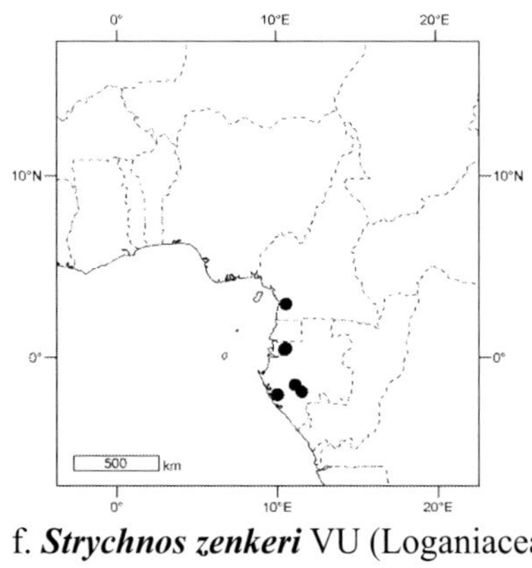

f. ***Strychnos zenkeri*** VU (Loganiaceae)

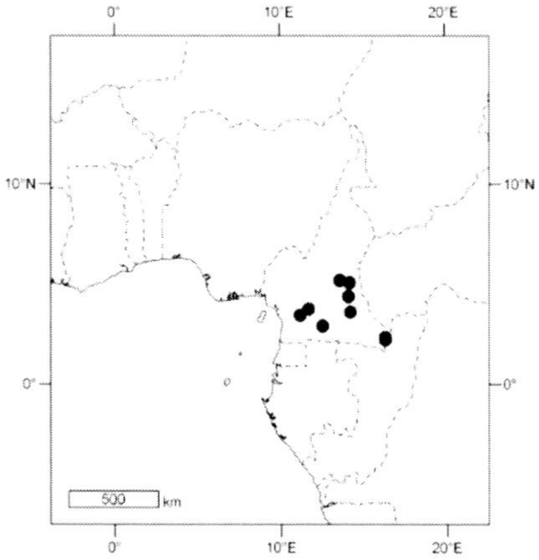

a. ***Agelanthus dichrous*** VU (Loranthaceae)

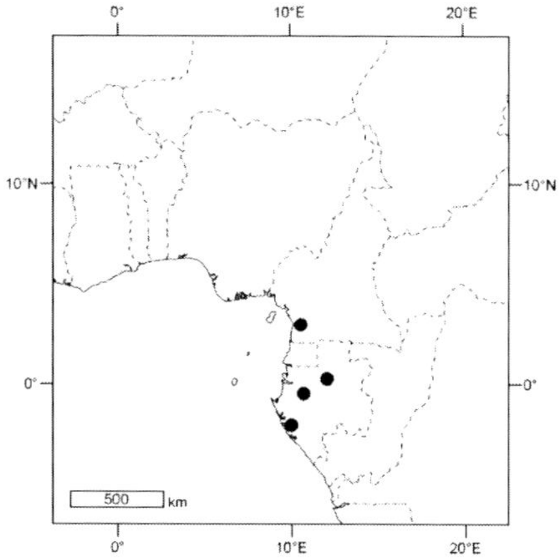

b. ***Agelanthus glaucoviridis*** EN (Loranthaceae)

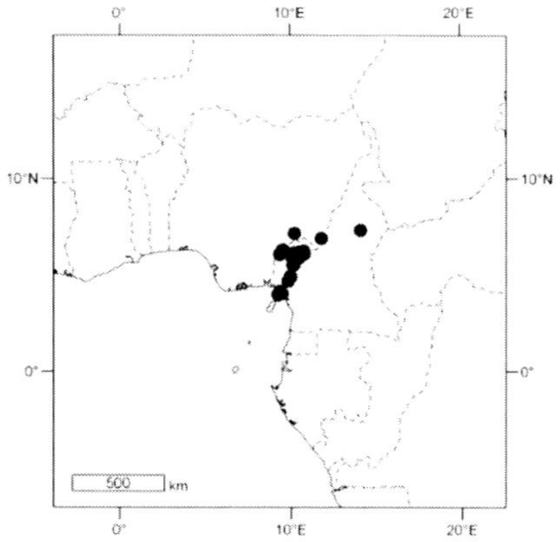

c. ***Globimetula oreophila*** VU (Loranthaceae)

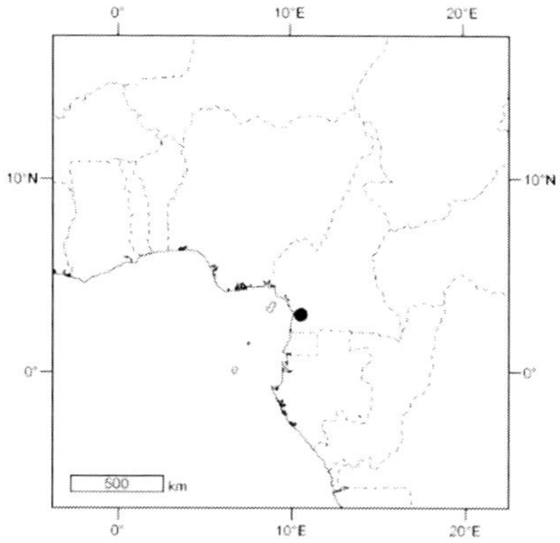

d. ***Helixanthera periclymenoides*** CR (Loranthaceae)

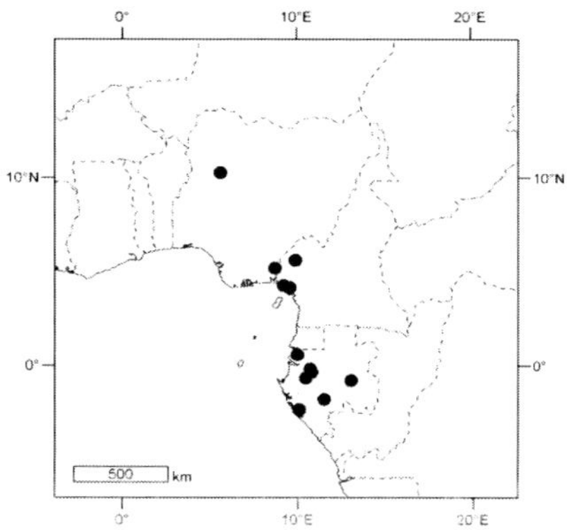

e. ***Phragmanthera kamerunensis*** VU (Loranthaceae)

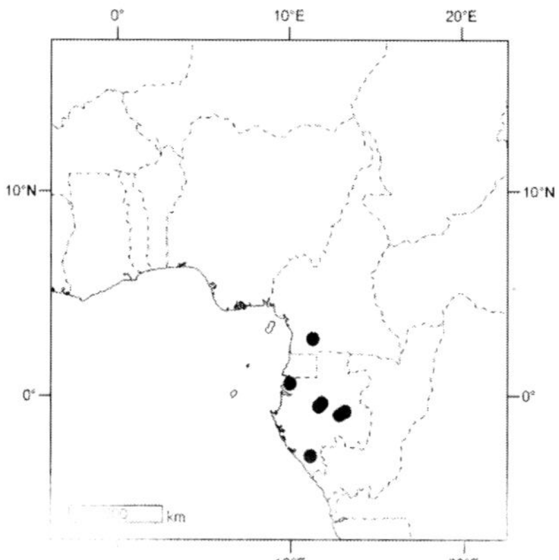

f. ***Phragmanthera longiflora*** subsp. ***longiflora*** EN (Loranthaceae)

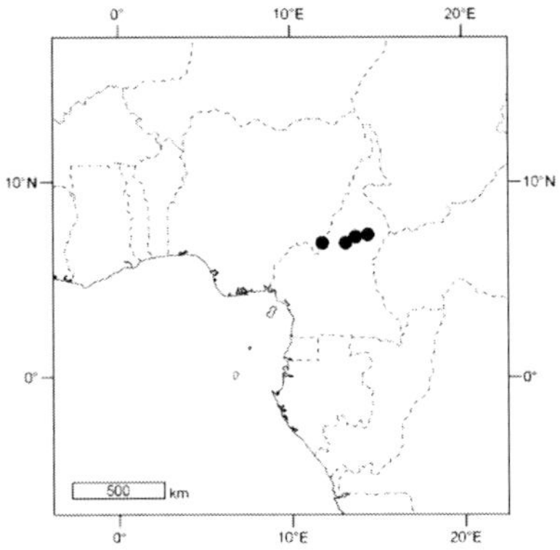

a. ***Phragmanthera raynaliana*** EN (Loranthaceae)

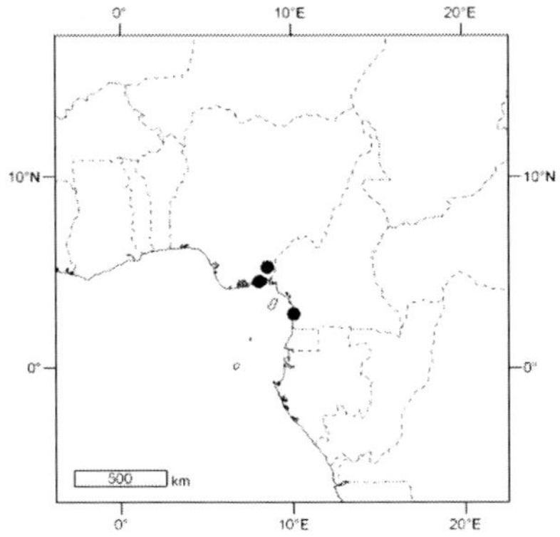

b. ***Phragmanthera talbotiorum*** EN (Loranthaceae)

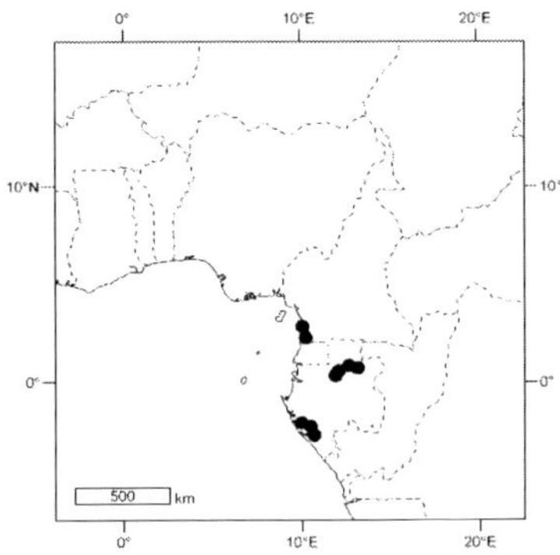

c. ***Acridocarpus camerunensis*** VU (Malpighiaceae)

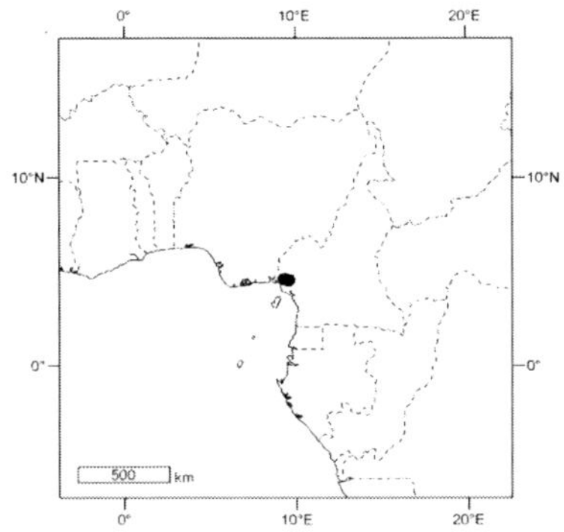

d. ***Acridocarpus staudtii*** CR (Malpighiaceae)

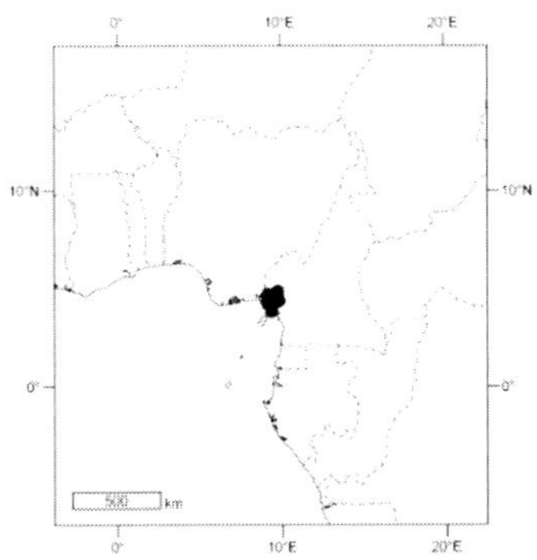

e. ***Medusandra richardsiana*** VU (Medusandraceae)

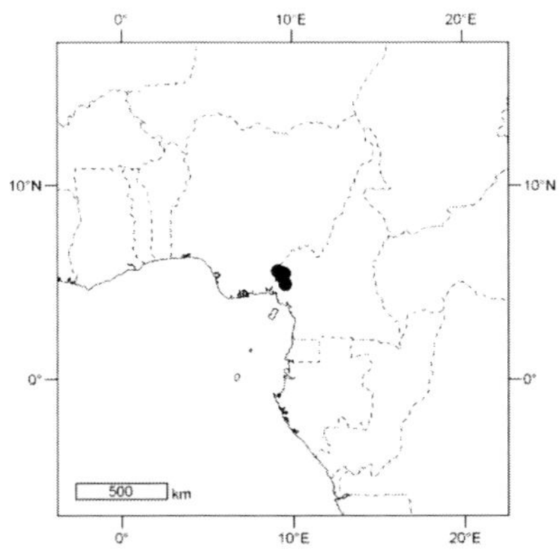

f. ***Amphiblemma amoenum*** EN (Melastomataceae)

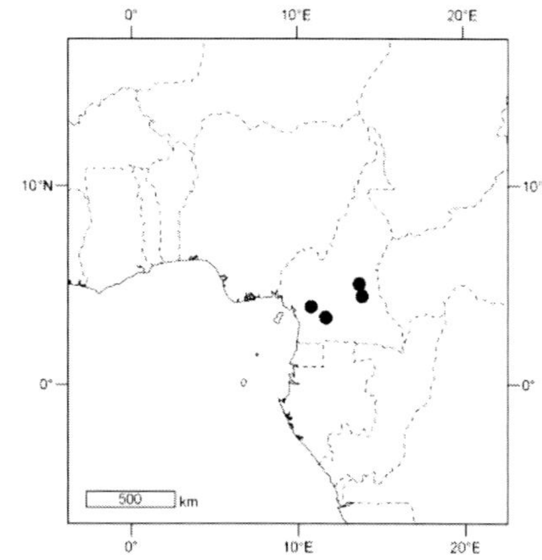

a. ***Amphiblemma lanceatum*** EN (Melastomataceae)

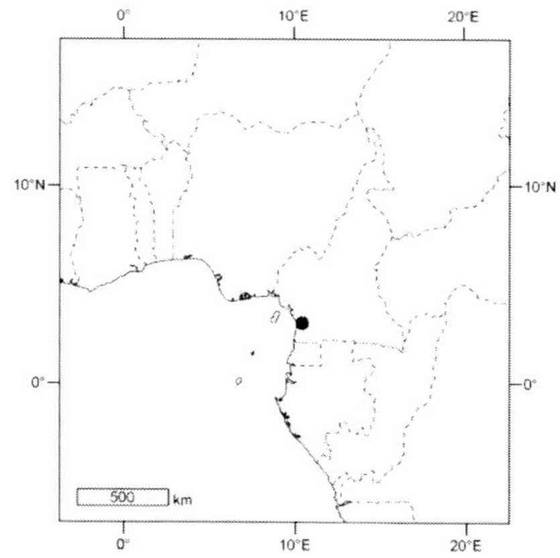

b. ***Amphiblemma letouzeyi*** CR (Melastomataceae)

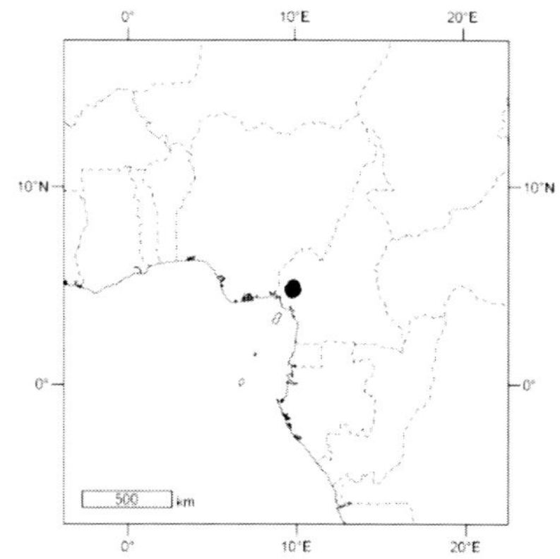

c. ***Amphiblemma monticola*** VU (Melastomataceae)

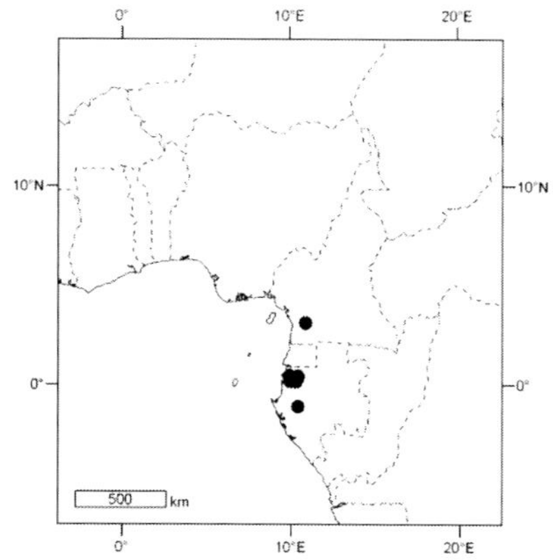

d. ***Amphiblemma soyauxii*** VU (Melastomataceae)

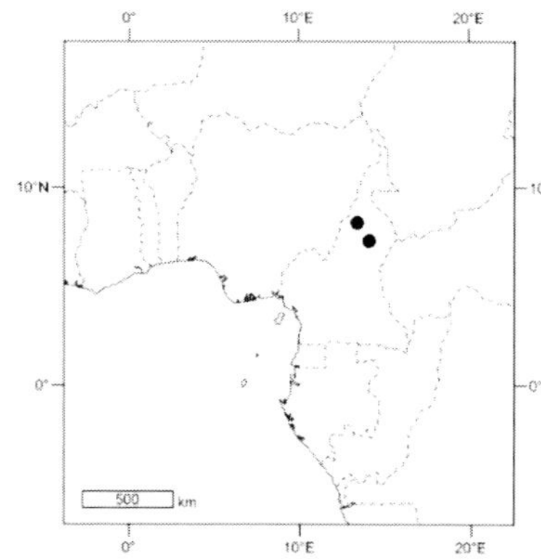

e. ***Antherotoma clandestina*** EN (Melastomataceae)

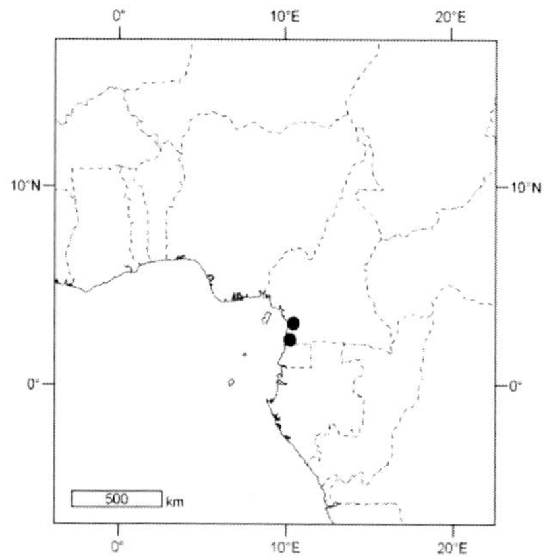

f. ***Calvoa calliantha*** EN (Melastomataceae)

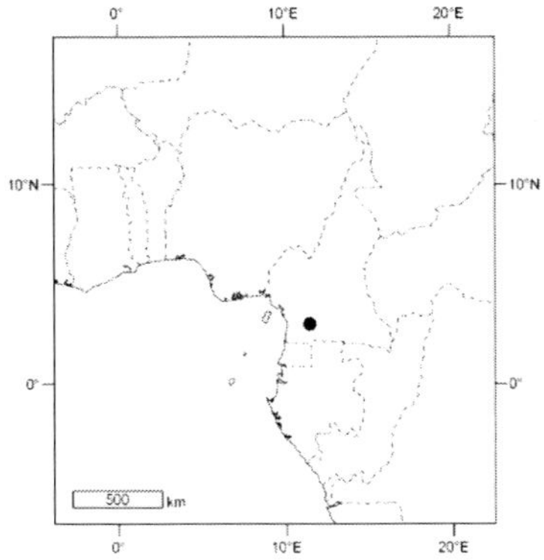

a. ***Calvoa stenophylla*** CR (Melastomataceae)

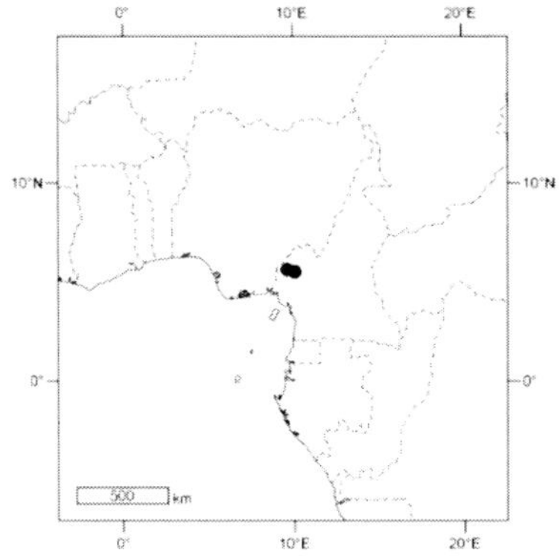

b. ***Cincinnobotrys letouzeyi*** EN (Melastomataceae)

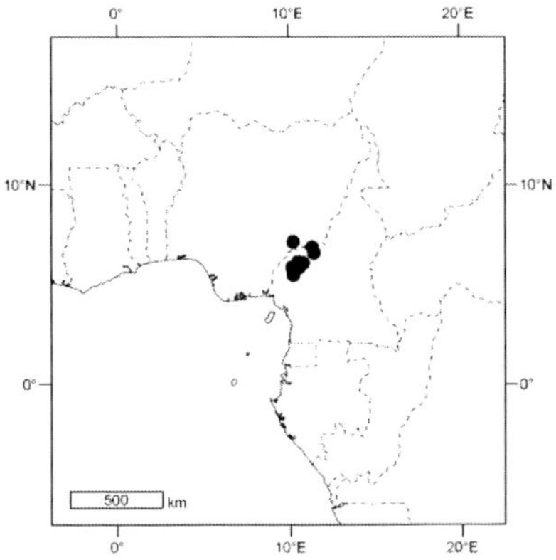

c. ***Dissotis bambutorum*** VU (Melastomataceae)

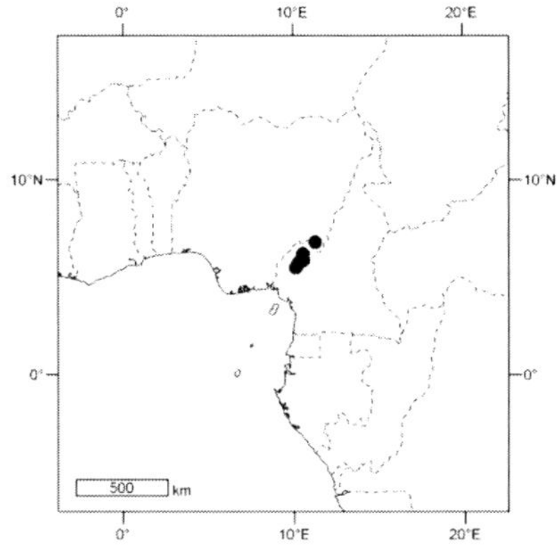

d. ***Dissotis bamendae*** VU (Melastomataceae)

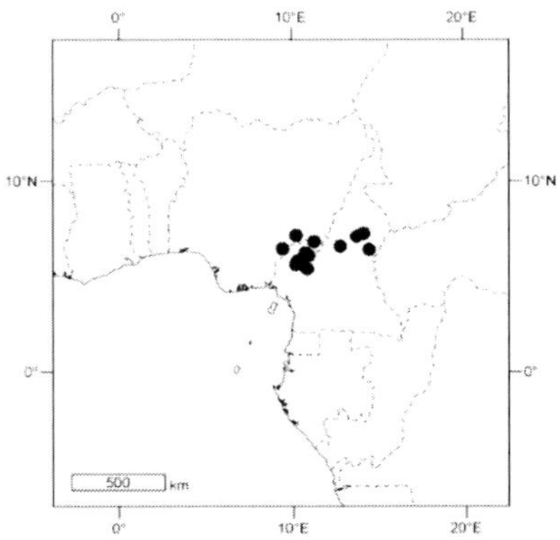

e. ***Dissotis longisetosa*** VU (Melastomataceae)

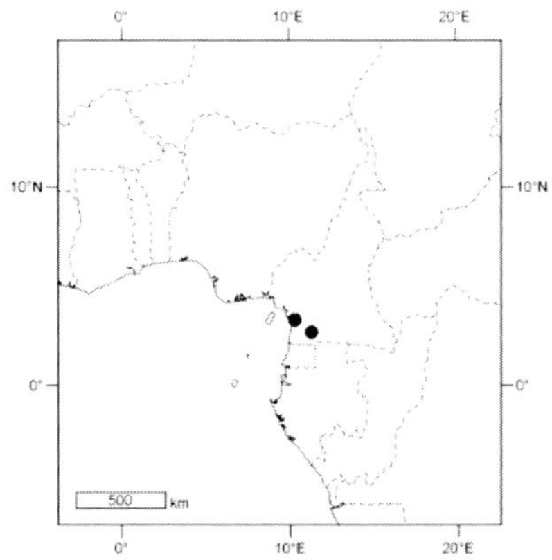

f. ***Tristemma camerunense*** EN (Melastomataceae)

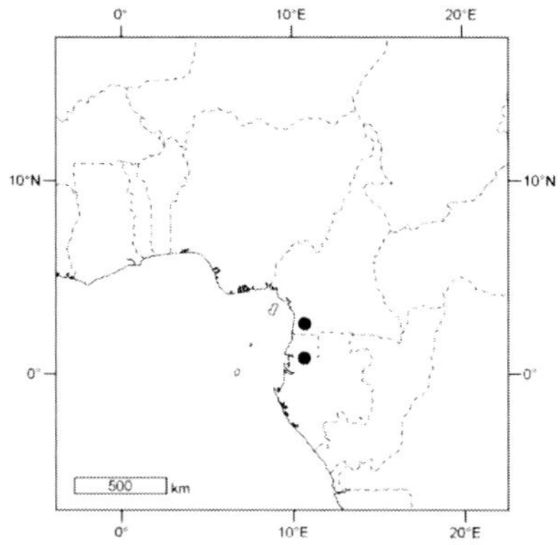

a. ***Memecylon alipes*** EN (Melastomataceae)

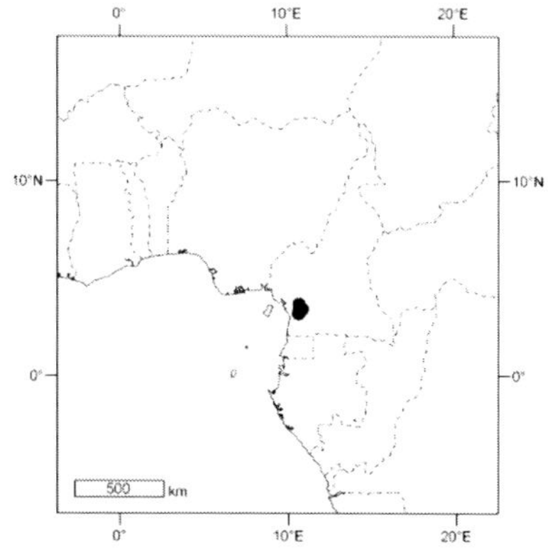

b. ***Memecylon amshoffiae*** EN (Melastomataceae)

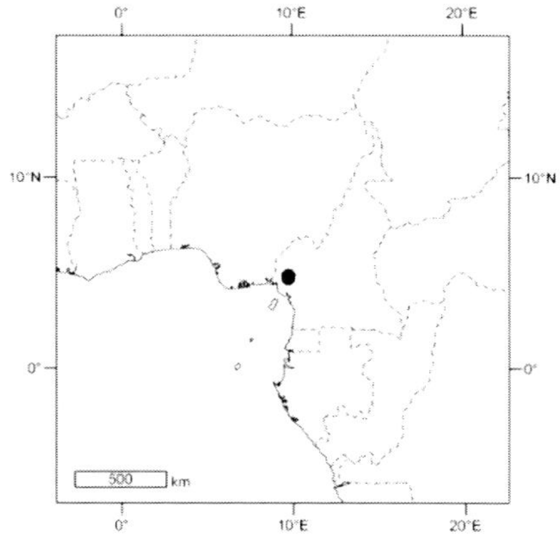

c. ***Memecylon bakossiense*** EN (Melastomataceae)

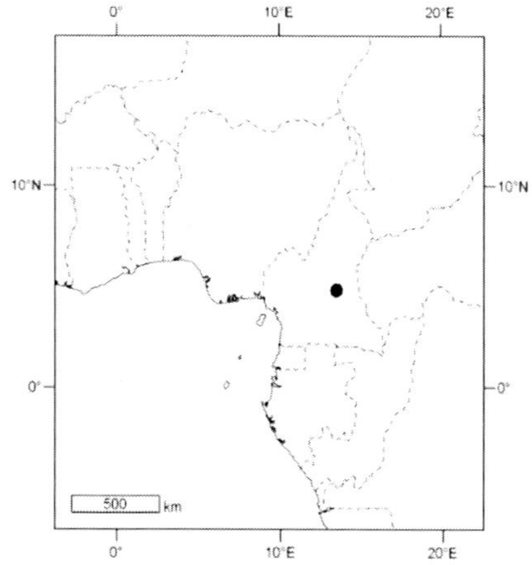

d. ***Memecylon breteleranum*** CR (Melastomataceae)

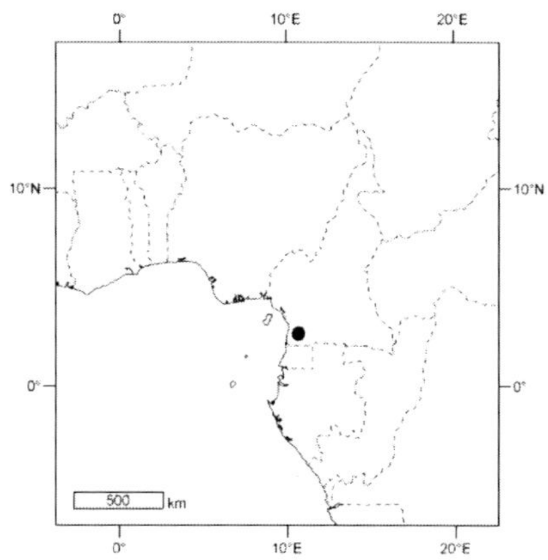

e. ***Memecylon fugax*** EN (Melastomataceae)

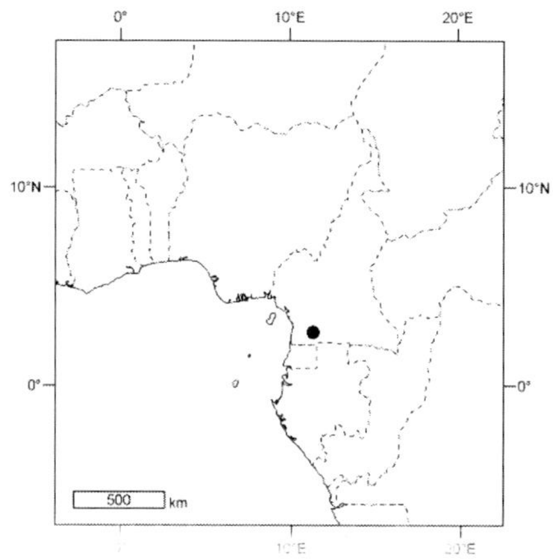

f. ***Memecylon [illegible]*** CR (Melastomataceae)

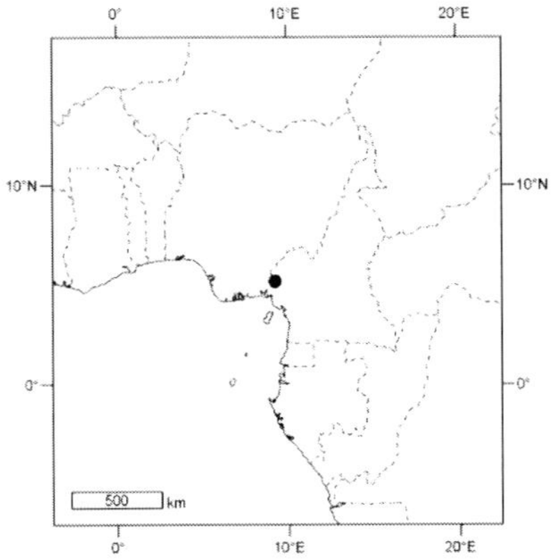

a. ***Memecylon korupense*** CR (Melastomataceae)

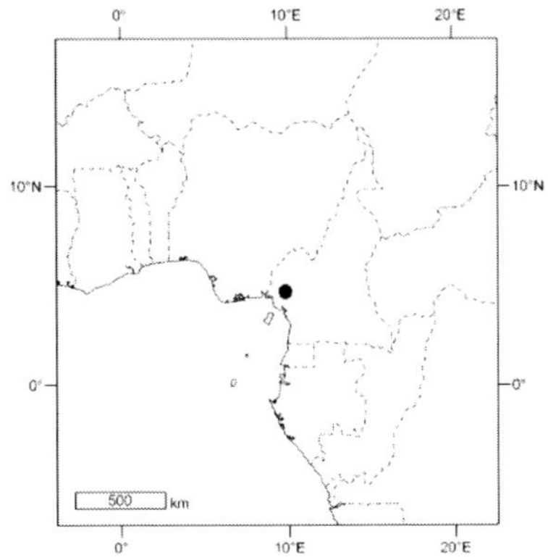

b. **Memecylon kupeanum** EN (Melastomataceae)

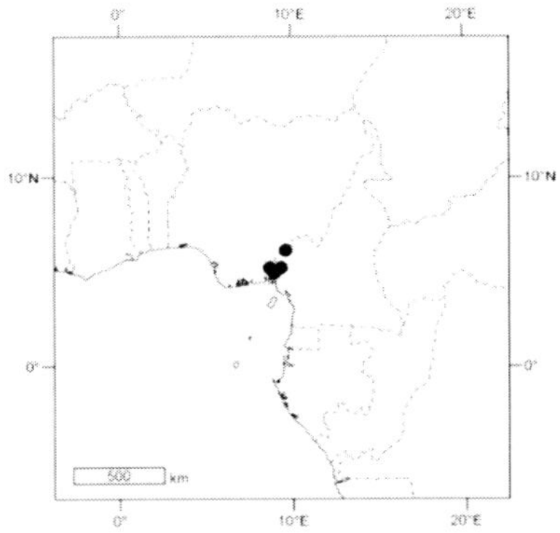

c. ***Memecylon mamfeanum*** EN (Melastomataceae)

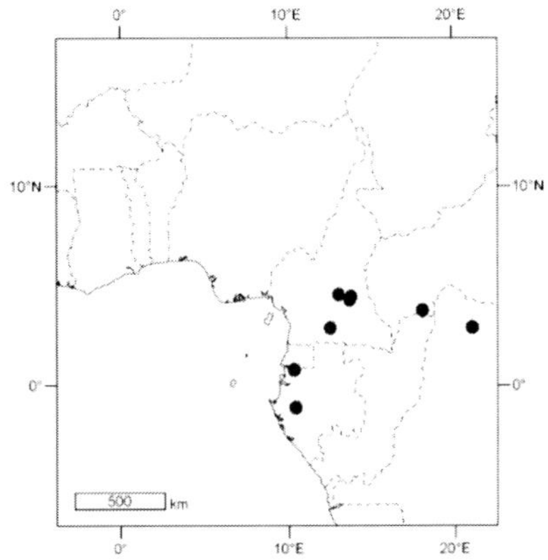

d. ***Memecylon oubanguianum*** VU (Melastomataceae)

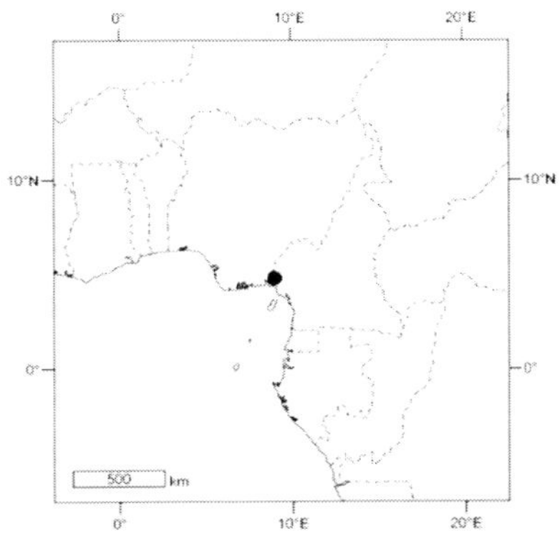

e. ***Memecylon rheophyticum*** VU (Melastomataceae)

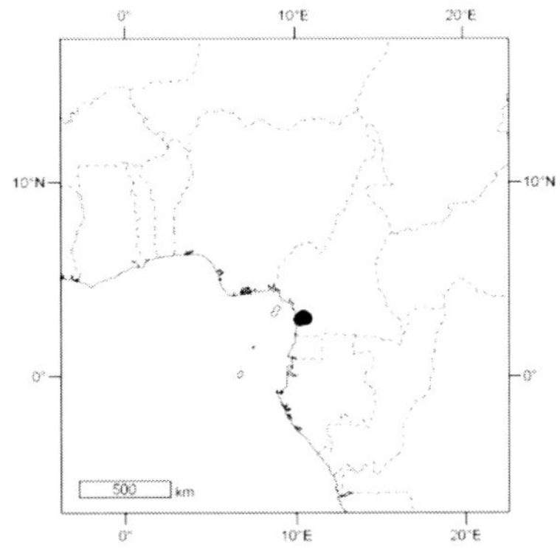

f. ***Memecylon simulans*** EN (Melastomataceae)

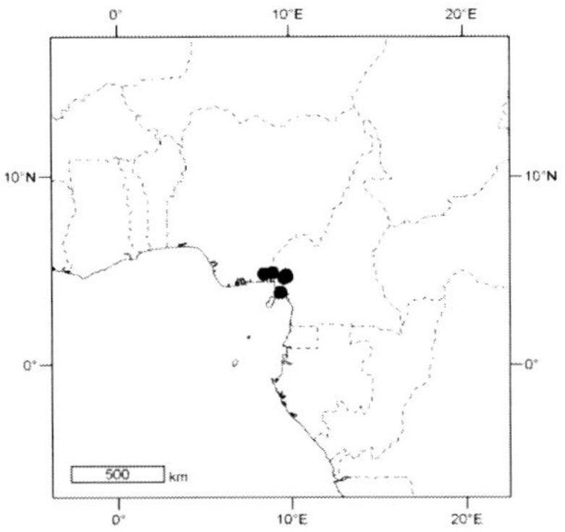

a. ***Warneckea austro-occidentalis*** EN (Melastomataceae

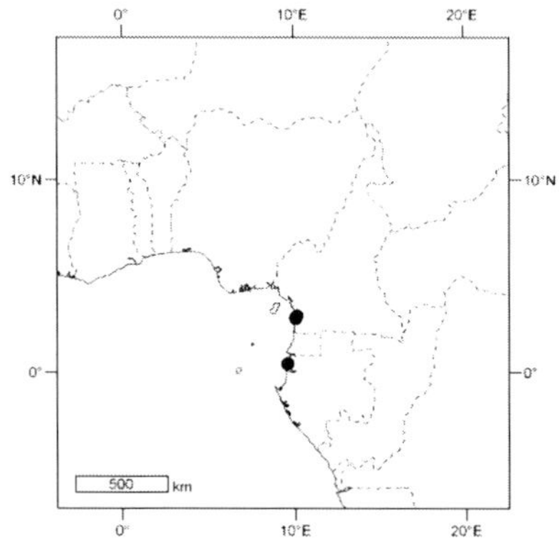

b. ***Warneckea mangrovensis*** EN (Melastomataceae)

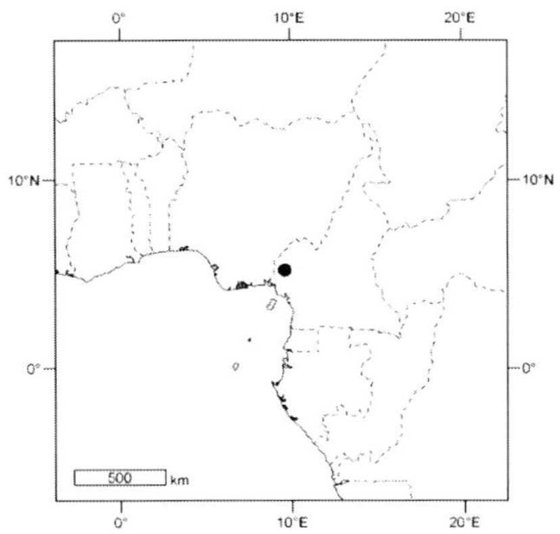

c. ***Warneckea ngutiensis*** CR (Melastomataceae)

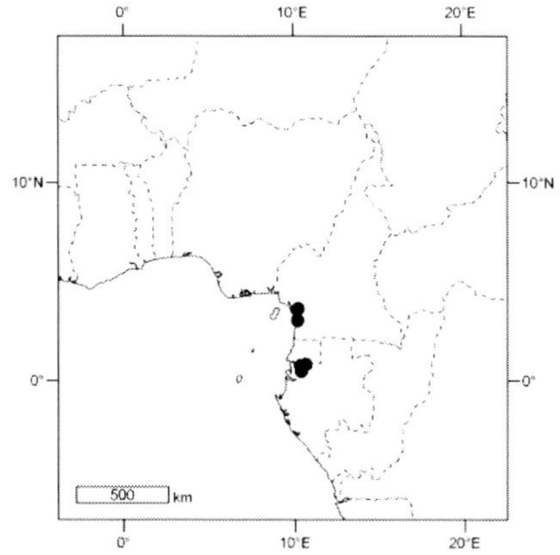

d. ***Warneckea wildeana*** EN (Melastomataceae)

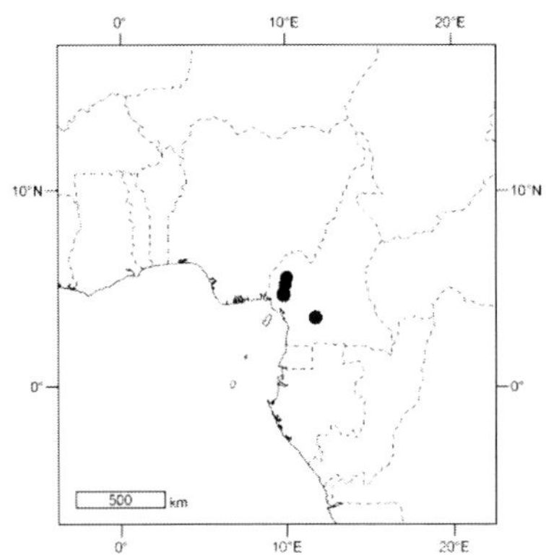

e. ***Heckeldora ledermannii*** EN (Meliaceae)

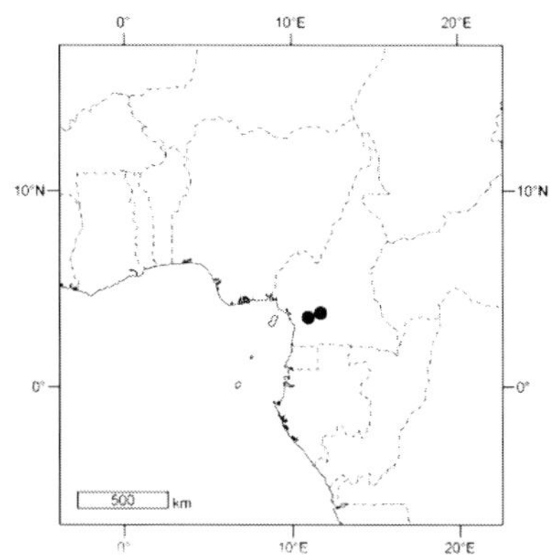

f. [illegible] VU (Meliaceae)

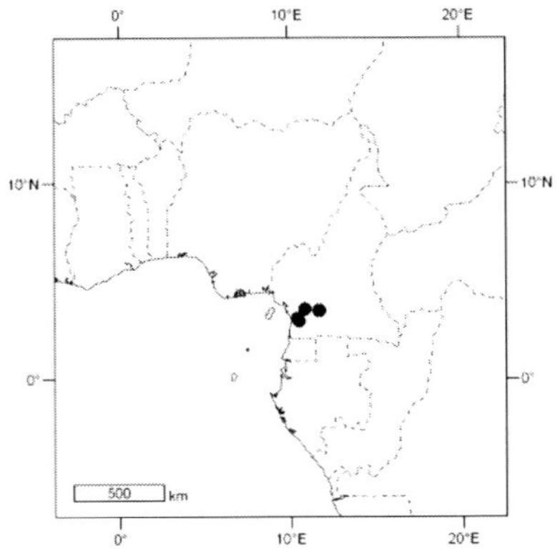

a. ***Trichilia zewaldae*** EN (Meliaceae)

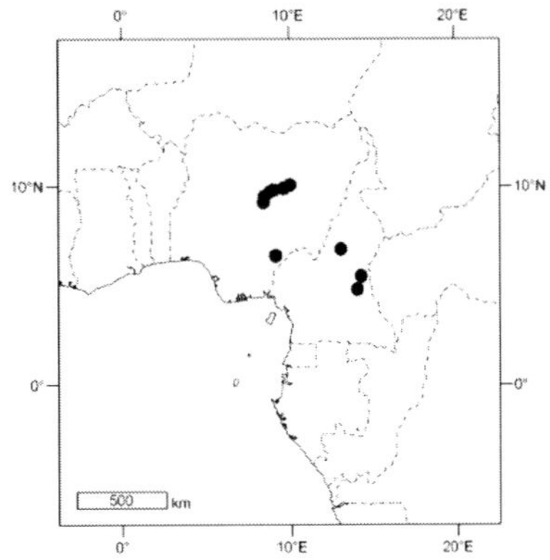

b. ***Turraea pellegriniana*** VU (Meliaceae)

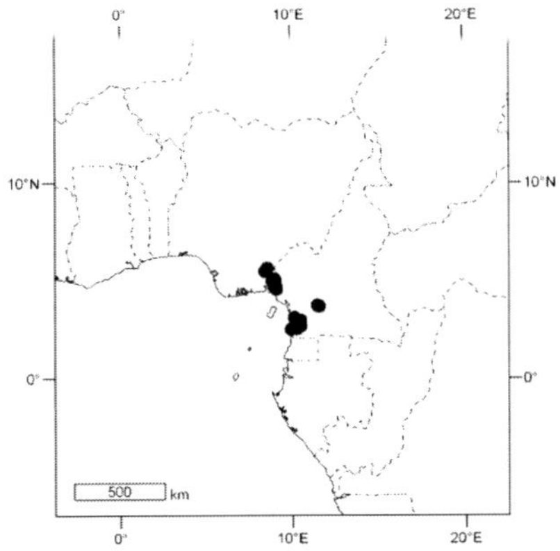

c. ***Turraeanthus mannii*** VU (Meliaceae)

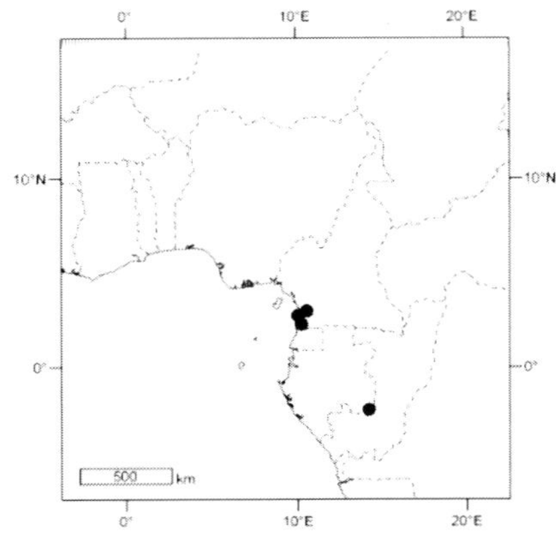

d. ***Albertisia capituliflora*** EN (Menispermaceae)

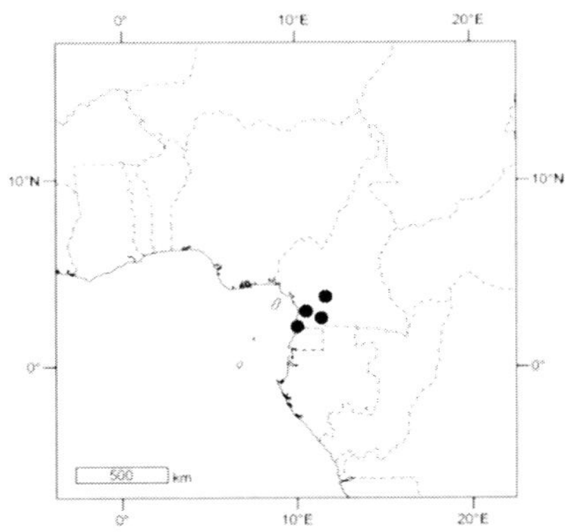

e. ***Albertisia glabra*** EN (Menispermaceae)

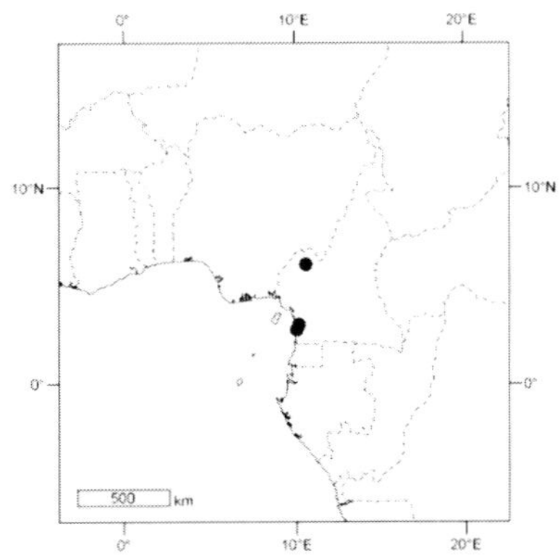

f. ***Platytinospora bucholzii*** var ***bucholzii*** EN (Menispermaceae)

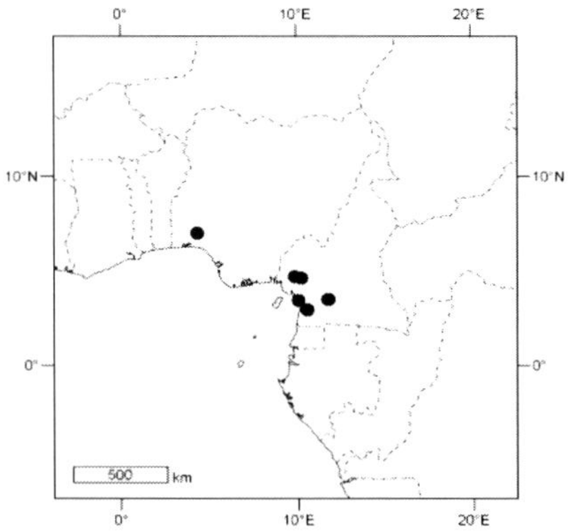

a. ***Platytinospora buchholzii*** var. ***macrophylla*** EN (Menispermaceae)

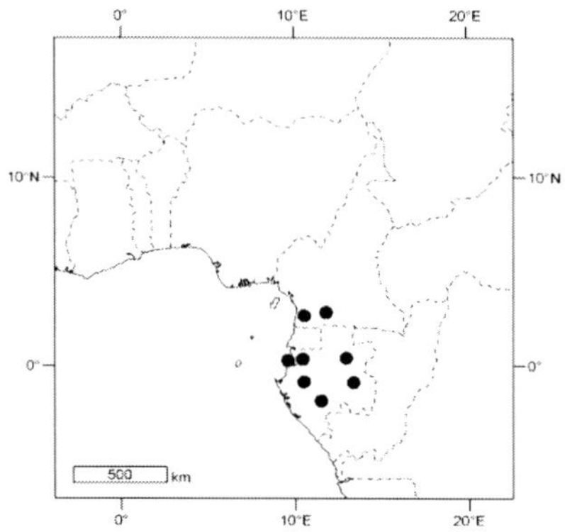

b. ***Sarcolophium suberosum*** VU (Menispermaceae)

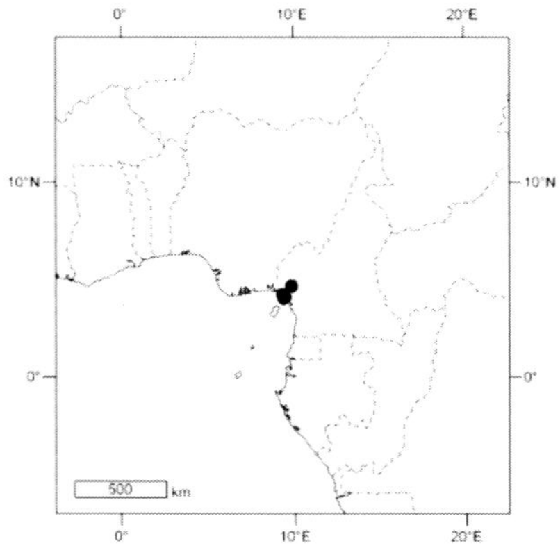

c. ***Tiliacora lehmbachii*** EN (Menispermaceae

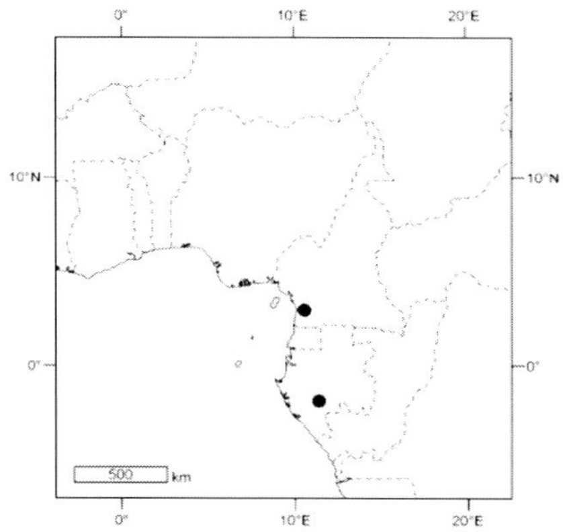

d. ***Tiliacora odorata*** EN (Menispermaceae)

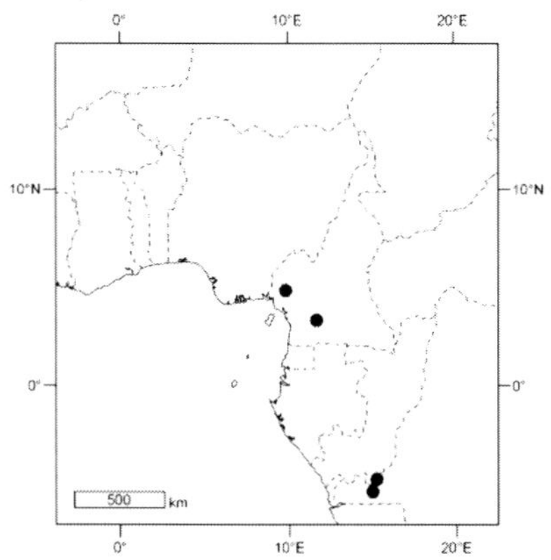

e. ***Triclisia lanceolata*** EN (Menispermaceae)

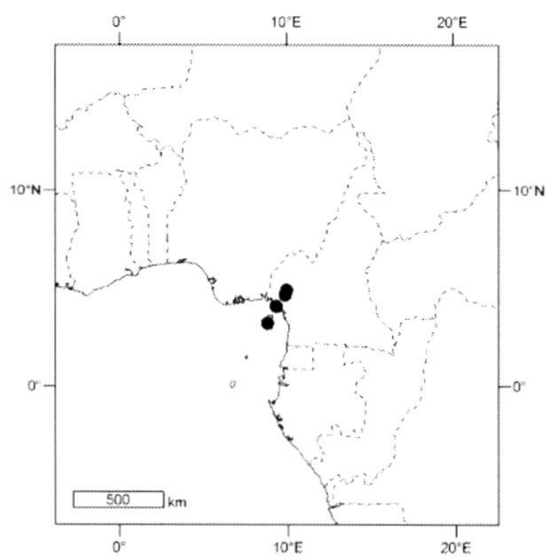

f. ***Triclisia macrophylla*** EN (Menispermaceae)

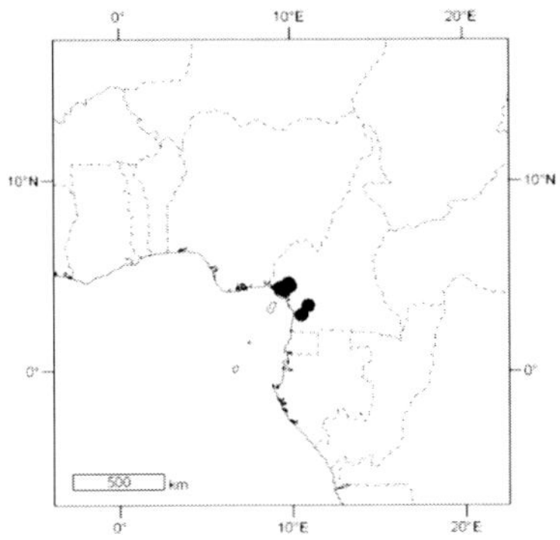

a. ***Dorstenia angusticornis*** VU (Moraceae)

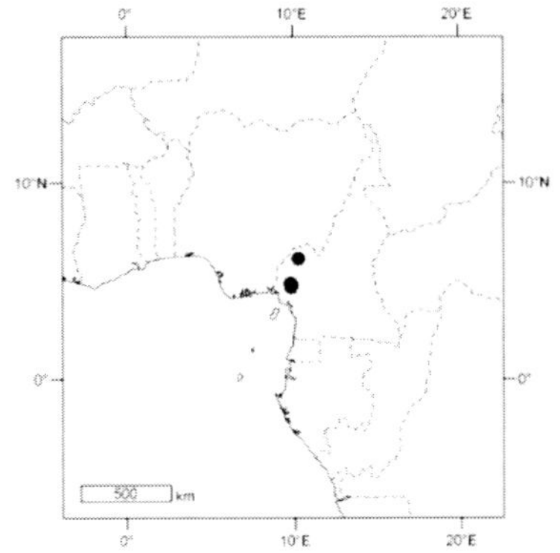

b. ***Dorstenia astyanactis*** VU (Moraceae)

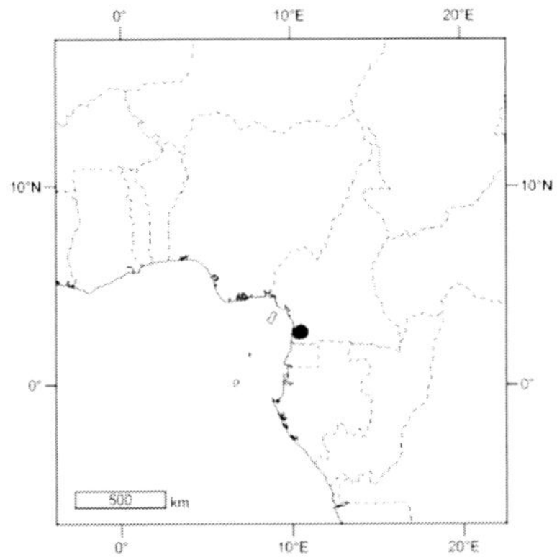

c. ***Dorstenia dorstenioides*** EN (Moraceae)

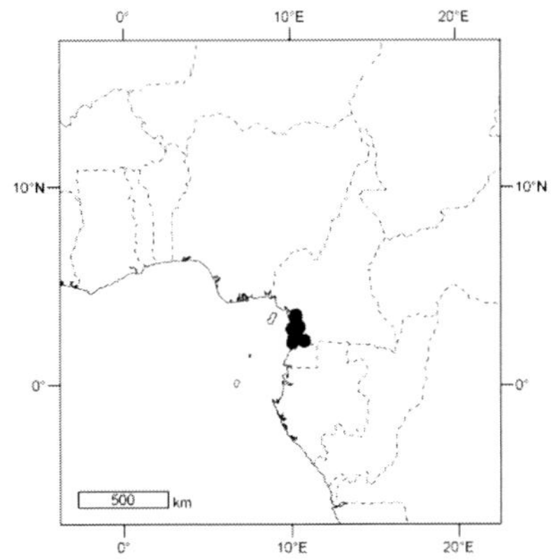

d. ***Dorstenia involuta*** VU (Moraceae)

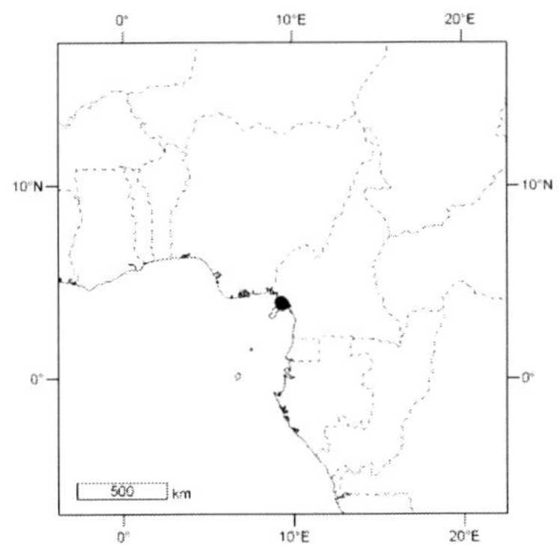

e. ***Dorstenia poinsettiifolia*** var. ***achoundongiana*** EN (Moraceae)

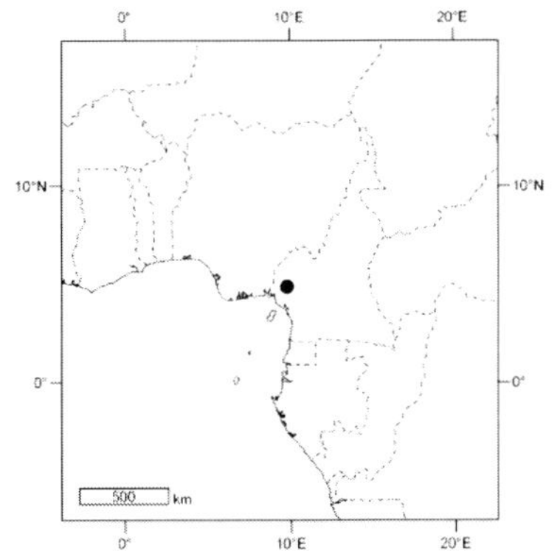

f. ***Dorstenia poinsettifolia*** var. ***etugeana*** EN (Moraceae)

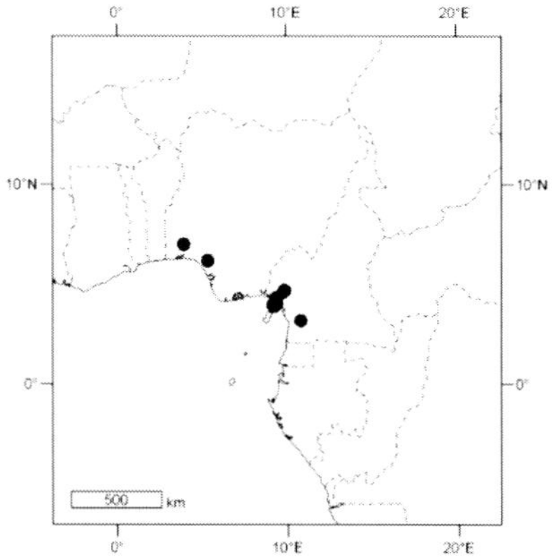

a. ***Dorstenia prorepens*** VU (Moraceae)

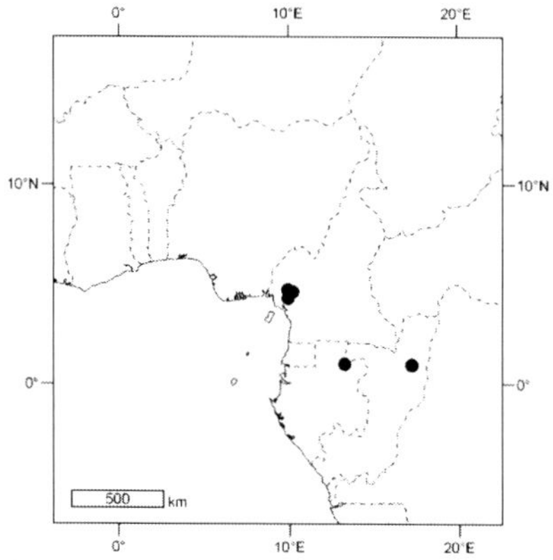

b. ***Ficus abscondita*** VU (Moraceae)

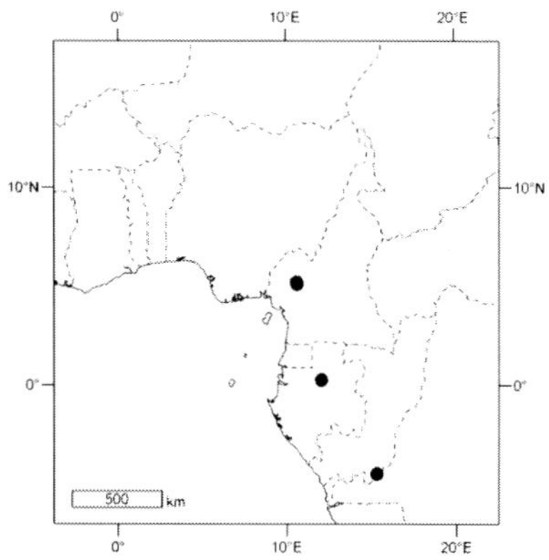

c. ***Ficus jansii*** EN (Moraceae)

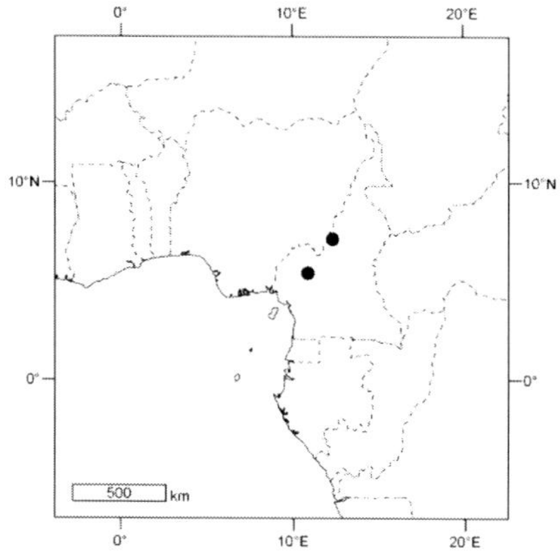

d. ***Ficus oresbia*** CR (Moraceae)

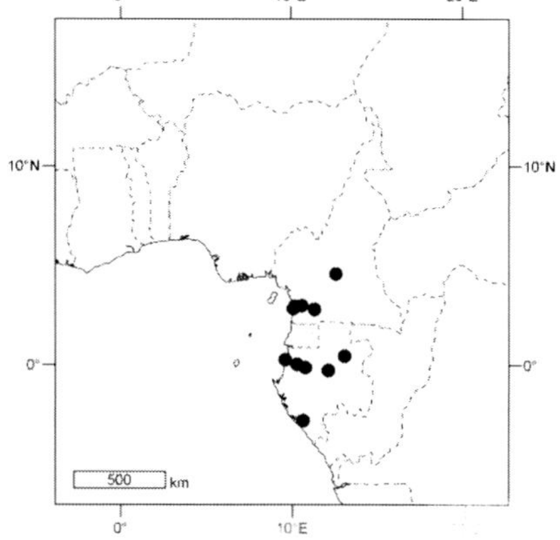

e. ***Ficus subsagittifolia*** VU

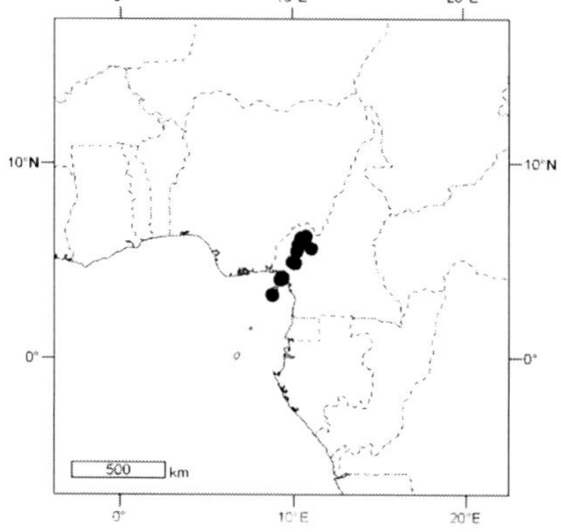

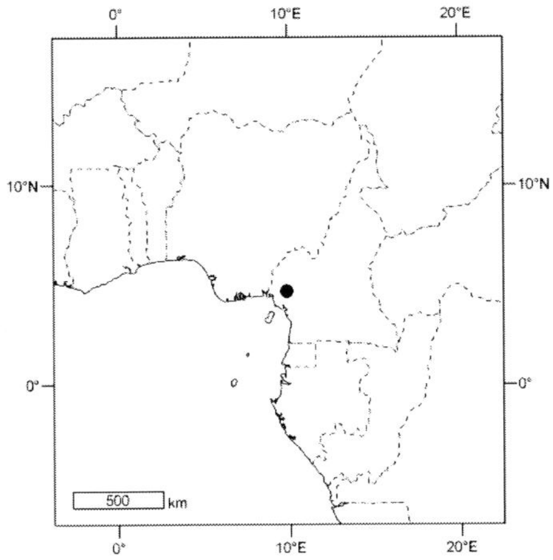

a. ***Ardisia alabastro-alata*** VU (Myrsinaceae)

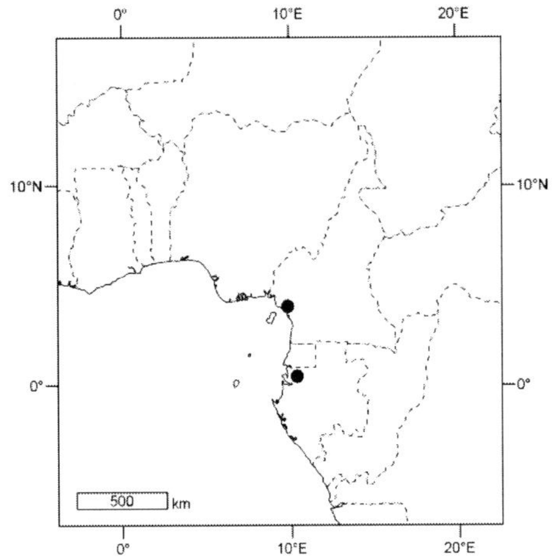

b. ***Ardisia atrobullata*** EN (Myrsinaceae)

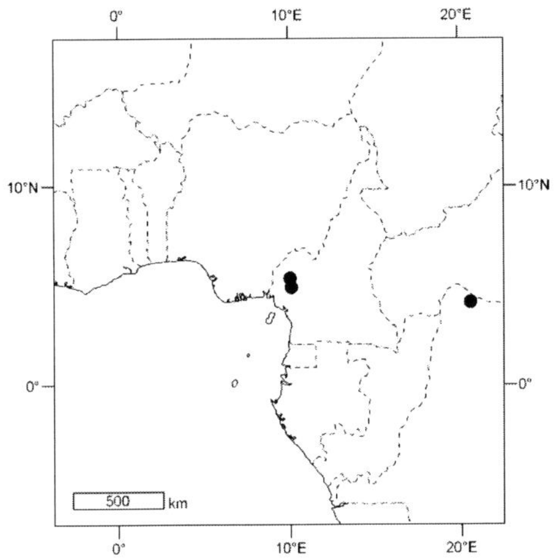

c. ***Ardisia conraui*** EN (Myrsinaceae)

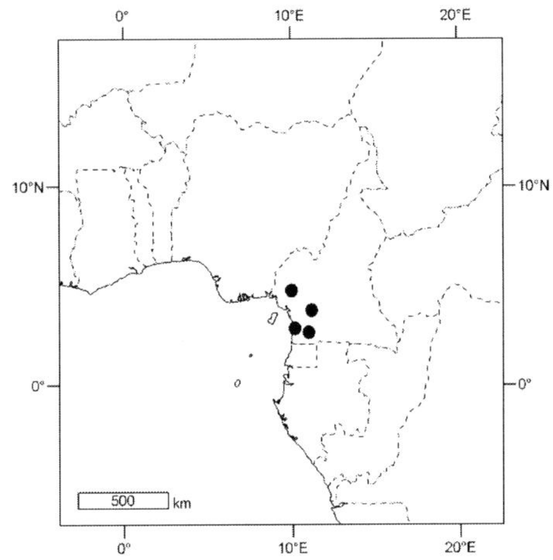

d. ***Ardisia dewitiana*** EN (Myrsinaceae)

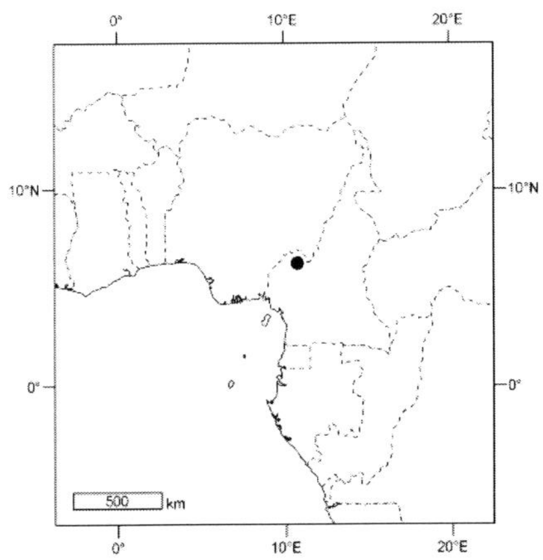

e. ***Ardisia dom*** CR (Myrsinaceae)

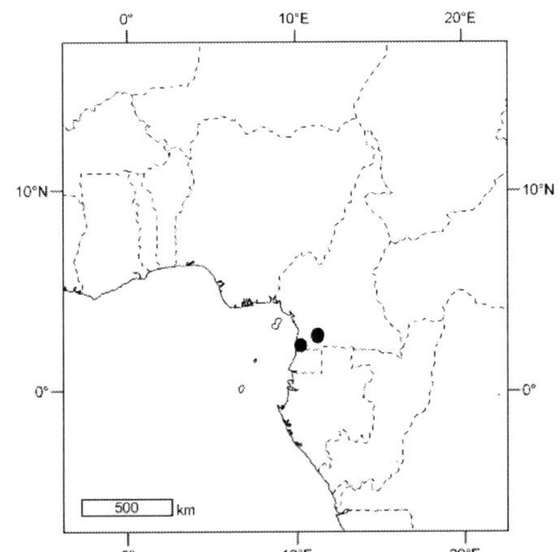

f. ***Ardisia ebolowensis*** EN (Myrsinaceae)

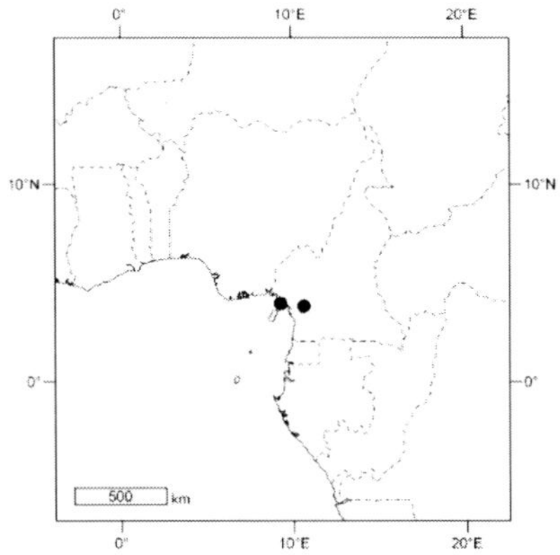

a. ***Ardisia etindensis*** EN (Myrsinaceae)

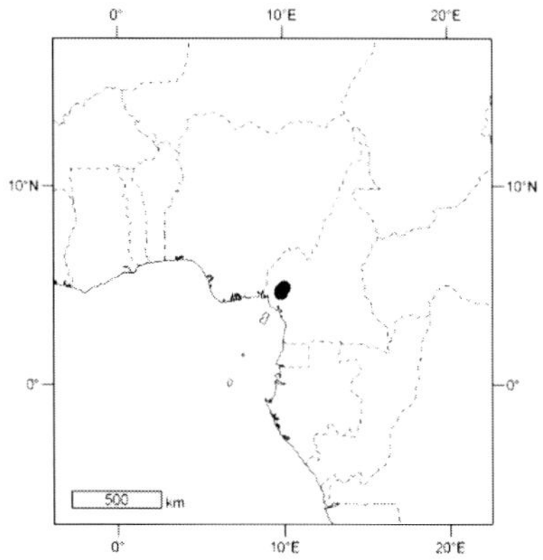

b. ***Ardisia koupensis*** EN (Myrsinaceae)

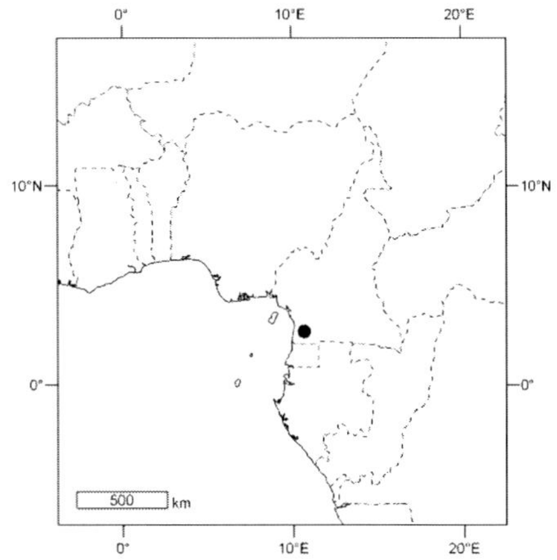

c. ***Ardisia letouzeyi*** CR (Myrsinaceae)

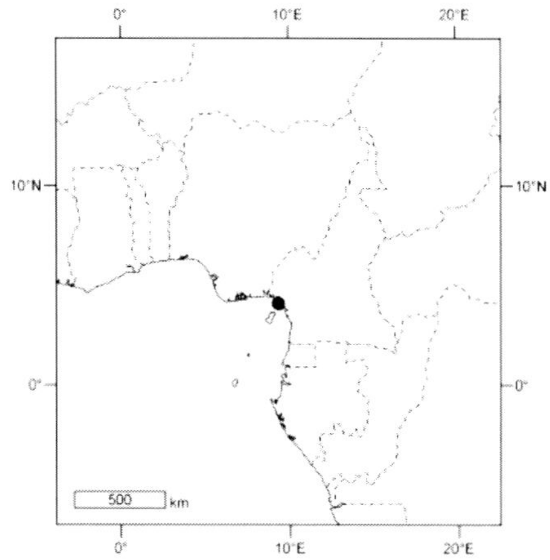

d. ***Ardisia oligantha*** CR (Myrsinaceae)

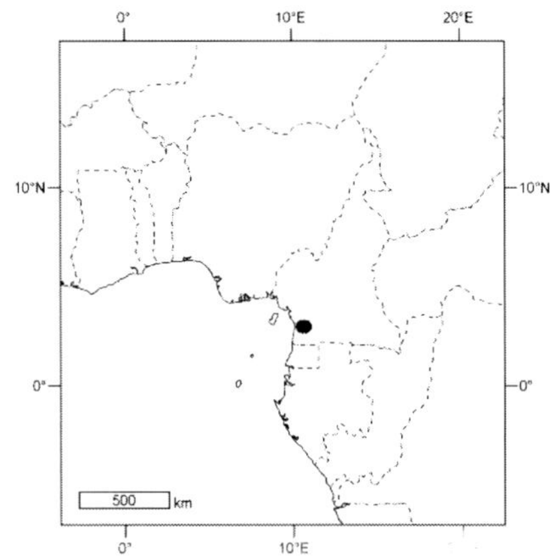

e. ***Ardisia platyphylla*** EN

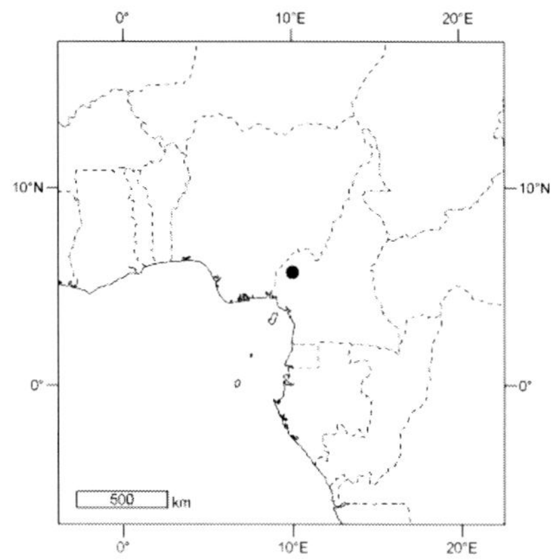

adenia CR (Myrsinaceae)

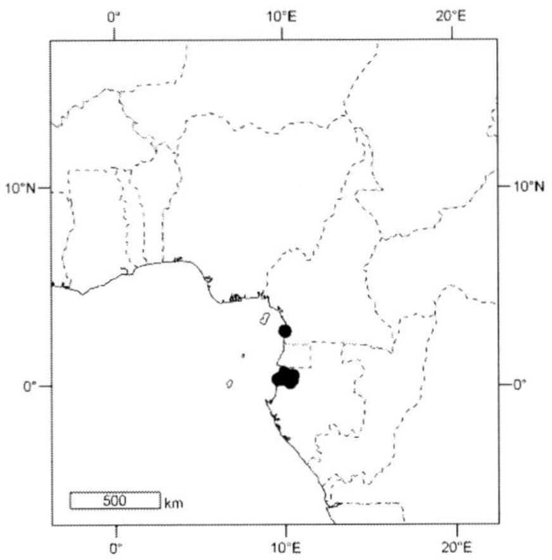

a. ***Ardisia sadebeckiana*** EN (Myrsinaceae)

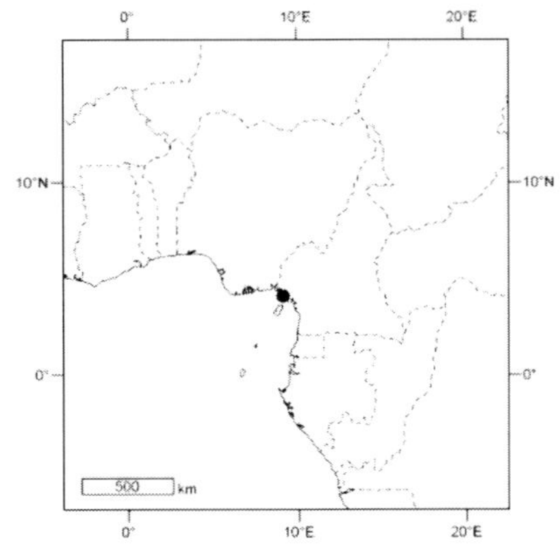

b. ***Ardisia schlechteri*** CR (Myrsinaceae)

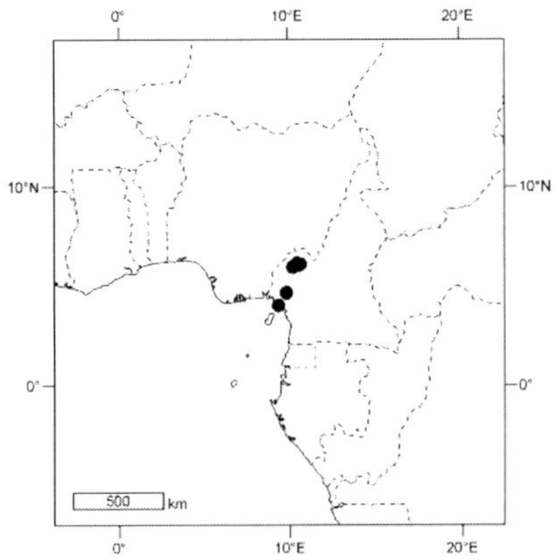

c. ***Embelia mildbraedii*** VU (Myrsinaceae)

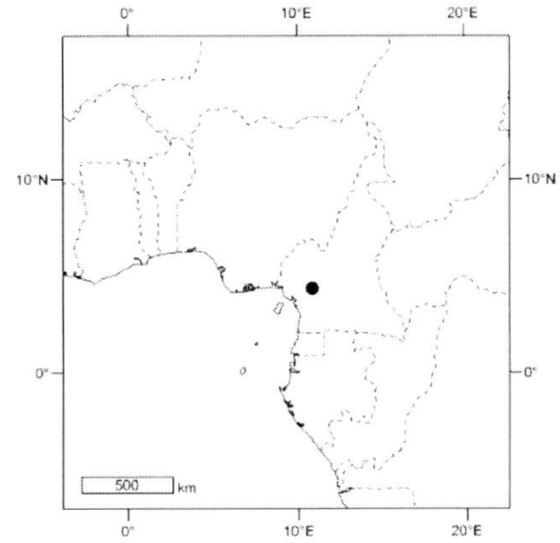

d. ***Eugenia ancorifera*** CR (Myrtaceae)

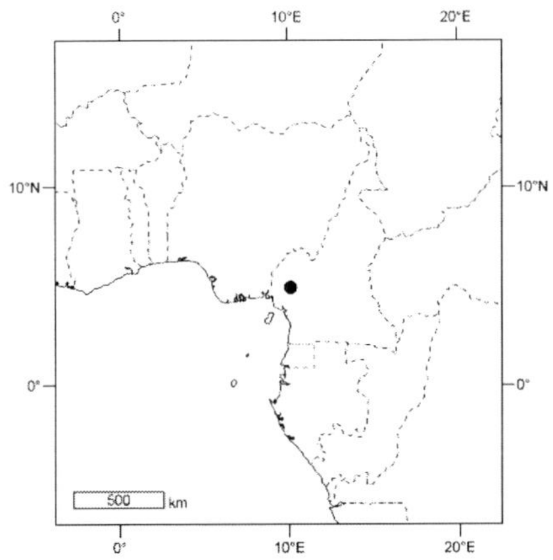

e. ***Eugenia bucholzii*** CR (Myrtaceae)

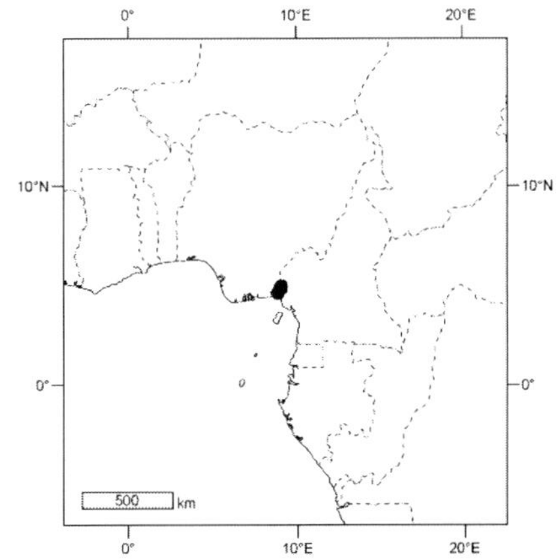

f. ***Eugenia dusenii*** VU (Myrtaceae)

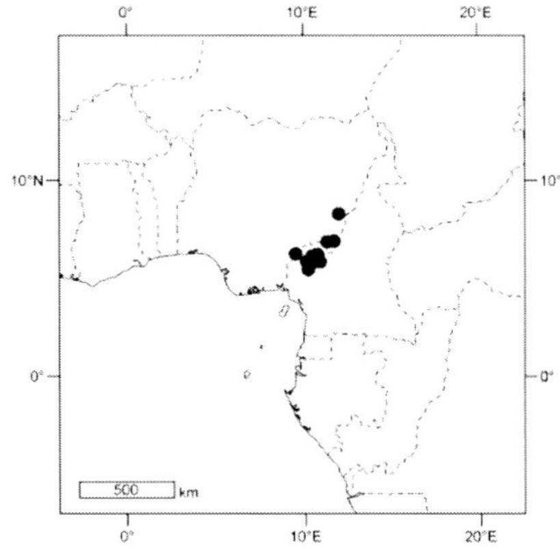

a. ***Eugenia gilgii*** CR (Myrtaceae)

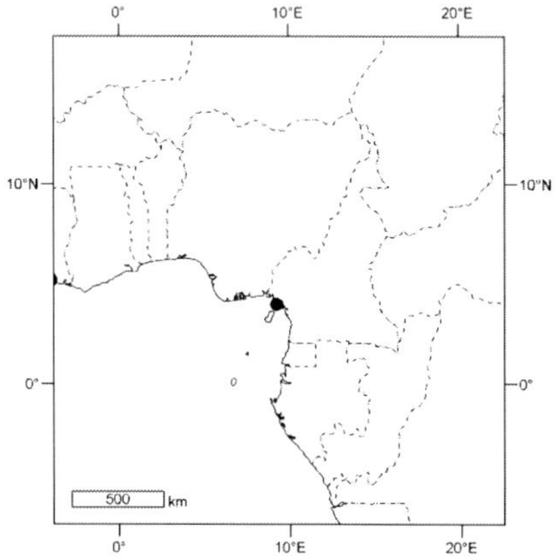

b. ***Eugenia kalbreyeri*** VU (Myrtaceae)

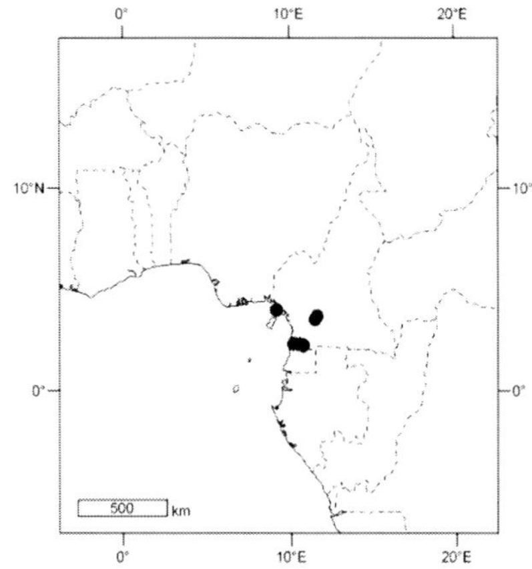

c. ***Eugenia kameruniana*** VU (Myrtaceae)

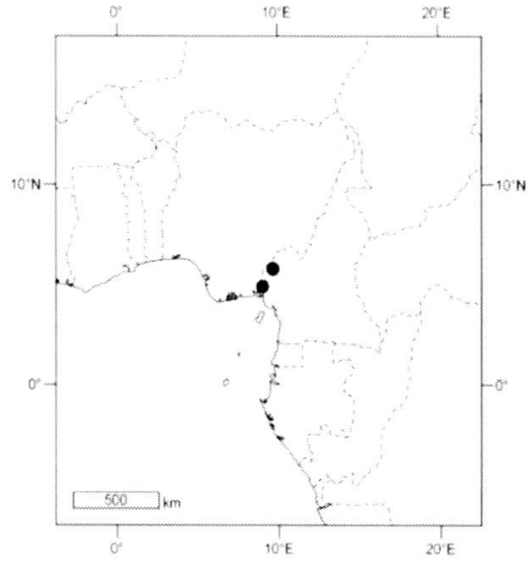

d. ***Campylospermum dusenii*** EN (Ochnaceae)

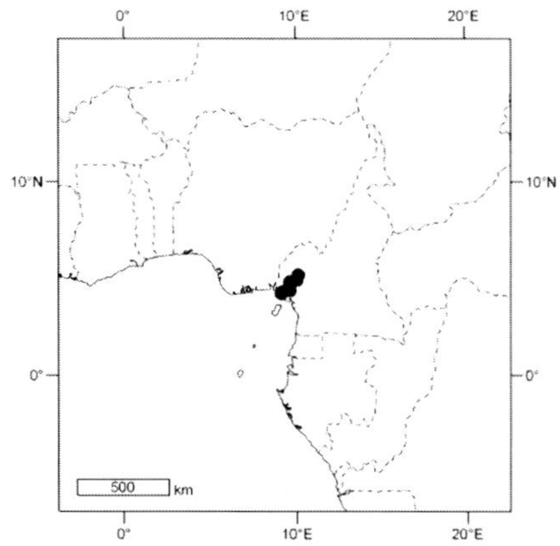

e. ***Campylospermum letouzeyi*** VU (Ochnaceae)

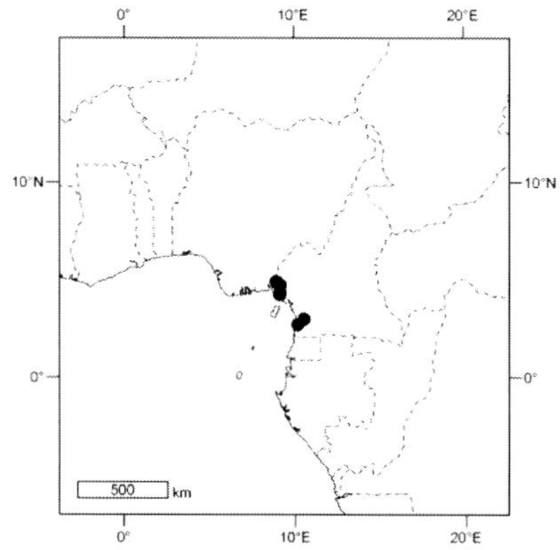

f. ***Campylospermum umbricola*** EN (Ochnaceae)

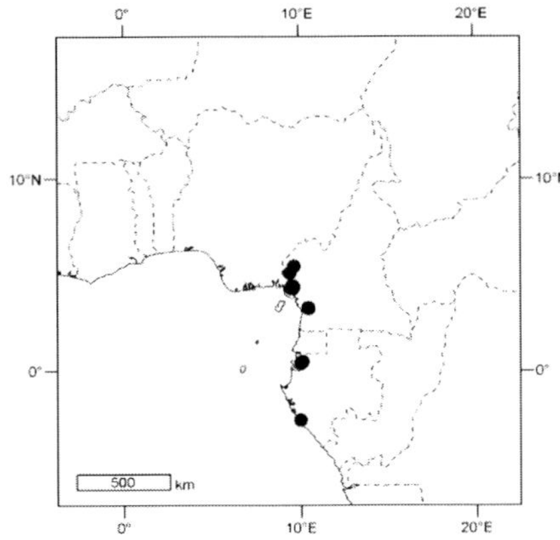

a. ***Idertia axillaris*** VU (Ochnaceae)

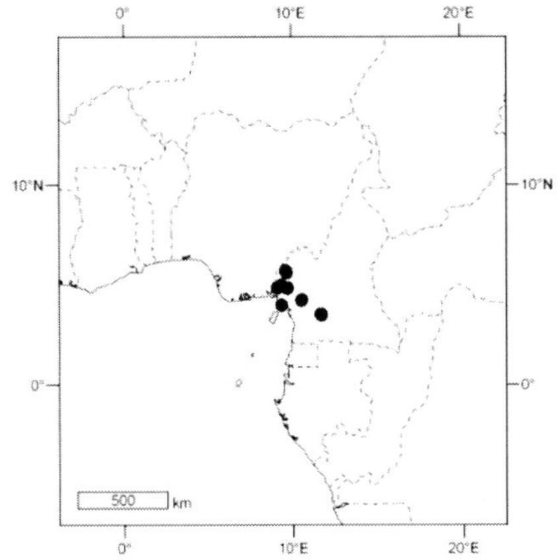

b. ***Lophira alata*** VU (Ochnaceae)

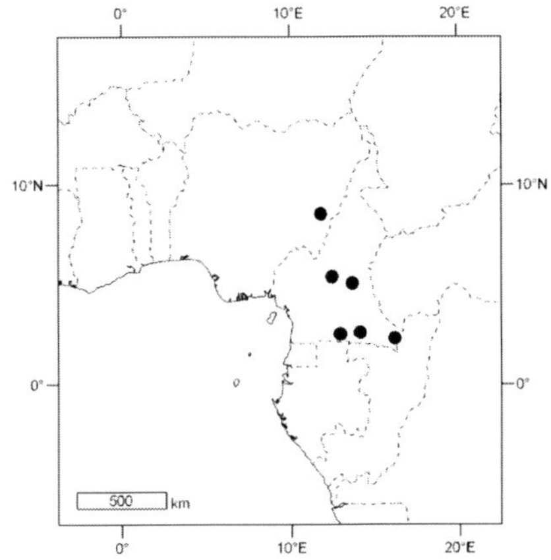

c. ***Ochna calodendron*** VU (Ochnaceae)

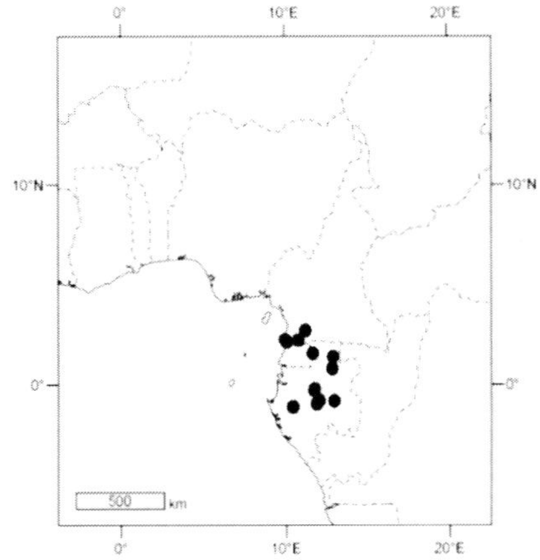

d. ***Testuluea gabonensis*** VU (Ochnaceae)

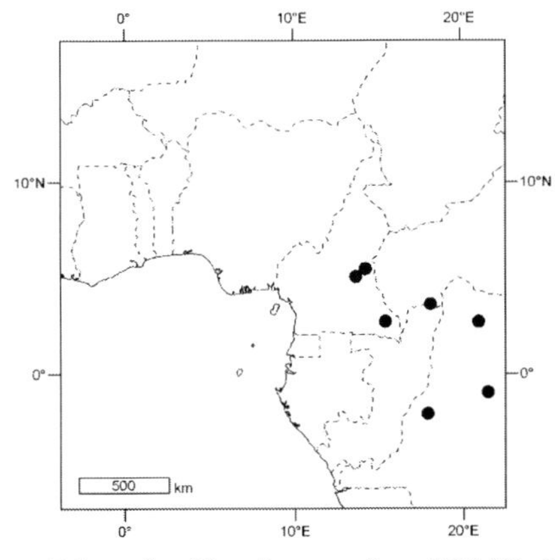

e. ***Rhopalopilia altescandens*** VU (Opiliaceae)

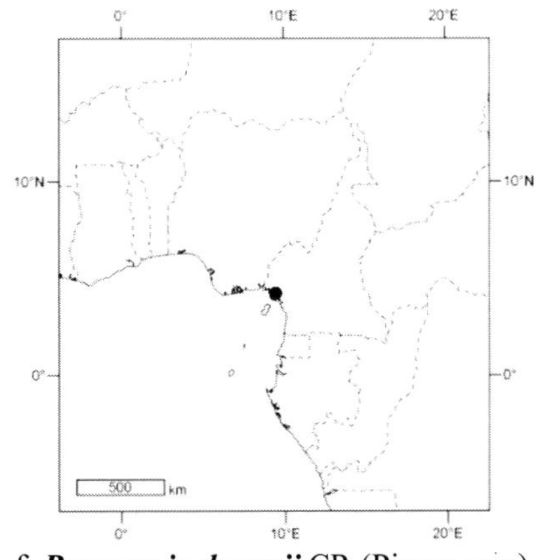

f. ***Peperomia dusenii*** CR (Piperaceae)

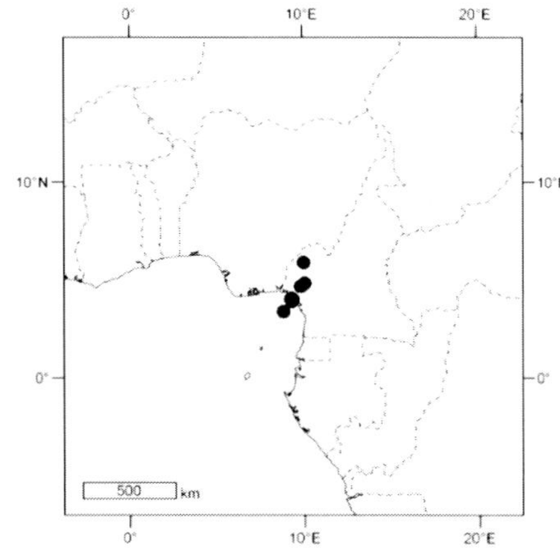

a. ***Peperomia kamerunana*** EN (Piperaceae)

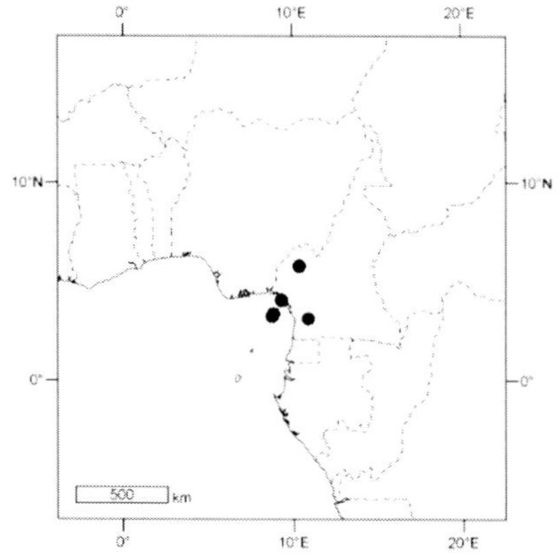

b. ***Peperomia laeteviridis*** EN (Piperaceae)

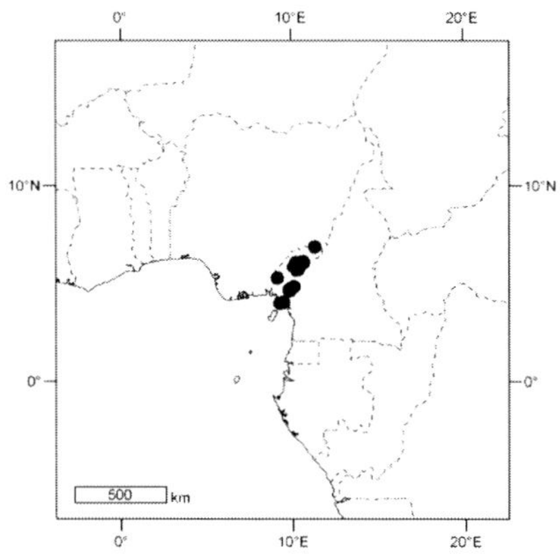

c. ***Peperomia thomeana*** VU (Piperaceae)

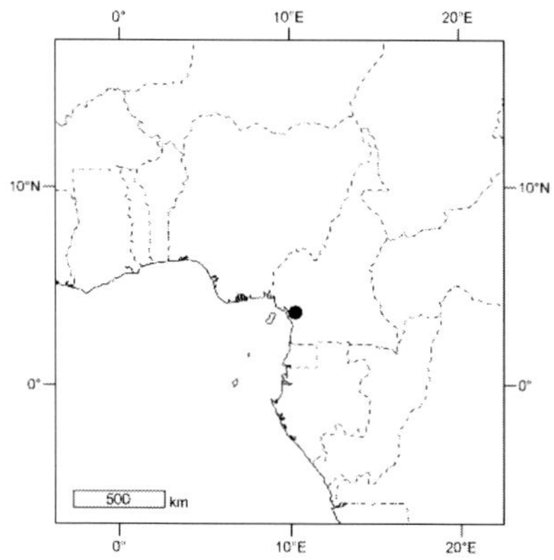

d. ***Dicraeanthus zehnderi*** CR (Podostemaceae)

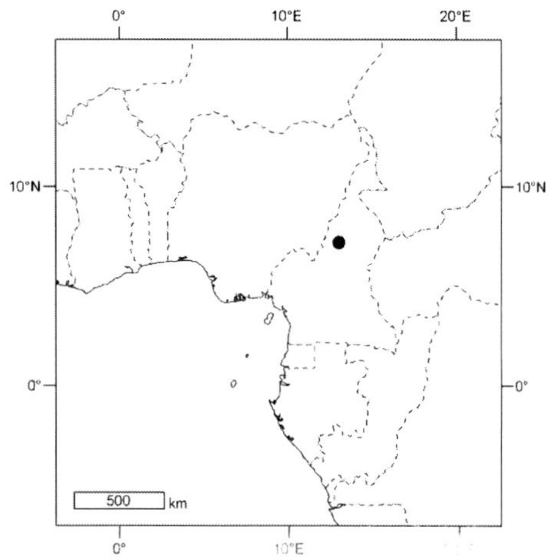

e. ***Djinga felicis*** CR (Po[illegible]

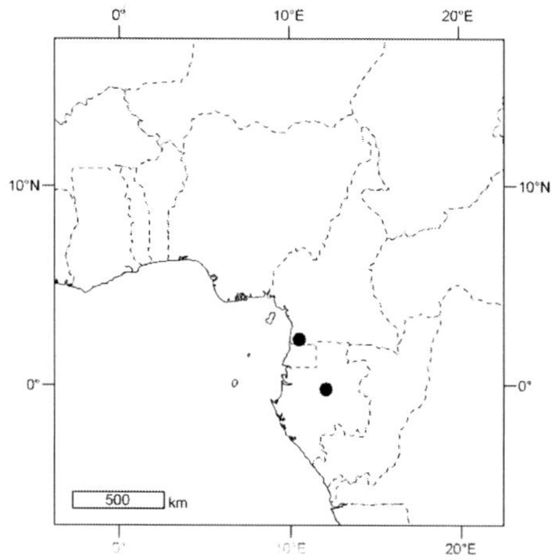

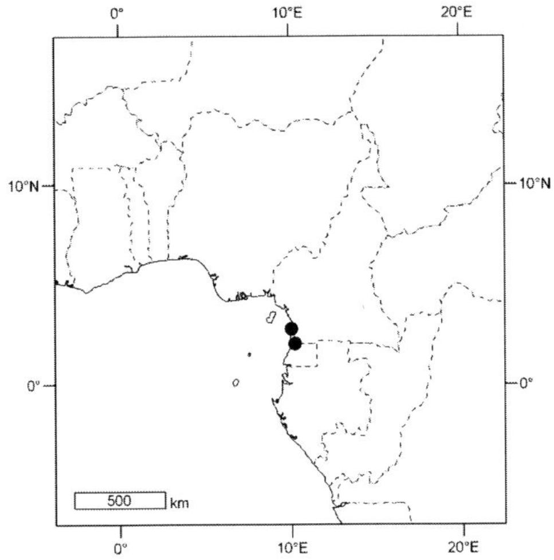

a. ***Ledermanniella bosii*** EN (Podostemaceae)

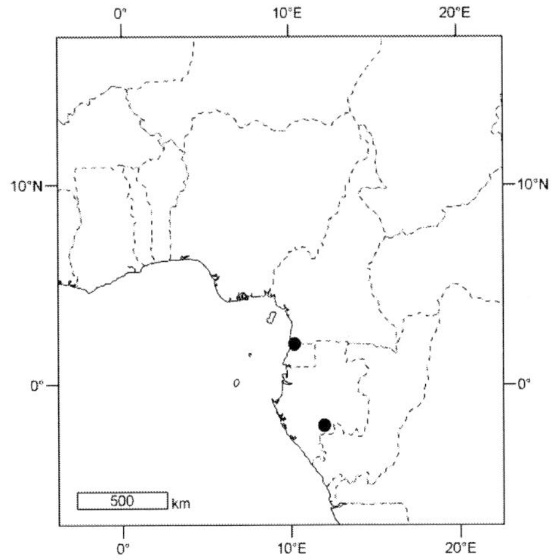

b. ***Ledermanniella boumiensis*** EN (Podostemaceae)

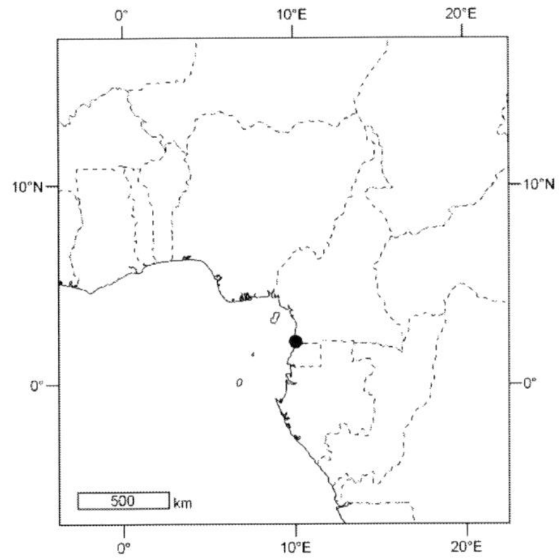

c. ***Ledermanniella kamerunensis*** CR (Podostemaceae)

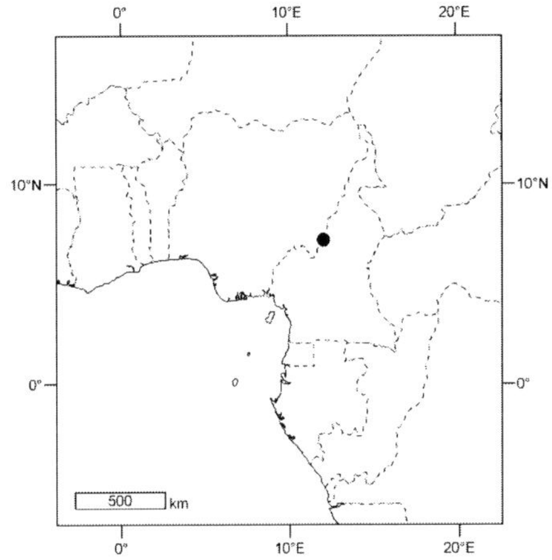

d. ***Ledermanniella aloides*** EN (Podostemaceae)

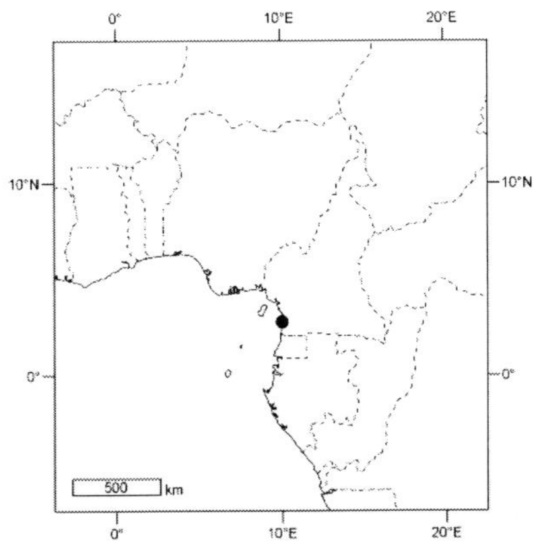

e. ***Ledermanniella batangensis*** CR (Podostemaceae)

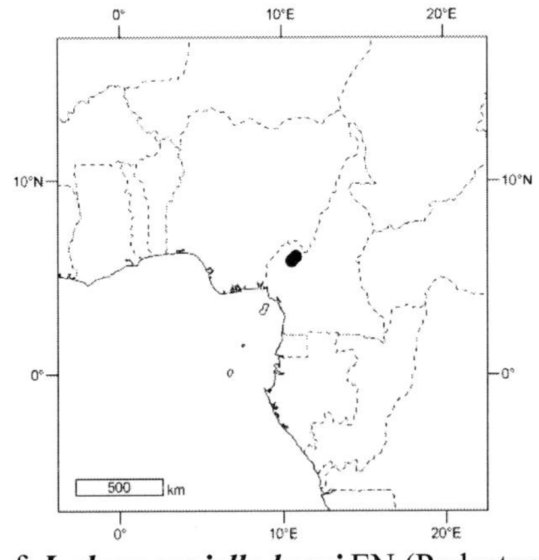

f. ***Ledermanniella keayi*** EN (Podostemaceae)

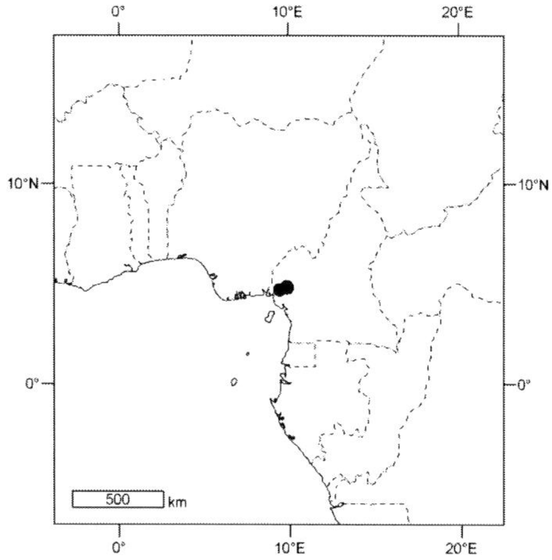

a. ***Ledermanniella letouzeyi*** EN
(Podostemaceae)

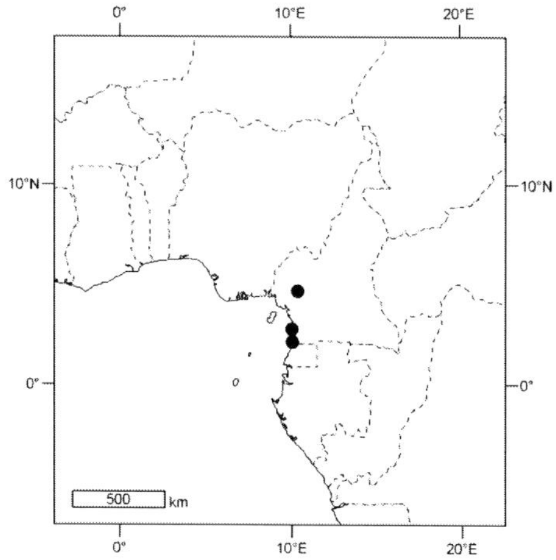

b. ***Ledermanniella linearifolia*** EN
(Podostemaceae)

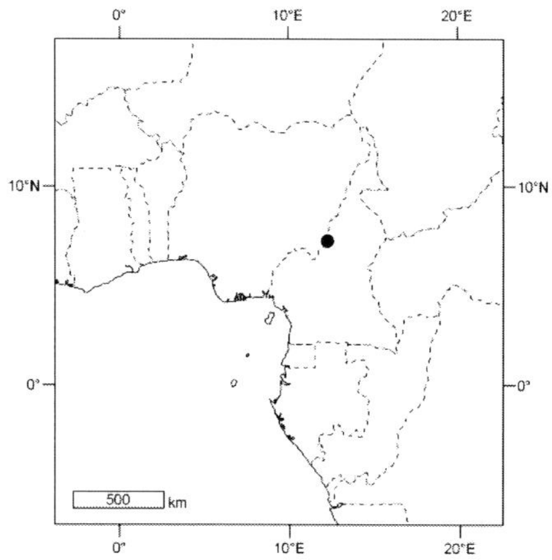

c. ***Ledermanniella monandra*** CR
(Podostemaceae)

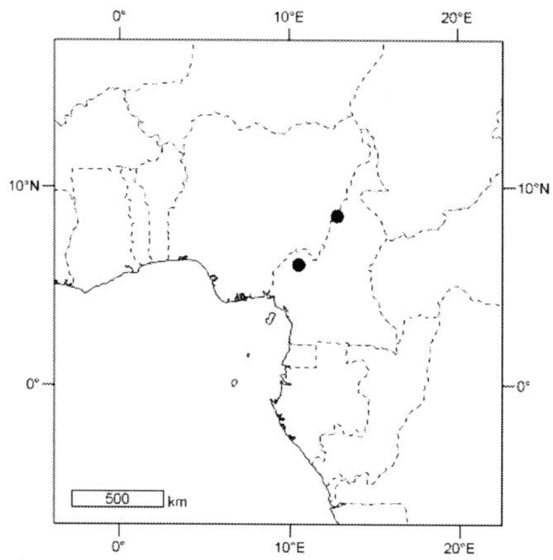

d. ***Ledermanniella musciformis*** EN
(Podostemaceae)

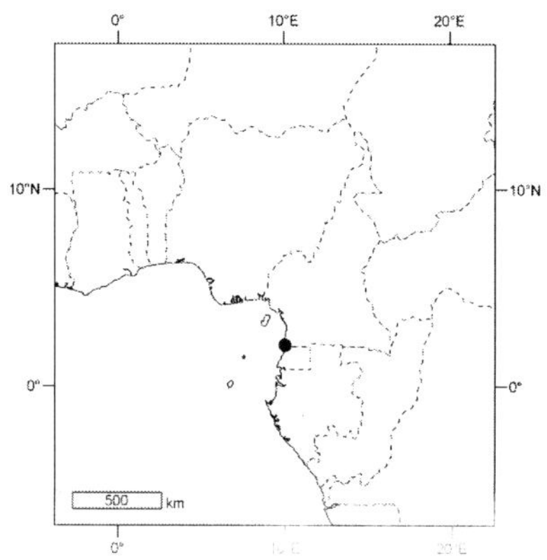

e. ***Ledermanniella ntemensis*** CR
(Podostemaceae)

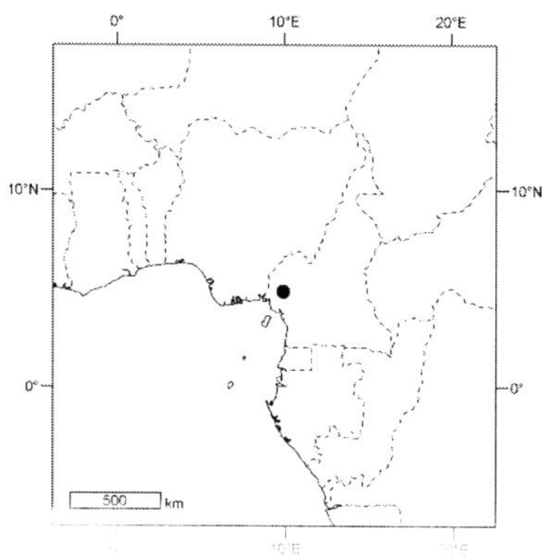

f. ***Ledermanniella onan***[illegible]
(Podostemaceae)

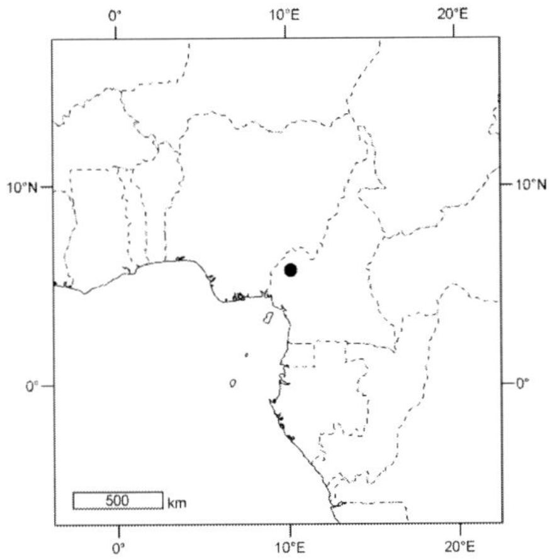

a. ***Ledermanniella pollardiana*** CR (Podostemaceae)

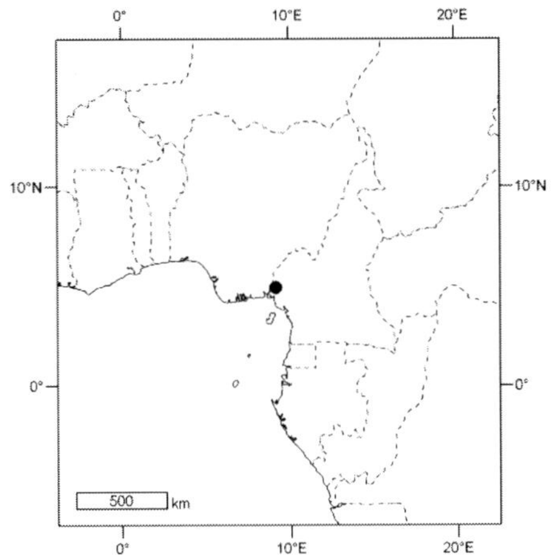

b. ***Ledermanniella prasina*** VU (Podostemaceae)

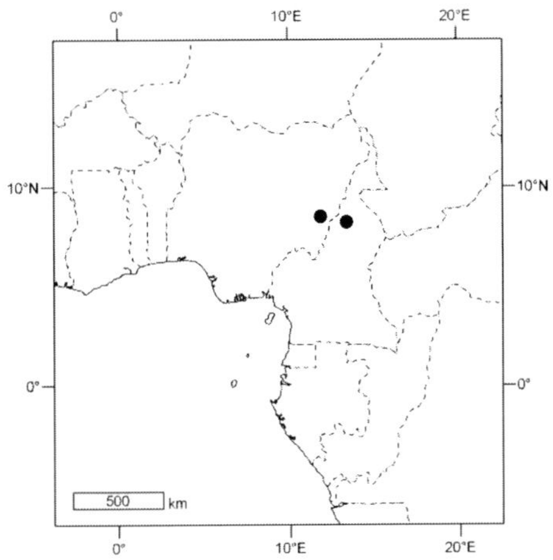

c. ***Ledermanniella raynaliorum*** EN (Podostemaceae)

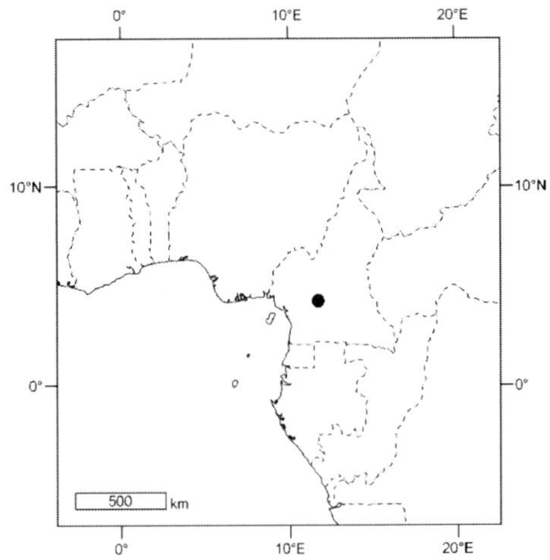

d. ***Ledermanniella sanagaensis*** VU (Podostemaceae)

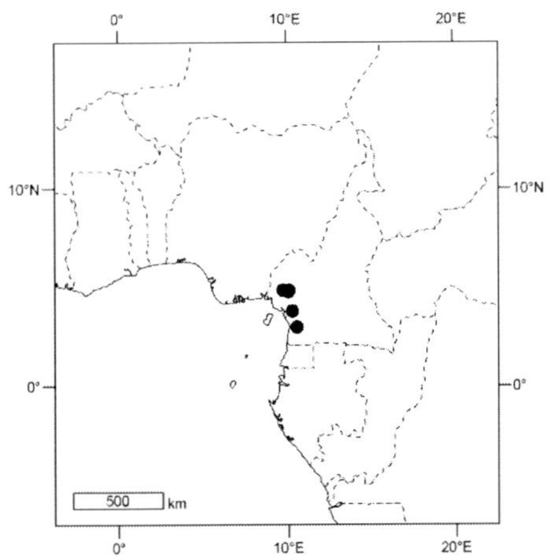

e. ***Ledermanniella thalloidea*** VU (Podostemaceae)

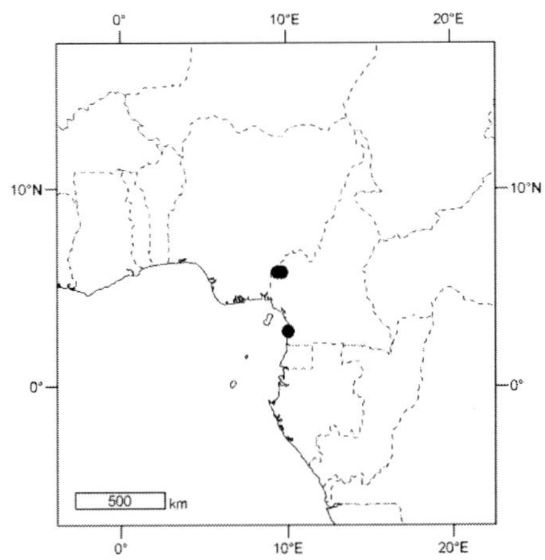

f. ***Ledermanniella variabilis*** EN (Podostemaceae)

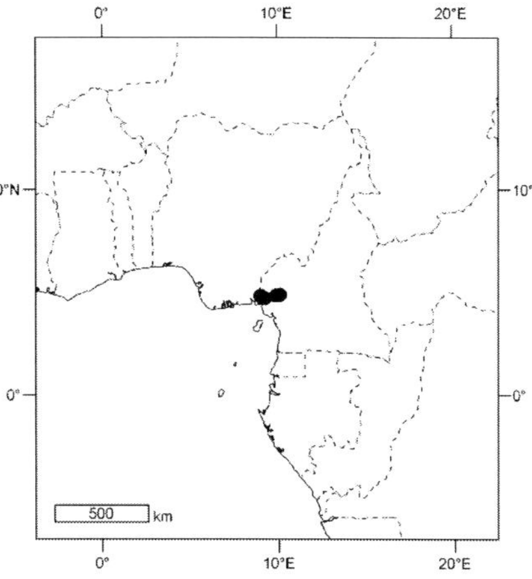

a. ***Macropodiella pellucida*** EN (Podostemaceae)

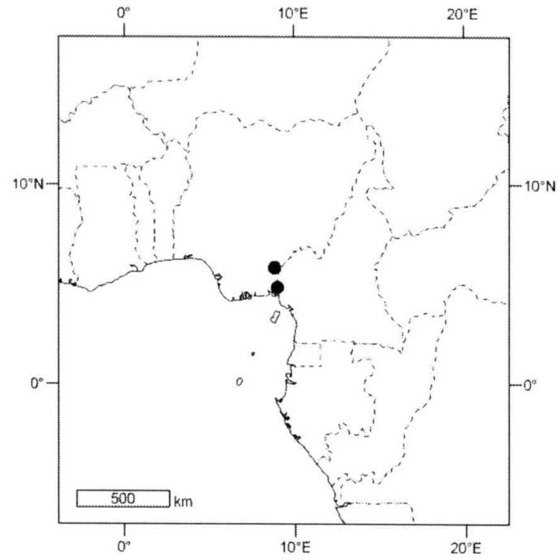

b. ***Saxicolella flabellata*** EN (Podostemaceae)

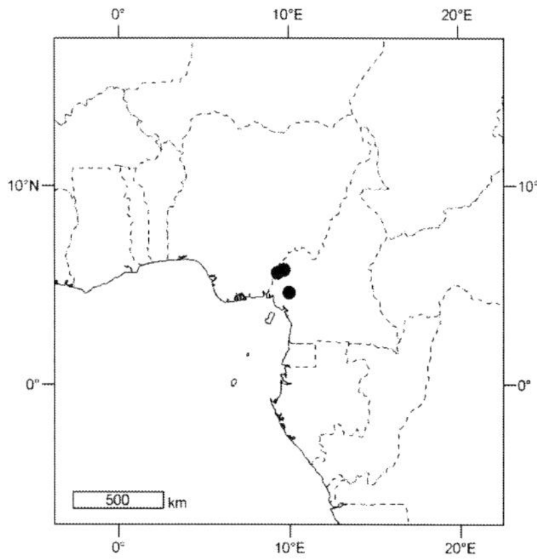

c. ***Saxicolella laciniata*** EN (Podostemaceae)

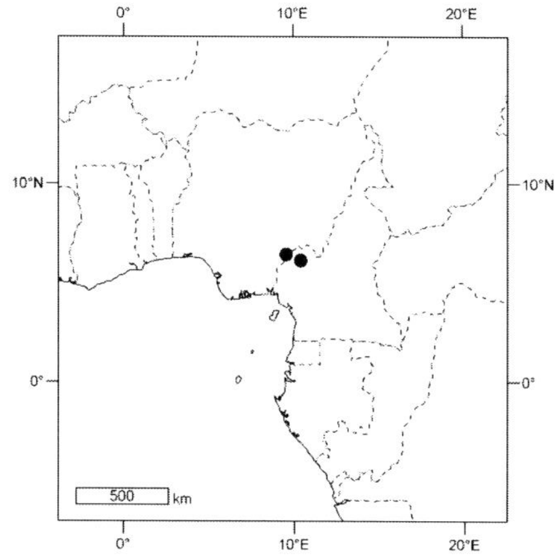

d. ***Saxicolella marginalis*** EN (Podostemaceae)

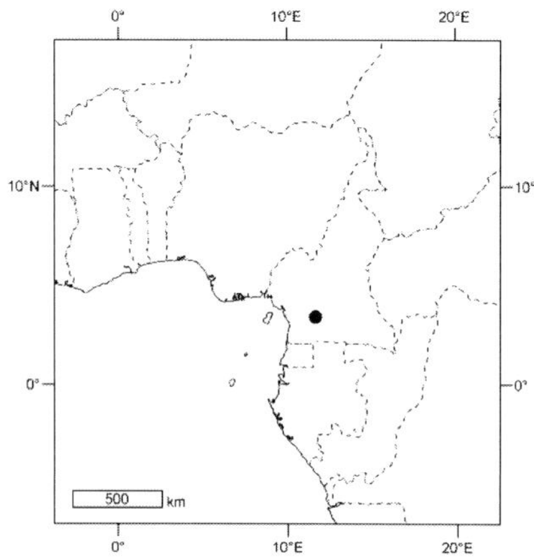

e. ***Saxicolella nana*** CR (Podostemaceae)

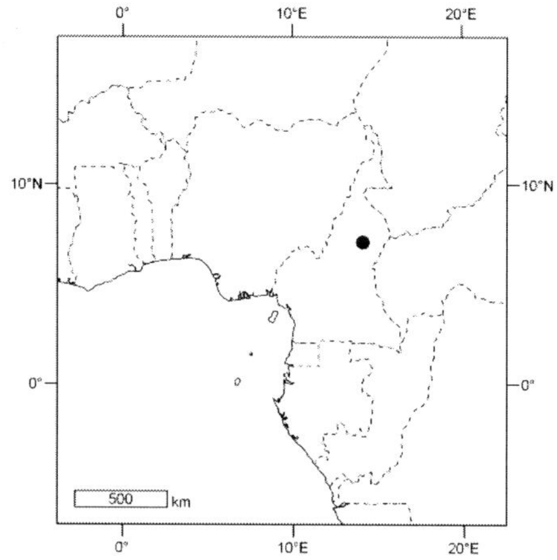

f. ***Stonesia ghoguei*** VU (Podostemaceae)

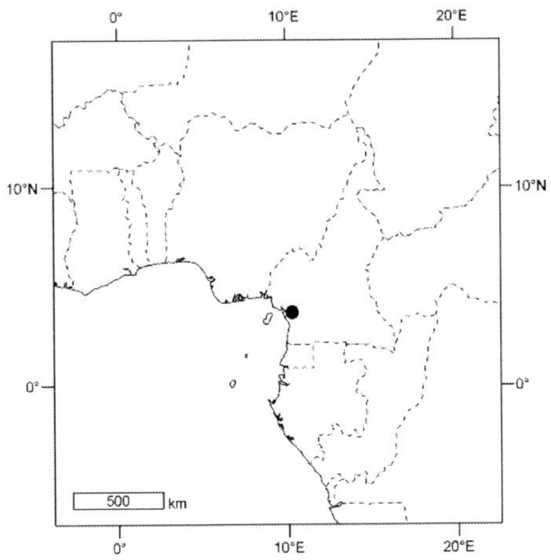

a. ***Winklerella dichotoma*** CR (Podostemaceae)

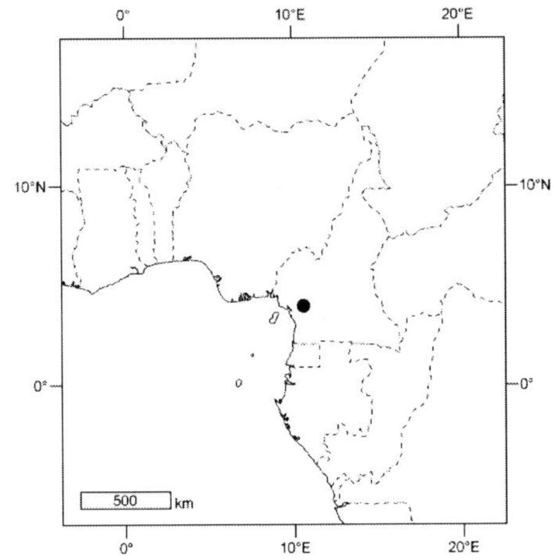

b. ***Zehnderia microgyna*** CR (Podostemaceae)

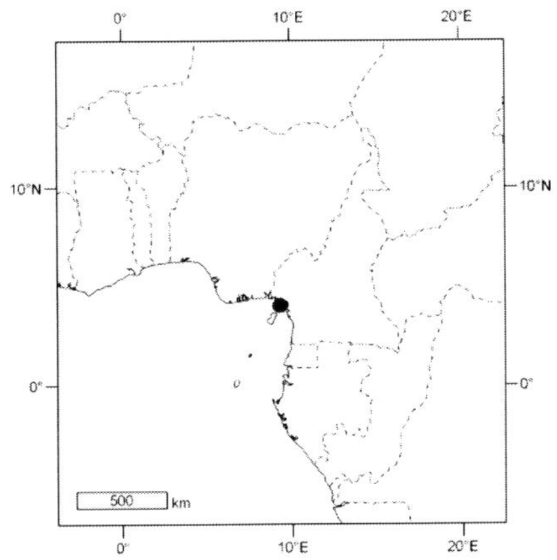

c. ***Polygala tenuicaulis*** subsp. ***tenuicaulis*** VU (Polygalaceae)

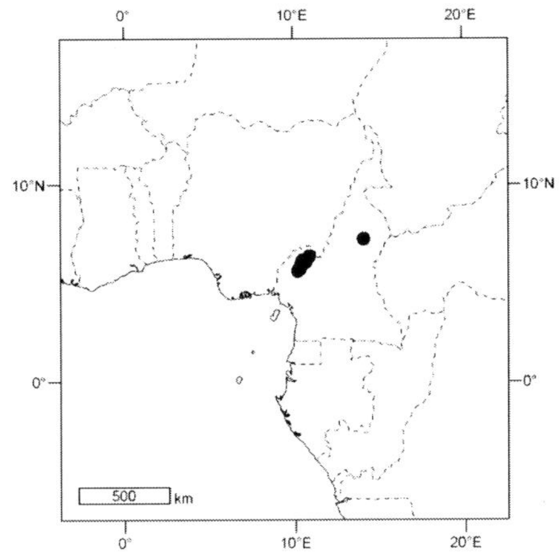

d. ***Polygala tenuicaulis*** subsp. ***tayloriana*** VU (Polygalaceae)

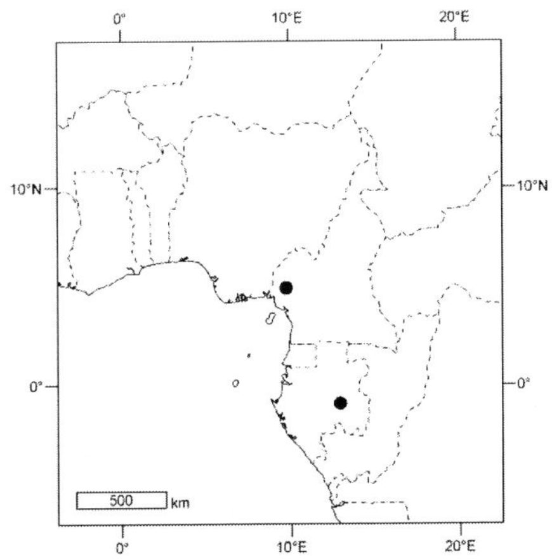

e. ***Cassipourea acuminata*** EN (Rhizophoraceae)

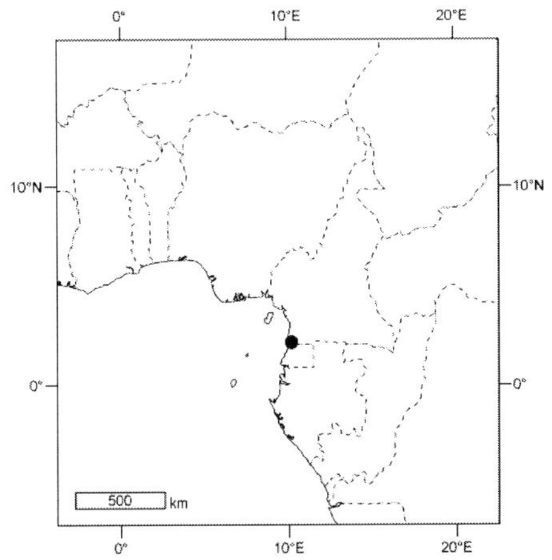

f. ***Cassipourea alternifolia*** CR (Rhizophoraceae)

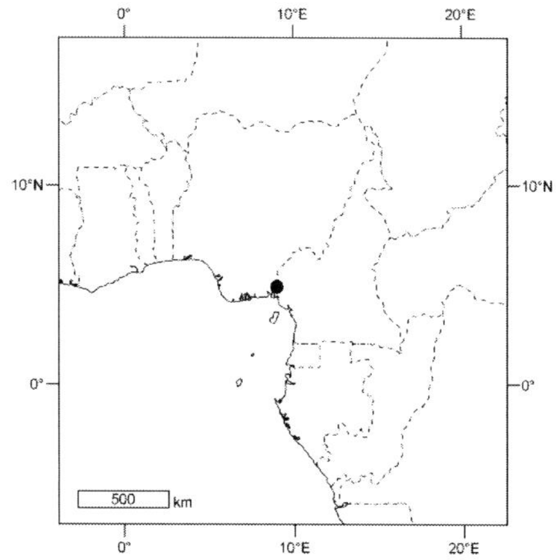

a. ***Cassipourea korupensis*** CR (Rhizophoraceae)

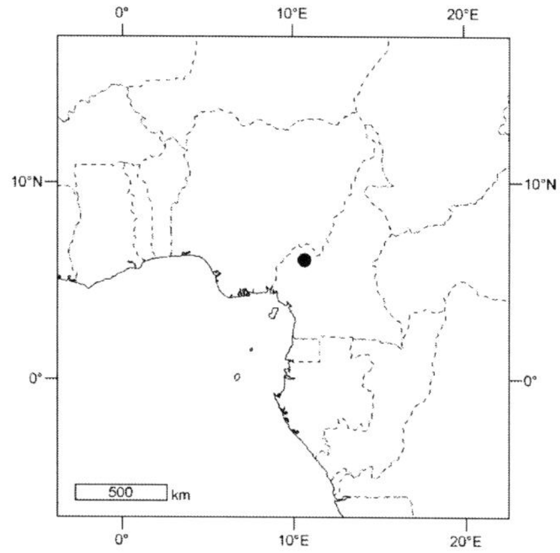

b. ***Alchemilla fischeri*** subsp. ***camerunensis*** CR (Rosaceae)

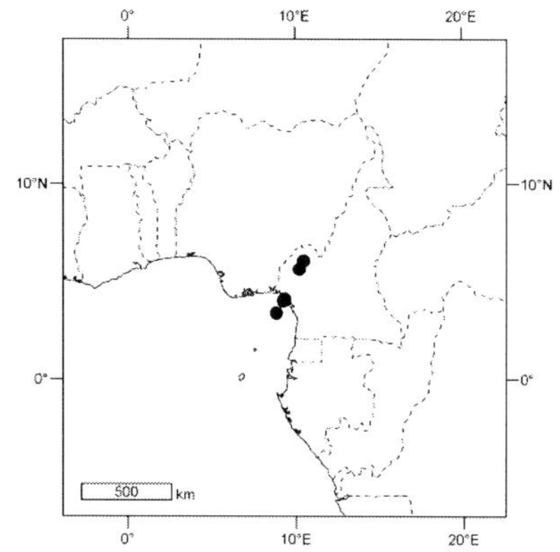

c. ***Anthospermum asperuloides*** EN(Rubiaceae)

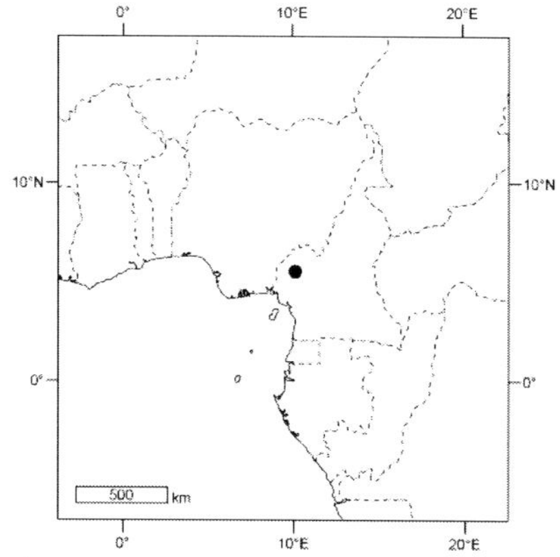

d. ***Argocoffeopsis fosimondi*** CR (Rubiaceae)

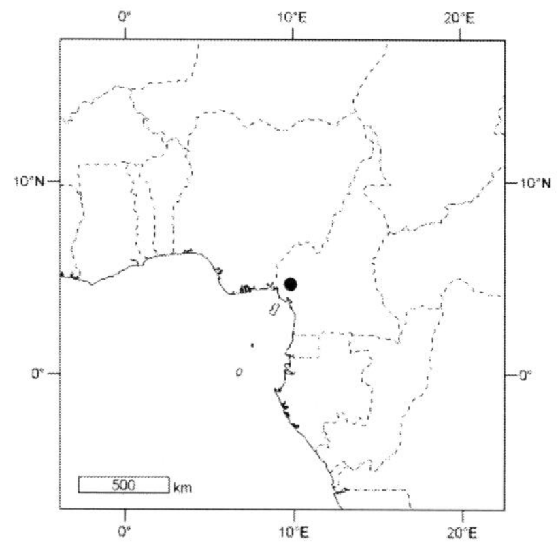

e. ***Argocoffeopsis kupensis*** CR (Rubiaceae)

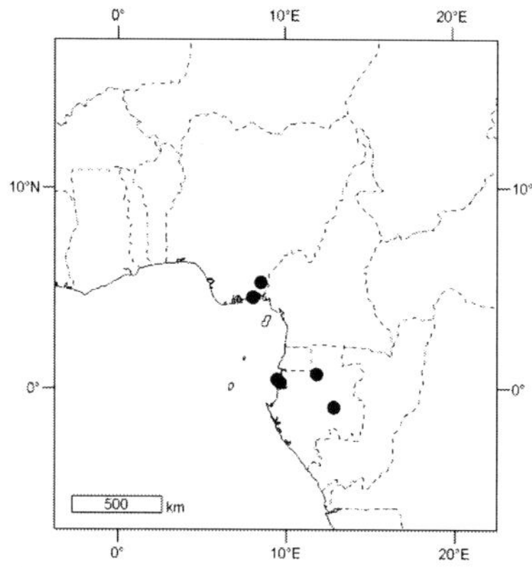

f. ***Argocoffeopsis pulchella*** EN (Rubiaceae)

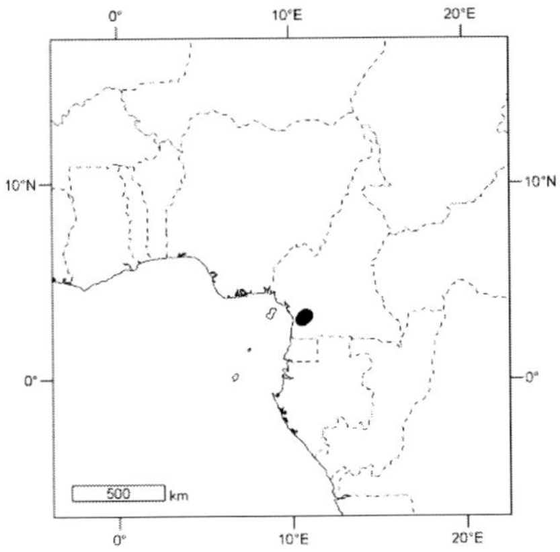

a. ***Argocoffeopsis spathulata*** VU (Rubiaceae)

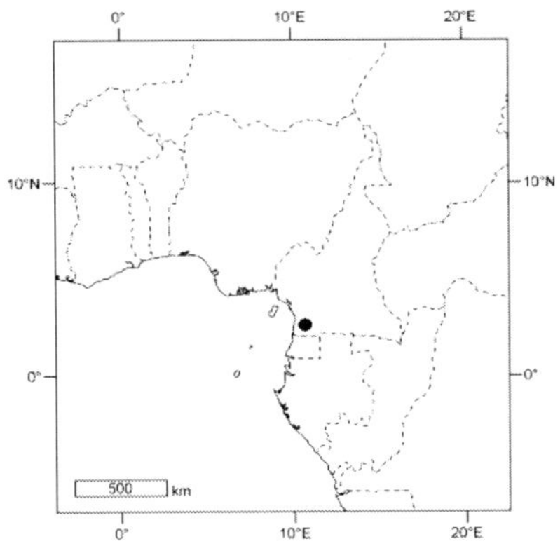

b. ***Aulacocalyx camerooniana*** CR (Rubiaceae)

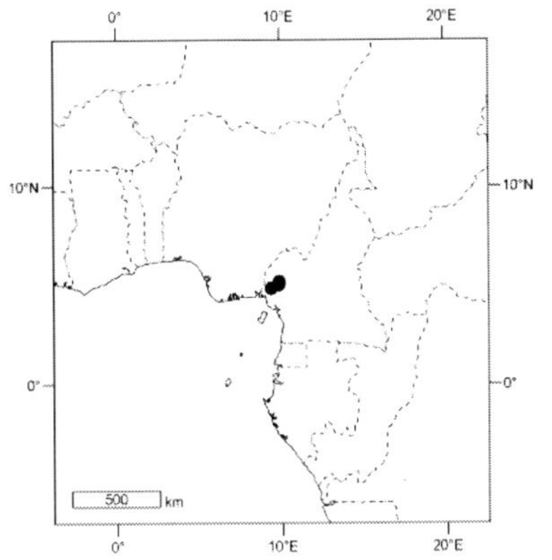

c. ***Aulacocalyx mapiana*** EN (Rubiaceae)

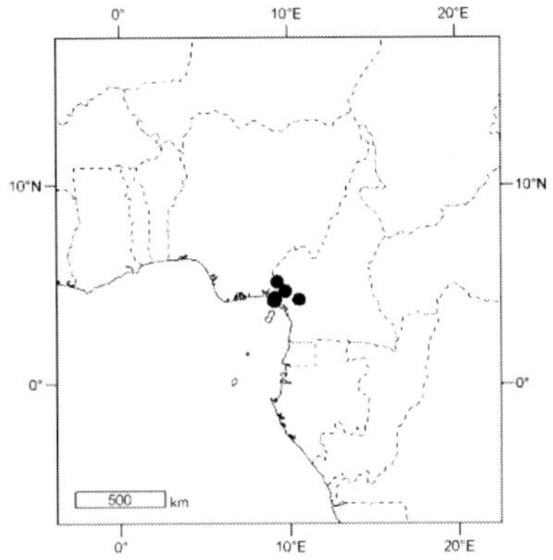

d. ***Belonophora ongensis*** CR (Rubiaceae)

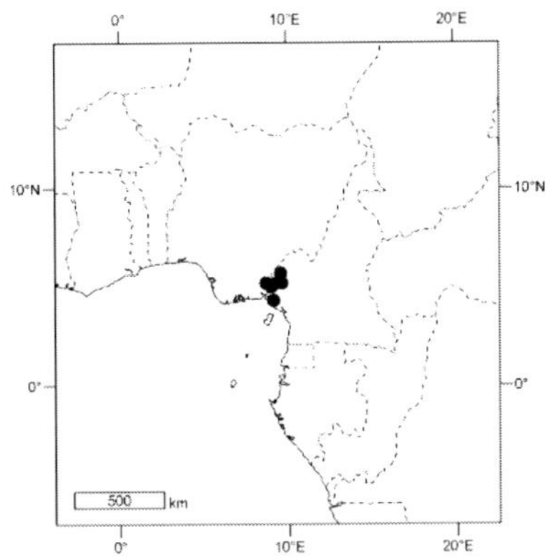

e. ***Belonophora talbotii*** CR (Rubiaceae)

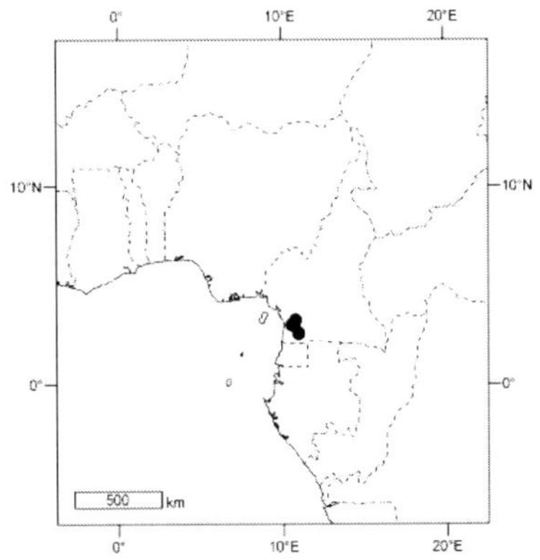

f. ***Bertiera heterophylla*** EN (Rubiaceae)

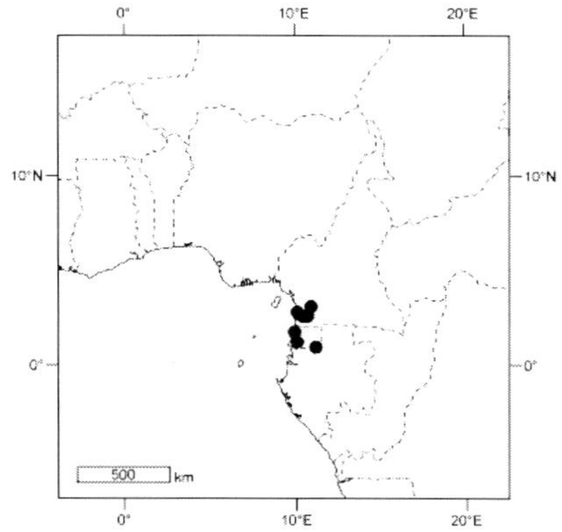

a. ***Bertiera rosseeliana*** VU (Rubiaceae)

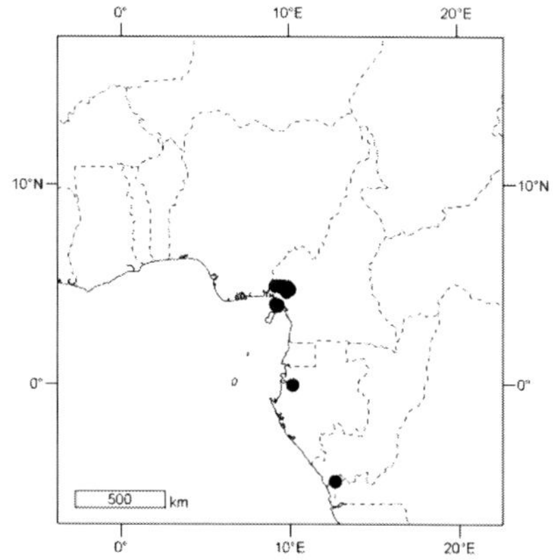

b. ***Calochone acuminata*** VU (Rubiaceae)

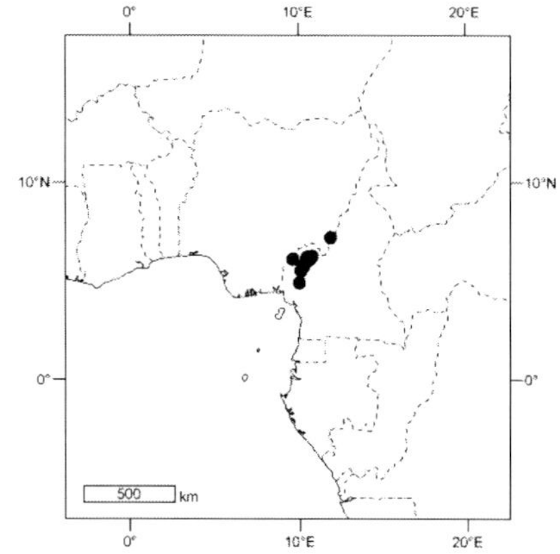

c. ***Chassalia laikomensis*** CR (Rubiaceae)

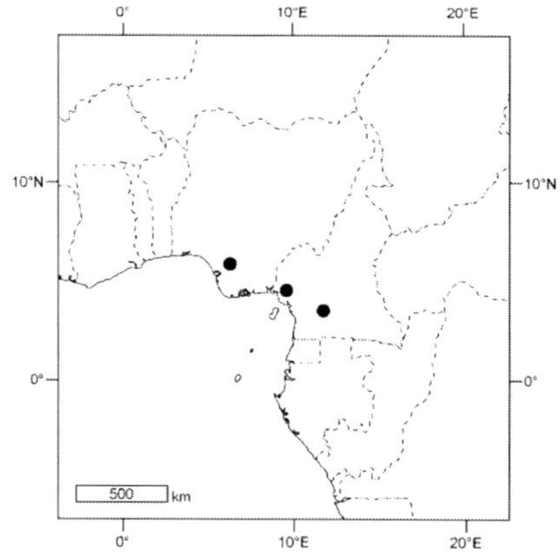

d. ***Chassalia manningii*** EN (Rubiaceae)

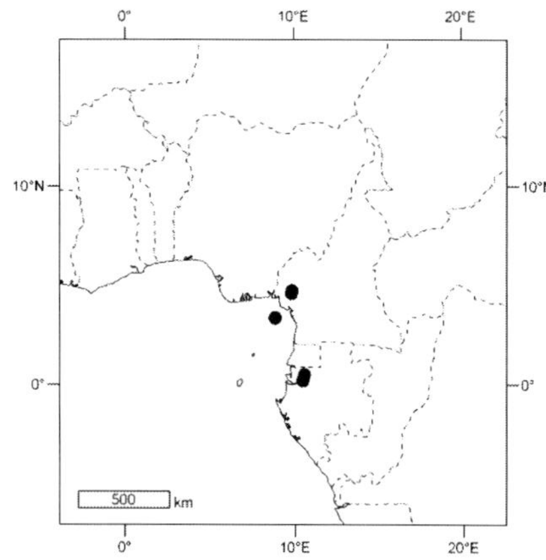

e. ***Chassalia petitiana*** VU (Rubiaceae)

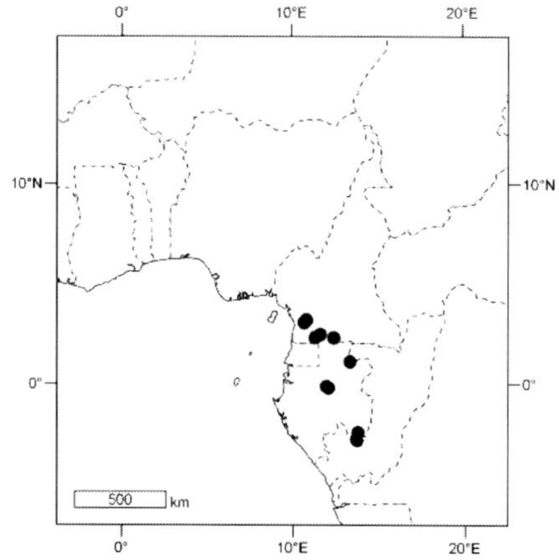

f. ***Chazaliella letouzeyi*** VU (Rubiaceae)

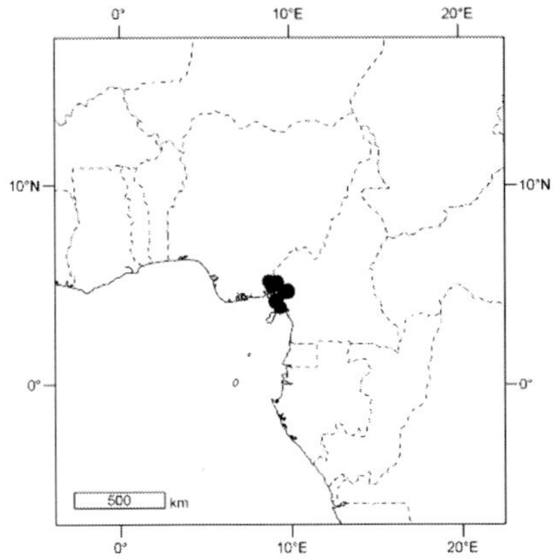

a. ***Chazaliella obanensis*** VU (Rubiaceae)

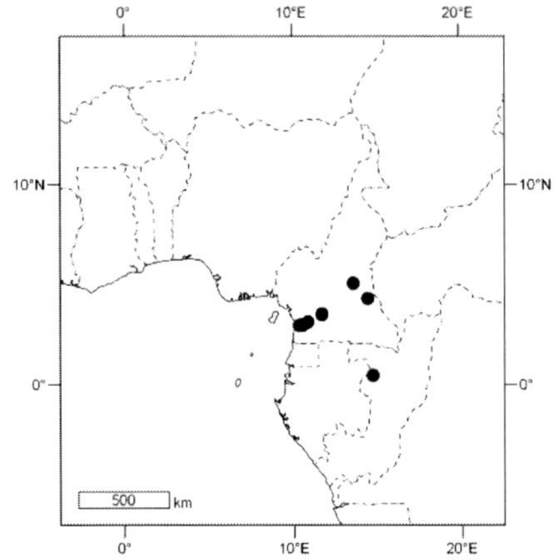

b. ***Chazaliella obovoidea*** subsp. ***villosistipula*** VU (Rubiaceae)

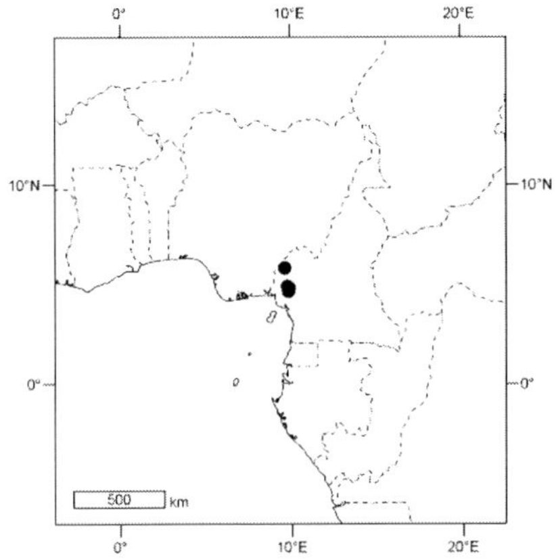

c. ***Coffea bakossii*** EN (Rubiaceae)

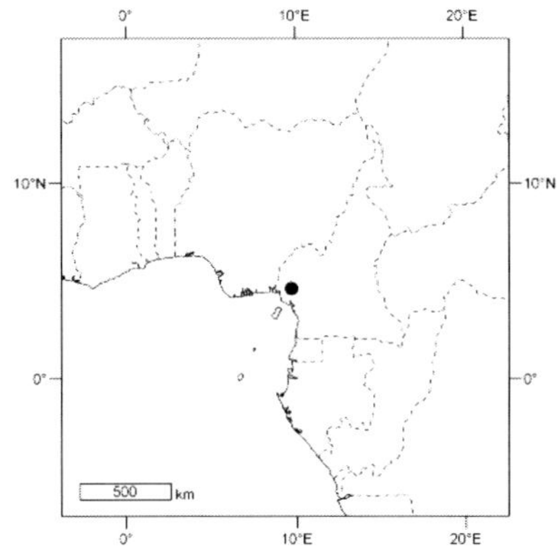

d. ***Coffea charrieriana*** CR (Rubiaceae)

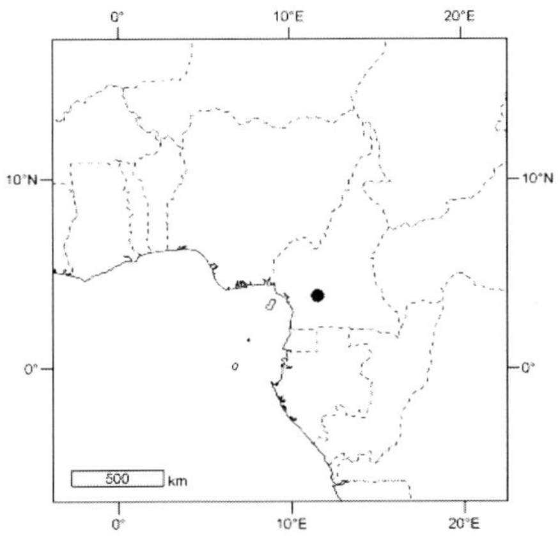

e. ***Coffea fotsoana*** CR (Rubiaceae)

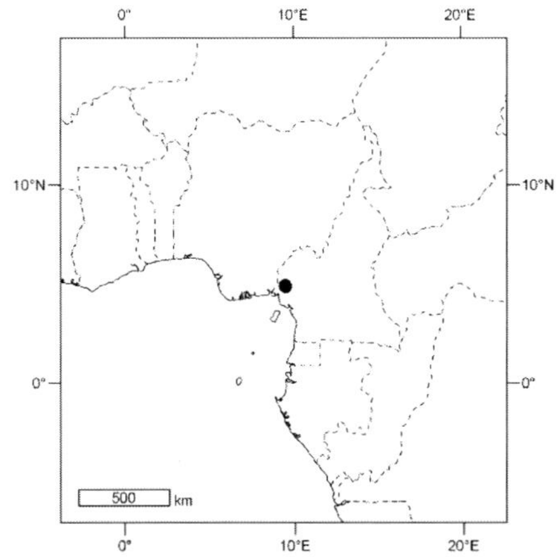

f. ***Coffea leonimontana*** CR (Rubiaceae)

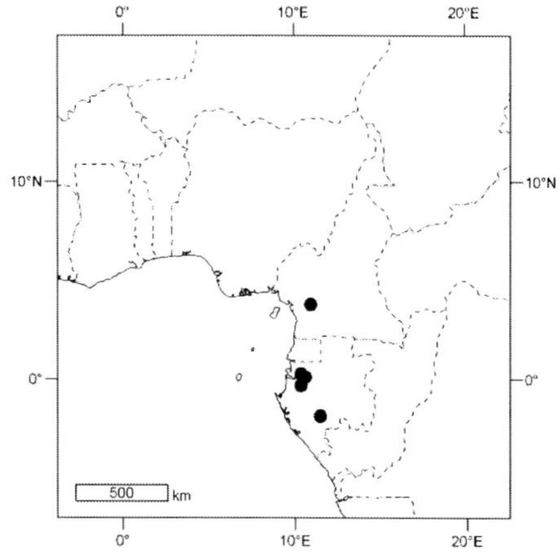

a. ***Coffea magnistipula*** EN (Rubiaceae)

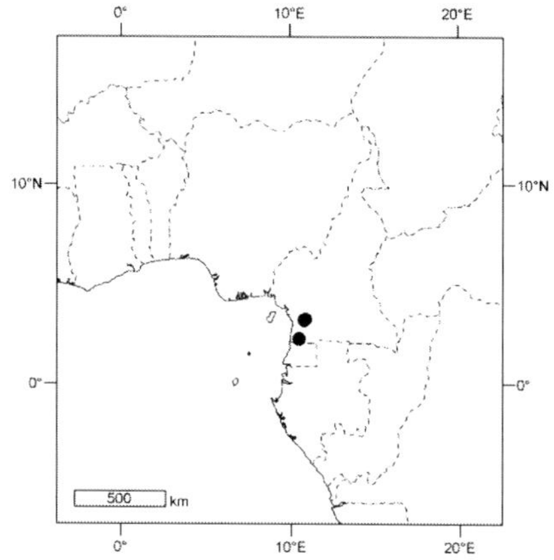

b. ***Coffea mapiana*** EN (Rubiaceae)

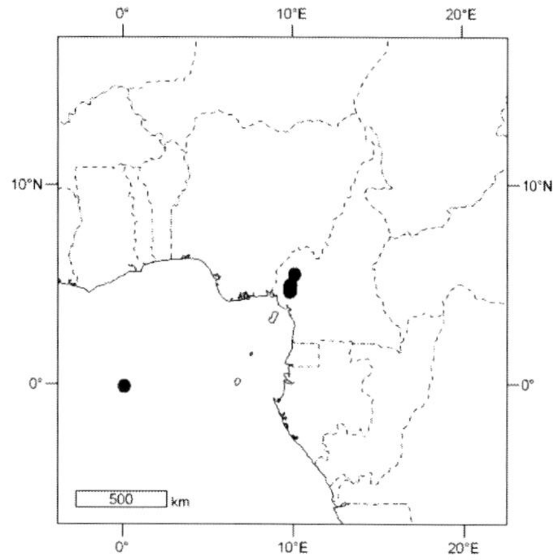

c. ***Coffea montekupensis*** VU (Rubiaceae)

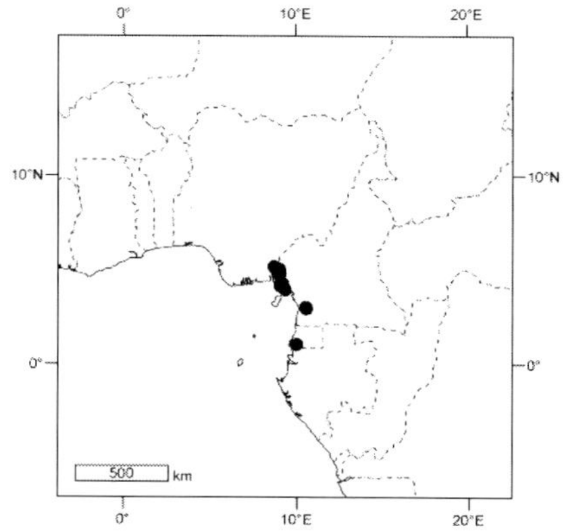

d. ***Craterispermum aristatum*** EN (Rubiaceae)

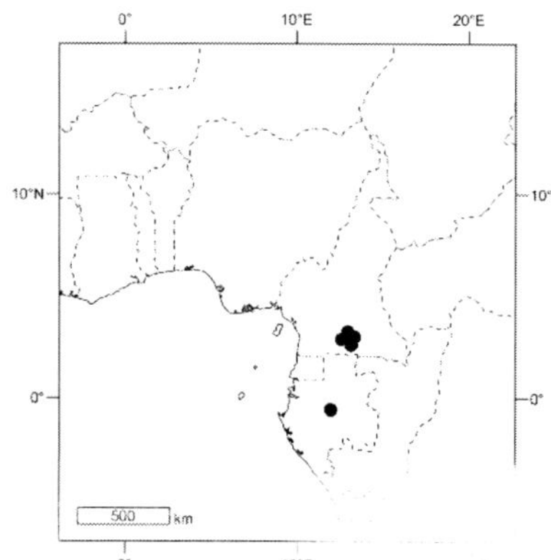

e. ***Cuviera cuvieroides*** EN (Rubiaceae)

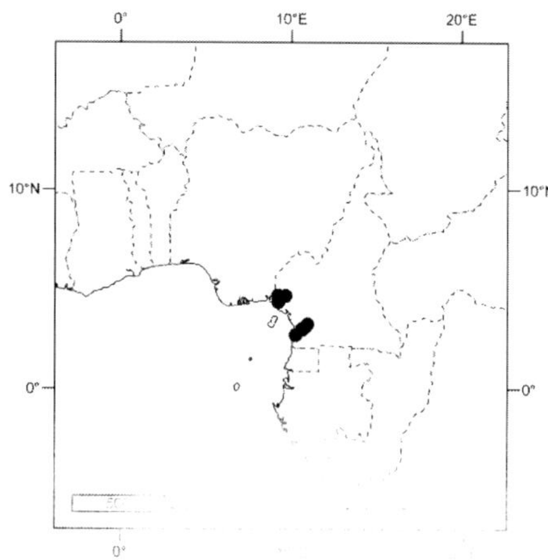

f. ***Cuviera leniochlamys*** EN (Rubiaceae)

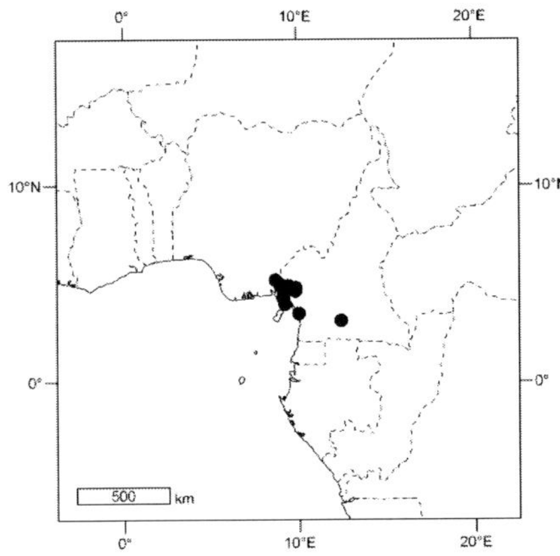

a. ***Cuviera talbotii*** VU (Rubiaceae)

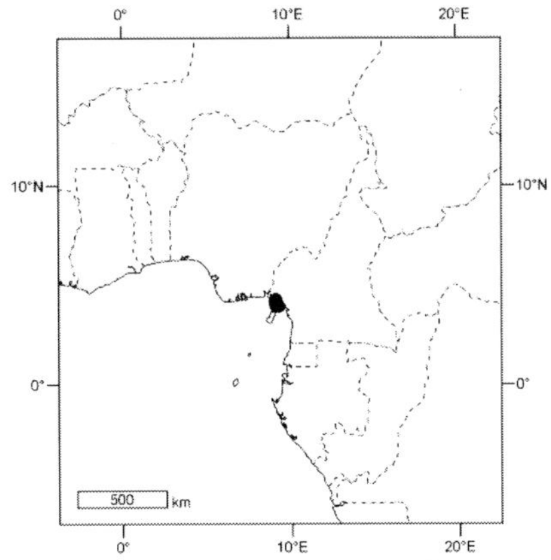

b. ***Cuviera wernhamii*** EN(Rubiaceae)

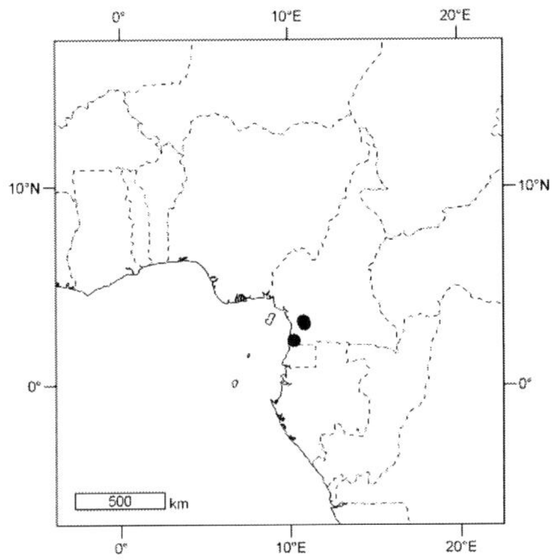

c. ***Ecpoma apocynaceum*** EN (Rubiaceae)

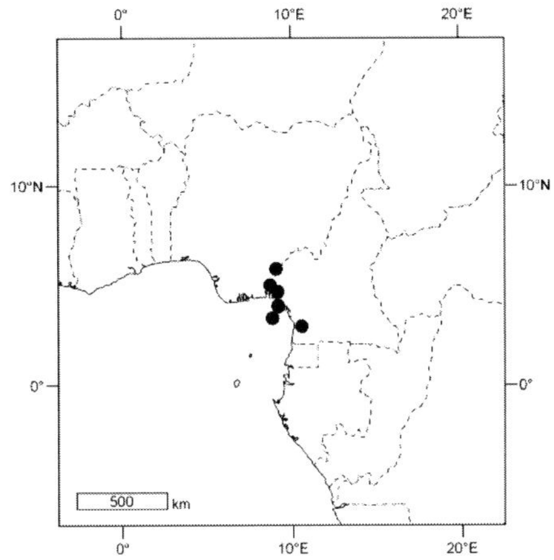

d. ***Ecpoma gigantostipulum*** EN (Rubiaceae)

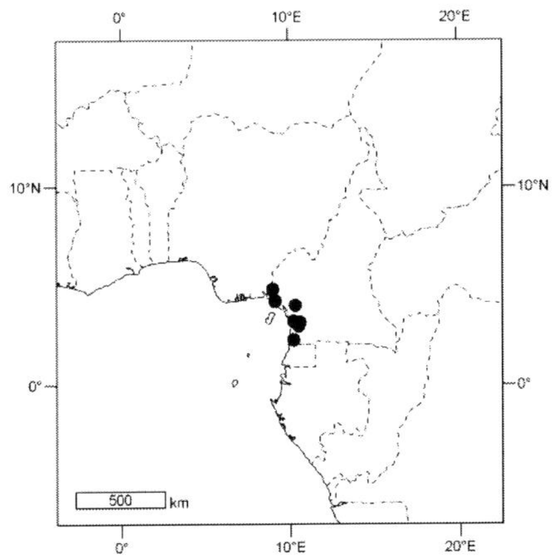

e. ***Gaertnera fissistipula*** VU (Rubiaceae)

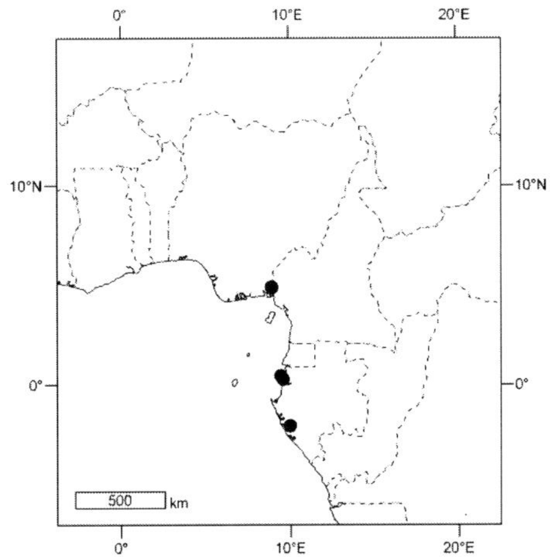

f. ***Gardenia epiphytica*** EN (Rubiaceae)

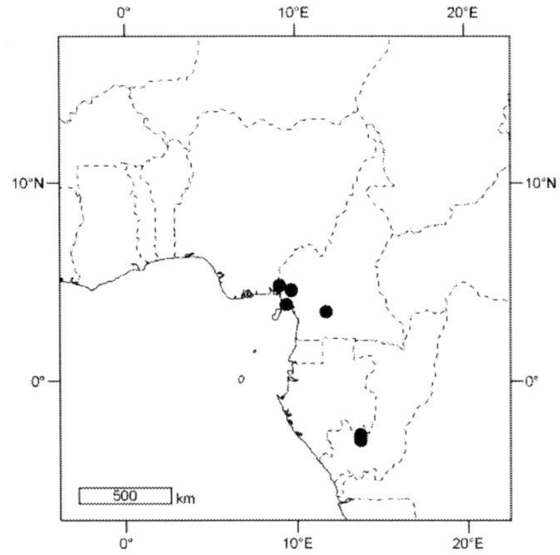

a. ***Hallea stipulosa*** VU (Rubiaceae)

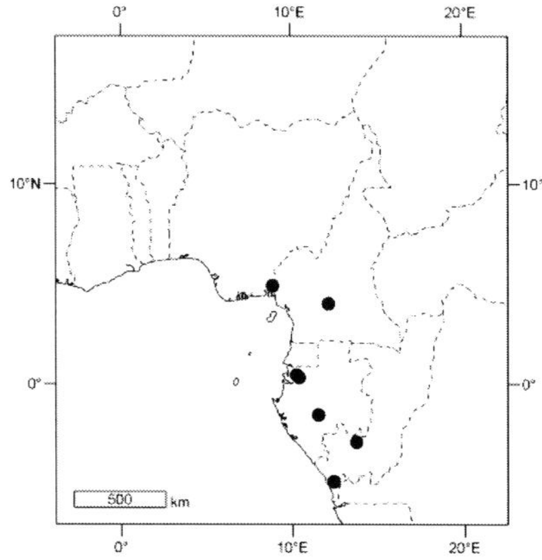

b. ***Hymenocoleus axillaris*** VU (Rubiaceae)

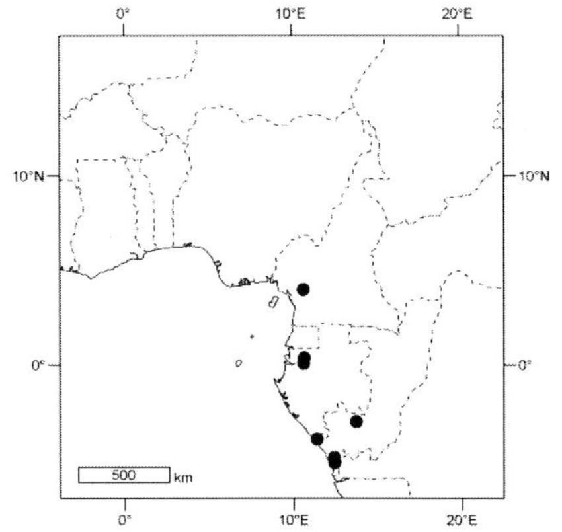

c. ***Hymenocoleus barbatus*** VU (Rubiaceae)

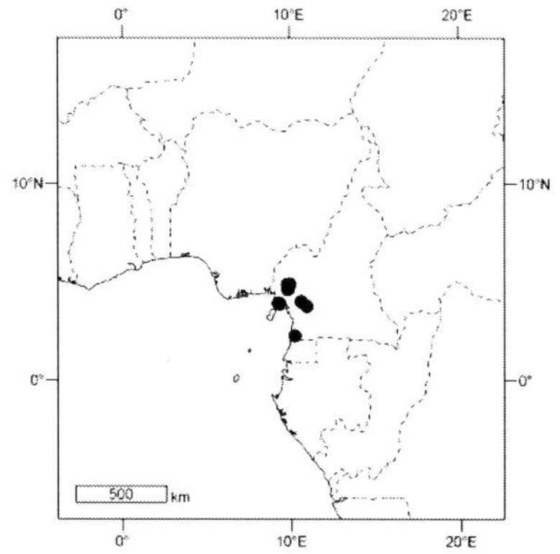

d. ***Hymenocoleus glaber*** VU (Rubiaceae)

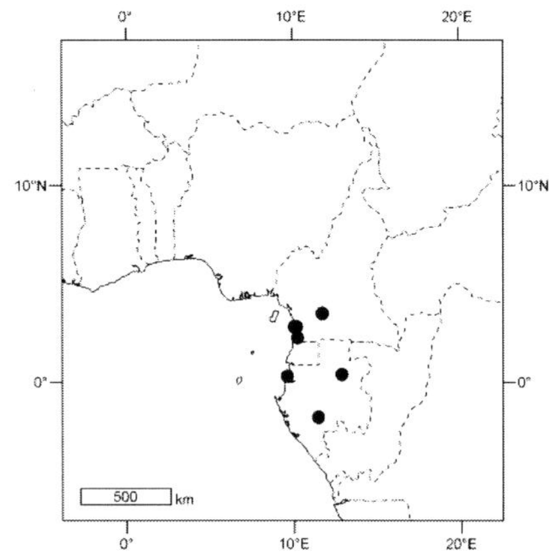

e. ***Hymenocoleus nervopilosus var orientalis***
VU (Rubiaceae)

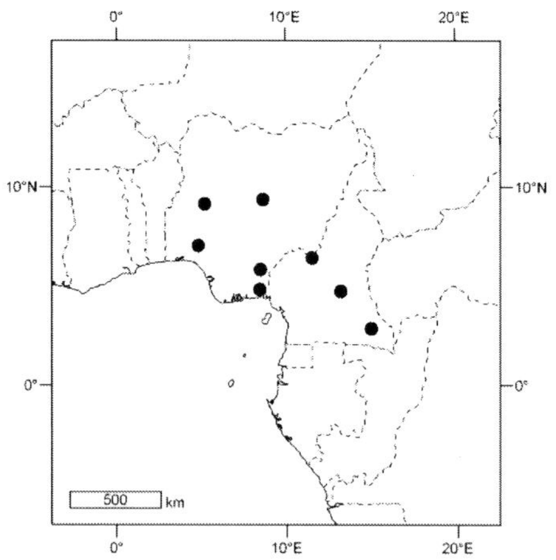

f. ***Hymenodictyon pachyantha*** EN

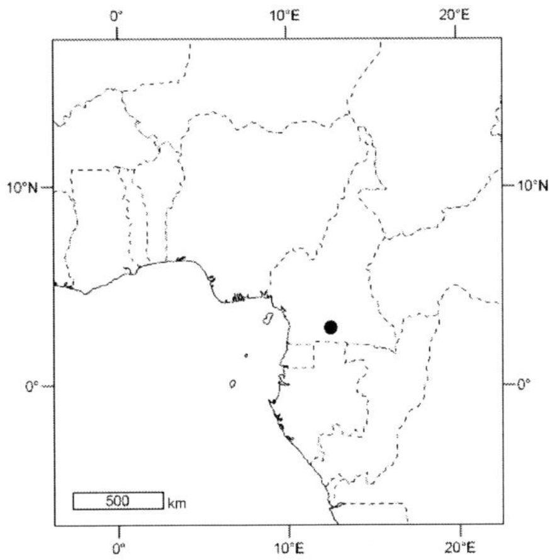

a. ***Ixora batesii*** CR (Rubiaceae)

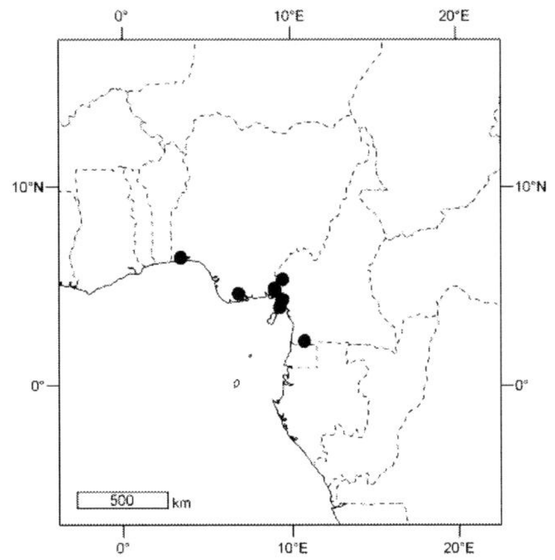

b. ***Ixora delicatula*** VU (Rubiaceae)

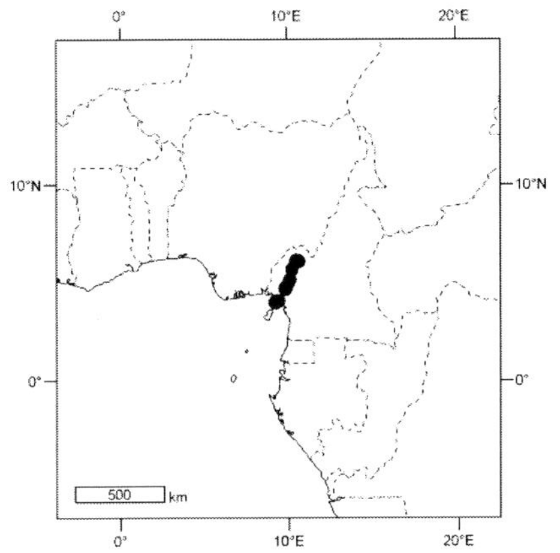

c. ***Ixora foliosa*** VU (Rubiaceae)

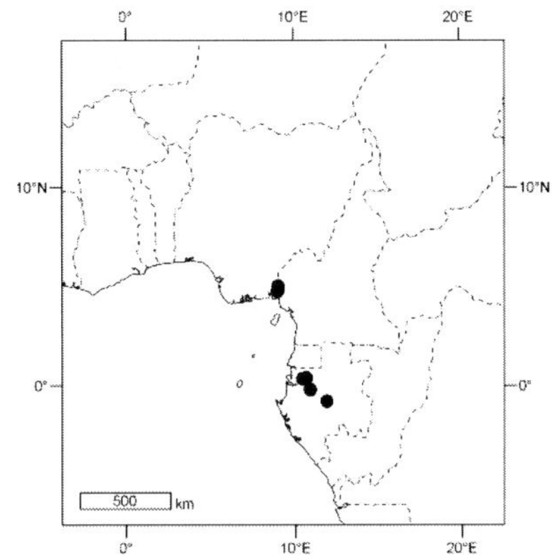

d. ***Ixora inundata*** EN (Rubiaceae)

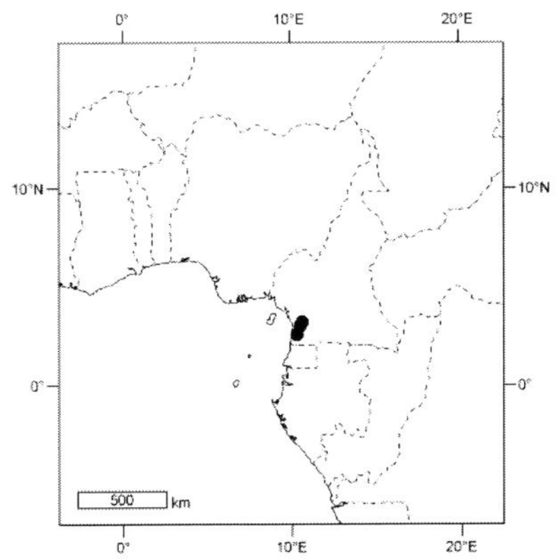

e. ***Ixora synactica*** EN (Rubiaceae)

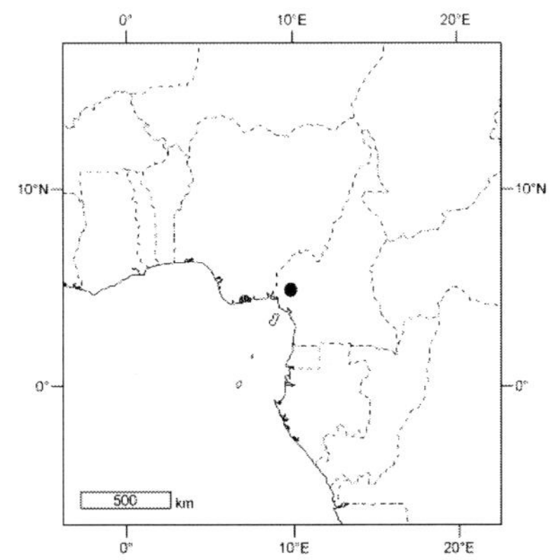

f. ***Keetia bakossiorum*** CR (Rubiaceae)

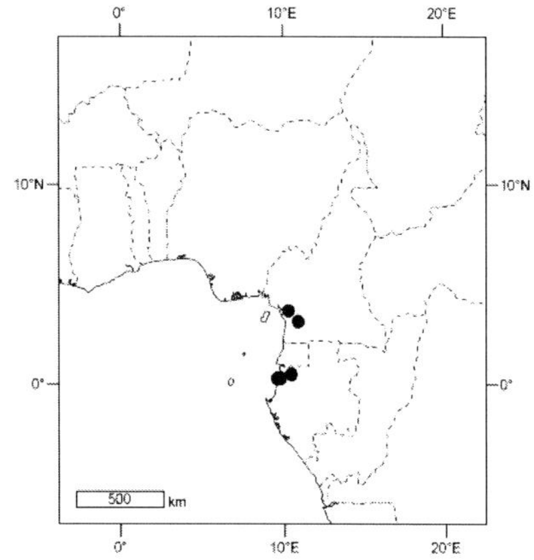

a. ***Leptactina latifolia*** EN (Rubiaceae)

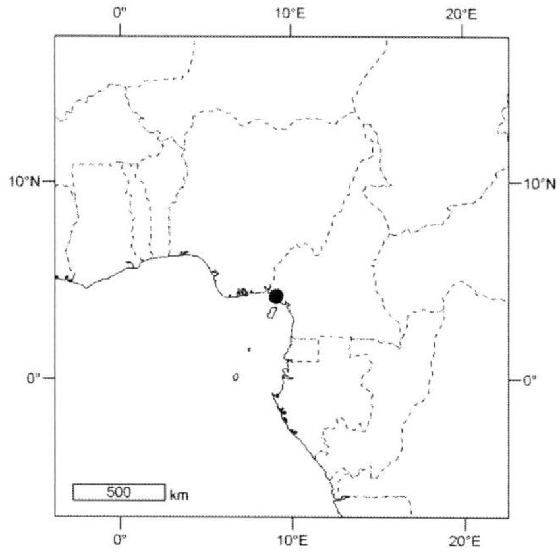

b. ***Mitriostigma bakweri*** CR (Rubiaceae)

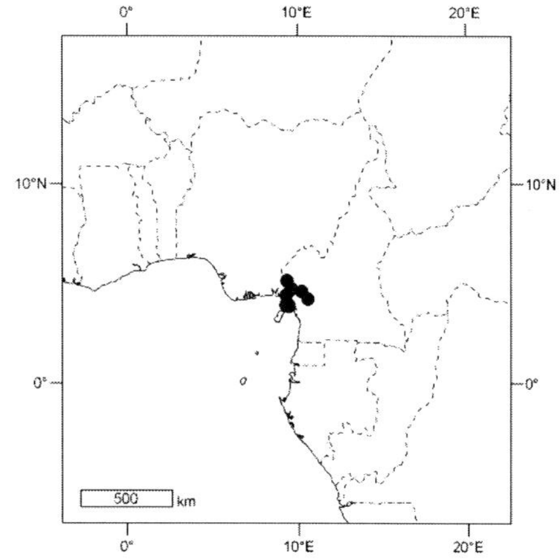

c. ***Mitriostigma cameroonense*** EN (Rubiaceae)

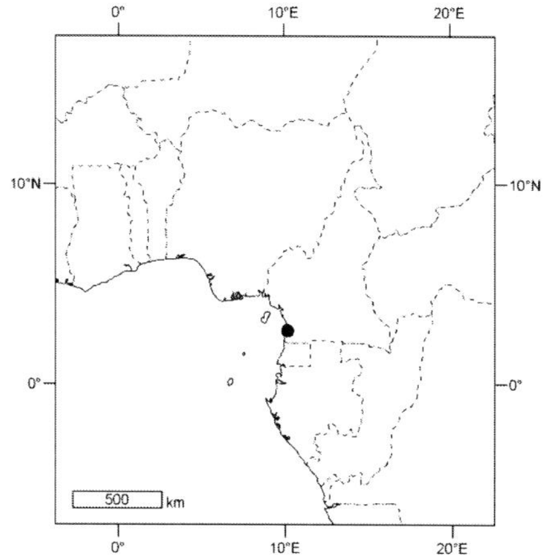

d. ***Mitriostigma monocaule*** CR (Rubiaceae)

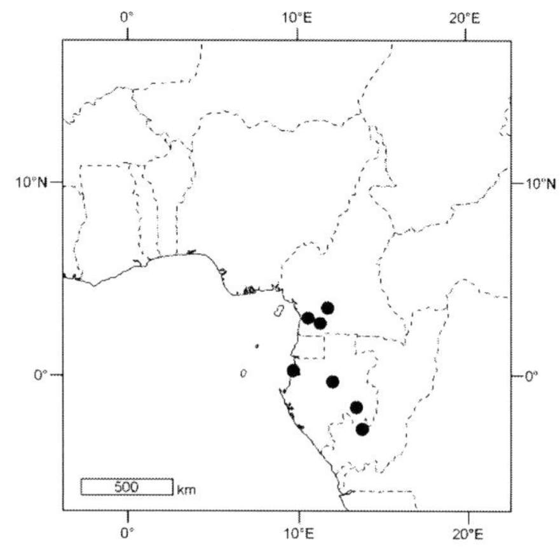

e. ***Morinda mefou*** VU (Rubiaceae)

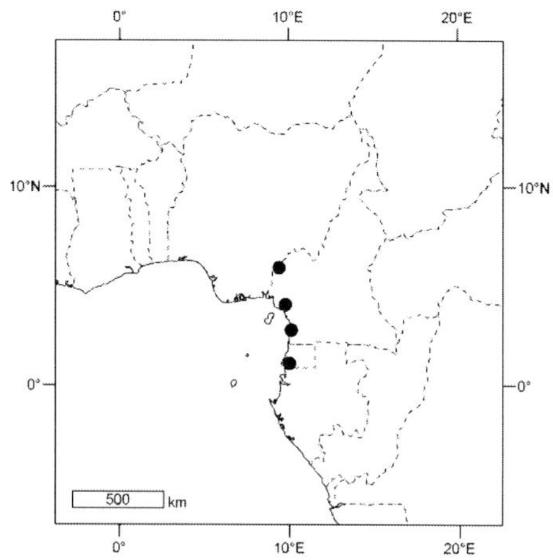

f. ***Mussaenda conopharyngifolia*** VU (Rubiaceae)

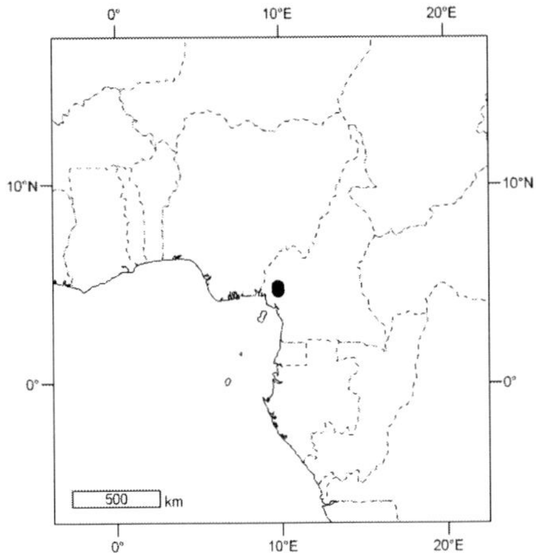

a. **Mussaenda epiphytica *VU*** (Rubiaceae)

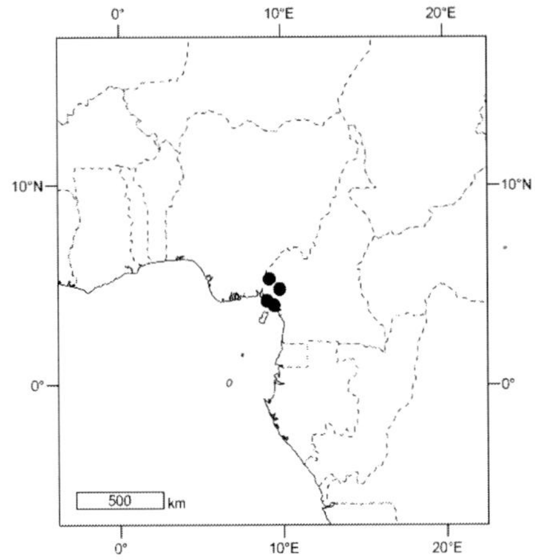

b. ***Nauclea diderrichii*** VU (Rubiaceae)

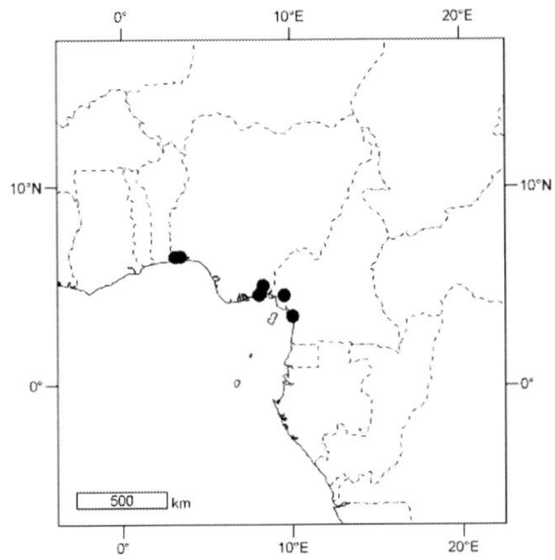

c. . ***Oligocodon cunliffeae*** EN (Rubiaceae)

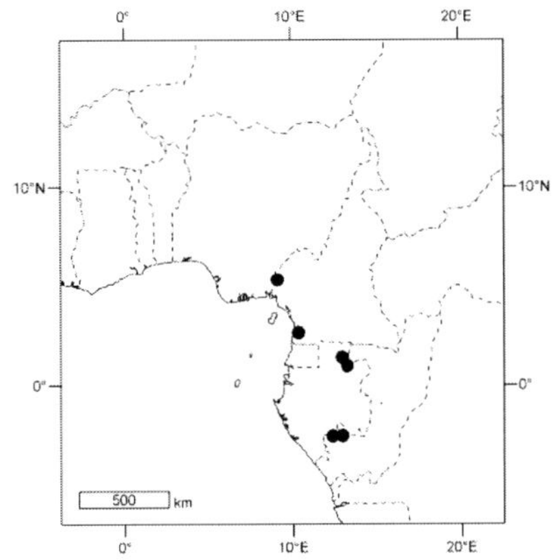

d ***Oxyanthus brevicaulis*** VU (Rubiaceae)

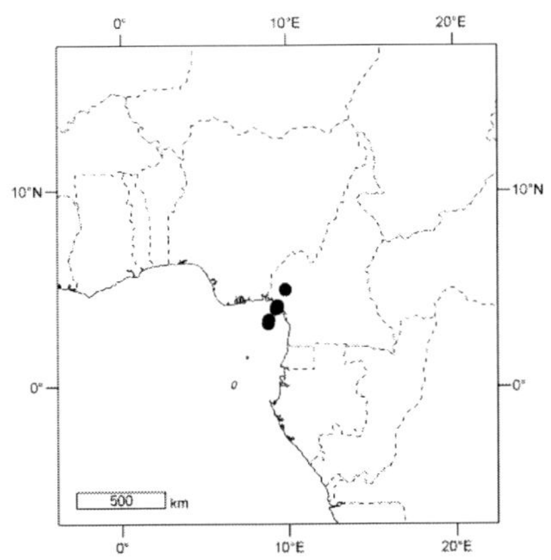

e. ***Oxyanthus montanus*** VU (Rubiaceae)

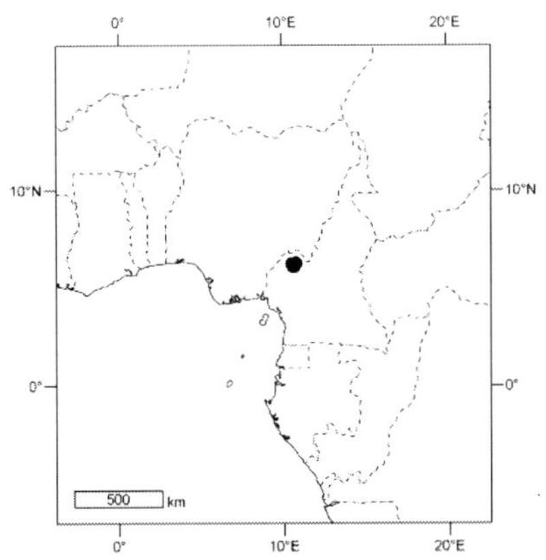

f. ***Oxyanthus okuensis*** CR (Rubiaceae)

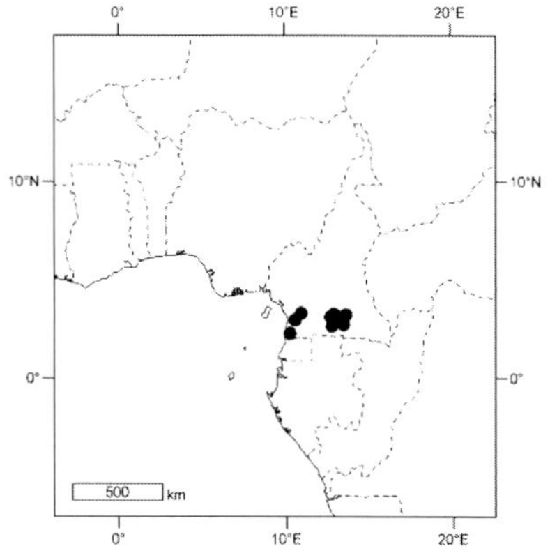

a. ***Oxyanthus oliganthus*** VU (Rubiaceae)

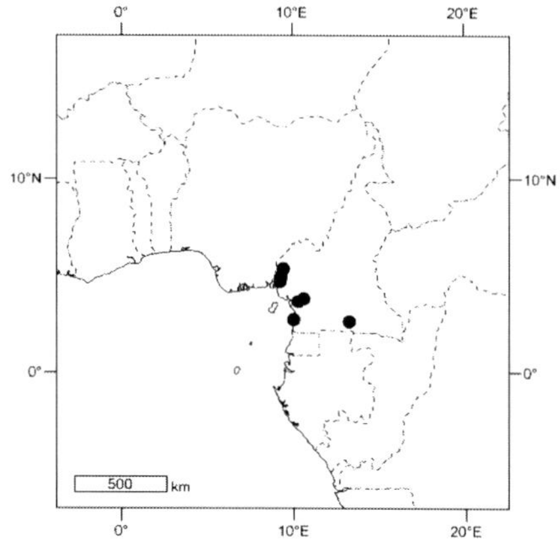

b. ***Pauridiantha multiflora*** VU (Rubiaceae)

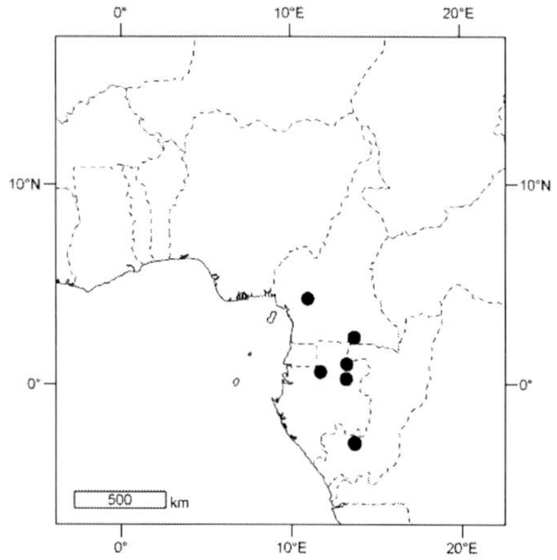

c. ***Pauridiantha siderophila*** VU (Rubiaceae)

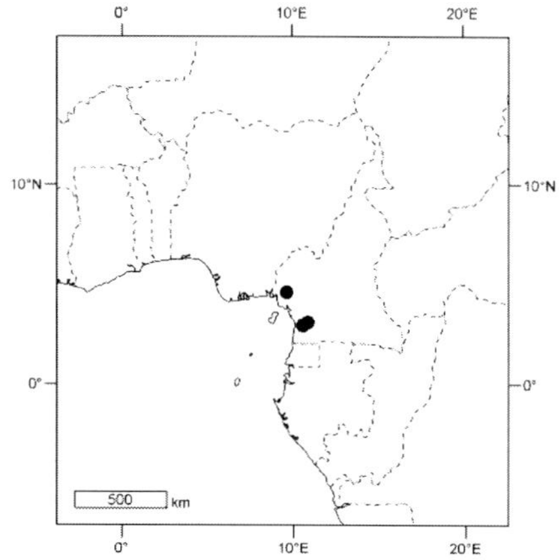

d. ***Pausinystalia brachythyrsum*** CR (Rubiaceae)

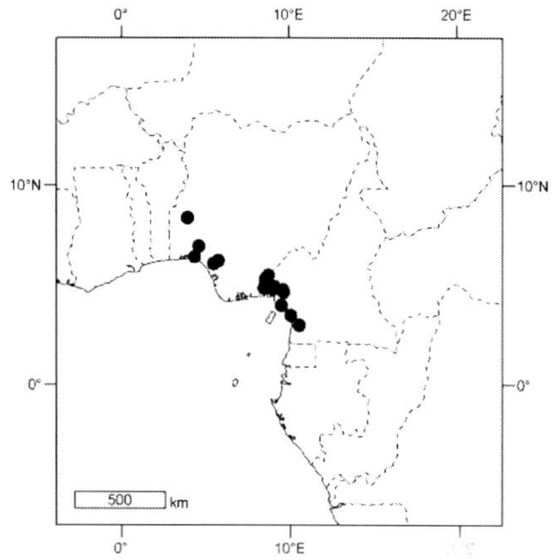

e. ***Pausinystalia talbotii*** VU (Rubiaceae)

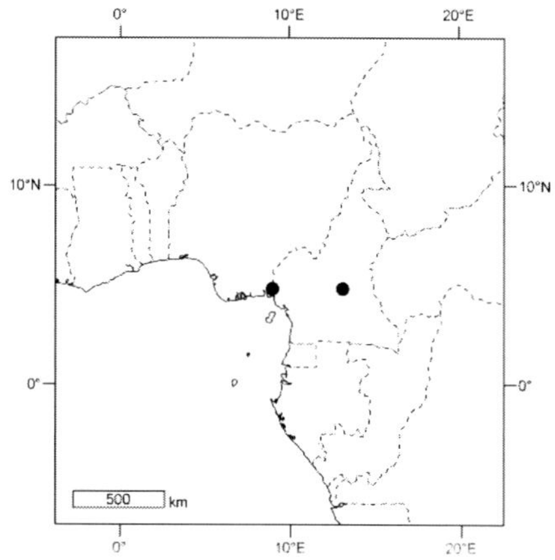

f. ***Pavetta baconiella*** VU (Rubiaceae)

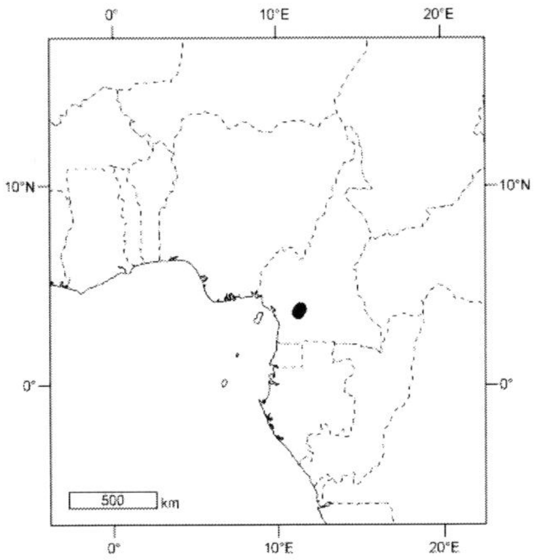

a. ***Pavetta bidentata var sessilifolia*** EN (Rubiaceae)

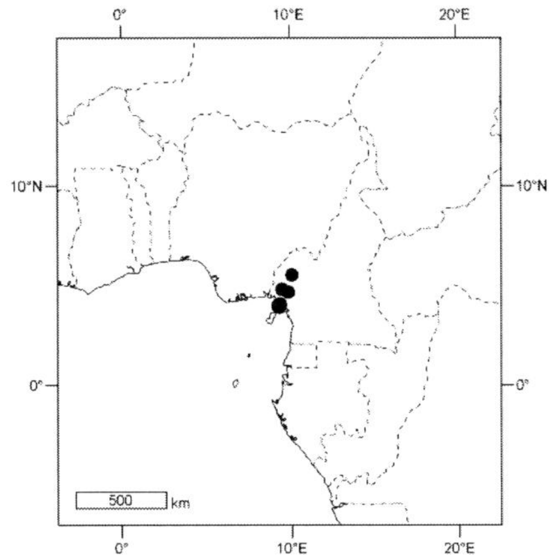

b. ***Pavetta brachycalyx*** EN (Rubiaceae)

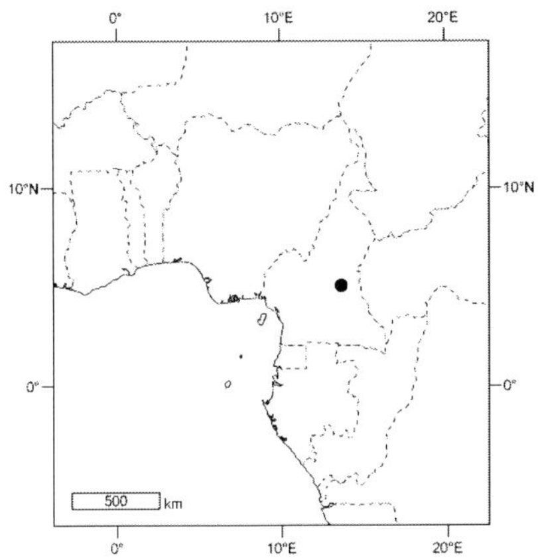

c. ***Pavetta brachysiphon*** VU (Rubiaceae)

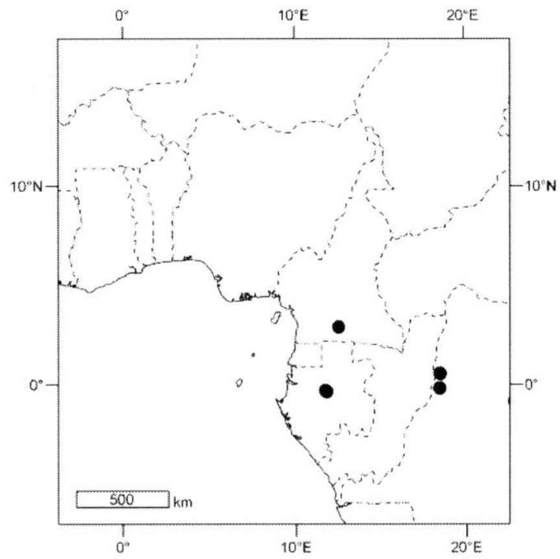

d. ***Pavetta cellulosa*** EN (Rubiaceae)

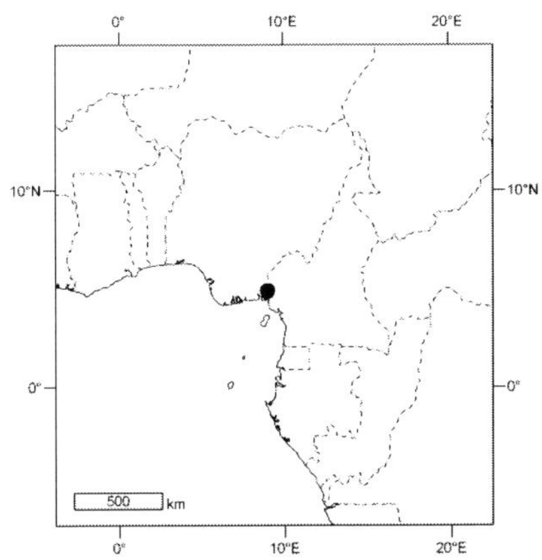

e. ***Pavetta grossissima*** EN (Rubiaceae)

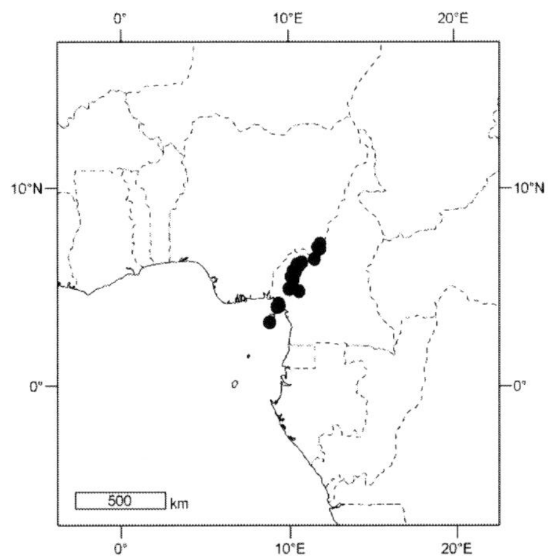

f. ***Pavetta hookeriana*** var. ***hookeriana*** VU (Rubiaceae)

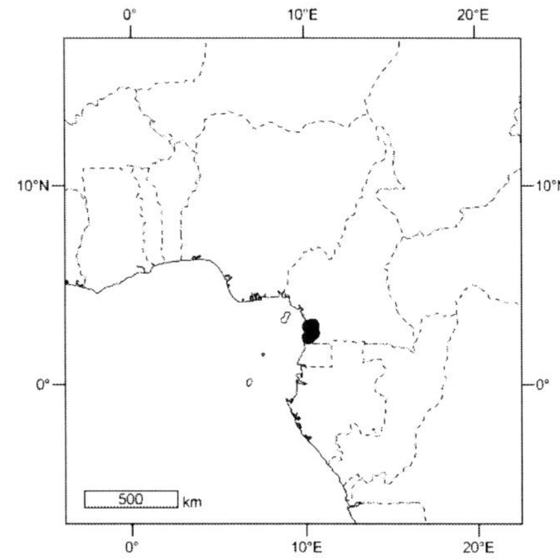

a. ***Pavetta kribiensis*** EN (Rubiaceae)

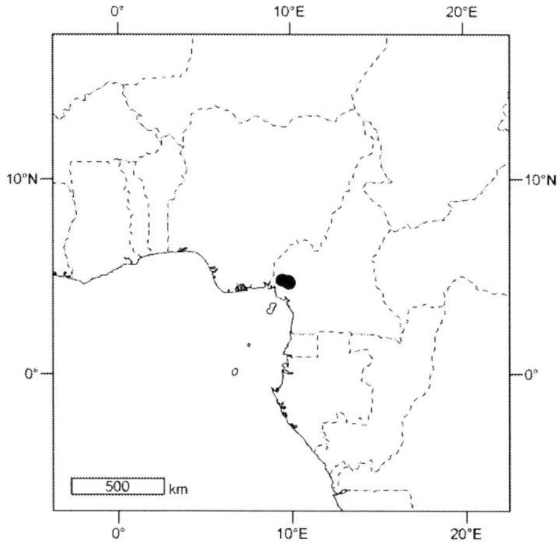

b. ***Pavetta kupensis*** CR (Rubiaceae)

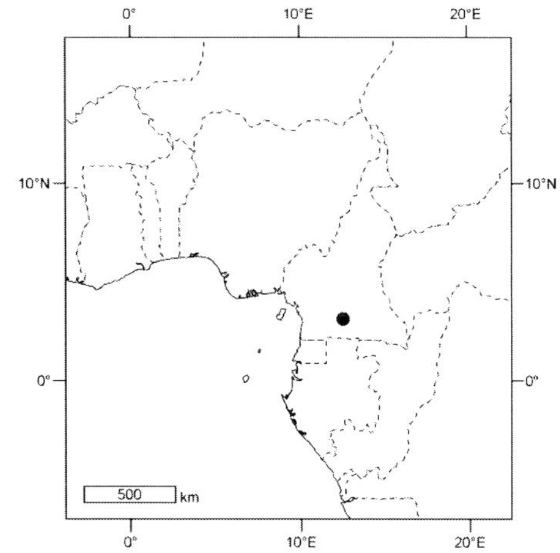

c. ***Pavetta laxa*** CR (Rubiaceae)

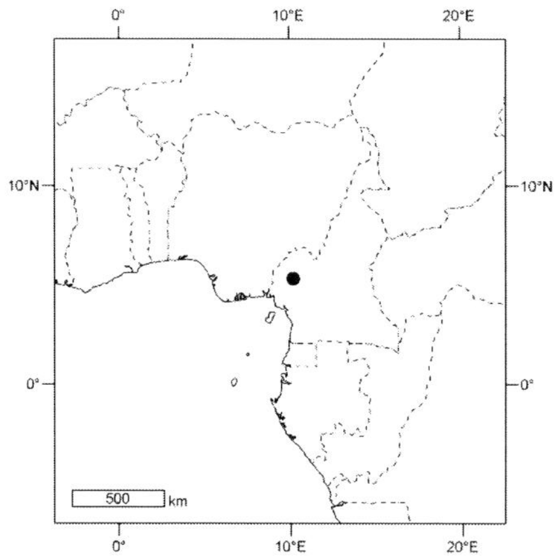

d. ***Pavetta longistyla*** CR (Rubiaceae)

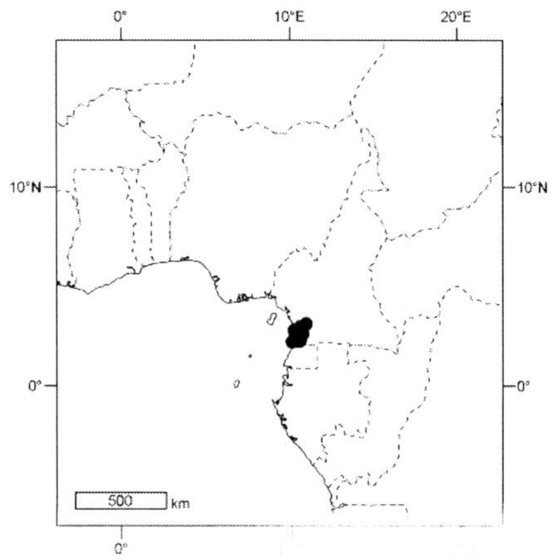

e. ***Pavetta mpomii*** EN (Rubiaceae)

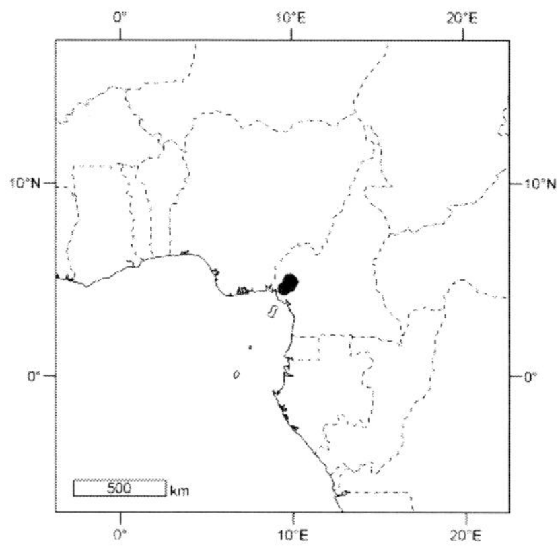

f. [illegible] EN (Rubiaceae)

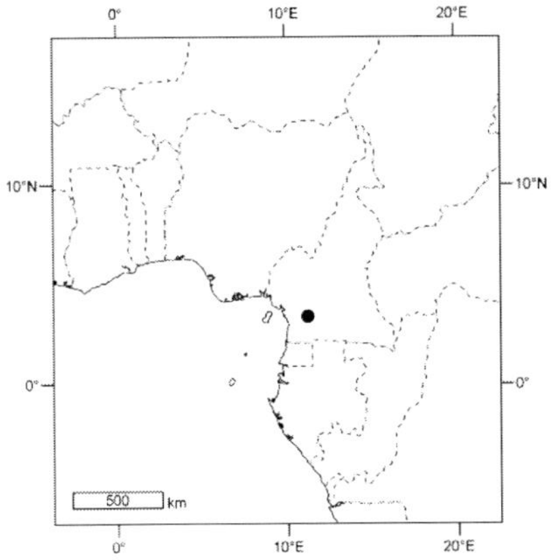

a. ***Pavetta namatae*** CR (Rubiaceae)

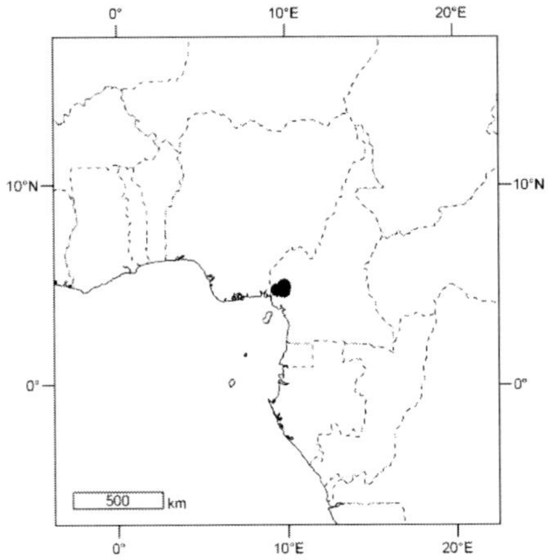

b. ***Pavetta owariensis*** *var* ***satabiei*** EN (Rubiaceae)

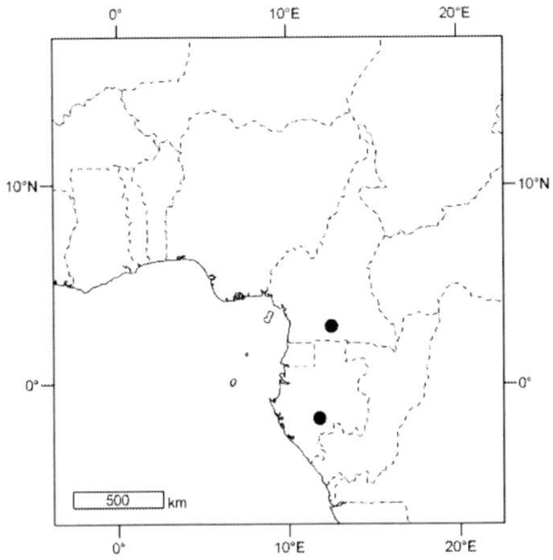

c. ***Pavetta robusta*** EN (Rubiaceae)

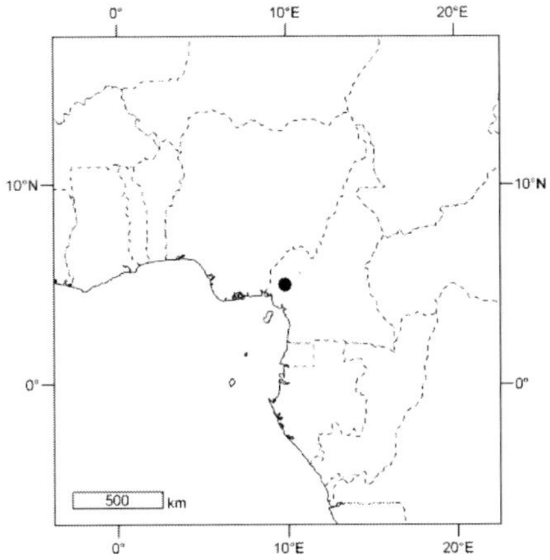

d. ***Pavetta rubentifolia*** CR (Rubiaceae)

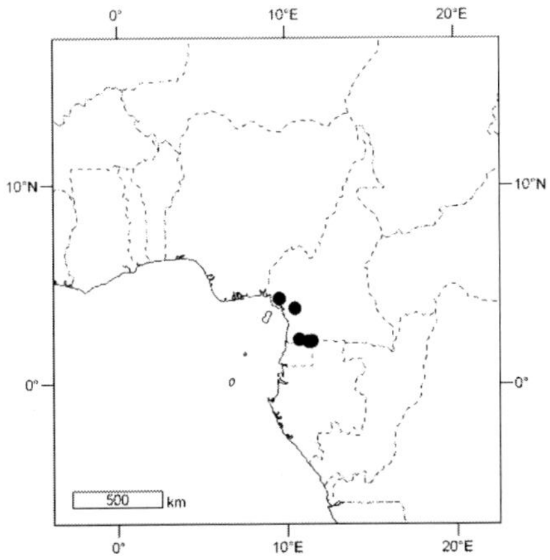

e. ***Pavetta tenuissima*** EN (Rubiaceae)

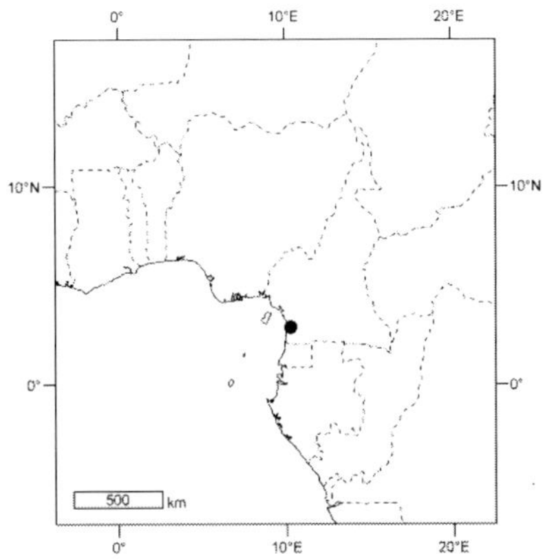

f. ***Pavetta urophylla*** *subsp* ***bosii*** CR (Rubiaceae)

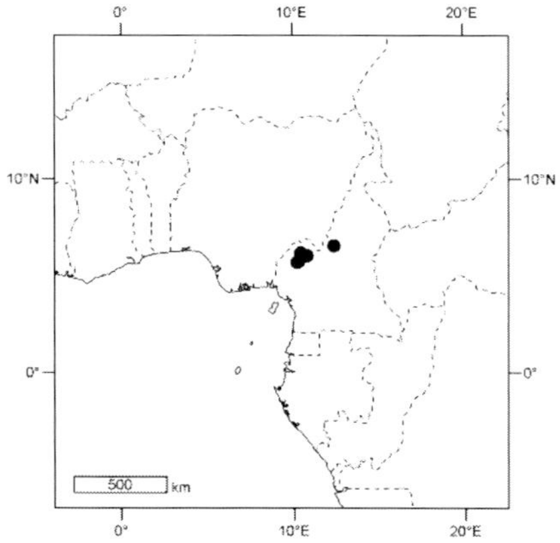

a. ***Pentas ledermannii*** VU (Rubiaceae)

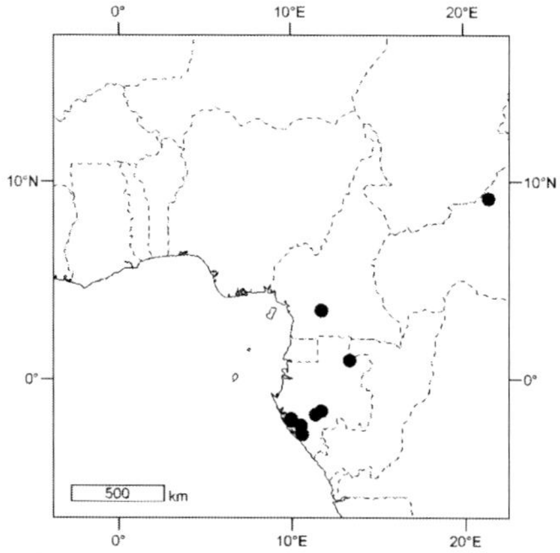

b. ***Poecilocalyx setiflorus*** VU (Rubiaceae)

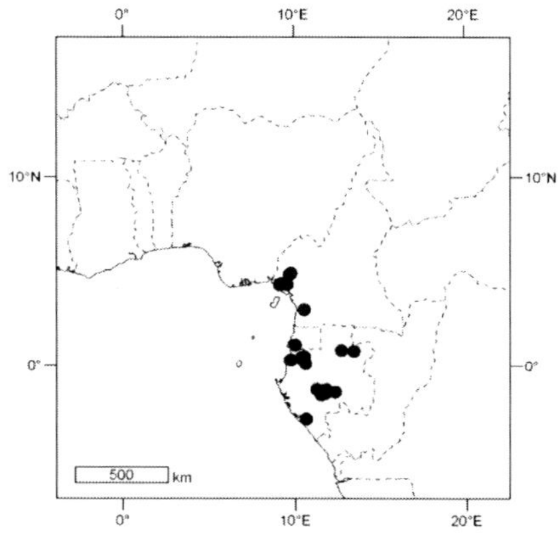

c. ***Pseudosabicea batesii*** VU (Rubiaceae)

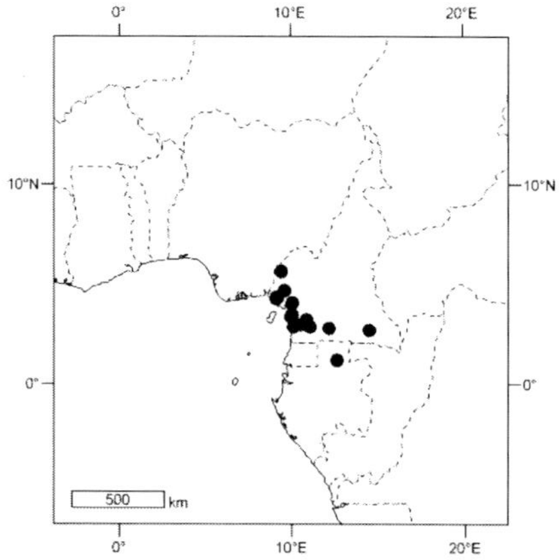

d. ***Pseudosabicea medusula*** VU (Rubiaceae)

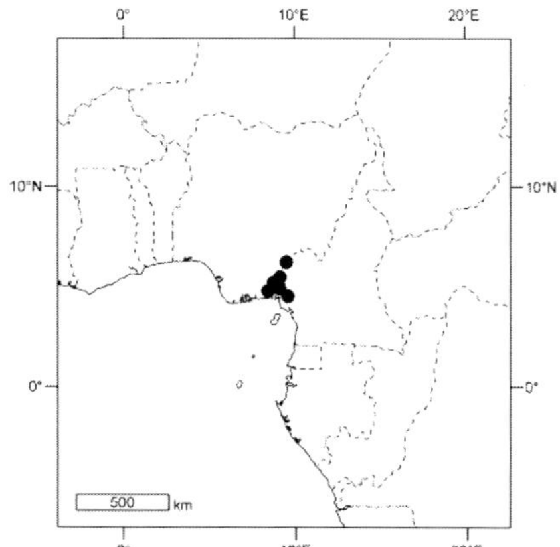

e. ***Pseudosabicea pedicellata*** VU (Rubiaceae)

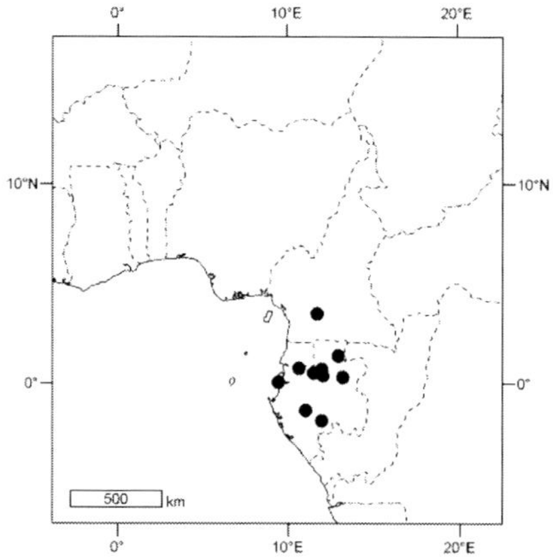

f. ***Pseudosabicea sthenula*** VU (Rubiaceae)

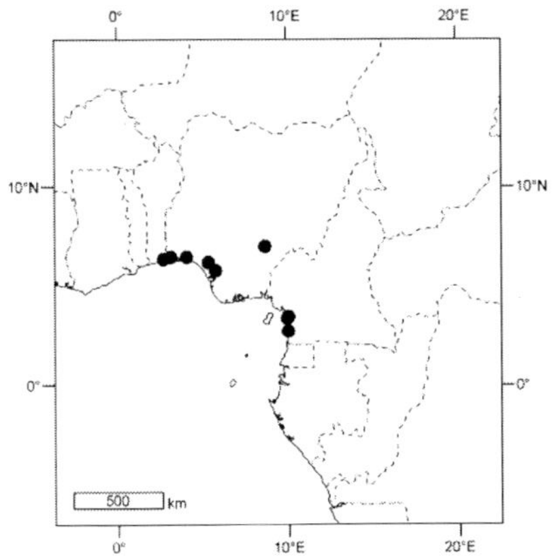

a. ***Psychotria articulata*** VU (Rubiaceae)

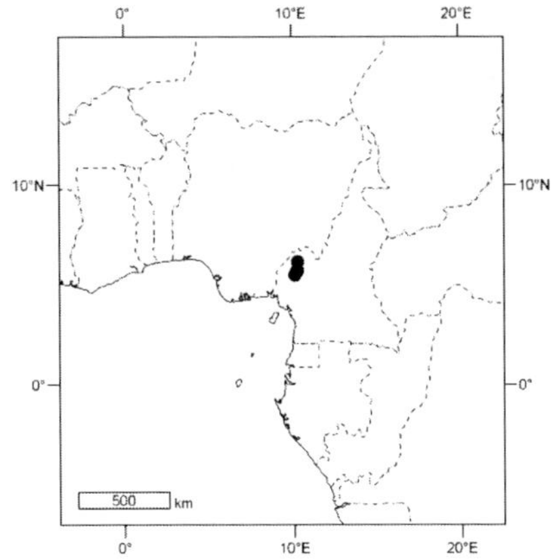

b. **Psychotria babatwoensis** EN (Rubiaceae)

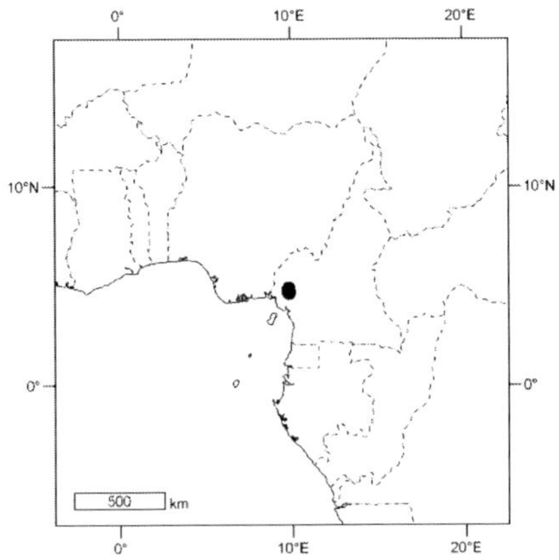

c. ***Psychotria bakossiensis*** EN (Rubiaceae)

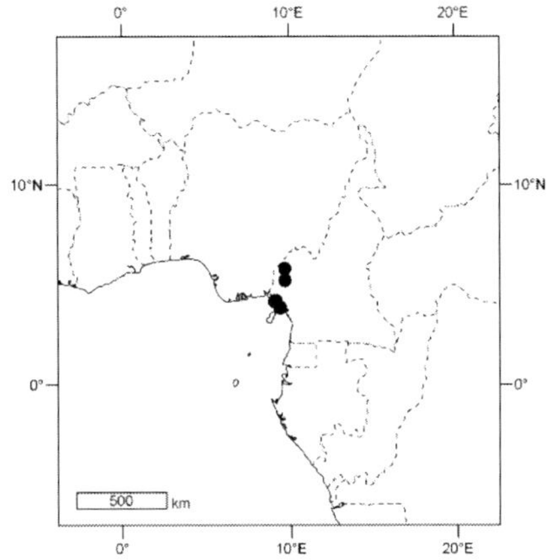

d. **Psychotria bimbiensis** CR (Rubiaceae)

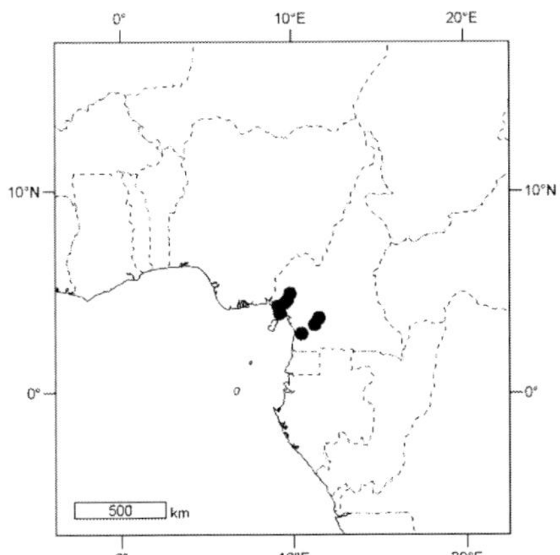

e. ***Psychotria camerunensis*** VU (Rubiaceae)

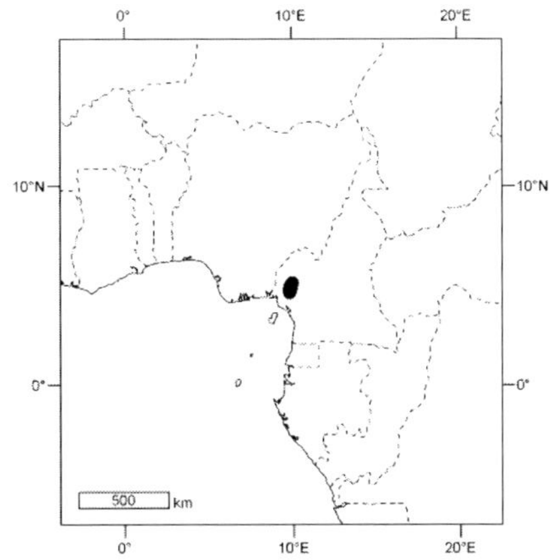

f. **Psychotria darwiniana** VU (Rubiaceae)

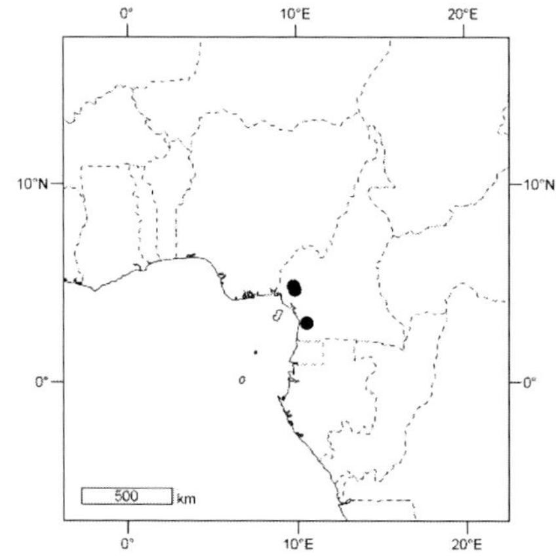

a. ***Psychotria densinervia*** EN (Rubiaceae)

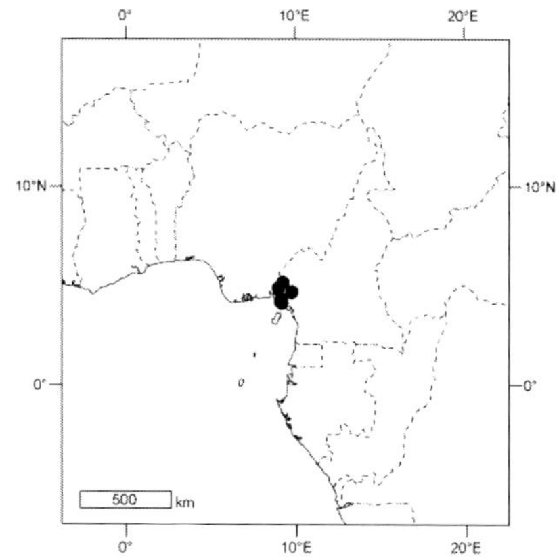

b. **Psychotria elephantina** EN (Rubiaceae)

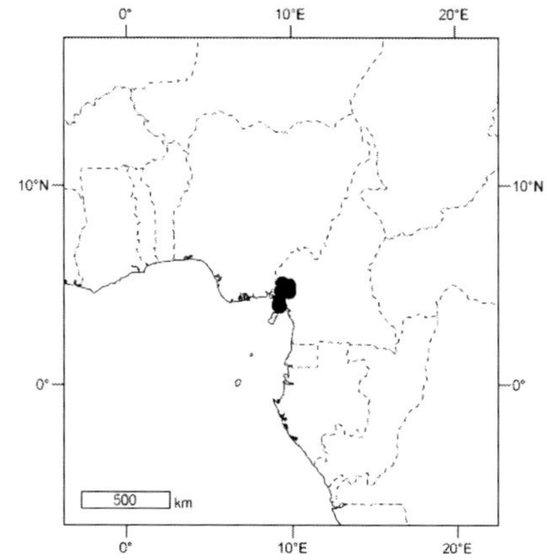

c. ***Psychotria geophylax*** VU (Rubiaceae)

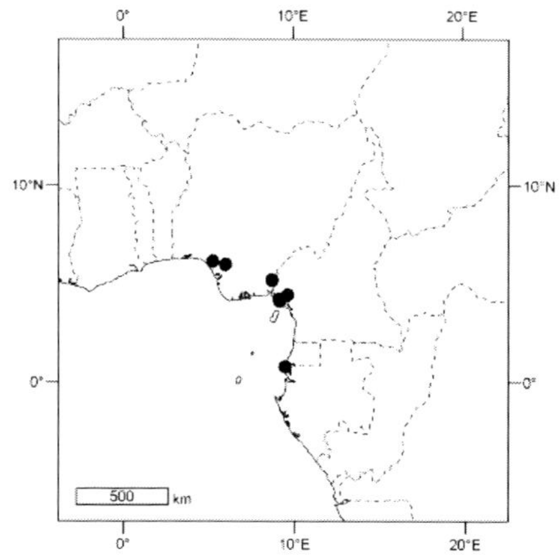

d. ***Psychotria humilis*** var. ***humilis*** VU (Rubiaceae)

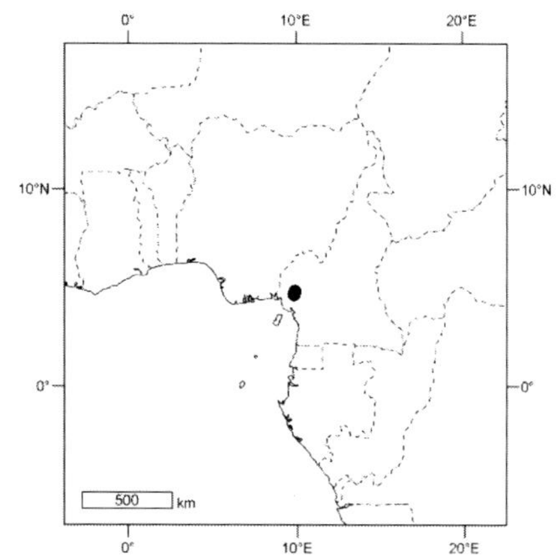

e. ***Psychotria kupensis*** EN (Rubiaceae)

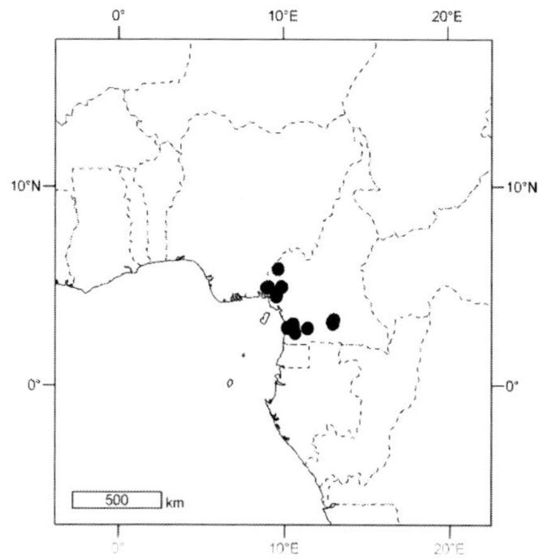

f. [illegible] ***…ifolia*** VU (Rubiaceae)

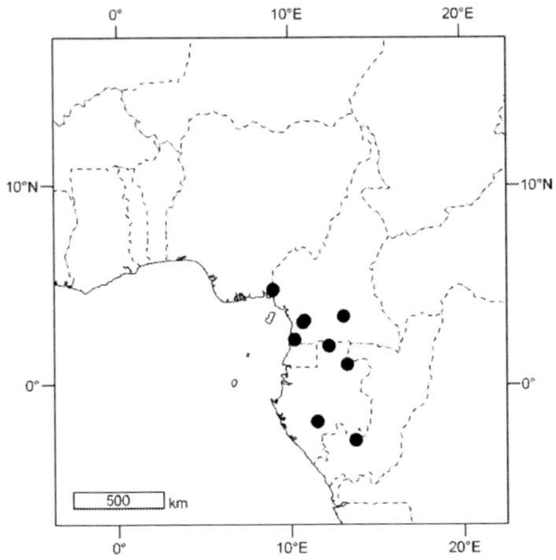

a. ***Psychotria letouzeyi*** VU (Rubiaceae)

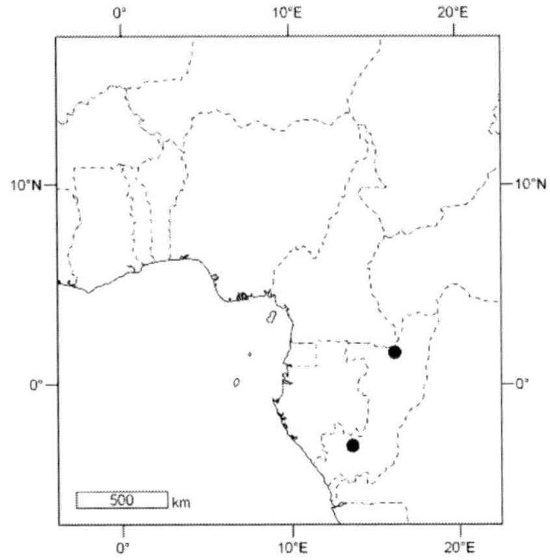

b. ***Psychotria microthyrsa*** CR (Rubiaceae)

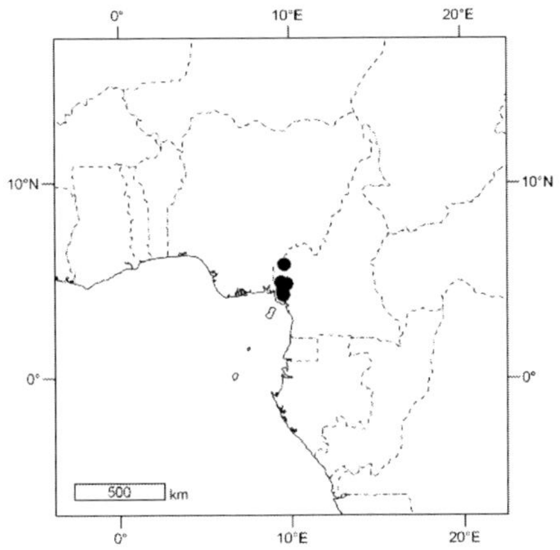

c. ***Psychotria minimicalyx*** CR (Rubiaceae)

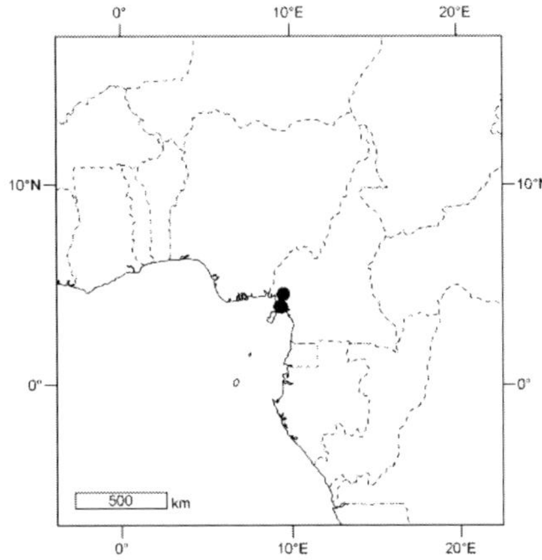

d. ***Psychotria moliwensis*** CR (Rubiaceae)

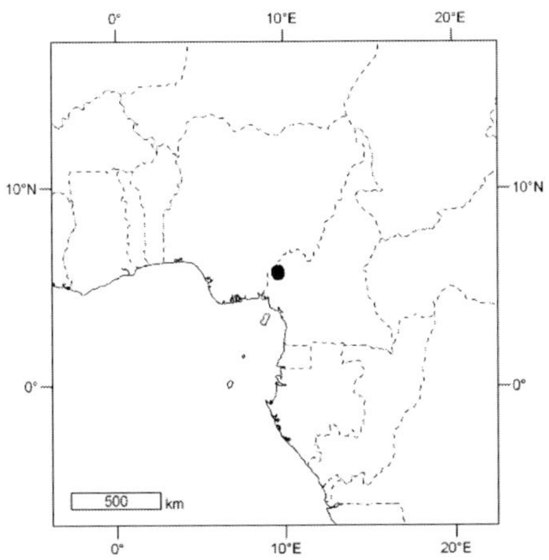

e. ***Psychotria monensis*** CR (Rubiaceae)

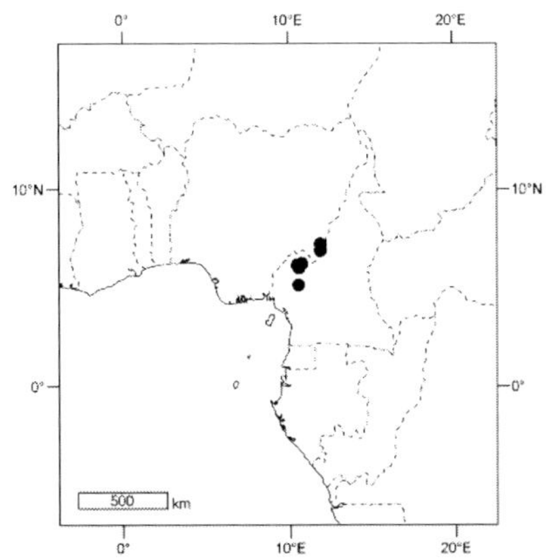

f. ***Psychotria moseskemei*** CR (Rubiaceae)

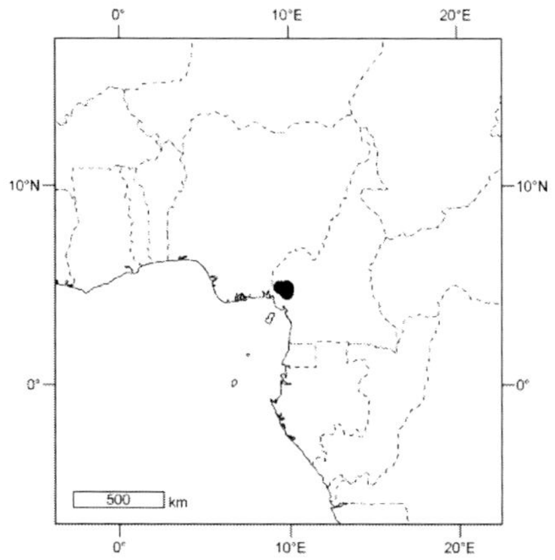

a. ***Psychotria ngollengollei*** VU (Rubiaceae)

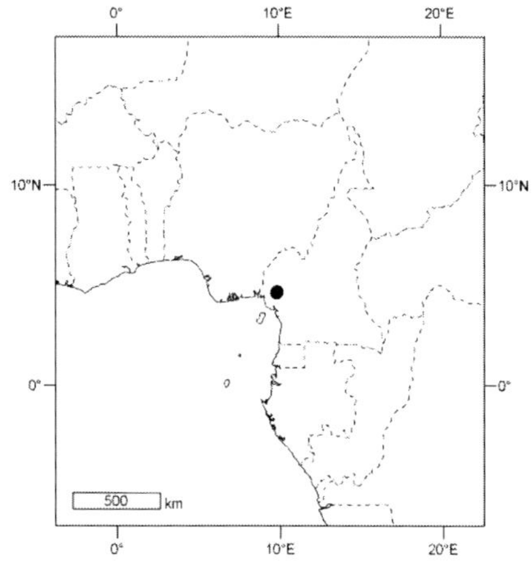

b. ***Psychotria njumei*** CR (Rubiaceae)

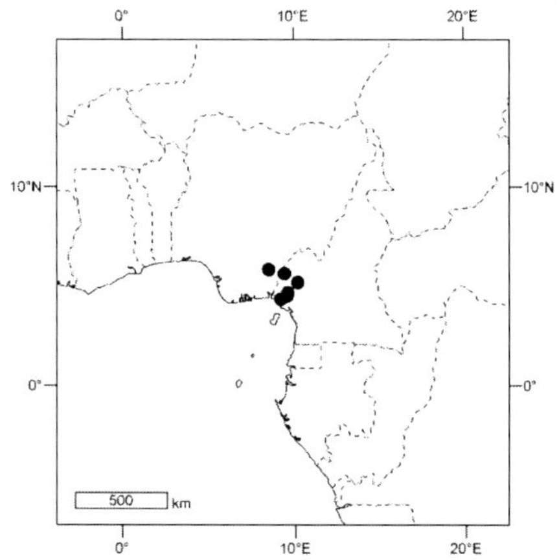

c. ***Psychotria piolampra*** VU (Rubiaceae)

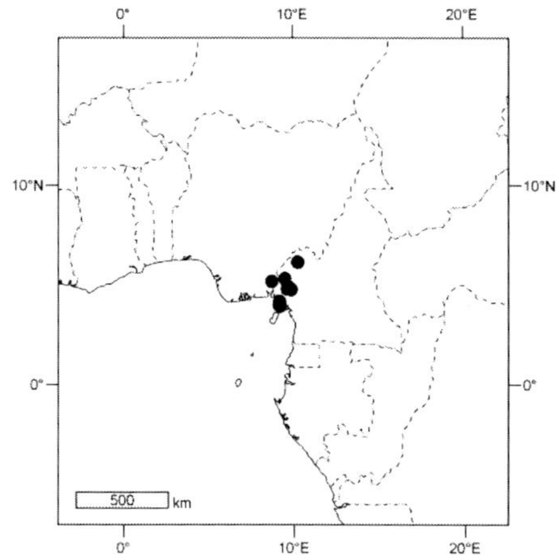

d. ***Psychotria podocarpa*** VU (Rubiaceae)

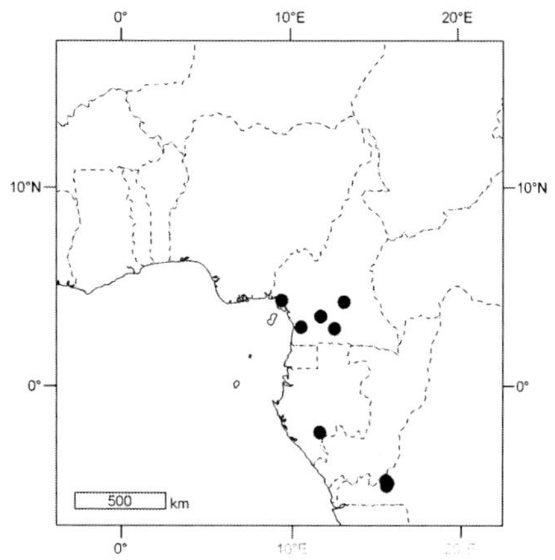

e. ***Psychotria rub*** [illegible]

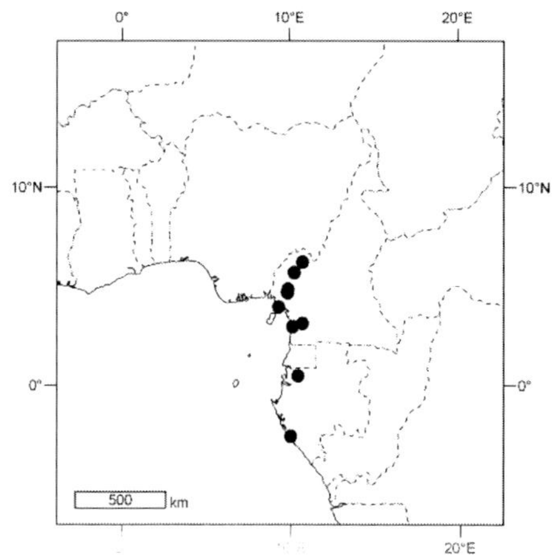

[illegible] ubiaceae)

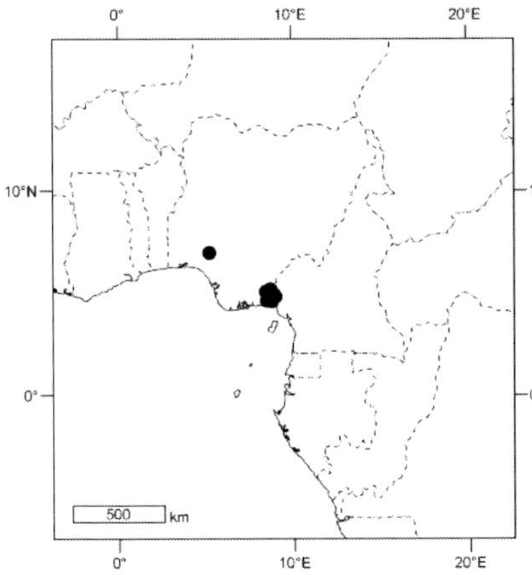

a. ***Psychotria talbotii*** EN (Rubiaceae)

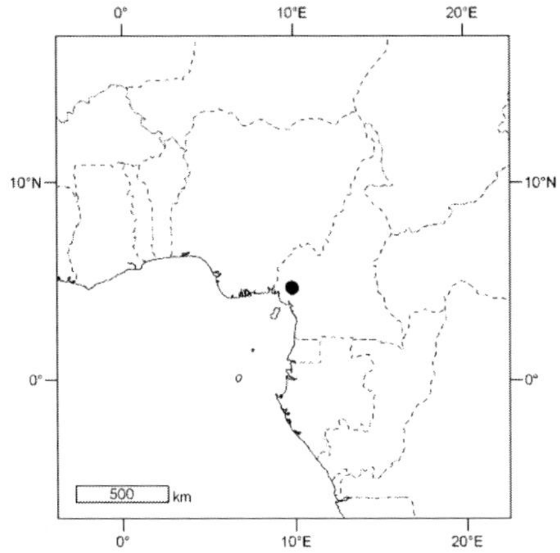

b. ***Psydrax bridsoniana*** EN (Rubiaceae)

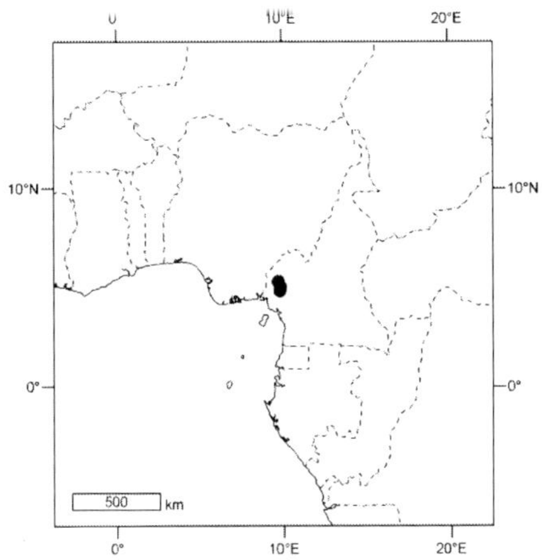

c. ***Rothmannia ebamutensis*** EN (Rubiaceae)

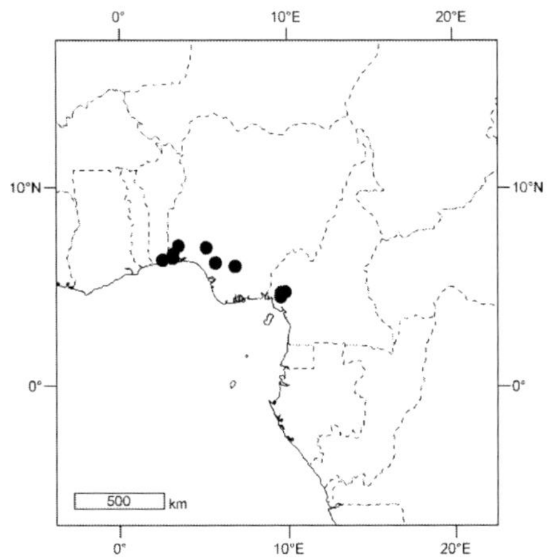

d. ***Rutidea nigerica*** VU (Rubiaceae)

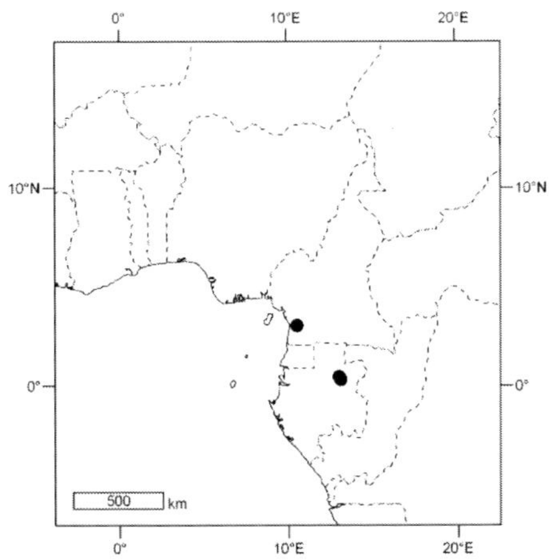

e. ***Sabicea bigerrica*** EN (Rubiaceae)

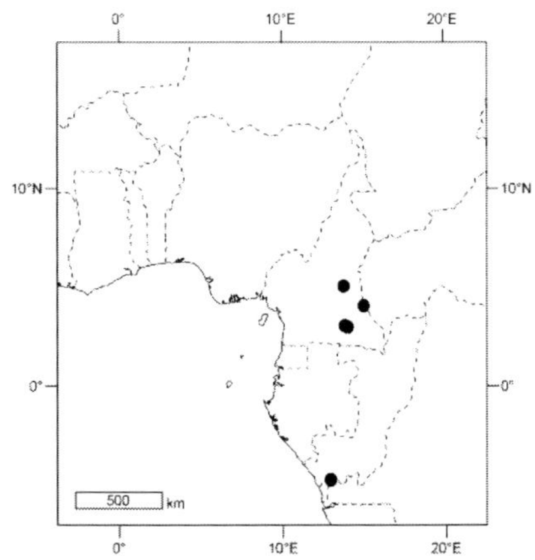

f. ***Sabicea cameroonensis*** VU (Rubiaceae)

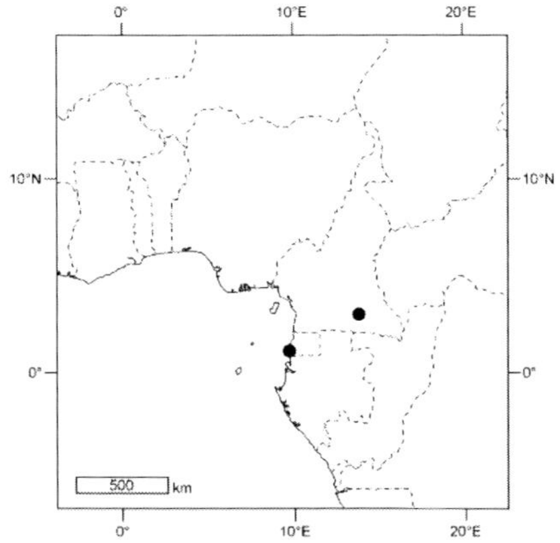

a. ***Sabicea cruciata*** EN (Rubiaceae)

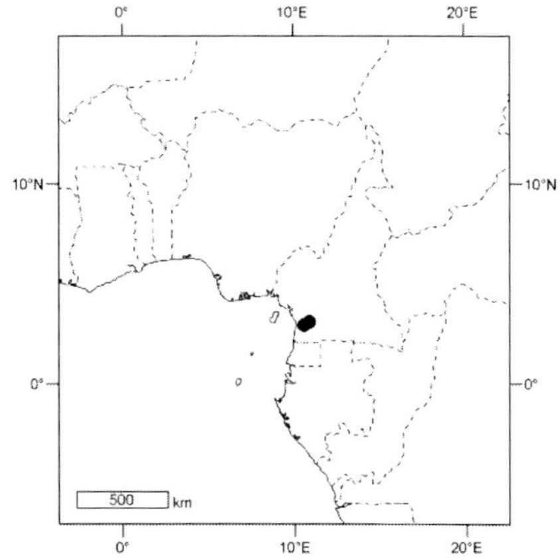

b. ***Sabicea laxa*** EN (Rubiaceae)

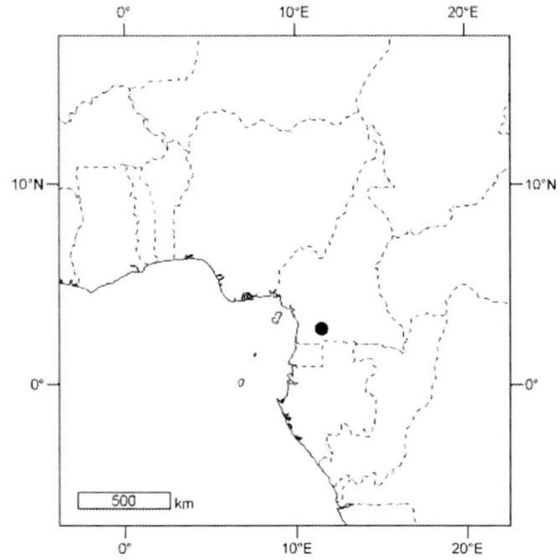

c. ***Sabicea leucocarpa*** CR (Rubiaceae)

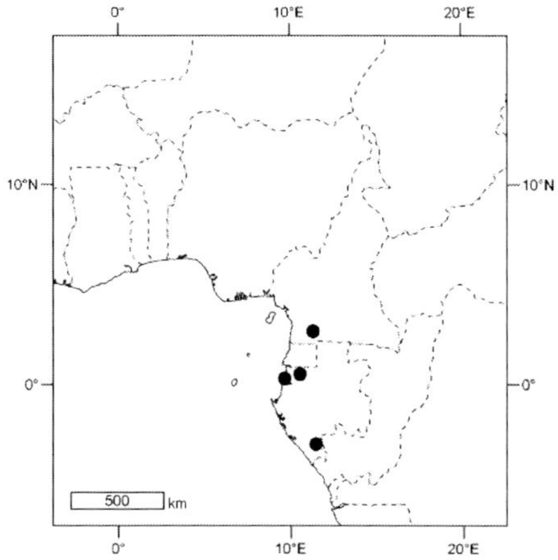

d. ***Sabicea najatrix*** EN (Rubiaceae)

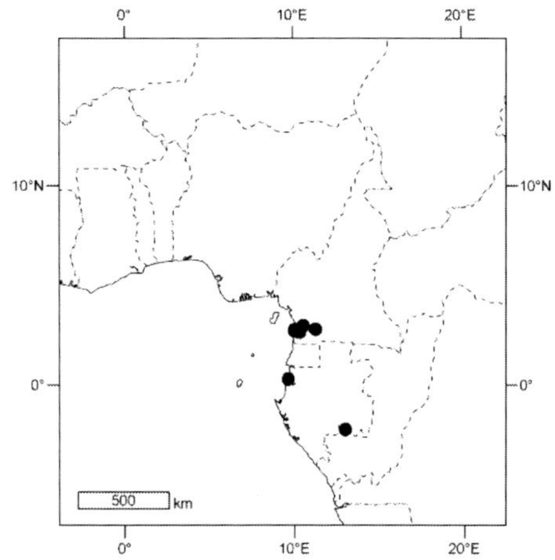

e. ***Sabicea rufa*** VU (Rubiaceae)

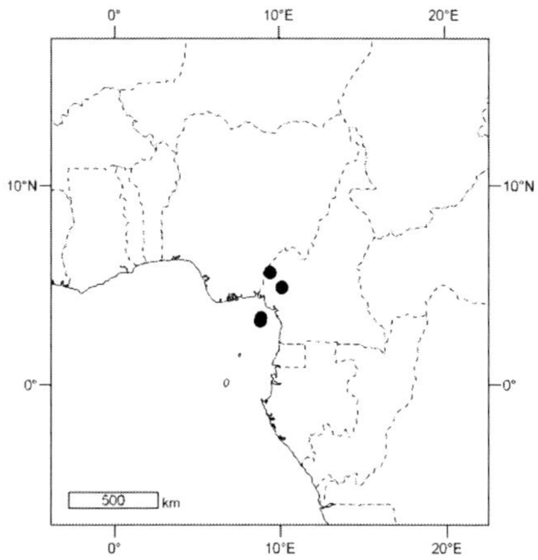

f. ***Sabicea schaefferi*** EN (Rubiaceae)

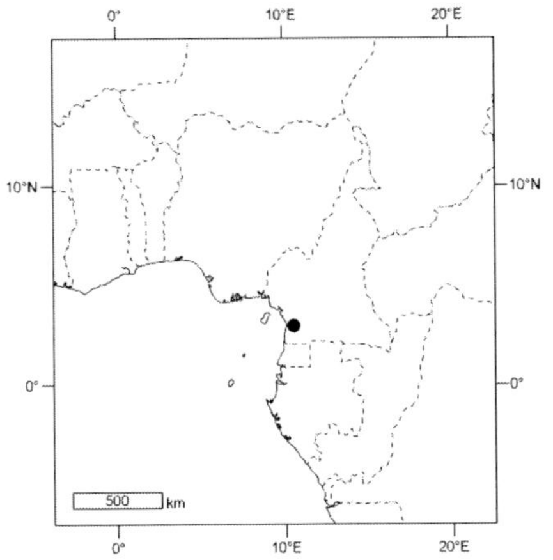

a. ***Sabicea trigemina*** CR (Rubiaceae)

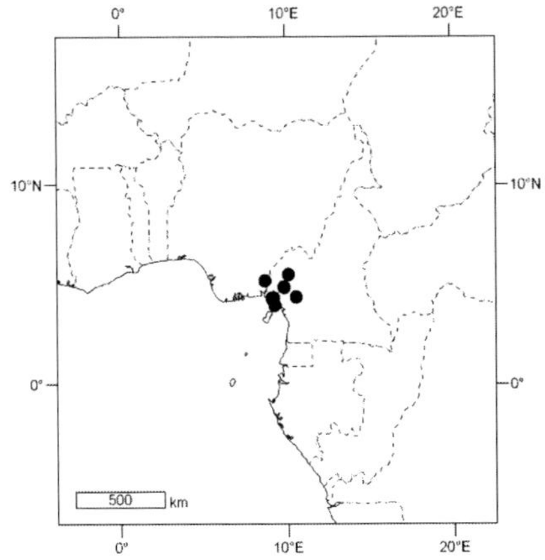

b. ***Sabicea xanthotricha*** VU (Rubiaceae)

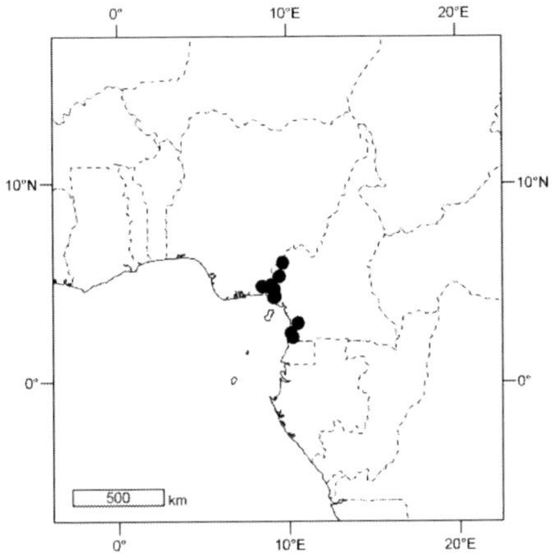

c. ***Sericanthe auriculata*** VU (Rubiaceae)

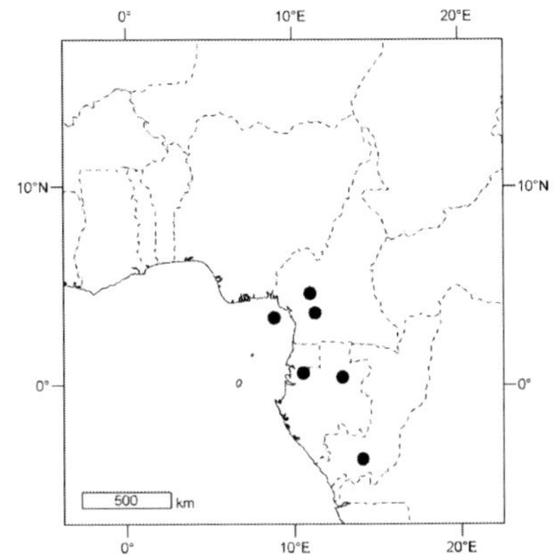

d. ***Sericanthe jacfelicis*** VU (Rubiaceae)

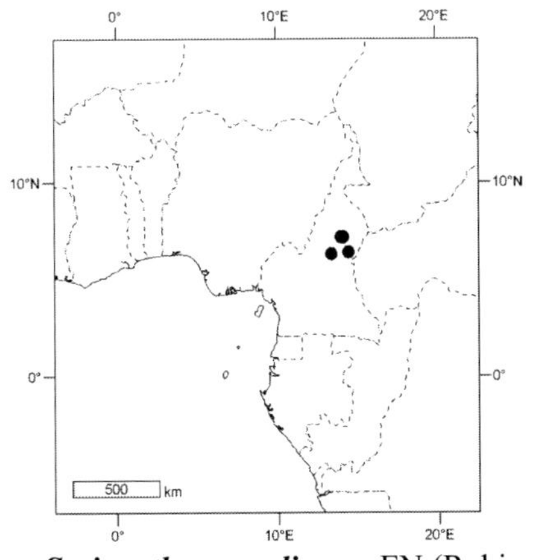

e. ***Sericanthe raynaliorum*** EN (Rubiaceae)

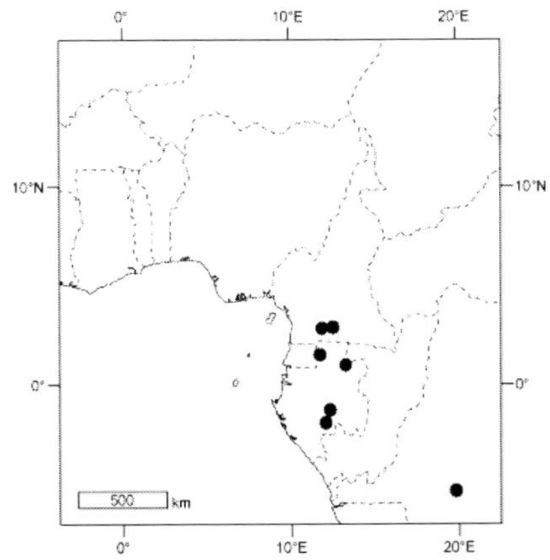

f. ***Sericanthe testui*** var. ***testui*** VU (Rubiaceae)

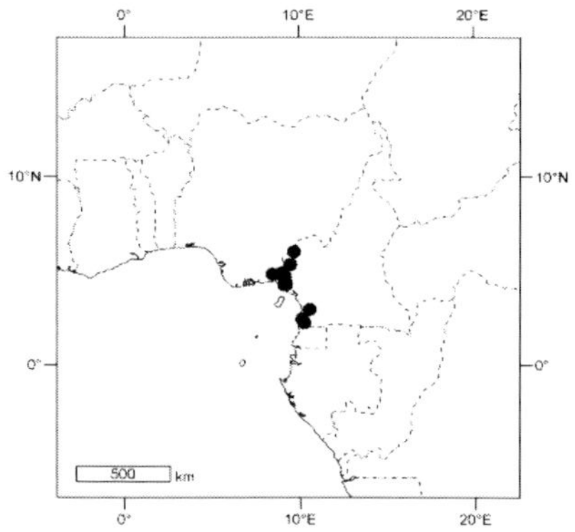

a. ***Stelechantha arcuata*** CR (Rubiaceae)

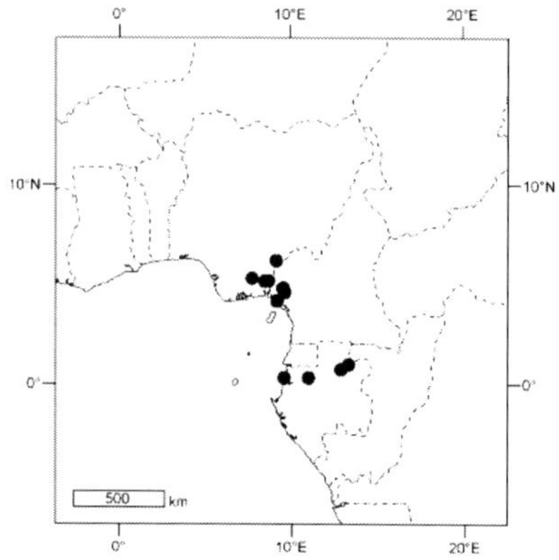

b. ***Tarenna baconioides*** EN (Rubiaceae)

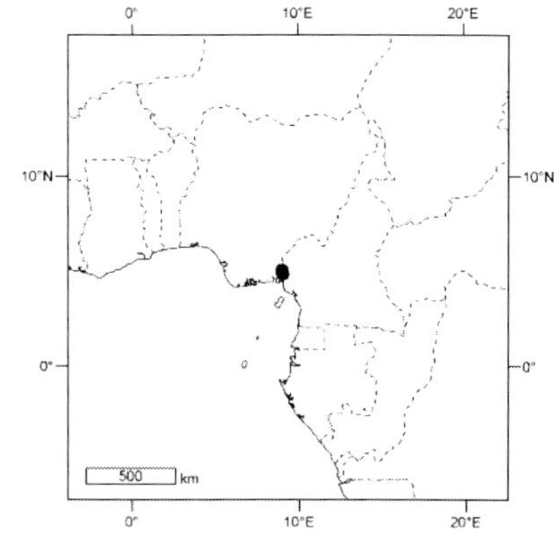

c. ***Tricalysia achoundongiana*** EN (Rubiaceae)

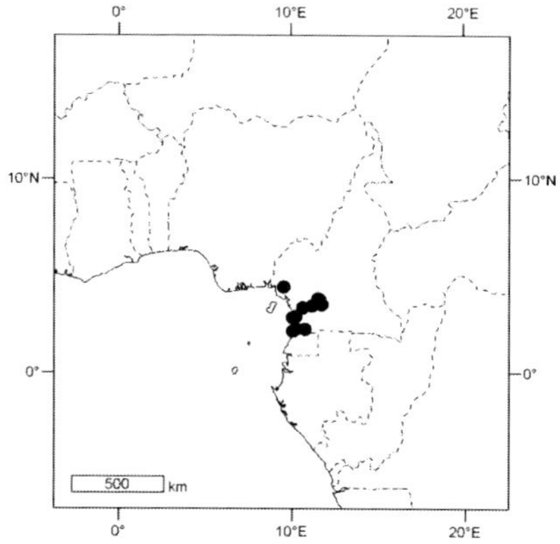

d. ***Tricalysia amplexicaulis*** VU (Rubiaceae)

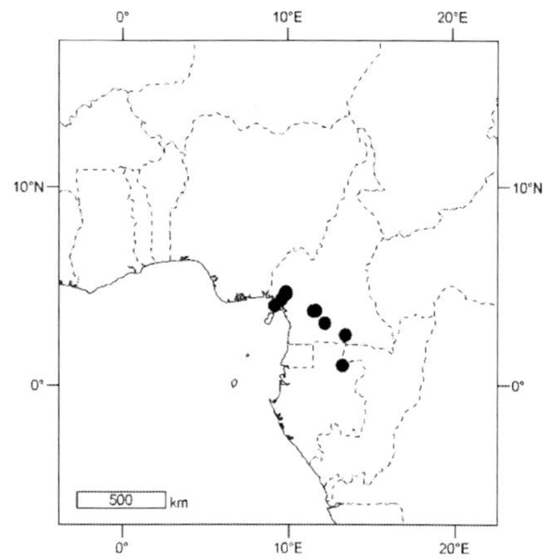

e. ***Tricalysia atherura*** EN (Rubiaceae)

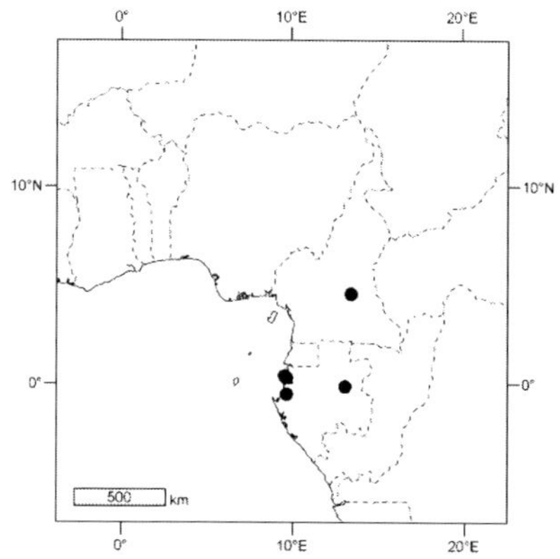

f. ***Tricalysia fangana*** EN (Rubiaceae)

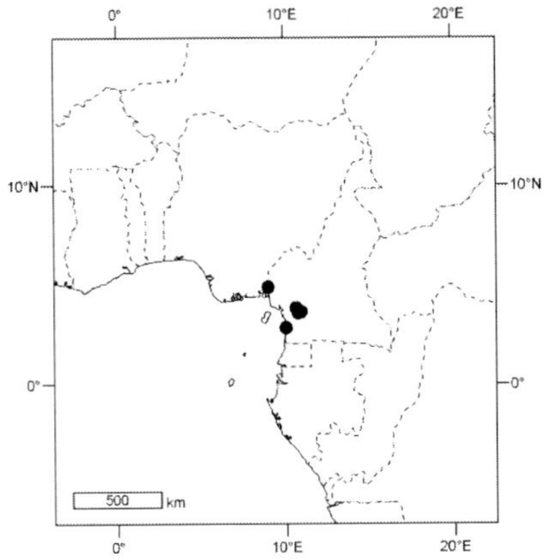

a. ***Tricalysia ferorum*** EN (Rubiaceae)

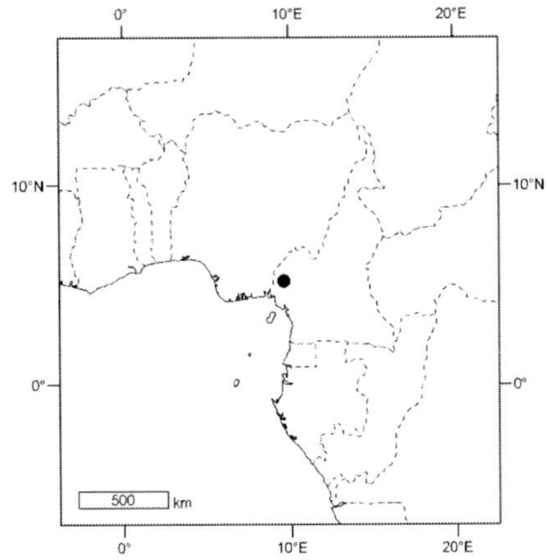

b. ***Tricalysia lejolyana*** CR (Rubiaceae)

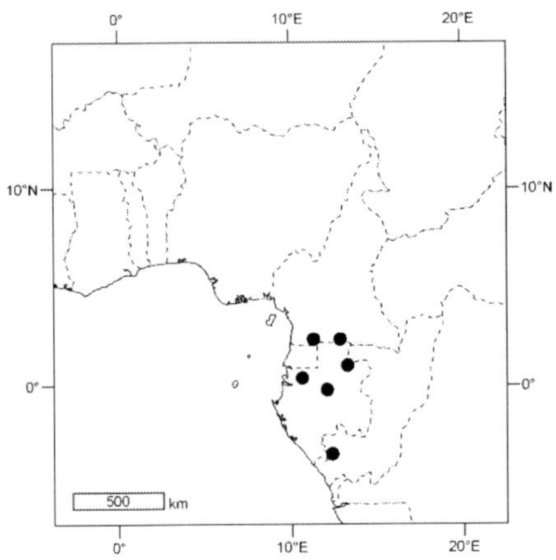

c. ***Tricalysia obstetrix*** VU (Rubiaceae)

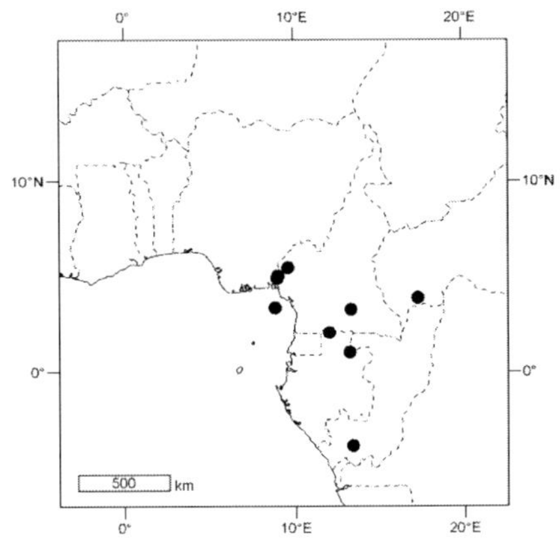

d. ***Tricalysia pangolina*** VU (Rubiaceae)

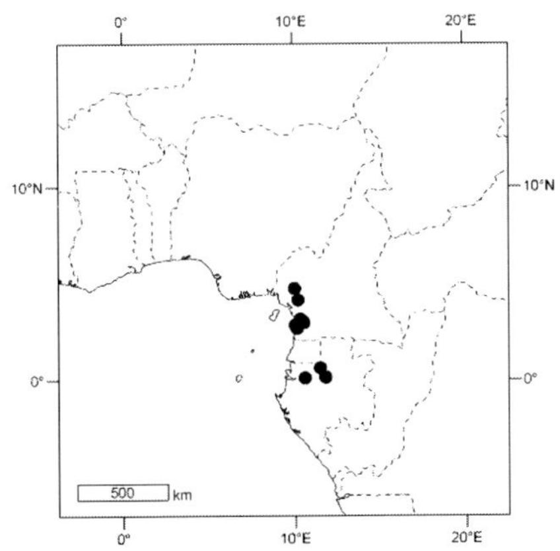

e. ***Tricalysia sylvae*** VU (Rubiaceae)

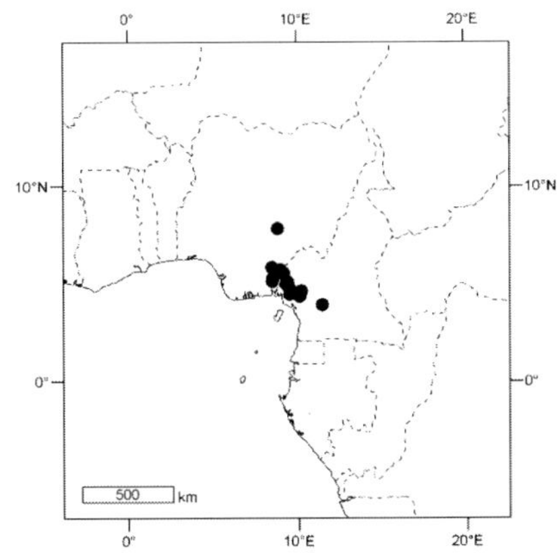

f. ***Tricalysia talbotii*** EN (Rubiaceae)

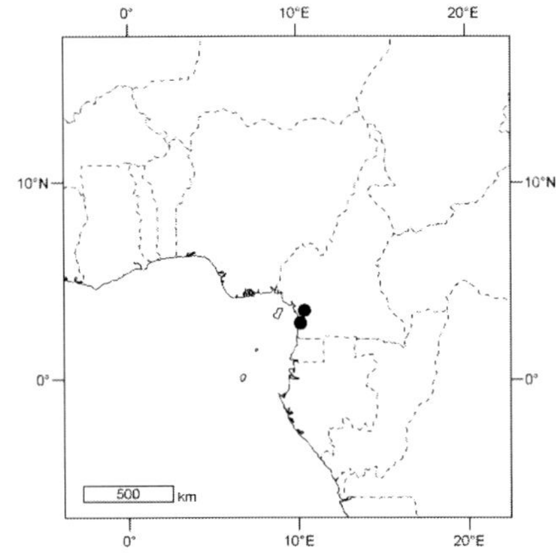

a. ***Tricalysia vadensis*** EN (Rubiaceae)

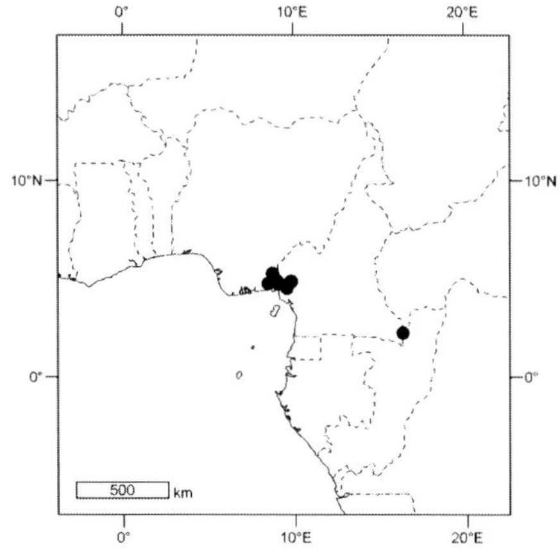

b. ***Trichostachys interrupta*** VU (Rubiaceae)

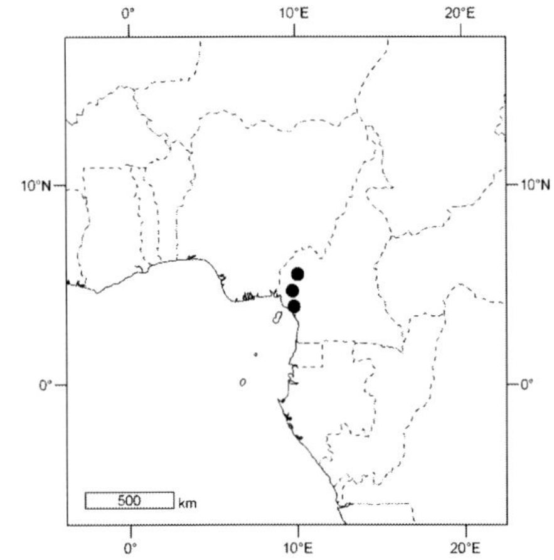

c. ***Trichostachys petiolata*** EN (Rubiaceae)

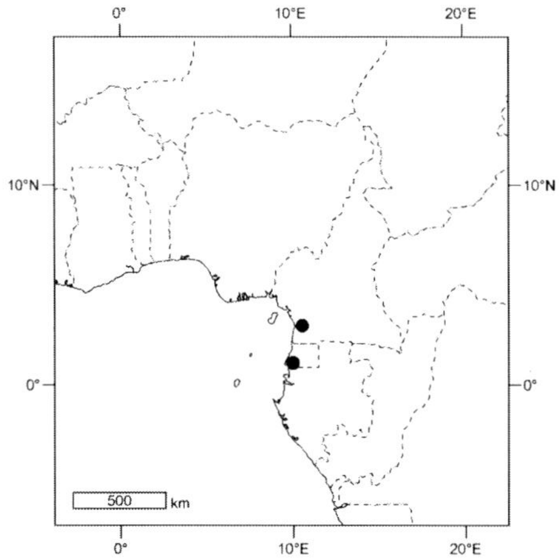

d. ***Trichostachys zenkeri*** EN (Rubiaceae)

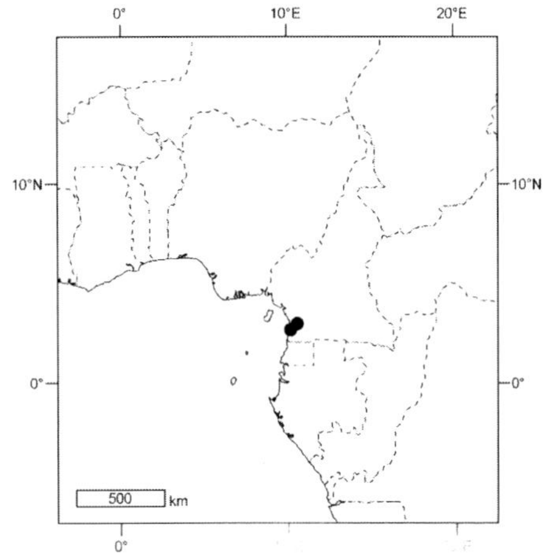

e. ***Vangueriella ze*** [illegible]

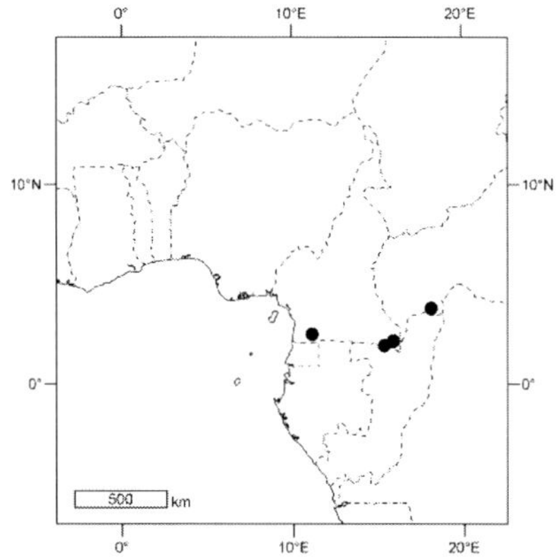

f. ***Afra*** [illegible] EN (Rutaceae)

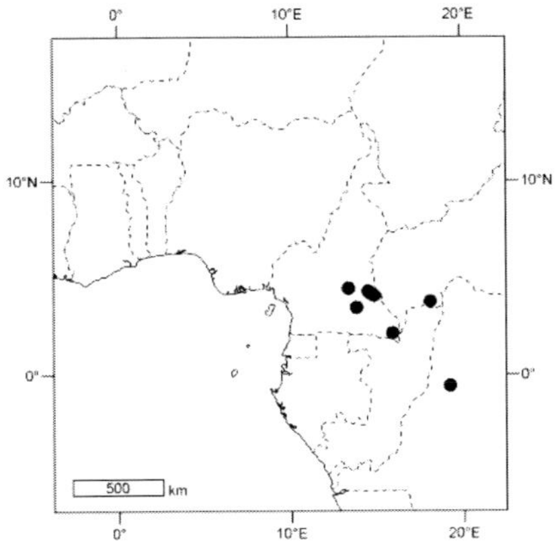

a. ***Balsamocitrus camerunensis*** VU (Rutaceae)

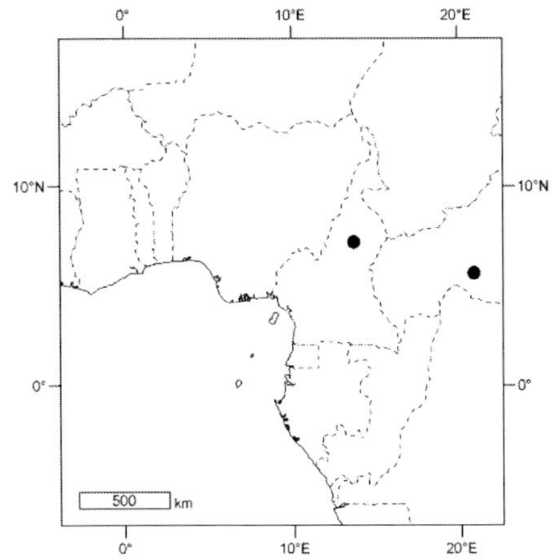

b. ***Vepris oubanguiensis*** EN (Rutaceae)

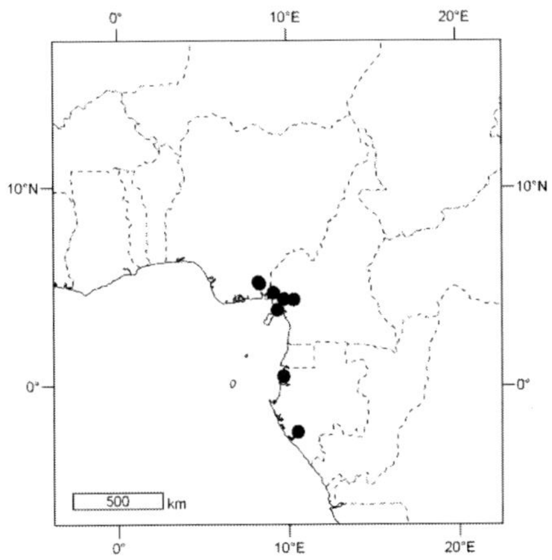

c. ***Vepris soyauxii*** VU (Rutaceae)

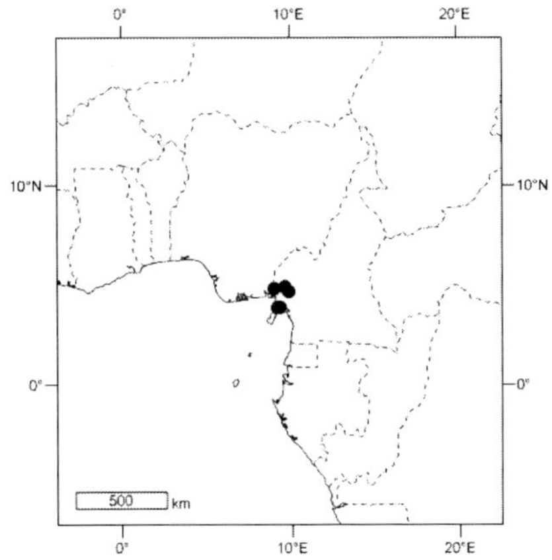

d. ***Vepris trifoliolata*** EN (Rutaceae)

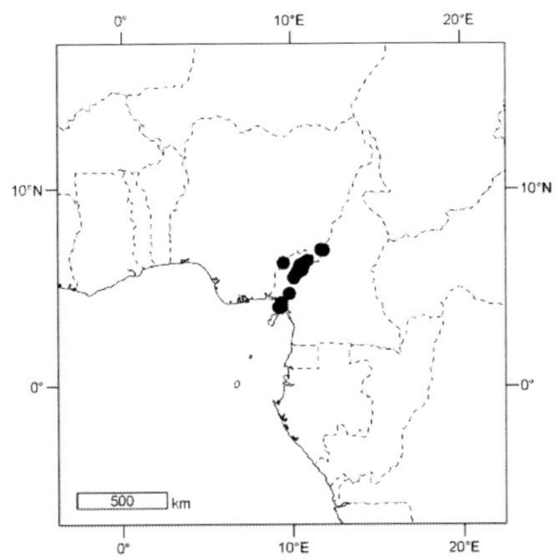

e. ***Allophylus bullatus*** VU (Sapindaceae)

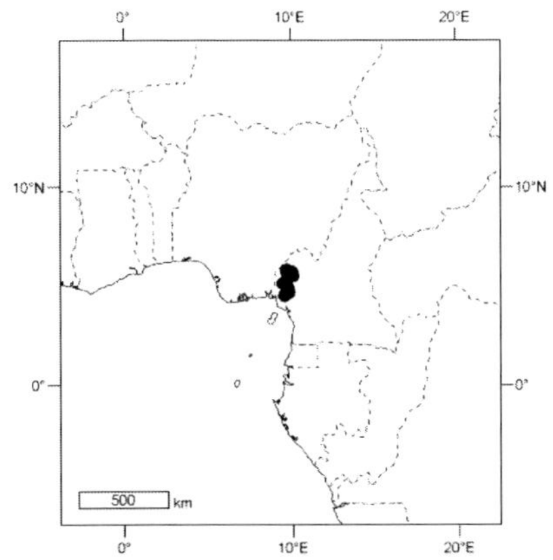

f. ***Allophylus conraui*** EN (Sapindaceae)

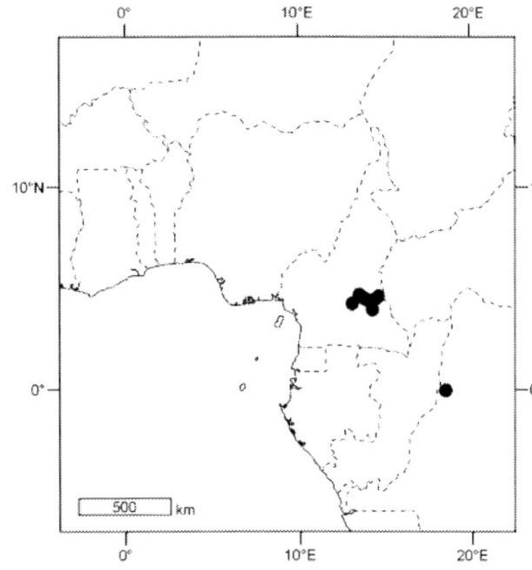

a. ***Allophylus hamatus*** VU (Sapindaceae)

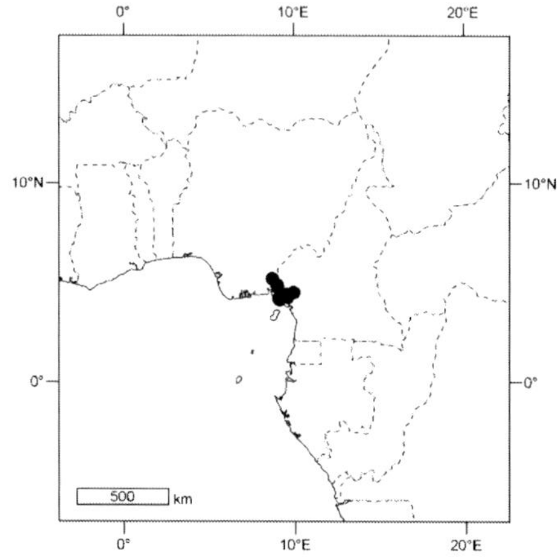

b. ***Allophylus megaphyllus*** EN (Sapindaceae)

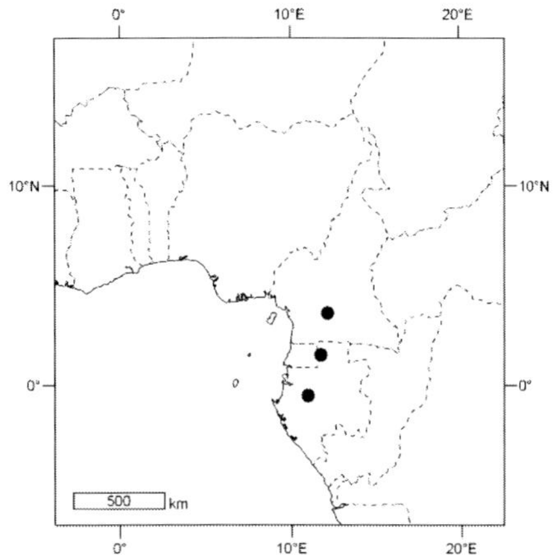

c. ***Allophylus oyemensis*** EN (Sapindaceae)

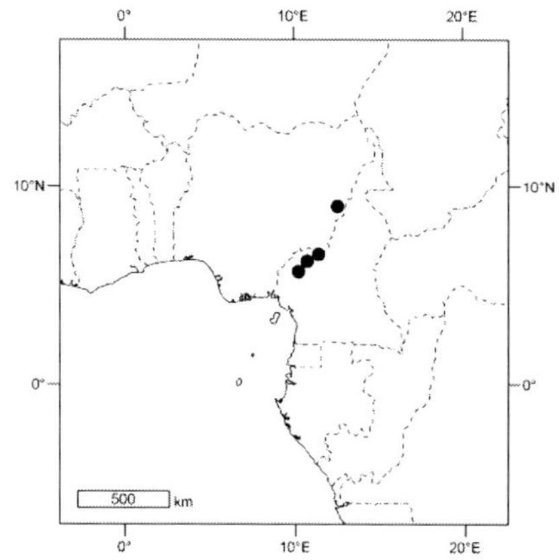

d. ***Allophylus ujori*** EN (Sapindaceae)

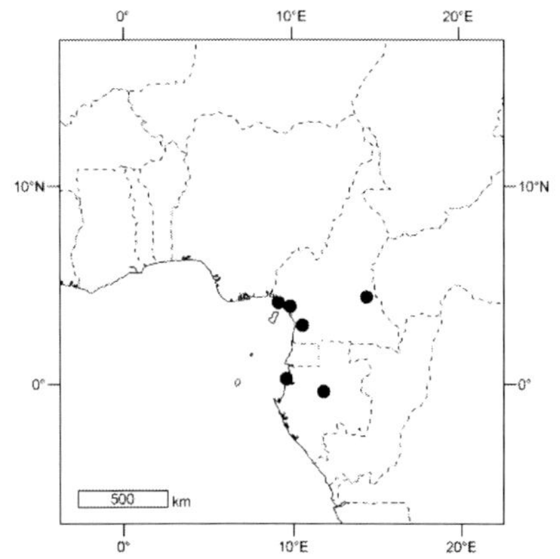

e. ***Allophylus zenkeri*** VU (Sapindaceae)

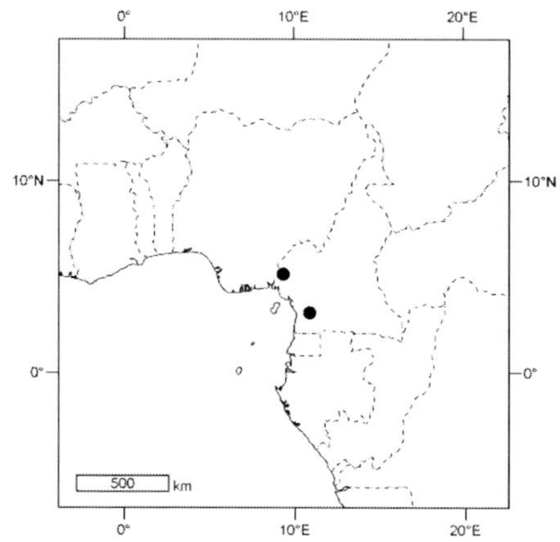

f. ***Aporrhiza multijuga*** CR (Sapindaceae)

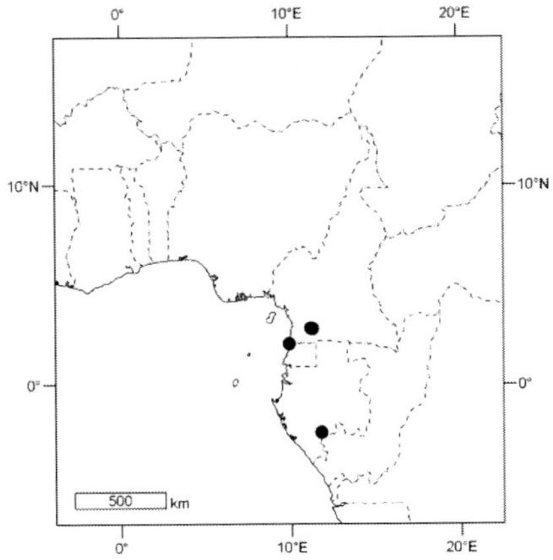

a. ***Aporrhiza tessmannii*** EN (Sapindaceae)

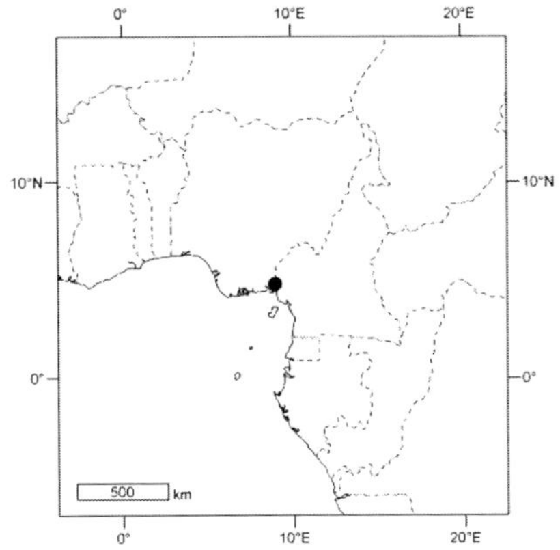

b. ***Deinbollia angustifolia*** VU (Sapindaceae)

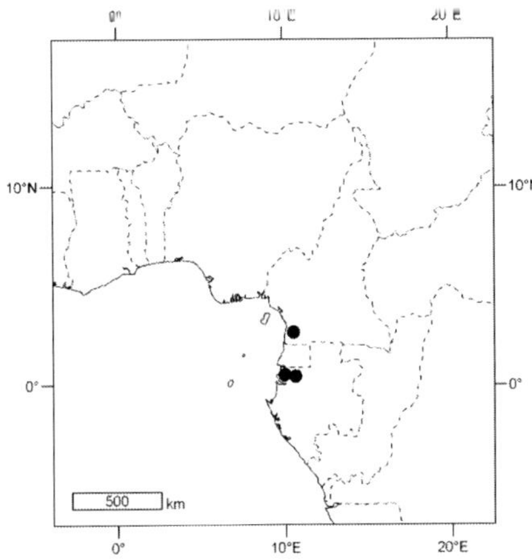

c. ***Deinbollia dasybotrys*** EN (Sapindaceae)

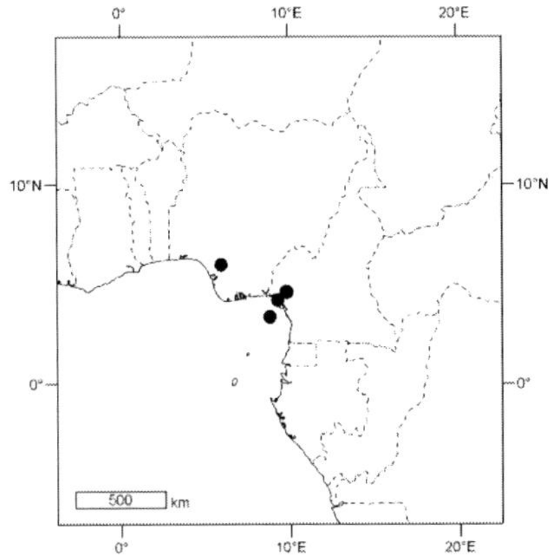

d. ***Deinbollia insignis*** VU (Sapindaceae)

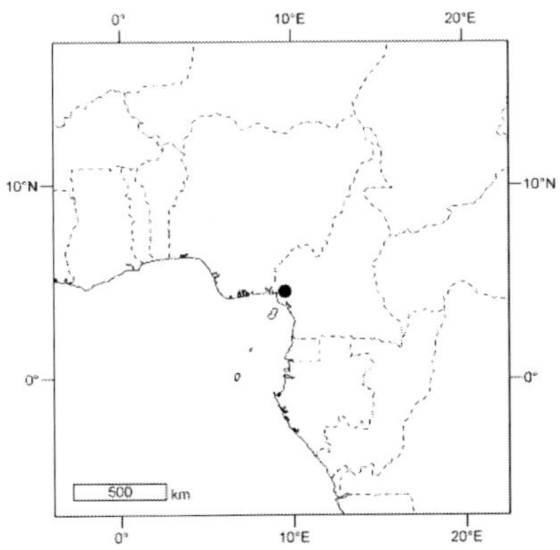

e. ***Deinbollia macrantha*** CR (Sapindaceae)

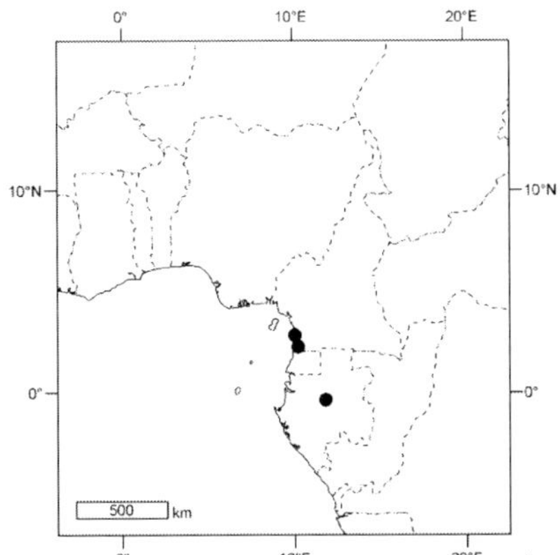

f. ***Deinbollia macroura*** EN (Sapindaceae)

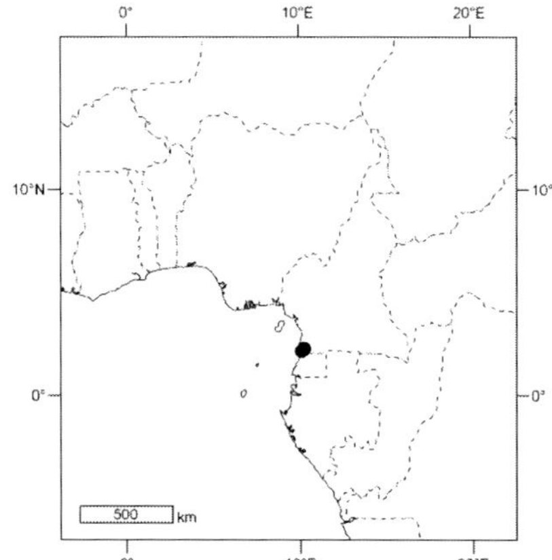

a. ***Deinbollia mezilii*** EN (Sapindaceae)

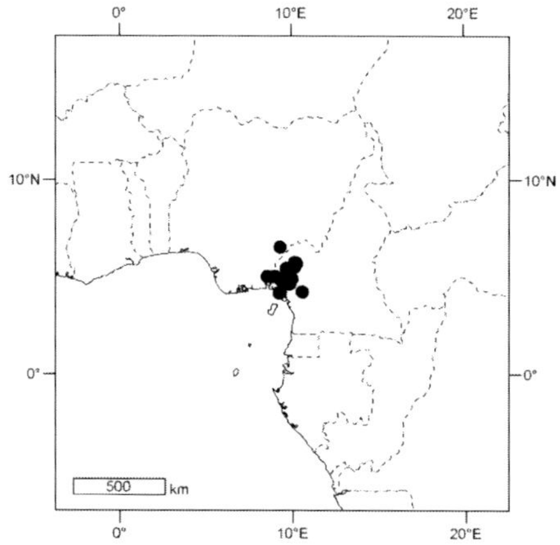

b. ***Deinbollia oreophila*** VU (Sapindaceae)

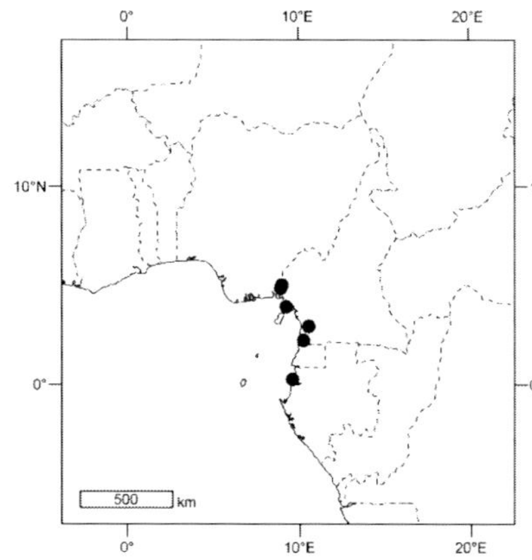

c. ***Deinbollia pycnophylla*** EN (Sapindaceae)

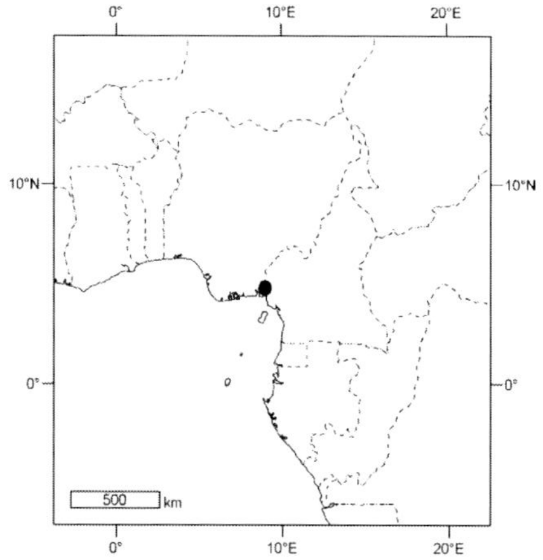

d. ***Deinbollia saligna*** VU (Sapindaceae)

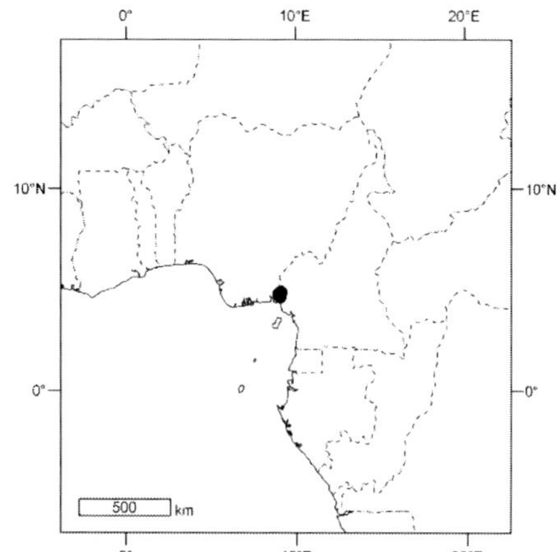

e. ***Deinbollia unijuga*** EN (Sapindaceae)

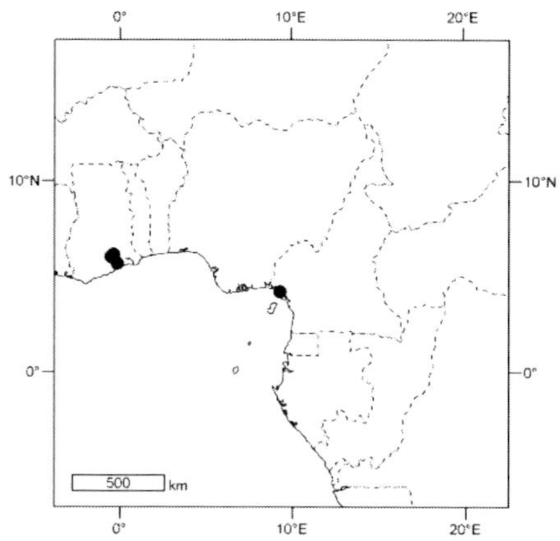

f. ***Leconiodiscus punctatus*** EN (Sapindaceae)

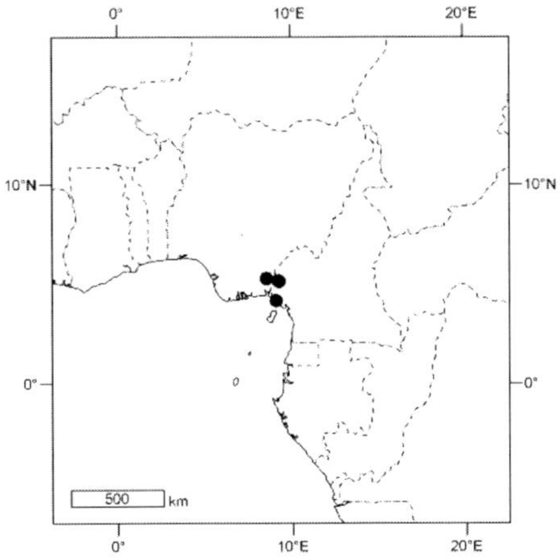

a. ***Namatea simplicifolia*** EN (Sapindaceae)

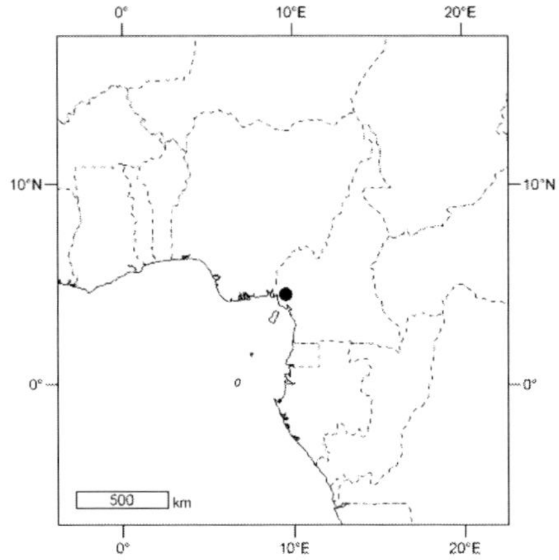

b. ***Pancovia polyantha*** CR (Sapindaceae)

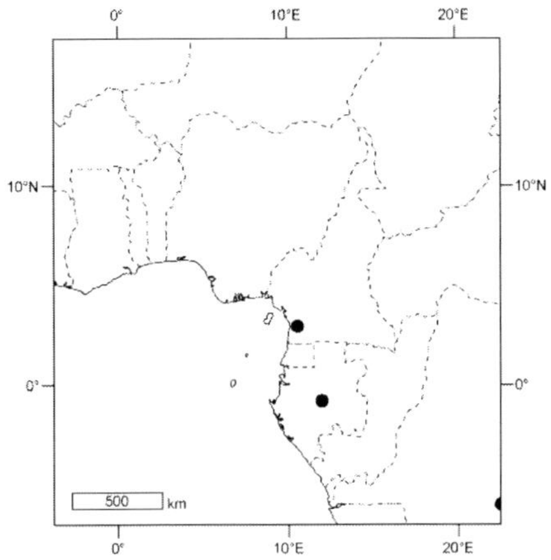

c. ***Placodiscus angustifolius*** EN (Sapindaceae)

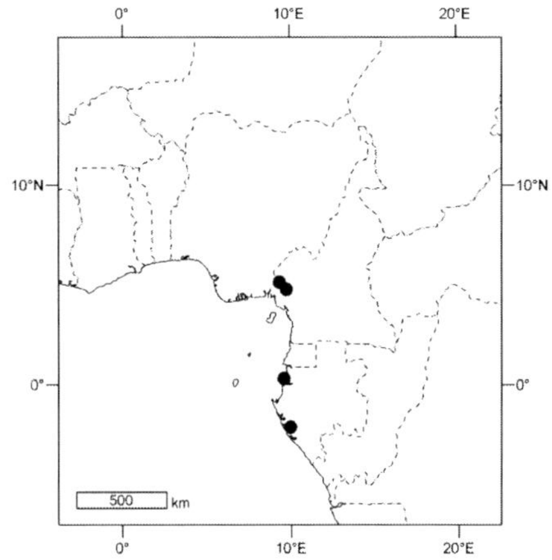

d. ***Placodiscus caudatus*** EN (Sapindaceae)

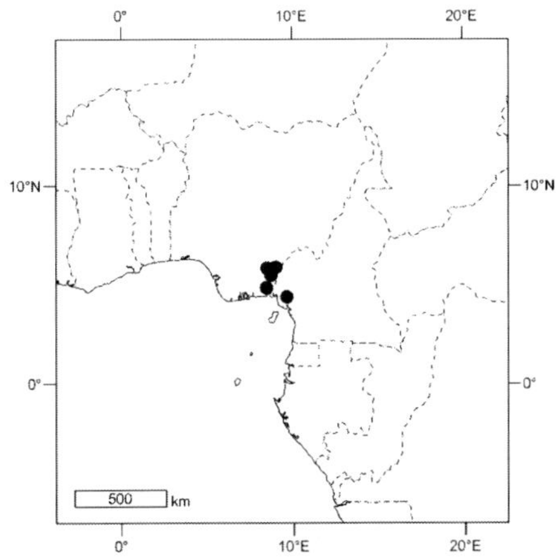

e. ***Placodiscus turbinatus*** EN (Sapindaceae)

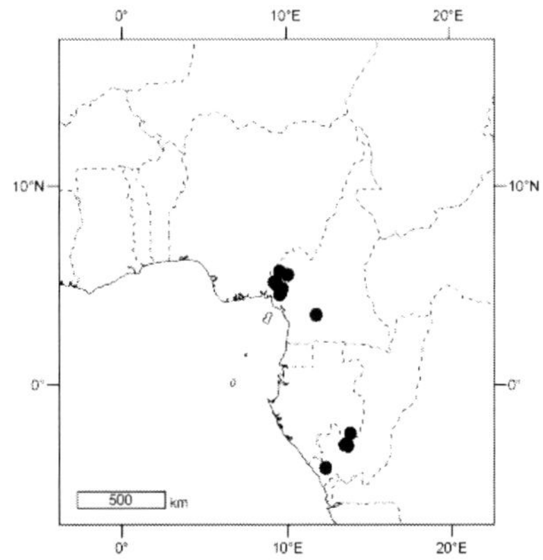

f. ***Baillonella toxisperma*** VU (Sapotaceae)

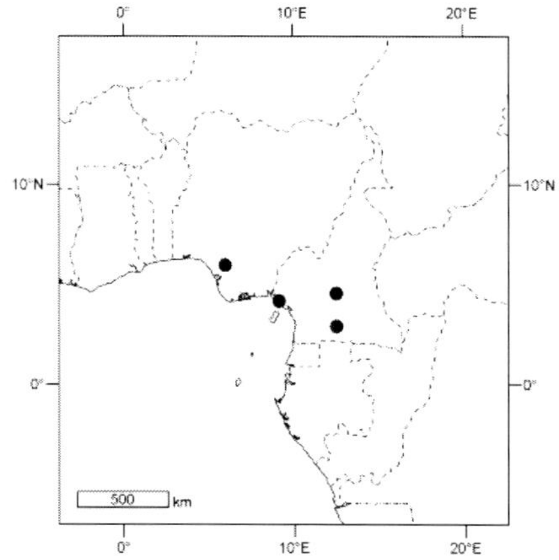

a. ***Englerophytum oubanguiense*** EN (Sapotaceae)

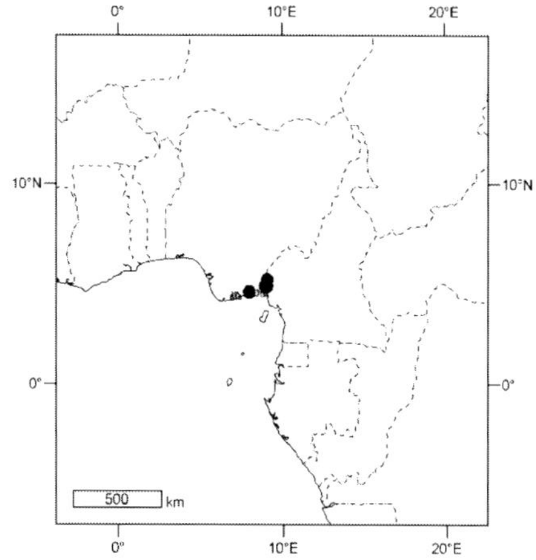

b. ***Gluema korupensis*** EN (Sapotaceae)

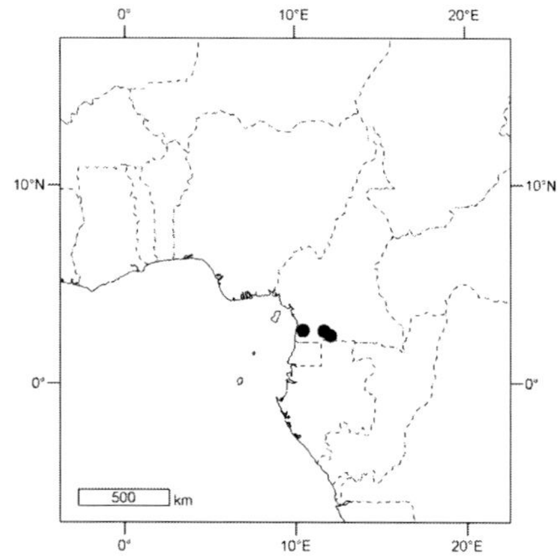

c. ***Inhambanella guereensis*** VU (Sapotaceae)

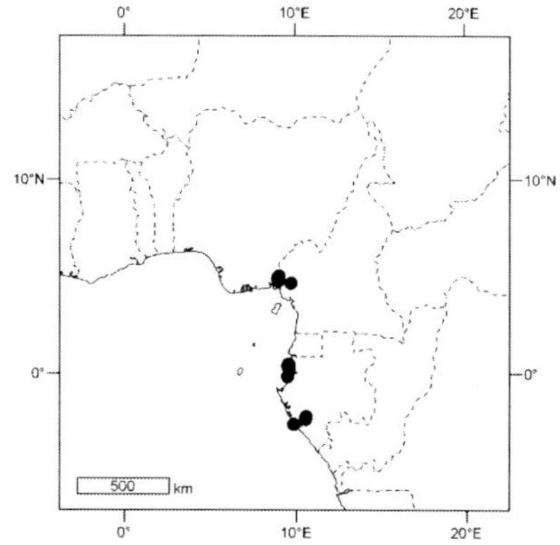

d. ***Lecomtedoxa klaineana*** VU (Sapotaceae)

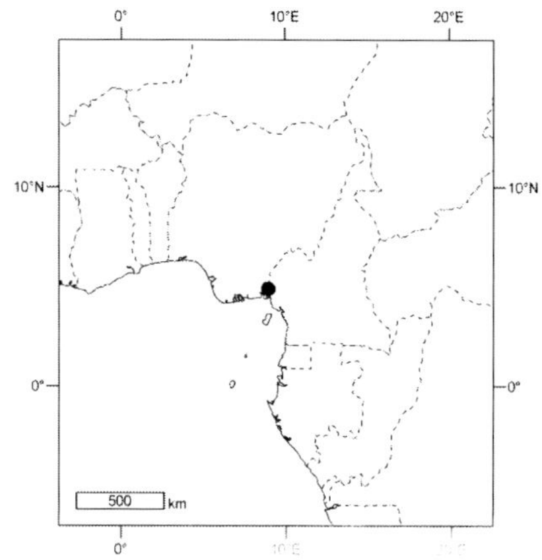

e. ***Lecomtedoxa p*** [illegible]

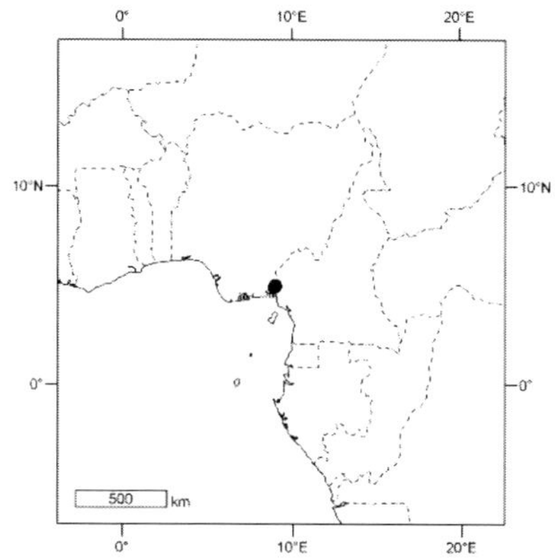

f. ***Manilkara lososiana*** CR (Sapotaceae)

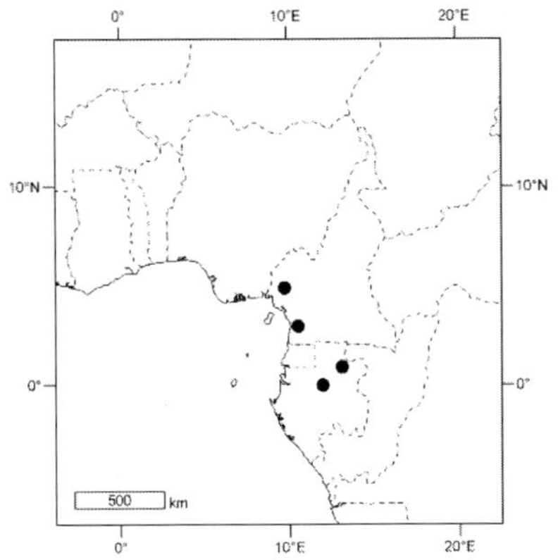

a. ***Manilkara zenkeri*** EN (Sapotaceae)

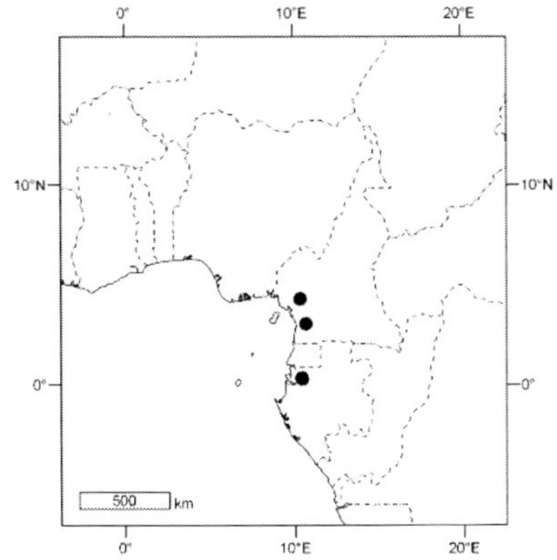

b. ***Neolemonniera batesii*** EN(Sapotaceae)

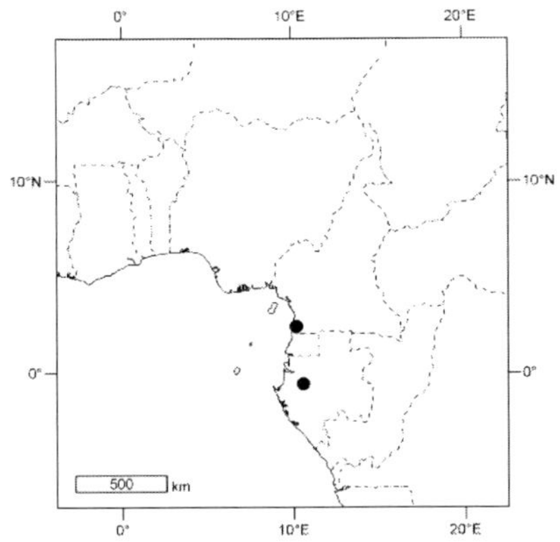

c. ***Neolemmoniera ogouensis*** EN (Sapotaceae)

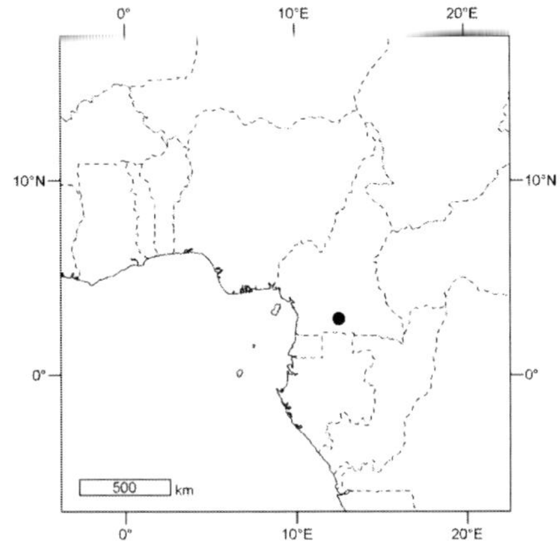

d. ***Synsepalum batesii*** CR (Sapotaceae)

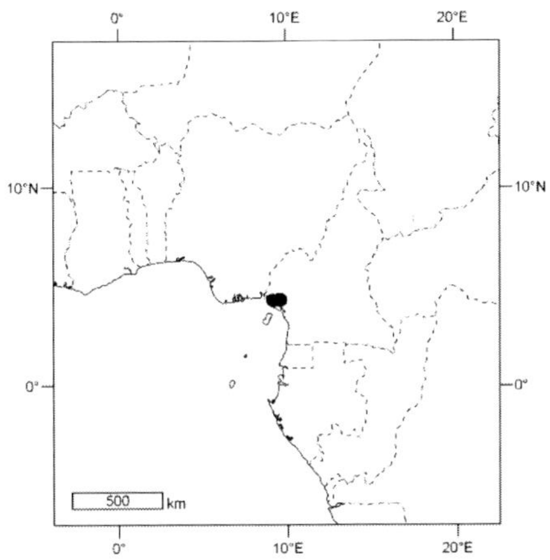

e. ***Synsepalum brenanii*** CR (Sapotaceae)

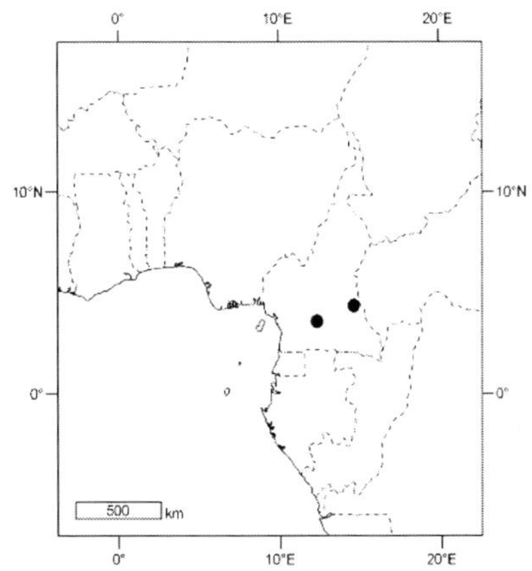

f. ***Synsepalum letouzei*** EN (Sapotaceae)

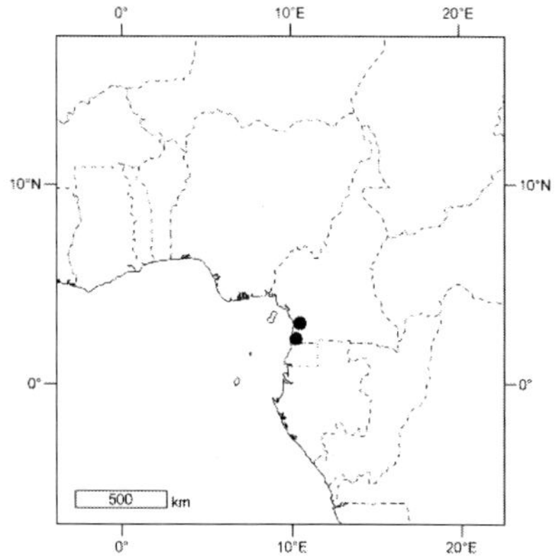

a. ***Synsepalum zenkeri*** EN (Sapotaceae)

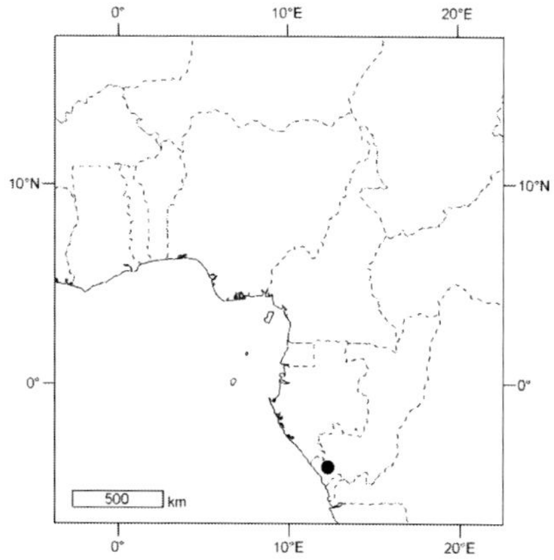

b. ***Tieghemella africana*** EN (Sapotaceae)

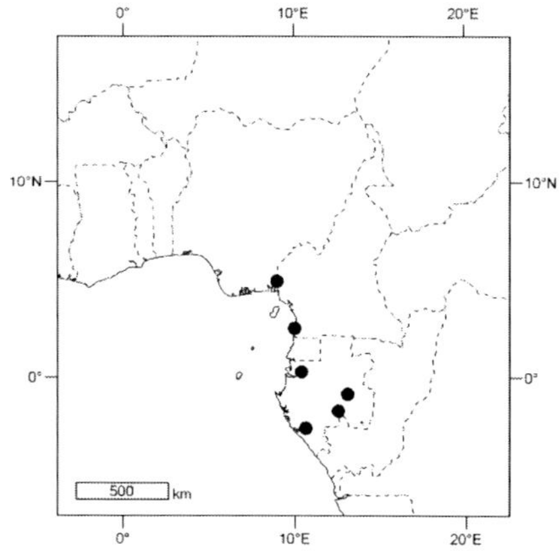

c. ***Zeyherella letestui*** VU (Sapotaceae)

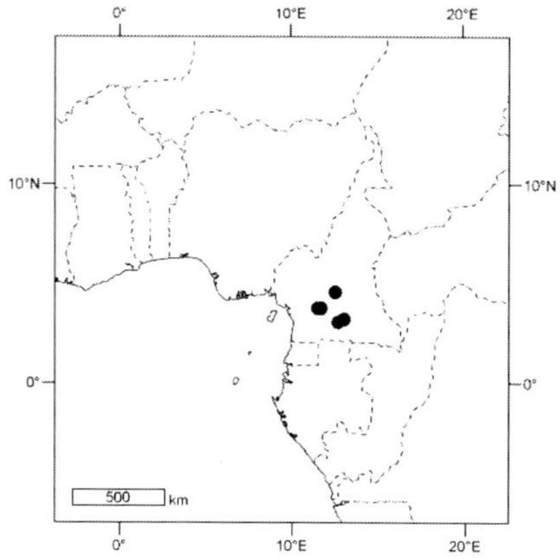

d. ***Lindernia yaundensis*** EN (Scrophulariaceae)

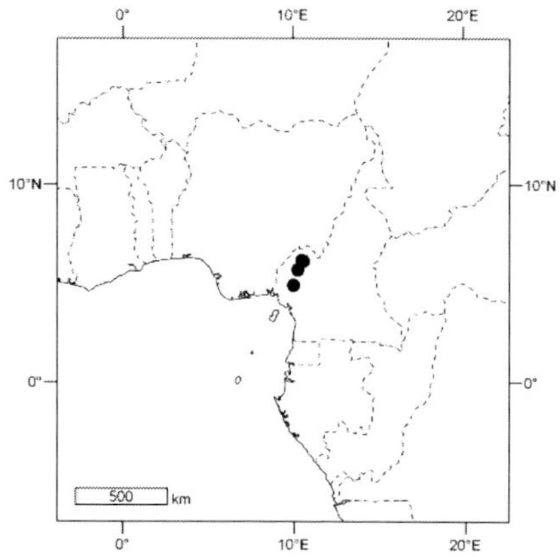

e. ***Rhabdotosperma densifolia*** VU (Scrophulariaceae)

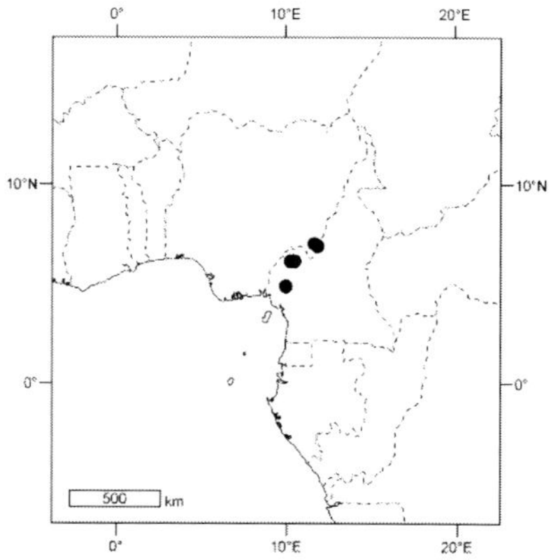

f. ***Rhabdotosperma ledermannii*** VU (Scrophulariaceae)

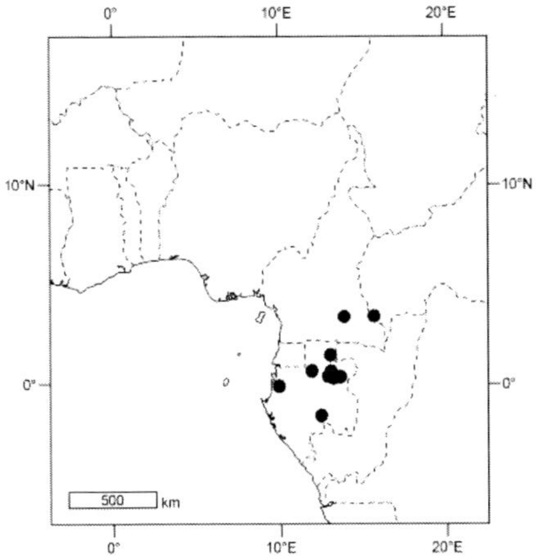

a. ***Torenia silvicola*** VU (Scrophulariaceae)

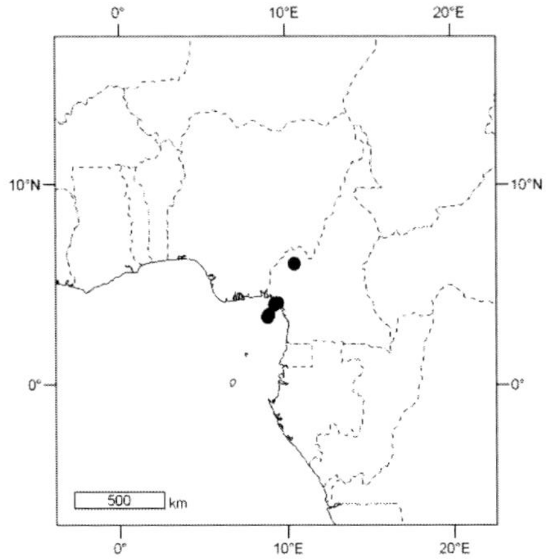

b. ***Veronica mannii*** EN (Scrophulariaceae)

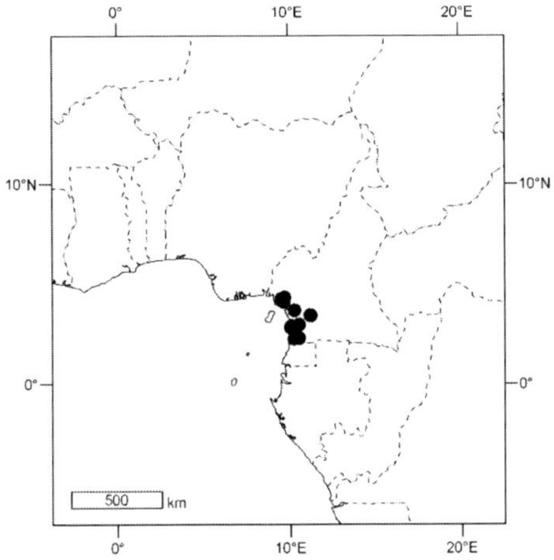

c. ***Pierrina zenkeri*** VU (Scytopetalaceae)

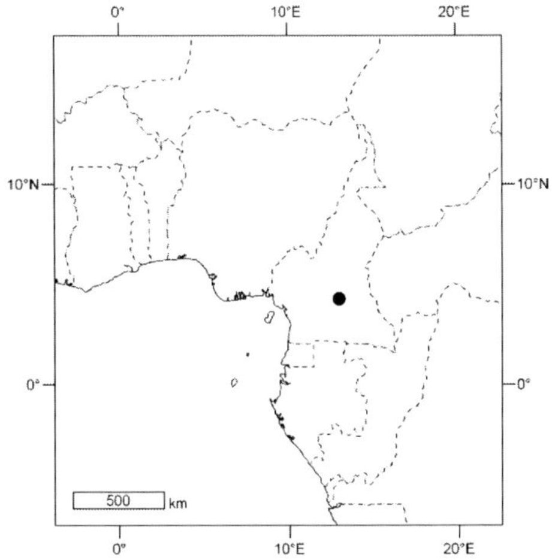

d. ***Rhaptopetalum breteleri*** CR (Scytopetalaceae)

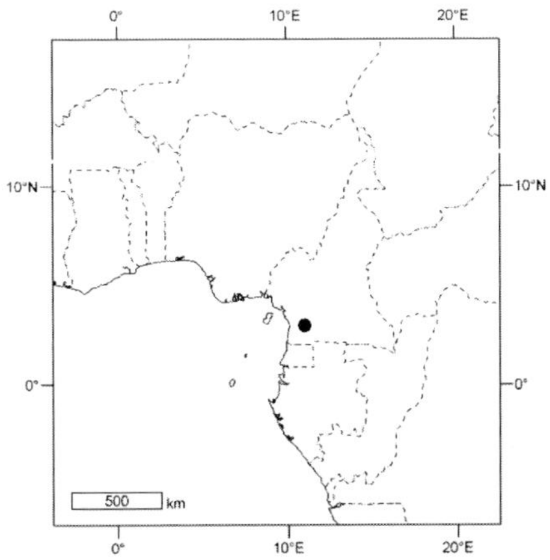

e. ***Rhaptopetalum depressum*** CR (Scytopetalaceae)

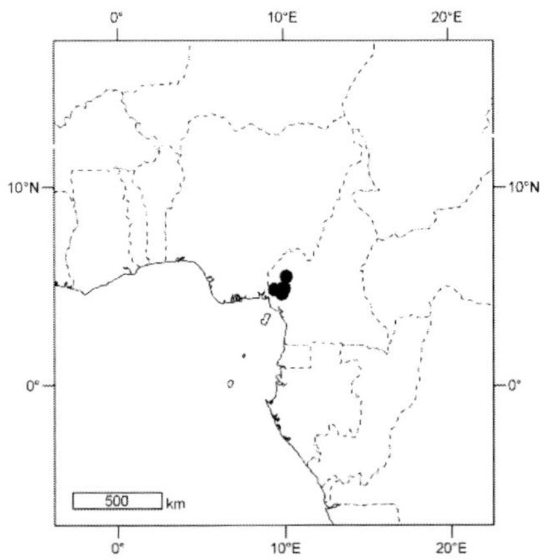

f. ***Rhaptopetalum geophylax*** EN (Scytopetalaceae)

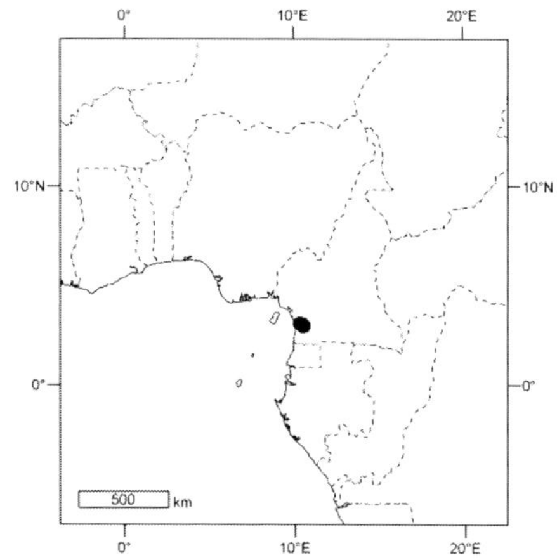

a. ***Rhaptopetalum sessilifolium*** EN (Scytopetalaceae)

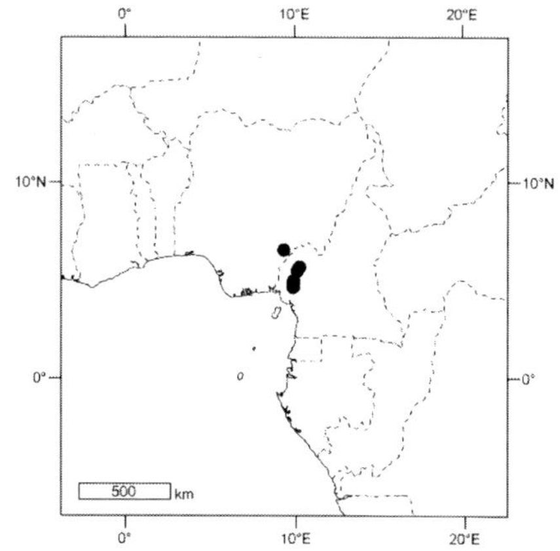

b. ***Quassia sanguinea*** VU (Simaroubaceae)

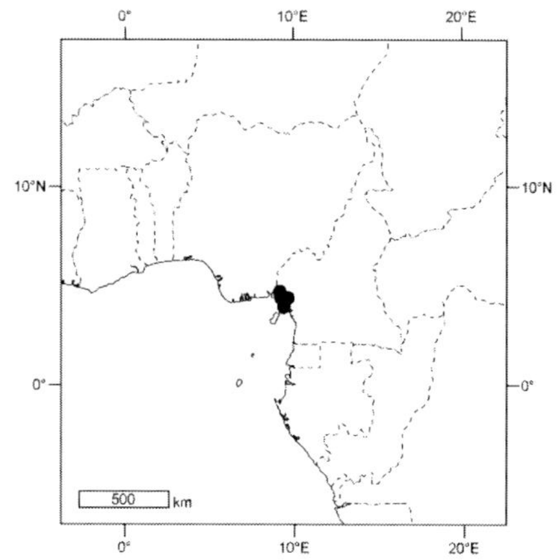

c. ***Cola cecidifolia*** CR (Sterculiaceae)

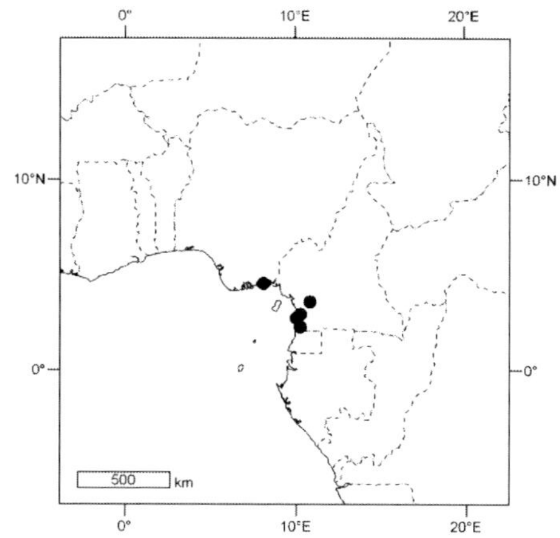

d. ***Cola hypochrysea*** EN (Sterculiaceae)

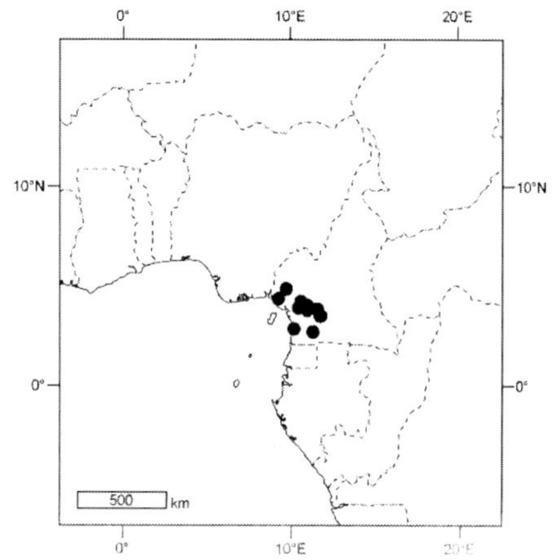

e. ***Cola letouzeyana*** VU (Sterculiaceae)

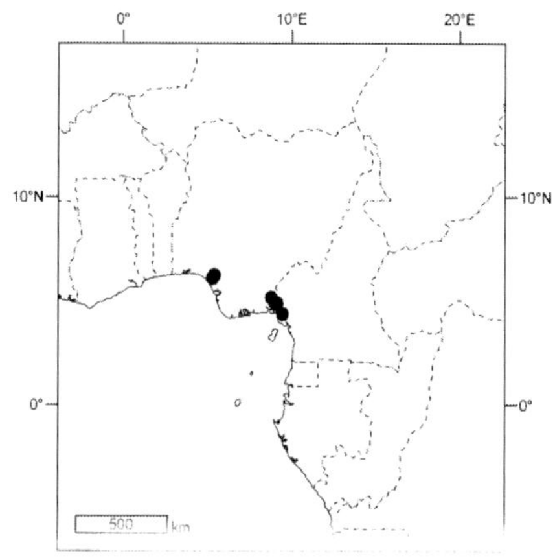

f. ***Cola*** [illegible]

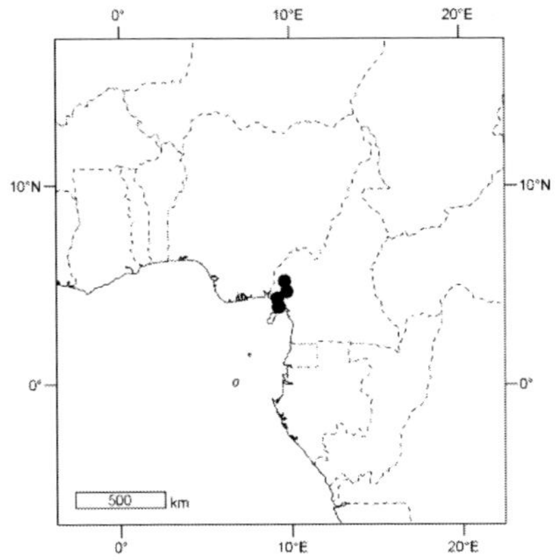

a. ***Cola metallica*** CR (Sterculiaceae)

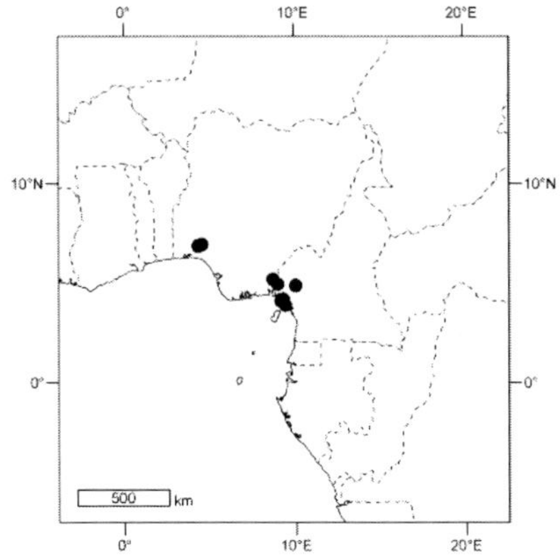

b. ***Cola nigerica*** CR (Sterculiaceae)

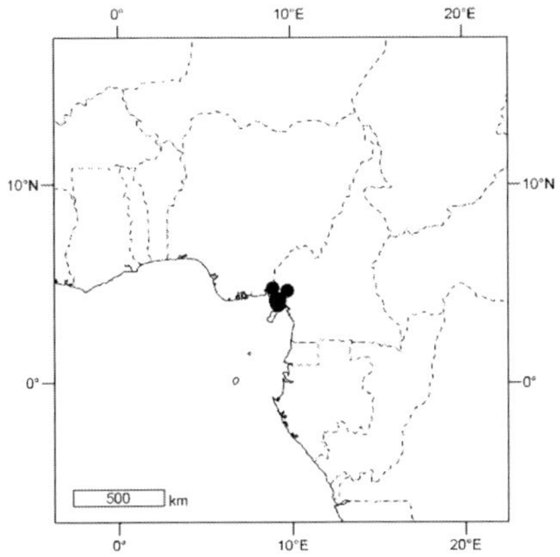

c. ***Cola praeacuta*** CR (Sterculiaceae)

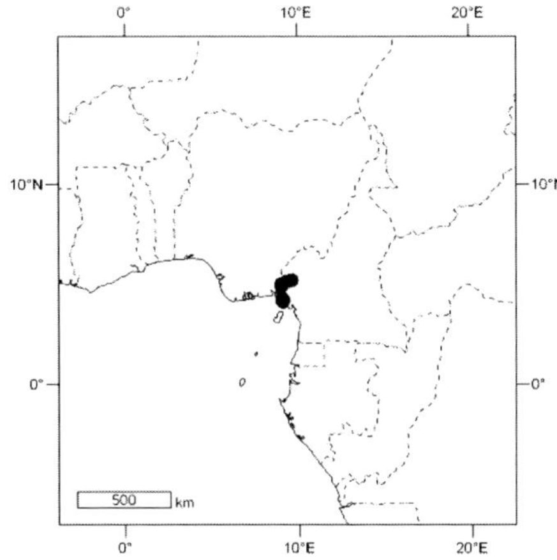

d. ***Cola suboppositifolia*** EN (Sterculiaceae)

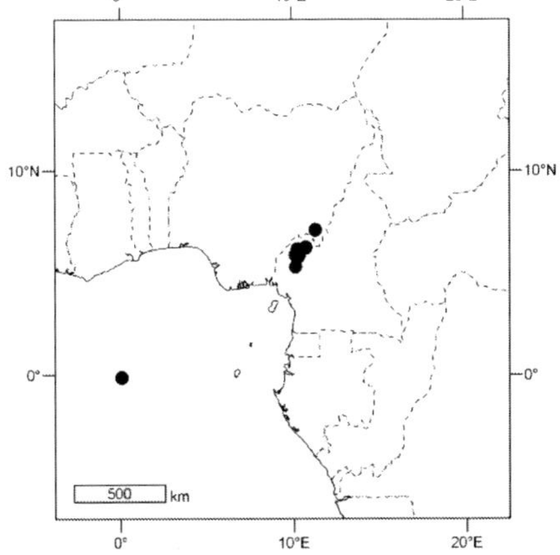

e. ***Dombeya ledermannii*** CR (Sterculiaceae)

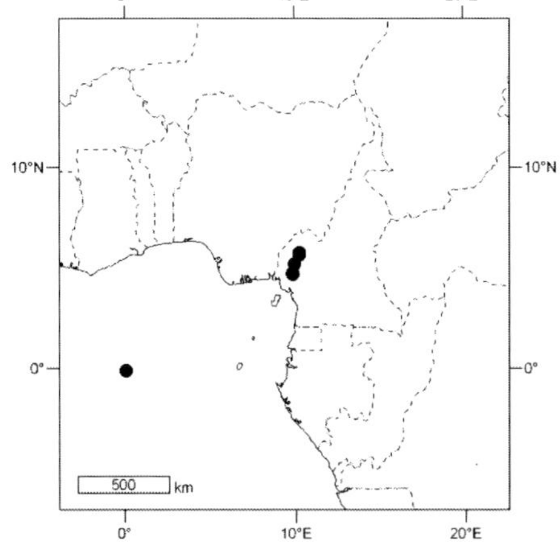

f. ***Leptonychia kamerunensis*** EN (Sterculiaceae)

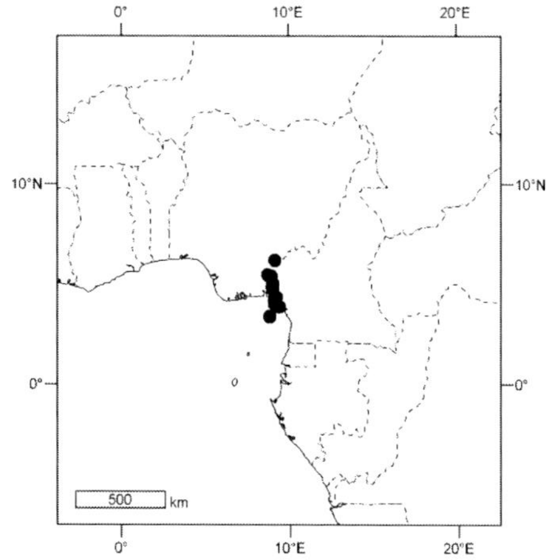

a. ***Leptonychia moyesiae*** VU (Sterculiaceae)

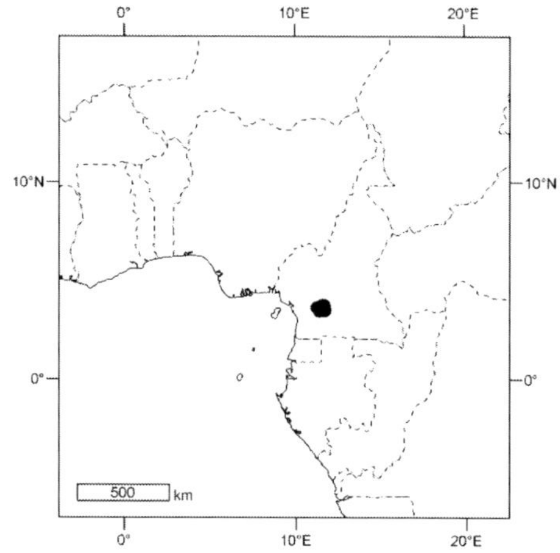

b. ***Leptonychia subtomentosa*** EN (Sterculiaceae)

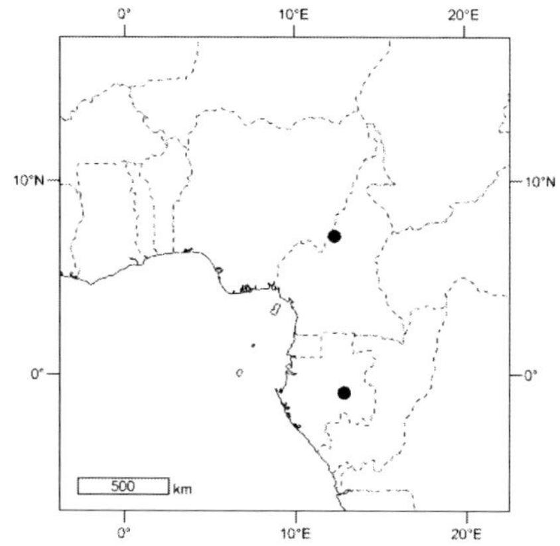

c. ***Craterosiphon montanum*** EN (Thymelaeaceae)

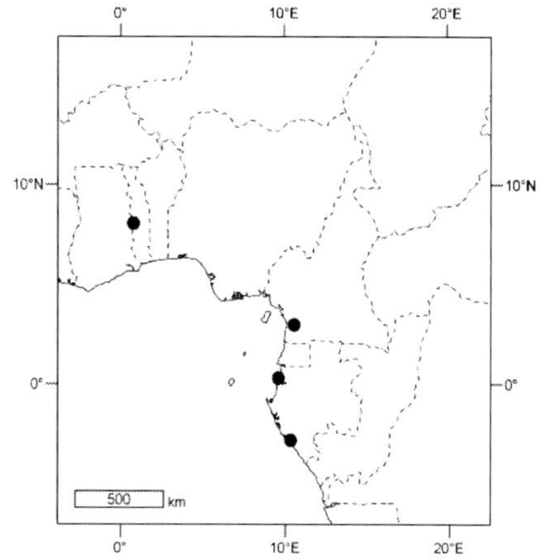

d. ***Craterosiphon pseudoscandens*** EN (Thymelaeaceae)

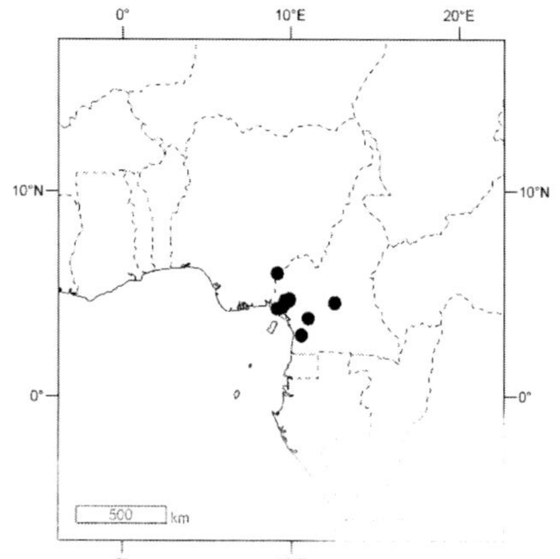

e. ***Dicranolepis polygaloides*** VU (Thymelaeaceae)

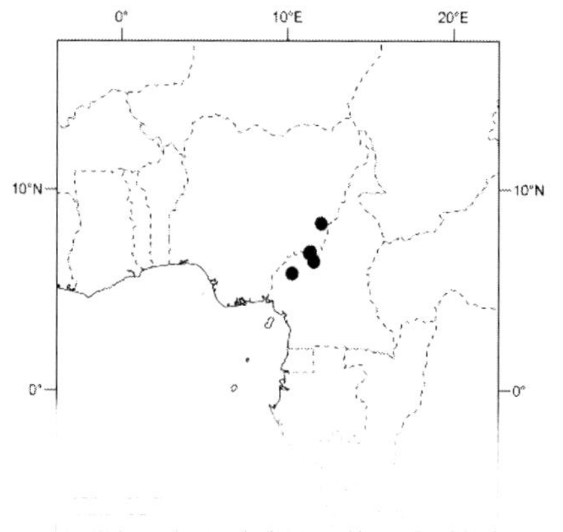

f. ***Gnidia bambutana*** EN (Thymelaeaceae)

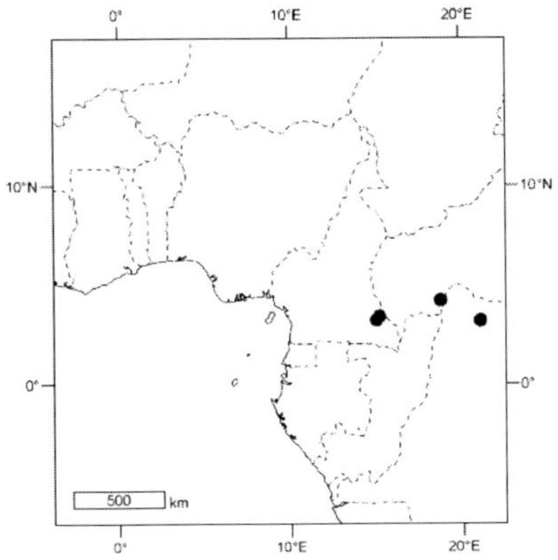

a. ***Synandrodaphne paradoxa*** VU (Thymelaeaceae)

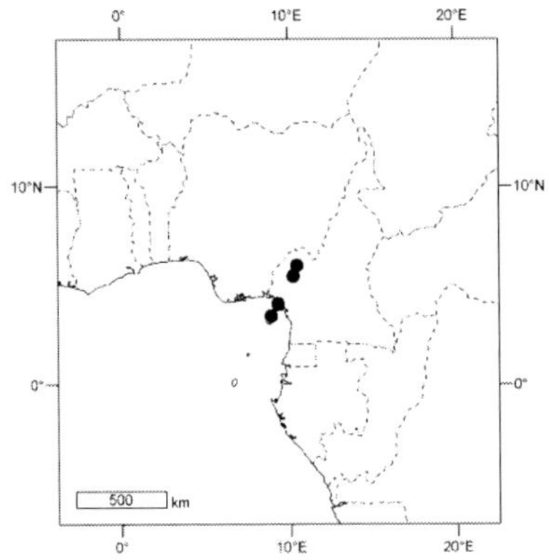

b. ***Afroligusticum townsendii*** EN (Umbelliferae)

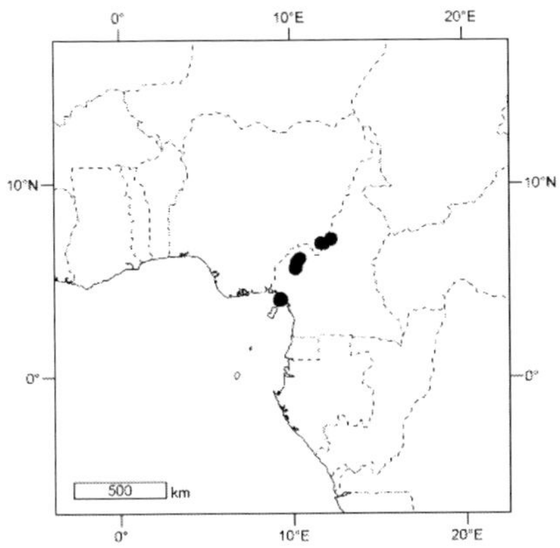

c. ***Lefebvrea angustisecta*** VU (Umbelliferae)

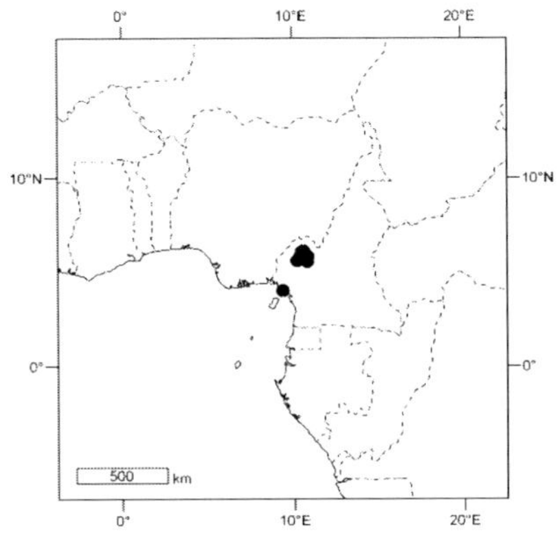

d. ***Lefebvrea camerunensis*** VU (Umbelliferae)

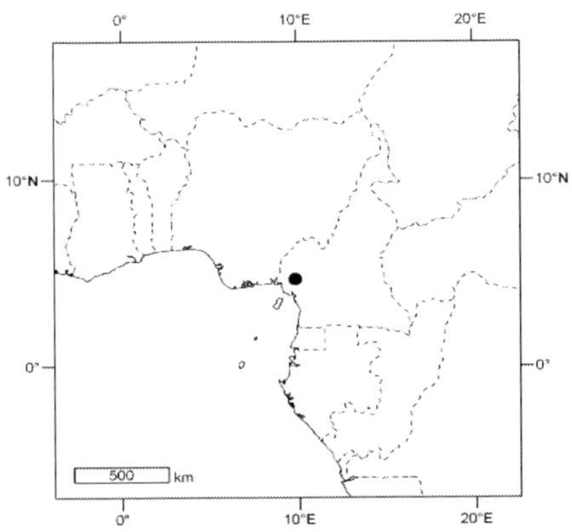

e. ***Lefebvrea kupense*** VU (Umbelliferae)

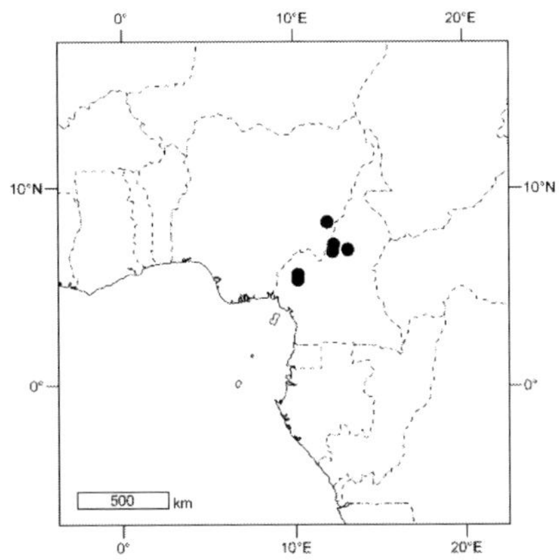

f. ***Pimpinella ledermannii*** *subsp.* ***ledermannii*** VU (Umbelliferae)

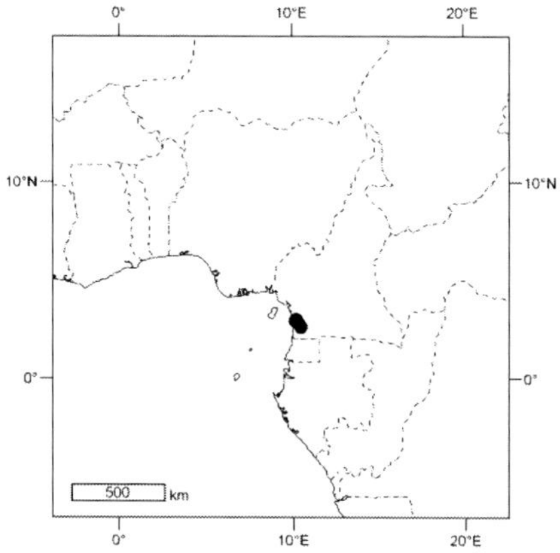

a. ***Allexis zygomorpha*** VU (Violaceae)

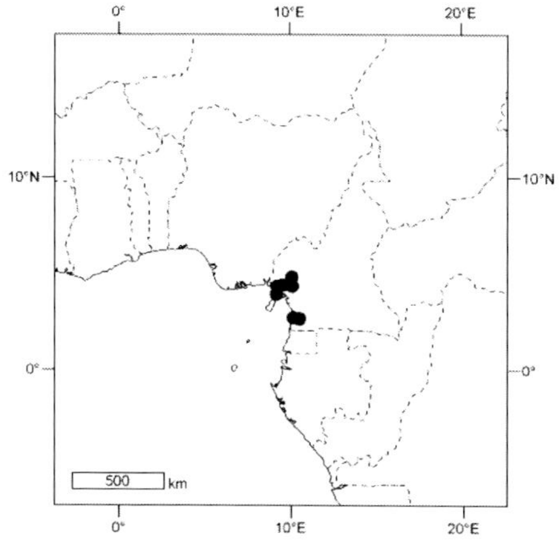

b. ***Rinorea dewitii*** VU (Violaceae)

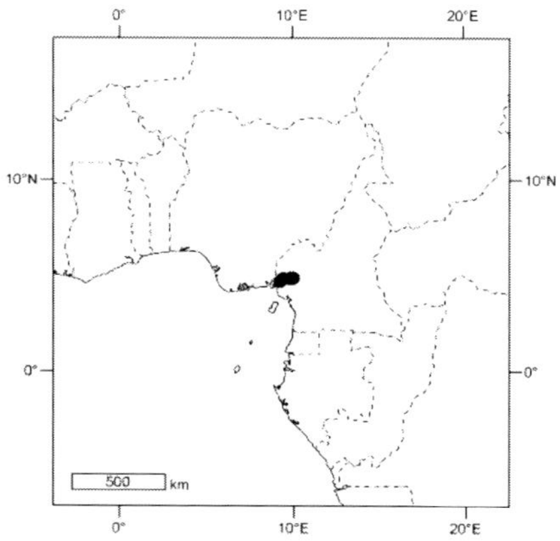

c. ***Rinorea fausteana*** EN (Violaceae)

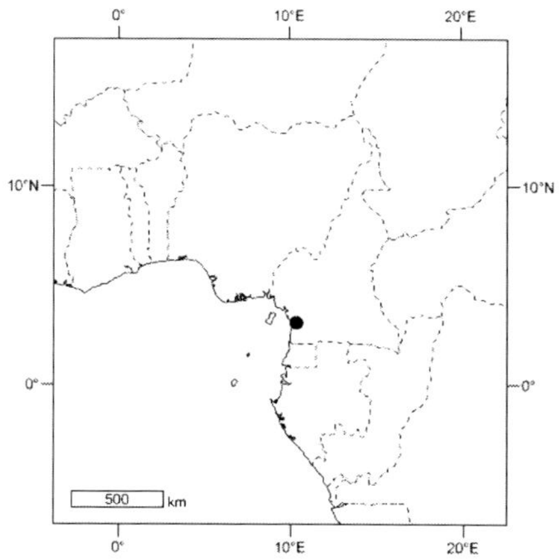

d. ***Rinorea simoneae*** EN (Violaceae)

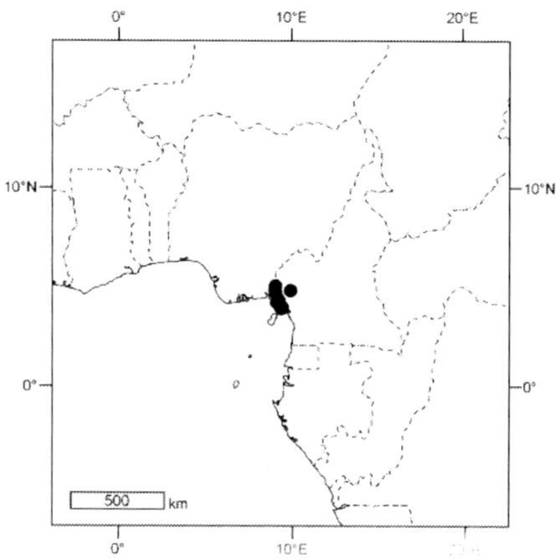

e. ***Rinorea thomasii*** VU (Violaceae)

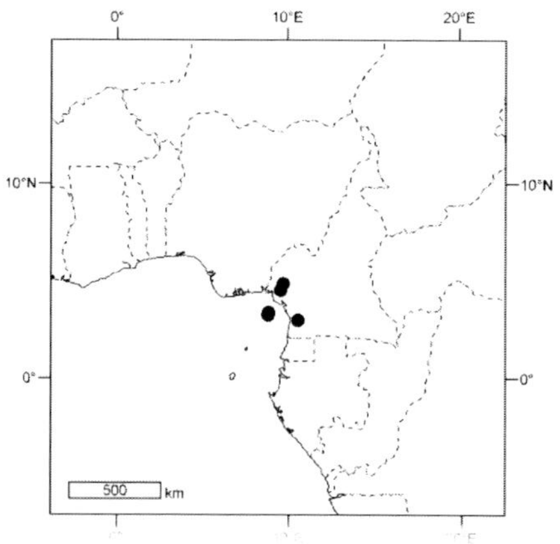

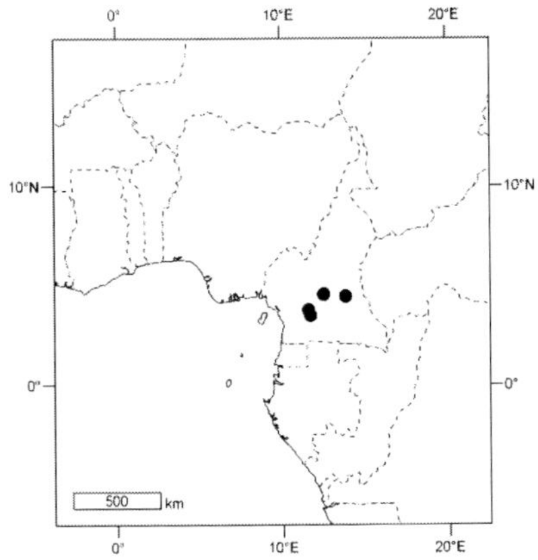

a. ***Cyphostemma cameroonense*** EN (Vitaceae)

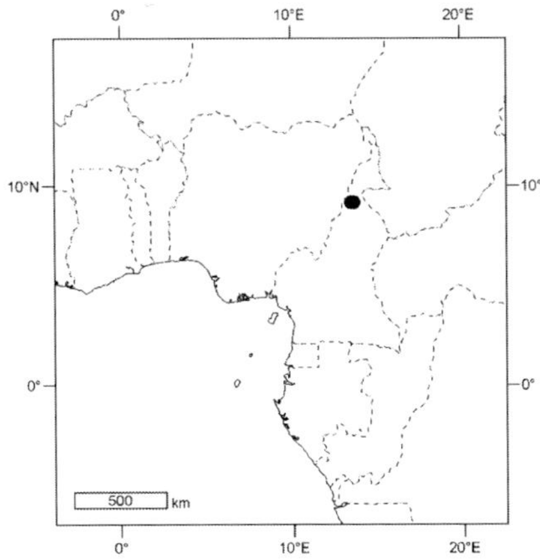

b. ***Cyphostemma cuneatum*** EN (Vitaceae)

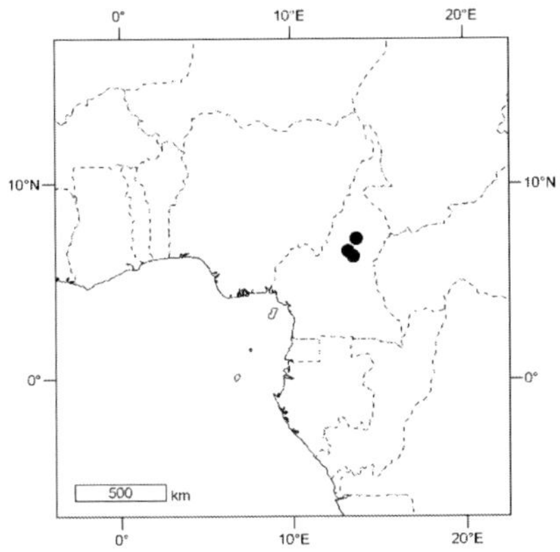

c. ***Cyphostemma letouzeyanum*** EN (Vitaceae)

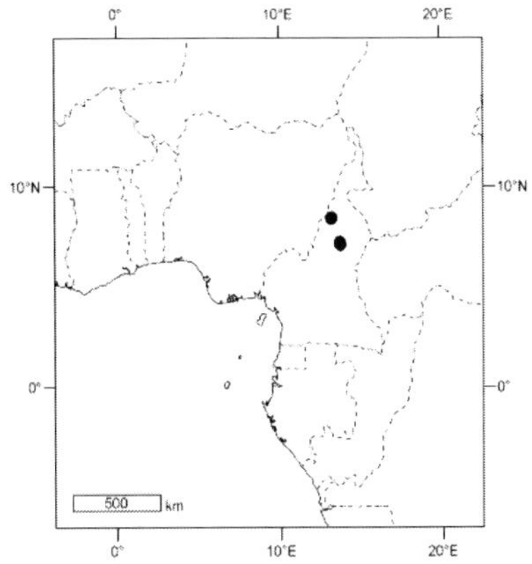

d. ***Cyphostemma leucotrichum*** EN (Vitaceae)

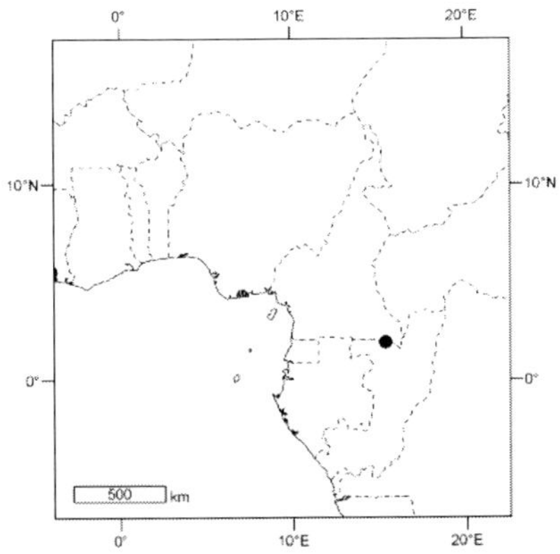

e. ***Cyphostemma ornatum*** EN (Vitaceae)

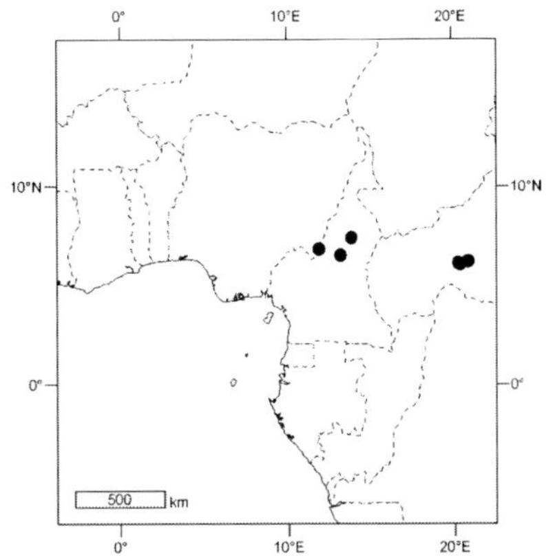

f. ***Cyphostemma ouakense*** EN (Vitaceae)

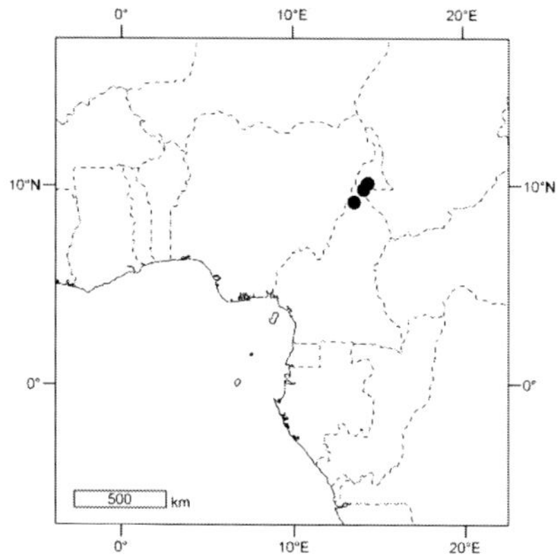

a. ***Cyphostemma rupicolum*** EN (Vitaceae)

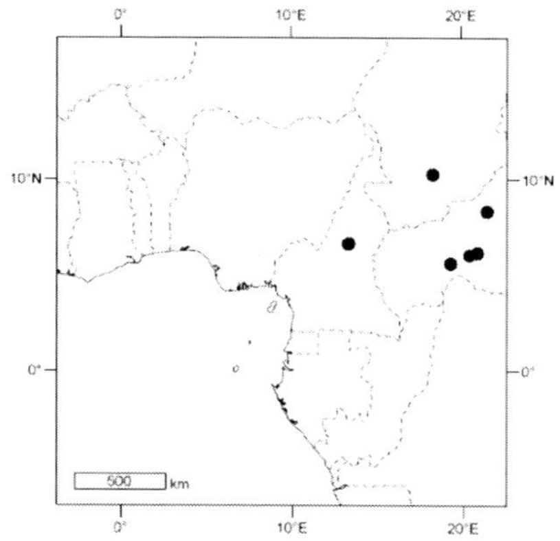

b. ***Cyphostemma tisserantii*** VU (Vitaceae)

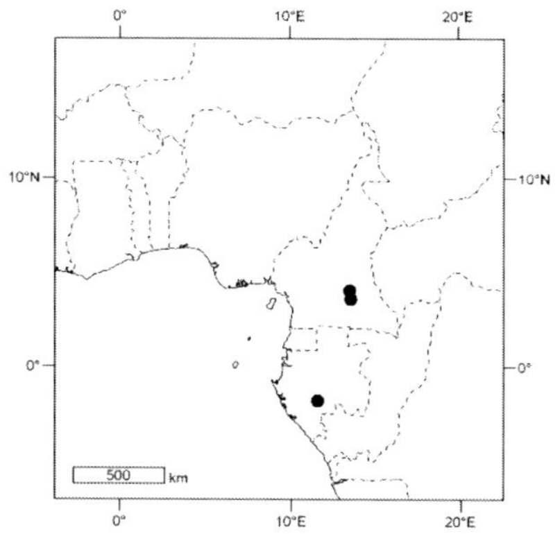

c. ***Erismadelphus sessilis*** VU (Vochysiaceae)

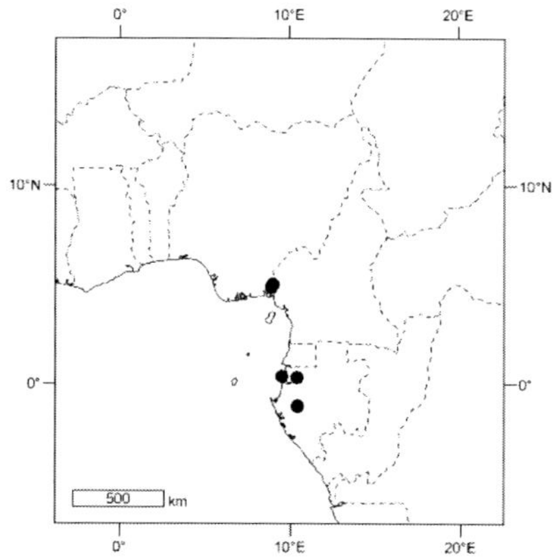

d. ***Korupodendron songweanum*** EN (Vochysiaceae)

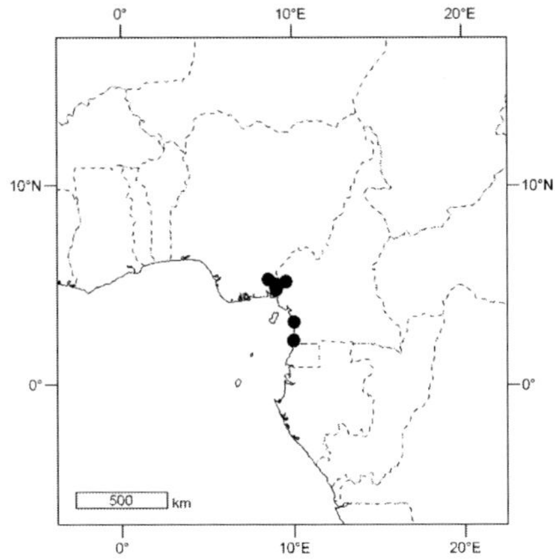

a. ***Limnophyton fluitans*** EN (Alismataceae)

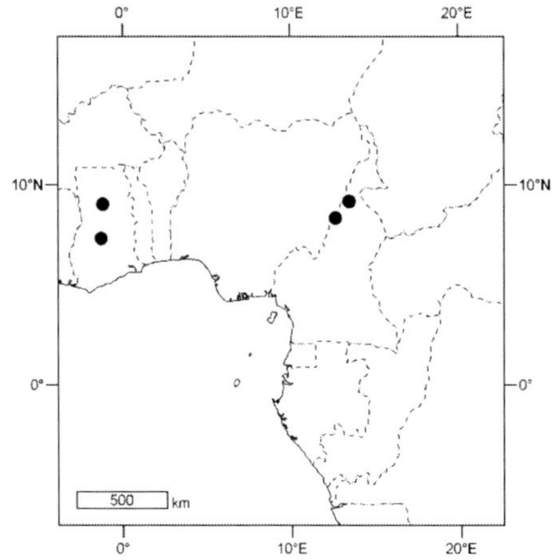

b. ***Chlorophytum altum*** EN (Anthericaceae)

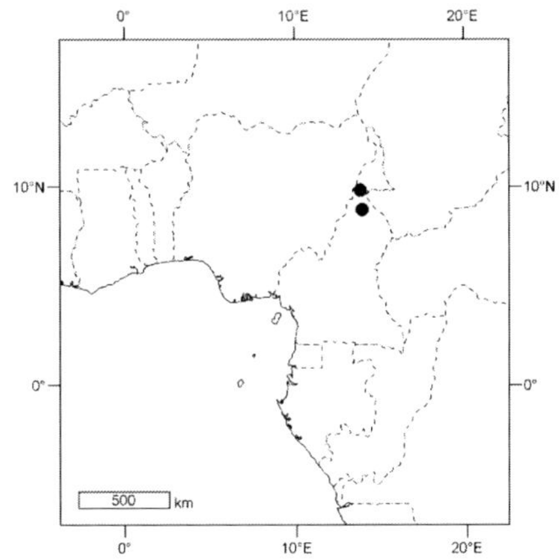

c. ***Chlorophytum benuense*** EN (Anthericaceae)

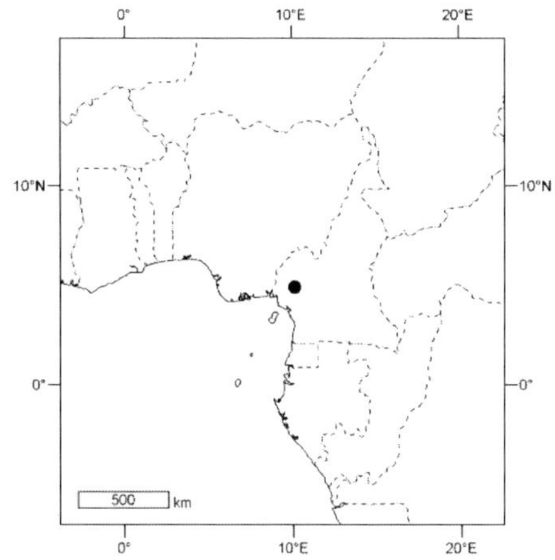

d. ***Chlorophytum camporum*** CR (Anthericaceae)

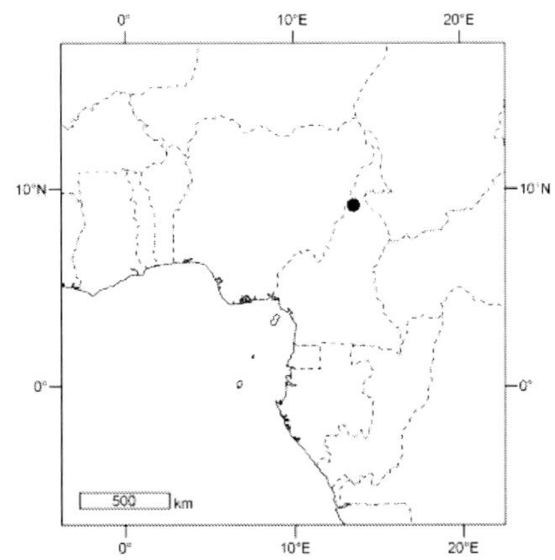

e. ***Chlorophytum caudatibracteatum*** CR (Anthericaceae)

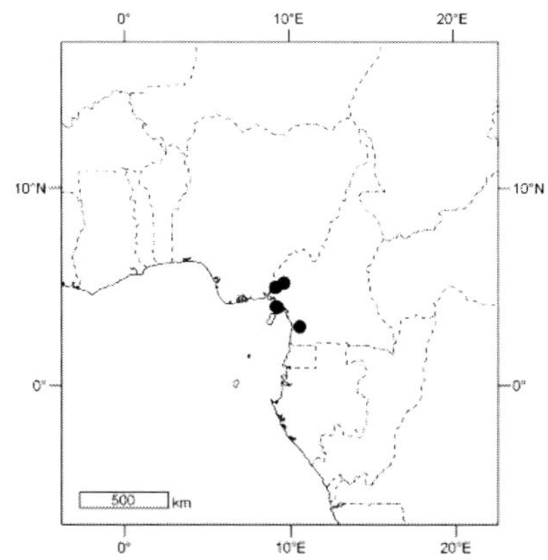

f. ***Chlorophytum petrophilum*** EN (Anthericaceae)

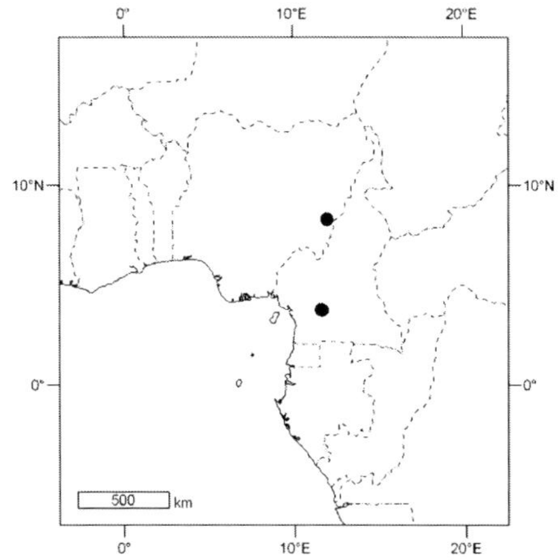

a. ***Chlorophytum staudtii*** EN (Anthericaceae)

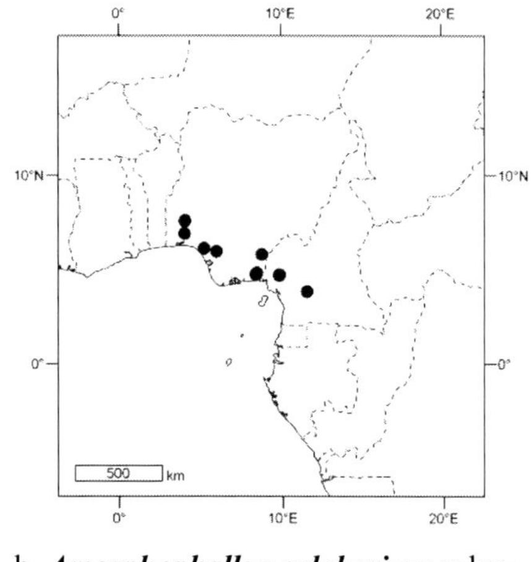

b. ***Amorphophallus calabaricus*** subsp. ***calabaricus*** VU (Araceae)

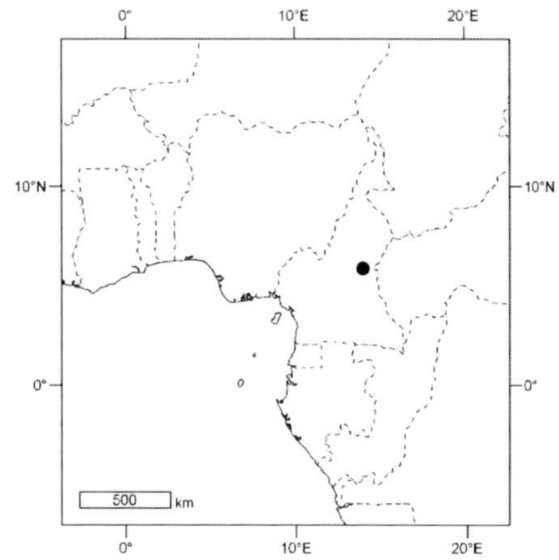

c. ***Amorphophallus mildbraedii*** CR (Araceae)

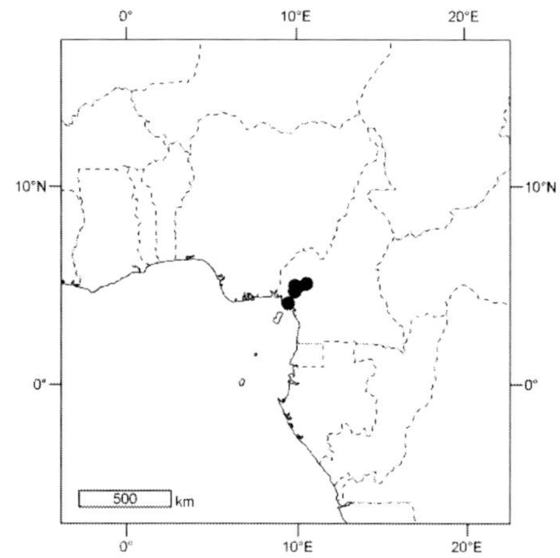

d. ***Amorphophallus preussii*** EN (Araceae)

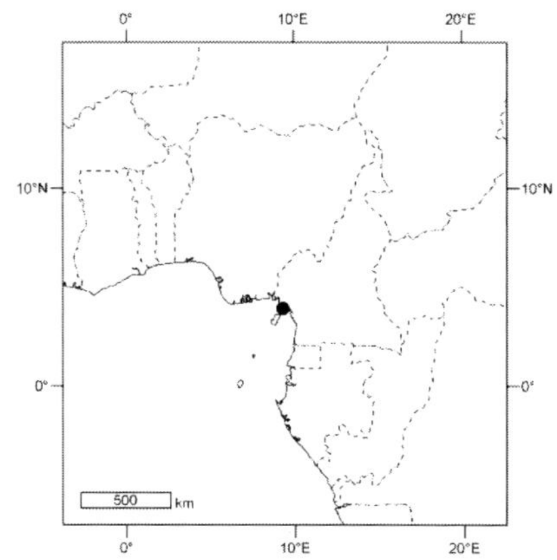

e. ***Amorphophallus zenkeri*** EN (Araceae)

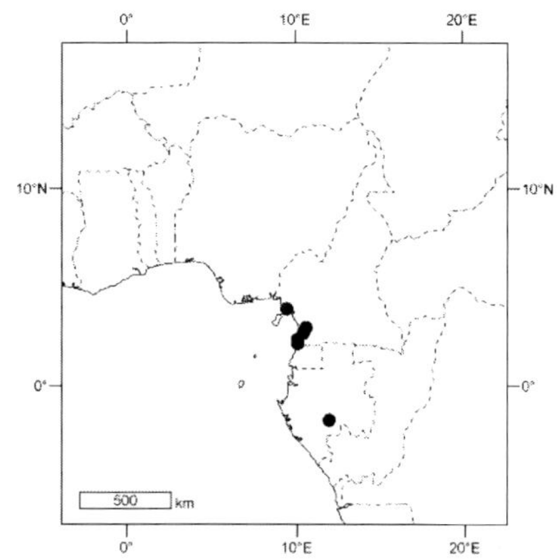

f. ***Culcasia bosii*** VU (Araceae)

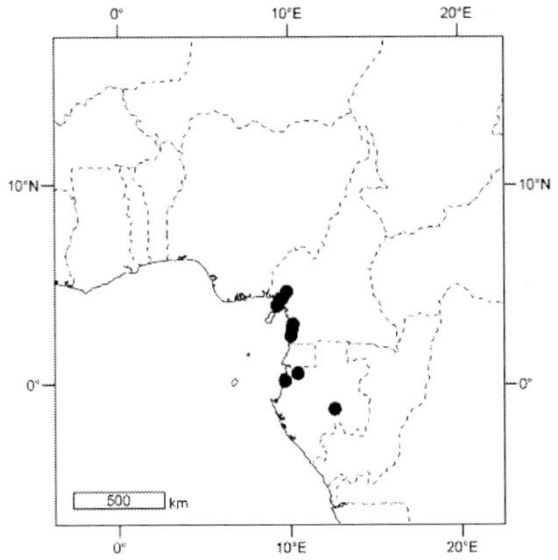

a. ***Culcasia obliquifolia*** VU (Araceae)

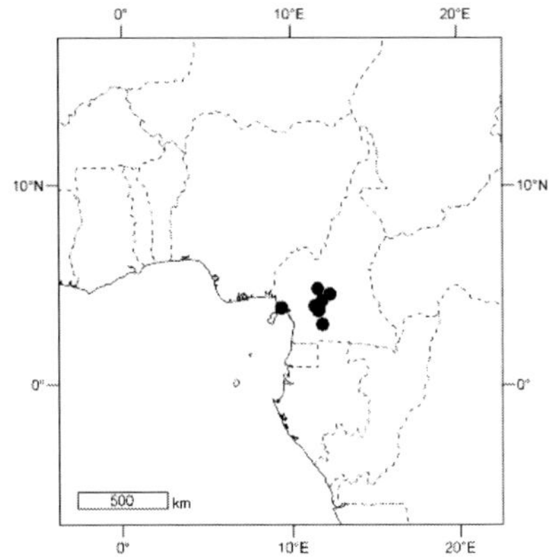

b. ***Culcasia sanagensis*** VU (Araceae)

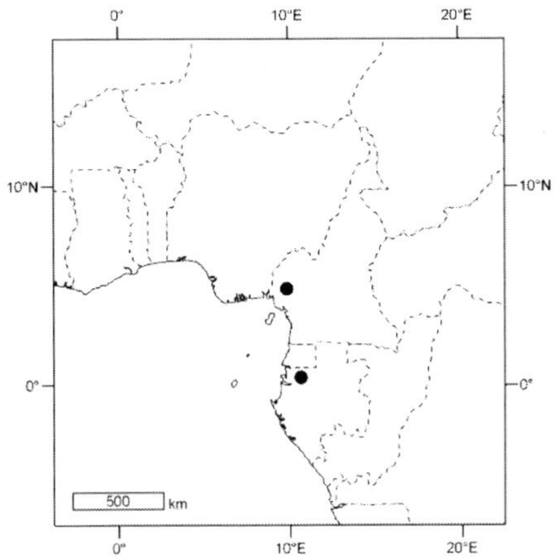

c. ***Rhaphidophora pusilla*** VU (Araceae)

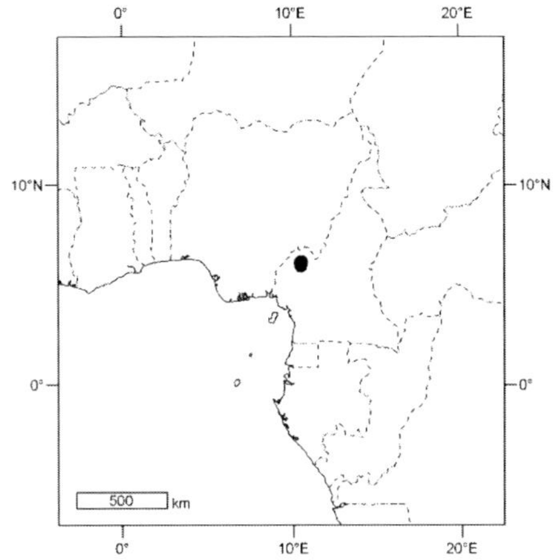

d. ***Kniphofia reflexa*** EN (Asphodelaceae)

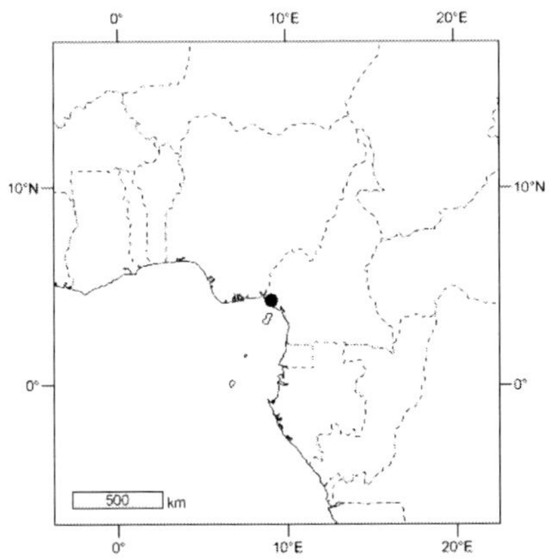

e. ***Afrothismia foertheriana*** CR (Burmanniaceae)

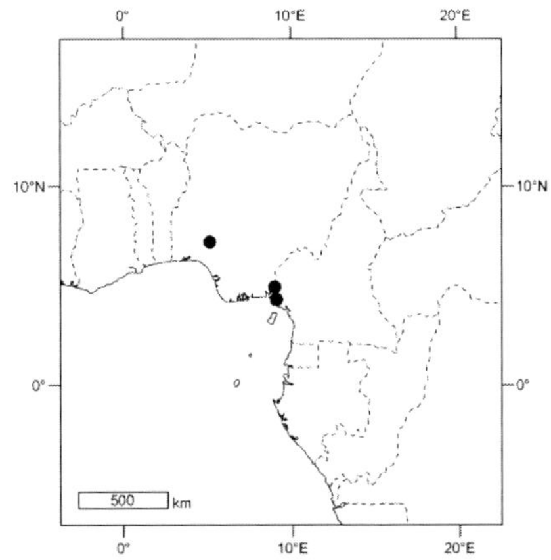

f. ***Afrothismia hydra*** EN (Burmanniaceae)

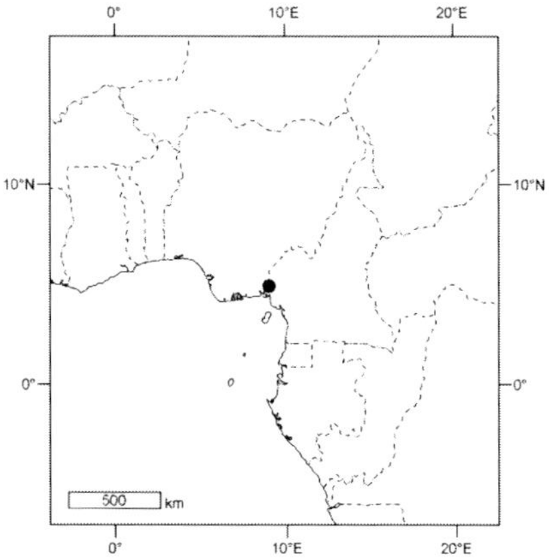

a. ***Afrothismia korupensis*** CR (Burmanniaceae)

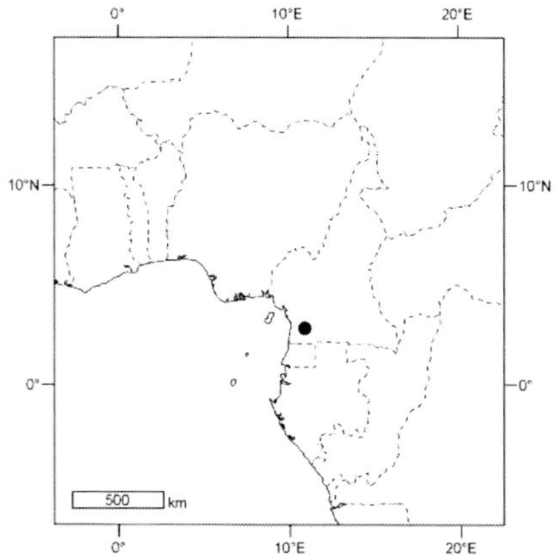

b. ***Afrothismia gesnerioides*** CR (Burmanniaceae)

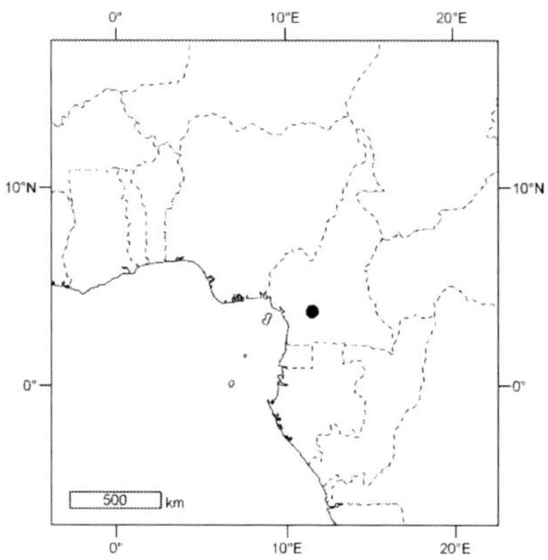

c. ***Afrothismia amietii*** CR (Burmanniaceae)

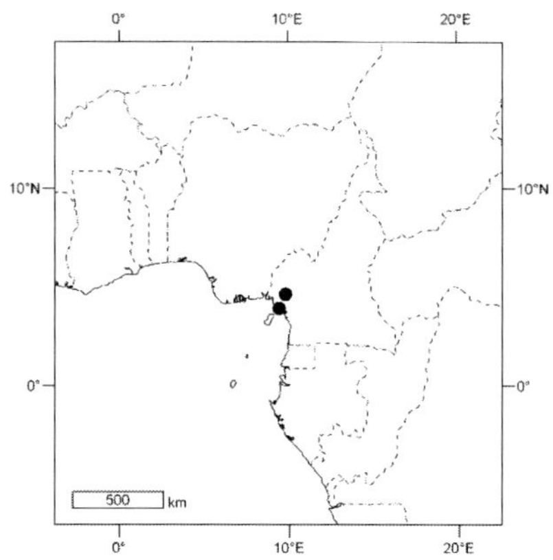

d. ***Afrothismia pachyantha*** CR (Burmanniaceae)

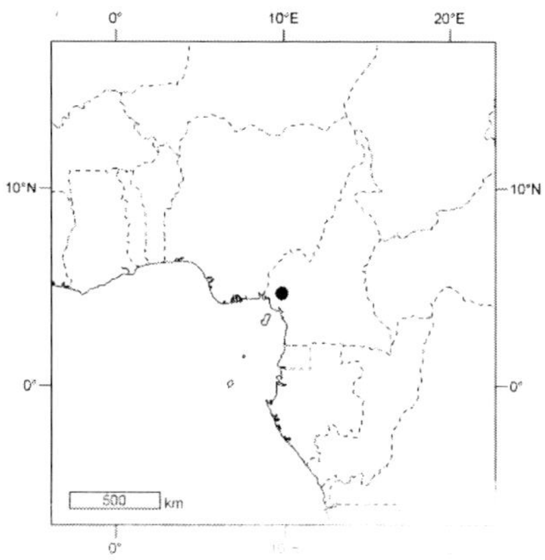

e. ***Afrothismia sai***

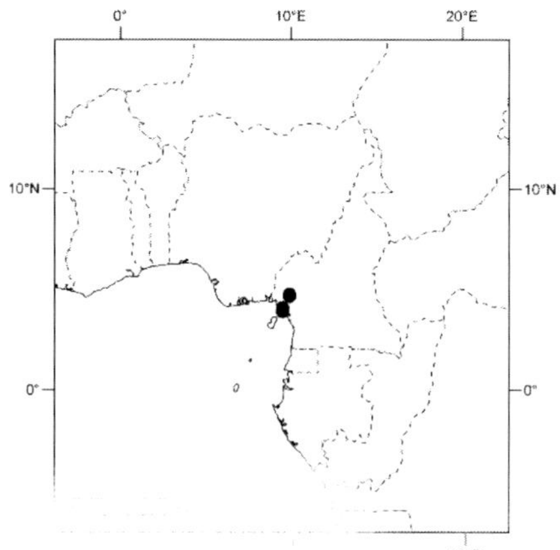

N (Burmanniaceae)

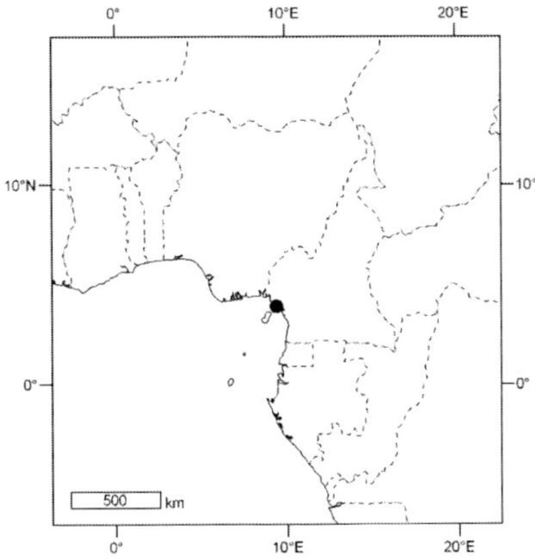

a. ***Oxygyne triandra*** CR (Burmanniaceae)

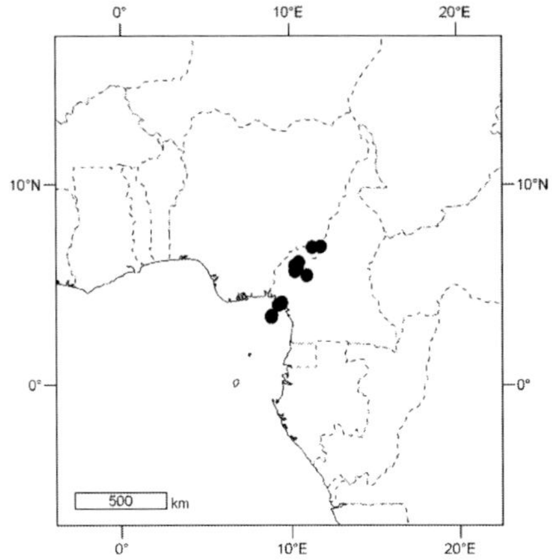

b. ***Wurmbea tenuis*** subsp.***tenuis*** VU (Colchicaceae)

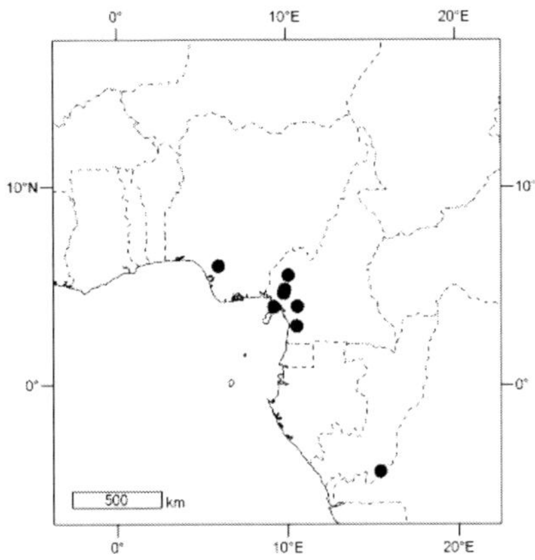

c. ***Aneilema silvaticum*** VU (Commelinaceae)

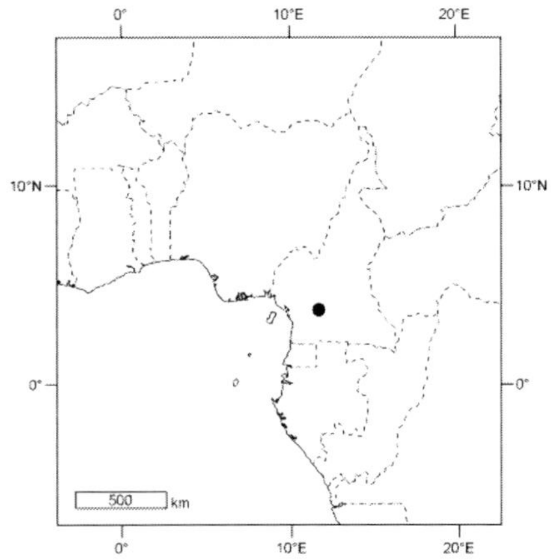

d. ***Commelina zenkeri*** CR (Commelinaceae)

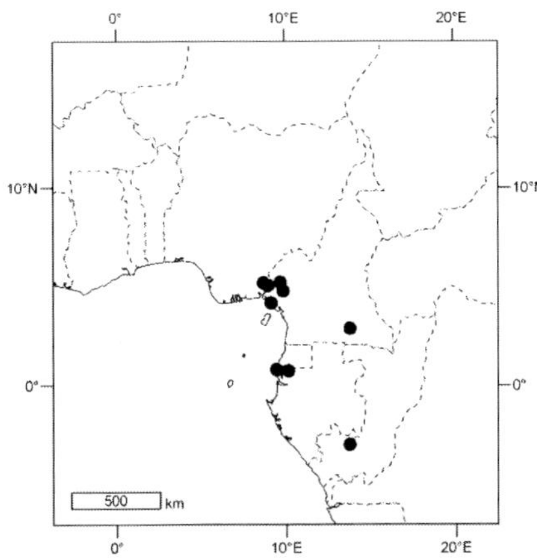

e. ***Floscopa mannii*** VU (Commelinaceae)

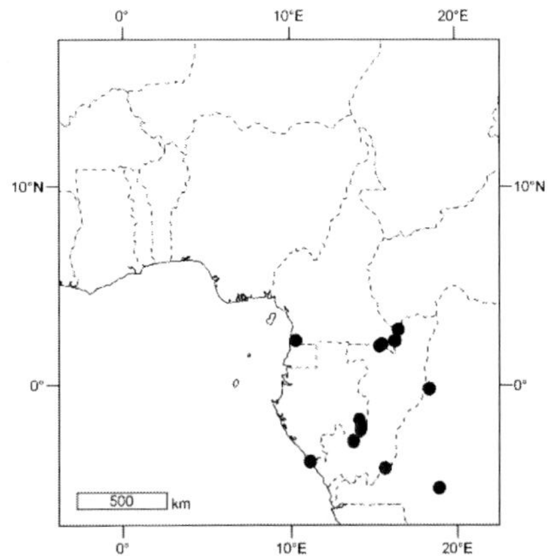

f. ***Palisota brachythyrsa*** VU (Commelinaceae)

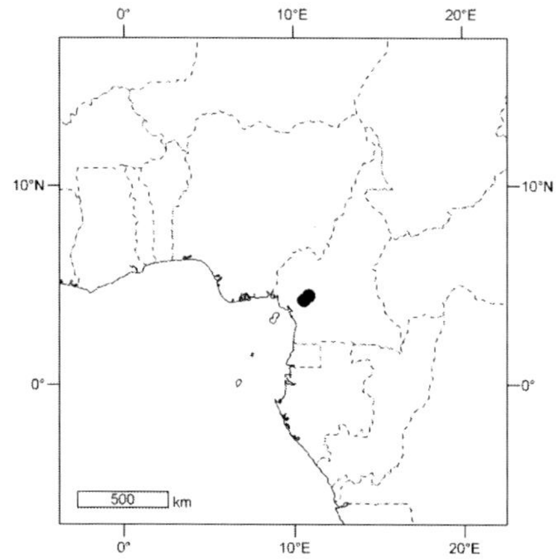

a. ***Palisota ebo*** EN (Commelinaceae)

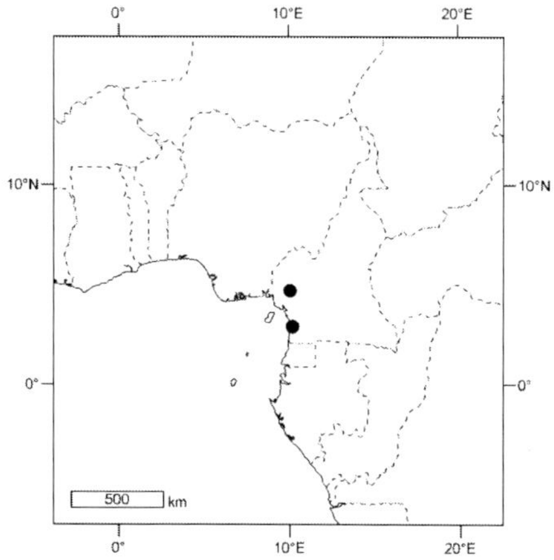

b. ***Palisota flagelliflora*** EN (Commelinaceae)

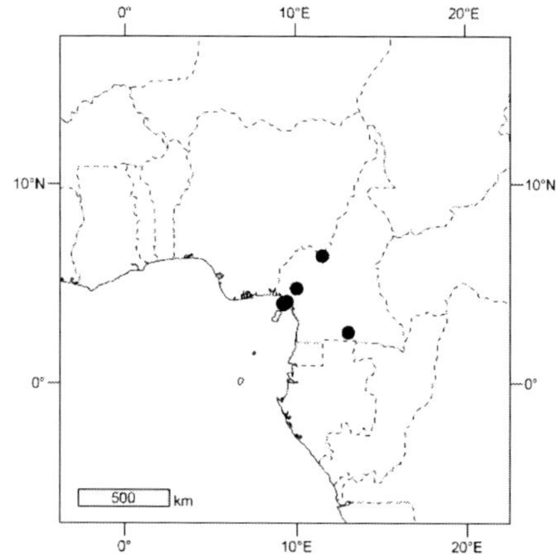

c. ***Palisota preussiana*** EN (Commelinaceae)

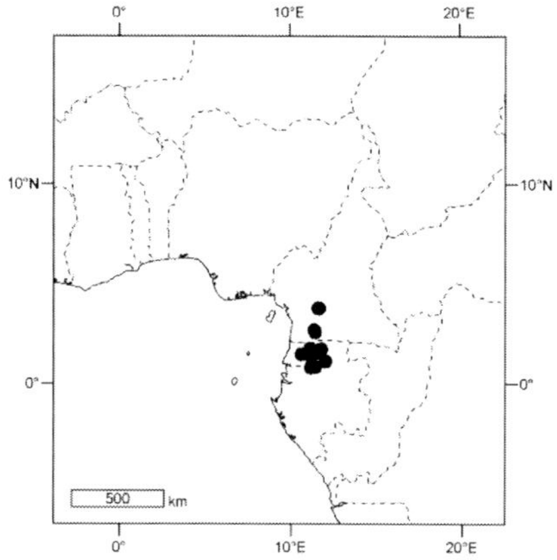

d. ***Tricarpelema africana*** VU (Commelinaceae)

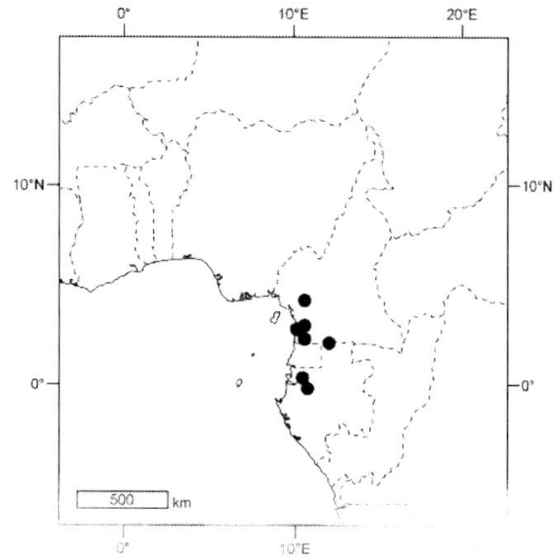

e. ***Costus phaeotrichus*** V[illegible]

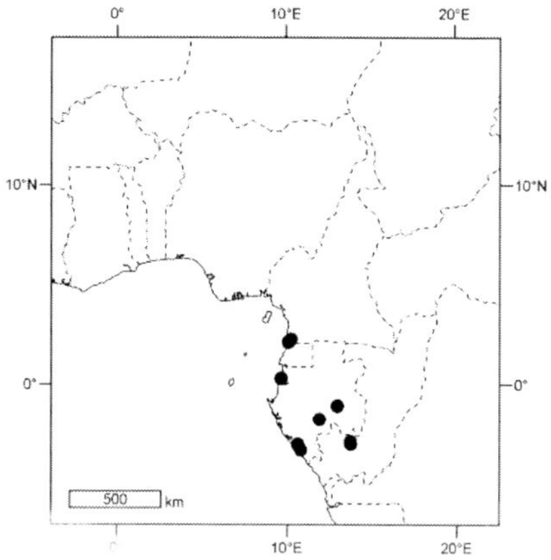

[illegible]***ppenbeckianus*** VU (Costaceae)

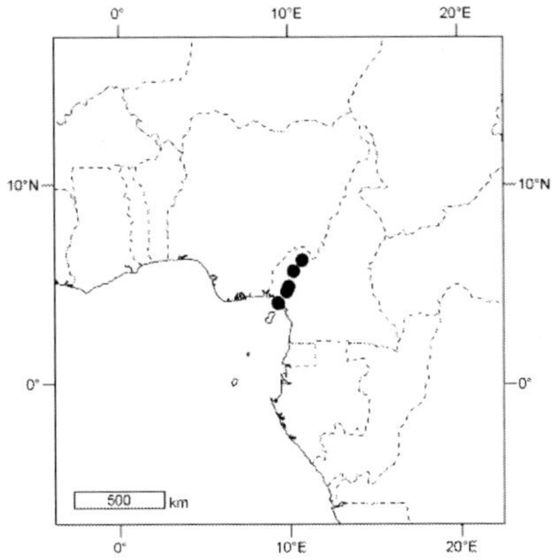

a. ***Bulbostylis densa*** var. ***cameroonensis*** EN (Cyperaceae)

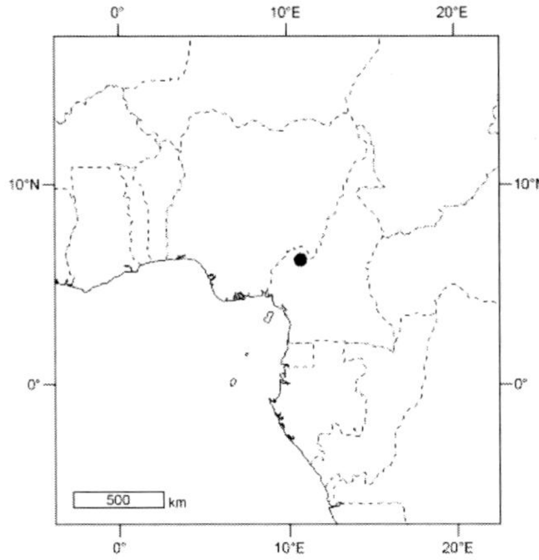

b. ***Coleochloa domensis*** CR (Cyperaceae)

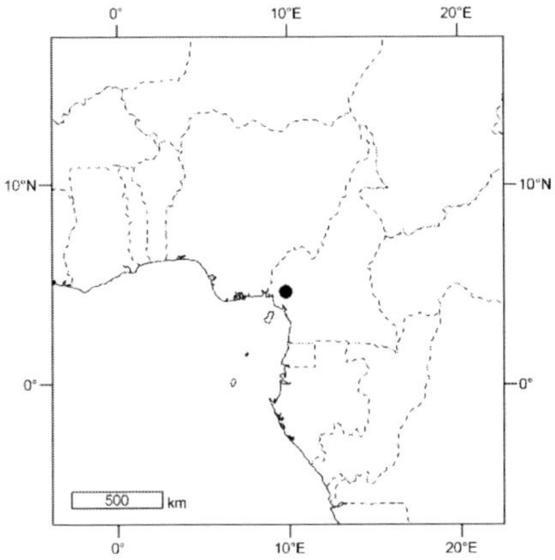

c. ***Cyperus microcristatus*** CR (Cyperaceae)

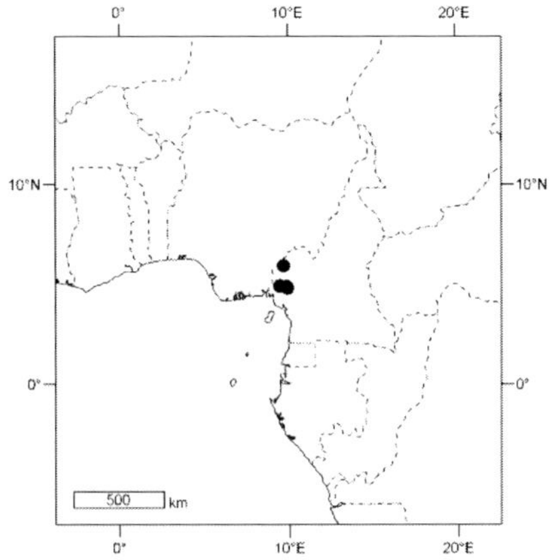

d. ***Cyperus rheophyticus*** EN (Cyperaceae)

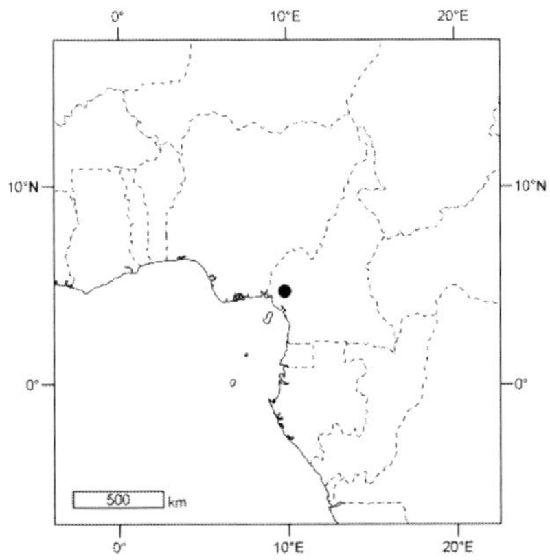

e. ***Cyperus tenuiculmis*** subsp. ***mutica*** CR (Cyperaceae)

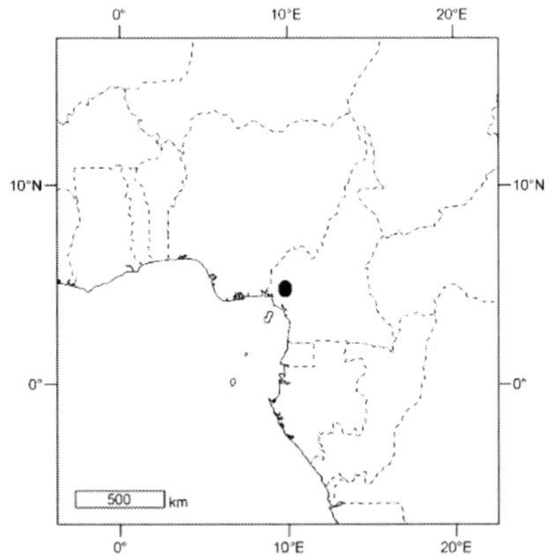

f. ***Hypolytrum pseudomapanioides*** EN (Cyperaceae)

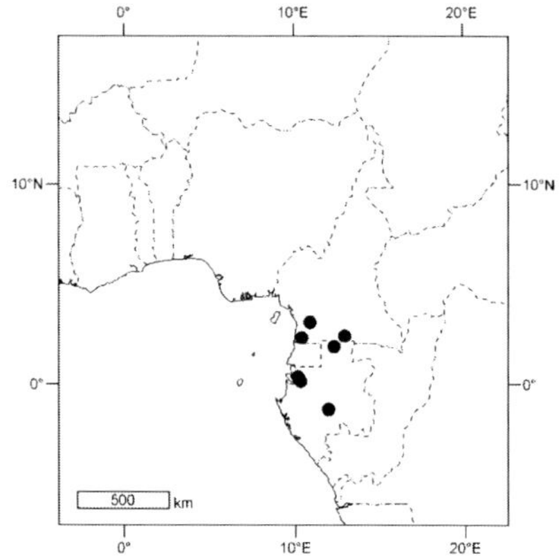

a. ***Hypolytrum secans*** VU (Cyperaceae)

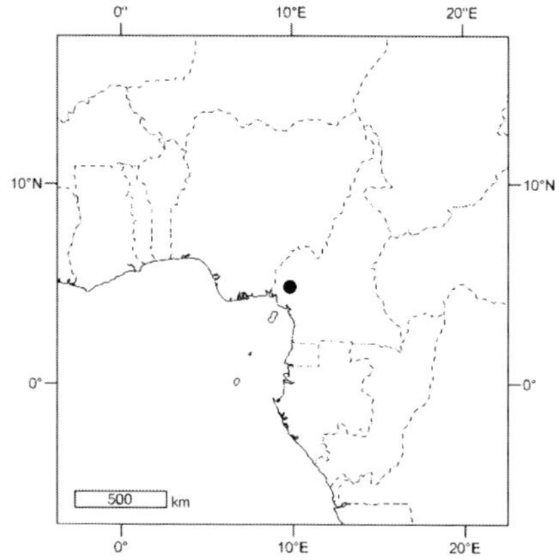

b. ***Hypolytrum subcompositus*** CR (Cyperaceae)

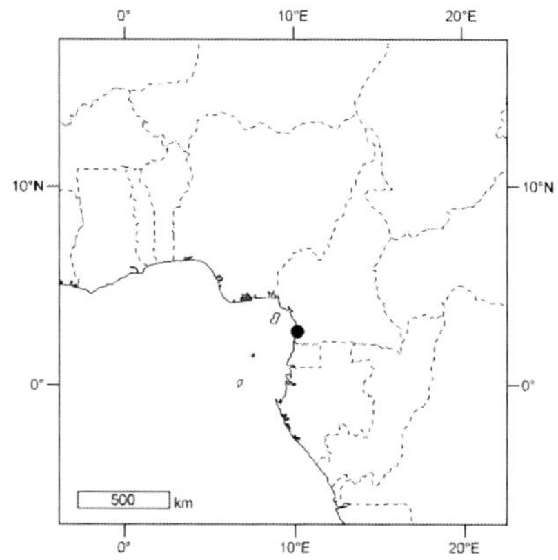

c. ***Hypolytrum unispicatum*** CR (Cyperaceae)

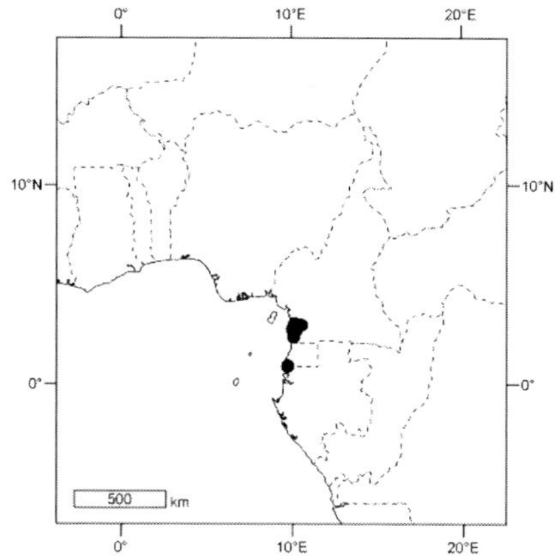

d. ***Mapania africana*** subsp. ***africana*** VU (Cyperaceae)

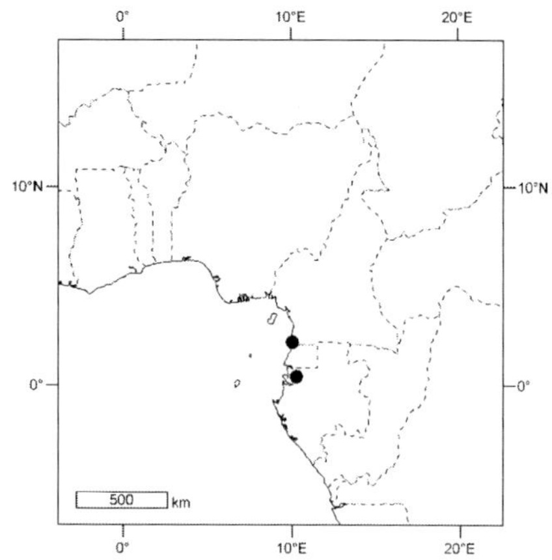

e. ***Mapania africana*** subsp. ***filipes*** EN (Cyperaceae)

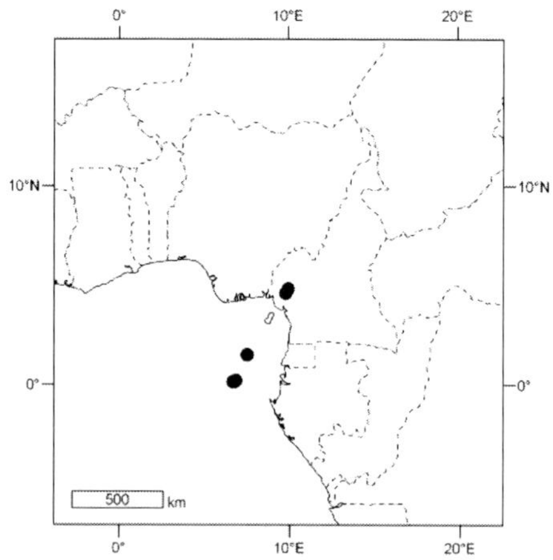

f. ***Mapania ferruginea*** VU (Cyperaceae)

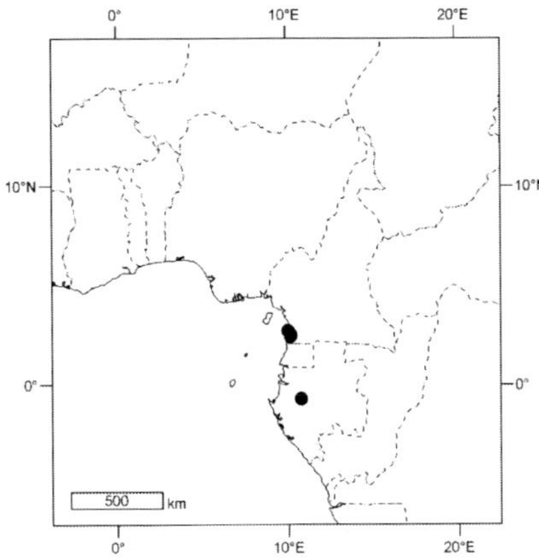

a. ***Mapania raynaliana*** EN (Cyperaceae)

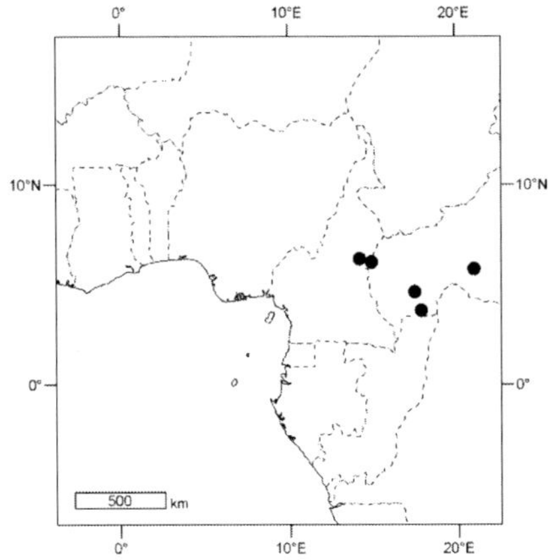

b. ***Nemum megastachyum*** EN (Cyperaceae)

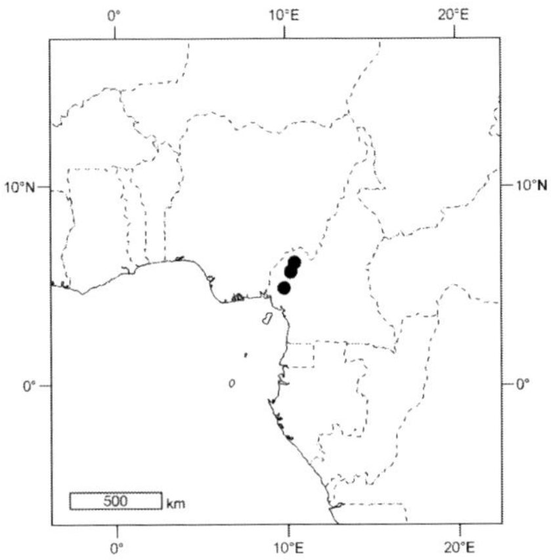

c. ***Scleria afroreflexa*** EN (Cyperaceae)

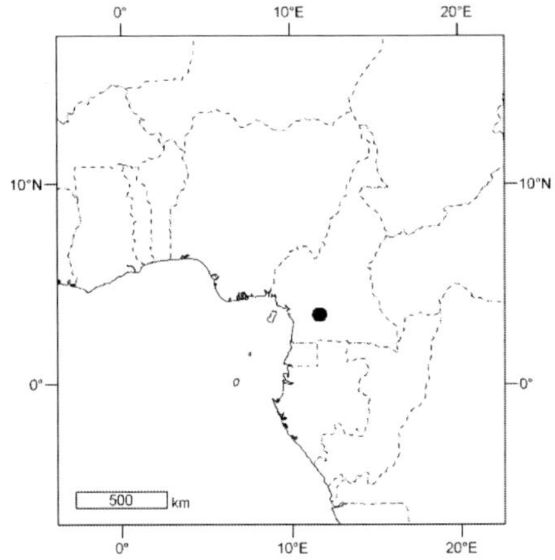

d. ***Scleria shielae*** CR (Cyperaceae)

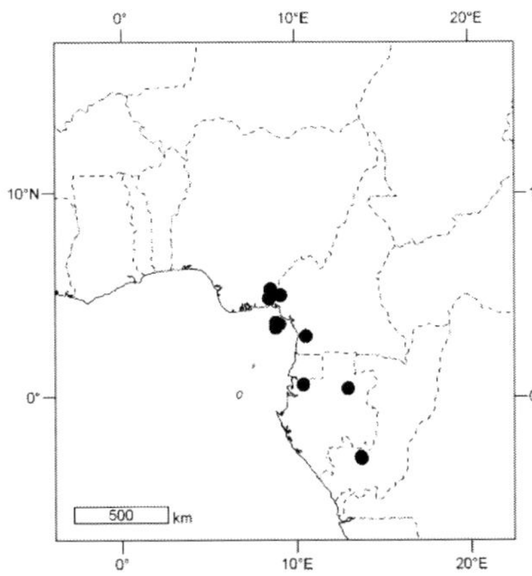

e. ***Dracaena bicolor*** VU (Dracaenaceae)

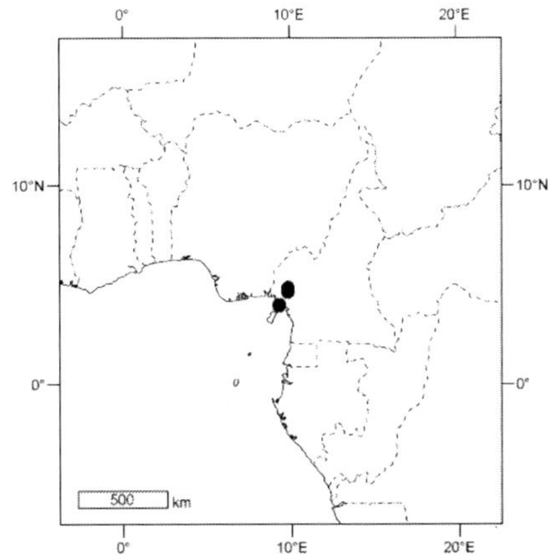

f. ***Dracaena kupensis*** EN (Dracaenaceae)

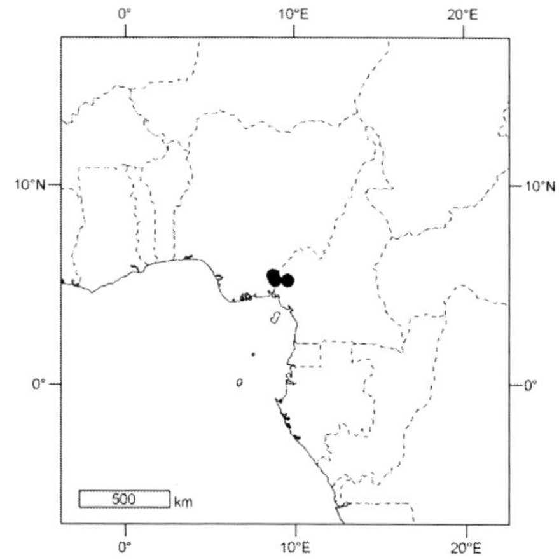

a. ***Dracaena goldieana*** EN (Dracaenaceae)

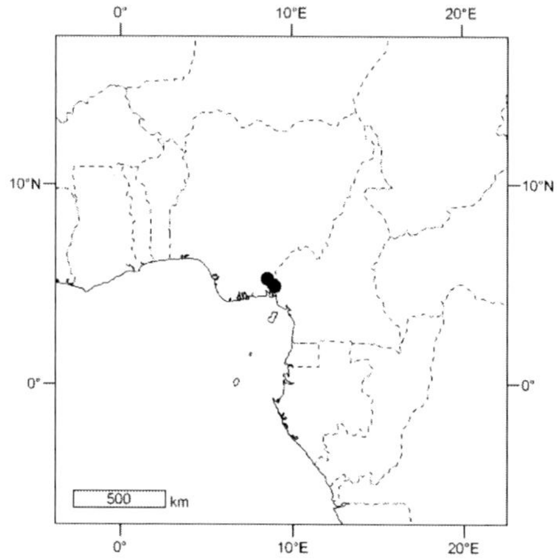

b. ***Dracaena talbotii*** EN (Dracaenaceae)

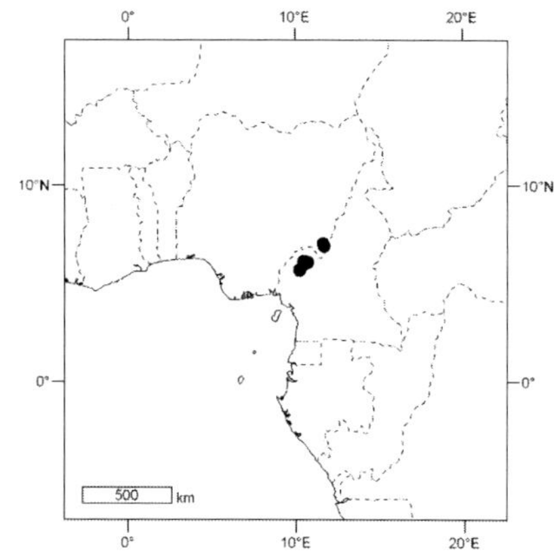

c. ***Eriocaulon asteroides*** VU (Eriocaulaceae)

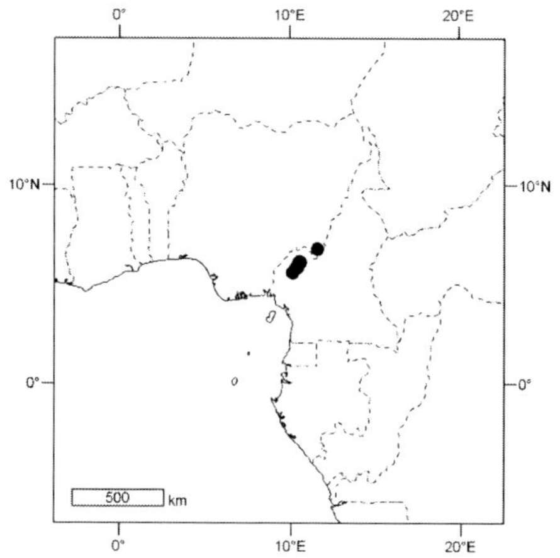

d. ***Eriocaulon bamendae*** VU (Eriocaulaceae)

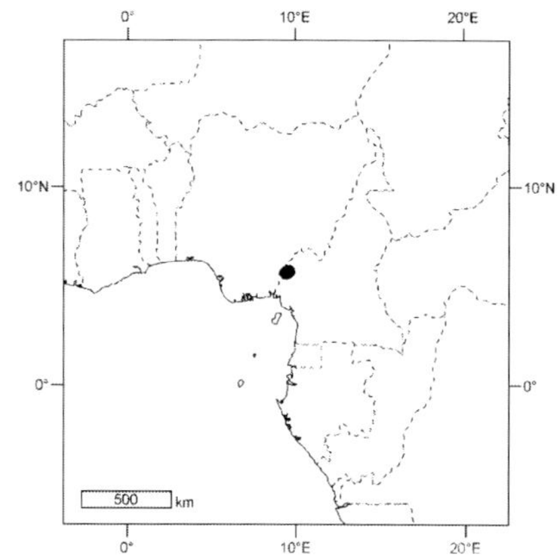

e. ***Eriocaulon mamfeense*** EN (Eriocaulaceae)

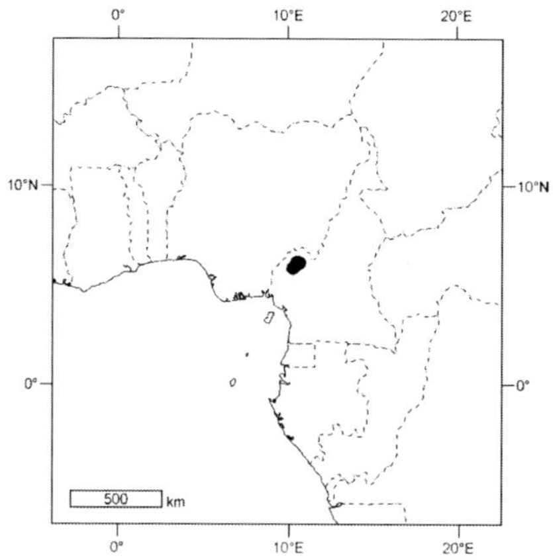

f. ***Eriocaulon parvulum*** VU (Eriocaulaceae)

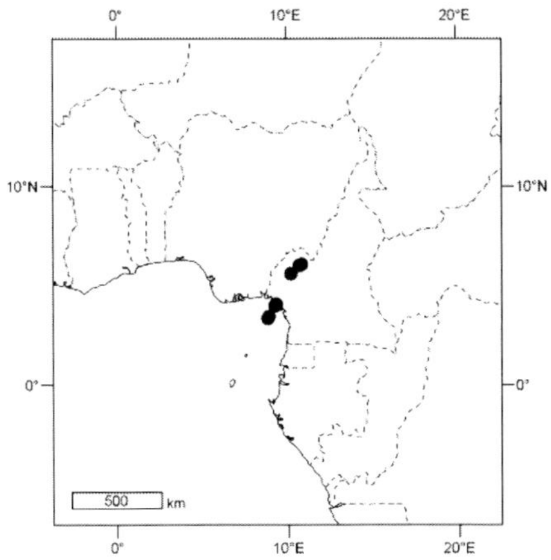

a. ***Agrostis mannii*** EN (Gramineae)

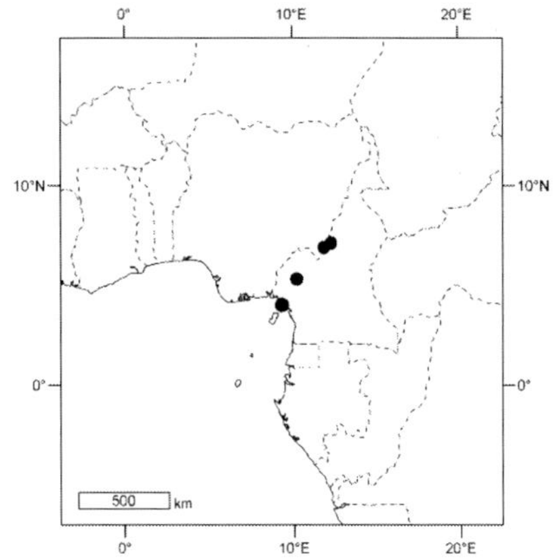

b. ***Andropogon pusillus*** EN (Gramineae)

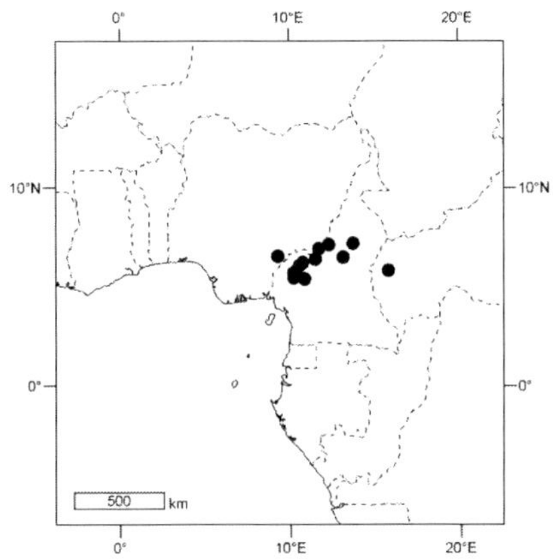

c. ***Ctenium ledermannii*** VU (Gramineae)

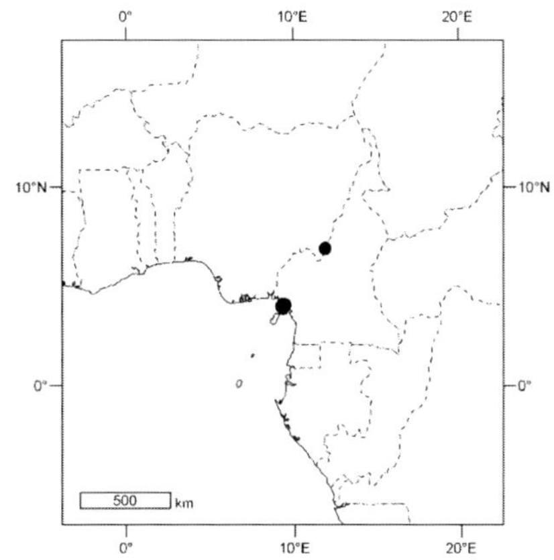

d. ***Deschampsia mildbraedii*** EN (Gramineae)

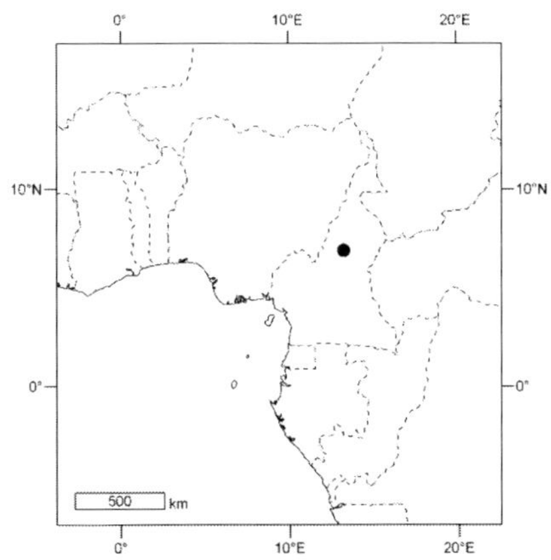

e. ***Digitaria adamouensis*** CR (Gramineae)

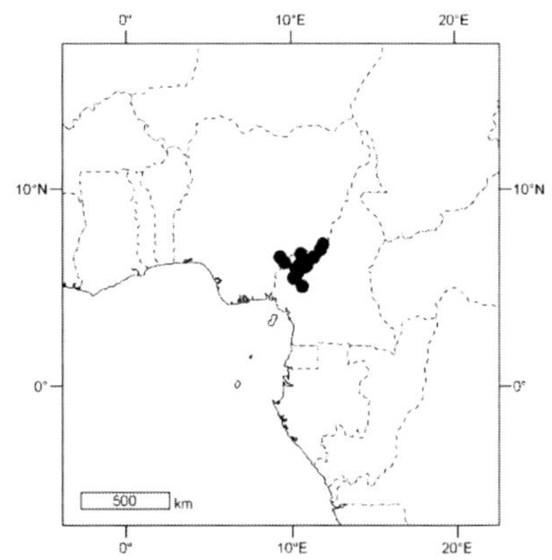

f. ***Eragrostis camerunensis*** VU (Gramineae)

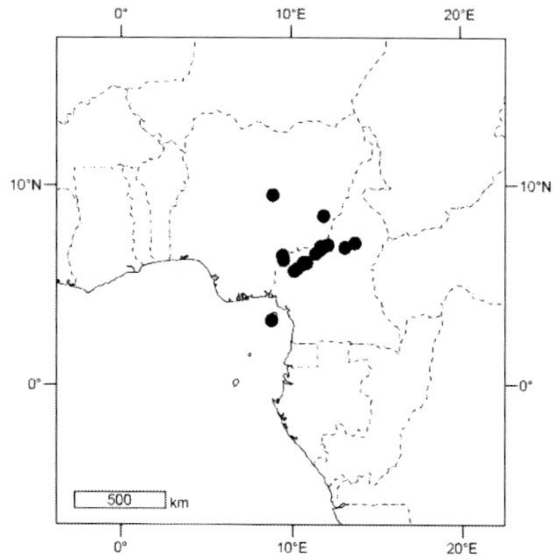

a. ***Eragrostis mokensis*** VU (Gramineae)

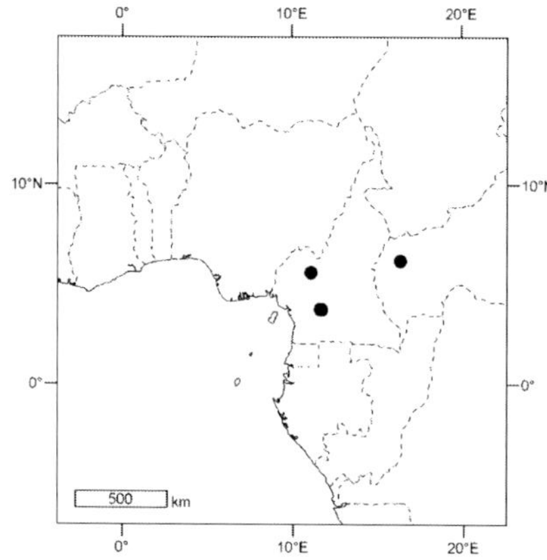

b. ***Eragrostis raynaliana*** EN (Gramineae)

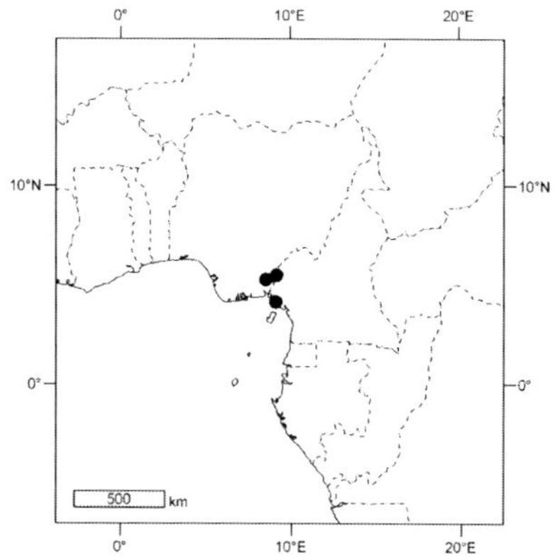

c. ***Guaduella humilis*** EN (Gramineae)

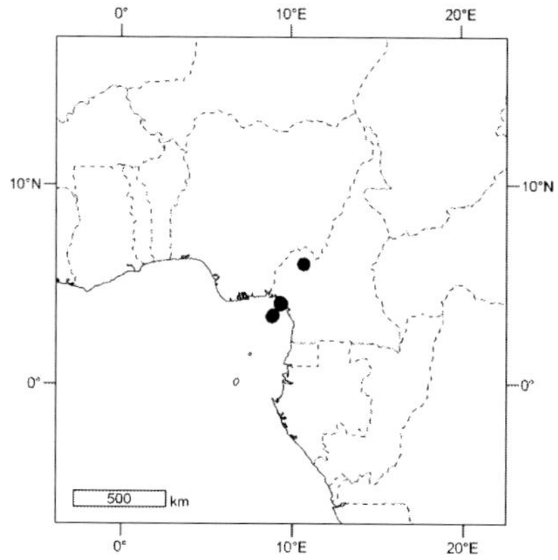

d. ***Helictotrichon mannii*** EN (Gramineae)

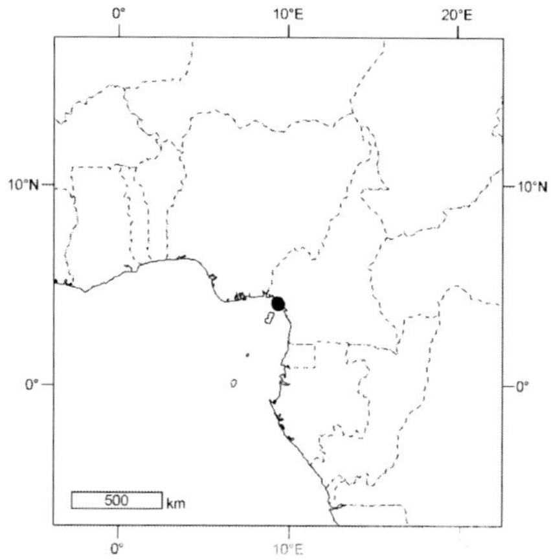

e. ***Hypseochloa camer***

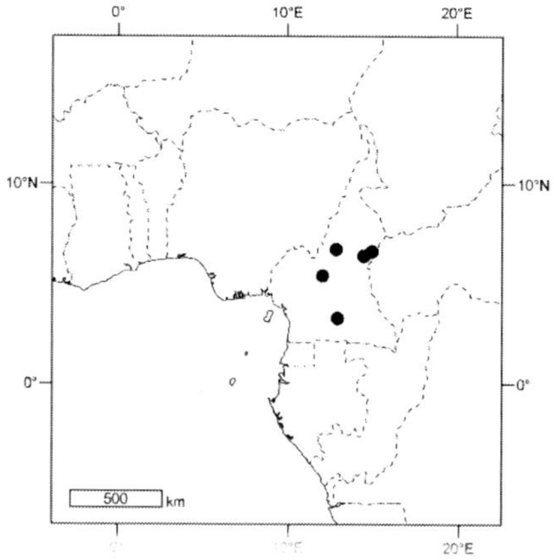

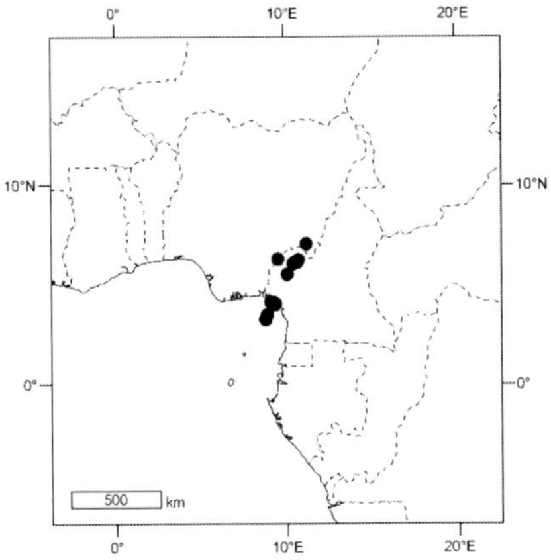

a. ***Panicum acrotrichum*** VU (Gramineae)

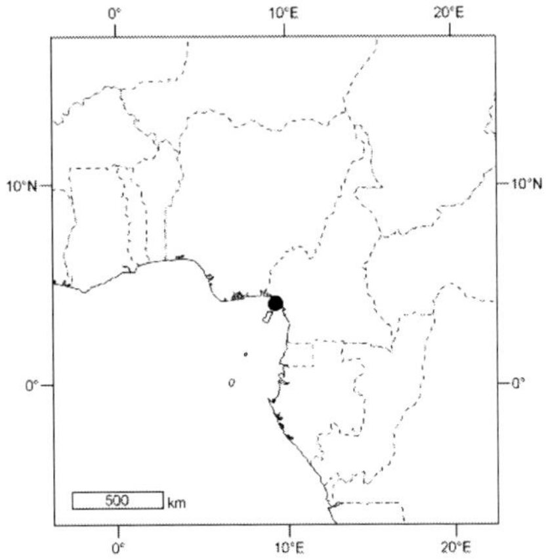

b. ***Pentaschistis mannii*** subsp. ***mannii*** VU (Gramineae)

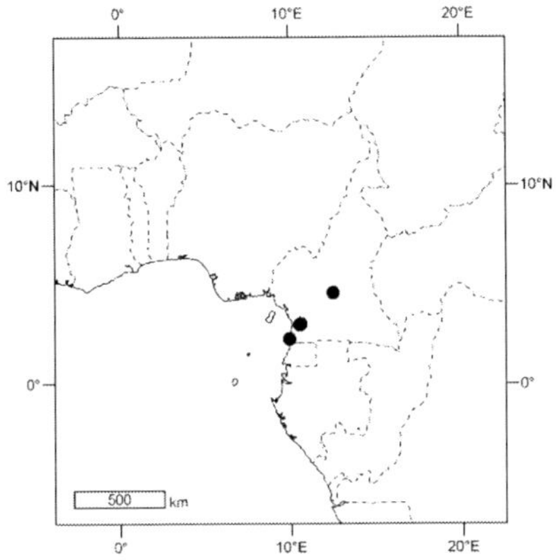

c. ***Puelia schumanniana*** EN (Gramineae)

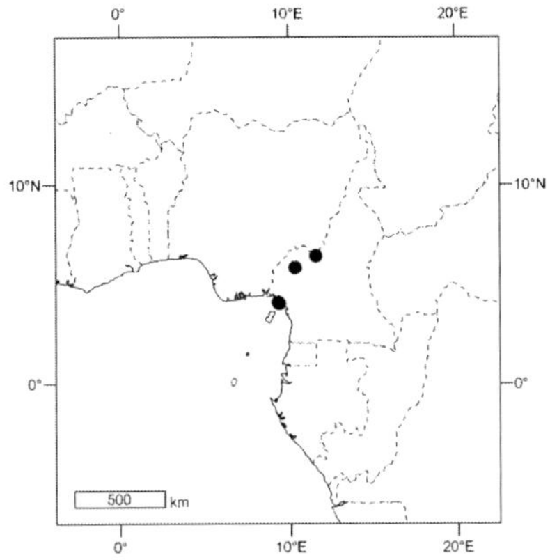

d. ***Sporobolus montanus*** EN (Gramineae)

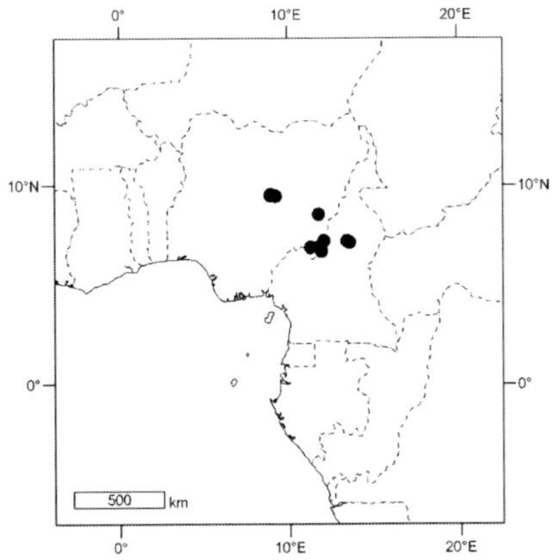

e. ***Hypoxis suffruticosa*** VU (Hypoxidaceae)

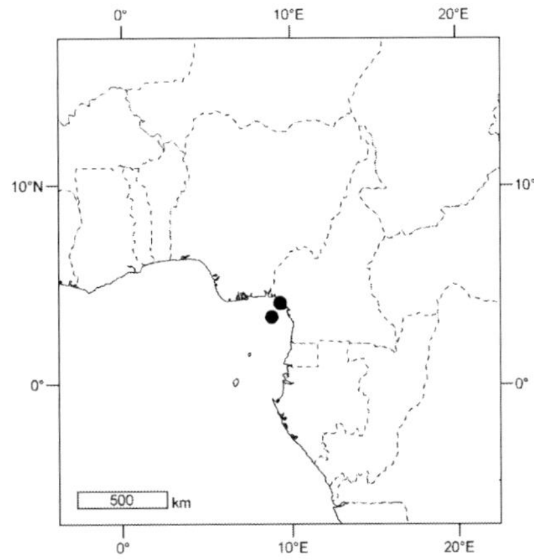

f. ***Luzula mannii*** subsp. ***mannii*** VU (Juncaceae)

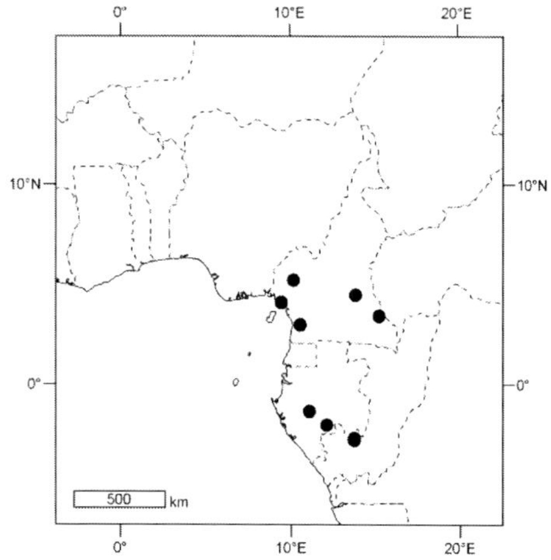

a. ***Hypselodelphys zenkeriania*** VU (Marantaceae)

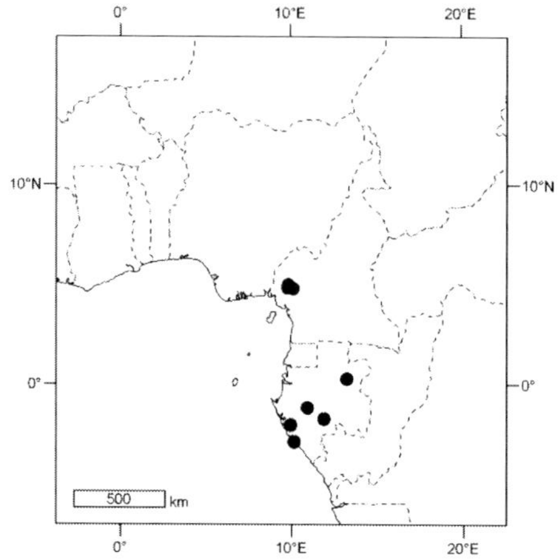

b. ***Sarcophrynium villosum*** VU (Marantaceae)

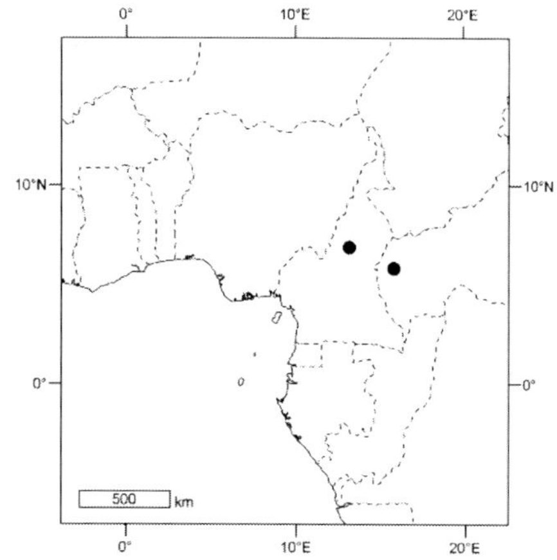

c. ***Aërangis bouarensis*** EN (Orchidaceae)

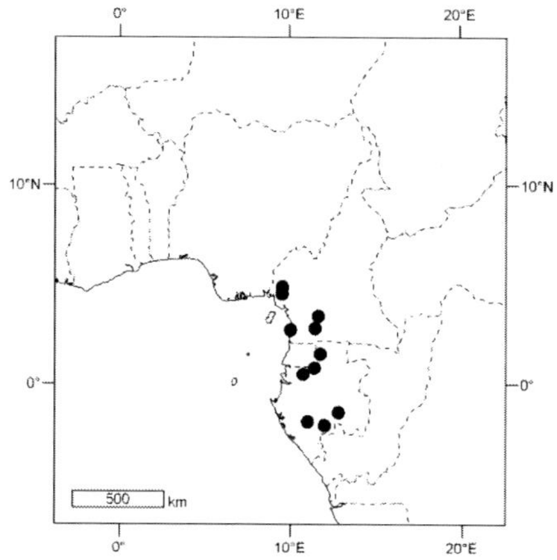

d. ***Aërangis gracillima*** VU (Orchidaceae)

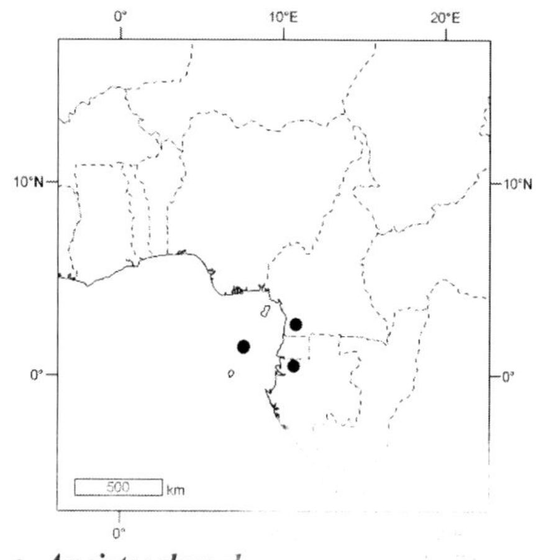

e. ***Ancistrorhynch*** [illegible] (Orchidaceae)

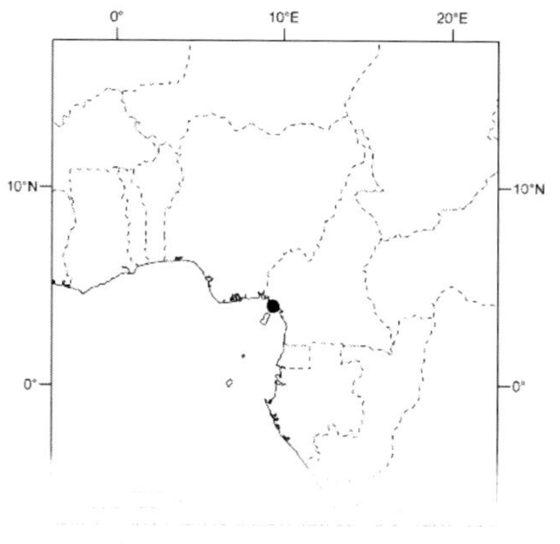

[illegible]

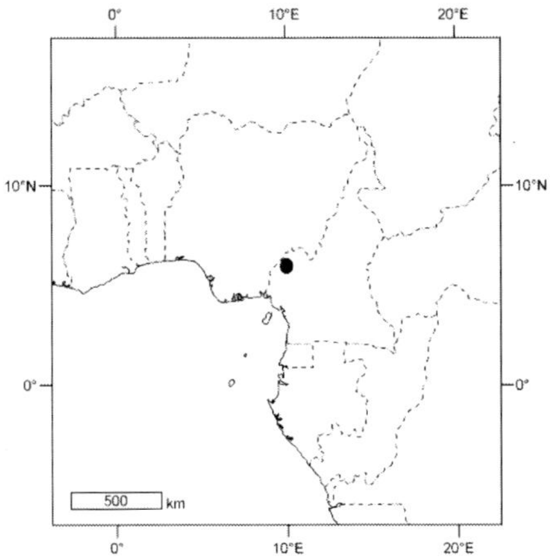

a. ***Angraecopsis lisowskii*** EN (Orchidaceae)

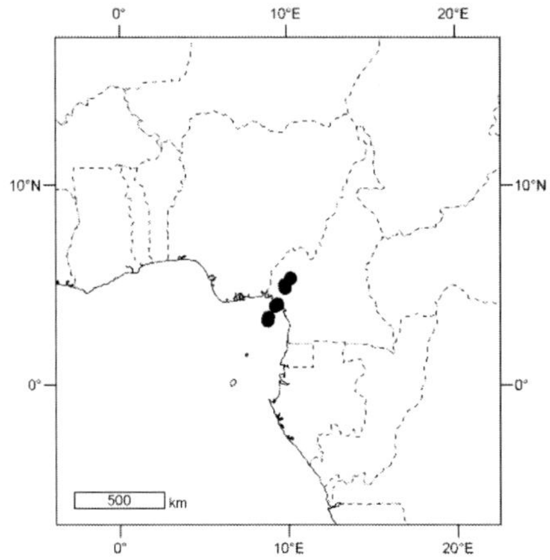

b. ***Angraecopsis tridens*** VU (Orchidaceae)

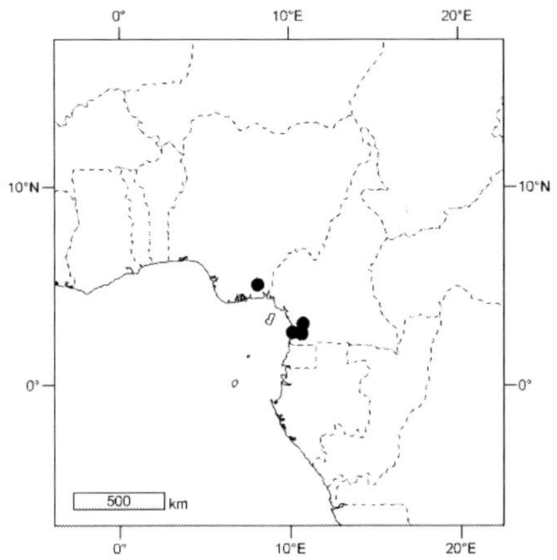

c. ***Angraecum angustum*** EN (Orchidaceae)

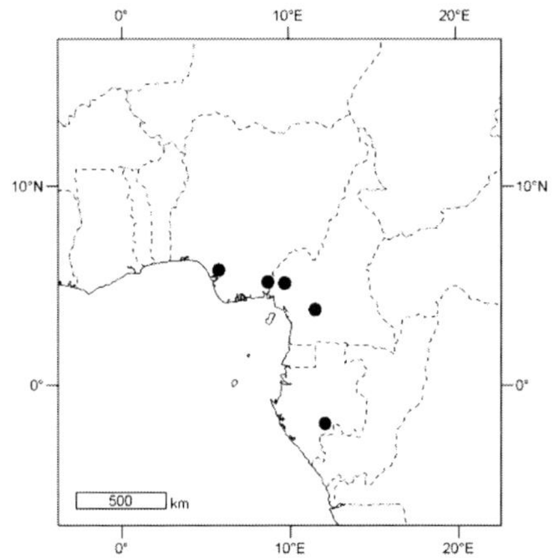

d. ***Angraecum egertonii*** EN (Orchidaceae)

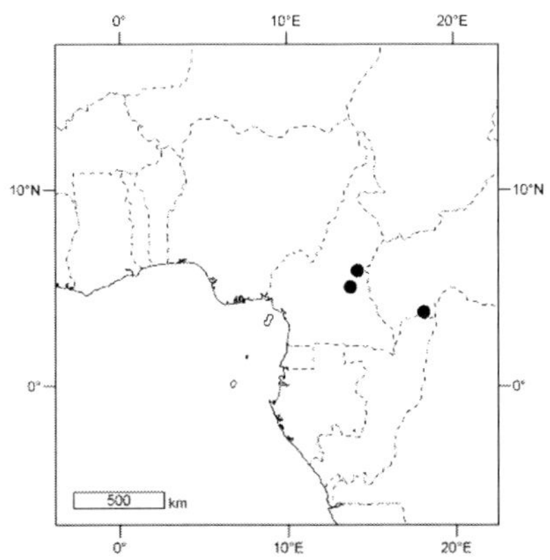

e. ***Angraecum firthii*** VU (Orchidaceae)

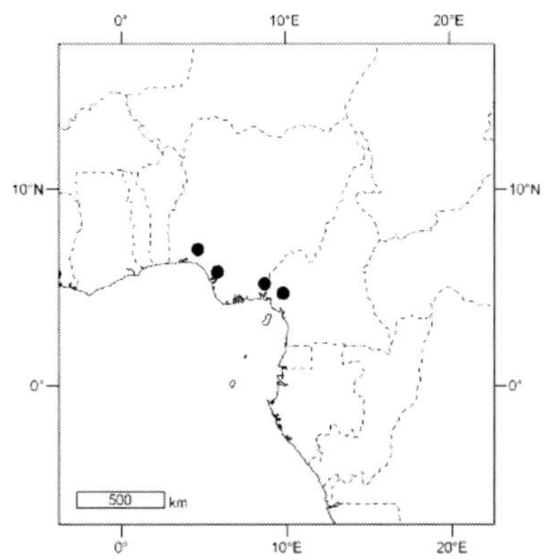

f. ***Angraecum pyriforme*** VU (Orchidaceae)

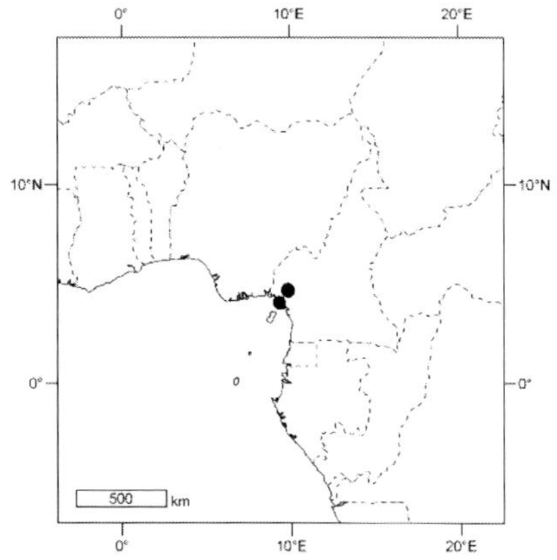

a. ***Angraecum sanfordii*** EN (Orchidaceae)

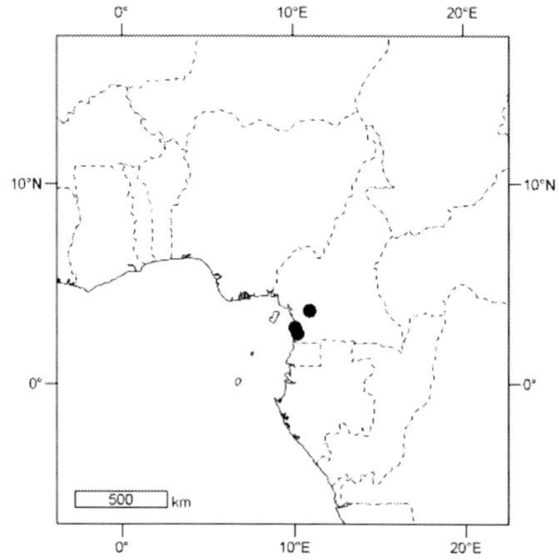

b. ***Bulbophyllum alinae*** EN (Orchidaceae)

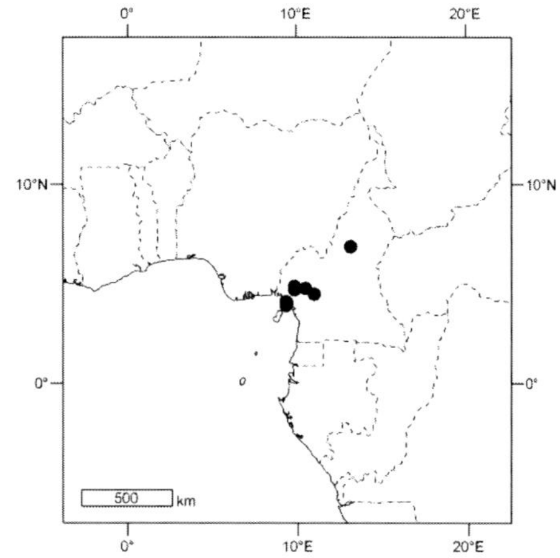

c. ***Bulbophyllum bifarium*** VU (Orchidaceae)

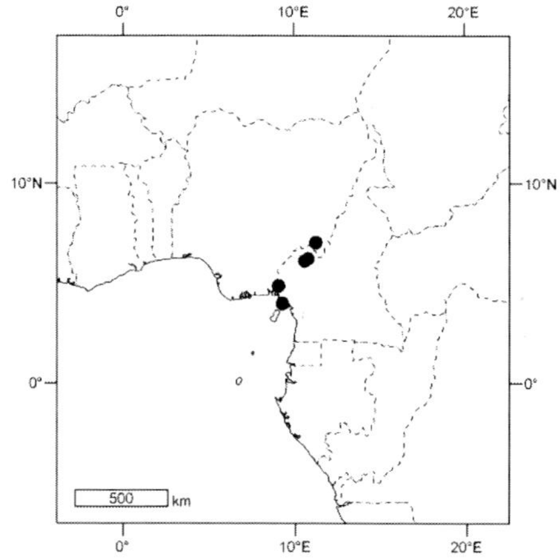

d. ***Bulbophyllum calvum*** EN (Orchidaceae)

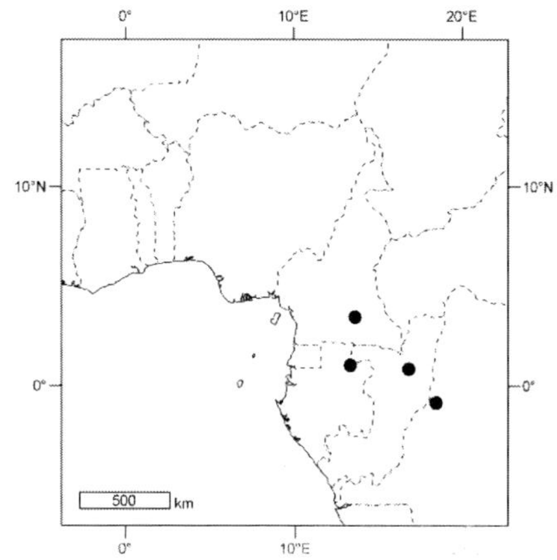

e. ***Bulbophyllum capi*** [illegible]
(Orchidaceae)

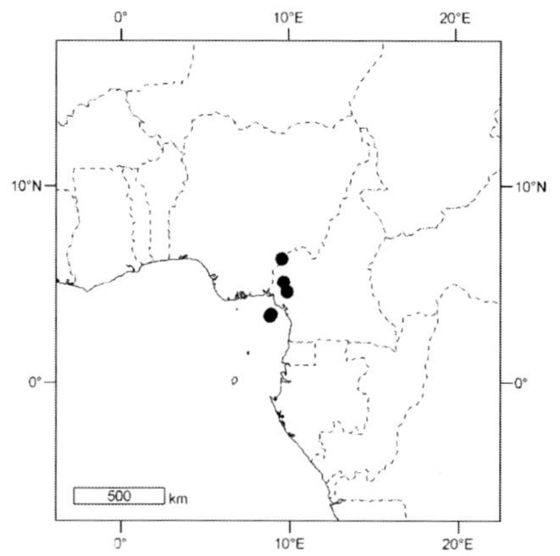

[illegible] ***omatum*** var. ***comatum*** EN
[illegible]

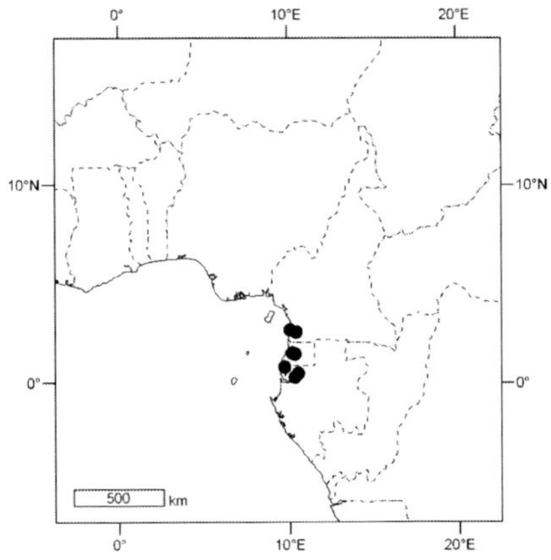

a. ***Bulbophyllum coriscense*** VU (Orchidaceae)

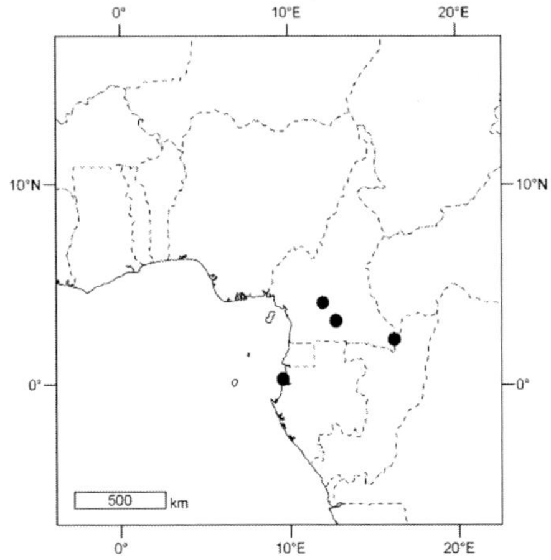

b. ***Bulbophyllum fayi*** EN (Orchidaceae)

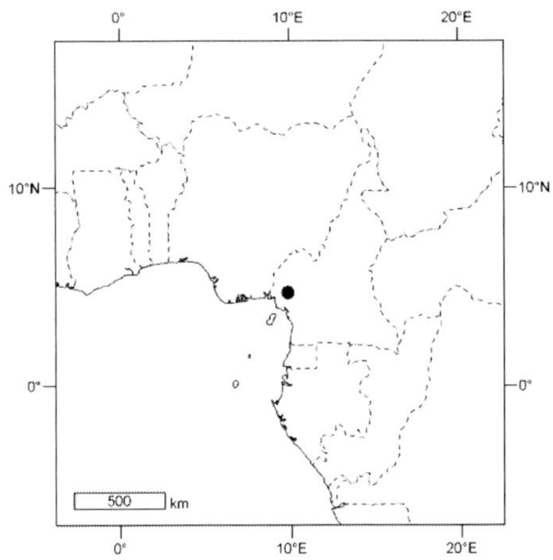

c. ***Bulbophyllum jaapii*** VU (Orchidaceae)

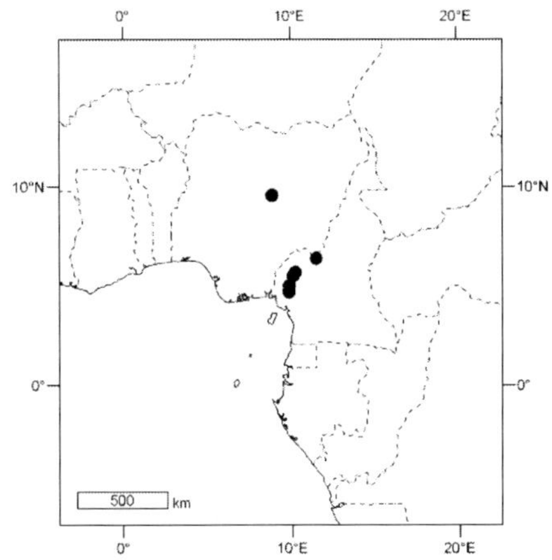

d. ***Bulbophyllum nigericum*** VU (Orchidaceae)

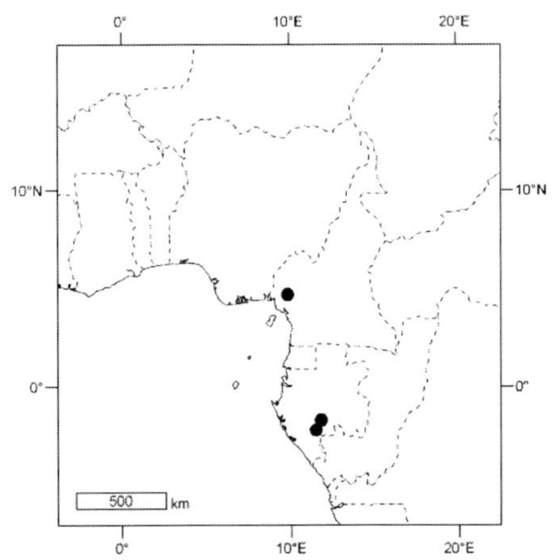

e. ***Bulbophyllum pandanetorum*** EN (Orchidaceae)

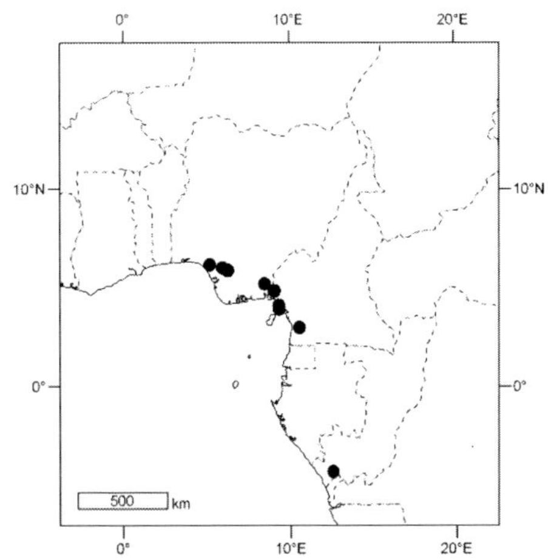

f. ***Bulbophyllum porphyrostachys*** VU (Orchidaceae)

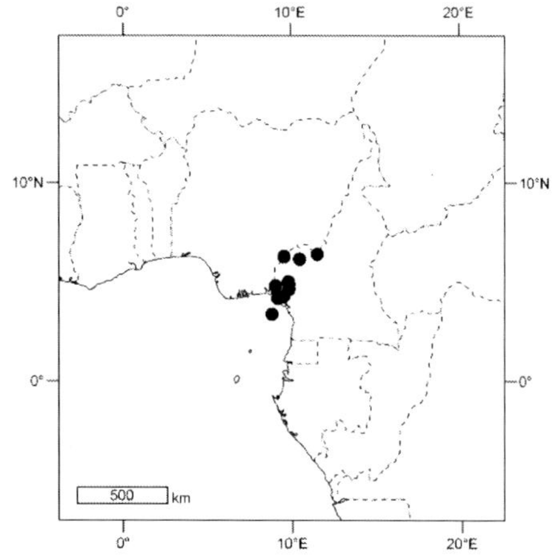

a. ***Bulbophyllum scaberulum*** var. ***fuerstenbergianum*** VU (Orchidaceae)

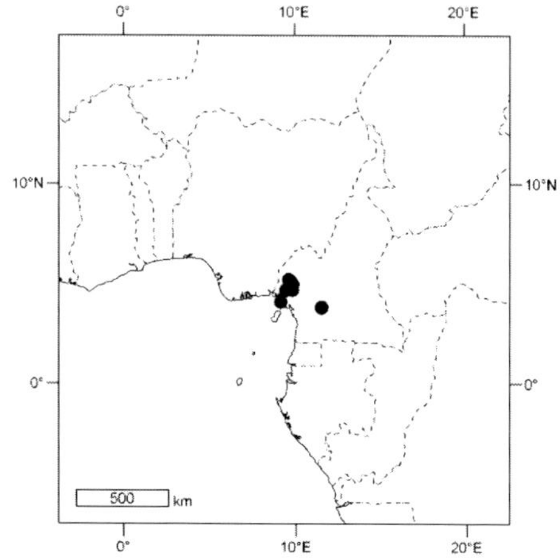

b. ***Bulbophyllum teretifolium*** VU (Orchidaceae)

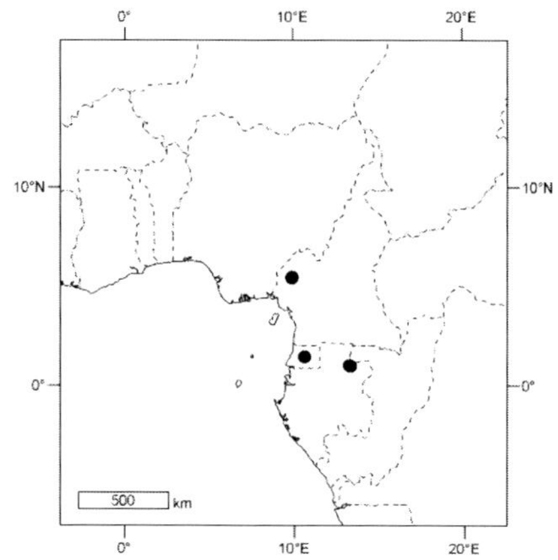

c. ***Bulbophyllum vanum*** EN (Orchidaceae)

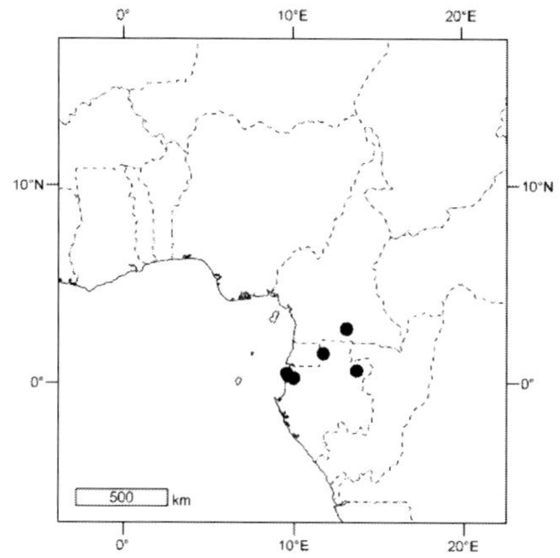

d. ***Chamaeangis gabonensis*** EN (Orchidaceae)

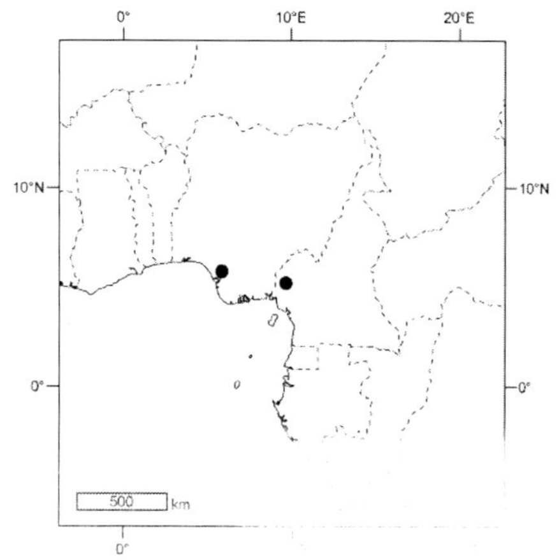

e. ***Chamaeangis lanceolata*** EN (Orchidaceae)

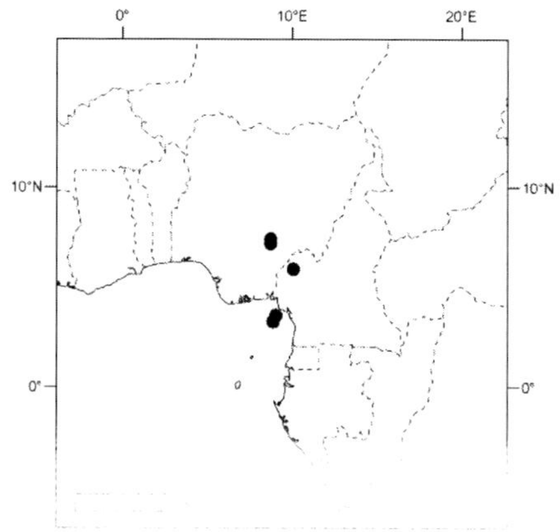

f. ***Chamaeangis letouzeyi*** VU (Orchidaceae)

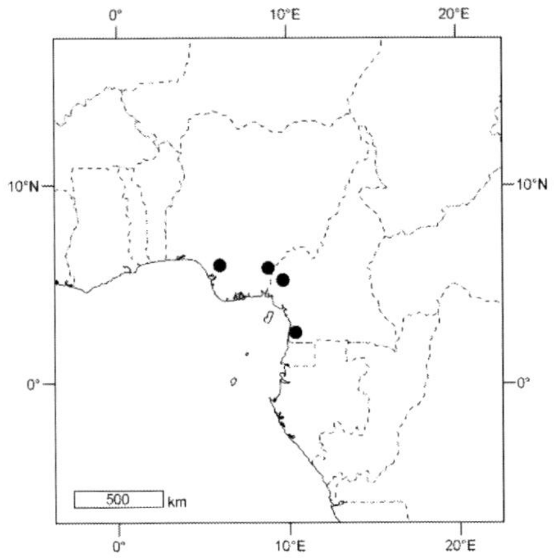

a. ***Chamaeangis spiralis*** EN (Orchidaceae)

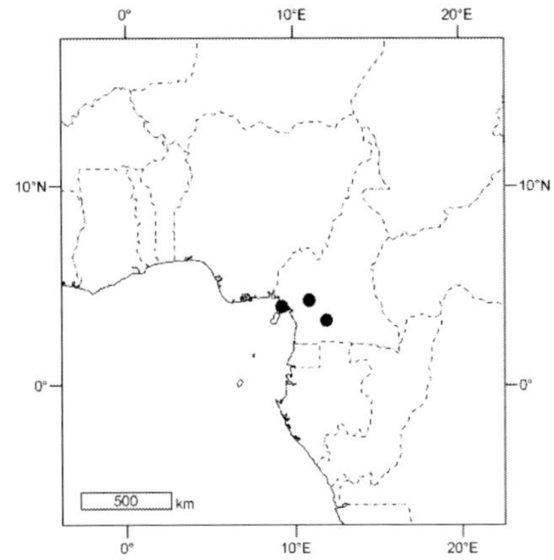

b. ***Cheirostylis divina*** var. ***ochyrae*** EN (Orchidaceae)

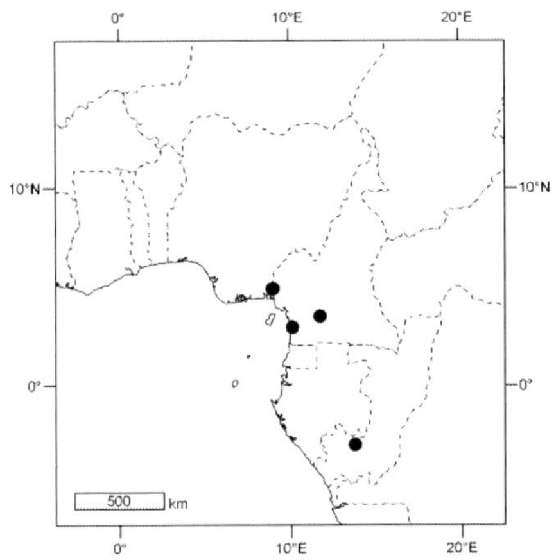

c. ***Corymborkis minima*** VU (Orchidaceae)

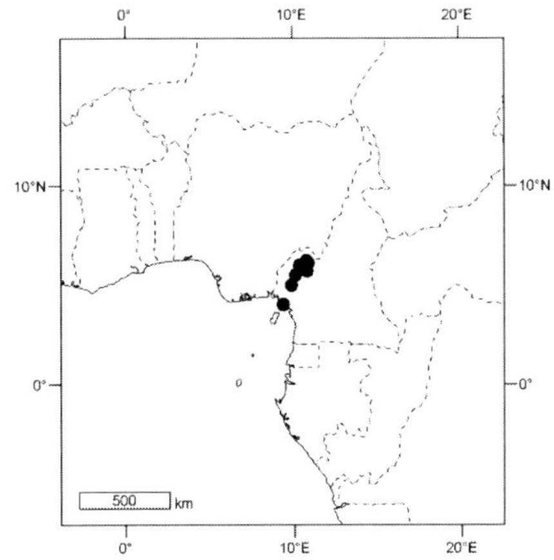

d. ***Diaphananthe bueae*** VU (Orchidaceae)

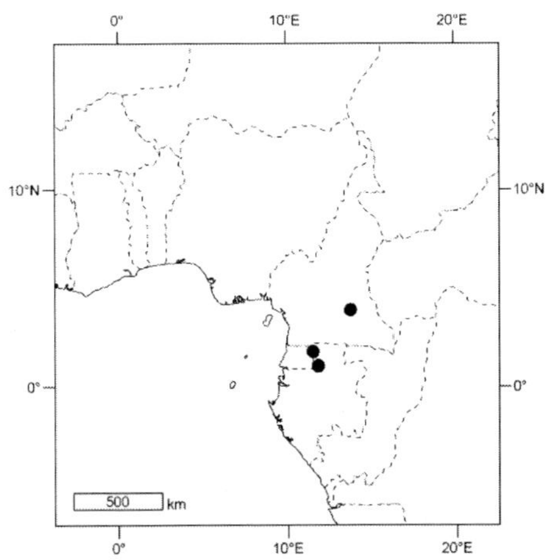

e. ***Diaphananthe ceriflora*** EN (Orchidaceae)

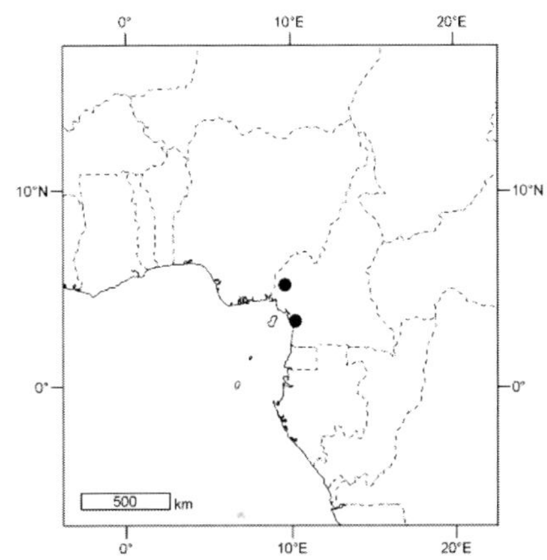

f. ***Diaphananthe garayana*** EN (Orchidaceae)

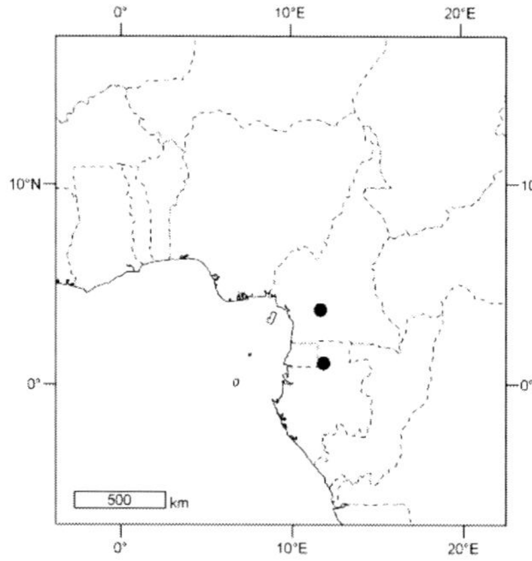

a. ***Diaphananthe sanfordiana*** EN (Orchidaceae)

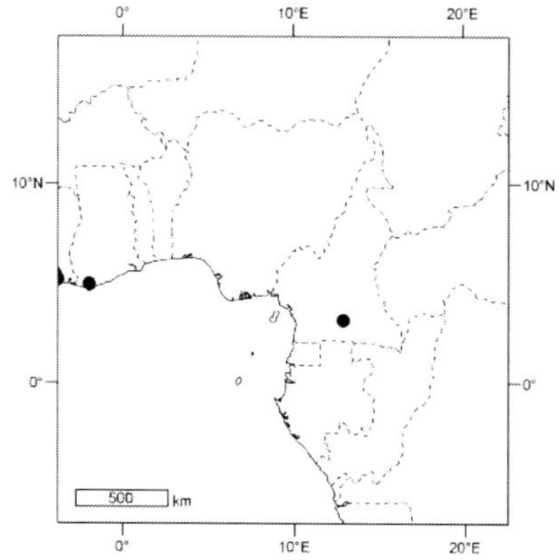

b. ***Diaphananthe sarcorhynchoides*** EN (Orchidaceae)

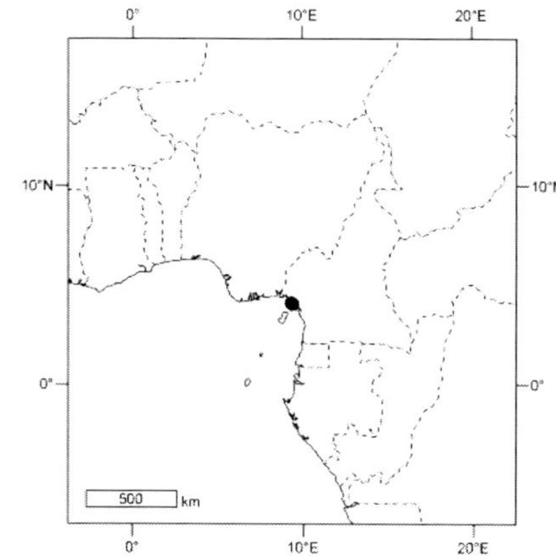

c. ***Disperis kamerunensis*** EN (Orchidaceae)

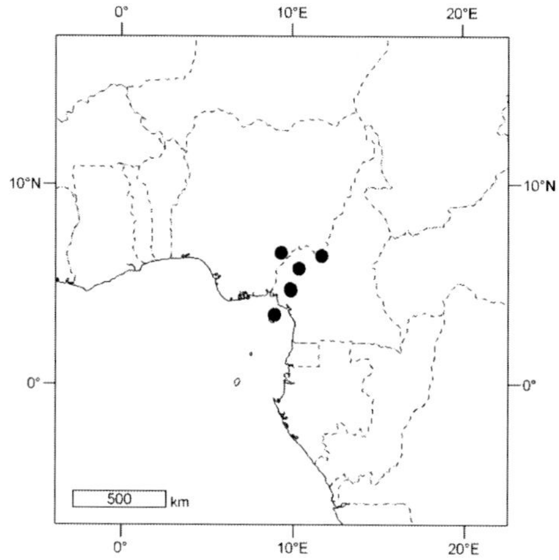

d. ***Disperis mildbraedii*** VU (Orchidaceae)

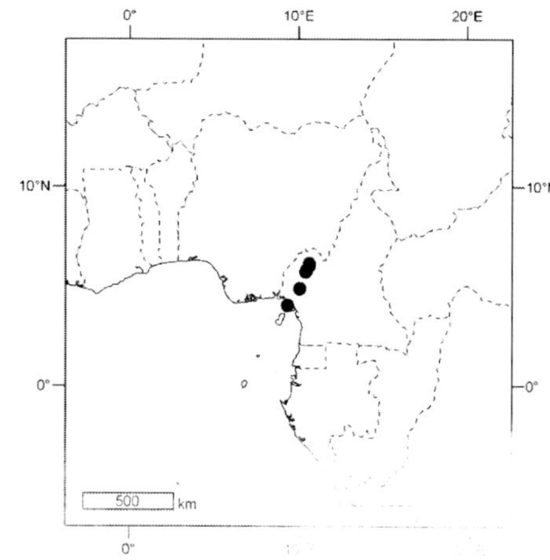

e. ***Disperis nitida*** VU (Orchidaceae)

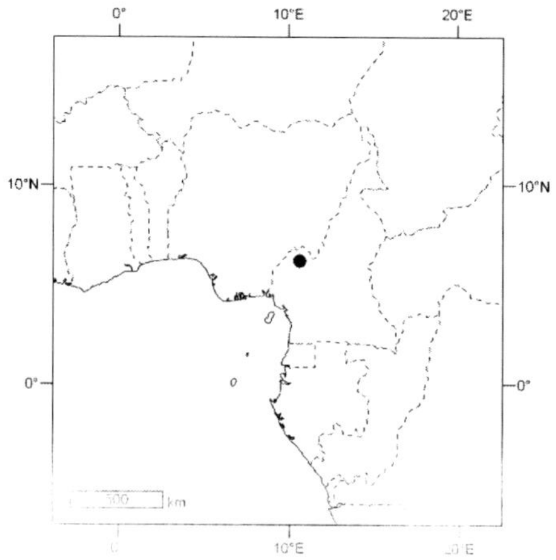

f. ***Disperis parvifolia*** EN (Orchidaceae)

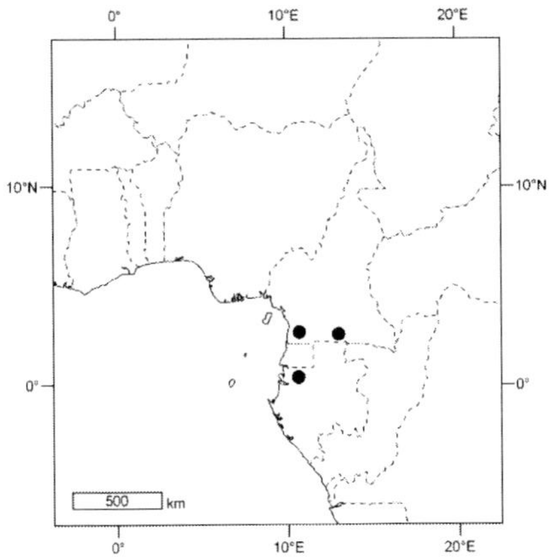

a. ***Eggelingia gabonensis*** EN (Orchidaceae)

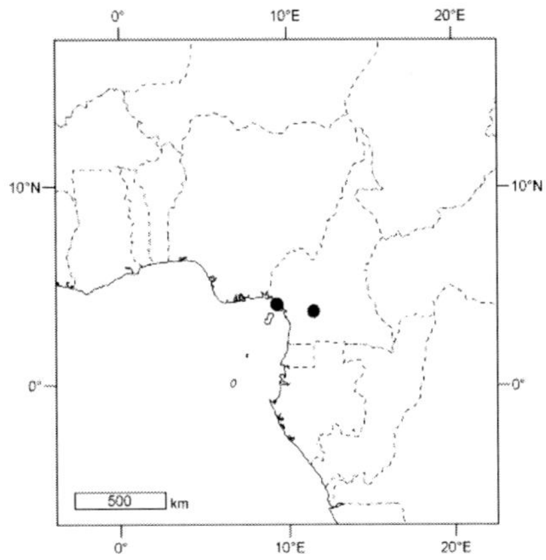

b. ***Gastrodia africana*** CR (Orchidaceae)

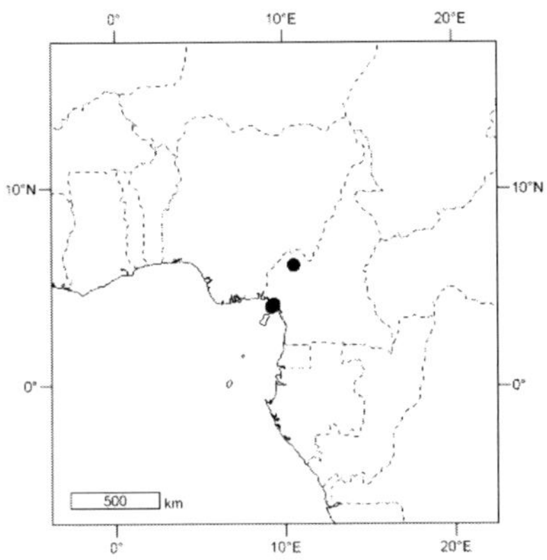

c. ***Genyorchis macrantha*** EN (Orchidaceae)

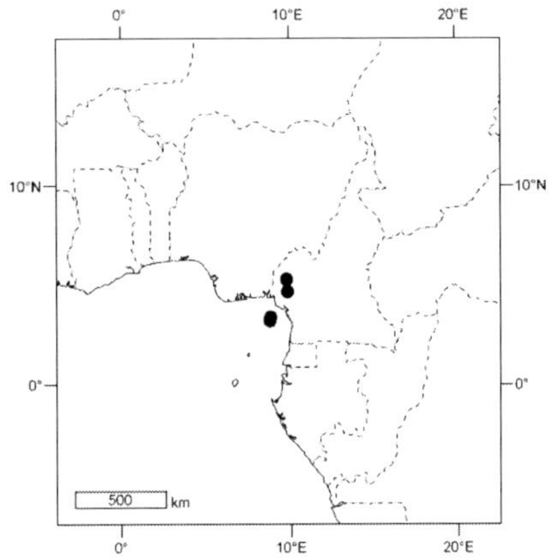

d. ***Genyorchis micropetala*** EN (Orchidaceae)

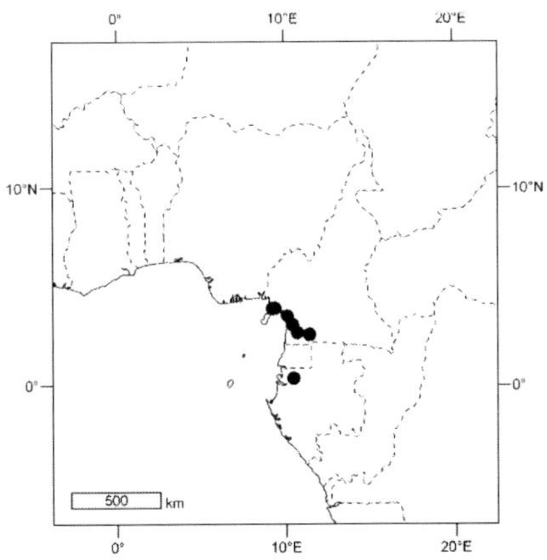

e. ***Genyorchis platybulbon*** VU (Orchidaceae)

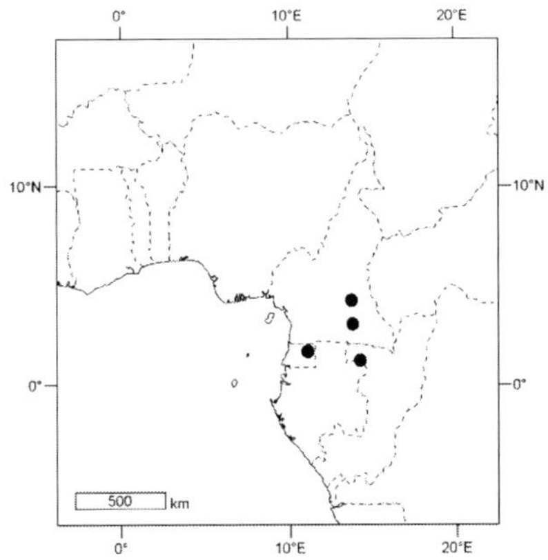

f. ***Genyorchis sanfordii*** EN (Orchidaceae)

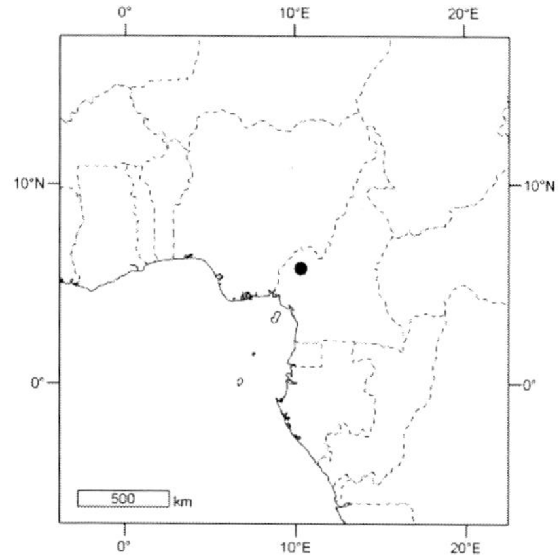

a. ***Habenaria alinae*** CR (Orchidaceae)

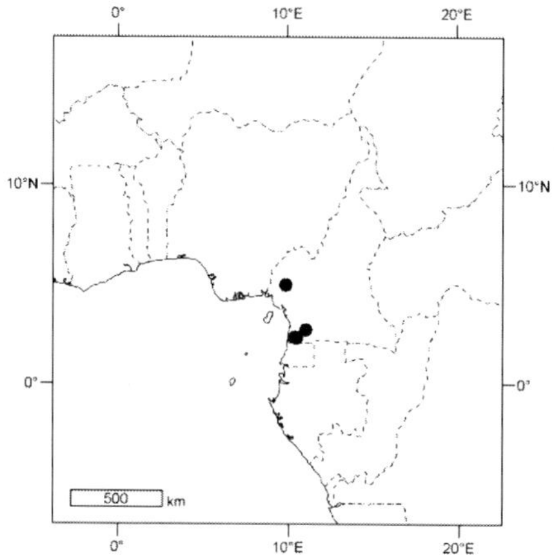

b. ***Habenaria batesii*** EN (Orchidaceae)

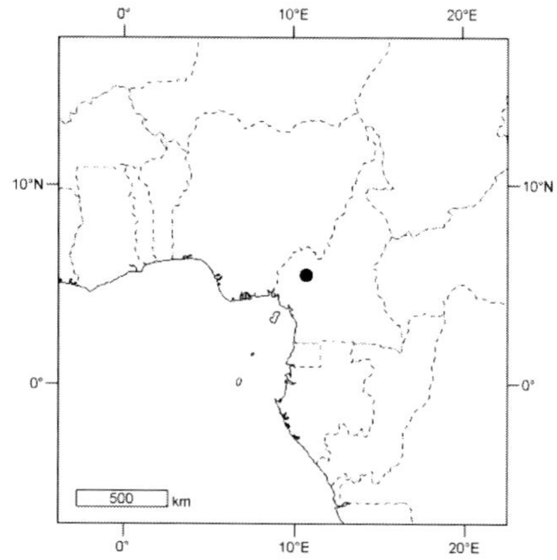

c. ***Habenaria bosseriana*** CR (Orchidaceae)

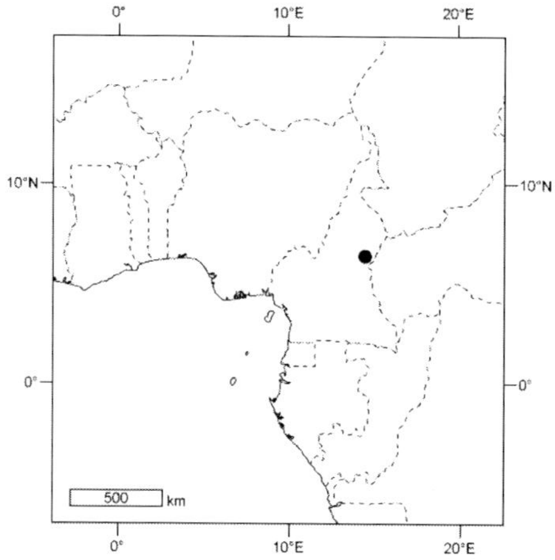

d. ***Habenaria egregia*** EN (Orchidaceae)

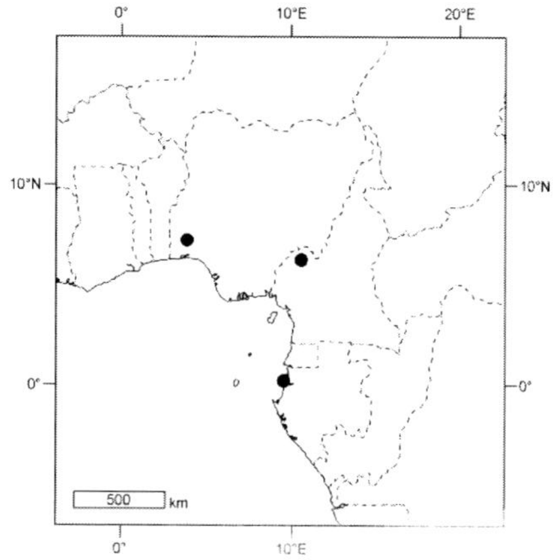

e. ***Habenaria maitlandii*** EN (Orchidaceae)

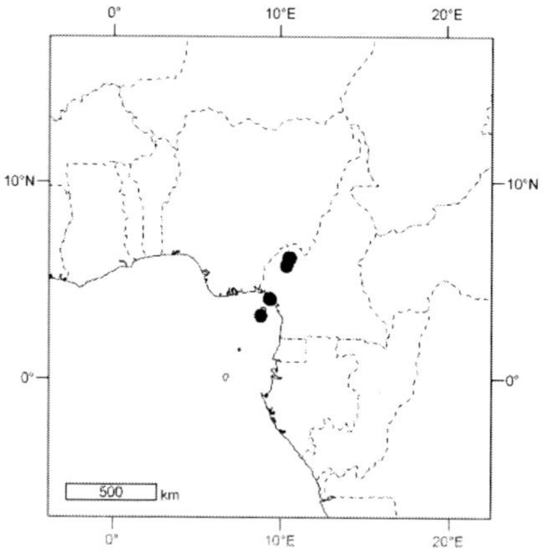

f. ***Habenaria microceras*** VU (Orchidaceae)

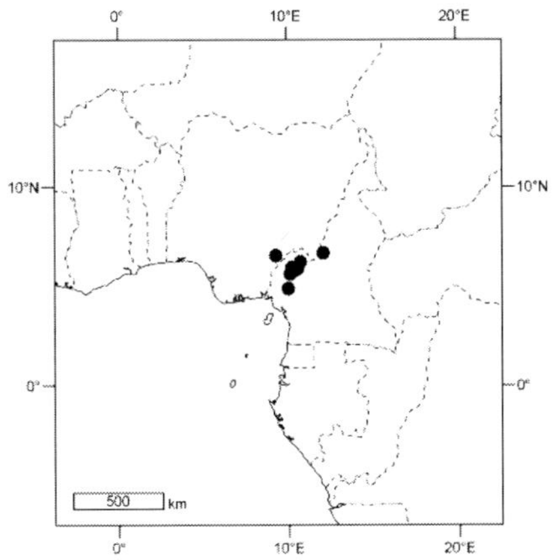

a. ***Habenaria nigrescens*** VU (Orchidaceae)

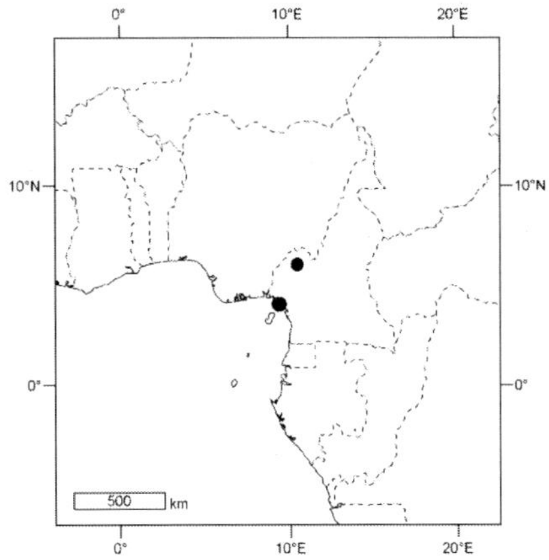

b. ***Habenaria obovata*** EN (Orchidaceae)

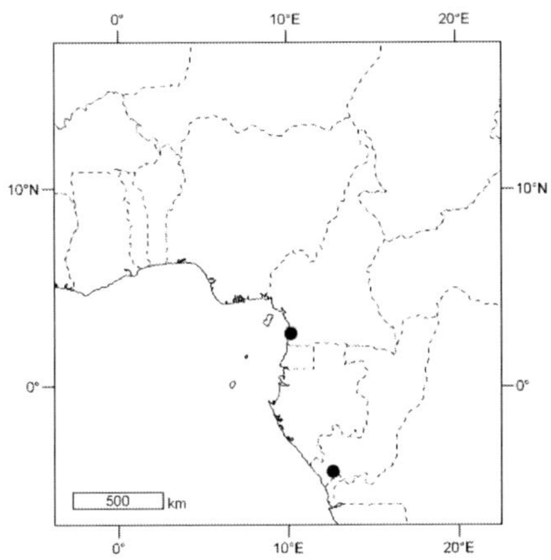

c. ***Habenaria phantasma*** EN (Orchidaceae)

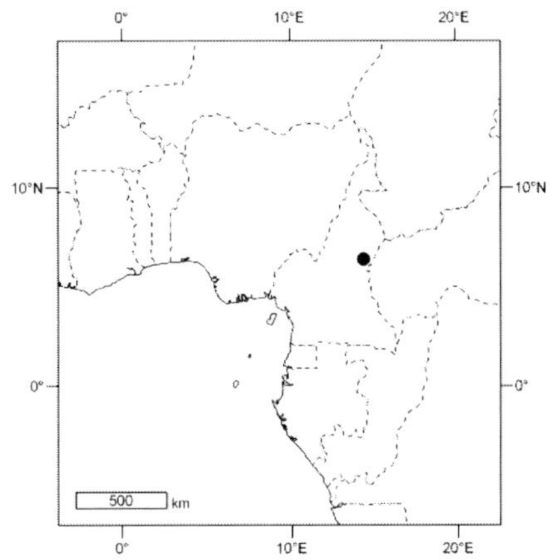

d. ***Habenaria sanfordiana*** EN (Orchidaceae)

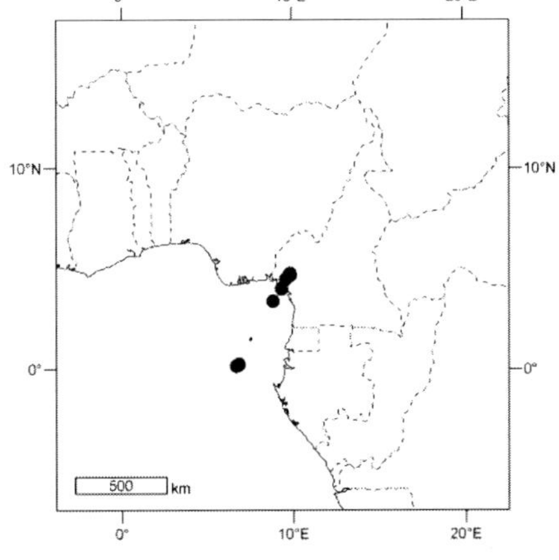

e. ***Habenaria thomana*** VU (Orchidaceae)

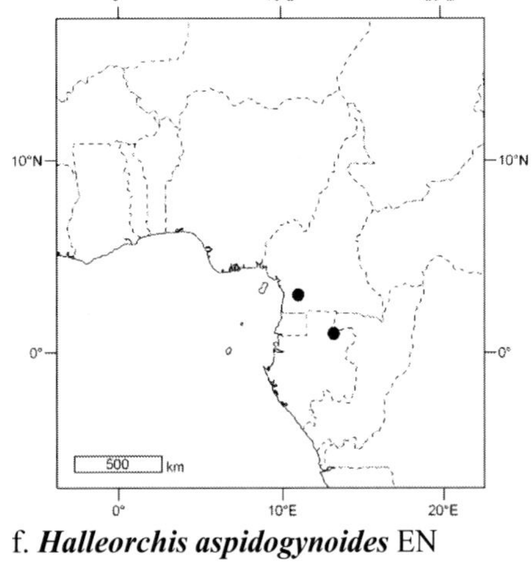

f. ***Halleorchis aspidogynoides*** EN (Orchidaceae)

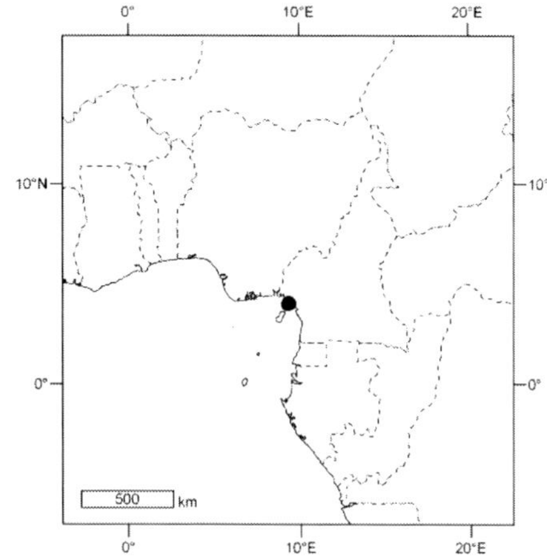

a. ***Holothrix tridentata*** VU (Orchidaceae)

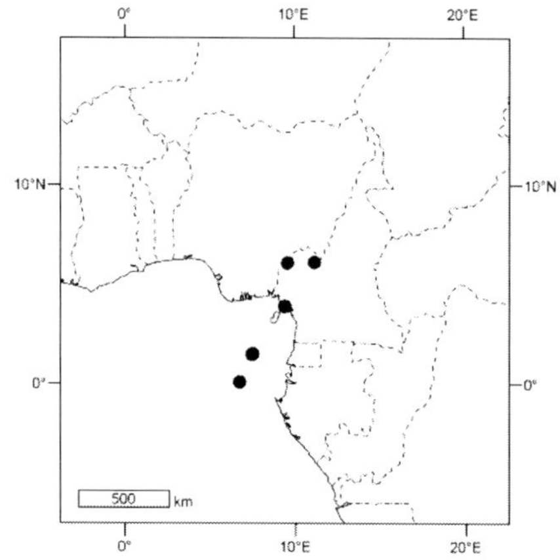

b. ***Liparis goodyeroides*** VU (Orchidaceae)

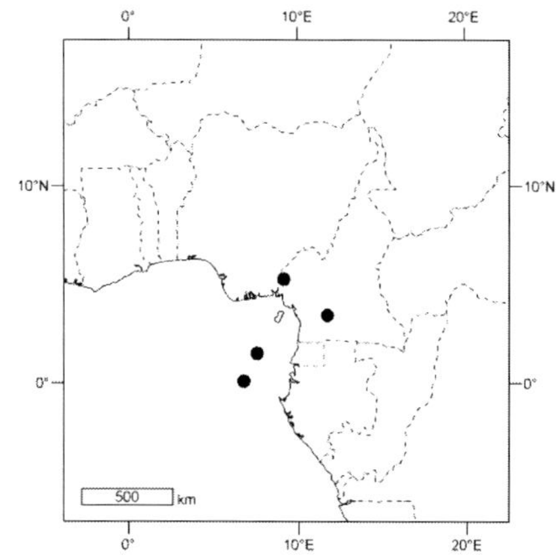

c. ***Liparis gracilenta*** EN (Orchidaceae)

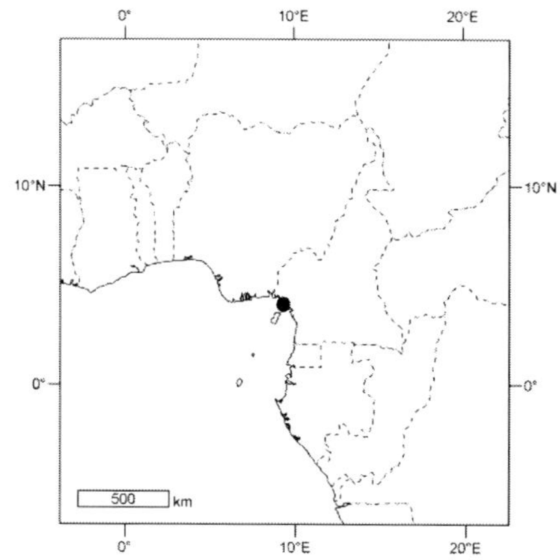

d. ***Liparis kamerunensis*** CR (Orchidaceae)

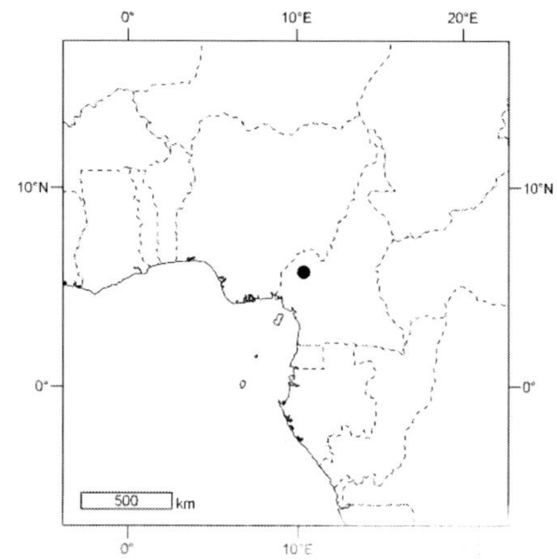

e. ***Liparis letouzeyana*** CR (Orchidaceae)

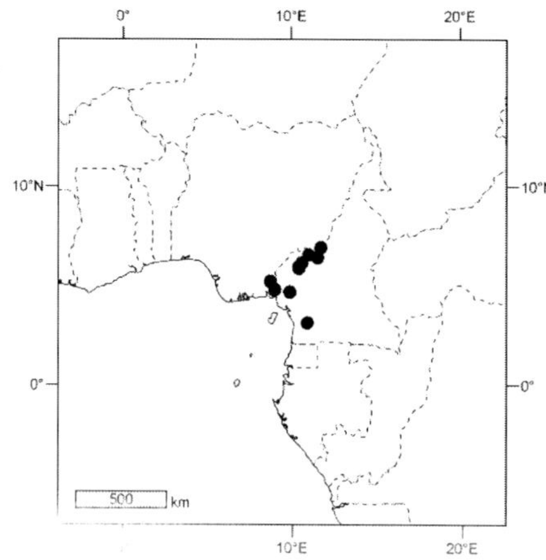

f. ***Liparis suborbicularis*** VU (Orchidaceae)

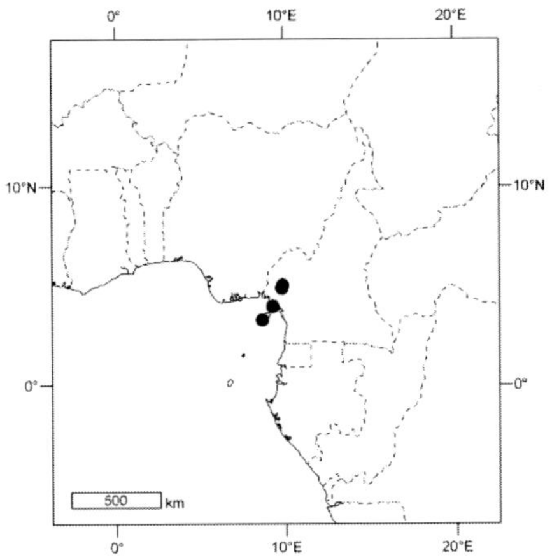

a. ***Manniella cypripedioides*** EN (Orchidaceae)

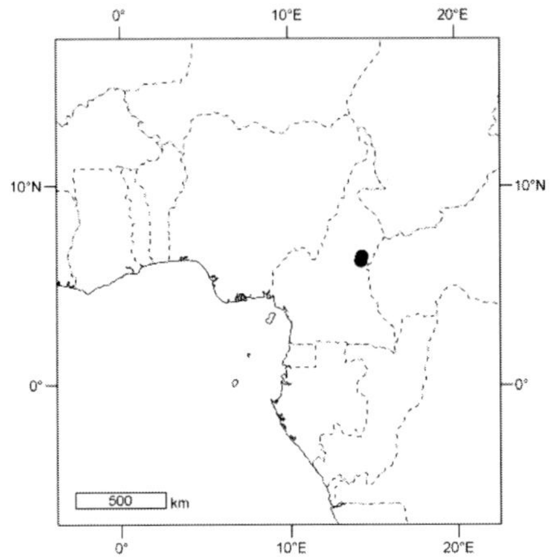

b. ***Microcoelia sanfordii*** EN(Orchidaceae)

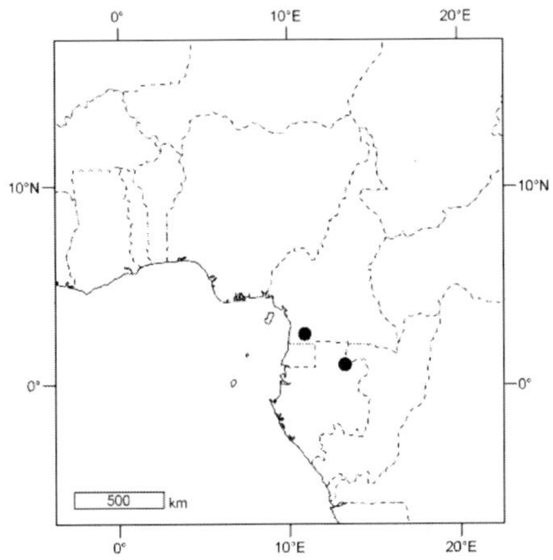

c. ***Nephrangis bertauxiana*** CR (Orchidaceae)

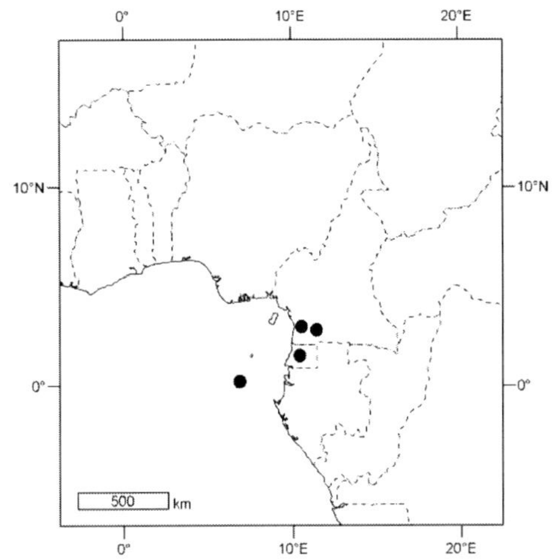

d. ***Orestias micrantha*** EN (Orchidaceae)

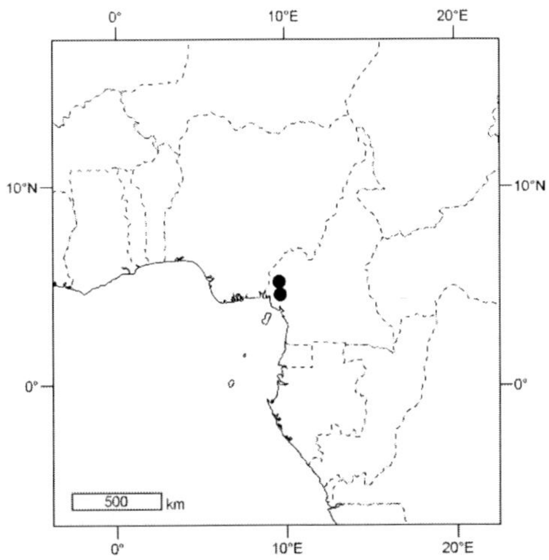

e. ***Ossiculum aurantiacum*** CR (Orchidaceae)

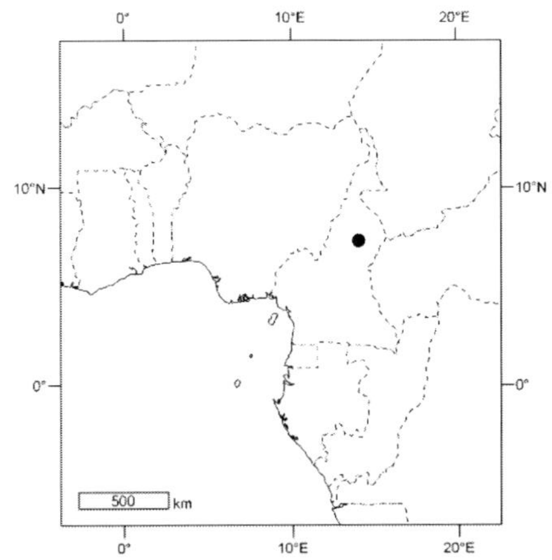

f. ***Platycoryne alinae*** CR (Orchidaceae)

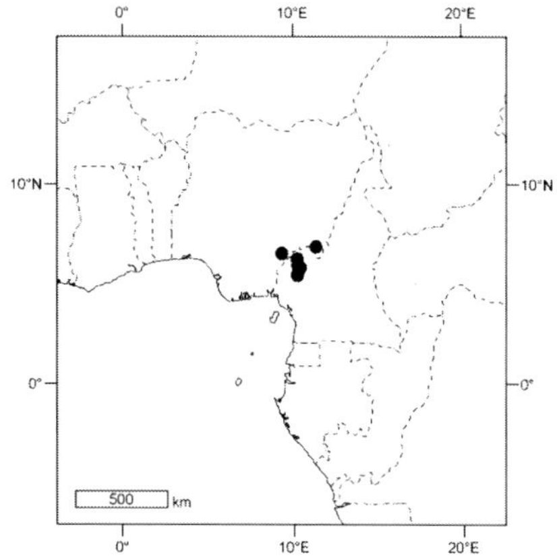

a. ***Platycoryne megalorrhyncha*** VU (Orchidaceae)

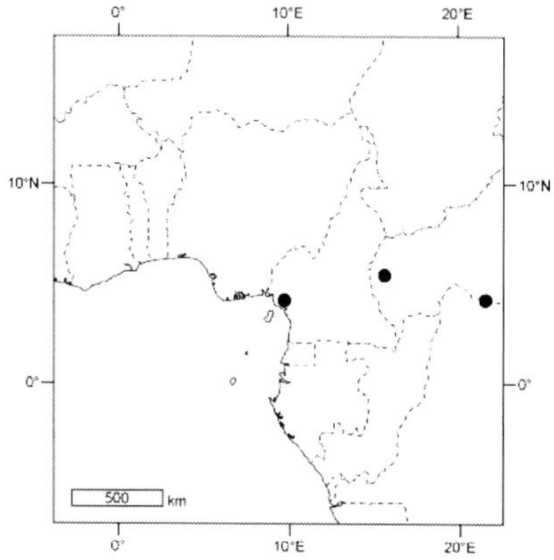

b. ***Plectrelminthus caudatus*** var. ***trilobatus*** EN (Orchidaceae)

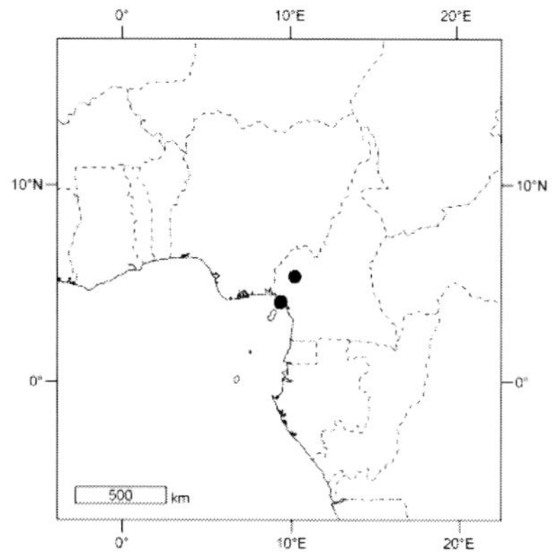

c. ***Polystachya albescens*** subsp. ***angustifolia*** EN (Orchidaceae)

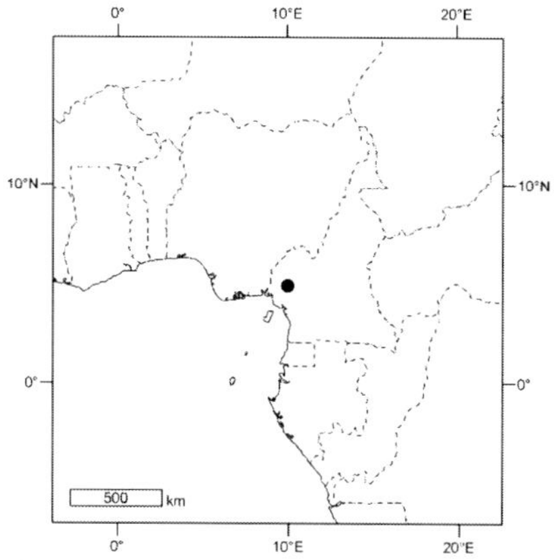

d. ***Polystachya albescens*** subsp. ***manengouba*** CR (Orchidaceae)

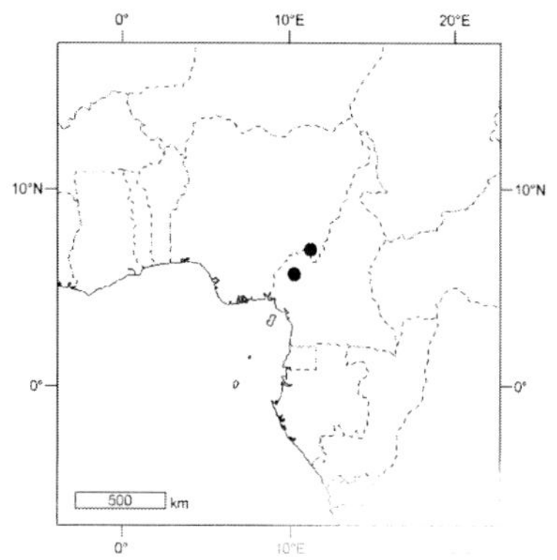

e. ***Polystachya anthoceros*** EN (Orchidaceae)

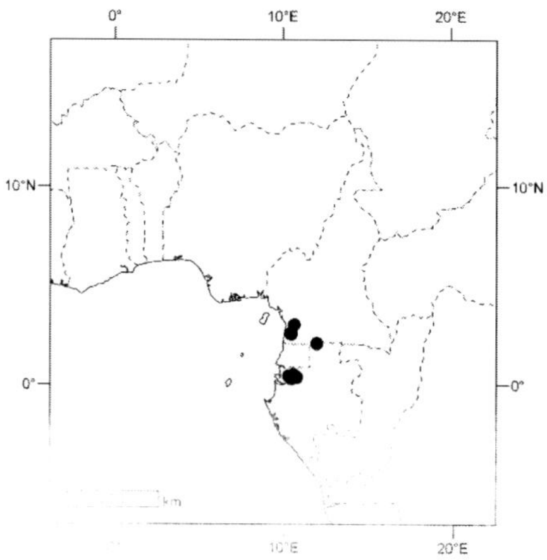

f. ***Polystachya batkoi*** VU (Orchidaceae)

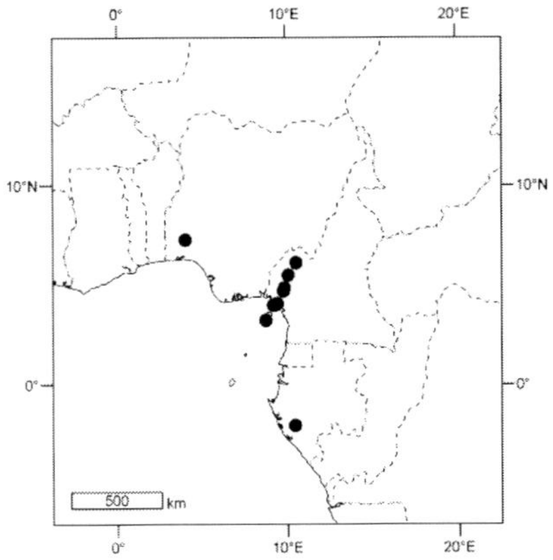

a. ***Polystachya bicalcarata*** VU (Orchidaceae)

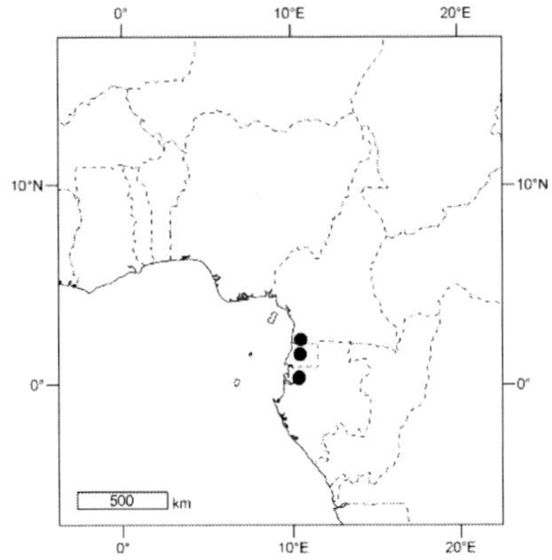

b. ***Polystachya bipoda*** EN (Orchidaceae)

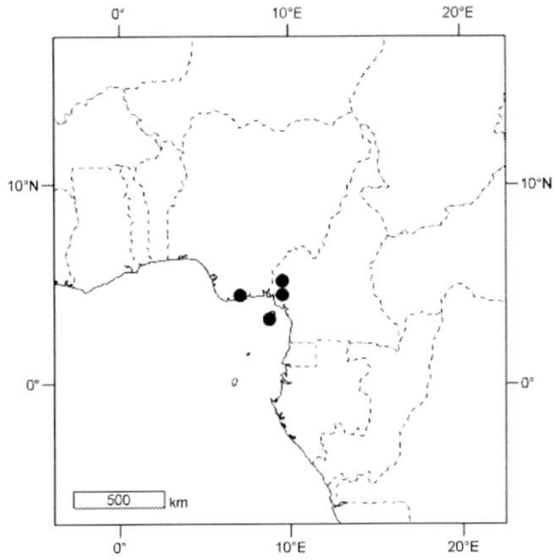

c. ***Polystachya camaridioides*** EN (Orchidaceae)

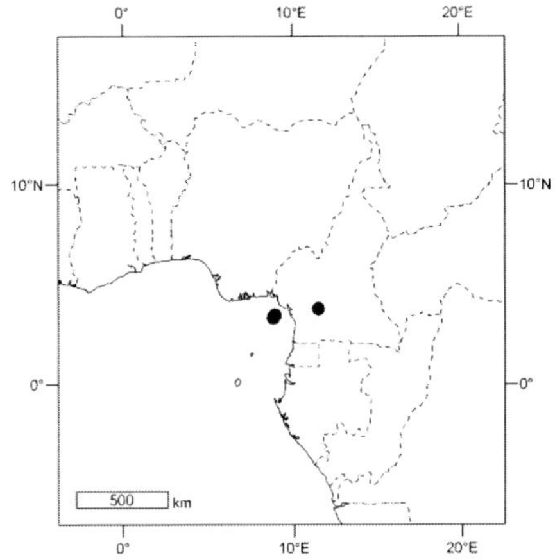

d. ***Polystachya carnosa*** EN (Orchidaceae)

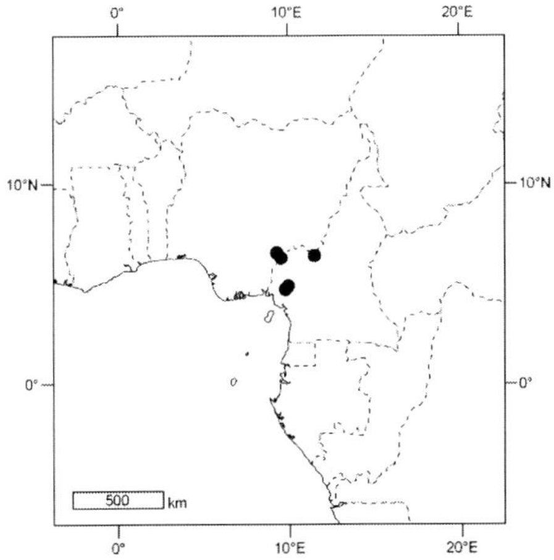

e. ***Polystachya cooperi*** EN (Orchidaceae)

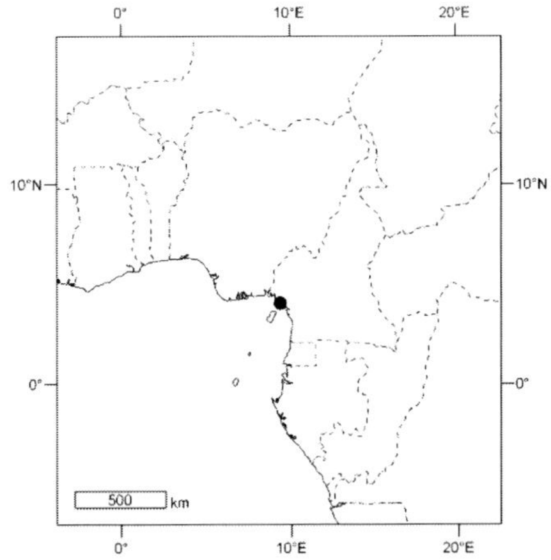

f. ***Polystachya crassifolia*** EN (Orchidaceae)

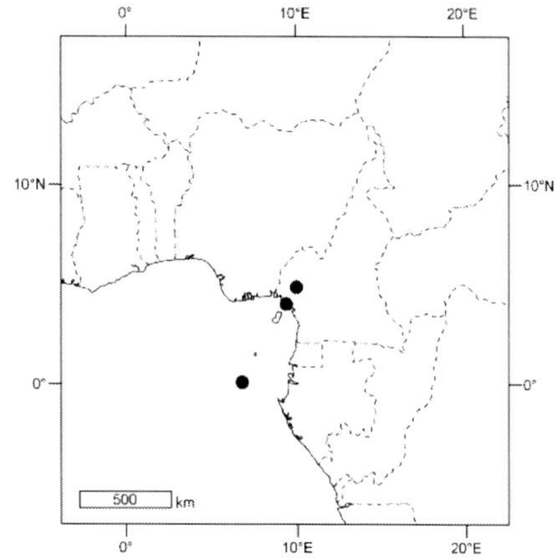

a. ***Polystachya farinosa*** EN (Orchidaceae)

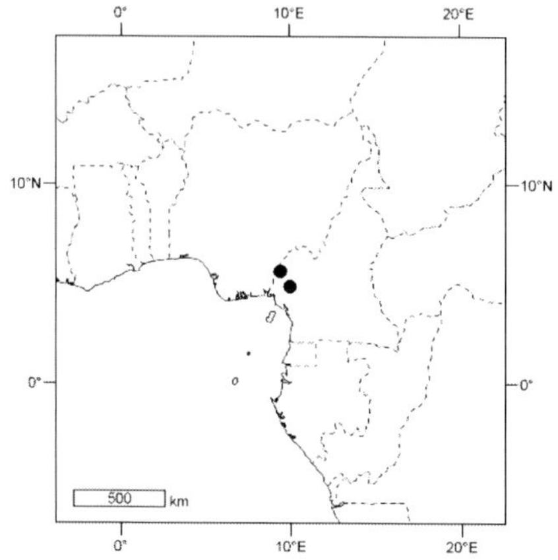

b. ***Polystachya geniculata*** EN (Orchidaceae)

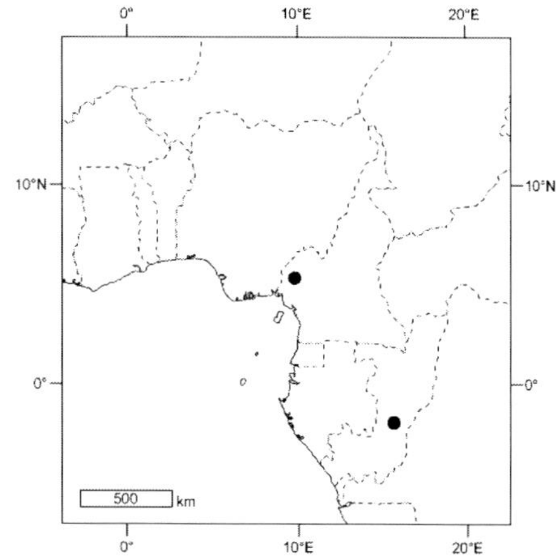

c. ***Polystachya kornasiana*** EN (Orchidaceae)

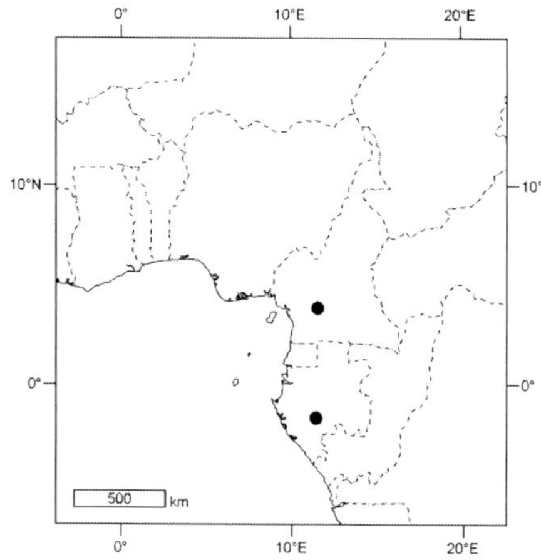

d. ***Polystachya kubalae*** EN (Orchidaceae)

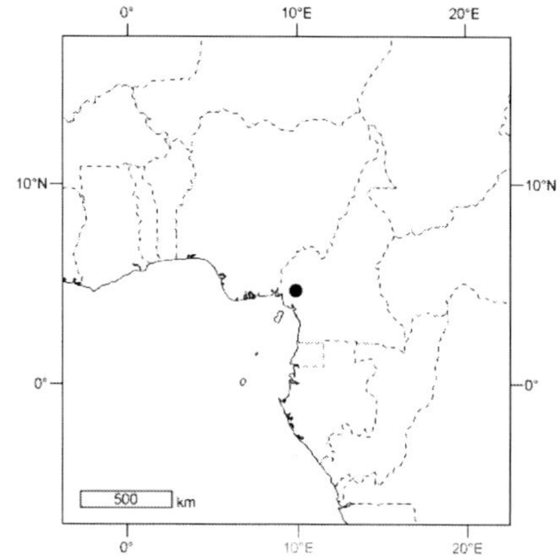

e. ***Polystachya kupensis*** CR (Orchidaceae)

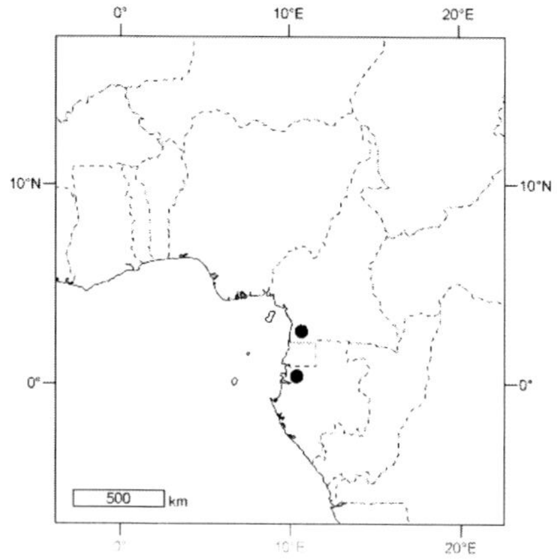

f. ***Polystachya lejolyana*** EN (Orchidaceae)

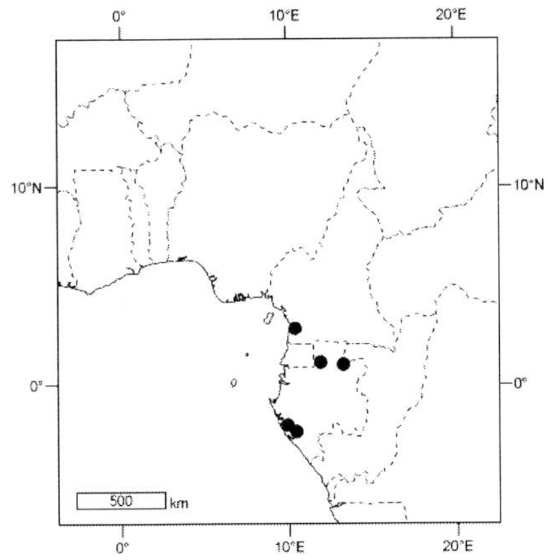

a. ***Polystachya letouzeyana*** EN (Orchidaceae)

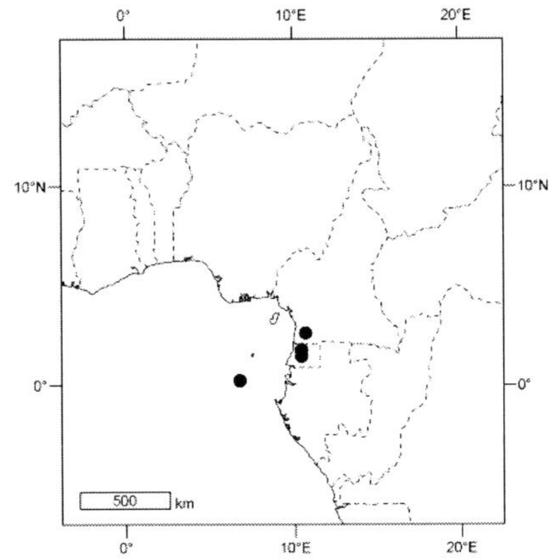

b. ***Polystachya moniquetiana*** EN (Orchidaceae)

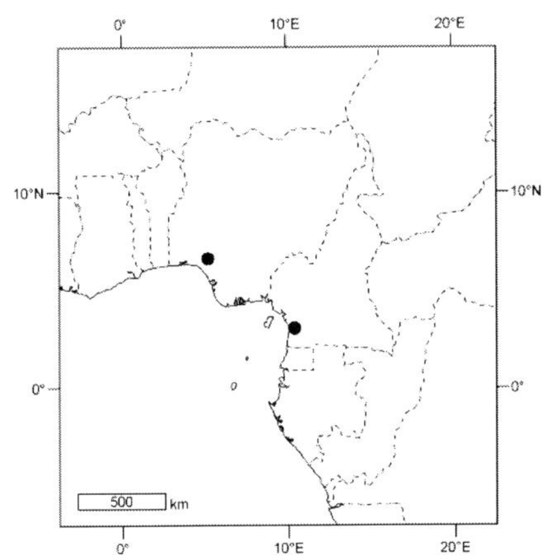

c. ***Polystachya odorata*** subsp. ***trilepidis*** VU (Orchidaceae)

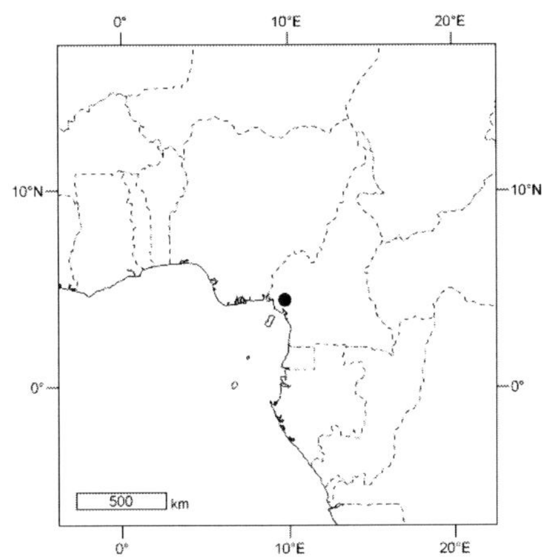

d. ***Polystachya retusiloba*** EN (Orchidaceae)

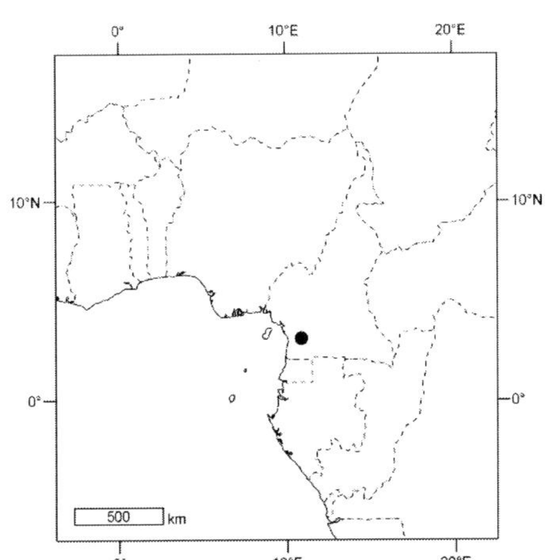

e. ***Polystachya stodolnyi*** EN (Orchidaceae)

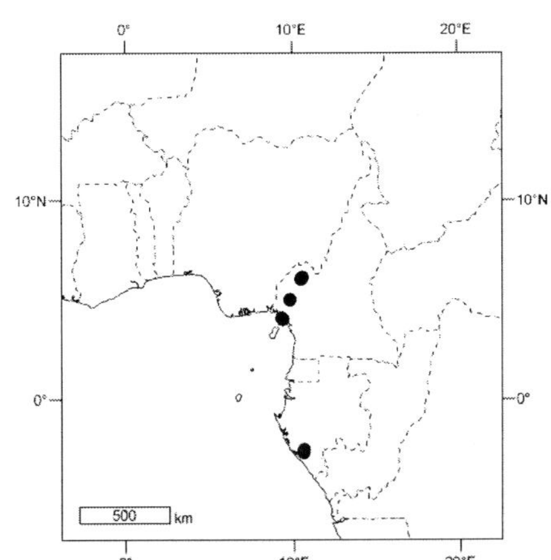

f. ***Polystachya superposita*** EN (Orchidaceae)

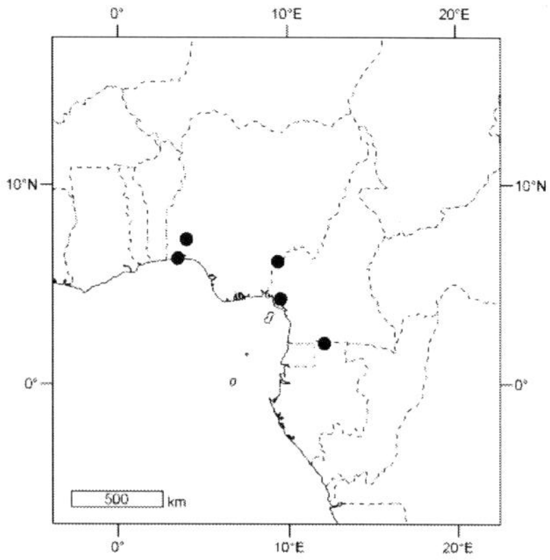

a. ***Rangaeris longicaudata*** VU (Orchidaceae)

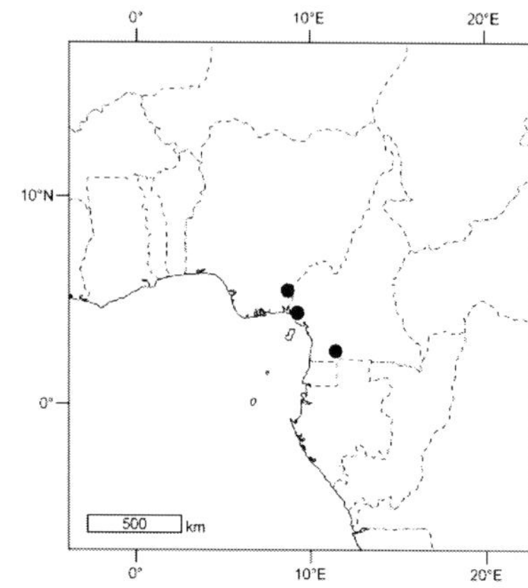

b. ***Rhipidoglossum obanense*** EN (Orchidaceae)

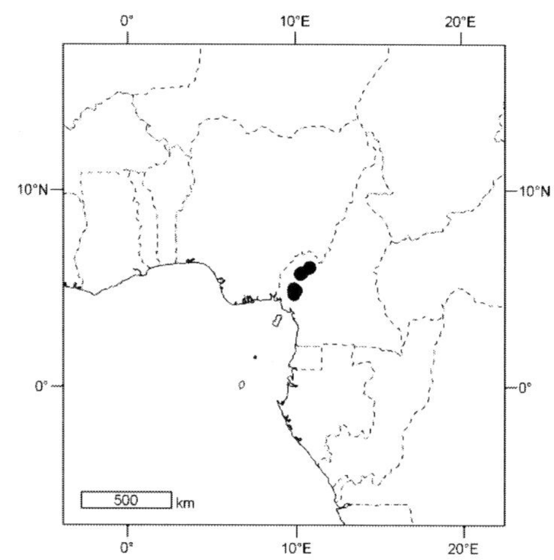

c. ***Rhipidoglossum polydactylum*** VU (Orchidaceae)

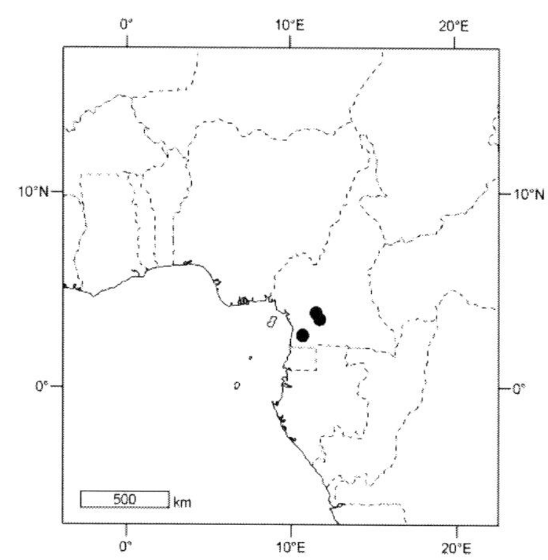

d. ***Stolzia repens*** var. ***cleistogama*** EN (Orchidaceae)

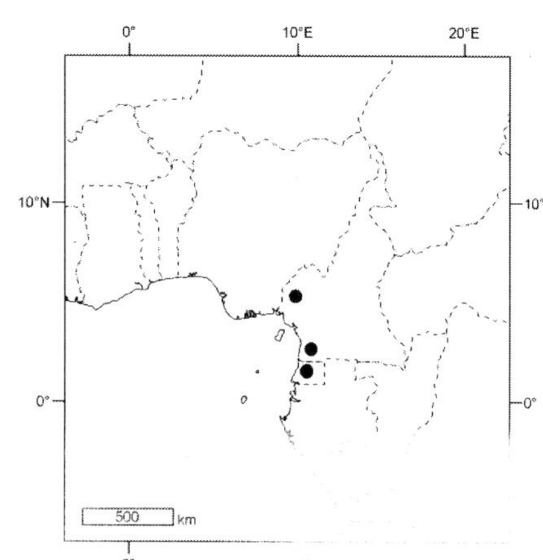

e. ***Tridactyle anthomaniaca*** subsp. ***nana*** EN (Orchidaceae)

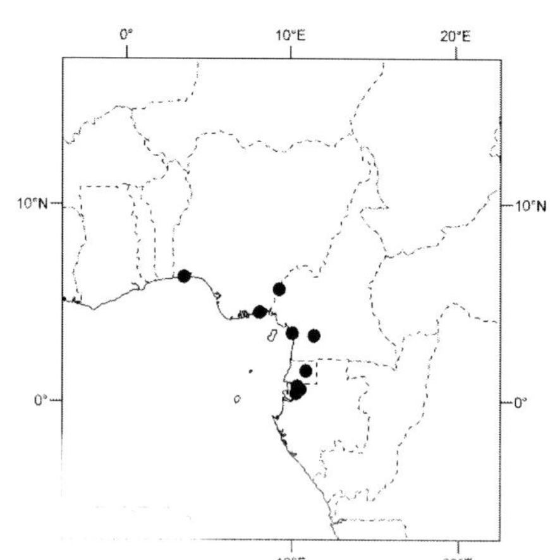

f. ***Tridactyle lagosensis*** VU (Orchidaceae)

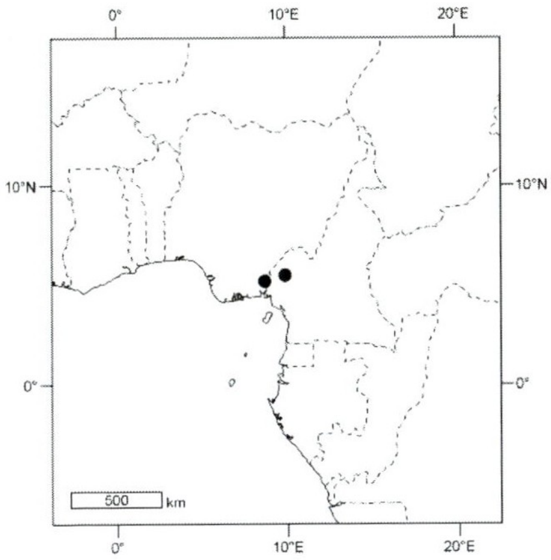

a. ***Tridactyle muriculata*** EN (Orchidaceae)

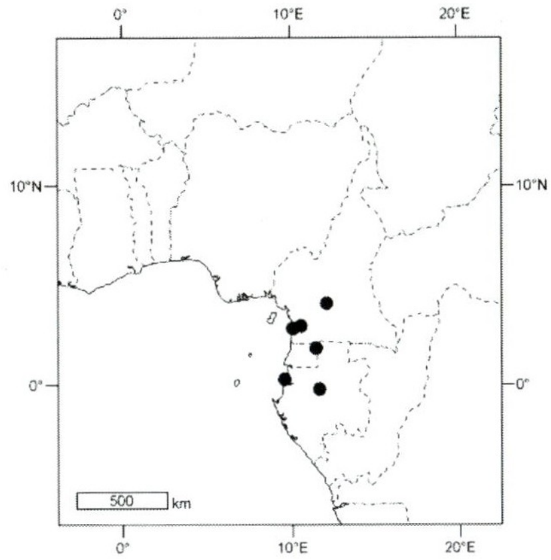

b. ***Vanilla africana*** subsp. ***cucullata*** EN (Orchidaceae)

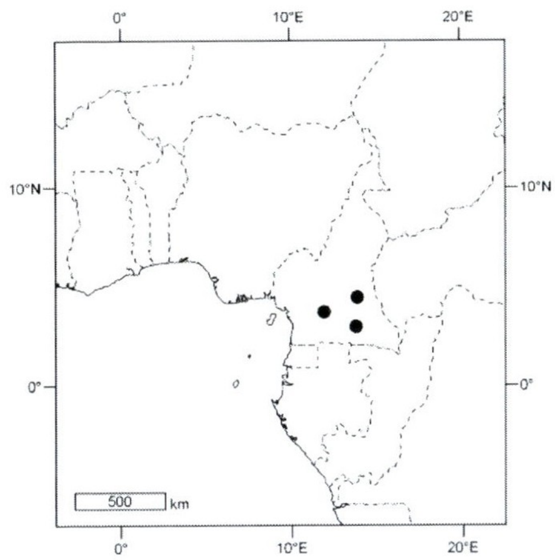

c. ***Vanilla ochyrae*** EN (Orchidaceae)

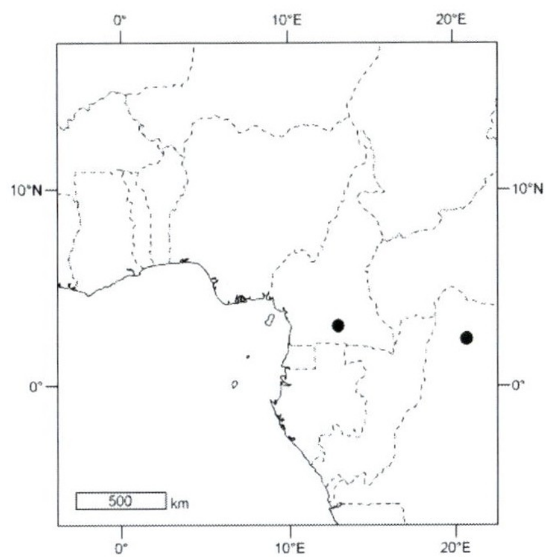

d. ***Vanilla seretii*** VU (Orchidaceae)

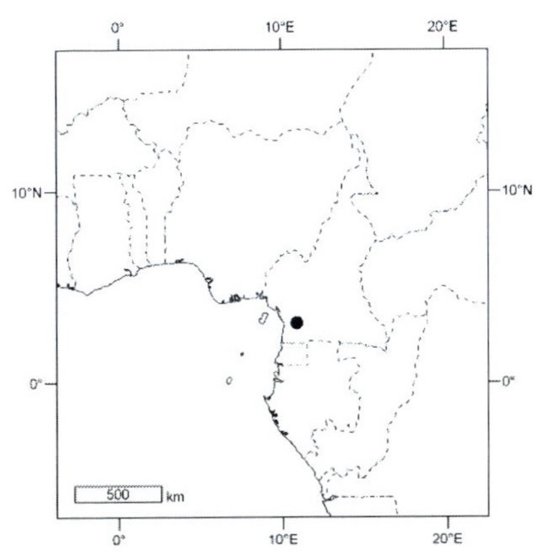

e. ***Eremospatha barendii*** CR (Palmae)

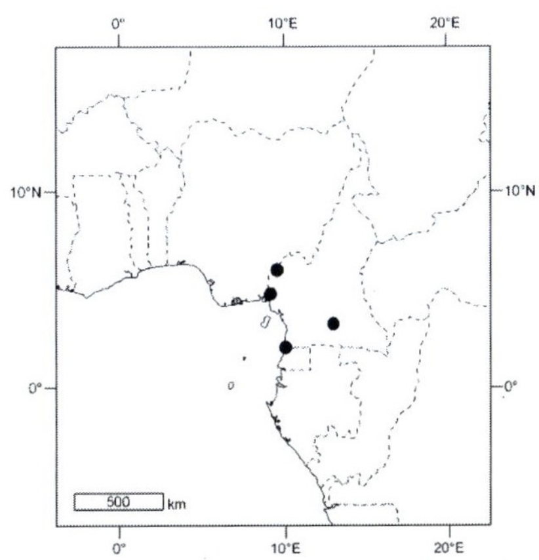

f. ***Eremospatha quinquecostulata*** VU (Palmae)

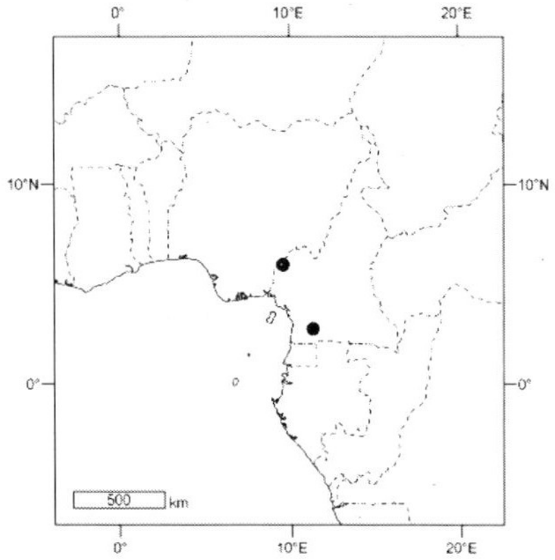

a. ***Eremospatha tessmanniana*** EN (Palmae)

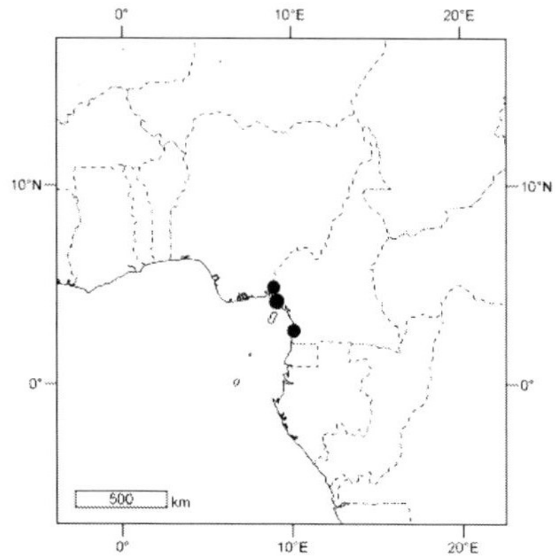

b. ***Laccosperma korupensis*** EN (Palmae)

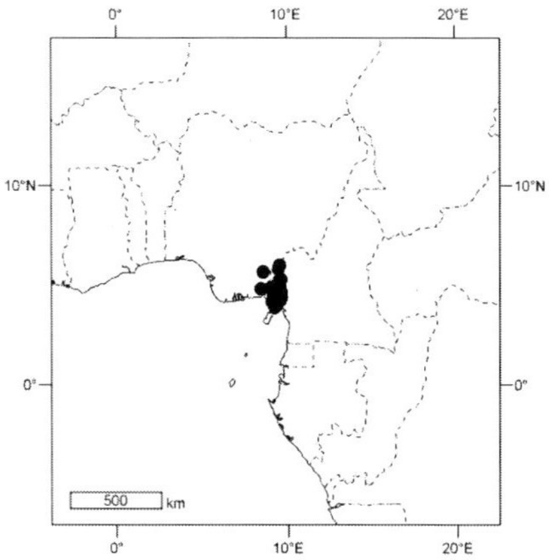

c. ***Oncocalamus tuleyi*** VU (Palmae)

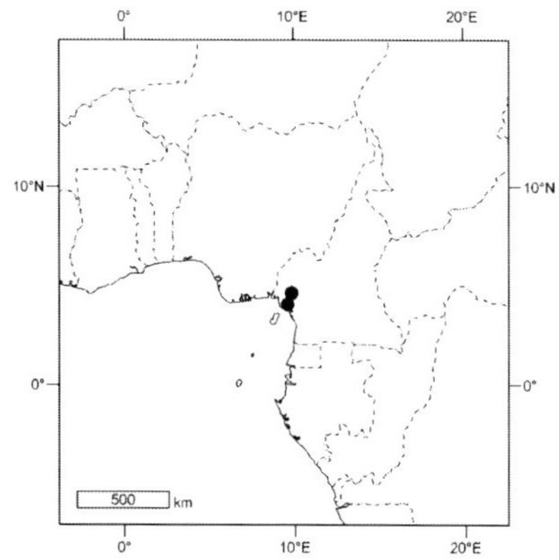

d. ***Kupea martinetugei*** CR (Triuridaceae)

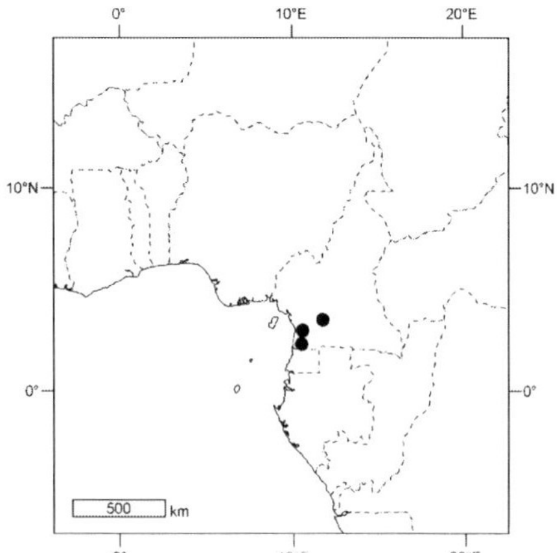

e. ***Aulotandra kamerunensis*** EN (Zing[illegible]